Advances in Intelligent Systems and Computing

Volume 436

Series editor

Janusz Kacprzyk, Polish Academy of Sciences, Warsaw, Poland
e-mail: kacprzyk@ibspan.waw.pl

About this Series

The series "Advances in Intelligent Systems and Computing" contains publications on theory, applications, and design methods of Intelligent Systems and Intelligent Computing. Virtually all disciplines such as engineering, natural sciences, computer and information science, ICT, economics, business, e-commerce, environment, healthcare, life science are covered. The list of topics spans all the areas of modern intelligent systems and computing.

The publications within "Advances in Intelligent Systems and Computing" are primarily textbooks and proceedings of important conferences, symposia and congresses. They cover significant recent developments in the field, both of a foundational and applicable character. An important characteristic feature of the series is the short publication time and world-wide distribution. This permits a rapid and broad dissemination of research results.

More information about this series at http://www.springer.com/series/11156

Millie Pant · Kusum Deep
Jagdish Chand Bansal · Atulya Nagar
Kedar Nath Das
Editors

Proceedings of Fifth International Conference on Soft Computing for Problem Solving

SocProS 2015, Volume 1

Springer

Editors
Millie Pant
Department of Applied Science and Engineering
Saharanpur Campus of IIT Roorkee
Saharanpur
India

Kusum Deep
Department of Mathematics
Indian Institute of Technology Roorkee
Roorkee
India

Jagdish Chand Bansal
Akbar Bhawan Campus
South Asian University
Chankyapuri, New Delhi
India

Atulya Nagar
Department of Mathematics and Computer Science
Liverpool Hope University
Liverpool
UK

Kedar Nath Das
Department of Mathematics
National Institute of Technology Silchar
Silchar, Assam
India

ISSN 2194-5357 ISSN 2194-5365 (electronic)
Advances in Intelligent Systems and Computing
ISBN 978-981-10-0447-6 ISBN 978-981-10-0448-3 (eBook)
DOI 10.1007/978-981-10-0448-3

Library of Congress Control Number: 2016930058

Printed on acid-free paper

This Springer imprint is published by SpringerNature
The registered company is Springer Science+Business Media Singapore Pte Ltd.

Preface

It is a matter of pride that the Annual Series of International Conference, called 'Soft Computing for Problem Solving' is entering its fifth edition as an established and flagship international conference. This annual event is a joint collaboration between a group of faculty members from institutes of repute like Indian Institute of Technology Roorkee; South Asian University, Delhi; NIT Silchar and Liverpool Hope University, UK. The first in the series of SocProS started in 2011 and was held from 20 to 22 December at the IIT Roorkee Campus with Prof. Deep (IITR) and Prof. Nagar (Liverpool Hope University) as the General Chairs. JKLU Jaipur hosted the second SocProS from 28 to 30 December 2012. Coinciding with the Golden Jubilee of the IIT Roorkee's Saharanpur Campus, the third edition of this international conference, which has by now become a brand name, took place at the Greater Noida Extension Centre of IIT Roorkee during 26–28 December 2013. The fourth conference took place at NIT Silchar during 27–29 December 2014. Like earlier SocProS conferences, the focus of SocProS 2015 is on Soft Computing and its applications to real-life problems arising in diverse areas of image processing, medical and healthcare, supply chain management, signal processing and multimedia, industrial optimisation, cryptanalysis, etc. SocProS 2015 attracted a wide spectrum of thought-provoking articles. A total of 175 high-quality research papers were selected for publication in the form of this two-volume proceedings.

We hope that the papers contained in this proceeding will prove helpful towards improving the understanding of Soft Computing at the teaching and research levels and will inspire more and more researchers to work in the field of Soft Computing. The editors express their sincere gratitude to SocProS 2015 Patron, Plenary Speakers, Invited Speakers, Reviewers, Programme Committee Members, International Advisory Committee and Local Organising Committee, without whose support the quality and standards of the conference could not be maintained. We express special thanks to Springer and its team for this valuable support in the publication of this proceedings.

Over and above, we express our deepest sense of gratitude to the 'Indian Institute of Technology Roorkee' to facilitate the hosting of this conference. Our sincere thanks to all the sponsors of SocProS 2015.

Saharanpur, India
Roorkee, India
New Delhi, India
Liverpool, UK
Silchar, India

Millie Pant
Kusum Deep
Jagdish Chand Bansal
Atulya Nagar
Kedar Nath Das

Conference Organising Committee

Patron
Prof. Pradipta Banerji, Director, Indian Institute of Technology Roorkee

Conference Chair
Dr. Y.S. Negi, Indian Institute of Technology Roorkee

Honaray Chair
Prof. Dr. Chander Mohan, Retired, Indian Institute of Technology Roorkee

Conveners
Prof. Kusum Deep, Indian Institute of Technology Roorkee
Dr. Millie Pant, Indian Institute of Technology Roorkee
Prof Atulya Nagar, Liverpool Hope University, Liverpool. United Kingdom

Organising Secretary
Dr. J.C. Bansal, South Asian University, New Delhi
Dr. Kedar Nath Das, NIT Silchar
Dr. Tarun Kumar Sharma, Amity University Rajasthan

Joint Organising Secretary
Dr. Dipti Singh, GBU, Greater Noida
Dr. Sushil Kumar, Amity University, Noida

Treasurer
Prof. Kusum Deep, Indian Institute of Technology Roorkee
Dr. Millie Pant, Indian Institute of Technology Roorkee

Best Paper and Best Ph.D. Thesis Chair
Prof V.K. Katiyar, IIT Roorkee
Prof S.G. Deshmukh, ABV-IITM Gwalior
Prof Atulya Nagar, Liverpool Hope University, UK

Publicity Chair
Dr. Musrrat Ali, Sungkyunkyan University, Korea
Dr. Radha Thangaraj, Indian Institute of Technology Roorkee
Dr. Rani Chinnappa Naidu, Northumbria University, UK

Social Media Chairs
Dr. Kedar Nath Das, NIT Silchar
Dr. Anupam Yadav, NIT Uttarakhand

Conference Proceedings and Printing and Publication Chair
Prof. Kusum Deep, IIT Roorkee
Dr. J.C. Bansal, South Asian University, New Delhi

Technical Sessions Chair
Dr. Kedar Nath Das, NIT Silchar, Assam
Dr. Manoj Thakur, IIT Mandi
Dr. Anupam Yadav, NIT Uttarakhand

Special Session Chair
Dr. Musrrat Ali, Sungkyunkyan University, Korea
Dr. Sushil Kumar, Amity University, Noida

Hospitality Chair
Dr. Millie Pant, Indian Institute of Technology Roorkee
Dr. Tarun Kumar Sharma, Amity University Rajasthan
Dr. Sushil Kumar, Amity University, Noida

Cultural Program and Registration Chair
Dr. Divya Prakash, AUR
Dr. Abha Mittal,CSIR-CBRI, Roorkee

Local Organising Committees

Amarjeet Singh, Indian Institute of Technology Roorkee
Asif Assad, Indian Institute of Technology Roorkee
Bilal Mirza, Indian Institute of Technology Roorkee
Garima Singh, Indian Institute of Technology Roorkee
Hira Zaheer, Indian Institute of Technology Roorkee
Kavita Gupta, Indian Institute of Technology Roorkee
Neetu Kushwaha, Indian Institute of Technology Roorkee
Renu Tyagi, Indian Institute of Technology Roorkee
Sunil Kumar Jauhar, Indian Institute of Technology Roorkee
Tushar Bharadwaj, Indian Institute of Technology Roorkee
Vanita Garg, Indian Institute of Technology Roorkee
Vidushi Gupta, Indian Institute of Technology Roorkee

Contents

About the Editors

Dr. Millie Pant is an Associate Professor with the Department of Paper Technology, Indian Institute of Technology Roorkee, Roorkee, India. She has to her credit several research papers in journals of national and international repute and is a well-known figure in the field of Swarm Intelligence and Evolutionary Algorithms.

Prof. Kusum Deep is working as a full-time professor in the Department of Mathematics, Indian Institute of Technology Roorkee, Roorkee, India. Over the last 25 years, her research is increasingly well cited making her a central International figure in the area of Nature-Inspired Optimisation Techniques, Genetic Algorithms and Particle Swarm Optimisation.

Dr. Jagdish Chand Bansal is an Assistant Professor with the South Asian University, New Delhi, India. Holding an excellent academic record, he is an excellent researcher in the field of Swarm Intelligence at the National and International Level, having several research papers in journals of national and international repute.

Prof. Atulya Nagar holds the Foundation Chair as Professor of Mathematical Sciences and is the Dean of Faculty of Science, at Liverpool Hope University, Liverpool, UK. Prof. Nagar is an internationally recognised scholar working at the cutting edge of theoretical computer science, applied mathematical analysis, operations research, and systems engineering and his work is underpinned by strong complexity-theoretic foundations.

Dr. Kedar Nath Das is now working as Assistant Professor in the Department of Mathematics, National Institute of Technology Silchar, Assam, India. Over the last 10 years, he has a good contribution towards to research in 'soft computing'. He has many papers to his credit in many journal in national and international level of repute. His area of interest is on Evolutionary and Bio-inspired algorithms for optimisation.

About the Book

The proceedings of SocProS 2015 will serve as an academic bonanza for scientists and researchers working in the field of Soft Computing. This book contains theoretical as well as practical aspects using fuzzy logic, neural networks, evolutionary algorithms, swarm intelligence algorithms, etc., with many applications under the umbrella of 'Soft Computing'. The book will be beneficial for young as well as experienced researchers dealing across complex and intricate real-world problems for which finding a solution by traditional methods is a difficult task.

The different application areas covered in the proceedings are: Image Processing, Cryptanalysis, Industrial Optimisation, Supply Chain Management, Newly Proposed Nature-Inspired Algorithms, Signal Processing, Problems related to Medical and Health Care, Networking Optimisation Problems, etc.

Optimization of Nonlocal Means Filtering Technique for Denoising Magnetic Resonance Images: A Review

Nikita Joshi and Sarika Jain

Abstract Magnetic resonance images are affected by noise of various types, which provide a hindrance to accurate diagnosis. Thus, noise reduction is still an important and difficult task in case of MRI. The objective behind denoising of images is to effectively decrease the unwanted noise by retaining the image features. Many techniques have been proposed for denoising MR images, and each technique has its own advantages and drawbacks. Nonlocal means (NLM) is a popular denoising algorithm for MR images. But it cannot be applied in its original form to different applications. The goal of this paper is to present the various optimization techniques for NLM filtering approach to reduce the noise present in MRIs. The original NLM filters along with its various advancements and mathematical models have been included.

Keywords Nonlocal means filtering · Similarity weights · Magnetic resonance imaging

1 Introduction

Magnetic resonance imaging (MRI) is a significant imaging technique used for producing specific and comprehensive images of organs and tissues of human body. The image quality in MRIs is of great clinical importance for correct and accurate diagnosis. However, these images are generally degraded through the noise present in the image. The noise in MRIs can be caused by varying factors such as stochastic

Nikita Joshi (✉)
Amity School of Engineering & Technology, Amity University Uttar Pradesh, Noida, Uttar Pradesh, India
e-mail: nikitajoshi502@gmail.com

Sarika Jain
Amity Institute of Information Technology, Amity University Uttar Pradesh, Noida, Uttar Pradesh, India
e-mail: ashusarika@gmail.com

M. Pant et al. (eds.), *Proceedings of Fifth International Conference on Soft Computing for Problem Solving*, Advances in Intelligent Systems and Computing 436, DOI 10.1007/978-981-10-0448-3_1

variation, various physiological processes and eddy currents, artifacts from the magnetic susceptibilities between neighboring tissues, rigid body motion, nonrigid motion, image acquisition process, and other sources. For details, refer Zhu et al. [1] and Mohan et al. [2]. The noise present in images leads to uncertainties in the measurements of various parameters and inaccurate automatic computer analysis such as segmentation, image reconstruction, and registration, which ultimately results in erroneous diagnosis. Therefore, removing noise from images is an important preprocessing step.

In MR images, all the signal frequencies are affected by noise and the noise is independent for each source. When noise is not dependent on the signal, the noise can be considered as complex additive white Gaussian noise (AWGN) having zero mean [3]. The spatial distribution of noise (in real and imaginary parts) in MR images generally becomes signal dependent and follows a Rician distribution, see Gudbjartsson and Patz [4] and Macovski [5], when the noise signal intensities have signal-to-noise ratio (SNR) less than 2. The Rician distribution follows a Gaussian distribution when SNR is quite high. The popular approaches for removing noise present in MR images include spatial and temporal filter, anisotropic diffusion filtering, nonlocal means algorithm, bilateral and trilateral filters, wavelet transformation, curvelet and the contourlet transformation, maximum likelihood approach, linear minimum mean square error estimation, nonparametric neighborhood statistics/estimation, and singularity function analysis. For details, refer Mohan et al. [2]. Finally, it was found that the nonlocal means filter introduced by Buades et al. [6], when applied to MR images, offered better results both qualitatively and quantitatively. But NLM filter could not be applied to different kinds of MR images in its original form. In this paper, we present a framework in which modifications to the original NLM filter have been made to make it suitable to apply under different conditions.

The outline of the paper is as follows: Section 2 presents the mathematical model for NLM filter. Section 3 discusses the various techniques in which optimization of NLM filter have been performed for various applications in medical imaging. Strength and weakness of all techniques have been revealed. Section 4 concludes about the applicability of discussed techniques and their medical importance.

2 NLM Filtering Approach for Denoising

Buades et al. [6] introduced a robust denoising technique known as NLM filter, which is built on the principle of nonlocal averaging of all pixels in the image. It compares the gray level in a single point in the image as well as compares the geometrical configuration in the entire neighborhood. As proposed by Buades et al. [6], for a discrete noisy image $v = \{v(i) | i \in I\}$ the approximate nonlocal means value $\mathrm{NL}(v(i))$, for a pixel i, is calculated as the weighted average of all the pixels in the image.

$$\mathrm{NL}(v(i)) = \sum_{j \in I} w(i,j)v(j)$$

where the set of weights $\{w(i,j)\}_j$ is dependent on the similarity between the two pixels i and j and satisfy the condition $0 \le w(i,j) \le 1$ and $\sum_j w(i,j) = 1$. The intensity gray level vectors $v(N_i)$ and $v(N_j)$ decide the similarity between the two pixels i and j, where N_k denotes a square neighborhood having fixed size and centered at a pixel k. The similarity between the pixels is measured as a decaying function of the weighted Euclidean distance d and is given by $d = \|v(N_i) - v(N_j)\|_{2,a}^2$ where $a > 0$ is the standard deviation of the Gaussian kernel. The pixel which has a similar gray level neighborhood as that of $v(N_i)$ has larger weight in the weighted average compared to others. The weights are defined as

$$w(i,j) = \frac{i}{Z(i)} e^{-\left(\frac{\|v(N_i)-v(N_j)\|_{2a}^2}{h^2}\right)}$$

where $Z(i)$ is the normalizing constant $Z(i) = \sum_j e^{-\left(\frac{\|v(N_i)-v(N_j)\|_{2a}^2}{h^2}\right)}$ and parameter h denotes the degree of filtering. The decay of exponential function is controlled by parameter h. Symbol $\|$ denotes the norm function. However, a pixel is equivalent to a volumetric pixel in 3-D, called as a voxel. It is the smallest distinguishable element of a 3-D object. Voxels contain multiple scalar values such as density, opacity, volumetric flow rate, and color. Therefore, voxels are used for analyzing medical data from ultrasound and MRI machines. According to the original NL means filter, every voxel in the image can be related to all the other voxels in the image, but since this procedure will greatly increase the computational speed, so the number of voxels that are considered in the weighted average can be constrained to a "search volume" V_i, which is centered at the current voxel x_i. In [7], Buades et al. demonstrated that NLM filter returned promising results for 2-D natural images.

3 NLM Based Denoising Methods

The major snag with NLM algorithm is its computational complexity. Various advancements have been proposed in the traditional NLM algorithm to overcome the computational complexity and thus increasing the speed. Some of them are discussed below.

3.1 An Optimized 3-D Blockwise NLM Denoising

The original nonlocal means filter introduced by Buades et al. [6] denoises 2-D images, but computational complexity increases when extended to 3-D images. Coupe et al. [8] focus on this critical aspect for 3-D medical images, specifically MRI by introducing a new variant of nonlocal means filter. The adaptation of this filter is based on the following:

(a) Automated modification of the smoothing parameter h: According to Buades et al. [6], for 2-D images, the smoothing parameter h is dependent on the standard deviation of the noise σ. But in 3-D imaging h should also depend upon $|N_i|$. So the value of h needs to be modified in order to get an equivalent filter as in 2-D imaging. The smoothing parameter h follows the relation $h^2 = f(\sigma^2, |N_i|, \beta)$, where β is a constant.
(b) Selection of voxel in the search volume: The computations are reduced by preselecting a subset of relevant voxels j in V_i. This avoids inadequate weight computations, as proposed by Mahmoudi and Sapiro [9], and skips unnecessary computation of all the Euclidean distances between $v(N_i)$ and $v(N_j)$. The selection of the voxels is based on the mean and variance of $v(N_i)$ and $v(N_j)$.
(c) Blockwise implementation of NLM: Blockwise implementation of NLM consists of

   - division of the complete volume into blocks, which have overlapping supports
   - implementing restoration of each of these blocks using NL-means
   - the intensity value of each voxel is restored based on the restored values of their respective blocks to which they belonged.

(d) Parallelized Computation: The operations are distributed on several processors using a grid. The entire volume is subdivided into further sub-volumes, each of which is computed by one processor.

Strengths and weaknesses. The above experiment was performed on the BrainWeb database and this technique significantly reduces the complexity of the NLM algorithm and demonstrates better results compared to the classical NL means filter and other techniques, such as the AD approach [10] and the TV minimization process [11]. However, the effect of this technique needs to be further explored on specific pathologies. The effect of NLM filter on segmentation and registration schemes should be investigated further.

3.2 Nonlocal Means Filter with Combined Patch and Pixel Similarity

In the original NLM filter, the contrast between small particles and their neighboring pixels gets reduced, thereby causing them to blur. To remove the Rician noise while using NLM, Daessle et al. [3] develop a Rician–NLM filter (RNLM) in which the self-weight of the pixel is the maximum weight of the noncentral pixels.

$$\mathrm{RNLM}(x_i) = \sqrt{\max\left(\left(\sum\nolimits_{j\epsilon V_i} w(i,j)\cdot y_j^2\right) - 2\sigma^2, 0\right)}$$

where σ denotes the standard deviation of Gaussian noise. In order to retain the high contrast details of the particles in MR images, the original RNLM algorithm is modified by using a new weight method which combines the patch/block and pixel similarity [12]. RNLM-combined patch and pixel algorithm (RNLM-CPP) which removes both Gaussian and Rician noise is given as

$$\mathrm{RNLM} - \mathrm{CPP}(x_i) = \sqrt{\max\left(\left(\sum\nolimits_{j\epsilon V_i} w(i,j)^{\mathrm{CPP}}\cdot y_j^2\right) - 2\sigma^2, 0\right)}$$

where $w(i,j)^{\mathrm{CPP}}$ is the weight after normalizing $\lambda(i,j)^{\mathrm{CPP}}$.

$$\lambda(i,j)^{\mathrm{CPP}} = \eta_{ij}\cdot\lambda(i,j), \quad \text{for} \quad j \neq i \quad \text{and} \quad \eta_{ij} = \frac{1}{1 + \left[\frac{|y_i - y_j|}{D_0}\right]^{2\alpha}}$$

where $\lambda(i,j)$ depicts the patches similarity between different patches, which are located at pixels i and j and η_{ij} denotes pixel similarity which is defined as a decreasing function of the intensity difference $|y_i - y_j|$. The position and slope of the transition are controlled by parameter D_0 and α. The pixel similarity η_{ij} ranges from 0 to 1.

Strengths and weaknesses. The above method is performed on synthetic data and real MR data. The details of small high contrast particles are retained better compared to the original RNLM filter. The weakness is that the expansion of this algorithm to 3-D would increase the computation cost as well as computation time.

3.3 Optimized Multicomponent MR Image Denoising

An MR image may comprise various types of images of the same patient. Various approaches such as gradient information [13], wavelet thresholding [14], and partial volume modeling-based approach [15] have been used to deal with multicomponent images. When multiple MR images are available, the additional related information in the images can be used to improve the filtering process. A new extension of NLM filter is introduced by Manjon et al. [16] which decreases random noise in multicomponent MR images. In order to remove noise, this algorithm uses a local principal component analysis decomposition, in which spatial information is taken not only from the spatial domain, but also from the intercomponent domain. According to the algorithm, when multiple MR images are accessed, having different acquisition times, then additional correlated information in these images can be used for improving the filtering process. This denoising method gives better results in both spatial and intercomponent domains. The similarity measure is enumerated by combining information from the surrounding pixels and from different components of the image. The similarity function for the multicomponent images is

$$w(i,j) = \frac{1}{Z(i)} e^{-\left(\sum_{k=1}^{C} \left(d(i^k, j^k)/h^{k2}\right)/C\right)}$$

where

$$\sum_{\forall j} Z(i) = e^{-\left(\sum_{k=1}^{C} \frac{\left(\frac{d(i^k, j^k)}{h^{k2}}\right)}{C}\right)}$$

C denotes the number of components and h^k denotes the standard deviation of noise of each image. Similar to the original NLM, the multicomponent-NLM (MNLM) consists of three parameters: first, the radius for search window R_{search}, second, the radius of similarity window R_{sim} and third, the degree of filtering h. In MNLM, the size of search window was decreased to a local window having smaller size. The size of similarity window increased the similarity measure between the neighborhoods. Smaller values of h remove less noise, whereas larger values of h introduce blurriness in the image. The following optimizations have been done to MNLM to increase accuracy.

(a) Similarity function: The MNLM optimization is useful on images with low noise. The weight associated with two equal noisy patches should be close to 1. To fulfill this, the normalized distance is calculated.
(b) Multicomponent preselection: In this work, pixels that have their first local moment smaller than the $k\,\sigma_i/n$ are selected, where σ_i is the standard deviation of noise in image i and n denotes the number of pixels that were used to find out the mean.
(c) Principal component analysis denoising (PCA): In this approach, the square window of the local surrounding (of a specific radius), of different images in the corresponding components is taken, discarding the less significant ones.

Each pixel in the image is denoised by decomposing this local surrounding square window. The PCA decomposition is carried on a local matrix of the size $N \times K$ (where N is the number of pixels belonging to the local window and K is the number of components). The components obtained after decomposing are processed before they are recomposed into the original matrix. After this, the intensity value for each pixel (filtered value) is computed by finding the average value of several overlapping windows. This averaging helps in reducing extra noise.

Strengths and weaknesses. As a result of increased data redundancy, this method presents advantages over single image techniques. Results are significantly improved when the number of images increases in contrast. However, in modern MRI, parallel imaging results in the generation of noise that is spatially dependent as well as variable. In such cases, evaluation of the local image noise is needed and hence it is required to use an automated local noise estimator.

3.4 Dynamic Nonlocal Means (DNLM)

Dynamic contrast enhanced magnetic resonance imaging (DCE-MRI) is a technique in which the imaging of tissues is performed numerous times after administering the contrast agent. Its applications include detection and characterization of tumors, cardiac obstruction, and MR angiography [17]. The analysis or explanation of DCE-MRI data is a typical task due to huge amount of data. The task is further complicated by induction of other noise such as geometric dissemination, non-consistency of intensity, and various artifacts due to motion. Yaniv et al. [17] present a novel variant of the NLM algorithm. The redundant information present in the temporal sequence of images is exploited, hence dipping the tendency of the nonlocal property, and thereby creating geometric objects. Nonlocal means algorithm takes into account the redundant information existing in the image, but at the same time, while handling medical images, it must be noted that similar patterns present in two different regions of the image may not always have the same meaning. If the NLM algorithm is simply extended, it does not consider local changes in intensity due to contrast enhancement. DNLM overcomes this drawback and suggests the following:

(a) Similarity metric: A similarity metric is considered that deals with local enhancement. The similarity metric is redefined as

$$\begin{aligned} S(V_i, V_j) &= S(V_i, C(V_i, V_j) \cdot V_j) \\ &= \left\| V_i - C(V_i, V_j) \cdot V_j \right\|_{2,a}^2 \end{aligned}$$

Weights can be computed as

$$w(i,j) = \frac{(C(v(\mathrm{N_i}), v(N_j))}{z(i)} e^{-S(v(N_i),v(N_j))/h^2}$$

where

$$C(V_i, V_j) = \begin{cases} \frac{E(V_i)}{E(V_j)} & t_1 \neq t_2 \text{ and } |E(V_i) - E(V_j)| > \sigma \\ 1 \end{cases}$$

t_1, t_2 are temporal components which helped in the selection of vectors V_i and V_j, σ is the approximate noise level in image and $E(V)$ denotes the expected value.

(b) Limited search: The search for the similarity window is limited to a small area in the spatial domain, but the search is not limited in the temporal domain. The DNLM algorithm filters the image in the spatial domain by utilizing the similarity over the temporal axis. This process did not affect the contrast enhancement or blurring of the image.

Strengths and weaknesses. The performance of the DNLM algorithm is empirically evaluated over seven other denoising methods and better results are obtained using DNLM. DNLM algorithm results in smallest mean square error between the denoised image (corrected of bias) and the corresponding original noise free image. In DCE-MRI data, this DNLM algorithm proves to be useful for improving the resolution of the image without degrading the image quality. However, its working is limited to exploit the redundant spatial information in the image and it also avoids the elimination of diagnostically important details/structures.

3.5 *Adaptive Nonlocal Means Denoising of MR Images with Spatially Varying Noise Levels*

When MR images are acquired, noise is introduced in the image, which varies spatially. Parallel MRI is a way of acquiring multiple images simultaneously, but at the same time the SNR is decreased by at least square root of the acceleration factor and nonhomogeneous spatial noise is also generated. Manjon [18] proposes a new scheme which takes into account both the Rician nature of MR data and noise patterns which vary spatially. The authors have followed the preselection approach used in Coupe et al. [8], which is based on finding the local mean and variance of patches/blocks, with some minor alteration. The task of preselection is accomplished by making use of the original mean and the inverted mean. The inverted mean is calculated as $\mathrm{Inv}(v(B_{ik})) = \max(v) - v(B_{ik})$, where B_{ik} refers to a block centered at x_{ik}. The most important parameter in NLM denoising is h^2 which regulates the smoothing strength of the filter. The most favorable value for h^2 is found to be σ^2, where σ is the noise standard deviation for blockwise NL means version. But when the noise is spatially varying in nature, local noise estimation should be introduced instead. As proposed by Buades et al. [19], the local noise estimation can be obtained by finding the Euclidean distance between two noisy patches as

$$d(N_i, N_j) = E\|v(N_i) - v(N_j)\|_2^2 = \|u_0(N_i) - u_0(N_j)\|_2^2 + 2\sigma^2$$

where u_0 denotes the noise free image. If each patch in the image is similar to itself, the variance of the local noise present in the image can be approximated as

$$\sigma^2 = \min(d(R_i, R_j))\forall j \neq i \quad \text{and} \quad R = u - \Psi(u)$$

where u is the original noisy image and $\Psi(u)$ is the low-pass filtered volume. This approach helps to find more number of similar patches that have a similar structure but varying mean level.

Strengths and weaknesses. This technique handles both Rician and Gaussian distributed noise. The interesting feature about this filter is that it adjusts itself according to the amount of noise existing in the image locally. Thus it can be utilized on images containing noise that is spatially homogeneous as well as non-homogeneous. The weakness of this approach is that while using this technique, there is a slight decrease in filtering accuracy.

3.6 DCT-Based Nonlocal Means Filter

Hu et al. [20] propose a new approach for denoising MR images by merging the NLM filter with the Discrete Cosine Transform (DCT). DCT inhibits useful properties like decrease in correlation among data and increase in level of energy compaction. In this method, the Euclidian distance $\|v(N_i) - v(N_j)\|_{2,a}^2$ is replaced by the distance obtained from DCT subspace of $v(N_i)$ and $v(N_j)$. First, transformation from time domain to frequency domain (using DCT) is performed on the blocks that lie in the image neighborhood. The lower dimension frequency coefficient subspace of DCT is achieved using the zigzag scan. The Euclidian distance is now $\| D_c(N_i) - D_c(N_j) \|^2 = \sum_{k=1}^{d} \left(D_c(N_i)_k - D_c(N_j)_k\right)^2$ where $D_c(N_i)$ is the coefficient in DCT subspace of neighborhood N_i and $D_c(N_i)_k$ is the kth coefficient in $D_c(N_i)$. This DCT based NLM filter is as follows:

$$\mathrm{NL}_{\mathrm{NLM-DCT}}(v(i)) = \sum_{j \in I} w_d(i,j)v(j) \quad \text{and} \quad w_d(i,j) = \frac{1}{Z_d(i)} e^{-\sum_{k=1}^{d} \frac{\left(D_C(N_i)_k - D_C(N_j)_k\right)^2}{h^2}}$$

where

$$Z_d(i) = \sum_{j \in I} e^{-\sum_{k=1}^{d} \frac{\left(D_C(N_i)_k - D_C(N_j)_k\right)^2}{h^2}}$$

is the normalizing term. MR images may contain both Gaussian and Rician noise. The MR image contains a noise bias of value $2\sigma^2$, which is signal independent [4]. In an MR magnitude image v, $E(v^2) = u_0^2 + 2\sigma^2$, where u_0 represents that image of v which is not contaminated with noise; v^2 and u_0^2 denote the squared images of v and u_0 respectively. The correction of this noise bias was illustrated by Manjon et al. [21]. The unbiased intensity value suggested by Manjon is

$$\mathrm{NL}_{\mathrm{UNLM}}(v(i)) = \sqrt{\max\left((\mathrm{NL}_{\mathrm{NLM}}(v_i(i)))^2 - 2\sigma^2, 0\right)}$$

where $\mathrm{NL}_{\mathrm{UNLM}}(v(i))$ represents the operation of unbiased-NLM (UNLM) filter on pixel i of a noisy image v. As described by Manjon et al. [21] above, the unbiased intensity value for the UNLM-DCT filter is given as

$$\mathrm{NL}_{\mathrm{UNLM-DCT}}(v(i)) = \sqrt{\max\left((\mathrm{NL}_{\mathrm{NLM-DCT}}(v(i)))^2 - 2\sigma^2, 0\right)}$$

Strengths and weaknesses. In this method, the similarity weights are calculated in DCT subspace so that noise can be reduced. The computation complexity is lower than the original NLM filter. Experimental results conclude that by using UNLM-DCT filter, the noise bias is successfully corrected and the original image gets returned. The weakness of this algorithm is that the computation time of UNLM-DCT is increased compared to NLM and UNLM because in UNLM-DCT, the discrete cosine transform is calculated for every neighborhood patch.

3.7 NLM Based on Salient Features Matching

Vega et al. [22] aim to reduce the computation load of NLM filtering technique while calculating the weights for the pixels. It is suggested to perform the similarity assessment by examining only a limited number of pixels. This limited number of pixels is a subset of the salient features of the associated pixels. This gives an estimation of the actual difference between the pixels, as computed by Buades et al. [6] in original NLM technique. The following features are proposed in this algorithm:

(a) Local structure: The local structure of the nontextured patches is retained.
(b) Reduced computation: The computation for calculating the distance between the pixels is reduced by considering simply a small subset of the features for all the pixels in the patch.
(c) Statistical characterization: The statistical characterization of the patches is preserved.

Strengths and weaknesses. This algorithm is developed to check merely a subset of the salient features that are associated to the pixels, which are similar to the task of computing the difference in the original NLM filter. This reduces the computational load of comparison. Moreover, the denoising outcomes are also improved. It is observed that in the process of computing the suitable salient features of the pixels, the assessment of the NLM weights becomes a bit stronger.

3.8 Adaptive Nonlocal Means Filtering

The structures present in the MR images may show different blurring levels according to the value of the decay control parameter h. Kang et al. [23] propose a new technique in which different weights are applied on the image based on edgeness. Edgeness per unit area measures the busyness of the texture, which is calculated by dividing the gradient magnitude of each pixel p by a region of N pixels in the image. The technique possesses the following features.

(a) Using adaptive decay control parameter: The edgeness of the image is estimated and then the NLM filter is applied. In this method, an adaptive decay control parameter h_i is used, which is dependent on edgeness and represents the amount of complete information in pixel i.

$$\sigma_i = \sigma\left(1 - \frac{E_i - E_{\mathrm{MIN}}}{E_{\mathrm{MAX}} - E_{\mathrm{MIN}}}\right)$$

where edgeness E_i is computed by a simple edge detector, and E_{MIN} and E_{MAX} represent the minimum and maximum edgeness value respectively. The calculated weight is dependent on patch similarity. Due to high edgeness, if σ_i is small, the original structure of image is preserved.
(b) Preserves image details: This technique retained the details present in high contrast regions and at the same time reduced noise in low contrast regions.
(c) Examining neighboring slice: To make denoising more effective, adjacent slices are examined in the search windows since the occurrence of similar patches is more in the neighboring slices.

Strengths and weaknesses. Experimental results show that by using this method, important structures in MR images are preserved. This algorithm is more accurate than conventional methods such as Gaussian averaging, total variation, anisotropic diffusion, bilateral filtering, and NLM. It also provides better restoration accuracy. The weakness of this method lies in the fact that the contrast-to-noise ratio has to be exploited for those images that do not have any ground-truth data.

3.9 Wavelet Mixing in NLM

Aksam et al. [24] suggested an improved adaptive nonlocal means algorithm used for brain MRI. The algorithm consists of the following extended features to NLM:

(a) Improved adaptive NLM filtering (IANLM): The original NLM calculates the similarity weights among the pixels in the search window only. IANLM provides an automatic selection of the size of search window for a particular pixel.
(b) Rician noise correction: The Rician noise is removed by taking the square of magnitude image for performing filtering, instead of the image itself and then subtracting the noise from the repaired value of each pixel, refer Nowak [25].
(c) Wavelet coefficients mixing: For denoising, the high and low frequency components of the image are split away. As a result, over-smoothed and under-smoothed images are obtained after filtering. Various subbands contained in the two images are then united so that the final image consists of the positive features of each image. The over-smoothed and under-smoothed images are disintegrated into four subbands each, by making use of discrete wavelet transform. Inverse discrete wavelet transform is used to combine the subbands. Peak signal-to-noise ratio gets improved in this method.

Strengths and weaknesses. This algorithm is more reliable and robust to noise and outperforms other denoising methods and preserves small image structures. The weakness of the algorithm is that the background area in the brain MR image needs to be removed so that no bias is created due to background tissues.

3.10 Filter Bank-Based Nonlocal Means

NLM filter works on the basis of patch similarity. But often, it is seen that these patches do not provide very accurate results. Therefore, some former information regarding the image is required. MRIs consist of many flat regions and edges which help in providing some prior information. Guo et al. [26] develop a NLM optimization technique in which the weights are calculated using the similarity of image features instead of similarity of patches. The image features are retrieved by convolving to Leung-Malik (LM) filters. Texton method by Leung and Malik [27] is an image texture analysis method. Using the Texton method, a set of filters is used which extracts various different texture features of the image. Guo mentioned that the LM filter bank consist of 48 filters partitioned as (i) first and the second order of Gaussian differential filters at 3 scales and 6 orientations, (ii) 8 Laplacian of Gaussian (LoG) filters, and (iii) 4 Gaussian filters. The responses of all the filters are combined as a vector. All the vectors are then clustered using a clustering algorithm like K-means clustering algorithm. The resulting cluster centers are called Textons. In this technique, first the LM filter bank is generated and then the image is filtered to get the feature vector string of each pixel $p : \{\psi_k(p), k = 1, 2, 3, \ldots, T\}$.

Secondy, based on all pixels in a particular search window, the weights are calculated as $w(i,j) = \frac{1}{Z(i)} e^{-\left[\sum_{k=1}^{T} \frac{\|\psi_k(i) - \psi_k(i)\|_2^2}{h^2}\right]}$, where $Z(i) = \sum_j e^{-\left[\sum_{k=1}^{T} \frac{\|\psi_k(i) - \psi_k(j)\|_2^2}{h^2}\right]}$

Finally the denoised intensity value of pixel i is defined as

$$\mathrm{NLM} - \mathrm{LMF}(v(i)) = \sum_{j \in I} w(i,j) v(j)$$

Strengths and weaknesses. The method is carried out on Rician noise, which is reduced effectively and the image details and weak edges are also preserved. This method proves to be efficient for highly noisy MR images. The major weakness is that the Texton method in the LM filter bank is very sensitive to the changes in the orientation of the texture.

4 Conclusion

This paper focuses on the various developments in the original nonlocal means filtering technique, specifically in MR images. There are different ways of capturing the MR images and the NLM filter in its original form cannot be applied to all types of MR images. The techniques used in optimizing the NLM filter make it suitable to apply NLM under various conditions. The strengths and weaknesses of all methods have also been included. Applications like segmentation and tactography can benefit from the enhanced denoised data.

References

1. Zhu, H., Li, Y., Ibrahim, J.G., Shi, X., An, H., Chen, Y., Gao, W., Lin,W., Rowe,D.B., Peterson, B.S.: Regression models for identifying noise sources in magnetic resonance imaging. J. Am. Stat. Assoc. **104**, 623–637 (2009)
2. Mohan, J., Krishnaveni, V., Guo. Y.: A survey on the magnetic resonance image denoising methods: Biomed. Signal Process. Control **9**, 56–69 (2014)

3. Daessle, N.W., Prima, S., Coupe, P., Morrissey, S.P., Barillot, C.: Rician noise removal by non-Local Means filtering for low signal-to-noise ratio MRI: applications to DT-MRI. In: 11th International Conference on Medical Image Computing and Computer-Assisted Intervention, pp. 171–179 (2008)
4. Gudbjartsson, H., Patz, S.: The rician distribution of noisy mri data. Magn. Reson. Med. **34**, 910–914 (1995)
5. Macovski, A.: Noise in MRI. Magn. Reson. Med. **36**, 494–497 (1996)
6. Buades, A., Coll, B., Morel J.M.: A non local algorithm for image denoising. In IEEE Computer Society Conference on Computer Vision and Pattern Recognition, pp. 60–65. IEEE Press, San Diego (2005)
7. Buades, A., Coll, B., Morel J.M.: A review of image denoising algorithms, with a new one. Multiscale Model. Simul. **4**, 490–530 (2005)
8. Coupe, P., Yger, P., Prima, S., Hellier, P., Kervrann, C., Barillot, C.: An optimized blockwise nonlocal means denoising filter for 3-D magnetic resonance images. IEEE Trans. Med. Imaging **27**(4), 425–441 (2008)
9. Mahmoudi, M., Sapiro, G.: Fast image and video denoising via nonlocal means of similar neighborhoods. IEEE Signal Process. Lett. **12**, 839–842 (2005)
10. Perona, P., Malik, J.: Scale-space and edge detection using anisotropic diffusion. IEEE Trans. Pattern Anal. Mach. intell. **12**(7), 629–639 (1990)
11. Rudin, L.I., Osher, S., Fatemi, E.: Nonlinear total variation based noise removal algorithms. Physica D **60**, 259–268 (1992)
12. Zhang, X., Hou, G., Ma,J., Yan, W., Lin, B.: Denoising MR images using non-local means filter with combined patch and pixel similarity. PLoS ONE **9** (2014)
13. Gerig, G., Kubler, O., Kikinis, R., Jolesz, F.A.: Nonlinear anisotropic filtering of MRI data. IEEE Trans. Med. Imaging **11**(1), 221–232 (1992)
14. Scheunders, P., Backer, S.D.: Wavelet denoising of multicomponent images using a Gaussian scale mixture model and a noise-free image as priors. IEEE Trans. Image Process. **16**(7), 1865–1872 (2007)
15. Thacker, N.A., Pokri, M.: Noise filtering and testing for MR using a multi-dimensional partial volume model. In: Proceedings of the Medical Image Understanding and Analysis, pp. 21–24 (2004)
16. Manjon, J.V., Thacker, N.A., Lull, J.J., Marti, G.G., Bonmati, L.M., Robles, M.: Multicomponent MR image denoising. Int. J. Biomed. Imaging (2009). doi:10.1155/2009/756897
17. Gal, Y., Mehnert, A.J.H., Andrew, P.B., B, Macmohan, K., Kennedy, D., Crozier, S.: Denioising of dynamic contrast enhanced MR images using dynamic non local means. IEEE Trans. Med. Imaging **29**(2), 302–310 (2010)
18. Manjon, J., Coupe, P., Bonmati, L.M., Collins, D.L., Robles, M.: Adaptive non-local means denoising of MR images with spatially varying noise levels. J. Magn. Reson. Imaging **31**, 192–203 (2010)
19. Buades, A., Coll, B., Morel, J.M.: Nonlocal image and movie denoising. Intern. J. Comput. Vis. **76**, pp. 123–139 (2008)
20. Hu, J., Pu, Y., Wu, X., Zhang, Y., Zhou, J.: Improved DCT-based nonlocal means filter for MR images denoising. Comput. Math. Methods Med. (2012). doi:10.1155/2012/232685
21. Manjon, J.V., Caballero, J.C., Lull, J.J., Marti, G.G., Bonmati, L.M., Robles, M.: MRI denoising using nonlocal means. Med. Image Anal. **12**(4), 514–523 (2008)
22. Vega, A.T., Perez, V.G., Fernandez, S.A., Westin, C.F.: Efficient and robust nonlocal means denoising of MR data based on salient features matching. J. Comput. Methods Program Biomed. **105**(2), 131–144 (2012)
23. Kang, B., Choi, O., Kim, J.D., Hwang, D.: Noise reduction in magnetic resonance images using adaptive non local means filtering. Elect. Lett. **49**(5) (2013)
24. Aksam, I.M., Jalil, A., Rathore, S., Ali, A., Hussain, M.: Brain MRI denoising and segmentation based improved adaptive non-local means. Int. J. Imaging Syst. Technol. **23**, 235–248 (2013)

25. Nowak, R.D.: Wavelet- based Rician noise removal for magnetic resonance imaging. IEEE Trans. Image Proc. **8**(10), 1408–1419 (1999)
26. Guo, T., Liu, Q., Luo, J.: Filter bank based nonlocal means for denoising magnetic resonance images. J. Shanghai Jiaotong Univ. (Sci.) **19**(1), 72–78 (2014)
27. Leung, T., Malik, J.: Representing and recognizing the visual appearance of materials using three-dimensional texton. Int. J. Comput Vis. **43**, 29–44 (2001)

A Production Model with Stock-Dependent Demand, Partial Backlogging, Weibull Distribution Deterioration, and Customer Returns

Chaman Singh, Kamna Sharma and S.R. Singh

Abstract In this paper, we derive an economic production model having two-parameter Weibull distribution deterioration. In this model, we considered a demand rate that depends on price stock and indirectly on time. Shortage is allowed and partially backlogged. We assume that customer return is a factor of quantity sold, price, and inventory level. Time horizon is finite. Production is also dependent on demand. The goal of this production is to maximize the profit function. An illustrative example, sensitivity analysis, and a graphical representation are used to interpret the usefulness of this model.

Keywords Production model · Two parameter Weibull distribution deterioration · Shortage · Partial backlogging · Customer return

1 Introduction

In every supply chain model, maintaining of deteriorating inventories is a major issue for almost all business organizations. Most of the goods decay over time. In general, some products deteriorate in a certain fixed period of storage like seasonal goods fruits, vegetables, etc., but certain goods lose their potentiality when the time passes, such as electronic items, radioactive substances, etc. Certain inventories like highly volatile liquids as ethanol, gasoline, etc., undergo depletion due to evaporation, so that deterioration is one of the most influential factors that affect the

Chaman Singh (✉)
Acharya Narendra Dev College, University of Delhi, New Delhi, India
e-mail: chamansingh07@gmail.com

Kamna Sharma
Department of Computer Science, D.N. College, CCS University, Meerut 250001, Uttar Pradesh, India
e-mail: shivrajpundir@gmail.com

S.R. Singh
Department of Mathematics, CCS University, Meerut 250001, Uttar Pradesh, India
e-mail: anushka.gautam17@gmail.com

M. Pant et al. (eds.), *Proceedings of Fifth International Conference on Soft Computing for Problem Solving*, Advances in Intelligent Systems and Computing 436, DOI 10.1007/978-981-10-0448-3_2

decision related to production and inventory management. Each business organization considers it quite seriously. With regard to all these issues, deterioration function is of various types that may be constant and time dependent. In our production model, we consider Weibull distribution as a deterioration function. Weibull distribution is one of the most reliable deterioration functions because it presents a perfect view of deteriorating inventory level. Covert and Philip [1] established an inventory model for deteriorating items having variable rate of deterioration. In their model, they use two-parameter Weibull deterioration. Misra [2] also presents a production model with two-parameter Weibull deterioration to show inventory depletion. Choi and Hwang [3] present an optimization of product planning problem with continuously distributed time lags. Aggarwal and Bahari-Hashani [4] synchronized production policies for deteriorating items in a declining market. Pakkala and Achary [5] present a deterministic inventory model for deteriorating items with two warehouses and finite replenishment rate. Jong et al. [6] developed an EOQ inventory model with time-varying demand and Weibull deterioration with shortages. Wu [7] presented an EOQ inventory model for items with Weibull distribution deterioration, ramp type demand rate, and partial backlogging. Lee and Wu [8] formulate an EOQ model for items with Weibull distributed deterioration, shortages, and power demand pattern. Banerjee and Agrawal [9] analyzed a two-warehouse inventory model for items with three-parameter Weibull distribution deterioration, shortages, and linear trend in demand. Roy and Chaudhuri [10] scheduled a production inventory model under stock-dependent demand, Weibull distribution deterioration, and shortage. Begum et al. [11] worked on an EOQ model for varying items with Weibull distribution deterioration and price-dependent demand. Konstantaras and Skouri [12] dealt a note on a production inventory model under stock-dependent demand, Weibull distribution deterioration, and shortage. Shilpi et al. [13] introduced an EPQ model of ramp type demand with Weibull deterioration under inflation and finite horizon in crisp and fuzzy environment.

In any production model, demand is a reliable factor on which the whole working of inventory model depends. Most researchers assume that demand depends on time as well as other factors. Stock-dependent demand is another way to look at practical situations. Many of the factors affect demand on a serious mode, but stock affects it in the most powerful manner. It may influence the production directly or indirectly, such as low stock raises the price of commodity in the market which decreases the demand and, if the stock level increases, then the price goes down and as a result demand increases. Therefore, it is observed that the stock level affects the demand in many ways. For example, if there are a large pile of goods available in the stock then the vendor announces a large discount to clear the stock. Many practitioners and researchers have analyzed this issue very seriously. Many researchers consider this as a realistic assumption, such as Datta et al. [14], Balki and Benkherouf [15], Teng and Chang [16], Wu et al. [17], Singh et al. [18], Singh and Singh [19], and finally, Sarker and Sarkar [20], Yang [21].

Customer return is also one of the most important factors that affect the production model. Customer returns are the products that may be returned by the customer after purchase. Customer may return these products due to several reasons such as defect in the product, customer is not satisfied with the product, some money-back guarantee, or maybe to replace the product, etc. Nowadays, customer returns occur in many different ways. Many researchers working in the stream like Hess and Mayhew [22] proposed a return of modeling merchandise in direct marketing. It is useful for the future studies of many researchers. Pasterneck [23] proposed a model for return policies of deteriorating items. In the same field, Anderson et al. [24] developed a relation between return and demand. Further, Ahmed et al. [25] introduced an inventory model for production as well as remanufacturing for quality and price-dependent return rate. In the same field, Hani et al. [26] derived an advertising policy customer's disadoption and subscriber services cost learning. Now Jiang and Chan [27] establish a lot of sizing polices for expiry date deteriorating items and partial trade credit risk customers.

In this proposed model, we considered a production inventory model with shortage, partial backlogging, and customer returns. Two-parameter Weibull deterioration is considered here. In this model, production is dependent on demand and demand depends on stock and price. Customer return is a function of price, quantity sold, and inventory level. To match the illustrated model with realistic situations, we discussed three cases of Weibull deterioration as constant, linear, and quadratic. To illustrate the model utility numerical example, sensitivity analysis, and concavity of the profit functions are shown here.

2 Notations and Assumptions

2.1 *Notations*

c_h Holding cost per unit per unit time
c_d Deterioration cost per unit per unit time
c_l Cost of lost sale per unit
p Selling price per unit, where $p > c$
θ Two-parameter Weibull deterioration rate
Q Order quantity
T Length of replenishment cycle time
B Backlogging rate
P Production rate
SV Salvage value per unit item
A Setup cost
$I_1(t)$ Inventory level at the time $t \in [0, t_1]$
$I_2(t)$ Inventory level at the time $t \in [t_1, t_2]$
$I_3(t)$ Inventory level at the time $t \in [t_2, t_3]$
$I_4(t)$ Inventory level at the time $t \in [t_3, t_4]$

2.2 Assumptions

1. Two-parameter Weibull distribution deterioration is considered here. $\theta = \alpha\beta t^{\beta-1}$.
2. Time horizon is finite.
3. The demand rate is $D(p,t) = (a - \text{bp} + \text{cI}(t))$ (where $a > 0$, $b > 0$) is a linearly decreasing function of the price but for the shortage and partial backlogging period demand depends on price only.
4. Shortage is allowed. The unsatisfied demand is backlogged, and the fraction of shortage back ordered is B, $(B > 0)$, and $0 \leq B \leq 1$.
4. We assume that the customer returns increase with both the quantity sold and price using the following general form: $R(p,t) = \text{AD}(p,t,I(t)) + \text{Bp}$ $(B \geq 0, 0 \leq A < 1)$.
5. Production is demand dependent, where $P(t) = \text{KD}(t)$.

3 Model Formulation

For the mathematical formulation of presented model, we solve the different inventory level as well as different costs. Firs, we can see that production starts when $t = 0$ then the inventory level goes up, but at the same time inventory goes down due to demand and deterioration. After time t_1, inventory decreases due to demand and deterioration. At the time interval $t_2 < t < t_3$, shortage occurs and the inventory level becomes negative and at the same time backlogging starts. In the fourth phase, production again starts and the backlogged demands get fulfilled partially.

$$\frac{dI_1(t)}{dt} = P - D(p,t,I(t)) - \theta I_1(t), \quad I_1(0) = 0, \quad 0 \leq t \leq t_1 \tag{1}$$

$$\frac{dI_2(t)}{dt} = -D(p,t,I(t)) - \theta I_2(t), \quad I_2(t_2) = 0, \quad t_1 \leq t \leq t_2 \tag{2}$$

$$\frac{dI_3(t)}{dt} = -D(p)B, \quad I_3(t_2) = 0, \quad t_2 \leq t \leq t_3 \tag{3}$$

$$\frac{dI_4(t)}{dt} = P - D(p), \quad I_1(t_4) = 0, \quad t_3 \leq t \leq t_4 \tag{4}$$

As we see in Fig. 1.

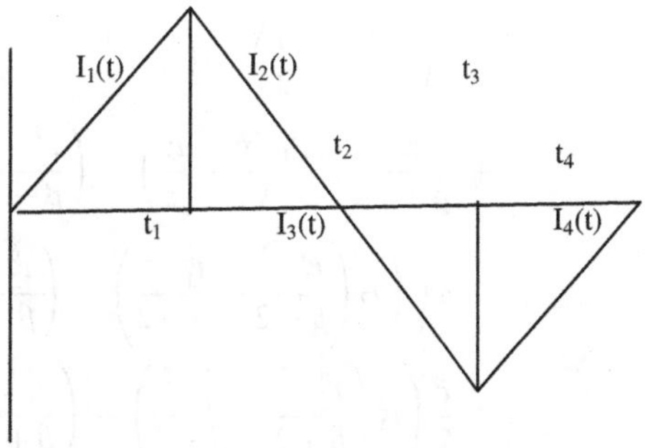

Fig. 1 Inventory level at time t

Now solving the above equations, we get

$$I_1(t) = (1-k)(a-\mathrm{bp})\left\{t + \frac{\alpha t^{\beta+1}}{\beta+1} - \frac{c(k-1)t^2}{2} + c(k-1)t^2 + \frac{\alpha c(k-1)t^{\beta+1}}{\beta+1} + \frac{c^2(k-1)^2t^2}{2} - \alpha t^{\beta+1} - \frac{\alpha^2 t^{2\beta+1}}{\beta+1} + \frac{c\alpha(k-1)t^{\beta+2}}{2}\right\} \quad (5)$$

$$I_2(t) = (1-\mathrm{ct}-\alpha t^{\beta})\left\{(a-\mathrm{bp})\left[(t_2-t) + \frac{c}{2}(t_2^2-t^2) + \frac{\alpha}{\beta+1}\left(t_2^{\beta+1} - t^{\beta+1}\right)\right]\right\} \quad (6)$$

$$I_3(t) = -B(a-\mathrm{bp})(t_2-t) \quad (7)$$

$$I_4(t) = (k-1)(a-\mathrm{bp})(t-t_4) \quad (8)$$

Now using the above equations, we can find the following cost:

The deterioration cost for the period (0, t_2)

$$= \theta c_d\left[\int_0^{t_1} I_1(t)\mathrm{d}t + \int_{t_1}^{t_2} I_2(t)\mathrm{d}t\right] \quad (9)$$

$$= C_d\alpha\beta t^{\beta-1}\left\{\left[\frac{t_1^2}{2} + \frac{\alpha t_1^{\beta+2}}{(\beta+1)(\beta+2)} - \frac{c(k-1)t_1^3}{6} + \frac{c(k-1)t_1^3}{3} + \frac{\alpha c(k-1)t_1^{\beta+2}}{(\beta+3)(\beta+2)} + \frac{c^2(k-1)^2t_1^4}{8} - \frac{\alpha t_1^{\beta+2}}{(\beta+2)} - \frac{\alpha^2 t^{2(\beta+1)}}{(\beta+1)(\beta+2)} + \frac{c\alpha(k-1)t^{(\beta+3)}}{2(\beta+3)}\right]\right.$$
$$+ (a-bp)\left[t_2(t_2-t_1) - \left(\frac{t_2^2}{2} - \frac{t_1^2}{2}\right) + \frac{c}{2}\left(t_2^2(t_2-t_1) - \left(\frac{t_2^3}{3} - \frac{t_1^3}{3}\right)\right) + \frac{\alpha}{\beta+1}\left(t_2^{(\beta+1)}(t_2-t_1) - \left(\frac{t_2^{\beta+2}}{\beta+2} - \frac{t_1^{\beta+2}}{\beta+2}\right)\right)\right.$$

$$- c\left(t_2\left(\frac{t_2^2}{2}-\frac{t_1^2}{2}\right)-\left(\frac{t_2^3}{3}-\frac{t_1^3}{3}\right)+\frac{c}{2}\left(t_2^2\left(\frac{t_2^2}{2}-\frac{t_1^2}{2}\right)-\left(\frac{t_2^4}{4}-\frac{t_1^4}{4}\right)\right)\right.$$

$$+\frac{\alpha}{\beta+1}\left(t_2^{\beta+1}\left(\frac{t_2^2}{2}-\frac{t_1^2}{2}\right)-\left(\frac{t_2^{\beta+3}}{\beta+3}-\frac{t_1^{\beta+3}}{\beta+3}\right)\right)\Bigg)\Bigg)$$

$$-\alpha\left(\left(t_2\left(\frac{t_2^{\beta+2}}{\beta+2}-\frac{t_1^{\beta+2}}{\beta+2}\right)-\left(\frac{t_2^{\beta+2}}{\beta+2}-\frac{t_1^{\beta+2}}{\beta+2}\right)\right)\right.$$

$$+\frac{c}{2}\left(t_2^2\left(\frac{t_2^{\beta+2}}{\beta+2}-\frac{t_1^{\beta+2}}{\beta+2}\right)-\left(\frac{t_2^{\beta+3}}{\beta+3}-\frac{t_1^{\beta+3}}{\beta+3}\right)\right)$$

$$\left.+\frac{\alpha}{\beta+1}\left(t_2^{\beta+1}\left(\frac{t_2^{\beta+1}}{\beta+1}-\frac{t_1^{\beta+1}}{\beta+1}\right)-\left(\frac{t_2^{2(\beta+2)}}{2(\beta+1)}-\frac{t_1^{2(\beta+2)}}{\beta+3}\right)\right)\right)\Bigg] \tag{10}$$

Holding cost for the inventory

$$=c_h\left[\int_0^{t_1} I_1(t)\mathrm{d}t+\int_{t_1}^{t_2} I_2(t)\mathrm{d}t\right]$$

$$=C_h\Bigg\{(1-k)(a-\mathrm{bp})\left[\frac{t_1^2}{2}+\frac{\alpha t_1^{\beta+2}}{(\beta+1)(\beta+2)}-\frac{c(k-1)t_1^3}{6}+\frac{c(k-1)t_1^3}{3}\right.$$

$$\left.+\frac{\alpha c(k-1)t_1^{\beta+2}}{(\beta+3)(\beta+2)}+\frac{c^2(k-1)^2t_1^4}{8}-\frac{\alpha t_1^{\beta+2}}{(\beta+2)}-\frac{\alpha^2 t^{2(\beta+1)}}{(\beta+1)(\beta+2)}+\frac{c\alpha(k-1)t^{(\beta+3)}}{2(\beta+3)}\right]$$

$$+(a-\mathrm{bp})\left[t_2(t_2-t_1)-\left(\frac{t_2^2}{2}-\frac{t_1^2}{2}\right)\right.$$

$$+\frac{c}{2}\left(t_2^2(t_2-t_1)-\left(\frac{t_2^3}{3}-\frac{t_1^3}{3}\right)\right)+\frac{\alpha}{\beta+1}\left(t_2^{(\beta+1)}(t_2-t_1)-\left(\frac{t_2^{\beta+2}}{\beta+2}-\frac{t_1^{\beta+2}}{\beta+2}\right)\right)$$

$$-c\left(t_2\left(\frac{t_2^2}{2}-\frac{t_1^2}{2}\right)-\left(\frac{t_2^3}{3}-\frac{t_1^3}{3}\right)+\frac{c}{2}\left(t_2^2\left(\frac{t_2^2}{2}-\frac{t_1^2}{2}\right)-\left(\frac{t_2^4}{4}-\frac{t_1^4}{4}\right)\right)\right.$$

$$+\frac{\alpha}{\beta+1}\left(t_2^{\beta+1}\left(\frac{t_2^2}{2}-\frac{t_1^2}{2}\right)-\left(\frac{t_2^{\beta+3}}{\beta+3}-\frac{t_1^{\beta+3}}{\beta+3}\right)\right)\Bigg)\Bigg)$$

$$-\alpha\left(\left(t_2\left(\frac{t_2^{\beta+2}}{\beta+2}-\frac{t_1^{\beta+2}}{\beta+2}\right)\right.\right.$$

$$\left.-\left(\frac{t_2^{\beta+2}}{\beta+2}-\frac{t_1^{\beta+2}}{\beta+2}\right)\right)+\frac{c}{2}\left(t_2^2\left(\frac{t_2^{\beta+2}}{\beta+2}-\frac{t_1^{\beta+2}}{\beta+2}\right)-\left(\frac{t_2^{\beta+3}}{\beta+3}-\frac{t_1^{\beta+3}}{\beta+3}\right)\right)$$

$$\left.\left.+\frac{\alpha}{\beta+1}\left(t_2^{\beta+1}\left(\frac{t_2^{\beta+1}}{\beta+1}-\frac{t_1^{\beta+1}}{\beta+1}\right)-\left(\frac{t_2^{2(\beta+2)}}{2(\beta+1)}-\frac{t_1^{2(\beta+2)}}{\beta+3}\right)\right)\right)\right]\Bigg\} \tag{11}$$

Return cost for the inventory

$$(p-\mathrm{SV})\{A(a-\mathrm{bp})t_2+Ac\left[\frac{t_1^2}{2}+\frac{\alpha t_1^{\beta+2}}{(\beta+1)(\beta+2)}-\frac{c(k-1)t_1^3}{6}+\frac{c(k-1)t_1^3}{3}\right.$$
$$+\frac{\alpha c(k-1)t_1^{\beta+2}}{(\beta+3)(\beta+2)}+\frac{c^2(k-1)^2t_1^4}{8}-\frac{\alpha t_1^{\beta+2}}{(\beta+2)}-\frac{\alpha^2 t^{2(\beta+1)}}{(\beta+1)(\beta+2)}+\left.\frac{c\alpha(k-1)t^{(\beta+3)}}{2(\beta+3)}\right]$$
$$+\mathrm{Ac}(a-\mathrm{bp})\left[t_2(t_2-t_1)-\left(\frac{t_2^2}{2}-\frac{t_1^2}{2}\right)+\frac{c}{2}\left(t_2^2(t_2-t_1)-\left(\frac{t_2^3}{3}-\frac{t_1^3}{3}\right)\right)\right.$$
$$+\frac{\alpha}{\beta+1}\left(t_2^{(\beta+1)}(t_2-t_1)-\left(\frac{t_2^{\beta+2}}{\beta+2}-\frac{t_1^{\beta+2}}{\beta+2}\right)\right)-c\left(t_2\left(\frac{t_2^2}{2}-\frac{t_1^2}{2}\right)-\left(\frac{t_2^3}{3}-\frac{t_1^3}{3}\right)\right.$$
$$+\frac{c}{2}\left(t_2^2\left(\frac{t_2^2}{2}-\frac{t_1^2}{2}\right)-\left(\frac{t_2^4}{4}-\frac{t_1^4}{4}\right)\right)+\frac{\alpha}{\beta+1}\left(t_2^{\beta+1}\left(\frac{t_2^2}{2}-\frac{t_1^2}{2}\right)-\left(\frac{t_2^{\beta+3}}{\beta+3}-\frac{t_1^{\beta+3}}{\beta+3}\right)\right)\Bigg)$$
$$-\alpha\left(\left(t_2\left(\frac{t_2^{\beta+2}}{\beta+2}-\frac{t_1^{\beta+2}}{\beta+2}\right)-\left(\frac{t_2^{\beta+2}}{\beta+2}-\frac{t_1^{\beta+2}}{\beta+2}\right)\right)\right.$$
$$+\frac{c}{2}\left(t_2^2\left(\frac{t_2^{\beta+2}}{\beta+2}-\frac{t_1^{\beta+2}}{\beta+2}\right)-\left(\frac{t_2^{\beta+3}}{\beta+3}-\frac{t_1^{\beta+3}}{\beta+3}\right)\right)$$
$$\left.+\frac{\alpha}{\beta+1}\left(t_2^{\beta+1}\left(\frac{t_2^{\beta+1}}{\beta+1}-\frac{t_1^{\beta+1}}{\beta+1}\right)-\left(\frac{t_2^{2(\beta+2)}}{2(\beta+1)}-\frac{t_1^{2(\beta+2)}}{\beta+3}\right)\right)\Bigg)\right]+\mathrm{Bpt}_2\Bigg\} \tag{12}$$

Lost sale cost for the inventory

$$=c_l\int_{t_2}^{t_3}(1-B)\mathrm{D}dt=c_l(1-B)\{(a-\mathrm{bp})(t_3-t_2)\} \tag{13}$$

Production cost for the inventory

$$=c_p\left[\int_0^{t_1}\mathrm{Pdt}+\int_{t_3}^{t_4}\mathrm{Pdt}\right]$$
$$=c_p\Bigg\{k(a-\mathrm{bp})t_1+ck\left[\frac{t_1^2}{2}+\frac{\alpha t_1^{\beta+2}}{(\beta+1)(\beta+2)}-\frac{c(k-1)t_1^3}{6}+\frac{c(k-1)t_1^3}{3}\right.$$
$$+\frac{\alpha c(k-1)t_1^{\beta+2}}{(\beta+3)(\beta+2)}+\frac{c^2(k-1)^2t_1^4}{8}-\frac{\alpha t_1^{\beta+2}}{(\beta+2)}-\frac{\alpha^2 t^{2(\beta+1)}}{(\beta+1)(\beta+2)}+\left.\frac{c\alpha(k-1)t^{(\beta+3)}}{2(\beta+3)}\right]$$
$$+k(a-\mathrm{bp})(t_4-t_2)+\mathrm{ck}\left[-B(a-\mathrm{bp})(t_2(t_4-t_2)-\left(\frac{t_4^2}{2}-\frac{t_2^2}{2}\right)\right.$$
$$\left.\left.+(k-1)(a-\mathrm{bp})\left(\frac{t_4^2}{2}-\frac{t_2^2}{2}-t_4(t_4-t_2)\right)\right]\right\} \tag{14}$$

Sales revenue for the inventory

$$
\begin{aligned}
&= p\Bigg\{ ((a-\mathrm{bp})(t_1+t_2+B(t_4-t_1))) + c\left[\frac{t_1^2}{2} + \frac{\alpha t_1^{\beta+2}}{(\beta+1)(\beta+2)} - \frac{c(k-1)t_1^3}{6}\right.\\
&+ \frac{c(k-1)t_1^3}{3} + \frac{\alpha c(k-1)t_1^{\beta+2}}{(\beta+3)(\beta+2)} + \frac{c^2(k-1)^2 t_1^4}{8} - \frac{\alpha t_1^{\beta+2}}{(\beta+2)} - \frac{\alpha^2 t^{2(\beta+1)}}{(\beta+1)(\beta+2)}\\
&\left.+ \frac{c\alpha(k-1)t^{(\beta+3)}}{2(\beta+3)}\right] + c(1+B)(a-\mathrm{bp})\left[t_2(t_2-t_1) - \left(\frac{t_2^2}{2}-\frac{t_1^2}{2}\right)\right.\\
&+ \frac{c}{2}\left(t_2^2(t_2-t_1) - \left(\frac{t_2^3}{3}-\frac{t_1^3}{3}\right)\right) + \frac{\alpha}{\beta+1}\left(t_2^{(\beta+1)}(t_2-t_1) - \left(\frac{t_2^{\beta+2}}{\beta+2}-\frac{t_1^{\beta+2}}{\beta+2}\right)\right)\\
&- c\left(t_2\left(\frac{t_2^2}{2}-\frac{t_1^2}{2}\right) - \left(\frac{t_2^3}{3}-\frac{t_1^3}{3}\right) + \frac{c}{2}\left(t_2^2\left(\frac{t_2^2}{2}-\frac{t_1^2}{2}\right) - \left(\frac{t_2^4}{4}-\frac{t_1^4}{4}\right)\right)\right.\\
&\left.+ \frac{\alpha}{\beta+1}\left(t_2^{\beta+1}\left(\frac{t_2^2}{2}-\frac{t_1^2}{2}\right) - \left(\frac{t_2^{\beta+3}}{\beta+3}-\frac{t_1^{\beta+3}}{\beta+3}\right)\right)\right)\\
&- \alpha\left(\left(t_2\left(\frac{t_2^{\beta+2}}{\beta+2}-\frac{t_1^{\beta+2}}{\beta+2}\right) - \left(\frac{t_2^{\beta+2}}{\beta+2}-\frac{t_1^{\beta+2}}{\beta+2}\right)\right)\right.\\
&+ \frac{c}{2}\left(t_2^2\left(\frac{t_2^{\beta+2}}{\beta+2}-\frac{t_1^{\beta+2}}{\beta+2}\right) - \left(\frac{t_2^{\beta+3}}{\beta+3}-\frac{t_1^{\beta+3}}{\beta+3}\right)\right)\\
&\left.\left.+ \frac{\alpha}{\beta+1}\left(t_2^{\beta+1}\left(\frac{t_2^{\beta+1}}{\beta+1}-\frac{t_1^{\beta+1}}{\beta+1}\right) - \left(\frac{t_2^{2(\beta+2)}}{2(\beta+1)}-\frac{t_1^{2(\beta+2)}}{\beta+3}\right)\right)\right)\right]\\
&+ \mathrm{Bc}(a-\mathrm{bp})\left[(t_2(t_3-t_2) - \left(\frac{t_3^2}{2}-\frac{t_2^2}{2}\right))\right]\\
&+ \mathrm{Bc}(a-\mathrm{bp})\left[(k-1)\left(\frac{t_4^2}{2}-\frac{t_2^2}{2}-t_4(t_4-t_2)\right)\right]\Bigg\}
\end{aligned}
\tag{15}
$$

Shortage cost for the inventory

$$
\begin{aligned}
&= c_s\left[-\int_{t_2}^{t_3} I_3(t)\mathrm{d}t - \int_{t_3}^{t_4} I_4(t)\mathrm{d}t\right]\\
&= c_s\left\{-(a-\mathrm{bp})B\left[\left(t_2(t_3-t_2) - \left(\frac{t_3^2}{2}-\frac{t_2^2}{2}\right)\right)\right] - (a-\mathrm{bp})(k-1)\left[\left(\frac{t_4^2}{2}-\frac{t_3^2}{2}-t_4(t_4-t_3)\right)\right]\right\}
\end{aligned}
\tag{16}
$$

4 Profit Function

PT = sales revenue (shortage cost–deterioration cost–production cost–lost sale cost–return cost–holding cost).

5 Numerical Example for All Three Cases

We use the following parameters to illustrate the numerical example for the described model.

$a = 24$; $b = 0.2$; $c_s = 0.03$; $c_h = 0.3$; $c_d = 0.05$; $c_l = 0.03$; $c_p = 100$; $B = 0.001$; $A = 0.01$; $p = 110$; $P = 10$; $\text{SV} = 100$; $\alpha = 0.005$; $k = 3$;

To solve the numerical example for all the three deterioration cases, we use the software mathematica 7 and the optimal results are presented as follows:

Case 1: When $\beta = 1$ the value of profit function and other variables is $\text{PT} = 31921.6$; $t_1 = 12.436$; $t_3 = 59.4748$.

Case 2: When $\beta = 2$ the value of profit function and other variables is $\text{PT} = 17560.4$; $t_1 = 5.64778$; $t_3 = 35.5941$.

Case 3: When $\beta = 3$ the value of profit function and other variables is $\text{PT} = 3719.8$; $t_1 = 3.5438$; $t_3 = 9.72077$.

6 Sensitivity Analysis for Different Parameters

To study the behavior of profit function w.r.t different parameter, see below.

Parameters	Change in values	When $\beta = 1$			When $\beta = 2$			When $\beta = 3$		
		TP	t_1	t_3	TP	t_1	t_3	TP	t_1	t_3
c_h	0.3	31912.6	12.436	59.4748	17560.4	5.6477	35.5941	13055.5	3.84339	26.278
	0.4	31931.1	12.439	59.4767	17561.3	5.6380	35.5810	13054.2	3.81433	26.2429
	0.5	31904.6	12.4462	59.4785	17562.1	5.6294	35.5684	13053.0	3.78729	26.2101
	0.4	31950.1	12.4453	59.4797	17563.0	5.6202	35.5563	13052.1	3.81433	26.1796
c	0.01	31912.6	12.436	59.4748	17560.4	5.6477	35.5941	13055.5	3.84339	26.278
	0.02	17586.2	10.4011	35.2377	9196.28	5.1735	20.7431	6751.02	3.6896	15.1687
	0.03	12577.4	9.2922	27.0972	6438.04	4.9260	15.8394	4701.31	3.5464	11.4945
	0.04	9882.97	8.5089	23.0866	5092.38	4.8097	13.5562	3719.8	3.5438	9.7207
B	0.001	31912.6	12.436	59.4748	17560.4	5.6477	35.5941	13055.5	3.84339	26.278
	0.002	31932.8	12.4364	59.4917	17564.0	5.6487	35.5997	13057.5	3.8444	26.2815
	0.003	31944.0	12.4365	59.5074	17576.5	5.6497	35.6054	13059.5	3.8454	26.2849
	0.004	31955.2	12.4368	59.528	17571.1	5.6508	35.6112	13061.4	3.8463	26.2881
c_s	0.03	31912.6	12.436	59.4748	17560.4	5.6477	35.5941	13055.5	3.84339	26.278
	0.04	31904.0	12.4364	59.4762	17560.1	5.6488	35.6073	13056.8	3.84431	26.289
	0.05	31886.4	12.4365	59.4769	17559.8	5.6498	35.6207	13058.2	3.84499	26.301
	0.06	31868.8	12.4367	59.4776	17559.6	5.6512	35.6343	13059.5	3.84621	26.312
θ	0.91	31912.6	12.436	59.4748	17560.4	5.6477	35.5941	13055.5	3.84339	26.278
	0.92	31868.8	12.4366	59.4776	17560.2	5.6477	35.5940	13055.2	3.8431	26.276
	0.93	31864.6	12.4368	59.4777	17560.1	5.6476	35.5938	13055.1	3.8429	26.274
	0.94	31860.4	12.4369	59.4774	17559.4	5.6478	35.5936	13054.4	3.8428	26.272

7 Observations

In this paper, we discussed the three cases of Weibull deterioration where we considered the different values of β such as $\beta = 1$, $\beta = 2$, and $\beta = 3$ in case first, second, and third case, respectively. For all these cases the values of profit function and decision variable are different. Now, we see the effect of change of different parameters on profit function and decision variables.

Case 1: When $\beta = 1$ (constant deterioration)

I. If we increase the value of parameter c_h the value of profit function is fluctuated up and down but the value of t_1 and t_3 increases regularly.
II. If we increase the value of c the value of profit function t_1 and t_3 decreases continuously.
III. If the value of B increases there is a continuous increase in the value of profit, as well as in t_1 and t_3.
IV. When there is increase in the value of c_s the profit function decreases but the value of t_1 and t_3 increases regularly.
V. On increasing the value of θ, profit decreases but the value of t_1 and t_3 increases.

Case 2: When $\beta = 2$ (linear deterioration)

I. When we increase the value of c_h the value of profit function increases but the value of t_1 and t_3 decreases.
II. If we increase the value of c the value of profit as well as t_1 and t_3 decreases vastly.
III. On increasing the value of B, value of profit function t_1 and t_3 increases simultaneously.
IV. If we increase the value of c_s the value of total profit decreases and the value of t_1 and t_3 increases.
V. When we increase the value of θ the value of total profit and t_1 and t_3 decreases.

Case 3: When $\beta = 3$ (quadratic deterioration)

I. After increasing the value of c_h, the values of TP, t_1, and t_3 decrease.
II. On increasing the value of c again, the values of TP, t_1, and t_3 decrease regularly.
III. When we increase the value of B the values of TP, t_1, and t_3 increase.
IV. On increasing the value of c_s the values of TP, t_1, and t_3 increase.
V. When we increase the value of θ the values of TP, t_1, and t_3 decrease.

8 Concavity of Profit Functions for Different Cases

See Figs. 2, 3, and 4.

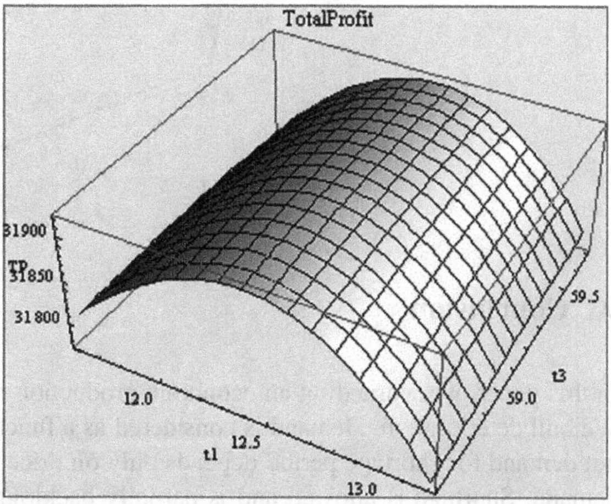

Fig. 2 Concavity of graph function for constant deterioration

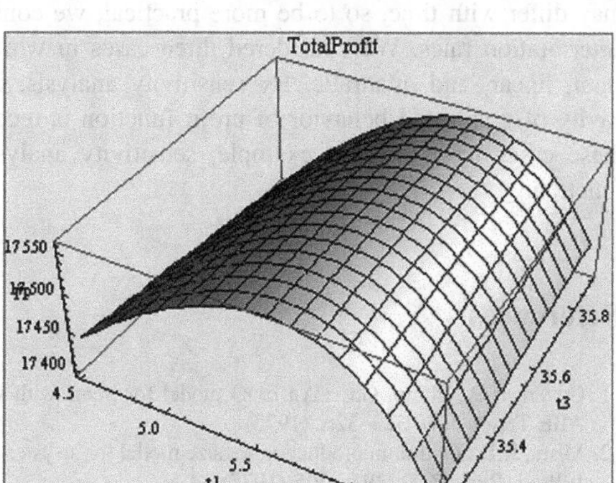

Fig. 3 Concavity of graph function for linear deterioration

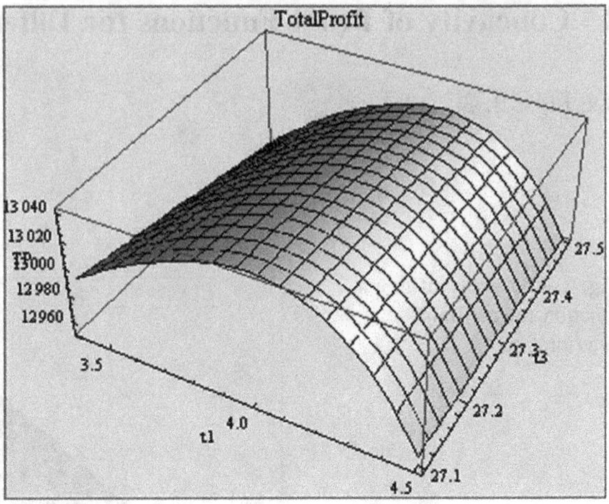

Fig. 4 Concavity of graph function for quadratic deterioration

9 Conclusion

In this paper, we worked on an economic production model having two-parameter Weibull deterioration. Demand is considered as a function of stock, price, and time but demand for shortage period depends only on price. Production also depends on demand. Shortage is allowed and is partially backlogged. To frame this model in real-life situations, we also considered customer return as a factor of quantity sold, price, and inventory level. As we know that in a realistic situation, deterioration may differ with time, so to be more practical, we consider three types of Weibull deterioration rates. We considered three cases in which deterioration rate is constant, linear, and quadratic. By sensitivity analysis, the difference between concavity of graph and behavior of profit function is recognizable. We also compare these cases by numerical example, sensitivity analysis, and concavity of profit function.

References

1. Covert, R.P., Philip, G.C.: An EOQ model for items with weibull distribution deterioration: AIIE Trans. **5**(4), 323–326, (1973)
2. Misra, R.B.: Optimum production lot size model for a system with deteriorating inventory. Int. J. Prod. Res. **13**(5), 495– 505 (1975)
3. Choi, S., Hwang, H.: Optimization of production planning problem with continuously distributed time-lags. Int. J. Syst. Sci. **17**(10), 1499–1508 (1986)
4. Aggarwal, V., Bahari-Hashani, H.: Synchronized production policies for deteriorating items in a declining market. IIE Trans. Oper. Eng. **23**(2), 185–197 (1991)
5. Pakkala, T.P.M., Achary, K.K.: A deterministic inventory model for deteriorating items with two warehouses and finite replenishment rate. Eur. J. Oper. Res. **57**(1), 157–167 (1992)

6. Wu, J.-W., Lin, C., Tan, B., Wen-Chuan.: An EOQ inventory model with time-varying demand and Weibull deterioration with shortages. Int. J. Syst. Sci. **31**(6), 677–683 (2000)
7. Wu, K.S.: An EOQ inventory model for items with Weibull distribution deterioration, ramp type demand rate and partial backlogging. Prod. Planning Control **12**(8), 787–793, (2001)
8. Lee, W.C., Wu, J.-W.: An EOQ model for items with Weibull distributed deterioration, shortages and power demand pattern. Int. J. Inf. Manag. Sci. **13**(2), 19–34 (2002)
9. Banerjee, S., Agrawal, S.: A two-warehouse inventory model for items with three-parameter Weibull distribution deterioration, shortages and linear trend in demand. Int. Trans. Oper. Res. **15**(6), 755–775 (2008)
10. Roy, T., Chaudhuri K.S.: A production-inventory model under stock-dependent demand, Weibull distribution deterioration and shortage. Int. Trans. Oper. Res. **16**(3), 325–346 (2009)
11. Begum, R., Sahoo, R.R., Sahu, S.K., Mishra, M.: An EOQ model for varying items with Weibull distribution deterioration and price-dependent demand. J. Sci. Res. **2**(1), 24–36 (2010)
12. Konstantaras, I., Skouri, K.: A note on a production-inventory model under stock-dependent demand, Weibull distribution deterioration, and shortage. Int. Trans. Oper. Res. **18**(4), 527–531 (2011)
13. Shipi, P., Mahapatra, G.S., Samanta G.P.: EPQ model of ramp type demand with weibull deterioration under inflation and finite horizon in crisp and fuzzy environment. Int. J. Prod. Econ. **156**, 159–166 (2014)
14. Datta, T.K., Paul, K., Pal, A.K.: Demand promotion by up-gradation under stock-dependent demand situation–a model. Int. J. Prod. Econ. **55**, 31–38 (1998)
15. Balkhi, T.Z., Benkherouf, L.: On an inventory model for deteriorating items with stock dependent and time varying demand rates. Comput. Oper. Res. **31**, 223–240 (2004)
16. Teng, J.T., Chang, C.T.: Economic production quantity models for deteriorating items with price and stock dependent demand. Comput. Oper. Res. **32**, 297–308 (2005)
17. Wu, K.S., Ouyang, L.Y., Yang, C. T.: An optimal replenishment policy for non-instantaneous deteriorating items with stock-dependent demand and partial backlogging. Int. J. Prod. Econ. **101**, 369–384 (2006)
18. Singh, S.R., Kumari, R., Kumar, N.: Replenishment policy for non-instantaneous deteriorating items with stock-dependent demand and partial back logging with two-storage facility under inflation. Int. J. Oper. Res. Optim. **1**(1), 161–179 (2010)
19. Singh, C., Singh, S.R.: An EPQ model with stock dependent demand under imprecise and inflationary environment using genetic algorithm. Int. J. Eng. Res. Technol. **1**(4), (2012)
20. Biswajit, S., Sumon, S.: An improved inventory model with partial backlogging time varying deterioration and stock dependent demand. Econ. Model. **30**, 924–932 (2013)
21. Yang, C.T.: Inventory model with both stock dependent demand and stock dependent holding cost rate. Int. J. Prod. Econ. **155**, 214–221 (2014)
22. Hess, J.D., Mayhew, G.E.: Modeling merchandise returns in direct marketing. J. Interact. Mark. **4**(2), 347–385 (1997)
23. Pasternack, B.A.: Optimal pricing and returns policies for perishable commodities. Mark. Sci. **4**(2), 166–176 (1985)
24. Anderson, E.T., Hansen, K., Simister, D., Wang, L.K.: How are demand and return related? Theory and empirical evidence: Working paper, Kellogg school of management, Northwestern University (2006)
25. Ahmed, M.A., Saadany, E.L., Jaber Mohomad, Y.: A production\remanufacturing inventory model with price and quality dependent return rate. Comput. Ind. Eng. **58**(3), 352–362 (2010)
26. Mesak Hani, I., Abdullahal, B., Babin Barry, J., Biron Laura, M., Antony, J.: Optimum advertising policy over time for subscriber services cost learning and customer's disadoption. Eur. J. Oper. Res. **211**(3), 642–649 (2011)
27. Wu, J., Ya-Lan, C.: Lot sizing policies for deteriorating items with expiration dates and partial trade credit risk customers. Int. J. Prod. Econ. **155**, 292–301 (2014)

Chemo-inspired Genetic Algorithm and Application to Model Order Reduction Problem

Rajashree Mishra and Kedar Nath Das

Abstract During the past three decades, evolutionary computing techniques have grown manifold in tackling all sorts of optimization problems. Genetic algorithm (GA) is one of the most popular EAs because it is easy to implement and is conducive for noisy environment. Similarly, amongst several swarm intelligence techniques, bacterial foraging optimization (BFO) is the recent popular algorithm being used for many practical applications. Depending on the complexity of the problem concerned, there is need for hybridized techniques which help in balancing exploration and exploitation capability over the search space. Many hybridized techniques have been developed recently to tackle such problems. This paper proposes a hybridization of GA and BFO to solve a real-life unconstrained electrical engineering problem. This unconstrained optimization problem is a model order reduction (MOR) problem of linear time invariant continuous single input and single output (SISO) system.

Keywords Genetic algorithm · Bacterial foraging optimization · Hybridization · Integral squared error · Impulse response energy · Model order reduction problem

1 Introduction

The present-day technology, societal, and environmental processes bring along a large number of problems that are complex due to the involvement of larger dimension and nonlinear functions. Model order reduction (MOR) is such a technique by which large-scale complex systems can be reduced to simpler models. The basic idea behind these techniques is to produce the input–output relation in

Rajashree Mishra (✉)
KIIT University, Bhubaneswar, Odisha, India
e-mail: rajashreemishra011@gmail.com

K.N. Das
NIT Silchar, Silchar, Assam, India
e-mail: kedar.iitr@gmail.com

M. Pant et al. (eds.), *Proceedings of Fifth International Conference on Soft Computing for Problem Solving*, Advances in Intelligent Systems and Computing 436, DOI 10.1007/978-981-10-0448-3_3

acceptable time, minimizing the error between the reduced model and the original model. Nowadays, real-life problems and large-scale complex systems can be time variant, time invariant, linear, nonlinear, parametric, or stochastic which are posing huge challenges for MOR problem. The system may be single input and single output (SISO) or multiple inputs and multiple outputs (MIMO). The goal of the model order reduction (MOR) problem includes the following points: (i) the approximation of the output of large-scale system by a reduced model should be evaluated in significantly less time, (ii) the computable error bound for the reduced model should exist, and (iii) the preservation of physical properties of the original system, such as stability, minimum phase, and passivity must be followed during the MOR process.

Many researchers have done excellent work in this regard. Deepa and Sugumaran proposed modified particle swarm optimization for MOR problem [1]. Minimization of error between the higher order and lower order system has been considered in the above algorithm. The algorithm is also compared with integral of time of squared error (ITSE) and integral of time of absolute error (ITAE). Bansal et al. [2] applied artificial bee colony (ABC) algorithm to solve SISO system. The results have been compared with Pade approximation, Routh approximation, and other conventional methods. They have considered the objective function for MOR problem considering minimization of both integral square error (ISE) and IRE. Bansal and Sharma [3] also applied differential evolution (DE) and its variant cognitive learning in differential evolution (CLDE) algorithm for obtaining the reduced models. In CLDE algorithm, cognitive learning factor (CLF) is introduced in the mutation operator in differential evolution (DE). Further, Bansal et al. [4] proposed fitness-based differential evolution (FBDE) for solving MOR problem. Kumar and Tiwari [5] used factor division method with clustering technique for finding stable, reduced, and ordered models of SISO system. Mondal and Tripathy [6] proposed a mixed method for SISO system where the numerator is reduced by Pade approximation and the denominator polynomial is reduced preserving the basic characteristics of higher order original system. This method tries to minimize the ISE. Desai and Prasad [7] considered Big Bang Big Crunch (BBBC) optimization procedure incorporating the stability equation (SE). The numerator is obtained by Big Bang Big Crunch (BBBC) algorithm and the denominator is obtained by stability equation (SE). Panda et al. proposed two evolutionary methods GA and PSO for reducing higher order large-scale linear system to lower order system [8]. Alsmadi et al. proposed artificial neural network for discrete model order reduction with substructure preservation [9].

This paper is organized as follows: Brief review of components considered in the algorithm and the pseudo code are discussed in Sect. 2. The MOR problem definition is described in Sect. 3. The experimental setup of the numerical example and the result discussion is given in Sect. 4. The conclusion of the paper is drawn in Sect. 5.

2 Brief on GA and BFO

This paper deals with the hybridization of some operators of GA and BFO. Although both GA and BFO are well-known optimizers, it is still essential to present a brief outline of them. GA is a population-based method consisting of four major steps, namely Selection, Crossover, Mutation, and Elitism. Based on Darwin's principle of survival of the fittest, GA selects the better individuals from the population by letting the worse die off. Similarly, BFO also works with a population of individuals through its major operators like Chemotaxis, Reproduction, and Elimination-Dispersal steps. The detailed mechanisms of both GA and BFO are presented in our latest paper [10, 11]. Collectively, the objectives of both the mechanisms are the same so as to explore the search space and come up with a near-optimal solution. Each of them has some pitfalls in their inherent mechanism. Researchers [12, 13] have tried to overcome them by the process of hybridization to solve unconstrained optimization problems. In this paper, hybridization of GA and BFO is done in a novel fashion that is implemented for being capable of handling unconstrained real-life optimization problems. The motivation and proposition of the method is presented in the next section.

2.1 Proposed Chemo-inspired Genetic Algorithm for Unconstrained Optimization

2.1.1 Motivation

This method of hybridization makes a mechanism more effective. Recently, Kim et al. tried to hybridize GA and BFO [12]. They have hybridized GA with BFO by taking entire steps of GA with entire steps of BFO. However, visualizing both GA and BFO at a time, the following observations are noted.

- Selection, Crossover, Mutation and Elitism are the four major operators in GA, whereas Chemotaxis, reproduction, and elimination-dispersal are the three principal mechanisms in BFO. The selection operation in GA and reproduction step in BFO are equivalent because both are retaining multiple copies of the strings.
- *Elitism* first combines the populations before and after the GA cycle and then eliminates the worse half in order to select the better half for the next generation. Elitism in GA and elimination-dispersal in BFO are equivalent because the elimination of the worst strings and insertion of new strings takes place in both the mechanisms.

From the above observations it is realized that while hybridizing to BFO with GA, probably the *reproduction* and *elimination-dispersal steps* of BFO become inefficient as it is just like a repetition of few of the existing operators in GA.

Hence, only the *Chemotaxis* step of BFO that plays a major role, has been picked up from BFO to hybridize in GA cycle as an additional operator.

2.1.2 Proposal

In our recent earlier work [10], the efficiency of the hybridization of Chemotaxis step with GA, namely chemo-inspired GA (CGA) over the GA-BF (a hybridization of GA and BFO) is well verified over the unconstrained problems. This paper is an application of the designed algorithm for MOR problem. The detailed mechanism of CGA is as follows: Selection, Crossover, Mutation, Elitism, and Chemotaxis are the five major components of the designed CGA. To improve the solution quality further, three productive properties namely Adaptive step size [10, 12], Squeezed search space [10, 12], Fitness function criterion [10, 12] and Modified Chemotactic step size [10] are incorporated in the algorithm which was followed in the improved BFO [12]. The detailed explanation of the proposed algorithm CGA can be referred from our earlier paper [10].

2.1.3 Solution Methodology of the Proposed CGA for the MOR Problem

Step 1: Initialize the CGA parameters. Evaluate k_i's and λ_i's of the unit step response of the unit step response of $G_1(s)$ as Eq. (7)
Step 2: Generate randomly $k_1^{'}, k_2^{'}, \mu_1, \mu_2$ as Eq. (10)
Step 3: Evaluate the Error Index as Eq. (13)
Step 4: Apply Selection operator
Step 5: Apply Cross over operator
Step 6: Apply Mutation operator
Step 7: Apply Elitism operator
Step 8: Evaluate IRE as Eq. (4)
Step 9: Select $R_1(s)$, where IRE is minimum
Step 10: *Repeat Step 2 until the maximum iteration is reached*

3 Model Order Reduction Problem

This section aims at verifying the efficiency of CGA over some recent algorithms like a variant of Cognitive Learning in Differential Evolution (CLDE), named as Linearly Increasing Cognitive Learning in Differential Evolution (LICLDE) [3] and Fitness-Based position update process in Differential Evolution (FBDE) [4]. One case of model order reduction (MOR) problem has been picked from the literature [14] and solved by CGA.

3.1 Problem Definition of Model Order Reduction Problem

A nth order linear time invariant dynamic single-input and single-output system given by

$$G_1(s) = \frac{N_1(s)}{D_1(s)} = \frac{\sum_{i=0}^{i=n-1} a_i s^i}{\sum_{i=0}^{n} b_i s^i}\text{, where} \tag{1}$$

a_i and b_i are constants. Here rth order reduced model in the transfer function form $R_1(s)$, where $r < n$ is constructed by Eq. (2) keeping the important characteristics of the original model. The minimization of Integral square error (ISE) as well as IRE of $G_1(s)$ and $R_1(s)$ are considered.

$$R_1(s) = \frac{N_r(s)}{D_r(s)} = \frac{\sum_{i=0}^{r-1} a'_i s^i}{\sum_{i=0}^{r} b'_i s^i}\text{, where} \tag{2}$$

The unknown constants in the reduced model are a'_i and b'_i respectively. Let $y(t)$ be the unit step response of the original system and $y_r(t)$ be the unit step response of the reduced system. The Integral Square Error index J given by Eq. (3) is as follows.

$$J = \int_0^\infty [y(t) - y_r(t)]^2 \mathrm{d}t\text{, where} \tag{3}$$

The error index is the function of unknown coefficients of reduced order model so that the error index is minimized.

The IRE for the original and the various reduced models is given by Eq. (4)

$$\mathrm{IRE} = \int_0^\infty g(t)^2 \mathrm{d}t\text{, where} \quad g(t) \tag{4}$$

is the Impulse response of the system. In this paper, the objective is to minimize the objective function based on both ISE and IRE.

3.2 Proposed Method to Solve MOR Problem

Mukherjee and Mishra [15] in their paper designed a reduced order model retaining the dominant eigenvalue of the original model. The designed reduced model contains the distinct and real eigenvalues. $R_1(s)$ is considered minimizing ISE discussed above.

Motivated by the above, in this paper, CGA (discussed in details [10]) has been used to solve MOR problem of linear continuous time invariant Single Input and Single Output (SISO) system where the original system is having real and distinct eigenvalues. Also in this proposed method, the steady state part of the unit step responses of the original system and reduced order models are closely matched. The minimization of the Integral Square Error (IRE) as well as IRE of unit step response of the original high order transfer function and the reduced low order transfer function for transient parts are considered. In the proposed method, the reduced model is designed considering the real and distinct eigenvalues.

The mth order original high order system transfer function is given by Eq. (5)

$$G_1(s) = \frac{b_n s^n + b_{n-1}s^{n-1} + \cdots + b_1 s + b_0}{s^m + a_{m-1}s^{m-1} + \cdots + a_1 s + a_0} \tag{5}$$

where, $m > n$.

$$G_1(s) = \frac{b_n s^n + b_{n-1}s^{n-1} + \cdots + b_1 s + b_0}{(s+\lambda_1)(s+\lambda_2)(s+\lambda_3)\cdots(s+\lambda_m)} \tag{6}$$

where-$\lambda_1 < -\lambda_2 < -\lambda_3 < \cdots < -\lambda_{m-1} < -\lambda_m$ are distinct real eigenvalues of the system.

The unit step responses of Eq. (6) is as follows.

$$\begin{aligned} T(s) &= \frac{G_1(s)}{s} = \frac{b_n s^n + b_{n-1}s^{n-1} + \cdots + b_1 s + b_0}{s(s+\lambda_1)(s+\lambda_2)(s+\lambda_3)\cdots(s+\lambda_m)} \\ &= \frac{k_0}{s} + \frac{k_1}{s+\lambda_1} + \frac{k_2}{s+\lambda_2} + \cdots + \frac{k_{m-1}}{s+\lambda_{m-1}} + \frac{k_m}{s+\lambda_m} \end{aligned} \tag{7}$$

where k_i^s are real constants. Taking inverse Laplace transformation,

$$y(t) = k_0 + k_1 e^{-\lambda_1 t} + k_2 e^{-\lambda_2 t} + \cdots + k_m e^{-\lambda_m t} \tag{8}$$

The steady state response is given by the term k_0, whereas k_1, k_2, k_3, ..., k_m are considered for transient response of the system.

Let the proposed reduced order system constructed be of the 2nd order, where

$$R_1(s) = \frac{a_0 s + a_1}{a_2 s^2 + a_3 s + a_4} = \frac{a_0 s + a_1}{(s+\mu_1)(s+\mu_2)} \tag{9}$$

where, μ_1 and μ_2 are distinct and real eigenvalues and $-\mu_1 < -\mu_2$.

The unit step response of (9) can be determined as

$$T_1(s) = \frac{R_1(s)}{s} = \frac{k'_0}{s} + \frac{k'_1}{s+\mu_1} + \frac{k'_2}{s+\mu_2} \tag{10}$$

where, $k'_0, k'_1, k'_2, \mu_1, \mu_2$ are real constants. Let the inverse Laplace transformation be

$$y_r(t) = k'_0 + k'_1 e^{-\mu_1 t} + k'_2 e^{-\mu_2 t} \tag{11}$$

The condition for matching the steady-state responses of $G_1(s)$ and $R_1(s)$ is defined as follows.

$$k_0 = k'_0 \tag{12}$$

Thus, Integral Square Error (ISE) of the transient responses of the system of (6) and (9) is given by

$$\begin{aligned} &= \int_0^\infty [y(t) - y_r(t)]^2 \mathrm{d}t \\ &= \int_0^\infty \left\{ \sum_{i=1}^{m} k_i e^{-\lambda_i t} - \sum_{j=1}^{2} k'_j e^{-\mu_j t} \right\}^2 \mathrm{d}t, \text{since } k_0 = k'_0 \end{aligned} \tag{13}$$

where k_i; s and λ_i; s are known and k'_1, k'_2, μ_1, μ_2 are all unknown constants, which are randomly generated in CGA and the value of J are calculated. The reduced model $R_1(s)$ is obtained for 30 independent runs so that the integral square error J is minimized. In the next step the reduced model $R_1(s)$ is taken, where IRE given by Eq. (4) is also minimized in Matlab7.0.

4 Experimental Setup and Numerical Example of MOR Problem

Example The MOR problem proposed by **Lucas** [14] is defined as follows.

$$G_1(s) = \frac{8169.13s^3 + 50664.97s^2 + 9984.32s + 500}{100s^4 + 10520s^3 + 52101s^2 + 10105s + 500} \tag{14}$$

In this paper the reduced model is constructed minimizing both the IRE and the ISE. The proposed CGA program code is designed in C++ and the experiment is carried out on a P-IV, 2.8 GHz machine with 512 MB RAM under WINXP platform. Parameter settings of CGA to solve MOR problem is defined as follows. Probability of Crossover ($P_c = 0.9$). Probability of Mutation ($P_m = 0.01$). Total Chemo tactic steps taken = 40. Number of swim steps taken = 4. Minimum step size taken ($C_{\min} = 0.008$). Maximum step size taken ($C_{\max} = 0.1$).

Bansal et al. [4] solved the above problem recently by suing a new Fitness Based position update process in Differential Evolution (FBDE) and also Bansal and Sharma [3] adopted a new variant of Cognitive learning in Differential evolution (CLDE) named as Linearly increasing cognitive learning factor in Differential evolution (LICLDE) to solve the problem. They claimed that LICLDE and FBDE outperform many other order reduction methods. This Paper reconsidered all these methods along with LICLDE and FBDE to compare with the proposed CGA.

At first hand a set of solutions have been collected for 30 runs where ISE is minimum given by Eq. (13). Then the emphasis is given in minimizing the IRE of the system given by Eq. (4). The solution where ISE and IRE are both minimum is considered as the global optimum. The reported solutions are plotted in form of impulse responses and step responses of the system in Matlab 7.0.

4.1 Reduced Model

Reduced Model for given problem is found out to be $R_1(s)$, where

$$R_1(s) = \frac{78.03187648s + 201.4534442}{s^2 + 92.318454s + 201.4534442} \tag{15}$$

The result comparison of ISE and IRE is shown in Table 1. The step and impulse responses are plotted in Figs. 1 and 2 respectively.

Table 1 Comparison of the methods for MOR problem [14]

Order reduction	Reduced models $R_1(s)$	ISE	IRE
Original	$G_1(s)$	——	**34.069**
CGA	$\frac{78.03187648s+201.4534442}{s^2+92.318454s+201.4534442}$	**0.001538260**	**34.06923470**
FBDE	$\frac{85.33529245s+462.3004006}{s^2+113.6582937s+462.3004006}$	0.0017826566	**34.06884**
LICLDE	$\frac{101.3218182s+867.893179}{s^2+169.4059231s+867.893179}$	0.0036228741	**34.069918**
DE	$\frac{220.8190s+35011.744}{s^2+1229.450s+35011.744}$	0.004437568	**34.069218**
Singh	$\frac{93.7562s+1}{s^2+100.10s+10}$	0.008964	43.957
Pade approximation	$\frac{23.18s+2.36}{s^2+23.75s+2.36}$	0.0046005	11.362
Routh approximation	$\frac{0.1936s+0.009694}{s^2+0.1959s+0.009694}$	2.3808	0.12041
Gutman et al.	$\frac{0.19163s+0.00959}{s^2+0.19395s+0.00959}$	2.4056	0.11939
Chen et al.	$\frac{0.38201s+0.05758}{s^2+0.58185s+0.05758}$	1.2934	0.17488
Marshall	$\frac{83.3333s+499.9998}{s^2+105s+500}$	0.00193	35.450

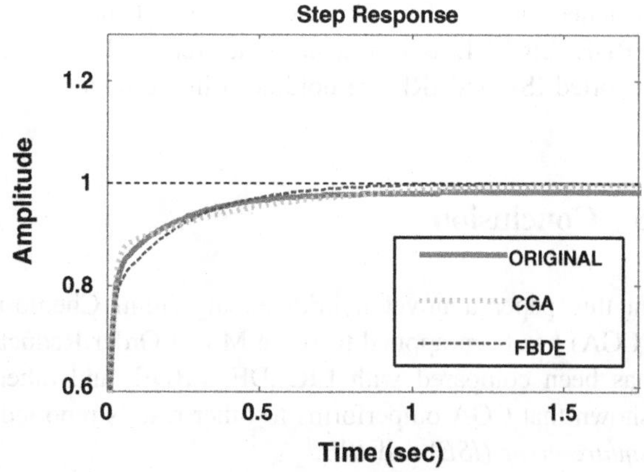

Fig. 1 Comparison of step responses for MOR problem

4.2 Result Analysis

The reduced ordered model by CGA and other reported algorithms for the corresponding original model are represented in Table 1 for the given problem. In the corresponding table, the errors (ISE and IRE) obtained by CGA are also compared with that of all the earlier reported results in [3, 4] last two columns. The unit step responses of CGA have been compared only with the original system and the best reported result so far which is given in [4] and are shown in Fig. 1. The corresponding impulse response is plotted in Fig. 2. It can also be visualized that for the given MOR problem in (Fig. 1), the step response is approximating the original polynomial more closely in comparison to FBDE [4]. Also, IRE of the reduced model obtained by CGA is the most close to the original model (Fig. 2). It can be viewed that the transient step responses as well as the steady state responses are

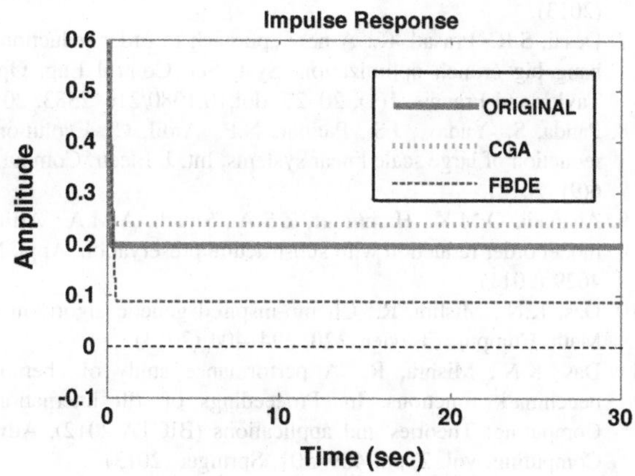

Fig. 2 Comparison of impulse responses for MOR problem

matched closely. Thus this example establishes the superiority of CGA over the FBDE, LICLDE as well as all other reported results for MOR problem. The best reported ISE and IRE are boldfaced in the table.

5 Conclusion

In this paper a novel hybridized algorithm Chemo-inspired Genetic Algorithm (CGA) has been applied to solve Model Order Reduction problem. The algorithm has been compared with LICLDE, FBDE, and other stochastic algorithm. It is shown that CGA outperforms to other results reported so far, in terms of *Integral square error (ISE)* and *IRE*.

References

1. Deepa, S.N., Sugumaran, G.: MPSO based model order formulation technique for SISO continuous system. World Acad. Sci. Eng. Technol. **51**, 838–843 (2011)
2. Bansal, J.C., Sharma, H., Arya, K.V.: Model order reduction of single input and single output systems using Artificial Bee Colony optimization Algorithm. In: Nature Inspired Co-operative Strategies for Optimization (NICSO 2011), Studies in Computational intelligence, vol. 387, pp. 85–100, Springer link (2011)
3. Bansal, J.C., Sharma, H.: Cognitive learning in differential evolution and its application to model order reduction problem for single-input and single–output systems, Meme tic computing, doi:10.1007/s12293–012-0089-8, Springer (2012)
4. Bansal, J.C., Sharma H., Arya, V.K.: Fitness based Differential Evolution, Memetic Computing, doi:10.1007/s12293-012-0096-9, 2012
5. Kumar, V., Tiwari, J.P.: Order reducing of linear system using Clustering method factor division algorithm. Foundation of Computer science FCS, New York, USA, vol. 3, no. 5, IJAIS, ISSN: 2249–0868, www.ijais.org, 2012
6. Mondal, S., Tripathy, P.: Model order reduction by mixed mathematical Methods, IJCER **3**(5) (2013)
7. Desai, S.R., Prasad, R.: A new approach to order reduction using stability equation and big bang big crunch optimization. Syst. Sci. Control Eng. Open Access J., Published online, Taylor and Francis, **1**(1), 20–27, doi:10.1080/21642583. 2013. 804463, (2014)
8. Panda, S., Yadav., J.S., Patidar, N.P., Ardil, C.: Evolutionary techniques for model order reduction of large scale linear systems. Int. J. Electr. Comput. Energ. Electron. Commun. Eng. **6**(9) (2012)
9. Alsmadi, O.M.K., Hammour, Z.S.A., Smadi, A.M.A.: Artificial neural network for discrete model order reduction with substructure preservation. Appl. Math. Model., Elsevier **35**, 4620–4629 (2011)
10. Das, K.N., Mishra, R.: Chemo-inspired genetic algorithm for function optimization. Appl. Math. Comput. Elsevier, **220**, 394–404 (2013)
11. Das, K.N., Mishra, R.: A performance study of chemo inspired genetic algorithm on benchmark functions. In: Proceedings of 7th international conference on Bio-inspired Computing: Theories and applications (BICTA-2012), Advances in Intelligent System and Computing, vol. 2, pp. 489–501, Springer (2013)

12. Kim, D.H., Abraham, A., Cho, J.H.: A hybrid genetic algorithm and bacterial foraging approach for global optimization. Information Sciences, **177**(18), 3918–3937 (2007)
13. Chen, Y., Lin, W.: An improved bacterial foraging optimization. In: Proceedings of the IEEE International Conference on Robotics and Biomimetics, Guilin, China, 19–23 Dec 2009
14. Lucas, T.N.: Continued- fraction expansion about two or more points: a flexible approach to linear system reduction. J. Franklin Inst. **323**(1), 49–60 (1986)
15. Mukherjee, S., Mishra, R.N.: Order reduction of liner systems using an error minimization technique. J. Franklin Inst., Pergamon Journals Ltd, **323**(1), 23–32 (1987)

Grammatical Evolution Using Fireworks Algorithm

Tapas Si

Abstract Grammatical Evolution generates computer programs automatically in any arbitrary language using Backus-Naur Form of Context-free Grammar in automatic programming. Variable-length Genetic Algorithm is used as a learning algorithm in Grammatical Evolution. Fireworks algorithm is a recently developed new Swarm Intelligent algorithm used for function optimization. This paper proposes Grammatical Fireworks algorithm which uses Fireworks algorithm as a learning algorithm in place of variable-length Genetic Algorithm in Grammatical Evolution to evolve computer programs automatically. Grammatical Fireworks algorithm is applied on three well-known benchmark problems such as Santa Fe ant trail, symbolic regression and 3-input multiplexer problems. A comparative study is made with Grammatical Evolution, Grammatical Swarm, Grammatical Artificial Bee Colony, and Grammatical Differential Evolution. The experimental results demonstrate that the proposed Grammatical Fireworks algorithm can be applied in automatic computer program generation.

Keywords Automatic programming · Grammatical evolution · Fireworks algorithm · Machine learning · Swarm intelligence

1 Introduction

Automatic Programming (AP) [1] is an important area of Machine Learning (ML) for automatic computer programs generation. Automatic Programming enables machines to evolve programs by themselves. Genetic Programming (GP) [1] is a popular algorithm in automatic programming. GP is a variant of

Tapas Si (✉)
Department of Computer Science and Engineering, Bankura Unnayani
Institute of Engineering, Bankura 722146, West Bengal, India
e-mail: c2.tapas@gmail.com

M. Pant et al. (eds.), *Proceedings of Fifth International Conference on Soft Computing for Problem Solving*, Advances in Intelligent Systems and Computing 436, DOI 10.1007/978-981-10-0448-3_4

Genetic Algorithm (GA) in which each individual with linear or nonlinear genomes are used to represent the computer programs. Grammatical Evolution (GE) [2, 3] is a variant of grammar-based GP in which linear genomes having set of integer codons known as *genotypes* in the range [0, 225] are used. The *phenotypes* are the evolved computer programs in any arbitrary language. Computer programs are evolved by a *genotype-to-phenotype* mapping process using Backus-Naur Form (BNF) of Context-free Grammar (CFG). In GE, variable-length binary strings are used to represent an individual. O'Sullivan et al. [4] used Simulated Annealing (SA), Hill Climbing (HC), and Random Search (RS) strategies as search engine in GE and the performances of these algorithms are investigated. O'Neill et al. [5] proposed Grammatical Swarm (GS) algorithm in which Particle Swarm Optimization (PSO) is used as a search engine for *genotype-to-phenotype* mapping process in GE instead of GA. O'Neill et al. [6] proposed another version of GS in which variable-length PSO was used. In variable-length PSO, particles with variable dimensions are used. O'Neill et al. [7] presented Grammatical Differential Evolution (GDE) in which different versions of Differential Evolution (DE) [8] algorithm were used as search engines in GE. As a proof of concept, Si et al. [9] have proposed Grammatical Bee Colony (GBC) algorithm in which Artificial Bee Colony (ABC) [10] algorithm is used to generate computer programs through grammatical evolution. In GS, GDE, and GBC algorithms, the individuals are represented as the set of integer codons in the range [0, 225] and Backus-Naur Form of CFG is used in *genotype-to-phenotype* mapping process. In this paper, Grammatical Fireworks algorithm (GFWA) is presented for generation of computer programs. The GFWA algorithm is applied on three well-known benchmark problems such as Santa Fe ant trail, symbolic regression, and 3-input multiplexer problems. A comparative study has been made with GE, GS, GBC, and GDE/rand/1/bin [7].

The remaining of this paper is organized as follows: Grammatical Evolution and its different learning algorithms are discussed in Sect. 2. Grammatical Fireworks algorithm is presented in Sect. 3. Experimental setup is given in Sect. 4. Results and discussion are given in Sect. 5. Finally a conclusion with future enhancement is given in Sect. 6.

2 Grammatical Evolution and Its Learning Algorithm

Genetic Programming (GP) was invented by Koza [1] for evolving the computer programs automatically. GP is a variant of Genetic Algorithm (GA) in which tree structure-based genomes are used to represent computer programs. GE is a variant of grammar-based genetic programming [11] in which Context-free Grammar (CFG) is combined with GP to restrict the search space according to problems of interest and CFG is used to construct the derivation trees. In GE, genomes are linear arrays of variable length in the form of binary strings. Bankus-Naur Form (BNF) of CFG is used in mapping from genotype to phenotype for generating programs.

Simulated Annealing, Hill Climbing, Random Search, Particle Swarm Optimiser, Differential Evolution, and Artificial Bee Colony algorithms are used as search engines or learning algorithms in Grammatical Evolution till date.

O'Sullivan et al. [4] investigated Simulated Annealing, Hill Climbing, Random Search *meta–heuristics* as search engines in GE. These algorithms can not compete with GA for Santa Fe ant trail, symbolic regression, and symbolic integration problems.

In Grammatical Swarm (GS), particle swarm optimisation (PSO) [12] algorithm is used as learning or searching algorithm instead of GA in GE for generating computer programs through *genotype-to-phenotype mapping* process. Each particle in PSO represents a set of integer codons from which computer programs are generated using grammatical rules. Each particle's position is a real vector rounded up to its nearest integers to form the genotype. GS is compared with traditional GE for Santa Fe ant, multiplexer, symbolic regression, and mastermind problems. GE outperforms GS in Santa Fe ant and symbolic regression. GS outperforms GE in multiplexer problem, whereas there is a tie between GS and GE on the Mastermind problem. Grammatical Swarm is a new form of automatic programming in which social learning or swarm intelligent algorithm is used and Grammatical Swarm is termed as *Social Programming* or *Swarm Programming* [5]. In this base line GS, fixed-length particles are used in PSO. O'Neill et al. [6] proposed another version of GS in which variable-length PSO was used instead of fixed-length PSO. In variable-length PSO, dimensions of positions of particles are of variable-length though this version is outperformed by base line GS [5] algorithm.

O'Neill et al. [7] proposed Grammatical Differential Evolution (GDE). In GDE, Differential Evolution (DE) [8] algorithm is used as a search engine in GE. Each individual in DE is represented as genotype. The real valued vectors of fixed length in DE are rounded up to form integer valued codons used to map from genotype to phenotype using BNF of production rules in CFG. Different versions of GDE such as GDE/rand/1/bin, GDE/best/1/exp, GDE/rand-to-best/1/bin, and GDE/rand-to-best/1/exp are examined in solving Santa Fe ant, multiplexer, symbolic regression and mastermind problems, and compared with GS and GE. GDE outperforms GE only for multiplexer problem, whereas GE outperforms GDE for other three problems. GDE performs better than GS for multiplexer and symbolic regression problems.

Si et al. [9] have proposed Grammatical Bee Colony (GBC) in which Artificial Bee Colony (ABC) algorithm is used as a search engine to generate computer programs through genotype-to-phenotype mapping using BNF of Context-free Grammar. In ABC algorithm, a solution is represented as food source's location. Obviously, a genotype in GBC is represented by a food source's location. The programs (i.e., phenotypes) are generated from the food source's location using BNF of Context-free Grammar. GBC is not compared with GE and GBC cannot compete with GS and GDE in [9].

3 Grammatical Fireworks Algorithm

3.1 Fireworks Algorithm

In Fireworks Algorithm (FWA) [13], the solutions of the problem are represented by a set of fireworks in the search space. The fireworks are let-off in the search space and a number of explosion sparks with different amplitudes is generated by each firework. A $\mathcal{N} \times \mathcal{D}$ matrix $\mathcal{X}$ represents the locations of $\mathcal{N}$ fireworks with $\mathcal{D}$ dimension as following:

$$\mathcal{X} = \begin{bmatrix} \mathcal{X}_1 \\ \mathcal{X}_2 \\ \vdots \\ \mathcal{X}_{\mathcal{N}} \end{bmatrix}_{\mathcal{N}\times 1} = \begin{bmatrix} x_{11} & x_{12} & \cdots & x_{1\mathcal{D}} \\ x_{21} & x_{22} & \cdots & x_{2\mathcal{D}} \\ \vdots & \vdots & \ddots & \\ x_{\mathcal{N}1} & x_{\mathcal{N}2} & \cdots & x_{\mathcal{N}\mathcal{D}} \end{bmatrix}_{\mathcal{N}\times\mathcal{D}} \tag{1}$$

The basic Fireworks algorithm has three main steps and they are as following: (i) explosion sparks generation, (ii) Gaussian sparks generation, and (iii) selection of best locations from all fireworks, explosion, and Gaussian sparks.

Explosion Sparks Generation The number of explosion sparks of ith firework is calculated as following:

$$\mathcal{S}_i = \mathcal{M} \times \frac{f_{\max} - f(\mathcal{X}_i) + \epsilon}{\sum_{i=1}^{\mathcal{N}} (f_{\max} - f(\mathcal{X}_i)) + \epsilon}, \tag{2}$$

where $f_{\max} = \max(f(\mathcal{X}_i)), i = 1, 2, \ldots, \mathcal{N}$, $\mathcal{M}$ is the maximum number of explosion sparks and ε is a very small number used to avoid 'division by zero' error.

The amplitude of the explosion sparks of ith firework is calculated as following:

$$\mathcal{A}_i = \mathcal{A} \times \frac{f(\mathcal{X}_i) - f_{\min} + \epsilon}{\sum_{i=1}^{\mathcal{N}} (f(\mathcal{X}_i) - f_{\min}) + \epsilon} \tag{3}$$

where $f_{\min} = \min(f(\mathcal{X}_i)), i = 1, 2, \ldots, \mathcal{N}$, $\mathcal{A} = (\mathcal{X}_{\max} - \mathcal{X}_{\min})$, is the maximum amplitude and $[\mathcal{X}_{\min}, \mathcal{X}_{\max}]$ is the search space range.

The maximum number of explosion sparks having less amplitude is generated by the best fireworks, Whereas the minimum number of explosion sparks having higher amplitude is generated by the worst fireworks. A good balance between exploration and exploitation is maintained by the fireworks during the search process. The exploitation is performed by the good fireworks using higher number of explosion sparks with less amplitude. The exploration is performed by the bad fireworks using less number of explosion sparks with higher amplitude.

Gaussian Sparks Generation Another type of sparks called as Gaussian Sparks are generated from the original fireworks to keep diversity in the search space. The original fireworks are multiplied with Gaussian distributed random number generated with mean 1 and standard deviation 1 to generate this type of sparks.

Selection of Locations In FWA, the current best location $\mathcal{X}^*$ having optimal solution $f(\mathcal{X}^*)$ in the current iteration t is kept for the next iteration $(t + 1)$. The remaining $(\mathcal{N} - 1)$ locations from $(\mathcal{K} - 1)$ number of locations are selected using Roulette Wheel Selection operator where $\mathcal{K}$ is the set of all fireworks and sparks in current iteration. The selection probability of a location $\mathcal{X}_i$ is defined as follows:

$$\mathcal{P}(\mathcal{X}_i) = \frac{\mathcal{R}(\mathcal{X}_i)}{\sum_{j\in\mathcal{K}} \mathcal{R}(\mathcal{X}_j)}, \tag{4}$$

where

$$\mathcal{R}(\mathcal{X}_i) = \sum_{j\in\mathcal{K}} \text{dist}(\mathcal{X}_i, \mathcal{X}_j) = \sum_{j\in\mathcal{K}} \| \mathcal{X}_i - \mathcal{X}_j \|_{1/\ell}^{\ell} \tag{5}$$

The fireworks or sparks having higher selection probability in low crowded regions are selected for next iteration over the fireworks or sparks having lower selection probability in high crowded regions. It guides the search process in the unexplored area of the search space.

The complete Fireworks algorithm is given in Table 1.

3.2 Genotype and Phenotype

In Grammatical Fireworks Algorithm, firework's position is represented as genome, a set of integer codons in the range [0, 255] In general, the firework's position is real valued vector in the range [0, 255] and rounded up to its nearest integer values to form the integer codons. A representation of genotype is given in Fig. 1 for an example.

The Backus-Naur Form (BNF) of Context-free Grammar (CFG) is used for genotype to phenotype mapping. BNF is a metasyntax and it is used to express Context-free Grammar. An example of CFG in BNF is given below:

```
1. <Expr> := (<Expr><Op><Expr>) (0) | <Var> (1)
2. <Op> :=  + (0) | - (1) | * (2) | / (3)
3. <Var> := x1 (0) | x2 (1)
```

A *mapping process* is used in mapping from codon to the rule number while deriving the expression using CFG as following:

rule number = (codon value) mod (number of choices for the current nonterminal)

Table 1 Fireworks algorithm

Algorithm	Fireworks algorithm
1	Initialize $\mathcal{N}$ fireworks $\mathcal{X}$ with dimension $\mathcal{D}$
2	Calculate the objective function $f(\mathcal{X}_i)$ of each firework
3	***while*** (<termination criteria>)
4	Calculate the number of explosion sparks $\mathcal{S}_i$ and the amplitude $\mathcal{A}_i$ of explosion sparks for *i*th firework using Eqs. (2) and (3) respectively. // Generate Explosion Sparks
5	***for*** *i*: = 1 to $\mathcal{N}$
6	***for*** *e*: = 1 to $\mathcal{S}_i$
7	Initialize the spark's location: $\dot{x}_e = \mathcal{X}_i$
8	Select the number of position randomly: $\mathcal{L} = \text{round}(\mathcal{D} \times \text{rand}(0,1))$
9	Calculate the displacement: $\delta x = \mathcal{A}_i \times \text{rand}(-1,1)$
10	***for*** $j = 1$ to $\mathcal{L}$
11	Select the index $k \in [1, \mathcal{D}]$ randomly
12	$\dot{x}_{ek} = \dot{x}_{ek} + \delta x$
13	***if*** $\dot{x}_{ek} < \mathcal{X}_{\min}$ or $\dot{x}_{ek} > \mathcal{X}_{\max}$ ***then***
14	$\dot{x}_{ek} = \mathcal{X}_{\min} + \lvert \dot{x}_{ek} \rvert \% (\mathcal{X}_{\max} - \mathcal{X}_{\min})$
15	***End if***
16	***End for***
17	***End for***
18	***End for*** // Generate Gaussian Sparks
19	***for*** $i = 1$ to $\mathcal{N}$
20	Select the firework $\mathcal{X}_r, \quad r \in [1, \mathcal{N}]$ randomly as following: $r = \text{round}(\mathcal{N} \times \text{rand}(0,1))$
21	Initialize the spark's location: $\dot{x}_g = \mathcal{X}_r$
22	Select the number of positions randomly: $\mathcal{L} = \text{round}(\mathcal{D} \times \text{rand}(0,1))$
23	Calculate the coefficient of Gaussian explosion: $\mathcal{E} = \mathcal{G}(1,1)$
24	***for*** $j = 1$ to $\mathcal{L}$
25	Select the index $k \in [1, \mathcal{D}]$ randomly
26	$\dot{x}_{gk} = \dot{x}_{gk} \times \mathcal{E}$
27	***if*** $\dot{x}_{gk} < \mathcal{X}_{\min}$ or $\dot{x}_{gk} > \mathcal{X}_{\max}$ ***then***
28	$\dot{x}_{gk} = \mathcal{X}_{\min} + \lvert \dot{x}_{gk} \rvert \% (\mathcal{X}_{\max} - \mathcal{X}_{\min})$
29	***End if***
30	***End for***
31	***End for***
32	Select the $\mathcal{N}$ best locations from all fireworks, explosion, and Gaussian sparks
33	***End while***

156	**61**	**170**	**98**	**223**	**203**

Fig. 1 Genotypic representation

If <Expr> is the current nonterminal in the derivation, then the rule number is generated as following:

rule number = (156 mod 2) = 0

Then, <Expr> is replaced by (<Expr> <Op> <Expr>). A derivation process is carried out from the genotype given in Fig. 1.

```
<Expr> :=(<Expr><Op><Expr>)     (156 mod 2)=0
       :=(<Var><Op><Expr>)      ( 61 mod 2)=1
       :=(x1<Op><Expr>)         (170 mod 2)=0
       :=(x1*<Expr>)            ( 98 mod 4)=2
       :=(x1*<Var>)             (223 mod 2)=1
       :=(x1*x2)                (203 mod 2)=1
```

It may happen that the derivation process runs out of codons. Then the derivation again starts from the beginning point of the genome. This process is known as *wrapping*. It is often found that the same rule number is generated repeatedly for a nonterminal (for an example, <Expr> is replaced by (<Expr> <Op> <Expr>)). As a result, *wrapping* takes indefinite time and it may become a failure. Therefore, in this work, *wrapping* is done for predefined number of times. After wrapping, if non-terminals exist in the derived expression, it is indicated as invalid. It is then assigned a very small fitness value so that better valid individual can replace it later during the search process.

4 Experimental Setup

4.1 Benchmark Problems

4.1.1 Santa Fe Ant Trail Problem

This is a standard benchmark problem in GP and GE research. The objective of this problem is to evolve a program by which an artificial ant can discover all 89 pieces of food in a noncontinuous trail in predefined time steps. The trail is placed on a 32×32 grid. The ant can take actions such as left, right, move one square forward, and look ahead one square in its front to determine whether that square contains any food or not. All actions excluding look ahead of food, take one time step. The ant starts from grid's top-left corner and it faces the first piece of food. The fitness function is calculated as the difference between the total number of food pieces in the grid before and after the runs:

$$\text{fitness} = 89 - F,$$

where F is the total number of food pieces eaten by the ant at the end of the run. The BNF of CFG for this problem is defined as follows:

```
1. <Code> := (<Code><Line>) | <Line>
2. <Line> :=  <Condition>    | <Op>
3. <Condition> := if(foodahead()) {<Line>} else {<Line>}
4. <Op> := left(); | right(); | move();
```

The evolved program using the above grammar is placed in a loop with following termination criteria: (i) time steps = 600 or (ii) ant eats all 89 food pieces placed in the grid.

4.1.2 Symbolic Regression

The target function is $f(x) = x + x^2 + x^3 + x^4$ and 100 fitness cases are generated randomly in the range $[-1, 1]$. The fitness function is calculated as the sum of absolute errors over 100 fitness cases. The BNF of CFG for symbolic regression problem is given below:

```
1. <Expr> := (<Expr><Op><Expr>)   | <Var>
2. <Op> :=  + | -  | * | /
3. <Var> := x
'/' is protected division.
```

4.1.3 3-Multiplexer

3-input multiplexer problem has eight fitness cases and the fitness function is calculated as the sum of absolute errors over eight fitness cases. The BNF of CFG for 3-Multiplexer problem is given below:

```
1. <Expr> := (<Expr><Op><Expr>) | <Var>
2. <Op> :=  or  | and   | nor   | nand
3. <Var> :=   x1  | x2  | x3
```

4.2 Parameters Settings

The parameters of GFWA are set as following: number of total fireworks ($\mathcal{N}$) = 10, dimensionm, i.e., number of codons in genotype = 100, maximum number of explosion sparks ($\mathcal{M}$) for each firework = 40, boundary constraints ($\mathcal{M}_{\min}, \mathcal{M}_{\max}$) on number of explosion sparks for each firework = [2, 32], magnitude of explosion ($\mathcal{A}$) = 255, number of Gaussian sparks ($\mathcal{N}'$) = 10.

The parameters of GE are set as following: population size = 500, maximum number of codons = 25, minimum number of codons = 15, Cross-over type: Single point, Cross-over probability = 0.9, Mutation probability = 0.1, gene duplication probability = 0.01, gene pruning probability = 0.01, Selection: Tournament Selection, Replacement: Steady state replacement with negative tournament selection, normalized fitness $\mathrm{fit}' = \frac{1}{1+\mathrm{fitness}}$.

The parameters of GS are set as following: population size = 30, dimension = 100, $V_{\max} = 255$, $c_1 = c_2 = 1.0$, $\{w_{\max}, w_{\min}\} = \{0.9, 0.4\}$.

The parameters of GDE/rand/1/bin are set as following: population size = 500, dimension = 100, Scale Factor = 0.8, Cross-over rate = 0.8.

The parameters of GBC are set as following: population size = 30, dimension = 100, limit = 10.

As the functions are not evaluated for invalid individual in the above algorithms, the maximum number generations or cycles are not fixed for the comparative study. Therefore, each of GFW, GE, GS, GDE, and GBC algorithms is allowed to run for maximum 30,000 number of Function Evaluations (FEs) in a single run. All algorithms are terminated when they reach the maximum FEs or target error. The target errors are set to 0, 0.01, and 0 for ant trail, symbolic regression, and 3-multiplexer problems respectively. The numbers of wrapping are set to 3, 2, and 1 for ant trail, symbolic regression, and 3-multiplexer problems respectively.

4.3 PC Configuration

System: Windows 7, CPU: AMD FX—8150 Eight-Core 3.60 GHz, RAM: 16 GB, Software: Matlab 2010b

5 Results and Discussion

The experiments have been carried out for each problem with 30 independent runs. Mean and standard deviation (in parenthesis) of best-run-errors of each problem for GFWA, GE, GS, GBC, and GDE are given in Table 2. The number of successful runs and success rates (in parenthesis) over 30 independent runs for each problem are given in Table 3. Success rate is the ratio of number of successful runs and total

Table 2 Mean and standard deviation of best-run-errors over 30 independent runs

Algorithms	Santa Fe Ant	Symbolic regression	3-Multiplexer
GFWA	24.57(16.9516)	**6.65(7.246)**	0.93(0.2537)
GE	**0.77(0.4302)**	12.83(4.353)	0.90(0.3051)
GS	14.67(16.0158)	7.99(6.7969)	0.87(0.3457)
GBC	32.57(17.8609)	10.35(7.1404)	**0.7(0.4661)**
GDE/rand/1/bin	20.00(18.8607)	7.99(6.5183)	0.93(0.2537)

Results in boldface indicates better

Table 3 Number of successful runs and success rates (in %) over 30 independent runs

Algorithms	Santa Fe Ant	Symbolic regression	3-Multiplexer
GFWA	1(3.33 %)	**15(50.00 %)**	2.(6.67 %)
GE	**7(23.33 %)**	3(10.00 %)	3(10.00 %)
GS	1(3.33 %)	12(40.00 %)	3(10.00 %)
GBC	0(0.00 %)	7(23.33 %)	**9(30.00 %)**
GDE/rand/1/bin	0(0.00 %)	11(36.67 %)	2(6.67 %)

Results in boldface indicates better

Table 4 Mean and standard deviation of FEs over 30 independent runs

Algorithms	Santa Fe Ant	Symbolic regression	3-Multiplexer
GFWA	29917(453.88)	23943(8657.80)	29062(4415.30)
GE	**27815.40(4471.95)**	28891.13(4247.20)	28366(5209.80)
GS	29813(1022.80)	**22808(10539.00)**	26321(9569.00)
GBC	30000(0.00)	27076(6935.20)	**23549(10420.00**
GDE/rand/1/bin	30000(0.00)	23735(9763.20	29522(2167.10)

Results in boldface indicate better

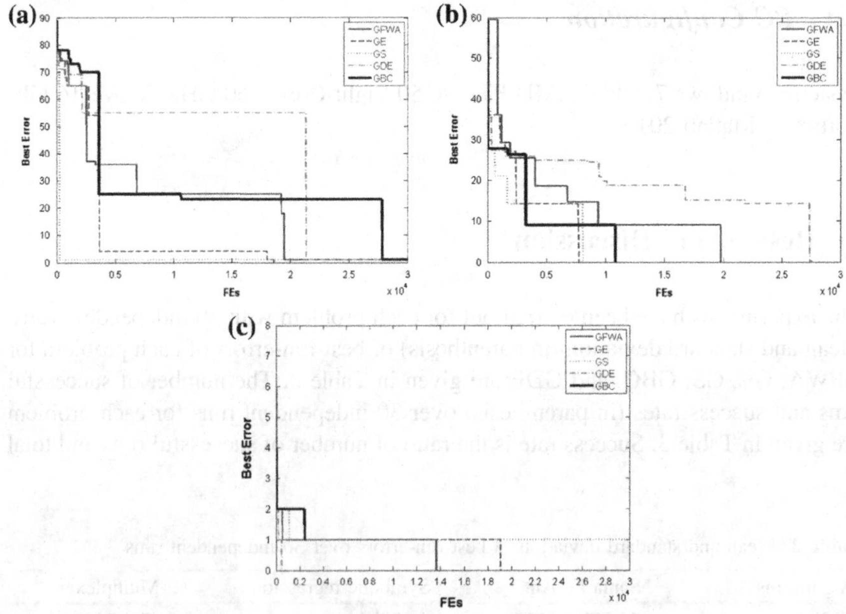

Fig. 2 Convergence graphs

number of independent runs. Mean and standard deviation (in parenthesis) of FEs over 30 independent runs are tabulated in Table 4. The convergence graphs for ant trail, symbolic regression, and 3-multiplexer problems are given in Fig. 2.

From Table 2, it has been seen that GE performs better than other algorithms in Santa Fe ant trail Problem. GFWA performs better than GE, GS, GBC, and GDE for only symbolic regression problem. GE, GS, GBC, and GDE perform better than GFWA for ant trail and 3-multiplexer problems.

It can be observed from Table 3 that GFWA has higher success rate than GBC for ant trail problem but lower success rate than GE, whereas GFWA and GS have the equal success rates for the same problem. GFWA has the higher success rate than GE, GS, and GBC for symbolic regression problem. For 3-input multiplexer problem, GFWA and GDE has the same success rates but GE, GS and GBC have the higher success rates than GFWA for the same problem.

From Table 4, it is noted that mean value of FEs of GFWA is lower than that of GBC and GDE algorithms for Santa Fe ant trail problem. The mean value of FEs of GFWA is lower than that of GE and GBC algorithms for symbolic regression problem. The mean FEs of GFWA is lower than that of GDE algorithm for multiplexer problem.

A series of pair-wise *t-tests* [14] has been carried out among GFWA, GE, GS, GBC, and GDE for statistical significance test with 95 % confidence interval and degree of freedom = 58 for each problem. The *p*-values and significances of *t-test* statistics have been given in Table 5. In the columns of significance of this table, '≈' indicates that there is no statistical difference in the performances of the algorithms in the pairs, '+' indicates that GFWA has the higher significant performance over its opposite algorithms in the pairs, '−' indicates that GFWA has the lower significant performance, '××' indicates that opposite algorithms of GFWA in the pair provide higher significant performance and '×' indicates that opposite algorithms of GFWA in the pair provide lower significant performance.

GE performs better than GFWA with higher significance. GS performs better than GFWA with lower significance, whereas there is no significant difference in the performances of GFWA and GBC for Santa Fe ant trail problem. For symbolic regression problem, GFWA provides higher significant performance over GE,

Table 5 t-test statistics

Pairs	Santa Fe Ant		Symbolic regression		3-Multiplexer	
	p-value	Significance	*p*-value	Significance	*p*-value	Significance
GFWA versus GE	1.7622e-008	'××'	2.1483e-004	'+'	0.6472	'≈'
GFWA versus GS	0.0236	'×'	0.4616	'≈'	0.3983	'≈'
GFWA versus GBC	0.0804	'≈'	0.0511	'≈'	0.0202	'×'
GFWA versus GDE	0.3281	'≈'	0.4556	'≈'	1.00	'≈'

whereas there are no significant differences in performances of GFWA, GS, GBC, and GDE. There are no statistical differences in the performances among GFWA, GE, GS, and GDE for 3-input multiplexer problem, whereas GBC performs better than GFWA with lower significance. The computer programs evolved by GFWA successfully, are given below:

The successful ant program evolved by GFWA (ant eats all 89 pieces of food):

```
if(foodahead()) if(foodahead()) move(); else left();  end;
else  if(foodahead()) move(); else right(); end;   end;
if(foodahead()) move(); else if(foodahead())
if(foodahead()) left(); else left(); end; else
right();  end; end; if(foodahead()) if(foodahead())
move(); else move(); end; else if(foodahead())
if(foodahead()) left(); else if(foodahead()) left();
else if(foodahead()) right(); else if(foodahead())
move(); else move(); end; end; end; end; else move();
end; end; right(); if(foodahead()) if(foodahead())
left(); else right(); end; else right(); end;
```

A successful program evolved by GFWA for symbolic regression problem (absolute error = $1.7837e-15$):

```
plus(times(plus(times(plus(x,times(x,x)),x),x),x),x)
```

A successful program evolved by GFWA for 3-multiplexer problem (absolute error = 0):

```
and(nor(and(x1,x2),and(and(x2,x3),x2)),or(or(x1,x3),  or
(x2,x3)))
```

From the above discussion of the results, FWA as a search engine, outperforms GA with higher success rate in solving symbolic regression problem only. FWA performs equal with GA for 3-multiplexer problem, whereas GA outperforms FWA for ant trail problem with higher success rate. The key finding of the present study is that GFWA can be applied in automatic computer program generation as GS, GBC, and GDE.

6 Conclusion

This paper presents Grammatical Fireworks algorithm for automatic programming in any arbitrary language. In Grammatical Fireworks algorithm, basic Fireworks algorithm is used as a learning algorithm in *genotype-to-phenotype* mapping process using Backus-Naur Form of Context-free grammar to generate computer programs. Grammatical Fireworks algorithm is applied on three well-known benchmark problems obtained from GP literatures. The experimental results establish the proof of concept that Grammatical Fireworks can be applied to generate computer programs in any arbitrary language. In future, this work will be extended and examined by applying different variants of Fireworks algorithm as learning algorithms in Grammatical Evolution.

References

1. Koza, J.R.: Genetic Programming: On the Programming of Computers by Means of Natural Selection. MIT Press (1992)
2. Ryan, C., Collins, J.J., O'Neill, M.: Grammatical evolution: evolving programs for an arbitrary language. In: Banzhaf, W., Poli, R., Schoenauer, M., Fogarty, T.C. (eds.) EuroGP 1998. LNCS, vol. 1391, pp. 83–95. Springer, Heidelberg (1998)
3. O'Neill, M., Ryan, C.: Grammatical evolution. IEEE Trans. Evol. Comput. **5**(4), 349–358 (2001)
4. O'Sullivan, J., Ryan, C.: An investigation into the use of different search strategies with grammatical evolution. In: Foster, J.A. et al. (eds.) EuroGP 2002, LNCS, vol. 2278, pp. 268–277. Springer, Heidelberg (2002)
5. O'Neill, M., Brabazon, A.: Grammatical swarm. In: Genetic and Evolutionary Computation Conference (GECCO), pp. 163–174 (2004)
6. O'Neill, M., Leahy, F., Brabazon, A.: Grammatical swarm: a variable-length particle swarm algorithm. Swarm Intelligent Systems. Studies in Computational Intelligence. Springer, pp. 59–74 (2006)
7. O'Neill, M., Brabazon, A.: Grammatical differential evolution. In: International Conference on Artificial Intelligence (ICAI 2006), CSEA Press, Las Vegas, pp. 231–236 (2006)
8. Storn, R., Price, K.: Differential evolution–a simple and efficient heuristic for global optimization over continuous spaces. J. Global Optim. **11**, 341–359 (1997)
9. Si, T., De, A., Bhattacharjee, A.K.: Grammatical bee colony. In: Panigrahi, B.K. et al. (eds.) SEMCCO 2013, Part I, LNCS, vol. 8297, pp. 436–445. Springer, Heidelberg (2013)
10. Karaboga, D.: An idea based on honey bee swarm for numerical optimization. In: Technical Report-TR06, Erciyes University, Engineering Faculty, Computer Engineering Department (2005)
11. Mckay, R.I., Hoai, N.X., Whigham, P.A., Shan, Y., O'Neill, M.: Grammar-based genetic programming: a survey. Genet. Programm. Evolvable Mach. **11**, 365–396 (2010)
12. Kennedy, J., Eberhart, R.: Particle swarm optimization. In: IEEE International Conference on Neural Networks. Perth, Australia (1995)
13. Tan, Y., Zhu, Y.: Fireworks algorithm for optimization. In: Tan, Y. et al. (eds.) ICSI 2010, Part I, LNCS, vol. 6145, pp. 355–364. Springer, Heidelberg (2010)
14. Derrac, J., Garcia, S., Molina, D., Herrera, F.: A practical tutorial on the use of nonparametric statistical tests as a methodology for comparing evolutionary and swarm intelligence algorithms. Swarm Evol. Comput. **1**, 3–18 (2001)

Gaussian Function-Based Particle Swarm Optimization

Priyadarshini Rai and Madasu Hanmandlu

Abstract This paper presents the Gaussian function-based particle swarm optimization (PSO) algorithm. In canonical PSO, potential solutions, called particles, are randomly initialized in the beginning. The proposed method uses the solutions of another evolutionary computation technique called genetic algorithm (GA) for initializing the particles in order to provide feasible solutions to start the algorithm. The method replaces the random component of the velocity update equation of PSO with the Gaussian membership function. The Gaussian function-based PSO is applied on eight benchmark functions of optimization and the results show that the proposed method achieves the same quality solution in significantly fewer fitness evaluations. This proposed modification of PSO will be useful to optimize efficiently.

Keywords Optimization · Particle swarm optimization · Genetic algorithm · Gaussian membership function

1 Introduction

Optimization is a process of finding the conditions that give the maximum or minimum value of a function. In recent years, some optimization methods such as genetic algorithm [1] and particle swarm optimization [2, 3] have been developed. These methods are conceptually different from traditional mathematical programming techniques and are known as modern or non-traditional methods of optimization.

Priyadarshini Rai (✉)
Apaji Institute of Mathematics & Applied Computer Technology, Banasthali University, Vanasthali, India
e-mail: priya.rai.lis@gmail.com

Madasu Hanmandlu
Department of Electrical Engineering, Indian Institute of Technology Delhi, New Delhi, India
e-mail: mhmandlu@gmail.com

M. Pant et al. (eds.), *Proceedings of Fifth International Conference on Soft Computing for Problem Solving*, Advances in Intelligent Systems and Computing 436, DOI 10.1007/978-981-10-0448-3_5

Particle swarm optimization (PSO) was originally proposed by Kennedy and Eberhart [2, 3] in 1995. PSO is based on swarm intelligence. 'Swarm' is a group of organisms or animals like group of flocking birds or fish schools. They live together and help each other to make their living. We model the artificial systems based on the lives of swarms or the properties they exhibit in order to get extra benefit. The swarm of birds inhabits 'intelligence' as a group. Because of this collective intelligence they are able to accomplish tasks like finding a path to a destination or achieving a goal. PSO is evolutionary as the solution path in the search space is arrived at in a piecemeal manner by maximizing some criterion function.

PSO is inspired from the behaviour of a group of flocking birds. They move in groups and keep track of the group movements. Each bird knows about its own group as well as other groups so that it can sustain in its own group in addition to competing with other groups to achieve a leading role. We use this endeavour of each bird in the biological world to guide us in learning complex systems.

In the literature, PSO has been modified in different ways for different optimization problems. Shi and Eberhart [4] introduced the inertia weight '*w*' in the velocity equation of PSO to procure different balances between exploration and exploitation for solving different problems. Naka et al. [5] utilized the mechanism of 'natural selection' in PSO. Particles with low function values were replaced by those with high function values. Da and Xiurun [6] had modified PSO-based artificial neural network (ANN) in order to train ANN in an improved manner. PSO accepts the personal best solution as the new global best solution when the personal best solution is greater than the global best solution with probability as 1. PSO accepts the personal best solution as the new global best solution even when the personal best solution is less than the global best solution with certain probability which gradually decreases to zero as the algorithm approaches towards the end. Victoire and Jeyakumar integrated Sequential Quadratic Programming (SQP) with PSO to fine-tune the improvements made by PSO run. SQP is used to solve the economic dispatch problem when the current global best solution is better than the global best solution of the previous iteration. Suganthan [7] replaced the global best component of the velocity equation with the neighbouring best component. Time varying values of inertia weight and acceleration constants are used. Hu et al. [8, 9] have modified the global as well as local version of PSO. Particles are repeatedly initialized until they satisfy all constraints. Personal best fitness of a particle is updated when the particle is in feasible space and the current fitness value of particle is better than the best fitness value in history.

Genetic algorithm (GA) was developed by Professor John Holland, his colleagues and students at the University of Michigan around 1975. GA is a powerful non-traditional optimization technique. It mimics the process of evolution and is based on the mechanics of natural selection and natural genetics. Though the algorithm was figured out in 1975 it gained popularity only in the past 15–20 years. It is based on the "survival of the fittest" concept of the Darwinian Theory. GA is not calculus-based search technique. It is a search algorithm and does not use derivative knowledge, like the first derivative or the second derivative.

In this paper, we propose some improvements for the canonical PSO algorithm. Instead of randomly initializing the particles, we have used the results of GA for the same to give feasible starting points to the algorithm. The randomness of the velocity equation is replaced with the Gaussian membership function in order to provide fast converging rate.

We present our modified PSO in Sect. 2. In Sect. 3, the benchmark functions used to test the modified PSO are given. Results are discussed in Sect. 4. The conclusions are given in Sect. 5.

2 Gaussian Function-Based PSO

In PSO, there is a search space with many peaks and valleys. The aim is to find the global optimum and to escape from the local optima. PSO is a multi-agent approach, where multiple particles or agents or birds search for global optimum in the search space. Every particle has a position (x_i) and fitness value. The particles in the search space are not static. They move somewhere in the fitness landscape in search of global optimum. The particles move in the search space with certain velocity. So, each particle has velocity (v_i) along with the position and fitness value. For any particle '*i*', we have a position vector, velocity vector and a fitness value.

The algorithm works in iterative manner. In iteration, each particle would get one chance to move. The particles will move with the magnitude of their velocities. If the velocity is very high, the particles will take bigger steps and very small steps if low. Initially, the particles will be located inside the fitness landscape at random positions. Therefore, they will have random fitness value in the beginning. We need to move the particles in the fitness landscape with some velocity. So, we will also randomly assign some velocity to the particles. The following equation is used to move particle in every iteration

$$x_i(\mathrm{itr}+1) = x_i(\mathrm{itr}) + v_i(\mathrm{itr}) \tag{1}$$

where

$x_i(\mathrm{itr}+1)$ position of particle in next iteration
$x_i(\mathrm{itr})$ position of particle in current iteration
$v_i(\mathrm{itr})$ velocity of particle in current iteration.

From Eq. (1), we infer that the position of the particle is updated according to its velocity. As we do not know where the local optimum lies, we make a prediction that the local optimum will lie at a position where the individual particle has the best fitness value so far. Each particle memorizes the best position or fitness value it has achieved so far. This gives us an idea about the local optimum. Global optimum is the best fitness achieved so far by all the particles.

We can never be sure that the global optimum reached after iteration is even near the actual global optimum. So, in order to attain the global optimum, there are two fundamental principles. According to one principle, the particle tries to converge towards its local optimum and according to another principle, it tries to converge towards the global optimum. The velocity update of a particle is expressed as

$$v_i(\mathrm{itr}+1) = v_i(\mathrm{itr}) + c_1 * r * [x_i - L\mathrm{best}] + c_2 * r * [x_i - G\mathrm{best}] \quad (2)$$

where

$v_i(\mathrm{itr}+1)$	Velocity of particle in next iteration
$v_i(\mathrm{itr})$	Current velocity of particle
'c_1 and c_2'	Constants to control the amount of variation in velocity
'***r***'	Random number to give randomness to the algorithm
'***Lbest***'	Local best position of particle
'***Gbest***'	Global best position
$[x_i - L\mathrm{best}]$	Intention to move towards local optimum
$[x_i - G\mathrm{best}]$	Intention to move towards global optimum

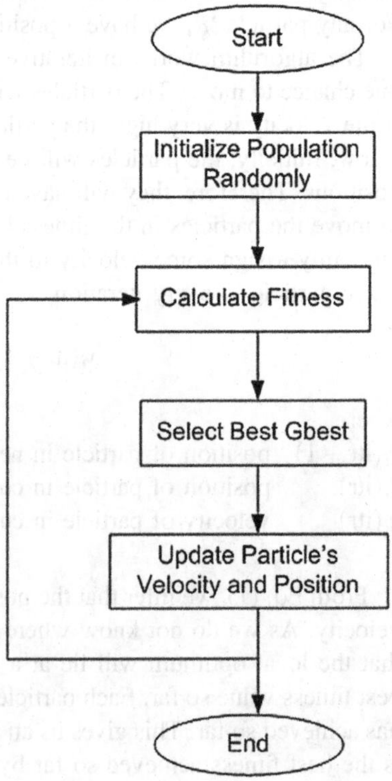

Fig. 1 Flowchart of canonical PSO

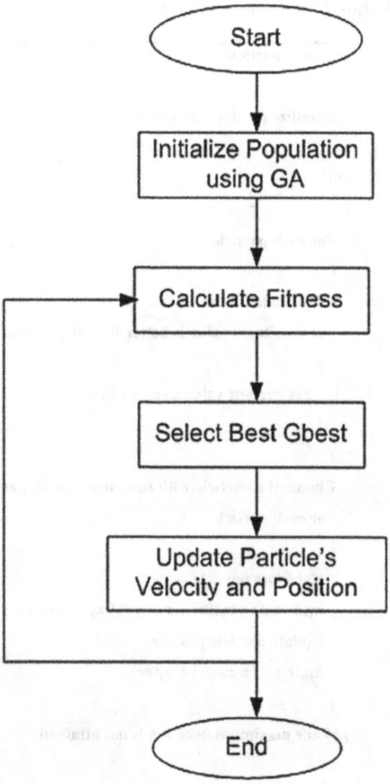

Fig. 2 Flowchart of Gaussian function-based PSO

Owing to the above two principles, changes in the velocity vector take place. Therefore, we shall add a historical parameter, $v_i(\text{itr})$, which tries to make the particle to move in the same direction as it was moving in history. 'c_1' and 'c_2' control the amount of variation in velocity. 'c_1' controls the intention of the particle to move towards the local optimum, whereas 'c_2' controls the intention of the particle to move towards the global optimum. Random number 'r' drives the facility to explore the solution space. The velocity 'v_i' is always restricted to lie in between the minimum and maximum values. When the constant 'c_2' becomes more dominant than the constant 'c_1' the particles start moving towards their global optimum thus indicating the convergence (Figs. 1, 2 and Tables 1, 2).

Modification 1: In the original PSO Eq. (2), the velocity update has component 'r' which gives randomness to the algorithm thus facilitating exploration. In the Gaussian function-based PSO, we have replaced the component 'r' of the velocity update with the Gaussian membership (MATLAB) function to achieve high convergence rate (as shown in Table 3).

Table 1 The original PSO algorithm

```
For each particle
{
 Initialize particle randomly
}
Do
{
 For each particle
 {
  Calculate fitness value
  If the fitness value is better than the best fitness value (Lbest) in history
  {
   Set current value as new Lbest
  }
 }
 Choose the particle with best fitness of all particles as 'Gbest'
 For each particle
 {
  Calculate particle velocity:
  v_i(itr + 1) = v_i(itr) + c_1 * r * [ x_i - Lbest] + c_2 * r * [x_i - Gbest]
  Update particle position:
  x_i(itr + 1) = x_i(itr) + v_i(itr)
 }
} While maximum iteration is not attained
```

$$y = \text{gaussmf}\ (x, [\text{sig}\ c]) \tag{3}$$

where

'x' function value of the global best particle
'sig' standard deviation of the function value of the particles
'c' mean of the function value of the particles.

The Gaussian function is given as

$$f(\mathrm{x}; \sigma, \mathrm{c}) = e^{-((x-c)^2)/2*\sigma^2} \tag{4}$$

where σ and c are the parameters of the Gaussian function.

Modification 2: In the original PSO, the initial values of particles are chosen randomly. Instead of randomly initializing the particles in the Gaussian function-based PSO, we initialize the particles using the initial population of GA in order to start the algorithm with feasible solutions.

Table 2 The Gaussian function based PSO algorithm

```
For each particle
{
  Initialize particle using GA
}
Do
{
  For each particle
  {
    Calculate fitness value
    If the fitness value is better than the best fitness value (Lbest) in history
    {
      Set current value as new Lbest
    }
  }
  Choose the particle with best fitness of all particles as Gbest
  For each particle
  {
    Calculate particle velocity
    v_i(itr + 1) = v_i(itr) + c_1 * gaussmf (x, [sig c])* [ x_i – Lbest] + c_2 * r * [x_i – Gbest]
    Update particle position
    x_i(itr + 1) = x_i(itr) + v_i(itr)
  }
} While maximum iteration is not attained
```

Table 3 Converging characteristics of original and Gaussian function-based PSO on benchmark functions

Benchmark function	Original PSO converges at *i*th iteration	Gaussian function-based PSO converges at *i*th iteration	Optimum values obtained using original PSO	Optimum values obtained using Gaussian function-based PSO
Ackley function	50	20	1.0467e-05, 6.7111e-06	−6.6199e-11, −5.4355e-11
Beale's function	40	10	2.6261, −0.4547	2.6261, −0.4547
Bukin function N.6	70	30	2.0620, 0.0425	−11.2347, 1.2622
Matyas function	55	15	2.7384e-05, 2.4934e-05	2.0745e-10, 1.2635e-11
Levy function N.13	35	10	1.0000, 0.9999	0.9999, 1.0000
Easom function	65	0	3.1415, 3.1416	2.0177, 3.7685
Eggholder function	50	5	512, 4.0423e+02	−4.6569e+02, 3.8571e+02
McCormick function	30	10	−0.5471, −1.5471	−0.5471, −1.5471

Table 4 Enhanced converging characteristics of Gaussian function-based PSO with GA

Benchmark function	Gaussian function-based PSO with modification 1 converges at *i*th iteration	Gaussian function-based PSO with modification 1 and modification 2 converges at *i*th iteration	Optimum values obtained using Gaussian function-based PSO with modification 1	Optimum values obtained by initializing Gaussian function-based PSO with modifications 1 and 2
Beale's function	10	1	2.6261, −0.4547	2.7464, −0.3849
Bukin function N.6	30	1	−11.2347, 1.2622	−15, 2.2535
Levy function N.13	10	4	0.9999, 1.0000	1, 1
McCormick function	10	1	−0.5471, −1.5471	−0.7667, −2.0067

Our experiments indicate that GA-based initialization yields very high convergence rate as GA provides feasible solutions (as shown in Table 4).

3 Function Optimization

In our investigation, we have employed Gaussian function-based PSO on the benchmark functions for optimization. The test functions used here are commonly used in the evolutionary computation literature. The following test functions are used to test the velocity of convergence, precision, robustness and general performance of the modified PSO.

- Ackley function:

$$f(x) = -a\exp\left(-b\sqrt{\frac{1}{d}\sum_{i=1}^{d} x_i^2}\right) - \exp\left(\frac{1}{d}\sum_{i=1}^{d}\cos(cx_i)\right) + a + \exp(1)$$

- Beale's function:

$$f(x) = (1.5 - x_1 + x_1x_2)^2 + \left(2.25 - x_1 + x_1x_2^2\right)^2 + \left(2.625 - x_1 + x_1x_2^3\right)^2$$

- Bukin function N.6:

$$f(x) = 100\sqrt{|x_2 - 0.01x_1^2|} + 0.01|x_1 + 10|$$

- Matyas function:

$$f(x) = 0.26\left(x_1^2 + x_2^2\right) - 0.48x_1x_2$$

- Levy function N.13:

$$(x) = \sin^2(3\pi x_1) + (x_1 - 1)^2\left[1 + \sin^2(3\pi x_2)\right] + (x_2 - 1)^2\left[1 + \sin^2(3\pi x_2)\right]$$

- Easom function:

$$f(x) = -\cos(x_1)\cos(x_2)\exp(-(x_1 - \pi)^2 - (x_2 - \pi)^2$$

- Eggholder function:

$$f(x) = -(x_2 + 47)\sin(\sqrt{\left|x_2 + \frac{x_1}{2} + 47\right|}) - x_1 \sin(\sqrt{|x_1 - (x_2 + 47)|}$$

- McCormick function:

$$f(x) = \sin(x_1 + x_2) + (x_1 - x_2)^2 - 1.5x_1 + 2.5x_2 + 1$$

4 Experimental Results

Here the minimization problems are dealt with, hence the results pertain to the minimum values. In our experiments, 50 particles are used. Maximum number of iterations (max_iteration) is fixed at 100. The optimum results of the original PSO as well as the modified PSO on each optimization test function correspond to the best results obtained out of 20 runs. We set: Min_Inertia_weight = 0.4, Max_Inertia_weight = 0.9.

The weight adaptation: The inertia weight 'w(itr)' is adapted as follows:

$$W(\text{itr}) = ((\text{max\_iterationitr}) * (\text{Max\_Inertia\_weight} - \text{Min\_Inertia\_weight}))/(\text{max\_iteration} - 1) + \text{Min\_Inertia\_weight}.$$

The parameters, c_1 and c_2 are set to 2. The canonical PSO and the Gaussian function-based PSO algorithms are run using the above parameter values. In the Gaussian function-based PSO algorithm, the 'random' component in the velocity update equation is replaced with the Gaussian membership function. Moreover, the initial populations of GA are used for initializing the position of the particles instead of random values.

Table 3 shows the comparison of the original and Gaussian function-based PSO when modification 1 is incorporated into the Gaussian function-based PSO, that is, the random component is replaced with the Gaussian membership function. The table shows the optimum values of the original and Gaussian function-based PSOs, when applied on the benchmark functions for optimization. It is evident from the table that the Gaussian function-based PSO converges faster than the original PSO without compromising the solution quality. The original PSO when applied on Ackley function converges at 50th iteration whereas the Gaussian function-based PSO converges at 20th iteration and gives the improved result. Similarly, the original and the Gaussian function-based PSOs yield convergence at 40th iteration and 10th iteration respectively on Beale's function. The converging characteristics of two PSOs are shown in Table 3.

Table 4 shows the comparison of Gaussian function-based PSO with only modification 1 (that is, random component of velocity update equation is replaced with the Gaussian membership function) and that with both modification 1 and modification 2 (that is, random component of velocity update equation is replaced with the Gaussian membership function and particles are initialized using GA populations instead of random values). The table shows the enhanced converging characteristics of Gaussian function-based PSO when initialized using GA populations and provides the same quality results in a very few fitness evaluations. When Gaussian function-based PSO with only modification 1 is used to optimize Bukin function N.6, it converges in 30 iterations but PSO with both the modifications converges in a single iteration. Similarly, for all other test functions for optimization, Gaussian function-based PSO with both the modifications converge within the first five iterations.

5 Conclusions

Optimization is the process of minimizing the cost and maximizing the profit. In recent years, modern or non-traditional methods of optimization have been developed. These methods are search algorithms and do not use derivative knowledge. They are based on certain characteristics and behaviour of biological, molecular, swarm of insects and neurobiological systems.

In this paper, particle swarm optimization (PSO) is modified to achieve high solution quality and fast convergence rate. In PSO's velocity equation, the randomness 'r' has been replaced with the Gaussian membership function. When this Gaussian function-based PSO is applied on eight test optimization functions, it provides better quality results in significantly fewer fitness evaluations. This Gaussian function-based PSO is then initialized using the initial populations of GA and it provides even more high convergence rate. However, time complexity of Gaussian function-based PSO increases when it is initialized using GA as the execution of GA requires more time.

References

1. Goldberg, D.E.: Genetic algorithms in search, optimisation and machine learning. Addison-Wesley, MA (1989)
2. Eberhart, R., Kennedy, J.: A new optimizer using particle swarm theory. In: Proceedings of sixth International Symposium on Micro Machine and Human Science, Nagoya, Japan, October 1995
3. Kennedy, J., Eberhart, R.: Particle swarm optimization. In: Proceedings of IEEE International Conference on Neural Networks, Perth, Australia, December 1995
4. Shi, Y., Eberhart, R.: A modified particle swarm optimizer. In: 1998 IEEE International Conference on Evolutionary Computation Proceedings. IEEE World Congress on Computational Intelligence (Cat.No.98TH8360), 1998
5. Naka, S., Genji, T., Yura, T., Fukuyama, Y.: A hybrid particle swarm optimization for distribution state estimation. IEEE Power Eng. Rev. **22**(11), 57–57 (2002)
6. Da, Y., Xiurun, G.: An improved PSO-based ANN with simulated annealing technique. Neurocomput. **63**, 527–533 (2005)
7. Suganthan, P.N.: Particle swarm optimiser with neighbourhood operator. In: Proceedings of the 1999 Congress on Evolutionary Computation-CEC99 (Cat. No. 99TH8406), 1999
8. Hu, X., Eberhart, R.C., Shi, Y.: Engineering optimization with particle swarm. In; Proceedings of the 2003 IEEE Swarm Intelligence Symposium. SIS'03 (Cat. No.03EX706), 2003
9. Hu, X., Eberhart, R.C.: Solving constrained nonlinear optimization problems with particle swarm optimization. In: Proceedings of the Sixth World Multi Conference on Systemics, Cybernetics and Informatics, 2002

Reinforcing Particle System Effects Using Object-Oriented Approach and Real-Time Fluid Dynamics for Next-Generation 3D Gaming

Rajesh Prasad Singh, Rashmi Dubey and Sugandha Agarwal

Abstract The focus of this paper is to use fluid dynamics and object-oriented approach in particle subsystems for interactive real-time visualization in 3D gaming. By exploiting all the areas of fluid dynamics and exercising it in the particle subsystem of gaming engine, we create an environment where we can map an artificial object to a real-world entity to such an extent that it can add realism to the virtual gaming environment. To do so, particle integration and particle rendering play an important part. Also, graphics accelerators and graphics processor units add a mammoth advantage calculating the particle coordinates followed by rendering. The current graphic processor though fast, fails to give the exact trajectory of particles when collision, sliding, outburst, explosion or stabbing by one character to another character occurs. They lack in tracing the exact path followed by the particles when certain forces are applied on a heavenly body or on liquidus body. The particle that emerges is random and has no correlation to reality as they follow random path. All these must be considered to enhance the user interaction in 3D gaming to enrich the real-time virtual environment gaming experience. Many particle subsystems tools are available that can create extraordinary particle effects in gaming, but they all lack in giving directions to the particle. Thus in this paper, we create a particle subsystem that is not only stable but also follows the law of physics so that each particle in the gaming environment can be advanced with random time steps. The principal motivation behind the paper is to examine the flow of particles when they are embedded with the laws of fluid dynamics and to calculate the rendering complexity which makes it impossible to implement it on the large scale on conventional graphic processing unit without using quantum technology.

R.P. Singh (✉) · Rashmi Dubey · Sugandha Agarwal
Computer Science and Engineering Department, Amity University,
Noida, Uttar Pradesh, India
e-mail: rj.singh@live.in

Rashmi Dubey
e-mail: rash.monu@gmail.com

Sugandha Agarwal
e-mail: aga.sugandha@gmail.com

M. Pant et al. (eds.), *Proceedings of Fifth International Conference on Soft Computing for Problem Solving*, Advances in Intelligent Systems and Computing 436, DOI 10.1007/978-981-10-0448-3_6

Keywords Object-oriented approach · Particle subsystem · Real-time fluid dynamics · Game engine · GPU

1 Introduction

"A particle system is a collection of many many minute particles that together represents a fuzzy object. Over a period of time, particles are generated into a system, move and change from within the system, and die from the system" [1]. The true nature of particle is that it flows freely under the forces acting on it. Particle system finds its application in many video games, digital art pieces, animations and models in which it represents the phenomena of nature like waterfall, bubbles, fire, smoke, etc. In the early 1980s, particle system spread widely in 3D gamings where it found some limitations when it comes to render millions of particles at a time. For e.g., When a character of game stabs someone with a knife or with some other weapon, then the blood coming out of another character is random, instead they must follow some pattern based on the laws of physics. The amount of blood flowing from the body part, exact trajectory of all the blood particle emerging from the point of contact between the weapon and body part were not taken into consideration in early gaming videos. As particle system is the only way in computer graphics for producing realistic animation and live effects, these problems need to be solved early. Many algorithms were proposed at that time and many advancements have been done to graphics processing unit so that they can render millions of particles at a time, but they all lack in giving the exact trajectory of particles emerging from a body. The only solution left is to introduce the real-time fluid dynamics, so that it can give exact directions to the particles and realism to the virtual environment. Superior options are available to use the already existing physical equation developed since the time of Euler, Naiver–Stokes. We generally use Navier–Stokes equations as it is the best suited model till date available for particle system. These physical equation statements only accept analytical solution in extremely straightforward cases. These equations make progress but lack in precision and are genuinely perplexing and tedious.

2 Background

This section reviews particle systems, their needs, previous work and their previous methodologies.

2.1 Particle System

William T. Reeves, a researcher at Lucas film Ltd., used Genesis effect in the movie called Star Trek: Wrath of Khan, it was the first computer generated particle system in commercial computer graphics [2].

A particle system is an advanced technique to produce a series of real-time individual particles that consist of particle velocity, position, colour and lifetime. It is heavily used to model complex systems such as fires, planets, smoke, fog, etc. Particle system is a way to create fuzzy objects and those objects whose behaviour varies with time. Its trade-off between the animator control and physical simulation demands motion to be physically correct and with complete availability of control. One way to implement the particle system in game engine is to first perform the complete physical simulation and then determine all the needed force to be applied to achieve the exact desired motion and trajectory of particle. But this is difficult and both are non-trivial. The practical way to approach this implementation is to combine dynamic simulation with kinematic control by giving a set of several levels of operations.

2.2 Need for Particle System

A particle system in game engine is represented as a dot and is considered to be the collection of independent objects. The principal aim of modelling the phenomena like explosions, collisions of objects, etc., is astonishing and useful but in reality there are far better reasons to use particle system. Particle system in computational fluid dynamics helps to find the position of objects in space and are used in medical applications where the amount of blood outflowing from blood vessels is predicted.

Modelling fuzzy objects with the help of particle system has numerous advantages over conventional surface-oriented technique. First, the particle itself is much simpler and primitive than the polygon, which is a simplest representation of surface. At this point, the particle itself is considered as a point in 3D space. So the amount of calculation time required in polygon representation can be utilized effectively to process more of the basic primitive objects like particle and can produce more complex image [3]. Particle system is also considered as one of the simplest as in particle system motion blurring is easy. Second, the whole particle system is based on procedural-oriented mechanism and is fully controlled by random generated numbers. In procedural implementation, there is no need to add minute details of existing surface-based area. Third, particle systems behave like "alive" objects as their value varies over time. Particle objects constantly change their formation over a regular interval of time. This complex dynamicity of particle systems is hard to achieve in surface-based modelling.

2.3 Particle Generation

Generation of particle in a complex molecular structure of particle system is a control stochastic procedure. The number of processes is executed at given time of frame. Each particle system process resolves the number of particles entering the system throughout a given interval of time. The density of the fuzzy object thus created is highly influenced by the number of particles generated in frame [4]. There are numerous ways to control the amount of particle. The first method controls the mean number of particles created at a given frame and also controls the total number of particle variance. Particles generated per frame is given by the formula

$$N_P = M_P + \text{Rand}\,() + V_P$$

where

N_P number of particles generated
M_P mean of particles generated
Rand () gives uniformly distributed number of particles
V_P variance of particle.

The second method relies on the size of the area under the effect of the object. In this methodology, the total number of particles emitted per unit work area size is considered. In this type of procedural model, viewing parameters determine the specific frame and calculate approximate screen area size and according to that set the number of new particles. The mathematical formula is given as (Fig. 1).

$$N_P = (M_P + \text{Rand}\,() + V_P)^{*} S_A$$

where S_A = screen size

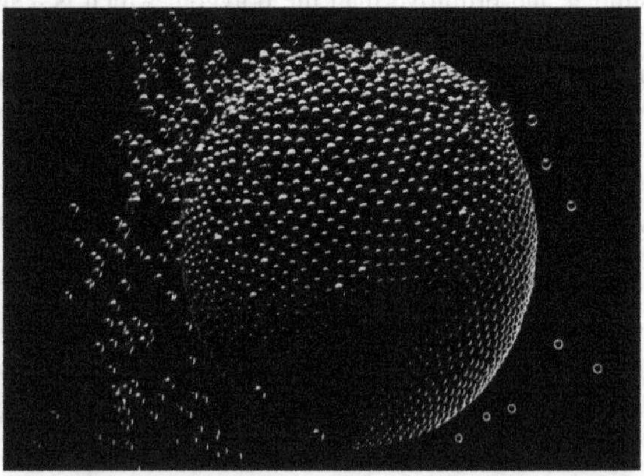

Fig. 1 Particle system example: an object is set to burst in particles, with all the generating particles moving in space free from gravity in a 3D virtual environment [4]

3 Simulation of Particles Using Fluid Dynamics

Simulation of particles using fluid dynamic provides strong premise to the advancement of real-time visualization strategies and completely exemplifies the implementation of particle simulation for interactive applications. Thus, the principal focus is not only on the physical correctness but also on the simple, clear and highly efficient implementation. Fluid dynamics harnesses macroscopic behaviour at time and length scales where intermolecular effects are not observable. In such situations, every point in a space holds discrete value having the same property. Property contains the character of observable macroscopic behaviours that characterize the particles.

3.1 Imposing Fluid Properties on Particles

When imposing fluid properties on particles, relevant properties to be considered are density, velocity, mass and pressure. The density ρ measures mass per volume, velocity v measures the direction and speed of particles passing in free space, mass m tells the amount of matter involved. Pressure p states the force acting on the particle and the whole formula is denoted by

$$\rho = \lim_{\Delta V \to L^3} \frac{\Delta m}{\Delta V} \tag{1}$$

where L is very small in length but greater than molecule spacing; V Volume.

If we consider Newton's second law of motion then usually its interpretations come from Lagrangian point of view in which the moving object is thoroughly observed. With particles possessing fluid properties, it means that the area under the scrutiny follows the particle flow so that same amount of particle is being watched, while this is not the case in Euler observation because in Euler the area of observation is locally fixed [4]. By doing so the amount of particle flowing from one point to another can be constantly monitored and at each time step their value will vary. In Euler observer, it not only notices the change in the variance in the recently watched amount of particles but also notices the change due to the fact that the watched amount of particle may be a different one every moment [5, 6].

The most conventional description of the fluid dynamics by Euler is that the acceleration is the special time derivative of the velocity that takes account of all the moving particles in both forms: advection and convection. Together they are called convection. It is represented by the formula (Fig. 2):

$$\frac{D\emptyset}{Dt} = \frac{\partial\emptyset}{\partial t} + v \cdot \nabla\emptyset = \frac{\partial\emptyset}{\partial t} + u\frac{\partial\emptyset}{\partial x} + v\frac{\partial\emptyset}{\partial y} + w\frac{\partial\emptyset}{\partial z} \tag{2}$$

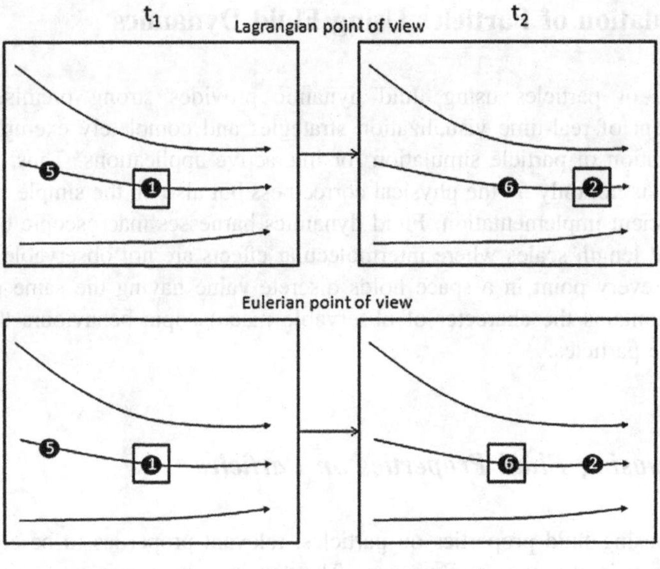

Fig. 2 Comparison of Euler and Lagrangian point of view [5]

The above formula is written in Cartesian coordinated in three dimensions where ∇: del operator; *x*, *y*, *z*: components of position; *u*, *v*, *w*: Components of velocity; Φ: an arbitrary quantity; $v \cdot \nabla \Phi$: it represents the change in advection, i.e. properties of transported matter. $\frac{\partial \Phi}{\partial t}$ represents the local change in currently observed amount of particle [4, 6]. By putting acceleration a in the above equation we get

$$F = m\frac{Dv}{Dt} = m\left(\frac{\partial v}{\partial t} + v \cdot \nabla v\right) \tag{3}$$

∇v: Gradient of the velocity.

Equation (3) states that the density depends on the mass of the particle inside the observed control volume. Therefore, we rewrite the equation as

$$\left(\frac{\partial v}{\partial t} + v \cdot \nabla v\right) = F_{\text{Fluid}} + F_{\text{External}} \tag{4}$$

The forces acting on the bunch of particles can be distinguished by internal forces that are produced by the particle itself and external forces like gravity and electromagnetic forces. They can be given as

$$\left(\frac{\partial v}{\partial t} + v \cdot \nabla v\right) = F_{\text{Fluid}} + F_{\text{External}} \tag{5}$$

Gravity is the most important external force which in fact is stated as gravitational acceleration that gives pull to the particles in virtual environment.

4 Proposed Technique in Particle System

4.1 By Imposing Euler Fluid Model in Particle System

The model that we proposed here considers each particle a portion of fluid. Thus, the particle possesses all the properties of fluid like velocity, position and mass, etc. Using the Eulerian fluid grid-based model an equation is derived for conservation of momentum like the Navier–Stokes equation and is represented as

$$\frac{\partial p}{\partial t} + \nabla.(\rho u) = 0 \tag{6}$$

Count of particle and mass of each particle are constant, so there is a guarantee of automatic mass conversation [7]. Hence, all that we needed is the momentum of equation that describes the movement of particles. So, the momentum equation of a single fluid particle is given as

$$\rho(r_i)\frac{\mathrm{d}v_i}{\mathrm{d}t} = F_i = -\nabla p(r_i) + \eta\nabla.\nabla v(r_i) + \rho(r_i)g(r_i) \tag{7}$$

F_i: force acting on particle i; $\rho(r_i)$: density at position of particle i; $\nabla\rho(r_i)$: pressure gradient at position of particle i; $\nabla\cdot\nabla v(r_i)$: velocity Laplacian at position of particle i.

For the acceleration of particle, we get therefore,

$$a_i = \frac{\mathrm{d}v_i}{\mathrm{d}t} = \frac{F_i}{\rho(r_i)} \tag{8}$$

The rest of the description of the particle model is based on the Naiver–Stokes equation of pressure and viscosity [8]. Applying pressure on a fluid particle will be given by the equation

$$F_i^{\text{pressure}} = -\nabla p(r_i) = -\sum_j p_j \frac{m_j}{\rho_j} \nabla W(r_i - r_j, h) \tag{9}$$

But in this equation, force is not symmetric so it could be seen only when two particles collide. As the gradient is zero at the centre, particle i only uses the pressure of particle j and vice versa [9, 10]. As the pressure is different in different areas, pressure force would be different for two particles. Balancing of forces by calculating the mean arithmetic of two particle is given as

$$F_i^{\text{pressure}} = -\sum_j m_j \frac{p_i + p_j}{2\rho_j} \nabla W(r_i - r_j, h) \tag{10}$$

Till now, the pressure at the particle position was unknown. Muller suggested the idea to use ideal gas equation to derive the pressure directly from the density and is given by [2].

4.2 By Implementing Objected Oriented Approach While Coding Particles in Game Engine

Using object-oriented approach, the set of properties can be assigned to huge number of particles generating in the virtual environment. In this approach, no single particle is referenced but the result will be full of particles flying on the screen. Example of object-oriented approach while designing the particle system is shown in Fig. 3.

Implementing particle system using object-oriented approach helps us to tackle two most advanced OOPs techniques, polymorphism and inheritance. By using inheritance and polymorphism, we can create a single list to store particles that can contain similar particle objects of different types. The fluid dynamics mechanism that is added to the particle subsystem, when used with the object-oriented programming approach, brings the virtual reality closer to the real world. It not only serves the purpose but also provides an interactive platform for game users to provide them the virtual real environment [11–13]. We have created few particle

Fig. 3 Designing particle system using object-oriented approach

```
Class Particle{
PVector location;
PVecor velocity;
PVector acceleration;

Particle(PVector 1){
Location=1.get();
Acceleration=new PVector();
Velocity=new PVector();
}

Void update(){
Velocity.add(acceleration);
Location.add(velocity);
}
Void display(){
Stroke(0);
Fill(175);
Ellipse(location x,location.y,8,8);
}
}
```

objects using object-oriented approach and using real-time fluid dynamics which is demonstrated in experiment section

$$p = k(\rho - \rho_0) \quad (11)$$

ρ: rest density; k: gas constant.

4.3 Experiment Results

The early stage results are the mere experiment to give each particle a particular direction so that they can behave according to the events happening in the virtual environment. Particles responding on the basis of event driven activities add more realistic effect in virtual environment. For e.g. blood pattern of stabbed character, explosion effect on the surroundings, etc., early method of implementation of particle systems are procedure-based and thus have no correlation with their surroundings while the experiment results shown above are based on object-oriented approach and include basic fluid dynamics mechanism. Using the analysis of early stage experiments, further advancement has been done on particle systems in unity 3D Game engine to create virtually realistic environment where particle can interact with each other when any event occurs [7]. In the experiment, it can be showed that each particle is entangled with another particle. When forces are applied on the particles, the others respond according to them. This property of particle

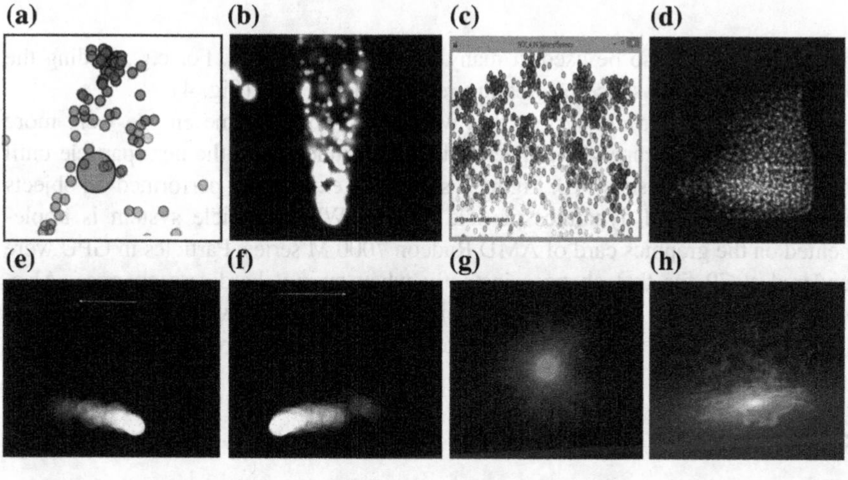

Fig. 4 Early stage results of particle system experiments using object-oriented approach. **a** Collision detection of particles and finding their directions after collision. **b** Dense particle system smoke. **c** Particle emitter based on mouse click event. **d** Sprite rendering. **e** Smoke creation using particle system and giving it a direction according to the mouse pointer, in this case smoke emerges to the *left side*. **f** Smoke creation to *right side*. **g** Explosion effect. **h** Surface to air smoke effect based on object-oriented approach [7]

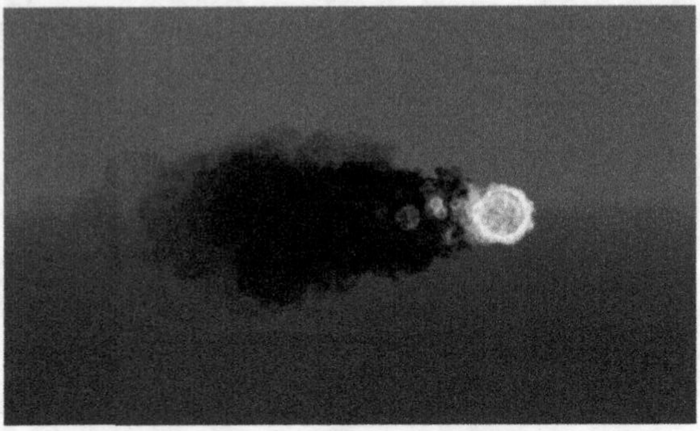

Fig. 5 Fireball with a *black smoke* burning effect (color online)

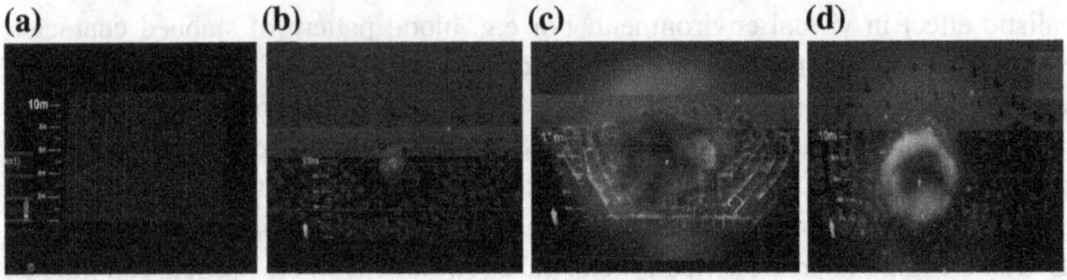

Fig. 6 Experiment result of particle system in unity after adding real-time fluid dynamics mechanism. **a** Wall of long vertical bricks in 3D space for testing. **b** Impact of explosion on bricks when forces are applied. **c** Brick particle creating mushroom effect using particle entanglement. **d** Effect of explosion in middle is transcended to each brick particle presented in 3D space

entanglement can also be used in many ways to add realism. For e.g., finding the blood splatter in 3D space or creating realistic fire effects (Fig. 4).

The results of particle system that we tested in unity game engine were more astonishing as they carry the impact of their surroundings to the next particle until their life span diminishes to 0. Figure 5 shows the experiment performed on objects using real-time fluid dynamics particle system. Whole particle system is implemented on the graphics card of AMD Radeon 7000 M series. Particles in GPU were rendered at 70 fps that changes instantly when current hardware changes. Also, better results can be achieved if rendering technique of hardware keeps pace with the emerging particles in the environment. Below are some results (Fig. 6).

5 Conclusion

There are many ways to stimulate the particle and finding their trajectories but the effective one is to use object-oriented approach using real-time fluid dynamics so that the particle associated with environment possesses the same behaviour. Also, for finding the trajectories of each particle with respect to other particles requires huge computational space and also requires complex quantum algorithm, so implementing this phenomena of adding realism to virtual environment is restricted to small splash games. Current GPU's process whole rendering phenomena of game and adding minute details of particles that work on fluid dynamics outnumber the computing and rendering capacity of graphics processor, which further results in lag. Though the particle system following fluid dynamics mechanism can be practiced on individual objects implementing it on large scale is still a tough task. Recently, a new technique called Ageia, which supports fluid-based particle simulation will be soon included in the PhysX SDK by Nvidi,a which wants to develop the same using CUDA programming framework. Demand for realistic and sophisticated visualization in games may rise soon, if fluid-based particle simulation is added in the current graphics processing technology with the support of quantum algorithm and quantum hardware.

Acknowledgments The author thanks Rashmi Dubey and Sugandha Aggarwal, Faculty of Computer Science and Engineering in Amity University of Science and Technology, for her ideas about particle systems that were insightful and helpful.

References

1. Reeves, W.T.: Particle system-a technique for modelling a class of fuzzy objects. Comput. Graph. **17**(3) (1983)
2. Hasting, E.J., Guha, R.K., Stanley, K.O.: Interactive evolution of particle systems for computer graphics and animation. IEEE Press (2009)
3. Amada, M., et al.: Particle-based fluid simulation on GPU. In: ACM Workshop on General Purpose Computing on Graphics Processors (2004)
4. Kolb, A., Cuntz, N.: Dynamic particle coupling for GPU-based fluid simulation. In: Proceedings 18th Symposium on Simulation Technique, pp. 722–727 (2005)
5. Kipfer, P., Westerman, R.: Realistic and interactive simulation of rivers. In: ACM International Conference Proceedings Series, p. 137 (2006)
6. Teitzel, C., Hopf, M., Ertl, T.: Scientific visualization on sparse grids. In: Nielson, H.H.G.M., Post, F. (eds.) Proceedings of Scientific Visualization - Dagstuhl '97, Heidelberg. IEEE Computer Society, IEEE Computer Society Press, pp. 284–295 (2000)
7. Li, G.-S., Bordoloi, U., Shen, H.-W.: Chameleon: an interactive texture-based rendering framework for visualizing three-dimensional vector fields. In: Visualization 2003. IEEE, pp. 241–248 (2003)
8. Hultquist, J.P.M.: Constructing stream surfaces in steady 3D vector fields. In: Proceedings of the 3rd Conference on Visualization '92, IEEE. IEEE Computer Society Press, pp. 171–178 (1992)
9. Max, N., Crawfis, R., Grant, C.: Visualizing 3D velocity fields near contour surfaces. In: Proceedings IEEE Visualization 94, pp. 248–255 (1994)
10. van Wijk, J.J.: Image based flow visualization. In: Proceedings Visualization. IEEE (2002)
11. Muller, M., et al.: Interaction of fluids with deformable solids. Comput. Animation Virtual Worlds **15**, 159–171
12. Bruckschen, R., Kuester, F., Hamann, B., Joy, K.I.: Real-time out-ofcore visualization of particle traces. In: Proceedings of the IEEE 2001 Symposium on Parallel and Large-Data Visualization and Graphics, pp. 45–50 (2001)
13. Bryson, S., Levit, C.: The virtual wind tunnel: an environment for the exploration of three-dimensional unsteady flows. In: Proceedings IEEE Visualization, pp. 17–24 (1991)

Searchless Fractal Image Compression Using Fast DCT and Real DCT

Preedhi Garg, Richa Gupta and Rajesh K. Tyagi

Abstract Growing need for pictorial data in information era makes image storage and transmission very expensive. Fast algorithms to compress visual information without degrading the quality are of utmost importance. To overcome this problem, this paper proposes new methods to reduce the encoding time for no search fractal image compression in DCT domain by curtailing the computational complexity of the discrete cosine transform (DCT) equations. Fast DCT and real DCT are the techniques, which are employed for the purpose of increasing the performance of searchless DCT. FDCT uses the concept of fast Fourier transform (FFT) which acts as fast discrete cosine transform (FDCT). Real DCT performs only real calculations and omits the imaginary complexity of the DCT. Proposed methods perform the calculations involved in DCT to be computed faster while keeping the quality of the image as much as nearly possible. Furthermore, the experimental results specified show the effectiveness of the methods being proposed for grayscale images.

Keywords Fractal image compression · Fast DCT · FFT · Real DCT · No search · Adaptive FIC

1 Introduction

Digital image compression is gaining high importance in the era of information and technology. Databases containing thousands of images are becoming an integral part of industry and academia. Fractal image compression is a highly potential

Preedhi Garg (✉) · Richa Gupta
CSE Department, Amity University, Noida, Uttar Pradesh, India
e-mail: preedhigarg@gmail.com

Richa Gupta
e-mail: richagupta@amity.edu

R.K. Tyagi
IT Department, KIET, Ghaziabad, Uttar Pradesh, India
e-mail: profrajeshkumartyagi@gmail.com

M. Pant et al. (eds.), *Proceedings of Fifth International Conference on Soft Computing for Problem Solving*, Advances in Intelligent Systems and Computing 436, DOI 10.1007/978-981-10-0448-3_7

image compression technique. Fractal image compression can serve the increasing demands of efficient transmission and storage of pictorial data. Self-similarity is the key property of fractals on which the fractal image compression is based [1].

The idea of fractal image compression was first introduced and executed by Barnsley and Sloan [2]. Jacquin [3] proposed a more practical approach for fractal image compression by introducing partitioned iterative system called PIFS. Jacquin achieved good image quality that too at higher compression rates but came across some drawbacks of high computational time for encoding procedures. The high computational demand as well as the existence of best match between domain and range block makes the execution of fractal image compression arduous. Matching of domain and range blocks is an extensively used operation which is very time-consuming and restricts the usage of fractal image compression in various applications.

Many researchers focused on decreasing the complexity of encoding algorithm such as Chong Fu et al. introduced the concept of DCT-based fractal image compression to increase the efficiency of exploiting the self similarities in an image [1, 4]. Discrete cosine transform (DCT) is a lossy technique which works by splitting an image into parts with differing frequencies. Thus the data present in the image called image pixels are transformed into sets of frequencies. DCT coefficients of low significance are excluded based on permissible threshold [5]. DCT-based image compression decreases the data needed for representing an image. Researchers have presented many algorithms for faster DCT calculations [6–8]. Boussakta and Alshibami [9] introduced the 3D vector-radix decimation in frequency algorithm, which calculates 3D DCT II directly. This regular structured algorithm can be implemented in place for efficient use of memory and faster results. Characteristics of DCT coefficient distribution and Parseval energy conservation theorem-based accelerated method is presented in [10]. This algorithm is designed to detect all zero blocks before transformation and quantization. Simulation result verified 47 % higher detection ratio of zero blocks.

Wohlberg and Jager [11] exploited the advantages of DCT in fractal image compression to achieve better results. DCT possess high energy packing properties, which makes it suitable for image compression. DCT's efficient compaction leads to lower memory requirement to store requirements for domain blocks in fractal image compression. Curtis et al. [12] presented a hybrid fractal and DCT-based image compression technique. In this communication DCT is used only when the quality of decoding block is permissible. In alternate case fractal image compression is used.

Cheng et al. [13] suggested the use of split radix DCT procedure whose fundamentals are based on recursive sparse matrix decomposition. DCT provides a more accurate distance measure for comparative analysis and contractivity factor may be checked rapidly without any inner computations required. DCT domain allows the application of symmetry operations more rapidly. The usage of DCT is beneficial in compression algorithm to cut down the encoding time by maintaining the image quality.

Fractal-based image compression has been combined with adaptive techniques in spatial domain for decimating searching time [14, 15]. Accelerating fractal encoding using adaptive image compression is broadly classified into two streams. Reduction in number of candidates for mapping between domain and range blocks called adaptive paradigm. Alternative approach taken by the researcher is reducing the computation complexity of fractal image compression to achieve faster encoding.

Hasan and Wu [16] suggested the use of adaptive approaches to minimize the long encoding time by reducing the complexity or number of matching operations. The various adaptive techniques used include adaptive quadtree partitioning, zero mean intensity level compression, reducing domain size, reducing number of range block, or reducing the number of domain block. Adaptive techniques attempt to reduce encoding time and increase the compression ratio by keeping the quality of image intact.

Adaptive-based fractal image compression executed in frequency domain using DCT is a novel approach [17]. Range exclusion, an adaptive technique in DCT domain performs a faster fractal image encoding comparatively to spatial domain adaptive method. Homogeneous range blocks are excluded from the range pool before transforming the domain using DCT. Reduced encoding procedure time and better compression ratio have been noted using the above approach.

A DCT-based fractal image encoding scheme helps in faster exploitation of self-similarities within images [4]. A faster approach for computing DCT via FFT is used which separates the even and odd part of DC coefficient. Substitution is performed on DCT via fast Fourier transform equations which give results much faster than the DCT without FFT.

Fractal image compression using DCT over grayscale images computes DC coefficient using cosine values which contain real as well as imaginary part. The maximum information storage capacity resides with real coefficients of DCT. These coefficients can save the encoding time by avoiding the imaginary DCT coefficients. Real DCT algorithm is applied for utilization in fractal image compression to get benefited in terms of encoding time. This correspondence attempts to use real DCT to decrease the computation time.

In this paper, a comparative analysis in terms of compression ratio (CR), peak signal–to-noise ratio (PSNR) and encoding time (ET) is performed. The fractal image compression is performed using fast DCT technique and real DCT method which is combined with searchless fractal image compression. The analysis proves that fractal image compression using these algorithm is competent with other techniques like, no search fractal image compression using DCT and adaptive (range exclusion) fractal image compression, for compressing the images.

The paper is organized into sections where Sect. 2 briefs the discrete cosine transform and techniques to make DCT computation faster. Further, Sect. 3 describes the procedure for searchless fractal image compression using DCT. Steps for performing the adaptive searchless FIC using DCT are mentioned in Sect. 4. Section 5 lists all the steps for the proposed methods which is followed by comparisons and results in Sect. 6.

2 Discrete Cosine Transform

The discrete cosine transform for a two-dimensional discrete function *f*(*x*, *y*) where $x = 0, 1, \ldots, M-1$ and $y = 0, 1, \ldots, N-1$ is

$$F(u,v) = \alpha(u)\alpha(v)\sum_{x=0}^{M-1}\sum_{y=0}^{N-1} f(x,y)\cos\left[(2x+1)\frac{\pi u}{2M}\right]\cos\left[(2y+1)\frac{\pi u}{2N}\right] \quad (1)$$

Inverse transform for *F*(*u*, *v*) where $u = 0, 1, \ldots, M-1$ and $v = 0, 1, \ldots, N-1$ is

$$f(x,y) = \sum_{x=0}^{M-1}\sum_{y=0}^{N-1} \alpha(u)\alpha(v)F(u,v)\cos\left[(2x+1)\frac{\pi u}{2M}\right]\cos\left[(2y+1)\frac{\pi u}{2N}\right] \quad (2)$$

where $\alpha(u) = \begin{cases} \frac{1}{\sqrt{2}}, & |u = 0 \\ 1, & |u \neq 0 \end{cases}$.

Equation (1) converts image defined in spatial domain into frequency domain. Equation (2) is inverse transform for mapping frequencies back to intensities of the image. These equations possess a property of energy compacting.

A fast discrete cosine transform is used to improve the computational time for encoding procedures using DCT via FFT [18]. The fast Fourier transform (FFT) reduces the multiplication and addition computations to the order of $MN \log_2 MN$ which becomes less complex when combined with DCT, and hence act as a fast discrete cosine transform [19, 20]. FFT uses Eqs. (3) and (4) to perform faster calculations.

$$F(u) = \frac{1}{2}\{F_{\text{even}}(u) + F_{\text{odd}}(u)W_{2M}^{u}\} \quad (3)$$

And we have $W_M^{u+M} = W_M^u$ and $W_{2M}^{u+M} = -W_{2M}^u$, thus we can also write

$$F(c+M) = \frac{1}{2}\{F_{\text{even}}(u) - F_{\text{odd}}(u)W_{2M}^{u}\} \quad (4)$$

The odd and even parts are calculated using two M/2 point transformations and substituted in Eq. (3) and the other is computed using Eq. (4). It is observed that usage of DCT via FFT increases the computing speed, and consecutively reduces the computational time.

A real discrete cosine transform algorithm involves calculation of only real operations and ignores the imaginary part thus fosters the image compression. Real function in MATLAB shows the real part of the complex number. DCT computes the dc coefficients which contain imaginary as well as real parts, real function cuts off the imaginary coefficients, saving the computation time.

3 Searchless FIC Using DCT

Fractal coding in fast DCT domain have an information compacting capability as additional advantage of faster computations. Fast DCT transform has been described in the preceding section using Eqs. (1), (2), (3), and (4). In fractal image compression, if fast DCT is applied on a block of size $A \times A$ then it will give $A \times A$ dc coefficients. The number of coefficients will increase with the increase of the block size, which will take more computational time. Applying DCT on the whole image takes more time and also it round off the small magnitude values near edges to zero [21].

The outline of searchless fractal image compression method using DCT domain is depicted in the Procedure 3.1.

Procedure 3.1 Searchless fractal image compression in DCT domain procedure

- Image is segmented into range and domain blocks of sizes $A \times A$ and $2A \times 2A$, respectively.
- DCT is applied on the blocks formed in the above step.
- Calculate range block mean.
- Perform the mean of four neighboring pixels of domain blocks for obtaining domain mean.
- Choose a threshold for each decomposition level.
- For each range block.

  - Fix the position of the range block on the domain block at location ($row_R - A/2$, $col_R - A/2$)
  - Calculate error using Eq. (5)

$$E(R,D) = \sum_{i=1}^{n} (sd_i + o - r_i) \tag{5}$$

    where R and D are the mean value of range and domain block, respectively, s is scaling factor and o is contrast scaling.
  - If $E(R,D) < \text{TH}$ then save the dc coefficient of the block which is $F_r(1,1)\,F_r(1,1)$.
  - Save the best scaling parameter by taking into account $F_r(1,1) = F_d(1,1) = 0$.
  - Otherwise apply the quadtree partitioning to divide the range blocks.

The dc coefficient of range block, $F_r(1,1)$ and domain block, $F_d(1,1)$ are calculated using Eq. (1). No search algorithm is applied, while performing the fractal image encoding procedure to reduce the computational time by fixing the position of the range block on the domain block at location ($row_R - A/2$, $col_R - A/2$) as it gives better reconstruction fidelity than other locations [22].

4 Adaptive Fractal Image Compression

The searching operation in fractal image compression utilizes most of the time, and thus produces the long encoding time during encoding phase. Range exclusion is an adaptive technique which focuses on less searching instead of faster searching, and hence attempts to reduce the range blocks by removing the range blocks which are homogeneous in nature [16]. The Procedure 4.1 outlined below depicts the working of range exclusion method in searchless fractal image compression in DCT domain.

Procedure 4.1 Adaptive fractal image compression using DCT procedure

- Choose a threshold limit called homogeneous permittivity (HP).
- For each range block,
  - Calculate range mean $\bar{r}$ and variance Vr using Eqs. (6) and (7).

$$\bar{r} = \frac{1}{\mathrm{MN}} \sum_{x=0}^{M-1} \sum_{y=0}^{N-1} M_{ij} \tag{6}$$

$$Vr = \frac{1}{\mathrm{MN}} \sum_{i=0}^{M-1} \sum_{j=0}^{N-1} \left[M_{ij} - \bar{r}\right] \tag{7}$$

  - If $Vr < \mathrm{HP}$ then exclude that range block and encode only mean $\bar{r}$ for these bocks which uses up 6 bits to store.
  - Otherwise include the range block for the pool.
  - Perform DCT transform on each range block.

Homogeneous permittivity (HP) is a factor which decides how much homogeneity is allowed for a range block to have. A block having low variance depicts high homogeneity, whereas blocks with greater variance are tending to show heterogeneous property. Thus, range exclusion accounts for less encoding time and higher compression ratio by reducing the number of search operations.

5 Proposed Algorithm

An improved algorithm for no search fractal image compression based on fast DCT and real DCT is proposed and presented in Algorithm 5.1. Encoding phase includes the partitioning of image into range and domain blocks, and then applying the fast DCT to the blocks with no search algorithm and thus forms the codebook. During the decoding phase, the codebook is taken as input and converted into the decoded image.

Algorithm 5.1 Searchless fractal image compression using (a) fast DCT and (b) real DCT algorithm

- Load the original image of size $M \times N$.
- Partition original image into nonoverlapping range blocks of size $A \times A$.
- Partition the original image into domain blocks of size $2A \times 2A$ with some integral steps. Size of a domain is twice the size of range blocks.
- Define a collection of contractive affine transformations mapping from domain block to the range block.
- Select a value for threshold for each decomposition level.
- For each range block of size $A \times A$ which is contained at the center of a domain block of size $2A \times 2A$ is taken.
  - Compute the mean of four neighbor pixels of domain block.
  - Apply
    (a) Fast DCT transform.
    (b) Real DCT transform.
  - Choose a corresponding domain block and symmetry so that error is within the threshold limit for particular decomposition level.
  - If threshold limit crosses then apply quadtree partition and repeat this loop.
- Save the best matched dc coefficient of range block as depicted in Procedure 3.1 and scaling factor s.
- Thus IFS code is generated which is compressed using Huffman encoding algorithm.
- At decoding phase, the IFS code is taken and Huffman decoding algorithm is applied and image blocks are generated using the codebook.

6 Experimental Results

To verify the effectiveness of the proposed method, a simulation is created by programming the same on MATLAB 7 implemented on A8 Vision AMD processor and Windows 7 operating system. Test images are of sizes 512×512 with 8-bit gray scale. The results are compared with no search fractal image compression in DCT domain and range exclusion searchless FIC in DCT Domain.

For all the ranges, the maximum range size is 16×16 with a quadtree depth of 4 and minimum range size is 2×2. Scaling factor is stored using 2 bits and dc coefficient is quantized using 8, 7, 7, 7 bits corresponding to each quadtree level.

The comparative analysis is done using three measures which are encoding time in seconds (ET(s)), compression ratio (CR), and peak signal to noise ratio (PSNR). All the values for these measures for the searchless DCT method, adaptive searchless DCT method, searchless FDCT method, and searchless RDCT method are summarized in Table 1.

Table 1 Comparison of the proposed method with other techniques

		Lena	Jet	Scene	Lady	Pepper	Baby	Cartoon	Boat	Baboon	Bridge
Searchless DCT	CR	20.32	13.33	10.19	12.12	17.85	36.39	5.74	12.08	4.92	6.54
	PSNR	30.01	29.27	27.11	23.89	29.46	32.71	25.32	27.33	22.19	24.69
	ET(s)	7.72	11.16	12.48	10.66	6.83	3.89	22.47	13.07	26.47	18.70
Adaptive searchless DCT	CR	23.56	12.99	13.45	14.67	17.98	38.68	7.43	15.90	5.52	8.31
	PSNR	33.45	32.76	29.33	25.98	30.11	34.79	27.45	28.71	25.67	26.69
	ET(s)	8.32	12.34	12.99	11.23	7.16	4.67	22.99	14.55	27.22	19.22
Searchless FDCT	CR	21.20	11.33	10.89	12.69	18.34	36.78	7.42	13.32	5.14	7.02
	PSNR	30.68	29.23	28.40	22.99	30.17	32.97	26.50	28.22	21.67	25.45
	ET(s)	6.96	10.23	10.90	13.65	8.41	5.79	23.56	13.99	27.11	18.99
Searchless RDCT	CR	31.40	19.44	14.39	18.69	30.75	58.50	7.23	17.10	5.73	9.44
	PSNR	30.25	29.58	27.60	23.86	29.69	33.14	27.76	27.86	22.44	24.91
	ET(s)	5.52	8.37	11.21	9.82	5.86	3.33	21.23	10.44	26.37	16.75

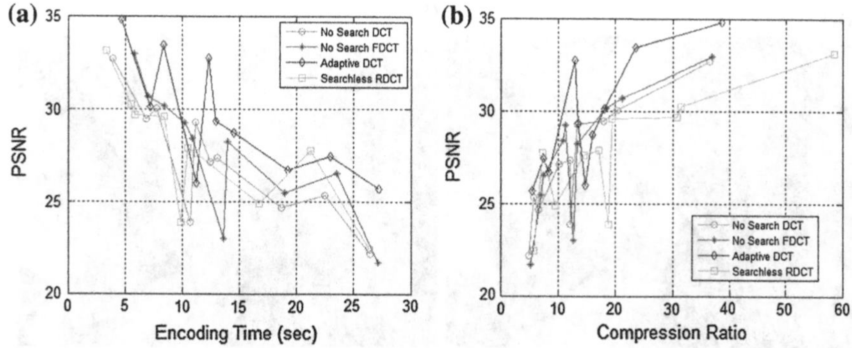

Fig. 1 **a** The PSNR versus encoding time (seconds) and **b** the PSNR versus compression ratio between various techniques

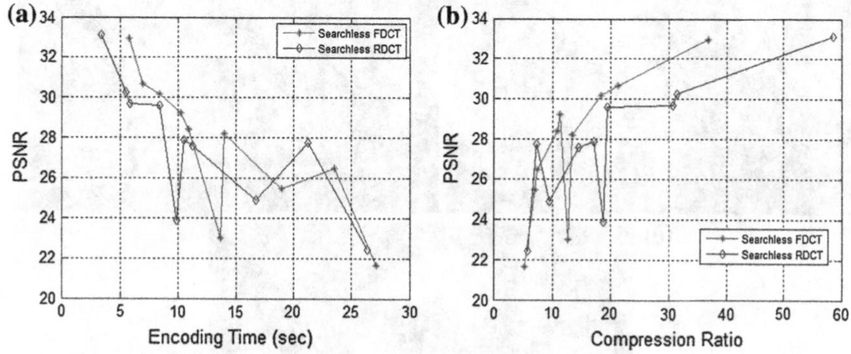

Fig. 2 **a** The PSNR versus encoding time (seconds) and **b** the PSNR versus compression ratio for fast DCT and Real DCT

The comparison of the PSNR and encoding time measured on various 512 × 512 images with all the four methods is shown in Fig. 1a and similarly Fig. 1b gives the comparison of compression ratio with the PSNR values for the various 512 × 512 images using all the four methods. Further, Fig. 3 shows the original image and also the decoded image using no search DCT, no search using Fast DCT, no search using real DCT, and range exclusion method, respectively.

The comparative analysis of the two DCT methods proposed in this paper, i.e., fast DCT and real DCT, are shown in Fig. 2a, b. PSNR versus encoding time (seconds) is depicted in Fig. 2a and PSNR versus compression ratio is showcased in Fig. 2b. The analysis shows that real DCT achieves much less encoding time and higher compression ratio in comparison with the fast DCT (Fig. 3).

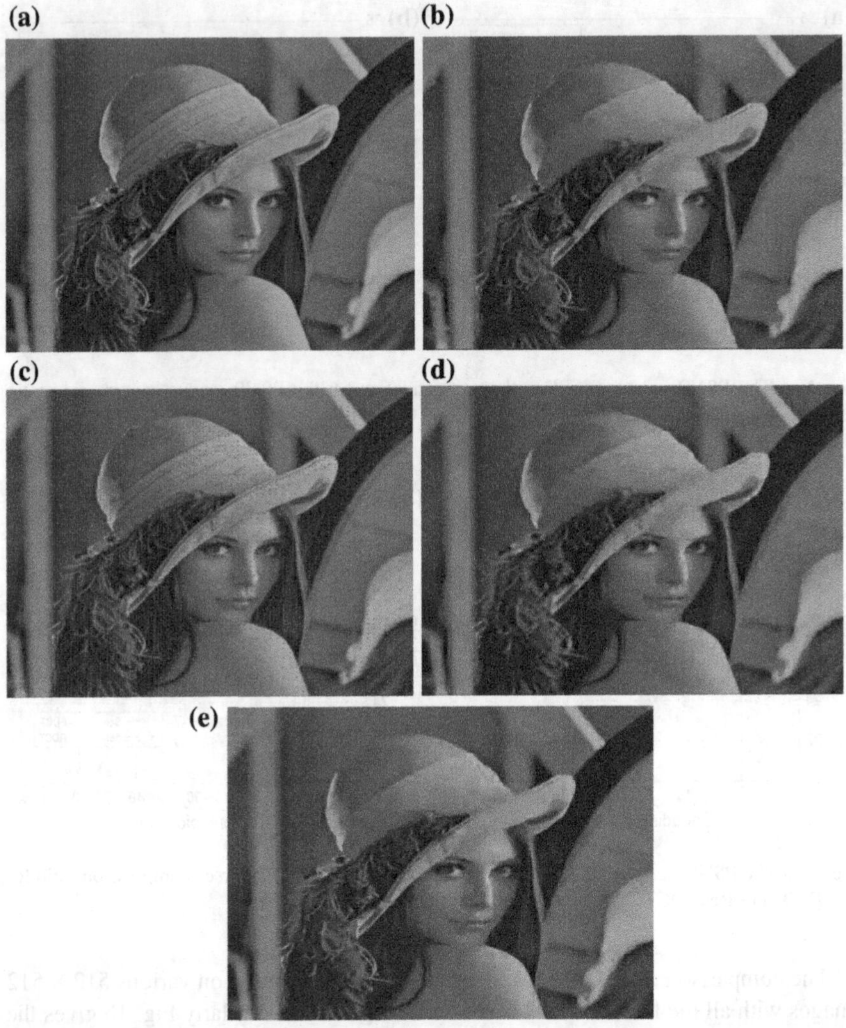

Fig. 3 Experimental results of image encoding: **a** original image **b** is image decoded using no search in DCT domain **c** decoded image using no search using fast DCT **d** image decoded using range exclusion method **e** image decoded using no search real DCT method

7 Conclusion

In this paper, we proposed a new approach to decrease the encoding time by implementing fast DCT and real DCT in no search fractal image compression. The experimental results shown prove that proposed method is better than searchless DCT and range exclusion method. Fast DCT method computes DC coefficient much faster by applying FFT where as real DCT only computes the real part of the complex DCT and ignores the imaginary one. The PSNR values are also better in fast and real DCT method than in other methods. The comparison between fast and real DCT depicts that the latter technique is a better image compressor with much lesser encoding time than the prior technique. In future, algorithms can be developed to perform the DCT calculations much faster to improve the performance.

References

1. Wohlberg, B., De Jager, G.: A review of the fractal image coding literature. IEEE Trans. Image Process. **8**(12), 1716–1729 (1999)
2. Barnsley, M.F., Sloan, A.D.: A better way to compress images. Byte **13**(1), 215–223 (1988)
3. Jacquin, A.E.: Fractal image coding: a review. Proc. IEEE **81**(10), 1451–1465 (1993)
4. Fu, C., Zhu, Z-L.: A DCT-based fractal image compression method. In: International Workshop on Chaos-Fractals Theories and Applications, 2009. IWCFTA'09. IEEE (2009)
5. Rawat, C., Meher, S.: A hybrid image compression scheme using DCT and fractal image compression. Int. Arab J. Inf. Technol. **10**(6), 553–562 (2013)
6. Dai, Q., Chen, X., Lin, C.: Fast algorithms for multidimensional DCT-to-DCT computation between a block and its associated subblocks. IEEE Trans. Signal Process. **53**(8), 3219–3225 (2005)
7. Nagaria, B., et al.: An optimized fast discrete cosine transform approach with various iterations and optimum numerical factors for image quality evaluation. In: 2011 International Conference on Computational Intelligence and Communication Networks (CICN), IEEE (2011)
8. Feig, E., Winograd, S.: Fast algorithms for the discrete cosine transform. IEEE Trans. Signal Process. **40**(9), 2174–2193 (1992)
9. Boussakta, S., Alshibami, H.O.: Fast algorithm for the 3-D DCT-II. IEEE Trans. Signal Process. **52**(4), 992–1001 (2004)
10. Xie, Z., et al.: A general method for detecting all-zero blocks prior to DCT and quantization. IEEE Trans. Circuits Syst. Video Technol. **17**(2), 237–241 (2007)
11. Wohlberg, B.E., De Jager, G.: Fast image domain fractal compression by DCT domain block matching. Electron. Lett. **31**(11), 869–869 (1995)
12. Curtis, K.M., Neil, G., Fotopoulos, V.: A hybrid fractal/DCT image compression method. In: 2002 14th International Conference on Digital Signal Processing, 2002. DSP 2002, vol. 2. IEEE (2002)
13. Cheng, L.Z., Xu, H., Luo, Y.: Integer discrete cosine transform and its fast algorithm. Electron. Lett. **37**(1), 64–65 (2001)
14. Xing, C.: An adaptive domain pool scheme for fractal image compression. In: International Workshop on Education Technology and Training, 2008. And 2008 International Workshop on Geoscience and Remote Sensing. ETT and GRS 2008, vol. 2. IEEE (2008)
15. Tong, C.S., Pi, M.: Fast fractal image encoding based on adaptive search. IEEE Trans. Image Process. **10**(9), 1269–1277 (2001)
16. Mohammed Hasan, T., Wu, X.: An Adaptive Fractal Image Compression (2013)
17. El Khamy, S.E., Hamdy, N.A., Shatila, H.: Adaptive fractal image compression using wavelet sub-tree coefficients. In: IEEE 46th Midwest Symposium on Circuits and Systems, 2003, vol. 2. IEEE (2003)
18. Chen, W.-H., Smith, C. H., Fralick, S.: A fast computational algorithm for the discrete cosine transform. IEEE Trans. Commun. 1004–1009 (1977)
19. Gonzalez, R.C.: Digital Image Processing. Pearson Education India (2009)

20. Rao, K.R., Yip, P.: Discrete Cosine Transform: Algorithms, Advantages, Applications. Academic press (2014)
21. Salarian, M., Hassanpour, H.: A new fast no search fractal image compression in DCT domain. In: International Conference on Machine Vision, 2007, ICMV 2007. IEEE (2007)
22. Furao, S., Hasegawa, O.: A fast no search fractal image coding method. Signal Process. Image Commun. **19**(5), 393–404 (2004)

Study of Neighborhood Search-Based Fractal Image Encoding

Indu Aggarwal and Richa Gupta

Abstract Fractal image encoding is one of the famous lossy encoding techniques ascertain high compression ratio, higher peak signal-to-noise ratio (PSNR), and good quality of encoded image. Fractal image compression uses the self-similarity property present in the natural image and similarity measure. The main drawback fractal image encoding suffers from is significant time consumption in search of appropriate domain for each range of image blocks. There have been various researches carried out to overcome the limitation of fractal encoding and to speed up the encoder. Initially, various classification and partitioning schemes were used to reduce the search space. A remarkable improvement was made by neighborhood region search strategies, which classify the image blocks on the basis of some feature vectors of image to restrict the region for best matching pair of domain and range, and also reduces the search complexity to logarithmic time. In this paper, three image block preprocessing approaches using neighborhood search method are explained in different domains and all these approaches are compared on the basis of their simulation results.

Keywords Fractal image encoding · IFS · Image block preprocessing · Neighborhood search · DCT · DWT

1 Introduction

In the past few years, development of multimedia products has increased. The need of high storage capacity of storage devices to store the graphics files can be accomplished by encoding of graphics files, which reduces the size of number of bits as much as possible. Image encoding reduces the size of file without hampering

Indu Aggarwal (✉) · Richa Gupta
CSE Department, ASET, Amity University, Noida, India
e-mail: induaggarwal1311@gmail.com

Richa Gupta
e-mail: rgupta6@amity.edu

M. Pant et al. (eds.), *Proceedings of Fifth International Conference on Soft Computing for Problem Solving*, Advances in Intelligent Systems and Computing 436, DOI 10.1007/978-981-10-0448-3_8

the quality of images in order to store and transmit in an efficient manner. Image encoding is broadly classified in two kinds: lossy and lossless encoding. A lossless method gives an identical image to the original image after decoding, whereas lossy will provide an image that somewhat resembles the original one after decoding process. Fractal image encoding is a lossy image compression technique. This approach uses self-similarity characteristic in natural image to perform the encoding. It improves the compression ratio and PSNR value with some acceptable loss of data to maintain the quality of an image. However, original process of coding is very time-consuming. The computation of similarity measure can be reduced using a robust search method that reduces computations and also coding time.

There have been various approaches to overcome the limitation of fractal image encoding in the past few decades. Some of the methods were based on classification of image blocks [1–3]. Many algorithms have been designed with different search strategy to reduce the region of interest [4–6]. Techniques developed to determine the best domain block for the range blocks of image for further transformation to encode and decode the image [7–9]. Antonioni et al. [10] proposed a hybrid approach using wavelet transform and vector quantization to give progressive transmission of an image to receiver to obtain the picture at a faster rate with minimum cost. Fisher [11] described the whole fractal encoding implementation in detail. Saupe [4] gave the nearest neighbor-based fractal encoding algorithm which speed up the encoder. Fu [1] introduced the fractal encoding based on classification in wavelet domain to overcome the limitation of basic fractal algorithm. Truong [12] developed a fast encoding algorithm by reducing MSE computations. Tong [5] made some improvements in Saupe's [4] nearest neighbor approach to obtain high PSNR value and compression ratio with less encoding time. Iano et al. [2] presented a fast hybrid fractal-wavelet approach which used Fisher's domain classification to apply at low-pass subband of wavelet-transformed image and modified SPIHT reduces 94 % of encoding and decoding time. Zhou et al. [7] anticipated an improved search scheme using image feature for best matching of domain and range blocks with a bit reduction in PSNR value. Fu and Zhu [3] introduced a discrete cosine transform (DCT) lower coefficient-based classification of range and domain blocks with high compression ratio and good reconstruction fidelity. Yu et al. [13] gave a segmentation-based approach to overcome the drawback of high encoding time. Wu and Lin [14] derived a genetic algorithm-based hybrid select approach to speed up the fractal encoder with little decay in image quality. Lin and Wu [15] improved the speed of fractal encoder using edge property-based neighborhood search in DCT domain with 0.9 dB decay in image quality. Further, he [8] used particle swarm optimization as search method under discrete wavelet transform (DWT) classification and dDihedral transformations to reduce the mean square error (MSE) computations. Wei et al. [16] improved the traditional fractal encoding by using only diagonal data to calculate MSE to speed up the encoder. de Quadros Gomes [17] introduced parallel modeling of fractal image encoding using dual or quadcore machines to get the encoded image in feasible time. Shiping and Lijing [9] used ortho difference sum-based method to restrict the domain for best matching and to reduce the encoding time. Zhang et al. [6] also used neighborhood search-based approach to save plenty of time to encode by using sub-block subtraction to limit the region for search. Nodehi et al. [18] gave an intelligent fuzzy approach to improve the encoding speed. There have been various approaches introduced by researchers to overcome the time consumption of fractal encoder and Du et al. [19] made a contribution using different a approach based on quantum mechanics to reduce the computational complexity.

1.1 Basic Fractal Image Encoding

Fractal image encoding is based on the natural image property of self-similarity. It decomposes the image using various techniques to determine the range blocks and domain blocks and then map these blocks on the basis of self-similarity of image blocks. M. F. Barnsley and Demko introduced the theory of iterated function systems IFS to define the natural objects. IFSs (iterated function systems) are the unified mathematical model to generate the parts of image that are similar to the whole image. Decoding of image is faster when compared to encoding.

The basic theory of fractal encoding based on IFS was introduced by Barnsley and Demko [20]. It was followed by the development of automated algorithm for fractal image encoding based on partitioned IFS (PIFS). Jacquin [21] exploits the restrictions of local self-similarity of image by dividing it into a number of sub-blocks. He suggested fractal block-based encoding technique for compressing the image using lesser memory and faster encoding. The algorithm for basic fractal encoding is as follows:

1. Partition the given image into mutually disjoint portions R_i, i.e., $R_i \cap R_j = \emptyset (i \neq j)$.
2. An equal number of other image portions are generated known as domain blocks D_j, which can overlap each other and double in size than that of range blocks to follow the contractive mapping fixed point theorem.
3. For each range block R_i, affine transformations are made to find the most similar domain block D_j, and store affine transformation parameters as the iterative parameter.
4. The image can be reconstructed using iterative parameters. In general, iterate for 5–8 times to produce the encoded image.

2 Image Block Transformation Techniques

In this section, three approaches based on neighborhood search are explained. All the algorithms describe three different domains for the search method. The first method uses spatial domain to find the best matching pair of range and domain blocks using orthogonal projection and pre-quantization, which reduces the problem to nearest neighbor search of transformed domain blocks for transformed blocks. The DCT-based approach provided frequency domain to gather all range and domain blocks with similar edge properties to reduce the area of search. However, DWT reduces the dihedral transformations of blocks by classifying them into different regions of same class to which they belong.

2.1 Neighborhood Search Strategy in Spatial Domain

In traditional fractal image encoding, the encoding scheme is time-consuming due to the search of appropriate domain block for each range block of an image. It uses linear search to determine the best domain blocks. This strategy suffers from redundancy present in the search method. To accelerate the search for domain, Saupe introduced a novel approach to find the best domain–range pair by using Euclidean space of feature vectors of domains and ranges. It reduces the search complexity from linear time to logarithmic time [4].

Saupe [4] derived the theorem stating that least square error *E*(*R*, *D*) is proportional to function *g* of Euclidean distance between normalized projections $\emptyset(R)$ and $\emptyset(D)$ is defined as:

$$E(R,D) = \|R - \bar{r}I\|^2 g(\Delta(R,D)) \quad (1)$$

where $\emptyset(.)$ is the normalized projection operator, $g(\Delta) = \Delta\sqrt{1 - \Delta^2/4}$ and $\Delta(R,D) = \min(\|\emptyset(R) + \emptyset(D)\|, \|\emptyset(R) - \emptyset(D)\|)$.

Since the function $g(\Delta)$ is monotonically increasing in the domain interest and has the essence of the new form for the distortion, in the domain pool Ω minimisation of *E*(*R*, *D*) is equivalent to minimizing Δ (*R*, *D*). Now nearest neighbor search is done $\emptyset(R)$ in the set $\{\pm\emptyset(D) : D \in \Omega\}$. In between the transformed domain blocks $\emptyset(D)$ and transformed range block $\emptyset(R)$ problem is now converted from domain range matching to nearest neighbor problem.

The brute-force linear search has $O(N_D)$, there are many algorithms and data structures of nearest neighbor research. Many variants of the k-d tree algorithm proposed by Arya et al. are adopted by Saupe [4]. $O(N_D \log N_D)$ Operations are required to build a tree data structure where domain blocks after transformation is done, before search. There is a degradation in the time complexity from $O(N_D)$ to $\mathrm{O}(\log N_D)$ [5].

2.2 Neighborhood Search Method Using Discrete Cosine Transform (DCT)

Lin and Wu [15] introduced a novel and efficient approach based on neighborhood region search method using the edge property in DCT frequency domain. In this approach, the two-dimensional coordinate system is prepared using lowest frequency coefficients of DCT to gather all domain and range blocks with same edge property. The two lowest DCT coefficients are used to build coordinate system of frequency domain, i.e., $F(1, 0)$ and $F(0, 1)$. The quantity $F(p, q)$ can be defined as

$$F(p,q) = \frac{2}{N} C_p C_q \sum_{i=0}^{N-1} \sum_{j=0}^{N-1} f(i,j) \cos\left(\frac{(2i+1)p\pi}{2N}\right) \cos\left(\frac{(2j+1)q\pi}{2N}\right) \quad (2)$$

where $f(i, j)$ is the image block of size $N \times N$ and $p, q = 0, 1, 2, 3, \ldots, N-1$ and,

$$C_m = \begin{cases} \frac{1}{\sqrt{2}}, & \text{if } m = 0 \\ 1, & \text{otherwise} \end{cases} \quad (3)$$

The algorithm for the neighborhood search-based fractal encoding in DCT domain is explained as follows:

For each image block, both the DCT coefficients must be calculated and their absolute values constitute a pair $(|F(1,0)|, |F(0,1)|)$ represents an image block. It was assumed that $(|F^*(1,0)|, |F^*(0,1)|)$ is the element with maximal norm. In Fig. 1, $(|F^*(1,0)|s, |F^*(0,1)|s)$ denote the scaled element with maximum norm. Therefore, all the ranges and domain blocks are further mapped into this unit loop in the first quadrant. All the similar edge-shaped blocks are gathered together in this quadrant and the two nearest blocks will have a highly similar edge shape.

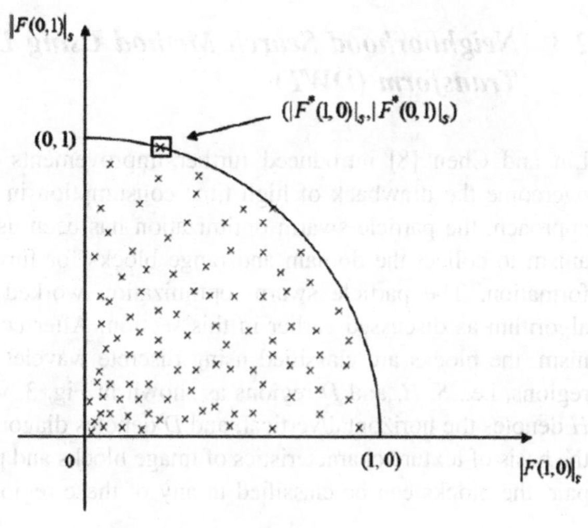

Fig. 1 Classification of image blocks in frequency domain (*Source* Computers and mathematics with applications 62, p. 313)

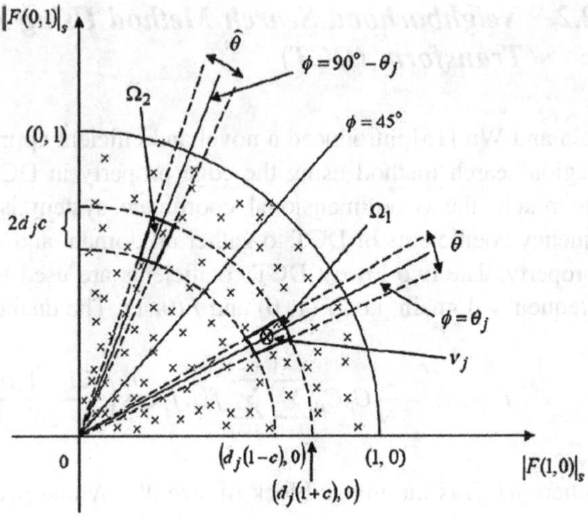

Fig. 2 Search space for fractal encoding (*Source* Computers and mathematics with applications 62, p. 314)

After completion of the nearest neighborhood search for image blocks, these blocks can be classified into two different regions, Ω_1 and Ω_2 as shown in Fig. 2. Begin with the initialization of spanned phase angle $\hat{\theta}$ of $\emptyset = \theta_j$ and positive constant value c. In Fig. 2, the d_j is the distance between the origin and any range block v_j, the straight line connecting origin and d_j is given as $\emptyset = \theta_j$, where $\theta_j = \tan^{-1}(|F(0,1)|s/|F(1,0)|s), |F(1,0)|s > 0$. All these notations must be calculated and should be used further to check whether search space contains any domain block or not. If it does not have any domain block then $\hat{\theta} = \hat{\theta} + \Delta\theta$ and $c = c + \Delta c$, otherwise determine the optimal domain from search space and store it as fractal code for range block v_j.

2.3 Neighborhood Search Method Using Discrete Wavelet Transform (DWT)

Lin and Chen [8] introduced further improvements in their previous works to overcome the drawback of high time consumption in the encoding phase. In this approach, the particle swarm optimization has been used as a block search mechanism to collect the domain and range blocks for further classification and transformation. The particle swarm optimization worked as a neighborhood search algorithm as discussed earlier in this section. After completing the search mechanism, the blocks are classified using discrete wavelet transform in different three regions, i.e., *S*, *H*, and *D* regions as shown in Fig. 3, where *S* denotes the smooth, *H* denotes the horizontal/vertical, and *D* denotes diagonal/sub-diagonal regions. On the basis of texture characteristics of image blocks and position of normalized DWT pair, the blocks can be classified in any of these regions.

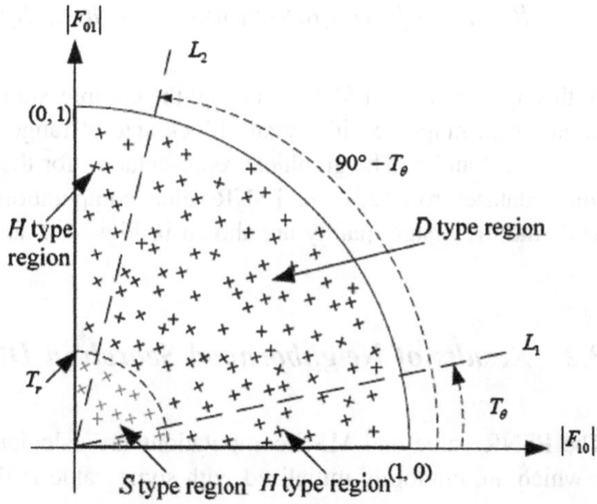

Fig. 3 Partition search space for DWT classification (*Source* JISE 28, p. 23)

After classification of all the blocks, the dihedral transformations are performed on domain blocks for each range block to find the best matching pair using dihedral property shown in Fig. 3. The MSE computations are completed after the transformations are performed according to the classification. The algorithm used for fractal image encoding using neighborhood search in DWT domain can be summarized as follows:

1. Initialize the swarm size and all the parameters of particle.
2. For each image block, check whether they belong to same class or not. If yes, then go to step 3. Otherwise, go to step 5.
3. For each block of same class, check whether they are in the same region (*S*, *H*, or *D*) or not. If yes, then apply the first four dihedral transformations on them. Otherwise, the last four transformations will be performed.
4. Calculate MSEs after performing the transformations.
5. Update the position and velocity of each particle.
6. If stopping criterion has been reached, then stop. Otherwise, go to step 2.

3 Experiments and Results

In this section, all the three neighborhood search-based approaches discussed in previous section are compared. All these approaches are analyzed on the basis of four parameters, i.e. PSNR value, compression ratio, MSE computations and CPU time for encoding and decoding. The results of comparison are as follows:

3.1 Results of Neighborhood Search in Spatial Domain

In this approach, the PSNR value and the compression ratio varies according to the variation in range sensitivity and block size of range cells which is shown in the Tables 1, 2 and 3. The graphical representation for the factors using miscellaneous image dataset from [22], i.e., PSNR value, computation time, and compression ratio to define the image quality are shown in Figs. 4, 5 and 6.

3.2 Results of Neighborhood Search in DCT Domain

The PSNR value and MSE computations are calculated using the variable $\hat{\theta}$ and c which are randomly initialized with some value is illustrated in Table 4.

Table 1 Analysis of four parameters with respect to block size of minimum range cells

Block size of range cells		Computation time	PSNR (dB)	Compression ratio
Min	Max			
2	64	0.0239	33.251	0.0167
4	64	0.0233	29.125	0.0140
8	64	0.0305	26.094	0.014
16	64	0.0254	23.926	0.0157

Table 2 Analysis of four parameters with respect to block size of maximum range cells

Block size of range cells		Computation time	PSNR (dB)	Compression ratio
Min	Max			
2	4	0.0292	33.41	0.0167
2	8	0.0197	33.29	0.0140
2	16	0.0258	33.25	0.014
2	64	0.0222	33.24	0.0150

Table 3 Analysis of three different images using evaluation parameters

Images	Size of compressed image	Computation time	PSNR (dB)	Compression ratio
Lena	5400	0.026	32.95	0.0158
Baboon	6232	0.025	34.75	0.0118
Flower	6056	0.022	34.81	0.0115

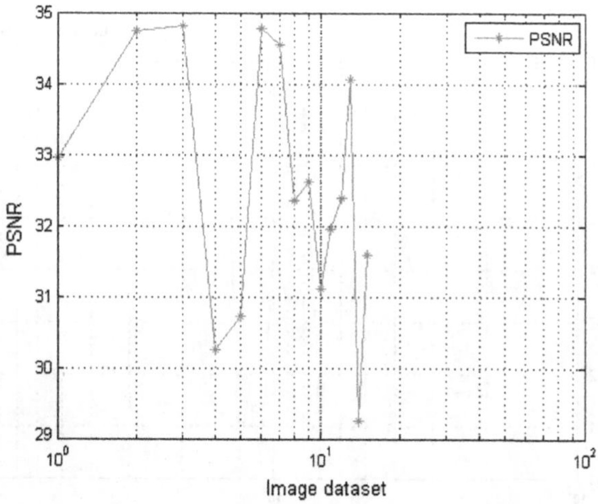

Fig. 4 PSNR values corresponding to image dataset

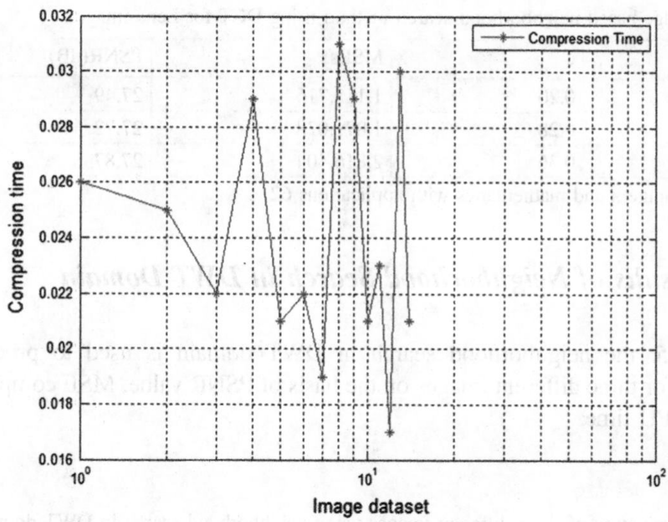

Fig. 5 Compression time values corresponding to image dataset

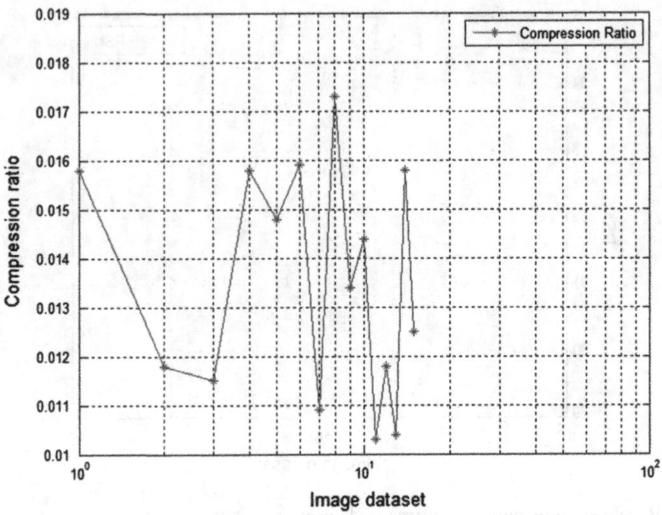

Fig. 6 Compression ratio values corresponding to image dataset

Table 4 Edge-based neighborhood search method using DCT for Lena image

$\hat{\theta}$ (°)	c	MSEs	PSNR(dB)
2.0	0.20	1,115,716	27.49
2.4	0.28	1,871,028	27.72
2.8	0.36	2,816,404	27.87

Source Computers and mathematics with applications 62

3.3 *Results of Neighborhood Search in DWT Domain*

In Table 5, the neighborhood search in DWT domain is used to process the encoding for three different images on the basis of PSNR value, MSE computations and the CPU time.

Table 5 Evaluation of three different images using neighborhood search in DWT domain

Images	CPU time	MSEs	PSNR (dB)
Baboon	8.125	2,062,584	19.715
Barche	8.266	2,115,532	25.867
Donna	10.125	2,090,948	31.134

Source JISE 28

4 Conclusion

In this paper, three major neighborhood fractal image compression approaches are observed and analyzed. Parameters used to compare and analyze are mentioned below:

1. PSNR value
2. Compression ratio
3. MSE computations
4. Computational time.

The neighborhood search strategy used in spatial domain reduces the encoding time from linear time to logarithmic time. This method reduces the search space by using Euclidean space of feature vectors of domain. It eventually provided multi-dimensional neighborhood search. The same principle was followed by neighborhood search in DCT-based lowest frequency domain by reducing the search space to gather image blocks with similar edge shapes. In addition to spatial domain-based approach, it take less time to map the range blocks and domain blocks which consecutively consumes lesser time to encode image. The discrete wavelet-based classification used particle swarm optimization to find the best domain block in global domain which considers only the same type of range and domain blocks. After classification, only four dihedral transformations are made for domain blocks which reduce the encoding time. All the three algorithms gave high PSNR value, high compression ratio and reduced MSE computations with little decay in image quality which is acceptable. For references, follow the given guidelines.

References

1. Fu, P., Tang, X., Zhu, Y., Wu, X.: A new fractal block coding scheme based on classification in the wavelet domain. In: Proceedings of IEEE Vehicle Electronics Conference, pp. 315–318 (1999)
2. Iano, Y., Da Silva, F.S., Cruz, A.L.M.: A fast and efficient hybrid fractal-wavelet image coder. In: IEEE Transactions on Image Processing, vol. 15, pp. 98–105 (2006)
3. Fu, C., Zhu, Z.L.: A DCT-based fractal image compression method. In: IEEE International Workshop on Chaos-Fractals Theories and Applications, pp. 439–443 (2009)
4. Saupe, D.: Fractal image compression via nearest neighbor search. Univ., Inst. für Informatik (1996)
5. Tong, C.S., Wong, M.: Adaptive approximate nearest neighbor search for fractal image compression. In: IEEE Transactions on Image Processing, vol. 11, pp. 605–615 (2002)
6. Zhang, A.H., Sheng, F., Sun, X.: A fast fractal encoding algorithm based on sub-block subtraction. In: 9th International Conference on Natural Computation (ICNC), pp. 1204–1208 (2013)
7. Zhou, Y., Zhang, C., Zhang, Z.: Fast fractal image encoding using an improved search scheme. Tsinghua Sci. Technol. **12**, 602–606 (2007)

8. Lin, Y.L., Chen, W.L.: Fast search strategies for fractal image compression. J. Inf. Sci. Eng. **28**, 17–30 (2012)
9. Shiping, L., Lijing, L.: The algorithm of fractal image block coding based on the ortho difference sum. In: 25th Chinese Control and Decision Conference (CCDC), pp. 655–659 (2013)
10. Antonini, M., Barlaud, M., Mathieu, P., Daubechies, I.: Image coding using wavelet transform. IEEE Trans. Image Process. **1**, 205–220 (1992)
11. Fisher, Y.: Fractal image compression. World Scientific, Fractals **2**, 347–361 (1994)
12. Truong, T.K., Jeng, J.H., Reed, I.S., Lee, P.C., Li, A.Q.: A fast encoding algorithm for fractal image compression using the DCT inner product. IEEE Trans. Image Process. **9**, 529–535 (2000)
13. Yu, H., Li, L., Liu, D., Zhai, H., Dong, X.: Based on quadtree fractal image compression improved algorithm for research. In: International Conference on E-Product E-Service and E-Entertainment, pp. 1–3 (2010)
14. Wu, M.S., Lin, Y.L.: Genetic algorithm with a hybrid select mechanism for fractal image compression. Digital Signal Process. **20**, 1150–1161 (2010)
15. Lin, Y.L., Wu, M.S.: An edge property-based neighborhood region search strategy for fractal image compression. Comput. Math. Appl. **62**, 310–318 (2011)
16. Wei, T.G., Shuang, W., Yan, Z.: An improved fast fractal image coding algorithm. In: 2nd International Conference on Computer Science and Network Technology (ICCSNT), pp. 730–732 (2012)
17. de Quadros Gomes, R., Guerreiro, V., da Rosa Righi, R., da Silveira, L.G., Yang, J.: Analyzing Performance of the Parallel-based Fractal Image Compression Problem on Multicore Systems. AASRI Procedia, vol. 5, pp. 140–146 (2013)
18. Nodehi, A., Sulong, G., Al-Rodhaan, M., Al-Dhelaan, A., Rehman, A., Saba, T.: Intelligent fuzzy approach for fast fractal image compression. EURASIP J. Adv. Signal Process. 1–9 (2014)
19. Du, S., Yan, Y., Ma, Y.: Quantum-accelerated fractal image compression: an interdisciplinary approach. IEEE Signal Process. Lett. **22**, 499–503 (2015)
20. Barnsley, M.F., Demko, S.: Iterated function systems and the global construction of fractals. Proc. Roy. Soc. Lond. A: Math. Phys. Eng. Sci. **399**(1817), 243–275 (1985)
21. Jacquin, A.E.: Image coding based on a fractal theory of iterated contractive image transformations. IEEE Trans. Image Process. **1**, 18–30 (1992)
22. USC-SIPI Image Database, http://sipi.usc.edu/database

An Optimized Color Image Watermarking Technique Using Differential Evolution and SVD–DWT Domain

Priyanka and Sushila Maheshkar

Abstract We present a new image watermarking technique for color images. Tradeoff between imperceptibility, robustness, and security is achieved by this technique. This technique is based on discrete wavelet transform (DWT) and singular valued decomposition (SVD). Advantage of both DWT and SVD has been utilized to achieve more robustness. Differential evolution (DE) is used for optimization of scaling factors. DE is incorporated to find optimal scaling factors to scale down the watermark and increase the visual quality and robustness of watermarked image. Color cover image is split into three color channels using RGB color space. To improve the security, watermark is scrambled before embedding using random Bit-Plane and XOR operation. Experimental results show that the watermarked image is imperceptible and resistant to various geometric attacks and image processing.

Keywords DWT · SVD · DE · RGB color space · PSNR

1 Introduction

With the advancements in computer and internet technology, transmission of digital multimedia contents has become very easy. At the same time advent of various image processing tools has made modification, tampering, or duplication effortless. Wide availability and easy manipulation has resulted in a serious problem of copyright protection, forgery, misuse, or violation. Digital watermarking has come up as a solution to deal with these problems. Digital image watermarking can be defined as a process where the watermark is embedded into the cover image.

Priyanka · Sushila Maheshkar (✉)
Department of Computer Science and Engineering, Indian School of Mines, Dhanbad, India
e-mail: sushila_maheshkar@yahoo.com

Priyanka
e-mail: priyankasingh401@gmail.com

M. Pant et al. (eds.), *Proceedings of Fifth International Conference on Soft Computing for Problem Solving*, Advances in Intelligent Systems and Computing 436, DOI 10.1007/978-981-10-0448-3_9

Multimedia object in which watermark is embedded is called host. A host can be image, audio, or video. A watermark can be an image, logo, noise etc. Watermarking techniques can be classified on the basis of domain, perceptivity, and robustness [1].

Watermarking techniques can be classified into spatial and frequency domain techniques on the basis of work domain. In spatial domain, the watermark is embedded by directly modifying the pixel intensity values of the cover image. Least significant bit (LSB) is simplest and the most popular methods used in spatial domain. This technique modifies the least significant bits (LSB) of the cover image pixels by watermark image pixels [2, 3]. Frequency domain is also called transform domain. In frequency domain, watermark is embedded by altering the frequency coefficients. DFT, DCT [2], SVD [4], and DWT are the popular transforms used in image watermarking.

Basic characteristics of watermarking techniques are imperceptibility, robustness, security, and capacity or payload [5]. Often there is tradeoff between these characteristics in achieving the watermarking objectives. In our proposed technique, we have tried to establish a balance between imperceptibility, robustness, and security. We have proposed a technique for color images as color images are practically more in use.

Organization of the paper is as follows: In Sect. 2, we discuss the related work. Section 3 describes the preliminaries (DWT, SVD, DE, RGB color space, and watermark scrambling by Random Bit-Plane and XOR operation) in brief. Section 4 explains the proposed algorithm. Simulation results are provided in Sect. 5. Section 6 concludes the paper.

2 Related Works

Watermarking has been very active research area in the past few decades. Various conventional and intelligence-based watermarking techniques have been proposed. Conventional techniques have high computing speed but it is unable to establish automatic balance between robustness and fidelity. Intelligence-based algorithms overcome the limitations of conventional techniques but takes more computational time. Computation time can be reduced by using parallelism. Some intelligence-based techniques have also been proposed such as genetic algorithm [6–9], particle swarm optimization [10–12], and differential evolution [13–15]. Aslantas [15] proposed a technique based on SVD using DE (DE + SVD). Ali and Ahn [14] proposed a watermarking technique based on DWT–SVD and incorporates self-adaptive DE(SDE). SDE adjusts the mutation factor and the crossover rate. They transformed the cover image into sub-bands of different frequencies using two-level DWT. Then SVD was applied to each sub-band at second level. Ali et al. [13] proposed DCT-SVD based watermarking technique.

3 Preliminaries

Here various methodologies used in proposed watermarking algorithm are discussed in brief.

3.1 Singular Value Decomposition (SVD)

SVD is a useful numerical analysis technique to extract geometric features from an image. It may be considered as a process which transforms correlated data set into uncorrelated one. Liu and Tan [16] proposed SVD-based watermarking scheme. SVD decomposes a rectangular matrix C into three matrices U, S, and transpose of V. SVD representation of image matrix (C) of size $M \times N$, where $M \geq N$, is as follows:

$$C = USV^T \tag{1}$$

where $UU^T = I_M$ and $VV^T = I_N$;

$U_{M \times M}$ and $V_{N \times N}$ are orthogonal matrices and $S_{M \times N}$ is The matrix S is square diagonal matrix. It is known as a singular value matrix. It contains square roots of the eigen values from U or V in descending order. These values are called singular values. The columns of U are called the left singular vectors, and that of V are called as right singular vectors. The brightness of the image is specified by the singular values. The geometry of the image is reflected by the corresponding pair of singular vectors.

3.2 Overview of Discrete Wavelet Transform (DWT)

DWT is based on wavelets. Wavelets are small waves of varying frequency. DWT has good energy compaction; therefore, it finds application in image processing and watermarking. In DWT, the original image is decomposed into four sub-bands. Frequency sub-bands are formed by low-pass filtering in both the directions row and column [14]. There are three high frequency sub-bands, i.e., HL, LH, and HH which is called as detail sub-images containing the fringe information. There is one low-frequency sub-band LL called approximate sub-image. The approximate sub-band contains a rough description of the image. Further decomposition of sub-bands can be done to get higher level of decomposition. This process can be carried out to more levels until the required number of levels is obtained. Watermark is embedded in the approximate sub-band to provide better robustness as it is more stable than the detail sub-images, and maximum of image energy is concentrated here.

3.3 Differential Evolution (dE)

DE was introduced by Storn and Price [17] in 1997. DE is a time efficient, simple yet robust evolutionary approach. DE is based on 'greedy' selection and is similar to evolutionary algorithm, but differs in generation of new candidate solutions. Figure 1 shows the main steps and operators applied in DE algorithm.

A terminating criterion is predefined which can be maximum number of generation. Operators are repeated for each generation, until the terminating criteria is satisfied. DE operators are as follows:

Mutation: mutation process begins by randomly selecting three distinct individuals $\{X_{r1}, X_{r2}, X_{r3}\}$ from the existing population and $X_{i,G}$ (i.e., $r1 \neq r2 \neq r3 \neq i$). Corresponding to each target individual $X_{i,G}$, it produces perturbed individual $V_{i,G}$. Vector difference between individuals of current population determines the degree and direction of the variation. The difference between two individuals after scaling by a scaling factor $F \in [0, 1]$, is added to the third individual. Mathematically, it is given as:

$$V_{i,G} = X_{r1,G} + F \times (X_{r2,G} - X_{r3,G}) \tag{2}$$

Fig. 1 Steps of DE algorithm

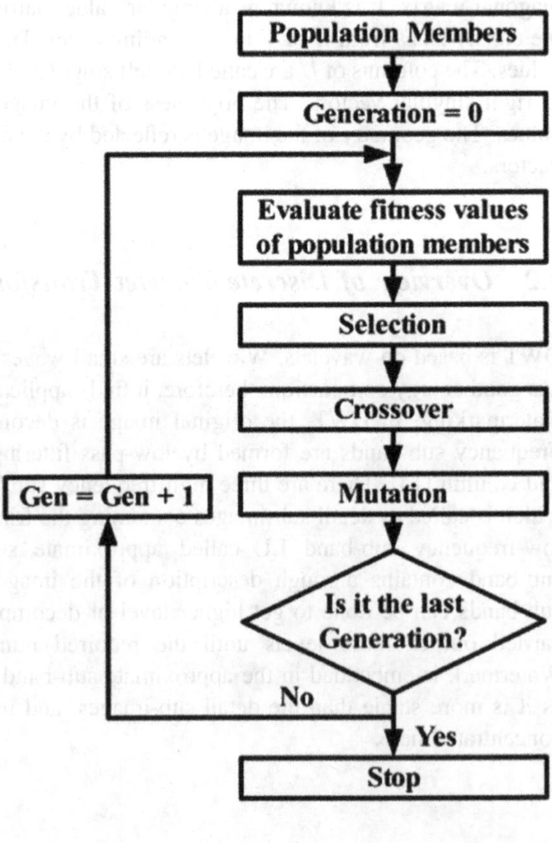

where, i ranges over the number of individuals.

Crossover: the trial individual, $T_{i,G}$ is generated by applying crossover operator. Trial individual is generated between $V_{i,G}$ (perturbed individual) and $X_{i,G}$ (target individual). Crossover probability $C_r \in [0, 1]$ decides the components of trial individual. Generation of trial individual can be mathematically written as:

$$t_{j,i,G} = \begin{cases} v_{j,i,G} & \text{if} \quad \text{rand}_j \leq C_r \vee j = k \\ x_{j,i,G} & \end{cases} \tag{3}$$

where j ranges over the dimension of the problem.

Selection: fitness is evaluated after reproduction of the trial individual. Trial individual is compared to its corresponding target individual. Selection operation selects the best individual from the target and trial individuals. Mathematically, it is defined by the following equation:

$$X_{i,G+1} = \begin{cases} T_{i,G} & \text{if} \quad f(T_{i,G}) \leq f(X_{i,G}) \\ x_{i,G} & \text{otherwise} \end{cases} \tag{4}$$

The trial individual replaces target individual if it is better in the next generation; otherwise, it will continue target individual.

3.4 RGB Color Space

This color space model is based on the 3D Cartesian coordinate system as shown in Fig. 2. It is a widely used model. All the colors appear in three basic components of red, green and blue.

3.5 Random Scrambling and XOR Operation

Image pixels are highly correlated to the pixels in its neighborhood. This fact can be utilized to predict the pixel value from its neighbors. To improve security, the watermark image can be scrambled using scrambling technique. In scrambling, the image pixels are shuffled to obtain the transformed image. Then the transformed image is divided into 2 × 2 pixel nonoverlapping blocks. Encryption of each individual block is done by applying XOR operation by keys. The length of key is 32 bit which is sufficiently strong [7, 18]. In the encryption process, an image is decomposed into l bit-plane. The l bit-plane image $X(l)$ is transformed into a one-dimensional vector $V(l)$. Two random sequences called R_1 and R_2 are generated having same length as V. R_1 and R_2 are generated by a random number generator by choosing two different seeds to produce and scrambles the $1 - D$ vector V. For encryption, XOR operation by four 8-bit keys is done with each block.

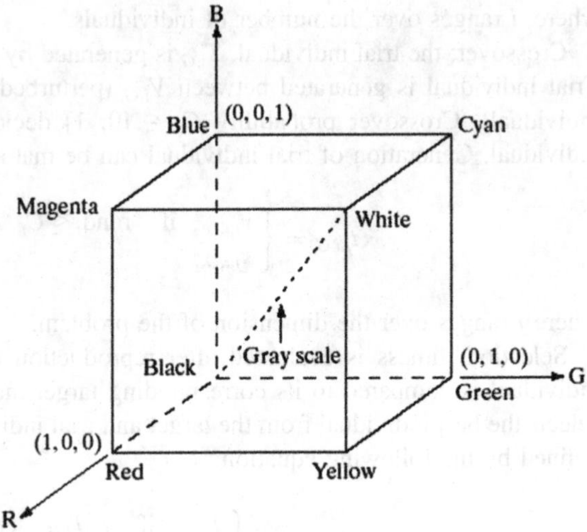

Fig. 2 3D Cartesian coordinate system of RGB color space

4 Proposed Algorithm

In this section, we describe the proposed watermarking scheme. Embedding, extraction, scrambling of watermark, and finding optimal scaling factors using DE have been discussed. Figures 3 and 4 show the embedding and extraction processes, respectively.

4.1 Embedding Process

Step 1. Split the color cover image into three color channels and select the blue channel for embedding the watermark.

Step 2. Decompose the blue channel of the cover image A by applying three-level DWT.

Step 3. Select the LL sub-band from level three and apply SVD operation to factorize these into matrices *U*, *D*, and *V* such that

$$A = UDV^T \tag{5}$$

Step 4. Scramble and encrypt the watermark by using scrambling technique

Step 5. DE Initialization: pseudorandom numbers ranging between 0 and 1 are used to populate the initial population of scaling factors.

Step 6. (a) Mutant individual is obtained using mutation operator as shown in Eq. (2)

(b) Crossover operator is applied to obtain trial individual, Eq. (3).

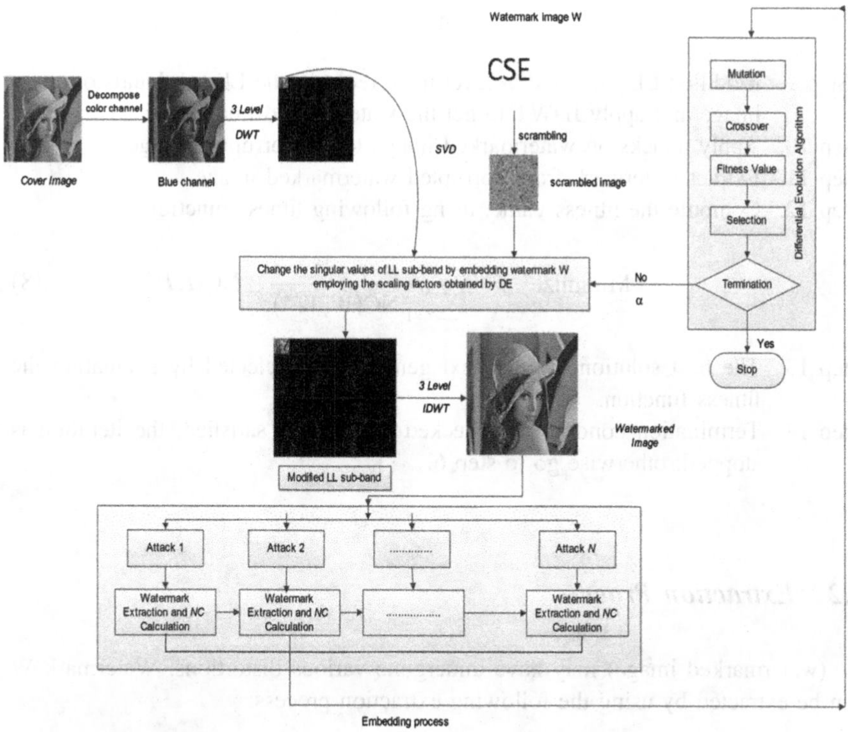

Fig. 3 Embedding process

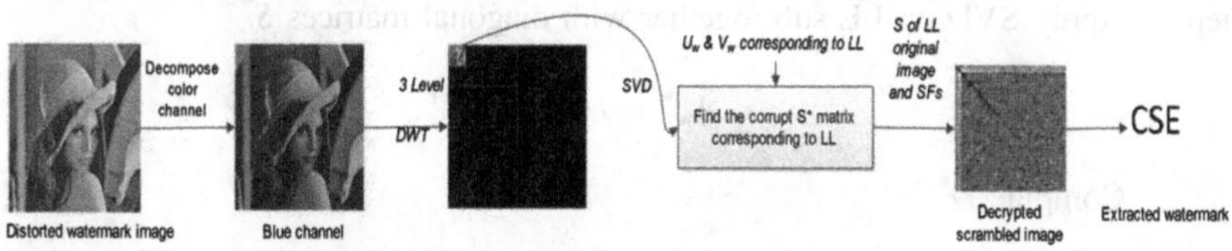

Fig. 4 Extraction process

Step 7. Singular values of LL sub-band are modified with the watermark image W and then SVD is applied to it.

$$S + \alpha W = U_w S_w V_w^T \tag{6}$$

where α is scaling factor obtained from DE algorithm.

Step 8. Obtain the modified LL sub-bands such that

$$A' = US_w V^T \tag{7}$$

Step 9. Modified LL sub-band at level three replaces the LL sub-bands of cover image and apply IDWT to get the watermarked image A_W.
Step 10. Apply attacks on watermarked image to get corrupted image.
Step 11. Extract watermark from corrupted watermarked image.
Step 12. Compute the fitness value, using following fitness function.

$$\text{Minimize} \quad f = \frac{N}{\sum_{i=1}^{N} \mathrm{NC}(W, W_i^*)} - \mathrm{NC}(I, I_w) \tag{8}$$

Step 13. The best solution for the next generation is selected by evaluating the fitness function.
Step 14. Termination condition is checked and if it is satisfied, the iteration is stopped; otherwise go to step 6.

4.2 Extraction Process

A_w (watermarked image) may have undergone various distortions. Watermark W can be extracted by using the following extraction process:

Step 1. Split the color cover image into three color channels and select the blue channel for embedding the watermark.
Step 2. Three-level DWT is applied to decompose the cover image A.
Step 3. Apply SVD on LL sub together with diagonal matrices S_w^*

$$A_w{}^* = U_w{}^* S_w{}^* V_w{}^{*T} \tag{9}$$

Compute D^*

$$D^* = U_w S_w{}^* V_w{}^T \tag{10}$$

Extract the scrambled and encrypted watermark W^*

$$W^* = (D^* - S)/\alpha \tag{11}$$

Step 4. Decrypt the obtained watermark and scramble it to obtain the extracted watermark.
DE parameters used in the proposed technique are given in Table 1.

Table 1 DE parameters

Parameters	Values
Population size NP	30
Scaling factor F	0.5
Crossover rate C_r	0.5
Maximum generations	300

Fig. 5 Cover images and watermark image

5 Results and Discussions

To assess the performance of the proposed technique, various simulations were performed on MATLAB. Five standard color test images Lena, Airplane, Tiffany, Mandrill, and Sailboat of size 512 × 512 have been used as cover image. Watermark image of size 64 × 64 as shown in Fig. 5.

PSNR (peak signal-to-noise ratio) is used for analyzing the visual quality of watermarked image. Mathematically PSNR can be represented as (12). The empirically tested threshold value of PSNR is greater than 30–35 dB, for an image without any perceptible degradation.

$$\mathrm{PSNR} = 10\log_{10}\left(\frac{(X_{\mathrm{MAX}})^2}{1/(n \times n)\sum_i \sum_j (X(i,j) - \hat{x}(i,j))^2}\right) \quad (12)$$

where $n \times n$ is the dimension of original image and $X(i, j)$ and $\hat{x}(i, j)$ are pixel values of original and watermarked image, respectively.

PSNR values of the proposed technique have been shown in Table 2. From the table, we can conclude that PSNR values are greater than threshold; so, the proposed technique has good visual quality of watermarked images.

Table 2 PSNR values of various cover images

S. no.	Cover images	PSNR
1	Lena	35.93
2	mandrill	33.65
3	Sailboat	32.68
4	Airplane	34.56
5	Tiffany	32.25

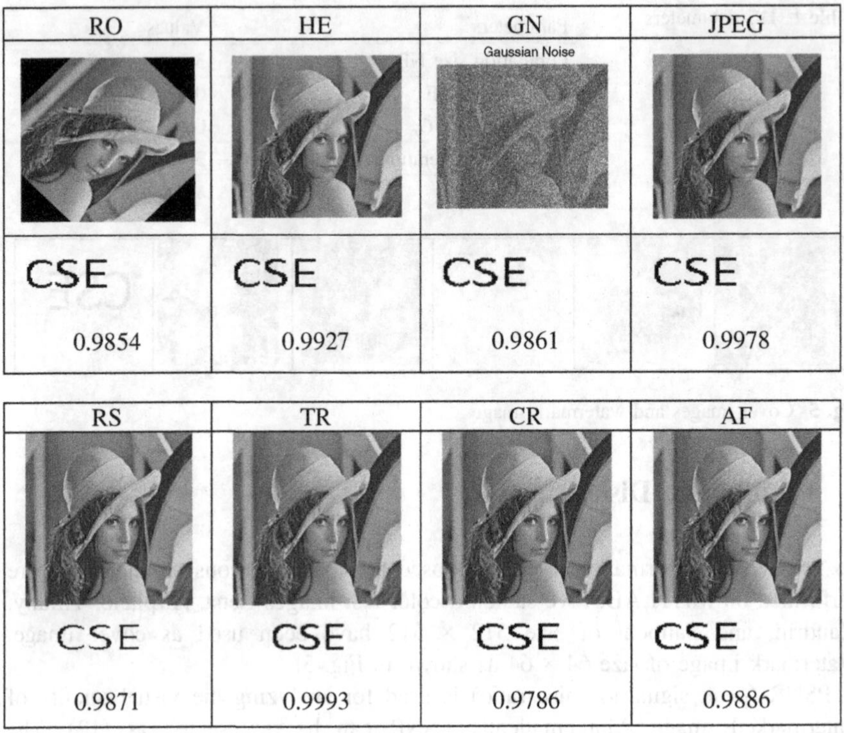

Fig. 6 Attacked image, extracted watermark, and NC values

Normalized correlation (NC) is a metric used to evaluate the similarity between original and extracted watermark. Mathematical equation for NC is given in Eq. 13.

$$\mathrm{NC}(X,\hat{X}) = \frac{\sum_i \sum_j X(i,j)\hat{X}(i,j)}{\sqrt{\sum_i \sum_j X(i,j)^2}\sqrt{\sum_i \sum_j \hat{X}(i,j)^2}} \tag{13}$$

NC value after applying attacks is given in Fig. 6. We can observe that the extracted watermarks are clearly visible and have appreciable NC values.

We have compared experimental results of the proposed technique with three similar existing schemes: DE + SVD, PSO, and tinyGA in terms of NC value. Figure 7 compares the results for the test image "sailboat"(color cover image) with its available counterparts. The proposed technique has better performance in comparison to the other algorithms in case of histogram equalization, JPEG, Gaussian noise, and translation. However, performances in few attacks are not as good as its counterparts, but the results are very near to the best one. Further improvement in proposed technique can be done to overcome these limitations.

Fig. 7 Comparison of techniques in terms of NC values

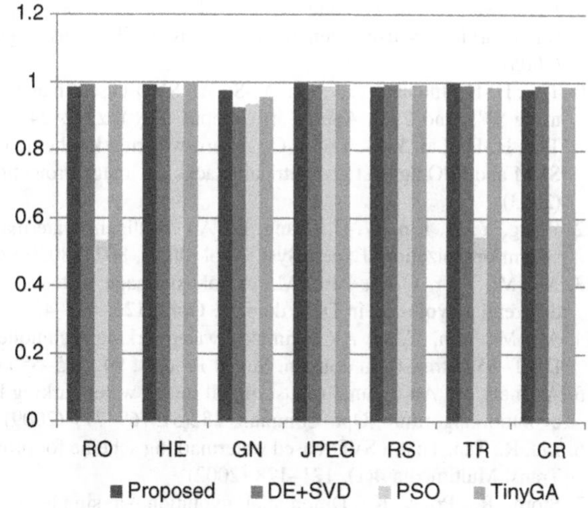

6 Conclusion

The proposed image watermarking technique based on DWT-SVD using DE achieves the balance between transparency, robustness, and security. Extensive analysis of the experimental results shows that multiple SFs estimated by DE are feasible and superior over the use of a single SF. Various image processing attacks were applied to test the robustness of the proposed scheme. Computational efficiency and improvement of robustness in some attacks are some of the future work.

References

1. Potdar, V.M., Han, S., Chang, E.: A survey of digital image watermarking techniques. In: Proceedings of IEEE Third International Conference on Industrial Informatics, INDIN'05, pp. 709–716 (2005)
2. Cox, I.J., Miller, M.L., Bloom, J.A.: Digital Watermarking. Morgan Kaufmann Publishers, San Francisco, CA (2002)
3. Celik, M.U., Sharma, G., Tekalp, A.M., Saber, E.: Lossless generalized-LSB data embedding. IEEE Trans. Image Process. 253–266 (2005)
4. Cao, L.: Singular value decomposition applied to digital image processing. In: CISSE Proceedings (2006)
5. Cox, I.J., Kilian, J., Leighton, F.T., Shamoon, T.: Secure spread spectrum watermarking for multimedia. IEEE Trans. Image Process. 1673 (1997)
6. Ali, M., Ahn, C.W., Siarry, P.: Differential evolution algorithm for the selection of optimal scaling factors in image watermarking. Eng. Appl. Artif. Intell. **31**, 15–26 (2014)
7. Pakshwar, R., Trivedi, V.K.: Image encryption using random scrambling and XOR operation. Int. J. Eng. Res. Technol. **2**(3) (March-2013). ESRSA Publications (2013)
8. Lai, C.C.: A digital watermarking scheme based on singular value decomposition and tiny genetic algorithm. Digit. Signal Process. **21**(4), 522–527 (2011)

9. Kumsawat, P., Attakitmongcol, K., Srikaew, A.: A new approach for optimization in image watermarking by using genetic algorithms. IEEE Trans. Signal Process. **53**(12), 4707–4719 (2005)
10. Tsai, H.-H., Jhuang, Y.-J., Lai, Y.-S.: An SVD-based image watermarking in wavelet domain using SVR and PSO. Appl. Soft Comput. **12**(8), 2242–2453 (2012)
11. Tsai, H.-H., Lai, Y.-S., Lo, S.-C.: A zero-watermark scheme with geometrical invariants using SVM and PSO against geometrical attacks for image protection. J. Syst. Soft. **86**(2), 335–348 (2013)
12. Wang, Y.-R., Lin, W.-H., Yang, L.: An intelligent watermarking method based on particle swarm optimization. Expert Syst. Appl. **38**(7), 8024–8029 (2011)
13. Ali, M., Ahn, C.W., Pant, M.: A robust image watermarking technique using SVD and differential evolution in DCT domain. Optik **125**, 428–434 (2014)
14. Ali, M., Ahn, C.W.: An optimized watermarking technique based on self-adaptive DE in DWT–SVD transform domain. Signal Process. **94**, 545–556 (2014)
15. Aslantas, V.: An optimal robust digital image watermarking based on SVD using differential evolution algorithm. Opt. Commun. **282**(5), 769–777 (2009)
16. Liu, R., Tan, T.: An SVD-based watermarking scheme for protecting rightful ownership. IEEE Trans. Multimedia **4**(1), 121–128 (2002)
17. Storn, R., Price, K.: Differential evolution—a simple and efficient heuristic for global optimization over continuous spaces. J. Global Optim. 341–359 1997
18. Sun, Q.D., Ma, W.X., Yan, W.Y., Dai, H.: A random scrambling method for digital image encryption: comparison with the technique based on Arnold transform. J. Shanghai Second Polytech. Univ. **25**(3), 159–163 (2008)

One Day Ahead Forecast of Pan Evaporation at Pali Using Genetic Programming

Narhari Dattatraya Chaudhari and Neha Narhari Chaudhari

Abstract Forecasting of pan evaporation is important for management of water resources. Evaporation process is highly nonlinear and complex. Hydrologists therefore try another technique in place of traditional deterministic and conceptual models to forecast pan evaporation with relative simplicity and accuracy. The present work uses genetic programming (GP) and model tree (MT) to forecast pan evaporation one day ahead at Pali in the Raigad district of Maharashtra, India. Daily minimum and maximum humidity, minimum and maximum temperature, wind speed, pan water temperature, and sunshine were the seven input parameters. Both models performed well for 2 years data with accuracy of prediction. Excellence of GP model is proved with correlation coefficient between GP forecasted and observed pan evaporation ($r = 0.97$), least error (MSRE = 0.012 mm/day), and high index of agreement ($d = 0.98$). These models can be useful for hydrologists and farm water managers.

Keywords Forecasting · Genetic programming · Pan evaporation

1 Introduction

Hydrological analysis for management of water resource in a particular region is influenced by pan evaporation at a specific site. Accuracy in forecasting pan evaporation plays an important role in optimal allocation of water supplies for local governing bodies, irrigation, river flow forecasting, and management of water

N.D. Chaudhari (✉)
Department of Civil Engineering, K.K. Wagh Institute of Engineering
Education and Research, Nasik, Maharashtra, India
e-mail: chaudhari_nd@rediffmail.com

N.N. Chaudhari
Department of Civil Engineering, Matoshri College of Engineering
and Research, Nasik, Maharashtra, India
e-mail: chaudhari_nn1994@rediffmail.com

M. Pant et al. (eds.), *Proceedings of Fifth International Conference on Soft Computing for Problem Solving*, Advances in Intelligent Systems and Computing 436, DOI 10.1007/978-981-10-0448-3_10

resources in forest regions. The knowledge of crop water need is important and can be achieved by forecasting evapotranspiration. Correct forecasting of climatic parameters, like temperature, relative humidity, solar radiation, and wind speed, yield better forecast of evapotranspiration. Forecast of evapotranspiration is governed by forecast of pan evaporation [1]. Cost of the irrigation water to farmers in most of the agro ecosystems is very important and plays a vital role in planning and management of water resources in the present scenario. Free water loss from a Class A type evaporation pan relates the estimates of daily consumptive use. Records at a specific site give the free water loss for the earlier day for an area whereas the free water loss for the next day should be forecasted based on forecasts of relative humidity, bright sunshine, wind, and rain. The process of evaporation is very difficult and nonlinear. At specific site, climatic parameters like humidity, wind speed, sunshine, temperatures of air and water govern the evaporation at a specific site. There are empirical methods available for prediction of evaporation; but, these methods require measurement of lot many climatic parameters, such as mean wind speed, actual vapor pressure, saturation vapor pressure at mean air temperature, mean air temperature in degrees Kelvin, maximum possible sunshine in hours, and actual duration of sunshine in hours and heat radiations to and from the water body [2].

Hydrologists, therefore, try to build different simple techniques to forecast evaporation accurately. Data-driven techniques relate input and output variables of physical process. Interconnection between the system state variables is the basis of working of data-driven techniques. Also they do not consider the physics of the phenomenon. These techniques work with only a limited knowledge of the behavior of the physical phenomenon [3]. Nowadays, there are various data centers established by government. As a result, data is available to the user from these data centers. Availability of data made these data-driven techniques popular since the last 20 years. These data-based modelings throw light on the use of machine learning technique for framing models which have the ability to replace the models driven by understanding the knowledge of physical process [4].

The present work uses genetic programming (GP) and model tree (MT) to forecast evaporation a day ahead at Pali station in the Raigad district of Maharashtra, India. Current evaporation and previous values of meteorological variables, such as sunshine, wind speed, maximum and minimum humidity, pan water temperature, and maximum and minimum air temperature at Pali station were used to build both the models. There are numerous papers available on evaporation modeling employing artificial neural networks (ANN) with climatic variables as the input for some models. The references of GP and MT for hydrological modeling are few. The references of GP and MT for evaporation forecasting are rare.

Recently, Ref. [5] noticed in his work noticeable results of MT technique for forecasting of evapotranspiration. Reference [6] in his study at Vaipar Basin used GP technique for studying the effect of climatic change on evaporation losses.

This study is an effort in that area which highlights the ability of this technique to yield better results with smaller data set. Information of data and area under study are discussed followed by short details of data-driven techniques. Later, model building and its assessment are discussed followed by discussions on results. At the end, concluding remarks are presented.

2 Data and Area Under Study

Current study is carried out at Pali (18.54°N, 73.22°E) in Raigad district of Maharashtra, India. Pali is a holy place with one of the eight ashtavinayaka temple. The aim of the present study is to forecast evaporation values one day in advance at Pali. The climate of this place resembles the climate of the west coast of India (Fig. 1). There is regular and huge amount of rainfall, highly humid weather due to vicinity of sea coast. Hot weather prevails from March till the end of May. Winter is fairly cold. Temperature variation is between 16.1 and 40.4 °C. In spite of average annual rainfall of 3884.3 mm, the nature of soil does not permit storage of water due to which there is acute shortage of water.

Data used for Pali station is from January 6, 2006 to May 31, 2008 (2 years). The input parameters were daily sunshine hours, pan water temperature, average wind speed, minimum temperature, maximum temperature, and minimum and maximum humidities.

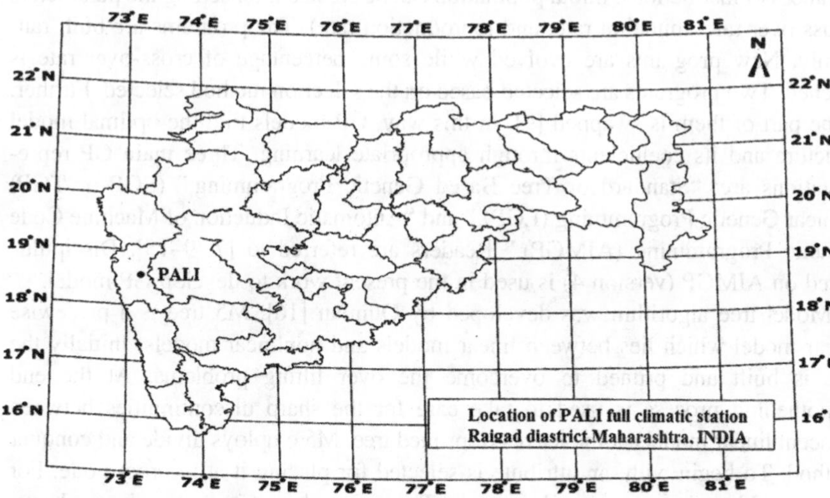

Fig. 1 Pali, Maharashtra, India (http://www.mapsofindia.com/maps/maharashtra/#)

3 Genetic Programming and Model Tree

Darwin's principle of survival of the fittest is the basis of genetic programming. This technique derives inspiration from nature and produces the best individual (program) with the help of mutation, cross-over, and reproduction processes [7]. In the beginning, GP population of individuals (programs or equations) is created, then evaluation of fitness of individual is completed, and then the fittest individual is selected as parent. Using cross-over, mutation, and reproduction, children are produced. Only stronger children survive. This is an iterative process and repeats till the criteria defined by user are satisfied. Criteria to terminate the process can be either achieving least error or completion of a specified number of generations. Knowledge of underlying physical process is not necessary for GP models. These models do not need huge climatic data sets. Initially, the user using this technique need not specify the overall function form of the model. This technique yields better approximations of the complicated natural phenomenon relating input variables into functional form [8]. The strength of GP in finding the solution depends on how instruction set express. For getting optimal solution, sufficient number of variables, registers, and range of constants is necessary. Search space includes instructions sets with the help of which it can construct all possible programs. The dimension of the search space is dependent on number instructions and registers and increases exponentially with their number. The chance of getting an optimal solution is reduced with lot many useless basic program elements. This reduction of a chance of getting optimal solution is due to assumption that initial population represents a small fraction of entire search space. To improve this, the instruction set may contain multiple instances of instructions. Initial population can be created after setting the parameters (cross-over rate, mutation rate, and reproduction rate). The programs are built randomly. New programs are evolved while some percentage of cross-over rate is reached. Two programs are selected based on the selection method selected. Further, some part of them is swapped [9]. In this way, GP models find the optimal model structure and its coefficients through appropriate learning. Three main GP representations are: "Standard or Tree Based Genetic Programming," (SGP or TGP) "Linear Genetic Programming (LGP)," and "Automatic Induction of Machine Code Genetic Programming (AIMGP)." Readers are referred to [7, 9–15]. Discipulus based on AIMGP (version 4) is used in the present work to develop GP model.

Model tree algorithm was developed by Quinlan [16]. M5 tree is a piecewise linear model which lies between linear models and nonlinear models. Initially the tree is built and pruned to overcome the over fitting problems. At the end smoothening process is used to take care for the sharp discontinuities between adjacent linear models at the leaves of pruned tree. M5 employs divide and conquer method. To begin with, an attribute is selected for placing it at the root node. For each possible value, one branch is made. Thereafter, the set is split up into subsets. There is one subset for every value of the attribute. Process is repeated for each branch. The splitting process terminates when the output values vary slightly [3]. Weka software is employed to build MT models [17].

GP is a relatively new technique in the field of hydrology with majority of applications is toward modeling of rainfall-runoff process [14, 18–20]. Lately, Refs. [21, 22] used GP for hydraulic modeling. Reference [23] assessed effort of meteorological variables to estimate pan evaporation using self-organizing map neural network. Reference [24] modeled pan evaporation using linear GP technique. References [25, 26] estimated daily pan evaporation using adaptive neural-based fuzzy inference system. Reference [5] employed M5 model trees and neural network to predict evapotranspiration in Ankara, Turkey.

4 Model Formulation and Assessment

Present study uses linear GP and MT technique to develop evaporation models. Performance of the networks was checked with respect to testing sets by calculating coefficient of correlation between forecasted and observed pan evaporation values. The scatter plots between these two values were plotted for both data-driven techniques. The present model was tried with data set of total daily records 730 values out of which 511 (70 %) were used for training the model and remaining 219 (30 %) were withheld for testing purpose, seven input climatic variables were maximum and minimum humidity, maximum and minimum air temperature, wind speed, daily sunshine, and pan water temperature. Data used was was from January 6, 2006 to May 31, 2008, i.e., 2 years. The initial control parameters like size of the population 2048, frequency mutation of mutation 95 %, and the frequency of cross-over 53 %, with a mean squared error as the fitness function.

The data-driven model discussed uses mean square error function to assess whether the set goal is achieved or not along with the coefficient of correlation to test the strength of the model during testing. Even if this is a common function to assess the model strength it fails to find the regions where the model is not effective [27]. To cater for this deficiency, other assessment functions like mean squared relative error (MSRE), mean absolute error (MAE), root mean square error (RMSE), index of agreement (d), and coefficient of efficiency (CE) are performed. Equations 1–5 give the expressions of these measures. MAE gives absolute differences disregarding over- or underestimation; similarly, RMSE and MSRE also neglect the distinction between over- or underestimation as the square of the differences. Balanced measure is given by relative error. Higher value of accuracy is indicated by their lower values. CE is independent of the scale of data and therefore, it is suitable when the scales of variables in data are different. CE value ranges from$-\infty$ to +1 and is an indicative of capabilities of estimation different from the mean. Satisfactory results are indicated by CE of 0.90 or above and unsatisfactory if below 0.8. Index of agreement (*d*) is a measure ranging between 0 and 1. It allows cross comparisons. This assess model is regardless of units. There is perfect agreement for value of this measure equal to 1. A value of 1 indicates a perfect agreement between the observed and estimated values, while 0 denotes complete disagreement [27].

4.1 Model Assessment Equations

$$\mathrm{MSRE} = \frac{\sum_{i=1}^{n} \frac{(E_i - \hat{E}_i)^2}{E_i^2}}{n} \tag{1}$$

$$\mathrm{RMSE} = \sqrt{\frac{\sum_{i=1}^{n} (E_i - \hat{E}_i)^2}{n}} \tag{2}$$

$$\mathrm{MAE} = \frac{\sum_{i=1}^{n} \left| E_i - \hat{E}_i \right|}{n} \tag{3}$$

$$\mathrm{CE} = 1 - \frac{\sum_{i=1}^{n} (E_i - \hat{E}_i)^2}{\sum_{i=1}^{n} (E_i - \overline{E})^2} \tag{4}$$

$$d = 1 - \frac{\sum_{i=1}^{n} (E_i - \hat{E}_i)^2}{\sum_{i=1}^{n} \left(\left| \hat{E}_i - \overline{E} \right| + \left| E_i - \overline{E} \right| \right)^2} \tag{5}$$

where,

$\hat{E}_i$ Model forecasted evaporation

E_i Observed evaporation

n Number of data

$\overline{E}$ Mean of observed evaporation.

5 Results and Discussions

GP model performed excellent as far as accuracy of the forecasting is concerned. Figure 2 shows balanced scatter plot between GP forecasted and observed evaporation values. A similar effect is observed in Fig. 3 indicating the scatter plot between MT model forecasted and observed evaporation values. From Figs. 2 and 3, it is clear that the GP model works slightly better than the MT model. Time series plots (Fig. 4) for GP and MT forecasted and observed pan evaporation indicate that GP model is slightly better than the MT model but with a marginal difference. Coefficient of correlation ($r = 0.97$) between GP forecasted and observed pan evaporation is excellent as well as error statistics are appreciable (Table 1) with least MSRE (0.011 mm/day) and other two values of error are also minimum. It is interesting to note high value of coefficient of efficiency (CE = 0.93) which is a

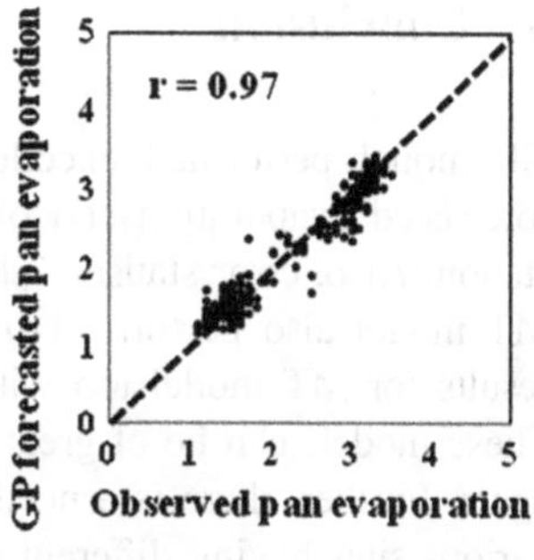

Fig. 2 Scatter plot of GP forecasted and observed evaporation

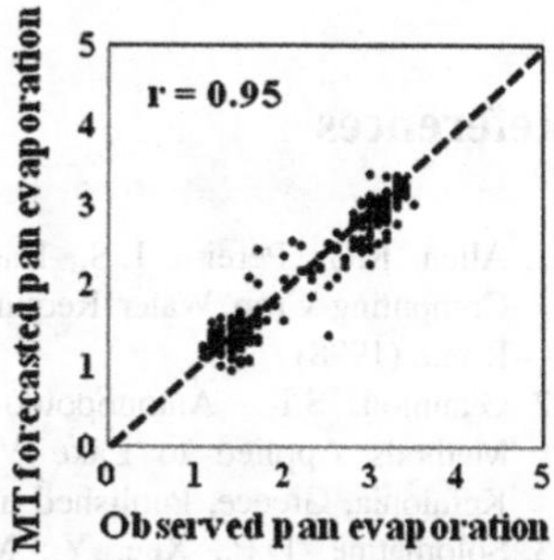

Fig. 3 Scatter plot of MT forecasted and observed evaporation

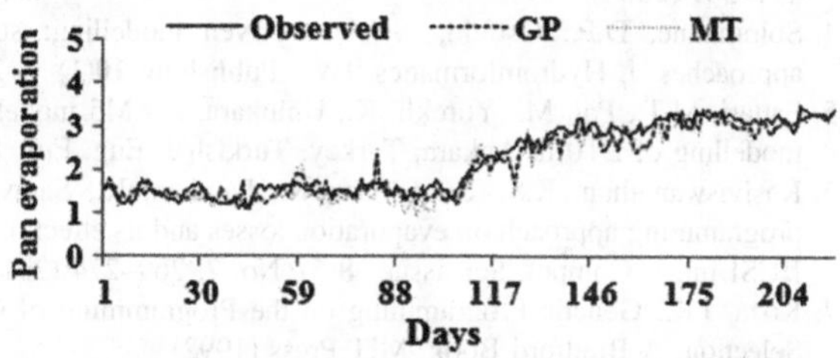

Fig. 4 Time series plot for observed, GP forecasted, and MT forecasted

symbol of satisfactory model; similarly, index of agreement ($d = 0.98$) agrees with the accuracy of the model, although higher values are underpredicted. Table 1 shows values for MT model which are also good and close to GP. These models can be of great use for hydrologists and irrigation farm water managers.

Table 1 Details of the GP and MT models

	MSRE	RMSE	MAE	CE	*d*	*r*
GP	0.012	0.21	0.17	0.93	0.98	0.97
MT	0.016	0.26	0.21	0.89	0.97	0.95

MSRE (mm/day) mean squared relative error, *RMSE* (mm/day) root mean square error, *MAE* (mm/day) mean absolute error, *CE* coefficient of efficiency, *d* index of agreement, and *r* coefficient of correlation

6 Conclusion

GP model performed excellently well with coefficient of correlation between forecasted evaporation and observed evaporation (r = 0.97) at Pali full climatic station. Also, error statistics show very good results with least error in forecasting. MT model also performed (r = 0.95) better with error statistics showing better results for MT model too with marginal difference if compared with GP results. These models can be of great use for hydrologists and irrigation farm water managers. Further, there is a need to explore these techniques for building models at various sites having different climate, different elevations, and larger data sets.

References

1. Allen, R.G., Pereira, L.S., Raes, D., Smith, M.: Crop evapotranspiration—Guidelines for Computing Crop Water Requirements. FAO Irrigation and Drainage Paper, No. 56, FAO, Rome (1998)
2. Gianniou, S.K., Antonopoulos, V.Z.: Comparison of Different Evaporation Estimation Methods Applied to Lake Vegoritis. Conference at Steven's Institute of Technology, Kefalonia, Greece. Published in PROTECTION (2008)
3. Solomatine, D.P., Xue, Y.: M5 model trees and neural networks: application to flood forecasting in the upper reach of the Huai River in China. J. Hydrol. Eng., ASCE. **9**(6), 491–501 (2004)
4. Solomatine, D.P., Ostfeld, A.: Data-driven modelling: some past experiences and new approaches. J. Hydroinformatics, IWA Publishing **10**(1), 3–22 (2008)
5. Sattari, M.T., Pal, M., Yurekli, K., Unlukara, A.: M5 model trees and neural network based modelling of ET0 in Ankara, Turkey. Turkish J. Eng. Env. Sci. **37**, 211–219 (2013)
6. Kasiviswanathan, K.S., Soundhara, R., Pandian, R., Saravanan, S., Agarwal, A.: Genetic programming approach on evaporation losses and its effect on climate change for vaipar basin. IJCSI Int. J. Comput. Sci. Issues **8**(5), No. 2, 269–274 (2011)
7. Koza, J.R.: Genetic Programming on the Programming of Computers by Means of Natural Selection. A Bradford Book, MIT Press (1992)
8. Giustolisi, O., Savic, D.A.: A symbolic data-driven technique based on evolutionary polynomial regression. J. Hydroinformatics **8**(3), 207–222 (2006)
9. Brameier, M.: On Linear Genetic Programming, PhD thesis, University of Dortmund (2004)
10. Banzhaf, W., Nordin, P., Keller, R.E., Francone, F.D.: Genetic Programming. Morgan Kauffman, San Francisco, CA (1998)
11. Banzhaf, W., Nordin, P., Keller, R.E., Francone, F.D.: Genetic Programming: An Introduction on the Automatic Evolution of Computer Programs and Its Applications. Morgan Kaufmann, Germany, ISBN: 978-1-55860-510-7 (1997)
12. Francone, F.D.: Discipulus: Fast Genetic Programming Based on AIM Learning Technology Owner's Manual. Register Machine Learning Technologies. Inc, Littleton, Colorado (2004)
13. Francone, F.D., Markus, C., Banzhaf, W., Nordin, P.: Homologous crossover in genetic programming. In: Proceedings of Genetic and Evolutionary Computation Conference, vol. 2, pp. 1021–1026 (1999)
14. Guven, A.: Linear genetic programming for time-series modeling of daily flow rate. J. Earth Syst. Sci. **118**(2), 137–146 (2009)
15. Nordin, J.P.: Evolutionary Program Induction of Binary Machine Code and its Application. Krehl Verlag, Muenster, Germany (1997)

16. Quinlan, J.R.: Learning with continuous classes. In: Adams, A., Sterling, L. (eds.) Proceedings of AI'92 Fifth Australian Joint Conference on Artificial Intelligence, pp. 343–348. World Scientific, Singapore (1992)
17. Witten, I.H., Frank, E.: Data Mining: Practical Machine Learning Tools and Techniques with Java Implementations. Morgan Kaufmann, San Francisco (2005)
18. Whigham, P.A., Crapper, P.F.: Modeling rainfall-runoff using genetic programming. Math. Comput. Model. **33**(6–7), 707–721 (2001)
19. Savic, D.A., Walters, G.A., Davidson, J.W.: A genetic programming approach to rainfall-runoff modeling. Water Res. Manag. **13**(3), 219–231 (1999)
20. Babovic, V., Keijzer, M.: Rainfall runoff modelling based on genetic programming. Nord. Hydrol. **33**(5), 331–346 (2002)
21. Azamathulla, H. Md., Ahmad, Z.: Gene-expression programming for transverse mixing coefficient. J. Hydrol. **434–435C**(1), 142–148 (2012a)
22. Azamathulla, H. Md., Ahmad, Z.: GP approach for critical submergence of intakes in open channel flows. J. Hydroinformatics **14**(4), 937–943 (2012b)
23. Chang, F.J., Chang, L.C., Kao, H.S., Wu, G.R.: Assessing the effort of meteorological variables for evaporation estimation by self-organizing map neural network. J. Hydrol. **384** (1–2), 118–129 (2010)
24. Guven, A., Kisi, O.: Daily pan evaporation modeling using linear genetic programming technique. Irrigation Sci. **29**, 135–145 (2011)
25. Chung, C.H., Chiang, Y.M., Chang, F.J.: A spatial neural fuzzy network for estimating pan evaporation at un-gauged sites. Hydrol. Earth Syst. Sci. **16**(1), 255–266 (2012)
26. Eslamian, S.S., Amiri, M.J.: Estimation of daily pan evaporation using adaptive neural-based fuzzy inference system. Int. J. Hydrol. Sci. Technol. **1**(3/4), 164–175 (2011)
27. Dawson, C.W., Wilby, R.L.: Hydrological modelling using artificial neural networks'. Progr. Phys. Geogr. **25**(1), 80–108 (2001)
28. Survey of India. http://www.mapsofindia.com/maps/maharashtra/#

Optimized Scenario of Temperature Forecasting using SOA and Soft Computing Techniques

Amar Nath, Rajdeep Niyogi and Santanu Kumar Rath

Abstract Weather forecasting at a given instant of time and location is a challenging activity as its data are continuous, highly intensive, multidimensional, and dynamic in nature. This paper presents an approach for maximum temperature forecasting over a given period of time using the service-oriented architecture (SOA) and soft computing techniques. SOA is used for collecting data of a particular location using the principles of SOA, i.e., reusability, interoperability, and composability. Large number of attributes of weather dataset gathered with the help of SOA concept can be curtailed sing one of the soft computing techniques, i.e., rough set theory (RST). The RST technique works by finding the relevant attributes and eliminating the irrelevant attribute which are not essential. The residual attributes are used to forecast temperature based on artificial neural network (ANN) technique. RST technique has been applied to improve the performance of ANN computationally as well as by its accuracy.

Keywords Artificial neural network (ANN) · Java business integration (JBI) · Rough set theory (RST) · Service-oriented architecture (SOA) · Temperature forecasting

Amar Nath (✉) · Rajdeep Niyogi
Indian Institute of Technology Roorkee, Roorkee 247667, Uttarakhand, India
e-mail: amardishilva@gmail.com

Rajdeep Niyogi
e-mail: rajdpfec@iitr.ac.in

S.K. Rath
National Institute of Technology, Rourkela, Rourkela 769008, Odisha, India
e-mail: skrath@nitrkl.ac.in

M. Pant et al. (eds.), *Proceedings of Fifth International Conference on Soft Computing for Problem Solving*, Advances in Intelligent Systems and Computing 436, DOI 10.1007/978-981-10-0448-3_11

1 Introduction

Weather forecasting is the analysis of the state of the weather at a given instance of time and in an area [1]. In old days, people used to forecast the weather mainly based upon the changes in barometric pressure, current weather conditions, and sky condition. But now, weather forecasting is carried out by the massive use of computer-based models.

In this study, SOA is being implemented by the technology of web service for collecting information. After gathering the information, soft computing techniques are applied for temperature prediction. The performance of ANN can be enhanced by giving it less number of inputs and RST helps in curtailing the number of attributes of actual weather dataset. In this study, the process of maximum temperature prediction is being discussed to study the impact of SOA coupled with soft computing technique.

The rest of the paper is organized as follows. Section 2 presents the review of the literature w.r.t. activeness of weather forecasting using ANN. Section 3 gives a brief description of techniques used for temperature forecasting. Section 4 presents the proposed approach of prediction for temperature forecasting scenario. Section 5 highlights the obtained results for temperature forecasting. Section 6 concludes the paper with the future work.

2 Review of Related Work

Weather forecasting is difficult to measure in concrete terms. Moreover, measuring weather in a way useful to social scientists requires careful attention. To map weather or climate to social phenomena of interest, careful measurements must occur across a large geographical areas. Weather constantly changes and climate can only be measured over long span of time. More than dozens of weather elements including temperature, sunlight, cloud, wind velocity, atmospheric pressure, humidity, and dew point can be measured in concrete terms using different equipments. Many researches have made an attempt to predict change in weather condition using neural network model.

French et al. used neural network model to predict the rainfall [2]. The authors have used backpropagation algorithm for training purpose. They trained their neural network model with training set of 1000 datasets, with 100–1000 iterations, input and output nodes as 625 and hidden nodes varying from 15 to 100. With 15 hidden nodes, 500 training iterations and their model consumes around 2 h to train and 1 s to forecast.

Luk et al. have applied artificial neural network to predict the rainfall and they developed three different types of artificial neural network: multilayer feed forward neural network (MLFN), Elman partial recurrent neural network (Elman) and time delay neural network (TDNN) for rainfall forecasting [3]. Finally, they compared the performance of result of these three artificial neural network models.

3 Details of Methodologies Used

3.1 *Service-Oriented Architecture*

An SOA is essentially a collection of services that communicate to each other, either by simple data passing or it could involve two or more services coordinating some activity. It also facilitates software to communicate with each other [4, 5]. The benefits of these SOA principles have been utilized for collecting the weather information of different locations by using services provided by some other service providers, i.e., http://www.webservicex.net.

Java business integration (JBI)-based implementation of SOA uses simple object access protocol (SOAP) [6, 7]. SOAP allows machines running in different environments to communicate by using the world wide web's hypertext transfer protocol (HTTP) and its extensible mark-up language (XML) as the mechanism for the information exchange [8].

3.2 *Soft Computing Techniques Used for Temperature Forecasting*

In this paper, two techniques from soft computing field namely, rough set theory and artificial neural network have been applied.

3.2.1 Rough Set Theory (RST)

The fundamental concept behind rough set theory is the approximation of lower and upper spaces of a set, the approximation of spaces being the formal classification of knowledge regarding the interest domain [9]. RST can be applied for the reduction of data to a minimal representation. In this technique, dataset is represented as a table known as information table. In this table, each row indicates an object (in this case, weather data for an instance of time) [10]. The RST technique works by finding the relevant attributes and eliminating the irrelevant attribute which are not essential. Every column represents an attribute.

3.2.2 Artificial Neural Networks

Artificial neural networks refer to a computing technique whose central theme is borrowed from the analogy of biological neural networks [11, 12]. In this soft computing technique, previously known data are used to train the system and then, the prediction is done using the trained system. In this paper, feed forward neural network (FFNN) is used to predict the temperature of the next day. A FFNN consists of neurons that are ordered in layers. The first, last, and layer in between

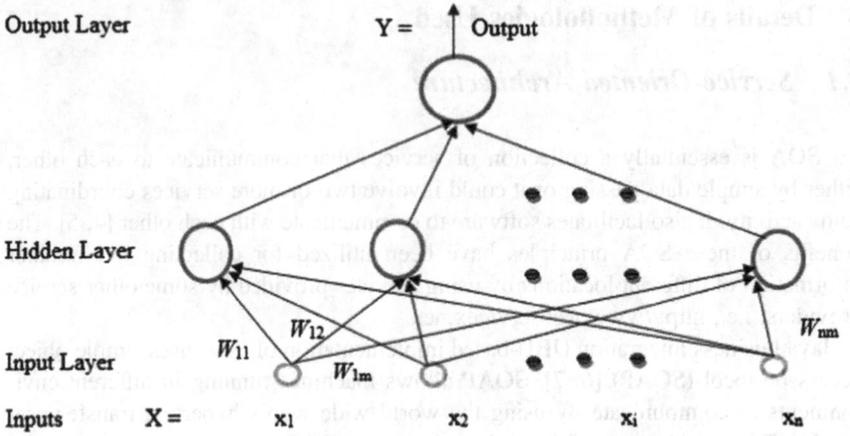

Fig. 1 A typical single layer FFNN

these two are called the input, output, and hidden layer, respectively. Each neuron in a particular layer is connected with all neurons in the next layer [13].

The interconnections between artificial neurons are called weights. The backpropagation learning algorithm is applied for the training of the FFNN. The graphical depiction of ANN is shown in Fig. 1. In backpropagation algorithm, computed error is used to update the weights between the layers to minimize the error which indicates the difference between the actual and computed value.

Different phases of ANN

1. **Training Phase** In this phase, the neural network is trained. The procedure are as follows.
   (a) **Input Layer Computation** In this layer, input data that we have are fed into the input layer of FFNN. The output of the input layer is the input to next level hidden layer of the network.
   (b) **Hidden Layer Computation** The weights between the input and hidden layer are multiplied with the output data from the input layer and outputs from all of the neurons are summed, and fed into the activation function as shown in Eq. 1. In this paper, sigmoid function is considered as the activation function represented in Eq. 2.

$$a = \sum x_i w_{ij} \tag{1}$$

where x_i represents the inputs fed to neuron and w_{ij} represents the neuron's weight.

$$f(a) = \frac{1}{1 + e^{-a}} \tag{2}$$

(c) **Output Layer Computation** The sigmoid function is used as the activation function in this layer to produce the output. Output from the activation function in the output layer is the expected output.

(d) **Computation of Error** This computed output from the neural net is compared with the actual output to compute the relative error. Calculation of the error is shown in Eq. 3.

$$\text{error}_i = |(\text{actual Output}_i - \text{computed Output}_i)| \tag{3}$$

where

error_i	error for the *i*th data point
actual Output_i	actual output for the *i*th data point
computed Output_i	output computed from the neural net for the *i*th data point

(e) **Updating of Weights** For every data point, error is calculated and then the weights are updated based on this relative error. The weights are updated according to the Eqs. 4, 5 and [14].

$$W_{\text{new}} = W_{\text{old}} + [\Delta W]_{\text{new}} \tag{4}$$

$$[\Delta W]_{\text{new}} = -\eta \frac{\partial E}{\partial W} + \alpha[\Delta W]_{\text{old}} \tag{5}$$

where η: learning rate, α: momentum coefficient and E: error

For every data mentioned, whole process is repeated. One epoch completed: when all data of the data set is considered; In every epoch mean square error (MSE) is computed as per Eq. 6. Epochs are repeated until MSE reaches a threshold value or the number of epochs exceed the limit of epochs.

$$\text{MSE} = \frac{\sum_{i=0}^{N} \text{error}_i^2}{N} \tag{6}$$

where N: Total number of data points.

2. **Testing Phase** Testing is performed using the known dataset; that is the corresponding output for the given input is known. For testing phase, weights will not be updated. After the training phase is finished, the last updated weights of the net are stored as the final weights. For testing, the inputs are fed and corresponding outputs are stored.

4 Temperature Forecasting Implementation: A Case

As the weather is very chaotic and unstable by nature, forecasting of weather becomes important. Basically, two methods are used for forecasting of weather: empirical approach and dynamic approach. The RST approach is used for predicting local scale weather. The second is applied in the form of computer modeling and is used to predict dynamically, the large scale weather phenomena [15]. In this paper, one attribute of weather, i.e., maximum temperature is predicted using SOA, RST, and ANN. Steps of implementation procedure are described below.

4.1 Collection and Analysis of Weather Data

The collected weather data are multidimensional in nature and possess the data on temperature, dew point, humidity, sea level pressure, visibility, wind speed, precipitation, etc. SOA implementation is carried out using JBI framework in OpenESB toolbox [16, 17]. This tool helps to build, integrate, and run the SOA applications.

Weather data are collected using SOA application which is created, deployed, and tested using OpenESB toolbox. In this paper, two locations, i.e., New Delhi and Calcutta are considered as case study. Deploying and testing of a composite application are done using the well-integrated NetBeans, glassfish, and OpenESB toolbox. A composite application is created that contains the composite application service assembly (CASA) editor that represents the relationship between WSDL ports, JBI modules, and external modules. The final collected weather data are shown in Fig. 2.

Attributes		Values					
Date		1-Jan-12	2-Jan-12	3-Jan-12	...	28-July-13	29-July-13
Temp (0C)	Max	18	17	19	...	36	35
	Min	16	14	13		31	31
	Avg	13	13	8		27	27
Dew Point (0C)	Max	16	15	11	...	28	30
	Min	13	13	8		27	27
	Avg	11	11	6		25	21
Humidity (%)	Max	100	100	100	...	100	100
	Min	85	94	80		77	81
	Avg	68	77	43		56	55
Sea Pres. (hpa)	Max	1016	1018	1018	...	1001	1001
	Min	1015	1016	1015		999	1000
	Avg	1013	1015	1013		997	998
Visibility (Km)	Max	3	1	2	...	4	4
	Min	1	0	1		2	3
	Avg	0	0	0		1	1
Wind(Km)	Max	13	10	13		11	14
	Avg	5	3	3		3	6
Event		Fog	Fog	Fog		Fog	Rain

Fig. 2 Final data collected for New Delhi

4.2 Data Discretization

Data analysis technique is carried out to check whether the data are in numerical form or in another form. In this paper, collected and stored weather data have a total of 18 attributes, where one of them, i.e., events are not in numerical form. This type of values are assigned to the corresponding number value in order to convert them into numerical form. The data discretization is carried out in order to get the discrete dataset. The observation for any attribute in weather data has a large number of possible outcomes, thus it is quantized into smaller number of discrete values. First, the maximum, b and minimum, a values for the attribute are computed. Then the overall interval is subdivided into a discrete set of subintervals. Number of intervals is computed using *Sturges* formula given in Eq. 7 [18].

$$k = \log_2(n + 1) \tag{7}$$

where n is number of observations. The width of subintervals w is calculated as Eq. 8.

$$w = \frac{b - a}{k} \tag{8}$$

The interval boundaries are at $a + w$, $a + 2w$, $a + (k - 1)w$. The technique is shown in the Table 1 with $b = 49$; $a = 9$; $k = \log_2(578) \approx 9$; $w = [(49–9)/9] \approx 4$.

4.3 Application of Dimension Reduction Technique

Dimension reduction technique is required for reducing the number of attributes. All the attributes that collected data may not be relevant to predict the desired attribute, so some of the attributes are reduced using RST algorithm; where all the irrelevant attributes are deducted. Initially the weather data collected have 18

Table 1 Discretization of the observable

Observation (Temp.)	Interval	Discrete value
O1	9–13	1
O2	14–17	2
O3	18–22	3
O4	23–26	4
O5	27–31	5
O6	32–35	6
O7	36–40	7
O8	41–44	8
O9	45–49	9

attributes. After applying the RST algorithm, number of attributes are reduced to nine and ten (including the target attribute) for New Delhi and Calcutta datasets, respectively. In this paper, attempt has been made to predict maximum temperature. So it is used as the target attribute as it should not be eliminated at the time of reduction.

4.4 Normalization

As in ANN technique, input and output values range between (−1, 1) and the data need to be normalized within this range. Normalization process maps range of original data into another scale. In this case, data are normalized in the range of (0, 1). The minimax normalization algorithm is used for normalizing the dataset [19]. The algorithm works as follows:

For a data vector $x(x_1; x_2; x_3; \ldots; x_n)$

- The maximum value, max is found out
- The minimum value, min is found out
- The maximum and minimum value of the normalized data are termed as NewMax and NewMin, respectively, which indicate the new range of the data
- For any value x_i the normalized $\bar{x}_i$ computed as Eq. 9

$$\bar{x} = \text{NewMin} + \frac{(x_i - \text{Min})(\text{NewMax} - \text{NewMin})}{\text{Max} - \text{Min}} \tag{9}$$

4.5 Temperature Prediction Using ANN Technique

ANN is applied to the normalized data for the prediction of temperature of the next day based on trained model with previous weather information. Different steps for predicting the weather are described below.

4.5.1 Training and Testing with ANN

Feed-forward neural network with a single hidden layer has been considered. The number of neurons in input layer, hidden layer, and output layer is shown in Table 2. The normalized data are fed to the neural network for training. 80 % of data is taken for training and the rest 20 % for testing the model. After the training the neural network, it is tested to find the accuracy of the training algorithm. Details of the parameters for the training are given Table 2.

Table 2 Testing and training details of neurons

Parameters	New Delhi	Calcutta
Neurons in input layer	8	9
Neurons in hidden layer	11	12
Neurons in output layer	1	1
Learning rate	0.6	0.6
Momentum coefficient	0.3	0.3
Maximum number of epochs	2000	2000

4.5.2 Prediction

When the neural net is trained, it is expected to give a correct output for a given input. For prediction if inputs are given for which output is not known; the net is expected to predict accurate output. The maximum temperature for the next day is predicted based on trained model with previous dataset, i.e., previous day's weather information is given as the input.

The following criteria are considered to evaluate the performance of algorithm used for prediction [20].

- **Root Mean Square Error (RMSE)** Root mean square error is used as the performance criterion defined in Eq. 10.

$$\text{RMSE} = \sqrt{\frac{\sum_{i=0}^{N} \text{error}_i^2}{N}} \tag{10}$$

where N: number of data points

- **Mean Absolute Error (MAE)** The mean absolute error (MAE) is calculated as follows [21]. The mean absolute error is given by Eq. 11.

$$\text{MAE} = \frac{1}{N}\sum_{i=1}^{N} |\text{actual}_i - \text{predicted}_i| \tag{11}$$

- **Mean Percentage Error (MPE)** Mean percentage error (MPE) is the mean of percentage errors and is calculated using Eq. 12.

$$\text{MPE} = \frac{1 * 100\,\%}{N}\sum_{i=1}^{N} \frac{|\text{actual}_i - \text{predicted}_i|}{\text{actual}_i} \tag{12}$$

- **Correlation Coefficient (*r*)** The Correlation coefficient, r measures relationship between two variables [22]. The formula to compute r is given in Eq. 13.

$$r = \frac{N\sum xy - (\sum x \sum y)}{\sqrt{[N\sum x^2 - (\sum x)^2][N\sum y^2 - (\sum y)^2]}} \tag{13}$$

where x is the predicted value and y is actual value.

- **Mean Magnitude of Relative Error (MMRE)** MMRE is also used to evaluate the performance of the prediction technique. MMRE is the common measure of the average estimation accuracy and calculated as per Eq. 14.

$$\text{MMRE} = \frac{1}{N}\sum_{i=1}^{N}\frac{|\text{actual}_i - \text{predicted}_i|}{\text{actual}_i} \tag{14}$$

- **Accuracy** Accuracy is the degree of matching between the predictions and the actual data. It is calculated as Eq. 15.

$$\text{Accuracy} = (1 - \text{MMRE}) * 100 \tag{15}$$

5 Result and Accuracy

In this paper, maximum temperature for two metro cities of India, i.e., New Delhi and Calcutta has been forecasted. The weather data for these cities are collected using SOA concept and then curtailed attribute set is obtained by applying RST. Finally, this reduced data have been fed as requisite input to ANN in order to obtain better prediction rate. The obtained results are shown in Figs. 3 and 4 representing the actual maximum temperature and the corresponding forecasted maximum temperature, respectively. In this paper, only 80 % of result is plotted for convenience. The accuracy and errors for the prediction are shown in Table 3.

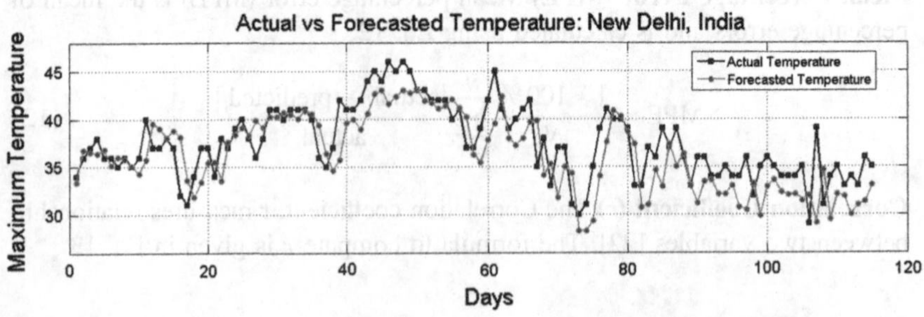

Fig. 3 Actual versus forecasted value for maximum temperature of New Delhi

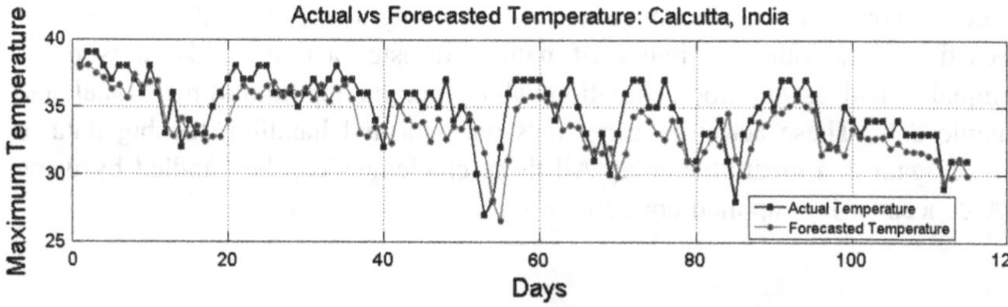

Fig. 4 Actual versus forecasted value for maximum temperature of Calcutta

Table 3 Result, Errors, and Accuracy

Accuracy parameters	New Delhi	Calcutta
Threshold MSE	0.001	0.001
RMSE	0.0705	0.0743
MAE	0.0560	0.0599
MMRE	0.0815	0.0834
MPE	8.1463	8.3360
Accuracy	91.85	91.66
r	0.7642	0.7556

The authors of papers [2, 3] have used neural network for predicting the different aspects of weather. Luk et al. [3] have used different models of neural network for weather prediction. This study has used SOA for data collection, RST for dimension reduction, and then ANN for prediction, so by applying all these three methodologies overall process of weather forecast can be optimized.

6 Conclusion and Future Work

In this paper, service-oriented architecture concept has been applied for collecting the weather data using existing service. The motivation behind application of SOA was to collect data automatically by using existing service which is located at any remote site [23] and access to that service in heterogeneous environment. Data collected are fed as input to application based on rough set theory and artificial neural network for dimension reduction and prediction, respectively. The main objective of this paper is to collect weather data using service-oriented architecture concept automatically and then applying dimension reduction technique for better prediction using artificial neural network. This paper has given emphasis on using SOA, RST, and ANN to make maximum temperature forecasting process optimal.

Future work may be extended toward applying the proposed approach for forecasting some other attributes of natural disaster activities like a tsunami, earthquake, and assess flood situation which are having multidimensional and dynamic data. These activities carry massive data and handling the big data to analyze again is a great challenge. All these challenges can be handled by using SOA concept if it is applied correctly.

References

1. Campbell, S.D., Diebold, F.X.: Weather forecasting for weather derivatives. J. Am. Stat. Assoc. **100**(469), (2005)
2. French, M.N., Krajewski, W.F., Cuykendall, R.R.: Rainfall forecasting in space and time using a neural network. J. Hydrol. **137**(1), 1–31 (1992)
3. Luk, K.C., Ball, J.E., Sharma, A.: An application of artificial neural networks for rainfall forecasting. Math. Comput. Model. **33**(6), 683–693 (2001)
4. Wang, Y.H., Chen, S.C., Peng, P.H.: Applying service-oriented architecture to construct the banking letter of credit system integration. Int. J. Manag. Res. Bus. Strateg. **2**(3), 68–82 (2013)
5. Dsilva, A.N., Maity, S., Rath, S.K.: Stock price prediction using service oriented architecture and soft computing techniques. Int. Conf. Bus. Anal. Intell. **1**(11), 1–13 (2013)
6. Griffin, D., Pesch, D.: A survey on web services in telecommunications. IEEE Commun. Mag. **45**(7), 28–35 (2007)
7. Christudas, B.A., Binildas, C.A.: Service Oriented Java Business Integration: Enterprise Service Bus integration solutions for Java Developers. Packt Publishing Ltd. (2008)
8. Ambler, C., Wallace, A.: U.S. Patent No. 6,393,456. Washington, DC: U.S. Patent and Trademark Office. Google Patent (2002)
9. Rissino, S., Lambert-Torres, G.: Rough set theory—fundamental concepts, principals, data extraction, and applications. Data mining and knowledge discovery in real life applications, pp. 35–58 (2009)
10. Ruzgar, B., Ruzgar, N.S.: Rough sets and logistic regression analysis for loan payment. Int. J. Math. Models Methods Appl. Sci. **2**(1), 65–73 (2008)
11. Soni, S.: Applications of ANNs in stock market prediction: a survey. Int. J. Comput. Sci. Eng. Technol. **2**(3), 71–83 (2011)
12. Hecht-Nielsen, R.: Theory of the backpropagation neural network. In: International Joint Conference on Neural Networks, 1989. IJCNN., pp. 593–605. IEEE (1989, June)
13. Maqsood, I., Khan, M.R., Abraham, A.: An ensemble of neural networks for weather forecasting. Neural Comput. Appl. **13**(2), 112–122 (2004)
14. Rajasekaran, S., Pai, G.V.: Neural networks, fuzzy logic and genetic algorithm: synthesis and applications (with cd). PHI Learning Pvt. Ltd. (2003)
15. Latha, B.C., Paul, S., Kirubakaran, E., Sathianarayanan, A.: A service oriented architecture for weather forecasting using data mining. Int. J. Adv. Netw. Appl. **2**(2), 608–613 (2010)
16. The open enterprise service bus, https://today.java.net/pub/a/today/2008/10/06/sending-sms-messages-with-jbi.html (2010)
17. Ip2geo, http://ws.cdyne.com/ip2geo/ip2geo.asmx?op=ResolveIP (2010)
18. Nina, F.R.C.: On Applications of Rough Sets Theory to Knowledge Discovery (Doctoral dissertation, University of Puerto Rico Mayaguez Campus) (2007)
19. Han, J., Kamber, M., Pei, J.: Data Mining: Concepts and Techniques: Concepts and Techniques. Elsevier (2011)
20. Menzies, T., Chen, Z., Hihn, J., Lum, K.: Selecting best practices for effort estimation. Software Engineering, IEEE Transactions on, 32(11), 883–895 (2006)

21. Willmott, C.J., Matsuura, K.: Advantages of the mean absolute error (MAE) over the root mean square error (RMSE) in assessing average model performance. Clim. Res. **30**(1), 79 (2005)
22. Person, K.: On lines and planes of closest fit to system of points in space. Philos. Mag. **2**, 559–572 (1901)
23. Web servicex net, http://webservicex.net/ws/WSDetails.aspx?CATID=12&WSID=56 (2013)

Software Reliability Prediction Using Machine Learning Techniques

Arunima Jaiswal and Ruchika Malhotra

Abstract Software reliability is an indispensable part of software quality. Software industry endures various challenges in developing highly reliable software. Application of machine learning (ML) techniques for software reliability prediction has shown meticulous and remarkable results. In this paper, we propose the use of machine learning techniques for software reliability prediction and evaluate them based on selected performance criteria. We have applied ML techniques including adaptive neuro fuzzy inference system (ANFIS), feed forward backpropagation neural network (FFBPNN), general regression neural network (GRNN), support vector machines (SVM), multilayer perceptron (MLP), bagging, cascading forward backpropagation neural network (CFBPNN), instance-based learning (IBK), linear regression (Lin Reg), M5P, reduced error pruning tree (reptree), and M5Rules to predict the software reliability on various datasets being chosen from industrial software. Based on the experiments conducted, it was observed that ANFIS yields better results and it predicts the reliability more accurately and precisely as compared to all the above-mentioned techniques. In this study, we also made comparative analysis between cumulative failure data and inter failure time data and found that cumulative failure data give better and more promising results as compared to inter failure time data.

Keywords Software reliability · Assessment · Prediction · Machine learning techniques

---

Arunima Jaiswal (✉) · Ruchika Malhotra
Delhi Technological University, Shahbad Daulatpur, Main Bawana Road, Delhi 110042, India
e-mail: ajaiswal34@amity.edu

Ruchika Malhotra
e-mail: ruchikamalhotra@dce.edu

M. Pant et al. (eds.), *Proceedings of Fifth International Conference on Soft Computing for Problem Solving*, Advances in Intelligent Systems and Computing 436, DOI 10.1007/978-981-10-0448-3_12

1 Introduction

Software reliability is defined as "The ability of the software to perform its required function under stated conditions for a stated period of time" [1]. Reliability is one of the important attributes of the software quality. According to ANSI, software reliability is "The probability of failure-free operation of a computer program for a specified period of time in a specified environment" [2]. Software reliability growth models (SRGM) have been used for predicting and estimating the probability of a system failing in given time interval or the expected time span between successive failures.

ML is an approach which is focused on learning automatically and allows computers to evolve and predict the system behavior based on past and the present failure data. Thus, it is quite natural for software practitioners and researchers to know that which particular method tends to work well for a given failure dataset and up to what extent quantitatively [3–7].

In this study, we present an empirical study of above-mentioned ML techniques for predicting software reliability based on five industrial datasets and investigate about the accuracy and performances of these models in predicting the software reliability when applied to past failure week data. We also performed a comparative analysis between cumulative failure data and inter failure time data to investigate the type of failure data more appropriate for reliability prediction.

2 Study Objective

Business applications which are critical in nature require reliable software, but developing such reliable software is a key challenge which our software industry faces today. With the increasing complexity of the software these days achieving software reliability is hard. In our study, we attempt to empirically assess the use of ANFIS for predicting the software failures. Although ANN, SVM etc., have been previously used in the literature [8] but for the first time ANFIS has been applied to cumulative week failure dataset. The background of using ANFIS was that if it had proven empirically to predict the software failures with least errors in comparison to the above-mentioned techniques, then it may possibly be used as a sound alternative to other mentioned existing techniques for software reliability predictions. Also, above-mentioned ML techniques were empirically analyzed for the first time together on five different types of datasets taken altogether.

3 Related Work

Several ML techniques have been proposed and applied in the literature for software reliability modeling and forecasting. Some of the techniques are genetic programming, gene expression programming, artificial neural network, decision trees, support vector machines, feed forward neural network, fuzzy models, generalized neural network, etc. [8–16]. Karunanithi et al. [11] carried out analysis of detailed study to explain the use of connectionist models in the reliability growth prediction for the software. Cai et al. [15] focused on the development of fuzzy software reliability models instead of probabilistic software reliability models as he says that reliability is fuzzy in nature. Ho et al. [17] carried out a comprehensive study of connectionist models and their applicability to software reliability prediction and inferred that these are better as compared to traditional models. Su and Huang (2006) [18] had applied neural network for predicting software reliability. Madsen et al., 2006 [19] focused on the application of soft computing techniques for software reliability prediction. Pai and Hong (2006) [20] performed experiments using SVMs for forecasting software reliability. Despite of recent advances in this field, it was observed that different models have varied predictive reliability capabilities.

4 Research Background

In this section, we summarize empirical data collection and independent and dependent variables.

4.1 Empirical Data Collection

In this paper, we have used software failure data from various projects given in (Project Data) [20, 21], data collected from tandem computers software data project (Project Data) (release 1 and 2) [22]. Other datasets include telecommunication system data (Project Data) (phase 1 and 2) [23]). Also, we have used data from the project on-line data entry IBM software package (Project Data) [24].

4.2 Independent and Dependent Variables

The dependent variable which we have used in this study is failure rate and the independent variable used is time interval in terms of weeks.

5 Research Methodology

In this study, we explored the above-mentioned ML techniques for predicting failures. We have divided entire dataset into two parts: training and testing datasets. The training and testing datasets selection are being employed using k-fold cross-validation. Here, 10 cross-validation is used where nine parts are used for training and one part is used for validation taken randomly 10 times and results are recorded for each of the 10 runs. This process is applied to each of the five datasets taken into consideration. Thus, cross-validation is used so as to maximize the utilization of failure past datasets by repeatedly resampling the same dataset randomly by reordering it and then splitting it into tenfolds of equal length [25].

A number of modeling techniques, both statistical and intelligent, were used by us to predict reliability.

5.1 Machine Learning Techniques

5.1.1 Adaptive Neuro Fuzzy Inference System (ANFIS)

ANFIS is a hybrid of two intelligent system models. It combines the low-level computational power of a neural network with the high-level reasoning capability of a fuzzy inference system [26]. The basic steps of the FIS model are (i) identification of input variables (failure time) and output variables (cumulative failures) (ii) development of fuzzy profile of the input/output variables (iii) defining relationships between input and output variables using fuzzy inference system. Thus, FIS is capable of making decisions under uncertainty which can be applied for reliability prediction when applied to unknown failure datasets [27].

5.1.2 Cascade Forward Backpropagation Neural Networks (CFBPNN)

It has more than one layer of neurons. The weights of each of the subsequent layer originate from the input and all the layers prior to the layer are in question. Each layer has biases. First, adaption is performed to create the model. Then, the model is trained using the stipulated number of epochs. Performance is computed according to the stipulated performance function. These networks have a weight connection from input to every successive layer and from every layer to all the following layers which sometimes improve the speed at which the model is trained and they only require past failure data as input for prediction analysis and no assumptions.

5.1.3 Feed Forward Backpropagation Neural Networks (FFBPNN)

It contains more than one layer of neurons and is carried out in order to build the model and train it. Single-layered feed forward neural networks have one layer of sigmoid neurons which are then followed by an output layer of linear neurons. The input layer with transfer functions that may be sigmoid (or of any other type except linear) permits the network model to learn both nonlinear as well as linear relationships between input variable vectors and output variable vectors.

5.1.4 Generalized Regression Neural Networks (GRNN)

A generalized regression neural network (probabilistic neural network) consists of a radial basis layer along with a special linear layer. Spread is associated with it whose value generally lies close to 1. A large spread leads to a bigger area from the input vector where the input layer will respond with a number of significant outputs.

5.1.5 Multilayered Perceptron (MLP)

It is a type of FFBPNN, where the backpropagation learning algorithm is in the form of gradient descent. It maps vectors of input data to a vector of appropriate outputs and contains more than one layer of nodes where each of the layers is completely connected to the next layer. In this type of networks, except for the input nodes, each node has a nonlinear activation function like sigmoid function associated with it.

5.1.6 Linear Regression (Lin Reg)

This technique envisages creation of the simplest model using a single input variable and a single output variable. In this model, there are some independent variables between which a linear relationship is found out to yield result in the form of a dependent variable.

5.1.7 M5P

It is a technique based on Quinlan's M5 algorithm and generates M5 model trees. Initially in order to build a tree, a decision tree induction algorithm is executed first. A splitting criterion is then applied till the class values of each of the instance that reaches a node vary slightly or when just a few instances are left. Then, the backward pruning from each of the leaf is being applied.

5.1.8 M5Rules

This technique uses divide and conquer method to generate a decision list. With every iteration, it builds a model tree and converts the best leaf into a separate rule.

5.1.9 Support Vector Machine (SVM)

It is also called as support vector regression (SVR). Application of this technique yields a model which is usually affected only by a subset of the data used while training the model. This is made possible since the cost function used for creating the model has no impact of the training points lying beyond the margin.

5.1.10 Bagging (Bootstrap Aggregating)

It is a ML ensemble which is employed to improve the accuracy as well as stability of the ML algorithms which are generally used in classification. It reduces the variance and helps avoiding over fitting issue.

5.1.11 Reduced Error Pruning Tree (REPTree)

It is a fast decision tree learner which builds a decision or a regression tree and performs the pruning with back over fitting. It sorts values for numeric attributes only. REPTree produces a suboptimal tree under the constraint that a subtree can only be pruned if it does not contain a subtree with a lower classification error than itself.

5.1.12 IBk (Instance-Based k-Nearest Neighbor)

It employs k-nearest neighbor algorithm in order to classify data. It stores the instances in memory while performing training and then compares new instances with these stored instances. It predicts the failure by looking at the k-nearest neighbors of a test instance.

5.2 *Statistical Efficacy Measures*

In order to validate the proposed reliability prediction models, we have used several efficacy measures which are summarized in Table 1.

The precise view of overall training and prediction process is being illustrated through a flow diagram in Fig. 1.

Table 1 Efficacy measures used for evaluating performance criteria

Efficacy measure	Definition
Correlation coefficient	Correlation coefficient measures how closely actual and predicted values are correlated with each other. The correlation coefficient lies between −1 and +1. No linear relationship is there if the correlation coefficient is 0
MARE	Mean absolute relative error is computed as the mean of the absolute difference of corresponding actual value and the corresponding predicted value divided by the corresponding actual value
MRE	Mean relative error is computed as the mean of the difference of corresponding actual value and the corresponding predicted value divided by the corresponding actual value
MSE	Mean squared error is the mean of the square of the difference of corresponding actual value and the corresponding predicted value, i.e., mean of the square of errors

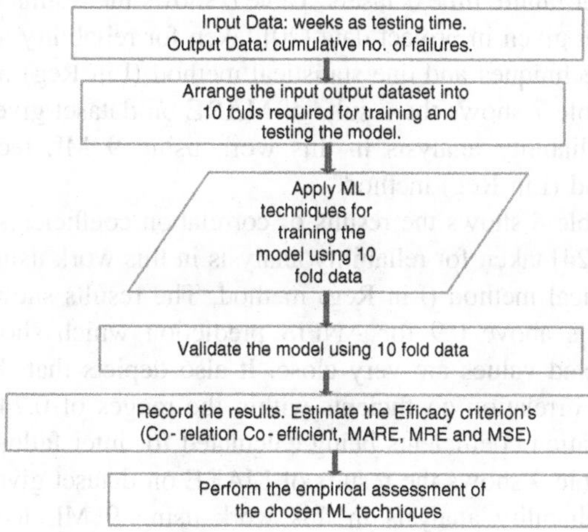

Fig. 1 Overview of software reliability prediction process

In this study, we predict the dependent variable (failure rate) based on the number of failures to be detected using various ML techniques and envisaging the application of tenfold cross-validation method required for training (training the model using training data) and testing the model (using the testing data). Next step includes recording of failures predicted by applying above-mentioned ML techniques. Then we estimate the statistical efficacy measures for all the chosen datasets. Finally, we perform the empirical assessment of the chosen ML techniques.

6 Result Analysis

In this section, we present the summary of results obtained for predicting reliability using five datasets which we have taken for comparison using ML techniques in terms of efficacy measures which are correlation coefficient, MARE, MRE, and

MSE. The results are being summarized in the following Tables 2, 3, 4, 5, 6, 7, 8, 9, and 10. Sometimes, MARE, MRE, and MSE are expressed in percentage (%). However, this paper follows the definition given in Table 1 and does not express MARE in percentage (%) [28].

Table 2 shows the results of correlation coefficients on five datasets taken for reliability analysis in this work using 11 ML techniques and one statistical method (Lin Reg) method.

Table 3 shows the values of MSE on five datasets taken for reliability analysis in this work using 11 ML techniques and one statistical method (Lin Reg) method.

Table 4 shows the values of MARE on five datasets taken for reliability analysis in this work using 11 ML techniques and one statistical method (Lin Reg) method.

Table 5 shows the values of MRE on five datasets taken for reliability analysis in this work using 11 ML techniques and one statistical method (Lin Reg) method.

These performance criteria were also being observed for the cumulative as well as inter failure time datasets. Table 6 shows the results of correlation coefficients on dataset given in project data [20] taken for reliability analysis in this work using 9 ML techniques and one statistical method (Lin Reg) method.

Table 7 shows the results of MARE on dataset given in project data [20] taken for reliability analysis in this work using 9 ML techniques and one statistical method (Lin Reg) method.

Table 8 shows the results of correlation coefficients on dataset given in project data [24] taken for reliability analysis in this work using 9 ML techniques and one statistical method (Lin Reg) method. The results show that the correlation coefficient is above 0.9 for ANFIS prediction which shows that the actual and the predicted values are very close. It also depicts that the cumulative failures yield high correlation coefficients within the ranges of 0.74–0.99 in comparison to the correlation coefficients being calculated for inter failure time data.

Table 9 shows the results of MARE on dataset given in project data [24] taken for reliability analysis in this work using 9 ML techniques and one statistical method (Lin Reg). The results show that the value of MARE is 0.04 for ANFIS prediction. It also depicts that the cumulative failures yield low MARE within the ranges of 0.04–0.48 in comparison to the MAREs being calculated for inter failure time data.

Table 10 shows the pred(0.25) for different datasets. It also depicts that most of the values of MRE lies below 0.25. Hence, the results are very high.

Figure 2 depicts the graphical representation of the analysis of the performances of above-mentioned ML techniques for predicting software reliability based on MARE for the industrial datasets taken into consideration.

Figure 3 depicts the graphical representation of the analysis of the performances of above-mentioned ML techniques for predicting software reliability based on correlation coefficient for the industrial datasets taken into consideration.

Figure 4 depicts the graphical representation of the analysis of the performances of above-mentioned ML techniques for predicting software reliability based on MRE for the industrial datasets taken into consideration.

Table 2 Summary of correlation coefficient predictions for different datasets

Project data	ANFIS	GRNN	FFBPNN	CFBPNN	Bagging	IBK	Linear regression	MSP	M5Rules	Multilayered perceptron	REPTree	SVM
Project data	0.999	0.998	0.978	0.99	0.969	0.992	0.891	0.967	0.984	0.994	0.967	0.866
Project data (release 2)	0.999	0.998	0.997	0.984	0.976	0.986	0.966	0.982	0.988	0.998	0.945	0.955
Project data (release 1)	0.999	0.998	0.984	0.987	0.965	0.99	0.961	0.982	0.994	0.998	0.951	0.96
Project data (phase 1)	0.997	0.986	0.978	0.986	0.98	0.978	0.982	0.985	0.984	0.995	0.957	0.984
Project data (phase 2)	0.998	0.997	0.987	0.993	0.98	0.983	0.985	0.986	0.994	0.997	0.94	0.983

Table 3 Summary of MSE for different datasets

Project data	ANFIS	GRNN	FFBPNN	CFBPNN	Bagging	IBK	Linear regression	MSP	M5 Rules	Multilayered perceptron	REPTree	SVM
Project data	15.838	22.077	231.55	116.534	438.628	85.902	1061.5	352.199	178.3	70.021	352.199	1636.6
Project data (release 2)	1.176	2.717	11.107	67.484	111.949	38	88.726	47.438	12.774	4.069	151.623	118.156
Project data (release 1)	1.658	4.857	43.193	39.131	61.866	15.875	61.834	28.861	12.774	3.286	91.562	64.69
Project data (phase 1)	0.39	2.792	2.607	1.649	3.401	3.321	2.681	2.277	2.381	0.852	10.663	2.781
Project data (phase 2)	0.548	0.618	4.431	3.043	9.001	6.81	5.912	5.516	4.784	1.109	24.313	7.557

Table 4 Summary of MARE for different datasets

Project data	ANFIS	GRNN	FFBPNN	CFBPNN	Bagging	IBK	Linear regression	M5P	M5Rule5	Multilayered perceptron	REPTree	SVM
Project data	0.145	0.293	0.442	0.151	1.031	0.194	1.34	0.832	0.188	0.155	0.832	2.166
Project data (release 2)	0.027	0.053	0.108	0.21	0.266	0.105	0.161	0.109	0.117	0.031	0.256	0.223
Project data (release 1)	0.031	0.059	0.194	0.11	0.162	0.075	0.125	0.083	0.067	0.032	0.158	0.131
Project data (phase 1)	0.14	0.55	0.638	0.219	0.397	0.147	0.246	0.222	0.213	0.145	0.449	0.214
Project data (phase 2)	0.073	0.09	0.249	0.142	0.268	0.141	0.177	0.161	0.141	0.074	0.318	0.194

Table 5 Summary of MRE for different datasets

Project data	ANFIS	GRNN	FFBPNN	CFBPNN	Bagging	IBK	Linear regression	MSP	M5RULES	Multilayered perceptron	REPTree	SVM
Project data	0.025	−0.249	−0.077	−0.005	−0.974	−0.115	−1.191	−0.755	−0.114	−0.074	−0.755	−2.112
Project data (release 2)	−0.012	−0.035	0.066	0.107	−0.22	−0.012	−0.074	−0.04	−0.011	0.006	−0.154	−0.116
Project data (release 1)	−0.015	−0.045	0.036	−0.064	−0.093	−0.015	−0.046	−0.028	−0.023	−0.084	−0.103	−0.058
Project data (phase 1)	0.019	−0.458	0.239	−0.121	−0.307	−0.045	0.077	0.076	0.062	0	−0.304	0.056
Project data (phase 2)	−0.008	−0.044	0.158	0.048	−0.201	−0.036	0.029	0.027	0.008	−0.002	−0.169	−0.013

Table 6 Summary of correlation coefficient predictions for cumulative versus inter failure time data

ML techniques	Inter failure data	Cumulative data
ANFIS	0.307	0.995
GRNN	0.36	0.979
FFBPNN	0.324	0.939
CFBPNN	0.158	0.975
Bagging	0.126	0.946
IBK	−0.147	0.951
Linear regression	0.264	0.92
Multilayered perceptron	0.333	0.975
REPTree	−0.093	0.856
SVM REG	0.392	0.87

Table 7 Summary of MARE predictions for cumulative versus inter failure time data

Mare	Inter failure data	Cumulative data
ANFIS	91.963	0.285
GRNN	104.925	0.314
FFBPNN	88.761	1.192
CFBPNN	94.627	0.504
Bagging	73.151	1.123
IBK	152.986	0.318
Linear regression	61.497	1.504
Multilayered perceptron	91.745	1.384
REPTree	84.267	2.255
SVM REG	32.054	0.798

Table 8 Summary of correlation coefficient predictions for cumulative versus inter failure time data

ML techniques	Inter failure data	Cumulative data
ANFIS	0.709	0.999
GRNN	0.897	0.996
FFBPNN	0.645	0.983
CFBPNN	0.635	0.984
Bagging	0.772	0.918
IBK	0.653	0.971
Linear regression	0.868	0.976
Multilayered perceptron	0.807	0.997
REPTree	0.609	0.743
SVM REG	0.817	0.979

Figure 5 depicts the graphical representation of the analysis of the performances of above-mentioned ML techniques for predicting software reliability based on MSE for the industrial datasets taken into consideration.

Table 9 Summary of MARE predictions for cumulative versus inter failure time data

Mare	Inter failure data	Cumulative data
ANFIS	0.798	0.046
GRNN	0.216	0.122
FFBPNN	0.657	0.45
CFBPNN	0.672	0.489
Bagging	0.506	0.225
IBK	0.288	0.186
Linear regression	0.386	0.188
Multilayered perception	0.175	0.123
REPTree	0.393	0.35
SVM REG	0.289	0.188

Figure 6 depicts the graphical representation of the analysis of the performances of above-mentioned ML techniques for predicting software reliability based on MARE for the project data [20] taken into consideration.

Figure 7 depicts the graphical representation of the analysis of the performances of above-mentioned ML techniques for predicting software reliability based on correlation coefficient for the project data [20] taken into consideration.

Figure 8 depicts the graphical representation of the analysis of the performances of above-mentioned ML techniques for predicting software reliability based on MARE for the project data [24] taken into consideration.

Figure 9 depicts the graphical representation of the analysis of the performances of above-mentioned ML techniques for predicting software reliability based on correlation coefficient for the project data [24] taken into consideration.

Based on the results obtained from rigorous experiments being conducted, we have made following observations regarding this study:

i. ANFIS yields better results as compared to other techniques in predicting failures in terms of statistical efficacy measures undertaken.
ii. ANFIS produces correlation coefficient nearest to +1 which depicts that it shows positive correlation coefficient as compared to all other techniques. The higher the correlation the better the reliability. Hence, we can say that ANFIS predicts failure more accurately than other mentioned techniques.
iii. The techniques GRNN and MLP follow ANFIS in predicting reliability. GRNN also showed to be encouraging and sound. GRNN and MLP produce positive correlation coefficient nearer to 1 and thus can also be used for predicting reliability efficiently and accurately after ANFIS.
iv. From the above results, we also found that ANFIS produces lowest MARE, MRE, and MSE scores as compared to rest of the techniques which again proves it to be better in terms of predicting failures. It shows that ANFIS yields least failure discrepancy between the actual and the predicted failures.

Table 10 Summary of prediction values pred(0.25) for MRE

Project data	ANFIS	GRNN	FFBPNN	CFBPNN	Bagging	IBK	Lin Reg	MLP	M5P	REPTree	M5Rules	SVM
Project data	0.964	0.964	0.929	0.964	0.964	0.964	1	0.964	1	1	0.964	1
Project data (release 2)	1	1	0.947	0.947	1	0.947	1	1	1	O.895	0.947	1
Project data (release 1)	1	1	0.95	1	1	1	1	1	1	0.95	1	1
Project data (phase 1)	0.905	1	0.905	1	1	0.952	0.905	0.905	0.905	0.952	0.952	0.905
Project data (phase 2)	1	1	0.81	0.952	1	0.952	0.905	0.952	0.905	0.905	0.905	0.905

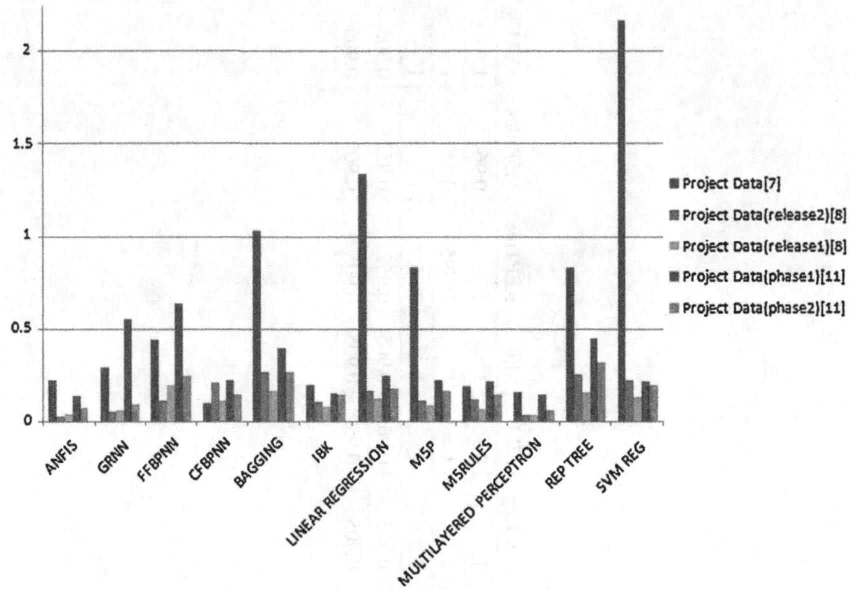

Fig. 2 Comparison of MARE for five different datasets

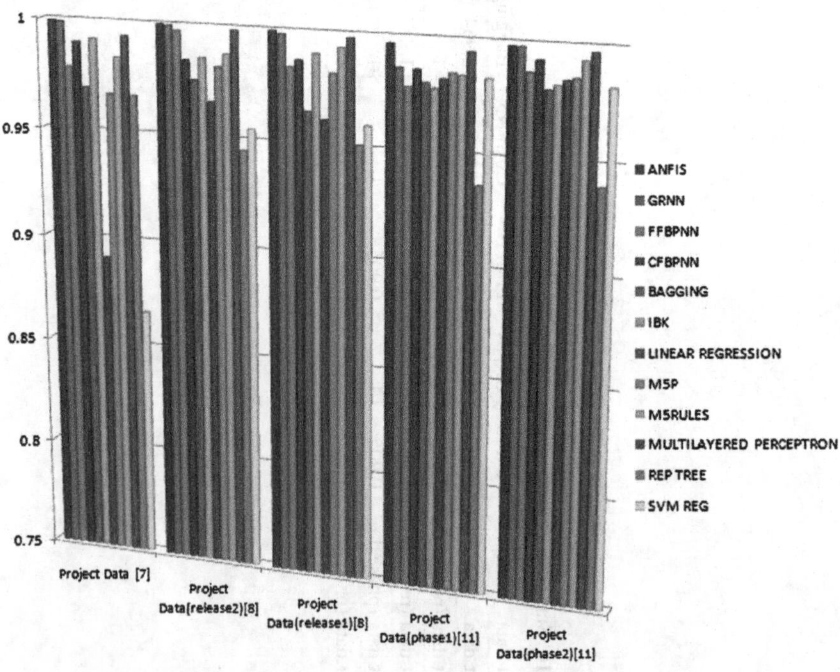

Fig. 3 Comparison of correlation coefficient for five different datasets

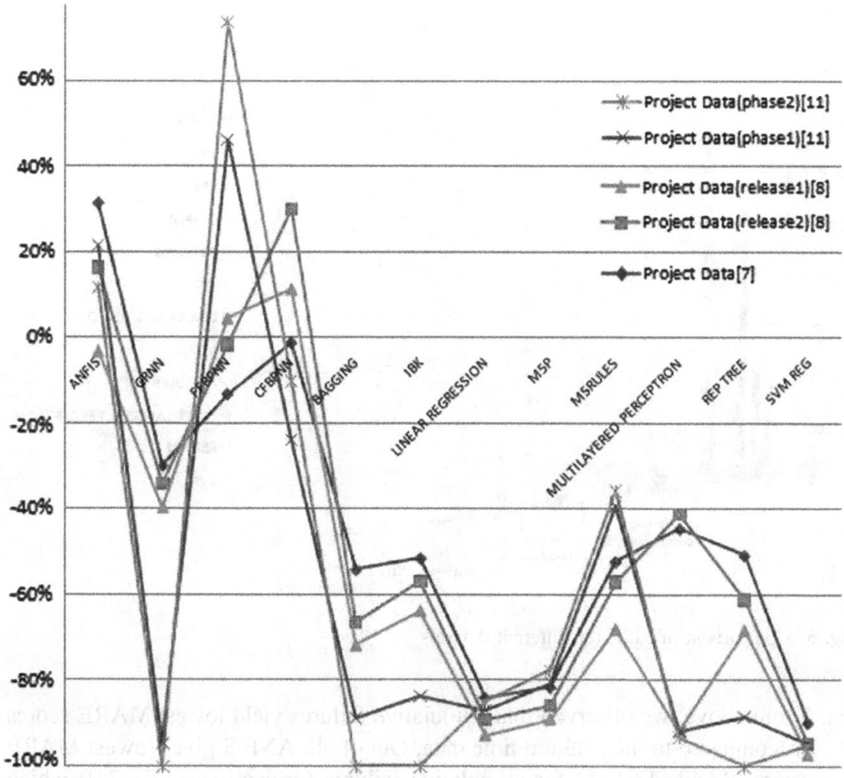

Fig. 4 Comparison of MRE for five different datasets

v. From the results obtained, we also infer that cumulative failures produce more accurate and precise results as compared to inter failure time data. Cumulative failures yield correlation coefficient nearer to 1 unlike inter failure time failures whose correlation coefficient lie more close to −1. Also, we observed that ANFIS yields correlation coefficient nearer to 1 (i.e., 0.9948 and 0.9989) for cumulative failures (project data [20, 24]) as compared to rest of the other mentioned techniques. Based on our experimental evaluation, we also observed that the predicted and the actual values are more closely related for cumulative failures in comparison to inter failure time failures. For cumulative failures, the correlation coefficient ranges from 0.743 to 0.999, which again proves that there is less deviation from actual and predicted values in case of cumulative failures compared to inter failure time failure whose correlation coefficient ranges from −0.043 to 0.897. Thus, it shows that values of correlation coefficient for cumulative failures are more closely related to +1 and hence yields better and accurate results with less deviation between actual and predicted values.

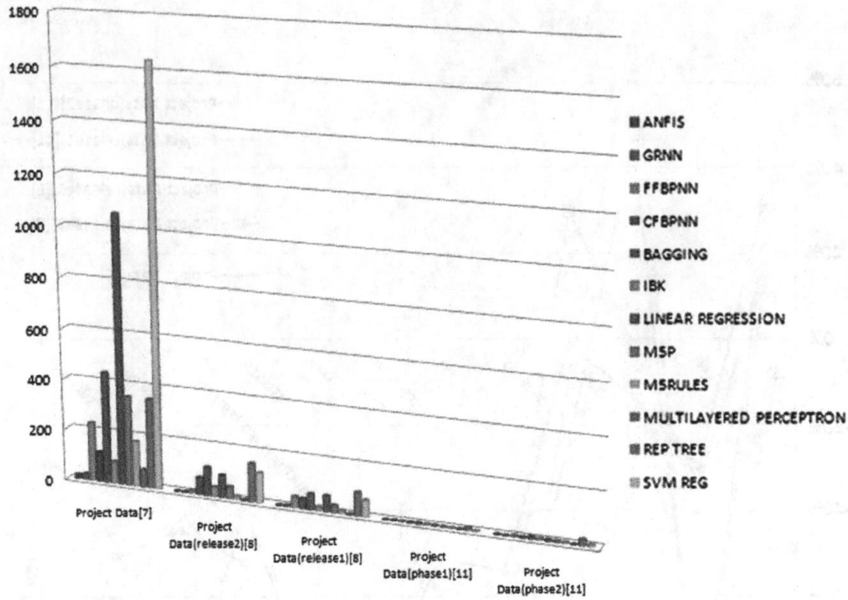

Fig. 5 Comparison of MSE for different datasets

vi. From above, we observed that cumulative failures yield lowest MARE scores as compared to inter failure time data. Out of all, ANFIS gives lowest MARE score (28.49, 4.64 %) for cumulative failures (project data [20, 24]) which again proves it to be the best method among the other mentioned techniques. Based on our experimental evaluation, we also inferred that MARE results ranges from 0.188 to 2.255 for cumulative failures in comparison to inter failure time failure which ranges from 32.054 to 152.986. We have compared MARE results for both the cumulative and inter failure time failure by analyzing it for two different datasets and comparing over ten different techniques. For each of our analysis, we observed that cumulative failures produce better and accurate results with very less percentage of deviation from the actual and the predicted values for software reliability and hence, they are being preferred over inter failure time data as per our study.

7 Threats to Validity

A few of the limitations confronted during the current study are given as follows: One is, the threats to internal validity are present due to the degree to which conclusions can be drawn between independent and dependent variables [28]. The data may not be cumulative and there may be lagging between the values which

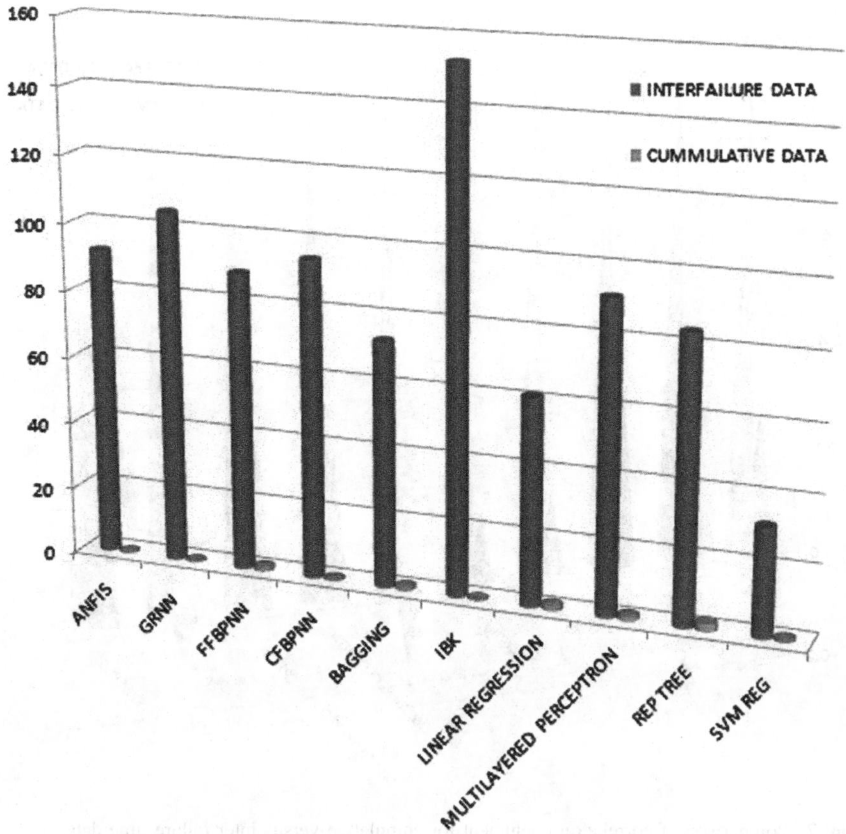

Fig. 6 Comparison of MARE for cumulative versus inter failure time data

need to be addressed. The other is the threats to external validity which are associated with the generalizability of the predicted models. The results in this paper are obtained from open source software WEKA and MATLAB tool, and hence these may not be applicable to other systems. In other terms, the effectiveness of the reliability prediction models depend on the operational environment. The size of the dataset is also not very large. These threats can be minimized by conducting more number of replicated studies across the various systems. Eventually, in spite of all these constraints and limitations, the findings of our work provide the guidance for the future research in order to assess the impact of past failure datasets for the prediction of software reliability using ML techniques.

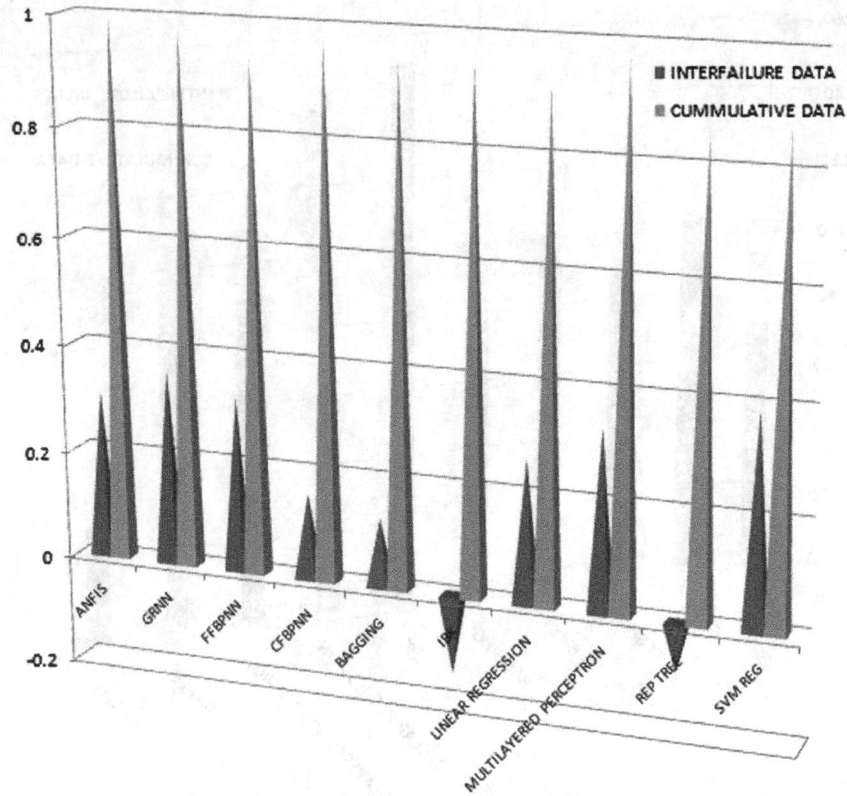

Fig. 7 Comparison of correlation coefficient for cumulative versus inter failure time data

8 Conclusions

Software reliability is dynamic and stochastic in nature so we may say that reliability is a probabilistic measure that assumes that the occurrence of failure of software is a random phenomenon [2]. In this paper, we have applied various machine learning techniques for predicting software reliability based on past failures of software products. The performance of these ML techniques has been evaluated using five different types of datasets being extracted from industrial data to predict the failure intensity of the softwares in use. For each of the five datasets taken into consideration, results show that the correlation coefficient is above 0.99 in most of the predictions for ANFIS which signifies that the actual and the predicted values are very close. Also, we found that the MARE is between the ranges of 0.025–1.5 in most of the predictions for ANFIS and is quite lowest in comparison to the MAREs and is being calculated by other mentioned techniques. The results also depict that the MSE ranges between 0.5 and 16.0 in most of the predictions for ANFIS and MRE and is quite lowest in comparison to the other

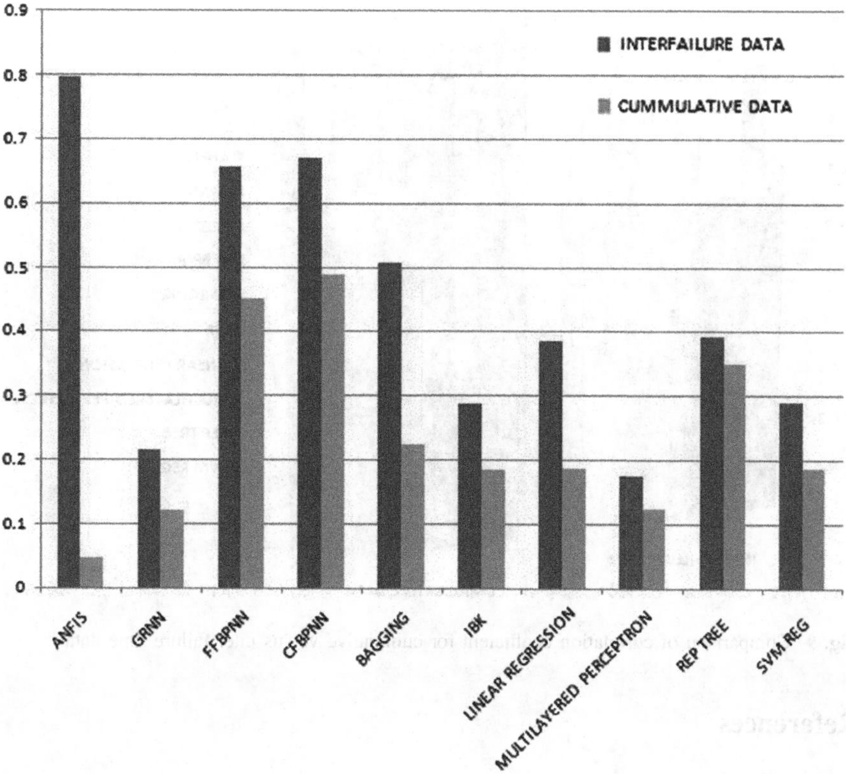

Fig. 8 Comparison of MARE for cumulative versus inter failure time data

mentioned techniques. The result shows that the cumulative failures yield high correlation coefficients within the ranges of 0.74–0.99 in comparison to the correlation coefficients being calculated for inter failure time data. The results also depict that the cumulative failures yield low MARE within the ranges of 0.04 to 1.38 in comparison to the MAREs being calculated for inter failure time data. This is the reason that cumulative failure data is always chosen for failure prediction experiments.

For the further work, more techniques like dynamic neuro fuzzy inference system (DENFIS), group method of data handling (GMDH), and probabilistic neural networks (PNN) can be applied to the data. More datasets can be collected and validated by applying various other machine learning techniques for failure predictions. Further research is planned in an attempt to combine above-mentioned models with other machine learning techniques so as to develop prediction models which can predict the reliability of software more accurately with least precision errors.

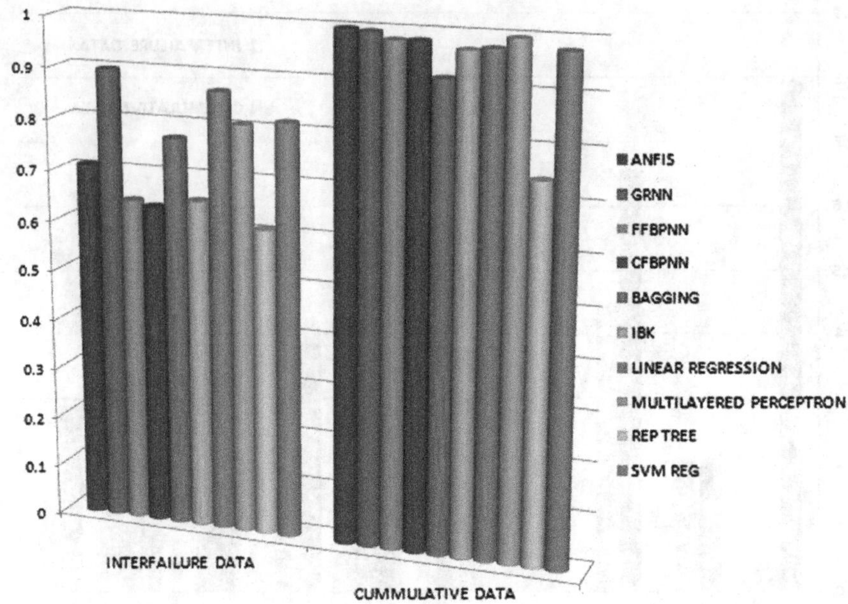

Fig. 9 Comparison of correlation coefficient for cumulative versus inter failure time data

References

1. Standards Coordinating Committee of the IEEE Computer Society, IEEE Standard Glossary of Software Engineering Terminology, IEEE-STD-610.12-1990, IEEE, New York (1991)
2. Quyoum, A., Din Dar, UdM., Quadr, S.M.K.: Improving software reliability using software engineering approach—a review. Int. J. Comput. Appl. **10**(5), 0975- 8887 (2010)
3. Aggarwal, K.K., Singh, Y., Kaur, A., Malhotra R.: Investigating the effect of coupling metrics on fault proneness in object-oriented systems. Softw. Qual. Prof. **8**(4), 4–16 (2006)
4. Goel, B., Singh, Y.: An empirical analysis of metrics. Softw. Qual. Prof. **11**(3), 35–45 (2009)
5. Singh, Y., Kumar, P.: A software reliability growth model for three-tier client–server system. Int. J. Comp. Appl. **1**(13), 9–16, doi:10.5120/289-451 (2010a)
6. Singh, Y., Kumar, P.: Determination of software release instant of three-tier client server software system. Int. J. Comp. Appl. **1**(3), 51–62 (2010b)
7. Singh, Y., Kumar, P.: Application of feed-forward networks for software reliability prediction. ACM SIGSOFT, Softw. Eng. Notes 35(5), 1–6 (2010c)
8. Xingguo, L., Yanhua, S.: An early prediction method of software reliability based on support vector machine. In: Proceedings international conference on wireless communications, networking and mobile computing (WiCom'07), pp. 6075–6078 (2007)
9. Malhotra, R., Kaur, A., Singh, Y.: Empirical validation of object-oriented metrics for predicting fault proneness at different severity levels using support vector machines. Int. J. Syst. Assur. Eng. Manag. **1**(3), 269–281. doi:10.1007/s13198-011-0048-7 (2011)
10. Hua Jung, L.: Predicting software reliability with support vector machines. In: Proceedings of 2nd International Conference on Computer Research and Development (ICCRD'10), Kuala Lumpur, Malaysia, pp. 765–769 (2010)
11. Karunanithi, N., Whitley, D., Malaiya, Y.: Prediction of software reliability using connectionist models. IEEE Trans. Softw. Eng. **18**(7), 563–574 (1992)

12. Singh, Y., Kumar, P.: Prediction of software reliability using feed forward neural networks. In: Proceedings of Computational Intelligence and Software Engineering (CiSE'10), Wuhan, China, pp. 1–5. doi:10.1109/CISE.2010.5677251 (2010d)
13. Singh, Y., Kumar, P.: Application of feed-forward networks for software reliability prediction. ACM SIGSOFT, Softw. Eng. Notes **35**(5), 1–6 (2010c)
14. Eduardo, OC., Aurora, TR., Silvia, RV.: A genetic programming approach for software reliability modeling. IEEE Trans. Reliab. **59**(1), 222–230 (2010)
15. Cai, Y.K., Wen, Y.C., Zhang, L.M.: A critical review on software reliability modeling. Reliab. Eng. Syst. Saf. **32**(3), 357–371 (1991)
16. Specht, F.D.: A general regression neural network. IEEE Trans. Neural Netw. **2**(6), 568–576 (1991)
17. Ho, S.L., Xie, M., Goh, T.N.: A study of connectionist models for software reliability prediction. Comput. Math. Appl. **46**(7), 1037–1045 (2003)
18. Su, S.Y., Huang, Y.C.: Neural network-based approaches for software reliability estimation using dynamic weighted combinational models. J. Syst. Softw. **80**(4), 606–615 (2006)
19. Madsen, H., Thyregod, P., Burtschy, B., Albeanu, G., Popentiu, F.: On using soft computing techniques in software reliability engineering. Int. J. Reliab. Qual. Saf. Eng. **13**(1), 61–72 (2006)
20. Pai, F.P., Hong, C.W.: Software reliability forecasting by support vector machines with simulated vector machines with simulated annealing algorithms. J. Syst. Softw. **79**, 747–755 (2006)
21. Hu, Q.P., Dai, Y.S., Xie, M., Ng, S.H.: Early software reliability prediction with extended ANN model. In: Proceedings of the 30th Annual International Computer Software and Applications Conference (COMPSAC '06), vol. 2, pp. 234–239, Sept 2006
22. Wood, A.: Predicting software reliability. IEEE Tandem Comput. **29**(11), 69–77 (1996)
23. Zhang, X., Jeske, D.R., Pham, H.: Calibrating software reliability models when the test environment does not match the user environment. Appl. Stoch. Models Bus. Ind. **18**, 87–99 (2002)
24. Ohba, M.: Software reliability analysis models. IBM J. Res. Dev. **21**(4) (1984a)
25. Kohavi, R.: The power of decision tables. In: The Eighth European Conference on Machine Learning (ECML-95), Heraklion, Greece, pp. 174–189 (1995)
26. Aljahdali, S.H., Buragga, K.A.: Employing four ANNs paradigms for software reliability prediction: an analytical study. ICGST-AIML J. **8**(2), 1687–4846 (2008)
27. Kumar, P., Singh, Y.: An empirical study of software reliability prediction using machine learning techniques. Int. J. Syst. Assur. Eng. Manag. **3**(3), 194–208. doi:10.1007/s13198-012-0123-8 (2012)
28. van Koten, C., Gray, A.R.: An application of Bayesian network for predicting object-oriented software maintainability. In: The Information Science Discussion Paper, Series Number 2005/02, pp. 1172–6024 (2005)

A Genetic Algorithm Based Scheduling Algorithm for Grid Computing Environments

Poonam Panwar, Shivani Sachdeva and Satish Rana

Abstract A grid computing environment is a parallel and distributed environment in which various computing capabilities are brought together to solve large size computational problems. Task scheduling is a crucial issue for grid computing environments; so it needs to be addressed efficiently to minimize the overall execution time. Directed acyclic graphs (DAGs) can be used as task graphs to be scheduled on grid computing systems. The proposed study presents a genetic algorithm for efficient scheduling of task graphs represented by DAG on grid systems. The proposed algorithm is implemented and evaluated using five real datasets taken from the literature. The result shows that the proposed algorithm outperforms other popular algorithms in a number of scenarios.

Keywords Directed acyclic graph (DAG) · Scheduling · Genetic algorithm (GA) · Makespan

1 Introduction

Scheduling is an important and computationally complex problem in grid computing environment. A grid computing environment is a parallel and distributed system that brings together various computing capacities to solve large complex problems [1]. Parallel program can be decomposed into a set of smaller tasks that generally have dependencies. This is usually done to load balance and share system

Poonam Panwar · Shivani Sachdeva (✉)
Department of Computer Science & Engineering, Ambala College of Engineering and Applied Research, Ambala 133101, India
e-mail: shivani.sachdeva91@gmail.com

Poonam Panwar
e-mail: rana.poonam1@gmail.com

Satish Rana
Department of Biotechnology & Engineering, Ambala College of Engineering and Applied Research, Ambala 133101, India
e-mail: satishrana.biotech@gmail.com

M. Pant et al. (eds.), *Proceedings of Fifth International Conference on Soft Computing for Problem Solving*, Advances in Intelligent Systems and Computing 436, DOI 10.1007/978-981-10-0448-3_13

resources effectively or achieve a target quality of service [2]. Scheduling algorithms are typically used to schedule all the subtasks on a given number of available processors in order to minimize the makespan without violating precedence constraints [3]. A task scheduling problem can be represented as directed acyclic task graph (DAG), for execution on multiprocessors with communication costs [4]. A parallel and distributed computing system could be a homogeneous system or heterogeneous system. Homogeneous computing refers to the system that uses same kind of processors to minimize possible time required to schedule all jobs on similar type of processors [5]. Heterogeneous computing refers to the system that uses different kind of processors to minimize possible time required to schedule all jobs on different types of processors [6–8]. The task scheduling problems are classified into categories which are static scheduling (all the information about tasks such as execution and communication costs for each task and relationship with other tasks is known beforehand) and dynamic scheduling (information is not available and decisions are made at runtime) [7]. In the literature, researchers have proposed various algorithms for solving static task scheduling problems which are deterministic (also known as heuristic-based algorithms) and nondeterministic approaches (also known as guided random search-based algorithms). Deterministic approaches are classified into three subcategories: list scheduling [7, 9], clustering scheduling [10], and task duplication scheduling [11].

Nondeterministic approaches are classified into various categories: particle swarm optimization (PSO) [12], ant colony optimization (ACO) [13], Tabu Search (TS) [14], simulated annealing [15], random search [16], and genetic algorithm (GA) [17]. In the proposed work, GA has been used because genetic algorithms (GAs) stand up as a powerful tool for solving search and optimization problems. GA is based on the principle of evolution and genetics. It is used to resolve complicated optimization problems. In the literature, various GA-based methods have been proposed to solve task scheduling problems but none can be applied to all types of grid environments with homogeneous as well as heterogeneous systems [18–21]. So, a new genetic algorithm that uses a specific way to assign priorities to tasks is proposed in this study that can be applied to both homogeneous as well as heterogeneous multiprocessor scheduling problems [22, 23]. An overview to DAG which is used as task graphs for scheduling is presented in Sect. 2. The proposed algorithm is given in Sect. 3. The algorithm has been applied to five heterogeneous multiprocessor scheduling problems and the results are presented in Sect. 4.

2 Directed Acyclic Graph

A directed acyclic graph (DAG) is a graph with no cycles. DAGs may be used to model many different kinds of information. A DAG is formed by a collection of vertices and directed edges, $G = (V, E)$, where V is set of vertices representing subtasks $T = \{ T_j : j = 1, \ldots, n\}$ and edges between vertices representing execution precedence between subtasks. For any two dependent subtasks T_i and T_j, if the

Table 1 Computation cost matrix for DAG given in Fig. 1

Task	P1	P2	P3
1	22	21	36
2	22	18	18
3	32	27	43
4	7	10	4
5	29	27	35
6	26	17	24
7	14	25	30
8	29	23	36
9	15	21	8
10	13	16	33

execution of T_j depends on the output from the execution of T_i, then T_i, is the predecessor of T_j, and, T_j is the successor of T_i. There is an entry subtask and an exit subtask in a DAG. The entry subtask T_{entry} is the starting subtask of the application without any predecessor, while the exit subtask T_{exit} is the final subtask with no successor. A weight is associated with each vertex and edge. The DAG is complemented by a matrix W that is a $n \times m$ computation cost matrix, where n is the total number of tasks and m is the total number of processors in the system. W_{ij} gives the computation time to execute task T_i on machine P_j. Communication cost is zero when the dependent tasks are executed on the same processor, otherwise they incur the communication cost equal to the edge weight, denoted as C_{ij}, represents the communication cost between subtask T_i and subtask T_j [3]. Table 1 represents the computation cost matrix of size $n \times m$, taken from the literature for DAG given in Fig. 1 [7].

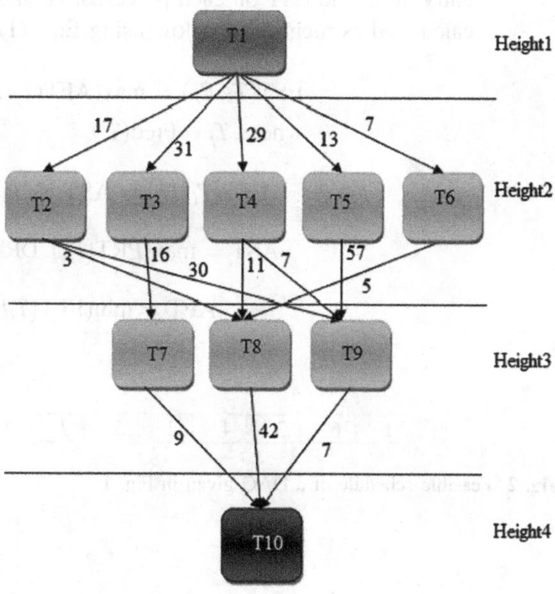

Fig. 1 Application DAG consisting of ten tasks at different levels of height

3 Proposed Algorithm

In the proposed algorithm, DAG is divided in different levels of height as shown in Fig. 1. The crossover and mutation operators are allowed only in the tasks of same height to maintain the precedence relationship of tasks. The proposed genetic algorithm works as follows:

Step 1 **Initialization**

Inputs:

Number of tasks in a program = n;
Number of chromosomes to generate = c;
Number of iterations = it;
Height of a DAG = h;
Number of tasks in each height;

Output:

To produce a feasible schedule, a random function is used to generate unique random numbers from 1 to 'n', where n is total number of tasks. The unique random function is used in such a way that precedence relations between tasks are not violated. Then a population of 1 to 'c' chromosomes is created, where each integer number generated by proposed algorithm represents a gene that corresponds to the task ID. An example of feasible schedule (chromosome encoding) is given in Fig. 2 for DAG shown in Fig. 1.

Step 2 **Deciding Processors for Feasible Schedule:**

After encoding, proposed algorithm calculates the earliest start time EST (T_i, P_k) and earliest finish time $\mathrm{EFT}(T_i, P_k)$ of task T_i on processor P_k. For entry node, the EST on each processor is taken as zero and for others it is calculated as mentioned below using Eqs. (1)–(4).

$$\begin{aligned} &\mathrm{EST}(T_i, P_k) = \max(\mathrm{AFT}(T_j, P_k) + \mathrm{CM}_{ij}) \\ &\text{where, } T_j \in \mathrm{Pred}(T_i) \end{aligned} \tag{1}$$

$$\mathrm{EFT}(T_i, P_k) = \mathrm{AST}(T_i, P_k) + \mathrm{CP}_{ik} \tag{2}$$

$$\mathrm{AST} = \max(\mathrm{PRT}(P_k), \mathrm{DRT}(T_i, P_k)) \tag{3}$$

$$\mathrm{AFT} = \min(\mathrm{EFT}(T_i P_k)) \tag{4}$$

1	6	3	4	2	5	7	9	8	10

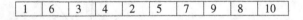

Fig. 2 Feasible schedule of a DAG given in Fig. 1

Indices

i, j: Task index; $i, j = 1, 2, \ldots, n$
k: Processor index; $k = 1, 2, \ldots, n$

Notations

AFT(T_j, P_k) is the actual finish time of task T_j on processor P_k.
CM_{ij} is the communication cost from parent task T_i on child task T_j.
AST(T_i, P_k) is the actual start time of task T_i on processor P_k.
CP_{ik} is the computation cost from task T_i on processor P_k.
PRT(P_k) is the processor ready time of processor P_k.
DRT(T_i, P_k) is the data ready time of task T_i on processor P_k.

Step 3 **Fitness Function**
Fitness function of each chromosome c in the gene pool is calculated using Eq. (5).

$$\mathrm{FT}(S) = \max(\mathrm{AFT}(P_k)) \tag{5}$$

where,
AFT(P_k) is the actual finishing time for the last task in processor P_k.
S is the feasible schedule generated by proposed algorithm.

Step 4 **Selection**
This algorithm makes use of tournament selection. In tournament selection, chromosomes are arranged in ascending order according to fitness function.

Step 5 **Crossover**
In crossover, randomly, height in good chromosomes is selected and copied unaltered in the worst chromosomes. In this, the second half of chromosome having worst fitness value gets replaced with the first half of chromosome. Select randomly level 2 of a chromosome to create a feasible solution, which consists of tasks from task 2 to task 6, and then copied selected tasks from parent 1 to parent 2 and generate two offsprings as shown in Fig. 3.

Parent 1	1	6	5	4	2	3	8	9	7	10
Parent 2	1	4	5	3	2	6	8	7	9	10

After Crossover

Offspring1	1	6	5	4	2	3	8	9	7	10
Offspring2	1	6	5	4	2	3	8	7	9	10

Fig. 3 Offspring generated after crossover

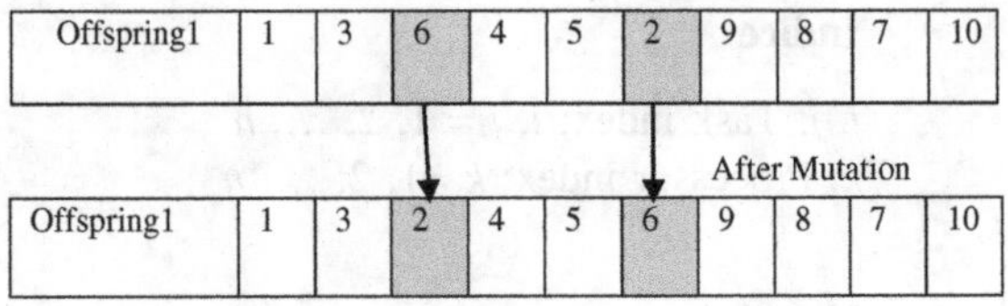

Fig. 4 Offspring generated after mutation

Step 6 **Mutation**
In mutation, randomly two positions of legal offspring are selected and then swap operator is used to swap the contents of that selected positions. But if mutation operator generates worst result of new offspring then mutation operation is not allowed. First, select the height 2 of a chromosome to create a feasible solution, and then randomly select the two positions in level 2. After that, swap the elements on a selected position as shown in Fig. 4.

Step 7 Repeat steps 2–6 until the number of iterations 'it' is achieved or stopping criteria is met.

Step 8 Exit and store the makespan, and execution schedule for each task.

4 Implementation and Experimental Results

The proposed algorithm has been coded in MATLAB using MATLAB801, version R2013a, and implemented on a DELL laptop with an Intel(R) Core(TM) i3-2310M CPU @ 2.10 GHz, x-64 based processor with the following parameters.

- Population size for experimental problem given in Fig. 1 is taken as = 4
- Maximum number of iterations used = 1000
- Stopping Condition: Number of iterations as specified by user
- Total number of trials = 10 for each problem

In this section, the results obtained using proposed genetic algorithm are compared with the results presented in the literature. Application on one of the problem is shown in some detail for DAG in Fig. 1 that consists of ten number of tasks $T = \{T_1, T_2, \ldots, T_{10}\}$ and computation cost matrix consists of "m = 3" heterogeneous computing processors $P = \{P_1, P_2, P_3\}$ in grid environment.

The encoding of tasks is done using encoding process as discussed in Sect. 3. After encoding the EST, EFT of each task is computed and finally the AFT of T_{exit} is computed, which is considered as the makespan of DAG; the chromosomes are evaluated on the basis of fitness function and after a specified number of iterations the best makespan is obtained. It is considered as the optimal schedule for the given task graph.

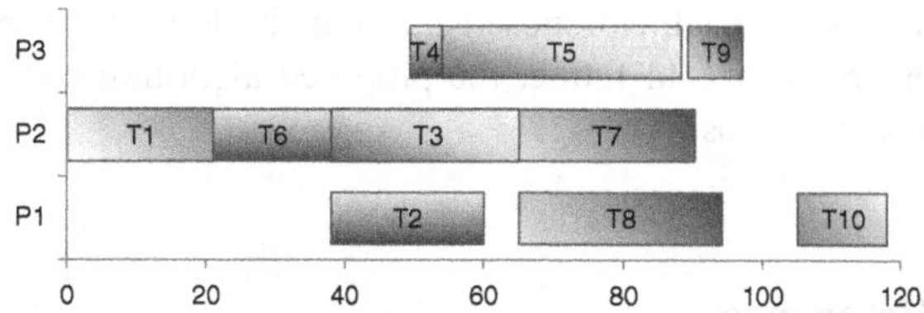

Fig. 5 The schedule of problem generated by proposed genetic algorithm (makespan = 118)

Table 2 Comparison of results obtained by our proposed algorithm with previously known results in the literature

Problem no.	No. of tasks	Makespan obtained using proposed algorithm		Previous best known results	Maximum iterations needed	Execution time (s)
		Best result	Worst result			
1	10	118	161	122 [7]	100	1.411980
2	9	16	19	23 [1]	50	0.260032
3	10	10	11	11 [22]	10	0.28507
4	8	66	88	66 [3]	50	0.241090
5	10	73	99	73 [4]	100	1.469137

The problem was previously solved by many algorithms, and the optimal result found till now was 122 time units [7], the proposed algorithm gives a schedule of 118 units of time presented as given in Fig. 5.

The algorithm has been applied to some more problems taken from the literature and the results obtained are shown in Table 2.

The algorithm has been applied to five heterogeneous problems taken from the literature for three processors. Each problem was solved ten times. The algorithm was made to stop when total number of iterations had been performed as specified by the user. The best and the worst results of these ten runs and comparison of obtained results with best available results in the literature and the execution time needed for the problem to get solved is shown in Table 2. The results shows that out of the five problems, in three cases the obtained makespan is less than the results present in the literature, in two cases the results are similar, and in no case the results are worst than the results present in the literature.

5 Conclusion

In this paper, a multiprocessor task scheduling algorithm for grid environments is proposed which can be used to schedule various tasks represented by DAG. Results in Table 2 shows that the proposed algorithm is able to achieve the optimal assignment schedule in 50–100 iterations in general. Finally, one can conclude from the results that the proposed algorithm is efficient to schedule the task graphs in grid environments. In future, the proposed algorithm can be modified to solve bigger size problems.

References

1. Jiang, Y.S., Chen, W.M.: Task scheduling for grid computing systems using a genetic algorithm. J. Supercomput., (2014). doi:10.1007/s11227-014-1368-6
2. Jin, S., Schiavone, G., Turgut, D.: A performance study of multiprocessor task scheduling algorithms. J. Supercomput. **43**(1), (2008). doi:10.1007/s11227-007-0139-z
3. Xu, Y., Li, K., Hu, J., Li, K.: A genetic algorithm for task scheduling on heterogeneous computing systems using multiple priority queues. Inf. Sci. **270**, 255–287 (2014)
4. Panwar, P., Lal, A.K., Singh, J.: A Genetic algorithm based technique for efficient scheduling of tasks on multiprocessor system. In: Proceedings of the International Conference on Soft Computing for Problem Solving (SocProS 2011), pp. 911–919. Springer India (2012)
5. Dhingra, S., Gupta, S.B., Biswas, R.: comparative analysis of heuristics for multiprocessor task scheduling problem with homogeneous processors. Adv. Appl. Sci. Res. **5**(3), 280–285 (2014)
6. Topcuoglu, H., Hariri, S., Wu, M.Y.: Performance-effective and low-complexity task scheduling for heterogeneous computing. IEEE Trans. Parallel Distrib. Syst. **13**(3), 260–274 (2002)
7. Arabnejad, H., Barbosa, J.G.: List scheduling algorithm for heterogeneous systems by an optimistic cost table. IEEE Trans. Parallel Distrib. Syst. **25**(3), 682–694 (2014)
8. Xu, Y., Li, K., He, L., Zhang, L.: A hybrid chemical reaction optimization scheme for task scheduling on heterogeneous computing systems. IEEE Trans. Parallel Distrib. Syst. (2014)
9. Tang, X., Li, K., Liao, G., Li, R.: List scheduling with duplication for heterogeneous computing systems. J. Parallel Distrib. Comput. **70**(4), 323–329 (2010)
10. Liou, J.C., Palis, M.A.: An efficient task clustering heuristic for scheduling DAGs on multiprocessors. In: Workshop on Resource Management, Symposium on Parallel and Distributed Processing, pp. 152–156 (1996)
11. Park, C.I., Choe, T.Y.: An optimal scheduling algorithm based on task duplication. In: Eighth International Conference on Parallel and Distributed Systems, ICPADS 2001, Proceedings, pp. 9–14. IEEE (2001)
12. Robinson, J., Rahmat-Samii, Y.: Particle swarm optimization in electromagnetics. IEEE Trans. Antennas Propag. **52**(2), 397–407 (2004)
13. Sim, K.M., Sun, W.H.: Ant colony optimization for routing and load-balancing: survey and new directions. IEEE Trans. Syst. Man Cybern. Part A: Syst. Hum. **33**(5), 560–572 (2003)
14. Rolland, E., Schilling, D.A., Current, J.R.: An efficient tabu search procedure for the P-median problem. Eur. J. Oper. Res. **96**(2), 329–342 (1997)
15. Romero, R., Gallego, R.A., Monticelli, A.: Transmission system expansion planning by simulated annealing. IEEE Trans. Power Syst. **11**(1), 364–369 (1996)
16. Price, W.L.: Global optimization by controlled random search. J. Optim. Theory Appl. **40**(3), 333–348 (1983)
17. Sivanandam, S.N., Deepa, S.N.: Introduction to Genetic Algorithms. Springer Science & Business Media (2007)
18. Lee, Y.H., Chen, C.: A Modified genetic algorithm for task scheduling in multiprocessor systems. In: Proceedings of the Ninth Workshop on Compiler Techniques for High-Performance Computing (CTHPC) (2003)

19. Khajemohammadi, H., Fanian, A., Gulliver, T.A.: Efficient workflow scheduling for grid computing using a leveled multi-objective genetic algorithm. J. Grid Comput. **12**(4), 637–663 (2014)
20. Adekunle, Y.A., Ogunwobi, Z.O., Jerry, A.S., Efuwape, B.T., Ebiesuwa, S., Ainam, J. P.: A comparative study of scheduling algorithms for multiprogramming in real-time systems. Int. J. Innov. Sci. Res. **12**(1), 180–185 (2014)
21. Iturriaga, S., Sergio, N., Francisco, L., Enrique, A.: A parallel local search in CPU/GPU for scheduling independent tasks on large heterogeneous computing systems. J. Supercomput. **71** (2), 648–672 (2015)
22. Heidari, H., Chalechale, A.: Scheduling in multiprocessor system using genetic algorithm. Int. J. Adv. Sci. Technol. **43**, 81–93 (2012)
23. Gupta, B., Dhingra, S.: Analysis of genetic algorithm for multiprocessor task scheduling problem. Int. J. Adv. Res. Comput. Sci. Soft. Eng. **3**(7), 339–344 (2013)

Effect of Imperfect Debugging on Prediction of Remaining Faults in Software

Poonam Panwar and Ravneet Kaur

Abstract Software reliability growth models have been used in the literature for estimating the remaining faults in the software based on the failure data collected during its testing phase. Most of the software reliability growth model (SRGMs) are based on the assumption of perfect debugging. However, in reality, this assumption may not be reasonable because imperfect debugging can occur in software development. Therefore, it will be interesting to study the effect of imperfect debugging on the prediction of remaining faults in the software. In this paper, an approach to estimate remaining faults in software using perfect and imperfect software reliability growth models is proposed. The proposed approach is applied on five distinct real datasets for the demonstration of how well these SRGMs predicted the expected total number of faults in case of perfect and imperfect debugging.

Keywords Imperfect debugging · Non-homogenous poisson process · Software reliability growth model

List of Symbols

$\alpha(t)$ The rate at which the faults may be introduced during the debugging process
$a(t)$ Expected number of initial faults
$b(t)$ Fault detection rate per fault in the steady state
$m(t)$ Total number of failures observed at time t according to the actual data
$\hat{m}(t)$ Expected number of failures at a time t estimated by a model

Poonam Panwar · Ravneet Kaur (✉)
Department of Computer Science & Engineering, Ambala College of Engineering and Applied Research, Ambala 133101, India
e-mail: rdhillon453@gmail.com

Poonam Panwar
e-mail: rana.poonam1@gmail.com

M. Pant et al. (eds.), *Proceedings of Fifth International Conference on Soft Computing for Problem Solving*, Advances in Intelligent Systems and Computing 436, DOI 10.1007/978-981-10-0448-3_14

1 Introduction

Over the last two decades, the demand for developing a robust, high-quality, and reliable software products has increased. In practice, project managers always want an assessment of the software reliability to determine when the desired level of reliability has been reached. This leads to the fact that software reliability is now an important research area and it is imperative to have sound methodologies to measure, quantify, and improve software reliability. According to the ANSI definition, software reliability is the probability of failure-free software operation for a specified period of time in a specified environment [1]. To assess software reliability quantitatively, one of the fundamental techniques is to apply software reliability growth models (SRGMs). SRGMs can provide useful information in predicting and improving the reliability of software products. Further, SRGMs can help the software engineers to measure the defect levels, failure rates, and reliability during the coding and testing phases. Non-homogenous poisson process (NHPP) forms one of the main classes of existing SRGMs due to its mathematical tractability and wide applicability. NHPP models are useful in describing failure processes, providing trends such as reliability growth and fault content. NHPP models determine an appropriate mean value function to denote the expected number of failures experienced up to a certain time [2, 3].

Most of these SRGMs assume that a software fault is fixed immediately upon detection with no debugging time delay and no new faults are introduced during the debugging process. These assumptions help to reduce the complexity of modeling software reliability growth. However, software debugging is a complex and labor-intensive process because the fault may not be immediately removed due to the complexity of the software. Therefore, imperfect debugging could occur in the real world. In reality, software developers experience cases when the faults are not removed perfectly and new faults may be introduced in the system while removing the original faults [4, 5]. So it will be quite interesting to study the effect of imperfect debugging models to predict the fault content of software so that the remaining number of faults can be calculated that may affect the software release time.

In this paper, both perfect and imperfect debugging NHPP SRGMs are used for estimating the number of software faults. Five distinct real datasets are taken from the literature and used for demonstration of how well these SRGMs predict the expected total number of failures. The rest of the paper is organized as follows. In Sect. 2, perfect and imperfect debugging are discussed in some details. Section 3 discussed proposed approach for estimating the faults in software. Section 4 gives some numerical examples with real software failure data employing the proposed approach. Finally Sect. 5 concludes the paper.

2 Perfect Debugging Versus Imperfect Debugging

Software debugging is the process to identify and rectify faults that interrupt the desired operation of software according to set specifications. There are two type of debugging process (i) Perfect and (ii) Imperfect debugging. Perfect debugging assumes that a detected fault is corrected with certainty with no debugging time delay. However, practical experiences shows that the time to fix an identified fault depends on a number of factors such as the complexity of the fault, the efficiency, and the skill level of the personnel, the size of the debugging team, the development environment, and the techniques being used [6]. In some practical situations, software developers also experience cases when fixing a fault and may introduce new faults which leads to imperfect debugging. Imperfect debugging occurs when root cause of failure remains undetected, or detected fault is not repaired perfectly. Therefore, to obtain realistic estimates it is desirable to incorporate the effect of imperfect debugging into the software reliability growth modeling. A number of perfect debugging models as well as imperfect debugging models are proposed in the literature by several authors to predict the remaining number of faults in software [7–15].

The existing perfect and imperfect debugging models are used in the literature to predict the faults in the software but which model to be used for a specific dataset is still a topic for research. So, in this study, a method is proposed to select the best perfect and imperfect debugging model for a given dataset. The selected model is then used to predict the remaining faults in software. A comparison of predicted number of remaining faults is done to validate that the assumption of imperfect debugging is necessary or not. The proposed approach is given in Sect. 3.

3 Proposed Approach

Many number of techniques exists in the literature to predict the faults remaining in software [7, 9, 14, 16]; we found that the proposed approach is simple and easy to use. It makes a use of sum of squared error (SSE), adjusted R-square, and root mean squared error (RMSE) which can be defined as follows:

- SSE: The sum of squared error (SSE) sum up the squares of the residuals of the actual data and the mean value function of each model in terms of the number of actual faults at any time points [17]. It can be expressed as given in Eq. (1):

$$\mathrm{SSE} = \sum_{i=1}^{k} (m(t) - \hat{m}(t))^2 \tag{1}$$

where k represents the sample size of the dataset.

- Adjusted R-square: This statistic uses the R-square statistic and adjusts it based on the residual degrees of freedom. It can be expressed as given in Eq. (2):

$$\text{Adjusted R square} = 1 - \frac{\text{SSE}(n-1)}{\text{SST}(v)} \tag{2}$$

where v is the residual degrees of freedom and n is number of response values.
- RMSE: The root mean squared error is an estimate of the standard deviation of the random component in the data [18]. It can be expressed as given in Eq. (3):

$$\text{RMSE} = \sqrt{\frac{1}{k}\sum_{i=1}^{k}[m(t) - \hat{m}(t)]^2} \tag{3}$$

where i represents a component of dataset; $i = 1, 2, 3, \ldots, k$ and k represents the sample size of the dataset.
The proposed approach works as follows:
- Identify various imperfect and perfect debugging NHPP SRGMs for the proposed study.
- Collect experimental data from the literature/real-life software applications.
- Fit selected models to the available failure data and use a software tool such as genetic algorithm and curve fitting tool of Matlab software to determine unknown parameters for each model using LSE parameter estimation technique, SSE, Adjusted R-Square, RMSE, and expected number of cumulative failures based on each of these models.
- Calculate the rank index of the selected models using Eq. (4):

$$\text{Rank Index} = \frac{1}{3}\left(\frac{\min_j^n(\text{SSE})}{\text{SSE}_j} + \text{Adjusted}\,R\,\text{Square}_j + \frac{\min_j^n(\text{RMSE})}{\text{RMSE}_j}\right) \tag{4}$$

where j is the index of models, $j = 1, 2, 3, \ldots, n$.
- Rank the models according to the value of rank index (i.e., model with maximum rank index is ranked 1).
- Estimate the $\hat{m}(t)$ using mean value functions equations given in Tables 1 and 2 for each model, respectively.
- Next, estimate the value of $a(t)$ using unknown parameters and calculate the remaining faults in software using Eq. (5):

$$\text{Remaining faults} = |\text{Total faults}\,a(t) - \text{Estimated Faults}\,\hat{m}(t)| \tag{5}$$

- Finally, plot graphs for each real dataset to see how well perfect and imperfect debugging SRGMs predicted the expected total number of faults.
In this paper, we considered 14 NHPP SRGMs. Tables 1 and 2 summarizes the perfect and imperfect debugging SRGMs with their mean value function, respectively, used in this study.

Table 1 Perfect debugging SRGMs used in proposed study

Model name	Mean value function
Delayed S-shaped [24]	$m(t) = a(1 - (1 + bt)e^{-bt}$
Generalized Goel [25]	$m(t) = a(1 - e^{-bt^c})$
Goel-Okumoto [26]	$m(t) = a(1 - e^{-bt})$
Gompert [25]	$m(t) = a(k^{b^t})$
Inflection S-shaped [27]	$m(t) = \frac{a(1-e^{-bt})}{1+\beta e^{-bt}}$
Logistic growth [25]	$m(t) = \frac{a}{1+ke^{-bt}}$
Pham Zhang IFD [28]	$m(t) = a - ae^{-bt}(1 + (b + d)t + bdt^2)$

Table 2 Imperfect debugging SRGMs used in the proposed study

Model name	Mean value function
P-N-Z model [29]	$m(t) = \frac{a\left(1-e^{-bt}\right)\left(1-\frac{\alpha}{b}\right)+\alpha at}{1+\beta e^{-bt}}$
P-Z model [30]	$m(t) = \frac{1}{(1+\beta e^{-bt})}\left((c+a)\left(1-e^{-bt}\right) - \frac{ab}{b-\alpha}\left(e^{-\alpha t} - e^{-bt}\right)\right)$
Yamada exponential [31]	$m(t) = a(1 - e^{-r\alpha\left(1-e^{-\beta t}\right)})$
Yamada Rayleigh [31]	$m(t) = a\left(1 - e^{-r\alpha\left(1-e^{\left(-\frac{\beta t^2}{2}\right)}\right)}\right)$
Yamada imperfect debugging model 1 [32]	$m(t) = \frac{ab(e^{\alpha t}-e^{-bt})}{\alpha+b}$
Yamada imperfect debugging model 2 [32]	$m(t) = a\left(1 - e^{-bt}\right)\left(1 - \frac{\alpha}{b}\right) + \alpha at$
Zhang-Teng-Pham [17]	$m(t) = \frac{a}{p-\beta}[(1 - \frac{(1+\alpha)e^{-bt}}{1+\alpha e^{-bt}})^{\frac{c}{b}(p-\beta)}]$

4 Application of Proposed Approach

To illustrate how well the perfect and imperfect SRGMs predict the expected remaining failures in software, a data analysis have been carried out on following five real software datasets.

i. Dataset 1 (DS-1): The DS-1 was collected from a subset of products for software releases at Tandem Computers Company [19]. The DS-1 reported 100 defects during 20 weeks.
ii. Dataset 2 (DS-2): Dataset 2 consists of the failure dataset [20] that came from a large medical record system of release-1. The dataset-2 shows the test time for 18 weeks and their cumulative failures.
iii. Dataset 3 (DS-3): The dataset 3 was obtained from a real software project [21] in which during an 81-week software execution period, 461 faults were recorded.

Table 3 Parameter estimation for perfect debugging SRGMs (DS-1)

Model Name	LSEs
Delayed S-shaped	a = 104, b = 0.2654
Generalized Goel	a = 118.6, b = 0.07651, c = 1.11
Goel-Okumoto	a = 130.2, b = 0.08317
Gompert	a = 110, b = 0.8328, k = 0.08877
Inflection S-shaped	a = 110.8, b = 0.1721, β = 1.205
Logistic growth	a = 103.9, b = 0.2849, k = 6.62
Pham Zhang IFD	a = 104, b = 0.2654, d = 1.529e−10

iv. Dataset 4 (DS-4): The dataset 4 was obtained from a wireless network product [22], 203 defects were recorded during 51-week software execution period.
v. Dataset 5 (DS-5): The dataset 5 was collected from a bug tracking system on the website of Xfce [23] which reported 167 cumulative numbers of failures within 21 weeks.

The application of proposed approach on DS-1 is explained below in detail. First, the unknown parameters for both type of debugging models were estimated using curve fitting tool of Matlab as given in Tables 3 and 4.

Tables 5 and 6 summarize the values of SSE, adjusted R-square and RMSE for each perfect and imperfect debugging SRGMs, respectively, for DS-1. The ranking of perfect and imperfect debugging SRGMs based on rank index is also given in Tables 5 and 6 for each model, respectively.

Tables 7 and 8 summarize the fault estimation for perfect and imperfect debugging SRGMs, respectively, for DS-1. Here one can see that, generally perfect debugging estimates are more close to actual data available as compared to imperfect debugging.

Figures 1, 2, 3, 4 and 5 shows the actual faults, estimated faults for both perfect debugging models, and imperfect debugging models for DS-1, DS-2, DS-3, DS-4, and DS-5, respectively. Results show that in most cases perfect debugging SRGMs

Table 4 Parameter estimation for imperfect debugging SRGMs (DS-1)

Model name	LSEs
P-N-Z model	a = 110.8, α = 2.65e−13, b = 0.1721, β = 1.204
P-Z model	a = 130.8, α = 0.07941, b = 38.29, β = 0.4786, c = 1.731
Yamada exponential	a = 117.2, α = 0.9997, β = 6.912e−05, r = 1461
Yamada Rayleigh	a = 115.8, α = 0.984, β = 0.0172, r = 2.018
Yamada imperfect debugging model 1	a = 130.2, α = 1.995e−11, b = 0.08317
Yamada imperfect debugging model 2	a = 130.2, α = 2.164e−09, b= 0.08317
Zhang-Teng-Pham	a = 10.42, α = 0.9937, b = 0.1509, β = 0.6603, c = 1.571, p = 0.7517

Table 5 Ranking of perfect debugging SRGMs (DS-1)

Model name	SSE	Adjusted R-square	RMSE	Rank index	Rank
Delayed S-shaped	505.1276	0.9672	5.2974	0.4308	6
Generalized Goel	217.4152	0.9851	3.5762	0.5070	4
Goel-Okumoto	232.3422	0.9849	3.5928	0.5031	5
Gompert	78.8286	0.9946	2.1534	0.6828	2
Inflection S-shaped	179.5843	0.9877	3.2502	0.5312	3
Logistic growth	32.4758	0.9978	1.3822	0.9993	1
Pham Zhang IFD	505.1276	0.9672	5.2974	0.4308	7

Table 6 Ranking of imperfect debugging SRGMs (DS-1)

Model name	SSE	Adjusted R-square	RMSE	Rank index	Rank
P-N-Z model	179.5843	0.9877	3.2502	0.9900	1
P-Z model	229.6817	0.9821	3.9131	0.8603	5
Yamada exponential	313.1996	0.9771	4.4244	0.7583	6
Yamada Rayleigh	790.7322	0.9422	7.0300	0.5426	7
Yamada imperfect debugging model 1	232.3422	0.9849	3.5928	0.8829	3
Yamada imperfect debugging model 2	232.3422	0.9849	3.5928	0.8829	4
Zhang-Teng-Pham	176.4086	0.9853	3.5497	0.9670	2

Table 7 Estimation of faults for perfect debugging SRGMs (DS-1)

Model name	Actual faults at time t ($m(t)$)	Estimated faults at time t ($\hat{m}(t)$)	Total faults in s/w ($a(t)$)	Remaining faults in s/w ($a(t) - \hat{m}(t)$)
Delayed S-shaped	100	100.7514	104	3.2486
Generalized Goel	100	104.4701	118.6	14.1299
Goel-Okumoto	100	105.5279	130.2	24.6721
Gompert	100	103.3495	110	6.6505
Inflection S-shaped	100	103.2721	110.8	7.5279
Logistic growth	100	101.6440	103.9	2.2560
Pham Zhang IFD	100	100.7514	104	3.2486

estimated faults closer to actual faults found in the software as compared to imperfect debugging SRGMs. So one can say that imperfect debugging assumption does not always effect the estimation of remaining faults in software.

Table 8 Estimation of faults for imperfect debugging SRGMs (DS-1)

Model name	Actual faults at time t ($m(t)$)	Estimated faults at time t ($\hat{m}(t)$)	Total faults in s/w ($a(t)$)	Remaining faults in s/w ($a(t) - \hat{m}(t)$)
P-N-Z model	100	103.2753	110.8000	7.5247
P-Z model	100	105.7539	105.8095	0.0555
Yamada exponential	100	101.6168	117.2000	15.5832
Yamada Rayleigh	100	98.8575	115.8000	16.9425
Yamada imperfect debugging model 1	100	105.5279	130.2000	24.6721
Yamada imperfect debugging model 2	100	105.5279	130.2000	24.6721
Zhang-Teng-Pham	100	103.8951	114.0044	10.1092

Fig. 1 Comparison of faults for DS-1

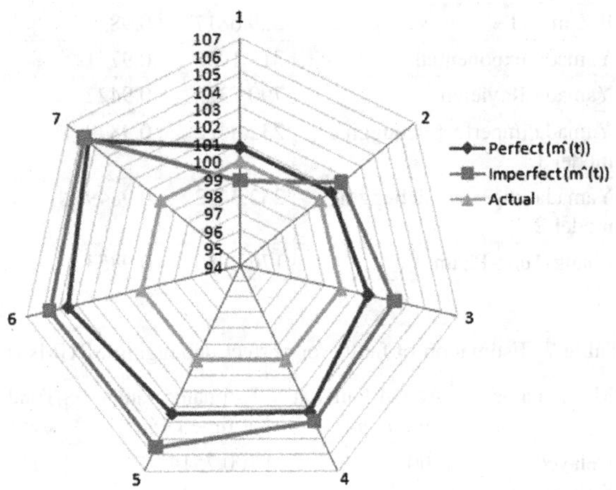

Fig. 2 Comparison of faults for DS-2

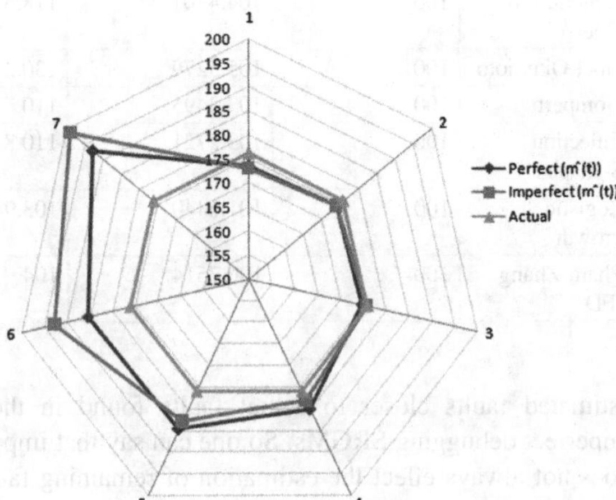

Fig. 3 Comparison of faults for DS-3

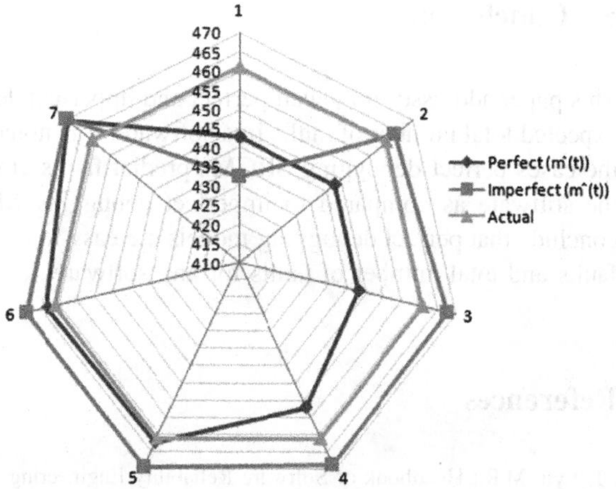

Fig. 4 Comparison of faults for DS-4

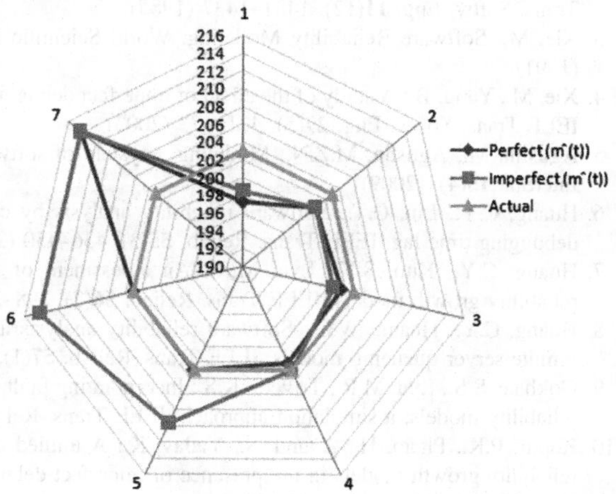

Fig. 5 Comparison of faults for DS-5

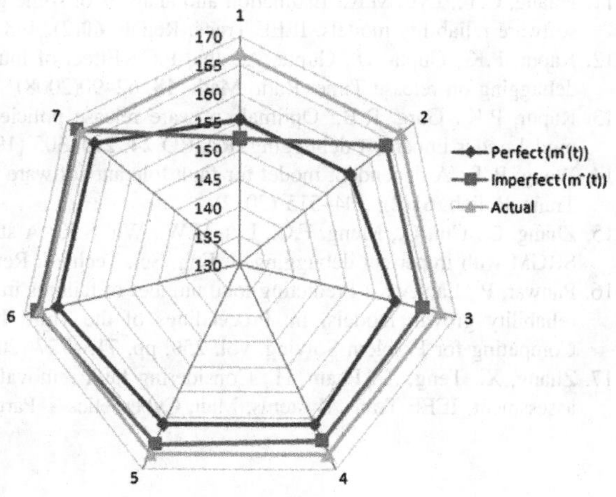

5 Conclusion

This paper addresses how well perfect and imperfect debugging SRGMs predict the expected total number of faults in a software. It is noted in the study that in most of the cases perfect debugging SRGMs predict faults closer to actual faults found in the software as compared to imperfect debugging SRGMs. So, finally, one can conclude that perfect debugging models are easy to use in estimating the remaining faults and total number of faults for any software.

References

1. Lyu, M.R.: Handbook of Software Reliability Engineering. McGraw Hill, New York (1996)
2. Yamada, S., Osaki, S.: Software reliability growth modeling: models and applications. IEEE Trans. Softw. Eng. **11**(12), 1431–1437 (1985)
3. Xie, M.: Software Reliability Modeling. World Scientific Publishing Company, New York (1991)
4. Xie, M., Yang, B.: A study of the effect of imperfect debugging on software development cost. IEEE Trans. Softw. Eng. **29**(5), 471–473 (2003)
5. Agustin, M., Agustin, M.Z.N.: Modeling a system of softwares under imperfect debugging. InterStat **15**(4), (2009)
6. Huang, C.Y., Lin, C.T.: Software reliability analysis by considering fault dependency and debugging time lag. IEEE Trans. Reliab. **55**(3), 436–450 (2006)
7. Huang, C.Y., Kuo, S.Y., Lyu, M.R.: An assessment of testing-effort dependent software reliability growth models. IEEE Trans. Reliab. **56**(2), 198–211 (2007)
8. Huang, C.Y., Huang, W.C.: Software reliability analysis and measurement using finite and infinite server queueing models. IEEE Trans. Reliab. **57**(1), 192–203 (2008)
9. Gokhale, S.S., Lyu, M.R., Trivedi, K.S.: Incorporating fault debugging activities into software reliability models: a simulation approach. IEEE Trans. Reliab. **55**(2), 281–292 (2006)
10. Kapur, P.K., Pham, H., Anand, S., Yadav, K.: A unified approach for developing software reliability growth models in the presence of imperfect debugging and error generation. IEEE Trans. Reliab. **60**(1), 331–340 (2011)
11. Huang, C.Y., Lyu, M.R.: Estimation and analysis of some generalized multiple change-point software reliability models. IEEE Trans. Reliab. **60**(2), 498–514 (2011)
12. Kapur, P.K., Gupta, D., Gupta, A., Jha, P.C.: Effect of introduction of faults and imperfect debugging on release Time. Ratio Math. **18**, 62–90(2008)
13. Kapur, P.K., Garg, R.B.: Optimal software release policies for software reliability growth model under imperfect debugging. RAIRO **24**, 295–305 (1990)
14. Wang, R.T.: A dependent model for fault tolerant software systems during debugging. IEEE Trans. Reliab. **61**(2), 504–515 (2012)
15. Zhang, C., Cui, G., Meng, F.C., Liu, H.W., Wu, S.X.: A study of optimal release policy for SRGM with imperfect debugging. J. Eng. Sci. Technol. Rev. **6**(3), 111–118 (2013)
16. Panwar, P., Lal, A.K.: Predicting total number of failures in a software using NHPP software reliability growth models. In: Proceedings of the Third International Conference on Soft Computing for Problem Solving, vol. 259, pp. 715–727 (2014)
17. Zhang, X., Teng, X., Pham, H.: Considering fault removal efficiency in software reliability assessment. IEEE Trans. Systems, Man, Cybernetics— Part A, **33**(1), 114–120 (2003)

18. Ehrlich, W.K., Emerson, T.J.: Modeling software failures and reliability growth during system testing. In: Proceedings of the 9th International Conference on Software Engineering (ICSE'87), Monterey, CA, pp. 72–82 (1987)
19. Wood, A.: Predicting software reliability. IEEE Comput. **11**, 69–77 (1996)
20. Stringfellow, C., Andrews, A.A.: An empirical method for selecting software reliability growth models. Empirical Softw. Eng. 319–343 (2002)
21. Kanoun, K., Martini, M., Souza, J.: A method for software reliability analysis and prediction application to the TROPICO-R switching system. IEEE Trans. Softw. Eng. **17**(4), 334–344 (1991)
22. Jeske, D.R., Zhang, X., Pham, L.: Adjusting software failure rates that are estimated from test data. IEEE Trans. Reliab. **54**(1), 107–114 (2005)
23. Tamura, Y., Yamada, S.: Comparison of software reliability assessment methods for open source software. In: Proceedings of the 11th International Conference on Parallel and Distributed Systems (ICPADS 2005), Los Almitos, CA, pp. 488–492 (2005)
24. Yamada, S., Ohba, M., Osaki, S.: S-shaped reliability growth modeling for software error detection. IEEE Trans. Reliab. **12**, 475–484 (1983)
25. Huang, C.Y., Lyu, M.R., Kuo, S.Y.: A unified scheme of some nonhomogenous poisson process models for software reliability estimation. IEEE Trans. Softw. Eng. **29**(3), 261–269 (2003)
26. Goel, A.L., Okumoto, K.: Time-dependent error-detection rate model for software and other performance measures. IEEE Trans. Reliab. **R-28**, 206–211 (1979)
27. Ohba, M.: Inflexion S-shaped software reliability growth models. In: Osaki, S., Hatoyama, Y. (eds.) Stochastic Models in Reliability Theory, pp. 144–162. Springer, Merlin, (1984)
28. Pham, H.: System Software Reliability. Springer, London (2006)
29. Pham, H., Nordmann, L., Zhang, X.: A general imperfect software debugging model with s-shaped fault detection rate. IEEE Trans. Reliab. **48**, 169–175 (1999)
30. Pham, H., Zhang, X.: An NHPP software reliability models and its comparison. Int. J. Reliab. Qual. Saf. Eng. **14**(3), 269–282 (1997)
31. Pham, H.: Software reliability and cost models: perspectives, comparison and practice. Eur. J. Oper. Res. **149**, 475–489 (2003)
32. Yamada, S., Tokuno, K., Osaki, S.: Imperfect debugging models with fault introduction rate for software reliability assessment. Int. J. Syst. Sci. **23**(12), 2241–2252 (1992)

Multiple Document Summarization Using Text-Based Keyword Extraction

Deepak Motwani and A.S. Saxena

Abstract The main focus of the paper is on the comparison between the proposed methodology keyword-based text extraction using threading and synchronization just like multiple files input as batch processing and previously used technologies for text extraction from research papers. Keyword-based summary is defined as selecting important sentences from actual text. Text summarization is the condensed form of any type of document whether pdf, doc, or txt files but this condensed form should preserve complete information and meaningful text with the help of single input file and multiple input file. It is not an easy task for human being to maintain the summary of large number of documents. Various text summarizations and text extraction techniques are being explained in this paper. Our proposed technique creates the summary by extracting sentences from the original document with the font type and pdf font or keyword extractor.

Keywords Document summarization · Information retrieval · Information extraction · Keyword extraction · Threading · Synchronization batch processing

1 Introduction

In the present era, most of the automated summarization systems create extracts only.

This section explains the various number of problems that we face during retrieval of full text document. There are so many problems associated with full text analysis, but two of them seem to be the most important one these days. First one is the bad quality of search engine mainly the internet search engines and there is a

Deepak Motwani (✉)
Department of CSE, Mewar University, Chittorgarh, Rajasthan, India
e-mail: dmotwani20005@gmail.com

A.S. Saxena
Faculty of Engineering & Technology, Mewar University, Chittorgarh, Rajasthan, India
e-mail: anand.saxena42@gmail.com

M. Pant et al. (eds.), *Proceedings of Fifth International Conference on Soft Computing for Problem Solving*, Advances in Intelligent Systems and Computing 436, DOI 10.1007/978-981-10-0448-3_15

Summarization = Topic Identification + Interpretation + Generation

Fig. 1 General overview of summarization

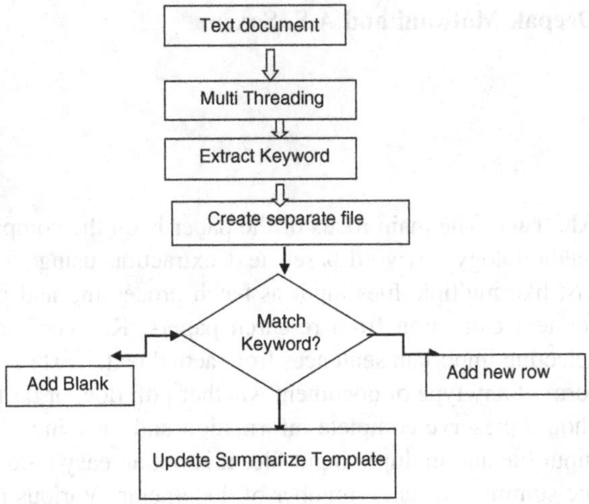

Fig. 2 Flowchart representing algorithm 1

shortage of automatic text categorization tools which are able to do fast assessment of large collection of documents. Many approaches are being used for the extraction in the last 10 years (Figs. 1, 2, 3, 4, 5 and 6).

Abstractive summarization writes the novel sentences which are not seen in the original file. Huge amount of information is present in the form of research publications on the web. Extracted header information is being used for all tasks related to data mining. For the extraction of header information from the research paper, author generates the hybrid approach by combining the given three ParsCit, GROBID, and Mendeley. The proposed methodology is tested on 75 research papers by the author and the precision rate was 95.97 %. Mendeley takes the help of

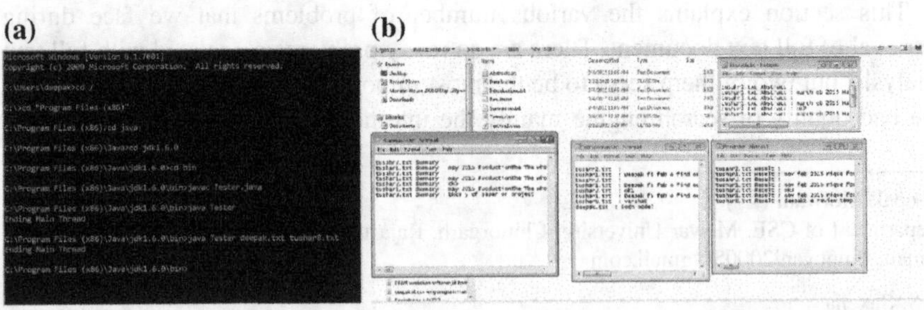

Fig. 3 **a** Command line argument for multiple files and **b** the output in separate files based on keywords

Fig. 4 Flowchart of algorithm 2

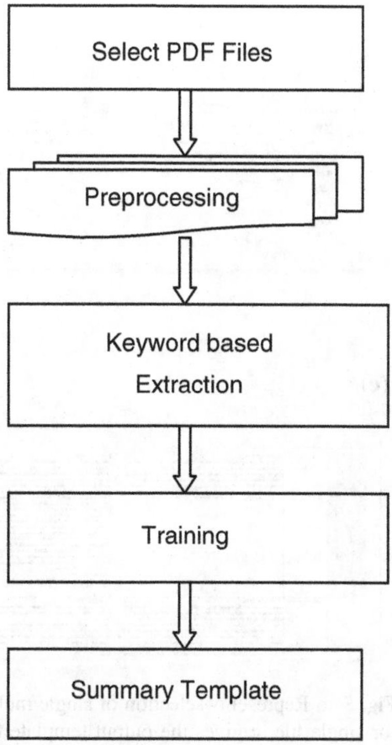

SVM and web-based lookup [1] for the citation extraction and embedded metadata from those research papers. Windows-based application is being provided by Mendeley which helps organizing the 111 research papers.

Another advance technique is being used from knowledge base named as Web-based lookup. Trained machine learning algorithm is being used for the authentication of an element in paper, and header information is being extracted. These methods are mainly known as HMM, CRF, and SVM [2]. In this paper, we proposed windows-based application (keyword-based extraction method) for the extraction of title, author, year, and references from research papers so that we can make the summary of research papers.

An artificial intelligence tool which is knowledge-based works in a narrow domain, to give smart decisions, with proper justification. KBS has their own command in the field of artificial intelligence [3].

Example 1 A classic patent analysis scenario

1. Categorization of Task: here one has to define the objective of analysis of task or scope and concept of task.
2. Searching: iteratively filter, seek, and download the linked patents.
3. Segmentation: normalizing the formless and prepared parts, clean, and segment.

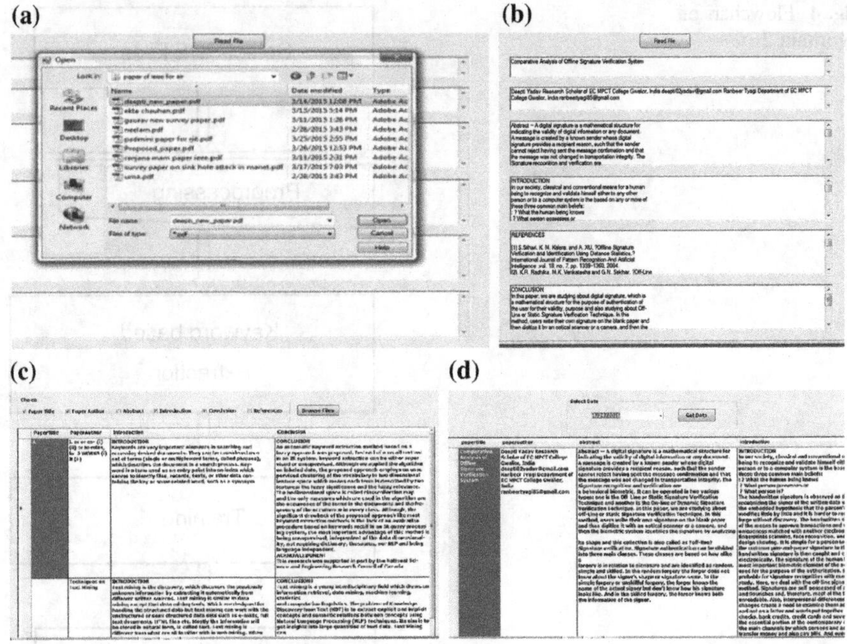

Fig. 5 **a** Represents selection of single/multiple file as an input and **b** represents output template for single file, **c** gives the output template for multiple file and **d** shows the date wise summary report

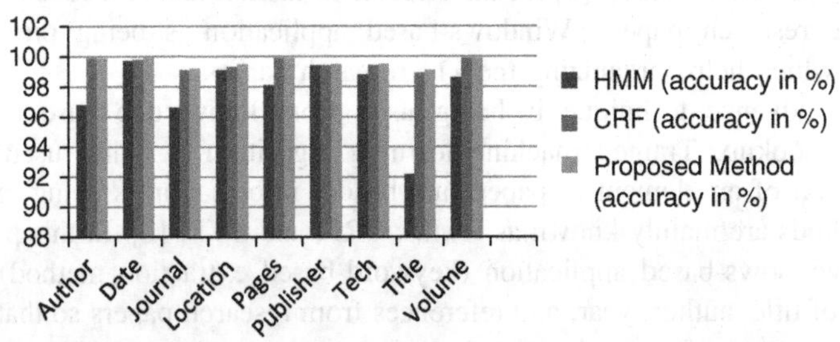

Fig. 6 Graph shows the comparison of accuracy for previous methods and proposed method

4. Abstraction: evaluating the patents content for summarizing their technologies, topics, claims, etc.
5. Clustering: categorizing or gathering of evaluated patent on the basis of extracted attributes.
6. Hallucination: generating the matrices of technology effects.
7. Interpretation: prediction of business trends and relations or technology.

Therefore, on the basis of patent analysis scenario [4] explained above, a text mining methodology dedicated to the analysis of full-text patent. First of all, collect the full patent document which is relevant for analysis purpose.

2 State of Art

In the present time, researchers are again interested in automatic text summarization as they were in fifties. During our literature survey, we came to know that, the paper in which researchers are using the concept of the paper which is published in 1958, the concept used in this paper is that the weight of the sentences of a document is being suggested as a task of high frequency words [5], disregard common words of very high frequency. In 1969, for determining the weight of the sentence "Automatic text summarization system" [6] uses the following three methods:

1. Title Method: this technique evaluates the weight of the sentence; the weight of any sentence is being calculated by summing up the contented words that appear in the title or heading of the text.
2. Cue Method: This technique calculates the weight of sentence by the presence or absence of cue of words.
3. Position technique: the working of this method is done on the basis of some hypothesis that, if a sentence occured in both text and individual paragraph at initial position then there is a high probability of relevance. In 1995, sentence extracting job was done with the help of trainable document summarizer on the basis of number of weighting heuristics [7] following features were used and evaluated:

    a. Sentence Length Cut-O Feature: this represents that those sentences which are containing less than predefined number of words are not considered as the part of abstract.
    b. Fixed-phrase attribute: here sentences are having few cue of words and phrases are also taken.
    c. Paragraph attribute: works as location method attribute [8]
    d. Thematic Word attribute: the most recurrent words are named as thematic words. Thematic words are having functions named as scores of words.

The ANES text extraction system [9], in 1995, is a system that performs automatic, domain-independent compression of news data. The process of summary generation has three major constituents:

1. Sentence weighting, 2. Corpus analysis, and 3. Selection of sentences

Hidden markov model (HMM) [5]: is the powerful mathematical statistical tool for document retrieval. IR can employ the subparts of the document like building links [10] and/or clusters between index terms and so on. Although this is not an issue in any of the above mentioned abstracting systems, it seems to be worth of consideration when building such systems (Table 1).

Table 1 Literature survey

Year	Author	Title	Methodology	Objective
1997	Thorsten Joachims	Text categorization with supports vector machines: learning with many relevant features [11]	Support vector machines (SVMs)	Text categorization
2004	Masoud Makrehchi	Fuzzy set approach to extract keywords [12]	Fuzzy set approach	Keyword extraction
2009	Rasim Alguliev	Evolutionary algorithm for extracting text summarization [13]	Discrete differential equation	Optimization of objective function
2012	Shu-Hsien Liao	Data mining techniques and applications [14]	Data mining techniques	Ability to continually change and provide new understanding by DMT
2012	Ozair Saleem	Information extraction from research papers by data integration and data validation from multiple header extraction sources [15]	(HybridMethod) GROBID, ParsCit, and Mendeley	To achieve accurate header extraction
2012	Hao Lul	Research on intelligent scientific research collaboration platform and taking journal intelligence system [16]	Intelligent journal prototype system	Prototype system can be changed to reliable, available, and effective scientific research collaboration platform
2012	Yogan Jaya Kumar	Automatic multi-document summarization approaches [17]	feature-based method, cluster-based method, graph-based method and knowledge-based method	To generate a better summary this is well suited for an informative type summary generation
2013	LI Yan-min	Applying information retrieval technology in analyzing the journals [18]	Variable of index term operators	CNKI's database searching
2013	Joeran Beel	Docear's PDF inspector: title extraction from PDF files [19]	Docear's PDF inspector	To extract titles from academic PDF files by applying a simple heuristic: the largest text on the first page of a PDF is assumed to be the title
2014	Xianfeng Yang and Liming Lian	A new data mining algorithm based on map reduce and Hadoop [20]	Map-reduce programming model and Hadoop, Newman algorithm	Discovering hidden information from large databases with respect to scalability

3 Problem Statement

In full text analysis, two problems seem to be the most important these days. They are firstly the bad quality of search engines mainly internet search engines and second one is the shortage of text categorization tools which permit the fast assessment for large number of documents. To deal with information explosion and to text, categorization will be the best technique. Text categorization would be much easier to deal with "information explosion" and assimilate all data that is going to flood us, if we are able to find out the main subject from the document and then arrange it into some sequential manner or say in some structured form, preferably hierarchical. The most traditional technique used for this problem is building of handcrafted index. Building a handcrafted index would be the traditional approach to this problem; in fact these indexes are of extensive use amid of the juridical communities and Internet. Unluckily, they merely cannot deal with number of new documents formed everyday. Due to the growing increment of the availability of large amount of information it seems incomplete and index creator is able to classify and analyze it. So, automatic text categorization is strongly required here. In this paper, we will build an extraction keyword-based text summarizer based on windows application that will provide a multiple text file using command line, create a summary of the original text in different files; run tester Java file. To build an automated text summarizer, assign the variable keywords to the set of significant words in the document.

4 Proposed Methodology

In this paper, we proposed keyword-based text extraction technique using threading and synchronization just like multiple files input as batch processing. Whole process is explained with the help of flowcharts and algorithms.

Experiment 1 Text files as an input (single document or multiple documents)

Algorithm 1 For .txt file

1. The Main module: Calls the thread (multithread) module to produce the summary of a text in different file using extract keyword. Main module (input f1, f2, f3, …, fn)
2. The Generator module contains the following methods: set Keywords generate summary: prints out the most significant sentences in the document. Create module (set keyword, limit summary)
3. Create Introduction, abstract, conclusion, and separate summary file.

Switch module
If == match keyword (for multiple files)
Then
Create separate abstract file
Create separate Introduction file
Create separate conclusion file

4. Input files like tushar.txt applied as command line argument
5. Separate (keyword, font, and style) text file created on the basis of keyword matching from abstract, introduction, result, and summary after program execution using thread and synchronization
6. Output summary file public class details which contains String intro, abstracts, result, and summary; and make it public class tester implements runnable.

   Create separate summary file

Experiment 2 pdf or doc files as input (single document or multiple documents)

Algorithm 2

The main module: reader application and class PDF Form: Form and doc or word document form
Main module(input file type doc, pdf, other)

The generator module contains the following methods: Choices for keyword: paper title, paper author, abstract, introduction, conclusion, and references. Read (keyword1, keyword2, keyword3, …)

If word == match
Match (file1, file2, file3.)
Else
No match ("word not match")
Then
pattern (title, abstract, introduction, reference, summary…..)
If pattern == title (Title extract: using font-size)
{
abstract paragraph fetch and store in data base
}
Else if pattern == reference(Match keyword and extract contents stored into specified template)
{
References fetch and store in data base
}
Else if pattern = introduction
{
Introduction paragraph fetch and store in data base
}
5. Display in template and view and exit

5 Experiment Results and Analysis

We have taken the datasets of 1000 pdf, doc, txt files for the extraction purpose and coding is done in Java, .Net, and C# programming. We compare our result with the result of other three methods used in previous papers. Out of these 1000 pdfs, SciPlore Xtract cannot extract data from 307 pdfs. These 307 pdfs consist of scanned images and OCR is being applied on them so SciPlore Xtract found difficulty in extraction. So, SciPlore Xtract works with 693 pdfs for the analysis. While comparing the result, we see that from 693 pdfs titles could not be extracted for 54 pdfs by using SciPlore or CiteSeer's SVM. All three approaches are able to identify the titles of only 160 pdfs. In final analysis, we observe that SciPlore Xtract can extract the titles of 540 pdfs correctly (77.9 %). CiteSeer's SVM can identify 481 titles correctly. SciPlore Xtract takes 8:19 min for extracting the titles and SVM needed 57 and our proposed methodology can extract titles from 640 pdfs correctly, 1.4 % is the error in the extraction and time taken for it is very less around 6:40 min. So, our results show the efficiency of the proposed methodology. Table 2 shows the result comparison of all the four methods for extracting the title and Table 3 represents the accuracy comparison chart extraction results for paper references from previous method and proposed system and figure shows the graphical representation of both the tables result. Overall accuracy for the results of HMM is 75 %, CRF is 99.5 %, and keyword-based extraction method (proposed method) 99.7 %.

Table 2 Title Extraction of 693 pdfs, (doc, txt, pdf: three format of file added in our work)

Methods	Correct		Slight errors		Total	
SciPlore Xtract	528	76.2 %	12	1.7 %	540	77.9 %
CiteSeer SVM + pdftotext	406	58.6 %	75	10.8 %	481	69.4 %
CiteSeer SVM + PDFBox	370	53.4 %	78	11.3 %	448	64.6 %
Keyword based extraction (proposed method)	630	90.9 %	10	1.4 %	640	92.3 %

Table 3 Extraction results for paper references from previous method and proposed system

Keywords	HMM (accuracy in %)	CRF (accuracy in %)	Proposed method (accuracy in %)
Author	96.8	99.9	99.9
Date	99.7	99.8	99.9
Journal	96.6	99.1	99.2
Location	99.1	99.3	99.5
Pages	98.1	99.9	99.9
Publisher	99.4	99.4	99.5
Tech	98.8	99.4	99.5
Title	92.2	98.9	99.1
Volume	98.6	99.9	99.9
Average accuracy	75	99.5	99.7

6 Conclusion and Future Scope

This paper made a clear and a simple overview of working of text extraction from pdf document in step by step process. We know that many text extraction systems are available in the market but researchers are working in this area to improve the efficiency because till now we find difficulty in extracting the text from the document containing tables, images, and so on. So the expectation of researchers is to extract the text from complex document very smoothly. Our method is able to work on these parameters and the accuracy is much better as compared to previous results. The accuracy percentage for the extraction for paper references from pdfs is approximately 99.7 % and for title extraction its 92.3 % and time taken for the extraction is very less. In future, we try to work using some hybrid techniques.

References

1. Mendeley is a desktop and web program for managing and sharing research papers, discovering research data and collaborating online
2. Accurate Information Extraction from Research Papers using Conditional Random Fields
3. Lin, C.-J., Lin, Y.-I.: Text mining techniques for patent analysis. Int. J. Inf. Proc. Manag., ACM, USA, **43**, 1216–1247 (2007)
4. Tu, Y.-N., Seng, J.-L.: Research intelligence involving information retrieval—an example of conferences and journals. Int. J. Expert Syst. Appl. 12151–12166 (2009)
5. Luhn, H.P.: The automatic creation of literature abstracts. Int. J. IBM J. Res. Dev., ACM, USA, vol. 2, pp. 159–165, 1958.
6. Edmundson, H.P.: New methods in automatic extracting. J. ACM, USA **16**, 264–285 (1969)
7. Kupiec, J., Pedersen, J., Chen, F.: A trainable document summarizer. In: Proceedings of the 18th ACMSIGIR Conference on Research and Development in Information Retrieval, USA, pp. 68–73 (1995)
8. Mittendorf, E., Schauble, P.: Document and passage retrieval based on hidden markov models. In: Proceedings of the 17th ACM-SIGIR Conference on Research and Development in Information Retrieval, New York, pp. 318–327 (1994)
9. Brandow, R., Mitze, K., Rau, L.F.: Automatic condensation of electronic publications by sentence selection. In: International Journal on Information Processing and Management, ACM, USA, vol. 31, pp. 675–685 (1995)
10. Bookstein, A., Klein S.T., Raita, T.: Detecting content-bearing words by serial clustering. In: Proceedings of the 18th ACM-SIGIR Conference on Research and Development in Information Technology, New York, pp. 319–327 (1995)
11. Joachims, T.: Text categorization with support vector machines: learning with many relevant features. In: Proceedings of European Conference on Machine Learning, ACM, London, pp. 137–142 (1998)
12. Makrehchi, M., Kamel, M.: A fuzzy set approach to extracting keywords from abstracts. IEEE Int. Conf. Fuzzy Inf. **2**, 528–532 (2004)
13. Alguliev, R., Aliguliyev, R.: Evolutionary algorithm for extractive text summarization. Int. J. Intell. Inf. Manag. **1** (2), 128–138 (2009).
14. Liao, S.-H., Chu, P.-H., Hsiao, P.-Y.: Data mining techniques and applications– A decade review from 2000 to 2011. J. Expert Syst. Appl., Elsevier **39**, 11303–11311 (2012)

15. Saleem, O., Latif, S.: Information extraction from research papers by data integration and data validation from multiple header extraction sources. In: World Congress on Engineering and Computer Science (WCECS), San Francisco, USA (2012)
16. Lu, H., Zheng, X., Sun, X., Zhang, N.: Research on intelligent scientific research collaboration platform and taking journal intelligence system as example. In: International Conference on Service Operations and Logistics, and Informatics (SOLI), IEEE, Suzhou, pp. 138–143 (2012)
17. Kumar, Y.J., Salim, N.: Automatic multi document summarization approaches. Int. J. Comput. Sci.
18. Xie, W.-L., Li, Y.-M., Zhang, Y.: Applying information retrieval technology in analyzing the journals. In: Fourth International Conference on Emerging Intelligent Data and Web Technologies (EIDWT), Xi'an, pp. 88–94 (2013)
19. Beel, J., Langer, S., Genzmehr, M., Müller, C.: Docear's PDF inspector: title extraction from PDF files. In: Proceedings of 13th ACM/IEEE-CS joint Conference on Digital Libraries, ACM, USA, pp. 443–444 (2013)
20. Yang, X., Lian, L.: A new data mining algorithm based on map reduce and Hadoop. Int. J. Signal Process. Image Process. Pattern Recogn. **7**, 131–142 (2014)

Implementation of the Principle of Jamming for Hulk Gripper Remotely Controlled by Raspberry Pi

Seema Rawat, Praveen Kumar and Geetika Jain

Abstract The Hulk Gripper has been constructed by replacing the fingers of a robotic hand with a mass filled with granular material, e.g., grounded coffee. This mass applies pressure on the article due to which the gripper adapts to the surface and engrosses it. Using the vacuum pump attached at the other end, air is pumped out making the granules contracted and hardened. It has been discovered that the volume change of approximately 0.5 is quite adequate to grab the object infallibly and lift them with a very large force. The ability of the granules to jam against each other in vacuum and unjam with air around is called as the principle of operation. The grip of the hand is based on three different mechanisms: friction, suction, and interlocking, which contributes to the engrossing force. With the help of all the mechanisms involved, it becomes possible in lifting heavy objects, exposing new prospects of design, with the ability to stand out quick engrossing of complicated objects. Our gripper is controlled mechanically with the help of radio frequencies. We are controlling the gripper with an android application. The mobile application will give commands to the microcontroller via Wi-Fi or Bluetooth, which in turn controls the movement of the gripper. The microcontroller being used turns the vacuum pump on and off. The existing gripper requires number of small and large joints which are to be controlled individually, in order to lift objects of different sizes, shapes, and delicacies whereas, our gripper uses a single point of contact to form the grip and do its task.

Keywords Jamming · Granular material · Raspberry pi · H-bridge

---

Seema Rawat (✉) · Praveen Kumar · Geetika Jain
Amity University, Noida, India
e-mail: srawat1@amity.edu

Praveen Kumar
e-mail: pkumar3@amity.edu

Geetika Jain
e-mail: geet.geet2577@gmail.com

M. Pant et al. (eds.), *Proceedings of Fifth International Conference on Soft Computing for Problem Solving*, Advances in Intelligent Systems and Computing 436, DOI 10.1007/978-981-10-0448-3_16

1 Introduction

One of the key tasks of robots is to grip, hold, and lift objects. Hulk grippers are one which can get a hold of objects of different shapes and sizes [1]. While the current designs have made it possible with multi-fingered hand, it has also introduced many software and hardware complexities. The existing robotic hands have sensory feedback that senses the object and computes the stress required to grip the object and involves a number of controllable joints, which makes it very complex and costly [2]. Instead of this, we declare a fairly new approach by replacing the fingers with a mass filled with granular material, e.g., grounded coffee [3]. As soon as the mass applies pressure on the article, it adapts to the surface and engrosses it. Using the vacuum pump attached at the other end, air is pumped out making the granules contracted and hardened, this mechanism helps the robot to grip the article without the use of any external sensory feedback (Fig. 1).

Hulk gripper has a wide scope of use from industrial to excavation sites; it can be used anywhere due to its capability of lifting objects of various shapes and sizes. In industries one of the major advantages of hulk gripper over others is that, one does not need to design specialized grippers and change between the gripping heads for grabbing [4, 5] different objects. A single gripper will be able to lift objects without any hassle. This reduces the downtime and increases the return on robot investment [6]. Apart from its industrial use, one can use it in excavation site for lifting and transporting the minerals from one site to another. Due to its property of adapting its grip according to the object it will be able to lift any extracted mineral, no matter how small or disproportionate it may be. The most challenging application of robots is helping people in their day-to-day life such as household works. Designing a robot for such purpose is tricky as they interact with objects of various sizes, shapes, and weights. The existing robots are not capable of adapting themselves with the objects, hence limiting their use [7]. With the introduction of hulk gripper, a bot will be able to interact with objects of various sizes such as books, glasses, pins; shapes like spheres, blocks, cones; and delicacies such as eggs, papers, etc. Such a bot can help us in our daily lives and make it better.

But this hulk gripper can have some disadvantages too. It can be very expensive. Its parts can be damaged easily and changing the damaged parts will increase the overall cost. We may call this hulk gripper as underactuated. This means that the hulk gripper falls between the active and passive grippers (Fig. 2).

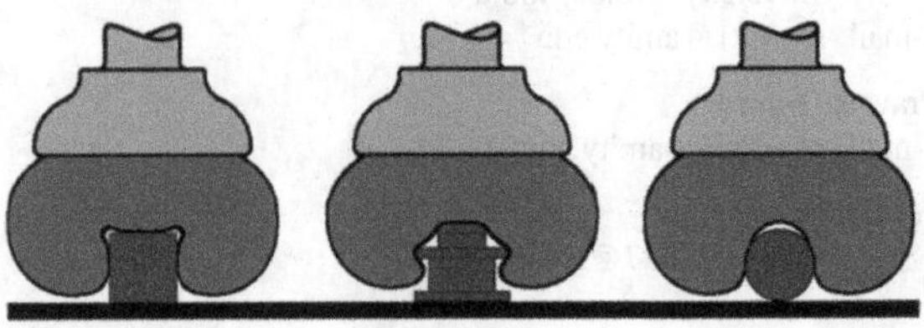

Fig. 1 Showing how the same gripper forms the shape of different objects to be picked

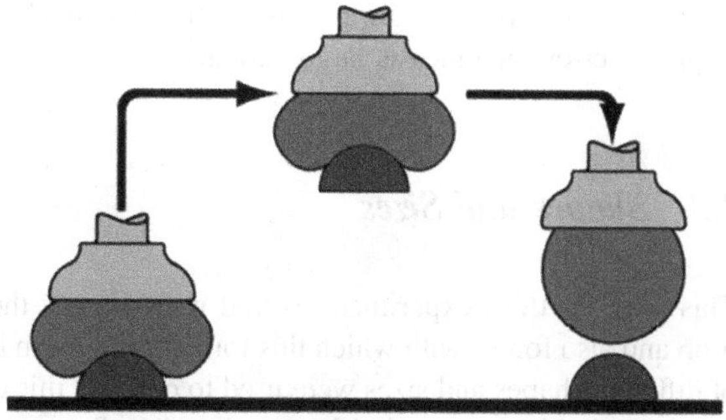

Fig. 2 Showing the gripper holding the object, picking it up and the releasing

To use this gripper in the best and easiest way possible we controlled it by an android application, so you can just take out your mobile phone out of your pocket and can pick up any object of your wish [8–10]. You do not manually need to control the gripper.

Some of the tests have been performed on the gripper and the characteristics we found are shown below.

1.1 Reliability

For this, a small experiment was performed. Few objects of different shapes and sizes were placed on a wooden table [11]. The aim was to pick up those objects by giving the instructions on the android application. As soon as the application gave the command the motor started doing its task. The motor creates the much needed vacuum due to which the balloon material takes the shape of the object to be picked and forms a tight grip around it [12]. Because of the grip formed, the hulk gripper was able to easily pick up all the objects placed on the table. After this, we instructed the hulk gripper to release the object by giving the instructions on the android application. The motor stopped creating the vacuum and hence, the balloon deformed. As a result, the object dropped on the table. By this experiment, we may say that this hulk gripper is reliable enough to be also used for bigger tasks.

1.2 Error Tolerance

We used the same setup as before to check for the error tolerance in the hulk gripper [11]. This time we placed the object in a way that it is unaligned with the central axis of the hulk gripper. What we found was pretty amazing as the gripper could still pick up the object. Though initially it took time to form the grip as a very small portion of the object was there for the gripper to form the grip. By switching off the pump, it was quite easy for the hulk gripper to release the object as for the same reason that the portion gripped was quite small [12]. So, we may say that the gripper's error tolerance is large enough.

1.3 Shapes and Sizes

This was our third experiment to find what shapes the hulk gripper can pick and drop and also forces with which this formed shape can be retained [11]. Five objects of different shapes and sizes were used to conduct this experiment. The shapes were sphere, cuboid, cube, cylindrical, and spring. To test the strength with which the object was held, we tried measuring the force with which the held object was removed. The results showed that to lift an object of a larger shape a large amount of jamming granular particles were displaced, whereas for an object of a smaller shape less amount of granular particles were displaced [9]. The strength to pick up a larger object was more than that of a smaller object. This is clearly understandable by the difference in the weights of the objects.

1.4 Speed

The speed with which a hulk gripper does its tasks of picking or lifting depends entirely on the vacuum created by the motor and the rate of flow of the granular particles [11]. These are the factors that are responsible for setting the time that will be taken by the griper to perform its intended task. Initially, we were not very satisfied with the speed of our hulk gripper [13]. It took around 20 s to form the grip of the object and then pick it up. Although the speed with which it released the object was pretty well. To solve this problem of the gripper formed by us we used a motor of a higher power. This solution was thought as a hit and trial method, as there was no surety whether this method will work in the way we want or not. But it did. The speed of picking up of the objects dropped down to 2–4 s. This was definitely a success for us.

2 Methodology Adopted

The hulk gripper built on the principle of jamming has a wide range of applications where a number of high ability of an individual's hand is required however not offered, or wherever a feedback is tough to get or costly, example, embodying things where varied objects are grabbed dependably and in quick series [14].

This approach will easily adapt to the shape of the article, from lifting a book to handling a raw egg. Such a gripper provides a vital benefit over the existing styles in situations where the available information is marginal, for example, lift articles whose materialistic properties are unknown beforehand. Because of the material used in the gripper, it becomes easy to adapt automatically to the shape of the article, and then a grip on the item with the help of jamming principle of the granules.

Out results have demonstrated how small changes within the packing density ($|\delta V/V| < 0.5$ %) related to a jamming/unjamming vacuum induced transition, modifies the hulk gripper to adapt its form to a good variety of objects, and grab them accurately without the necessity of a sensory feedback; the gripper achieves its skillfulness and noteworthy holding strength through a mix of mechanisms-friction [15], suction, and interlocking, solely a fraction of the entire surface has to be gripped to carry it firmly [2].

There is one unique thing that we have done with this gripper. We have used the L293 IC. We tried to make this gripper as easy as possible. This is why we preferred this basic IC than any other complicated one. L293D is a dual H-bridge integrated [16] circuit for motor drivers. In fact, this IC contains two inbuilt H-bridge driver circuits. A H-bridge is an electronic circuit that enables a voltage to be applied across a load in either direction. These help in running the motors in forward as well as backward direction (Figs. 3 and 4).

To simplify the use of L293D circuit we made a special circuit (Figs. 5 and 6).

The following circuit diagrams will show how the motor has been connected with the L293 IC (Fig. 7).

Raspberry pi is a low cost, affordable, small (credit card sized) computer that can be plugged into any monitor and used with the help of normal keyboard and mouse. Raspberry pi was basically designed to explore computing, learn new languages, however it is being extensively used in projects especially for automation [8]. It is the same in our project too, the raspberry [17] pi has been used to control and

Fig. 3 Circuit diagram of H-bridge

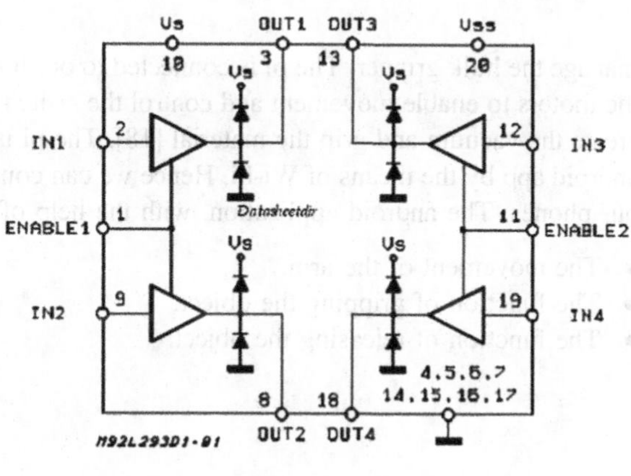

Fig. 4 PIN diagram of L293D

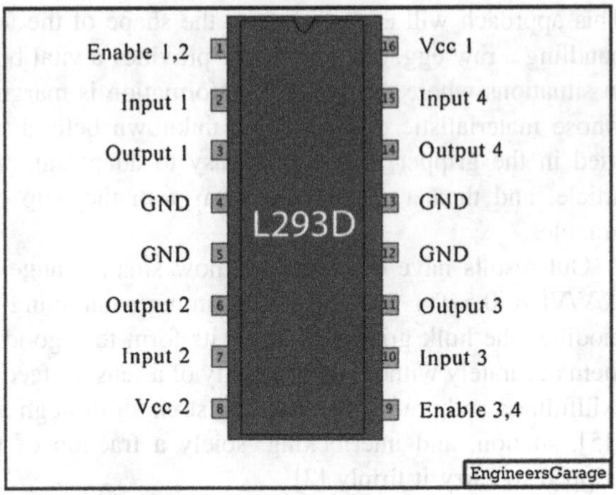

Fig. 5 Real image of the special circuit made

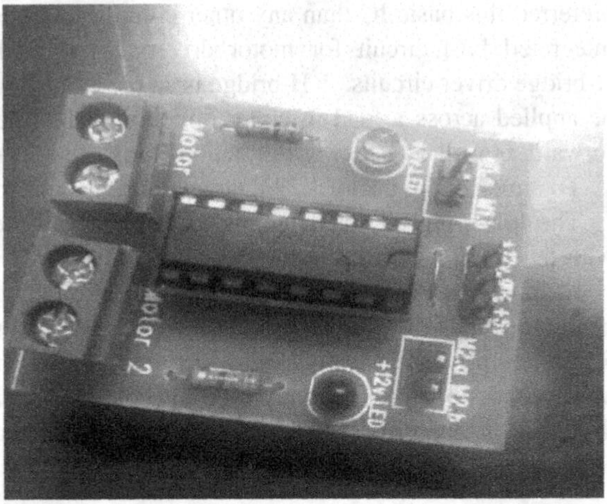

manage the hulk gripper. The pi is connected to our hulk gripper which can turn on the motors to enable movement and control the switch of vacuum pump in order to create the vacuum and grip the material [18]. The pi is wirelessly connected to the android app by the means of Wi-Fi. Hence we can control the gripper directly from our phones. The android application, with the help of pi, controls the following:

- The movement of the arm.
- The function of gripping the object.
- The function of releasing the object.

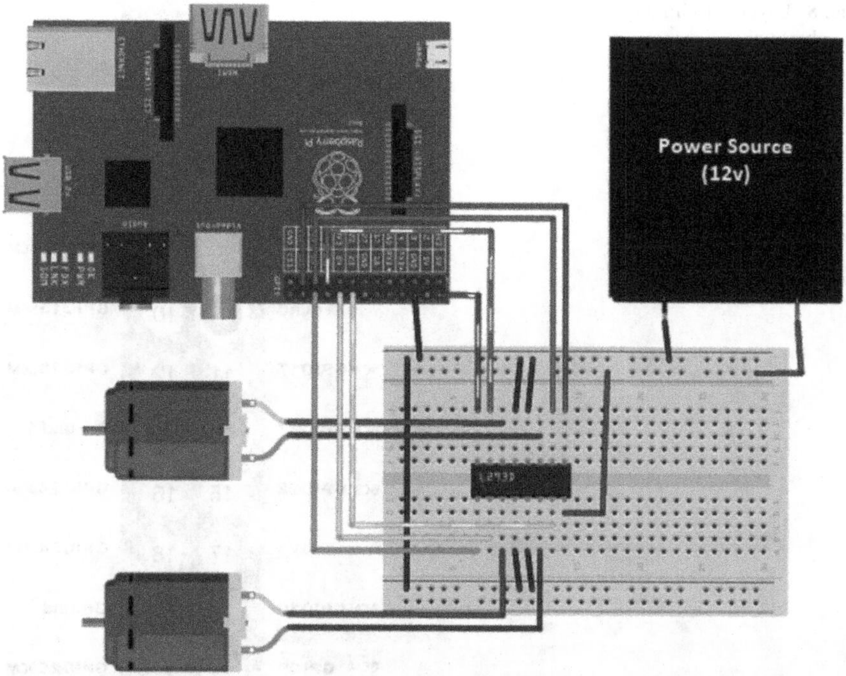

Fig. 6 Circuit diagram of the above-mentioned circuit

Fig. 7 Circuit diagram of L293D with motor

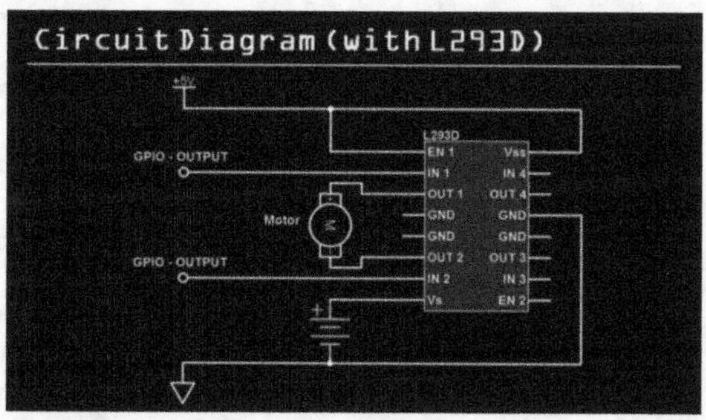

One of the most interesting feature of raspberry pi is number of pins present [19] on it. These are the general purpose input output (GPIO) pins. These pins provide a physical interface between pi and the outside world (Fig. 8).

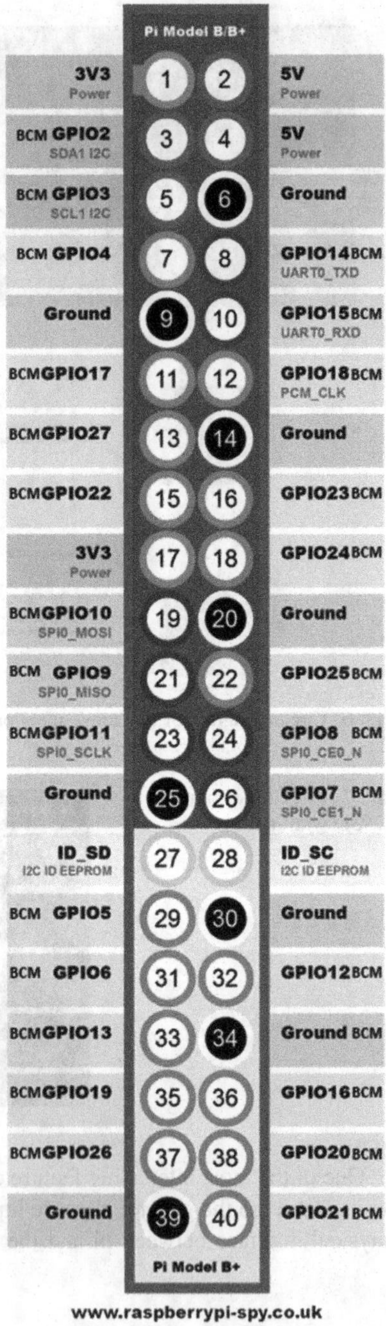

Fig. 8 Layout of GPIO of Raspberry pi

3 Conclusion

In this paper, we have tried representing the effects of applying and removing pressure (in order to lift and drop objects) on a hulk gripper and we also have tried evaluating its performance by conducting several tests to check it on the basis of five parameters, viz., reliability, error tolerance, shape, size, and speed. The prototype developed by us was used to test the above mentioned characteristics. This prototype was developed to check the real-time applications of a hulk gripper. After performing all the tests, the hulk gripper was proved capable of picking up the objects on applying a positive pressure and releasing the objects easily on applying a negative pressure. The principle of jamming of granular materials serves as the basis of gripper. On applying positive pressure the motor jams the particle to make it rigid and hence, forms the grip of object [20]. With this particulate principle, the objects of different shapes, sizes, weights, and fragility can be picked and dropped. This diverse nature of the gripper makes it suitable for scenarios like industrial tasks, military environments, and even at homes. Regarding its thermal environment, the limit of the gripper can only be decreased because of the material used inside the hulk gripper.

References

1. Fink, A.: Universal gripper apparatus for robotic device. United States Patents (1986)
2. Birglen, L., Lalibert′e, T., Gosselin, C.: Under actuated robotic hands. Springer, Berlin, Germany (2008)
3. Esquivel, P., Jimenez, V.M.: Food Res. Int. **46**(2) (2012)
4. Scott, P.B.: Omnigripper: a form of robot universal gripper. Robotics **3**(3) Cambridge Journals, (1985)
5. Forcella, P.W., Meek, D.S., French, G.D.: Universal gripper assemble for frozen confections. United States patent (1976)
6. Harvey, I., Husbands, P., Cliff, D.: Issues in evolutionary robotics (1992)
7. O'Neil, J.P.: Drive mechanism for pick-and-place unit. United State Patents (1980)
8. Enck, W., Octeau, D., McDaniel, P., Chaudhary, S.: A study of android application security. In: USENIX Security Symposium (Aug 2011)
9. Brown, E., Rodenberg, N., Amend, J., Mozeika, A., Steltz, E., Zakin, M.R., Lipson, H., Jaeger, H.M.: Universal robotic gripper based on jamming of granular material, editorial board (2010)
10. Shabtai, A., Fledel, Y., Kanonov, U., Elovici, Y., Dolev, S., Glezer, C.: Google android: a comprehensive security assessment. IEEE Secur. Policy **8**(2), 35–44 (2010)
11. Amend, Jr., J.R., Student member, IEEE, Brown, E., Rodenberg, N., Jaeger, H.M., Lipson, H., Member, IEEE: a positive pressure universal gripper based on the jamming of granular material. In: IEEE Transactions on Robotics, vol. 28, pp. 341–350 (Jan 2012)
12. Sniderman, A.: Automatic vacuum gripper. United States Patent (1988)
13. Sawdon, E.G., Krugar, D.J., Sprotberry, S.J.: Universal gripper. United States Patent (1998)
14. Corwin, E.I., Jaeger, H.M., Nagel, S.R.: Structural signature of jamming in granular media. Nature **435**, 1075–1078 (2005)

15. Massa, B., Roccella, S., Carrozza, M.C., Dario, P.: Design and development of an underactuated prosthetic hand. Proc. IEEE Int. Conf. Rob. Autom. **4**, 3374–3379 (2002)
16. Dollar, A.M., Howe, R.D.: Simple, robust autonomous grasping in Unstructured environ ments. In Proceedings of IEEE International Conference Robotic Automation, pp. 4693– 4700 (Apr 2007)
17. Paramanathan, A., Pahlevani, P., Thorsteinsson, S., Hundebøll, M., Lucani, D.E., Fitzek, F.H. P.: Sharing the pi. In: IEEE VTC Vehicular (2014)
18. Tella, R., Birk, J., Kelley, R.: A contour-adapting vacuum gripper. In: Pham T., Heginbotham, W.B. (eds.) Robot Grippers, pp. 86–100. Springer, New York (1986)
19. Jones, T.J., Duke, B.R.: Gripper and wrist joint for a robotic arm. United States Patent (1988)
20. Fikes, R.E., Hart, P.E., Nilsson, N.J.: Artificial intelligence. **3**, 251–288 (1972)

Differential Evolution: An Overview

Amritpal Singh and Sushil Kumar

Abstract Differential evolution (DE) is one of the most influential optimization algorithms up-to-date. DE works through analogous computational steps as used by a standard evolutionary algorithm. Nevertheless, not like traditional Evolutionary Algorithms, the DE-variants agitate the current generation populace members with the scaled differences of indiscriminately preferred and dissimilar population members. Consequently, no discrete probability dissemination has to be utilized for producing the offspring. Ever since its commencement in 1995, DE has dragged the interest of numerous researchers around the globe ensuing in a lot of alternative of the fundamental algorithm with enhanced working. This paper introduces a comprehensive review of the basic conception of a DE and an inspection of its key alternatives and the academic studies carried out on DE up to now.

Keywords Differential evolution (DE) · Evolutionary algorithms (EAs) · Genetic algorithms (GAs)

1 Introduction

In past few years, global optimization has got attention from a lot of researchers worldwide. As optimization problems are all over the place in the field of engineering, finance, and other areas of science, researchers require vigorous optimization method which can decipher optimization problems and should not be very difficult to implement. Every species needs to revise the physical structures to stay fit in atmosphere where they live. The association between optimization and

Amritpal Singh (✉)
Department of Computer Science, Lovely Professional University, Phagwara, Punjab, India
e-mail: apsaggu@live.com

Sushil Kumar
Department of Computer Science, Amity University, Noida, Uttar Pradesh, India
e-mail: kumarsushiliitr@gmail.com

M. Pant et al. (eds.), *Proceedings of Fifth International Conference on Soft Computing for Problem Solving*, Advances in Intelligent Systems and Computing 436, DOI 10.1007/978-981-10-0448-3_17

evolution guides to the expansion of evolutionary computing techniques. Evolutionary algorithms are one of the most attractive types of guided random search techniques enthused by Darwin's theory of evolution. An Evolutionary algorithm became accustomed during the search process, using the information it will find out to break the nuisance of dimensionality that makes nonrandom and comprehensive search methods computationally hard to deal with.

The optimization technique should meet following requirements:

i. Ease of use
ii. Parallelizability
iii. Reliable and adaptable function optimizer
iv. Robustness
v. Accuracy and convergence speed,

The differential algorithm comes out as a reasonable form of evolutionary computing. This technique was first initiated by Price and Storn [1]. Differential evolution (DE) started to locate applications to optimize problems in different domains of computer science like image processing, big data, and other scientific fields.

Differential evolution (DE) is an optimization method that provides following advantages:

i. Differential evolution is very simple to implement as compared to other evolutionary algorithms.
ii. Differential evolution displays better performance in terms of correctness, convergence speed, and robustness.
iii. Main body of algorithm takes very few lines to code in C or any other programming language.
iv. Space complexity is low as compared to other algorithms.
v. Control parameters in differential evolution are only a handful which do not put any burden on performance.

2 Advances in Differential Evolution

Differential evolution (DE) is a dependable and adaptable global optimizer. Many diverse improved DE alternatives have been projected to improve the functioning of standard DE. The DE was projected by Price and Storn [1]. Many researchers focus on tuning Cr to build up the performance of DE as the crossover operator plays an important role in the performance of DE. Yiqiao Cai and Jiahai Wang presented a novel linkage utilization technique, called hybrid linkage crossover (HLX). The HLX automatically extracts the linkage information of a specific problem and then uses the linkage information to guide the crossover process [2]. By incorporating HLX into DE, the resulting algorithm, named HLXDE, is presented. The HLXDE differs from the original DE algorithm only in the crossover operator.

The HLX consists of three main operators: (1) constructing the linkage matrix to extract and store the linkage information; (2) adaptively grouping the problem variables to detect building blocks; (3) applying groupwise crossover to explicitly use building blocks for guiding the crossover process. In summary, HLXDE can employ the linkage learning technique to enhance the search ability of DE. Therefore, it is expected that HLXDE can achieve a good balance between exploration and exploitation. Carlos Segura et al. provide a better insight into reasons of the curse of dimensionality (i.e., performance of DE deteriorates rapidly when dealing with large scale problems) and to propose techniques to lessen this problem. Authors have proposed new DE scheme with two modifications which change the way of perturbing solutions in DE [3]. The principle of first modification is to increase the number of potential trial solutions while at the same time preserving the basic behavior of DE. The aim of second modification is to improve the exploration capabilities of DE so as to correctly deal with large search spaces. Computational results from large set of scalable problems with various complexities show that new proposal is much better than the original DE scheme. The number of resources that can be saved with new scheme is significant. In addition, scalability study demonstrates that as the dimensionality of problem involved increases, advantages of proposed scheme are more significant.

Dynamic economic dispatch (DED) is a vital job in the functioning of power system, it permits more superior treatment of economic dispatch (ED), and the major aim is to drive electric power systems at smallest possible total fuel cost at the same time as fulfilling system disparity and fairness restriction. Emission dispatch becomes one more critical goal in the power dispatch, and DED has been extended to environmental/economic dispatch (DEED) problem, which requires optimizing the fuel cost and pollutant emission at the same time. On the other hand, the factor of pollutant emission will add to the fuel cost, DEED can be taken as a multiobjective problem. Several methodologies have been projected to resolve this problem. Huifeng Zhang projected an improved multiobjective differential evolutionary algorithm named multiobjective hybrid differential evolution with simulated annealing technique (MOHDE-SAT) to solve dynamic economic emission dispatch (DEED) problem [4].

Rammohan Mallipeddi and Minho Lee presented an evolving surrogate model-based differential evolution (ESMDE) method, wherein a surrogate model concept based on the population members of the current generation is used to help out the DE algorithm in order to produce spirited offspring by means of the suitable parameter setting during various phases of the evolution [5]. As the population goes forward over generations, the surrogate model also goes forward over the iterations and better represents the basin of search by the DE algorithm. The projected method makes use of a simple Kriging model to build the surrogate. The performance of the proposed algorithms was compared to the state-of-the-art self-adaptive DE algorithms such as JADE, CoDE, SaDE, and EPSDE on a set of 17 bound-constrained problems. From results, ESMDE showed better performance due to better convergence speed caused by use of the surrogate model.

Josef Tvrdíka and Ivan presented new algorithm combining DE and k-means [6]. The reason of choosing k-means is due its fast convergence to a local minimum and having low time complexity. Clustering is an important exploratory technique used for splitting a collection of objects into relatively homogeneous groups (called clusters) based on object similarities. There are two main approaches to solve the clustering problem, hierarchical and nonhierarchical. The nonhierarchical clustering algorithms try to decompose the data sets directly into a set of disjoint clusters via optimizing a chosen function. Authors have applied their algorithm on nonhierarchical clustering and tested on eight well-known real-world data sets. The problem of optimal nonhierarchical clustering is addressed. Two criteria (clustering validity indexes), namely TRW and VCR, were used in the optimization of classification. The experimental results illustrate that hybrid variants with k-means algorithm are more effective than the non-hybrid ones. Compared to a standard k-means algorithm with restart, the new hybrid algorithm was found more reliable and more efficient, especially in challenging jobs.

The unit commitment (UC) problem is one of the challenging problems in power system operation. The UC problem is a scheduling problem which consists of two jobs to be performed: one is decisive the on/off status of the thermal units; the second job is the power send off which necessitates to distribute the system load demand to the committed thermal units. The most favorable UC involves carrying out the above two jobs to meet the estimated load demand over a specific time horizon, fulfilling a large set of units and system restrictions and meeting the aim of lessening the system operation cost. It is a nonlinear and link to the set of NP-hard problems. Anupam Trivedi et al. proposed a hybrid of genetic algorithm and differential evolution (DE), termed hGADE, to resolve the power system optimization problems known as the unit commitment (UC) scheduling [7]. The proposed hGADE algorithm is generic and it can easily be put into practice for solving other challenging single-objective mixed-integer optimization problems.

Mohamed et al. presented an alternative DE algorithm for work out unimpeded global optimization problems [8]. Authors have proposed four modifications in order to considerably improve the by and large operation of the standard DE algorithm. Authors have proposed modifications in control parameters such as mutation scheme, scaling factor, crossover rate, and population size. They projected new mutation rule based on weighted difference vector between the finest and the nastiest individuals of a specific generation. The adapted mutation scheme is as follows:

$$v_i^{G+1} = x_r^G + F_l \cdot \left(x_b^G - x_w^G\right) \tag{1}$$

where x_r^G is a random chosen vector and x_b^G and x_w^G are the best and worst vectors in the total population, in that order. On the other hand, authors have proposed two scaling factors F_l and F_g for the two different mutation rules, where F_l and F_g designate scaling factor for the local mutation scheme and the scaling factor for global mutation scheme, in that order. Authors have done modification in crossover rate in order to provide steadiness among the diversity and the convergence rate or among global exploration ability and local exploitation tendency, a new crossover scheme is planned as follows:

$$\mathrm{CR} = \mathrm{CR}_{\max} + (\mathrm{CR}_{\min} - \mathrm{CR}_{\max}) \cdot (1 - \mathrm{G}/\mathrm{GEN})^{k} \quad (2)$$

where G is the current generation number, GEN is the maximum number of generations, $\mathrm{CR}_{\min}$ and $\mathrm{CR}_{\max}$ denote the minimum and maximum value of the CR, respectively, and k is a positive number. This alteration is revealed to build up the local search power of the basic DE and to boost the convergence rate.

Wenyin Gong et al. proposed two DE alternatives with adaptive tactic selection namely, probability matching and adaptive pursuit to separately choose the most apt tactic at the same time as solving the problem [9]. Investigational outcome verifies that they are able to select the most appropriate tactic for a specific problem in an effective way. The authors have addressed some important issues in their paper. The first issue which they addressed is implementation of adaptive tactic selection, credit assignment, i.e., how to allot a reward to a tactic following its application. Second important issue in adaptive strategy selection is the tactic selection method itself, i.e., how to choose the next tactic to be applied based on the rewards lately received by each of the available ones.

XIN Bin et al. proposed an adaptive hybrid optimizer based on particle swarm and differential evolution for global optimization [10]. The authors have proposed novel technique of arithmetical learning approach which chooses the evolution method for each entity in accordance with the relative success ratio of another method in a preceding learning period. If the discovered best objective value is improved by PSO, the success number of PSO is raised. If not, the failure number of PSO is raised. This rule is same for DE. The probability of choosing PSO as an evolution method is as follows:

$$\mathrm{Pr} = \frac{\mathrm{N}_{\mathrm{s}}^{\mathrm{PSO}}\left(\mathrm{N}_{\mathrm{s}}^{\mathrm{DE}} + \mathrm{N}_{\mathrm{f}}^{\mathrm{DE}}\right)}{\mathrm{N}_{\mathrm{s}}^{\mathrm{PSO}}\left(\mathrm{N}_{\mathrm{s}}^{\mathrm{DE}} + \mathrm{N}_{\mathrm{f}}^{\mathrm{DE}}\right) + \mathrm{N}_{\mathrm{s}}^{\mathrm{DE}}\left(\mathrm{N}_{\mathrm{s}}^{\mathrm{PSO}} + \mathrm{N}_{\mathrm{f}}^{\mathrm{PSO}}\right)} \quad (3)$$

Here $\mathrm{N}_{\mathrm{s}}^{\mathrm{PSO}}$ and $\mathrm{N}_{\mathrm{f}}^{\mathrm{PSO}}$ are success number and failure number of PSO in a learning period. Denoted by $\mathrm{N}_{\mathrm{s}}^{\mathrm{DE}}$ and $\mathrm{N}_{\mathrm{f}}^{\mathrm{DE}}$ are success number and failure number of DE in a learning period.

Ahmet Bedri Ozer projected new DE, named as chaotically initialized differential evolution based on chaotic maps for initial population generation [11]. Author has embedded different chaotic maps to create the initial population of DE algorithm. This has been done by means of chaotic number generators whenever a random number is desirable in the initial population generating step of the classical DE algorithm. Seven chaotic maps have been investigated in the benchmark functions. It has been perceived that coupling emergent results in different areas, like those of DE and complex dynamics, can enrich the quality of results in some optimization problems and also that chaos may be a preferred method. This technique helps in enhancing the global probing competence.

Ferrante Neri proposed a new approach to get in to the bottom of continuous optimization problems typify by memory restrictions [12]. The projected Algorithm employed multiple unequal searches inside the main framework and carried out a numerous step global search using a randomized perturbation of the virtual population related to a regular randomization of the look for the manipulative Operators. Mathematical outcome showed the supremacy of the anticipated algorithm in connection with other contemporary algorithms present in text.

Yong Wang et al. recommended a basic structure for using an Orthogonal Crossover in DE alternatives to develop the search power of DE and projected OXDE, a mixture of DE/rand/1/bin and Orthogonal Crossover [1, 13]. Authors proposed to utilize the QOX operator to investigate the hyper-rectangle defined by the mutant vector and the target vector and, as a result, to perk up the search power of DE. In framework, authors applied QOX operator just once at each generation to preclude the computational cost and to keep the execution simple. By doing this, the expense of QOX in setting up trials at each generation is fairly insignificant, particularly when the population magnitude is huge or the cost of the assessment of objective function is costly. In addition, the sampling solutions produced by QOX make DE more efficient in investigating the search space. For example, DE/rand/1 mutation has been altered to the following procedure for the mutant vector which will take part in QOX:

$$\vec{v}_{i,G} = \vec{x}_r 1, G + \text{rand}(0,1) \cdot \left(x_{r2,G} - x_{r3,G}\right) \tag{4}$$

The only key differentiation between OXDE and DE/rand/1/bin is that, in the former, for one indiscriminately chosen individual member $_{k,G}$ at each generation and its mutant vector $_{k,G}$ to produce its trial vector $_{k,G}$ (see Fig. 1).

Xiaoting Ma et al. put forward an improved DE variant by making use of hybrid strategies for solving multimodal problems [14]. The projected approach is called

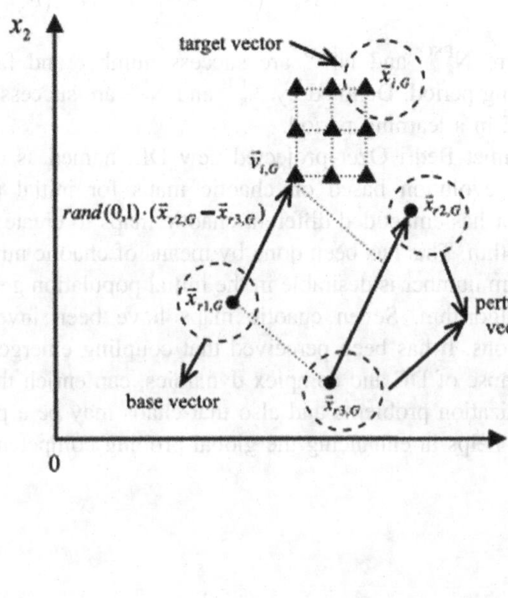

Fig. 1 Demonstration of OXDE. $_{i,G}$ is the target vector, $_{r1,G}$; $_{r2,G}$ and $_{r3,G}$ are mutually exclusive vectors indiscriminately selected from the population and the triangle points symbolize the trial vectors obtained by QOX [13]

Table 1 Results of DE and HDE on *f1–f6* [14]

Functions	DE mean	DE standard deviation	HDE mean	HDE standard deviation
f1	–5961.4	1328.2	–6654.7	325.1
f2	155.6	42.7	32.3	21.7
f3	2.19 e–14	3.26e–14	4.14 e–15	5.20 e–15
f4	0	0	0	0
f5	6.38 e–17	1.42 e–19	0	0
f6	2.45 e–15	5.16 e–16	0	0

HDE, which makes the most of opposition-based learning concept and optional external archive. Experiments on a multimodal benchmark set showed that HDE attains good solutions in many test cases. Authors have compared the standard DE with HDE. In order to measure the performance of HDE, six benchmark problems are chosen in the experiments. The population size is set as 100 (see Table 1).

Yiqiao Cai et al. proposed a new technique called learning-enhanced DE (LeDE) that encourages the individuals to swap information methodically [15]. LeDE adopted an unique learning tactic, specifically clustering-based learning strategy (CLS). In CLS, there are two types of learning strategies, intracluster learning strategy and intercluster learning strategy. They are implemented for swapping information inside the same cluster and between dissimilar clusters, in that order (see Fig. 2).

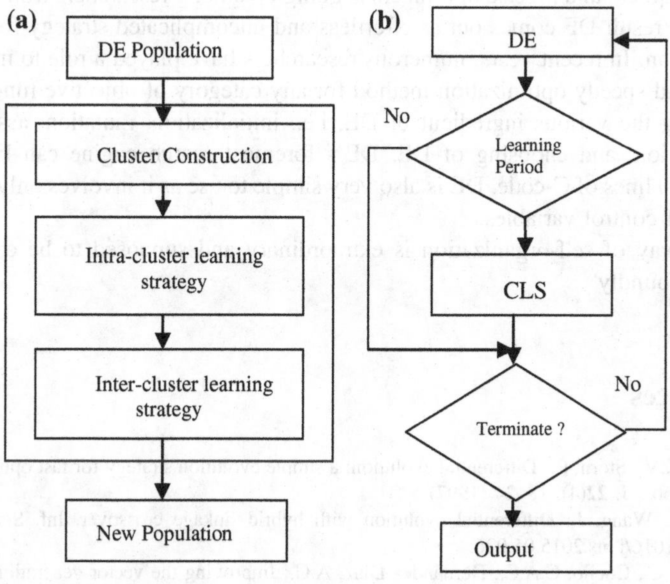

Fig. 2 Working of **a** CLS **b** LeDE [15]

Karthik Sindhya et al. put forward a new fusion mutation operator composed of a polynomial-based operator with nonlinear curve tracking potential and the differential evolution's original mutation operator, for the effective treatment of various interdependencies amid decision variables [16]. The resultant fusion operator is simple to put into practice and can be used within the majority evolutionary algorithms. Particularly, it can be used as a substitute in all algorithms makes most of the fundamental mutation operator of differential evolution.

3 Applications

Differential evolution (DE) has effectively been attached to several artificial and existent optimization troubles such as decision making/decision support systems, automated mirror design, protein structure analysis, neural network-training algorithm based on DE [17], maximize the hydropower production [18], engineering component/equipment design, effectiveness of radial active magnetic bearings, and effectiveness of fermentation using high ethanol tolerance yeast.

4 Conclusion

With the growing complication of physical world optimization problems, insist for vigorous, quick, and precise optimizers is going up among researchers from various fields as a result DE comes out as effortless and uncomplicated strategy for global optimization. In recent years, numerous researchers have played a role to make it a general and speedy optimization method for any category of objective function by fine-tuning the various ingredient of DE, i.e., initialization, mutation, assortment augmentation, and choosing of DE. DE's foremost search engine can be jotted down in 30 lines of C-code. DE is also very simple to use as it involves only a small number of control variables.

DE's way of self-organization is extraordinary and supposed to be examined more profoundly.

References

1. Price, K.V., Storn, R.: Differential evolution: a simple evolution strategy for fast optimization. Dr. Dobb's J. **22**(4), 18–24 (1997)
2. Cai, Y., Wang, J.: Differential evolution with hybrid linkage crossover. Inf. Sci. (2015). doi:10.1016/j.ins.2015.05.026.
3. Segura, C., Coello, C.A.C., Hernández-Díaz, A.G.: Improving the vector generation strategy of differential evolution for large-scale optimization. Inf. Sci. (2015). doi:10.1016/j.ins.2015.06.029.

4. Zhang, H., Yue, D., Xie, X., Hu, S., Weng, S.: Multi-elite guide hybrid differential evolution with simulatedannealing technique for dynamic economic emission dispatch. Appl. Soft Comput. (2015). doi:10.1016/j.asoc.2015.05.012.
5. Mallipeddi, R., Lee, M.: An evolving surrogate model-based differential evolution algorithm. Appl. Soft Comput. (2015). doi:10.1016/j.asoc.2015.06.010.
6. Tvrdík, J., Krivy, I.: Hybrid differential evolution algorithm for optimal clustering. Appl. Soft Comput. (2015). doi:10.1016/j.asoc.2015.06.032.
7. Trivedi, A., Srinivasan, D., Biswas, S., Reindl, T.: Hybridizing genetical gorithm with differential evolution for solving the unit commitment scheduling problem. Swarm Evol. Comput. (2015). doi:10.1016/j.swevo.2015.04.001.
8. Mohamed, A.W., Sabry, H.Z., Khorshid, M.: An alternative differential evolution algorithm for global optimization. J. Adv. Res. (2011)
9. Gong, W., Fialho, A., Cai, Z., Li, H.: Adaptive strategy selection in differential evolution for numerical optimization: an empirical study. Inf. Sci. **181**, 5364–5386 (2011)
10. Xin, B., Chen, J., Peng, Z.H., Pan, F.: An adaptive hybrid optimizer based on particle swarm and differential evolution for global optimization. Sci. China Inf. Sci. (2010). doi:10.1007/s11432-010-0114-9.
11. Ozer, A.B.: CIDE: chaotically initialized differential evolution. Exp. Syst. Appl. 4632–4641 (2010)
12. Neri, F., Iacca, G., Mininno, E.: Disturbed exploitation compact Differential Evolution for limited memory optimization problems. Inf. Sci. 2469–2487 (2011)
13. Wang, Y., Cai, Z., Zhang, Q.: Enhancing the search ability of differential evolution through orthogonal crossover. Inf. Sci. **185**, 153–177 (2012)
14. Maa, X., Chen, C.: Improving differential evolution using hybrid strategies for multimodal optimization. Energy Procedia **11**, 850–856. (2011)
15. Cai, Y., Wang, J., Yin, J.: Learning-enhanced differential evolution for numerical optimization. Soft Comput. (2011) doi:10.1007/s00500-011-0744-x.
16. Sindhya, K., Ruuska, S., Haanpa¨a, T., Miettinen, K.: A new hybrid mutation operator for multiobjective optimization with differential evolution. Soft Comput. (2011). doi:10.1007/s00500-011-0704-5.
17. Si, T., Hazra, S., Jana, N.D.: Artificial neural network training using differential evolutionary algorithm for classification. Adv. Intell. Soft Comput. (2012)
18. Regulwar, D.G., Choudhari, S.A., Anand, P.R.: Differential evolution algorithm with application to optimal operation of multipurpose reservoir. J. Water Res. Prot. (2010). doi:10.4236/jwarp.2010.26064.

Multi-objective Parametric Query Optimization for Distributed Database Systems

Vikram Singh

Abstract A classical query optimization compares solutions on single cost metric, not capable for multiple costs. A multi-objective parametric optimization (MPQ) approach is potentially capable for optimization over multiple cost metrics and query parameters. This paper demonstrated an approach for multi-objective parametric query optimization (MPQO) for advanced database systems such as distributed database systems (DDBS). The query equivalent plans are compared according to multiple cost metrics and query related parameters (modeled by a function on metrics), cost metrics, and query parameters are semantically different and computed at different stage of optimization. MPQO also generalizes parametric optimization by catering the multiple metrics for query optimization. In this paper, performance of MPQO variants based on nature-inspired optimization; 'Multi-Objective Genetic Algorithm' and a parameter-less optimization 'Teaching-learning- based optimization' are also analyzed. MPQO builds a parametric space of query plans and progressively explores the multi-objective space according to user tradeoffs on query metrics. In heterogeneous and distributed database system, logically unified data is replicated and distributed across multiple distributed sites to achieve high reliable and available data system; this imposed a challenge on evaluation of Pareto set. An MPQO attempt exhaustively determines the optimal query plans on each end of parametric space.

Keywords Distributed query processing · Genetic algorithm · Multi-objective optimization · Parametric query optimization · Top-K query plans · Teacher-Learner based optimization

Vikram Singh (✉)
National Institute of Technology, Kurukshetra 139119, Haryana, India
e-mail: viks@nitkkr.ac.in

M. Pant et al. (eds.), *Proceedings of Fifth International Conference on Soft Computing for Problem Solving*, Advances in Intelligent Systems and Computing 436, DOI 10.1007/978-981-10-0448-3_18

1 Introduction

In the era of information processing, end users are principally attracted toward best query answers in potentially vast solution space [1]. A database is an abstract view over various data form and logical layers, as distributed databases systems (DDBS). DDBS encompasses coherent data, spread across various sites of a computer network. A distributed database management system (DDBMS) deals with managing such distributed data and provides a simple and unified interface for this purpose [2]. In DDBS, relevant data is distributed over multiple sites to ensure higher reliability and availability of data. The distribution is based on either partially replicated or fully replicated with the effect of which instances of a relation or sub relations can be found in multiple sites. On the arrival of user query, query optimizer parses the query and identifies the relevant relations for the result retrieval. In DDBS, for a user query multiple equivalent query plans (QEPs) are possible, as the relation or sub relation is stored in multiple sites. Each QEP is potentially capable to generate desired result and thus selecting best alternative/alternatives become a computationally complex task for a query optimizer. Query optimizer selects optimal QEPs based on design objective function or cost function. In this paper, cost function/design objectives such as query affinity cost (QAC), query localization cost (QLC), local processing cost (LPC), and predicate selectivity (PS) are proposed for query optimization in DDBS.

In multi-objective query optimization (MQ), solution involves finding not only one but Pareto set of solutions or the best possible trade-offs among the objective functions [3–5]. Parametric query optimization (PQ) is an evolution on this course, which associates query metrics value with each QEP. PQ performs optimization over a pruned population (using metrics values) on parametric space [6–8]. PQ is potentially unable to optimize a problem with multiple metrics, which is handled by multi-objective query optimization (MPQO). MPQO identifies QEPs set that realize a good comparison between conflicting design objectives, such as minimizing the affinity and maximizing the predicate selectivity. The MPQO generates near-optimal plans, applying traditional optimization on pruned QEPs on parametric space, and further imposing post-optimization according to the metrics values [6, 8, 9].

Two trivial approaches for query optimization in parameterized space are viable. In first approach, query optimizer generates a new set of QEPs iteratively. MPQO variant based on the multi-objective genetic algorithm (MOGA) [6] is used for optimization of QEPs. This algorithm is principally based on 'Optimize-Always' approach. Another trivial approach is principally based on the 'Optimize-Once' approach. In 'Optimize-Once' a physical QEP goes through optimization once with some set of parameters and metrics values and reuses the resulting physical plan for subsequent set of parameters [6, 10]. MPQO variant based on teaching-learning-based optimization (TLBO) is adopted for the query optimization which is principally according to 'Optimize-Once' approach. TLBO is an optimization parameter-less technique, in which a QEP physically gone through only once, as subsequent optimization takes place on the modified values of the query parameters. TLBO also minimizes the optimization computation, as it does not take any algorithm-specific parameters from user. Both the above mentioned variant of MPQO have clear disadvantages, as 'Optimize-Once' consumes lot of memory resources and limits the concurrent queries in the systems. 'Optimize-Always' returns single QEP, that is used for all points in the parametric space. However, in actuality the cost functions of physical plans and regions of optimality are not so well behaved.

1.1 Related Work

An CQ approach on DDBS is an exhaustive search and a combinatorial optimization problem, which is addressed by various techniques based on heuristics like greedy, evolutionary, and randomized for single-objective problems [2–4]. MQ algorithms compare QEPs according to several cost tradeoffs with primary goal to find a plan that represents the best conciliation among conflicting cost metrics. In [5, 11, 12] similar approaches are discussed, which are effective for problem with multiple cost metrics but does not support parameters. PQ algorithms overcome the constraints capability of MQ by associating each QEP with cost functions instead of cost values [6–8]. MPQO is a generalization of MQ and PQ approaches while it is not possible to apply existing MQ or PQ algorithms to MPQO. Since a PQ algorithm supports only one cost metric and MQ algorithms do not support multiple parameters [6]. It may at first seem possible to model cost metrics as parameters; if all but one cost metric could be represented as parameters then PQ algorithms could be applied. Query optimizer considered that query processing over lesser number of sites would be more efficient and thus query plans involving fewer sites preferred [13–15] for result generation. In some existing research work, query parameters are used to generate the optimal solution from initialized QEPs, while in the proposed algorithm query specific parameters are used to optimize the build parametric space metrics that are used for post-optimization.

1.2 Contribution and Outline

The main contribution of this paper includes, MPQO algorithms for DDBS according based on multiple cost metrics and multiple query parameters,

- First key contribution is MPQO algorithm that considers ranges of query-related parameters and metrics to evaluate Pareto optimal set. Teaching-learning-based optimization (TLBO) and genetic algorithm variants, such as aggregation genetic algorithm (GA) and vector evaluated genetic algorithm (VEGA) are used to simulate the performance of MPQO.

- Another contribution is the validation of various cost models for query optimization on MPQO model. Approach exploits query parameters and metrics to design efficient optimization approach and believes that our annotations can hand out as preliminary point for the application of MPQO algorithms in distributed computing scenarios.

The fundamental of distributed query processing and related cost models and classification in metrics/ parameters is explained in Sect. 2. The fundamental of proposed MPQO variants is explained in Sect. 3. In Sect. 3.1 the parametric front initiation is shown and a TLBO-based MPQO is discussed in Sect. 3.1.1. The optimization performance among TLBO, MPQO, and various GA-based MPQO is illustrated in result section.

2 Distributed Query Processing

Distributed query processing involves multiple logical sites, thus multiple processing costs are associated in query processing among them inter-site communication cost is dominant. This cost directly affected by the amount data transfer from various sites involves in query processing, which means fewer sites in QEP lead to less cost as relevant data are stored at the less number of sites. In order to process a query, the data required may have to be obtained from several local sites (locally processed, e.g., selection, projection, etc.) distributed over a network. Furthermore, as the number of database sites in distributed database network are increased (to achieve higher degree of reliability and availability of data closer to user), relations accessed by the query increase lead to increased number of possible query equivalent plans (QEPs). Hence, identification of optimal QEP that entails an optimal cost for query processing becomes critical [9, 11. Thus, a large parametric search space comprising all possible query plans needs to be explored in order to derive the optimal query plans.

2.1 Query Equivalent Plans (QEPs) in DDBS

In distributed computing environment, a user query is parsed before arriving at an effective query processing strategy, this strategy is called distributed query processing (DQP) [4]. A query is represented by a set of database tables or relations that need to be joined and it specifies the join order and the operators executing scan and join operations [12]. DQP strategy comprises of efficient QEPs that would decompose the global query into local sub-queries to be executed at their respective local sites. Also, the order and the site at which the results of the sub-queries are integrated part of this plan. Finally, integrated result is provided as the answer to the query user. Thus, the DQP strategy aims to generate query processing plans that reduce the amount of data transfer between sites and thereby reduces the distributed query response time [4]. Query processing in such environment is critical task for a query processor, as multiple query equivalents are possible due to distributed and replicated logical data. Selecting optimal query plans from pool of alternatives is a critical task to a query optimizer, whose primary objective is to choose optimal (best) solutions [9], as entire activity of optimization lies on the various query parameters and tradeoff values of query metrics. In the build query parametric space, the trade off values of query metrics are used to select near-optimal QEPs which are the part of post-optimization work. Following exemplify the scenario,

Example A given user query and relation-site matrix (RSM) as in Table 1(a) shows, relation R_1 is replicated among S_1, S_2, S_3, S_5, S_6, S_8, S_{10}. Multiple QEPs are possible and all the QEP alternatives are initialized based on the RSM, Table 1(b) shows some QEPs for below mentioned user query.

SELECT *a*, *m*
FROM $R_1, R_2, R_3, R_4, R_5, R_6, R_7, R_8$
WHERE P_1 **AND** P_2

There are two important aspects namely, the content of the query plans and the length of the QP. The content is the sites of each relation. Query plans represent the set of data sites on which relevant data are stored and result will be aggregated, e.g., in query plans [1–3, 5], relation R_1 and R_2 from S_1, relation R_3, R_4 and R_5, from S_2, R_6 and R_8 from S_3, and finally relation R_7 from S_5.

2.2 *Query Parameters and Query Metrics*

It may first seem possible to model cost metric as query parameter, while optimizing for optimal QEPs in DDBS. If all but one cost metric could be represented as parameters then PQ algorithms could be applied. There are two strong arguments stated, first said in some scenario, semantic of some query parameters are different and cannot be treated in same way, e.g., LPC and QPC are time-based while PS, affinity is a degree of site heterogeneity-based parameter. Another argument is based on the computation of query parameters (values) known in advance or values can be taken form user along with query, e.g., values of predicate selectivity (PS) and affinity, while some are evaluated at execution time. Query Parameters that can be evaluated during execution time and used for post-optimization are considered as metrics, while others are considered parameters. The parametric solution space is build according to the pre-computed values (supplied values by user) of query parameters along with query.

Query Parameters: query parameters (QPs) values are supplied along with QEPs to the optimization algorithm and values are pre-computed. The parametric

Table 1 (a) Relation-Site Matrix (RSM) (b) Initialized QEPs (c) Pareto QEPs

(a) Relation-site matrix (distribution of relations among sites)

Site/Rel.	R_1	R_2	R_3	R_4	R_5	R_6	R_7	R_8
S1	1	1	1	1	1	0	0	0
S2	1	1	1	1	1	0	0	0
S3	1	1	0	0	1	1	1	1
S4	1	0	0	0	0	1	0	1
S5	1	1	1	0	0	1	1	1
S6	1	1	0	0	1	1	1	0
S7	0	1	1	1	0	1	0	0
S8	1	1	1	1	0	0	1	1
S9	1	1	0	0	0	0	0	1
S10	1	0	0	0	0	0	0	0
S11	1	1	1	1	0	1	1	1
S12	0	1	0	0	0	0	0	1
S13	0	0	1	0	0	0	0	0
S14	0	0	0	1	0	0	0	1
S15	1	0	1	1	1	1	0	0
S16	0	1	1	0	1	0	1	0

(b) Initial solution space (SS)

QEP	R_1	R_2	R_3	R_4	R_5	R_6	R_7	R_8
1	2	2	2	1	1	3	3	5
2	3	5	7	8	15	4	6	8
3	6	7	8	11	16	6	8	11
4	4	11	11	15	16	11	16	14
5	8	8	1	14	16	11	11	14
6	11	12	13	11	15	15	11	12
7	1	2	5	7	2	4	6	8
8	3	5	7	8	15	3	5	3
9	2	2	2	2	2	4	5	4
10	2	1	13	2	15	3	5	3
11	9	9	16	14	16	5	16	14
12	3	7	7	8	16	7	16	3
13	2	2	2	2	2	5	16	14
14	1	1	8	8	2	7	8	8
15	8	8	8	8	2	7	8	8
16	5	6	15	15	6	6	8	9
17	1	1	1	1	15	15	16	14
18	10	11	11	8	6	3	5	3
19	15	16	15	15	15	15	16	14
20	3	3	8	8	3	7	8	8

(continued)

Table 1 (continued)

(c) Pareto Optimal Set (POS) with 0.1 ≤ QAC ≤ 72 && 0.1 ≤ PS ≤ 0.93 (After pruning)								
QEP	R_1	R_2	R_3	R_4	R_5	R_6	R_7	R_8
9	2	2	2	2	2	4	5	4
13	2	2	2	2	2	5	16	14
14	1	1	8	8	2	7	8	8
15	8	8	8	8	2	7	8	8
17	1	1	1	1	15	15	16	14
19	15	16	15	15	15	15	16	14
20	3	3	8	8	3	7	8	8

space is build according to the span of imposed values of QP's, following are the formulation of query parameters:

(i) **Query Affinity Cost (QAC)**: This quantifies the degree of heterogeneity of QEP, in which 'affinity' refers sites referred for final result retrieval. Formulated as follows:

$$\text{QPC} = \sum_{i=1}^{M} \frac{K_i}{N}\left(1 - \frac{K_i}{N}\right) \tag{1}$$

where M is the number of sites accessed by the query plan, K_i is the number of times the ith site used in the query plan; N is the number of relation accessed by the QEP [13, 14].

(ii) **Selectivity of Predicate (SOP)**: the number of database objects or tuples retrieved by query predicate and query user provides the ranges of selectivity. In DDBS, logical data is partitioned or replicated between multiple logically related sites which imposed challenges to evaluate the predicate selectivity.

Query metrics: the values of the metrics are computed by optimizer for each of QEP on compilation time. In the proposed solution, following query parameters are modeled as query metrics for query optimization:

(i) **Query localization cost (QLC)**: this combines two aspects, first amount of data communicated between participating sites in a QEP and another it identifies the control site for query execution. QLC between S_1 to S_2 is a ratio of actual contribution of S_1 or S_2 and formulated as follows [15]:

$$QP_{Cost} = MIN_{i=1}^{N_r}\left[\sum_{j=1\&\&j\neq i}^{N_r} \frac{Size(R_{S_j})}{\sum_{k=1}^{N_r} Size(R_k)}\right] \quad (2)$$

where QP_{cost} communication/ localization cost, MIN is a function to evaluate minimum for i = (1 to N_r), N_r number of relations in given RSM or number of relations in FROM clause in query, size(R_{sj}) is number of tuples in relation present at site S_j, size(R_k) is number of tuples in relation R_k.

(ii) **Local processing cost (LPC)**: this cost concerns with predicate selectivity of stored relations and depends on two sub-costs, first at remote sites and second on control sites. (Assumptions) N_t is number of tuples, N_r is number of relations, S_{qp} is total number of sites in the QP, R_s is total number of relations on local site, S_r is selectivity of relation R on local site, S_j is selectivity of join, and N_j is number of joins for a query plan. CLPC is control site LPC [15].

$$\textbf{Relation Processing Cost}\ (RPC) = N_t * \left(S_r \Big/ \sum_{(k=1)}^{N_r} N_t(k) * k\right) \quad (3)$$

(a)
$$RLPC = Max_{(i=1\ to\ R_s)}[RPC(i)] \quad (3a)$$

(b)
$$CLPC = Max_{(i,j=1\ to\ N_r)}\left[N_t(JOIN(R_i, R_j)) * S_j((JOIN(R_i, R_j)) \Big/ \sum_{k=1}^{N_r} N_t(k)k)\right] \quad (3b)$$

3 Multi-objectives Parametric Query Optimization (MPQO)

Evolution of various advanced database architectures over the span of decades imposed an optimization challenge on query processing, as additional objectives which are of interest that should be considered during query optimization are added [16]. Query optimization becomes a complex and multi-objective inspired scenario, where the goal is to find a query plan that realizes the best compromise between conflicting objectives or pareto-optimal set. A classical query optimization (CQ), approach's focus is on to reduce the query execution time during the result retrieval, while of a MPQO is to find query plans that minimize a weighted effect over different cost metrics while respecting all cost bounds. This means that multiple cost metrics are finally combined into a single metric (the weighted sum); it is still not possible to reduce MPQO to single-objective query optimization and use classic

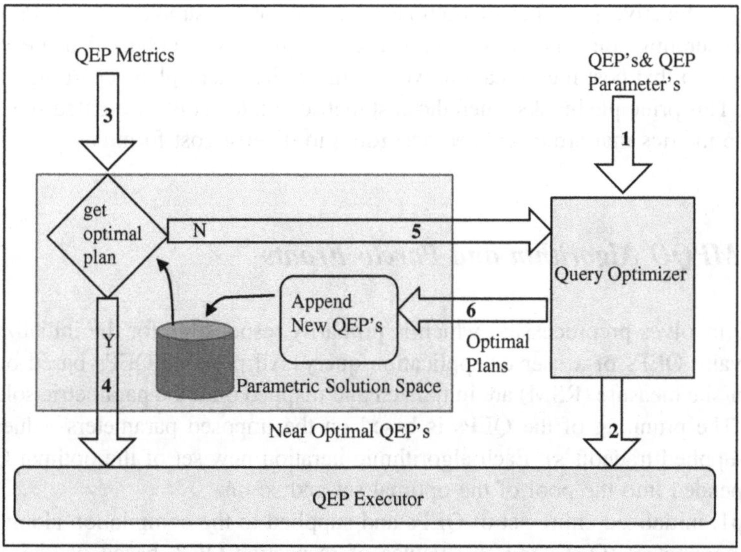

Fig. 1 MPQO Schematic diagram

optimization algorithms such as the approach in [10]. A schematic diagram of MPQO is shown in Fig. 1.

MPQO starts before run time; so inputs to an MPQO are user query and coupled query parameters. A parameter may represent any quantity that influences the cost of query plans and is unknown at optimization time. The goal of MPQO is to generate a complete set of relevant QEPs plans, meaning a set that contains a plan *p* for each possible plan *p* and each point in the parameter space *x* such that *p* has at most the same cost as *p* at *x* according to each cost metric [6]. Formulated differently, the goal is to find a set of Pareto optimal query plans for all points in the parametric space. As in PQ, all relevant query plans are generated in advance so that no query optimization is required at run time [17, 18]. The parametric space is generated based on the query parameters supplied along with user query, the parameters values are computed post at runtime and based on the computed values parametric space is build and the QEPs are optimized in post-optimization. The resultant QEPs are considered near-optimal set and further based on which results are generated for the user [6, 19].

The Post-optimization is based on the provided tradeoff by the user or decision maker. The cost metric values are imposed on the parametric solution space of each of the query parameter and QEPs with poor or lesser optimal are pruned from the parametric space. In the next section, the MPQO variant-based algorithm is explained, In which QEPs are pruned based on multi cost metric which must rely on the multi-objective principle of optimality: by replacing subplans or QEPs (e.g., plans generating join operands) within a query plan by subplans that are better according to that cost metric cannot worsen the entire query plan according to that metric. This principle breaks when the cost metric of interest is a weighted sum over multiple metrics that are calculated according to diverse cost formulas.

3.1 MPQO Algorithm and Pareto Fronts

MPQO involves preprocessing which is primarily responsible for the initialization of all valid QEPs of a user or application query. All possible QEPs based on the relation-site measure (RSM) are initialized and mapped onto the parametric solution space. The prunning of the QEPs is based on the imposed parameters values (or users supplied tradeoff's). Each algorithmic iteration new set of the optimal QEPs are appended into the pool of the optimal set and so on.

Step1, initialized entire set of QEPs and supplied to the compilation along with query parameters (QAC and PS) values of respective QEP, based on the query parameter domain parametric solution space is build. Each QEPs attain certain position on the parametric solution space, which indicates the fitness of QEP. Next step is prunning of QEPs based on the users tradeoff on the parametric space, QEPs, assumed less optimal, are sent to preprocessing and reconciliation cycle while better solution referred towards near-optimal pool and included for post-optimization. Subsequently, on the selected QEP, post-optimization tradeoff is applied according to the values of metrics (QLC and LPC) to identify the Pareto-optimal set, submit pareto optimal to the query executor for final result generation. In final step, resubmission of less fitter QEPs to the preprocessing for further generation and reoptimized QEPs are again submitted to the parametric space for the reconsideration and again the tradeoff values are applied on the set of QEPs for and so on.

Pareto Front: based on the metrics

For the given user query (Q_1), relations [R_1, R_2, …, R_8] used are distributed among logical sites [S_1, S_2, …, S_{16}], shown in Table 1(a)–(c).

Pareto front: in parametric solution space, a QEP considers Pareto optimal than other QEP, if former is better parametric value of each parametric plans (see Fig. 2).

3.1.1 MPQO Using Teaching-Learning-Based Optimization (TLBO)

Teaching-learning-based optimization (TLBO) is a parameter-less optimization technique for the optimization of the multi-objective. TLBO, firstly introduced in [20, 21], which primarily preferred for the optimization in classical engineering problem in mechanical, civil, etc. Optimization computation in most of the evolutionary and swarm intelligence-based algorithms are probabilistic and requires controlling parameters, like the search space size, number of generations, elite size, etc. In addition to the common control parameters, algorithm-specific control-parameters are required, such as in GA rate of mutation and crossover rate, similarly, inertia weight and social parameters in PSO [21]. The proper regulation of algorithm explicit parameters is a crucial factor, as it affects the overall performance of the abovementioned algorithms improper regulation of algorithm-specific parameters may lead to increases in the computational effort or yields a localized solution. Therefore, Rao et al. in [20] recently introduced the teaching-learning-based optimization (TLBO) algorithm, which requires only the common control parameters and does not require any algorithm-specific control-parameters [20].

Other evolutionary algorithms require the control of common control parameters as well as the control of algorithm-specific parameters. The burden of tuning control parameters is comparatively less in the TLBO algorithm. Thus, the TLBO algorithm is simple and effective and involves comparatively less computational effort. TLBO algorithm is designed in two phases, teacher phase and learner phase, teacher phase starts by evaluating the best learner among the pool of QEPs according to the calculated mean values of the metrics values (QLC and LPC). The metrics values are considered as performance of a learner (QEP) on the subject (metric). Subsequently, entire population is upgraded based on the best learners (T_f, teaching factor) metrics values, this step represents the knowledge sharing by the best student to the rest of learners in the pool initialized.

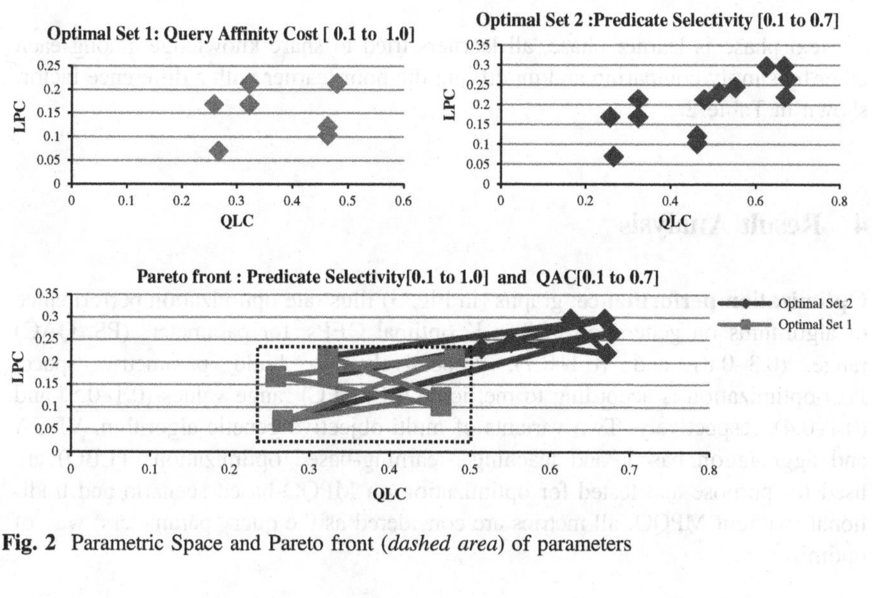

Fig. 2 Parametric Space and Pareto front (*dashed area*) of parameters

Algorithm

Input: *Initialized QEPs set and evaluation objectives such as: Query Metrics and Query Parameters*
Output: *Pareto set of QEPs*

Step 1: Identify Query Parameters and Metrics, e.g. QLC and LPC as Query Parameters and QA and PS as Query Metrics.

Step 2: Initialize, Solution Space (SS) of Query Equivalent Plans(QEPs), according to RSM.

Step 3: Pruning (QAC[a,b], PS[c,d]) & build Parametric Space(PS) of Pareto Optimal set of QEPs

```
{
  POS =Prun ( QPi(QACi,SSi), QPi+1(QAC i+1,SS i+1i)); for i=0 to n, where n is total no of QEPs;
}
```

Step 4: Supply POS to query executor to compute Query Parameters Vector (QPV) of LPC & QLC values.

Step 5: Supply POS and QPV to TLBO, each QEP from POS is Initialized as learner and QPV values as score,

```
Teacher Phase,
{
  Calculate means ( LPC, QLC);
  Best Teacher = QPbest, where QPBest is the QP with best mean value for each of Query Parameters
  Difference Meani = Ri (Mnew-Tf *Mj), Where Tf =round(1+ r and(0,1) in the range {2-1}
}
Xnew,i = Xold,i+Difference_Meani
New POS = ( QPi(QACi,SSi), QPi+1(QAC i+1,SS i+1i)); for i=0 to n, where n is total no of QEPs
```

Step 6: Student Phase

```
{
  Upgrade knowledge(fitness) among learner,
    Xnew,i = Xold,i+ri(Xi-Xj) if f(Xi) < f(Xj)
    Xnew,i = Xold,i+ri(Xi-Xj) if f(Xi) < f(Xj),
  Accept Xnew with better function value.
}
```

Step 7: Repeat (**Steps 4**), until termination criteria as Pareto Optimal set.

Next phase is learner phase, all learners tried to share knowledge among each other by simply comparing and modifying the poor learner with a difference factor, shown in Table 2.

4 Result Analysis

Optimization performance: graphs (in Fig. 3) illustrate optimization performance of algorithms on generation of Top-K optimal QEPs, for parameters (PS, QAC) range (0.3–0.8) and (0.1–0.7), respectively, to build parametric space. Post-optimization is according to metrics (QLC, LPC) range values (0.1–0.8) and (0.1–0.4), respectively. Two variants of multi-objective genetic algorithm, VEGA and aggregation based, and teaching—learning-based optimization (TLBO) are used for purpose and tested for optimization on MPQO-based scenario and traditional (without MPQO, all metrics are considered as the query parameters) way of optimization.

Table 2 TLBO results (a) Teacher phase (b) Learner phase

(a) Teacher phase

Query metric cost (QMC)			Fitness 1
QEP	QLC (F_1)	LPC (F_2)	$F_1^2 + F_2^2$
9	0.3228	0.1682	0.1325
13	0.3228	0.2132	0.1496
14	0.4634	0.1210	0.2294
15	0.2665	0.0694	0.0758
17	0.4802	0.2132	0.2760
19	0.2581	0.1682	0.0949
20	0.4634	0.1031	0.2254
Mean	0.36822	0.1509	
tf = 2, Best QP_No. = 19			
Difference factor		−0.20336	−0.1630

(b) Learner phase

Query metric cost			Fitness 3	After	
QEP	QLC (F_1)	LPC (F_2)	$F_1^2 + F_2^2$	Predefined iterations	Final value
9	0.119446	0.0052074	0.0142945		0.0030309
13	0.119446	0.0501512	0.0167825	15	0.0127646
14	0.1860071	0.0031337	0.0694304	9	0.0142945
15	0.0631722	−0.093669	0.0127646	13	0.0167825
17	0.2696389	0.0501512	0.0791668	14	0.0694304
19	0.0548072	0.0052074	0.0030309	20	0.0712631
20	0.2601304	−0.049067	0.0712631	17	0.0791668

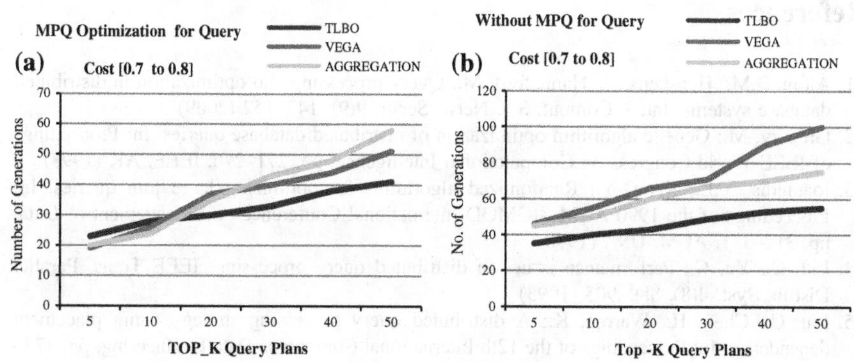

Fig. 3 Optimization performance by TLBO and GA-based MPQO using a parametric space. **b** Similar optimization without parametric

5 Conclusion

MPQO is a novel variant of multi-objective optimization based on multiple cost metrics and query parameters for optimization on parametric space. The parametric space is built according to the various pre-computed values of query parameters provided with user query to query processor. The identification of query-related cost metrics and parameters are purely based on the semantics and evaluation approach. In this paper, a MPQO variant is proposed over the multiple cost heuristics and multiple query parameters. There are two basic approaches for query optimization, 'Optimized-Once' and 'Optimized-Always'. Our approach is principally based on 'Optimized-Once' strategy and optimization performance is compared with variants based on the 'Optimized-Always' strategy. TLBO-based MPQO is designed for optimized one, as specific query plans physically pass through optimization cycle only once and subsequently updated optimized query instance is used by the approach. Genetic algorithm-based variant for the optimized always approach as GA optimizes solution iteratively and without modifying the QEPs. The performance over multiple metrics and query parameters of both algorithms is evaluated. TLBO generates Pareto front with lesser number of algorithmic runs as lesser computational effort is required.

Acknowledgments The author was inspired (in different ways) by discussion at the AEOTIT 2014, a workshop on advanced optimization techniques had the discussions that ultimately led to this research work. Dr. R.V. Rao, Professor at SVNIT Surat, Gujrat, India. The TLBO played key role on the development of optimized solution in various fundamental engineering problems.

References

1. Alom, B.M., Henskens, F., Hannaford, M.: Query processing and optimization in distributed database systems. Int. J. Comput. Sci. Netw. Secur. **9**(9), 143–152 (2009)
2. Gregory, M.: Genetic algorithm optimization of distributed database queries. In: Proceedings of IEEE World Congress on Computational Intelligence, pp. 271–276. IEEE, AK (1998)
3. Ioannidis, Y.E., Kang, Y.: Randomized algorithms for optimizing large join queries. In: Proceedings of the 1990 ACM SIGMOD International Conference on Management of Data, pp. 312–321. ACM, USA (1990)
4. Liu, C., Yu, C.: Performance issues in distributed query processing. IEEE Trans. Parallel Distrib. Syst. **4**(8), 889–905 (1993)
5. Liu, C., Chen, H., Warren, K.: A distributed query processing strategy using placement dependency. In: Proceedings of the 12th International Conference Data Engineering, pp. 477–484. New Orleans, Louisiana (1996)
6. Trummer, I., Koch, C.: Multi-objective parametric query optimization. In: Proceedings of 41st International Conference on VLDB, pp. 221–232, Coast, Hawaii (2015)
7. Ganguly, S.: Design and analysis of parametric query optimization algorithms. In: Proceedings of 24th International Conference on VLDB, pp. 228–238. Morgan Kaufman, USA (1998)
8. Ganguly, S., Hasan, W., Krishnamurthy, R.: Query optimization for parallel execution. In: Proceedings of the 1992 ACM SIGMOID International Conference on Management of Data, pp. 9–18. ACM, USA (1992)

9. Bruno, N.: Polynomial heuristics for query optimization. In: Proceedings of International Conference of Data Engineering, pp. 589–600. IEEE, USA (2010)
10. Trummer, I., Koch, C.: Approximation schemes for many-objective query optimization. In: Proceedings of the 2014 ACM SIGMOID International Conference on Management of Data, pp. 1299–1310. ACM, USA (2014)
11. Rho, S., March, S.T.: Optimizing distributed join queries: a genetic algorithmic approach. Ann. Oper. Res. **71**, 199–228 (1997)
12. Swami, A., Gupta, A.: Optimization of large join queries. In: Proceedings of the 1988 ACM SIGMOD International Conference on Data Management, vol. 17, no 3, pp. 8–17. ACM, USA (1998)
13. Kumar, T.V., Singh, V., Verma, A.K.: Generating distributed query processing plans using genetic algorithm. In: Proceedings of the 2010 International Conference on Data Storage and Data Engineering, pp. 173–177. IEEE, USA (2010)
14. Singh, V., Mishra, V.: Distributed query plan generation using aggregation based multi-objective genetic algorithm. In: Proceedings of International Conference on TCS, pp. 20–29. ACM, USA (2014)
15. Mishra, V., Singh, V.: Generating optimal query plans for distributed query processing using teacher-learner based optimization. In: Proceedings of 11th International Conference on Data Mining and Warehouse, vol. 54, pp. 281–290. Elsevier, India (2015)
16. Kambhampati, S., Nambiar, S., Nie, Z., Vaddi, S.: Havasu: A multi-objective, adaptive query processing framework for web data integration. Technical Report (2002)
17. Hulgeri, A., Sudarshan, S.: Parametric query optimization for linear and piecewise linear cost functions. In: Proceedings of 28th International Conference on VLDB, pp. 167–178, VLDB (2002)
18. Bizarro, P., Bruno, N., DeWitt, D.: Progressive parametric query optimization. Trans. Knowl. Data Eng. **21**(04), 582–594 (2009)
19. Xu, Z., Tu, Y.C., Wang, X.: PET: Reducing database energy cost via query optimization. In: Proceedings of VLDB endowment, vol. 5, no. 12, pp. 1954–1957 (2012).
20. Rao, R.V., Savsani, V.J., Vakharia, D.P.: Teaching–learning-based optimization: an optimization method for continuous non-linear large scale problems. Inf. Sci.: Int. J. **183**(1), 1–15 (2012)
21. Rao, R.V., Patel, V.: An elitist teaching learning-based optimization algorithm for solving complex constrained optimization problems. Int. J. Ind. Eng. Comput. **3**(4), 535–560 (2012)

Energy Efficient Routing Protocol for MANET Using Vague Set

Santosh Kumar Das and Sachin Tripathi

Abstract In the prevailing epoch, the application of mobile ad hoc networks (MANET) has risen quickly. All the nodes of the network communicate directly with each other to carve up information within the assortment. The network is vigorous and infrastructureless. So the topology of this network is able to amend very commonly. MANET nodes are power-driven through narrow capacity battery and due to this sometimes nodes cannot successfully broadcast data packets from source node to destination node. In this chapter, we propose a new energy efficient routing protocol for MANET using vague set. The main aim of this proposed protocol is to choose an energy efficient route that diminishes energy expenditure of MANET based on the scheme of vague set. The proposed scheme is primarily used for interval-based membership where each parameter of energy efficient routing (i.e. energy and distance) is characterized by true and false membership functions. Therefore, this approach helps to determine the energy efficient route. The simulation of proposed protocol by NS2 and relative study with available protocol AODV is scrutinized wherein proposed routing protocol improves the performance of MANET based on throughput, average end-to-end delay, packet delivery ratio and packet loss.

Keywords Mobile ad hoc network · Vague set · Fuzzy logic · Expert system

S.K. Das (✉) · Sachin Tripathi
Department of Computer Science and Engineering, Indian School of Mines, Dhanbad 826004, Jharkhand, India
e-mail: sunsantosh2007@rediffmail.com

Sachin Tripathi
e-mail: var_1285@yahoo.com

M. Pant et al. (eds.), *Proceedings of Fifth International Conference on Soft Computing for Problem Solving*, Advances in Intelligent Systems and Computing 436, DOI 10.1007/978-981-10-0448-3_19

1 Introduction

MANETs [1–3] are the theme of many research works in the field of computer networks. It is infrastructureless and the fastest growing network [4, 5]. A mobile node can be eligible for notice of arrival and departure of different devices and achieve essential arrangement to ease communique and distribution of data and services among devices. MANET permits the devices to maintain links to the network in addition to easy joining and separating of devices to and from the network. The battery power expenditure transpires due to distributing a packet, getting a packet and when the node is inactive. Therefore, energy efficient routing point out to select the route that necessitates a shortest distance with the highest energy. Therefore, to expand life of the network, mobile-nodes should choose the finest path based on the remaining battery power. In this paper, energy of batteries and distances between nodes can be deliberated by a vague set.

The remainder of the paper is organized as follows: Sect. 2 gives a survey of energy competent routing protocol for MANET. Section 3 talks about the preliminaries based on proposed protocol. Details of proposed protocol are described in Sect. 4. The illustrations of the simulation results of the proposed protocol appear in Sect. 5. Finally, the conclusion of this paper is drawn in Sect. 6 and the paper is completed with several relevant references.

2 Related Works

In the past decade, a lot of research has been done with energy efficient routing such as Wang et al. [6] introduced a fuzzy logic-oriented power-aware protocol to select stable manager from mobile hosts based on parameter speed, battery power, and location. In this protocol, the authors select a proper power-aware manager to take care of the intra-cell and inter-cell tasks. This protocol also uses the multi-mobile agent to split the manager workload. However, it has a limitation that fuzzy logic work with point-based membership function which is less expressive in capturing vagueness of the data. Su et al. [7] introduced a fuzzy logic-oriented multicast routing protocol FMAR for MANET. The prime goal of this FMAR is to dynamically assign the active path based on multi-value logic. However, the proposed protocol has a drawback that it does not calculate rating value of all possible routes and selects some specific routes only. Therefore, it does not determine which paths are intensely usable. Zussman and Segall [8] proposed a method to create a MANET of wireless astute badges for obtain knowledge from attentive survivors. Here, the authors examine the energy proficient routing problem which occurs in this network. It is an inadequate emergency situation due to limited power. So, the subject is prepared as all types of routing problems for aiming to increase the lifetime of battery awaiting the initial battery drains out. Xu et al. [9] proposed an algorithm GAF that decreases energy expenditure in an ad hoc network. It safeguards energy through recognizing nodes that are correspondent from a routing viewpoint and then exits pointless nodes. Many topological-control, energy-efficient routing protocols have been proposed [10, 11]. Chen et al. [12] proposed an algorithm that adaptively selects coordinators within the network. It rotates the coordinator role between nodes to equilibrium the energy savings. Thus, coordinators proceed as backbone routers for the complete network and provide definite connectivity by ensuring that at least one energetic node is in the coordinator's range. The coordinators are chosen with their remaining energy and the deployment of the node [13]. Yu et al. [14] surveyed and grouped various energy-conscious routing protocols for MANETs. Here the authors brought down either the dynamic communique energy essential to send out or collect data or the motionless energy consumed at the time of inactive mobile node. Zarifzadeh et al. [15] proposed a new topology control strategy where all nodes energetically maintain their devolution energy in each packet and try to reduce energy expenditure by allowing parameters such as transmission power and traffic load. In this method, the authors first talk about the difficulty of the problem, then present an estimate algorithm to solve it.

However, none of the above techniques addresses the energy efficient routing and network lifetime matters jointly based on vague set. Therefore, this paper considers both these matters based on vague set.

3 Preliminaries: Concepts and Definitions

In this section, some preliminaries are discussed that serve an important role in designing this protocol. Short descriptions of these preliminaries are given below.

3.1 Vague Set

In mathematics, fuzzy set theory [16–18] is used to established logic between two Boolean value such as true and false. Its work based on point-based membership function. Vague set [19] is an extension of fuzzy sets. It is work on interlude-based membership as an alternative of point based. In vague set interlude-based true membership function is denoted by α and false membership function is denoted by β.

3.2 Expert System

Expert system [20] is a program used to solve exacting unpredictable problem in a specific area. Basically it contains two parts, knowledge-based system and inference engine. The knowledge-based system is used to solve any problem with the help of intelligence techniques. The inference engine is used to process data by using a rule-based system.

4 Proposed Protocol

In this section the proposed protocol is illustrated using interval-based membership function driven by an expert system. Expert system uses a vague set to evaluate energy efficient route. In this protocol, two input parameters and one output parameter are considered. The input variables are energy and distance while rating of the route has been taken as an output variable. The control flow illustration of proposed scheme is given in Fig. 1. One hypothetic instance is taken to visualize the representation of proposed protocol. In this example energy is considered as 5 J and distance is considered as 50 m. Membership functions for each input parameter are given in Tables 1 and 2.

A vague set V [21] in a universe of discourse U is discriminated by true and false membership function as αv and βv given as: $\alpha v : U \rightarrow [0, 1]$, $\beta v : U \rightarrow [0, 1]$, and

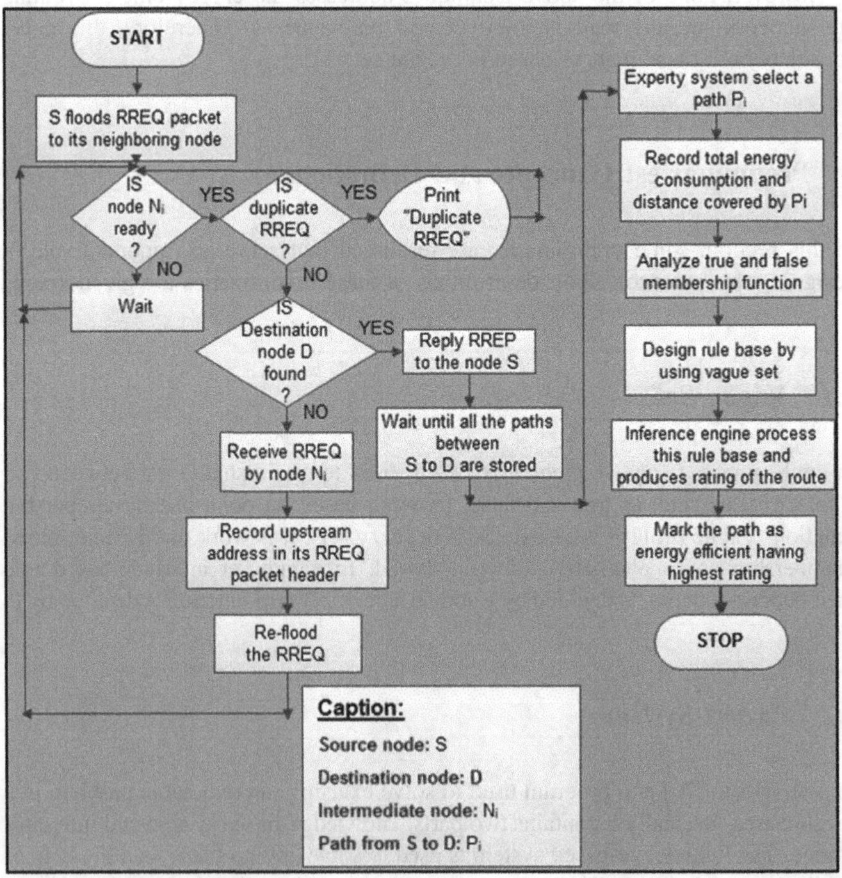

Fig. 1 Control flow diagram

Table 1 Membership function of energy

Linguistic values	Notation	Range	Base value
Low	E_L	$[E_{L-}, E_{L+}]$	(0, 1.8)
Medium	E_M	$[E_{M-}, E_{M+}]$	(1.8, 3.5)
High	E_H	$[E_{H-}, E_{H+}]$	(3.5, 5)

Table 2 Membership function of distance

Linguistic values	Notation	Range	Base value
Short	D_S	$[D_{S-}, D_{S+}]$	(0, 17)
Medium	D_M	$[D_{M-}, D_{M+}]$	(17, 34)
Long	D_L	$[D_{L-}, D_{L+}]$	(34, 50)

$\alpha v(u) + \beta v(u) \leq 1$. It provides an analysis of membership ranking. For example, the vague predicate related to energy such as energy is lower is given by a unique interval [0.3, 0.9], which means the true membership α is 0.3 and false membership β is 0.1, that is, $1 - \beta = 0.9$. So the vague membership value is [0.3, 0.9] for a node having low energy. It may be written as $[0.3, 0.9]/E_L$ which stands for the inferior bound of membership of E_L whose least fuzzy membership is 0.3 and the superior bound is the maximum of fuzzy membership 0.9. Thus, the following can be written as for other linguistic values (such as energy is low, medium and high while distance is short, medium and long respectively) as: $\alpha(E_L) + \beta(E_L) \leq 1$, $\alpha(E_M) + \beta(E_M) \leq 1$, $\alpha(E_H) + \beta(E_H) \leq 1$, $\alpha(D_S) + \beta(D_S) \leq 1$, $\alpha(D_M) + \beta(D_M) \leq 1$ and $\alpha(D_L) + \beta(D_L) \leq 1$.

Thus we have a corresponding vague set $V_t = \{V_1, V_2, V_3, V_4, V_5, V_6\}$ where V_1, V_2, V_3 are elements of the first input parameter energy and V_4, V_5 and V_6 are elements of the second input parameter distance. The value of each element is given below: $V_1 = (E_L, (\alpha(E_L), 1 - \beta(E_L)))$, $V_2 = (E_M, (\alpha(E_M), 1 - \beta(E_M)))$, $V_3 = (E_H, (\alpha(E_H), 1 - \beta(E_H)))$, $V_4 = (D_S, (\alpha(D_S), 1 - \beta(D_S)))$, $V_5 = (D_M, (\alpha(D_M), 1 - \beta(D_M)))$ and $V_6 = (D_L, (\alpha(D_L), 1 - \beta(D_L)))$.

Each element of vague set consists of two elements; the first is a true membership function which is αv and another is false membership function which is βv. The value of true and false membership functions are taken from Tables 1 and 2. In this paper, for the most suitable and ideal solution false membership function is discarded and only true memberships function α and $1 - \beta$ are chosen by expert system. These α and $1 - \beta$ of input parameters energy and distance are help to design fuzzy implications which are given in Table 3. Expert system processes these implications and produces ratings of different routes by using Eq. (1) given below:

$$R_{ij} = \frac{Avg(\alpha(E_i), 1 - \beta(E_i))}{Avg(\alpha(D_j), 1 - \beta(D_j))} \tag{1}$$

where $i \in 1 \leq i \leq 3$ for E_L, E_M, E_H and $j \in 1 \leq j \leq 3$ for D_S, D_M, D_L.

Table 3 Fuzzy implication for route selection

Rules	Energy (E)	Distance (D)	Rating (R)
Rule 1	E_L	D_L	R_{VB}
Rule 2	E_L	D_M	R_B
Rule 3	E_L	D_S	R_S
Rule 4	E_M	D_L	R_M
Rule 5	E_M	D_M	R_{LG}
Rule 6	E_M	D_S	R_G
Rule 7	E_H	D_L	R_{VG}
Rule 8	E_H	D_M	R_E
Rule 9	E_H	D_S	R_{VE}

Table 4 Illumination of rating for different route

Route No.	Abbreviation	Illumination of rating	Rating of route
R1	R_{VB}	Very bad	1.058824
R2	R_B	Bad	0.352941
R3	R_S	Satisfactory	0.214286
R4	R_M	Medium	3.117647
R5	R_{LG}	Less good	1.039216
R6	R_G	Good	0.630952
R7	R_{VG}	Very good	5.000000
R8	R_E	Excellent	1.666667
R9	R_{VE}	Very excellent	1.011905

Therefore, every route has a unique rating in MANET. In Table 4, Route R7 is the best choice for route selection because it has the highest rating. The sequences of different route based on their descending order rating are R7 > R4 > R8 > R1 > R5 > R9 > R6 > R2 > R3. Hence, expert system chooses R7 as an optimal energy efficient route because it has higher energy and shortest distance and R3 as a worst choice route because it has lower energy and longest distance.

5 Simulation Results and Analysis

In this section, to demonstrate the benefits of the proposed scheme, the presentations of the scheme with respect to packet loss, average end-to-end delay, packet delivery ratio and throughput are simulated using NS2. Here, we used AODV protocol for comparison. This protocol is not an energy efficient protocol, so first it is made an energy efficient routing protocol by enabling energy to all mobile nodes and some changes are made in aodv.cc and aodv.h files. To make the energy efficient routing protocol using a vague set we again use aodv.cc and aodv.h file and

Table 5 The simulation environment

Simulation parameter	Value
NS-2 simulator version	NS 2.25
Topology size	1000 × 1000
MAC layer type	IEEE 802.11
Number of nodes	50
Protocols under test	AODV
Initial energy	1000 J
Transmission power (W)	1.4
Receive power (W)	1.0
Idle power (W)	0.83
Mobility pause time	0 s
Packet size	512 Bytes
Packet interval	0.1 ms
Node movement model	Random waypoint
Number of source	1
Number of destination	1

compare it with energy enable AODV protocol. At the time, comparing we use a TCL script for testing the proposed protocol with the existing AODV protocol within a predefined 1000 m × 1000 m grid area. Details of general simulation environment are depicted in Table 5, and software requirement specification and hardware system configuration are given in Tables 6 and 7.

Table 6 Software requirement specification

Software	Specification
NS-2 version	NS 2.25
Windows OS	Windows 7 professional (service pack 1)
Linux OS	Ubuntu 14.100
MS office	2013

Table 7 Hardware system configuration

Hardware	Specification
Processor	Intel (R) core i5
Speed	3.20 GHz
RAM	2 GB
Hard disk	300 GB
Key board	HCL
Mouse	HCL
Monitor	HP

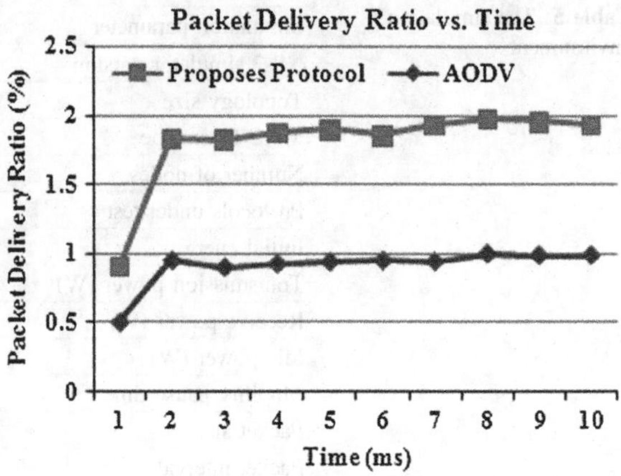

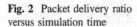

Fig. 2 Packet delivery ratio versus simulation time

5.1 *Packet Delivery Ratio*

The relation of the number of data packets are fruitfully distributed to the destination of those created by CBR sources. Figure 2 illustrates a comparison among the proposed protocol and AODV based on packet delivery ratio and simulation time. Equation (2) represents the formula of packet delivery.

$$\text{Packet delivery ratio} = \frac{\text{Received packets}}{\text{Sent packets}} * 100 \tag{2}$$

5.2 *Average End-to-End Delay*

The average end-to-end delay indicates time between data transmission from source node to destination. It is calculated by the formula in Eq. (3). Figure 3 shows a comparison among the proposed protocol with existing protocol AODV based on the average end-to-end delay and simulation time.

$$\text{Average end-to-end delay} = \frac{\sum(\text{arrive time} - \text{send time})}{\sum \text{Number of connections}} \tag{3}$$

5.3 *Throughput*

Throughput is the amount of successfully received packets in a unit time and it is represented in bps. In this paper, throughput needs to be measured in bytes per millisecond. Throughput is measured by the formula in Eq. (4). Figure 4 shows an

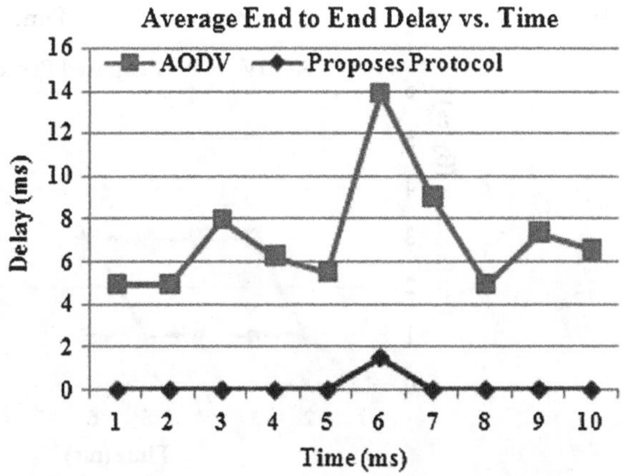

Fig. 3 Average end-to-end delay versus simulation time

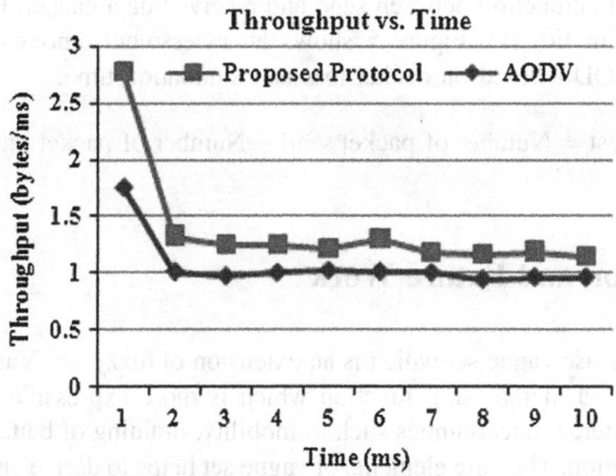

Fig. 4 Throughput versus simulation time

assessment among the proposed protocol with AODV based on throughput and simulation time.

$$\text{Throughput} = \frac{\text{Number of delivered packet} * \text{Packet size}}{\text{Total duration of the simulation}} \quad (4)$$

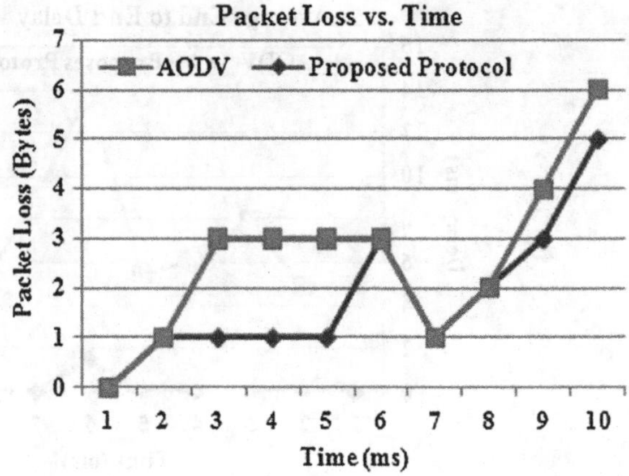

Fig. 5 Packet loss versus simulation time

5.4 *Packet Loss*

Packet loss is a malfunction between send and received data packet. It is measured by the formula in Eq. (5). Figure 5 shows an assessment among the proposed protocol with AODV based on packet loss and simulation time.

$$\text{Packet lost} = \text{Number of packet send} - \text{Number of packet received} \tag{5}$$

6 Conclusion and Future Work

In this paper, we use vague set which is an extension of fuzzy set. Vague set works with interval-based membership function which is more expressive for capturing vagueness of different uncertainties such as mobility, draining of battery power and energy consumption. The core elements of vague set helps to derive energy efficient route in MANET which has the highest energy with the shortest distance.

Future enrichment scopes include but are not limited to (i) the mathematical analysis of the proposed algorithm in which we wish to compare the analytical results with simulation results and (ii) the plan to implement a new energy efficient routing protocol which contains the proposal of route maintenance.

References

1. Das, S.K., Tripathi, S., Burnwal, A.P.: Intelligent energy competency multipath routing in wanet. In: Information Systems Design and Intelligent Applications, pp. 535–543. Springer (2015)
2. Das, S.K., Tripathi, S., Burnwal, A.P.: Fuzzy based energy efficient multicast routing for ad-hoc network. In: 2015 Third International Conference on Computer, Communication, Control and Information Technology (C3IT), pp. 1–5. IEEE, (2015)

3. Das, S.K., Tripathi, S., Burnwal, A.P.: Design of fuzzy based intelligent energy efficient routing protocol for wanet. In: 2015 Third International Conference on Computer, Communication, Control and Information Technology (C3IT), pp. 1–4. IEEE, (2015)
4. Yadav, A.K., Tripathi, S.: DLBMRP: Design of load balanced multicast routing protocol for wireless mobile ad-hoc network. Wireless Pers. Commun. 85(4), 1–15 (2015)
5. Yadav, A.K., Tripathi, S.: Load balanced multicast routing protocol for wireless mobile ad-hoc network. In: 2015 Third International Conference on Computer, Communication, Control and Information Technology (C3IT), pp. 1–6. IEEE, (2015)
6. Yan, K.-Q., Wang, S.-C., Chiang, M.-L., Tseng, L.-Y.: A fuzzy-based power-aware management for mobile ad hoc networks. Comput. Stand. Interfaces 31(1), 209–218 (2009)
7. Su, B.-L., Wang, M.-S., Huang, Y.-M.: Fuzzy logic weighted multi-criteria of dynamic route lifetime for reliable multicast routing in ad hoc networks. Expert Syst. Appl. 35(1), 476–484 (2008)
8. Zussman, G., Segall, A.: Energy efficient routing in ad hoc disaster recovery networks. Ad Hoc Netw. 1(4), 405–421 (2003)
9. Xu, Y., Heidemann, J., Estrin, D.: Geography-informed energy conservation for ad hoc routing. In: Proceedings of the 7th Annual International Conference on Mobile Computing and Networking, pp. 70–84. ACM, (2001)
10. Guodong, W., Gang, W., Jun, Z.: Elgr: an energy-efficiency and load-balanced geographic routing algorithm for lossy mobile ad hoc networks. Chin. J. Aeronaut. 23(3), 334–340 (2010)
11. Srisathapornphat, C., Shen, C.-C.: Coordinated power conservation for ad hoc networks. In: IEEE International Conference on Communications, 2002. ICC 2002, vol. 5, pp. 3330–3335. IEEE, (2002)
12. Chen, B., Jamieson, K., Balakrishnan, H., Morris, R.: Span: an energy-efficient coordination algorithm for topology maintenance in ad hoc wireless networks. Wireless Netw. 8(5), 481–494 (2002)
13. Saravanan, K., Velmurugan, T., Bagubali, A.: Increasing the lifetime of manets by power aware protocol–span. J. Theor. Appl. Inf. Technol. 54(2), (2013)
14. Yu, C., Lee, B., Yong Youn, H.: Energy efficient routing protocols for mobile ad hoc networks. Wireless Commun. Mob. Comput. 3(8), 959–973 (2003)
15. Zarifzadeh, S., Yazdani, N., Nayyeri, A.: Energy-efficient topology control in wireless ad hoc networks with selfish nodes. Comput. Netw. 56(2), 902–914 (2012)
16. Das, S.K., Tripathi, S., Burnwal, A.P.: Some relevance fields of soft computing methodology. Int. J. Res. Comput. Appl. Rob. 2, 1–6, (2014)
17. Das, S.K., Kumar, A., Das, B., Burnwal, A.P.: Ethics of reducing power consumption in wireless sensor networks using soft computing techniques. Int. J. Adv. Comput. Res. 3(1), 301–304, (2013)
18. Das, S.K., Kumar, A., Das, B., Burnwal, A.P.: On soft computing techniques in various areas. Comput. Sci. Inf. Technol. 3, 59 (2013)
19. Das, S.K., Tripathi, S.: Energy efficient routing protocol for manet based on vague set measurement technique. Procedia Comput. Sci. 58, 348–355 (2015)
20. Kim, K.-S., Roh, M.-I., Ha, S.: Expert system based on the arrangement evaluation model for the arrangement design of a submarine. Expert Syst. Appl. 42(22), 8731–8744 (2015)
21. Zhang, Q., Wang, J., Wang, G., Hong, Y.: The approximation set of a vague set in rough approximation space. Inf. Sci. 300, 1–19 (2015)

Data Storage Security in Cloud Paradigm

Prachi Deshpande, S.C. Sharma and Sateesh K. Peddoju

Abstract The advent of social networking has given rise to the huge data processing in terms of image and video streams. This, in turn, increased the use of cloud computing services by the users. Secure data storage and access are the main challenges in front of the cloud scenario. This paper reports a novel method of multimedia data security in the cloud paradigm. The proposed method watermarks and compresses the data before its storage in the cloud. This approach not only safeguards the data storage but also reduces the storage requirement and allied monitory overheads. The simulation result shows a 7 and 36 % use of CPU and memory capacity, which overrules the additional hardware requirement for the proposed module in the cloud paradigm.

Keywords Cloud · Data · Image · Multimedia · Security · Watermark

1 Introduction

Cloud computing has emerged as a landmark information processing technology in the past few years. The National institute of science and technology (NIST) had defined cloud computing as "Cloud computing is a model for enabling ubiquitous, convenient, on-demand network access to a shared pool of configurable computing resources (e.g., networks, servers, storage, applications, and services) that can be

Prachi Deshpande (✉) · S.C. Sharma
Department of Applied Science and Engineering, Indian Institute of Technology Roorkee, Roorkee 247667, Uttarakhand, India
e-mail: psd17dpt@iitr.ac.in

S.C. Sharma
e-mail: scs60fpt@iitr.ac.in

S.K. Peddoju
Department of Computer Science and Engineering, Indian Institute of Technology Roorkee, Roorkee 247667, Uttarakhand, India
e-mail: drpskfec@iitr.ac.in

M. Pant et al. (eds.), *Proceedings of Fifth International Conference on Soft Computing for Problem Solving*, Advances in Intelligent Systems and Computing 436, DOI 10.1007/978-981-10-0448-3_20

rapidly provisioned and released with minimal management effort or service provider interaction" [1]. The cloud computing paradigm has become attractive due to its inherent capacities like pay-per-use, scalability, accessibility, and ubiquity. The past decade witnessed the rapid increase in the use of digital multimedia contents such as images and videos in the social networking applications. Due to limited storage and processing capacity, users and enterprises were forced to upload their data onto the cloud servers [2–4]. The need of the speedy processing and data storage had been solved by the advent of cloud computing paradigm. However, the users are absolutely unaware of the level of security and confidentiality of their data in the cloud. The data may be transferred and stored in a cloud system wirelessly. Owing to this aspect, the data may become vulnerable to unauthorized access, modifications, and disclosures. Hence, security and privacy are the critical issues of such out sourcing of data [5, 6]. Figure 1 depicts the user behavior model in view of cloud computing. Generally, the streaming-level authentication methods such as media authentication code (MAC) or content-level method such as digital watermarking may be used to protect the multimedia data over a wireless network. MAC methodology adds a hash value to each data packet. This feature may cause a high computational overhead. Watermarking techniques consider the characteristics of media data for authentication and thus may cause lower computational overheads. Hence, in the proposed work, digital watermarking technique is used to secure the user data in the cloud environment.

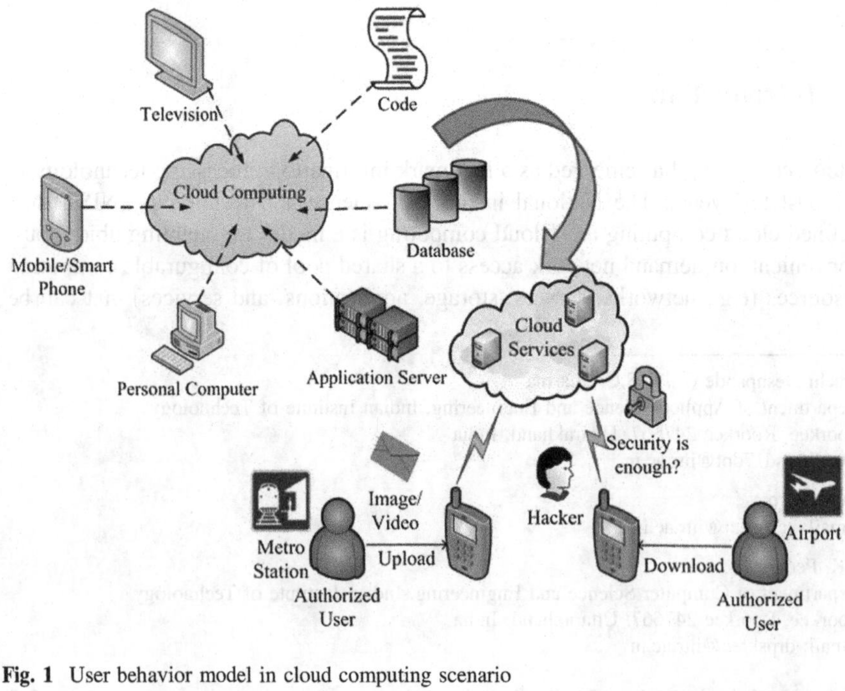

Fig. 1 User behavior model in cloud computing scenario

In the proposed work, a two-prong strategy is adopted to secure the multimedia content in the cloud. The proposed module, first, adds a digital watermark to the original image/video to be stored. Then, the newly formed image is compressed using the block matching algorithm. Due to this strategy, even if, the information has been tapped; it can not be decoded easily. The reduction in the storage requirement is also another advantage of the proposed methodology.

The paper is organized as: Sect. 2 provides the overview of the related work, whereas the proposed architecture and its performance are discussed in Sect. 3. The paper concludes in Sect. 4 with the discussion of the future research.

2 Overview of the Related Work

2.1 Image/Video Watermark Scenario

Digital watermarking of the multimedia content had drawn attention of academicians and researchers over the past several years. Industries are also coming forward with watermarking products like Digimarc's *Media Bridge* [7]. Encryption, steganography and watermarking are the most used techniques for data protection. Encryption techniques are useful for transmission processes. However, examination of the original data in its encrypted (protected) form is never possible. In steganography, the methodology for hiding the message may be confidential and the message is secret, which must be protected by intelligent hiding techniques. Here, the means for hiding the message is not important. The watermarks generally remain in the content in its original form and do not avert a user to listen, view, examine, or to operate the content. The watermark embedding process is known and the messages need not to be confidential. However, the effective coupling of the message and watermark is critical as watermarking is a direct embedding of additional content in the original signal. Hence, there should not be any noticeable difference between the original signal and the watermark and it should not be possible to remove the watermark without damaging the original signal [8, 9]. Discrete cosine transform (DCT)-based watermarking, embedding of watermark in the variable DCT blocks, watermarking using discrete wavelet transform (DWT), qualified significant wavelet tree (QSWT) were used by the researchers and academicians for enforcing the watermarks [10–14]. Spatial domain techniques are also very attractive due to the ease in generation of watermark [15, 16]. Many watermarking techniques had used the discrete Fourier transform (DFT) techniques due to its immunity to the noise [17, 18]. Fractal compression technique based on block-based local iteration function system coding was also used for watermarking [19, 20].

2.2 Cloud Computing Scenario

In recent years, various commercial cloud-based models such as 'X as a service (XaaS)' were developed. The 'X' may be storage, hardware or software [21]. Microsoft Azur, Amazon EC2, and Google App Engine are some of the cloud infrastructure and service providers (CSP) [22–24]. Owing to the distributed nature of operation, the data owners are furious about the safety of the data. Hence, there has been an increase in the demand for data authentication, confidentiality, and access control. Figure 2 depicts the basic cloud model for the data access.

A variety of approaches have been adopted by the researchers to safeguard the user generated multimedia data. A modified Diffie–Hallman key exchange protocol has been proposed by Sanka et al. [25].

The authors proposed to maintain a list of authorized users on the owner server. The access may be granted to the users who are listed in the access control list (ACL). However, in this approach, the owner has to maintain a server and manage the ACL; it is not applicable in the personal storage scenario. This method also suffers from the periodic requirement of user credentials for safety purpose. Segmentation of image data approach was proposed by Nourin and Mahesweran [26]. In this method, the image data to be stored on the cloud is segmented and each segment is arbitrarily stored on different anonymous servers in the cloud. This method is not attractive due to the complexities that may arise due to segmentation process and selection of anonymous servers. A cryptography-based encryption technique is reported by Kester et al. [27]. In this method, the data security has been achieved in two stages. First, the advanced encryption standard is used to generate the shared secrete key and this key is then used by the RGB pixel displacement algorithm for encryption process. However, this process is computationally complex when a huge data is to be processed. Data encryption using reversible data hiding (RDH) techniques has been proposed in [28]. This approach requires a pre-allocation of memory before the initiation of image encryption process. A data

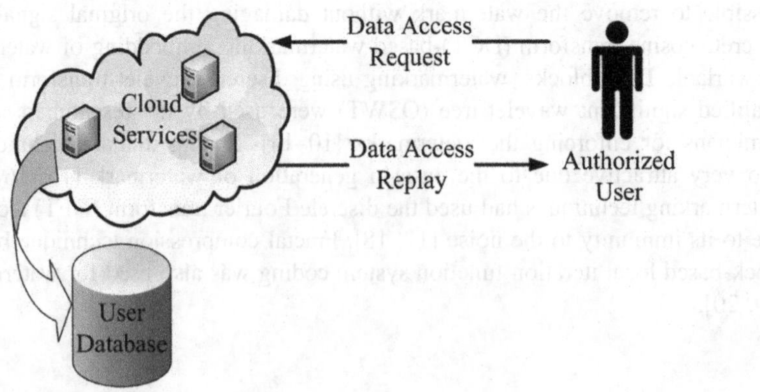

Fig. 2 Data access model in cloud paradigm

Table 1 State-of-the-art data access security in cloud paradigm

Contribution	Methodology	Disadvantage
Sanka et al. [25]	Cryptography	Periodic verification of user credentials
Nourin and Mahesweran [26]	Segmentation	Complex process and requirement of independent servers for storage
Kester et al. [27]	Cryptography and RGB pixel motion estimation	Computationally complex for huge database
Markandey et al. [28]	Reversible data hiding	Pre-allocation of memory before the initiation of image encryption process
Kamara and Lauter [29]	Data sharing algorithm	Fixed data access policy
Dai and Zhou [30]	Identification of authorized user by an access control matrix	Separate access control matrix definition for each user

sharing and searching architecture between the data owners and other users was proposed by Kamara and Lauter [29]. However, this method suffered from a fixed access policy. An access control matrix (ACM) method was proposed by Dai and Zhou [30]. In this approach, each user needs to define an access control matrix (ACM) which is sent to cloud storage provider (CSP). The CSP may check this matrix against each user request. However, in the enterprise scenario, it is difficult to maintain such a huge matrix for each user.

All these methodologies suffers from attributes such as a huge computational overhead, degraded image quality after its retrieval, separate arrangements for authentication and pre-allocation of memory, etc. The storage requirement aspect is also never considered along with the data security by any of the reported methods. Most of the reported methods were dealing with the data storage requirement and security separately. Owing to all these facts, a new methodology is proposed for the security of data access in cloud scenario. The proposed method adopts a two-prong strategy to safeguard the data. First it adds a watermark in the user-generated image/video and then it compresses it before storage. This approach not only secures the user data but also reduces the storage overhead of the multimedia data. Table 1 provides a brief comparison of data access security techniques.

3 The Proposed Methodology

As per the definition of NIST [1], a cloud may be structured with three main layers of operation such as—infrastructure as a service (IaaS); platform as a service (PaaS), and software as a service (SaaS). Each of them is meant for providing some specialized services to the cloud users. The IaaS layer is responsible for storage of data in cloud mechanism. Hence, we suggest a watermarking and a compression/decompression module (WCDM) at each virtual machine (VM) as a software abstraction in IaaS layer of the cloud to store the video data in secured and compressed form. In the proposed approach, CSP will provide a secret key to its registered users. This key may be used for the authorized access of cloud-based data. Whenever, a user or a third-party wish to access the data stored in the cloud, it needs to send a request to the CSP. Upon receipt of the request, CSP will ask the secret key to the user. The access to the requested data may be granted to the user after verification of the required credentials. In the proposed analysis, a fragile image/video watermarking scheme based on DCT along with H.264 based interframe predictive coding is used.

Figure 3 depicts the architecture of the proposed concept. H.264-based interframe predictive coding is used in the proposed analysis for eliminating the temporal and spatial redundancy in video sequences for its effective compression. At the encoder side, the first frame of the image/video sequence is initially considered as the reference frame. The next frame is considered as the incoming frame. A watermark (secret key) is added to each incoming frame before the further processing. In other words, a watermarked image is processed for achieving compression and decompression. The individual watermarked image may be divided into macroblocks of desired dimensions (i.e., 16 × 16, 8 × 8, 4 × 4). In this study, the block matching algorithm (BMA) [31] with a macroblock of dimension 16 × 16 is used as a motion compensation (MC) tool for interframe predictive coding. BMA works on block by block basis. It finds a suitable match for each block in the current frame from the reference frame. The quality of reconstruction of

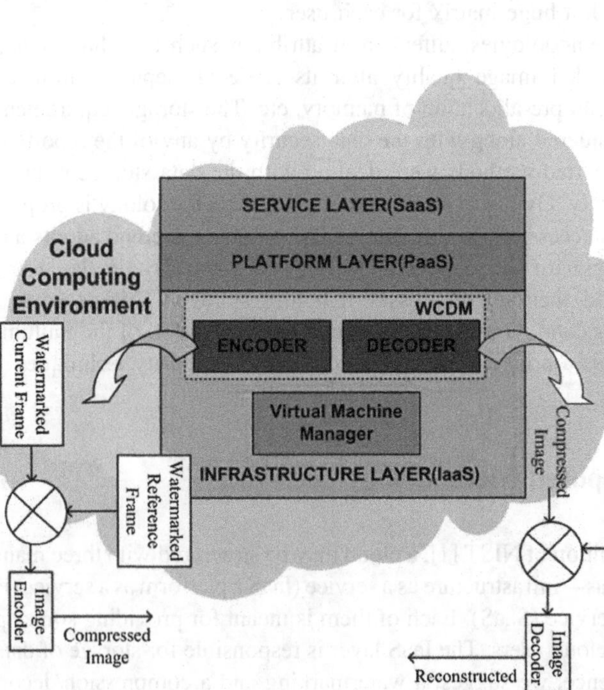

Fig. 3 Proposed data access security model in cloud paradigm

image is evaluated by its PSNR value. According to [32], a reconstructed image with a PSNR value less than 30 dB is considered as worst match.

The proposed mechanism may run as a software part with each VM in the cloud. Whenever, image/video data is to be processed, this module will perform the encoding (addition of watermark and compression) or decoding (decompression and removal of watermark) as per the use requirement. This module will not incur any monitory or tactical burden on the existing mechanism as no additional hardware is required. In this way, it may be the best alternative for securing the data access as well as reduction of data storage size and cost for the image/video data.

4 Results and Discussion

Proposed module adopts two-prong strategy with the image/video data to be stored. First it safeguards the image/video data to be stored by the addition of the watermark and second, it compresses the data on frame by frame basis. For this purpose, the video to be stored is converted into equal-sized frames. Figure 4a depicts the basic process of watermark generation and insertion in an image frame. The watermark is generated by calculating the hash value H of the frame. H is a 128-bit hash value generated by message-digest-5 (MD5) algorithm. This hash function is computed for each block of the image frame. When all the blocks are taken into account, the digital signature algorithm (DSA) is applied on the final H value along with the secrete key (S_k) to generate the watermark. After addition of a watermark, the image frames are further compressed using BMA. For generation of watermark as well as for compression, a 16 × 16 pixel (pels) block size is considered. The purpose of the particular choice is to ensure uniformity as well as to achieve the minimal computational complexity. Figure 4b shows the basic concept of a BMA. It consists of a macroblock of '$M \times N$' size, which is to be searched within a search window of size 'd' in all directions of the macro block.

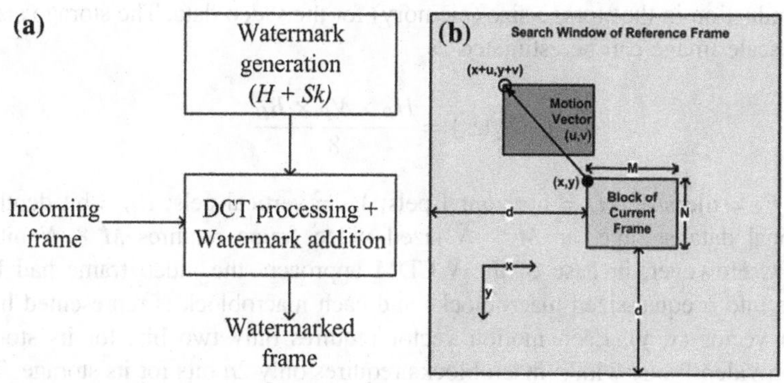

Fig. 4 **a** Generation of watermark. **b** Concept of macroblock matching

The output of a cost function influences the macroblock matching with each other. A macroblock with the least cost is the most excellent equivalent to current block under search. Equations 1–3 provides the computationally efficient cost functions such MAE; MSE; and PSNR. 'CPU TIME,' an inbuilt function of MATLAB, is used to estimate the computational time for the proposed method. In block matching approach, the WCDM divides the incoming frame into a matrix of macroblocks. These macroblocks are compared with the equivalent block and its neighboring blocks in the reference (previous) frame. The process forms a two-dimensional vector, which indicates the motion of a macroblock from one position to another in the reference frame. The motion in the incoming frame is anticipated based on the movement of all the macroblocks in a frame. In order to have an accurate macroblock match, the search range is confined to '*d*' pels (pixels) on all four directions of the equivalent macroblock in the reference frame. This '*d*' is the search range restriction and is proportional to the nature of motion.

$$\mathrm{MAE}(dx, dy) = \frac{1}{M \times N} \sum_{i=0}^{M-1} \sum_{j=0}^{N-1} |C(x, y) - R(x, y)| \tag{1}$$

$$\mathrm{MSE}(dx, dy) = \frac{1}{M \times N} \sum_{i=0}^{M-1} \sum_{j=0}^{N-1} |C(x, y) - R(x, y)|^2 \tag{2}$$

$$\mathrm{PSNR\ (dB)} = 10 \log_{10} \left(\frac{255}{\sqrt{\mathrm{MSE}}} \right) \tag{3}$$

Here $C(x, y)$ = the pels in the current macroblock and $R(x, y)$ = the pels in the reference block. M and N = the size of the macroblock. The experimentations were carried out in private cloud setup as reported in [33]. Figure 5 shows the software abstraction used for the experimentation of the proposed method. Table 2 provides the performance of BMA for the proposed method.

The reconstructed image at the decoder side provides a PSNR value of 33.75 dB which well above the desired value. The principle advantage of the WCDM module is the reduction in the storage size (memory) for the video data. The storage size of a gray scale image can be estimated as

$$F_S(\text{bytes}) = \frac{H_P \times V_P \times B_D}{8} \tag{4}$$

where F_S = file size; H_P = horizontal pels; V_P = vertical pels; B_D = bit depth. In traditional data storage, an $M \times N$ sized video frame requires $M \times N$ bits of memory. However, in case of the WCDM approach, the video frame had been divided into n equal-sized macroblocks and each macroblock is represented by its motion vector (x, y). Each motion vector requires only two bits for its storage. Hence a video frame with n macroblocks requires only $2n$ bits for its storage. Thus the storage requirement has been reduced significantly. Table 3 summarizes the

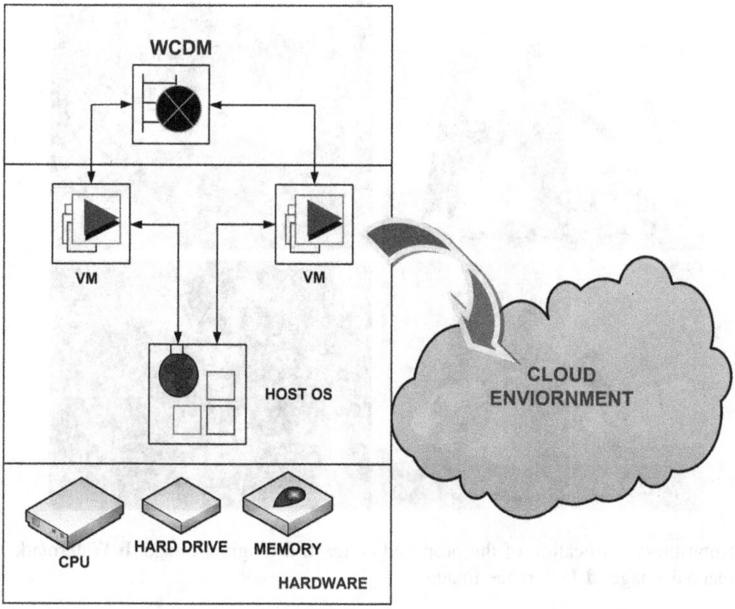

Fig. 5 Cloud environment setup of the proposed concept

Table 2 BMA performance

Algorithm	Block size	CPU time (s)	PSNR (dB)	MSE	MAE
Full Search	16 × 16	18.04	33.75	10.90	3.30

Table 3 Storage requirement

Approach	Block size (pels)	No. of macroblocks	Memory requirement (bytes)
Traditional	–	–	76800
Full search	16 × 16	300	75

storage requirement with the proposed approach for an image/frame size of 320 × 240 pels with 8 bit depth.

As far as security is concerned, the proposed method adopts a two-way approach. First it computes the secure H function for each block which is then combined together for the final H value for a particular frame. Second, it may allow the access only when the secure key has been correctly provided by the user. The principle advantage of the proposed method is its ability to provide the highest security to the stored data. The eavesdropper may not get access or temper the data easily without damaging the original information. This has been possible due to the block by block calculation of the H value. Figure 6 depicts the original image (Fig. 6a), watermark image (Fig. 6b), watermarked image (Fig. 6c) and the difference image of the

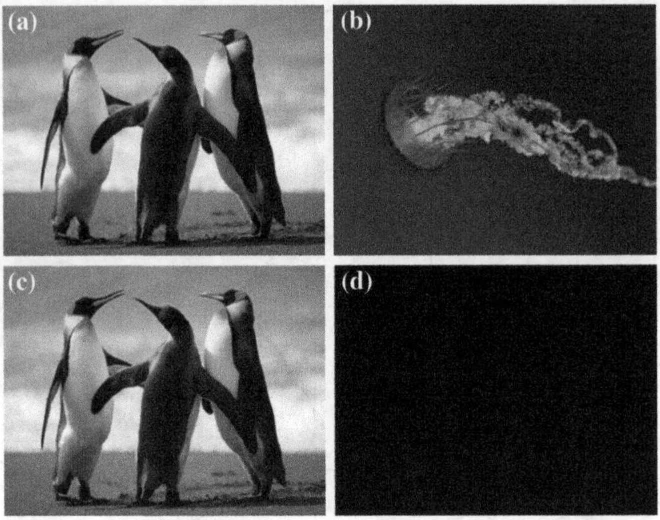

Fig. 6 Robustness verification of the proposed concept. **a** Original image. **b** Watermark image. **c** Watermarked image. **d** Difference image

original, and watermarked image (Fig. 6d). The robustness of the proposed approach is verified by taking the difference between the original and watermarked image. No significant variation is observed between the original image and the watermarked image. Thus the proposed approach may be very useful in the highly insecure cloud environment. Table 4 provides limitations of the existing methodologies of data security in cloud paradigm.

Table 4 Cloud specific limitations of the existing methodologies

Contribution	Data compression	Cloud specific attribute
Sanka et al. [25]	No	Additional mechanism for continuous verification of user credentials
Nourin and Mahesweran [26]	No	Requirement of independent servers for storage
Kester et al. [27]	No	Approach fails against huge database
Markandey et al. [28]	No	Huge pre-allocation of memory and allied monitory overheads
Kamara and Lauter [29]	No	Stucked up due to fixed data access policy
Dai and Zhou [30]	No	Computational overhead is proportional to the number of user
Proposed approach	Yes	Requires minimum storage with data tamper protection

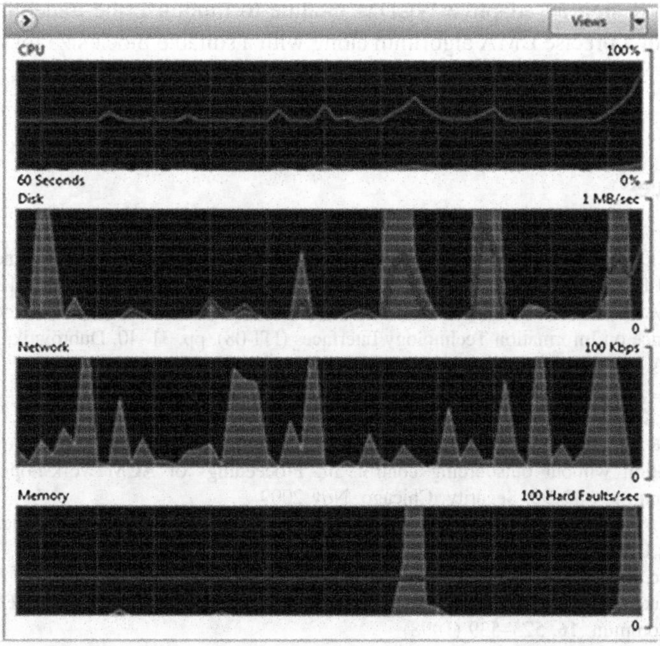

Fig. 7 CPU and memory performance for the proposed module for BMA (16 × 16 block)

Figure 7 shows the cloud performance for the proposed WCDM module with the help of '*resource manager.*' The performance has been analyzed in terms of CPU and memory usage. It has been observed that only 7 % of the CPU capacity and 36 % of the memory capacity has been used by the WCDM module. This indicates that the proposed module works without causing an additional burden on the available resources.

5 Conclusion

The paper reports WCDM, a watermarked, compression, and decompression module for cloud-based video data storage security. This module may be placed in each VM as a software abstraction at IaaS (*SasIaas*) layer of the cloud architecture. The novelty of the present work is the two-prong strategy for storage data security, i.e., addition of watermark on the basis of block by block computation and then the data compression. With the proposed module, the requirement of a dedicated cloud for data processing and storage has been overruled and the multimedia data storage requirement may be reduced with minimal overheads. The proposed module demonstrated a fair QoS in terms of the reconstructed images. Hence, this approach may be the best candidate for the future generation information processing technology. In future, efforts may be initiated to design an adaptive WCDM module to minimize the tradeoff between selections of a precise BMA algorithm along with a suitable block size for a specific application.

References

1. Brown, E.: NIST issues cloud computing guidelines for managing security and privacy. National Institute of Standards and Technology Special Publication 800-144 (2012)
2. Vouk, M.: Cloud computing—issues, research and implementations. In: 30th International Conference on Information Technology Interfaces (ITI-08), pp. 31–40. Dubrovnik, June 2008
3. Weiss, A.: Computing in the clouds. NetWorker **11**(4), 16–25 (2007)
4. Wang, L., Laszewski, G., You, A., et al.: Cloud computing: a perspective study. New Gen. Comput. **28**(2), 137–146 (2010)
5. Chow, R., Golle, P., Jakobsson, M., et al.: Controlling data in the cloud: outsourcing computation without outsourcing control. In: Proceedings of ACM workshop on Cloud Computing, pp. 85–90. Security, Chicago, Nov 2009
6. Modi, C., Patel, D., Borisaniya, B., et al.: A survey of intrusion detection techniques in cloud. J. Netw. Comput. Appl. **36**, 42–57 (2013)
7. Digimarc. http://www.digimarc.com
8. Podilchuk, C., Zeng, W.: Image-adaptive watermarking using visual models. IEEE J. Sel. Areas Commun. **16**, 525–539 (1998)
9. Podilchuk, C., Delp, E.: Digital watermarking algorithms and its applications. IEEE Signal Proc. Mag. **18**(4), 33–46 (2001)
10. Wu, C., Hsieh, W.: Image refining technique using digital watermarking. IEEE Trans. Consumer Elect. **46**(1), 1–5 (2000)
11. Kwon, O., Kim, Y., Park, R.: A variable block-size dot based watermarking method. IEEE Trans. Consumer Elect. **45**(4), 1221–1229 (1999)
12. Langelaar, G., Lagendijk, R.: Optimal differential energy watermarking of DCT encoded images and video. IEEE Trans. Image Proc. **10**(1), 148–158 (2001)
13. Huang, J., Yang, C.: Image digital watermarking algorithm using multi-resolution wavelet transform. In: Proceedings of IEEE International Conference on Systems, Man and Cybernetics, pp. 2977–82, 2004
14. Hsieh, M., Tseng, D., Huang, Y.: Hiding digital watermarks using multiresolution wavelet transform. IEEE Trans. Ind. Elec. **48**(5), 875–882 (2001)
15. Megalingam, R., Nair, M., Srikumar et al.: Performance comparison of novel, robust spatial domain digital image watermarking with the conventional frequency domain watermarking techniques. In: International Conference on Signal Acquisition and Processing (ICSAP-10), pp. 349–353. Bangalore, India, Feb 2010
16. Lancini, R., Mapelli, F., Tubaro, S.: A robust video watermarking technique in the spatial domain. 4th EURASIP-IEEE Region 8 International Symposium on Video/Image Processing and Multimedia Communications, pp. 251–256, 2002
17. Ruanaidh, J., Pun, T.: Rotation, scale and translation invariant spread spectrum digital image watermarking. Signal Process. **66**(3), 303–317 (1998)
18. Fallahpour, M., Shirmohammadi, S., Semsarzadeh, M., Jiying, Z.: Tampering detection in compressed digital video using watermarking. IEEE Trans. Instrum. Meas. 63(5), 1057–1072 (2014)
19. Keyvanpour, M., Farnoosh, M.: A new encryption method for secure embedding in image watermarking. In: 3rd International Conference on Advanced Computer Theory and Engineering (ICACTE-10), vol. 2, pp. 403–407. Chengdu, Aug 2010

20. Cox, I., Kilian, J., Leighton, T., Shamoon, T.: Secure spread spectrum watermarking for multimedia. NEC Research Institute, Princeton, NJ, Tech. Rep. 95–10, 1995
21. Armbrust, M., Fox, A., Griffith, R. et al.: Above the Clouds: A Berkeley View of Cloud Computing. University of California, Berkeley, Tech. Rep. USB-EECS-2009-28, Feb 2009
22. Chappell, D.: Introducing the Azure Service Platform. White paper, Oct 2008
23. Amazon EC2 and S3. http://aws.amazon.com/
24. Google App Engine. http://code.google.com/appengine/
25. Sanka, S., Hota, C., Rajarajan, M.: Secure data access in cloud computing. In: IEEE Fourth International Conference on Internet Multimedia Systems Architecture and Application (IMSAA-10), pp. 1–6. Bangalore, India, Dec 2010
26. Nourian, A., Maheswaran, M.: Using segmentation for confidentiality aware image storage and retrieval on clouds. In: Communication and Information System Security Symposium (Globecom-12), pp. 758–763. Anaheim, CA (2012)
27. Kester, Q., Nana, L., Pascu, A.: A novel cryptographic encryption technique for securing digital images in the cloud using AES and RGB pixel displacement. European Modeling Symposium, pp. 293–298, 2013
28. Markandey, A., Moghe, S., Bhute, Y., Honale, S.: An image encryption mechanism for data security in clouds. In: IEEE Global Humanitarian Technology Conference—South Asia Satellite (GHTC-SAS-14), pp. 227–231. Trivandrum, India, Sept 2014
29. Kamara, S., Lauter, K.: Cryptographic cloud storage. Financial Cryptography and Data Security. Lecture Notes in Computer Science, vol. 6054, pp. 136–149 (2010)
30. Dai, L., Zhou, Q.: A PKI-based mechanism for secure and efficient access to outsourced data. In: Proceedings of Second International Conference on Networking and Digital Society (ICNDS-10). Wenzhou, vol. 1, pp. 640–643 (2010)
31. Barjatya, A.: Block matching algorithms for motion estimation. DIP 6620 Spring 2004 Final Project Paper, pp. 1–6 (2004)
32. Bruckmann, A., Hammerle, J., Reichl, M., Uhl, A.: Hybrid fractal/wavelet image compression in a high performance computing environment. In: High-Performance Computing and Networking, Lecture Notes in Computer Science, vol. 1225, pp. 115–126 (1997)
33. Deshpande, P., Sharma, S., Peddoju, S.: Implementation of a private cloud: a case study. In: Advances in Intelligent Systems and Computing, vol. 259, pp. 635–647 (2013)

Axisymmetric Vibrations of Variable Thickness Functionally Graded Clamped Circular Plate

Neha Ahlawat and Roshan Lal

Abstract The axisymmetric vibrations of functionally graded clamped circular plate have been analysed on the basis of classical plate theory. The material properties, i.e. Young's modulus and density vary continuously through the thickness of the plate, and obey a power law distribution of the volume fraction of the constituent materials. A semi-analytical technique, i.e. differential transform method has been employed to solve the differential equation governing the equation of motion. The effect of various plate parameters, i.e. volume fraction index g and taper parameter γ have been studied on the first three modes of vibration. Three-dimensional mode shapes for the first three modes of vibration have been presented. A comparison of results with those available in the literature has been given.

Keywords Functionally graded circular plates · Differential transform method · Axisymmetric vibrations

1 Introduction

The wide applications of functionally graded materials (FGMs) in space vehicles, nuclear reactor, defence industries and chemical plants have attracted many researchers throughout the world. FGMs are microscopically inhomogeneous materials whose mechanical properties vary continuously in one or more directions [1]. In a metal-ceramic FGM, the metal-rich side is placed in regions where mechanical properties, such as toughness need to be high whereas the ceramic-rich side which has low thermal conductivity and can withstand high temperatures is placed in regions of large temperature gradients. Due to these characteristics, FGM

Neha Ahlawat (✉) · Roshan Lal
Indian Institute of Technology Roorkee, Roorkee, India
e-mail: ahlawatneha@gmail.com

Roshan Lal
e-mail: rlatmfma@iitr.ernet.in

M. Pant et al. (eds.), *Proceedings of Fifth International Conference on Soft Computing for Problem Solving*, Advances in Intelligent Systems and Computing 436, DOI 10.1007/978-981-10-0448-3_21

plate-type components of different geometries are extensively used as structural elements in various fields of modern science and technology.

A lot of studies have been concerned dealing with the vibration characteristics of FGM plates and reported in Refs. [2–12], to mention a few. Out of these, Jha et al. [2] have presented a critical review of recent research on functionally graded plates till 2012. Ferreira et al. [3] used collocation method to analyse the free vibrations of functionally graded rectangular plates of various aspect ratios. Zhao et al. [4] used element-free *kp*-Ritz method for free vibration analysis of rectangular and skew plates with different boundary conditions taking four types of functionally graded materials on the basis of first-order shear deformation theory. Liu et al. [5] have analysed the free vibration of FGM rectangular plates with in-plane material inhomogeneity using Fourier series expansion and a particular integration technique on the basis of classical plate theory. Free vibration analysis of functionally graded thick annular plates with linear and quadratic thickness variation along the radial direction is investigated by Tajeddini and Ohadi [6] using the polynomial-Ritz method. The vibration behaviour of rectangular FG plates with non-ideal boundary conditions has been studied by Najafizadeh et al. [7] using Levy method and Lindstedt–Poincare perturbation technique. The free vibrations of FGM circular plates of variable thickness under axisymmetric condition have been analysed by Shamekhi [8] using a meshless method in which the point interpolation approach is employed for constructing the shape functions for Galerkin weak form formulation. Chakraverty and Pradhan [9] have applied Rayleigh–Ritz method to study the free vibrations of exponentially graded rectangular plates subjected to different combinations of boundary conditions in thermal environment using Kirchhoff's plate theory. Recently, Dozio [10] has derived first-known exact solutions for free vibration of thick and moderately thick FGM rectangular plates with at least on pair of opposite edges simply supported on the basis of a family of two-dimensional shear and normal deformation theories with variable order. Very recently, the natural frequencies of FGM nanoplates are analysed by Zare et al. [12] for different combinations of boundary conditions by introducing a new exact solution method.

The aforementioned survey of the literature reveals that there is almost no work on the vibration analysis of functionally graded circular plate of variable thickness using classical plate theory. Keeping this in view, the present paper analyses the axisymmetric vibrations of FGM circular plate of linearly varying thickness based on classical plate theory. Differential transform method (DTM) which is a semi-analytical technique has been employed to obtain the frequency equation. This resulting equation has been solved using MATLAB to get the frequencies. The material properties, i.e. Young's modulus and density are assumed to vary in the thickness direction according to a power law distribution. The effect of various parameters such as volume fraction index g and taper parameter γ on the natural frequencies have been illustrated for the first three modes of vibration. Three-dimensional modes shapes for a specified plate and for the first three modes of vibration have been plotted. For the validation of the present results, a comparison of results with the existing literature has been made which ensure the versatility of the present technique.

2 Mathematical Formulation

Consider a two-directional functionally graded circular plate of radius a, thickness h, mass density ρ and subjected to hydrostatic in-plane tensile force N_0. Let the plate be referred to a cylindrical polar coordinate system (R, θ, z), $z = 0$ being the middle plane of the plate. The top and bottom surfaces are $z = +h/2$ and $z = -h/2$, respectively. The line $R = 0$ is the axis of the plate. The equation of motion governing transverse axisymmetric vibration of the present model (Fig. 1) is given by [13]

$$\begin{aligned} & Dw_{,RRRR} + \frac{2}{R}\left[D + R\,D_{,R}\right]w_{,RRR} + \frac{1}{R^2}\left[-D + R\,(2+\nu)D_{,R} + R^2 D_{,RR}\right]w_{,RR} \\ & + \frac{1}{R^3}\left[D - R\,D_{,R} + R^2\nu\,D_{,RR}\right]w_{,R} + \rho\,h\,w_{,tt} = 0, \end{aligned} \tag{1}$$

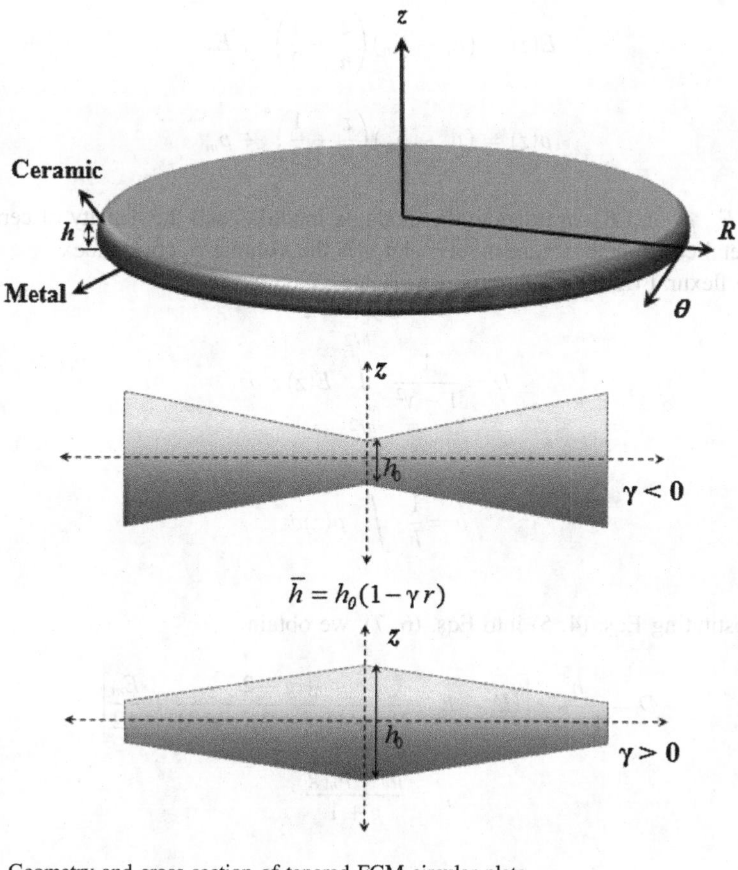

Fig. 1 Geometry and cross-section of tapered FGM circular plate

where w is the transverse deflection, D the flexural rigidity and ν the Poisson's ratio and a comma followed by a suffix denotes the partial derivative with respect to that variable.

For a harmonic solution, the deflection w can be expressed as

$$w(R,t) = W(R)e^{i\omega t}, \tag{2}$$

where ω is the radian frequency. Equation (1) reduces to

$$\begin{aligned} &DW_{,RRRR} + \frac{2}{R}\left[D + R D_{,R}\right] W_{,RRR} + \frac{1}{R^2}\left[-D + R(2+\nu)\, D_{,R} + R^2 D_{,RR}\right] W_{,RR} \\ &+ \frac{1}{R^3}\left[D - R D_{,R} + R^2 \nu\, D_{,RR}\right] W_{,R} - \rho\, h\, \omega^2 W = 0. \end{aligned} \tag{3}$$

Assuming that the top and bottom surfaces of the plate are ceramic and metal-rich, respectively, for which the variations of the Young's modulus $E(z)$ and the density $\rho(z)$ in the thickness direction are taken as

$$E(z) = (E_c - E_m)\left(\frac{z}{h} + \frac{1}{2}\right)^g + E_m \tag{4}$$

$$\rho(z) = (\rho_c - \rho_m)\left(\frac{z}{h} + \frac{1}{2}\right)^g + \rho_m \tag{5}$$

where E_c, ρ_c and E_m, ρ_m denote the Young's modulus and the density of ceramic and metal constituents, respectively, and g is the volume fraction index.

The flexural rigidity and mass density are given as

$$D = \frac{1}{1-\nu^2} \int_{-h/2}^{h/2} E(z)\, z^2 dz \tag{6}$$

$$\rho = \frac{1}{h} \int_{-h/2}^{h/2} \rho(z) dz \tag{7}$$

Substituting Eqs. (4, 5) into Eqs. (6, 7), we obtain

$$D = \frac{h^3}{1-\nu^2}\left[(E_c - E_m)\frac{g^2 + g + 2}{4(g+1)(g+2)(g+3)} + \frac{E_m}{12}\right] \tag{8}$$

$$\rho = \frac{\rho_c + \rho_m g}{g+1} \tag{9}$$

Introducing the non-dimensional variables $r = R/a, f = W/a, \bar{h} = h/a$, Eq. (3) now reduces to

$$D f_{,rrrr} + \frac{2}{r}\left[D + r D_{,r}\right] f_{,rrr} + \frac{1}{r^2}\left[-D + r(2+\nu) D_{,r} + r^2 D_{,rr}\right] f_{,rr} + \frac{1}{r^3}\left[D - r D_{,r} + r^2 \nu D_{,rr}\right] f_{,r} = \rho\, a^4 \omega^2 f\, \bar{h} \tag{10}$$

Assuming the linear variation in the thickness, i.e. $\bar{h} = h_0(1 - \gamma r)$, γ being the taper parameter and h_0 is the non-dimensional thickness of the plate at the centre. Substituting the values of D and ρ from Eqs. (8, 9) into Eq. (10), we get

$$\begin{aligned} & r^3(1-\gamma r)^3 B f_{,rrrr} + 2r^2\left((1-\gamma r)^3 - 3\gamma r(1-\gamma r)^2\right) B f_{,rrr} \\ & + rB\left(-(1-\gamma r)^3 - 3\gamma r(2+\nu)(1-\gamma r)^2 + 6r^2\alpha^2(1-\gamma r)\right) f_{,rr} \\ & + B\left((1-\gamma r)^3 + 3r\alpha(1-\gamma r)^2\right) f_{,r} = r^3 \Omega^2 A(1-\gamma r) f \end{aligned} \tag{11}$$

where

$$D = D^* B(1-\gamma r)^3 a^3, \quad \Omega^2 = \frac{\rho_c h_0 a^4}{D^*}\omega^2,\ D^* = \frac{E_c h_0^3}{12(1-\nu^2)},$$
$$A = \left(\frac{\rho_c + \rho_m g}{\rho_c(g+1)}\right), \quad B = \left[3\left(1 - \frac{E_m}{E_c}\right)\frac{g^2+g+2}{(g+1)(g+2)(g+3)} + \frac{E_m}{E_c}\right]$$

Equation (11) is a fourth-order differential equation with variable coefficients whose exact solution is not possible. The approximate solution with appropriate boundary and regularity conditions has been obtained employing differential transform method.

2.1 *Boundary Conditions:* Clamped Edge

$$f(1) = 0,\ \frac{df}{dr}\Big|_{r=1} = 0 \tag{12}$$

Regularity conditions at the centre (r = 0) of the plate-

$$\frac{df}{dr}\Big|_{r=0} = 0, Q_r\big|_{r=0} = \left[D\left(\frac{d^3 f}{d r^3} + \frac{1}{r}\frac{d^2 f}{d r^2} - \frac{1}{r^2}\frac{df}{dr}\right) + D_{,r}\left(\frac{d^2 f}{d r^2} + \frac{\nu}{r}\frac{df}{dr}\right)\right]_{r=0} = 0 \tag{13}$$

where Q_r the radial shear force.

3 Method of Solution: Differential Transform Method

Following the description of the method given in Ref. [11], the transformed form of the governing differential Eq. (11) around $r_0 = 0$ will be written as

$$\begin{aligned} F_{k+1} = \frac{1}{(k^2-1)^2} \cdot & [3\gamma k(k-1)(k^2-k-1+\nu) F_k \\ & + \{3\gamma(k-4)(k-3)(k-2)(k-1) - 6\nu\gamma^2(k-2)(k-1) \\ & - 3\gamma^2(k-1)(6k^2-25k+2\nu k-2\nu)\} F_{k-1} \\ & + \gamma^3(k-2)\{k^3-4k^2+(2+3\nu)k-3\nu+1\}F_{k-2} \\ & + \frac{\Omega^2 A}{B} F_{k-3} - \gamma\Omega^2 \frac{A}{B} F_{k-4}] \end{aligned} \quad (14)$$

The transformed form of boundary and regularity conditions will be

$$\textit{Clamped edge condition}: \sum_{k=0}^{n} F_k = 0, \sum_{k=0}^{n} k F_k = 0 \quad (15)$$

$$\textit{Regularity condition}: F_1 = 0, F_3 = \frac{2}{3}\gamma(1+\nu)F_2 \quad (16)$$

4 Frequency Equation

Now, applying the boundary condition and regularity condition (15, 16) on the resulted F_k expressions (14), we get the following equations:

$$\begin{aligned} \Phi_{11}^{(m)}(\Omega)F_0 + \Phi_{12}^{(m)}(\Omega)F_2 = 0 \\ \Phi_{21}^{(m)}(\Omega)F_0 + \Phi_{22}^{(m)}(\Omega)F_2 = 0 \end{aligned} \quad (17)$$

where $\Phi_{11}^{(m)}$, $\Phi_{12}^{(m)}$, $\Phi_{21}^{(m)}$ and $\Phi_{22}^{(m)}$ are polynomials in Ω of degree m where $m = 2n$. Equation (17) can be expressed in matrix form as follows:

$$\begin{bmatrix} \Phi_{11}^{(m)}(\Omega) & \Phi_{12}^{(m)}(\Omega) \\ \Phi_{21}^{(m)}(\Omega) & \Phi_{22}^{(m)}(\Omega) \end{bmatrix} \begin{Bmatrix} F_0 \\ F_2 \end{Bmatrix} = \begin{Bmatrix} 0 \\ 0 \end{Bmatrix} \quad (18)$$

For a non-trivial solution of Eq. (18), the frequency determinant must vanish and hence

$$\begin{vmatrix} \Phi_{11}^{(m)}(\Omega) & \Phi_{12}^{(m)}(\Omega) \\ \Phi_{21}^{(m)}(\Omega) & \Phi_{22}^{(m)}(\Omega) \end{vmatrix} = 0 \quad (19)$$

5 Numerical Results and Discussion

The frequency Eq. (19) provides the values of the frequency parameter Ω. The lowest three roots of this equation have been obtained using MATLAB to investigate the influence of taper parameter γ and volume fraction index g on the frequency parameter Ω. In the present analysis, the values of Young's modulus and density for aluminium as metal and alumina as ceramic constituents are taken from [11], as follows:

$$E_m = 70\,\text{GPa},\ \rho_m = 2{,}702\ \text{kg/m}^3 \quad \text{and} \quad E_c = 380\,\text{GPa},\ \rho_c = 3{,}800\ \text{kg/m}^3$$

The variation in the values of Poisson's ratio is assumed to be negligible all over the plate and its value is taken as $\nu = 0.3$. From the literature, the values of parameters are taken as

$$\text{Volume fraction index}\ g = 0, 1, 2, 3, 4, 5;$$
$$\text{Taper parameter}\ \gamma = -0.5,\ -0.3, -0.1,\ 0.1,\ 0.3,\ 0.5.$$

In order to choose an appropriate value of the number of terms 'n', a computer program has been developed and run for various values of g and γ. The convergence of frequency parameter Ω for the first three modes of vibration for a specified plate taking $g = 5, \gamma = -0.5$ is shown in Table 1, as maximum deviations were

Table 1 Convergence study for first three modes of vibration for $g = 5$, $\gamma = -0.5$

No. terms n	I	II	III
10	10.6033	38.0750	82.9112
11	**10.6033**	38.0686	83.5110
12	–	38.0678	83.5261
13	–	38.0683	83.5013
14	–	**38.0683**	83.5058
15	–	–	83.5061
16	–	–	83.5058
17	–	–	**83.5058**
18	–	–	–

Table 2 Values of frequency parameter Ω

g	Modes	γ					
		−0.5	−0.3	−0.1	0.1	0.3	0.5
0	I	14.3021	12.6631	11.0301	9.4027	7.7783	6.1504
	II	51.3480	46.7813	42.1337	37.3763	32.4610	27.3002
	III	112.6360	103.4123	93.9486	84.1680	73.9467	63.0611
1	I	11.8983	10.5347	9.1762	7.8223	6.4710	5.1166
	II	42.7176	38.9184	35.0520	31.0941	27.0050	22.7117
	III	93.7044	86.0310	78.1579	70.0212	61.5179	52.4620
2	I	11.3730	10.0696	8.7711	7.4770	6.1853	4.8907
	II	40.8316	37.2002	33.5045	29.7214	25.8128	21.7090
	III	89.5675	82.2329	74.7074	66.9299	58.8020	50.1458
3	I	11.0756	9.8063	8.5418	7.2815	6.0235	4.7628
	II	39.7640	36.2275	32.6284	28.9442	25.1378	21.1413
	III	87.2254	80.0827	72.7539	65.1798	57.2645	48.8346
4	I	10.8269	9.5861	8.3500	7.1180	5.8883	4.6559
	II	38.8712	35.4141	31.8959	28.2944	24.5735	20.6667
	III	85.2671	78.2847	71.1205	63.7164	55.9788	47.7382
5	I	10.6033	9.3881	8.1775	6.9709	5.7667	4.5597
	II	38.0683	34.6826	31.2370	27.7099	24.0659	20.2398
	III	83.5058	76.6676	69.6514	62.4003	54.8225	46.7521

Table 3 Comparison of frequency parameter Ω for $g = 0, \gamma = 0$

Ref.	First mode	Second mode	Third mode
Present	10.2158	39.7711	89.1041
Leissa [13]	10.2158	39.771	89.104
Wu et al. [14]	10.216	39.771	89.104

observed for this data. The frequency parameter Ω converges with the increasing value of n. The value of n has been fixed as 18, as there was no further improvement in the values of Ω even at the fourth place of decimal.

Numerical results have been given in Tables 2 and 3 and presented in Figs. 2, 3 and 4. The effect of volume fraction index g on the frequency parameter Ω for three different values of taper parameter γ has been demonstrated in Fig. 2. It has been observed that the value of frequency parameter Ω decreases with the increasing values of g whatever be the value of taper parameter γ. The corresponding rate of decrease is higher for smaller values of g (< 2) as compared to the higher values of g (> 3). Further, it increases with the increase in the number of modes.

To study the effect of taper parameter γ on the frequency parameter Ω, a graph has been plotted for two different values of volume fraction index $g = 0, 5$ in Fig. 3.

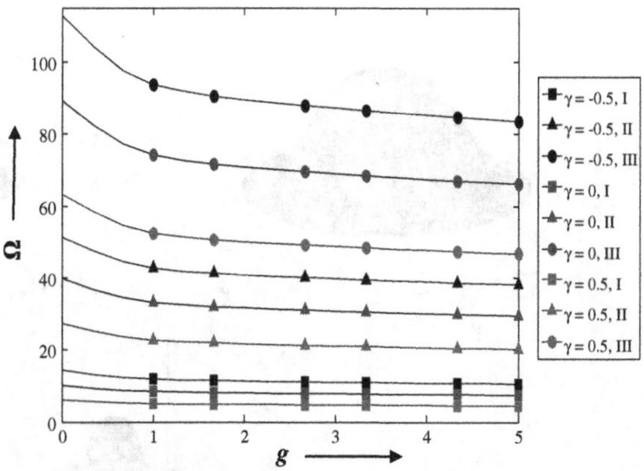

Fig. 2 Frequency parameter Ω versus volume fraction index *g*

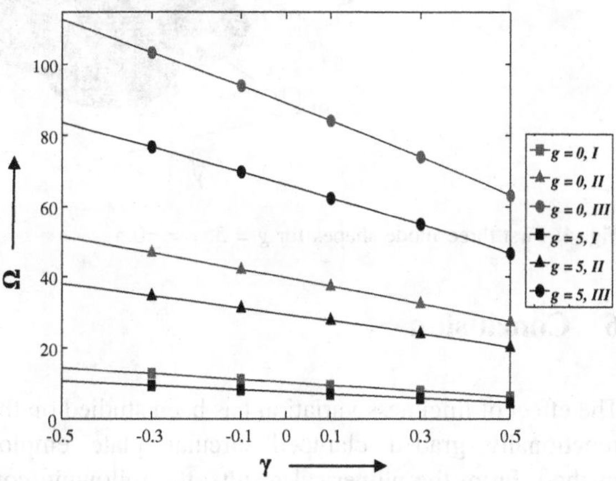

Fig. 3 Frequency parameter Ω versus taper parameter γ

It has been noticed that the frequency parameter Ω decreases as the plate becomes thinner and thinner towards the outer edge. This effect is more pronounced for isotropic plate ($g = 0$) as compared to FGM plate ($g = 5$) and increases with the increase in the number of modes. Three-dimensional mode shapes for a specified plate, i.e. $g = 5, \gamma = -0.5$ for the first three modes of vibration has been presented in Fig. 4. A comparison of frequency parameter Ω for an isotropic plate has been given in Table 3. A close agreement of the results shows the versatility of the present technique.

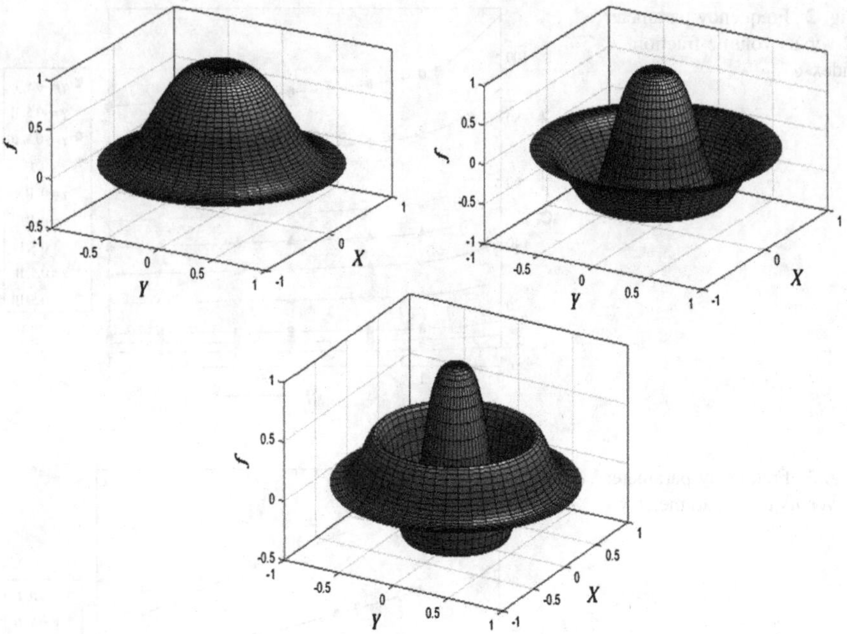

Fig. 4 First three mode shapes for $g = 5$, $\gamma = -0.5$

6 Conclusion

The effect of thickness variation has been studied on the axisymmetric vibrations of functionally graded clamped circular plate employing differential transform method. From the numerical results, the following conclusions can be made:

- The frequency parameter decreases with the increasing values of volume fraction index. From this fact, it can be observed that the frequencies for an isotropic plate ($g = 0$) are higher than that for the corresponding FGM plate ($g > 0$) which shows the superiority of the FGM plates over isotropic plate.
- The frequency parameter decreases with the increasing values of taper parameter, i.e. the frequency parameter increases as the plate become thicker and thicker towards the outer boundary of the plate.

Acknowledgements The authors wish to express their sincere thanks to the learned reviewers for their valuable suggestions in improving the paper. One of the authors, Neha Ahlawat, is thankful to University Grants Commission, India, for providing the research fellowship.

References

1. Suresh, S., Mortensen, A.: Fundamentals of Functionally Graded Materials. Maney, London (1998)
2. Jha, D.K., Kant, T., Singh, R.K.: A critical review of recent research on functionally graded plates. Compos. Struct. **96**, 833–849 (2013)
3. Ferreira, A.J.M., Batra, R.C., Roque, C.M.C., Qian, L.F., Jorge, R.M.N.: Natural frequencies of functionally graded plates by a meshless method. Compos. Struct. **75**(1), 593–600 (2006)
4. Zhao, X., Lee, Y.Y., Liew, K.M.: Free vibration analysis of functionally graded plates using the element-free *kp*-Ritz method. J. Sound Vib. **319**(3), 918–939 (2009)
5. Liu, D.Y., Wang, C.Y., Chen, W.Q.: Free vibration of fgm plates with in-plane material inhomogeneity. Compos. Struct. **92**(5), 1047–1051 (2010)
6. Tajeddini, V., Ohadi, A.: Three-Dimensional vibration analysis of functionally graded thick, annular plates with variable thickness via polynomial-Ritz method. J. Vib. Control (2011). 1077546311403789
7. Najafizadeh, M.M., Mohammadi, J., Khazaeinejad, P.: Vibration characteristics of functionally graded plates with non-ideal boundary conditions. Mech. Adv. Mater. Struc. **19**(7), 543–550 (2012)
8. Shamekhi, A.: On the use of meshless method for free vibration analysis of circular FGM plate having variable thickness under axisymmetric condition. Inter. J. Res. Rev. Appl. Sci. **14**(2), 257–268 (2013)
9. Chakraverty, S., Pradhan, K.K.: Free vibration of exponential functionally graded rectangular plates in thermal environment with general boundary conditions. Aerosp. Sci. Technol. **36**, 132–156 (2014)
10. Dozio, L.: Exact free vibration analysis of lévy fgm plates with higher-order shear and normal deformation theories. Compos. Struct. **111**, 415–425 (2014)
11. Lal, R., Ahlawat, N.: Axisymmetric vibrations and buckling analysis of functionally graded circular plates via differential transform method. Eur. J. Mech. A-Solid **52**, 85–94 (2015)
12. Zare, M., Nazemnezhad, R., Hosseini-Hashemi, S.: Natural frequency analysis of functionally graded rectangular nanoplates with different boundary conditions via an analytical method. Meccanica **50**, 1–18 (2015)
13. Leissa, A.W.: Vibration of Plates, vol. 160. NASA SP, Washington (1969)
14. Wu, T.Y., Wang, Y.Y., Liu, G.R.: Free vibration analysis of circular plates using generalized differential quadrature rule. Comput. Meth. Appl. Mech. Eng. **191**(46), 5365–5380 (2002)

Performance Evaluation of Geometric-Based Hybrid Approach for Facial Feature Localization

Sourav Dey Roy, Priya Saha, Mrinal Kanti Bhowmik and Debanjana Debnath

Abstract Nowadays, facial recognition technology (FRT) has come into focus because of its various applications in security and non-security perspective. It provides a secure solution for identification and verification of person identity. Accurate localization of facial features plays a significant role for many facial analysis applications including biometrics and emotion recognition. There are several factors that make facial feature localization a challenging problem. Facial expression is one of the influential factors of FRT. The paper proposes a new geometric-based hybrid technique for automatic localization of facial features in frontal and near-frontal neutral and expressive face images. A graphical user interface (GUI) is designed that could automatically localize 16 landmark points around eyes, nose, and mouth that are mostly affected by the changes in facial muscles. The proposed system has been tested on widely used JAFFE and Bosphorous database. Also, the system is tested on DeitY-TU face database. The performance of the proposed method has been done in terms of error measures and accuracy. The detection rate of the proposed method is 96.03 % on JAFFE database, 94.06 % on DeitY-TU database, and 94.21 % on Bosphorous database.

Keywords Biometrics · Face recognition technology (FRT) · Facial feature localization · Image processing

S.D. Roy (✉) · P. Saha · M.K. Bhowmik · D. Debnath
Department of Computer Science & Engineering, Tripura University
(A Central University), Agartala, Tripura, India
e-mail: souravdeyroy49@gmail.com

P. Saha
e-mail: priyasaha.cse@gmail.com

M.K. Bhowmik
e-mail: mkb_cse@yahoo.co.in

D. Debnath
e-mail: debanjanadebnath24@gmail.com

M. Pant et al. (eds.), *Proceedings of Fifth International Conference on Soft Computing for Problem Solving*, Advances in Intelligent Systems and Computing 436, DOI 10.1007/978-981-10-0448-3_22

1 Introduction

Facial feature localization is defined as the detection and localization of certain characteristic points that play a distinctive role on the face graph. Accurate localization of facial features plays a significant role in many face image analyses such as face recognition, facial expression understanding, animation, face tracking, etc. [1]. Among the successful applications of image analyses, face recognition technology (FRT) has been largely investigated for the past two decades. In the perspective of law and non-law enforcement, FRT is more convenient than other techniques like passwords, fingerprints, and others because of its individuality and noncontactness. The face recognition can be broadly classified into two different categories: holistic and feature-based approach [2]. The holistic-based approach of face recognition uses the whole image and considers the global patterns of the face. Whereas in feature-based approach of face recognition, the system works on some extracted features of the face instead of the whole face. Automatic localization of facial features is a difficult task, which faces all the difficulties of face recognition, such as occlusion, illumination, expression, pose, and camera resolution [3]. Some important landmark points that are mostly affected by the activity of muscles in human faces are eye corners, eyebrow corners, and mouth points. These landmark points are significant to identify the face that may provide applications for surveillance in criminal identification and also for finding missing peoples in public places.

The paper proposes a newly framed hybrid approach for facial feature localization. The proposed work is a combination of methods, which decreases the computational burden and also advances the accuracy by making the method flexible to changes in expression. Also graphical user interface (GUI)-based software is designed that could automatically localize 16 landmark points on the face. The strength of the system is assessed by testing on JAFFE database, DeitY-TU database, and Bosphorous database. This paper also explores a rigorous literature survey and comparison of our method with various techniques proposed by various researchers.

The whole paper is organized as; Sect. 2 describes the literature survey on facial landmark detection. Section 3 gives a brief description of the database used for the experiment. Section 4 explains the proposed methodology of automatic facial feature localization. In Sect. 5, a performance evaluation measure has been illustrated. Section 6 reports the experimental results with discussion of the proposed methodology. Section 7 describes the comparison of the proposed methodology with other techniques developed by various researchers. Section 8 gives an overview of graphical user interface (GUI) design of the proposed landmark methodology. And finally, Sect. 9 concludes the paper.

2 Related Work

Many approaches to facial analysis rely on robust and automatic facial landmarking to correctly function. But the localization of facial landmark is still an open and difficult problem. In the recent years, the research communities have sparked off thunder in their research for automatic localization of facial features that immensely varies with the changes in facial expressions. Table I shows some recent landmarking approaches in the literature. In [4], Li et al. proposed a multitask sparse representation-based fine-grained matching algorithm, which accounts for the average reconstruction error of probe face descriptors sparsely represented by a large dictionary of gallery descriptors in identification. In [5], Dibeklioglu et al. proposed a statistical method for automatic localization of facial landmark. The landmarking uses a Gabor wavelet features on the coarse scale is complemented with a structure analysis shape and fine tuning shape. In [6], Huang et al. used 2D Gabor filter to extract the features from a given face image. Next, face mapping is carried out using weighted warping procedure. Finally, proposed an improved ASM method to capture the features and locate the frontal face image in order to efficiently advance the convergence performance and accurately locate feature points. In [7], Valstar et al. developed a graph-based feature point detection system for detection of 22 landmark points. The method used support vector regression of Haar-like features and Markov random fields where the search space is constrained. The algorithm was tested on visual face images of FERET, MMI, and BioID database and obtained 94.75 % accuracy. Liu et al. [8] proposed an adaptive algorithm that uses generic active appearance model (AAM) and subject- specific appearance model together for detecting 72 facial feature points. Zhao et al. [9] used Gabor feature to align 13 control points on the face and further 83 points are generated by constrained profile and flexible shape models. In [10], Sohail et al. proposed a method for detection of 18 landmark points in the face based on Anthropometric Face Model. The method obtained success rate of 90.44 % on JAFFE database. Gizatdinova et al. [11] proposed a method for feature-based landmarking on extracting oriented edges and constructing edge maps at two resolution level. In [12], Vukadinovic et al. proposed a method that uses Gabor feature-based boosted classifiers. In this approach, the detected face is divided into region of interest (ROI). Then based on grayscale texture information and Gabor wavelet features individual GentleBoost templates are used to detect landmarks within the relevant ROI independently.

3 Database Description

In order to develop a system for automatic facial feature localization, a database is required which is full of facial images under various lightning conditions and camera positions. For this experiment, we used frontal images of two most widely used databases, i.e., JAFFE database and Bosphorous database. Also the experiment was conducted on DeitY-TU face database which was created in Biometrics Laboratory of Tripura University. The brief description of these three databases are given below.

3.1 DeitY-TU Face Database

The DeitY-TU face database is a visual face image database [13]. The database was created in Biometrics Laboratory of Computer Science and Engineering Department, Tripura University (A Central University), India. The database contains the face images of different tribal and nontribal people of Seven North Eastern states of India. These face images are taken under strictly control conditions. The database contains total 96,995 images of 524 subjects. The database contains a total of eight expressions; including neutral expression, closed eye, and six basic expressions; four different types of illumination variations and images with glasses; and each of these variations are being clicked concurrently from five different angles to provide pose variations. The face images collected from different states and from different communities illustrate the existence of facial structural differences.

3.2 Japanese Female Facial Expression (JAFFE) Database

The Japanese Female Expression (JAFFE) Database [14] is also a visual face image database. The JAFFE database was designed at the Department of Psychology in Kyushu University. The database was released in the year 1998. The database contains totally 213 images of 10 Japanese females. These face images are also taken under strictly controlled conditions. The resolution of each image is 256 × 256. The database contains six basic facial expressions + 1 neutral of each subject. And all the images are in TIFF (Tagged Image File Format) format and frontal.

3.3 Bosphorous Database

The Bosphorous database [15] consists of 3D faces and corresponding texture images, specially collected for expression analysis purpose. The database was designed by Yale University, United States in the year 2001. The subject variation in this database comprises not only various expressions and poses but also realistic occlusion. The database consists of total 4666 face images of 105 subjects. The size of each image is 1128 × 1368 and all the images are in PNG format.

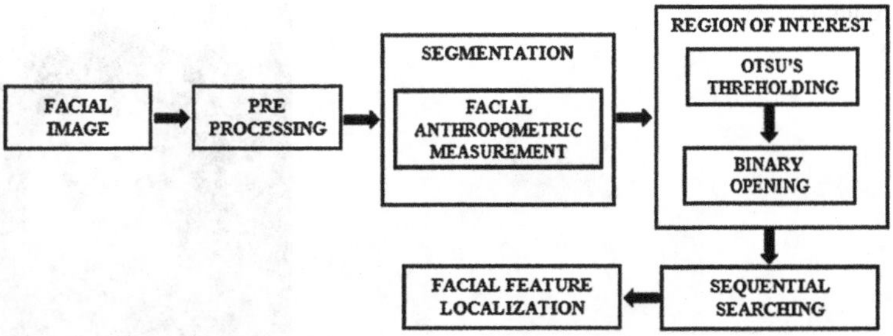

Fig. 1 Block-diagram of proposed facial feature localization method

4 Methodology

The proposed method has been illustrated in Fig. 1 through the block diagram. The theoretical details have been described below in different stages.

4.1 Preprocessing Using Elliptical Mask

The input image is considered as a two-dimensional array, that is, with X × Y size. In the preprocessing stage, the facial images are cropped with an elliptical mask by taking the nose as the centre and cropped to a standard resolution. It is applied over the image to remove all the unnecessary parts of the face image except the central face region. Then the preprocessed image has been taken for segmentation that has been described in the next section.

4.2 Segmentation

Segmentation is the method of partitioning an image into several segments having a significant effect of easier analysis. For region-based segmentation, we have taken two ideas, i.e., ideal facial proportions and the facial anthropometry.

4.2.1 Facial Anthropometric Measurement

Facial anthropometric measurement includes different distance measurements of facial landmarks. It begins with the identification of landmark points on the face. Farkas [16] proposed a system, where a large amount of anthropometric data is available for describing a face using 47 landmark points. The landmarks, which are used here, are as follows: tr for trichion (hairline), n for nasion, sn for subnasale,

Fig. 2 Anthropometric landmarks

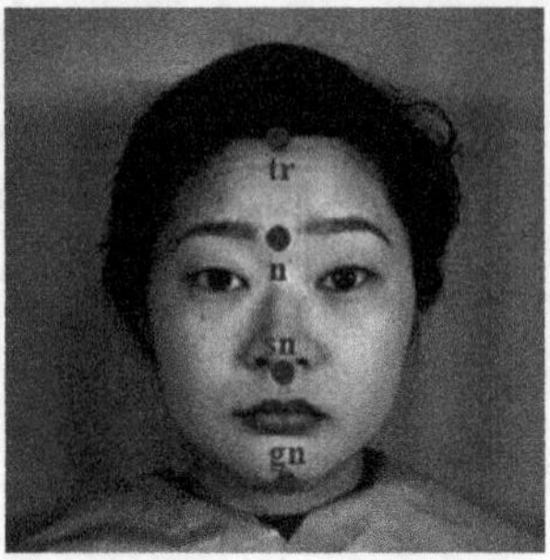

and gn for gnathion (the lowest point on the chin). It has been observed during the distance measurements of these landmarks on face images of JAFFE database, DeitY-TU database, and Bosphorous database, facial measures slightly differ between subjects. Based on the anthropological measurements, the preprocessed image is segmented into horizontal thirds, [17] i.e., the upper third contains only eyebrows; the middle third contains eyes and nose and the lower third contains lips. Figure 2 represents the face image containing four landmark points.

4.2.2 Region of Interest Formation

According to the previous measurement, it is now easier to move forward to detect the region of interest (ROI) for facial feature localization. For ROI formation we have taken two ideas, i.e., Otsu's thresholding and morphological opening. Same strategy has been maintained for three different facial proportions, i.e., the upper third, the middle third, and the lower third. Thresholding is a very important technique for converting a grayscale or color image into binary image based on the threshold criterion T. A threshold using Otsu's method [18] has been chosen on the resultant face images so that only the feature of interest remains.

After binarizing the three regions using Otsu's method, we can see that there are some background pixels in the foreground region of interest. To overcome this problem, binary opening is used to remove small objects from the foreground of the image and placing them in the background. If we deeply see the shape of the eyebrow, eye, and lip, it will be clearly seen that they all contain a disk-shaped structure. A disk-shaped binary mask has been created and using a mask; at first erosion was done and then expanded or dilated the three thirds separately based on anthropometric statistics [4, 19].

4.3 Sequential Searching and Facial Feature Localization

The sequential search considers the first element in the list and then examines each sequential element in the list until a match is found. Forward and reverse iterative sequential search procedures have been applied to the binarized thresholded facial parts.

4.3.1 Localization in Eyebrow Region

To detect the eyebrow region, four landmark points need to be localized, i.e., two inner and two outer eyebrow corners. Localization of eyebrow corners has been done using the reverse sequential search. To detect the inner corner of the left eyebrow, the search starts from the middle of the last row of the upper facial part to the landmark point (i.e., bottom to top). The search proceeds iteratively in the left direction of the facial part and carries on until a black pixel is found. In this way, left inner eyebrow corner is localized. Likewise right inner eyebrow corner is localized using iterative sequential searching in the right direction of the facial part. Then the algorithm proceeded to detect the outer corner of left and right eyebrow. Two search processes have been performed from the first and last of the upper facial part. And finally the four detected landmark points of the eyebrow are localized in the original image. Figure 3a–c shows the inner and outer eyebrow corners points in face image of JAFFE, DeitY-TU and Bosphorous database, respectively.

4.3.2 Localization in Eye Region

To detect the eye region, eight landmark points need to be localized, i.e., two inner eye corners, two outer eye corners, two upper eyelids, and two lower eyelids. Like the eyebrow corner localization, four eye corners have been localized using reverse sequential searching. Then the search proceeded to detect the upper and lower eyelid of left and right eye. The central point of the eye can be located using inner and outer corner of the eye. From the middle of the first and last row of the two eye

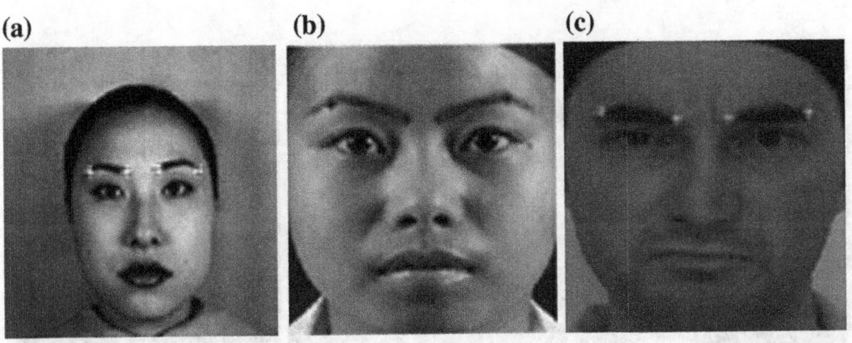

Fig. 3 Eyebrow corners localization **a** JAFFE database, **b** DeitY-TU database, **c** Bosphorous Database

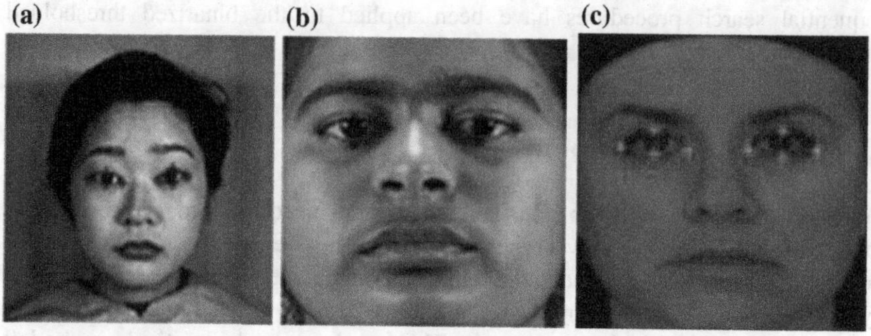

Fig. 4 Eye corners localization **a** JAFFE database, **b** DeitY-TU database, **c** Bosphorous Database

corner, iterative sequential searching (i.e., top to bottom for upper eyelid and bottom to top for lower eyelid) has been performed so that the upper and lower eyelid can be determined. The same process has been performed on both the eyes to detect the four landmark points. Thus eight detected landmark points of the eye are localized in the original image as shown in Fig. 4a–c.

4.3.3 Localization in Mouth Region

To detect the mouth region, we have to localize four facial feature points of the mouth, i.e., two lip corners, upper lip, and lower lip. Using forward and reverse sequential search in the lower facial part, two lip corners can be easily detected. The central point of the lip also can be located using two lip corners. From the middle of the first and last row of the two lip corners, iterative sequential searching has been performed so that the upper and lower lip point can be determined. That is to locate the lower and upper lip middle point, bottom-top and top-bottom search has been applied on lower facial part respectively. And finally the four detected landmark points of lip are localized in the original image. Figure 5a–c localize the mouth points in face image of JAFFE, DeitY-TU, and Bosphorous database, respectively.

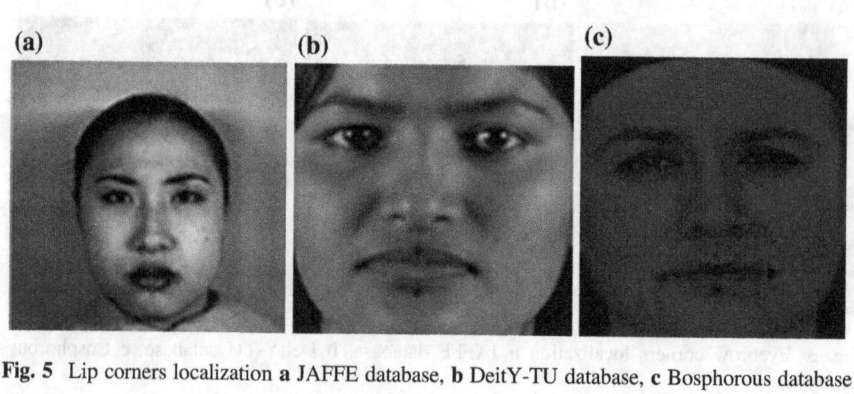

Fig. 5 Lip corners localization **a** JAFFE database, **b** DeitY-TU database, **c** Bosphorous database

5 Performance Evaluation Measures

The performance evaluation of facial feature localization methods proposed in the literature has been given either visual inspection of the detection result or error measure. The localization error can be measured using the distance between manually annotated and automatically detected feature points. The distance is calculated in terms of Euclidean pixel distance. According to Dibeklioglu [5], an error measure is calculated using interocular (d_{io}) distance, which is the distance between the left and right eye centers. A landmark location is correctly detected if the distance to the manually annotated area is less than a percentage of the interocular distance. This threshold was set at 10 % of the interocular distance. The error measure is mathematically represented as

$$\text{Error Measure} = \frac{\|\text{M} - \text{A}\|}{d_{\text{io}}} \quad (1)$$

The average detection rate of each landmark points is computed as

$$\text{Detection Rate} = \frac{\text{TP} \times 100}{\text{TN}} \quad (2)$$

where,

TP Number of images correctly detected
TN Total number of images
M Manually detected landmark points
A Automatically detected landmark points
d_{io} = Interocular distance.

6 Experimental Results and Discussions

In this section, performance analysis of our proposed landmark detection algorithm is made in terms of error measure and accuracy. For measuring the performance of the proposed algorithm, 75 individuals of DeitY-TU face database including both tribe and non tribe with variable image sizes, 100 images of JAFFE database and 300 images of Bosphorous database including male and female are used. The average success rates of the system on DeitY-TU face database, JAFFE database, and Bosphorous database are 94.06, 96.03, and 94.21 % respectively. The localization error and detection rate of the proposed method on these three databases are listed in Table 1.

The proposed methodology has 100 % detection rate of 4 corners of left and right eye. Whereas the average detection rate of all 16 landmark points for JAFFE

Table 1 Localization error and detection rate of the proposed landmark method on various databases

Landmarks	Mean error (% of IOD)			Detection rate (%)		
	JAFFE database (%)	DeitY-TU database (%)	Bosphorous database (%)	JAFFE database (%)	DeitY-TU database (%)	Bosphorous database (%)
Left inner eye corner	3	6	8	100	100	100
Left outer eye corner	2	8	6	100	100	100
Right inner eye corner	2	6	3	100	100	100
Right outer eye corner	3	5	4	100	100	100
Left inner eyebrow	3	7	9	96	95	91.5
Left outer eyebrow	13	15	17	78	75	83.3
Right inner eyebrow	2	7	2	94	92.5	95
Right outer eyebrow	9	11	15	80	80	82.5
Left upper eyelid	2	11	1	100	97.5	95
Left lower eyelid	3	7	2	100	95	100
Right upper eyelid	3	2	3	100	95	95
Right lower eyelid	2	5	6	100	92.5	95
Left lip corner	2	8	5	98	97.5	90
Right lip corner	3	5	2	100	95	95
Upper middle lip	3	2	5	94	95	95
Lower middle lip	1	5	8	96.5	95	90

database is more compared to DeitY-TU face database and Bosphorous database. Also in case of JAFFE database 100 % detection rate for left inner eye corner, left outer eye corner, right inner eye corner, right outer corner, left and right upper, and lower eyelid was achieved which justifies that the proposed work is effective in the three face databases. Experimental results on the face dataset demonstrate that the system works well for all fully frontal face images. The localization error rate for right and left outer eyebrow is more serious than other 14 detected landmark points. This amount of error happens mostly due to the shadow effects around the outer eyebrow corner.

7 Comparative Study

Most researches in facial feature localization have been performed based on error measure. In [20], S. Arca et al. reported the 95.1 % localization accuracy for 22 landmarks on JAFFE database. Sohail et al. [10] and Huang et al. [6] have also made their comparative study based on point error measure. So, by comparing success rate of our proposed method with techniques developed by other researchers as shown in Table 2, we can say that our proposed landmark detection method generates noticeable results on JAFFE face images.

8 Graphical User Interface Design

It is a graphical user interface (GUI)-based prototype system for localization of facial feature from frontal view image. For performing all the required steps a total of 16 interfaces are designed. The first interface contains the database list, i.e., for which database we want to see the landmark points as shown in Fig. 6a. If we click the push button among the three database in the “Database Menu Page” then it will open the second interface, “Load Image.” The second interface is to load the face image and after loading the face image, from the list of popup menu we have to select whether we want to detect the landmark around eye, eyebrow, lip, or whether we want to see all the 16 landmark points at once. After selecting the appropriate

Table 2 Comparison of the proposed methodology with other methods

Author	Number of landmarks	Database	Accuracy (%)
Sohail et al. [10]	16	JAFFE database	93.04
Huang et al. [6]	–	JAFFE database	63.92
Li et al. [4]	20	Bosphorous database	91.14
Dibeklioglu et al. [5]	22	Bosphorous database	97.62
Arca et al. [20]	22	JAFFE database	95.1
Our method	16	JAFFE database	96.03
		DeitY-TU database	93.05
		Bosphorous database	94.21

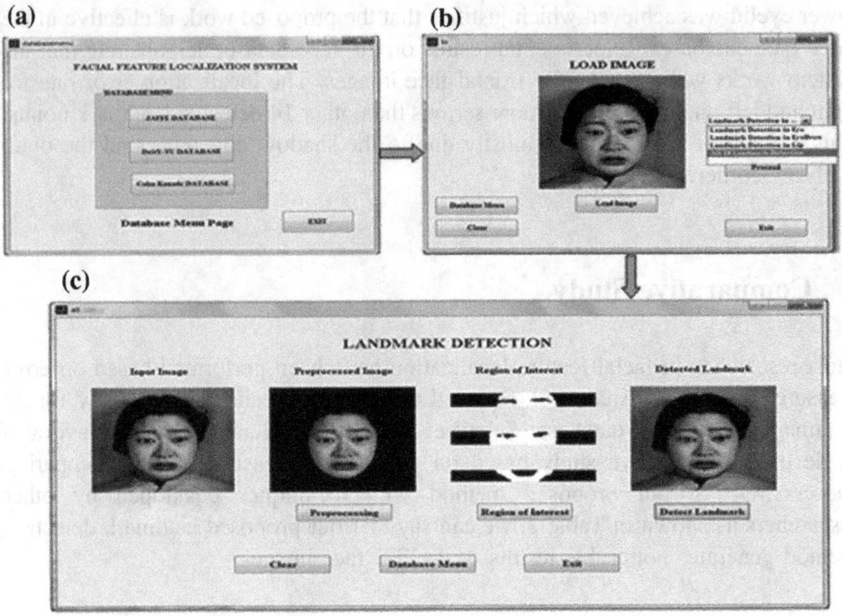

Fig. 6 Graphical user interface (GUI) based prototype system of the proposed method

choice from the popup menu we have to proceed for the landmark detection by clicking on "Proceed" button as shown in Fig. 6b. Then it will open the third interface based on the user's choice selected in the previous interface. Then by clicking the respective buttons in the interface as shown in Fig. 6c we can see the outputs of all the intermediatary steps of the proposed system. And finally if we click on "Detect Landmark" button then it will localize the facial features around eye, eyebrow, and lip on the basis of user's choice as selected in the second interface.

9 Conclusion

The paper presents an automatic facial feature localization method based on a geometric-based hybrid approach. A group of methods are integrated to develop an automatic system that detects 16 landmark points accurately. The experimental results also justify the proposed approach. The average success rate of the methodology is 94.78 % for three databases. The methodology contains some limitations in localization of outer eyebrows' corners. We will try to improve this method in future. Although the present study only focuses on frontal faces, in future, faces with varying poses, occlusion, and illumination will be taken into consideration.

Acknowledgments The work presented here was conducted in the Biometrics Laboratory of Department of Computer Science and Engineering of Tripura University (A central university), Tripura, Suryamaninagar-799,022. The research work was supported by the Grant No. 12(2)/2011-ESD, Dated 29/03/2011 from the DeitY, MCIT, Government of India.

References

1. Celiktutan, O., Ulukaya, S., and Sankur, B.: A comparative study of face landmarking techniques. EURASIP J. Image Video Process. 2013. **2013**(1), pp. 1–27 (2013)
2. Jafri, R., Arabnia, H.R.: A survey of face recognition techniques. JIPS. **5**(2), pp. 41–68 (2009)
3. Salah, A.A., Cinar, H., Akarun, L., Sankur, B.: Robust facial landmarking for registration. Ann. Des Telecommun. **62**(1–2), pp. 83–108. Springer (2007)
4. Li, H., Huang, D., Morvan, J.M., Wang, Y. and Chen, L.: Towards 3d face recognition in the real: a registration free approach using fine-grained matching of 3d keypoint descriptors. Int. J. Comput. Vis. 1–15 (2014)
5. Dibeklioglu, H., Salah, A. A., and Gevers, T.: A statistical method for 2-d facial landmarking. IEEE Trans. Image Process. **21**(2), 844–858 (2012)
6. Huang, H.Y., Hsu, S.H.: Improved active shape model for facial feature localization. In: WIAMIS 2011 (2011)
7. Valstar, M., Martinez, B., Binefa X. and Pantic, M.: Facial point detection using boosted regression and graph models. In: 2010 IEEE Conference on Computer Vision and Pattern Recognition (CVPR), pp. 2729–2736. IEEE (2010)
8. Liu, X.: Video-based face model fitting using adaptive active appearance model. Image Vis. Comput. **28** (7), 1162–1172 (2010)
9. Zhao, S., Gao, Y., Zhang, B.: Gabor feature constrained statistical model for efficient landmark localization and face recognition. Pattern Recog. Lett. **30** (10), pp. 922–930 (2009)
10. Sohail, A.S.M. Bhattacharya, P.: Detection of facial feature points using anthropometric face model. In: Signal Processing for Image Enhancement and Multimedia Processing, pp. 189–200. Springer, US (2008)
11. Gizatdinova Y., Surkka V.: Feature based detection of facial landmarks from neutral and expressive facial images. IEEE Trans. Pattern Anal. Mach. Intell. **28** (1), 135–139 (2006)
12. Vukadinovic, D., Pantic, M.: Fully automatic facial feature point detection using gabor feature based boosted classifier. In: 2005 IEEE International Conference on Systems, Man and Cybernetics, vol. 2, pp. 1692–1698. IEEE (2005)
13. Bhowmik, M.K., Saha, K., Saha, P., Bhattacharjee, D.: DeitY-TU face database: its design, multiple camera capturing, characteristics, and evaluation. In: Opt. Eng. **53** (10), 102106–102106 (2014)
14. Lyons, M., Akamatsu, S., Kamachi, M., Gyoba, J.: Coding facial expressions with Gabor wavelets. In: Third IEEE International Conference on Automatic Face and Gesture Recognition, 1998. Proceedings, pp. 200–205 (1998)
15. Savran, A., Alyuz, N., Dibeklioglu, H., Celiktutan, O., Gokberk, B., Sankur, B., Akarun, L.: Bosphorus database for 3d face analysis. In: Biometrics and Identity Management, pp. 47–56. Springer, Berlin (2008)
16. Farkas, L.G., Munro, I.R.: Anthropometric Facial Proportions in Medicine. Charles C. Thomas Publisher (1987)
17. Teck, S.R.S., Smith, J.D., Chan, A.S.Y.: Comparison of the aesthetic facial proportions of Southern Chinese and white women. In: Arch. Facial Plast. Surg. **2** (2), 113–120 (2000)
18. Otsu, N.: A threshold selection method from gray-level histograms. Automatica **11** (285–296), 23–27 (1975)

19. Jayaraman, S., Esakkirajan, S., Veerakumar, T.: Digital Image Processing. Tata McGraw Hill Education Private Limitedn (2009)
20. Arca, S., Campadelli, P., Lanzarotti, R.: A face recognition system based on automatically determined facial fiducial points. J. Pattern Recogn. **39** (3), 432–443 (2006)

Optimal Land Allocation in Agricultural Production Planning Using Fuzzy Goal Programming

Babita Mishra and S.R. Singh

Abstract Agricultural production is dependent on several imprecise factors and therefore the parameters used in defining fuzzy goals in agricultural production system should be imprecise rather than crisp. Thus for modeling of such systems, we take coefficients defining fuzzy goal as fuzzy number rather than crisp one. In this paper, we deal with the agricultural production planning problem as undertaken by Ghosh et al. (Opsearch 30(1):15–34, 1993) in the more realistic case of having fuzzy inequality with fuzzy coefficients. We transformed the problem as fuzzy goal programming problem and used the triangular possibility distribution for obtaining solution. The results obtained have been compared with the existing one to show its superiority.

Keywords Multiobjective fuzzy linear programming · Linear membership function · Possibility distribution · Fuzzy number · Land use planning

1 Introduction

Goal programming, a multi-criteria decision-making technique, was first used by Charnes and Cooper [9] but the actual power of its applications in decision making emerged with the works of Lee [17], Ignizio [14], and Cohan [10]. Since then, goal programming is being used as a powerful tool in the paradigm of multiobjective programming and variants of goal programming are presented in several potential areas of applications for decision making. The three major variants of classical goal programming in the literature for decision making are weighted goal programming

Babita Mishra (✉) · S.R. Singh
Department of Mathematics, Banaras Hindu University, Varanasi 221005, India
e-mail: babitamishra1983@gmail.com

S.R. Singh
e-mail: srsingh_mathbhu@rediffmail.com

M. Pant et al. (eds.), *Proceedings of Fifth International Conference on Soft Computing for Problem Solving*, Advances in Intelligent Systems and Computing 436, DOI 10.1007/978-981-10-0448-3_23

or non-preemptive goal programming, lexicographic goal programming or preemptive goal programming, and Chebyshev goal programming or minmax goal programming.

Application of decision-making tools in agricultural management system has always been a core area of study but the tools have been changing from time to time. It started with linear programming then multiobjective linear programming, and later, goal programming. With time, more variants of goal programming like stochastic goal programming and fuzzy goal programming developed and are getting strengthened due to the keen interest of research workers due to its suitability to handle the problem in stochastic and possibilistic situations. Fuzzy goal programming was initiated by Zimmermann [25, 26], who combined fuzzy set theory with goal programming to handle the imprecision in information and made it flexible to handle real-life problems by fuzzy goal programming with more accuracy.

In many problems of optimization arising in handling of real-life situations, we are encountered by situations where the information available is imprecise. Thus, fuzzy goal programming emerged as a natural extension of goal programming to model such situations of practical importance. In decision-making problems where coefficients defining objective and constraints need not be crisp, Dubois [11, 12] considered it as linear programming problem with coefficients as fuzzy numbers. The theory of fuzzy linear programming with fuzzy inequality was further enriched by Tanaka and Asai [22, 23] in consideration of possibility distribution of fuzzy coefficients. This concept was further generalized for multiobjective programming problem by several authors, a comprehensive view of various tools of fuzzy multiobjective programming in decision making can be seen in the work of Lai and Hwang [16]. Further, Li and Yu [19] extended the fuzzy multiobjective programming for a case when membership function is quasiconcave. Buckley [7, 8] considered the possibility distribution of uncertain parameters and solved such problems as possibilistic linear programming problem. Further, Arenas et al. [2–4] solved the multiobjective possibilistic programming problems and Jimenez et al. [15] solved a possibilistic linear programming problem through compromise programming. A detailed review of various optimization techniques under uncertainty can be obtained in the work of Sahinidis [20]. Angiz et al. [1] solved a possibilistic linear programming problem considering membership function for coefficients. In agricultural management system, Ghosh et al. [13] considered a problem of optimal land allocation for crop planning though goal programming with penalty functions and Biswas and Pal [6] applied fuzzy goal programming technique to land use planning in agriculture system. Further, Sahoo et al. [21] developed a fuzzy multiobjective and linear programming-based management model for optimal land–water crop system planning and recently, Xieting et al. [24] applied fuzzy multiobjective linear programming to crop area planning.

The objective of the present work is to primarily show the suitability of fuzzy goal programming in optimal land allocation for various crops in agricultural production system as the information available in agricultural system is imprecise in a natural way being field data. The paper describes a systematic approach to deal with decision-making problem in fuzzy environment in which aspiration level for goals have linear membership function. The work is organized as follows: Section 2 deals with stepwise procedure for converting the fuzzy goal programming problem into mixed 0-1 linear programming problem. The steps have been implemented on a numerical problem as undertaken by Ghosh et al. [13] and are illustrated in Sect. 3. The results obtained are presented and discussed in Sect. 3.

2 Multiobjective Fuzzy Goal Programming Problem

Let b_k be the aspiration level given by the decision maker for the kth objective $Z_k(X)$, then the multiobjective fuzzy goal programming problem is to find decision vector X such that

$$\begin{aligned} &Z_k(X) * b_k, \quad \text{where } k = 1, 2, \ldots, K \\ &\tilde{A}X \leq \tilde{b}, \\ &X \geq 0 \end{aligned} \tag{1}$$

where $\tilde{A}$ and $\tilde{b}$ are fuzzy parameters and $(*)$ can be one of fuzzy inequality $(\gtrsim, \approx, \lesssim)$.

2.1 Transformation of Fuzzy Goal Programming into Fuzzy Linear Programming

The fuzzy goal programming problem (1) can be written in its equivalent linear programming problem by the Maxmin model given by Zimmermann [25].

2.1.1 Goals with Fuzzy Coefficients

As the goals considered are of the form

$$Z_k(X) = \sum_{j=1}^{m} \widetilde{b_{kj}} x_j, k = 1, 2, \ldots, K \tag{2}$$

the triangular possibility distribution of goal Z_k using [22, 23] can be expressed as

$$\prod(Z_k) = 1 - \frac{\left|Z_k - \sum_{j=1}^{m} b_{kj} x_j\right|}{\sum_{j=1}^{m} p_{kj} x_j} \tag{3}$$

where fuzzy coefficient $\widetilde{b_{kj}}$ have symmetric triangular distribution, with b_{kj} as the most possible central value and p_{kj} are the most possible deviations from central values on both sides, for all $j = 1, 2, \ldots, m$.

Let us denote $h = \prod(Z_k)$, with $0 \leq \prod(Z_k) \leq 1$,

Now we have

$$h = 1 - \frac{\delta_k^+ + \delta_k^-}{\sum_{j=1}^m p_{kj}x_j} \tag{4}$$

where

$$Z_k - \sum_{j=1}^m b_{kj}x_j = \delta_k^+ - \delta_k^-, \quad \delta_k^+, \delta_k^- \geq 0, \quad 0 \leq h \leq 1 \tag{5}$$

Now using (4) and (5), the fuzzy linear programming problem according to Zimmarmann can be written as

$$\begin{aligned}
&\text{maximize} \quad \lambda \\
&\text{subject to} \quad \lambda - \mu_k(Z_k(X)) \leq 0, \\
&\lambda \leq \text{wh}, \\
&h\left(\sum_{j=1}^m p_{kj}x_j\right) = \left(\sum_{j=1}^m p_{kj}x_j\right) - \delta_k^+ - \delta_k^-, \\
&Z_k - \sum_{j=1}^m b_{kj}x_j = \delta_k^+ - \delta_k^-, k = 1, 2, \ldots, K \\
&AX \leq b, \\
&X \geq 0, \delta_k^+, \delta_k^- \geq 0, 0 \leq h \leq 1
\end{aligned} \tag{6}$$

where w is weight between the degree of preference for membership functions and degree of possibility for possibility distribution of which lower and upper bound can be obtained by algorithm given by Biswal [5]. A major difficulty of solving the above problem is the constraints containing product term hx_j, this nonlinear term can be linearized by the following proposition given by Li and Chang [18].

Proposition A product term hx_j, where $0 \leq h \leq 1$ and $x_j \geq 0$ can be approximately linearized as $v_j = hx_j$. The relationship among h, x_j, v_j are expressed as

$$v_j = \frac{1}{\sum_{r=1}^R 2^{r-1}} \left[\sum_{r=1}^R 2^{r-1} \alpha_{jr}\right], \tag{7}$$

$$h = \frac{1}{\sum_{r=1}^{R} 2^{r-1}} \left[\sum_{r=1}^{R} 2^{r-1} \theta_{jr} \right], \tag{8}$$

$$M(\theta_{jr} - 1) + x_j \le \alpha_{jr} \le 1 - \theta_{jr} + x_j, \tag{9}$$

$$0 \le \alpha_{jr} \le M\theta_{jr}, \alpha_{jr} \le x_j, r = 1, 2, \ldots R, \tag{10}$$

where θ_{jr} are 0–1 variables, α_{jr} are continuous variables such that (7) is true, M is the upper bound of x_j and R is an integer specified by the decision maker in view of maximal tolerable error arising due to linearization larger the integer R, smaller the linearization error.

2.2 Mixed 0–1 Linear Programming Problem

Now using the above proposition, (7), (8), (9) and (10) in (6), we obtain a mixed 0–1 linear programming problem as

$$\begin{aligned}
&\text{maximize} \quad \lambda \\
&\text{subject to} \quad \lambda - \mu_k(Z_k(X)) \le 0, \\
&\lambda \le wh, \\
&\left(\sum_{j=1}^{m} p_{kj} v_j \right) = \left(\sum_{j=1}^{m} p_{kj} x_j \right) - \delta_k^+ - \delta_k^-, \\
&Z_k - \sum_{j=1}^{m} b_{kj} x_j = \delta_k^+ - \delta_k^-, k = 1, 2, \ldots, K \\
&v_j = \frac{1}{\sum_{r=1}^{R} 2^{r-1}} \left[\sum_{r=1}^{R} 2^{r-1} \alpha_{jr} \right], \\
&h = \frac{1}{\sum_{r=1}^{R} 2^{r-1}} \left[\sum_{r=1}^{R} 2^{r-1} \theta_{jr} \right], \\
&M(\theta_{jr} - 1) + x_j \le \alpha_{jr} \le 1 - \theta_{jr} + x_j, \\
&0 \le \alpha_{jr} \le M\theta_{jr}, \alpha_{jr} \le x_j, r = 1, 2, \ldots, R, \\
&AX \le b, \\
&X \ge 0, \delta_k^+, \delta_k^- \ge 0, 0 \le h \le 1
\end{aligned} \tag{11}$$

Problem (11), a mixed 0–1 linear programming, can be easily solved by any optimization package.

3 Numerical Illustration

For numerical illustration of the above method presented in Sect. 2, we consider a land allocation problem as undertaken by Ghosh et al. [13] with slight modification of considering parameters as fuzzy in view of more realistic modeling of problem. Gosh et al. considered a problem of production of crops in Hooghly district of West Bengal, India.

With assumption that the land available for cultivation is to be fully utilized, the mathematical formulation of the above problem as fuzzy goal programming is as follows.

3.1 *Production Achievement Goal (S)*

$$Z_1(X) : \widetilde{21.62}x_{211} + \widetilde{25.78}x_{221} \gtrsim 3250 \tag{12}$$

$$Z_2(X) : \widetilde{44.45}x_{111} + \widetilde{26.67}x_{112} \gtrsim 10500 \tag{13}$$

$$Z_3(X) : \widetilde{282.79}x_{313} \gtrsim 31350 \tag{14}$$

$$Z_4(X) : \widetilde{5.28}x_{413} \gtrsim 170 \tag{15}$$

$$Z_5(X) : \widetilde{20.61}x_{513} \gtrsim 1250 \tag{16}$$

$$Z_6(X) : \widetilde{8.85}x_{613} \gtrsim 450 \tag{17}$$

$$Z_7(X) : \widetilde{44.45}x_{111} + \widetilde{26.67}x_{112} + \widetilde{20.61}x_{513} \gtrsim 11750 \tag{18}$$

3.2 *Man-Days Goal*

$$\begin{aligned} Z_8(X) : {} & \widetilde{340}x_{111} + \widetilde{195}x_{112} + \widetilde{350}x_{211} + \widetilde{375}x_{221} + \widetilde{95}x_{313} + \widetilde{47}x_{413} + \widetilde{156}x_{513} \\ & + \widetilde{93}x_{613} \lesssim 154215 \end{aligned} \tag{19}$$

3.3 *Machine-Hour Goal*

$$\begin{aligned} Z_9(X) : {} & \widetilde{7.3}(x_{111} + x_{112}) + \widetilde{5.15}(x_{211} + x_{221} + x_{413} + x_{513} + x_{613}) \\ & + \widetilde{12.15}x_{313} \lesssim 5112.5 \end{aligned} \tag{20}$$

3.4 Cash Expenditure Goal

$$\begin{aligned} Z_{10}(X) &: \widetilde{6215.78}x_{111} + \widetilde{3165.78}x_{112} + \widetilde{2900.42}x_{211} + \widetilde{3265.64}x_{221} \\ &\widetilde{5586.85}x_{313} + \widetilde{1291.74}x_{413} + \widetilde{2601.3}x_{513} + \widetilde{1465.7}x_{613} \lesssim 2767458.32 \end{aligned} \tag{21}$$

3.5 Profit Goal

$$\begin{aligned} Z_{11}(X) &: \widetilde{8445.5}\, x_{111} + \widetilde{6000.75}\, x_{112} + \widetilde{7245.51}\, x_{211} + \widetilde{6663.1}\, x_{221} \\ &\widetilde{31870.43}x_{313} + \widetilde{2376}\, x_{413} + \widetilde{5337.99}x_{513} + \widetilde{4566.6}x_{613} \gtrsim 7557976 \end{aligned} \tag{22}$$

and other constraints are given as below.

Land utilization in summer season

$$x_{111} + x_{211} + x_{221} = 253.432 \tag{23}$$

Land utilization in rainy season including crops of late variety jute and rainy paddy

$$x_{112} + x_{221} = 253.432 \tag{24}$$

Land utilization in winter season

$$x_{313} + x_{413} + x_{513} + x_{613} = 253.432 \tag{25}$$

Further, the condition that after harvesting the early variety of jute and boro paddy, attempt is to be done to utilize the total available land for cultivating the amon paddy, thus this relational goal is expressed as

$$x_{111} + x_{211} - x_{112} = 0 \tag{26}$$

As ravi pulse and mustard are generally cultivated at the beginning of winter, so after harvesting the late variety of jute at the end of rainy season, the available land can be used for cultivating ravi pulse and mustard, hence,

$$x_{221} - x_{413} - x_{613} = 0 \tag{27}$$

one can transform the fuzzy goals to its equivalent constraints to get an equivalent mixed 0–1 linear programming problem as

Max λ
s.t.

$$
\begin{aligned}
&975\lambda - Z_1 \leq -2275\\
&2625\lambda - Z_2 \leq -7875\\
&9405\lambda - Z_3 \leq -21945\\
&51\lambda - Z_4 \leq -119\\
&375\lambda - Z_5 \leq -875\\
&135\lambda - Z_6 \leq -315\\
&2937.5\lambda - Z_7 \leq -8812.5\\
&30843\lambda + Z_8 \leq 185058\\
&1278.125\lambda + Z_9 \leq 6390.625\\
&553491.664\lambda + Z_{10} \leq 3320949.984\\
&755797.6\lambda - Z_{11} \leq -6802178.4\\
&\lambda - wh \leq 0
\end{aligned}
$$

$$
\begin{aligned}
&2.162v_2 + 2.578v_3 - 2.162x_{211} - 2.578x_{221} + \delta_1^+ + \delta_1^- = 0\\
&4.445v_1 + 2.667v_4 - 4.445x_{111} - 2.667x_{112} + \delta_2^+ + \delta_2^- = 0\\
&28.279v_5 - 28.279x_{313} + \delta_3^+ + \delta_3^- = 0\\
&.528v_6 - .528x_{413} + \delta_4^+ + \delta_4^- = 0\\
&2.061v_7 - 2.061x_{513} + \delta_5^+ + \delta_5^- = 0\\
&.885v_8 - .885x_{613} + \delta_6^+ + \delta_6^- = 0\\
&4.445v_1 + 2.667v_4 + 2.061v_7 - 4.445x_{111} - 2.667x_{112} - 2.061x_{513} + \delta_7^+ + \delta_7^- = 0\\
&34v_1 + 19.5v_4 + 35v_2 + 37.5v_3 + 9.5v_5 + 4.7v_6 + 15.6v_7 + 9.3v_8 - 34x_{111}\\
&\quad - 19.5x_{112} - 35x_{211} - 37.5x_{221} - 9.5x_{313} - 4.7x_{413} - 15.6x_{513} - 9.3x_{613}\\
&\quad + \delta_8^+ + \delta_8^- = 0\\
&.73v_1 + .73v_4 + .52v_2 + .52v_3 + 1.22v_5 + .52v_6 + .52v_7 + .52v_8 - .73x_{111}\\
&\quad - .73x_{112} - .52x_{211} - .52x_{221} - 1.22x_{313} - .52x_{413} - .52x_{513} - .52x_{613}\\
&\quad + \delta_9^+ + \delta_9^- = 0\\
&621.578v_1 + 316.578v_4 + 290.042v_2 + 326.564v_3 + 558.685v_5 + 129.174v_6\\
&\quad + 260.13v_7 + 146.57v_8 - 621.578x_{111} - 316.578x_{112} - 290.042x_{211} - 326.56x_{221}\\
&\quad - 558.685x_{313} - 129.174x_{413} - 260.13x_{513} - 146.57x_{613} + \delta_{10}^+ + \delta_{10}^- = 0\\
&844.55v_1 + 600.07v_4 + 724.55v_2 + 666.31v_3 + 3187.04v_5 + 237.6v_6 + 533.8v_7\\
&\quad + 456.66v_8 - 844.55x_{111} - 600.07x_{112} - 724.55x_{211} - 666.31x_{221} - 3187.04x_{313}\\
&\quad - 237.6x_{413} - 533.8x_{513} - 456.66x_{613} + \delta_{11}^+ + \delta_{11}^- = 0
\end{aligned}
$$

$$
\begin{aligned}
&Z_1 - 21.62x_{211} - 25.78x_{221} - \delta_1^+ + \delta_1^- = 0\\
&Z_2 - 44.45x_{111} - 26.67x_{112} - \delta_2^+ + \delta_2^- = 0\\
&Z_3 - 282.79x_{313} - \delta_3^+ + \delta_3^- = 0\\
&Z_4 - 5.28x_{413} - \delta_4^+ + \delta_4^- = 0\\
&Z_5 - 20.61x_{513} - \delta_5^+ + \delta_5^- = 0\\
&Z_6 - 8.85x_{613} - \delta_6^+ + \delta_6^- = 0\\
&Z_7 - 44.45x_{111} - 26.67x_{112} - 20.61x_{513} - \delta_7^+ + \delta_7^- = 0\\
&Z_8 - 340x_{111} - 195x_{112} - 350x_{211} - 375x_{221} - 95x_{313} - 47x_{413} - 156x_{513}\\
&\quad - 93x_{613} - \delta_8^+ + \delta_8^- = 0\\
&Z_9 - 7.3x_{111} - 7.3x_{112} - 5.2x_{211} - 5.2x_{221} - 12.2x_{313} - 5.2x_{413} - 5.2x_{513}\\
&\quad - 5.2x_{613} - \delta_9^+ + \delta_9^- = 0\\
&Z_{10} - 6215.78x_{111} - 3165.78x_{112} - 2900.42x_{211} - 3265.6x_{221}\\
&\quad - 5586.85x_{313} - 1291.74x_{413} - 2601.3x_{513} - 1465.7x_{613} - \delta_{10}^+ + \delta_{10}^- = 0\\
&Z_{11} - 8445.5x_{111} - 6000.7x_{112} - 7245.5x_{211} - 6663.1x_{221} - 31870.4x_{313}\\
&\quad - 2376x_{413} - 5338x_{513} - 4566.6x_{613} - \delta_{11}^+ + \delta_{11}^- = 0
\end{aligned}
\tag{28}
$$

$$
\begin{aligned}
x_{111} + x_{211} + x_{221} &= 253.432\\
x_{112} + x_{221} &= 253.432\\
x_{313} + x_{413} + x_{513} + x_{613} &= 253.432\\
x_{111} + x_{211} - x_{112} &= 0\\
x_{221} - x_{413} - x_{613} &= 0
\end{aligned}
$$

$$
\left.
\begin{aligned}
v_1 &= \tfrac{1}{15}[\alpha_{11} + 2\alpha_{12} + 4\alpha_{13} + 8\alpha_{14}]\\
v_2 &= \tfrac{1}{15}[\alpha_{21} + 2\alpha_{22} + 4\alpha_{23} + 8\alpha_{24}]\\
v_3 &= \tfrac{1}{15}[\alpha_{31} + 2\alpha_{32} + 4\alpha_{33} + 8\alpha_{34}]\\
v_4 &= \tfrac{1}{15}[\alpha_{41} + 2\alpha_{42} + 4\alpha_{43} + 8\alpha_{44}]\\
v_5 &= \tfrac{1}{15}[\alpha_{51} + 2\alpha_{52} + 4\alpha_{53} + 8\alpha_{54}]\\
v_6 &= \tfrac{1}{15}[\alpha_{61} + 2\alpha_{62} + 4\alpha_{63} + 8\alpha_{64}]\\
v_7 &= \tfrac{1}{15}[\alpha_{71} + 2\alpha_{72} + 4\alpha_{73} + 8\alpha_{74}]\\
v_8 &= \tfrac{1}{15}[\alpha_{81} + 2\alpha_{82} + 4\alpha_{83} + 8\alpha_{84}]
\end{aligned}
\right\}
$$

$$
\left.
\begin{aligned}
h &= \tfrac{1}{15}[\theta_{11} + 2\theta_{12} + 4\theta_{13} + 8\theta_{14}]\\
h &= \tfrac{1}{15}[\theta_{21} + 2\theta_{22} + 4\theta_{23} + 8\theta_{24}]\\
h &= \tfrac{1}{15}[\theta_{31} + 2\theta_{32} + 4\theta_{33} + 8\theta_{34}]\\
h &= \tfrac{1}{15}[\theta_{41} + 2\theta_{42} + 4\theta_{43} + 8\theta_{44}]\\
h &= \tfrac{1}{15}[\theta_{51} + 2\theta_{52} + 4\theta_{53} + 8\theta_{54}]\\
h &= \tfrac{1}{15}[\theta_{61} + 2\theta_{62} + 4\theta_{63} + 8\theta_{64}]\\
h &= \tfrac{1}{15}[\theta_{71} + 2\theta_{72} + 4\theta_{73} + 8\theta_{74}]\\
h &= \tfrac{1}{15}[\theta_{81} + 2\theta_{82} + 4\theta_{83} + 8\theta_{84}]
\end{aligned}
\right\}
$$

$$253.432(\theta_{jr}-1)+x_j \leq \alpha_{jr} \leq 1-\theta_{jr}+x_j,$$
$$0 \leq \alpha_{jr} \leq 253.432\theta_{jr}, \alpha_{jr} \leq x_j, \text{where } j=1,2,\ldots,8 \text{ and } r=1,2,3,4$$
$$X \geq 0,$$

This mixed 0–1 LPP (28) has been solved using LINDO software for different values of w = 0.5, 0.7, 0.9 and 1 using $\lambda = 0.5, 0.7, 0.9, 1$ with h = 1.

4 Results and Discussion

The computed value of decision variables and of the goal achievement for different values of w are considered for problem (28). The results obtained provide a variety of options for decision maker to select a cropping model (decision variable X) in accordance to a suitable value of w, a compromise value between degree of preference for membership functions, and degree of possibility for possibility distribution. For example, if one decides to go for a cropping model so that all goals are completely satisfied, pick up the model solution with w = 1. Further, we compare results obtained with previous existing solution obtained by goal programming method as given in Table 1.

The results obtained by our proposed method placed in Table 1 clearly reveal the superiority over Ghosh et al. [13] results. This superiority is twofold, the first is allocating the land for various crops by proposed model fulfill goal achievements completely. Secondly, the above proposed model overcomes the demerits of Ghosh

Table 1 Comparison of optimal solution by proposed method with existing solution

Goal(s)	Aspiration level	Tolerance limit		Solution by Ghosh [13]	Solution by proposed method (for $w = \lambda = h = 1$)
		Lower	Upper		
Production achievement (qtls.)					
Jute	3250	2275	–	2927.695	3250.00
Paddy	10,500	7875	–	10500.44	10,500
Potato	31,350	21,945	–	31350.1	31,350
Ravi pulse	170	119	–	170	171.645
Wheat	1250	875	–	1226.913	1250
Mustard	450	315	–	449.996	450
Paddy+wheat	11,750	8812.5	–	11727.35	11,750
Man-days (in days)	154,215	–	185,058	142724.3	133496.11
Man-hour (h)	5112.5	–	6390.625	4918.32	4327.02
Cash-exposure (Rs.)	2767458.32	–	3320949.98	2639387	2293081
Profit goal (Rs.)	7,557,976	6802178.4	–	7130773.59	7,557,976
Ratio (profit/exp.)	2.73	2.05	–	2.702	3.296

et al. model in which productions of jute, wheat, paddy + wheat and profit goals are not achieved. Further, the proposed model for allocation of land to different crops provide better achievement of set goals, moreover the overall profit, i.e., ratio of profit with expenditure increases to 3.296 as against the existing 2.702. The cash expenditure needed for proposed cropping model is also lower.

Thus the significance of the present study lies in the fact that the fuzzy goal programming having aspiration levels for goals with upper and lower tolerance for goals provides better solution than goal programming with penalty function. Further, in agriculture production system it is unrealistic to set crisp goals and various parameters are also imprecise due to several nondeterministic factors responsible for affecting production of crops. Thus in view of this fact, we considered a more realistic case of having 10 % of variation in parameter used in the previous study. The study also reveals that fuzzy goal programming is versatile to accommodate the imprecision in data and is more suitable for modeling the problem of agricultural production planning problems.

References

1. Angiz, M.Z., Saati, S., Memariani, A., Movahedi, M.M.: Solving possibilistic linear programming problem considering membership function of the coefficients. Fuzzy Sets Syst. **1**(2), 131–142 (2006)
2. Arenas, M., Bilbao, A., Rodriguez, M.V.: Solving the multiobjective possibilistic linear programming problem. Eur. J. Oper. Res. **117**, 175–182 (1999a)
3. Arenas, M., Bilbao, A., Rodriguez, M.V.: Solution of a possibilistic multiobjective linear programming problem. Eur. J. Oper. Res. **119**, 338–344 (1999b)
4. Arenas, M., Bilbao, A., Gladish, B.P., Rodriguez, M.V.: Solving a multiobjective possibilistic problem through compromise programming. Eur. J. Oper. Res. 164, 748–759 (2005)
5. Biswal, M.P.: Use of projective and scaling algorithm to solve multi-objective fuzzy linear programming problems. J. Fuzzy Math. **5**, 439–448 (1997)
6. Biswas, A., Pal, B.B.: Application of fuzzy goal programming technique to land use planning in agricultural system. Omega **33**, 391–398 (2005)
7. Buckley, J.J.: Solving possibilistic linear programming problems. Fuzzy Sets Syst. **31**, 329–341 (1987)
8. Buckley, J.J.: Possibilistic linear programming with triangular fuzzy numbers. Fuzzy Sets Syst. **26**, 135–138 (1988)
9. Charnes, A., Cooper, W.W.: Management models of industrial applications of linear program 1 and 2. Wiley, New York (1961)
10. Cohon, J.L.: Multiobjective programming and planning. Academic Press, New York (1978)
11. Dubois, D., Prade, H.: Operations on fuzzy numbers. Int. J. Sys. Sci. **9**(6), 613–626 (1978)
12. Dubois, D., Prade, H.: Fuzzy Sets and Systems: Theory and Applications. Academic Press, New York (1980)
13. Ghosh, D., Pal, B.B., Basu, M.: Determination of optimal land allocation in agricultural planning through goal programming with penalty functions. Opsearch **30**(1), 15–34 (1993)
14. Ignizio, J.P.: Goal programming and extensions. Lexington, Mass: D.C. Heath (1976)
15. Jimenez, M., Arenas, M., Bilbao, A., Rodriguez, M.V.: Solving a possibilistic linear program through compromise programming. Mathware Soft Comput. **7**(2–3), 175–184 (2000)
16. Lai, Y.J., Hwang, C.L.: Fuzzy multiple objective Decision making. Springer: New York (1994)

17. Lee, S.M.: Goal programming for decision analysis. Auerbach, Philadelphia (1972)
18. Li, H.L., Chang, C.T.: An approximately global optimization method for assortment problems. Eur. J. Oper. Res. **105**, 604–612 (1998)
19. Li, H.L., Yu, C.S.: A fuzzy multiobjective program with quasiconcave membership functions and fuzzy coefficients. Fuzzy Sets Syst. **109**, 59–81 (2000)
20. Sahinidis, N.V.: Optimization under uncertainty: state-of-the-art and opportunities. Comput. Chem. Eng. **28**, 971–983 (2004)
21. Sahoo, B., Lohani, A.K., Sahu, R.K.: Fuzzy multiobjective and linear programming based management model for optimal land-water-crop system planning. Water Resour. Manag. **20**, 931–948 (2006)
22. Tanaka, H., Asai, K.: Fuzzy solution in fuzzy linear programming problems. IEEE Trans. Syst. Man Cybernet **14**, 325–328 (1984)
23. Tanaka, H., Asai, K.: Fuzzy linear programming problems with fuzzy numbers. Fuzzy Sets Syst. **13**, 1–10 (1984)
24. Zeng, X., Kang, S., Li, F., Zhang, L., Guo, P.: Fuzzy multi-objective linear programming applying to crop area planning. Agric. Water Manag. **98**, 134–142 (2010)
25. Zimmermann, H.J.: Fuzzy programming and linear programming with several objective functions. Fuzzy Sets Syst. **1**, 45–55 (1978)
26. Zimmermann, H.J.: Using fuzzy sets in operational research. Eur. J. Oper. Res. **13**, 201–206 (1983)

Quantitative Estimation for Impact of Genomic Features Responsible for 5′ and 3′ UTR Formation in Human Genome

Shailesh Kumar, Sumita Kachhwaha and S.L. Kothari

Abstract UnTranslated Regions (UTRs) are part of messenger ribonucleic acid (mRNA) that do not undergo protein translation mechanism but plays an important role in translation control. Various genomic and non genomic features are responsible for controlling the translation. We have attempted to find various genomic features and their information content that are contributing to the length of UTRs. With the increase in length of UTRs, the translation process becomes slower resulting into less protein output. In this study results revealed that as length of 5′ UTR and 3′ UTR increase the information content of the sequence also increase but it becomes stable at longer UTR. Trimeric features are having more information content as compared to Dimeric features. As length of UTR increase the entropy of the information increases but after certain length it becomes stable. As 5′ UTR length increases the GC content decreases while AT increases and it is opposite in 3′ UTRs. Some genomic features like CG, TAA, CGT, CGC, CCG, CGG, ACG are having correlation <0.70 where as features like CT, TC, AC, CA, GT, GA. ACT, CAT, CTT, TCA, TGA are having correlation >0.90.

Keywords Translation · Genome · Ribonucleic acid · Entropy of information

S. Kumar (✉) · S.L. Kothari
Amity Institute of Biotechnology, Amity University Rajasthan, Jaipur, Rajasthan, India
e-mail: shailesh_iiita@hotmail.com

S.L. Kothari
e-mail: slkothari@jpr.amity.edu

Sumita Kachhwaha
Bioinformatics Infrastructure Facility, University of Rajasthan, Jaipur, Rajasthan, India
e-mail: kachhwahasumita@rediffmail.com

M. Pant et al. (eds.), *Proceedings of Fifth International Conference on Soft Computing for Problem Solving*, Advances in Intelligent Systems and Computing 436, DOI 10.1007/978-981-10-0448-3_24

1 Introduction

A Gene of any organism after transcription and maturation is represented as messenger Ribonucleic Acid (mRNA) which may be composed of translated and untranslated regions. This mRNA helps in formation of proteins by protein translation mechanism. Protein Translational regulation comprises of many sis and trans regulatory factors that help in controlling the amount of protein synthesized in a biological system. These regulatory factors are classified as Genomic and Non genomic factors. Genomic factors are those which are present on the genomic sequence like specific pattern of sequence for secondary structure formation, self complementarities in the UTR region, etc. Non genomic factors are not the part of genome but act as regulatory elements. There are a number of studies representing the role of various genomic features of 5′ and 3′ UnTranslated Regions (UTR) in controlling the translational mechanism [1]. It was reported that if there are higher number of regulatory elements in the UTR then the translation process is compromised leading to reduced protein output [2]. These regulatory factors on 5′ UTR, mediated regulation by 7-methyl-guanine (cap), hairpin-like secondary structure, RNA-protein interactions, upstream open reading frames (uORFs), and internal ribosome entry sites (IRES) [3]. On 3′ UTR it regulates by, antisense RNA interactions, RNA-protein interactions, involving also multiprotein complexes, cytoplasmic polyadenylation elements (CPE), poly (A) tail, and variation of its size.

The presence of introns is well understood in gene transcripts at coding sequences (CDSs). A small but significant fraction of introns is also found to reside within the untranslated regions (5′ UTRs and 3′ UTRs) of expressed sequences. These introns tend to increase the length of UTRs, thus controlling the gene expression [4]. The alignment study of whole genome and expressed sequence tags (ESTs) of the model plant Arabidopsis thaliana has identified introns residing in both coding and non-coding regions of the genome [5, 6]. A bioinformatics analysis revealed some interesting observations: (1) Density of introns in 5′ UTRs is similar to that in CDSs but much higher than that in 3′ UTRs. (2) 5′ UTR introns are preferentially located close to the initiating ATG codon. (3) Introns in the 5′ UTRs are, on average, longer than introns in the CDSs and 3′ UTRs (4) 5′ UTR introns have a different nucleotide composition to that of CDS and 3′ UTR introns. Approximately, 35 % of human genes contain introns within the 5′ untranslated region (UTR). Introns in 5′ UTRs differ from those in coding regions and 3′ UTRs with respect to nucleotide composition, length distribution and density. In particular, genes with regulatory roles were surprisingly enriched in having 5′ UTR introns [7]. It is hypothesize that the selection pressure on a combination of genomic features is also important for 5′ UTR evolution [8]. The evolution was studied by find the alternative splicing sites in UTR regions which resulted in variations in UTR length [9, 10].

Recently, it has been deduced that there was about 10 % impact of GC content in UTR at 5′ Region on gene expression of chicken [11]. This study had taken only GC content into consideration which is one of the features that UTR is having in it, there are many more features that also influence the expression [12, 13]. One computational model was generated in 2013 for identification of splice site at 5′ UTR using sequence and structural specification. [14]. Classification and regression tree (CART) a database was developed for classification of 5′ UTR [15]. In this study, the main aim is to identify genomic features and to find quantitative impact of these genomic features with the length of UTR sequences.

2 Materials and Methods

2.1 UTR Data Set

The UTR data set used in the study was obtained from curated database UTRdb (http://utrdb.ba.itb.cnr.it/) which was published in 2010. This database contains information of 473,330 5′ UTR, 527,323 3′ UTR and 483,605 genes across 79 species. This is the largest database for UTR information on sequence and motifs. Out of these large number of UTR dataset, we used only UTR data of Homo sapience which consists of 124316 raw 5′ UTR and 194503 raw 3′ UTR sequences which represents nearly ∼5 % of whole UTR dataset among all species [16].

2.2 Validation of Data

The data validation is an important step before data analysis; here to validate the UTR data standalone blast program was used to verify UTR sequences. Whole data validation was not carried out, only randomly selected human gene sequences were chosen for blast search using all UTR sequences of Homo sapience as database. This process validate whether the UTR sequences are well defined or not or any data mutation had occurred during sequencing. This validation process is also to find whether the sequence is having any discrepancies from the original gene sequence. On validation, there were no mutational variation and discrepancies were found in the local search. Standalone BLAST+ [17] source code is downloaded from http://ftp.ncbi.nlm.nih.gov/blast/executables/blast+/LATEST/. The inbuilt program "formatdb" was used to create a local database for performing blast search. Parameters for blast search command are taken as default for all the genes.

2.3 Extracting Genomic Features

For extracting genomic features from the raw genomic sequence data of UTRs a PERL program was developed as UTR explorer which works on the DNA sequences. This program takes input in FASTA file format and gives output in tab delimited text for various genomic features in the sequence with their frequency like

A, T, G, C, GC%, AT%, dimer frequency like GC, GA, GT, GG, CG, CT, CA, CC, TG, TT, TA, TC, AC, AG, AT, AA, trimer count, ATG, TGA, TAA, TAG, TTT, TCT, TAT, TGT, TTC, TCC, TAC, TGC, TTA, TCA, TTG, TCG, TGG, CTT, CCT, CAT, CGT, CTC, CCC, CAC, CGC, CTA, CCA, CAA, CGA, CTG, CCG, CAG, CGG, ATT, ACT, AAT, AGT, ATC, ACC, AAC, AGC, ATA, ACA, AAA, AGA, ACG, AAG, AGG, GTT, GCT, GAT, GGT, GTC, GCC, GAC, GGC, GTA, GCA, GAA, GGA, GTG, GCG, GAG, GGG, Possible uORF, Entropy of the sequences, and Perplexity of the sequence. This txt file which was generated and imported to Microsoft excel and R Package for further analysis.

2.4 Data Normalization

Normalization of data is required to remove all the noises or duplicates in the data which causes problem in final data interpretation. For normalization, a unique strategy was used where UTR IDs which have all the above features identical were said to be duplicated or redundant data which was removed from raw dataset. By using this strategy, total number of 5′ UTR sequences was reduced from 124,316 > 55,059, resulting to removal 44.2895 % of redundancy and in 3′ UTR data were reduced from 194,503 > 86,948 resulting to reduction of 44.7026 % data redundancy. UTR IDs which are having sequence length less than 16 bp are also removed to make data more reliable, because probabilistically for repeating a dimer in a DNA sequence minimum 16 base pair are required. After data normalization, in 5′ UTR the unique sequences which are having length between 16 and 6170 bp are about 54,333, whereas in 3′ UTR had having length between 16 and 14,575 bp producing 8,6093 unique sequences.

2.5 Entropy of Information

Information content of a DNA sequence is important feature because DNA is information source all biological processes. Here, the entropy of information is calculated for all monomer, dimer and trimer separately for finding the information content of each features representing one sequence. This will also help in finding the information flow between the variables. The entropy of information $H(X)$ was calculated by Shannon entropy method [18–20] represented in Eq. (1). A R program was developed for entropy calculation, which takes PERL program generated text file as input and generated a file with entropy of each feature type.

$$H(X) = \sum_{i=1}^{n} p(x_i)I(x_i) = \sum_{i=1}^{n} p(x_i)\log_b = \frac{1}{p(x_i)} = -\sum_{i=1}^{n} p(x_i)\log_b p(x_i) \quad (1)$$

$H(X)$ is entropy of information; $p(x_i)$ is probability of each feature.

2.6 Data Analysis

The rectified UTR data (5′ and 3′ UTR) was analyzed by R package for quantitative estimation of each feature for contributing the length of the UTR sequences. This estimation is calculated by finding correlation between various genomic features and UTR length. The graphical plots were developed by scatterplot3D package; and plot packages of R.

3 Results and Discussion

3.1 Validation of UTR Sequences

Standalone Blast program is use for validation of UTR sequences, where UTR database is created by formatdb program and query is taken as randomly selected genes from human genome. This query gene when aligned by blast program with locally created UTR database. If the sequence after alignment is having differences like gaps, insertion, and deletion in the alignment then the UTR sequence will be rejected for further analysis and if no variations are found then the UTR data is accepted. It is assumed that by doing these exercises the effect of data noises will be minimized. Fortunately, the data used in the study do not have any variations as validated by blast program. Figure 1 is representing blast output of Homo sapiens breast cancer metastasis suppressor 1 (BRMS1) gene with UTR database. Here, there are variations are not found which results to validated UTR dataset.

3.2 Entropy of Sequences

Entropy of Information is a concept for finding randomness in a system by finding the frequency of each events occurring in the system. Here, the event in this data is

```
> SHSAA011068 BA011068
Length=161

 Score =  272 bits (147),  Expect = 4e-072
 Identities = 147/147 (100%), Gaps = 0/147 (0%)
 Strand=Plus/Plus

Query  1    AAGCACCGATAGGCTCTGCCTCCCGAAGAAAAGGGAGCCGCGCAGCGCCTACGGGAGTCC  60
            ||||||||||||||||||||||||||||||||||||||||||||||||||||||||||||
Sbjct  15   AAGCACCGATAGGCTCTGCCTCCCGAAGAAAAGGGAGCCGCGCAGCGCCTACGGGAGTCC  74

Query  61   GGCGGCAGCAGCCGGTACCGGCAACCACGGGCAGCTCTCAGGGAATCTCCGTCGTGAGGC  120
            ||||||||||||||||||||||||||||||||||||||||||||||||||||||||||||
Sbjct  75   GGCGGCAGCAGCCGGTACCGGCAACCACGGGCAGCTCTCAGGGAATCTCCGTCGTGAGGC  134

Query  121  CAGAGGCTCCAGTCCCCGCGAGTCCAG  147
            |||||||||||||||||||||||||||
Sbjct  135  CAGAGGCTCCAGTCCCCGCGAGTCCAG  161
```

Fig. 1 BLAST result of BRMS1 gene showing a 100 % identity with UTR ID: SHSAA011068

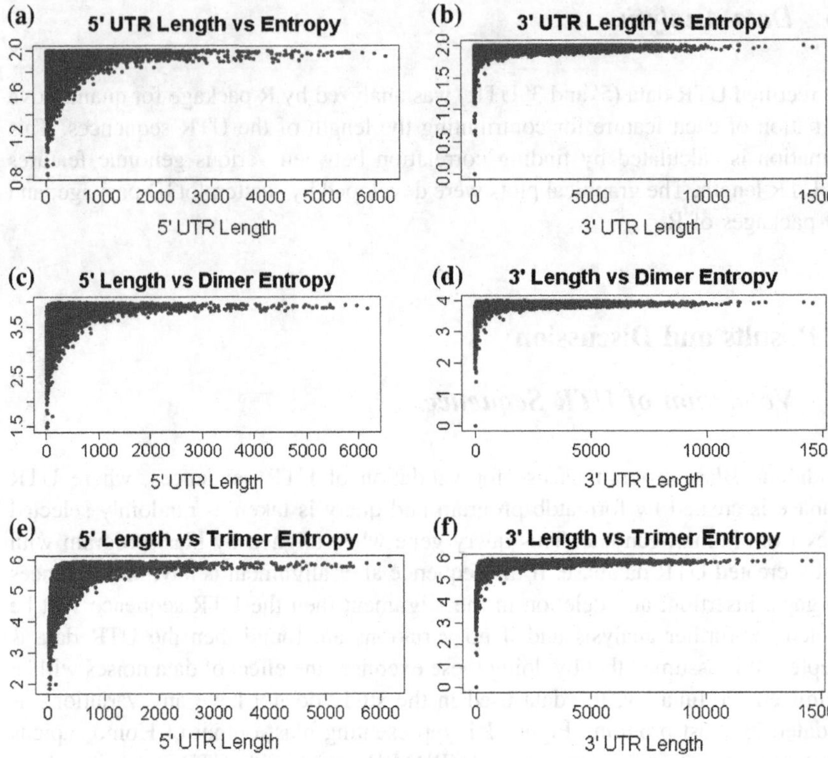

Fig. 2 The variation in the entropy *H*(*X*) with respect to change in length of UTR. Figure **a** and **b** is entropy by monomeric features for 5′ and 3′ UTRs. Figure **c** and **d** represents dimer entropy with length and E and F represents trimer entropy with length on 5′ and 3′, respectively

UTR length of the sequence and the entropy is calculated for different events which are cumulatively representing the UTR length. For example, for representing a UTR sequence monomer, dimer, and trimer are taken as event that contributes to total length of UTR. A comparative analysis of monomer, dimer, and trimer is carried out to find which types of events are representing a sequence in more informative manner. Here it is found that as the string length of an event in a sequences increases the information content of the sequence is also increasing. The genomic *H*(*X*) in monomer is up to 2, in dimer it is up to 4 and in trimer it is up to 6. The information content is also varies with length as represented in Fig. 2a–e. This entropy comparison reveals that as the length of the sequence is increasing the randomness is also increasing but after a certain length (~1000) the entropy of the sequences gets stabilized. It is concluded that the UTR length which have sequence length <1000 bp are more informative.

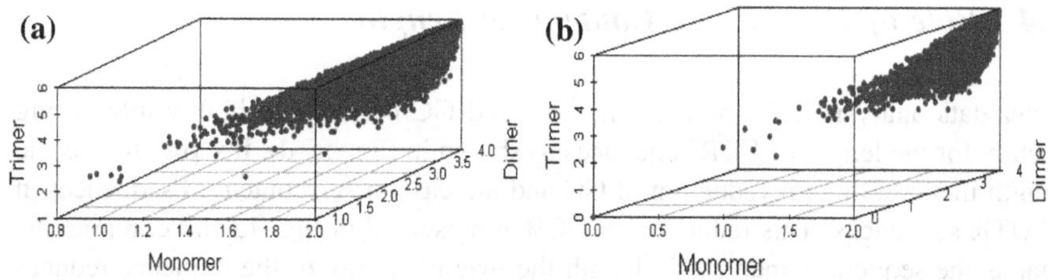

Fig. 3 3 D Scatter plot for information gain in monomeric, dimeric, and trimeric features in 5′ UTR (**a**) and in 3′ UTR (**b**)

3.3 *Information Gain in the Features*

The information gain is observed as the sequence grows in the length in both 5′ and 3′ UTR. If we compare the information stored in monomer, dimer, and trimer features of a sequence it reveals that information content is increasing as we move from monomer < dimer < trimer. This pattern of information growth is similar in each feature type. The information of 5′ UTR sequences are more versatile as compared to 3′ UTR as presented in Fig. 3a for 5′ UTR and b for 3′ UTR sequences.

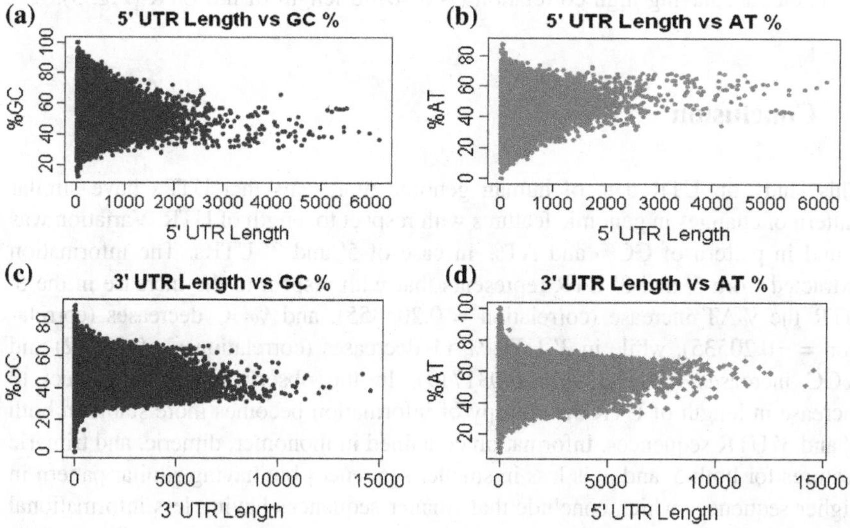

Fig. 4 **a**, **b** are showing scatter plot between GC (*blue*) and AT (*red*) content to 5′ UTR length, and **c**, **d** showing scatter plot between GC (*blue*) and AT (*red*) content to the 3′ UTR length, respectively

3.4 Role of GC and AT Content on Length

After data analysis it is observed that AT and GC content are highly influencing factors for the length of UTR sequences as shown in Fig. 4a–d). It shows that as the length increase there is reduction of GC and increase in AT content of sequences in 5′ UTR sequences. This reduction in GC% represented in Fig. 4a and c, represents that as the sequences increase in length the overall energy of the sequence reduces resulting as GC% is an important factor which stabilizes the UTR.

3.5 Correlation Between Various Genomic Features and Length of 5′ UTR and 3′ UTR

Simple Pearson correlation (Crln) was calculated with respect to various genomic features (GF) isolated from 5′ and 3′ UTR sequences as shown in Table 1. On analyzing the correlation table it is observed that in monomeric features have similar correlation with length but in dimer some features like GA, GT, TG, and AG are highly influencing the length with correlation >0.95 in 5′ and 3′ UTR while CG is having correlation <0.53 in 5′ and 0.68 in 3′ UTR said to be the least influencing factor. In trimeric features CGC, CCG, CGG, GCG are having least influence on length with correlation below 0.60 while TCA, TGG, CAT, AGA, CTT, etc. are having high correlation (~0.90) to length of the UTR (Fig. 5).

4 Conclusion

This study on UTR data of human genome represents that UTRs have similar pattern of changes in genomic features with respect to length of UTR. Variation was found in pattern of GC% and AT% in case of 5′ and 3′ UTRs. The information extracted from 5′ and 3′ UTR represents that with respect to the increase in the 5′ UTR the %AT increase (correlation = 0.205355), and %GC decreases (correlation = −0.20535), while in 3′ UTR %AT decreases (correlation = −0.03172) and %GC increases (correlation = 0.031719). In the observation, with respect to increase in length of UTR the entropy of information becomes more stable in both 5′ and 3′ UTR sequences. Information contained in monomer, dimeric, and trimeric features for both 5′ and 3′ is less in smaller sequences but having similar pattern in higher sequences which conclude that smaller sequences having less informational entropy and randomness, while the higher sequences become more vibrant. This information becomes stable in the sequences with sequence length more than 1000 bp. In correlation studies with length of the sequence to various genomic features, it is found that some features like CG, TAA, CGT, CGC, CCG, CGG, ACG are having less correlation with length (correlation < 0.70) while CT, TC, AC,

Table 1 Correlation table of all the genomic features isolated from 5′ and 3′ UTR sequences with respect to their length

GF	Crln 5′	Crln 3′	GF	Crln 5′	Crln 3′	GF	Crln 5′	Crln 3′
A	0.95584	0.97219	TTC	0.88504	0.94737	ACC	0.87027	0.90509
T	0.95136	0.96533	TCC	0.85521	0.90494	AAC	0.85508	0.92824
G	0.95763	0.96751	TAC	0.82268	0.90797	AGC	0.86797	0.90760
C	0.94472	0.95198	TGC	0.88958	0.92859	ATA	0.73293	0.77545
%GC	−0.20535	0.03172	TTA	0.73715	0.77943	ACA	0.80981	0.86267
%AT	0.20536	−0.03172	TCA	0.89200	0.96365	AAA	0.79623	0.83784
GC	0.83075	0.88482	TTG	0.85905	0.91728	AGA	0.87524	0.94055
GA	0.95139	0.97018	TCG	0.88958	0.92859	ACG	0.64689	0.71138
GT	0.95008	0.98077	TGG	0.90109	0.92596	AAG	0.87392	0.93462
GG	0.87658	0.89982	CTT	0.88657	0.94717	AGG	0.87378	0.90766
CG	0.52940	0.67223	CCT	0.87114	0.88509	GTT	0.84083	0.90397
CT	0.96093	0.97773	CAT	0.87351	0.95022	GCT	0.87856	0.91468
CA	0.95733	0.98278	CGT	0.62709	0.75478	GAT	0.85739	0.93420
CC	0.85939	0.87084	CTC	0.84097	0.89649	GGT	0.85875	0.93395
TG	0.96331	0.98805	CCC	0.71765	0.75874	GTC	0.85166	0.93444
TT	0.86683	0.89356	CAC	0.86263	0.91478	GCC	0.71828	0.77986
TA	0.82161	0.84327	CGC	0.39018	0.56400	GAC	0.85931	0.90252
TC	0.95527	0.98131	CTA	0.83888	0.91756	GGC	0.69939	0.78530
AC	0.95364	0.98421	CCA	0.88538	0.90588	GTA	0.80476	0.86182
AG	0.96092	0.97745	CAA	0.86790	0.94014	GCA	0.88252	0.93562
AT	0.86686	0.89898	CGA	0.60381	0.70043	GAA	0.86492	0.91738
AA	0.88294	0.91103	CTG	0.89688	0.90461	GGA	0.88015	0.91317
ATG	0.86027	0.94285	CCG	0.43742	0.58888	GTG	0.87972	0.93298
TGA	0.89652	0.96101	CAG	0.90142	0.92107	GCG	0.38268	0.58503
TAA	0.74224	0.77447	CGG	0.42406	0.59095	GAG	0.85562	0.89449
TAG	0.80981	0.86267	ATT	0.77717	0.81390	GGG	0.73381	0.79469
TTT	0.78359	0.83109	ACT	0.88212	0.95111	uORF	0.86027	0.94285
TCT	0.88284	0.95240	AAT	0.78006	0.81153	Entropy	0.27299	0.24040
TAT	0.73550	0.78628	AGT	0.88710	0.94604	Perplexity	0.28006	0.24819
TGT	0.86798	0.93534	ATC	0.86642	0.94250			

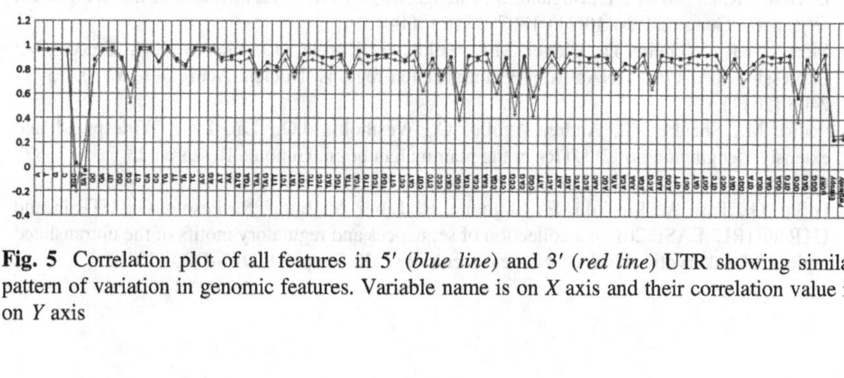

Fig. 5 Correlation plot of all features in 5′ (*blue line*) and 3′ (*red line*) UTR showing similar pattern of variation in genomic features. Variable name is on *X* axis and their correlation value is on *Y* axis

CA, GT, GA, ACT, CAT, CTT, and TCA are highly correlated (correlation > 0.90) to the length of sequences. This quantitative estimation of UTR features will be helpful in development of stable miRNA for the gene silencing to treat diseases.

References

1. Sangeeta, C., Pal, J.K.: Role of 5′- and 3′-untranslated regions of mRNAs in human diseases. Biol. Cell **101**(5), 251–262 (2009)
2. Francesca, C., Miller, W., Bouhassira, E.E.: Gene length and proximity to neighbors affect genome-wide expression levels. Genome Res. **13**(12), 2602–2608 (2003)
3. Chen, C.-H., Lin, H.-Y., Pan, C.-L., Chen, F.-C.: The genomic features that affect the lengths of 5′ untranslated regions in multicellular eukaryotes. BMC Bioinf. **12**(Suppl 9): S3 (2011). doi:10.1186/1471-2105-12-S9-S3
4. Chung, B.Y.W., Simons, C., Firth, A.E., Brown, C.M., Hellens, R.P.: Effect of 5′ UTR introns on gene expression in Arabidopsis thaliana. BMC Genomics **7**(1), 120 (2006)
5. Alexandrov, N.N., Troukhan, M.E., Brover, V.V., Tatarinova, T., Flavell, R.B., Feldmann, K.A.: Features of Arabidopsis genes and genome discovered using full-length cDNAs. Plant Mol. Biol. **60**(1), 69–85 (2006)
6. Chung, B.Y.W., Simons, C., Firth, A.E., Brown, C.M., Hellens, R.P.: Effect of 5′ UTR introns on gene expression in Arabidopsis thaliana. BMC Genomics (2006). doi:10.1186/1471-2164-7-120
7. Can, C., Chua, H.N., Zhang, H., Tarnawsky, S.P., Akef, A., Derti, A., Tasan, M., Moore, M.J., Palazzo, A.F., Roth, F.P.: Genome analysis reveals interplay between 5′ UTR introns and nuclear mRNA export for secretory and mitochondrial genes. PLoS Genet. **7**(4) (2011). doi:10.1371/journal.pgen.1001366
8. Shlomi, D., Velten, L., Sharon, E., Zeevi, D., Carey, L.B., Weinberger, A., Segal, E.: Deciphering the rules by which 5′-UTR sequences affect protein expression in yeast. In: Proceedings of the National Academy of Sciences 110.30 E2792–E2801 (2013)
9. Wegrzyn, J.L., Drudge, T.M., Valafar, F., Hook, V.: Bioinformatic analyses of mammalian 5′-UTR sequence properties of mRNAs predicts alternative translation initiation sites. BMC Bioinf. **9**(1) (2008). doi:10.1186/1471-2105-9-232
10. Piscuoglio, S., Kishore, S., Kovac, M., Gylling, A., Wenzel, F., Trapani, F., Heinimann, K.: 3′-UTR poly (T/U) tract deletions and altered expression of EWSR1 are a hallmark of mismatch repair–deficient cancers. Cancer Res. **74**(1), 224–234 (2014)
11. Rao, Y.S., Chai, X.W., Wang, Z.F., Nie, Q.H., Zhang, X.Q.: Impact of GC content on gene expression pattern in chicken. Genetics Sel. Evol. **45**(1) (2013). doi:10.1186/1297-9686-45-9
12. Jun-Jie, L., Ke-Jun, W., Wei-Xing, F., Xin, W., Xin-yan, X.: Identification of 5′ UTR splicing site using sequence and structural specificities based on combination statistical method with SVM. Appl. Math. Inf. Sci. **7** (1L), 91–97 (2013)
13. Davuluri, R.V., Suzuki, Y., Sugano, S., Zhang, M.Q.: CART classification of human 5′ UTR sequences. Genome Res. **10**(11) 1807–1816 (2000)
14. Mercer, T.R., Wilhelm, D., Dinger, M.E., Solda, G., Korbie, D.J., Glazov, E.A., Truong, V.: Expression of distinct RNAs from 3′ untranslated regions. Nucleic Acids Res. **39**(6), 2393–2403 (2011)
15. Pedro, M., Shenker, S., Andreu-Agullo, C., Westholm, J.O., Lai, E.C.: Widespread and extensive lengthening of 3′ UTRs in the mammalian brain. Genome Research **23**(5), 812–825 (2013)
16. Grillo, G., Turi, A., Licciulli, F., Mignone, F., Liuni, S., Banfi, S., Pesole, G.: UTRdb and UTRsite (RELEASE 2010): a collection of sequences and regulatory motifs of the untranslated regions of eukaryotic mRNAs. Nucleic Acids Res. **38**(Suppl 1), D75–D80 (2010)

17. Camacho, C., Coulouris, G., Avagyan, V., Ma, N., Papadopoulos, J., Bealer, K., Madden, T.L.: BLAST+: architecture and applications. BMC Bioinf. **10**, 421(2009). doi:10.1186/1471-2105-10-421
18. Daniel, H., Romashchenko, A., Shen, A., Vereshchagin, N.: Inequalities for Shannon entropy and Kolmogorov complexity. J. Comput. Syst. Sci. **60**(2), 442–464 (2000)
19. Schmitt, A.O., Herzel, H.: Estimating the entropy of DNA sequences. J. Theor. Biol. **188**(3): 369–377 (1997)
20. Machado, J.A.T.: Shannon entropy analysis of the genome code. Math. Probl. Eng. **2012**, 12 (2012). doi:10.1155/2012/132625

Comparison of Multilayer Perceptron (MLP) and Support Vector Machine (SVM) in Predicting Green Pellet Characteristics of Manganese Concentrate

Mohammad Nadeem, Haider Banka and R. Venugopal

Abstract A huge portion of available minerals and materials are in the form of fine powder that makes their management and utilization a tedious job. Pelletization, a size enlargement technique, is used to tackle aforementioned problems and considered as a combination of two subprocesses; wet or green pelletization and induration. Green pelletization is highly sensitive to the slightest variation in operating conditions. As a result, identification of the impact of varying parameters on the behaviour of the process is a challenging task. In this paper, we employ MLP and SVM, two soft computing methods, to exhibit their applicability in predicting pellet characteristics. The scarcity of training data is addressed by employing genetic algorithm. Results demonstrate the better accuracy of MLP over SVM in forecasting green pellet attributes.

Keywords Multilayer perceptron · Support vector machine · Genetic algorithm · Green pelletization

1 Introduction

Solving a problem with precision and certainty using conventional computing techniques has a serious drawback of associated cost. However, in practical situations, we are often interested in a low-cost 'good' solution rather than a costly 'optimum' one. Soft computing provides low-cost, low-precision, reasonable

Mohammad Nadeem (✉) · Haider Banka
Department of Computer Science and Engineering, Indian School of Mines, Dhanbad, India
e-mail: mail.mdnadeem@gmail.com

Haider Banka
e-mail: hbanka2002@yahoo.com

R. Venugopal
Department of Fuel and Mineral Engineering, Indian School of Mines, Dhanbad, India
e-mail: vrayasam@yahoo.com

M. Pant et al. (eds.), *Proceedings of Fifth International Conference on Soft Computing for Problem Solving*, Advances in Intelligent Systems and Computing 436, DOI 10.1007/978-981-10-0448-3_25

solutions by leveraging the tolerance for uncertainty, approximate reasoning, imprecision, and partial truth. It is a consortium of methodologies that works synergistically and provides in one form or another flexible information processing capabilities for handling real life ambiguous situations [1]. Core methodologies of soft computing include fuzzy logic (FL), artificial neural networks (ANNs), evolutionary computation (EAs), rough sets (RS) and support vector machines (SVMs). Soft computing tools are not used very extensively in pelletization literature but nevertheless found limited applications.

Pelletization is a size enlargement process where fine particles are transformed into larger aggregate particles. In many industries (such as drugs and pharmaceuticals, ore processing, fertilizers, iron and steel, etc.), size enlargement of fine material is a common practise. It enhances the productivity of industrial plants by providing desired size distributions and product geometry for increased functionality and protects users from hazards such as dust. The main purpose of agglomeration is to increase bulk density, better flowability and control of porosity and to reduce dustiness [2]. It finds variety of applications in mineral and material processing as well. Pelletization can be seen as a combination of two subprocesses; wet or green pelletization and induration. First subprocess produces green pellets which, after drying, are subjected to heat treatment to bring about surficial fusion for better strength in the later stage.

First step of pelletization process is wet pelletization in which fine particulates and water are mixed and rotated in a device, called pelletizer, to form spherical balls. The solid particles coated with liquid collide with other particles and because of surface tension effects of liquid, granules grow subsequently to form coarser particles under the influence of rotatory motion of pelletizer. Wet pelletization plays a crucial role in overall efficiency and productivity of the process since following induration subprocess relies heavily on quality of wet pellets. Besides other characteristics, pellet size, percentage of desired size pellets and pellet strength are of prime significance. Basic process parameters that affect pellet quality are: speed of pelletizer, moisture content, inclination of pelletizer, binder content, pelletization time, particle size of material, etc. $D50$ (median pellet diameter) is the commonly used measure of pellet size, % yield of +9.42 mm size pellets describes the quantity of useful products while weighted mean drop number interprets its strength.

Producing pellets with desired characteristics is a complicated task due to its sensitivity towards even the slight variation in affecting variables. This paper presents the attempt to predict three key quantity and quality measures of wet pellets using MLP and SVM for manganese concentrate. The adopted methodology is based on laboratory data but can be extended for industrial data as well. Since lab-data is limited, supply of synthetic data was fulfilled using GA so that MLP and SVM can be trained properly to produce good predictions. Objective functions of GA are devised using multiple regression. Finally, comparison of the predictive power of MLP and SVM is made and results are presented in graphical and tabular format.

The paper is organized in four sections. Section 2 provides basic preliminary on green pelletization process. We assume that readers are familiar with the working of

MLP and SVM. Related works are presented in Sect. 3. Section 4 outlines the methodology adopted for this work. Results are summarized in Sect. 5. Section 6 concludes with the findings.

2 Green Pelletization

Green pelletization is the process of obtaining green pellets by rotating a mixture of water and material into a pelletizer. Pelletizer's rotary motion is the reason that fine particles achieve a near spherical form. Water added during the rotating motion gives green pellets enough strength to withstand collision between them and with the surface of pelletizer. Sometimes, binder material is also added to provide pellets extra strength. The process is influenced by many variables that can be material-dependent, mechanical properties of pelletizer or some other factors. Some of these factors have been described below:

1. Particle size of feed: Growth of pellets is significantly influenced by initial feed size and important in determining their strength. It has been observed that the final pellet size is a function of the starting feed size fineness.
2. Moisture content: Water added in terms of weight percentage of the total feed is called moisture content of the feed and is one of the most important variables that affect green pelletization. If moisture is less, green balls formation is hindered whereas high moisture content diminishes their strength. Therefore, optimum moisture should be determined for a given feed material system.
3. Inclination of pelletizer: Inclination plays an important role in forming green pellets. As the angle of inclination is increased, compaction is more and pellets of better strength are obtained. It is measured in degrees.
4. Speed of rotation of pelletizer: With the increase in rotational speed of pelletizer, the rate of growth increases but higher speed may results in higher level of impact which break or deform the pellets. Its unit is revolution/minute.
5. Pelletization time: Time in minutes, up to which a pelletizer is rotated to form pellets.
6. Binder Content: Binders, like bentonite, starch, dextrine, etc., are added in terms of weight percentage of the feed to improve balling and strength of green pellets.

To estimate the percentage of desired size pellets and their quality, there are some measures, which are outlined in the following:

1. *D*50 (median pellet diameter): It is the value of the pellet diameter at 50 % in the cumulative weight percentage distribution and represents pellet diameter for a single experiment. Cumulative weight percentage passing is plotted against pellet size and the corresponding pellet size value against 50 % of cumulative weight percentage passing is *D*50.

2. Percentage yield of +9.42 mm pellets: It is a quantity measure of green pelletization that describes the percentage of total pellet quantity having size greater than 9.42 mm.
3. Compressive strength: It is determined by compressing green pellets between two steel plates till they break into fragments. Usually, about 20 pellets are tested for their compression strength. The unit of compressive strength of pellets is kg/pellet or gm/pellet.
4. Drop Number: It indicates the capacity of the pellets to sustain the impact by repeated drops from a fixed height. A minimum of 10 pellets are taken and are dropped one by one from a height of 30 or 45 cm on a plastered floor till they develop apparent cracks.

In the present study, moisture content, inclination and disc speed are taken as operating parameters and pellet properties such as; *D*50, % yield of +9.42 mm size pellets and drop number are predicted.

3 Some Existing Works

Sensitive nature of green pelletization process motivated the practitioners to develop model equations so that produced pellet attributes can be measured beforehand. Many researchers have worked towards the modelling of pelletization process because modelling provides an insight into its behaviour, studying its characteristics and improving its output. Some of these models were developed by Capes and Danckwerts [3], Kapur and Fuerstenau [4], Sastry and Fuerstenau [5], Ramabhadran [6] and Venugopal [7]. Most of these research attempts demonstrate the work carried out to develop model equations to predict the size distribution of pellets at any given level of influential variable parameters.

Artificial neural networks (ANNs) have been exercised at some places, especially in pharmaceutical industry [8–12] to predict pellet characteristics. Petrović et al. [12] applied various types of ANNs to model the influence of input parameters on output properties of pellets. Different variants of ANN, multilayered perceptron (MLP), generalized regression neural network (GRNN), modular neural network (MNN) and radial basis function neural network (RBFNN) were developed for the study. However, SVM has never been used so far in modelling pelletization process to the best of our knowledge. Zafari et al. [13] estimated density of the biomass based on process variables such as moisture content, speed of piston and particle size using four layered MLP pellet. They also reported the superiority of ANN based model over statistical model. Benkovic et al. [14] used MLP to predict the physical and chemical properties of agglomerated cocoa powder mixtures based on mixture composition, water content and duration of agglomeration. Sauter diameter, bulk density, porosity, chroma wettability and solubility were taken as physical attributes while chemical characteristics included phenolic content and antioxidant capacity. Mathew [15] employed ANN to accurately forecast cold compression strength of iron ore pellets for a given set of bentonite (binder content) and moisture weight percentage. Radial basis function neural network (RBFNN)-based models also found limited applications in agglomeration literature [16, 17]. Recently, Wang et al. [17] proposed a RBF neural network based modelling technique for rotary kiln pellet sintering process.

4 Method Description

MLP and SVM rely on quantity and quality of training set for producing acceptable predictions. The method of material preparation and experimental setup is outlined next.

4.1 Experimental

A total of 27 experiments were conducted on manganese concentrate as the material system with the help of disc pelletizer.

Material Preparation Manganese concentrate, prepared through magnetic separation, was stage crushed and ground to produce 100 % passing 150 μ fines. Table 1 shows the size distribution of −150 μ feed to pelletization.

Experimental setup For this experiment, feed samples of 1 kg were pelletized in a disc pelletizer of 50 cm diameter and 12.5 cm rim height. The disc was driven by a 735.5 W motor, having a variable speed drive to adjust the rotational speed. By adjusting gear arrangement, the disc inclination could be varied between 10 and 60°. We considered three operating conditions for this experiment as well, with each having three levels as shown in Table 2.

Table 1 Size distribution of manganese concentrate used in experiments

Size range (μ)	Weight %	Cumulative wt% passing
–150 to +105	7.85	100
–105 to +75	14.70	92.15
–75 to +53	8.62	77.45
–53	68.83	68.83

Table 2 Design of experiment for set of experiments

Variables	Level 1	Level 2	Level 3
Moisture content (% by wt. of feed)	11	11.5	12
Inclination (°)	30	35	40
Disc Speed (rev/min)	30	35	40

4.2 Generating Data Using Genetic Algorithm

Since data obtained from laboratory experiments was insufficient, concept of random population generation of genetic algorithm (GA) is exploited to serve the purpose. First, there was need to identify mathematical relationships between input parameters and pellet properties that could be supplied as fitness function to GA. Multiple regression is a suitable option to develop the relationships using available data and the same has been employed in the present study. For each experiment, *D*50 (median pellet diameter), % yield of +9.42 mm size pellets and weight mean drop number were estimated as a function of process parameters:

$$D50 = 6.767 \times 10^{-5} S^{1.4248} M^{3.3918} I^{-0.4157} \quad (1)$$

$$P = 4.8172 \times 10^{-13} S^{4.2045} M^{8.7001} I^{-1.1088} \quad (2)$$

$$Dr = 0.6074 \times S^{0.9533} M^{-0.4960} I^{0.1532} \quad (3)$$

where *D*50 denotes median pellet diameter in mm, *P* is the % yield of +9.42 mm pellets, *Dr* is weight mean drop number, *M* denotes moisture content in % weight of feed, *I* is disc inclination in degree and *S* denotes disc speed in rev/min.

To run GA for a given fitness function, range of different input parameters must be decided at first, which is described in Table 3. Once the range of various operating conditions was available, the next important step was to encode them. Binary encoding has been used for this task and chromosome length was taken as 15. GA was run for 10 generations with population size as 20, crossover probability as 0.8 and mutation probability as 0.01. The process followed to generate data has been described in Fig. 1.

We did not use GA to produce optimum solution. The primary goal was to select chromosomes by encoding the input variables, spanned in the pre-specified range, so that they can be treated as replacement of original data. Once several populations have been generated, encoded candidates were selected randomly according to our data requirement. Duplicate chromosomes were carefully removed. Finally, collection of chromosomes was decoded to obtain records that could be used further to train MLP. For each experiment, 100 such data samples were generated.

Computational procedure Training and testing of MLP and SVM was carried out using the publicly available software tool Weka [18]. Both prediction

Table 3 Range of input parameters used in GA

Input variable	Symbol used	Unit	Range
Disc speed	*S*	rev/min	25–45
Moisture content	*M*	% by weight of feed	10–13
Inclination	*I*	degree	25–45

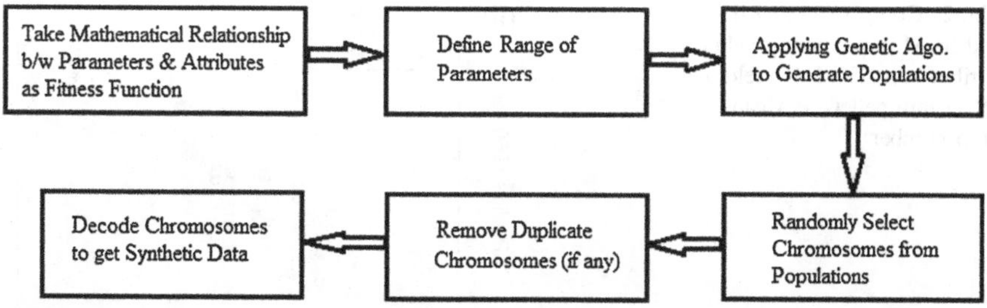

Fig. 1 Process used for synthetic data generation

algorithms were set up with three parameters as the input and particular property as the output. Out of available data sets, 60 % was used to train the network, 20 % as validation set and remaining 20 % for testing the performance of algorithms. The MLP architecture with one hidden layer consisting of two nodes was chosen, learning rate was set to 0.1 and number of epochs to 500. For SVM, complexity parameter was set to 1.0 and radial basis kernel function was employed.

5 Comparison of MLP and SVM

Estimating the product characteristics, based on certain operating condition setting, has substantial significance while modelling granulation process. Two soft computing techniques, MLP and SVM were used for this purpose as they have the ability to predict trends and patterns by learning from historical data without deriving any mathematical relationship between variables. Three pellet characteristics, namely *D*50, % yield of +9.42 mm pellets and mean drop number, were forecasted using MLP and SVM based on three process parameters—moisture content, disc inclination and disc speed. For better and instant comparative study, actual and predicted pellet attributes are presented graphically in Fig. 2.

Both tools predicted *D*50 reasonably well as evident from Fig. 2a and in doing so, they were also close to each other suggested by a lot of overlapping points. For % yield of +9.42 mm pellets, prediction results were not as satisfactory as in the case of *D*50 (Fig. 2b). However, MLP showed better prediction power than SVM since points were nearer to the perfect fit region for the former tool. Predictability for mean drop number was better than in the case of % yield of +9.42 mm pellets for MLP as well as SVM. As we can see in Fig. 2c that almost all predicted values by both tools are overlapped, suggesting same forecasting capabilities in this case. Prediction accuracies for all the three pellet characteristics have been summarized in Table 4.

The accuracy in predicting *D*50 and % yield of +9.42 mm pellets was higher for MLP than SVM. However, the difference was marginal for *D*50 (0.212 %) while for

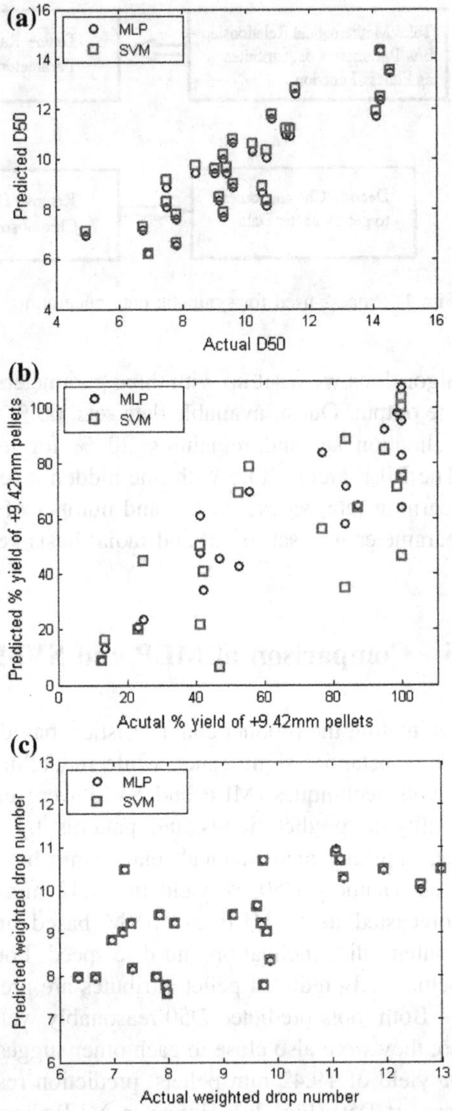

Fig. 2 Comparison of MLP and SVM in predicting pellet attributes. **a** *D*50. **b** % yield of +9.42 mm pellets. **c** Mean drop number

Table 4 Prediction accuracies of MLP and SVM for various pellet characteristics

Pellet characteristics	MLP (%)	SVM (%)
*D*50 (median pellet diameter)	89.564	89.352
Percentage yield of +9.42 mm pellets	83.204	68.448
Mean drop number	84.663	84.665

yield it was high enough (14.756 %). In the case of mean drop number, both tools showed almost same accuracy (a negligible difference of 0.002 %).

Figure 2 and Table 4 manifest the fact that pelletization is a highly complex non-linear process affected by several intangible parameters and prediction of granule characteristics with sufficient accuracy is not easy. For the present study, MLP outperformed SVM in above-stated purpose but the statement may not be true for all granulation processes.

6 Conclusion

We considered a green pelletization system for manganese concentrate and applied two soft computing techniques (MLP and SVM) to predict pellet properties under the influence of several process variables based on data generated in laboratory. MLP exhibited better prediction power than SVM in present study but this result can not be generalized for all pelletization systems.

Major points observed from the present study are:

- Soft computing techniques are capable of modelling complex non-linear processes.
- Synthetic data was generated using genetic algorithm.
- Comparison of the prediction power of MLP and SVM was made in forecasting green pellet characteristics.
- For the material system considered in the present study, MLP outperformed SVM in pellet attribute prediction.

A very few research works, especially in pharmaceutical industry, have been carried out to model pelletization process using oft computing methods and got comparable results. These techniques may serve as an alternative to conventional modelling approaches because of their effectiveness and potentiality.

References

1. Zadeh, L.A.: Making computers think like people. IEEE Spectr. **21**(8), 26–32 (1984)
2. Green, D.W., Perry, R.H.: Perrys Chemical Engineers Handbook. McGraw-Hill, New York, 7th edn. (2008)
3. Capes, C.E., Danckwerts, P.V.: Granule formation by the agglomeration of damp powders. Part II: the distribution of granule sizes. Trans. Inst. Chem. Eng. **43**, 125–130 (1965)
4. Kapur, P.C., Fuerstenau, D.W.: Size distributions and kinetic relationships in nuclei region of wet pelletization. Ind. Eng. Chem. Process Des. Dev. **5**(1), 5–10 (1966)
5. Sastry, K.V.S., Fuerstenau, D.W.: Size distribution of agglomerates in coalescing dispersed phase systems. Ind. Eng. Chem. Fund. **9**(1), 145–149 (1970)
6. Ramabhadran, T.E.: On the general theory of solid granulation. Chem. Eng. Sci. **30**(9), 1027–1033 (1975)

7. Venugopal, R.: Studies on wet pelletization characteristics of manganese concentrate and pyriteferous shales. PhD Thesis (1986)
8. Murtoniemi, E., Yliruusi, J., Kinnunen, P., Merkku, P., Leiviskä, K.: The advantages by the use of neural networks in modelling the fluidized bed granulation process. Int. J. Pharm. **108**(2), 155–164 (1994)
9. Watano, S., Takashima, H., Miyanami, K.: Scale-up of agitation fluidized bed granulation by neural network. Chem. Pharm. Bull. **45**(7), 1193–1197 (1997)
10. Behzadi, S.S., Klocker, J., Hüttlin, H., Wolschann, P., Viernstein, H.: Validation of fluid bed granulation utilizing artificial neural network. Int. J. Pharm. **291**(1), 139–148 (2005)
11. Behzadi, S.S., Prakasvudhisarn, C., Klocker, J., Wolschann, P., Viernstein, H.: Comparison between two types of artificial neural networks used for validation of pharmaceutical processes. Powder Technol. **195**(2), 150–157 (2009)
12. Petrović, J., Chansanroj, K., Meier, B., Ibrić, S., Betz, G.: Analysis of fluidized bed granulation process using conventional and novel modeling techniques. Eur. J. Pharm. Sci. **44**(3), 227–234 (2011)
13. Zafari, A., Kianmehr, M.H., Abdolahzadeh, R.: Modeling the effect of extrusion parameters on density of biomass pellet using artificial neural network. Int. J. Recycl. Org. Waste Agric. **2**(1), 1–11 (2013)
14. Benković, M., Tušek, A.J., Belščak-Cvitanović, A., Lenart, A., Domian, E., Komes, D., Bauman, I.: Artificial neural network modelling of changes in physical and chemical properties of cocoa powder mixtures during agglomeration. LWT-Food Sci. Technol. (2015)
15. Mathew, M.: Predicting the cold compressive strength of iron ore pellet using artificial intelligence technique. Int. J. Glob. Technol. Initiatives **4**(1), D33–D42 (2015)
16. Kusumoputro, B., Faqih, A., Sutarya, D., et al.: Quality classification of green pellet nuclear fuels using radial basis function neural networks. In: 2013 12th International Conference on Machine Learning and Applications (ICMLA), vol. 2, pp. 194–198. IEEE (2013)
17. Wang, J., Shen, N., Ren, X., Liu, G.: Rbf neural network soft-sensor modeling of rotary kiln pellet quality indices optimized by biogeography-based optimization algorithm. J. Chem. Eng. Jpn. **48**(1), 7–15 (2015)
18. Hall, M., Frank, E., Holmes, G., Pfahringer, B., Reutemann, P., Witten, I.H.: The weka data mining software: An update. SIGKDD Explor. **11** (2009)

Audio Pattern Recognition and Mood Detection System

Priyanka Tyagi, Abhishek Mehrotra, Shanu Sharma and Sushil Kumar

Abstract Music is and has been an integral part of our society since time immemorial. It is a subtle display of a person's emotions. Over the decades even though the way music is composed or heard has greatly evolved but what has remained constant is the entwined relationship it shares with mood. The kind of music one listens is to be governed solely by their mood at that instant. This paper proposes an automated and efficient method of classifying music on the basis of the mood it depicts, by extracting suitable features that show significant variation across songs. A database of 300 popular Bollywood songs was taken into consideration in which timbral and temporal features were extracted to classify songs into four moods: Happy, sad, relaxed and romantic. 200 songs were used to train the model by using Multilayer perceptron with backpropagation algorithm. The model exhibited an accuracy of 75 % when tested over a set of 100 songs.

Keywords Mood detection · Audio feature extraction · Multilayer perceptron

1 Introduction

Music is an expression of emotions in the form of lyrics strung together by rhythm and melody. The way music is composed, created, or even defined varies across cultures. Even after its varying forms the purpose for music for almost everyone

Priyanka Tyagi (✉) · Abhishek Mehrotra · Shanu Sharma · Sushil Kumar
Department of Computer Science and Engineering, Amity School of Engineering and Technology, Amity University, Noida, Uttar Pradesh, India
e-mail: pri_tyagi246@yahoo.co.in

Abhishek Mehrotra
e-mail: abhiscorpionster@gmail.com

Shanu Sharma
e-mail: ssharma6@amity.edu

Sushil Kumar
e-mail: kumarsushiliitr@gmail.com

M. Pant et al. (eds.), *Proceedings of Fifth International Conference on Soft Computing for Problem Solving*, Advances in Intelligent Systems and Computing 436, DOI 10.1007/978-981-10-0448-3_26

remains the same. Be it a hectic day at work, devotion of love, or getting over a heartbreak, music acts as an effective remedy for all such problems. With the fast paced technology and the invention of smart phones, music today has reached almost every person's personal gadget be it phones, iPods, mp3 players, or laptops. Thus, access to music is easily and readily available. With this wide variety of songs available, one never really likes to listen to all of them. People usually like listening to music emoting their current emotional state. Even after this strong association, music software's today are devoid of this facility of mood-aware song selection facility.

Building on the same inspiration this paper is an effort towards creating an efficient system to automatically analyze and classify songs based on their underlying mood. The proposed method would help reduce the time spent on choosing the appropriate song. However, classifying music as per its mood is a much harder task. The main reasons being:

- First, mood not just in context of music but itself is subjective in nature. Mood is influenced by many factors such as the state of mind, different cultural backgrounds, surrounding environment and various other things
- Second, a musical piece can be annotated to many similar meaning adjectives. For example, Sad or depressed can refer to the same song.

To study how music and mood are related, mood models like Hevner's and Thayer's were studied. Timbral and temporal features are extracted from music files. Multilayer perceptron with backpropagation was used for the classification of songs on the basis of mood.

The rest of the paper is organized as follows. The mood models studied for the paper are described in Sect. 2. Section 3 describes the related research work. A detailed description of the audio features extracted is given in Sect. 4. Section 5 explains the decision rule for detecting mood. Section 6 presents the results of the work carried out. The last section concludes the work and presents the opportunities for future work that can be carried out.

1.1 Mood Model

In order to map audio clips to different moods, the issue of mood taxonomy needs to be tackled. In the past, researchers have come up with models using adjective descriptor-based approach. Among the best known are that of Hevner's and Thayer's. Hevner [1, 2] designed a circle consisting of eight clusters of adjectives as depicted in Fig. 1. The adjectives with similar meaning were placed in the same cluster while adjacent clusters differed slightly in their meaning. The difference in the meaning between clusters increases with the distance. Hevner's model is a categorical model as the mood spaces comprise of categories of different moods. The problem with Hevner's model is that the use of a lot of adjectives increases the complexity of mood mapping. On the other hand, Thayer's model, [3, 4] as shown

Fig. 1 Hevner's mood model [1]

in Fig. 2, is a multidimensional model since the moods are positioned in a multidimensional space. Unlike other adjective-based models which collectively form a mood pattern, this dimensional approach adopts the theory that mood is entailed from following factors and divides music mood into four clusters: Contentment, Depression, Exuberance, and Anxious/Frantic.

Fig. 2 Thayer's model [3]

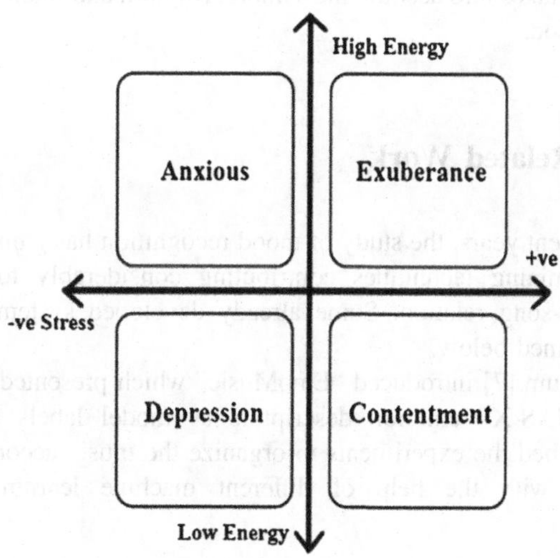

- Stress (happy/anxious)
- Energy (calm/energetic)

However, after studying the above mentioned models it was realized that a real time system application would require mood categories used/needed most by the people. Thus, a survey was held in order to attain the most sought after mood categories and taking the results into account audio clips have been classified into the following moods: Happy, Sad, Relaxed and Romantic.

1.2 Audio Features

Extraction of features is required in order to classify music on the basis of mood. McKay et al. [5] developed "jAudio," an open source audio feature extraction framework. It placed great emphasis on implementations of meta-features and aggregators that were used to generate many more features useful for musical analysis automatically (for example, running mean, instance, standard deviation, etc.). This tool is widely used for music analysis research and has been used for feature extraction in this paper as well.

Mitrovicet al. [6] dealt with the analysis and finding the relation between a wide set of latest audio features and low-level MPEG-7 audio descriptors. The analysis comprised of review and anomaly detection, which revealed the similarities between the two. The work employs Principal Components Analysis, depicting that there is low redundancy between the MPEG-7 descriptor groups. However, high redundancy was found in some groups of descriptors such as the basic spectral group and the timbre spectral group. Redundancy tells about the similarity of properties of the media objects and therefore, should not be used together. The paper provides a good insight on the choice of audio features for analysis. This paper takes into account the Timbre, Rhythm and Intensity features for the detection of mood.

2 Related Work

In recent years, the study of mood recognition has gained a lot of momentum, with data-mining techniques contributing considerably to analyze and identify the mood-song relation. Some already developed systems in this problem area are explained below.

Baum [7] introduced "EmoMusic" which presented a study of feasibility of the "PANAS-X" emotion descriptors as model labels for music classification. It described the experiments to organize the music according to its underlying emotions with the help of different machine learning methods. Naïve Bayes,

Self-Organizing Maps and Random Forest classifiers proved to be useful in predicting the mood of music with a good success rate.

Dewi and Harjoko [8] presented a mood classification system which was based on mood parameter systems using k-Nearest Neighbor classification method and Self-Organizing Maps. Its parameters were based on Thayer's Stress Energy model and used rhythm patterns of music. Classification of 120 songs with k-Nearest Neighbor had an accuracy of 73.33 % whereas it reached 86.67 % with Self-Organizing Maps.

Li and Ogihara [9] discussed SVM-based multi-label classification method for two problems: classification into thirteen sub-groups and into six super-groups. The model had a low performance as there were many borderline cases and it was difficult to make a decision. This was resolved by expanding the data sets and collecting labeling in multiple rounds.

Shirai and Dang [10] proposed the classification of moods of songs based on lyrics and meta-data. The model used three machine learning algorithms: SVM, Naïve Bayes and Graph-based methods. The results showed that the accuracy of this model was poor because mood is a subjective meta-data and lyrics cannot be considered an accurate measurement of the underlying mood of a song. Later, audio information was used with lyrics to improve upon the model.

Dan et al. [11] in their paper presented a hierarchical framework for detection of mood from acoustic data. Gaussian Mixture model and Bayesian classification were made use of for modeling the data set and classifying the features into different categories, respectively. Accuracy up to 86.3 % was achieved, about 5.7 % more than the non-hierarchical framework. The scope of this work though was restricted to classical western music.

Vyas et al. [12] in their paper have proposed an efficient method to detect mood from music by taking into account only three features-MFCC, peak difference and frame energy. K-means clustering and thresholding were used for detecting mood. An accuracy of 90 % was achieved when system was tested over a database of 100 songs. Even though satisfactory results were attained but the songs were classified into only two categories of mood: Happy and Sad.

Hampiholi [13] proposed a method to classify western as well as Indian music by extracting rhythm, intensity and timber features. The extracted features were mapped to moods in Thayer's model using C4.5 classifier. An accuracy of 60 % was achieved when tested over Indian music. It was found that accuracy fell during detection of western music.

Ujlambkar et al. [1] used Random forest followed by k-means clustering for classifying the features into four clusters-happy, sad, exciting and silent and an accuracy of 70 % was achieved. It was observed that accuracy fell by 30 % when tested with western music as music classification framework was entirely based on the Indian cultural context and the audiences classifying the western music through listening tests were all from Indian cultural background.

3 Proposed Methodology

The proposed approach aims at developing automatic audio pattern recognition and mood detection system. Figure 3 shows the proposed algorithm which is divided in two phases.

- Training
- Testing

200 songs have been taken to train the mood model then the accuracy of the model has been tested on 100 songs. Various steps of training the mood model are explained below.

3.1 Audio Preprocessing

Audio clips of mp3 format prior to extraction are first split into 30 s clips for reducing the complexity and increasing the processing time of the system. The first 30 s of clips have been used for classification as they contain the least amount of

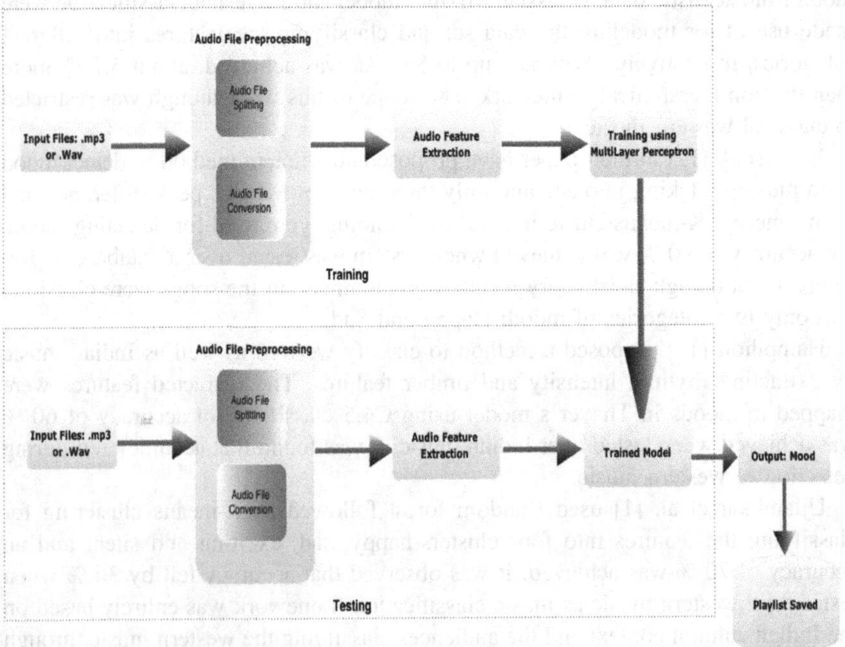

Fig. 3 Flow chart of proposed methodology

lyrics and greater amount of significant features required for extraction. The split files are then converted into .wav format sampled at 22050 Hz, mono 8 bit PCM.

3.2 Audio Feature Extraction

Features are extracted from clips using jAudio [5], an open source feature extraction application. Extraction has been performed on frame window size of 32 ms and 0.5 overlapping at 22,050 Hz. The recordings were normalized in order to ease the process of studying the extracted values. Mean and standard deviation was calculated over all recordings.

- Timbre Features: Timbre represents the quality of sound [14, 15]. It is used for differentiating between sounds having same loudness or pitch [13]. It helps judge the nature of a song as being negative or positive. Timbre can further be divided into spectral shape and spectral contrast features. Spectral shape features are known to contribute primarily for timbre in music. The spectral shape features used for extraction of timbre from music files are as follows:
  - Spectral centroid
  - Spectral roll-off point
  - Spectral flux
  - Zero crossings
  - Peak-based spectral smoothness
  - MFCC
- Intensity Feature: Intensity is useful primarily in distinguishing between Romantic and Happy songs. Even if they both are fast paced, the intensity would act as the deciding factor. Intensity is calculated by the extraction of following features:
  - Root mean square
  - Fraction of low energy windows

  RMS and low energy are inversely proportional. The higher the RMS value of a song, lower will be the low energy value [16].
- Rhythm Features: Rhythm of a song is measured significantly through the beats of a song. A happy song would have a higher value of beats per minute as compared to a sad song. In order to extract rhythm the following features have been extracted:
  - Strongest beat
  - Beat sum

  Table 1 comprises of a description of features extracted. Only those features showing significant variation across songs have been taken into consideration.

Table 1 Description of key features

Feature		Definition
Timbre	Spectral centroid	It is an indication of the position of "center of mass" of a spectrum
	Spectral roll-off	A measure of the skewness of a spectral shape
	Spectral flux	A measure of the amount of spectral change in a signal
	Zero crossings	A measure of the pitch as well as the amount of noise in a signal
	MFCC	It is a representation of the shape of the spectrum
Rhythm	Strongest beat	Measure of the strongest beat in a signal
	Beat sum	It is the sum of all beats in a histogram. Measure of the importance of regular beats in a piece of music
Intensity	Root mean square	Measure of the power of a signal
	Fraction of low energy windows	Measure of much of a signal is quiet with respect to other signals

3.3 *Mood Model Learning Using Multilayer Perceptron*

An ARFF file comprising of the extracted features is obtained post feature extraction process. Weka [17], an open source machine learning software has been used for training data, since it is written in Java and hence increases inter-hardware compatibility. The ARFF files belonging to different moods are clubbed into a single file and fed to Weka. Multilayer perceptron with backpropagation algorithm has been used for classification of music files. Multilayer perceptron is a feed-forward (data flows only in one direction) artificial neural networks model that is used for mapping a set of inputs onto their appropriate outputs. The network consists of interconnected layers, wherein each layer consists of nodes in a directed graph. MLP can distinguish data that is not linearly separable. Learning in MLP occurs by changing connection weights after each piece of data is processed, based on the amount of error in the output compared to the expected result. For training MLP, backpropagation algorithm has been made use of.

Backpropagation Algorithm:

- Architecture for the network is chosen and fixed containing the input, hidden and output layers. The inputs are in the form of sigmoid functions.
- Weights ranging between −0.5 and 0.5 are assigned randomly to all nodes.
- In order to re-train the weights, training examples are used one after the other.
- A termination condition is checked after running through all training examples. Thus, marking the end of the training of network.

3.4 Mood Detection

Post training of model, ARFF files of music clips of 30 s frame length are used for testing the data. The test files are annotated the most probable mood tag and then tested by applying the previously trained model. The number of correctly classified songs out of the total number of songs gives the accuracy of the model.

4 Results and Discussion

A database of 200 songs was used for training the data using Multilayer perceptron with backpropagation algorithm. The ROC statistics of the training data are depicted in Table 2.

As can be seen from Table 3, the songs belonging to mood—Happy and Sad are more correctly classified as compared to those belonging to mood—Romantic and Relaxed.

The final confusion matrix of the trained instances is depicted in Table 3.

Table 4 depicts the value of a few extracted features of test songs belonging to the four moods. As can be seen the values of features vary over different moods and hence help in classification.

Table 2 ROC statistics of trained data

Class	ROC area
Happy	0.963
Sad	0.949
Romantic	0.905
Relaxed	0.911

Table 3 Confusion matrix of training data

a	b	c	d	Classified as
45	0	1	2	Happy
0	42	0	8	Sad
4	3	33	10	Romantic
3	3	1	43	Relaxed

Table 4 Values of extracted features of test songs

S. No.	Name of the song	Mood	MFCC	Beat sum	RMS
1	Tu mere agal bagal hai	Happy	−87.56	1519	0.232
2	Aap ki nazron	Relaxed	−99.83	395.4	0.112
3	Sun le zara	Romantic	−82.57	1695	0.242
4	Ae kash kahin	Sad	−94.75	512.8	0.136

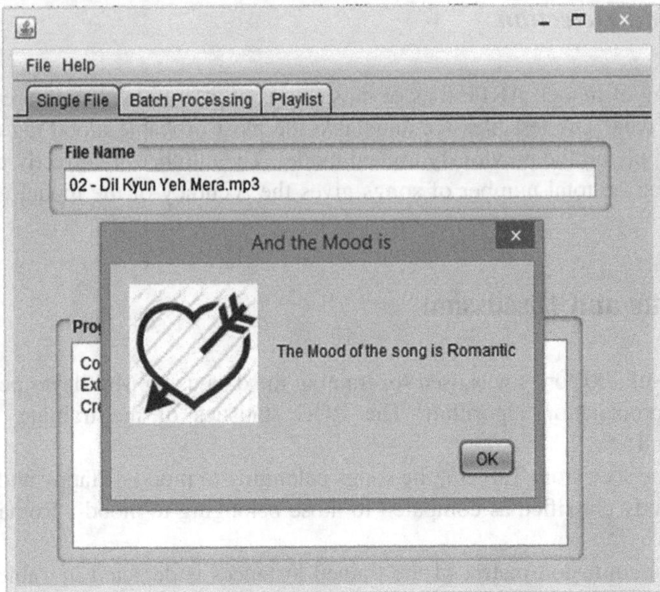

Fig. 4 Screenshot of the graphical user interface (GUI)

A Graphical User Interface (GUI) was developed for integrating the modules into a single system for testing. Figure 4 shows a screenshot of the GUI after predicting the most probable song of the chosen file. A file selected from the system is chosen and classified into one of the four moods: Happy, Sad, Relaxed, and Romantic. The GUI after internally splitting and converting the input file predicts the most probable mood.

A set of 100 songs were used for testing the application with the trained model. The results of a few test audio clips are depicted in Table 5. As can be seen from the

Table 5 Results of test audio clips

S. No.	Name of the song	Clip duration (s)	Expected mood tag	Actual mood tag
1	Tu mere agal bagal hai	30	Happy	Happy
2	Aap ki nazron	30	Relaxed	Relaxed
3	Mannat	30	Romantic	Relaxed
4	Ae kash kahin	30	Sad	Romantic
5	Ae kash ke hum	30	Relaxed	Relaxed
6	Dhating naach	30	Happy	Happy
7	Aankh hai bhari	30	Sad	Romantic
8	Humein tumse pyaarkitna	30	Relaxed	Romantic
9	Sun le zara	30	Romantic	Romantic
10	Chithi	30	Sad	Romantic
11	Chingam chabake	30	Happy	Happy
12	Tere hoke rahenge	30	Romantic	Romantic

table almost all songs belonging to mood Happy were annotated the right mood while some of the songs belonging to mood Romantic were predicted as Relaxed and vice versa. This is because the values of extracted features belonging to these moods have overlapping or close values. The overall accuracy of the model after testing with a set of 100 songs was achieved to be around 75–80 %.

5 Conclusions

The described system for Automatic Mood Detection System has numerous applications. It can be used as a part of mood sensitive music player or as an android application for classifying the user's existing playlist on the basis of the mood selected. Further, it can work as a back end application in an intelligent system. This paper was an effort towards designing an efficient real time mood detection system for classification of popular Indian Bollywood music on the basis of the mood. Multilayer perceptron with backpropagation algorithm generated satisfying results and a classification accuracy of 75 % was achieved. The completion of the paper has resulted in generation of a base model for classification of popular Indian Bollywood music. The current model can be improvised in future through experimentation and analysis of more classification algorithms and features for increasing the accuracy of mood detection. The scope of this paper can further be extended to different genres of Indian music like Classical music, Ghazals and Carnatic Music.

References

1. Ujlambkar, A., Upadhye, O., Deshpande, A., Suryawanshi, G.: Mood based music categorization system for bollywood music. Int. J. Adv. Comput. Res. **4**(1, 14) (2014). ISSN (print): 2249-7277 ISSN (online): 2277-7970
2. Hevner, K.: Experimental studies of the elements of expression in music. Am. J. Psychol. **48**, 246–268 (1936)
3. Thayer, R.E.: The Biopsychology of Mood and Arousal. Oxford University Press, New York (1998)
4. Thayer, R.E.: The Origin of Everyday Moods: Managing Energy, Tension, and Stress. Oxford University Press, New York (1996)
5. McEnnis, D., McKay, C., Fujinaga, I., Depalle P.: jAudio: a feature extraction library. In: Proceedings of the International Conference on Music Information Retrieval, p. 6003 (2005)
6. Mitrovic, D., Zeppelzauer, M., Eidenberger, H.: Analysis of the data quality of audio descriptions of environmental sounds. J. Digital Inf. Manag. **5**(2), 48 (2007)
7. Baum, D.: Emomusic—classifying music according to emotional. In: Proceedings of the 7th Workshop on Data Analysis (WDA), Kosice, Slovakia (2006)
8. Dewi, K.C., Harjoko, A.: Kid's song classification based on mood parameters using K-nearest neighbor classification method and self organizing map. In: International Conference on Distributed Frameworks for Multi-media Applications (DFmA) (2010)

9. Li, T., Ogihara, M.: Detecting emotion in music. In: Proceedings of the International Symposium on Music Information Retrieval, Washington D.C., USA (2003)
10. Dang, T.T., Shirai, K.: Machine learning approaches for mood classification of songs toward music search engine. In: International Conference on Knowledge and Systems Engineering (2009)
11. Liu, D., Lu, L., Zhang, H.J.: Automatic mood detection from acoustic music data. In: Conference: ISMIR 2003, 4th International Conference on Music Information Retrieval, Baltimore, Maryland, USA, October 27–30 (2003)
12. Vyas, G., Dutta, M.L.: Automatic mood detection of indian music using MFCCs and K-means algorithm. In: Seventh International Conference on Contemporary Computing (IC3) (2014)
13. Hampiholi, V.: A method for music classification based on perceived mood detection for Indian bollywood music. World Acad. Sci. Eng. Technol. **6**, 12–25 (2012)
14. Bhat, A.S., Amith, V.S., Prasad, N.S.: An efficient classification algorithm for music mood detection in Western and Hindi music using audio feature extraction. In: 2014 Fifth International Conference on Signals and Image Processing
15. Aucouturier, J.-J., Pachet, F., Sandler, M.: The way it sounds: timbre models for analysis and retrieval of music signals. IEEE Trans. Multimedia **7**(6) (2005)
16. Singh, P., Kapoor, A., Kaushik, V., Maringanti, H.B.: Architecture for automated tagging and clustering of song files according to mood. IJCSI Int. J. Comput. Sci. **7**(4, 2) (2010)
17. Hall, M., Frank, E., Holmes, G., Pfahringer, B., Reutemann, P., Witten, I.H.: The WEKA data mining software: an update. SIGKDD Explor. **11**(1) (2009)

Sumdoc: A Unified Approach for Automatic Text Summarization

Mudasir Mohd, Muzaffar Bashir Shah, Shabir Ahmad Bhat, Ummer Bashir Kawa, Hilal Ahmad Khanday, Abid Hussain Wani, Mohsin Altaf Wani and Rana Hashmy

Abstract In this paper, we focus on the task of automatic text summarization. Lot of work has already been carried out on automatic text summarization though most of the work done in this field is on extracted summaries. We have developed a tool that summarizes the given text. We have used several NLP features and machine learning techniques for text summarizing. We have also showed how WordNet can be used to obtain abstractive summarization. We are using an approach that first extracts sentences from the given text by using ranking algorithm, by means of which we rank the sentence on the basis of many features comprising of some classical features as well as some novel ones. Then, after extracting candidate sentences, we investigate some of the words and phrases and transform them into their respective simple substitutes so as to make the final summary a hybrid summarization technique.

Mudasir Mohd (✉) · M.B. Shah · S.A. Bhat · U.B. Kawa · H.A. Khanday · A.H. Wani · M.A. Wani · Rana Hashmy
Department of Computer Sciences, University of Kashmir, Srinagar, India
e-mail: mudie.mohammad@gmail.com

M.B. Shah
e-mail: shahmzfr21@yahoo.com

S.A. Bhat
e-mail: shabirmca1@gmail.com

U.B. Kawa
e-mail: ubksinceborn@gmail.com

H.A. Khanday
e-mail: hilalhyder@gmail.com

A.H. Wani
e-mail: abid.ku@gmail.com

M.A. Wani
e-mail: mohsin.strx@gmail.com

Rana Hashmy
e-mail: ranahashmy@gmail.com

M. Pant et al. (eds.), *Proceedings of Fifth International Conference on Soft Computing for Problem Solving*, Advances in Intelligent Systems and Computing 436, DOI 10.1007/978-981-10-0448-3_27

Keywords Automatic text summarization · Sentence extraction · Power factor · Stemming · Cosine similarity

1 Introduction

Today, internet contains vast amount of electronic collections that often contain high quality information. However, usually the Internet provides more information than is needed. User wants to select best collection of data for particular information need in minimum possible time. Text summarization is one of the applications of information retrieval, which is the method of condensing the input text into a shorter version, preserving its information content and overall meaning. The field of text summarization consists of two approaches, abstract summaries [1–5] and extracts summaries. Extract summarization is the process of extracting some important parts (sentences) from original text which contain the main idea of topic. While as abstract summarization is process of giving the main idea of text in own words. There has been a huge amount of work on extract-based summaries. Most of the previous methods on the sentence extraction-based text summarization take into account the features like term frequency [2], text length to assign indexing weight to terms, as it was suggested by LUHN that topic words are the most frequently occurring words in the topic and Edmundson [6] suggested that the larger sentences are most important sentences of the topic. Further researches proposed some other features like giving extra importance to, noun phrases [7], capital words [8], etc. Our proposed work generates summaries by extracting sentences from given topic by using these features and some new features for ranking sentences. To further hybrid approach [9], we in our proposed work use lesk algorithm and WordNet at the last stage of our algorithm so, that the summary generated is partially abstract.

2 Methodology

Our algorithm works in three phases that is preprocessing, feature extraction and post processing:

This has been shown diagrammatically in Fig. 1.

2.1 Preprocessing

After getting the raw text as input we first preprocess the text. It includes five steps as:

- **Segmentation:** The input text is tokenized into sentences and then into words so, that different rank can be assigned to different words and then sentences.

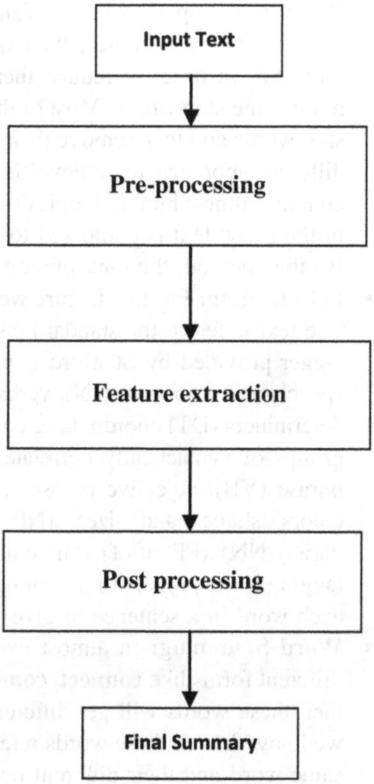

Fig. 1 Different phases of Sumdoc

- **Synonym removal:** The author often uses two or three synonyms in its text to refer to same word for example **story and tale**. Now, if we process the text as such these words will be treated as different words and will get different ranks but they refer to the same idea and should be treated as same words, this will improve the performance of our system. For example, if we rank words only on the basis of their performance then if our text contains the words like "*beast*"—5 times, "*animal*"—4 times and word "*plant*"—7 times, then these words will get rank as 5, 4, and 7, respectively, thus considering the word "*plant*" as more important than other two. But if we have changed the word "*beast*" and "*animal*" into same word (as they refer to the same idea) then they should have got higher and hence more chances to be selected in summary.
  This is achieved by making use of Lesk algorithm [10, 11] and WorldNet [12]; here, we change the meaning of all synonyms into same word.
  Lesk algorithm is used here in order to minimize word sense disambiguity.
- **Removing ambiguity initially:** Authors often write some parts of text more than one time. So, redundancy needs to be removed initially. Cosine Similarity measure is used for this purpose. Cosine similarity is calculated among all sentences, and out of two sentences whose cosine similarity is more than 80 %, out is discarded

- **Removing Stop Word:** According to Luhn if we assign rank to words on the basis of their occurrence, then words like the, is, of, are etc. will get highest rank. So, in order to reduce their rank (as they are not the topic words) we remove the stop words. Most of the systems proposed so far make use of a list of stop words and then remove them from the input text. But our system uses some different approach to achieve this goal [4], we use background corpus which contains almost hundred topics on different fields, then occurrence of each word in the given text is compared to its occurrence in all background corpus texts. By this method, the rank of stop words is decreased to almost zero.
- **POS tagging:** For this feature we use pos tagger [13–16] prepared by Stanford. The text is fed to the standard parser for generating tokens. A parser cum POS tagger provided by Stanford is used to tag the input text into various parts of speech such as nouns (NN), verbs (VBZ), adjectives (JJ), and adverbs (ADVB), determiners (DT) coordinating conjunction (CC), etc. It also divides the text into groups of syntactically correlated parts of words as Noun phrase (NP), verb phrase (VB), adjective phrase (AP), etc. Example: The rose has a variety of colors, shapes and sizes. (NP the/DT rose/NP) (VP has/VBZ) (NP a/DT variety/NN) (PP of/of) (NP colors/NN shapes/NN and/CC sizes/NN). Word tagging is the process of assigning P.O.S) like (noun, verb, and pronoun, etc.) to each word in a sentence to give word class.
- **Word Stemming:** In almost every text document there are words with their different forms like **connect, connected, connecting**. If we process them as such then these words will get different ranks according to considered features, but we know that all these words refer to the same idea, as they should be treated as same word and their different occurrences should be given same rank. This is achieved by stemming; removing suffixes by automatic means is called Stemming [7]. The proposed system employs the Porter stemming algorithm with some improvements on its rules for stem. Terms with a common stem will, for example: (CONNECT, CONNECTED, CONNECTING, CONNECTION, CONNECTIONS) are conflated into a single term. This may be done by removal of the various suffixes -ED, -ING, -ION; IONS to leave the single term CONNECT. In addition, the suffix stripping process will reduce the number of terms in the system, and hence reduce the size and complexity of the data in the system, which is always advantageous.

This can be shown diagrammatically in Fig. 2.

2.2 Assigning Ranks to Words and Sentences

This step is dived into two stages, in first stage we rank the words according to different features and in second stage we rank the sentences.

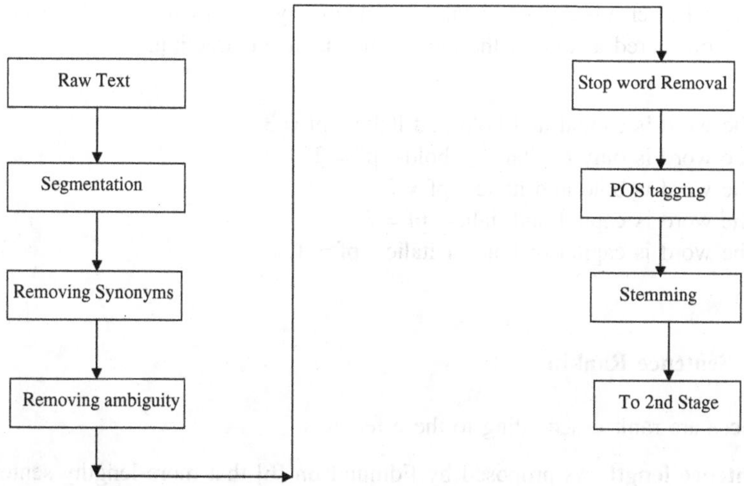

Fig. 2 Preprocessing

2.2.1 Word Ranking

All the words are ranked according to these features:

- **Tf * Idf ranking** Term frequency and inverse document frequency [4] of each word is calculated by using formulae as:

$$\text{Tf} * \text{Idf} = C(w) * \log\left(\frac{D}{d(w)}\right)$$

  where *C*(*w*) is the number of occurrences of current word in given text,
  D is number of background corpus documents taken. And
  d(*w*) is number of background documents containing that word.
  By using this formulae the stop words will got their Tf * Idf rank almost nearer to zero, hence reducing their importance and other words will get their rank according to their occurrences.
- **Noun and verb Phrase ranking** For every word it is checked weather it is noun or verb phrase. It is binary feature for each word its value is 1 or 0. Then, this feature is considered in sentence ranking giving the sentences weight age according to number of Noun or Verb Phrases it contains besides other features.
- **Power factor** Author most often writes the important words and phrases in bold and/or italic and/ or capital. We take this feature as one of the factors for extracting important words [7], this is also the binary feature, for each word we

check whether it is capital or italic or bold or any one of two or all three and it is then considered as one of the factors of sentence extraction.

if the word is capital and bold and italic—pf = 3
if the word is only capital and bold—pf = 2
if the word is bold and italic—pf = 2
if the word is capital and italic—pf = 2
if the word is capital or bold or italic—pf = 1.

2.2.2 Sentence Ranking

Sentences are ranked according to these features:

- **Sentence length** As proposed by Edmundson [6] that more lengthy sentences are more important.

Hence, for each sentence we assign it value *len* equal to the number of words contained in sentence.

- **Tf * Idf** For each sentence we calculate Sum of tf * idf of each word [2, 4]. Then, divide it by the tf * idf of the sentence with highest tf * idf value to make this value normalized (i.e., between 0 and 1).
- **Existence of Noun and verb phrases** For each sentence, we calculate the number of noun phrase and verb phrases. Hence, for each sentence we assign value nvf, where nvf is equal to number of Noun and verb phrases in given sentence [13, 17, 18]. This is then divided by the number of noun and verb phrases of the sentence with most noun and verb phrases to make this value normalized (i.e., between 0 and 1).
- **Summation of power factor** For each sentence we calculate sum of pf values. This is then divided by the summation of pf value of the sentence with highest summation of pf value to make this value normalized (i.e., between 0 and 1).

We assign weights to all the sentences by using below model which considers all the above factors as:

$$l_i = \alpha \frac{\sum_{w \epsilon s_i} \mathrm{tf}(w) \cdot \mathrm{idf}(w)}{\max\limits_{s_j \epsilon c}\left\{\sum_{w \epsilon s_i} \mathrm{tf}(w) \cdot \mathrm{idf}(w)\right\}} + \beta \frac{\mathrm{len}\, s_i}{\max\limits_{s_j \epsilon c}(\mathrm{len}\, s_j)} + \gamma \frac{\sum \mathrm{pf}(w)}{\max\limits_{s_j \epsilon C}(\sum \mathrm{pf}(w))}$$
$$+ \Phi \frac{\sum \mathrm{nvf}(w)}{\max\limits_{s_j \epsilon C}(\sum \mathrm{nvf}(w))} + \partial * \frac{\sum \mathrm{nvf}}{\max\limits_{s_j \epsilon C}(\{\sum \mathrm{nvf})}$$

where α, β, γ, ϕ and ∂ are constants used to vary the value of different factors. C is set of all the sentences.

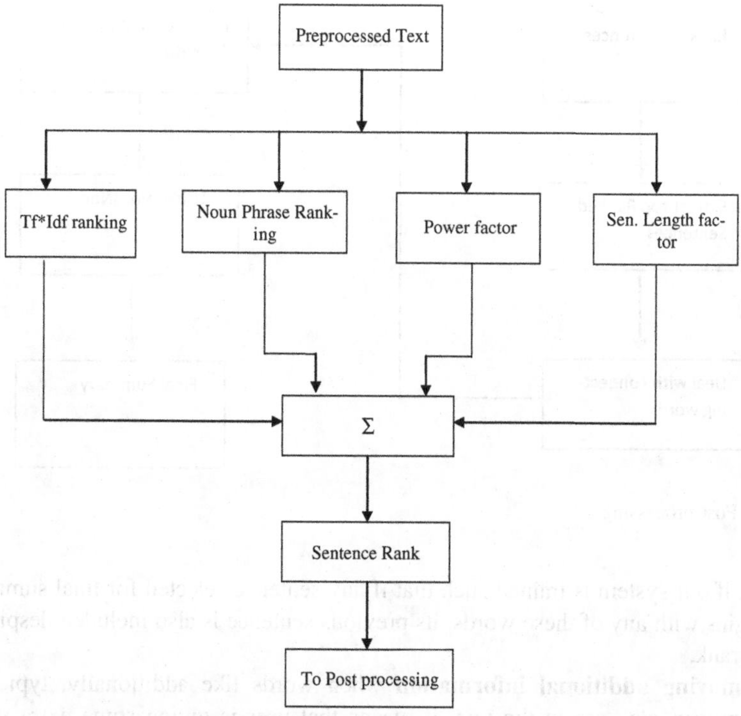

Fig. 3 Assigning ranks to words and sentences

Tf(*w*) * idf (*w*) is the term frequency multiplied by inverse document frequency calculated for each word,

Len is length of each sentence, pf(*w*) is the number value of power factor for each word (which varies between 0 and 3 as discussed above), nvf is the number of Noun and Verb phrases contained in that sentence.

Max{Σ} for each term is the maximum value of that factor for all sentences.

This stage can be shown diagrammatically in Fig. 3.

2.3 *Post Processing (Generating Summary)*

Post processing consists of four steps:

a. **Sentence extraction** After calculating Sen_{rank} for every sentence, we select the sentences with highest rank to form the summary such that summary is 30 % of original text.
b. **Dealing with connecting words** In natural languages there are some words called connecting words such as however, although, but, etc., and if any sentence begin with them, there meaning is incomplete without previous sentence.

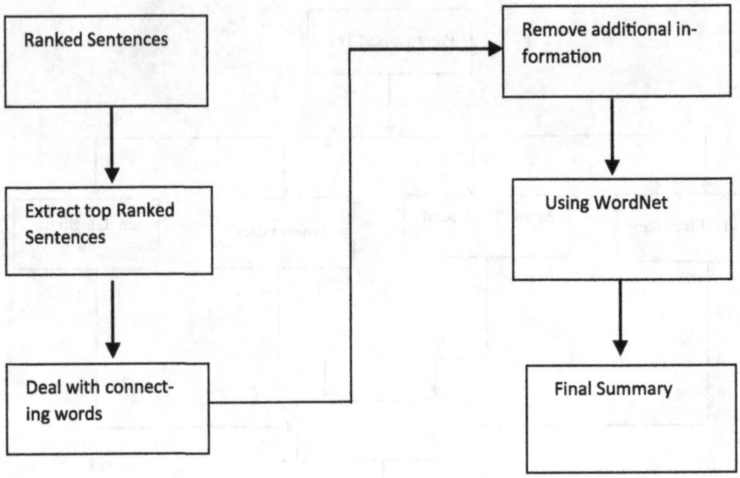

Fig. 4 Post processing

So, if our system is trained such that if any sentence selected for final summary begins with any of these words, its previous sentence is also included despite of its rank.

c. **Removing additional information** when words like additionally, typically, moreover, etc., are in the text, it means that user is giving some extra information which is not too much important. So, our system always, discards those parts of sentences which contain these words.
d. **Using WordNet** Here again, we make use of WordNet and lesk algorithm [19, 20] to change some words and phrases in final summary into their Simple Synonyms (meanings), This step helps us to make the summary somewhat abstract-based.

This stage can be shown diagrammatically in Fig. 4.

3 Result

We have developed an automatic Text Summarizer viz. Sumdoc. The methodological literature on Text Summarization is large and growing, and no one package can hope to implement all known measures and techniques. Sumdoc provides a collection of procedures which are diverse, and which cover many of the methods currently seeing wide use within the field. In Sumdoc in which we have used several NLP and Machine learning techniques like POS tagging, Stemming, WordNet, Cosine Similarity algorithm, and Lesk algorithm. We have tested our system with 10 documents. Here each document comprises of around 25 sentences.

For auto summarization we have fixed the percentage as desired by user, for example, if user desires his summary to be 50 %, i.e., it will shrink the summary to half of the original document. The Screen shots of our system and the graph for the results produced is given in Figs. 5, 6 and 7.

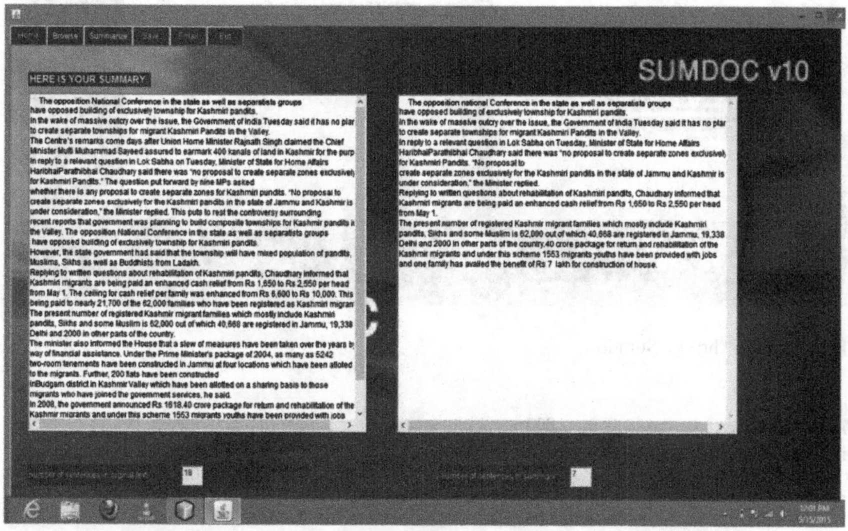

Fig. 5 Screen shot of Sumdoc

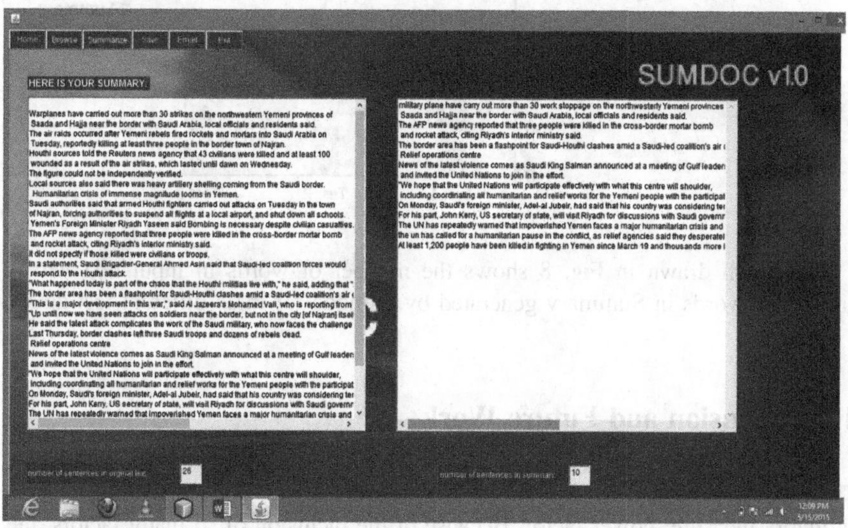

Fig. 6 Screen shot of Sumdoc

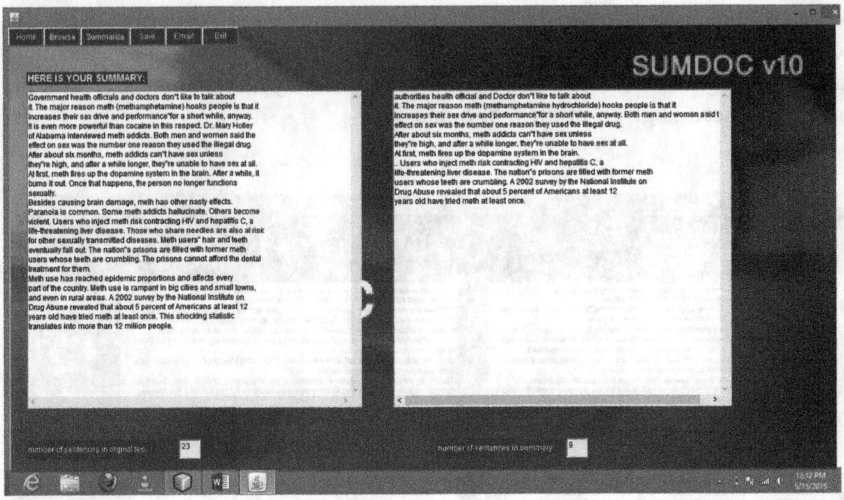

Fig. 7 Screen shot of Sumdoc

Fig. 8 Input text versus summary generated text

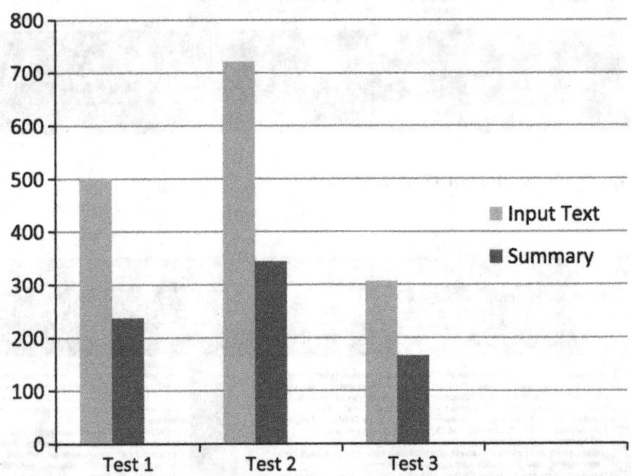

The graph drawn in Fig. 8 shows the number of words in input text versus number of words in Summary generated by Sumdoc for three different test data.

4 Conclusion and Future Work

The scheme proposed in this paper uses WordNet with lesk algorithm, Cosine similarity measure, power factor. Because of the inclusion of so many factors, the accuracy and effectiveness of our system is more than that of other automatic text summarization systems in most cases. Even though it is difficult to find meticulous abstraction of text but involvement of multiple factors in our newly developed tool has shown decent performance gains as is evident from performance studies. As future work, we intend to come up with a comprehensive abstract-based summarization model which shall be more accurate than the current conventional systems.

References

1. Nenkova. A., McKeown, K.: Found. Trends Inf. Retrieval **5**(2–3), 103–233 (2011)
2. Amini, M.R., Usunier, N., Gallinari, P.: Automatic text summarization based on word-clusters and ranking algorithms. Computer Science Laboratory of Paris 6, 8 Rue du Capitaine Scott, 75015 Paris, France
3. Patil, V., Krishnamoorthy, M., Oke, P., Kiruthika, M.: A statistical approach for document summarization. Department of Computer Engineering Fr. C. Rodrigues Institute of Technology, Vashi, Navi Mumbai, Maharashtra, India
4. Yu, L., Liu, M., Ren, F., Kuroiwa, S.: A Chinese automatic text summarization system for mobile devices. Int. Inf. Inst. **13**, 3(B) (2010)
5. Ren, F.: Automatic abstracting important sentences. Int. J. Inf. Technol. Decis. Making **4**(1), 141–152 (2005)
6. Edmundson, H., Wyllys, R.: Automatic abstracting and indexing—survey and recommendations. Commun. ACM **4**(5), 226–234 (1961)
7. Balabantaray, R., Sahoo, D., Sahoo, S.M.: Text summarization using term weights. Int. J. Comput. Appl. (0975–8887) **38**(1), 10–14 (2012)
8. Pal, A., Maiti, P.K., Saha, D.: An approach to automatic text summarization using simplified Lesk algorithm and Wordnet
9. Chang, T., Hsiao, W.: A hybrid approach to automatic text summarization. In: 8th IEEE International Conference on Computer and Information Technology (CIT 2008), Sydney, Australia, 2008
10. Banerjee, S.: Adapting the Lesk algorithm for word sense disambiguation to WordNet
11. Kulkarni, A.R.: An automatic Text summarization using feature terms for relevance measure (2002)
12. Carlos, B.: WordNet.Br An Exercise of Human Language Technology Research. Dias-da-Silva Universidad estadual, Paulista, Brazil
13. Das, D.: Unsupervised part-of-speech tagging with bilingual Graph-Based projections. Carnegie Mellon University, Pittsburgh, PA 15213, USA
14. Manne, S., Fatima S.S.: A feature terms based method for improving text summarization with supervised POS tagging
15. Stanford NLP Group.: Stanford log-linear part of speech tagger. http://nlp.stanford.edu/software/tagger.shtml. Accessed 15 June 2009
16. Brill, E.: A simple rule-based part-of-speech tagger. In: Proceedings of the Third Conference on Applied Computational Linguistics. Association for Computational Linguistics (1992)
17. Dalianis, H.: SweSum-a text summarizer for Swedish. Technical Report, TRITA-NA-P0015, IPLab-174, KTH NADA, Sweden, 2000
18. Edmundson, H.P.: New methods in automatic extracting. J. Assoc. Comput. Mach. **16**(2), 264–285 (1969)
19. Hull, D.A.: Information Retrieval Using Statistical Classification. Ph.D. dissertation, Stanford University, 1994
20. Marcu, D.: The Rhetorical Parsing, Summarization, and Generation of Natural Language Texts. Ph.D. dissertation, University of Toronto, 1997

Improvising and Optimizing Resource Utilization in Big Data Processing

Praveen Kumar and Vijay Singh Rathore

Abstract This paper is to improvising and optimizing the scenario of Big data processing in cloud computing. A homogeneous cluster setup supports static nature of processing which is a huge disadvantage for optimizing the response time towards clients. In order to avail utmost client satisfaction, the host server needs to be upgraded with the latest technology to fulfil all requirements. Big data processing is a common frequent event in today's Internet and the proposed framework improvises the response time. This will also make sure that the user gets its entire requirement fulfilled in optimal time. In order to avail utmost client satisfaction, the server needs to eliminate homogeneous cluster setup that is encountered usually in parallel data processing. The homogeneous cluster setup is static in nature and dynamic allocation of resources is not possible in this kind of environment. This will improve the overall resource utilization and, consequently, reduce the processing cost.

Keywords Data mining · Data warehousing · Parallel data processing

1 Introduction

In today's digital generation, a huge amount of data has been processed parallel in the Internet. Providing optimal data processing in least time improvises the output of parallel data processing. There are many users that try to access the same data over the Internet and it is a stimulating job for the server to provide optimal outcome. The large volume of data they have to deal with everyday has made old style data bank solutions prohibitively expensive. Instead, these companies have

Praveen Kumar (✉)
Department of Computer Science, NIMS University, Jaipur, India
e-mail: praveenvashisht07@gmail.com

V.S. Rathore
Shri Karni College, Jaipur, Rajasthan, India
e-mail: vijaydiamond@gmail.com

M. Pant et al. (eds.), *Proceedings of Fifth International Conference on Soft Computing for Problem Solving*, Advances in Intelligent Systems and Computing 436, DOI 10.1007/978-981-10-0448-3_28

promoted an architectural paradigm based on a large number of servers. Problems like processing crawled documents or restoring a web file are split into several independent subjobs, distributed among the available nodes, and computed in parallel.

Big data processing is a key feature in accessing and operating on huge set of data's [1]. There are several ways available to process data parallel which improvises time and response. Today's framework has a huge disadvantage that can be termed by a homogeneous cluster setup [2]. To be more precise, when a job manager is allocated with a job it then divides that job into many subjobs and it allocates to each task manager [3]. Now once this cluster is setup and the Big data processing begins, there can be no possible ways by which we can add more task managers or eliminate any executed task managers until all have executed. This is an ambiguous situation when there can be no data or resources allocation during the middle of data processing. This creates a problem for the server (Job Manager) to offer complete satisfaction to its users.

1.1 Overview

There are several amount data being processed in today's web. There are several challenges during parallel processing of this huge amount of data. During our research [4], there were many ambiguities we came across regarding parallel data processing.

The ambiguities are like:

(1) More delayed response time
(2) Resources cannot be allocated once but the numbers of task managers are allocated
(3) High traffic if data with many peers in action
(4) Data access used to be slower as per approximation, etc.

All these challenges being very general, the most important problem was if the data to be accessed by the user was of huge size than the processing becomes slower. Facing these challenges regarding parallel data processing, there was a straightforward approach that was deployed in our proposed paper.

The *primary task* that [5] was applied in our experiment is the use of map reduce algorithm. This map reduce algorithm stated a divide–conquer organization of working with dataset. This algorithm made it easy for the host server (job manager) to handle the job quite efficiently. It halts the job into many subjobs and executes them individually with the help of task managers.

The *second idea* was all about eliminating the homogeneous cluster setup of networks. This will allow the dynamically allocation of resources to the host server at any point of time. Dynamic allocation of resources allows optimizing the Big data processing in a new manner. This approach offers a new scope of viewing these given challenges and moulding it to operate in an efficient manner.

The experiment is analyzed by taking into account various time slots that allows imagining the whole operation between job manager, task manager, and user.

When the program has been fit in [2] the appropriate map, the task of splitting the job into subtask is taken care by the execution framework. The job to reduce a single Map always consist of reduce program with a distinct map. Client–server computing, also called networking is a distributed application architecture which divides the job or workload of the system between the component, i.e. the service requester (Client) and the service provider (Server).

The Clients and the servers work over separate hardware on a computer network. A server (Host) machine is high performance system. It shares the resources with the client. A client also shares its resources over the network by initiating the communication sessions with the servers that wait for the incoming requests form client. The execution of the theorem is based on networks search processes [6] for fast processing.

1.2 Proposed System Overview

A number of systems have been proposed in few years to facilitate MTC. Although Systems proposed shared common goals (e.g. Hiding issues of parallelism or fault tolerances), their field of application is differed. Whereas Map Reduce algorithm has been [7] design to analyze large datasets, i.e. in order to improvise the Big data processing between big number of nodes (users) and servers.

The proposed system also exhibits the isolated allocation of resources [8] that are not available on the server but are available on distant servers, parallel while handling the available data. The projected framework allows a platform for the server and users in efficient and optimized parallel data processing. It allows the task/job manager to distribute resources [2] at any given time and this improvises the response time. Therefore the problem of a homogeneous cluster network is removed and thus it is more optimized approach.

When a user has accepted the data for processing into the required map and reduce pattern [9], the execution framework takes care of splitting the job into subtasks, distributing and executing them. The proposed system also offers dynamic allocation of resources to any of the task managers during execution. The allotted resources are then available on the host server always and it can be operated later.

2 Objective of the Paper

The objectives and purpose for this paper are to improvise and optimize the scenario of Big data processing in data warehousing [6]. Millions of data are accessed in the web by the user and it is the utmost responsibility of the server to provide satisfaction to the user. It is always viable to be on the other side but dealing with such huge amount of data everyday makes the situation more complicated. Hence there are very few loopholes in the current framework but these are enough to degrade the performance in parallel data processing.

Therefore the main objective and purpose of this paper is to optimize [1] the parallel data processing. In order to avail utmost client satisfaction, the host server needs to be upgraded with the latest technology to fulfil all requirements. The proposed map reduce algorithm is a generic framework that can be deployed in this scenario. Another important goal of this paper [8] is to allocate resources or data dynamically to the host server (job manager) so that every requirement of resources can be fulfilled at any instant of time. This allows higher and sharper response time and avoiding delay in transfer.

The job manager should be aligned [10] with all its task managers to avail maximum optimization. The task managers are mapped by the job manager with many jobs and they all solve it individually which is later reduced to return it back to the client. This allows the load for the execution to be shared and the overall execution of huge sized data is more feasible in less time. Typically data of huge size [4] are the toughest challenge to be dealt in the web for parallel data processing. This paper makes sure that the user gets its entire requirement fulfilled in optimal time.

We discussed pros and cons of Map Reduce and classified its improvements. Map Reduce is simple [5] but provides good scalability and fault tolerance for massive data processing. The performance evaluation gives a first impression on how the ability to assign specific jobs to specific task manager of a processing job, as well as the possibility to automatically allocate/de-allocate virtual machines in the course of a job execution, can help to improve the overall resource utilization and, consequently, reduce the processing cost.

3 Architecture of Projected System

See Fig. 1.

3.1 Client

This module deals with the Client or the Customer whose needs are to be fulfilled. The client always requests [11] to the server for executing a particular operation and send a response back to it accordingly. Nevertheless a client is always volatile about its operation. In our proposed paper, the client selects the file that it wants to download. After the file is selected the client clicks on the download button. It is obvious [12] though that the client always tries to request to download a file in this scenario. After clicking the download button, it waits for the server to send a response back. The status of the downloading interface is shown to the client so as

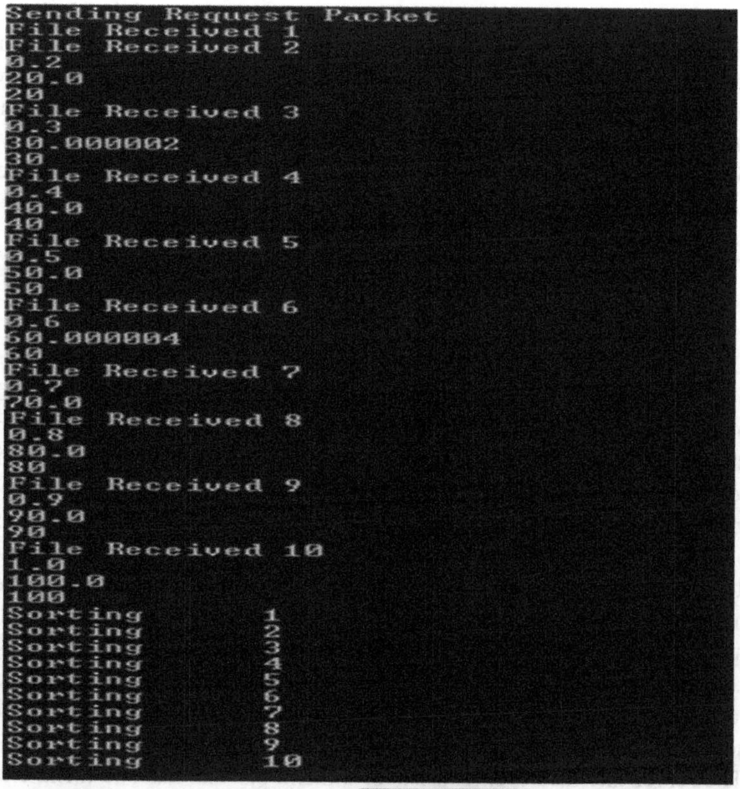

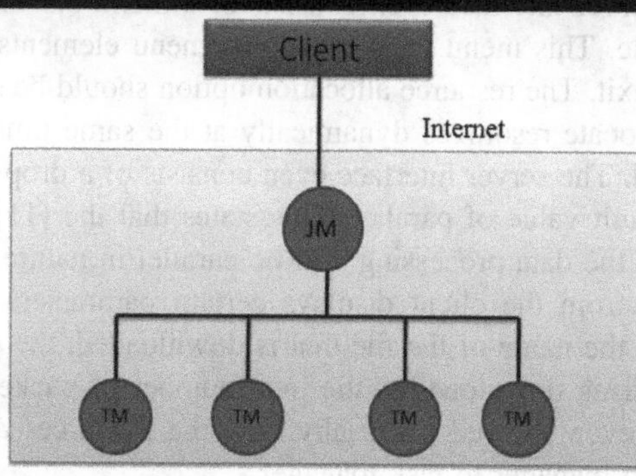

Fig. 1 Module of the proposed system

to it can check the status of the downloading. The client is demonstrated [6] by building a simple interface which consists of simple components. The client interface is event driven and the concept of swings in java is largely implemented. The client interface consists of a text area which displays all the files that are available presently in the database [3]. The text area is updated dynamically as per the uploading of resources dynamically. The name in the title bar of the interface is named as select file which basically states to select a file to download. There are two swing buttons included named as download and cancel. Both these buttons are event driven and upon clicking on the button a specified event takes place. The download button allows downloading a file and the cancel button closes the client interface.

3.2 Server

A server is a computer system that is responsible for servicing the request made by the client. The server is normally located remotely [13] and is used to service requests from multiple clients. The server is always responsible for maintaining resources and allocating them as required by the host clients. The server even manages and controls data processing between server and client. In today's modern multiprocessor architecture [2], Big data processing plays an important role. In order to have efficient Big data processing where many clients participate to execute certain tasks, the server needs to execute data processing faster and efficiently. In our proposed paper the server is an entity [3] that services the request made by the client. The client request for downloading a file and the server makes it sure that the file is downloaded and opened at the end of downloading for the clients. The server interface consists of two important criterions that are job manager and task manager [14]. The server interface even consists of a menu bar that has one menu element names as file. This menu consists of two menu elements named as resource allocation and exit. The resource allocation option should be selected only if the server needs to allocate resources dynamically at the same time when the file is getting downloaded. The server interface even consists of a drop down select menu which has the default value of parallel. This states that the [11] data distribution type is parallel and the data processing will be parallel in nature. The server upon getting the request from the client displays certain parameters in its command prompt output. It is the name of the file that is downloaded, the port numbers that will be involved during downloading, the total number of packets sent and received, etc. The server even showed pictorially how the resources are allocated dynamically from the job manager to task manager.

3.3 Job Manager

The job manager is an essential component of the server. It is like the master component of the entire client server layout. The job manager [1] accepts the request that comes from the client and is responsible for processing it. The job manager follows divide and conquer approach for executing the job that. The job

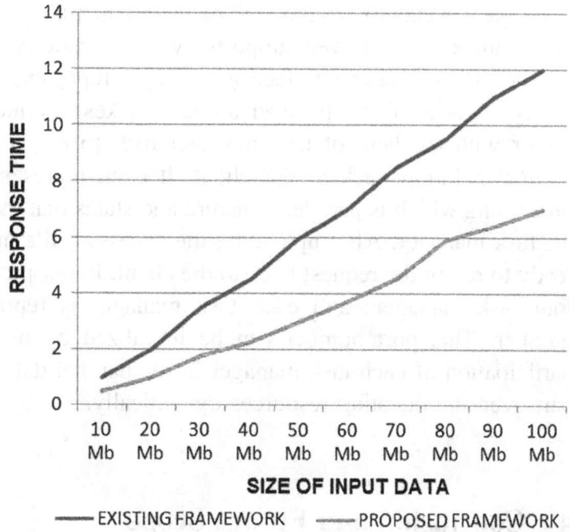

Fig. 2 Experimental analysis of existing versus proposed framework

manager has to schedule and control the execution of the jobs and returning back a valid response to the client as per its request.

In the first scenario upon achieving a request from the client the job manager divides the job into many subjobs or packets. It distributes evenly and randomly all the subjobs and allocates it to the task managers. After the task managers finish executing their individual subjobs, they return the resultant data to the job manager. All the subjobs or packets returned by the task manager to the job manager would not be in sorted order and hence the job manager sorts all the packets as they were allocated initially. After sorting all the packets the job manager tries to merge all the executed packets into one data so that it can return a single solution to the client.

The command prompt terminal output shows the time instances at which the job manager sends a packet to the task managers. It even shows the sorting and merging of packets accordingly so as to return a single solution to the client. During dynamic resource allocation to the task managers, the job manager itself uploads the file to the task manager. The job manager is responsible for uploading the files to the task manager whenever there is a need of allocation of resources (Fig. 2).

3.4 Task Manager

The task manager is an essential part of the server. It is like a basic block of execution that helps the job manager to execute the sub tasks and return it back to the job manager. The task manager's responsibility is to execute the individual packets allocated to them and return it back to the job manager. When a client requests for downloading a file, the task manager is the one which is responsible for executing the operation and performs efficient parallel data processing. Upon the use of task managers with the job manager, the time for Big data processing is much more efficient and it even supports dynamic resource allocation.

The task manager interface is a simple representation of the operation it performs. It shows that whenever a client makes a request to download the file, the server with the help of task manager tries to execute the request and return an optimal solution back to the client. It even represents the distributed type data processing which is parallel in nature and states that the file uploading is done with the task manager. After uploading the necessary file in the back end database, it is ready to return the request back to the client. In our proposed paper we have chosen four task managers and each task manager is represented with a unique port number. This port number can be initialized by us but moreover it shows the participation of each task manager in the parallel data processing. It plays a major part even in allocating resources dynamically.

4 Conclusion and Future Scope

Today's digital generation executes its key ingredient at a regular basis and that is data [5]. Everyday there are many probabilities of parallel data processing. There are many search engines like Google or Yahoo which has to process a lot of data simultaneously for returning a response [13] to its users and even at a faster rate. The reliability and feasibility should not be hampered during this parallel data execution [4]. Presently, the mechanisms used for parallel data execution creates a homogeneous cluster setup within the network. The homogeneous cluster setup states that when there is a parallel downloading environment [8] under processing between the client and server, if the client at the same time requests for downloading a particular file and the server does not have it currently in it back end database then it causes a huge problem. The file cannot be uploaded until all the downloading under progress stops its execution.

In order to avoid this kind of scenario and to decrease the delay in response time from the server, we propose a framework that represents efficient Big data processing with [11] no homogeneous cluster setup. It makes sure that when a client request for a file that is not present in the server, it can dynamically allocate that resource or file to the client even at the same time all the parallel downloading scenario is under progress. This improves the reliability [8] and response time since the client has to no more wait for its response. This framework defines a new level in Big data processing that is not encountered in today's world.

Eyeing for efficient future requirements, it is obvious that the proposal of this framework that can eliminate the possibility of homogeneous cluster setup is the best option introduced so far. The data is a key integrity [9] for processing many tasks in day-to-day lives and thus to make its execution easier will introduce a new level of data computation. In data mining where the data has to be mined and returned back as a response can follow this approach that will lead to better execution.

References

1. Wu, S., Li, F., Mehrotra, S., Ooi, B.C.: Query optimization for massively parallel data processing. School of Computing, National University of Singapore, March 2012
2. Parallel Data Processing. http://server-demo-ec2.cloveretl.com/clover/docs/clustering-parallel-processing.html
3. Dean, J., Ghemawat, S.: MapReduce: simplified data processing on large clusters. In OSDI'04: Proceedings of the 6th Conference on Symposium on Operating Systems Design & Implementation, pp. 10–10. Berkeley, CA, USA, 2004. USENIX Association
4. Chih Yang, H., Dasdan, A., Hsiao, R.-L., Parker, D.S.: Map-Reduce-Merge: simplified relational data processing on large clusters. In: SIGMOD'07: Proceedings of the 2007 ACM SIGMOD International Conference on Management of Data, pp. 1029–1040. New York, NY, USA, 2007. ACM
5. Lee, K.H., Lee, Y.J.: Big data processing with Map Reduce: A Survey. Department of Computer Science KAIST, December 2011
6. Gropp, W., Lusk, E., Skjellum, A.: Using MPI: Portable Parallel Programming with the Message-Passing Interface. MIT Press, Cambridge, MA (1999)
7. Deelman, E., Singh, G., Su, M.-H., Blythe, J., Gil, Y., Kesselman, C., Mehta, G., Vahi, K., Berriman, G.B., Good, J., Laity, A., Jacob, J.C., Katz, D.S.: Pegasus: a framework for mapping complex scientific workflows onto distributed systems. Sci. Program. **13**(3), 219–237 (2005)
8. Pike, R., Dorward, S., Griesemer, R., Quinlan, S.: Interpreting the data: parallel analysis with Sawzall. Sci. Program. **13**(4), 277–298 (2005)
9. Thain, D., Tannenbaum, T., Livny, M.: Distributed computing in practice: the condor experience. Concurrency Comput.: Pract. Exp. (2004)
10. Pike, R., Dorward, S., Griesemer, R., Quinlan, S.: Interpreting the data: parallel analysis with Sawzall. Sci. Program. **13**(4), 277–298 (2005)
11. Jiang, D., et al.: Map-join-reduce: towards scalable and efficient data analysis on large clusters. IEEE Trans. Knowl. Data Eng. (2010)
12. Li, B., et al.: A platform for scalable one-pass analytics using MapReduce. In: Proceedings of the 2011 ACM SIGMOD, 2011
13. Babu, S.: Towards automatic optimization of map reduce programs. In: Proceedings of the 1st ACM Symposium on Cloud Computing, pp. 137–142 (2010)
14. Arpaci-Dusseau, A.C., Arpaci-Dusseau, R.H., Culler, D.E., Hellerstein, J.M., Patterson, D.A.: High-performance sorting on networks of workstations. In: Proceedings of the 1997 ACM SIGMOD International Conference on Management of Data, Tucson, Arizona, May 1997
15. Thusoo, A., Sarma, J.S., Jain, N., Shao, Z., Chakka, P., Anthony, S., Liu, H., Wychoff, P., Murthy, R.: Hive—a warehousing solution over a map-reduce framework. In: VLDB, 2009

Region-Based Prediction and Quality Measurements for Medical Image Compression

P. Eben Sophia and J. Anitha

Abstract This paper presents a prediction-based compression algorithm for medical image containing region of interest. The medical image on the whole consumes a lot of memory space which makes them difficult for storage and transmission. In a medical image with only a particular part needed for diagnosis, the important decision that has to be made is whether to go for block compression or region-based compression. Here region-based compression plays a vital role since a particular region alone can be preserved and the other regions can be compressed in a lossy way. Such methods are of great interest in tele-radiology applications with large storage requirements. Here, the quality of compression can be measured by capturing the size of the selected regions. A new method for calculating total compression ratio and total bits per pixel is proposed for such selective image compression algorithms. Since selected area of medical images is compressed lossless, the performance of the proposed system is compared with other lossless compression algorithms. The results showed comparatively good performance.

Keywords Medical image · Region-based compression · Prediction · Tele-radiology · Wavelet transforms · Arithmetic coding · Thresholding

1 Introduction

Medical images such as MRI are rich in radiological information. Transmitting these images from one place to another is a tedious task. This makes tele-radiology impractical and paves the way for image compression. Compression enhances the

P. Eben Sophia (✉) · J. Anitha
Department of Electronics and Communication, Karunya University, Coimbatore, India
e-mail: ebensophia@gmail.com

J. Anitha
e-mail: anithaj@karunya.edu

M. Pant et al. (eds.), *Proceedings of Fifth International Conference on Soft Computing for Problem Solving*, Advances in Intelligent Systems and Computing 436, DOI 10.1007/978-981-10-0448-3_29

performance of a tele-radiology system by reducing the time and cost with or without loss of information. The main objective of this paper is to develop an image compression scheme which provides images with good visual quality and with fewer bits to represent them. The performance of the compression scheme is further enhanced by using region-based compression, where the diagnostically important information remains unaffected compared to other information. Several image compression algorithms and hybrid techniques which combine neural networks [1], discrete wavelet transform (DWT), and predictive coding are still under research to achieve good compression ratio. A lossless predictive technique which combines existing prediction algorithms were also investigated [2].

This paper uses the properties of spatial prediction along with wavelet transform and entropy encoding for selective compression of medical images. Prediction algorithms are simpler and more efficient for image lossless compression and several complex prediction algorithms are also available [3]. There are several lossless image compression algorithms available in the literature [3–6]. The images are split into region of interest (ROI) and non-region of interest (NROI) for selective compression [7–10]. Based on spatial prediction of these regions the error values are transformed and encoded in lossy and lossless mode using wavelet transform. Several wavelet transform approaches play an important role in the compression process [11–14]. Adaptive thresholding is another technique used to further improve compression. An optimized threshold value can be selected after multilevel thresholding. This thresholding plays an important role in reducing the source entropy [15]. The transformed coefficients are then coded using arithmetic coding technique. The results of the proposed approach indicate that there is an increase in compression performance for large image data. Two types of prediction algorithms are used. For ROI the hard prediction techniques that use some standard mathematical functions both at the encoder and decoder are used. The appropriate function was selected based on the entropy values. On the other hand, for NROI, a binary mask is generated based on the mean values of the wavelet coefficients on each subband. Then the appropriate prediction function is applied to the masked area alone. Masking before prediction greatly increased the compression ratio with acceptable visual quality.

Prediction is a method of capturing the correlation between image or video pixels. The prediction model is shown in Fig. 1. This method is widely utilized in image and signal processing applications. In the case of compression, the main idea is to encode and transmit to the decoder the unpredictable part in the prediction process, which is the prediction error. The decoder with the knowledge of the

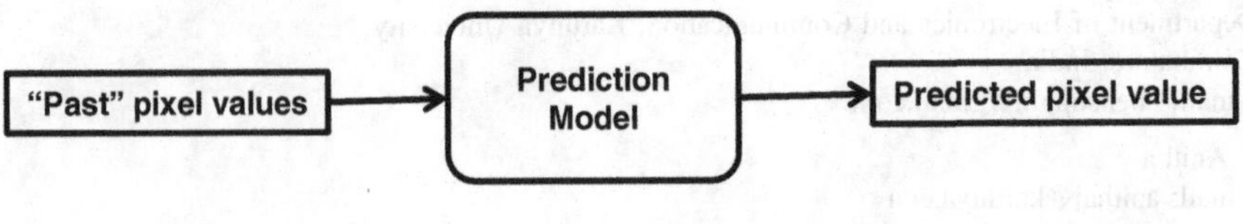

Fig. 1 The prediction model

prediction model performs the same procedure as the encoder and adds the transmitted prediction error for reconstructing the image. In the case of lossless compression the prediction error is coded error free and for lossy coding the error signal is first quantized and then coded. Prediction schemes are fully reversible and are used by several lossless encoding schemes [16].

2 Prediction-Based Compression Algorithm

Median edge detector (MED) is the basis for the proposed prediction algorithm. The prediction function which gives the best entropy value is chosen for ROI and NROI prediction. The main aim is to reduce the prediction error, so that the transmission process can be made simpler. The reconstructed value will be the sum of predicted value and the prediction error. Consider the equations below which show the predicted value of $x(n)$ which is $x_p(n)$.

$$x_p(n) = x(n-1) \tag{1}$$

The prediction error $e(n)$ is given as

$$e(n) = x(n) - x_p(n) \tag{2}$$

$$e(n) = x(n) - x(n-1) \tag{3}$$

Let the entropy calculation for encoding original image be H_1 and error image be H_2. Then,

$$H_1 > H_2 \tag{4}$$

This means that encoding original image requires more bits than encoding the prediction error. Hence the prediction error is encoded and transmitted instead of the original image. This enhances the performance of compression.

2.1 ROI Coding Methods

ROI part of medical image is the diagnostically important region and is compressed lossless. The structure of the ROI coder is shown in Fig. 2. For ROI compression, the region is partitioned from the original image and is cropped to its own size. Based on the size of the ROI the computation complexity of the prediction process gets increased or decreased. Then a linear mapping is done to map the error values in the positive range. This helps the compression algorithm to compress more efficiently. The prediction error is directly encoded using the lossless arithmetic

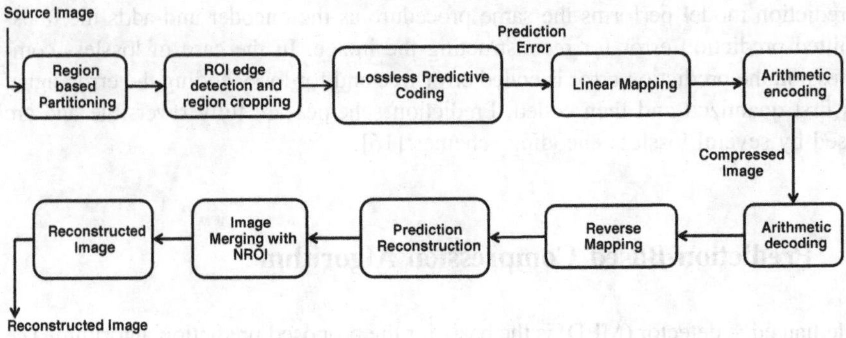

Fig. 2 Structure of the proposed ROI coder

coding technique. Prediction algorithm is a completely lossless method of compression and hence we can conclude that there is no loss of data at any point of ROI compression.

2.2 NROI Coding Methods

NROI part of the image is the region where more compression can be obtained. Hence, NROI coder consists of lossy predictive coding along with wavelet transform and lossy entropy coding as shown in Fig. 3. Discrete wavelet transform (DWT) technique decomposes the image into multilevel subbands. The NROI part of the image is passed through high pass and low pass filters, and then they are down-sampled to get the decomposed image at level 1. The same process is repeated using the high frequency band to get a multilevel decomposed image. The maximum level of decomposition is given as

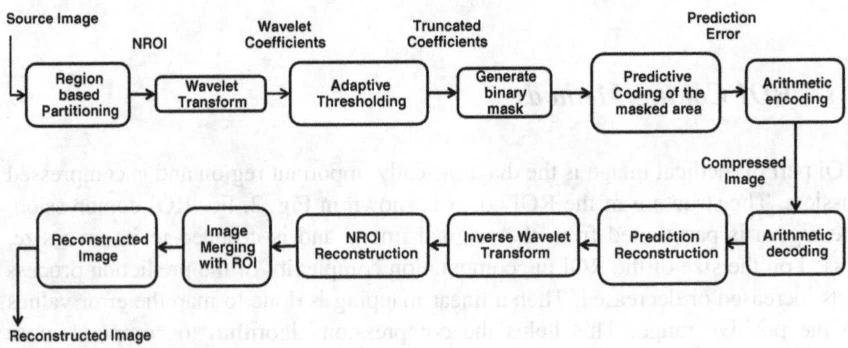

Fig. 3 Structure of the proposed NROI coder

$$\text{Maximum decomposition level} = \log_2(n) \tag{5}$$

where '*n*' is the size of the square image. As the thresholding level of the wavelet coefficients is increased the compression ratio also increases. Wavelet transformation as such does not produce any compression. Hence, quantization is done after subband decomposition. This process induces loss of image data and for this reason they are not included in ROI regions. The truncated coefficients are then subjected to lossy prediction. For lossy prediction, the mean value of the wave data for each subband is calculated and based on this value a binary mask is generated. Then the masked area alone is predicted using an appropriate prediction function selected based on entropy calculation. The other areas are truncated and do not undergo prediction. This reduces the number of wave data to be predicted with loss of information. Then the error coefficients are entropy encoded using arithmetic coding technique. The decoding procedure is just the reverse process.

3 Quality Measurement for Region-Based Compression

The compressed ROI and NROI part of the images are transmitted one after another. Let the original bytes for ROI and NROI be represented as orig_roi and orig_nroi respectively. Then the encoded bytes for ROI and NROI are represented as enc_bytes_roi and enc_bytes_nroi respectively. For compression ratio calculation, initially the size of ROI has to be calculated in percentage.

$$\begin{aligned} O_{\text{img}} &= n * m \text{ (Size of original image in bytes)} \\ \text{ROI}_{\text{cr}} &= \text{nr} * \text{mr} \text{ (Size of cropped ROI in bytes)} \\ S_{\text{roi}} &= \text{nr} * \text{mr} * 100/(n * m)\text{(Percentage of size ROI in original Image)} \\ \text{CR}_{\text{roi}} &= \text{ROI}_{\text{cr}}/\text{enc\_bytes\_roi(CR of ROI)} \\ \text{CR}_{\text{nroi}} &= \text{O}_{\text{img}}/\text{enc\_bytes\_nroi(CR of NROI)} \\ \text{CR}_{\text{T}} &= S_{\text{roi}} * \text{CR}_{\text{roi}} + (1 - S_{\text{roi}}) * \text{CR}_{\text{nroi}}\text{(CR of the full image)} \end{aligned} \tag{6}$$

This is the formula for calculating the compression ratio of the full image using the compression obtained for ROI and NROI. The BPP can also be calculated using the percentage size of the ROI (S_{roi}) using the formula below.

$$\begin{aligned} \text{BPP}_{\text{roi}} &= \left(\text{enc\_bytes\_roi} * 8 * \text{size } \left(O_{\text{img}}, 3\right)\right)/\text{ROI}_{\text{cr}}\text{(BPP of the ROI)} \\ \text{BPP}_{\text{nroi}} &= \left(\text{enc\_bytes\_nroi} * 8 * \text{size } \left(O_{\text{img}}, 3\right)\right)/O_{\text{img}}\text{(BPP of the NROI)} \\ \text{BPP}_{\text{T}} &= S_{\text{roi}} * \text{BPP}_{\text{roi}} + (1 - S_{\text{roi}}) * \text{BPP}_{\text{nroi}}\text{(BPP of the full image)} \end{aligned} \tag{7}$$

MSE and PSNR calculations are done with the final reconstructed image and hence the usual calculation process is followed as shown below. The PSNR is given by

$$\mathrm{PSNR} = 10\log_{10}\left((255)^2/\mathrm{MSE}\right) \quad (8)$$

where, mean square error MSE is given as

$$\mathrm{MSE} = \frac{1}{mn}\sum\nolimits_{i=1}^{m}\sum\nolimits_{j=1}^{n}\left|x_{ij} - y_{ij}\right| \quad (9)$$

This equation gives the error value between the reconstructed and the original image. Then the structural information is extracted from an image to find its quality. This is calculated using SSIM (structural similarity index). If SSIM is 1, it means that the image has no loss of data and the loss increases with decreasing values of SSIM. The equation for SSIM is shown below.

$$\mathrm{SSIM} = \frac{\left(2\mu_f\mu_{\bar{f}} + C_1\right)\left(2\delta_{f\bar{f}} + C_2\right)}{\left(\mu_f^2 + \mu_{\bar{f}}^2 + C_1\right)\left(\delta_f^2 + \delta_{\bar{f}}^2 + C_2\right)} \quad (10)$$

where f is the original image and $\bar{f}$ is reconstructed image, μ represents the average gray value, δ represents the variance, and C_1 and C_2 are constants to prevent unstable results.

Initially, the aim is to determine the prediction function suitable for the given image. The source entropy value and the entropy value of the error sequences are calculated using the equation

$$H = -\mathrm{sum}(p. * \log_2(p)) \quad (11)$$

where 'p' is the probability of occurrence of each symbol in the image. The source entropy value gives the minimum number of bits needed for encoding the original image. This is calculated based on the redundancy values of that image. Then the entropy value of the error image is also calculated using the same formula as in Eq. 11 for comparison.

4 Simulation Results

The medical images used for testing the algorithm are shown in Fig. 4. The ROI is marked, for which results were obtained.

The prediction function used are $C, A, B, A + B - C, A + (B - C)/2, B + (A - C)/2$ and $(A + B)/2$. Here A, B and C are the pixel values of the image. To predict the value of the pixel at location say (m, n) we need $(m, n - 1)$ pixel value which is 'A' and $(m - 1, n)$ pixel value which is 'B' and $(m - 1, n - 1)$ pixel value which is 'C'. This is shown in detail in Fig. 5.

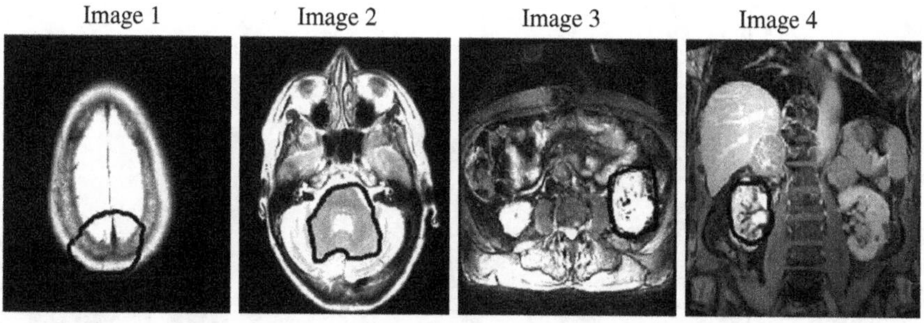

Fig. 4 Test images: *Image* 1, 2 MRI brain, *Image* 3, 4 MRI Abdomen

Fig. 5 Sample neighborhood pixels *A*, *B* and *C* for prediction of *X*

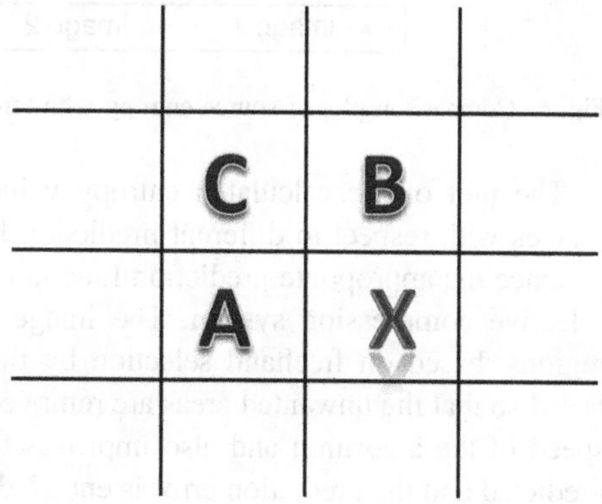

Then the entropy values are calculated for the test images and are listed in Table 1. *H*1 is the entropy of the original image and *H*2 is the entropy of the error image. Based on the entropy and MSE values the prediction function is selected for ROI and NROI. The prediction function that gives minimum entropy for that particular image is selected for further process.

Table 1 Entropy and MSE values calculated for the test images for various prediction functions

Prediction function	Image 1 (512 * 512) *H*1 = 3.6901		Image 2 (512 * 512) *H*1 = 5.1354		Image 3 (256 * 256) *H*1 = 6.9584		Image 4 (320 * 320) *H*1 = 7.5952	
	*H*2	MSE	*H*2	MSE	*H*2	MSE	*H*2	MSE
C	2.658	6.097	3.899	13.01	6.000	25.58	5.418	12.95
A	2.730	6.454	3.969	13.70	6.026	26.29	5.495	13.68
B	2.705	6.801	3.966	14.89	6.109	27.78	5.596	14.73
A + *B* − *C*	2.803	25.70	4.173	38.75	6.097	36.85	5.747	35.74
A + (*B* − *C*)/2	2.673	5.892	3.892	12.44	6.023	25.71	5.385	12.42
B + (*A* − *C*)/2	2.659	6.071	3.895	13.27	5.961	24.58	5.460	13.31
(*A* + *B*)/2	2.742	13,578	4.057	21.63	5.976	26.98	5.622	20.17

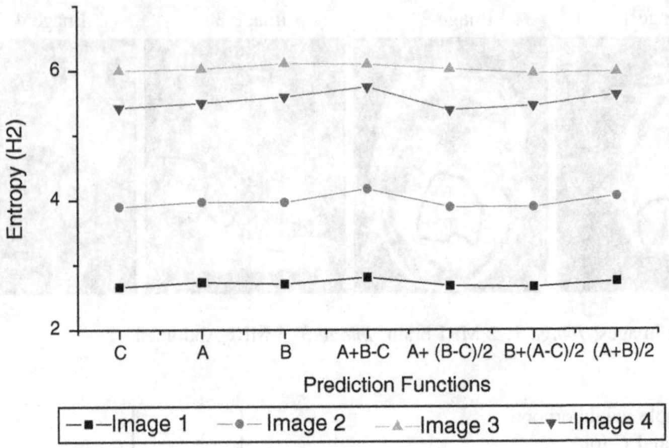

Fig. 6 Comparison plot of source entropy with error entropies of the test images

The plot of the calculated entropy value of the original image and the error images with respect to different prediction functions is shown in Fig. 6.

Once the appropriate prediction function is found the image can be subjected to selective compression system. The image is initially split into ROI and NROI regions, based on freehand selection by the user. The ROI part is then cropped based so that the unwanted areas are removed. This greatly increases the processing speed of the algorithm and also improves the compression ratio. The ROI part is predicted and the prediction error is encoded using the arithmetic coding technique. The NROI part initially undergoes 5-level wavelet decomposition and then the prediction algorithm is applied to the quantized wavelet coefficients. The performance analysis is done based on CR, BPP, PSNR and SSIM. The tabulations are shown below for the test images. Table 2 shows the result for test images 1 and 2. The compression ratio, PSNR and SSIM are calculated for various BPP.

Table 3 shows the result for test images 2 and 3. PSNR and SSIM show the quality of the reconstructed images.

The graphical representations of the results obtained are shown in Figs. 7 and 8. Comparisons are made between BPP_T versus CR_T and PSNR.

Table 2 CR_T, PSNR and SSIM values calculated for the test images 1 and 2 for various BPP_T

BPP_T	CR_T		PSNR		SSIM	
	Image 1	Image 2	Image 1	Image 2	Image 1	Image 2
3	2.7451	2.6759	61.2002	48.4649	0.9999	0.9976
2.5	3.0991	3.357	59.7269	42.5399	0.9998	0.9925
2	3.73	4.3753	54.8	37.4936	0.993	0.9833
1.5	5.3766	5.7128	47.3236	33.7191	0.9957	0.9722
1	7.7474	9.0105	42.22	28.989	0.9873	0.9421
0.5	16.95	19.6835	35.0831	24.1345	0.9713	0.8469

Table 3 CR_T, PSNR and SSIM values calculated for test images 1 and 2 for various BPP_T

BPP_T	CR_T		PSNR		SSIM	
	Image 3	Image 4	Image 3	Image 4	Image 3	Image 4
3	2.4765	2.5386	33.05	36.8973	0.9035	0.9476
2.5	3.07	3.219	30.0821	33.8257	0.8591	0.9066
2	4.2134	3.9439	26.7321	31.8329	0.7864	0.8682
1.5	5.9541	5.4692	24.2164	30.1681	0.7055	0.8133
1	9.4776	9.646	21.4884	25.7953	0.5971	0.6791
0.5	22.63	17.4819	18.05	22.9036	0.4044	0.5284

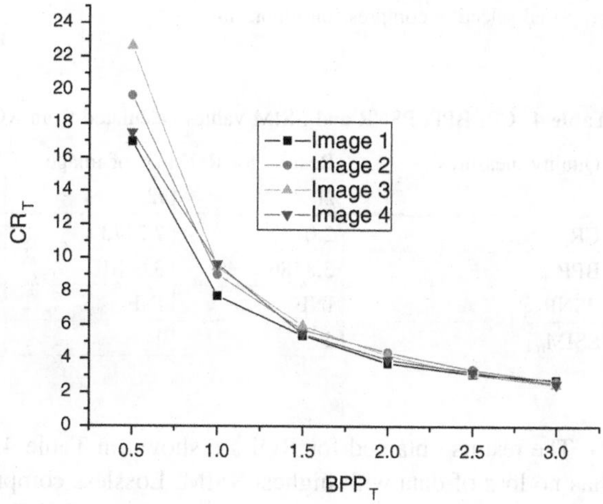

Fig. 7 Comparison plot of BPP_T with CR_T of the test images

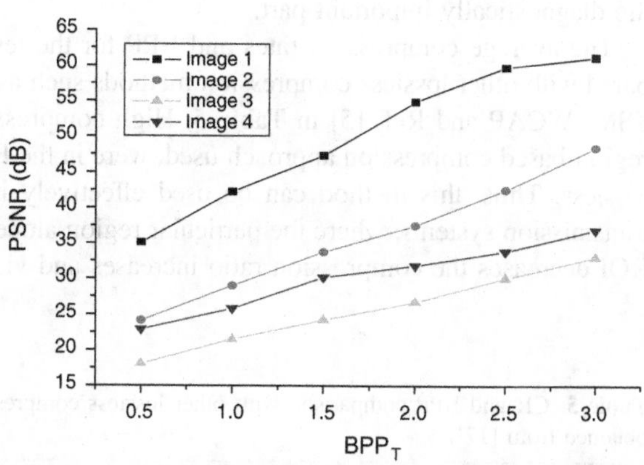

Fig. 8 Comparison plot of PSNR with BPP_T of the test images

The visual quality of the resultant image for fully lossy compression and ROI-based compression are shown in Fig. 9. Test image 2 is used for this comparison and the lossy compression technique used is the technique which is applied to the NROI part of the image.

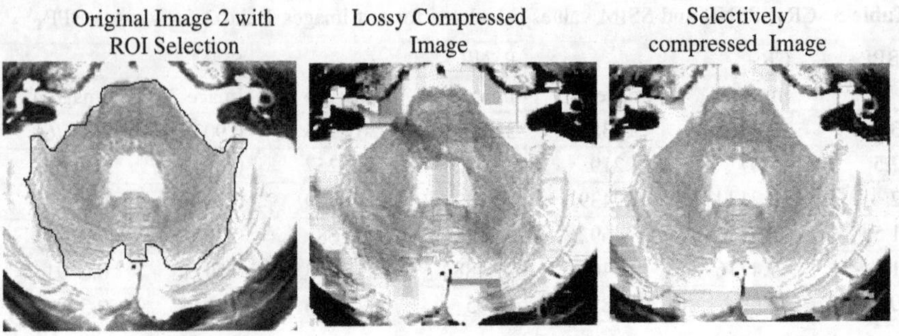

Fig. 9 Visual comparison of the test image 2 after lossy compression and after applying the proposed selective compression algorithm

Table 4 CR, BPP, PSNR and SSIM values calculated from ROI part of the test image

Quality measures	Results for ROI part of image			
	*I*1	*I*2	*I*3	*I*4
CR_{roi}	2.0787	2.2173	2.1995	2.0925
BPP_{roi}	3.8486	3.6081	3.6372	3.8233
$PSNR_{roi}$	INF	INF	INF	INF
$SSIM_{roi}$	1	1	1	1

The results obtained for ROI are shown in Table 4. The ROI part of the image has no loss of data with highest SSIM. Lossless compression ratio achieved is less than the total compression ratio with maximum PSNR. This confirms the quality of the diagnostically important part.

The average compression rates and BPP for the test images obtained are compared with other lossless compression methods such as CALC, SPIHT, JPEG2000, SSM, WCAP and Ref. [5] in Table 5. High compression ratio is obtained due to region-based compression approach used, were in the ROI part alone is compressed lossless. Thus, this method can be used effectively in medical image storage or transmission systems, where the particular region alone is of interest. As the size of ROI decreases the compression ratio increases and vise versa.

Table 5 CR and BPP comparison with other lossless compression techniques (Existing results obtained from [17])

Quality measures	CALC	SPIHT	JPEG2000	SSM	WCAP	Ref. [5]	Proposed
CR	3.50	3.61	3.53	3.45	3.31	3.30	4.20
BPP	2.28	2.21	2.26	2.32	2.42	2.42	2

5 Conclusion

The proposed prediction-based compression algorithm shows good performance for selective medical image compression as shown above. Medical image should always be compressed in a lossless format. Hence the results are compared with some of the lossless compression algorithms such as CALC, SPIHT, JPEG2000, SSM, WCAP and Ref. [5]. The performance of the proposed method can also be improved in several ways, such as using an optimization algorithm or zero tree coding. Segmentation algorithms can also be used for partitioning the ROI. But these techniques increase the complexity of the algorithm as well as computational time. The algorithm presented here is simple and less time-consuming. The ROI part of the image remains unaffected and also high compression is obtained using the NROI part of the image. Thus the overall compression achieved is satisfactory for tele-radiology where only limited bandwidth is available and the diagnostic details are also well preserved.

References

1. Hussain, A.J., Al-Jumeily, D., Radi, N., Lisboa, P.J.G.: Hybrid Neural Network Predictive-wavelet image compression system. Neurocomputing **151**, part 3, 975–984 (2014)
2. Shirsat, T.G., Bairagi, V.K.: Lossless medical image compression by integer wavelet and predictive coding. ISRN Biomed. Eng. Hindawi Publishing Corporation, Article ID 832527, 6 pp. (2013)
3. Martchenko, A., Deng, G.: Bayesian predictor combination for lossless image compression. IEEE Trans. Image Process. **12**, 5263–5270 (2013)
4. Avramović, A., Reljin, B.: Gradient edge detection predictor for image lossless compression. In: 52nd International Symposium, ELMAR, pp. 15–17 (2010)
5. Li, J.: An improved wavelet image lossless compression algorithm. Optik **124**(11), 1041–1044 (2013)
6. Miaou, S.-G., Ke, F.-S., Chen, S.-C.: A lossless compression method for medical image sequences using JPEG-LS and interframe coding. IEEE Trans. Inf. (2009)
7. Bairagi, V.K., Sapkal, A.M.: Automated region-based hybrid compression for digital imaging and communications in medicine magnetic resonance imaging images for telemedicine applications. IET Sci. Meas. Technol. **6**(4), 247–253 (2012)
8. Chavez, R.F.L., Iano, Y., Higa, R.S., Arthur, R., Saotome, O.: Generalized region of interest coding applied to SPIHT. J. Sel. Areas Telecommun. (JSAT), Special Issue, 23–31 (2012)
9. Garcia-Alvarez, J.C., Führ, H., Castellanos-Dominguez, G.: Evaluation of region-of-interest coders using perceptual image quality assessments. J. Vis. Commun. Image **24**(8), 1316–1327 (2013)
10. Sran, P.K., Gupta, S., Singh, S.: Content-based medical image coding with fuzzy level set segmentation algorithm. In: Proceedings of the Fourth International Conference on Signal and Image Processing, pp. 161–171. Springer, India (2013)
11. Jiang, H., Ma, Z., Hu, Y., Yang, B., Zhang, L.: Medical image compression-based on vector quantization with variable block sizes in wavelet domain. Comput. Intell. Neurosc. **5**, 1–8. Hindawi Publishing Corporation, Article ID 541890 (2012)
12. Kanumuri, T., Dewal, M.L., Anand, R.S.: Progressive medical image coding using binary wavelet transforms. SIViP **8**(5), 1–17 (2012)

13. Kathirvalavakumar, T., Ponmalar, E.: Self organizing map and wavelet-based image compression. Int. J. Mach. Learn. Cyber, **4**(4), 319–326 (2013)
14. Zadeh, P.B., Akbari, A.S., Buggy, T.: Wavelet-based image compression techniques. In: Chapter 18 in 'Advances in Wavelet Theory and Their Applications in Engineering, Physics and Technology'. Inntech Open Access Publisher, pp. 423–448, ISBN 979-953-307-385-8 (2012)
15. Ahmadi, K., Javaid, A.Y., Salari, E.: An efficient compression scheme based on adaptive thresholding in wavelet domain using particle swarm optimization. Signal Process.: Image Commun. Article in press, http://dx.doi.org/10.1016/j.image.2015.01.001 (2015)
16. European society of radiology: usability of irreversible image compression in radiological imaging. Insights Imaging **2**(2), 103–115 (2011)
17. Chen, Y.-T., Tseng, D.-C.: Wavelet-based medical image compression with adaptive prediction. Comput. Med. Imaging Graph. **31**, 1–8 (2006)

Recent Advancements in Energy Efficient Routing in Wireless Sensor Networks: A Survey

Gaurav Kumar Pandey and Amritpal Singh

Abstract Wireless sensor networks have set a new realm in the field of wireless transmission technology. Their applications have diversified over the years and now they cover various sophisticated areas of applications which include military applications, surveillance, agriculture, monitoring and control, etc. This paper covers various recent developments in the field of energy-aware routing techniques to minimize the energy consumption and extend the lifetime of wireless sensor networks.

Keywords Energy efficient routing · Balanced clustering · Learning automata · Restricted routing · Virtual backbone networks · Residual energy

1 Introduction

The supremacy of wireless sensor networks became prominent with the arrival of Internet of Things. WSN was able to reach and gather information from those places and environments that are not suitable for humans to visit. They have helped to mine information from deep seabeds to volcanic mountains. But their performance has always been hindered by their diminutive battery power. To minimize energy depletion of the nodes has always been the center of attraction for researchers. This gave rise to the concept of *energy-aware routing* protocols for data transmission in WSN. Over the years many energy efficient traditional routing protocols were developed such as LEACH, PEGASIS, TEEN, APTEEN, SPIN, etc., to lower the rate of energy dissipation and obtain an increase in the lifetime of wireless sensor networks. But with the advancement of time, many new and much

G.K. Pandey (✉) · Amritpal Singh
Department of CSE, Lovely Professional University, Phagwara, Punjab, India
e-mail: pandeygauravkumar10@gmail.com

Amritpal Singh
e-mail: apsaggu@live.com

M. Pant et al. (eds.), *Proceedings of Fifth International Conference on Soft Computing for Problem Solving*, Advances in Intelligent Systems and Computing 436, DOI 10.1007/978-981-10-0448-3_30

more efficient routing techniques have emerged which have outperformed their ancestors. Recent algorithms have made use of more dynamic approaches such as equalized clustering, dynamic cluster head categorization, elimination algorithms, learning automata, Scalable routing techniques, modified ABC (artificial Bee Colony) algorithms, fitness functions, virtual backbone networks, residual energy aware techniques, etc.

Recent advancements in field of energy efficient routing techniques have been explained and discussed in this paper from Sects. 2–9.

2 Equalized Cluster Head Election Routing Protocol

ECHERP (equalized cluster head election routing protocol) is an energy efficient novel routing protocol [1] which makes use of *balanced clustering* for conservation of energy in WSN. The energy of a sensor node is utilized for doing three major tasks: sensing the environment, processing the data, and transfer of data to next node or base station. So to mitigate energy losses, EHCERP makes use of well-adjusted equalized clustering approach whereby existing residual energy and expected future residual energy of the sensor node is considered so as to increase the lifetime of the wireless sensor network. The authors have used a linear model along with a Gaussian elimination algorithm to compute sensor nodes that may be elected as cluster heads. Compared to other protocols like LEACH, PEGASIS, TEEN, etc., which make use of stochastic or hierarchical mechanisms, EHCERP analyzes current and future residual energies of the sensor nodes. Also, EHCERP determines the number of rounds for which a sensor node can act as a cluster head.

Certain *assumptions* made in case of EHCERP:

- The base station (BS) has unrestrained communication power and residual energy.
- The position of BS is fixed which may be either outside the field or inside the field.
- All sensor nodes are assembled dynamically into clusters.
- The number of clusters are equivalent to the number of cluster heads.

The authors have categorized cluster heads as: *first level* cluster head, *second level* cluster head, and so on depending on their proximity to the base station. The first level cluster heads directly communicate with the base station. The second, third, and other level cluster heads transfer data to those cluster heads that are higher in hierarchy.

EHCERP is more efficient than other proposed algorithms because it takes into account residual energy as well as routing information, which are together considered as a linear system. If only residual energy is taken into consideration (in case of some other proposed algorithms), then it may cause disorganization. If we consider a scenario where a sensor node is elected as a cluster head since it has greater residual energy than other members of the cluster, its position is much far apart from BS compared to other members in the cluster. This will cause other sensor nodes in the cluster to transmit data in the opposite direction of BS, which in turn will increase the rate of energy dissipation. This problem is targeted by EHCERP.

Steps of EHCERP:

- A TDMA schedule is initiated by the BS according to which the sensor nodes send their advertisement.
- Every sensor node broadcasts its energy level, proximity from BS, and its unique id in the form of message to its nearby sensor nodes.
- Every sensor node builds up a *neighbor information table* containing energy level and location of its neighbor on the basis of information exchanged in the above step and sends this table in conjunction with its own information to its neighbors.
- The above step is continued until the BS receives information about every node in the network. At this step the BS obtains global statistics about all nodes and every node is treated as a potential cluster head candidate.
- Now Gaussian elimination algorithm is executed by the BS to calculate the number of rounds for which every node can act as a cluster head. Also, the BS categorizes cluster heads into first level and other levels depending on their proximity to base station.
- Then the BS broadcasts the ids of selected cluster heads along with their cluster demarcation and other information based on which every node knows the frequency of times they can act as cluster head.
- Every cluster head will build and broadcast its TDMA schedule to other nodes in its cluster so as to indicate the time slot during which its members can transmit data. Also, the transmission radio will be activated only at during this TDMA schedule in order to save energy of every node.
- Now the data transfer stage takes place in which all the sensor nodes send their respective sensed data to their cluster heads at the allotted TDMA schedule.
- Aggregation of data will be done by cluster heads which are at lower level from base station. Then the aggregated data will be sent to higher level cluster heads. This sequence will continue until BS receives the data.
- After the completion of a particular round, re-election of cluster heads will take place according to the procedure specified in the above steps.

Simulation results have proved that EHCERP outperform protocols such as LEACH, PEGASIS, TEEN, etc.

3 Balanced Energy Aware Routing Protocol

BEAR (Balanced Energy Aware Routing) [2] is a combination of energy management and energy saving tactics which make use of *learning automata concept* to discover a compromise between optimal distance and energy balancing. Many of the energy aware routing protocols aim to discover the shortest route between source and destination node which may result in network partition. On the contrary, considering only energy balancing to find the optimal route may result in a lengthy route along with increased latency which diminishes the lifetime of WSN. BEAR protocol tends to balance and reduce energy consumption by routing the data according to the proximity from the BS and residual energy of nodes. This protocol is an advancement of SEER (simple energy efficient routing protocol) [3]. Steps of BEAR:

- The sink floods the entire network with broadcast messages. Upon receiving the packets from sink node, every node add certain details which include hop count, energy level, and id of neighbor. Finally, the node calculates *selection probability* of nodes in neighborhood and inserts this entry based on the probabilistic value obtained from hop counts and energy level.
- After completion of the above step, every node in the network has information about their energy level, their hop count from BS, and probability of its neighbors.
- To forward the sensed data, the mechanism of *learning automata* is used. The node that needs to forward its data looks into its neighbor table for probability values of its neighbors.
- The sender node piggybacks its energy level to actual data packet and forwards it to the neighbor having highest value of selection probability.
- Then the neighbor node that receives the packet updates the value of energy level of sender node.
- In each hop, the sender node's probability is updated into the list of neighbors of preceding sender nodes.

The authors used *GloMoSim simulator* to compare and contrast BEAR with SEER protocol. Results have shown that BEAR outperforms SEER in terms of managing the energy because BEAR transfers data along a balanced route which helps to increase the lifetime of wireless sensor network.

4 Scalable Energy Aware Protocol

ScEP (Scalable Energy Aware Routing) protocol is an energy efficient protocol that makes use of MapReduce technique [4] along with clustering method to increase the lifetime of the network. All sensor nodes relay their sensed data to their respective cluster heads which finally transmit the data to the base station. Cluster heads are chosen based on their remaining energies and distance from the base station. Steps of mapping phase of ScEP are as follows:

- The BS sends a broadcast which requests the position of nodes and its energy level.
- The BS then determines a *key–value pair* based on a clustering method which denotes a set of primary cluster heads as "key" and other nodes position and energy values as "value."

- Then the BS appoints each node as cluster head or member of a particular cluster.
- The elected cluster heads then sends its information regarding energy level and one hop transmission to other sensor nodes.
- Finally the output *key–value pair* is generated.

Steps of reduce phase of ScEP:

- The output *key–value pair* from mapping phase is taken as input for reduce phase.
- Second cluster is generated based on "value" of key–value pair.
- Then based on the remaining energy of the node, a cluster head is found for every node.
- The remaining energy of initial cluster heads is compared with the minimum energy needed to be a cluster head and if the energy of initial cluster heads is found less, then another cluster head for the cluster is selected.
- Finally the new cluster heads are updated and a new output *key–value pair* is generated.

The main aim of the above algorithm is to elect the best possible cluster heads based on their remaining energy levels and transmission hops from the base station. Simulation results have shown that ScEP outperforms LEACH and MRKCP protocols based on energy consumption and improvement in residual energies of other sensor nodes in the network which altogether increases the network lifetime.

5 Cluster-Based Routing Using Artificial Bee Colony Algorithm

The proposed algorithm is a novel energy aware protocol based on artificial bee colony mechanism [5] in order to increase the lifetime of WSN. The use of artificial bee colony simulates the intellectual foraging aspects of the swarm of honey bees. The authors have proposed a central clustering mechanism for electing cluster heads. The cluster head selection algorithm is deployed at the base station. The clustering method incorporates ABC (artificial bee colony) algorithm [6] to provide an optimal routing mechanism for wireless sensor networks. The algorithm assumes that there are N sensor nodes and K clusters.

Steps of the protocol are as follows:

- The first step of is to initialize the network with information about distance and clusters.
- Then a fitness function is evaluated for every sensor node.
- The fitness function is obtained by computing the distance of every node from every cluster head.

- Assign the sensor node to that cluster head for which there is a minimum distance between node and cluster head.
- Update the position by incorporating optimization algorithm.
- Repeat the process until all the nodes are initialized.

In this algorithm communication energy is regarded as the essential factor. Energy consumption is defined by distance between the communication nodes.

The algorithm executes in three phases as follows:

- Initialization (computing cross-distances (using ABC—ant bee colony algorithm) for selection of specific roles (cluster heads and members))
- Setup (organizing the sensor nodes into clusters and assigning nodes to respective cluster heads)
- Data-gathering (relaying data to the cluster heads)

Simulation results have demonstrated that the above proposed algorithm was far better than LEACH and other similar kind of routing protocols.

6 Energy Efficient Routing Using Virtual Backbone

Wireless sensor networks deal with the problem of *broadcasting or flooding* which tends to increase the energy consumption of the entire network. Energy of every node decreases as it broadcasts the packets to other nodes. In the case of interference, the node has to retransmit the packets, which further causes energy dissipation. This makes way for a scenario termed as—*broadcast storm* [7, 8]. This problem can be solved using *backbone nodes* which are a subset of the required active nodes that help to minimize and exclude unrequired transmission links by shutting down the radio of extraneous nodes. Collection of such backbone nodes is called-*a backbone network.* This backbone network enhances the routing mechanism which in turn escalates efficiency of bandwidth and lowers the overall energy consumption.

Load balancing is achieved by backbone nodes which are operative when transmission occurs, else they go dormant by shutting down their radios. But having only a single backbone network does not properly boost up the network lifetime, hence it is required to have multiple backbone networks termed as disjoint CDS (connected dominated sets) that are adaptive to changes in topology [9]. The nodes that do not belong to CDS group are put into dormant state to save energy.

To further increase the efficiency through backbone routing, the concept of localized efficient backbone routing was proposed [10]. The authors made use of a *schedule transition graph,* which is a centralized approximation algorithm that maps the backbone node to a particular state. Further, the concept of *restricted region* was introduced to confine forwarding routes. Umesh et al. proposed the concept of *greedy method routing,* which chooses the next hop node on the basis of its distance from the base station. The steps include:

- The backbone node which needs to send a packet chooses a neighbor backbone node in the restricted region based on its distance from the neighbor. The nearest neighbor has the most optimal link.
- If the backbone node fails to find a nearby backbone neighbor it expands its *critical transmission radius* to discover the next best neighbor node.
- If after expansion of transmission radius the backbone node still does not find any best neighbor, the transmission takes place by greedy mechanism.

A flowchart depicting the stages of this algorithm is given in Fig. 1.

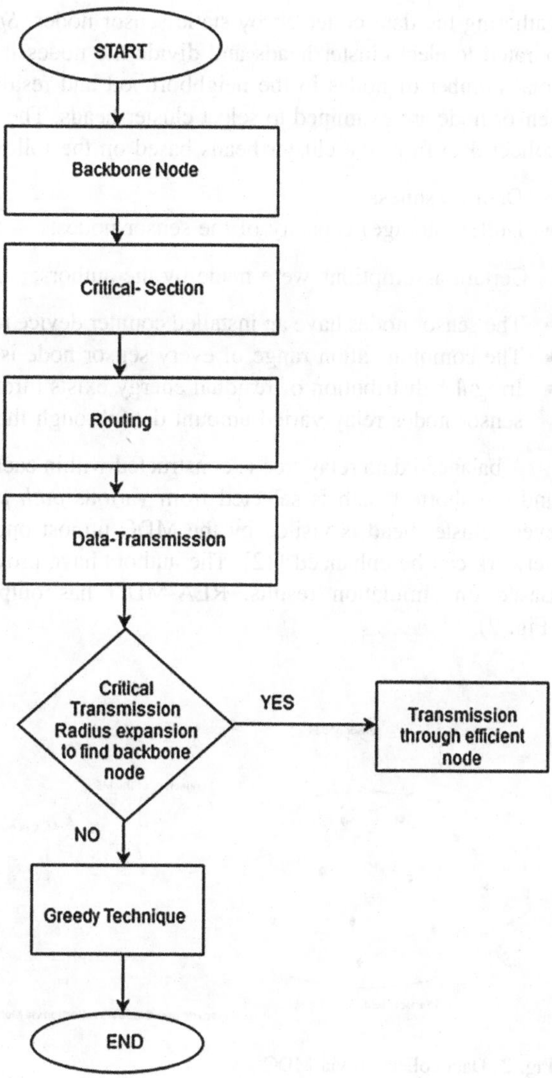

Fig. 1 A flowchart depicting steps of *greedy routing* in virtual backbone networks

The simulation results demonstrated that the above routing was able to conserve energy during the entire process of routing compared to other tradition routing methods of WSN.

7 Residual Energy Aware Mobile Data Gathering (REA-MDG)

The idea of *MDC* (mobile data collector) was used by the authors to maximize the lifetime of WSN. An approach involving *residual energy* [11] was coined for gathering the data collected by static sensor nodes. *Spectral clustering* was incorporated to elect cluster heads and divide the nodes into different cluster sets. The total number of nodes in the neighborhood and residual energy of each neighbor sensor node are examined to select cluster heads. The tour of MDC is organized to collect data from the cluster heads based on the following criteria:

- Data Freshness
- Buffer (storage) capacity of the sensor nodes.

Certain assumptions were made by the authors:

- The sensor nodes have an installed counter device to check their residual energy.
- The communication range of every sensor node is the same.
- Irregular distribution of residual energy exists throughout the sensor field since sensor nodes relay varied amount data through themselves.

A balanced data relay tree is constructed within each cluster for data aggregation and the shortest path is selected from various *path points* (cluster heads) so that every cluster head is visited by the MDC utmost once so that the lifetime of the network can be enhanced [12]. The authors have proved their algorithm *NP-hard*. Based on simulation results, REA-MDG has outperformed LEACH protocol (Fig. 2).

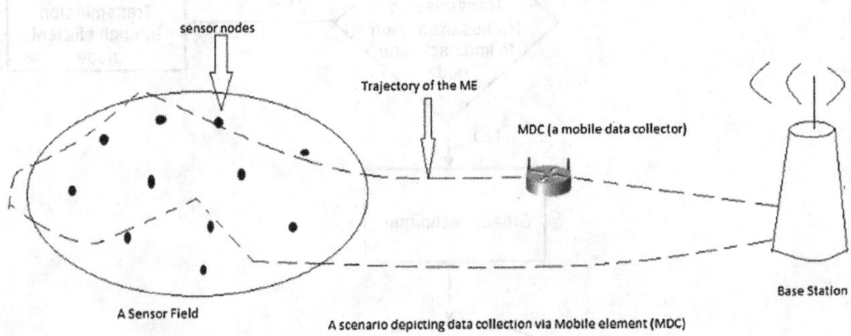

Fig. 2 Data collection via MDC

8 Comparison of the Above Discussed Algorithms

Protocol	Clustering	Mobility	Data aggregation	Proximity from base station	Technique used for maximizing lifetime of WSN
EHCERP	Yes	Limited	Yes	Considered	Balanced clustering
BEAR	No	No	No	Considered	Learning automata
REA-MDG	Yes	Mobile data collector, static sensor field	Yes	Not considered	Spectral clustering, MDC
SCEP	Yes	No	Yes	Considered	MapReduce technique
Cluster-based routing using ABC	Yes	No	Yes	Not considered	Fitness function
Virtual backbone	No	No	No	Not considered	Schedule transition graph, restricted region, greedy method

9 Conclusion

This paper has given extensive information about the recent developments in the field of energy efficient routing techniques. All existing famous energy aware routing techniques are covered which will help researchers to get an insight into the topic and support them to carry out further developments in the field of energy efficient routing. The basic idea conveyed by the paper is that energy should be minimized and balanced during data transmission. Future research problems will aim at improving QOS (quality of service) and finding more efficient ways for data transfer along with mitigation in transmission latency.

References

1. Nikolidakis, S.A., Kandris, D.: Energy efficient routing in wireless sensor networks through balanced clustering. Algorithms **6**, 29–42. ISSN 1999-4893. www.mdpi.com/journal/algorithms (2013). doi:10.3390/a6010029
2. Ahvar, E., Fathy, M.: BEAR: a balanced energy-aware routing protocol for wireless sensor networks. Wireless Sens. Netw. **2**, 793–800 (2010)
3. Hancke, G.P., Leuschner, C.J.: SEER: a simple energy efficient routing protocol for wireless sensor networks. S. Afr. Comput. J. **39**, 17–24 (2007)
4. Naderi, H., Kangavari, M.R., Okhovvat, M.: ScEP: a scalable and energy aware protocol to increase network lifetime in wireless sensor networks. Wireless Pers. Commun. **82**, 611–623, © Springer Science+Business Media, New York (2015)
5. Karaboga, D., Okdem, S., Ozturk, C.: Cluster based wireless sensor network routing using artificial bee colony algorithm. Springer Science+Business Media, LLC (2012)

6. Karaboga, D., Ozturk, C.: A novel clustering approach: Artificial bee colony (ABC) algorithm. Appl. Soft Comput. **11**, 652–657 (2011)
7. Akbari Torkestani, J., Meybodi, M.R.: An intelligent backbone formation algorithm for wireless ad networks based on distributed learning automata. Comput. Netw. **54**, 826–843 (2010)
8. Hussain, S., Shafique, M.I., Yang, L.T.: Constructing a CDS-based network backbone for energy efficiency in industrial wireless sensor network. In: Proceedings of HPCC, pp. 322–328 (2010)
9. Misra, C., Mandal, R.: Rotation of cds via connected domatic partition in ad hoc sensor networks. IEEE Trans. Mob. Comput. 488–499 (2009)
10. Umesh B.N., Dr Siddaraju: Energy efficient routing of wireless sensor networks using virtual backbone. Int. J. Wireless Mob. Netw. (IJWMN) (2013)
11. Rao, X., Huang, H., Tang, J., Zhao, H.: Residual energy aware mobile data gathering in wireless sensor networks. Springer Science+Business Media New York (2015). doi:10.1007/s11235-015-9980-1
12. Pandey, G.K., Singh, Amritpal: Energy conservation and efficient data collection in WSN-ME: a survey. Indian J. Sci. Technol. [S.l.]. ISSN 0974-5645 (2015). doi:10.17485/ijst/2015/v8i17/68648

A Novel Approach for Market Prediction Using Differential Evolution and Genetic Algorithm

Apoorva Gupta, Manoj Kumar and Sushil Kumar

Abstract A novel approach is proposed for the purpose of market analysis by optimizing the reviews of customers using differential evolutionary algorithm. The approach is further compared with the genetic algorithm for improved results analysis. The customer reviews are analyzed in terms of their hidden sentiments and these sentiments form the basis for the recommendation of a product in comparison to the other product reviews. The differential evolutionary and the genetic algorithms provide an advantage of optimized Sentiwords analysis and further enabling a more efficient product recommendation in terms of the reviews of that product, plus more.

Keywords Market prediction · Differential evolution · Genetic algorithm · Business intelligence · NLP · Text analysis · Recommender system · Big data

1 Introduction

Business intelligence and market predictions are one of the major technology trends in the 2000s as per the IBM tech trends report [1]. The two interrelated terms form a crucial part of today's business culture that primarily targets improved customer

---

Positive and negative words available at http://www.unc.edu/~ncaren/haphazard/.

Database available at http://snap.stanford.edu/data/web-Amazon.html.

Web data: Product_and_Accessories reviews size:20 M. Last accessed on 20/02/15.

Apoorva Gupta (✉) · Manoj Kumar · Sushil Kumar
Department of Computer Science and Engineering, Amity School of Engineering and Technology, Amity University, Noida, Uttar Pradesh, India
e-mail: apoorva.2901@gmail.com

Manoj Kumar
e-mail: mkumar7@amity.edu

Sushil Kumar
e-mail: skumar21@amity.edu

M. Pant et al. (eds.), *Proceedings of Fifth International Conference on Soft Computing for Problem Solving*, Advances in Intelligent Systems and Computing 436, DOI 10.1007/978-981-10-0448-3_31

satisfaction leading to enhanced company profits and better company reputation among their customers and service providers. There has been tremendous growth in these fields during the past decades with a handful of related academic and industry publications. It is believed by many market researchers that there exists a "conversation" between the business providers and the customers providing a unique opportunity for the business instead of the conventional long-accepted business-customer marketing. This supports a two-way conversation between the customer and the business providers wherein the customer reviews regarding any service, products, platform, etc., provided by the business lead are of prime importance. These further enable the business providers to provide recommendations and enhance their services as per the customer demands and needs. The improvement in any product/service, the likelihood of a product/service, the future sustainability of a product in the market majorly depends on the customer reviews that further influence and attract other customers, increasing the company profits. These reviews form a silent marketing strategy that formulates not only customer satisfaction but at the same time advertises the product/services for the other customers. Hence, there has been a tremendous need for devising a mechanism such that these reviews can benefit maximum to the company. Therefore, focusing on these reviews, the paper focuses on extracting the sentiments from the reviews by the customers, optimizing the reviews such that the reviews with huge difference in their positive and negative sentiments are ignored, and the same product is categorized as the less likely product as per the differential evolutionary algorithm in comparison to the genetic algorithm, and finally providing the recommendation of the products in terms of customer reviews and in resemblance to other product reviews analysis. The dataset used is provided by Amazon.com, one of the leading e-commerce sites. The reviews are categorized as positive and negative as per the customer sentiments exhibited in the review text. The dataset is handled using the Hadoop storage framework such that the data fetching is performed using the mapper and reducer that perform the job of NLP tokenization, NLP stemming, and NLP filtering making the dataset more suitable for analysis. Primarily, the "Products and the accessories" review dataset is taken into consideration. The positive and the negative sentiments in the reviews, hence extracted, are optimized in terms of the valid review and also provide recommendation with the help of the differential evolutionary and the genetic algorithm.

The fist section gives the brief overview of the paper. The second and third sections focuses on the current business intelligent techniques with its application areas and the step-wise proposed methodology respectively. The fourth section provides an introduction to the NLP module, sentiment words used and the database required for analysis. The fifth and sixth sections gives the algorithms used and the case study using the fitness function proposed respectively. The paper is concluded along with the future scope.

2 Business Intelligence and Market Prediction

Studying the current market statistics in order to predict the future market trends not only fetches benefits to the company but at the same time is also important for good customer–business relationship in order to maintain the customer trust and having enhanced profits with greater market sustainability. Keeping in mind such an objective market prediction has become one of the prime areas of research for market analysts. Now, there exist predictive models giving a link between the dependent and the explanatory variables (example: next likely customer based on their preference, fraud detection), descriptive models giving a collaborative data elements having similar characteristics (example: product preference, profitability, customer segmentation as per their social responses), and decision models which provide an optimal solution to make the desired decision (example: resource optimization and scheduling). The earlier prediction approaches directly focused on the future events but keeping in mind the feasibility and the acceptability of the results obtained decreasing the uncertainty factor, a better classification approach for future prediction has been used in predicting the present and shaping the future. The patterns of the present behavior of customers, business officials, and likeability of the products/services are used in predicting the present and hence shaping the future, that is, making assumptions about future events. The present prediction can be applied to many functional domains such as providing big data to big impact [2] (Table 1).

The future is hence shaped keeping in mind the realistic assumptions, priority-based assumptions, and categorizing the controlled drivers. Therefore, it improves the organization's excellence.

There have been many areas of emerging research such as [1, 2]

- Big data analytics—web mining, cloud computing, Hadoop, data mining, cloud security, parallel DBMS, MapReduce, spatial mining, temporal mining, machine learning.
- Text analytics—statistical NLP, sentiment analysis, multilingual analysis, speech analysis, relevance feedback, search engine optimizations, query processing, Hadoop, MapReduce.
- Web analytics—web crawling, mushups, cloud services, social marketing, web-based auctions, internet security.
- Network analytics—link mining, fraud detection, community detection, trust/reputation, criminal networks.
- Mobile analytics—web services, smartphone platform, games, mobile social networking.

Table 1 Functionality overview of business intelligent applications [1, 2]

	Data	Analytics	Impact	Functionality	Application
E-commerce and market intelligence	1. Logs 2. Transaction records 3. Social comments 4. Reviews 5. Content generated by customer 6. Responses 7. Feedback forms 8. Informal opinions	1. Sentiment analysis 2. Opinion mining 3. Association rule 4. Data segmentation 5. Text analysis 6. Web analysis 7. Social network analysis 8. Anomaly detection	1. Marketing 2. Recommender engines 3. Customer satisfaction 4. Increased sale 5. Enhanced profits 6. Advertising	1. Classification 2. Regression 3. Anomaly detection 4. Association rules 5. Clustering 6. Feature extraction	1. Recommender systems 2. Social media modeling 3. Virtual games 4. Market basket analysis 5. Fraud detection 6. Employee retention 7. Network intrusion
Science and technology	1. Sensor and network content 2. System generated data 3. Large scale records 4. Multiple modality	1. Analytical and mathematical models	1. Technological advances 2. Space observations 3. Interplanetary conclusions 4. Archeological advances 5. Natural calamities prediction	1. Association rules 2. Feature extraction 3. Attribute importance	1. Knowledge discovery 2. Hypothesis testing 3. Link analysis
Health	1. Health records 2. Genomics and sequential data 3. Patient social media	1. clustering 2. Genomics and sequence analysis 3. Ontology 4. Data mining 5. Patient social analysis 6. Drug analysis	1. Improved health care 2. Enhanced life expectancy 3. Quality services 4. Customer satisfaction	1. Classification 2. Clustering 3. Attribute importance	1. Human plant genomics 2. Healthcare decision support 3. Patient social analysis 4. Predicting future diseases

(continued)

Table 1 (continued)

	Data	Analytics	Impact	Functionality	Application
					5. Protein analysis 6. Surgery preparation 7. Customer profitability modeling
Security and safety [3]	1. Criminal records 2. Terrorism databases 3. Viruses, cyber-attacks, botnets 4. Crime maps	1. Cyber-attacks 2. Criminal network analysis 3. Multilingual text analysis 4. Sentiment analysis 5. Criminal association rule and clustering	1. Improved public safety 2. Enhanced security	1. Feature extraction 2. Anomaly detection 3. Classification 4. Association rules	1. Text analysis 2. Fraud detection 3. Credit default modeling 4. Link analysis
Government and politics	1. Citizen feedback and comments 2. Legacy systems 3. Citizen conversations	1. Content analytics 2. Sentiment analysis 3. Information collaboration 4. Price flu	1. Better decision making 2. Empowering citizens 3. Equality 4. Participation 5. Enhanced transparency	1. Classification 2. Attribute importance 3. Association Rules	1. E-polling 2.Citizen participation 3. Response modeling 4. Text analysis & search 5. Analogous government

In the paper, sentiment analysis is done on the Amazon review dataset which further enables prediction of the acceptance of the product in the market, customer satisfaction, and also recommending the products. The other application areas have been done using various soft computing techniques found in the literature [3–7].

3 Methodology

A novel computational intelligent framework is propounded for customer sentiment analysis using the textual reviews. The positive and negative sentiments are extracted from the reviews of the selected reviews such that these are assigned a numerical positive and negative value in reference to the positive and negative sentiments present in the review [8]. These reviews are further optimized using the soft computing approach, wherein the fitness function for analysis is propounded and the results are compared for the differential evolution and genetic algorithm approaches used. The learning phase comprises NLP, sentiwords, and the database module. The learning phase gives the number of reviews and the sentiwords present in the selected product reviews. The evaluation phase deals with the soft computing module such that the fitness function propounded gives an optimized review analysis. The overall analysis predicts the popularity of the products in comparison to the review of the products of the same database. Also, the significance of any review is computed in terms of the lesser difference between the positive and the negative reviews about the product such that in the case of a huge difference between the two, the product review is regarded as less significant since there may be certain customers who gave an extremely positive exceptional response about the product but there is no consistency in the positive feedback for the product. In other words, there is only one review that is extreme positive and the rest are coined positive reviews with very little positive values (Fig. 1).

4 Learning Phase

4.1 NLP Module

The natural language processing (NLP) module prepares the data for analysis. Many e-commerce portals are regularly accessed by billions of customers to leave their reviews about the business products/services which forms a major part of the database repositories. These are analyzed for extracting the sentiments and

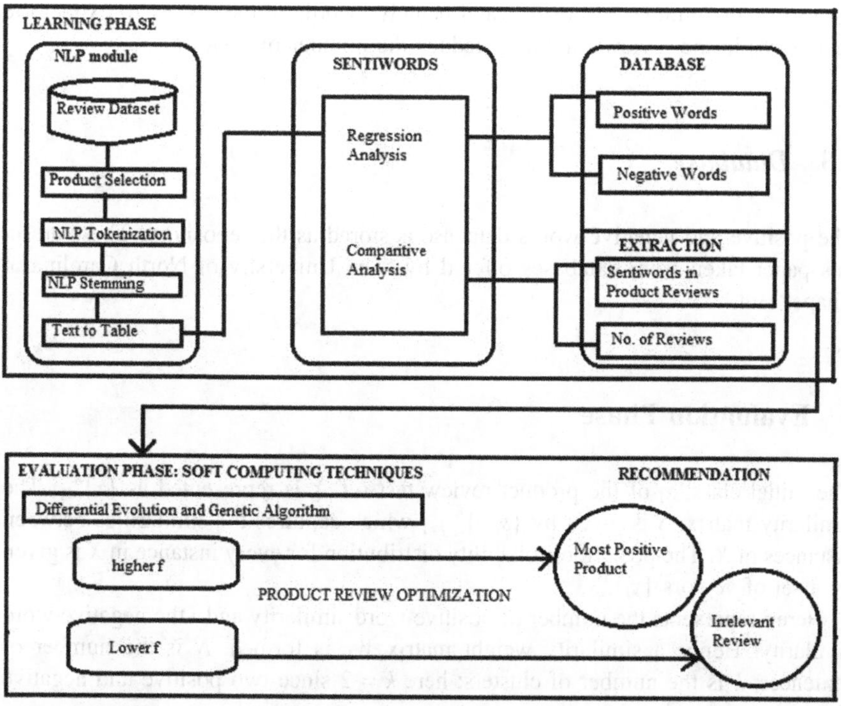

Fig. 1 Stepwise methodology adopted

deducing the popularity of a product among common man enabling enhanced profits. Now, the dataset size as well as the discrepancy in words is controlled by:

- NLP tokenization—the process of breaking a stem into words called tokens. The tokens are the input for text mining. Stemming is removing the suffixes and prefixes from the word. It is a part of lexical analysis [9].
- NLP stemming—the uppercase words are converted into lower case and the punctuations are removed to maintain consistency in the words for analysis [10].
- Text to table—the format of the data is converted from text to table for easy access and data analysis with a further predefined structure to the data.

4.2 Sentiwords

Regression analysis and a comparative study are performed over the reviews of the selected products for extraction of the positive and negative sentiments. Regression analysis is performed over the positive and negative words in the database and a comparative analysis is performed on the review and the positive negative words list to find the number of positive and negative words in the review to extract the sentiwords in the reviews and also deduce the number of reviews.

4.3 Database

The positive and negative words database is stored as the repository. These are in this paper taken from databases offered by 'The University of North Carolina at Chapel Hill.'

5 Evaluation Phase

The initial class, x_i of the product review test set, X is represented as $\{x\}_{i=1}^{n}$. The similarity matrix, S is given by $\{s_{i,\,j}\}_{i,\,j=1}^{n}$ which captures the similarities between instances of X. The posterior probability distribution for every instance in X is given by a set of vectors $\{y_i\}_{i=1}^{n}$.

Here i represents the number of positive word similarity and j the negative word similarity. Hence a similarity weight matrix W_{ij} is formed. N is the number of sentences, k is the number of clusters; here $k = 2$ since two positive and negative clusters are considered.

5.1 Differential Evolution

Evolutionary computing is developed from interaction between the optimization and the biological evolution and hence, the differential evolution (DE) algorithm emerges. DE was proposed by Storn and Price in [11, 12] sharing many features of the classical genetic algorithm and is summarized below. CR (crossover rate), SR (scaling factor), fitness function, k and n as input to algorithm are given as [13] (Fig. 2).

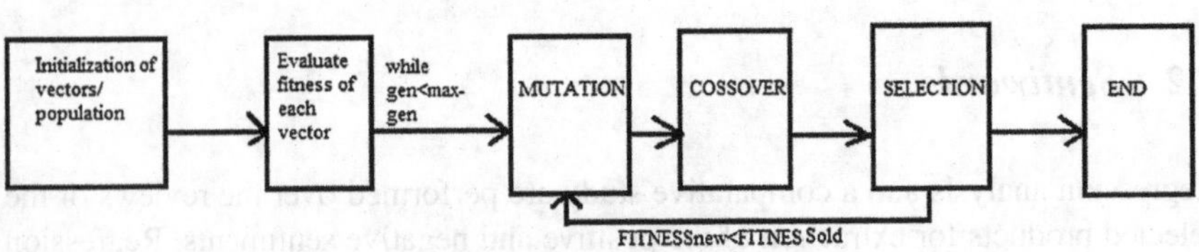

Fig. 2 Aspects of DE [16]

Algorithm 1- Differential Evolution Algorithm

Input: CR, k, n, SF, fitness
Output: f_{best}
//Initialize population
for i <- 1 to n
for j <- 1 to k
$X_{i,j}$ = randInt(1,k)
//iterations
while (!Stopping condition)
for r <- 1 to n
randomly select i, j, m from 1 to n such that i, j, l, m are different
for r <-1 to n
if rand(0,1)<=CR
{
//Generating new vector by adding weighted difference of 2 vectors to 3rd vector; X_{mr} mutation given by func DE/rand/1; SF is adjust usually between 0-2

$X'_{l,r} = SF(X_{i,r}-X_{j,r}) + X_{m,r}$ [MUTATION]
//Mix new vector with target vector to yield trial vector
$X'_{l,r}$ = int(abs($X'_{l,r}$)) [CROSSOVER]
//Replace target vector with trail vector if latter is strictly superior
if $X'_{l,r}$<1 or $X'_{l,r}$>k [SELECTION]
$X'_{l,r}$ = randInt(1,k)
}
if fitness(X'_l)>fitness(X_l)
nextX_l = X'_l
else
nextX_l = X_l
X = nextX
Return f_{best}

5.2 Genetic Algorithm

See Fig. 3.

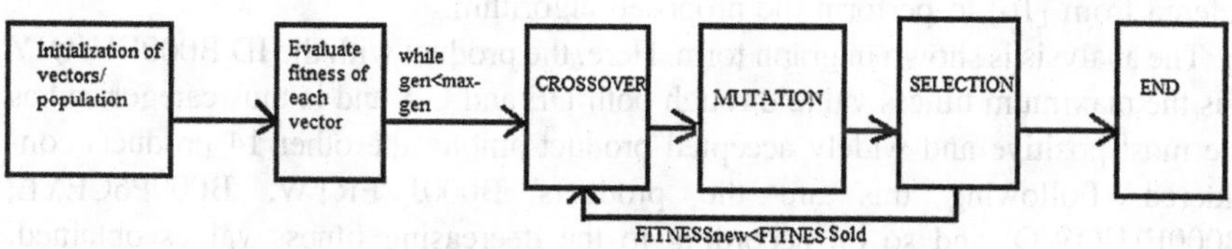

Fig. 3 The main aspects of genetic algorithm

5.3 *Objective Function*

The fitness is an important function responsible for the quality of the chromosomes (The chromosomes represent the sentiwords extracted from the review that are clustered between [1, *k*] where *k* is the number of clusters) produced. In this paper, the function is a combination of two criterion functions combined to balance both positive sentiments (intracluster similarity) and the negative sentiments (intercluster dissimilarity) [14].

$$\text{Intra-cluster similarity function} f1 = \sum\nolimits_{l=1}^{k} |c1| \sum\nolimits_{sisj} \text{sim}_x(s_i s_j) \rightarrow \max \quad (1)$$

where c is the cluster ($c1$ is positive cluster representing the total number of positive words that act as reference), l is the cluster number, k is the total number of clusters required, sim is similarity, x is the current selected similarity measure.

$$\text{Inter-cluster dissimilarity function} f2 = \sum\nolimits_{l=1}^{k-1} 1/|c1| \sum\nolimits_{m=l+1}^{k} 1/|c2| \sum\nolimits_{si\ \varepsilon\ c1} \sum\nolimits_{sj\ \varepsilon\ c2} \text{sim}x(si, sj) \rightarrow \min \quad (2)$$

where $c2$ is negative cluster representing the total number of negative words that act as reference.

The objective fitness function is computed from (1) and (2) as

$$f = f1/f2 \rightarrow \max$$

6 A Case Study: Evaluating Product Reviews

This section provides a case study performed on the Amazon dataset. In addition to the products_and_accessories review dataset (20 M), 15 products are selected and the customer reviews for them through the propounded algorithm provide the best liked product and also list the irrelevant review (Table 2). The Amazon Dataset is refered from [16] to perform the proposed algorithm.

The analysis is shown in graph form. Here, the product with the ID B000F1UQJY has the maximum fitness value through both DE and GA and is thus categorized as the most positive and widely accepted product among the other 14 products considered. Following this are the products B000JVERTW, B000P6CEYE, B000F1UQWQ, and so on according to the decreasing fitness values obtained. However, the product reviews with very low fitness values are observed to have a huge gap between the positive and negative sentiments marking a question on the

Table 2 Products comparison for value of *f*, NFE, time for DE, and GA

Product ID	No. of reviews	DE (*f*)	DE (NFE)	DE elapsed time (s)	GA (*f*)	GA (NFE)	GA elapsed time
B000JVERTW	5	6.473058 e+010	1000	13.069793	6.473058 e+010	1	7.380601
B000924R51	9	3.136944 e+010	1000	10.009981	3.136944 e+010	1	7.214541
B000F1UQJY	5	8.738629 e+010	1000	11.462348	8.738629 e+010	1	6.740137
B00004WIN0	218	2.912876 e+010	1000	8.295068	2.912876 e+010	1	8.741099
B00004WINT	9	2.912876 e+010	1000	8.412276	2.912876 e+010	1	12.858335
B0006J27C4	92	3.883835 e+010	1000	5.560718	3.883835 e+010	1	6.654296
B000M9N5GA	23	3.107068 e+010	1000	11.811993	3.107068 e+010	1	9.621744
B000JUV21 W	11	1.262246 e+010	1000	10.620127	1.262246 e+010	1	7.619004
B000F1UQWQ	69	4.766525 e+010	1000	8.272758	4.766525 e+010	1	6.319275
B000FV8S58	56	3.883835 e+010	1000	12.098943	3.883835 e+010	1	6.141944
B000NJGDUY	18	3.883835 e+010	1000	7.460046	3.883835 e+010	1	6.29608
B000FPGZTA	40	3.155616 e+010	1000	7.973906	3.155616 e+010	1	13.331299
B0002VQ3SU	19	3.530759 e+010	1000	7.816322	3.530759 e+010	1	11.962311
B000P6CEYE	13	5.825752 e+010	1000	6.474509	5.825752 e+010	1	7.76768
B0002DFW2Q	58	1.941917 e+010	1000	12.165299	1.941917 e+010	1	10.609424

wide acceptance of the customer reviews and whether the reviews are genuine or not. Hence, the products with ID(s) B000JUV21W and B0002DFW2Q are least recommended. The review optimization also takes place at the same time since the lower value of f recommends a non-genuine review about the product.

Also, with the same fitness values of f in GA and DE, it is observed that the NFE (number of function evaluation, i.e., generations) is 1000 in DE versus 1 in GA. Hence, it can be said that the same value is obtained in 1 evaluation versus 1000 evaluation but the net iteration time is more efficient in case of DE compared to GA

Fig. 4 Best fitness value versus elapsed time in differential evolutionary algorithm (DE)

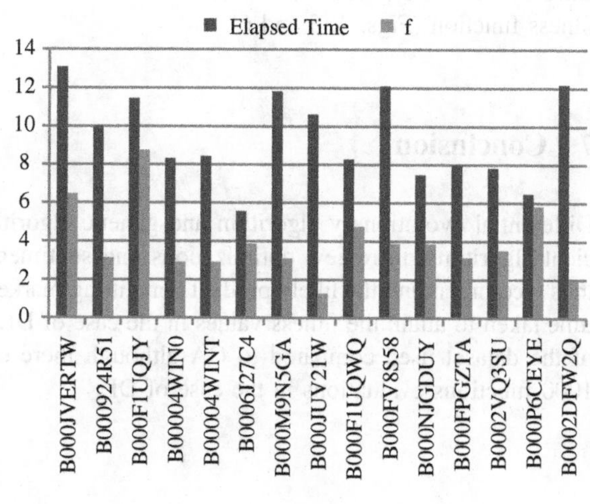

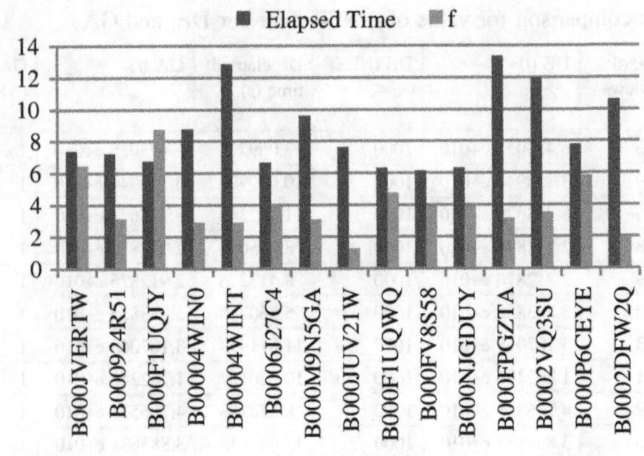

Fig. 5 Best fitness value versus elapsed time in genetic algorithm (GA)

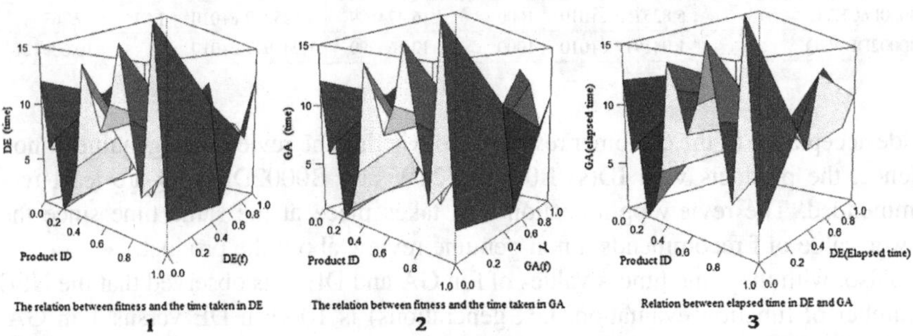

Fig. 6 The relation between fitness and time taken in DE(1), GA(2), and time elapsed in DE and GA(3)

since in GA for 2 products, it is observed to be more for 1 iteration to obtain the fitness function (Figs. 4, 5 and 6).

7 Conclusion

Differential evolutionary algorithm and genetic algorithm are observed to be efficient algorithms in review optimizations and sentiment analysis on the products, thus recommending the likely product benefitting market intelligence. However, the time taken to attain the fitness values in the case of DE is more consistent in terms of the dataset used compared to GA although there exists only one NFE versus 1000 functions evaluations in the case of DE.

8 Future Scope

The approach can be used for sentiment analysis and imparting business intelligence on huge chunks of dataset using Hadoop and MapReduce framework. Also, the sentiment analysis is not restricted to the review dataset used in this paper, it can be applied to other opinion sites like Twitter, Facebook, e-commerce sites like Flipkart, and many others such that the objective can be modified as per the required analysis agenda [15].

References

1. Chen, H., Chiang, R., Storey, V.: Busines intelligence and analytics: from big data to big impact. MIS Q. **36**(4), 1165–1188 (2012)
2. Predictive Analytics: Bringing The Tools To The Data, An Oracle White Paper, Sept 2010
3. Maheshwari, R., Gupta, A., Chandra, N.: Secure authentication using biometric templates in Kerberos. In: 2nd International Conference on Sustainable Global Development (INDIAcom), IEEE, 2015
4. Aggarwal, C.C., Procopiuc, C., Yu, P.S.: Finding localized associations in market basket data. IEEE Trans. Knowl. Data Eng. **14**(1) (2002)
5. Ravi, V., Kurniawan, H., Thai, P.N.K., Kumar, P.R.: Soft computing system for bank performance prediction. Appl. Soft Comput. Elsevier (2007)
6. Kiran, N.R., Ravi, V.: Software reliability prediction by soft computing techniques. Elsevier (2007)
7. Maqsood, I., Khan, M.R., Abraham, A.: An ensemble of neural networks for weather forecasting. Neural Comput. Appl. (2004)
8. Pratap, A., Kanimozhiselvi, C.S., Vijayakumar, R., Pramod, K.V.: Soft computing models for the predictive grading of childhood autism—a comparitive study. Int. J. Soft Comput. Eng. (IJSCE) **4**(3) (2014)
9. Acampora, G., Cosma, G.: A hybrid computational intelligence approach for efficiently evaluating customer sentiments in E-commerce reviews. In: IEEE Symposium on Intelligent Agents, 2014
10. Baiza-Yates, R., Ribeiro-Neto, B.: Modern Information Retrieval. ACM Press/Addison-Wesley (1999)
11. Porter, M.F.: An algorithm for suffix stipping. Program: Electron. Libr. Inf. Syst. **14**(3), 130–137 (1980)
12. Suresh, K., Ghosh, S., Kundu, D., Das, S.: Clustering with mulitobjective differential evolution—a comparative study. In: The International Conference on Advanced Computing Technologies (ICACT 2008), Hyderabad, 2008
13. Daoudi, M., Hamena, S., Benmounah, Z., Batouche, M.: Parallel differential evolution clustering algorithm based on MapReduce. In: International Conference on Soft Computing and Pattern Recognition, IEEE, 2014
14. Abbass, H.A., Sarker, R.: The pareto differential evolution algorithm. Int. J. Artif. Intell. Tools **11**(4), 531–552 (2002)
15. Karwa, S., Chatterjee, N.L: Discrete differential evolution for text summarization. In: International Conference on Information Technology, IEEE, 2014

16. McAuley, J., Leskovec, J.: Hidden factors and hidden topics: understanding rating dimensions with the review text. In: The Proceedings of the 7th ACM Conference on Recommender Systems, ser RecSys '13. New York, NY, USA: ACM, 2013, pp. 165–172. http://doi.acm.org/10.1145/2507157.2507163.
17. Gupta, A.: Big data analysis using computational intelligence and Hadoop: a study. In: 2nd International Conference on Computing for Sustainable Global Development (INDIAcom), IEEE, 2015

A Novel Approach for Actuation of Robotic Arm Using EEG and Video Processing

Saurin Sheth, Saurabh Saboo, Harsh Dhanesha, Love Rajai, Prakash Dholariya and Parth Jetani

Abstract In today's fascinating world of technology, much advancement is being done in various fields of science. However, the field which is showing the most rapid growth in modern times is brain–computer interface. One extremely effective tool for this purpose is Emotiv EPOC headset. The present research focuses on the process of creating a novel BCI that makes the use of the Emotiv EPOC system to measure EEG waves and then it consequently controls the robot. The experiments are performed on 30 different subjects and the obtained results have been analyzed to confirm the usage of the data for the actuation and the control of numerous actuators. The paper presents how video processing can be used to control the robotic arm. The use of video processing gives a new dimension to the variety of applications. The objective is also to provide a low-cost brain-controlled robotic arm.

Keywords Brain–computer interface · EEG · Actuation · Brain-controlled · Robotic arm · Video processing

1 Introduction

Robot is an integral part in automating the flexible manufacturing system which is greatly in demand these days. A step ahead from machines, robots are the solutions to the future increasing labor cost and customer demands. The cost of acquiring a robotic system is expensive but is required to overcome the everlasting demand in every field along with the requirement of maintaining and establishing the quality standard based on ISO grades. Robots and automation have been used extensively

Saurin Sheth (✉) · Saurabh Saboo · Harsh Dhanesha
Mechatronics Engineering Department, G. H. Patel College of Engineering and Technology, Vallabh Vidhyanagar, India
e-mail: saurinsheth@gcet.ac.in

Love Rajai · Prakash Dholariya · Parth Jetani
Vallabh Vidhyanagar, India

M. Pant et al. (eds.), *Proceedings of Fifth International Conference on Soft Computing for Problem Solving*, Advances in Intelligent Systems and Computing 436, DOI 10.1007/978-981-10-0448-3_32

for work which is dull or dangerous, or precision involvement is necessary as well as a replacement for the parts of body in the form of bionic limbs. Advancement in technology has greatly influenced the automation which has led to the enhancement of production capability, product quality, as well as brought a decrement in extent of production cost. Biopac kit is also one of the modern technologies that has capabilities of recording multiple signals. However, it has numerous limitations. These limitations have been stated in [1]. The alternative available is Emotiv EPOC headset which has been well explained and implemented in this paper.

1.1 Electroencephalography

In human brain, the flow of ionic current within the neurons leads to the fluctuation in voltage along the scalp and recording of these changes is termed as electroencephalography (EEG). In early 90s, a Ukrainian physiologist Vladimir Pravdich-Neminsky successfully acquired EEG signals by experimenting on dogs and termed it as "electrocerebrogram." Almost a decade later, German psychiatrist Hans Berger for the first time measured EEG waves being produced in human brain and then coined the name EEG. Multiple electrodes are placed along the scalp at points which are sensitive enough to detect cerebral cortex neural postsynaptic potential changes. Reference electrodes are placed along the scalp at positions of minimal cortical activity and use them as a base to reduce or eliminate unwanted signals but still because of the human errors like poor contact of electrodes with the scalp or eye movement and linking fall under some of the common artifacts. The signal flow diagram of the system is shown in Fig. 1.

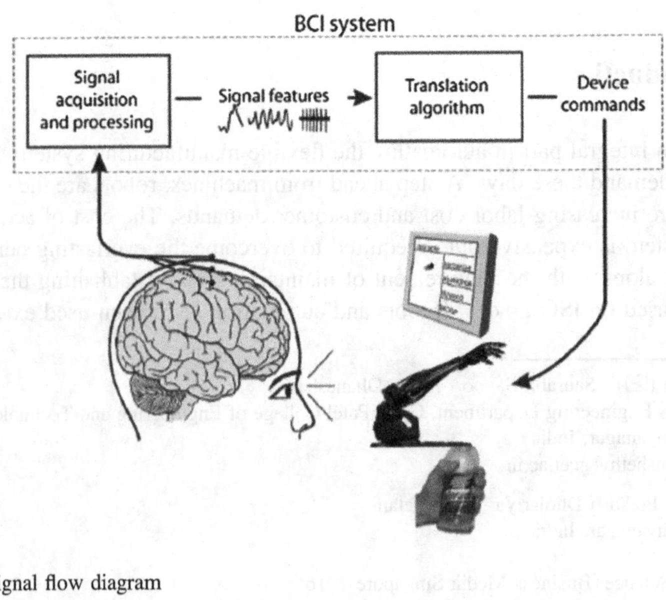

Fig. 1 Signal flow diagram

1.2 Literature Review

Vaish et al. in their study in Brainwave-based user identification system [2] show the use of brainwave computing for providing higher security. In their work, they have developed a brainwave-based user authentication system. As per their view, brainwaves are produced by the brain activities which depend upon the external environment. They have utilized this property which is unique among different individuals to become a new approach in user authentication. They extracted the raw data of the user and then obtained the corresponding values and hence developed a unique user authentication system.

Pinki et al. in their study in Brainwave's energy feature extraction using wavelet transform [3] have represented that the brainwaves have an extremely important security and communication implication. However, the reliability of the methods for differentiation between genuine and imposter is still underdeveloped. They have stated that the efforts are made to increase the reliability in having the exact identification using EEG. Considering the fact that EEG is a non-stationary signal, the utilization of the joint time–frequency feature may provide more reliable results.

Andrew Campbell et al. in their study in brain–mobile phone interface using a wireless EEG headset [4] stated that neural signals are everywhere. They proposed the use of neural signals to have the control over mobile phones for silent, hands-free, and effortless human–mobile interaction.

2 Brain–Computer Interface

The ultimate goal of BCI research is to create a system that is a "closed loop" system giving feedback to the user. The motor cortex of the brain is the area which controls muscle movements, and testing on animals showed that the natural learning behaviors of the brain could easily adapt to new stimuli as well as control the firing of specific areas of the brain. This was an invasive technique but slowly algorithms were developed that decode the responses of the motor neuron in real time and translate it to perform a robotic activity but invasive BCIs are costly and dangerous. Noninvasive alternatives for BCIs include EEG technology, magnetoencephalography (MEG), magnetic resonance imaging (MRI), as well as the "partially invasive," and electrocorticography in which the sensors are implanted within the skull but essentially outside the gray matter of the brain [5]. Noninvasive methods are limited due to the interference of unwanted signals, difficulties to record the inner brain activities, and a poor resolution of signals. However, more sophisticated systems [6] are constantly emerging to combat these difficulties and noninvasive techniques have the advantage of lower cost, greater portability, and the fact that they do not require any special surgery. The primary difficulty in creating an EEG-based BCI is the feature extraction [7] and classification of EEG data that must be done in real time if it is to have any use.

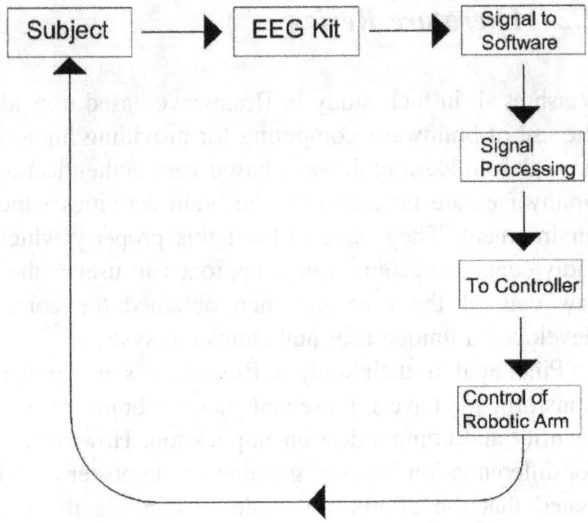

Fig. 2 System design

2.1 System Design

The system is categorized into three blocks as shown in Fig. 2, EEG signal acquisition, signal transmission and signal processing, each targeting at the acquisition of the EEG signal from user scalp. The scalp potentials are then amplified, digitized, and transmitted to a controller that further processes to map them to a robotic arm control [5, 6]. There can be several issues of signal interference because of various sources such as noise, electrical sparks, etc. Hence, there may be a need of signal filtering to obtain precise results. For this purpose, a number of filters are available which may be used. This filtered signal is to be sent to the controller which in turn controls the robotic arm and accordingly, the movements which are thought of can be obtained as desired. The feedback can also be provided in the system to confirm whether the desired movement has been achieved or not.

3 EPOC Suites

Emotiv EPOC headset is provided with the control panel which in turn consists of Expressiv suite, Cognitiv suite, Affectiv suite, and the Mouse Emulator. Of these, the Cognitiv suite can be used along with some other resources for the control of various equipments. The Cognitiv suite essentially consists of a three-dimensional object which can be moved around by proper calibration and perfect training provided by the subject. Once the training has been done and the headset has been calibrated, the subject can again put the headset on and the data that is then obtained is compared with the recorded or the trained data and if the value matches, then the

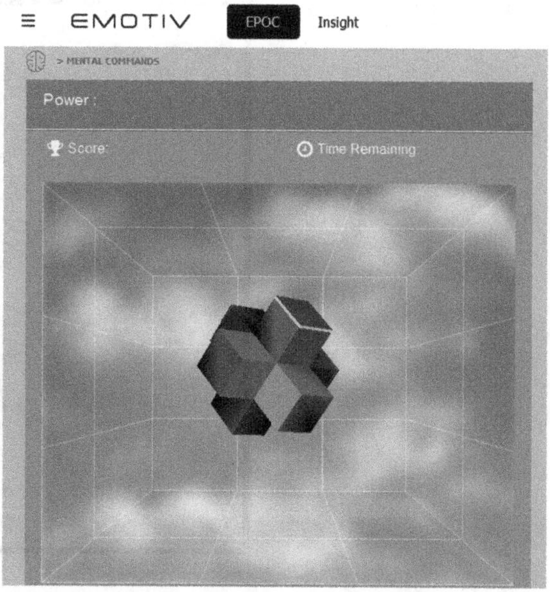

Fig. 3 Neutral action

three-dimensional object is shifted from its neutral position. The Cognitiv suite can be used for getting the training for numerous actions such as lift action, drop action, left action, right action, rotate clockwise action, rotate counter clockwise action, pull action, push action, etc. However, the same procedure can be followed for twelve different actions. The view of the three-dimensional object in its neutral position is shown in Fig. 3.

The headset can be trained in subsequent steps. At any point of time, only four actions can be recorded and used. However, more than one action can be observed at a time. The effect of the Drop Action, Right Action, Left Action, and Lift Action is shown in Fig. 4 which can be observed after a lot of training.

4 Use of Video Processing to Control the Robotic Arm

This process makes use of computer vision or image processing for the control purpose [7–9]. Hence, a camera is required which can capture the video and send it to the computer which further processes the captured video and it can then be sent to the controller. There are several steps involved in the use of video processing for the actuation and the control of the robotic arm. All the steps are performed in Python. These steps can be briefly described as follows:

Step 1: This is the basic step in which only the three-dimensional cube is to be detected which is shown in Fig. 5. In this step, only the frame of the cube along with the blue background is to be detected. One additional task in this step is to crop the image of the software and have only the frame of the cube.

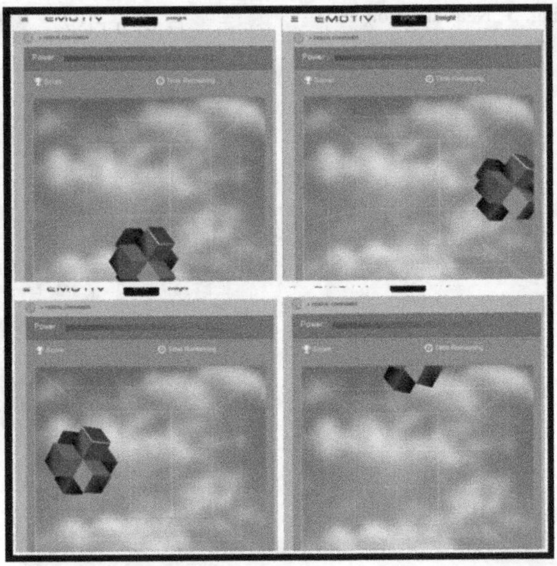

Fig. 4 Four different actions

Fig. 5 Original cube

Step 2: In this step, the original image is converted into the HSV form which separates color components from the intensity and the edges get highlighted. Hence, it can be interpreted as a color model that is used to describe colors in terms of their shade. Rather than using the value, the color model may make the use of brightness. Brightness in combination with saturation describes the intensity of the color. The sample HSV image is shown in Fig. 6.

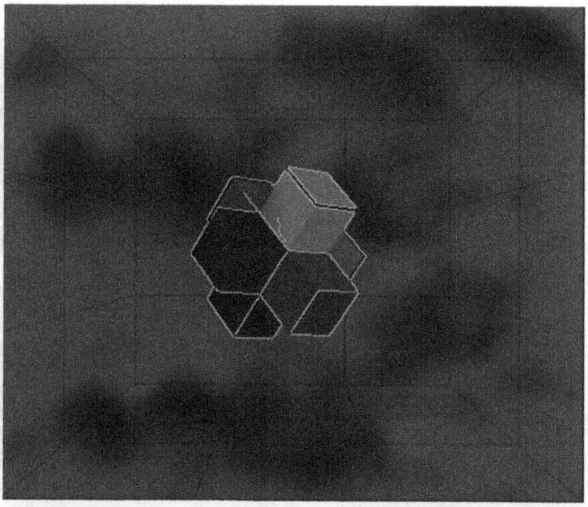

Fig. 6 HSV image

Fig. 7 Edge detection

Step 3: The third step is getting the edges of the cube detected. This can be done from the HSV image. The detection of the edges is important for determining the extreme points of the cube which is again important in obtaining the center of the cube. The center of the cube is then to be followed to get the movement of the robotic arm in this case. The image which shows the edge detection can be seen in Fig. 7.

Step 4: The fourth step involves the conversion of the HSV image to the threshold image. In this image, only the edges of the cube are retained and the rest of the cube is masked. This can be done through Python and the image is shown in Fig. 8.

Fig. 8 Threshold Image

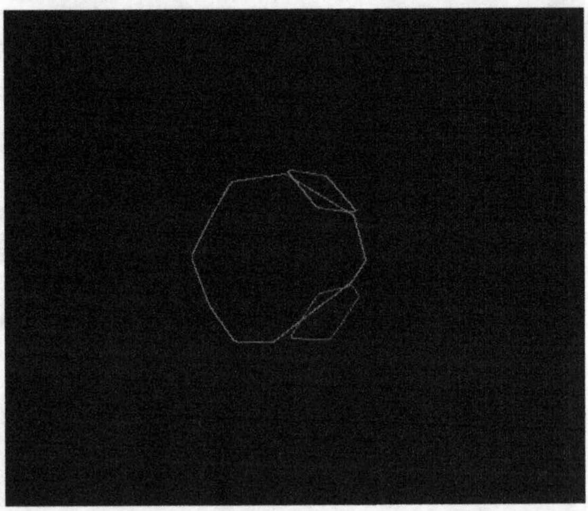

Fig. 9 Contour of the object

Step 5: In the fifth step, the contour of the cube is plotted from the masked image that is obtained in the fourth step. The contour of the cube in the present case is shown in Fig. 9. As it can be seen from the figure, three contours are generated from the cube. Hence, it becomes a cumbersome procedure to detect the center of the contour. Consequently, proper care has to be taken while determining the center of the contour, since all the three contours have to be taken into account. One of the methods for obtaining the control of the cube counterpart in this case is locating the four extreme points on the contour and then according to the pixels that the corner points occupy and correspondingly control the robotic arm.

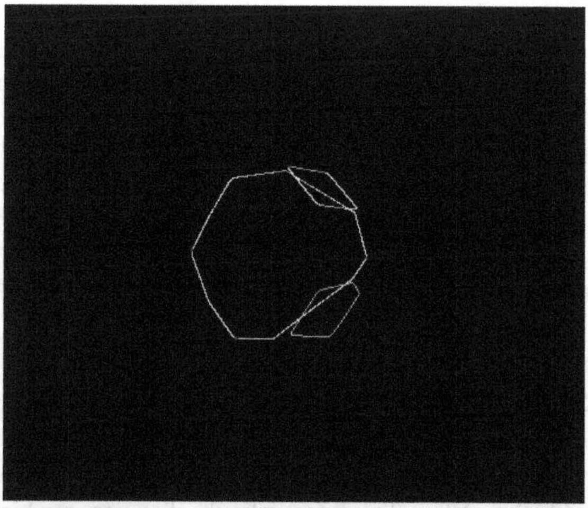

Fig. 10 Corner point determination

Fig. 11 Corner points of the cube

Step 6: Once the smooth contour of the cube is obtained, the corner points of the contour can then be located. This can be done by designating the outermost points as the extreme points of the cube. The extreme points can be seen by red dots in Fig. 10. This image can again be masked to plot only the corner points. These points can be evidently seen in Fig. 11.

Step 7: The seventh step is the final step in the video processing in which the cube counterpart movement can be controlled by knowing the pixels of the extreme points and observing the real-time movement of the cube and making a note of whether the pixels then increase or decrease from the reference values and accordingly the movement of the actuator can be controlled. In this case, if the value of x is less than 150, the arm moves toward the left; if the value of x is more than 315, the arm moves toward the right; if the value of y is less than 110, the arm moves upwards; and if the value of y is more than 292, the arm moves downwards. However, there is some tolerance kept in the pixels during the coding. This is because, even in the neutral position or in the center position, the cube tends to move up and down. Hence, it becomes difficult to detect the extreme points of the cube or the contour in this case. Consequently, the tolerance of five pixels can be taken into account while preparing the code for further processing for the actuation of the arm.

5 Experimental Results

The experimental results show that the cube can be readily detected with a negligible time delay. Simultaneously, the direction in which the cube moves can also be printed on the screen in words. This can again be taken on the serial monitor of the Arduino [5, 6] and accordingly, the codes can be framed for moving the actual object or the actual robotic arm by controlling the motors. The output of the experiments is shown in Fig. 12. The mechanical assembly of the robotic arm excluding the electrical components, which is to be controlled from the above experimental results, is shown in Fig. 13.

6 Conclusion and Future Scope

From a number of experiments that have been performed on 30 subjects in order to check the versatility of the headset and hence to observe its response, it can be concluded that using the brain–computer interface and video processing in

Fig. 12 Experimental output

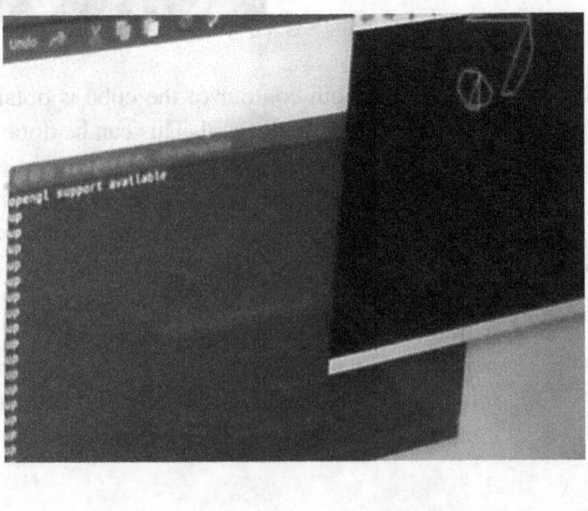

Fig. 13 Robotic arm

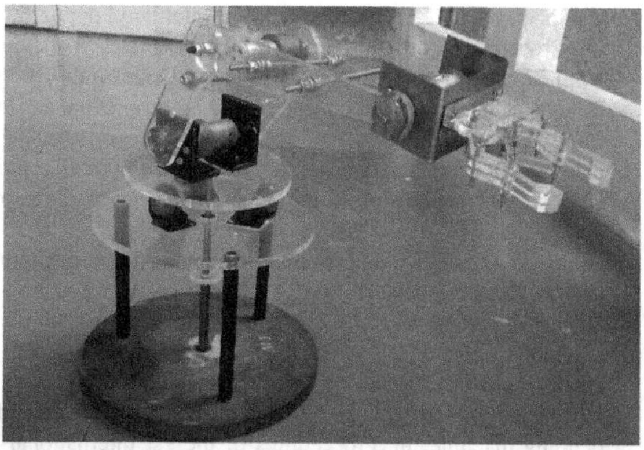

conjunction, any object can be controlled directly with the brain. As a lot of work is done in the field of image and video processing, none of them are reported as use of video processing to control the objects. In this case, the use of video processing to actuate the robotic arm is the novelty as it was achieved without the SDK license of the EPOC. Thus it leads to low-cost actuation. Hence, the headset only serves as a gaming device and there is no means of taking the signals from the software to the real world. Hence, the raw EEG cannot be analyzed, so filters cannot be used, even though fairly accurate response can be obtained. However, the same approach can be used for the number of applications such as follows:

- To control the wheelchair according to the user's desire.
- The control of the computer cursor by thoughts.
- Can be used to recognize the sensations which are generated in the brain when a patient is healed by acupressure or acupuncture. Knowing that, a patient can be directly healed by just creating those sensations in the patient, thus eliminating the need of any of the two above-mentioned methods.
- The brain–computer interface can also be used to dampen the vibrations that of the aged people experience when they take hold of the things.

References

1. Sheth, S., Rajai, L., Jetani, P., Dholariya, P.: Scope, study and experimentation of biopac kit and its implications on a brain controlled robotic arm. Discovery **43**(196), 1–7 (2015)
2. Kumari, P., Vaish, A.: Brainwave based user identification system: A pilot study in robotics environment. Robot. Auton. Syst. **65**, 15–23 (2015)
3. Kumari, P., Vaish, A.: Brainwave's energy feature extraction using wavelet transform. In: 2014 IEEE Students' Conference on Electrical, Electronics and Computer Science (SCEECS), IEEE (2014)

4. Campbell et al., T.: NeuroPhone: brain-mobile phone interface using a wireless EEG headset. In: Proceedings of the 2nd ACM SIGCOMM Workshop on Networking, Systems, and Applications on Mobile Handhelds, New Delhi, India, August 30, 2010
5. Levine, S.P.: A direct brain interface based on event-related potentials. IEEE Trans. Rehab. Eng. **8**, 180–185 (2000)
6. Kher, R., Pawar, T., Thakar, V., Shah, H.: Physical activities recognition from ambulatory ECG signals using neuro-fuzzy classifiers and support vector machines. J. Med. Eng. Technol. **39**(2): 138–152 (2015)
7. Penny, W.D.: EEG-based communication: a pattern recognition approach. IEEE Trans. Rehab. Eng. **8**, 214–215 (2000)
8. Gajjar, B.R., Sheth, S.: Design and automation in back plug press fitting process of ball pen assembly. Appl. Mech. Mach. **592–594**, 2596–2600 (2014). doi:10.4028/www.scientific.net/AMM.592-594.2596
9. Virani, M., Vekariya, J., Sheth, S., Tamboli, K.: Design and development of automatic stirrup bending machine. In: Proceedings of the 1st International and 16th National Conference on Machines and Mechanisms (iNaCoMM 2013), IIT Roorkee, India, pp. 598–606 (2013)
10. Parikh, P.A., Joshi, K.D., Sheth, S.: Color guided vehicle—an intelligent material handling mechatronics system. In: Proceedings of the 1st International and 16th National Conference on Machines and Mechanisms (iNaCoMM 2013), IIT Roorkee, India, pp. 628–635 (2013)
11. Chauhan, V., Sheth, S., Hindocha, B.R., Shah, R., Dudhat, P., Jani, P.: Design and Development of a Machine Vision System for Part Color Detection and Sorting, ICSSA (2011) doi:10.13140/2.1.1628.7526
12. Gupta, M., Patel, A., Dave, N., Goradia, R., Sheth, S.: Text based image segmentation methodology. Procedia Technol. **14**, 465–472 (2014). doi:10.1016/j.protcy.2014.08.059.

A Survey: Artificial Neural Network for Character Recognition

Mrudang D. Pandya and R. Patel Jay

Abstract Due to advancement in technology many recognition task have been automated. Optical Character Recognition (OCR) aims to convert the images of handwritten or printed text into a format that is capable for a machine to understand and process it. For the recognition to be precise various properties are calculated, on the basis of which characters are classified and recognized. Character recognition has been an attractive area for researchers using the Artificial Intelligence. Recognition is easy for humans, but what about machines? Advancement in Artificial Intelligence has led to the developments of various devices. The open issue is to recognize documents both in printed and written format. Character recognition is widely used for authentication of person as well as document. OCR is a technique where digital image that contains machine printed or handwritten input into software and translating it into a machine readable digital format. A Neural network can be designed the way in which the brain performs a particular task or function of interest. In this paper we present the survey of how efficient an Artificial Neural network can be utilized for character recognition process.

Keywords OCR · Artificial neural network · Neuron · Classifier · K-means · Epoch · Clustering

1 Introduction

Character recognition is the phenomenon to identify the character which may be in the form of scanned document or typed text in different fonts and effects. Classical methods for recognition face the problems due to [1]: (1) the characters that are

M.D. Pandya (✉) · R. Patel Jay
Department of Information Technology, CSPIT, Charotar University of Science and Technology, Changa 388421, India
e-mail: mrudangpandya.it@charusat.ac.in

R. Patel Jay
e-mail: engjaypatel@gmail.com

M. Pant et al. (eds.), *Proceedings of Fifth International Conference on Soft Computing for Problem Solving*, Advances in Intelligent Systems and Computing 436, DOI 10.1007/978-981-10-0448-3_33

same often differ in shapes as well as styles with person to person. (2) Characters are subject to damage due to presence of noise that might be interpreted as another character. (3) There are no fix formats that define the appearance of a character. Several recognition approaches have been applied including statistical, structural methods, and neural networks. Some methods identify strokes; others attempt to identify entire words or groups of characters [2].

2 Artificial Neural Networks

Artificial Neural Network (ANNs) is inspired by biological nervous systems, such as the neuron. Its architecture consists of highly interconnected processing elements i.e. neurons to solve specific problems. ANNs learn by example. ANN is also a robust classifiers and able to make decisions about input data. The cells in ANN are linked through weighted links. This way, the input to the ANN is more number of cells so that an output is in the form of one or more activated cells. The information is always stored in parameters that can be set or trained by providing input and expected output.

A typical ANN system has layered architecture as shown in Fig. 1. The first layer has input neurons which act as input to the system which send data to the second layer of neurons for further processing. After passing second layer, the active neurons output the results by applying activation functions. More complex systems have more layers of neurons with increased layers of input neurons and output neurons.

An ANN uses parameters defined like:

(1) The weight of neurons.
(2) The learning parameter i.e. α to update the weights.
(3) The activation function to transform the activation level of a neuron.

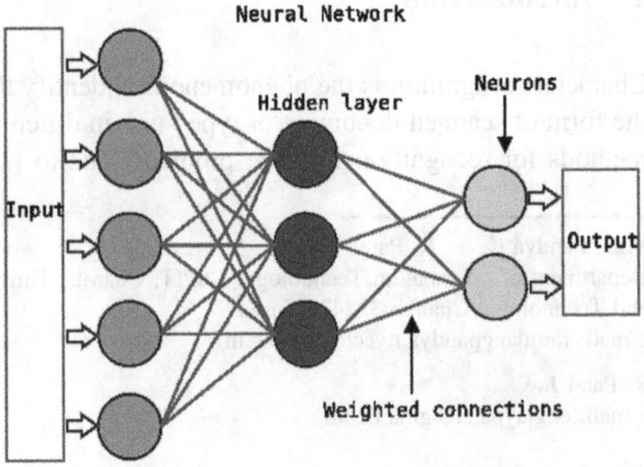

Fig. 1 Architecture of artificial neural network

2.1 *Advantages of Multilayer Perceptron Network*

(1) **Generalization** Neural networks are capable of generalization that is they recognize an unidentified pattern with other known patterns that have the same distinguishing features.
(2) **Fault Tolerance** Neural networks are highly fault tolerant which is known as graceful degradation. A neural network keeps on learning even if some interconnection between neurons fails.

2.2 *Limitations of Multilayer Perceptron's*

(1) Expensive learning process.
(2) No guarantee about solution: Remedies such as the momentum, add computational cost, estimates of transfer functions, estimates of error values.
(3) Extension problem: Do not extend up well from small systems to larger real systems.

3 Character Recognition Procedures

A classical method for recognition requires high computational cost. Figure 2 shows the classification of character recognition techniques and a brief description of all these methods are depicted in the further subsections.

3.1 *Online Recognition*

Online character recognition system is posed by the challenges to increase the accuracy and reduce the recognition time that requires the conversion of text, where a sensing plate identifies the pen point movement.

(1) **Direction Algorithm Based**
Directional information is used to identify a character [4], which assumes that orientation information is generated every time whenever a character is written. For e.g. four directions right, down, left and up helps in modelling strokes.
(2) **KNN Classifier**
Classification is done using the dissimilarity measures [4] by KNN classifier between the input character and the training sample. It's computationally heavy with large sets and complex similarity measures. For similarity measure it uses DTW measure between two strokes.

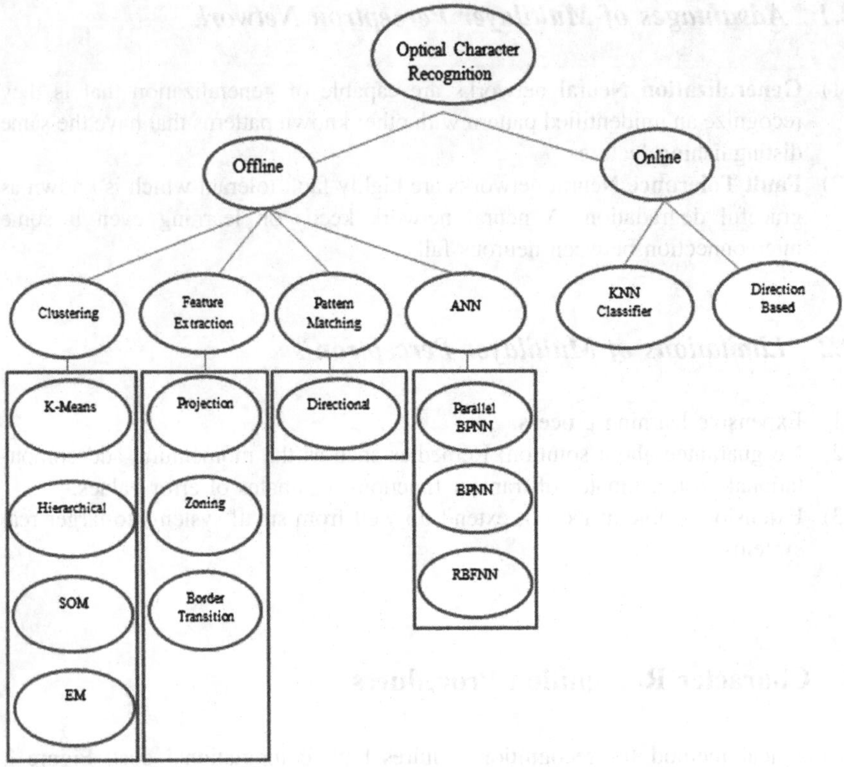

Fig. 2 Classification of character recognition techniques [3]

3.2 *Offline Recognition*

An Off-line recognition works on images that have been scanned. The data is space ordered and two dimensional which means that overlapping characters are not separated.

(1) **Clustering**
K-means is well known clustering algorithm [5] which used for data alignment when the number of the clusters are known. So this procedure or algorithm works for a fixed set of characters. Each point is assign to one of the initial set of given clusters and then each centre of the cluster is replaced by the mean point in that cluster.

(2) **Feature Extraction**
The aim of the feature extraction [6] is to recognize characters from properties which are similar to that of humans utilize to recognize letters [1, 7] such as percent of pixels above horizontal half point, aspect ratio, strokes, structural, directional feature etc., are used as features to gain high performance.

The projection method compresses data through a projection. Marginal distributions through image area are formed by counting black pixel along parallel lines [8].
The character gets divided into smaller fragment of areas (zone) in zoning. The features get extracted by counting each zone's black pixels and the profiles are averaged in each zone [8].
A graph matching method [9] uses structural properties of character. Such as end point is connected to only one pixel which has positional information, a branch point is connected to three pixels and a curve point is connected to the two pixels

(3) **Pattern Matching**
Directional planes are generated after normalizing the character into standard size which records the stroke in a different orientation. The dimension can be reduced to directional plane in some fixed size blocks and then strengths of every block are averaged. Low pass filtering and down-sampling are used to smooth the margins of blocks,

(4) **ANN**
An Artificial Neural Network is a model which processes information using a connectionist approach. ANN is an adaptive system that modifies its structure based on information during the learning phase. The ANN is trained using Back propagation algorithm for various inputs to get expected outputs for classifying characters. The 'Target' values are specified by user to accommodate for small recognition errors.

4 Character Recognition Procedure

(1) Image/character input [10]
Input can be handwritten character or image file of written character for recognition.

(2) Pre-processing [10]
The pre-processing phase gives a document with maximum compression and minimal noise on image.

(3) Segmentation
Segmentation [11] is a critical phase because the one can reach in separation of words, lines or characters directly degrade/increase the recognition rate of the input [12].

(4) Feature extraction
Extraction of feature like height, width, and horizontal, vertical line, top and bottom check is done.

(5) Recognition
For recognition back propagation algorithm [13] is implemented.

Table 1 Results for variation in number of epochs

Epoch (300)			Epoch (600)		Epoch (800)	
No. of characters error (%)			No. of characters error (%)		No. of characters error (%)	
Arial	34	4.4	34	3.33	34	1.11
Tahoma	34	1.11	34	0	34	0
Roman	34	0.00	34	0	34	1.00

Table 2 Results for variation in learning rate parameter

50			100		120	
No. of characters error (%)			No. of characters error (%)		No. of characters error (%)	
Arial	34	91.10	34	20	34	3.33
Tahoma	34	62.50	34	12.22	34	1.11
Roman	34	80.10	34	16.66	34	0.00

Number of characters = 90 Number of Epochs = 600, Sigmoid slope = 0.014

(6) Output
Output is saved.
For results the parameters are:
Learning rate = 150, Sigmoid Slope = 0.014,
Weight bias = 30,
Number of Epochs = 300–800,
Mean error threshold value = 0.0002,
Learning rate = 150, Sigmoid slope = 0.01 (Tables 1 and 2)
From the results it can be seen that if number of epochs and learning parameter gets increased, the error in character identification reduces.

5 Issues and Challenges

Though various techniques and tools for optical character recognition have been developed and implemented, still it faces the problem that lies into the input.

Some of them are depicted below:

(1) **Noise** sensible to disconnected gaps, broken lines, filled loops, etc.
(2) **Translations** Sensitivity to movement of the character or its constituent parts.
(3) **Distortions** Variations for e.g. corners, improper dilations, shrink and expansion of characters.
(4) **Uneven Lighting conditions** Certain discrepancies can remain into the input image that has been captured from the camera under uneven lighting conditions which causes flooding of light.

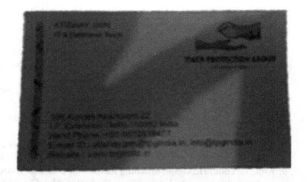

Fig. 3 Image under uneven lighting conditions

(5) **Skewness** When the image is taken from camera due to rotation, the image gets skewed that limits the OCR to identify the character which further results to reduction into accuracy.
(6) **Blur** Due to moving camera the image captured in it get blurred because of fast moving objects.
(7) **Overlapped** Characters may cause problem in identifying the actual character due to mixing of upper and lower region of the line in which the characters are placed. When two or more characters appear side by side in Devanagari, the header lines touches with each other and generate a bigger header line.
(8) **Deformations** This is caused due to broken characters, smeared characters and spot of color or patch (Fig. 3).
(9) **Variations in spacing** It occurs due to variable spacing, header lines, and superscript/subscript.

6 Conclusions

Over the decades the character recognition techniques have developed. In this paper, we have presented a survey on how different techniques can work for character recognition. Though neural network is used frequently in the research for character recognition, combination with fuzzy logic and genetic programming can be used to improve the performance or accuracy. There are various parameters or factors that affect the efficiency of recognition procedure. The issues and challenges are also mentioned in this paper with the reasons specified of why it may occur.

References

1. Dedgaonkar, S.G., Chandavale, A.A., Sapkal, A.M.: Survey of Methods for Character Recognition. Int. J. Eng. Innovative Technol. (IJEIT) **1**(5), 180–189 (2012)
2. Sharma, A., Chaudhary, D.R.: Character recognition using neural network. Int. J. Eng. Trends Technol. (IJETT) **4**(4), 662–667 (2013)

3. A. Rajavelu, Musavi, M.T., Shirvaikar, M. V.: A Neural Network Approach to Character Recognition, Neural Networks
4. Gupta, K., Rao, S.V., Viswanath, P.: Speeding up online character recognition. In: Proceedings of Image and Vision Computing New Zealand (2007)
5. Krasteva, R., Bulgarian hand-printed character recognition using fuzzy c-means clustering. Probl. Eng. Robot. **53**, 112–117
6. Patil, V.V., Sanap, R.V., Kharate, R.B.: Optical character recognition using artificial neural network, Int. J. Eng. Res. General Sci. **3**(1), (2015)
7. Ayshi, M.A., Jay Kimmel, M., Simmons, D.C.: Character recognition system using spatial and structural features. US 7,010,166B2
8. Khawaja, A., Tingzhi, S., Memon, N.M., Rajpar, A.: Recognition of printed Chinese characters by using neural network, 1–4244-0794-X/06/$20.00 ©2006 IEEE, pp 169–172
9. Kim, J., Yoon, H.: Graph matching method for character recognition in natural scene images. In: INES 2011, pp. 347–350, 978-1-4244-8956-5/11/$26.00 ©2011 IEEE
10. Singh, S.: Optical character recognition techniques: a survey. J. Emerg. Trends Comput. Inf. Sci. **4**(6), (2013) ISSN: 2079–8407
11. Shah, P., Karamchandani, S., Nadkar, T., Gulechha, N., Koli, K., Lad, K.: OCR-based chassis-number recognition using artificial neural net-works. In: ICVES (2009)
12. Vasudeva, N., Parashar, H.J., Vijendra, S.: Offline character recognition system using artificial neural network
13. Chung, Y.Y., Wong, M.T.: Handwritten character recognition by fourier descriptors and neural network. IEEE TENCON, pp 391–394 (1997)

Crosstalk Noise Voltage Analysis in Global Interconnects

Purushottam Kumawat and Gaurav Soni

Abstract The rapid growth of VLSI technology is generally due to the continuous reduction in the feature size of device. The work of an interconnect is to distribute the data signals and to make available power or ground to and among the different circuit functions on the chip. As the scaling of devices, the traditional transistor has thus far met the challenges like crosstalk, coupling, and noise margin. Due to this, scaling of interconnects has become one of the performance limiting factor for the new VLSI designs. As the advancements in process technology scaling are going on, the spacing between the adjacent interconnect wires keeps shrinking. This causes an increase in coupling capacitance between the interconnect wires. Hence, coupling noise became an important part which must be taken into account while performing timing verification for VLSI chips. The crosstalk generated due to switching of signals will induce noise onto nearby lines which can further deteriorate the signal integrity and reduce noise margins. These aspects of crosstalk make the system performance dependent on data patterns, switching rates, and line-to-line spacing.

Keywords Crosstalk · Parasitic capacitance · Signal integrity

1 Introduction

Interconnects are used to connect components on a VLSI chip, to connect chips on a multichip structure, and to connect multichip structures on a system board. As transistors become smaller in size, they dissipate lesser power, switch faster, and are cheaper to manufacture. Interconnects can be categorized as local, semi-global, and global. Local interconnects are the lowest level of interconnects. These are used in

Purushottam Kumawat (✉) · Gaurav Soni
Poornima University, Jaipur, India
e-mail: purushottamkumawatsesgi@gmail.com

Gaurav Soni
e-mail: gaurav.soni@poornima.edu.in

M. Pant et al. (eds.), *Proceedings of Fifth International Conference on Soft Computing for Problem Solving*, Advances in Intelligent Systems and Computing 436, DOI 10.1007/978-981-10-0448-3_34

localized regions of a chip to provide electrical path between nearby nodes. They use the most bottom layers of metals. Semi-global interconnects are usually used to connect devices within a block. The coupling phenomenon occurs in digital as well as in analog blocks. However, digital blocks are more densely interconnected to each other and for this reason, coupling of capacitance tends to be more significant in this type of circuit. Crosstalk owing to capacitive coupling wires can result in functional failure of the circuit. It is a well-known fact that signal integrity (SI) is robustly affected by both the nonlinear behavior of CMOS drivers and the transmission line behavior of interconnect. Therefore, accurate estimation of time delay and crosstalk noise should be done. Crosstalk in one line to another will occur when the electromagnetic fields from different structures come interact. In multiconductor systems, excessive line-to-line coupling can create some pernicious effects. The crosstalk is also one of them that will affect the performance of interconnect by modifying the propagation velocity and effective characteristic impedance. This will adversely affect signal integrity and system level timings. The crosstalk will also induce noise into other nearby lines which can further degrade the signal integrity of the integrated circuit. This aspect of crosstalk makes the system performance densely dependent on switching rates, data patterns, and line-to-line spacing [1].

2 Signal Transmission in Interconnect

Any electronic system is composed of functional blocks interconnected with each other; that is, capable of transmitting signal from one part to another. This information is in the form of a voltage or a current value. Ideally, the transmission of the signal between two interconnected blocks must be without any distortion and instantaneous. However, this ideal picture cannot be achieved in practice. The reason behind is that physically there is always a propagation time for transmitting the information from one point to the other. If the signals varies very slowly compared to this propagation time, the transmission may be considered instantaneous and the ideal picture is right. Hence, signal delay must be taken into account when designing the system to ensure that all the blocks receive the information at the expected time. If, on the other hand, the signal varies hastily compared to the propagation time, several effects may be observed:

- Reflections coming from the end of the interconnection and it interfered with the signal sent.
- Interference between nearby interconnections (Crosstalk).

All these effects may influence the behavior of the system and they represent a limitation on performance. If they are not properly addressed at the design stage, they may be adequate to completely distort the transmitted signal and therefore cause a system malfunction [2].

3 Physical Factors Affecting Interconnect Design

The signal transmission will eventually depend on some physical factors of the interconnections that determine the respective values of *R*, *L*, and *C*. There are three physical components influencing electrical parameters:

- Material properties of the conductors forming interconnects and the dielectrics between them. This affects *R* and *C* respectively.
- The Dimensions of conductor and dielectric materials: width, length and thickness of conductors, pitch and dielectric thickness, which determines the distance between stacked interconnections. This influences *R*, *L* and *C*.
- Connection of the wires carrying reference voltages, generally known as power supply distribution. This Influence on *L* and *C*.

The first depends exclusively on technology, the second one depends on technological design and restrictions, and the third one depends mostly on design, but is also restricted by the type of technology used [3].

4 Modeling an Interconnect

4.1 Transmission Line Model

The RLC transmission line model is used to represent the wire as interconnects. The model was developed to calculate equivalent circuit parameters for a copper-based interconnect geometry (Fig. 1).

According to the model, the thickness of interconnect is *t*, the width of interconnect is *w*, and *h* is the height of interconnect above the ground. The mathematical formulae for these are [4]:

Resistance

$$R = \frac{\rho L}{A} = \frac{\rho L}{\mathrm{HW}} \tag{1}$$

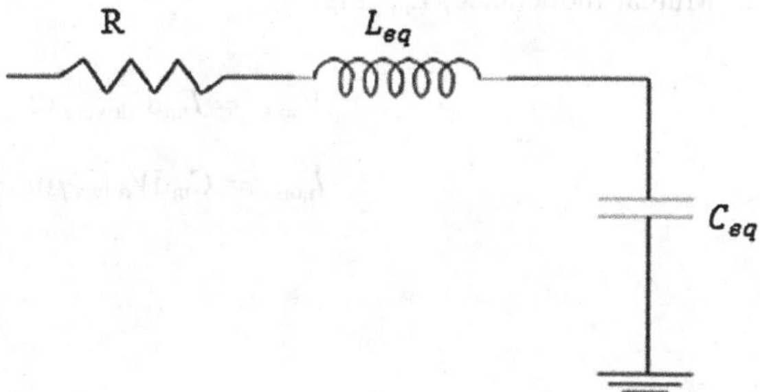

Fig. 1 Transmission line using RLC [4]

Capacitance

$$C_W = \varepsilon\left[1.15\frac{w}{t} + 2.80\left(\frac{h}{t}\right)^{0.222} + \left(0.66\frac{w}{t} + 1.66 - 0.14\left(\frac{h}{t}\right)^{0.222}\right)\cdot\left(\frac{t}{s}\right)^{1.34}\right] \quad (2)$$

$$C_c = \varepsilon\left[1.14\frac{t}{s}\left(\frac{h}{h+2.06s}\right)^{0.09} + 0.74\left(\frac{w}{w+1.59s}\right)^{1.14} + 1.16\left(\frac{w}{w+1.87s}\right)^{0.16}\left(\frac{h}{h+0.98s}\right)^{1.18}\right] \quad (3)$$

where

s is the interwire spacing (assumed $s = w$), ε is the dielectric permittivity, and C_c = Coupling Capacitance.

For global wires, we include inductance whose formula is:

Inductance

$$L_W = 2\times10^{-7} l\left(\ln\frac{2l}{w+h} + 0.5 + \frac{w+h}{3l}\right) \quad (4)$$

4.2 Crosstalk

Crosstalk is defined as the energy proclaimed to a transmission line due to signals in nearby lines. The magnitude of the induced crosstalk noise is a function of net configuration, rise time, and signal line geometry. More aggressive technology scaling will cause an increment in the contribution of the coupling capacitances to the total interconnects capacitance.

4.3 Interconnect Modeling Fundamentals

Crosstalk is caused by energy coupling from one line to another line via:

1. Mutual capacitance, C_m (Fig. 2)
2. Mutual inductance, L_m (Fig. 3)

$$V_{noise} = L_m \mathrm{d}I_{driver}/\mathrm{d}t \quad (5)$$

$$I_{noise} = C_m \mathrm{d}V_{driver}/\mathrm{d}t \quad (6)$$

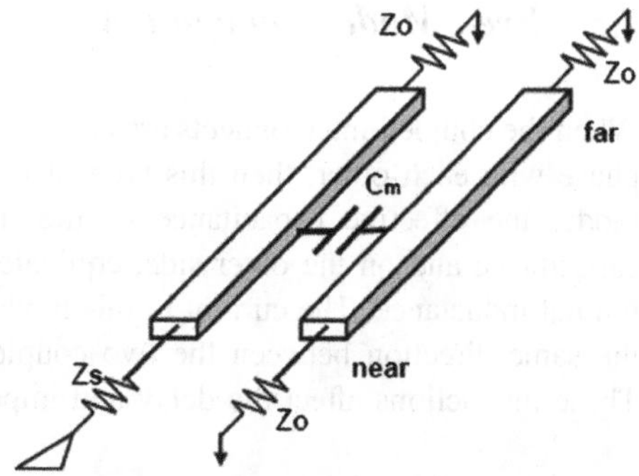

Fig. 2 Mutual capacitance (C_m) [5]

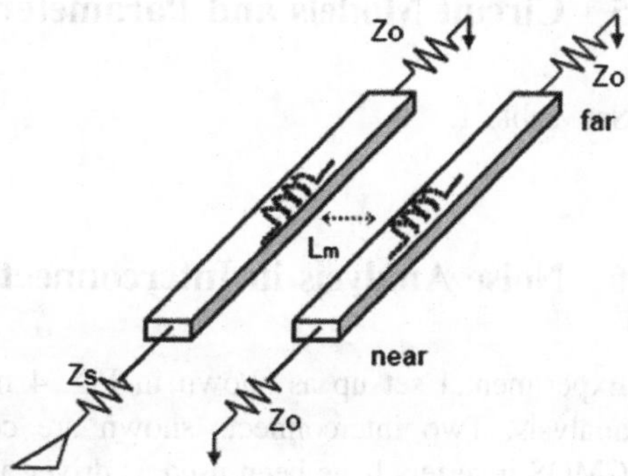

Fig. 3 Mutual inductance (L_m) [5]

Coupled currents on victim line is sum to produce near and far-end crosstalk noise which is given by

$$I_{near} = I_{Cm} + I_{Lm} \tag{7}$$

$$I_{far} = I_{Cm} - I_{Lm} \tag{8}$$

4.4 Odd Mode Propagation

When the two coupled interconnects are carrying the voltages of equal magnitude and phase difference of 180° with each other, then it is called odd mode propagation. The effective capacitance of interconnects will be augmented by twice the mutual capacitance and on the other side, the equivalent inductance will be decreased by the mutual inductance [1, 5].

4.5 *Even Mode Propagation*

When the coupled interconnects are carrying the voltages of same magnitude and in phase with each other, then this type of propagation is called even mode. In this mode, the effective capacitance of the line will be decreased by the mutual capacitance and on the other side, equivalent inductance will be increased by the mutual inductance. The current in this mode contains equal magnitude and flow in the same direction between the two coupled lines and interacts with each other. These interactions affect the delay and impedance of the transmission line [1, 5].

5 Circuit Models and Parameters Used

See Table 1.

6 Noise Analysis in Interconnect

Experimental set up as shown in Fig. 4 have been used for the crosstalk noise analysis. Two interconnects shown are coupled by a coupling capacitor (C_c). CMOS inverters have been used as driver and load. The input signal is transmitted through the first interconnect (known as aggressor interconnect) and the effect of switching is measured in the second interconnect (known as victim interconnect). When the signal goes from one state to another state in aggressor, a spike (crosstalk noise) is generated in the victim line which affects the information traveling through it. Input has been taken in the form of a pulse for transient analysis and the output waveforms are shown in Figs. 5 and 6. The experiment has been done for different lengths of interconnects, different power supply values, different frequencies of input signal, and also for different dielectric material. The RLC parameters calculated for different interconnect lengths is shown in Table 2.

Table 1 Hardware, software, and parameters used by various researchers

Circuit model used	Software used	Parameter	Results
Interconnect using RLCG	H-SPICE	$R = 8.829\ \Omega$	The run time off is reduced to 0.01–0.003 [6]
		l (pH/μm) = 1.538	
		c (fF/μm) = 0.18 (parasitic parameter)	
Interconnect using RLCG	H-SPICE	$R_s = 100\ \Omega$	The Time delay is reduced to 181.32–173.23 ps [7]
		$C_l = 1.75$ pF	
RLC interconnect network	H-SPICE	line resistance varying from 0 to 90 Ω	% error in delay is reduced to 7.5–6.3 % [8]

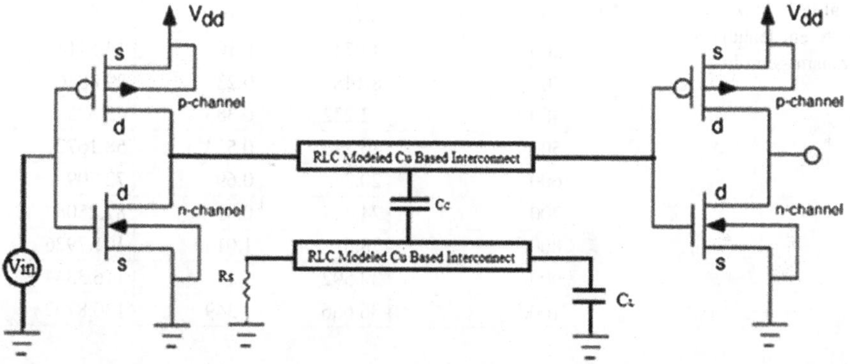

Fig. 4 Experimental set up

Fig. 5 Transient response of coupled interconnect at 200 μm length

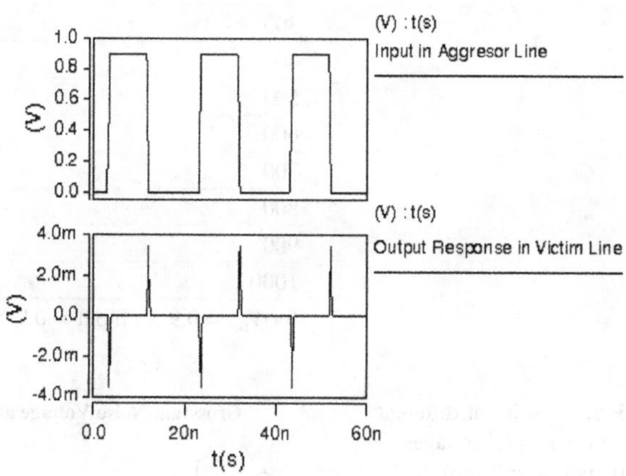

Fig. 6 Transient response of coupled interconnect at 300 μm length

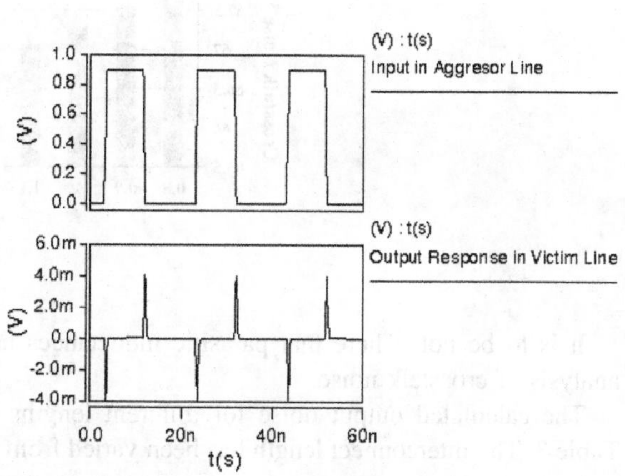

Table 2 RLC parameters for different lengths of transmission line

Length (μm)	R (Ω)	L (nH)	C_g (fF)
200	4.074	0.106	14.5418
300	8.148	0.239	29.0836
400	12.222	0.383	43.6254
500	16.296	0.534	58.1672
600	20.37	0.69	72.709
700	24.444	0.85	87.2508
800	28.518	1.014	101.7926
900	32.592	1.18	116.3344
1000	36.666	1.349	130.8762

Table 3 Crosstalk noise voltage at different lengths

Length of interconnect (μm)	Noise voltage (μV)
200	63.5230
300	64.4574
400	64.6542
500	64.9832
600	65.1891
700	65.5469
800	65.9544
900	66.7898
1000	66.9887

(for V_{dd} = 0.9 V; input = 0.9 V)

Fig. 7 Noise at different power supply voltages (length = 700 μm)

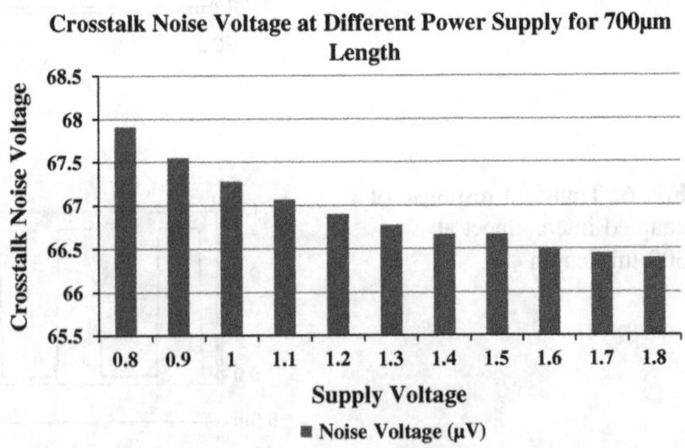

It is to be noted here that parasitic inductances have not been considered for analysis of crosstalk noise.

The calculated output noise for different lengths of interconnects is listed in Table 3. The interconnect length has been varied from 200 to 1000 μm for crosstalk analysis.

Table 4 Crosstalk noise voltage of 700 μm interconnect length with different power supply voltages

V_{dd} (V)	Noise voltage (μV)
0.8	67.9102
0.9	67.5469
1.0	67.2750
1.1	67.0662
1.2	66.9023
1.3	66.7710
1.4	66.6642
1.5	66.6642
1.6	66.5034
1.7	66.4418
1.8	66.3877

(Input = 0.9 V)

Table 5 Crosstalk noise voltage at different interconnect lengths with different dielectric materials

Interconnect length (μm)	Dielectric material		
	SiO_2 (μV)	Al (ceramic package) (μV)	HfO_2
200	64.6542	63.9926	62.5667
300	64.523	64.4494	63.4727
400	64.4574	63.4295	63.2951
500	64.255	63.7595	62.7733
600	64.1891	62.8861	62.8031
700	67.5469	64.2344	62.0253
800	63.9832	63.5778	61.8396

(for V_{dd} = 0.9 V; input = 0.9 V)

Power supply voltage is varied for 700 μm interconnect length from 0.8 to 1.8 V. The effect of power supply variation is shown in Fig. 7 (Table 4).

The noise voltages at different frequencies are listed in Table 6 (Fig. 9).

7 Results

The two capacitively coupled interconnects with CMOS driver and load circuit were used for analyzing crosstalk noise voltage. The interconnect length was varied from 200 to 1000 μm. Table 3 shows that as the length is increased the noise voltage is also increased for capacitively coupled interconnect. However it increases if parasitic inductances are also considered for crosstalk noise analysis [2, 9]. The

Table 6 Crosstalk noise voltage for 700 μm interconnect length with different frequencies

Frequency	Noise voltage (μV)
200 MHz	12.8216
400 MHz	18.0933
600 MHz	22.0887
800 MHz	25.4019
1 GHz	28.2622
1.2 GHz	30.7866
1.4 GHz	33.0451
1.6 GHz	35.0837
1.8 GHz	36.9352
2 GHz	38.6246
10 GHz	59.3642
20 GHz	62.0447

(for V_{dd} = 0.9 V; input = 0.9 V)

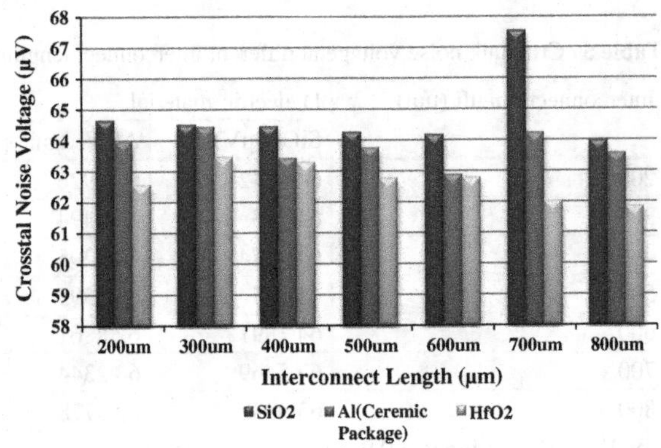

Fig. 8 Crosstalk noise voltage at different dielectric materials

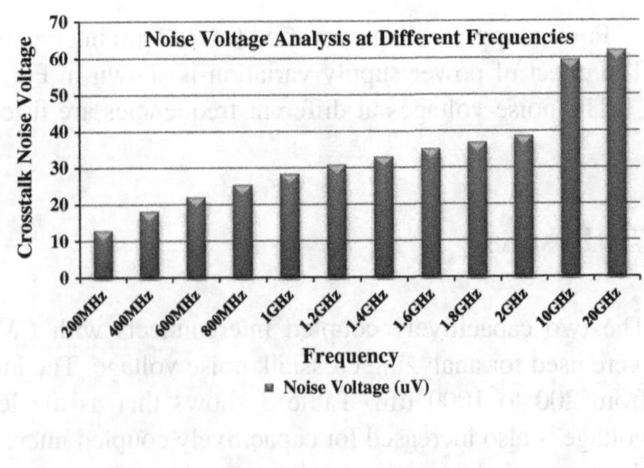

Fig. 9 Crosstalk noise voltage at different frequencies (length = 700 μm)

effect of power supply variation was also carried out in which supply voltage was varied from 0.8 volt to 1.8 volt. The result as shown in Fig. 7 indicates that as the power supply voltage is increased, the crosstalk noise voltage is decreased. At lower supply voltages, the noise voltage is high. Also, the crosstalk noise voltage was measured for different dielectric materials like SiO_2, Al (Ceramic Package), and HfO_2. Lower crosstalk noise voltage was observed with HfO_2 material as shown in Fig. 8. Figure 9 represents that as the frequency increases, the noise voltage also increases. At lower frequency, the noise is low but at higher frequencies it increases rapidly.

8 Conclusion

The experimental results show that copper is not a suitable material for interconnects in low power, high speed, and complex chips based on devices of nanometer regime. The ever-increasing density, performance, and reliability requirements of VLSI circuits create significant process integration challenges for the future interconnect systems. So facing to these issues, new materials such as CNT and optical interconnects can play a vital role in meeting the performance challenge.

References

1. Kumawat, P., Soni, G.: Crosstalk analysis in global interconnects. Int. Bull. Math. Res. (Alied Journals-IBMR) **2**(1), 257–263 (2015) ISSN: 2394–7802
2. Rabeay Jam, M., Chandrakasan, A., Nikolic, B.: Digital Integrated Circuits. Prentice Hall, New York (2003)
3. Moll, Sc., Roca, M.: Interconnection noise in VLSI circuits. Kluwer Academic Publishers, Norwell (2004)
4. http://www.intechopen.com/books/electronic-properties-of-carbon-nanotubes/carbon-nanotube-as-a-vlsiinterconnect (Online Chapter of book "Electronic Properties of Carbon Nano Tubes)
5. Hall, S.H., Hall, G.W., McCall J.A.: High-speed digital system. In: Design—A Handbook of Interconnect Theory and Design Practices
6. Li, X.-C., Mao, J.-F., Swaminathan, M.: Transient analysis of CMOS-Gate-Driven interconnects based on FDTD. IEEE Trans. Comput.-Aided Design Integr. Circ. Syst. **30** (4), 574, 583 (2011)
7. Kar, R., Maheshwari, V., Choudhary, A., Singh, A., Mal, A.K., Bhattacharjee, A.K.: Inductive coupling aware explicit cross-talk and delay formula for on-chip VLSI RLCG interconnects using difference model approach. In: 2010 International Conference on Computing Communication and Networking Technologies (ICCCNT), pp. 1, 6, 29–31 July 2010
8. Roy, A., Xu, J., Chowdhury, M.H.: Impacts of signal slew and skew variations on delay uncertainty and crosstalk noise in coupled RLC global interconnects. In: 15th IEEE International Conference on Electronics, Circuits and Systems, 2008. ICECS 2008. pp. 1055, 1058, 31 Aug 2008–3 Sept 2008

9. Agarwal, K., Sylvester, D., Blaauw, D.: Modeling and analysis of crosstalk noise in coupled RLC interconnects. IEEE Trans. Comput.-Aided Des. Integr. Circ. Syst. **25**(5), 892, 901 (2006)
10. Coulibaly, L.M., Kadim, H.J., Analytical crosstalk noise and its induced-delay estimation for distributed RLC interconnects under ramp excitation. IEEE Int. Symp. Circ. Syst. ISCAS **2**, 1254, 1257 (2005)
11. Choudhary, A., Maheshwari, V., Singh, A., Kar, R.: Wave propagation based analytical delay and cross talk noise model for distributed on-chip RLCG interconnects. In; 2010 IEEE International Conference on Semiconductor Electronics (ICSE), pp. 153, 157, 28–30 June 2010
12. You, H., Soma, M.: Crosstalk analysis of interconnection lines and packages in high-speed integrated circuits. IEEE Trans. Circ. Syst. **37**(8), 1019, 1026 (1990)
13. Ilumoka, A.A., Efficient prediction of crosstalk in VLSI interconnections using neural networks. In: 2000 IEEE Conference on Electrical Performance of Electronic Packaging, pp. 87, 90 (2000)
14. Kavicharan, M., Murthy, N.S., Rao, N.B.: An efficient delay estimation model for high speed VLSI interconnects. In: 2013 International Conference on Advances in Computing, Communications and Informatics (ICACCI), pp. 1358, 1362, 22–25 Aug 2013
15. Kar, R., Maheshwari, V., Agarwal, V., Choudhary, A., Singh, A., Mai, A.K., Bhattacharjee, A.K.: Accurate estimation of on-chip global RLC interconnect delay for step input. In: 2010 International Conference on Computer and Communication Technology (ICCCT), pp. 673, 677, 17–19 Sept 2010
16. Kar, R., Maheshwari, V., Choudhary, A., Singh, A.: Modeling of on-chip global RLCG interconnect delay for step input. In: 2010 International Conference on Computer and Communication Technology (ICCCT), pp. 318, 323, 17–19 Sept 2010
17. Maheshwari, V., Agarwal, S., Goyal, A., Jain, J., Kumar, S., Kar, R., Mandal, D., Bhattacharjee, A.K.: Elmore's approximations based explicit delay and rise time model for distributed RLC on-chip VLSI global interconnect. In: 2012 IEEE Symposium on Humanities, Science and Engineering Research (SHUSER), pp. 1135, 1139, 24–27 June 2012
18. Maheshwari, V., Baboo, A., Kumar, B., Kar, R., Mandal, D., Bhattacharjee, A.K.: Delay model for VLSI RLCG global interconnects line. In: 2012 Asia Pacific Conference on Postgraduate Research in Microelectronics and Electronics (PrimeAsia), pp. 201, 204, 5–7 Dec 2012
19. Mehran, M., Masoumi, N.: A tapered partitioning method for delay energy product optimization in global interconnects. In: 50th Midwest Symposium on Circuits and Systems, 2007. MWSCAS 2007. pp. 21, 24, 5–8 Aug. 2007
20. Zangeneh, M., Masoumi, N.: An analytical delay reduction strategy for buffer-inserted global interconnects in VDSM technologies. In: IEEE Conference, pp. 470–475 (2009)
21. Paul, C.R.: Incorporation of terminal constraints in the FDTD analysis of transmission lines. IEEE Trans. Electromagn. Compat. **36**(2), 85, 91 (1994)
22. Yu, Q., Kuh, E.S.: Moment computation of lumped and distributed coupled RC trees with application to delay and crosstalk estimation. Proc. IEEE **89**(5), 772, 788 (2001)
23. Roy, A., Chowdhury, M.H.: Global interconnect optimization in the presence of on-chip inductance. In: IEEE International Symposium on Circuits and Systems, 2007. ISCAS 2007. pp. 885, 888, 27–30 May 2007
24. Seki, S., Hasegawa, H.: Analysis of crosstalk in very high-speed LSI/VLSI's using a coupled multiconductor MIS microstrip line model. IEEE Trans. Microw. Theory Tech. **32**(12), 1715, 1720 (1984)
25. Semerdjiev, B., Velenis, D.: Optimal crosstalk shielding insertion along on-chip interconnect trees. In: 20th International Conference on VLSI Design, 2007. Held jointly with 6th International Conference on Embedded Systems, pp. 289, 294, 6–10 Jan 2007
26. Qi, X., Wang, G., Yu, Z., Dutton, R.W., Young, T., Chang, N.: On-chip inductance modeling and RLC extraction of VLSI interconnects for circuit simulation. In: Proceedings of the IEEE Custom Integrated Circuits Conference, 2000. CICC. 2000, pp. 487, 490 (2000)

27. Eo, Y., Eisenstadt, W.R., Jeong, J.Y., Kwon, O.-K.: A new on-chip interconnect crosstalk model and experimental verification for CMOS VLSI circuit design. IEEE Trans. Electron Dev. **47**(1), 129, 140 (2000)
28. Zangeneh, M., Masoumi, N.: An analytical delay reduction strategy for buffer-inserted global interconnects in VDSM technologies. In: European Conference on Circuit Theory and Design, 2009. ECCTD 2009. pp. 470, 475, 23–27 Aug 2009

A Comparative Analysis of Copper and Carbon Nanotubes-Based Global Interconnects in 32 nm Technology

Arti Joshi and Gaurav Soni

Abstract At a high-pace advancements in the technologies today and their ubiquitous use, speed and size, has been the important aspects in VLSI interconnect. Channel length of device decreases to tens of nanometers, as the technology is shifting to the deep submicron level. Hence, the die size and device density of the circuit increase rapidly. This increase makes the requirement of long interconnects in VLSI chips. Long interconnects lead to increase in propagation delay of the signal. In deep submicron meter VLSI technologies, it has become increasingly difficult for conventional copper-based electrical interconnects to gratify the design requirements of delay, power, and bandwidth. Promising candidate to solve this problem is carbon nanotube (CNT). In this paper, the prospects of carbon nanotubes (CNT) as global interconnects for future VLSI Circuits have been examined. Due to high thermal conductivity and large current carrying capacity, CNTs are favored over copper as VLSI future interconnects. The energy, power, propagation delay, and bandwidth of CNT bundle interconnects have been examined and compared with that of the Cu interconnects at the 32-nm technology node at two different global interconnects lengths. The simulation has been carried out using HSPICE circuit simulator with a transmission line model at 200 and 1000 μm lengths. The results show that power consumption and energy of CNT-based interconnects are reduced by 66.49 and 66.86 %, respectively, at 200 μm length in comparison with the Cu-based Interconnects. At 1000 μm length, a reduction of 43.90 and 44.04 % has been observed in power consumption and energy, respectively, using CNT interconnects. Furthermore, the propagation delay is reduced approximately 61.17 % for 200 μm and 69.13 % for 1000 μm length while the bandwidth increases

---

Arti Joshi (✉)
Poornima University, Jaipur, India
e-mail: arti.joshi@poornima.edu.in

Gaurav Soni
EEE, Poornima University, Jaipur, India
e-mail: gaurav.soni@poornima.edu.in

M. Pant et al. (eds.), *Proceedings of Fifth International Conference on Soft Computing for Problem Solving*, Advances in Intelligent Systems and Computing 436, DOI 10.1007/978-981-10-0448-3_35

up to 90 %. This work suggests single-wall carbon nanotubes (SWCNT) bundle interconnects for global interconnects in VLSI designing as they devour low energy and are faster when compared with conventional copper wires.

Keywords Copper · SWCNT · Global interconnects · Power · Energy

1 Introduction

Due to advances in technology scaling, today's ICs are composed of billions of transistors. Interconnects in these transistors were treated as optimal conductors that proliferated signals instantaneously [1]. They are used to associate components on a VLSI chip, connect chips on a multichip module, and connect multichip modules on a system. While device sizes were timiding with every technology, the multi-level metal structures raised superior above the surface of the silicon and began to overshadow the prospect of the chip. In the sub-100-nm scaling tenure, interconnect behavior curbs the operation and accuracy of VLSI systems. The wiring in today's IC forms a complicated geometry that introduces resistive, inductive, and capacitive parasitics. They can cause an upsurge in delay, power distribution, energy dissipation, and includes extra noise sources, which affect the functionality of the device. As scaling increases, the impact of interconnect in the VLSI circuits became even more important. It controls all these important electrical characteristics on the chip [2]. Also as the width of wires is decreased, the resistance rises. This increase in wire resistance causes RC delay to increase. The spacing among wires has been decreasing to the point where the coupling among wires is significant. This capacitive coupling offers additional delay and noise effects.

2 Limitations of Copper Interconnects

In the past, copper interconnects replaced aluminum interconnects due to the low resistance of copper wires when compared with aluminum and the resistance to electromigration was much higher in copper when compared with aluminum, now copper interconnects are going through similar problems due to the growing resistivity. As the processing technology is approaching the subnanometer regime, delay is becoming serious concern. From the report of International Technology Roadmap for Semiconductors (ITRS) devised below, we find that copper resistivity for future technologies is increasing at a very fast rate [3] (Fig. 1).

From 90 to 32 nm, resistivity does not increase much; but below 32 nm, a sharp resistivity increment can be observed due to scattering effects.

Apart from increment in resistivity, the wire width is also diminishing with new technologies. This further increases the overall resistance, since resistance of a wire is inversely proportional to its width. Therefore, even though the wire length is

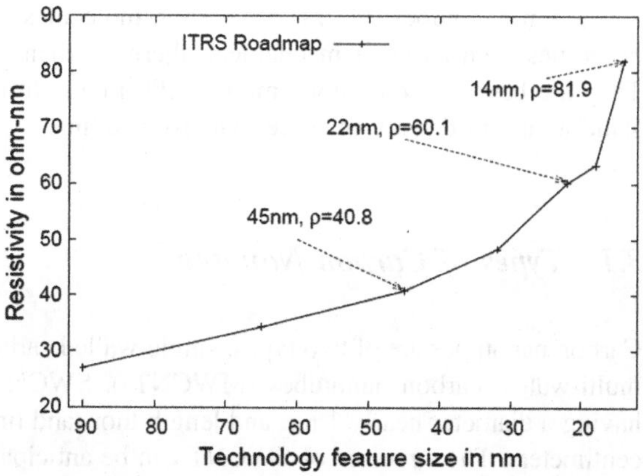

Fig. 1 Increase in Cu resistivity with technology scaling [3]

getting smaller, decreasing cross-sectional area and increasing resistivity result in higher interconnect delay.

So, as we extend from deep submicron technology to nanometer regimes, the conventional copper wires will not be able to continue, and we will have to look for new materials for interconnect design. So, surrogate approaches such as carbon nanotube (CNT) interconnects have been suggested to ignore the problems associated with global wires.

3 Carbon Nanotubes

Carbon nanotubes commonly called by the acronym CNTs were discovered by Sumio Lijima in 1991. Nanotubes are made of sp^2 bonds and they align themselves into ropes wrapped by van der Waals forces. These are allotropes of carbon being cylindrical in shape [4] (Fig. 2).

Fig. 2 Carbon nanotube [2]

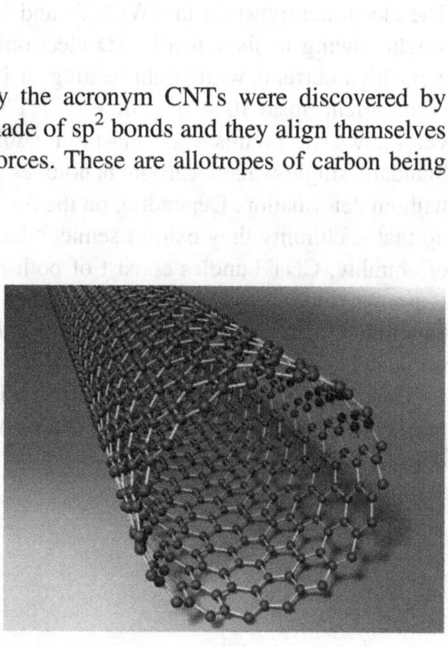

Carbon nanotubes (CNTs) are carbon molecules with cylindrical shape having properties such as great mechanical, thermal, chemical, and electrical properties. They are best field emission emitters, 100 times stronger than steel and can withstand a current density of more than 10^{10} A/cm^2.

3.1 Types of Carbon Nanotubes

Carbon nanotubes are of two types: single-walled carbon nanotubes (SWCNTs) and multi-walled carbon nanotubes (MWCNTs). SWCNTs were discovered in 1993, having a diameter nearly 1 nm and length thousand times larger and of the order of centimeters. The structure of SWCNT can be anticipated by bundling a thick layer of graphite into a consistent cylinder. The way the graphene sheets wrap can be denoted by a duo of indices (n, m) named as chiral vector. The relationship between n and m defines the three categories of CNTs namely, zigzag, chiral, and arm chair. MWCNTs consist of several layers of graphite rolled in to form a tube with an interlayer spacing of 3.4 Å. The outlying diameter of MWCNTs ranges from 1 to 50 nm while the inner diameter is of few nanometers.

3.2 General Properties of Carbon Nanotubes

The electronic transport in SWCNTs and MWCNTs occurs ballistically over long lengths owing to their nearly 1D electronic structure. This enables nanotubes to carry high currents with slight heating. It is reported that the MWCNTs can carry high current up to 10^9–10^{10} A/cm^2. CNT's are the stiffest and strongest materials yet known in premises of elastic modulus and tensile strength, respectively. Standard single-walled carbon nanotubes can tolerate a pressure up to 24 GPa without deformation. Depending on the direction in which the carbon sheet is rolled up that is chirality they exhibit semiconducting or metallic properties. Due to lack of chirality, CNT bundles consist of both metallic and semiconducting nanotubes. For a nanotube, if $n = m$, the nanotube is metallic; if n-m is multiple of 3, then the nanotube is semiconducting with a small band gap; otherwise, the nanotube is a semiconductor [1].

This work investigates the effect of using SWCNT bundle as an interconnect material as they absorb low energy and also have greater bandwidth as compared to conventional copper wires.

4 Circuit Parameters Modeling

To study the parasitic effects elaborated above, we need the introduction of electrical models that can estimate and approximate the real behavior of interconnect as a function of its parameters.

4.1 Modeling Parameters for Copper

The resistance of a wire is [3] (Fig. 3):

$$R = \frac{\rho L}{A} = \frac{\rho L}{HW} \quad (1)$$

where ρ is the resistivity of material (in Ωm).

The interconnect capacitance is modeled as follows [4]:

$$C_W = \varepsilon\left[1.15\frac{w}{t} + 2.80\left(\frac{h}{t}\right)^{0.222} + \left(0.66\frac{w}{t} + 1.66 - 0.14\left(\frac{h}{t}\right)^{0.222}\right).\left(\frac{t}{s}\right)^{1.34}\right] \quad (2)$$

where h is wire height, ε is the dielectric permittivity, t is intermetal layer spacing, and s is the interwire spacing (assumed $s = w$).

For global wires, inductance can be modeled as shown in the following equation [5]:

$$L_W = 2 \times 10^{-7} l\left(\ln\frac{2l}{w+h} + 0.5 + \frac{w+h}{3l}\right) \quad (3)$$

4.2 Modeling Parameters for CNT Bundles

The various parameters are (Fig. 4):

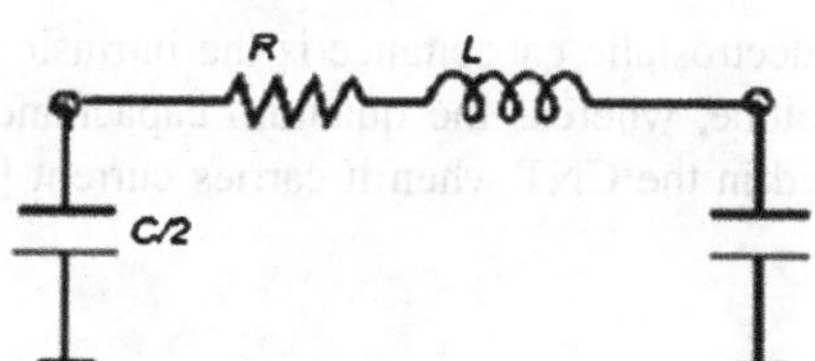

Fig. 3 RLC interconnect model [6]

Fig. 4 Equivalent RLC circuit model for an isolated SWCNT [7]

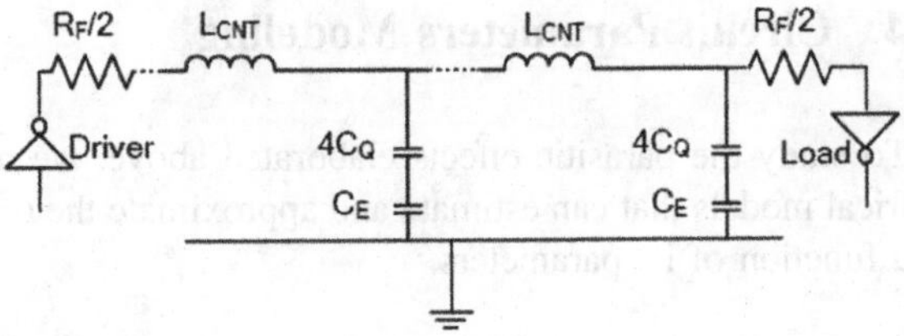

1. R_F is the fundamental resistance,
2. L_{CNT} is the total inductance,
3. C_E is electrostatic capacitance, and
4. C_Q is quantum capacitance.

The fundamental resistance R_F, is equally divided among the two contacts on both side of the nanotube. It can be given as [4]:

$$R_F = \frac{h}{4e^2} \tag{4}$$

The mean free path of electrons in CNT is 1 μm [8]. If CNT length is less than mean free path, electron transport is ballistic and the fundamental resistance does not depend on wire length. This resistance is given by $(h/4e^2) = 6.45$ kΩ. But, if wire length is longer than λ_{CNT}, an extra-ohmic resistance exists due to scattering [4]:

$$R_{CNT} = R_F \frac{1}{\lambda_{CNT}} \tag{5}$$

Here l is length of CNT.

This additional scattering resistance appears as distributed resistance per unit length in the equivalent circuit.

$$R_{CNT}(\text{p.u.l}) = \frac{R_F}{\lambda_{CNT}} = \left(\frac{h}{4e^2}\right) \cdot \frac{1}{\lambda_{CNT}} \tag{6}$$

The mean free path of a CNT is proportional to diameter [8].

The total capacitance of a nanotube consists of two components:

1. Electrostatic capacitance (C_E)
2. Quantum capacitance (C_Q)

Electrostatic capacitance is the intrinsic plate capacitance of an isolated carbon nanotube, whereas the quantum capacitance accounts for the electrostatic energy stored in the CNT when it carries current [3].

$$C_{\mathrm{E}} = \frac{2\pi\varepsilon}{\cosh^{-1}\left(\frac{2y}{d}\right)} \tag{7}$$

$$C_{\mathrm{Q}} = \frac{2e}{hv_{\mathrm{F}}} \tag{8}$$

Here y is distance of CNT from ground plane, v_{F} is Fermi velocity in graphite (8×10^5 m/s), ε is dielectric permittivity, and d is CNT diameter.

The total inductance of a CNT consists of two components [4]:

1. Magnetic inductance (L_{M})
2. Kinetic inductance (L_{K})

The magnetic inductance per unit length is given by:

$$L_{\mathrm{M}} = \frac{\mu}{2\pi}\cosh^{-1}\left(\frac{2y}{d}\right) \tag{9}$$

Apart from magnetic inductance, another inductive component exists due to the kinetic energy of the electrons. The kinetic inductance per unit length can be given as [4]

$$L_{\mathrm{K}} = \frac{h}{2e^2 v_{\mathrm{F}}} \tag{10}$$

5 Circuit Models and Parameters Used

See Table 1.

6 Performance Comparison

Experimentation is done using HSPICE and waveform is observed using Cosmos Scope. The output waveform is simulated under the process of transient analysis (Table 2).

Table 1 Circuit models and parameters used by researchers

Interconnect model used	Parameters	Result
RLC circuit [5]	TN = 22 nm	Larger DCNT increases CNT resistance
	d_{cnt} = 1 nm	
	Wire length = 1/10 mm	For (MFP, PD) = (1.6 μm, 1/3), CNT has 1.7 % delay advantage
	PMOS: NMOS = 2:1 (W/L ratio)	
	MFP = 1.6 μm/2.8 μm	For (MFP, PD) = (2.8 μm, 1), CNT has 5 % delay advantage
	PD = 1/3/1	
RLC wire model [9]	TN = 22 nm	CNT interconnects consume less power than Cu interconnects
	Wire length = 10 mm	
	Cu ρ = 1.9 μΩ cm	
	MFP = 1.6 μm	Increasing MFP results in decreasing Power density
	V_{bias} = 4.7 V	
	SA = 20 %	
RLC circuit [8]	TN = 22 nm	For MFP = 0.9 μm CNT's has 1.6× improvement in latency
	Contact resistance = 1.2 kΩ	
	MPF = 2.8 μm	
	Scattering coefficient = 0.5	For MFP = 2.8 μm CNT's are 1.2× energy efficient than Cu
	Grain boundary coefficient = 0.5	
	W/L ratio = 2:1	
Equivalent circuit metallic SWCNT's [10]	Length = 10 μm	For a metallic SWCNT density of one per 3 nm^2, the bandwidth density can potentially increase by up to 40 %
	R = 0.198 kΩ	
	L = 0.952 and 0.010 nH	While density of one per 6 nm^2 offers less than 10 % improvement
	C = 30.36 fF	
RLC equivalent circuit model [11]	TN = 32 nm	CNT bundle consumes 1.5 to twofolds smaller power
	V_{dd} = 0.9 V at 700 MHz	
	Length = 200 μm	
	R = 374.2 Ω	CNT bundle are 1.4–3 times faster than Cu interconnect
	L = 317 pH	
	C = 36.283 fF	
Equivalent transmission line model [12]	TN = 22/32 nm	CNT has an 80 % performance enhancement in delay reduction
	Length = 10 μm	
	W = 70 n	
	H = 168 nm	
	Cu ρ = 2.963 μΩ cm	

Table 2 RLC parameters for Cu and CNT interconnect

Interconnect length (μm)	Interconnect	Interconnect parameters		
		R (Ω)	*L* (pH)	*C* (fF)
200	Cu	374.2	317	36.283
	CNT	109.05	5270	28.3
1000	Cu	1870.8	1907	181.42
	CNT	546.44	26,700	142.0

Note All values have been taken from the IEEE Paper [11]

7 Results

The Cu and CNT global interconnects are modeled by RLC transmission line. Average and maximum power obtained are shown in Tables 3 and 4 (Figs. 5, 6, 7, 8, and 9).

Table 3 Analysis of power, delay, and energy of Cu and CNT interconnects at 200 μm length

Performance metrics	Interconnect		Percentage reduction (%)
	Cu	CNT	
Propagation delay (ns)	0.7272	0.2823	61.17
Average power (μW)	1.365	0.4573	66.49
Total energy (fJ)	6.76	2.24	66.86
Leakage power (μW)	0.1692	0.05604	66.87
Bandwidth (MHz)	2.398	38.22	93.72

Table 4 Analysis of power, delay, and energy of Cu and CNT interconnects at 1000 μm length

Performance metrics	Interconnect		Percentage reduction (%)
	Cu	CNT	
Propagation delay (ns)	2.488	0.7680	69.13
Average power (μW)	2.558	1.435	43.90
Total energy (fJ)	12.789	7.1562	44.04
Leakage power (μW)	0.319	0.178	44.20
Bandwidth (MHz)	2.365	42.12	94.38

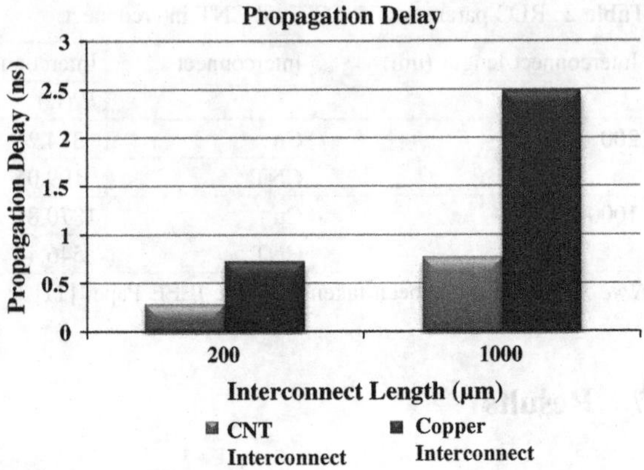

Fig. 5 Propagation delay of Cu and CNT interconnects at different lengths

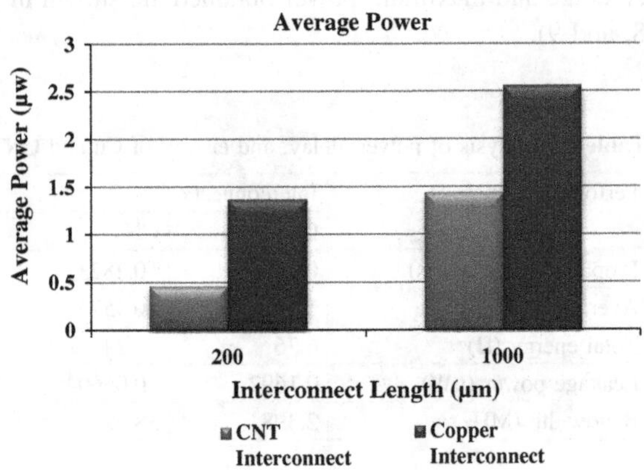

Fig. 6 Average power consumed by Cu and CNT interconnects at different lengths

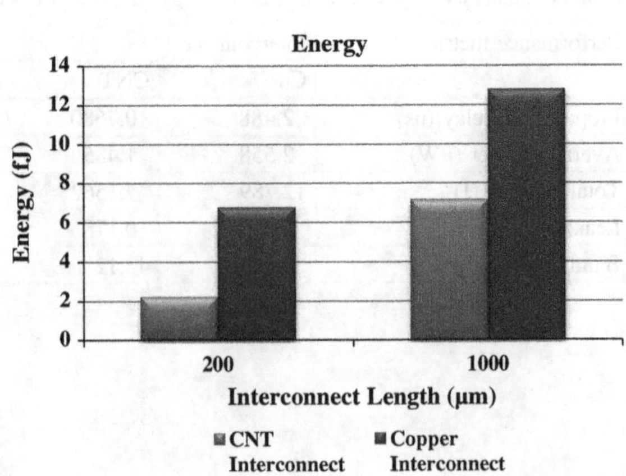

Fig. 7 Energy consumed by Cu and CNT interconnects at different lengths

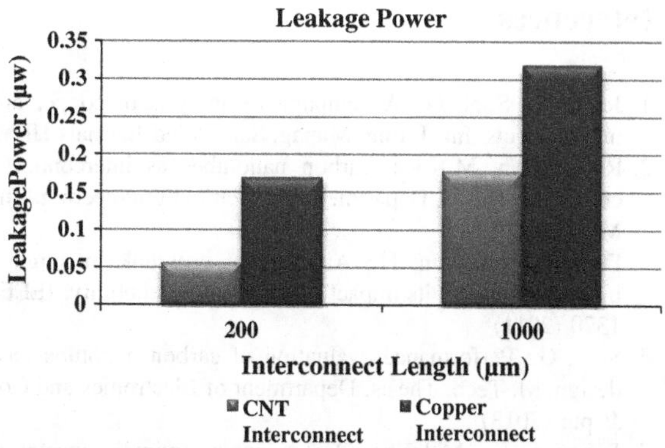

Fig. 8 Leakage power consumed by Cu and CNT interconnects at different lengths

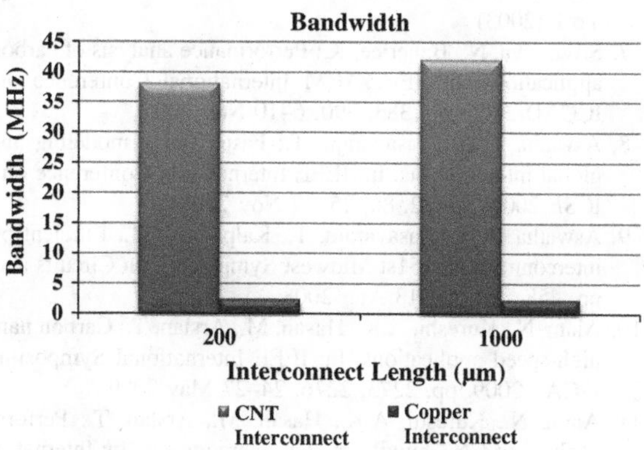

Fig. 9 Bandwidth of Cu and CNT interconnects at different lengths

8 Conclusion

This paper presents the limitations of copper-based technology and recommends the carbon nanotube technology as the possible future alternative. The performance analysis in terms of power, energy, propagation delay, and bandwidth at 200 and 1000 μm interconnect lengths is carried out. SPICE simulation is used to compare SWCNT interconnect performances with that of copper interconnects for the 32-nm technology node. Simulation results show that densely packed CNT bundle interconnects display significant improvement in performance as compared to copper interconnects, in spite of imperfect metal nanotube contacts. The simulation results obtained strengthened the fact that CNT technology is a possible and viable replacement for the present copper-based technology.

References

1. Joshi, A., Soni, G.: A comparative analysis of copper and carbon nanotubes based global interconnects. Int. J. Eng. Manag. Sci. (Alied Journals-IJEMS) ISSN-2348–3733, **2**(5), (2015)
2. Rajasekaran, M.K.R.: Carbon nanotubes as interconnect for next generation network on chip. M.S. Thesis, Department of Electrical Engineering, The University of New Mexico, New Mexico (2012)
3. Das, D., Rahaman, H.: Analysis of crosstalk in single- and multiwall carbon nanotube interconnects and its impact on gate oxide reliability. IEEE Trans. Nanotechnol. **10**(6), 1362, 1370 (2011)
4. Soni, G.: Performance evaluation of carbon nanotube based devices and circuits for VLSI design. M. Tech. Thesis, Department of Electronics and Communication Engineering, MNIT, Jaipur (2013)
5. Banerjee, K., Mehrotra, A.: A power-optimal repeater insertion methodology for global interconnects in nanometer designs. IEEE Trans. Electron Dev. **49**(11), 2001, 2007 (2002)
6. Rabeay, J.M., Chandrakasan, A., Nikolic, B.: Digital Integrated Circuits. Prentice Hall, New York (2003)
7. Srivastava, N., Banerjee, K.: Performance analysis of carbon nanotube interconnects for VLSI applications. In: IEEE/ACM International Conference on Computer-Aided Design, 2005. ICCAD-2005. pp. 383, 390, 6–10 Nov 2005
8. Aswatha, A.R., Basavaraju, T.: Faster delay modeling and power optimization for on-chip global interconnects. In: IEEE International Conference on Semiconductor Electronics, 2008. ICSE 2008. pp. 82, 86, 25–27 Nov 2008
9. Aswatha, A.R., Basavaraju, T., Kalpana, A.B.: Efficient power modeling for on-chip global interconnects. In: 51st Midwest Symposium on Circuits and Systems, 2008. MWSCAS 2008, pp. 458, 461, 10–13 Aug 2008
10. Alam, N., Kureshi, A.K., Hasan, M., Arslan, T.: Carbon nanotube interconnects for low-power high-speed applications. In: IEEE International Symposium on Circuits and Systems, 2009. ISCAS 2009, pp. 2273, 2276, 24–27 May 2009
11. Alam, N., Kureshi, A.K., Hasan, M., Arslan, T.: Performance comparison and variability analysis of CNT bundle and Cu interconnects. In: International Multimedia, Signal Processing and Communication Technologies, 2009. IMPACT '09. pp. 169, 172, 14–16 March 2009
12. Koo, K.-H., Cho, H., Kapur, P., Saraswat, K.C.: Performance comparisons between carbon nanotubes, optical, and Cu for future high-performance on-chip interconnect applications. IEEE Trans. Electron Dev. **54**(12), 3206, 3215 (2007)
13. Bartur, M., Nicolet, M.-A.: Utilization of NiSi2 as an interconnect material for VLSI. Electron Dev. Lett. IEEE **5**(3), 88, 90 (1984)
14. Bhatia, H.S.: A comparative study of delay analysis for carbon nanotube and copper based VLSI interconnect models, M. Tech. Thesis, Department of Electronics And Communication Engineering, Thapar University, Patiala (2011)
15. Li, H., Yin, W.-Y., Mao, J.-F.: Modeling of carbon nanotube interconnects and comparative analysis with Cu interconnects. In: Proceedings of the Asia-Pacific Microwave Conference, 2006. APMC 2006, pp. 1361, 1364, 12–15 Dec 2006
16. Cho, H., Koo, K.-H., Kapur, P., Saraswat, K.C.: The delay, energy, and bandwidth comparisons between copper, carbon nanotube, and optical interconnects for local and global wiring application. In: International Interconnect Technology Conference, IEEE 2007, pp. 135, 137, 4–6 June 2007
17. Cho, H., Koo, K.-H., Kapur, P., Saraswat, K.C.: Performance comparisons between Cu/low-κ, carbon-nanotube, and optics for future on-chip interconnects. Electron Dev. Lett. IEEE **29**(1), 122, 124 (2008)
18. International Technology Roadmap for Semiconductors. http://public.itrs.net/ (2007)
19. International Technology Roadmap for Semiconductors, 2013. [Online] Available: http://public.itrs.net/

20. Majumder, M.K., Das, P.K., Kaushik, B.K., Manhas, S.K.: Optimized delay and power performances for multi-walled CNT in global VLSI interconnects. In: 2012 5th International Conference on Computers and Devices for Communication (CODEC), pp. 1, 4, 17–19 Dec 2012
21. Majumder, M.K., Kaushik, B.K.; Manhas, S.K.: Comparison of propagation delay characteristics for single-walled CNT bundle and multiwalled CNT in global VLSI interconnects. In: Recent Advances in Intelligent Computational Systems (RAICS), 2011 IEEE, pp. 911, 916, 22–24 Sept 2011
22. Majumder, M.K., Pandya, B.D., Kaushik, B.K., Manhas, S.K.: Analysis of MWCNT and bundled SWCNT interconnects: impact on crosstalk and area. Electron Dev. Lett. IEEE **33**(8), 1180, 1182 (2013)
23. Murugeswari, P., Kabilan, A.P., Vaishnavi, M., Divya, C.: Performance analysis of single-walled carbon nanotube and multi-walled carbon nanotube in 32 nm technology for on-chip interconnect applications. In: International Conference on Computing, Communication and Networking Technologies (ICCCNT), 2014, pp. 1, 6, 11–13 July 2014
24. Sahoo, M., Rahaman, H.: Performance analysis of multiwalled carbon nanotube bundles. In: 2013 IEEE XXXIII International Scientific Conference Electronics and Nanotechnology (ELNANO), pp. 200, 204, 16–19 April 2013
25. Srivastava, N., Joshi, R.V., Banerjee, K.: Carbon nanotube interconnects: implications for performance, power dissipation and thermal management. In: IEEE International Electron Devices Meeting, 2005. IEDM Technical Digest. pp. 249, 252, 5–5 Dec 2005
26. Parihar, T., Sharma, A.: A comparative study of Mixed CNT bundle with Copper for VLSI Interconnect at 32 nm. Int. J. Eng. Trends Technol. **4**(4), 606–610 (2013)
27. Parihar, T., Sharma, A., Parihar, D.: A comparative delay analysis of copper interconnect with future candidate CNT. Int. J. Curr. Eng. Technol. **3**(2), 606–610 (2013)
28. Duksh, Y.S., Kaushik, B.K., Sarkar, S., Singh, R.: Analysis of propagation delay and power with variation in driver size and number of shells in multi walled carbon nanotube interconnects. J. Eng. Des. Technol. **11**(1), 19–33 (2013)

Comparative Analysis of Si-MOSFET and CNFET-Based 28T Full Adder

Rishika Sethi and Gaurav Soni

Abstract In this paper, 28T CNFET-based full adder circuit is proposed. With the increase in the number of transistors and speed per unit chip area, power consumption of VLSI circuits has also increased. Power has become an extremely important design constraint along with the area and speed in modern VLSI design. So, carbon nanotubes with their superior properties, high current drivability, and high thermal conductivities have emerged as potential alternative devices to the CMOS technology. In this paper, average power consumption, energy and delay of Si MOSFET and CNFET-based full adder have been analyzed. The simulation was carried out using HSPICE circuit simulator. The simulation results show that power consumption, energy, and PDP of CNFET-based full adder is 56, 54.74, and 59 % reduced, respectively, in comparison to the Si MOSFET-based full adder. Moreover, the delay is also reduced approx by 8.69 % for sum output and 8.63 % for carry output.

Keywords CNFET · CMOS · Full adder

1 Introduction

With the increase in the number of transistors and speed per unit chip area, power consumption of VLSI circuits has also increased. Power becomes an important design constraint along with the area and speed in modern VLSI design. Among the most important hindrance against further device scaling beyond 32 nm technology

Rishika Sethi (✉)
Poornima University, Jaipur, India
e-mail: rishika.sethi@poornima.edu.in

Gaurav Soni
EEE, Poornima University, Jaipur, India
e-mail: gaurav.soni@poornima.edu.in

M. Pant et al. (eds.), *Proceedings of Fifth International Conference on Soft Computing for Problem Solving*, Advances in Intelligent Systems and Computing 436, DOI 10.1007/978-981-10-0448-3_36

node is the performance variation introduced by increased process variations as the feature sizes reduce. Increased device density and device parametric variation, rising of subthreshold leakage current, and higher device temperatures all increase power consumption problems.

An effective path to reduce power consumption is to lower power supply voltage level of a circuit. However, the reduction in the supply voltage causes an increment in the circuit delay and, consequently, decrement in the throughput. Electronic industries have been searching for novel strategies to overcome the constraints of power consumption and chip size with optimal performance (speed). This has made scaling stringent requirement.

Moreover, as VLSI technology is progressing into the deep submicron (DSM) region, global interconnects play an increasingly important role in the overall performance and power consumption of IC (Integrated Circuit) chips. VDSM (Very Deep Submicron) technologies have led to largely complex, billion-transistor chips. This has resulted in a new circuit prototype system-on-chip (SoC). Then wires, not transistors, dominate performance and power. Interconnects are the key design objective in deep submicron SoC [1].

In order to sustain Moore's law and to ensure further improvements in performance of interconnects and field effect transistors, it is necessary to find an alternative to MOSFETs and copper interconnects. Carbon nanotubes are considered as the most promising candidate for future MOSFETs and interconnect because of their superior performance, ballistic transport, and size advantage [1, 2].

In this paper we have done comparative performance evaluation of CNFET and Si-MOSFET-based 1 bit full adder using HSPICE simulation.

2 Structure of Nanotubes

2.1 *Carbon Nanotubes (CNTs)*

Carbon nanotubes are considered the most suitable candidates for future interconnects and CMOS-based circuits because of their excellent performance and size advantage. A CNT is a hollow cylinder in shape. It is constructed by rolling up sheets of graphene [3]. Graphene is a single layer (atomic) of graphite which in turn is the crystalline form of carbon [1].

2.2 *Chirality*

Chirality is basically used to classify the electronic and physical structure of carbon nanotubes. It is primarily used to show the reflection symmetry among an object and its mirror image [3]. An object that is not superimposable on its mirror image is

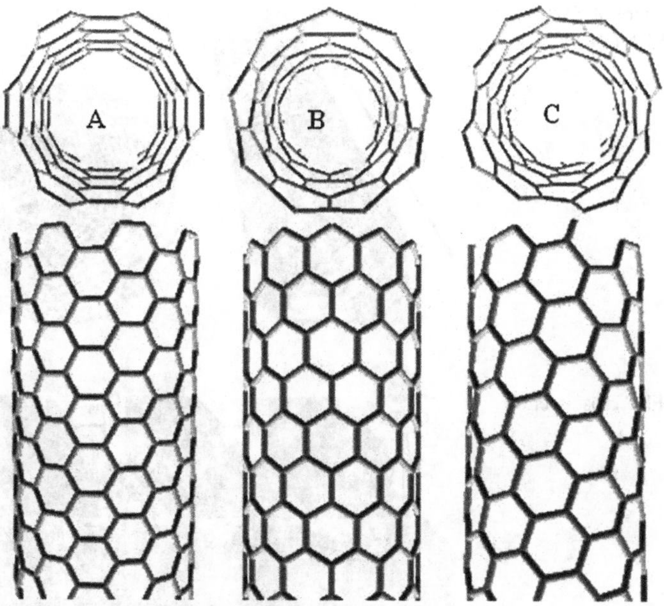

Fig. 1 Carbon nanotube types: *A* Armchair; *B* Zig-Zag; *C* Chiral [1, 3]

a chiral object. CNTs that are superimposable on their mirror images are defined as achiral CNTs. Achiral CNTs are further classified as [1]:

1. Armchair CNTs ($n = m$)
2. Zigzag CNTs ($n = 0$ or $m = 0$)

Carbon nanotubes that are not superimposable on their mirror images are defined as chiral CNTs [3]. The circumferential edge shaped of different types of carbon nanotubes is shown in Fig. 1.

2.3 Single-Walled and Multiwalled Nanotubes

Carbon nanotubes can be considered as single-walled or multi-walled depending on the number of shells forming the tubular structure [4]. Single-walled carbon nanotubes consist of a single cylinder of graphene that can be either metallic or semiconducting in nature. Moreover, multi-walled nanotubes (MWNTs) consist of metallic concentric cylinders of CNT which occur together with van der Waals forces [3]. They contain both types of semiconducting and metallic cylinders [4]. The structures of multi-walled carbon nanotubes (MWCNTs) or single-wall carbon nanotubes (SWCNTs) are shown in Fig. 2 [4].

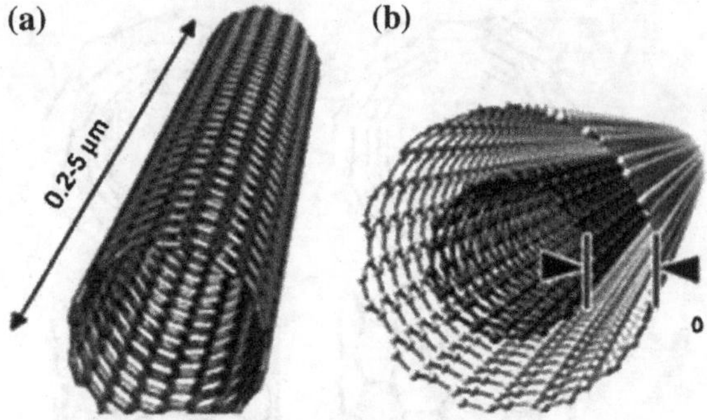

Fig. 2 Structure of SWCNT (**a**) and MWCNT (**b**) [3, 4]

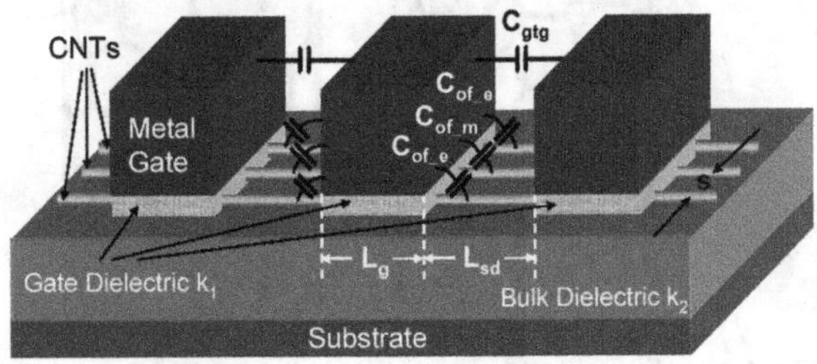

Fig. 3 CNFET cross section [3, 5]

2.4 *Carbon Nanotube Field Effect Transistor*

Carbon nanotube FET is the field effect transistor that is estimated to outperform the ITRS target of 11 nm node. Intrinsic delay (CV/I) of CNTFETs is very low. They show electron mobility higher than bulk silicon and provide better power-delay product. They have superior electrical properties such as compatibility with high-k dielectric layer, high speed, and reduced short channel effects (SCEs). Moreover, Si MOSFET circuit blocks can be realized using CNTFETs, since there is a similar operation principle. The schematic of CNFET is shown in Fig. 3 [4].

Carbon nanotube field effect transistor (CNTFET) is considered as a promising technology to replace the existent silicon technology due to the following reasons [3]:

- The structure and operation principle of CNFET devices are similar to CMOS devices.
- CNFETs have the fantabulous experimentally demonstrated device current carrying ability [3].

In future integrated circuits, to search the role of CNTFETs, their performance should be examined and compared with the existent metal oxide semiconductor field-effect transistor (MOSFET) [3]. Stanford University has developed a compact model that can be used to measure the various performances of carbon nanotube field-effect transistors (CNTFETs) [6].

3 Advantages of CNFETs over Traditional MOSFETs

The strengths of CNFET making them ideal for applications of Nanoscale are [3]:

- One-dimensional carbon nanotubes offer ballistic transport result in high-speed devices and also acting as channel reduces the scattering probability.
- CNFET power consumption is lesser than an equivalent Si MOSFET device.
- CNFET has the ability to carry large current. It can also carry current density of the order of 10 μA/nm.
- CNFET current levels do not depend on channel length. Thus, there is no channel length minimization problem [4].

All these unique properties suggest that CNFETs have the ability to be a successful replacement for MOSFETS in nanoscale devices [4].

4 Full Adder Design

Three one-bit numbers, i.e., *A*, *B* and *C* can be added using 1-bit full adder, where *A* and *B* are the operands and *C* is a bit imported from the next less significant stage. Full adder is basically constructed from a cascade of adders, which add 8, 16, 32, etc., binary numbers. The circuit basically used to generate a two-bit output sum is represented by the signals Carry and Sum [7]. In this paper a full adder is designed using two half-adders and generate Sum and Carry Output. The expression of sum and carry output is as inTable 1:

$$\text{Sum} = A \oplus B \oplus C \tag{1}$$

$$\text{Carry} = AB + BC + CA \tag{2}$$

The CMOS implementation of full adder is shown in Fig. 4.

The schematic diagram of CNFET-based 1 bit full adder is shown in Fig. 5.

Table 1 Truth table of full adder [8]

In			Out	
A	*B*	*C*	*Sum*	Carry
0	0	0	0	0
0	0	1	1	0
0	1	0	1	0
0	1	1	0	1
1	0	0	1	0
1	0	1	0	1
1	1	0	0	1
1	1	1	1	1

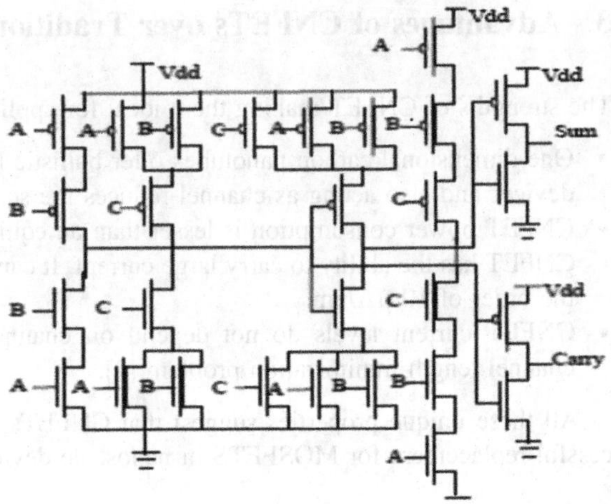

Fig. 4 Conventional 28T Si MOSFET-based full adder [8]

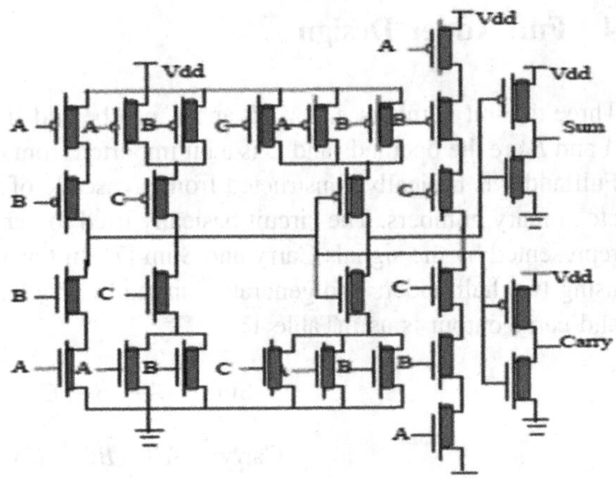

Fig. 5 Proposed 28T CNFET-based full adder

5 Important CNT Parameters

5.1 Chirality

- The chirality of CNT is explained using the vector C (the roll-up direction). The vector C is a combination of lattice vectors a_1 and a_2 [3]. It connects any two primitive lattice points of graphene such that when folded into a nanotube the two points are indistinguishable [1].

$$c = n\overrightarrow{a_1} + m\overrightarrow{a_2} \tag{3}$$

- where (*n*, *m*) are a pair of integers known as chiral indices. The vector *C* is given as [1, 3]:

$$\mathrm{C} = 3a_1 + 3a_2 = (3,3) \tag{4}$$

- Translational vector (*T*) defines the periodicity of the lattice along the tubular axis. This vector is the smallest graphene lattice vector perpendicular to *C*. It can be calculated from the orthogonality condition [1, 3].

$$C \cdot T = 0 \tag{5}$$

- Chiral angle (*θ*) explains the tilt angle of the hexagons w.r.t the tubular axis. It is the angle among the chiral vector *C* and the primitive lattice vector a1 [1, 3]

$$\cos\theta = \frac{c \cdot a_1}{|c||a_1|} \tag{6}$$

5.2 Diameter of CNT

Depending on their bandgap, the CNTs are either metallic or semiconducting. The bandgap of a CNT is inversely proportional to its diameter [3]. The diameter of the CNT is determined by the chiral indices (*n*, *m*) as [1, 9]:

$$D = \frac{|\overrightarrow{C}|}{\pi} = \frac{a}{\pi}\sqrt{n^2 + m^2 + nm} \tag{7}$$

Here, *a* is lattice constant of grapheme (0.246 nm).

5.3 Threshold Voltage of CNT

In CNTFETs, the threshold voltage of the transistor is accomplished by the diameter of the CNT. Hence a multiple threshold design can be achieved by employing carbon nanotubes with different diameters in the carbon nanotube FET. The threshold voltage of CNTFET is determined by the CNT diameter as [1, 3]:

$$V_{\mathrm{TH}} = \frac{aV_\pi}{\sqrt{3} * q \cdot D_{\mathrm{CNT}}} \tag{8}$$

Here V_π is carbon π to π bond energy (3.033 eV), DCNT is diameter of CNT, *q* is electronic charge.

6 Simulations and Results

The simulation of the proposed 28T CNFET-based full adder has been done using H-SPICE. Following are the results for the supply voltages of 0.9 V (32 nm) The functionality of the input and output of 28T full adder such as A, B, C, sum, and carry is shown in Fig. 6.

The proposed circuit, i.e., 28T Si MOSFET and CNFET-based full adder is simulated using HSPICE at 32 nm technology node. The simulated results show reduction of power consumption, delay, and energy. The obtained results are shown in Table 2.

Figure 7 shows the graphical representation of average power.

Average power consumption reduced by approx 56 %, leakage power, leakage current by 54.742 and 54.740 %, respectively, with respect to conventional CMOS full adder. Figure 8 shows the performance of both Si MOSFET and CNFET-based full adder.

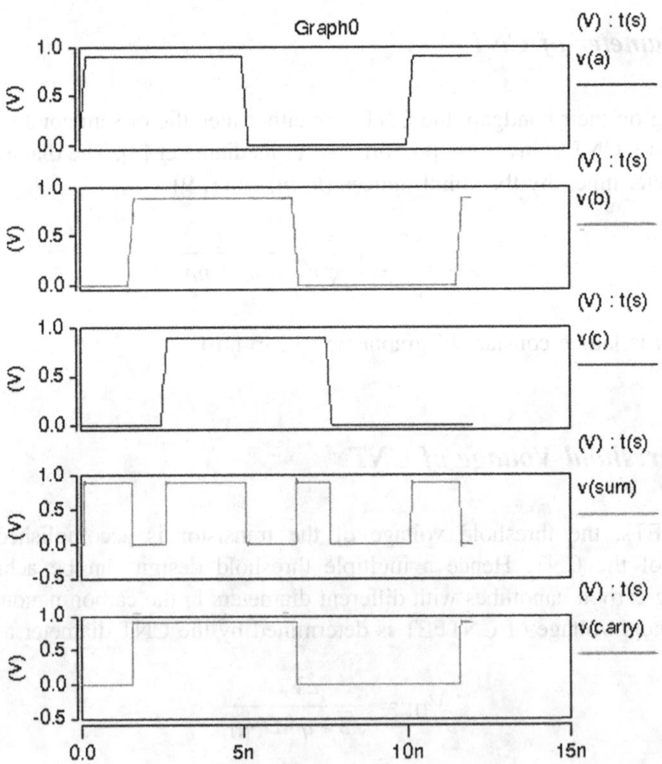

Fig. 6 According to the applied input signal sum output and carry output of 28T CNFET-based full adder

Table 2 Comparative analysis of Si-MOSFET and CNFET-based full adder

Parameter (32 nm)	Si MOSFET-based full adder	CNFET-based full adder	Percentage deduction (%)
Average power consumed	0.61754 μW	0.27173 μW	56
Delay sum	1.41385 ns	1.29095 ns	8.69
Delay carry	1.65875 ns	1.51545 ns	8.63
Total energy	2.6814 fJ	1.21355 fJ	54.74
Leakage power	0.13407 μW	0.060677 μW	54.742
Leakage current	0.14896 μA	0.067418 μA	54.740
Power delay product	0.873108 fJ	0.350531 fJ	59.85

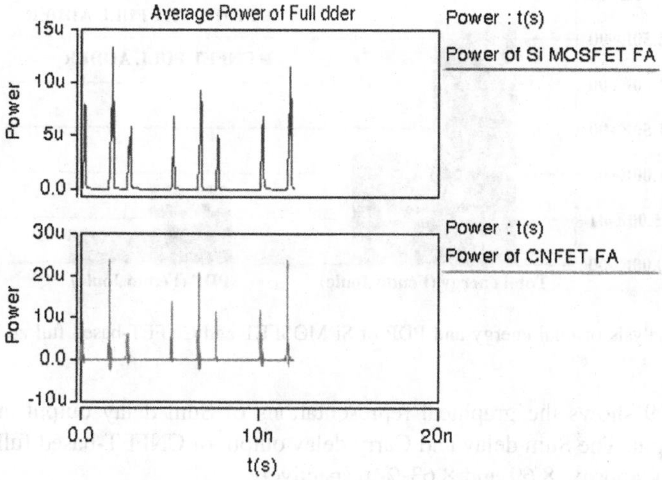

Fig. 7 Average power of Si MOSFET and CNFET-based full adder

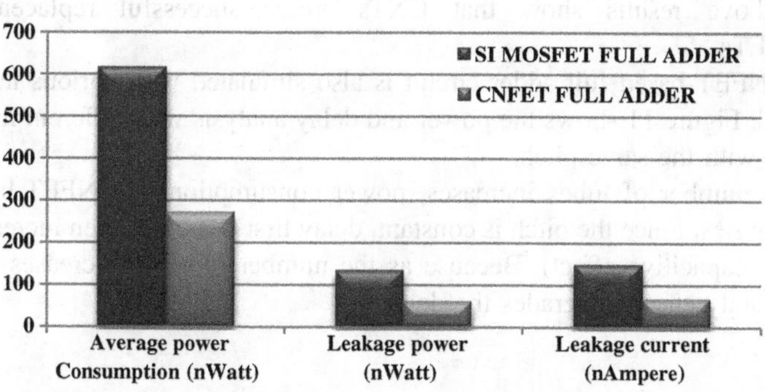

Fig. 8 Analysis of average power consumption, leakage power, and leakage current of Si MOSFET and CNFET-based full adder

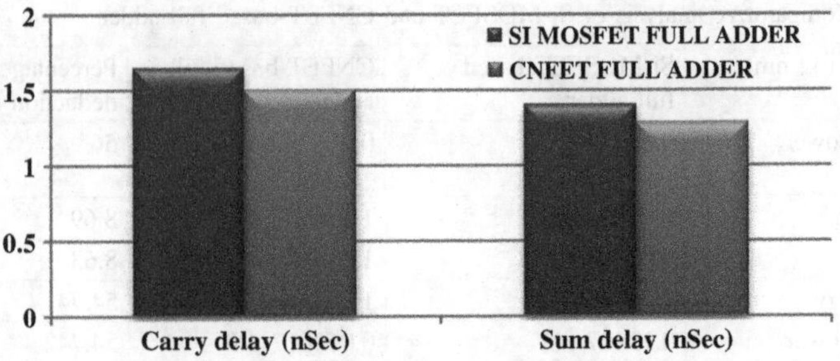

Fig. 9 Analysis of sum delay and carry delay of Si MOSFET and CNFET-based full adder

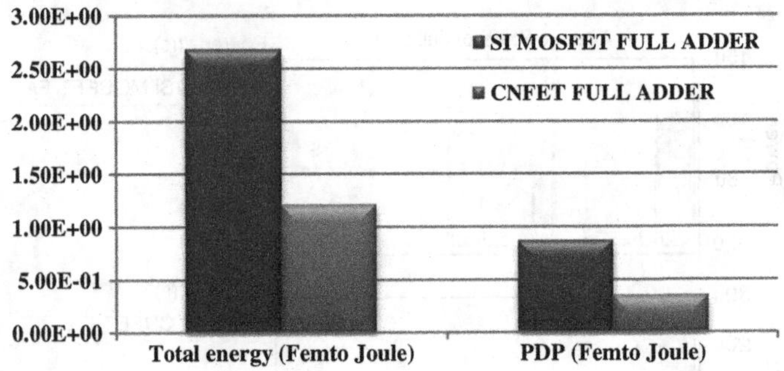

Fig. 10 Analysis of total energy and PDP of Si MOSFET and CNFET-based full adder

Figure 9 shows the graphical representations of Sum delay output and Carry Delay output. The Sum delay and Carry delay output of CNFET-based full adder is reduced by approx. 8.69 and 8.63 % respectively.

The graphical representation of total energy and power-delay product is shown in Fig. 10. The total energy and PDP of proposed design of full adder is reduced by approx 54.74 and 59 % respectively.

The above results show that CNTs are a successful replacement for Si-MOSFETs.

The CNFET-based full adder circuit is also simulated with various number of tubes used. Figure 11 shows the power and delay analysis with different number of tubes and with the same pitch.

As the number of tubes increases, power consumption of CNFET-based full adder increases. Since the pitch is constant, delay first decreases then increases (due inter-CNT capacitive effect). Because as the number of tubes increases crosstalk increases and hence it degrades the delay.

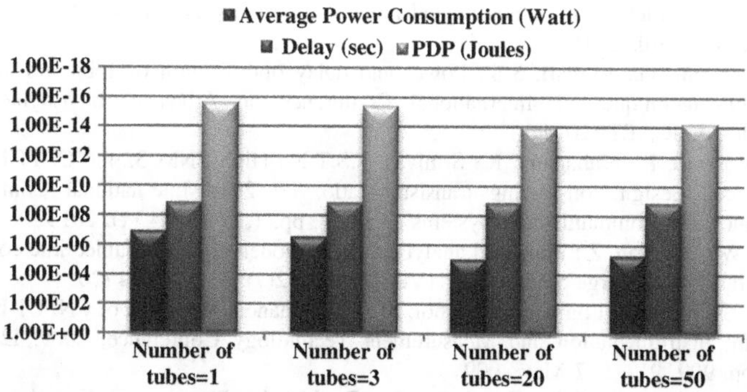

Fig. 11 Analysis of power, delay, and PDP in CNFET-based full adder with different number of tubes

7 Conclusion

In VLSI design, a digital system performance largely depends on the performance of full adder engaged in such systems. In this paper 28T CNFET-based full adder circuit is proposed. The proposed circuit is simulated using HSPICE and Stanford model of CNTFET (at 32 nm technology node). The performance analysis of full adder cell and its comparative studies show that CNT can obviously be a viable option for the future and a superior technology for applications to circuit designs. Thus it is a replacement for traditional MOSFET technology as it consumes less power and has high speed. It is emerging as an excellent candidate for building highly energy-efficient future electronic systems.

References

1. Soni, G.: Performance evaluation of carbon nanotube based devices and circuits for VLSI design. M.Tech. Thesis, Department of Electronics and Communication Engineering, Malaviya National Institute of Technology Jaipur, June 2013
2. Deng, J.: Device modeling and circuit performance evaluation for nanoscale devices: silicon technology beyond 45 nm node and carbon nanotube field effect transistors. Ph.D. thesis, Department of Electrical Engineering, Stanford University, Stanford (2007)
3. Sethi, R., Soni, G.: Power analysis of Si MOSFET and CNFET based logic gates. Int. J. Eng. Manag. Sci. (Alied Journals-IJEMS) **2**(5) (2015) ISSN-2348–3733
4. Saha, P.: Noise margin modeling and performance analysis of CNFET based SRAM. M.Tech. Thesis, Department of Electronics and Telecommunication Engineering, Jadavpur University (2013)
5. Shahidipour, H.: A study on the effects of variability on performance of CNFET based digital circuits, Ph.D. thesis, School of Electronics and Computer Science, University of Southampton, Mar, 2012
6. https://nano.stanford.edu/stanford-cnfet-model

7. Kang, S.M., Leblebici, Y.: Cmos Digital Integrated Circuits: Analysis and Design, 3rd edn. Tata McGraw Hill, 2003
8. Gaur, S., Soni, G., Kumari, S.S.: Power and delay optimization of 1 bit full adder using MTCMOS technique. In: International Conference on Advances in Engineering and Technology—ICAET (2014)
9. Dhilleswararao, P., Mahapatra, R., Srinivas, P.S.T.N.: High SNM 32 nm CNFET based 6T SRAM cell design considering transistor ratio. In: 2014 International Conference on Electronics and Communication Systems (ICECS), pp. 1, 6, 13–14 Feb 2014
10. Shi, F., Wu, X., Yan, Z.: Improved analytical delay models for RC-coupled interconnects. In: IEEE Trans. Very Large Scale Integr. (VLSI) Syst. **22**(7), 1639, 1644 (2014)
11. Cho, G., Kim, Y.-B., Lombardi, F., Choi, M.: Performance evaluation of CNFET-based logic gates. In: Instrumentation and Measurement Technology Conference, 2009. I2MTC '09. IEEE, pp. 909, 912, 5–7 May 2009
12. Halder, A., Maheshwari, V., Goyal, A., Kar, R., Mandal, D., Bhattacharjee, A.K.: Moment based delay modelling for on-chip RC global VLSI interconnect for unit ramp input. In: 2012 International Joint Conference on Computer Science and Software Engineering (JCSSE), pp. 164, 167, 30 May 2012–1 June 2012
13. http://www.sr.bham.ac.uk/yr4pasr/project06/GT/CNT.html
14. http://www.itrs.net/Links/2011ITRS/2011Chapters/2011SysDrivers.pdf
15. Shin, J.Y., Dutt, N., Kurdahi, F.: Vision-inspired global routing for enhanced performance and reliability. In: 2013 14th International Symposium on Quality Electronic Design (ISQED), pp. 239, 244, 4–6 March 2013
16. Kavicharan, M., Murthy, N.S., Rao, N.B.: An efficient delay estimation model for high speed VLSI interconnects. In: 2013 International Conference on Advances in Computing, Communications and Informatics (ICACCI), pp. 1358, 1362, 22–25 Aug 2013
17. Seokjoong, K., Guthaus, M.R.: SNM-aware power reduction and reliability improvement in 45 nm SRAMs. In: IEEE/IFIP 19th International Conference on VLSI and System-on-Chip (VLSI-SoC), 2011, pp. 204, 207, 3–5 Oct 2011
18. Lamberti, P., Tucci, V.: Impact of the variability of the process parameters on CNT-based nanointerconnects performances: a comparison between SWCNTs bundles and MWCNT. IEEE Trans. Nanotechnol. **11**(5), 924–933, (2012)
19. Sun, L., Mathew, J., Shafik, R.A., Pradhan, D.K., Li, Z.: A low power and robust carbon nanotube 6T SRAM design with metallic tolerance. In: Design, Automation and Test in Europe Conference and Exhibition (DATE), 2014, pp. 1, 4, 24–28 March 2014
20. Maheshwari, V., Agarwal, S., Goyal, A., Jain, J., Kumar, S., Kar, R., Mandal, D., Bhattacharjee, A.K.: Elmore's approximations based explicit delay and rise time model for distributed RLC on-chip VLSI global interconnect. In: 2012 IEEE Symposium on Humanities, Science and Engineering Research (SHUSER), pp. 1135, 1139, 24–27 June 2012
21. Maheshwari, V., Baboo, A., Kumar, B., Kar, R., Mandal, D., Bhattacharjee, A.K.: Delay model for VLSI RLCG global interconnects line. In: 2012 Asia Pacific Conference on Postgraduate Research in Microelectronics and Electronics (PrimeAsia), pp. 201, 204, 5–7 Dec 2012
22. Majumder, M.K., Das, P.K., Kaushik, B.K., Manhas, S.K.: Optimized delay and power performances for multi-walled CNT in global VLSI interconnects. In: 2012 5th International Conference on Computers and Devices for Communication (CODEC), pp. 1, 4, 17–19 Dec 2012
23. Majumder, M.K., Pandya, N.D., Kaushik, B.K., Manhas, S.K.: Analysis of MWCNT and Bundled SWCNT Interconnects: Impact on Crosstalk and Area. IEEE Electron Device Lett. **33** (8), pp. 1180, 1182, Aug 2012
24. Mehran, M., Masoumi, N.: A tapered partitioning method for "delay energy product" optimization in global interconnects. In: 50th Midwest Symposium on Circuits and Systems, 2007. MWSCAS 2007, pp. 21, 24, 5–8 Aug 2007

25. Moradi, M., Mirzaee, R.F., Moaiyeri, M.H., Navi, K.: An applicable high-efficient CNTFET-based full adder cell for practical environments. In: 2012 16th CSI International Symposium on Computer Architecture and Digital Systems (CADS), pp. 7, 12, 2–3 May 2012
26. Somorjit Singh, N., Madheswaran, M.: Simulation and analysis of 3t and 4t cntfet dram design in 32 nm technology. Int. J. Electron. Sign. Syst. (IJESS) **3**, 59–65 (2013) ISSN: 22315969
27. Pushkarna, A., Raghavan, S., Mahmoodi, H.: Comparison of performance parameters of SRAM designs in 16 nm CMOS and CNTFET technologies. In: 2010 IEEE International SOC Conference (SOCC), pp. 339, 342, 27–29 Sept 2010
28. Prasad, S.R., Madhavi, B.K., Kishore, K.L.: Design of a 32 nm 7T SRAM Cell based on CNTFET for low power operation. In: 2012 International Conference on Devices, Circuits and Systems (ICDCS), pp. 443, 446, 15–16 March 2012
29. Lin, S., Kim, Y.-B., Lombardi, F.: Design of a CNTFET-based SRAM cell by dual-chirality selection. IEEE Trans. Nanotechnol. **9**(1), 30, 37, (2010)
30. Kim, Y.B., Kim, Y.-B., Lombardi, F., Lee, Y.J.: A low power 8T SRAM cell design technique for CNFET. In: International SoC Design Conference, 2008. ISOCC '08. vol. 01, pp. I–176, 24–25 Nov 2008
31. Kim, Y.B.: Design methodology based on carbon nanotube field effect transistor (CNFET), Ph.D. thesis, Department of Electrical and Computer Engineering, Northeastern University Boston, Massachusetts, Jan 2011
32. Zhang, Z., Delgado-Frias, J.G.: Low power and metallic CNT tolerant CNTFET SRAM design. In: 2011 11th IEEE Conference on Nanotechnology (IEEE-NANO), pp. 1177, 1182, 15–18 Aug 2011
33. Zhang, Z., Delgado-Frias, J.G., CNTFET SRAM cell with tolerance to removed metallic CNTs. In: 2012 IEEE 55th International Midwest Symposium on Circuits and Systems (MWSCAS), pp. 186, 189, 5–8 Aug 2012

Cuckoo Search-Based Scale Value Optimization for Enhancement in Retinal Image Registration

Ebenezer Daniel and J. Anitha

Abstract Retinal image registration has significant role in various medical applications such as diabetic retinopathy, glaucoma, and many other retinal diagnosis applications. Contrast enhancement plays vital role in disease identification. In this paper, we proposed an enhancement method for intensity-based retinal image registration. In our approach, simulated images are blurred images using gaussian filter. Scale value for transformation is optimized using cuckoo search algorithm. The resultant enhanced images show better values in terms of PSNR (peak signal-to-noise ratio) and RMSE (root mean square error) which ultimately results in quality retinal image registration.

Keywords Retinal imaging · Image registration · Cuckoo search algorithm · Image restoration

1 Introduction

Retinal imaging has vital application in diabetic retinopathy and glaucoma identification. Image registration is the process of finding the correspondence image using transformation techniques. Image registration, combines different images may be taken in different time or using different imaging modalities. The combined image provides multiframe information in single image, which ease storage and communication. For example, Golabbakhsh proposed a retinal image registration

Ebenezer Daniel (✉) · J. Anitha
Department of Electronics and Communication Engineering,
Karunya University, Coimbatore 641114, India
e-mail: ebydaniel89@gmail.com

J. Anitha
e-mail: anithaj@karunya.edu

M. Pant et al. (eds.), *Proceedings of Fifth International Conference on Soft Computing for Problem Solving*, Advances in Intelligent Systems and Computing 436, DOI 10.1007/978-981-10-0448-3_37

technique in which he combined OCT (optical coherence tomography) image and Fundus Images [1]. Delibasis proposed a registration algorithm based on artificial immune system (AIS) optimization technique for retinal images and dental images. In which he generated 20 retinal images from 10 reference images based on sinusoidal transformation technique [2]. Suicheng performed a lungs image registration based on B-Spline affine transformation (BSAT) [3]. Contrast enhancement is necessary for all images for object identification. Human eyes are low sensitive to light; so, contrast enhancement has importance [4]. Histogram-based enhancement techniques are well-known approaches in preprocessing. There are various modifications available in histogram; Khan proposed segment-selective dynamic histogram equalization [5–8]. The green plane of retinal image has more information than the other two color planes. OGPM (Optimum Green Plane Masking] is an enhancement technique for retinal images in which contrast improvement is evaluated using PSNR, RMSE, and AMBE measures [9]. Retinal image registration is mainly classified into feature-based registration and intensity-based registration. In feature-based approach, segmentation and bifurcation are essential for registration [10]. Pixel-based techniques, contrast enhancement is the main advantage. Molodij proposed a nonlinear image registration for retinal image enhancement [11]. Optimum transformation parameters are necessary in image registration. Rouet proposed a Genetic Algorithm (GA) for MR and CT image registration [12]. Das compared GA and Particle Swarm Optimization (PSO) for MR and CT image registration, in which PSO provided better performance than conventional GA [13]. In this paper, we proposed an intensity-based registration approach, simulated images are generated using transformation techniques. Optimum scale value for transforms is selected using Cuckoo Search Optimization (CSO) algorithm [14, 15]. Quality of enhancement is evaluated using PSNR and RMSE values.

2 Proposed Intensity-Based Image Registration

Our proposed technique is cuckoo search optimization-based retinal image registration. In our registration, simulated images is generated using gaussian filter. Contrast of blurred image [16, 17] is improved using optimum masking techniques. The blurred image is filtered using median filter. Scale value (*k*) is multiplied with filtered image and it is used for mask formulation. The obtained mask is added to the simulated image and the resultant image is called masked image. Optimum scale value is updated based on RMSE metric value. Optimum resultant image is called registered image. Similarity calculation between reference image and registered images is calculated using RMSE value (Fig. 1).

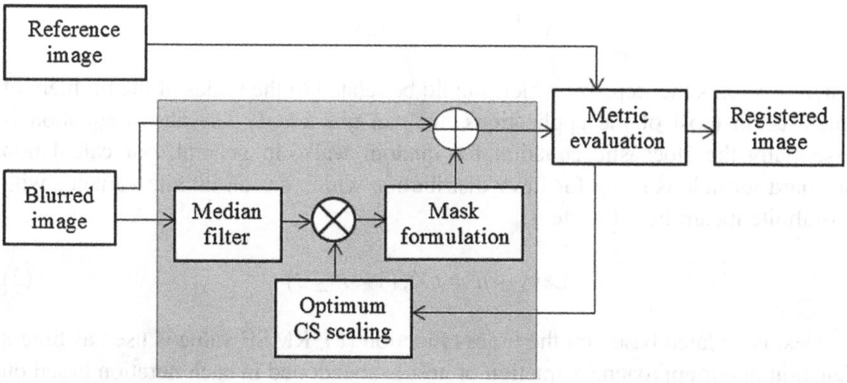

Fig. 1 Proposed block diagram for retinal image registration

3 Cuckoo Search Optimization (CSO)

Cuckoo search optimization algorithm is used here for finding the optimum scalar values for wavelet transform. Range of scalar values is obtained using trial and error values. In CSO, each solution is considered as each nest. In our application, we generated 50 nests within the range. Random cuckoo selecting a nest for laying eggs is performed based on Levy fight. After nest selection evaluated the fitness of nest, based on the fitness value cuckoo updates the rank of nests (Fig. 2).

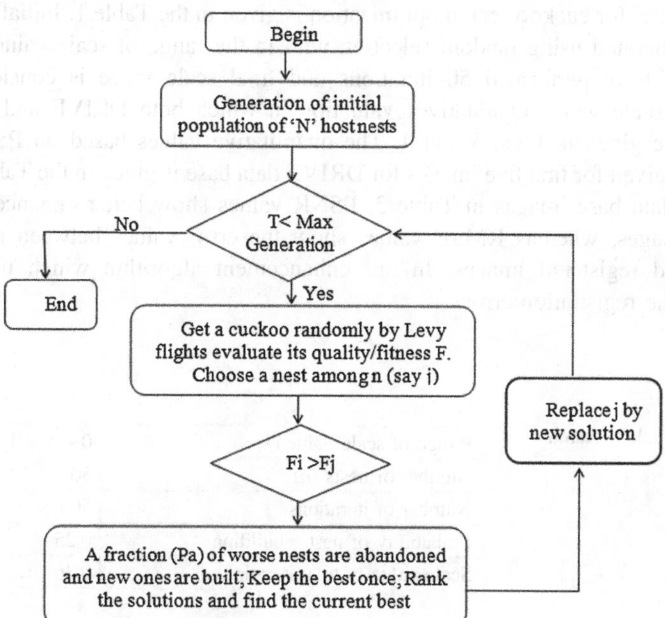

Fig. 2 Flowchart of cuckoo search optimization (CSO) algorithm

$$x_i^{t+1} = x_i^t + \alpha \otimes \text{Levy}(\lambda) \quad (1)$$

where $\alpha > 0$, is the step size which should be related to the scales of the problem of interests. In most of the applications, we can use $\alpha = 1$. The above equation is essentially the stochastic equation for random walk. In general, our calculation standard formula is using for Levy distribution which has an infinite variance with an infinite mean; here the steps.

$$\text{Levy} \sim u = t^{-\lambda}, (1 < \lambda \leq 3) \quad (2)$$

Nest is updated based on the fitness function (F), RMSE value is used as fitness function in our approach. A fraction of nest is abandoned in each iteration based on a portability value (Pa). In our optimization task, we selected a random probability value 0.25.

4 Results and Discussion

For result analysis five test images are taken from DRIVE and STARE database. Enhancement of retinal images based on our proposed registration technique is analyzed using both quantitatively and the qualitatively. Quantitative image enhancement is measured using PSNR, AMBE, and RMSE values. The higher value of PSNR gives the better performance and lower values of RMSE provides better enhancement.

Parameter for cuckoo search optimization is given in the Table 1. Initial 50 scale values generated using random selection with in the range of scale value. In our work, we have performed 50 iterations and final scale value is considered as optimum scale value. Qualitative evaluation in which both DRIVE and STARE images are given in Figs. 3 and 4. The quantitative values based on PSNR and RMSE is given for first five images for DRIVE data base is given in the Table 2 and STARE data base images in Table 3. PSNR values show better enhancement of retinal images, whereas RMSE values show the error values between reference image and registered images. In our enhancement algorithm which ultimately reduced the registration error.

Table 1 Cuckoo search parameters

Range of scale value (k)	$0 < k < 1$
Number of nests (n)	50
Number of iterations	50
Probability of nest rebuilding	0.25
Scale value of final iteration	0.36

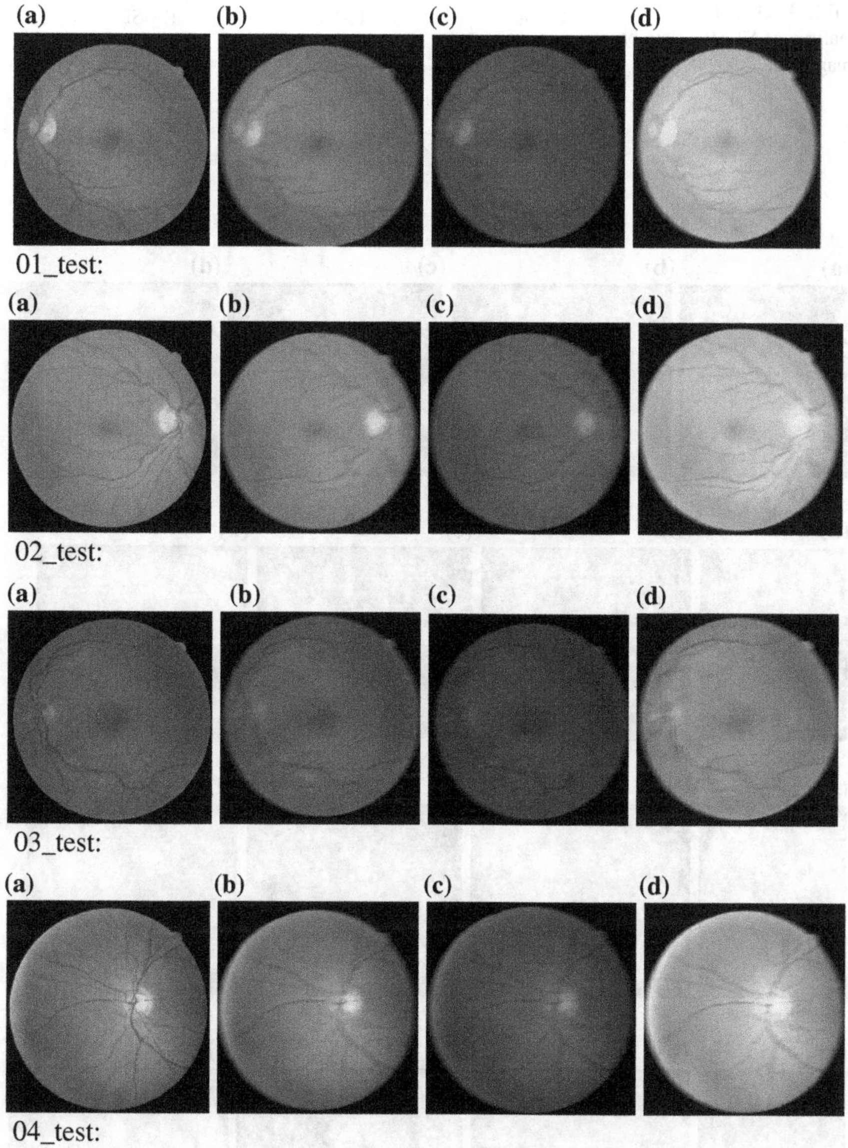

Fig. 3 DRIVE data base image: **a** reference image **b** blurred image **c** mask **d** registered image

Table 2 Quantitative analysis of DRIVE data base images

Image	PSNR	RMSE
01_test	60.02	0.2551
02_test	57.78	0.3303
03_test	61.84	0.2071
04_test	59.30	0.2773

Table 3 Quantitative analysis of STARE data base images

Image	PSNR	RMSE
im0001	59.98	0.2563
im0002	59.80	0.2617
im0003	58.24	0.3133
im0004	59.38	0.2748

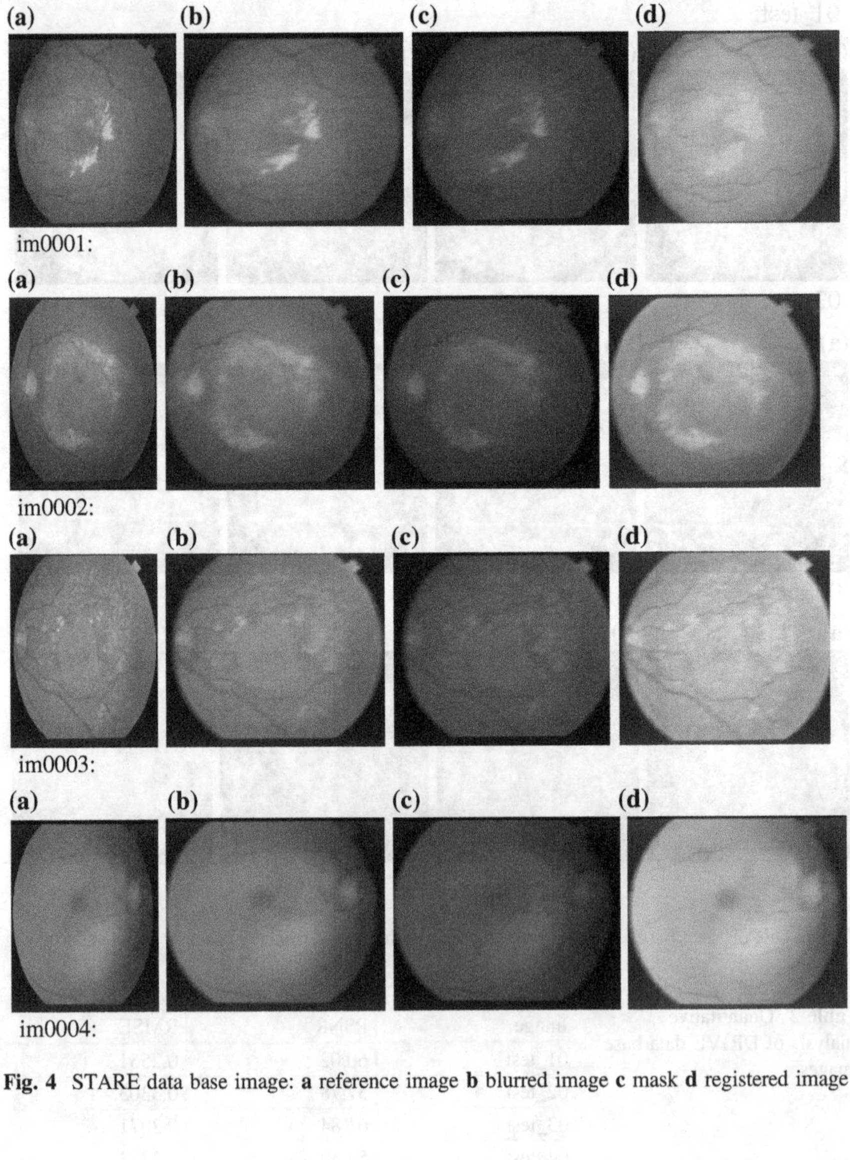

Fig. 4 STARE data base image: **a** reference image **b** blurred image **c** mask **d** registered image

5 Conclusion

In this paper, we presented an enhancement algorithm for retinal image registration. In the first section, we generated blurred images using gaussian filter. In second section, optimum mask is generated using Cuckoo Search Algorithm. Optimum mask is added to the simulated image. The resultant image is registered image. In this work, cuckoo search-based image registration is used for contrast enhancement of retinal images. The registered images shows contrast improvement as well as restoration of blurred noise. The performance of registration is evaluated both quantitatively and qualitatively. Performance is measured using PSNR and RMSE values.

References

1. Golabbakhsh, M., Rabbani, H.: Vessel-based registration of fundus and optical coherence tomography projection images of retina using a quadratic registration model. IET Image Process (2013). doi:10.1049/iet-ipr.2013.0116
2. Delibasis, K., Asvestas, P.A., Matsopoulos, G.K.: Automatic point correspondence using an artificial immune system optimization technique for medical image registration. Comput. Med. Imaging Graphics **35**, 31–41 (2011)
3. Gu, S., et.al.: Bidirectional elastic image registration using B-spline affine Transformation. Comput. Med. Imaging Graphics **38**, 306–314 (2014)
4. Daniel, J.A., Retinal image enhancement using wavelet domain edge filtering and scaling. In: International Conference on Electronics and Communication Systems (2014). doi:10.1109/ECS.2014.6892670
5. Khan, M.F., Khan, E., Abbasi, Z.A.: Segment selective dynamic histogram equalization for brightness preserving contrast enhancement of images. Optik **125**, 1385–1389 (2014)
6. Khan, M.F., Khan, E., Abbasi, Z.A.: Segment dependent dynamic multi-histogram equalization for image contrast enhancement. Digital Signal Process. **25**, 198–223 (2014)
7. Khan, M.F., Khan, E., Abbasi, Z.A.: Artifact suppressed image enhancement through Bi Histogram equalization. In: International Conference on Multimedia, Signal Processing and Communication Technologies (2013). doi:10.1109/MSPCT.2013.6782089
8. Khan, M.F., Khan, E., Abbasi, Z.A.: Weighted average multi segment histogram equalization for brightness preserving contrast enhancement. In: IEEE International Conference on Signal Processing, Computing and Control (2012). doi:10.1109/ISPCC.2012.6224340
9. Daniel, J.A.: Optimum green plane masking for the contrast enhancement of retinal images using enhanced genetic algorithm. Optik (2015). doi:10.1016/j.ijleo.2015.05.027
10. Chen, L., Huanga, X., Tiana, J.: Retinal image registration using topological vascular tree segmentation and bifurcation structures. Biomed. Signal Process. Control **16**, 22–31(2015)
11. Molodij, G., Ribak, E.N., Glanc, M., Chenegros, G.: Enhancing retinal images by nonlinear registration. Optics Commun. **342**, 157–166 (2015)
12. Rouet, J.M., Jacq, J.J., Roux, C.: Genetic algorithms for a robust 3-D MR-CT registration. IEEE Transaction On Information Technology in Biomedicine, Vol. 4, No. 2 (2000)
13. Das, A., Bhattacharya, M.: Affine-based registration of CT and MR modality images of human brain using multi resolution approaches: comparative study on genetic algorithm and particle swarm optimization. Neural Comput. Appl. **20**, 223–237 (2011)

14. Bhandari, A.K., Soni, V., Kumar, A., Singh, G.K.: Cuckoo search algorithm based satellite image contrast and brightness enhancement using DWT–SVD. ISA Trans. **53**, 1286–1296 (2014)
15. Panda, R., Agrawal, S., Bhuyan, S.: Edge magnitude based multilevel thresholding using Cuckoo search Technique. Expert Syst. Appl. **40**, 7617–7628 (2013)
16. Al-Ameen, Z., Sulong, G.: Deblurring computed tomography medical images using a novel amended landweber algorithm. Interdisc. Sci. Comput. Life Sci. **7**, 1–7 (2015)
17. Jiang, M., et.al.: Blind deblurring of spiral CT images. IEEE Trans. Med. Image **22**(7), (2003)

Optimal Demand-Side Bidding Using Evolutionary Algorithm in Deregulated Environment

Subhojit Dawn, Sadhan Gope, Prashant Kumar Tiwari and Arup Kumar Goswami

Abstract This paper presents an efficient and optimization proficiency for minimization of fuel cost and losses of an electrical system in a completely deregulated power system. Single-side bidding and double-side bidding both cases are considered in this paper with the help of sequential quadratic programming (SQP) and evolutionary algorithm like firefly algorithm (FA) and cuckoo search algorithm (CSA) for checking the effectiveness of the presented approach. Modified IEEE 14 bus test system and modified IEEE 30 bus test system are considered for validating and analyzing the impact of proposed approach.

Keywords Deregulation · Firefly algorithm · Cuckoo search algorithm · Sequential quadratic programming · Optimal power flow

1 Introduction

In the last 40–50 years, electricity companies have divided into some parts which are monopoly in nature, for which there is a competitive attitude build up among these parts of the electricity companies, which creates a market environment for electricity. The demand of electricity is increased day-by-day throughout the world in a rapid manner. Every consumer wants to purchase the electrical energy with a

Subhojit Dawn (✉) · Sadhan Gope · P.K. Tiwari · A.K. Goswami
National Institute of Technology Silchar, Silchar, Assam, India
e-mail: subhojit.dawn@gmail.com

Sadhan Gope
e-mail: sadhan.nit@gmail.com

P.K. Tiwari
e-mail: prashant081.in@gmail.com

A.K. Goswami
e-mail: gosarup@gmail.com

M. Pant et al. (eds.), *Proceedings of Fifth International Conference on Soft Computing for Problem Solving*, Advances in Intelligent Systems and Computing 436, DOI 10.1007/978-981-10-0448-3_38

minimum cost, which is possible only in the competitive electricity market. There are several contract models present in a competitive power market like bilateral contracts, multilateral contracts, and pool market. There are two types of bidding present in the electricity market: (1) Single-side bidding and (2) double-side bidding. In single-side bidding, single buyer and multiple sellers (or vice versa) are participated in the electricity auction. On the other hand, multiple sellers and multiple buyers are participated in the double-auction bidding of electricity market.

In deregulated market, the bidding price of the generated energy is not fixed by any governing body but the generation companies (GENCOS) decide it by themselves. In this market, consumer can choose their suppliers (DISCOS). But this full marketing service is controlled by independent system operator (ISO).

The Ref. [1] presents an approach for improving power quality and reliability of the distribution system using quantum firefly algorithm. In this paper, author used quantum theory integration with firefly algorithm and used quantum firefly algorithm as an optimization technique. In paper [2], authors present an approach for optimal placement of active power conditioner in distribution systems by dynamic discrete firefly algorithm. This paper describes that how the presented methods reduce the power quality of hybrid renewable energy-based generators in smart grids. Modified IEEE 16-bus test system has been used by authors in this paper for performance analysis. Reference [3] presents execution of firefly algorithm with online wavelet filter on AGC model for a three-area thermal power system. Reference [4] presents an optimization approach for optimal sizing and placement for active power conditioner to increase power quality in the distributed power system by improved discrete firefly algorithm. IEEE 16 and 69 bus test system is used to test the system performance. In paper [5], total energy capacity of any wind firm is calculated using the firefly algorithm and this result is compared with other approaches using genetic algorithm. In paper [6], reconfiguration methodology using cuckoo search algorithm has been used for voltage magnitude maximization and active power minimization of an electrical system. Reference [7] presents penalty-based cuckoo search algorithm to optimize the solution of redundancy and reliability allocation with nonlinear constraints. The redundancy and reliability involve in the selection of reliability of components in each subsystem. Paper [8] presents a modified cuckoo search algorithm to optimize the short-term hydro-thermal scheduling.

From the literature, it exhibits that optimization approach in fully deregulated power environment is a very efficient and beneficial process. This paper presents a simple optimization approach to minimize total generation cost and losses of the electrical system. Three types of optimal power flow (OPF) techniques have been used in this paper: (i) Sequential quadratic programming, (ii) firefly algorithm, and (iii) cuckoo search algorithm. In this presented approach, both single- and double-side biddings have been incorporated in the optimal power flow problem. MATLAB 2013 has been used for programming purpose in this work.

2 Presented Method

2.1 Pool Market Model

In this model, generation companies (GENCOS) participate in the pool with their cost function and maximum generation which they want to sell to the pool. Distribution companies (DISCOS) also bid their cost function and maximum demand which they want to take from the pool. After social welfare optimization, the generation and demand of all buses are known.

Let vector of pool real power generation be

$$\mathrm{RPG}^{\mathrm{P}} = [\mathrm{RPG}_i^{\mathrm{P}}, i = 1, 2, 3 \ldots N_{\mathrm{g}}] \tag{1}$$

and vector of pool real power demand be

$$\mathrm{RPL}^{\mathrm{P}} = [\mathrm{RPL}_j^{\mathrm{P}}, j = 1, 2, 3 \ldots N_l] \tag{2}$$

Now let vector of the total real power generation and demand be

$$\mathrm{RPG}^{\mathrm{T}} = [\mathrm{RPG}_i^{\mathrm{T}}, i = 1, 2, 3 \ldots N_{\mathrm{g}}] \tag{3}$$

$$\mathrm{RPL}^{\mathrm{T}} = [\mathrm{RPL}_j^{\mathrm{T}}, j = 1, 2, 3 \ldots N_l] \tag{4}$$

2.2 Firefly Algorithm

The firefly algorithm is a meta-heuristic algorithm, inspired by the flashing behavior of fireflies [9, 10]. The primary purpose of a firefly's flash is to act as a signal to attract other fireflies. Xin-She Yang was the inventor of this algorithm. The characteristics of fireflies are taken care in this algorithm.

1. All fireflies are unisexual, so that one firefly will be attracted to all other fireflies.
2. Attractiveness is proportional to their brightness, and for any two fireflies, the less bright one will be attracted to the brighter one; the brightness can be decreased as their distance increased.
3. If there are no fireflies brighter than a given firefly, it will move randomly. The brightness should be associated with the objective function. The variation of firefly attractiveness 'β' with the distance 'r' is determined by

$$\beta = \beta_0 e^{-\gamma r^2} \tag{5}$$

where 'β_0' is the attractiveness at $r = 0$.
The movement of a firefly is determined by

$$x_i^{t+1} = x_i^t + \beta_0 e^{-\gamma r_{ij}^2}\left(x_j^t - x_i^t\right) + \alpha_t \varepsilon_i^t \tag{6}$$

where the second term of the right-hand side of Eq. (6) is due to the attraction. The third term is randomization with a randomization parameter *αt* and ε_i^t is a vector of random numbers drawn from a Gaussian distribution or uniform distribution at time *t*.

Pseudocode of Firefly Algorithm

1. Objective function: $f(x)$, $x = (x_1, x_2 \ldots, x_d)$
2. Generate an initial population of fireflies x_i (for $i = 1, 2, \ldots, n$)
3. Formulate light intensity *I* so that it is associated with $f(x)$
4. Define absorption co-efficient γ
5. While (t < Max generation)
   For $i = 1$: n (all n fireflies)
   For $j = 1$: n (n fireflies)
   If ($I_j > I_i$)
   Move firefly i toward j
   End if
   Vary attractiveness with distance r via exp ($-\gamma r$)
   Evaluate new solutions and update light intensity
   End for j
   End for i
   Rank fireflies and find the current best;
   End while
   Post processing the results and visualization;
6. End

2.3 Cuckoo Search Algorithm

Cuckoo search is an optimization algorithm developed by Xin-She Yang and Suash Deb in 2009 [11]. Cuckoos have the female laying her fertilized eggs in the nest of another species so that the surrogate parents unwillingly raise her brood. Sometimes when the cuckoo's egg in the host nest is discovered (eggs are not its own), the surrogate parents either throw it out or abandon the nest and builds their own brood elsewhere.

Some cuckoo species have evolved in such a way that female parasite cuckoos are often much specified in the mimicry in color and pattern of the eggs of a few chosen host species. This reduces the probability of eggs being abandoned and increases their reproductively. Parasitic cuckoos often choose a nest where the host bird just laid its own eggs. In general, the cuckoo eggs hatch slightly earlier than their host eggs. Once the first cuckoo chick is hatched, the first instinct action it will take is to evict the host eggs by blindly propelling the eggs out of the nest, which increases the cuckoo chick's share of food provided by its host bird.

Pseudocode of Cuckoo Search Algorithm

1. Objective function: $f(x)$, $x = (x_1, x_2, \ldots, x_d)$
2. Generate an initial population of n host nests
3. While (t < Max Generation) or (Stop Criterion)
4. Get a cuckoo randomly (say i) and replace its solution by performing Levy flights
5. Evaluate its quality/fitness F_i
   [For maximization, $F_i \propto f(x_i)$]
6. Choose a nest among n (say j) randomly
7. If ($F_i > F_j$)
   Replace j by the new solution
8. End if
9. A fraction (p) of the worse nests is abandoned and new ones are built
10. Keep the best solutions/nests
11. Rank the solutions/nests and find the current best
12. Pass the current best solutions to the next generation
13. End while

3 Problem Formulation

3.1 Objective Function

The following mathematical expression is used to finding the objective of the proposed method:

$$\text{minimize}\, P = aC_i(P_i)^2 + bC_i(P_i) + c, \quad i = 1, 2, \ldots, N_g \tag{7}$$

where N_g is total number of generator; a, b, and c are the generator cost co-efficient; and $C_i(P_i)$ is the generation cost for each unit. Total generation cost P has been minimized, subjecting to the following constraints.

3.2 Constraints

Equality Constraints

The constraints are real power balance equation and power flow equation.

a. **Real Power Balance Equation**

$$\sum_{i=1}^{Ng} P_{Gi} - P_{\text{loss}} - P_D = 0 \qquad (8)$$

$$P_{\text{loss}} = \sum_{J=1}^{N_{\text{TL}}} G_J \left[|V_i|^2 + |V_j|^2 - 2|V_i||V_j| \cos(\delta_i - \delta_j) \right] \qquad (9)$$

where P_{Gi} is the generation capacity of ith generator, P_D is the total load; P_{loss} is the total transmission loss; N_{TL} is the number of transmission lines; G_J is the conductance of the line between buses i and j; $|V_i|$ is the voltage magnitude of bus i; and δ_i is the voltage angle of bus i.

b. **Power Flow Equations**

$$P_i - \sum_{k=1}^{N_b} |V_i V_k Y_{ik}| \cos(\theta_{ik} - \delta_i + \delta_k) = 0 \qquad (10)$$

$$Q_i + \sum_{k=1}^{N_b} |V_i V_k Y_{ik}| \sin(\theta_{ik} - \delta_i + \delta_k) = 0 \qquad (11)$$

where P_i and Q_i are the real and reactive powers injected into power system at bus i, N_b is the number of buses; Y_{ik} is the element in the ith row and kth column of bus admittance matrix; and θ_{ik} is the angle of the element in the ith row and kth column of bus admittance matrix.

Inequality Constraints

These are the set of discrete and continuous constraints that represent the system security and operational limits as follows:

a. **Voltage Magnitude Limits of Each PQ Bus**

$$V_i^{\min} \le V_i \le V_i^{\max} \quad i = 1, 2, 3 \ldots N_b \qquad (12)$$

where '$V_i^{\min}$' and '$V_i^{\max}$' are the voltage limits of bus 'i', and 'N_b' is the total number of buses.

b. **Phase Angle Limits of Voltage of Each PQ Bus**

$$\emptyset_i^{\min} \leq \emptyset_i \leq \emptyset_i^{\max} \quad i = 1, 2, 3 \ldots N_{\mathrm{b}} \tag{13}$$

where '$\emptyset_i^{\min}$' and '$\emptyset_i^{\max}$' are the phase angle limits of voltage at bus 'i'.

c. **MVA Limits of Lines**

$$\mathrm{TL}_l \leq \mathrm{TL}_l^{\max} \quad l = 1, 2, 3 \ldots N_{\mathrm{TL}} \tag{14}$$

where 'TL_l' is line flow at line 'l', '$\mathrm{TL}_l^{\max}$' is the maximum line flow limit of the line 'l', and 'N_{TL}' is the number of transmission lines.

d. **Real and Reactive Power Limits at PV Buses**

$$P_{Gi}^{\min} \leq P_{Gi} \leq P_{Gi}^{\max} \quad i = 1, 2, 3, \ldots, N_{\mathrm{b}} \tag{15}$$

$$Q_{Gi}^{\min} \leq Q_{Gi} \leq Q_{Gi}^{\max} \quad i = 1, 2, 3, \ldots, N_{\mathrm{b}} \tag{16}$$

where '$P_{\mathrm{Gi}}^{\min}$' and '$P_{\mathrm{Gi}}^{\max}$' are the real power limits of bus 'i', and '$Q_{\mathrm{Gi}}^{\min}$' and '$Q_{\mathrm{Gi}}^{\max}$' are the reactive power limits of bus 'i'.

3.3 Proposed Approach

This paper presents an optimization approach to minimize the total system generation cost and also minimize the system losses. The steps associated with the proposed approach are as follows:

(1) Take the bus data, line data, and generation data of modified IEEE 14 bus test system.
(2) Run optimal power flow program using sequential quadratic programming and read the value of total generation cost and power losses in the electrical system with single-side bidding.
(3) Run OPF problem using firefly algorithm and cuckoo search algorithm and take the value of system generation cost and system loss.
(4) Compare the results taken from firefly algorithm and cuckoo search algorithm with the results of sequential quadratic programming.
(5) If
Firefly algorithm and cuckoo search algorithm give the better result, compared with SQP, then continue the process,
Else
end.
(6) Repeat step [2–5] for double-side bidding.
(7) Repeat step [1–6] for modified IEEE 30 bus system.

4 Results

In this paper the optimization approach has been applied to the modified IEEE 14 and modified IEEE 30 bus system.

4.1 Optimization Approach for Modified IEEE 14 Bus System

In modified IEEE 14 bus system, 5 generators, 14 buses, 20 transmission lines, and 10 loads are present. The system data is acquired from [12, 13]. 100 MVA has been considered as base MVA rating and bus 1 as reference bus or slack bus throughout this paper. There are four cases that have been taken in this paper. They are as follows:

Case I. Optimal power flow (OPF) using sequential quadratic programming (SQP) in single-auction bidding in deregulated environment.
Case II. OPF using firefly algorithm and cuckoo search algorithm in single-auction bidding.
Case III. OPF using SQP in double-auction bidding in competitive power market.
Case IV. OPF using firefly algorithm and cuckoo search algorithm in double-auction competitive power market.
Case I In this case OPF has been done using SQP in deregulated environment with single-auction bidding. In this approach optimized generation cost is 899.09 $/h and the system active power loss is 3.02 MW.
Case II In this case firefly algorithm is used as OPF technique in deregulated environment with single-auction bidding. In this approach the OPF converges within 300 iterations. To check the robustness of the technique, it has been analyzed for 100 times. The optimum result for the generation cost is 878.1058 $/h and the system active power loss is 3.6132 MW. Using the cuckoo search algorithm the generation cost is minimized more, as generation cost is 876.9681 $/h and system active power losses are also minimized as 3.3714 MW. The generation cost using cuckoo search algorithm is considerably less than the others (shown in Table 1).
Case III OPF has been done using SQP in deregulated environment with double-auction bidding in this case. Demand-side bidding has been done in bus no. 4 in the test system. The objection function (Generation cost) in this case is 680.06 $/h The total generation cost is very less because of the generation rescheduling done here for double-auction bidding. The active power loss after optimization is 2.54 MW and the losses are also minimized for the double-auction bidding.
Case IV In this case first the firefly algorithm is used as OPF technique in double-action bidding in deregulated energy market. Within 300 iterations this method converges and gives optimized results. 671.7609 $/h is the optimized generation cost and 2.8769 MW is the system active power loss in this case. After

Table 1 Results in modified IEEE 14 bus system

	Control Variable	Single-side bidding			Double-auction bidding		
		SQP	Firefly	Cuckoo	SQP	Firefly	Cuckoo
Generation (MW)	P_{G1}	54.1930	58.550	58.7189	44.542904	50.3123	50.3259
	P_{G2}	67.3540	75.523	75.5938	57.148671	65.8865	65.9718
	P_{G3}	45.8448	28.806	28.7139	35.116639	25.9358	25.8000
	P_{G6}	63.9902	68.847	68.7979	51.631733	48.9606	48.4449
	P_{G8}	30.6387	30.884	30.5469	25.299563	22.9817	23.2395
	P_{L4}	–	–	–	−0.000015	−0.0001	−0.0011
Voltage (V)	V_1	1.050	1.0496	1.1000	1.011	1.0476	1.0961
	V_2	1.045	1.0440	1.0941	1.004	1.0427	1.0922
	V_3	1.023	1.0196	1.0709	0.982	1.0207	1.0721
	V_4	1.016	1.0242	1.0700	0.996	1.0363	1.0857
	V_5	1.019	1.0275	1.0728	0.997	1.0313	1.0796
	V_6	1.032	1.0676	1.1000	1.050	1.0476	1.0894
	V_7	1.019	1.0421	1.0817	1.021	1.0261	1.0884
	V_8	1.033	1.0670	1.1000	1.050	1.0200	1.1000
	V_9	1.001	1.0250	1.0649	1.005	1.0131	1.0706
	V_{10}	0.998	1.0243	1.0631	1.005	1.0111	1.0663
	V_{11}	1.011	1.0414	1.0772	1.023	1.0251	1.0739
	V_{12}	1.016	1.0509	1.0843	1.033	1.0311	1.0747
	V_{13}	1.009	1.0431	1.0772	1.025	1.0242	1.0691
	V_{14}	0.985	1.0143	1.0523	0.995	0.9992	1.0521
System generation cost ($/h)		899.09	878.10	876.968	680.06	671.760	670.559
System losses (MW)		3.02	3.6132	3.3714	2.54	2.8769	2.5821

using the cuckoo search algorithm the generation cost is minimized more with 670.5596 MW and system active power loss is also minimized with 2.5821 MW. Cuckoo search algorithm in double-auction bidding gives the most optimum cost and active power loss with comparison to other cases. System voltage has been also improved after using the firefly algorithm and cuckoo search algorithm in double-auction bidding in fully deregulated environment (Table 1).

In Table 1, P_{G1}, P_{G2}, P_{G3}, P_{G6}, and P_{G8} are represented as the power generated by the generator connected at bus no. 1, 2, 3, 6, and 8. P_{L4} is the generated power by load connected at bus no. 4. In deregulated market, load is considered as negative generator. So, in every case for P_{L4}, we see that the result is negative. V_1, V_2, …, V_{14} are the values of voltages at bus no. 1, 2, 3, …, 14.

4.2 Optimization Approach for Modified IEEE 30 Bus System

This system with 6 generators, 30 buses, 41 transmission lines, and 21 loads has been used to check the effectiveness of the presented approach. The bus data, line data, generator data, and generator cost co-efficient have been taken from [13]. System data and results are based on 100 MVA and the reference bus is bus-1. Like modified IEEE 14 bus experiment, four cases have been taken in order to verify the presented approach and illustrate the impacts of firefly algorithm and cuckoo search algorithm in fully deregulated environment—(1) OPF using SQP in single-auction bidding in deregulated environment, (2) OPF using firefly algorithm and cuckoo search algorithm in single-auction bidding in deregulated environment, (3) OPF using SQP in double-auction bidding in deregulated environment, and (4) OPF using firefly algorithm and cuckoo search algorithm in double-auction bidding. First, optimal power flow has been done using SQP with single-side bidding, and got total generation cost as 969.91 $/h, and system active power loss as 3.99 MW. Then, firefly algorithm is used with single-side bidding, and optimized total generation cost is minimized to 799.5296 $/h and system active power losses is minimized to 9.2539 MW. To check the robustness of the technique, firefly algorithm and cuckoo search algorithm have been tested for 100 times. Cuckoo search algorithm has been used in single-side bidding and generation cost has been minimized more as 798.9156 $/h and system active power loss is also minimized as 9.0549 MW. Again OPF has been done using SQP with double-auction bidding. In this case demand-side bidding has been done in bus no. 7. The OPF converges within 300 iterations in this approach. The objective function (total generation cost) in this case is 866.32$/h and system active power loss is 3.35 MW. The total generation cost is very less compared with single-side bidding, because of the rescheduling of generation has been done here for double-auction bidding. System losses have been also minimized after applying the double-auction bidding. Firefly algorithm has been used in double-action bidding in the fourth case. Within 300 iterations this method converges and gives optimized generation results as 718.5260 $/h and in this case active power loss of the system is 8.0658 MW. After this cuckoo search algorithm also has been used for checking the effectiveness of this algorithm in the double-action bidding. We get 717.3140 $/h as a generation cost, which is minimum among all three optimization techniques used in this paper. System active power loss has been also minimized with losses 7.8387 MW. Like IEEE 14 bus system, in this system also system voltage has been improved after using the firefly algorithm and cuckoo search algorithm (shown in Table 2).

In Table 2, P_{G1}, P_{G2}, P_{G5}, P_{G8}, P_{G11}, P_{G13} are represented as the power generated by the generator connected at bus no. 1, 2, 5, 8, 11, and 13. P_{L7} is the generated power by load connected at bus no. 7. V_1, V_2, …, V_{30} are the values of voltages at bus no. 1, 2, …, 30.

Table 2 Results in modified IEEE 30 bus system

	Control variable	Single-side bidding			Double-auction bidding		
		SQP	Firefly	Cuckoo	SQP	Firefly	Cuckoo
Generation (MW)	P_{G1}	61.758	177.816	177.570	55.704	164.630	166.318
	P_{G2}	79.254	48.793	48.548	72.117	45.357	45.167
	P_{G5}	30.011	21.417	21.392	27.734	20.184	20.670
	P_{G8}	54.999	20.874	20.989	54.999	16.374	14.281
	P_{G11}	29.999	11.751	11.934	27.124	10.119	10.000
	P_{G13}	31.370	12.000	12.019	26.273	12.000	12.000
	P_{L7}	–	–	–	−0.0001	−0.001	−0.001
Voltage (V)	V_1	1.063	1.1000	1.1000	1.062	1.0910	1.1000
	V_2	1.058	1.0843	1.0872	1.057	1.0720	1.0890
	V_3	1.050	1.0614	1.0678	1.050	1.0504	1.0708
	V_4	1.046	1.0550	1.0628	1.047	1.0433	1.0662
	V_5	1.034	1.0541	1.0587	1.036	1.0385	1.0620
	V_6	1.045	1.0567	1.0639	1.046	1.0449	1.0684
	V_7	1.033	1.0472	1.0534	1.043	1.0452	1.0676
	V_8	1.049	1.0645	1.0683	1.050	1.0493	1.0716
	V_9	1.047	1.0736	1.0925	1.048	1.0612	1.0946
	V_{10}	1.022	1.0667	1.0920	1.023	1.0524	1.0937
	V_{11}	1.100	1.0818	1.0999	1.100	1.0726	1.1000
	V_{12}	1.034	1.0797	1.1073	1.032	1.0650	1.1000
	V_{13}	1.055	1.0673	1.1000	1.051	1.0535	1.0998
	V_{14}	1.019	1.0678	1.0968	1.017	1.0525	1.0979
	V_{15}	1.014	1.0655	1.0954	1.013	1.0498	1.0964
	V_{16}	1.022	1.0684	1.0957	1.021	1.0540	1.0970
	V_{17}	1.017	1.0640	1.0908	1.017	1.0499	1.0924
	V_{18}	1.004	1.0568	1.0861	1.004	1.0402	1.0873
	V_{19}	1.002	1.0549	1.0838	1.002	1.0378	1.0850
	V_{20}	1.006	1.0589	1.0874	1.006	1.0416	1.0887
	V_{21}	1.009	1.0580	1.0850	1.010	1.0438	1.0865
	V_{22}	1.009	1.0585	1.0856	1.010	1.0442	1.0870
	V_{23}	1.002	1.0585	1.0909	1.001	1.0424	1.0911
	V_{24}	0.995	1.0516	1.0810	0.995	1.0366	1.0819
	V_{25}	0.999	1.0587	1.0822	1.000	1.0432	1.0829
	V_{26}	0.981	1.0418	1.0657	0.982	1.0261	1.0665
	V_{27}	1.010	1.0714	1.0910	1.012	1.0558	1.0917
	V_{28}	1.041	1.0542	1.0621	1.042	1.0413	1.0659
	V_{29}	0.990	1.0609	1.0843	0.992	1.0431	1.0823
	V_{30}	0.978	1.0465	1.0687	0.980	1.0293	1.0678
System generation cost ($/h)		969.91	799.52	798.91	866.32	718.52	717.31
System losses (MW)		3.99	9.2539	9.0549	3.35	8.0658	7.8387

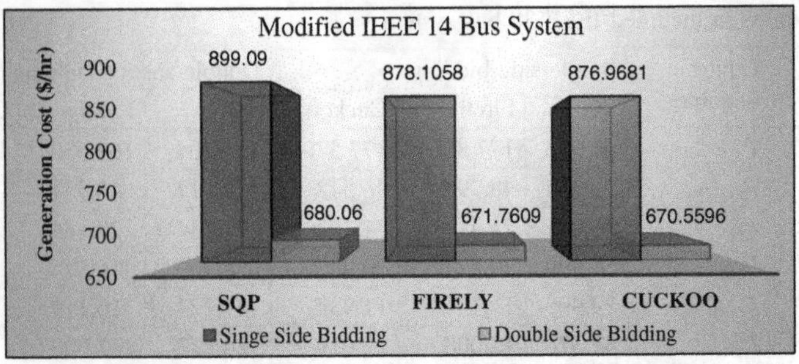

Fig. 1 Comparative study of generation cost in modified IEEE 14 bus system

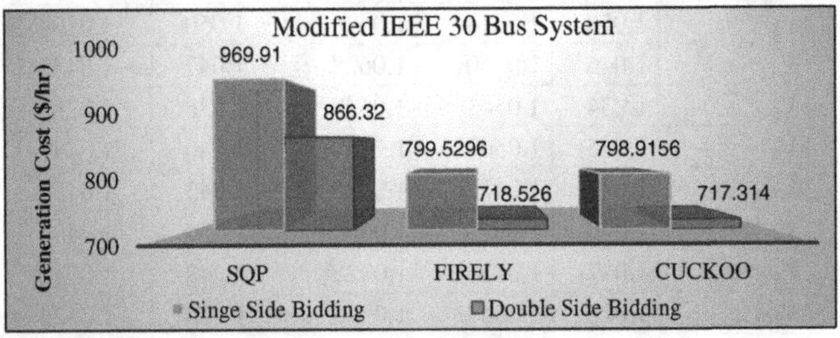

Fig. 2 Comparative study of generation cost in IEEE 30 bus system

Figures 1 and 2 show the comparative study of generation cost using the optimizing techniques: Sequential quadratic programming, firefly algorithm, and cuckoo search algorithm in singe-side bidding and double-side bidding. We can conclude from Figs. 1 and 2 that, after using the cuckoo search algorithm, generation cost is going to the minimum point in both modified IEEE 14 bus system and modified IEEE 30 bus system cases.

5 Conclusion

This paper presents an optimization strategy for minimization of generation cost of an electrical system in a fully deregulated power market. Firefly algorithm and cuckoo search algorithm have been incorporated to optimize the generation cost. Proposed techniques are applied to the modified IEEE 14 bus and modified IEEE 30 bus test systems. The test results prove the effectiveness of the proposed techniques. System losses are minimized after using the optimization strategy. Voltage has been also improved after applying firefly algorithm and cuckoo search algorithm in double-side bidding. This proposed approach is a generalized approach which can be applied to any small and large integrated as well as deregulated power system.

Appendix

Parameters of firefly and cuckoo search algorithm:

Sl. no.	Firefly algorithm		Sl. no.	Cuckoo search algorithm	
	Parameter	Value		Parameter	Value
1	No of firefly	20	1	No of nest	25
2	Randomness	0.25	2	Discovery rate	0.25
3	Firefly attractiveness	0.20	3	No of iteration	300
4	Absorption co-efficient	1			
5	No of iteration	300			

References

1. Shareef, H., Ibrahim, A.A., Salman, N., Mohamed, A., Ling, W.: Power quality and reliability enhancement in distribution systems via optimum network reconfiguration by using quantum firefly algorithm. Electr. Power Energy Syst. **58**, 160–169 (2014)
2. Farhoodnea, M., Mohamed, A., Shareef, H., Zayandehroodi, H., Optimum placement of active power conditioners by a dynamic discrete firefly algorithm to mitigate the negative power quality effects of renewable energy-based generators. Electr. Power Energy Syst. **61**, 305–317 (2014)
3. Naidu, K., Mokhlis, H., Bakar, A.H.A., Terzija, V., Illias, H.A.: Application of firefly algorithm with online wavelet filter in automatic generation control of an interconnected reheat thermal power system. Electr. Power Energy Syst. **63**, 401–413 (2014)
4. Farhoodnea, M., Mohamed, A., Shareef, H., Zayandehroodi, H.: Optimum placement of active power conditioner in distribution systems using improved discrete firefly algorithm for power quality enhancement. Appl. Soft Comput. **23**, 249–258 (2014)
5. Massan, S., Wagan, A.I., Shaikh, M.M., Abro, R.: Wind turbine micro-siting by using the firefly algorithm. Appl. Soft Comput. **27**, 450–456 (2015)
6. Nguyen, T.T., Truong, A.V.: Distribution network reconfiguration for power loss minimization and voltage profile improvement using cuckoo search algorithm. Electr. Power Energy Syst. **68**, 233–242 (2015)
7. Garg, H.: An approach for solving constrained reliability redundancy allocation problems using cuckoo search algorithm. Beni-Suef Univ. J. Basic Appl. Sci. **4**, 14–25 (2015)
8. Nguyen, T.T., Vo, D.N.: Modified cuckoo search algorithm for short-term hydrothermal scheduling. Electr. Power Energy Syst. **65**, 271–281 (2015)
9. Yang, X.-S., Hosseini, S.S.S., Gandomi, A.H.: Firefly Algorithm for solving non-convex economic dispatch problem with valve loading effect. Appl. Soft Comput. **12**, 1180–1186 (2012)
10. Yang, X.-S.: Multi objective firefly algorithm for continuous optimization. Eng. Comput. **29**, 175–184 (2013)
11. Yang, X.-S., Deb, S.: Engineering optimization by cuckoo search. Int. J. Math. Model. Numer. Optim. **1**(4) (2010)
12. Mahdad B., Srairi K., Bouktir T.: Optimal power flow for large-scale power system with shunt FACTS using efficient parallel GA. Electr. Power Energy Syst. **32**, 507–517 (2012)
13. Zimerman R.D., Murillo-Sanchez C.E., Gam D.: MATPOWER—A MATLAB Power System Simulation Package, Version 3. www.pserc.cornell.edu/matpower

Optimized Point Robot Path Planning in Cluttered Environment Using GA

Motahar Reza, Saroj K. Satapathy, Subhashree Pattnaik and Deepak R. Panda

Abstract In this paper, an optimized path planning for mobile robot by using genetic algorithm is analyzed. A hybrid method based on the visible midpoint and genetic algorithm is implemented for finding optimal shortest path for a mobile robot. The combination of both the algorithms provides a better solution in case of shortest and safest path. Here the visible approach is efficient for avoiding local minima and generates the paths which are always lying on free trajectories. Genetic algorithm optimizes the path and provides the shortest route from source to destination.

Keywords Genetic Algorithm (GA) · Robot path planning · Cluttered environment

1 Introduction

Path planning is an important and challenging problem in mobile robotics. A mobile robot must be able to search collision-free paths to navigate from a start position to a goal position in an environment with barriers. Paths generated by robot must be optimized under some certain criteria for the smooth running of the robot in the environment. These criteria may be shortest path, smoothest path, or safest path. From the mid-80s, different predictable methods have been developed to solve the robot navigation problem [1], such as global C-space methods [2], potential field

Motahar Reza (✉) · S.K. Satapathy · Subhashree Pattnaik · D.R. Panda
High Performance Computing Lab, School of Computer Science and Engineering,
National Institute of Science and Technology, Berhampur 761008, Odisha, India
e-mail: reza@nist.edu

S.K. Satapathy
e-mail: saroj@nist.edu

Subhashree Pattnaik
e-mail: pattnaiksubhashree91@gmail.com

D.R. Panda
e-mail: deepakpanda93@gmail.com

M. Pant et al. (eds.), *Proceedings of Fifth International Conference on Soft Computing for Problem Solving*, Advances in Intelligent Systems and Computing 436, DOI 10.1007/978-981-10-0448-3_39

methods [1], and neural networks approaches [3]. Path planning problem can be solved by different conservative algorithms, such as the potential field, Road Map cell decomposition [1]. Almost all approaches were based on the concept of configuration space. These approaches show lack of adaptation and nonrobust activities. So each method has its own strength over others in certain aspects, but also has some drawbacks. To get rid from the limitation of these procedures, researchers explored variety of solutions. In [4], Fuzzy Logic and Neural Network approach is implemented to handle the path planning problem. Genetic Algorithm (GA) is proposed in 1975 [5], and can be employed to control a robot moving in a configuration space which has static obstacles and/or dynamic obstacle [6, 7].

GA has a greater impact on this field and this algorithm (GA) with parallel search technique is used for finding the optimal solution of this kind of problem in dynamics environments [7, 8]. It is also applied to optimize the path of a mobile robot moving in an environment which has number of static barriers. Some of the proposed methodologies in [9] suffer from different type of problems. They take (1) high computational cost (2) large memory spaces for dealing the large-size dynamic environments (3) more time-consuming. In the last years, scientists are using the genetic algorithms widely for finding the optimum path by intriguing the advantage of its potential optimization capability. This work is aggravated by our earlier work presented in [10]. In this study, we provide an initial idea-based genetic algorithm along with midpoint algorithm to select the shortest path in cluttered environment which will be able to handle static obstacles.

As per above discussion, in this paper, we propose a procedure to find an optimized path for robot in a cluttered environment by using GA. This paper is organized as follows. Section 2 introduces problem statement for robot path planning (RPP). Section 3 discusses the proposed methodology for RPP. Section 4 presents the simulation results and demonstrates the solution for RPP. Finally, Sect. 5 discusses some conclusions and considers possible future work.

2 Problem Description

The robot navigation planning problem is usually described as follows: a mobile robot is taken as a point robot and a complete description of configuration space which is two-dimensional by nature. We must have to design an obstacle-free path between the two assigned locations, a source and a destination. The path should be collision free and satisfies certain optimization criteria (i.e., shortest path). According to the above discussion, path planning problem is categorized as an optimization problem. Researchers use various methods to solve the path planning problem as per the two common factors, the environment type (i.e., static or dynamic) and path planning algorithms (i.e., global or local). The Work space used for the problem is a static environment. It is considered that no objects other than robot are moving in the static environment. A complete awareness about the search space and that all terrains of the space should be static, is required for implementing

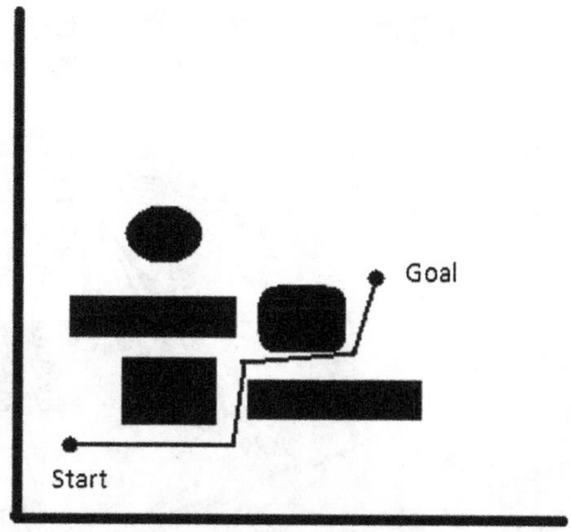

Fig. 1 A model for robot path planning

global path planning algorithms. Local path planning suggests that path planning is being implemented while the robot starts its navigation; in other words, these algorithms provide the collision-free movement in local region of robot. Figure 1 defines a model in which the black color objects are the obstacles in a two-dimensional area with a start and goal point. The robot must move from start point to goal point without collision.

3 Proposed Methodology

In our proposed methodology, we have followed two steps to provide an optimized result. (1) Implement visible midpoint approach and (2) implement genetic algorithm to optimize the path. First step is to generate a graph in which all paths lie on free trajectories available in configuration space. And in second, the optimized path can be found among the available path by using genetic algorithm.

3.1 Visible Midpoint Approach

Visible midpoint approach is a type of classical path planning which provides a graph consists of all collision-free paths in the configuration space. This approach provides two important activities:

1. This approach avoids the local minima in the configuration space. The local minima is such a case where no results can be found in the configuration space.
2. The paths generated in the configuration space are the safest path because it does not touch the obstacle lines.

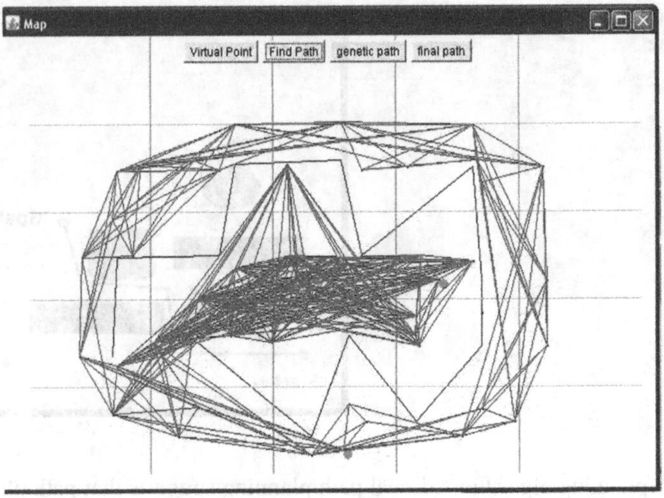

Fig. 2 Visible midpoint output of line obstacle

The cluttered terrains are the main reasons to put the robot in local minima situation. Local minima can be easily avoided by using visible midpoint algorithm because in each cluttered environment, there must be some paths which can help robot to overcome the cluttered environment. The main focus of this algorithm is the generation of midpoints in free trajectories. The midpoints are generated according to the obstacles present in the environment. The paths are generated by connecting all the visible midpoints available in configuration space and should not touch any obstacles present in the environment. Figure 2 represents the graph for line obstacle by using said algorithm and Fig. 3 represents the graph for square-type obstacles by the help of visible midpoint algorithm.

3.2 *Optimized Path by Genetic Algorithm*

This step gives the optimum result for path planning. The methodology takes the primary population from the result provided by midpoint algorithm. The population contains number of individuals (i.e., chromosomes). Each candidate from the population represents a solution for the given problem under study. In our case, each solution is in fact a path from source to destination in the search space. The entire graph is decomposed into group of cells. Each cell consists of a number of nodes and the information related to each cell is stored in a two-dimensional matrix. The data for a single cell consists of the number of nodes it is having along with the neighbor cells. Then the minimum edge connecting one cell to all other cells is stored in another adjacent matrix. Then the cell where the source and destination node is present is determined by traversing through the entire cell matrix.

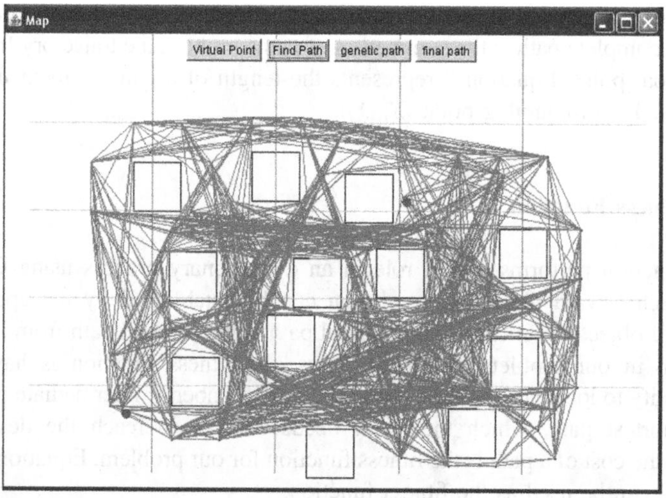

Fig. 3 Visible midpoint output of square obstacle

Then starting from the source node another node is determined by taking the minimum of all the nodes connected to it. When a node is selected for the path its duplicity is also checked with previously determined paths. If the node finds duplicity the node is discarded and other neighboring nodes are checked. Once the condition is satisfied the nodes are included in the path. The path is determined by connecting nodes to nodes and repeating the above procedure till the destination node is found. The entire process is repeated a number of times to get a desired number of paths (10–20 path) as initial population. Each path now represents a chromosome which is further optimized with the application of genetic algorithm.

3.2.1 Initialization

The size of initial population is n and can be taken for the implementation purpose. Therefore the chromosomes are represented as $\{C_1, C_2 \ldots C_n\}$. Each chromosome has its own fitness value and the fitness value is calculated as the sum of paths from start to goal. This means a path is the collection of paths from start to goal and can be represented as:

$$P = \sum_{i=1}^{n} pi \tag{1}$$

$$pi = \sqrt{(x_{i+1} - x_i)^2 - (y_{i+1} - y_i)^2} \tag{2}$$

The Eq. 1 represents the total length of path from a start point to goal point. Because a complete path is the summation of paths in the same trajectory from start point to goal point. Equation 2 represents the length of a path segment from one point (X_{i+1}, Y_{i+1}) to another point (X_i, Y_i).

3.2.2 Fitness Function

Fitness function performs a vital role in an evolutionary process using GA. The quick search toward the optimal solution can be established by the appropriate selection of objective function. The optimal path is the shortest path from source to destination in our problem statement. Thus, the fitness function is having the responsibility to locate the optimal path. The total number of intermediate path may be the shortest path, which the robot needs to take to reach the destination. Therefore the cost of a path is the fitness function for our problem. Equation 1 along with Eq. 2 can be used as the fitness function.

3.2.3 Crossover

Then the crossover algorithm is applied to the pair of chromosomes. The crossover is done by determining the crossover points for each pair. The crossover points are calculated by taking the cell common to both the chromosomes. The entire cell matrix is again traversed to find the cell common to both the chromosomes excluding the source cell and destination cell.

After determining the crossover point four nodes are determined as s1, d1, s2, d2 where s1 and d1 are the successor nodes and predecessor node, respectively, connected to the common cell's node of first chromosome and s2 and d2 are the successor node and predecessor node, respectively, connected to the common cell's node of second chromosome of a pair. The crossover node is determined to carry out the crossover procedure. The crossover node of the first child chromosome is determined by checking the existence of path from s1 as well as from d2 to the nodes of the common cell of second chromosome. If the crossover node is found then the first crossover path is calculated. The first crossover path is calculated by taking the source node till the successor node of first chromosome (s1) then crossover node and finally the predecessor node till the destination node of second chromosome (d2).

The crossover node of the second child chromosome is determined by checking the existence of path from s2 as well as from d2 to the node of the common cell of first chromosome. If the crossover node is found then the second crossover path is calculated. Then the second crossover is done by taking the source node till the successor node of second chromosome (s2) then crossover cell node and finally the predecessor node till the destination node of first chromosome (d1). Figure 4 represents the chromosome selected for crossover. Figure 5 represents the new chromosomes generated through the help of crossover step.

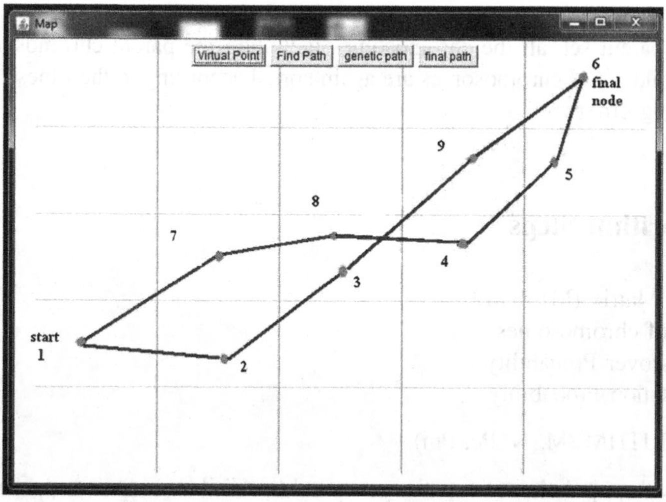

Fig. 4 Crossover step 1 output

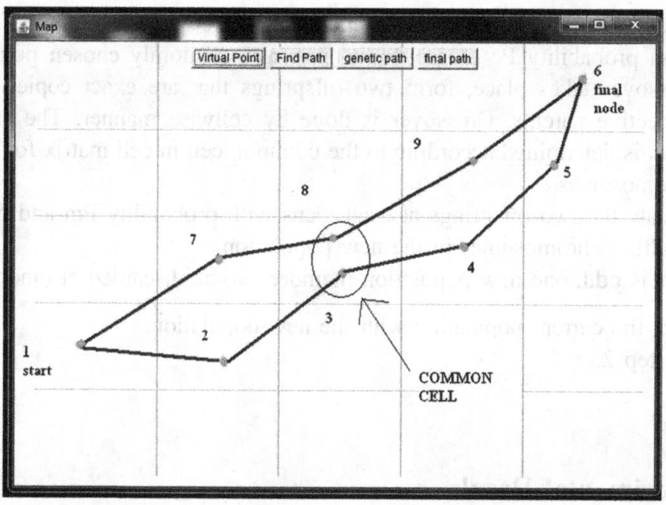

Fig. 5 Crossover step 2 output

3.2.4 Mutation

After getting the child chromosome pair the mutation operator is applied on the chromosomes. The nodes are interchanged in between the chromosomes to determine the new child chromosomes after mutation. Before the nodes are interchanged the existence of path for the new node is also checked. Then this crossover and mutation method is applied to rest other pairs of chromosomes. After crossover of the entire parent set, all the chromosomes including the parent chromosomes and the new child set of chromosomes are again sorted according to the fitness function in ascending order.

4 Algorithm Steps

\\M<-Cell Matrix (M1 * M2)
\\N<-No. of chromosomes
\\P_c<-Crossover Probability
\\P_m<-Mutation Probability

ALGORITHM (M, N, Pc, Pm)

1. Start with population of n paths generated by Visible midpoint approach
2. Calculate the path distance from source to destination using Eqs. 1 and 2 of each chromosome N_i in the population.
3. Repeat the following steps until N offsprings have been created:
   a. Select a pair of parent chromosomes from the current population.
   b. With probability Pc crossover the pair at a randomly chosen point. If no crossover takes place, form two offsprings that are exact copies of their respective parents. Crossover is done by cellwise manner. The crossover point is determined according to the common cell in cell matrix for both the chromosomes.
   c. Mutate the two offsprings at each locus with probability Pm and place the resulting chromosomes in the new population.
   If N is odd, one new population member can be discarded at random.
4. Replace the current population with the new population.
5. Go to step 2.

5 Experimental Result

We have implemented both the approach with the same machine configuration that is PC with Intel Core 2 CPU, 1.86 GHz. Movement space for robot is taken as 700 * 500 pixels. For our experiment, the graph is divided into group of 25 cells (5 * 5). Figure 6 represents the first step of proposed algorithm. In this experiment we have taken the obstacles as Square type. Figure 7 represents the second step that is optimized path from start point to goal point by using GA. Figure 8 represents midpoint based visibility graph. In this experiment we have taken the obstacles as

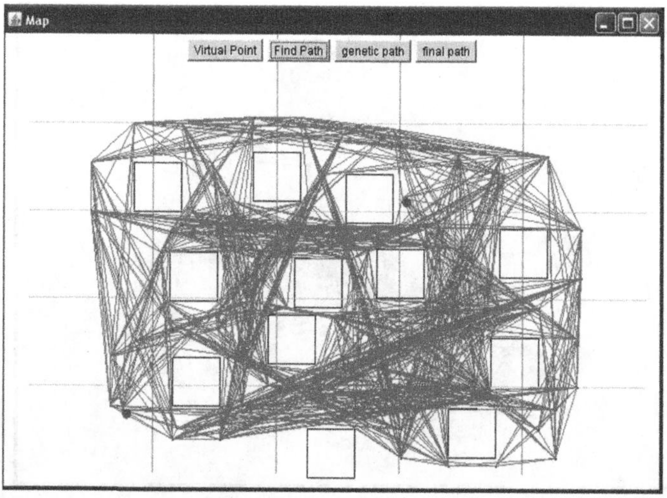

Fig. 6 Visible midpoint output for squared obstacles

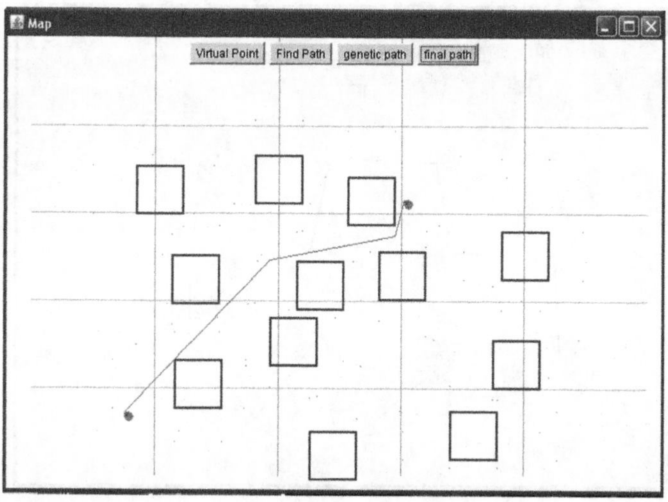

Fig. 7 Optimized path by GA for squared obstacles

Line which generates the Local Minima. Figure 7 represents the genetic algorithm path from start point to goal point. In our experiment we have taken only 20 chromosomes for 50th iterations where the chromosomes belong to different regions. Figure 10 shows the optimized value which is the distance from start to goal with respect to the number of iteration. After 17th iteration, the fitness value is not changed further (Fig. 9).

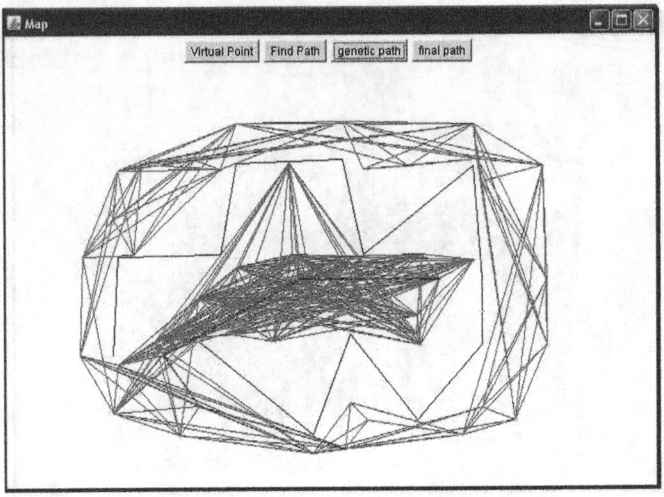

Fig. 8 Visible midpoint output for line obstacles

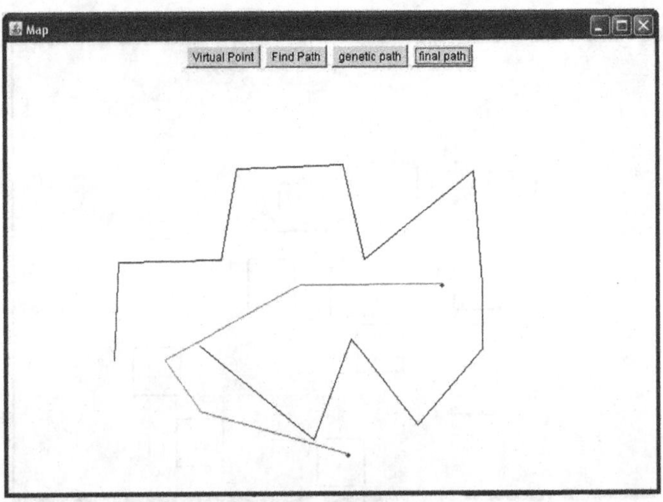

Fig. 9 Optimized path By GA for line obstacles

Fig. 10 Fitness value versus iteration

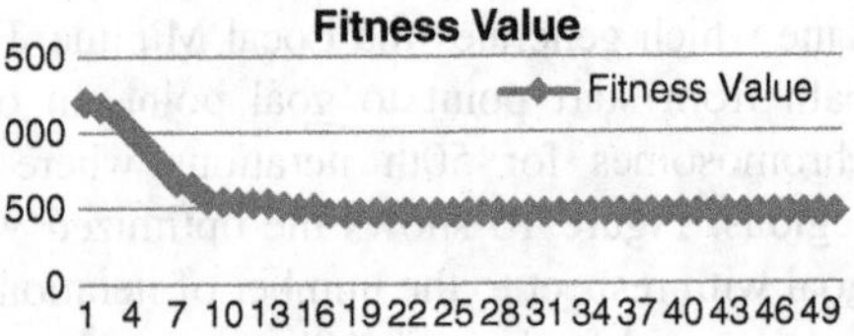

6 Conclusion

Genetic Algorithm is easy to apply on path planning problem because the path planning problem is a quite related to traveling Sales problem. The result, what we get from this implementation is effective as compared to classical path planning problem because the time complexity of classical path planning problem with DIJkstra's algorithm is very much high. The problem converges quickly within a less number of iterations. But we have seen that the mutation does not put any drastic changes on the result. The cell-to-cell crossover makes the algorithm significantly good. The ultimate objective of developing a fully independent robot is beyond the scope of the research carried out for this dissertation. This dissertation necessarily focuses on a more specific component of an independent robot.

References

1. Latombe, J.C.: Robot motion planning. Kluwer Academic Publisher, Boston (1991)
2. Sharir, M.: Algorithmic motion planning in robotics, Computer **22**(3), 9–19 (1989)
3. Zhong, Y., Shirinzadeh, B., Tian, Y.: A new neural network for robot path planning. In: Proceedings of Advanced Intelligent Mechatronics, IEEE/ASME International Conference on AIM 2008, pp. 1361–1366. Xian, 2–5 July 2008
4. Huan, Z., Zhiguo, S., Xiancui, W.: A rapid path planning adaptive optimization algorithm based on fuzzy neural network for multi-robot systems. IEEE Conference on CyberSpace Technology, pp. 32–38. Beijing, 23 Nov 2013
5. Holland, J.: Adaptation in natural and artificial systems. Ann Arbor, University of Michigan Press (1975)
6. Lee, J., Kang, B.-Y., Kim, D.-W.: Fast genetic algorithm for robot path planning. Electron. Lett. **49**, 1449–1451 (2013)
7. Davis, L.: Handbook of genetic algorithms, VNR Comp. Library, New York (1990)
8. Dorigo, M., Schnepf, U.: Genetic-based machine learning and behavior based robotics: a new synthesis, IEEE Trans. Syst. Man Cybern. **23**(1), 141–154 (1993)
9. Ismail, A.-T., Sheta, A., Al-Weshah, M.: A mobile robot path planning using genetic algorithm in static environment. J. Comput. Sci. **4**(4), 341–344 (2008)
10. Satapathy, S.K., Reza, M.: A visible midpoint approach for point robot path planning in cluttered environment, pp. 1–4. NCCCS, Durgapur, 21–22 Nov 2012

ITMS (Intelligent Traffic Management System)

Rahul Kumar and Kunal Gupta

Abstract In the present work, an ITMS (Intelligent Traffic Management System) is used for managing and controlling traffic lights based on photoelectric sensors placed on one side of the road. The suitable space between each sensor is selected by the traffic control authority. As a result, the traffic control authority can supervise vehicle that are running toward a particular traffic direction and can thereby manage transmission of information signal to microcontrollers which are fixed to the traffic control cabinet. An Arduino is a microcontroller which is capable of managing the traffic signals by using the information sent by the infrared sensors. In case of emergency, this system can pass the ministries vehicles, ambulance, and fire brigade buses that oblige urgent opening, i.e., clearance from traffic signal system by using RFID-based technology.

Keywords Microcontroller (Arduino) · IR sensor · RFID · Traffic · Traffic control cabinet

1 Introduction

In the present work, our endeavor is to present a blueprint of an ITMS (Intelligent Traffic Management System) for managing and run the traffic signals using Infrared Sensors (IR Sensors) positioned at before and subsequent to the traffic lights. The appropriate space between each sensor is selected by the traffic control authority. As a result, they can supervise vehicles that are running toward a particular traffic

Rahul Kumar (✉) · Kunal Gupta
Department of CS&E, Amity University, Noida, Uttar Pradesh, India
e-mail: rahul.rkverma@gmail.com

Kunal Gupta
e-mail: kgupta@amity.edu

M. Pant et al. (eds.), *Proceedings of Fifth International Conference on Soft Computing for Problem Solving*, Advances in Intelligent Systems and Computing 436, DOI 10.1007/978-981-10-0448-3_40

Fig. 1 Intelligent traffic management system

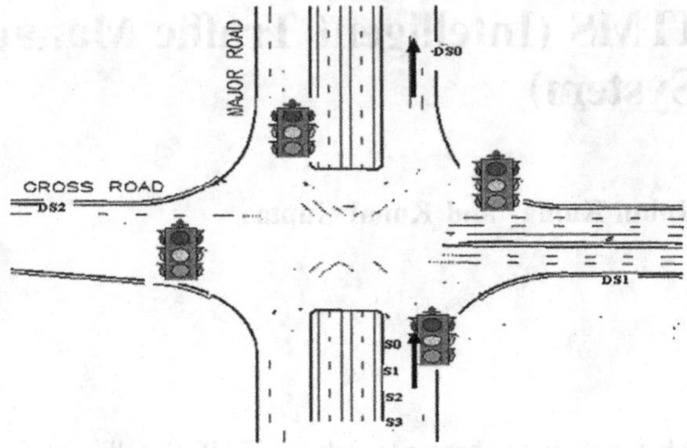

direction, and consequently, are able to transmit information signal to microcontrollers which are fixed to the traffic control cabinet. Microcontroller is capable of managing the traffic signal lights according to the information sent by the infrared sensors [1, 2].

This arrangement is in a position to release the vehicles that are jam-packed and thus provide it an extended time larger than the specified time to the traffics. This system is programmed in such a way that facilitates the user to have better judgment for smart automatic management of traffic signals. In case of emergency, this system can pass the ministries, ambulance, and fire brigade buses that oblige urgent opening, i.e., clearance from traffic signal system by using Radio-frequency identification (RFID) based technology [3]. The projected arrangement can be flow of traffic without any individual interference or be able to adjust to permit individual intercession at assured situation.

The traffic management authority sends this information to any commercial company to circulate the status of the traffic jams at particular location at minimum cost to the daily goers. ITMS automatically switches to the usual mode while there is no necessity for operating the smart traffic management arrangement, and vice versa. The suppleness of the projected arrangement to be programmed with changeable periods of time that goes with numerous levels of traffic blockage and requirements (Fig. 1).

2 Related Work

A systematic literature survey was done regarding the research done on this particular topic till date in order to get apprised and collect all the facts, data, and status as well as quality of existing system and problems encountered by the existing system. It was accomplished by the deep study of research papers, theses, and application-based books on the topic. In this connection, more than 45 research papers were searched, out of which 15 were reckoned as the most relevant and significant for our purpose. The analysis for literature survey with the selected papers may be specified as below:

In [4], Engelbrecht et al. focus on the conception of hardware in the loop traffic simulation and describe how it can be used to estimate the consequence of traffic signal management on traffic routine. Their paper concisely narrates the utilization of simulation-based traffic control assessment as an alternative to field survey-based estimation. The existing traffic lights are generally controlled by solid state microprocessor-based traffic signal controllers kept inside the cabinet on the side of the road near the intersection. The hardware in the loop simulation is a computer simulation in which some of the components can be replaced with actual hardware. It has been successfully used in aerospace and defense industries. This model runs in a real time and can produce a detector which is operated by modeling simulated vehicles crossing simulated detectors. This detector is then able to send information to the controller which responds to them by bringing up-to-date phase indications according to the phasing and timing plan programmed in the controller. The phase indications are read from the controller for the proposed simulation and allocated to the simulated traffic signals. The simulated traffic signals provide ways and means for stopping or allowing the simulated vehicles through the test traffic system. The methodology and guidelines in this paper allow the traffic engineer, analyst, or researcher to perform efficient evaluations of traffic controller characteristic.

In [5], Yang et al. focus over the modern roundabouts with traffic signal to resolve the disorder of the traffic causes when more than two lanes meet together at a place. Traffic signal controls the run of vehicles and takes a left turn traffic run on circulatory roadway. The left turn vehicle on roundabout would be stopped before the red signal of traffic system. This new method solves the serious traffic congestion problem in Xiamen City. A roundabout structure removes vehicle crossing clashes by changing all arrangements to right turn. Left turn vehicles would be stopped before red signal till the green signal is on in the circulatory roadway. In their paper, Traffic Flow of the whole day is divided into 3–6 periods of time for signal control. Sensor or manual accounting data can be acquired to measure the traffic volume. Then the sum of the run ratios of vehicles is computed, which is maximal for each one of the phases that have the green time interval. By using this new technique, the green signal time along with the operation path of circulatory roadway or roundabout is employed in the best possible ways. This technique is thus capable of improving the capacity and travel safety.

In [6], Ghaman et al. focus on the traffic lights simulation software development at Traffic Research Laboratory. A traffic simulation is incorporated to build up, test, remove the bugs, and estimate the software of traffic light system. In their paper, the possibility of the use of simulation software and traffic light control tool and interface between them have been explored. An ACS (Adaptive Control System) framework is used in Federal Highway Administration (FHWA) underneath the Intelligent Transportation System (ITS). This framework supports five prototypes for theoretical assessment. The first model is OPAC (Optimization Policies for Adaptive Control), second one is ARTS (Adaptive Real-Time Traffic Control System), third one is RHODES (Real-Time Hierarchical Optimization Distributed

Effective System), fourth one is RTACL (Real-Time Adaptive Control Logic), and fifth or the last one is AFT (Adaptive Fixed Time). In their paper, different methods of simulation such as CORSIM, NGSIM, VISSIM, and ACS-Lite have been used to fabricate the entire traffic control system. This arrangement is enviable to incorporate simulation of traffic and software to control traffic signal.

In [7], Quan et al. focus over the existing traffic signal system working in India. It only gives instructions for stopping during running of the vehicles. Smart traffic signal functioning based on the microcontroller and ultrasonic sensor have been proposed for use in this system in which Ultrasonic Sensors are placed at one side of road in such a way so as to cover required area of road from where the vehicles are restricted to pass. In case when the traffic signal is red and any vehicle is to cross the signal, then ultrasonic sensor detects it and microcontroller takes a quick action to buzzer alarm along with camera capture the image of that vehicle. The proposed system for smart traffic signal using ultrasonic sensor and controller is advantageous to existing systems. The sensor creates a secure zone at road to meet at junction and its interfacing with controller and MATLAB makes it easier to implement. It is an inexpensive and does not require a system to be installed inside the vehicle.

In [8], Kanungo et al. focus on the situation in which, by chance, many vehicles congregate on a constrained road. The traffic light system cycle is now bound to create an end to the confirmed delay time and startup wait time. The usual traffic cycle, 30–45 % does not match with current traffic cycle. Here the researchers propose electrosensitive traffic light using fuzzy look-up table technique in order to diminish the average vehicle wait time and get better average vehicle pace. Traffic simulation outcome demonstrates the dropping of average vehicle wait time. The passage of vehicles for most favorable traffic phase is thus improved than that in permanent signal technique. The proposed A.I. traffic simulation controller system has been put into action using look-up table technique and examined with numerous types of traffic states. With fuzzy controller, the average waiting time could be reduced by 15 % as compared to that measured with the usual controller. The fuzzy controller simulation very well matched with waiting time of T.O.D. signal light and fuzzy traffic light. Conclusively, traffic simulation substantiates that vehicle waiting time gets enhanced by 10–15 % even in case of large vehicles' unexpected entry.

3 Research Problems and Objectives

3.1 Problems

After a thorough review of the operational traffic light system in numerous places inside busy metropolitan cities [9–15], a number of encountered troubles have been acknowledged which are summarized as below:

- We have seen traffic congestion occurring at main crossings, particularly in peak time, that leads to prolonged wait time, and more fuel consumption.
- Manual control by the traffic official occasionally solves the traffic congestion problem particularly at the junction point.
- Existing traffic light management system does not provide longer duration for the green light to a certain traffic lane than to other ones to clear the over-crowded traffics.
- Till now, no data have so far been procured to provide priority to the traffic lane.

3.2 *Objective of the Present Work*

Our purpose is to:

- Provide a method for traffic management system to alleviate or overcome the problem of traffic jam by way of lessening the wait time at traffic signal.
- Eliminate the current manual traffic system and automate the control of traffic lights.
- Provide useful information to traffic control authority/department for assessment and better controlling of the traffic.
- Assure a clear path in case of emergency such as for fire brigade buses and ambulance cars.

4 Projected Intelligent Traffic Management System Model

The important modules of the projected system are as follows:

(a) *Central Traffic Control System* This component is set up at the traffic location with a microcontroller. A microcontroller is used as a central traffic control system. An Arduino is an open-source physical computing platform based on a simple I/O board and a development environment where the developer uses the Arduino IDE to implement the logic to perform the particular task. Arduino can be used to develop a stand-alone interactive object that is connected to the software on our computer. A physical I/O board with a programmable Integrated Circuit (IC).
The main function of this system is:

- To change the regular traffic light form to the smart traffic light form and conversely by using program logic.
- To handle an emergency case such as when an ambulance or a fire brigade bus is to cross the traffic signal.
- To receive the information signal from sensor and RFID reader to change the regular traffic light form to smart traffic light form by using controller (Fig. 2).

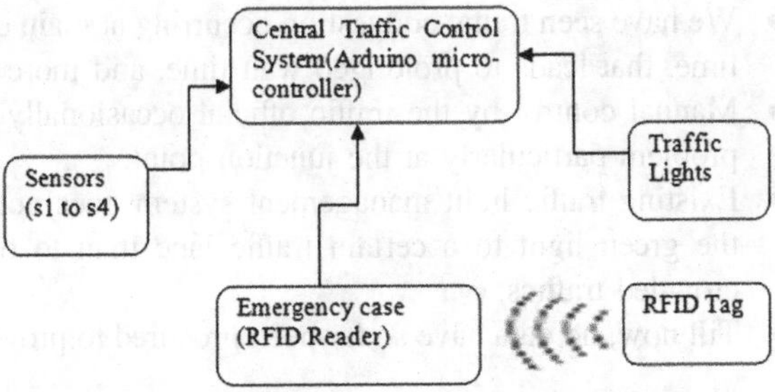

Fig. 2 The proposed intelligent traffic management system architecture

(b) *Sensors* There are two main types of infrared sensor. First is **thermal infrared detector** and the second one is **photodetector infrared detector**. In this model, photodetector sensors are used, which are positioned at suitable places to be determined by the traffic admin branch as per their need and knowledge-based suitability. Each one of the sensors (from S_1 to S_4) is allocated a proper space before traffic light. Each sensor owes only one lane side of the road.

(c) *Emergency Cases Detector* Emergency situations such as the obligation to give clearance to fire brigade buses, police force vehicles, ambulances can be managed in this projected system by one of two ways. First, setting up a Radio-Frequency Identification (RFID) tag in those particular vehicles which need such emergency exit services. Second, RFID reader that is located besides the sensor to detect the RFID tag vehicle comes into the region of RFID reader which sends the needed information to the microcontroller to take the correct accomplishment.

(d) *Traffic Light*(*s*) The traffic signal lights are coupled with the central traffic control cabinet and perform their function according to the acknowledged commands from the microcontroller.

5 Operation Technique of ITMS

The projected Intelligent Traffic Management System (ITMS) employs an infrared sensor lying on one side of road. The infrared sensors are positioned at varying distances in the subsequent order from S_1 to S_4 represent the feasible addition to a particular path. The foremost role of these sensors is to provide traffic information about the capabilities of releasing the traffic flow on these paths. Sensors of this ITMS model send their readings to the microcontroller in the traffic signal cabinet. The central traffic control system sends the message to the traffic management authority about the traffic congestion and wait time at traffic light. The traffic management authorities send this information to any commercial company to circulate the status of the traffic jams at particular location at minimum cost to the daily goers.

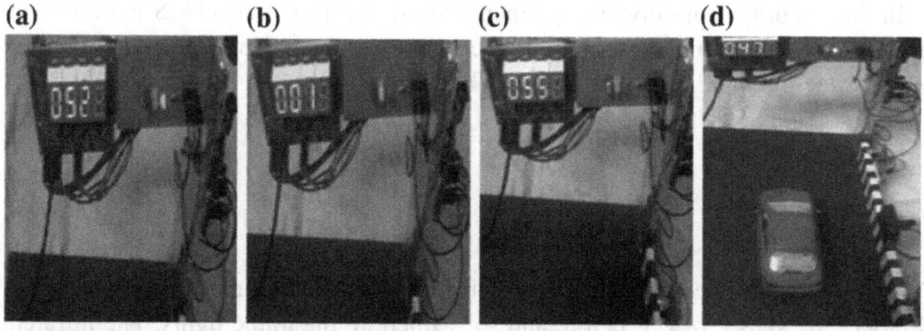

Fig. 3 Some screenshots of the ITMS operation: **a** the *green light* of traffic light glows when normal mode operation; **b** the *yellow light* of traffic light glows in under normal mode operation; **c** the *red light* of traffic light glows in under normal mode operation; **d** glow of *green traffic light* when sensor S_4 or RFID reader detects the special vehicle (color online)

Fig. 4 Emergency case activity diagram

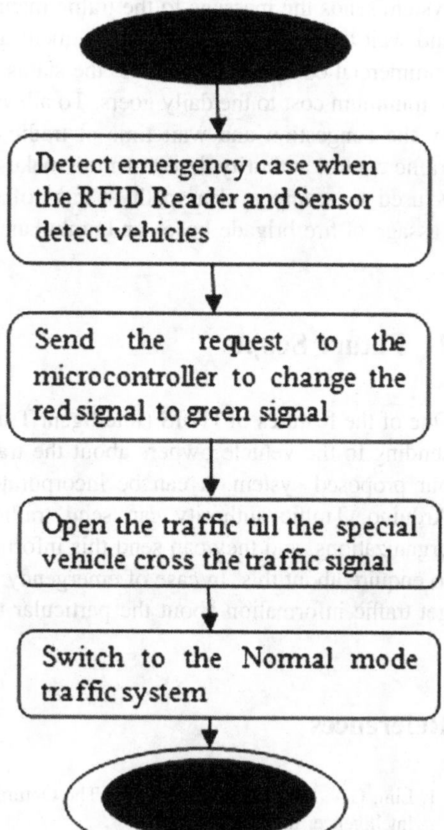

In this section, a prototype implementation of the proposed ITMS is described and validated. A simple proof of the concept has been developed in order to demonstrate the validity of our System (Figs. 3 and 4).

6 Conclusion

The present work encompasses the Intelligent Traffic Management System using a photoelectric sensor placed before and subsequent to the traffic lights. The infrared sensors are positioned at varying distances in the subsequent order from S_1 to S_4, thus representing the feasible addition to a particular path. The foremost role of these sensors is to provide traffic information about the capabilities of releasing the traffic flow on these paths. Sensors of this ITMS model send their readings to the microcontroller installed at the traffic signal cabinet. The central traffic control system sends the message to the traffic management authority about the congestion and wait time at traffic. The management authorities send this information to any commercial company to circulate the status of the traffic jams at particular location at minimum cost to the daily goers. To alleviate or circumvent the problems created by the congestion and wait time at traffic, ITMS provides useful information to traffic control authority department to make decisions and monitor the traffic. ITMS is used to assure a clear path in case of emergency such as the obligation for passage of fire brigade buses and ambulance cars, etc., through the traffic.

7 Future Scope

One of the features of ITMS (Intelligent Traffic management system), i.e., message sending to the vehicle owners about the traffic information is not implemented in our proposed system. It can be incorporated using GSM Shield connected with Arduino. Traffic authority can send traffic jam information to the commercial organizations, and they can send this information to the vehicle owner who wishes to enquire about this. In case of emergency, registered government official vehicles get traffic information about the particular traffic location automatically.

References

1. Lim, G.Y., Kang, J.J., Yousik, H.: The Optimization of Traffic Signal Light using Artificial Intelligence, IEEE (2001)
2. Srinivasan, D., Choy, M.C., Cheu, R.L.: Neural Networks for Real-Time Traffic Signal Control, IEEE (2006)
3. Karygiannis, A., Phillips, T., Tsibertzopoulos, A.: RFID Security: A Taxonomy of Risk, IEEE (2006)

4. R. Engelbrecht, Using Hardware-in-the-Loop Traffic Simulation to Evaluate Traffic Signal Controller Features, IEEE (2001)
5. Yang, X., Li, X.: New Method of Traffic Signal control for Modern Roundabout, IEEE (2003)
6. Ghaman, R.S., Zhang, L., McHale, G., Stallard, C.: The Role of Traffic Simulation in Traffic Signal Control System Development", IEEE, 2003.
7. Quan, Y., Jin-guang, L., Pei-hua, L., Jian, R., Xiaoming, L.: Dynamic optimization project study between the traffic organization and the traffic signal control of urban traffic. In: World Congress on Computer Science and Information Engineering (2009)
8. Kanungo, A., Sharma, A., Singla, C.: Smart Traffic Lights Switching and Traffic Density Calculation using Video Processing, IEEE (2014)
9. Lin, S., Chao, K., Lo, C.: Service-Oriented Dynamic Data Driven Application Systems to Traffic Signal Control, IEEE (2010)
10. Riley, D., Nellen, G., Barrera, R., Quevedo, J.: Crowd Souring Traffic Simulation to Improve Signal Timing, CCSC (2014)
11. Zhang, L., Haas, C., Tighe, S.L.: Evaluating weigh-in-motion sensing technology for traffic data collection. In: Annual Conference of Transportation Association of Canada (2007)
12. Aoyama, K.: Universal Traffic Management System (UTMS) in Japan, IEEE (1994)
13. Thanasis, Z., Ahmad, A.: A prototype of an integrated advanced traffic management system for the columbus area: an interface between the traffic signal control and the freeway management systems. IEEE (1995)
14. Ahmed, S.S., Bahaa, K.S., Mohamad, M.E.: Intelligent cross road traffic management system (ICRTMS), In: 2010 2nd International Conference on Computer Technology and Development (2010)
15. Jain, D.K., Sharma, S., Yadav, M.: An optimal and dynamic vehicular traffic management system, ICCCT (2011)

Intelligent Parking Management System Using RFID

Priyanka Singh and Kunal Gupta

Abstract In India the parking management system we have is manually controlled. This causes long queues for parking and traffic in the road in search for a free parking slot. So we need a system which saves people's time and reduces the emission from the vehicles. This can be done by the use of RFID technology in our parking management system. With RFID technology we can automate our parking management system and reduce long queues for parking vehicles. In this paper we have implemented the RFID technology into the parking management system using Arduino for connections and Visual Studio with SQL Server Management Studio for database.

Keywords Parking management · RFID technology · Arduino · Visual studio · SQL server

1 Introduction

RFID is Radio Frequency Identification (RFID) which is a wireless use of electromagnetic fields for transferring data. It is widely used in many fields for better and efficient performance. RFID is also used for automatically identifying and tracking tags attached to many objects. Currently in India the traditional parking system is implemented. In this system a lot of man power is used for identification, managing, and ticketing for parking. The traffic in India is too crowded. So an efficient and flawless system is required for managing the traffic which is caused by

Priyanka Singh (✉) · Kunal Gupta
Computer Science and Engineering, Amity University, Noida, Uttar Pradesh, India
e-mail: priyankasinghjan@yahoo.com

Kunal Gupta
e-mail: kgupta@amity.edu

M. Pant et al. (eds.), *Proceedings of Fifth International Conference on Soft Computing for Problem Solving*, Advances in Intelligent Systems and Computing 436, DOI 10.1007/978-981-10-0448-3_41

the search of a free parking slot. With the use of RFID in parking management system our system will become automatic. RFID is a fast, reliable, and does not require physical sight or contact between reader/sender and tagged object. RFID is similar to bar code identification, only difference in them is that bar code needs the tag to be in the line of sight of the reader. There are two modules for the implementation of RFID in parking management system: hardware module and software module.

Components used in hardware module are: RFID reader, RFID tag, Barriers (Motor Shield), Display device (LCD Shield), IR sensor, LED device (Basic Shield), a GSM Shield, and computer system. Components used in software module are: Visual Studio 2010 with SQL Server Management Studio.

Arduino: Arduino is an open source physical computing platform based on a simple microcontroller board and provides a development environment for writing software. The programming is done in Arduino IDE which is open source software. The board of Arduino can be assembled or purchased. An Arduino board consists of an 8-bit AVR microcontroller and other associated components.

RFID components: RFID reader and RFID tags. They are broadly classified into two parts, passive RFID and active RFID. The passive RFID range is smaller and the active RFID range is larger. For our project we need an active RFID reader and tags. The cost of active RFID is expensive so for our prototype we used passive RFID Reader and tags. The tags contain electronically stored information. Every tag has a unique number. This system can also be implemented for toll roads. It will maintain the flow of the traffic and will avoid the long queue for the toll payment. The payment can be done automatically when the vehicle is identified and the user will not have to stop.

2 Related Work

A literature review on Parking Management System using RFID is done. In [1, 2], Shilpa et al. proposed a method by which they can reduce queues at parking lots. There are two modes of payments—prepaid and postpaid. Vehicles have passive RFID Tags. Fees are deducted based on entry and exit timings. SMS or email is sent for alert for balance deduction and if low balance. In postpaid a bill is sent at user's home at the end of the month. If the user fails to pay their tags are barred. Drawback with this system is tags' collisions. To avoid this anti-collision protocols are used. This makes it scalable system. The system is managed by cloud computing. An application is made to access and manage the backend of the system.

There are many risks involved in using RFID in any system. Many risks are yet not known. In [3], Karygicmnis et al. discussed about RFID which is widely used in mostly every field and the key feature is the automation done by it. Until now risks in RFID are not completely known. Due to this it is tough to manage the risks involved with RFID systems. Risks are divided into three parts. They are network-based risks, business process risks, and business intelligent risks [4, 5]. Each part has many risk elements and contributing risk attributes. They gave security features for many attributes which are important, for example, data confidentiality, authentication of user, optimizing protocols.

In [3], Yuejun et al. discussed about the RFID Structure. It contains integrated circuit and antenna. When there is more than one tag in the range of RFID reader to read, it creates problems in the RFID-based system and it also reduces the performance of the system. In RFID technology there is no need for the line of sight of the tag with reader to read the ID from the tag. The passive RFID tags are used more commonly due to the lower costing. But it only responds once for the reader's signal. This limits the design of wireless communication protocol. When a large amount of tags are read simultaneously it causes collision. The reader is not able to differentiate between the unique numbers. ALOHA algorithm and Binary tree splitting algorithm are used to solve collision problem. In ALOHA algorithm if there is a collision the sender transmits wait signal and retransmission is done after some fixed time. In binary tree splitting algorithm the range of the tag is checked. The algorithm splits the collision tags into two subsets and further narrows down each part into two parts whenever collision is occurred. This algorithm is better than ALOHA.

In [4], Hong et al. discussed about the implementation of RFID System on the parking management for motorcycle. In this paper, they have implemented a system using visual basic language and MySQL for the database. PHPMyAdmin is used for integration of database. The database files manage update edit and store information about the user parking vehicle in the parking lot. A parking application is also available for checking of parking information. This feature is also on mobile application. The distance between the reader and the host computer is very large so a USB hub is used. This system helps in fast search for the free parking spot for the motorcycle.

3 Proposed Work

In the proposed system we have only two levels. First one is Entry level and second is Exit level. The registered tag values are stored in microcontroller of the Arduino board and in the database. In the database other details of the user are also stored. The RFID reader sends 1 bit of data at a time to Arduino serially. So to avoid tag collision the range of the RFID reader should be maintained. Only one vehicle with only one tag should be allowed in front of the RFID reader.

At the entry level, the vehicle with the RFID tag enters in the range of the reader to read the tag unique number [8]. If the user is already registered with the parking system he will be allowed to enter and park the vehicle [9]. A green color LED will glow to indicate that the user is allowed to enter. The same massage is also being displayed in the LCD screen along with the number of vehicles. Side by side in the computer system the tag value is read from the COM port where the serial value is read from the RFID tag [10]. It sends a mail to the user with the attachment of the receipt of details stored in database along with the time of the check-in in the parking lot. But if the user is not registered he can take a temporary card and enter the parking lot. They will not get any notification mail or message as they will be given a predefined temporary card and the payment option will be manually like in the traditional parking system.

At the exit level, the vehicle again will come in the range of RFID reader. The reader reads the tag unique number, opens the barrier and at the same time in the computer system the parking fees is deducted from the account, also a mail is generated and sent to the user. There will be same components at both levels.

4 Implementation

In this project we have done embedded coding in Arduino IDE, interfacing coding in Microsoft Visual Studios 2010 [11]. There are many modules in this project. At entry level first module is to read the value from the RFID reader. Next module is to compare it with the already stored tag values to match if the user is already registered or not. Last module is when match is found, the barrier opens, a green signal is showed and "Access Allowed" signal is displayed in the LCD shield. Algorithms for these modules are as follows.

4.1 Algorithm for Reading Tag Value

In this algorithm the data is read from the RFID tag through the serial port. In my system it reads from COM3 [12, 13]. The reader reads tag value bit by bit. The tag value is of 12 bits. Therefore, we have a counter to check if all the values are stored in the new variable (Fig. 1).

4.2 Algorithm for Comparing Registered Tag to the New Tag Value

In this algorithm the new value stored in previous algorithm is compared with the registered values. Until a match is found the algorithm compares the new value with every tag value stored in the microcontroller (Fig. 2).

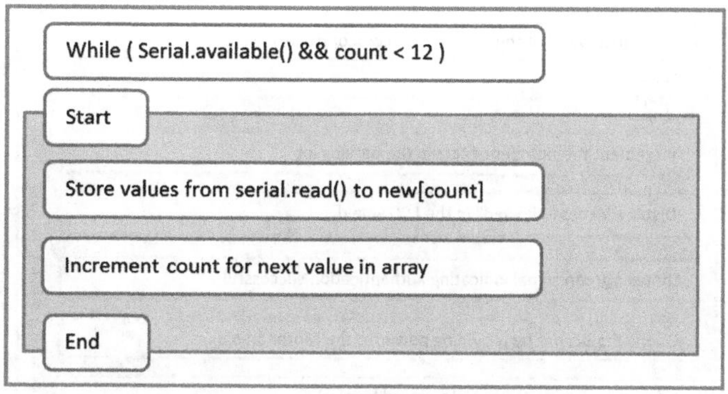

Fig. 1 Algorithm for reading tag value

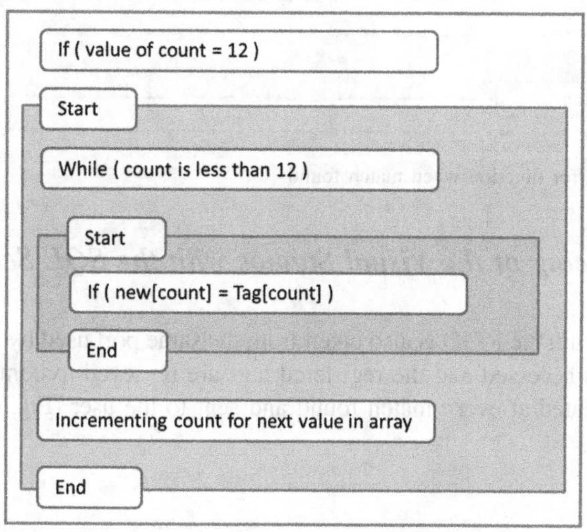

Fig. 2 Algorithm for comparing tags

4.3 Algorithm for Display and Opening Barrier When Match Found

The vehicle is authenticated and is to be allowed to pass the barrier into the parking lot. The interfacing is done with LCD shield, LEDs, and IR sensor in this (Fig. 3).

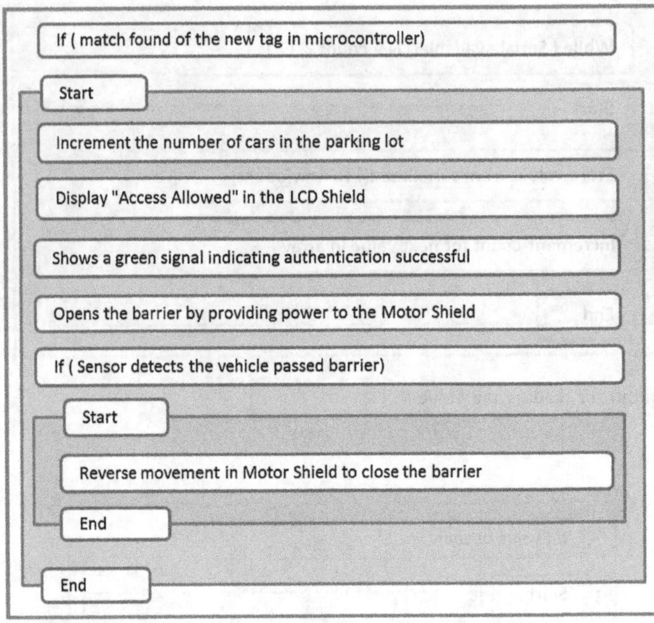

Fig. 3 Algorithm for function when match found

4.4 *Interfacing of the Visual Studios with the SQL Server*

The tag value from the RFID is also taken from the same port used by Arduino [14]. The database is accessed and the registered tags are retrieved to compare the tags. A receipt is created at every match found and sent to the user (Fig. 4).

5 Results

The proposed system is fully automated except at the time of recharging of account because someone is required to enter the amount into parking system's database. The Windows Application is automatic at both Entry and Exit level. When the COM3 sends the value to the application it checks if the tag is already registered. If it is a mail is sent to the user's account (Figs. 5 and 6).

The system automatically detects value and search the database for the match and perform function when match is found. This makes the system autonomous. This windows application is to be running at both Entry and Exit level.

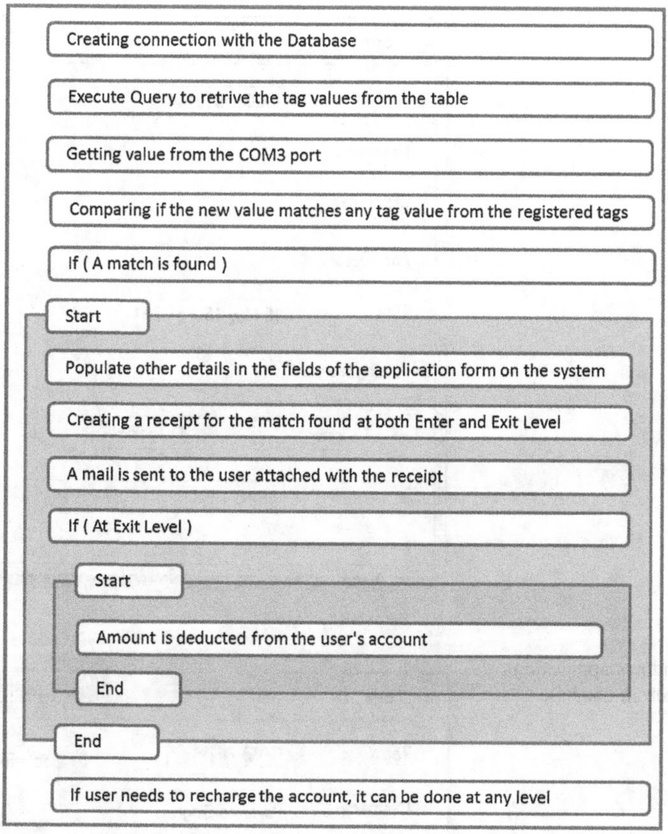

Fig. 4 Algorithm for interfacing with database

6 Conclusion and Future Scope

Intelligent parking management uses RFID technology to uniquely identify a vehicle. This helps avoiding the wait time at the Entry level and Exit level in the parking lot. This system can be implemented in the existing systems. Only a small costing is needed at the initial level. But as the personnel costing is reduced it will be beneficial after few years. The Active RFID readers and tags are to be used for the greater distance. But a larger range Passive RFID reader and tags can also be used depending on the requirement.

One of the parking features is not yet implemented, i.e., capturing images. Images of under the vehicle, of the driver, and of the license plate are to be taken to store more information in the database in order to maintain the security.

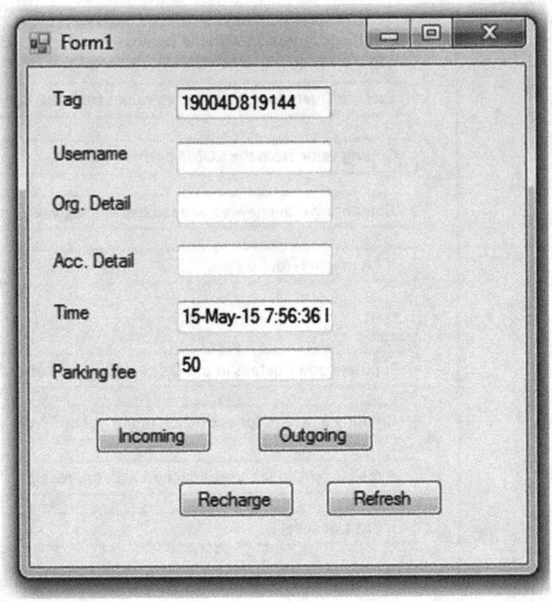

Fig. 5 Windows application reads tag from port

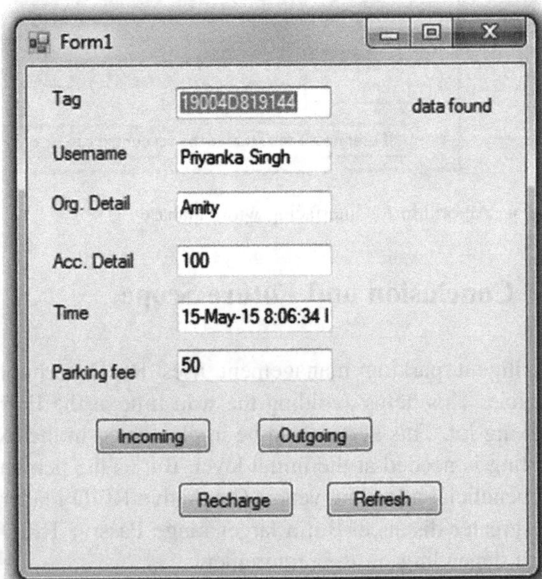

Fig. 6 Windows application—finds match in database

The payment in proposed system is done from the parking system database (credit system). In future the user's details will be connected to user's bank account and the parking fee will be deducted directly from the user's bank account.

References

1. Shilpa, B. et al.: Electronic parking lot payment using passive RFID and cloud computing. Int. J. Comput. Sci. Inf. Technol. **5**(4), 2014
2. Rahman, M.S., Youngil, P., Ki, D.K.: Relative location estimation of vehicles in parking management system. In: 11th International Conference on Advanced Communication Technology, vol. 1, pp. 729–732 (2009)
3. Karygicmnis, A., Phillips, T., Tsibertzopoulous, A.: RFID security: a taxonomy of risk. In: First International Conference on Communications and Networking in China, pp. 1–8 (2006)
4. Lv, A., Li, Y., Sun, Y., Cao, C.: Research on the integrated management of highway based on radio frequency identification technology. In: 3rd International Conference on Measuring Technology and Mechatronics Automation, vol. 3, pp. 116–119 (2011)
5. Srinivasan, S., Ranganathan, H., Vani, S.: An embedded system and RFID solution for transport related issues. In: The 2nd International Conference on Computer and Automaton Engineering, vol. 1, pp. 298–302 (2010)
6. Yuejun, D., Yan, C., Fangfang, S., Zhiyong, D.: The research and application in intelligent parking management system of RFID anti-collision algorithm. In: IEEE 3rd International Conference on Communication Software and Networks, pp. 468–471 (2011)
7. Horng, L.S., Wen, S.C., Shih, F.L., Syu, B.J.: A motorcycle parking lot management system based on RFID. In: International Conference on Fuzzy Theory and Its Applications, pp. 268–272 (2013)
8. Mainette, L., Palano, L., Patrono, L., Stefanizzi, M.L.: Integration of RFID and WSN technologies in smart parking system. In: 22nd International Conference on Software, Telecommunications and Computer Networks, pp. 04–110 (2014)
9. Wang, H., Tang, Y.: RFID technology applied to monitor vehicle in highway. In: Third International Conference on Digital Manufacturing and Automation, pp. 736–739 (2012)
10. Poorva, P., Snehal, T.: RFID-based parking management system. Int. J. Adv. Res. Comput. Commun. Eng. **3**(2) (2014)
11. Pala, Z., Inanc, N.: Smart parking applications using RFID technology. In: 1st Annual RFID Eurasia, pp. 1–3 (2007)
12. Manjusha, P., Vasant, N.B.: Wireless sensor network and RFID for smart parking system. Int. J. Emerg. Technol. Adv. Eng. **3**(4), (2013)
13. Du, S., Sun, S.: The research and design of intellectual parking system based on RFID. In: 9th International Conference on Fuzzy Systems and Knowledge Discovery, pp. 2427–2430 (2012)
14. Lanxin, W., Qisheng, W., Mei, Y., Wei, D.: Design and implementation of smart parking management system Based on RFID and internet. In: International Conference on Control Engineering and Communication Technology, pp. 17–20 (2012)

Performance Analysis of DE over *K*-Means Proposed Model of Soft Computing

Kapil Patidar, Manoj Kumar and Sushil Kumar

Abstract In real-world data increased periodically, huge amount of data is called Big data. It is a well-known term used to define the exponential growth of data, both in structured and unstructured format. Data analysis is a method of cleaning, altering, learning valuable statistics, decision-making, and advising assumption with the help of many algorithms and procedures such as classification and clustering. In this paper we discuss about big data analysis using soft computing technique and propose how to pair two different approaches like evolutionary algorithm and machine learning approach also try to find better cause.

Keywords Big data · *K*-means algorithm · DE (differential evolution) · Data clustering

1 Introduction

Day-by-day amount of data generation is accumulative in drastic manner. Wherein to describe the data, for zetta byte, popular term used is "Big data." The marvelous volume and mixture of real-world data surrounded in massive databases clearly overcome old-fashioned manual method of data analysis, such as worksheets and ad hoc inquiries. A new generation of tools and techniques with the capabilities of perceives and repeatedly promotes users in investigating elevations of data in warehouse in bits for useful knowledge. These procedures and tools are the issue of the field of Knowledge Discovery on Database (KDD), which is mining fascinating

Kapil Patidar (✉) · Manoj Kumar · Sushil Kumar
Department of CSE, Amity School of Engineering and Technology,
Amity University, Noida, Uttar Pradesh, India
e-mail: kpl.ptdr@gmail.com

Manoj Kumar
e-mail: manojbaliyan@gmail.com

Sushil Kumar
e-mail: kumarsushiliitr@gmail.com

M. Pant et al. (eds.), *Proceedings of Fifth International Conference on Soft Computing for Problem Solving*, Advances in Intelligent Systems and Computing 436, DOI 10.1007/978-981-10-0448-3_42

information or design from data in large databases [1]. As in the current situation, data mining tools are very expensive, only few companies have enough money to afford them. The techniques being used for data analysis are spontaneous cluster recognition. As doing online analysis the algorithm used should be fast, that is, in scientific duration it should not be calculated exhaustive but quite probably provide a good result, so the technique expresses worldwide finest discrete cluster. Clustering is a corporate data mining task that it has been examined for use in a number of different areas of data mining and statistics recovery. It is an important unsupervised classification technique, where set of design and frequent vector in a multidimensional space, are grouped into a cluster if pattern is same then it belongs to the same cluster but, if pattern is different, then the cluster is dissimilar. The aim of clustering techniques is to partition a heterogeneous multidimensional data set into group of more homogenous characteristics [2]. Unsupervised clustering may be generally classified into two types—'hierarchical' and 'partitional'. Hierarchical clustering produces a nested sequence of panels, with a single, all-encompassing cluster at the highest and singleton clusters of separable points at the lowest. The result of hierarchical clustering can be represented by graphically displaying it as a tree called a dendogram. Partitional technique produces an unnested, just opposite to hierarchical clustering. If K is the anticipated numeral of clusters, then partitional methodologies typically find all K clusters at once [3, 4]. The first suggested clustering algorithm is K-means which was published in 1957. K-means algorithm is distinctive clustering algorithm based on Euclidean distance as similarity measure and used clustering and evaluates fitness function of DE.

DE (differential evolution algorithm) is a stochastic, population-based optimization algorithm, it is introduced by Storn and Price in 1996. Why we use DE? It is used for global optimization and is compulsory in turfs such as engineering, measurements, and investment but many real-world problematic scenarios have unbiased function that are nondifferentiable, nonconstant, nonlinear, loud, horizontal, multidimensional or have many local minima, constraints or stochasticity such problems are difficult [5]. DE can be used to find estimate explanation to such problem. It is self-organizing pattern taking the modification vector of two casually selected population vector to disturb a prevailing vector [6].

2 Related Work

Clustering algorithm is broadly classified into four ways partitioning algorithm, hierarchical algorithm, density-based algorithm, and grid-based algorithm. Partitioning algorithm build a partition of N object into a set of k clusters. The problem of partition clustering has been approached from miscellaneous fields of knowledge, such as artificial neural network, graph theory, expectation maximization algorithms, evolutionary computing, and so on. For evolutionary computation between them; Differential Evolution (DE) algorithm is developed. It has been shared by many features of classical Genetic Algorithms (GA). In hierarchical algorithm, it creates a classified decomposition of the database that can be represented by dendrogram. It can be agglomerative (bottom-up) or divisive (top-down). Agglomerative algorithms initiate per element as a distinct cluster and merge them in sequentially large cluster. Disruptive algorithms initiate the entire set and continue to divide it into one after another minor clusters. Hierarchical clustering algorithms have two basic advantages. First, the numbers of classes requirement not to be quantified a priori, and second, they are independent of the initial conditions.

The main drawback of hierarchical clustering techniques is that it is static. Density-based clustering algorithm played important character in discovery non-linear outlines construction based on the density. It is used to search regions from data space that are denser than a threshold from dense states. Grid-based algorithms quantize the exploration planetary into a limited amount of cell and then activate on the quantized planetary [2]. Clustering can also be performed in two different modes first 'fuzzy' in this clustering; a pattern may belong to all the classes. Second 'crisp' in this clustering pattern may belong to one and only one class [7].

3 Soft Computing Techniques

There are four types of soft computing techniques as follows.

3.1 *Fuzzy System*

Fuzzy system is upgraded version of traditional system of set membership (in Fuzzy sets) and logic (truth value) originated from ancient Greek philosophy and for the application artificial intelligent. It is a mathematical system that evaluates analog input values between 0 and 1 to digital values either 0 or 1 (i.e., true or false). 'Fuzzy' term determines concepts evaluating true or false value rather than partial values. It is a new area for development.

3.2 *Neural Network*

"A computing system made up of a number of simple, highly interconnected processing elements, which process information by their dynamic state response to external inputs" as stated by Hecht-Nielsen [8]. Neural network is an information processing hypothesis which is motivated by human brain. It includes huge number of interconnected elements such as neurons and also includes activation function working together solve a problem. It is a classically systematized in layers. It consists of three layers namely: input layer, one or more then hidden layer(s) through system of weighted connections.

3.3 Evolutionary Computation

Evolution simulated on system through evolutionary computation. It results in series of optimization algorithms which is based on set of rules. Optimization is performed iteratively in order to get final optimal and feasible result. The evolutionary computation applied to those places where heuristic solution in not available. It is applied to the frequent problem from diverse domain such as social media, machine learning, operation research, automatic programming, optimization, and bioinformatics. Important advantage of evolutionary computation technique is that they can deal with multimodel function. Evolutionary computations are broadly classified into two categories, GA and genetic programming [9]. GA is briefly described in Sect. 4.

3.4 Machine Learning

Machine learning is the branch of measurements it is known as computation theory and allows a computer to learn from data. Learning does not essentially include realization but it is a problem of finding arithmetical predictabilities or other patterns in the data. Many machine learning algorithm learn from human activity or task. It is a subpart of computer science stream and research into artificial intelligence. Machine learning is broadly classified into four types; First, 'Supervised learning' in this learning algorithms class labeled are presented and it generates a function that map given input to desired output. Second, 'Unsupervised learning' classes labeled are not presented. Third, one is 'Semi-supervised learning' which is combination of supervised and unsupervised learning instance generate suitable utility or classifier. Last one is 'Reinforcement learning' these types of learning algorithms learn how to act assuming a reflection of real world. Every act influence on the environment and the environment delivers to the opinion that is usably to the learning algorithm [10]. In this paper we mainly focus on unsupervised learning algorithm such as *K*-means discussed in section.

4 Genetic Algoritham

Genetic algorithms (GA) are adaptive search algorithm based on the appliance of regular variety and normal inheritances. It is also global optimization algorithm which combines survival of the fittest among string structures with a structured, yet randomized information exchange to form a search algorithm with some of the innovative flair of human search [11]. In the Darwinian natural collection standard is used in GA for a biological development to search for the optimum or adjacent optimum clarification in multidimensional feature space. GA is inhabitants built parallel approach. GA is effectively applied to the many problems such as financial marketing, traveling salesmen problem, scheduling problems, network routing problems; these problems are hard to solve by predictable method, that is why we use GA.

A classic GA contains following steps:

1. Generate a random population P of chromosome.
2. Evaluate each chromosome.
3. Select the best chromosome.
4. After manipulation to create new population of chromosome.

First step of GA initialized random population P of chromosome. The P is represented as the population chromosome. Chromosome is commonly signified by string of variables, which is represented as the binary number or real number. The processes of modified component have some termination situation which influences to be all components of the population meet some minimum criteria. After generating random population evaluate a fitness function. This is an important link between GA and the system. The fitness values of all chromosomes are calculated by impartial function. After calculating fitness value selects the operator, it depends on fitness value of chromosome. The GA usually uses three types of operators such as selection, crossover, and mutation. In selection operator select population from chromosome according to the fitness value. The combination of evaluation and selection process is called reproduction. After selection is done second is crossover (sensual recombination) operator combine two specific information of chromosome and produce a new gene (offspring) that contains subparts of both parents. Third one is mutation in this operator mutates randomly any of the bit in the string and generate new genes. The genetic manipulation process uses standard 'crossover' and 'mutation' operator to produce new population of entities (offspring) by influencing the 'genetic information' mentioned to as genes overcome by the paternities of the present population [12].

5 Differential Evolution Algorithm

DE (differential evolution algorithm) is a stochastic, population based on global optimization algorithm, it is introduced by Storn and Price in 1996. It is applied to the composite problem such as Travelling salesmen problem, NP-hard problem, and so on. Differential evolution (DE) is alike to GA because both are used to comparable operator; like selection, crossover, and mutation. The main difference between GA and DE genetic algorithm result is dependency on crossover operator and in differential evolution algorithm result relies on mutation operator. DE is effectively useful to the many factual optimization and artificial problem such as magnetic bearing, automated mirror design, and mechanical engineering design [13, 14]. In the optimization technique, define an objective function for the minimization of task. To the end of the minimization problem objective function is more exactly i.e. is called 'cost' function. Nonlinear, nondifferentiable, and multimodal cost functions are easily handled by DE.

Three advantages of Differential evolution; first one is finding a true global minimum values. Second one is few control parameter direct minimization. In the minimization process self-association so that actual petite input is essential from handler. Third advantage of DE is fast convergence property in the global optimization [15, 16]. The DE used for non-uniform crossover that can take child vector from one close relative than is does from other. The weighted difference between two vectors is a third vector that is call mutation. The mutation vector parameter find alternative vector and the target vector to mix with the all parameter produce the so-called trail vector [17]. After the mutation process is done referred to crossover and the last operation of DE is selection.

Basic strategy of Differential evolution (DE) algorithm is as described below: DE is used for NP D-dimensional parameter vector

$$x_{i,G},\ i = 1, 2, \ldots, \mathrm{NP} \tag{1}$$

- Initialization: population is generate a randomly in distribution procedure.
- Mutation: Randomly select three vectors from population indexes r_1, r_2, r_3 mutually subtract between two vector and differences are applied weight by a factor and finally add difference to the third vector variance. Mutation vector is:

$$v_{i,G+1} = x_{r_1,G} + F.(x_{r_2,G-}x_{r_3}) \tag{2}$$

where r_1, r_2, r_3 random indexes are belonging to {1, 2, …, NP}, F is real and constant influence € [0, 2] which is magnification of difference variation $(x_{r_2,G} - x_{r_3,G})$ [18].

- Crossover: After mutation operation completion recombination is performed, taking of each and individual's population as the primary parent and other parent are selected randomly and generate a new chromosome. If generated child has a produced better cost of objective function then prime parent is replace it [19]. Crossover vector is:

$$u_{i,G+1} = (u_{1i,G+1}, u_{2i,G+1}, \ldots, u_{Di,G,+1},) \tag{3}$$

- Selection: all vectors are selected once as prime parent without contingent on objective function, check selected prime vector and generated child vector values. If it is greater than conserved that value otherwise value is replace by child [20].

In the last few years clustering algorithm is used for differential evolution algorithm and provide a better result. Using DE calculate fitness value of the objective function. So if we use *K*-means algorithm as an objective function of DE then i.e. provide a better result of the *K*-means.

6 *K*-means Clustering

K-means clustering deals with categorizing data objects having similar properties and nearest possible data object into one cluster as well as increasing the clusters with upcoming similar data. *K*-means depends on the distance of the data object and the similarity index is decided by the distance calculation. Let us say, two data values are closer to each other, it shows how similar they are and decides to which cluster it will belong. *K*-means similarity index is calculated by Euclidean distance. The algorithm consists of basic steps for clustering:

1. Basic step of *K*-means is simplest to initially make "*K*" clusters of all the data objects and also assume the center of these clusters.
2. Then iteration is formed until data object is finished with best clustering:
   (a) Determine the center i.e. centroid of each cluster.
   (b) Using Euclidean distance calculate the distance of each object to the centroid in a cluster.

$$d\left(X, \mu_i\right) = ||X - \mu_i||^2 \quad (4)$$

where $X_i \cdot i = 1, \ldots, n$ data point, $\mu_i \cdot i = 1, \ldots, k$ cluster minimize the distance from data point.

3. Make a group of all the objects whose distance is minimum.

This is how the clustering is done by *K*-means. Initially, we cluster the data objects in order to take the best clustering results by analyzing the distance and find accurate and appropriate cluster. This iteration continues until we get all the data objects in the cluster and if new data set is involved it continues for them also [21]. Since *K*-means is simplest but have many problems, one of which is the data set is in convex form where data objects in each cluster is very small whereas the data set is large, due to which the number of clusters increases, therefore we does not get any benefit by using the algorithm. Moreover, it increases the complexity. To overcome this convex data set clustering problems we can pair this algorithm with another algorithms for better results which reduces the complexity. DE is the best option to pair with the *K*-means and generate a best result. *K*-means is used for fitness function of DE and calculate fitness values [22]. After having the problems still *K*-means is the best practice algorithm for clustering because it is simple and is easily understandable.

7 Proposed Model

The proposed model implements performance analysis of DE over *K*-means using clustering algorithm by calculating fitness function. It consists of two levels. In the first level DE import data from txt or csv files after importing all data, fitness function is called i.e. function, calculate the best fitness value and transfer it to the next level. In the next level; mutation, crossover, and selection operations are performed individually. DE is based on mutation operator. Finally, fitness value is found and selection process is completed. The proposed model is shown in Fig. 1.

Procedure of this model:

1. Import data from machine learning repository and randomly initialize the position of the data point.
2. In the first level *K*-means algorithm evaluates the fitness for each data point.
3. Each data point creates different genes.
4. Calculate the fitness rate of the new genes.
5. If new fitness rate is better than the other, then it is conserved; otherwise replace it by parent.
6. Loop to first level until obtained good fitness value or reach maximum number of iteration.

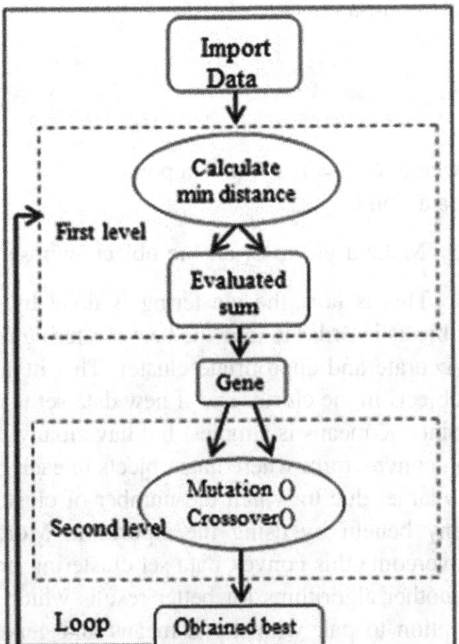

Fig. 1 Flowchart of the proposed model

7.1 First Level

In first level, algorithm import data and calls the objective function. First value checks their fitness rate after calculating first value check fitness rate of remaining values. In this model we used *K*-means as objective function of DE (Differential Evaluation). *K* means uses Euclidian distance, using this sum of minimum distance is calculated and transferred to the DE and produce new offspring. The fitness is calculated as follows:

$$\text{fitness} = \arg\min \sum_{i=1}^{k} \sum_{x \in c_i} ||X - \mu_i||^2 \quad (5)$$

where k represent number of cluster, ci is set of point i.e. belongs to cluster i, X_i $i = 1, \ldots, n$ data point and μ_i $i = 1, \ldots, k$ cluster reduce the distance from data point.

7.2 Second Level

In second level, DE performs all operations like mutation, crossover, and selection. These operations analyze each and every mutation and crossover vector. DE depends on mutation operator Storn and Price. Suggested five mutation schemes are as follows:

- Scheme 1-DE/best/1
- Scheme 2-DE/rand to best/1
- Scheme 3-DE/rand/1
- Scheme 4-DE/best/2
- Scheme 5-DE/rand/2

These are five mutation schemes [23, 24] in this experiment we use scheme no 1. The disposition mutation scheme is the following equation:

$$v_{i,G+1} = x_{r_1,G} + F.(x_{r_2,G-}x_{r_3}) \quad (6)$$

Randomly select three vectors from population indexes r_1, r_2, r_3 mutually subtract between two vector and differences are applied weight by a factor (F), F represent the mutation vector in our case we set the value 0.6 and finally add difference to the third vector variance. The crossover operation is use 0.8. DE check the entire fitness rate, if first generated rate is better than the other, then it is conserved; otherwise replace it by new generated rate. Finally, select best fitness value and result are saved in txt, csv, or exl file.

8 Data Sets and Experimental Result

8.1 Data Sets

We tested our methodology on well-known seven different publically available data sets from machine learning repository [25]. They are:

1. Wdbc (Wisconsin diagnostic breast cancer) dataset (n = 569, D = 31, c = 2) which contains three attribute, where n is number of sample categorized by D dimension (ID, diagnosis, 30 real-valued input features) some real values of nucleus like radius, texture, area, smoothness, perimeter, symmetry, concavity, fractal dimension, and so on and c represented the number of classes or clusters. There are two classes in this data.
2. Thyroid dataset (n = 215, D = 5, c = 2) which contains three attributes, where n is number of sample categorized by D dimension and c represented the number of classes. There are two classes in this data.
3. Fisher's iris dataset (n = 150, D = 4, c = 3) which contains three attributes, where n is the number of sample categorized by D dimension (sepal length, sepal width, petal length, petal width) all lengths are in cm and c represented the number of classesor clusters. There are three classes in this data: Iris versicolor (50), Iris setosa (50), Iris virginica (50).
4. Wine recognition data are the result of a chemical analysis (n = 178, D = 14, c = 3) which contains three attributes, where n is number of sample categorized by D dimension (Alcohol, Malicacid, Ash, Alcalinity of ash, Magnesium, Total phenols, Flavanoids, Nonflavanoid phenols, Proanthocyanins, Colorintensity, Hue, OD280/OD315 of diluted wines, Proline) and c represented the number of clusters. There are three classes in this data: class 1, class 2, and class3.
5. Glass dataset (n = 214, D = 9, c = 7) which contains three attributes, where n is the number of sample categorized by D dimension (Refractive index, Sodium, Magnesium, Aluminum, Silicon, Potassium, calcium, Barium, Iron) and c represented the cluster. There are seven clusters in this datasets: Building windows float and non-float processed Vehicle window float and non-float processed, Containers, Tableware, and Headlamps.
6. Seeds data are X-ray technique (n = 210, D = 7, c = 3) which contains three attributes, where n is number of sample categorized by D dimension (area, perimeter, compactness, length of kernel, width of kernel, asymmetry coefficient, length of kernel groove) and c represented to the cluster. There are three clusters in this datasets: Kama, Rosa, and Canadian.
7. Yeast dataset localization site of protein(n = 1484, D = 8, c = 10) which contains three attributes, where n is number of data point categorized by D dimension (McGeoch's, Heijne's, Alom, Mitochondrial, Hdel, peroxisomal, Vacuolar proteins, vacuolar proteins, nuclear proteins) and c represented the number of clusters. There are ten classes in this data.

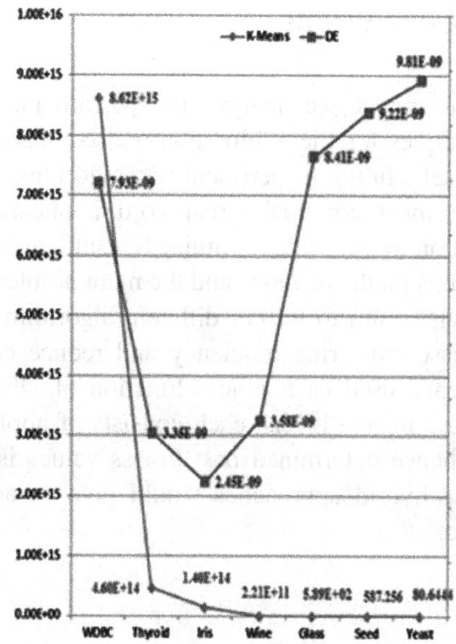

Fig. 2 Comparison of fitness values

For DE we have to set the value of crossover and mutation vector. In this experiment, we set the value for crossover vector as 0.8 and for mutation as 0.6.

8.2 *Experimental Result*

In this Fig. 2 shown graphical representations of both algorithms. We compared the clarity result of two algorithms, one is standard single *K*-means algorithm and second one is *K*-means used as a fitness function of DE. In Table 1, we show the obtained best values of fitness.

It is pragmatic from Table 1 we achieved the finest result in all cases when *K*-means used objective function of DE.

Table 1 Best fitness values obtained

Datasets	*K*-means	DE
WDBC	8.6185E+15	7.929101E−09
Thyroid	4.6031E+14	3.347882E−09
Iris	1.4003E+14	2.452023E−09
Wine	2.2067E+11	3.576548E−09
Glass	5.89E+02	8.411398E−09
Seed	587E+02	9.215005E−09
Yeast	80.6444	9.808668e−09

9 Conclusions

In this paper we analyze the performance using two different soft computing techniques for the exploration and compare fitness value (Table 1) on real-world data sets. In this experiment two algorithms are considered: DE and *K*-means. DE solve most composite real-word application and minimized continuous space function or provide optimized result, and *K*-means use for data clustering. It depends on the distance and the main problem is for convex data set but we can pair this algorithm to further different algorithm in order to get a better result and can increase clustering efficiency and reduce complexity In this work, we proposed *K*-means used as a fitness function of DE. The proposed algorithm was implemented in two levels, each consists of applied on differential evolution operation and hence determined best fitness values is more truthful compared to *K*-means. These hybrid approaches would give an advantage of a global search and high speed.

References

1. Harun, P., Burak, E., Andy, D.P., Cetin, Y.: Clustering of high throughput gene expression data. Comput. Oper. Res. **39**, 3046–3061 (2012)
2. Sarafis, I., Zalzala, A.M.S., Trinder, P.W.: A genetic rule-based data clustering toolkit. In: Proceedings of the 2002 Congress on Evolutionary Computation, 2002. CEC'02, vol. 2. IEEE, (2002)
3. Abraham, A., Das, S., Konar, A.: Document clustering using differential evolution. In: IEEE Congress on Evolutionary Computation, 2006. CEC 2006. IEEE (2006)
4. Song, W., et al.: A hybrid evolutionary computation approach with its application for optimizing text document clustering. Expert Syst. Appl. **42**(5), 2517–2524 (2015)
5. Hatamlou, A.: Black hole: a new heuristic optimization approach for data clustering. Inf. Sci. **222**, 175–184 (2013)
6. Coletta, L.F.S., et al.: A differential evolution algorithm to optimise the combination of classifier and cluster ensembles. In: International Journal of Bio-Inspired Computation (2014)
7. Forgy, E.W.: Cluster analysis of multivariate data: efficiency versus interpretability of classifications. Biometrics **21**, 768–769 (1965)
8. Sankaralingam, K.: A basic introduction to neural networks. Internet: http://pages.cs.wisc.edu/~bolo/shipyard/neural/local.html, 1 May 1996 (20 April 2015)
9. Georgieva, K.S., Engelbrecht, A.P.: Dynamic Differential Evolution Algorithm for Clustering Temporal Data. Large-Scale Scientific Computing, pp. 240–247. Springer, Berlin (2014)
10. Ayodele, T.O.: Types of machine learning algorithms. INTECH Open Access Publisher (2010)
11. Goldberg, D.E.: Genetic Algorithms in Search, Optimization and Machine Learning. Addison-Wesley (1989)
12. Rawat, M.K.: Cluster detection using GA-KNN conjunction approach. J. Global Res. Comput. Sci. **3**(5), 7–10 (2012)
13. Abdual-Salam, M.E., Abdul-Kader, H.M., Abdel-Wahed, W.F.: Comparative study between differential evolution and particle swarm optimization algorithms in training of feed-forward neural network for stock price prediction. In: 2010 the 7th International Conference on Informatics and Systems (INFOS), IEEE (2010)

14. Kumar, S., et al.: Colour image segmentation with histogram and homogeneity histogram difference using evolutionary algorithms. Int. J. Mach. Learn. Cybern. 1–21 (2015)
15. Storn, R., Price, K.: Differential evolution-a simple and efficient adaptive scheme for global optimization over continuous spaces, vol. 3. Berkeley: ICSI, (1995)
16. Kumar, S., Pant, M., Ray, A.: Differential evolution embedded Otsu's method for optimized image thresholding. In: World Congress on Information and Communication Technologies (WICT), IEEE (2011)
17. Paterlinia, S., Krink, T.: Differential evolution and particle swarm optimization in partitional clustering. Comput. Stat. Data Anal. **50**(5), 1220–1247 (2006)
18. Santana Quintero, L.V., Coello Coello, C.A.: Un Algoritmo Basado en Evoluci´on Diferencial para Resolver Problemas Multiobjetivo, Master's thesis, IPN (2004)
19. Storn, R., Price, K.: Differential evolution–a simple and efficient heuristic for global optimization over continuous spaces. J. Global Optim. **11**(4), 341–359 (1997)
20. Botía, J.A., Charitos, D.: Genetic algorithms and differential evolution algorithms applied to cyclic instability problem in intelligent environments with nomadics agents. In: Workshop Proceedings of the 9th International Conference on Intelligent Environments, vol. 17. IOS Press (2013)
21. Tian, Y., Liu, D., Qi, H.: K-harmonic means data clustering with differential evolution. BioMedical Information Engineering, 2009. FBIE 2009. In: International Conference on Future, IEEE (2009)
22. Abraham, A., Das, S., Konar, A.: Document clustering using differential evolution. In: IEEE Congress on Evolutionary Computation, 2006. CEC 2006. IEEE (2006)
23. Govardhan, A., Satapathy S.C.: Data Clustering Using Almost Parameter Free Differential Evolution Technique
24. Zaheer, H., et al.: A new guiding force strategy for differential evolution. Int. J. Syst. Assur. Eng. Manag. 1–14 (2014)
25. Lichman, M.: UCI Machine Learning Repository. Internet: https://archive.ics.uci.edu/ml/, 15 April 2015 (27 April 2015)

Fuzzy Controller for Flying Capacitor Multicell Inverter

P. Ponnambalam, M. Praveen Kumar, V. Surendar and G. Gokulakrishnan

Abstract This article provides the analysis of flying capacitor multicell inverter with different levels, which shows that the THD decreases with the increase in number of levels. The topology called flying capacitor multicell inverter is an inverter which has the property of natural balancing of capacitors available in between each cell, this property of natural balancing makes this type of converter to have more number of levels. In this article, five, seven, and nine levels flying capacitor multicell inverters are analyzed. PD-PWM technique is used to control the RMS voltage at the output. To control the RMS output voltage of these converters, suitable fuzzy controllers are designed and the controlled outputs were verified. The verification of the fuzzy controller is carried out by keeping the reference voltage constant for some time and then varying the same after some time. The waveforms concerned with this verification are provided for validation.

Keywords Fuzzy control · Fuzzy logic · Fuzzy inference systems · Inverters · Power conditioning · Power quality · Power conversion harmonics · Phase disposition pulse-width modulation · Total harmonic distortion

P. Ponnambalam (✉) · M. Praveen Kumar · G. Gokulakrishnan
SELECT, VIT University, Vellore, Tamil Nadu, India
e-mail: p.ponnambalam@gmail.com

M. Praveen Kumar
e-mail: praveen.m@vit.ac.in

G. Gokulakrishnan
e-mail: gokul.g@vit.ac.in

V. Surendar
EEE Department of Kongu Engineering College,
Perundurai, Erode, Tamil Nadu, India
e-mail: surendar136@gmail.com

M. Pant et al. (eds.), *Proceedings of Fifth International Conference on Soft Computing for Problem Solving*, Advances in Intelligent Systems and Computing 436, DOI 10.1007/978-981-10-0448-3_43

1 Introduction

The flying capacitor multicell inverter (FCMI) has the inherent property of natural balancing; the function of flying capacitors in a multicell inverter is to split the input voltage and clamp the voltage stress across the devices [1]. The primary requirement of functioning of FCMI is to have balanced flying capacitor voltages at their reference values, because it dictates both the safe and efficient operation of the inverter under all operating conditions [2, 3]. This FCMI is analyzed with different levels and their THDs are compared in this article, which gives an inference that the THD (total harmonic distortion) decreases with the increase in the number of levels.

There are many attractive properties possessed by FCMI for medium voltage applications, including the advantage of transformer less operation and the capacity to maintain the cell capacitor voltages at their target operating levels, this capacity is termed as natural balancing and this capacity helps in the construction of such converters with a large number of voltage levels [4, 5]. The safe and efficient operations of the converters are dictated by natural balancing property of flying capacitor [6].

Natural self-balancing phenomenon of the clamping capacitor's voltages is one of the advantages of FCM (flying capacitor multicell) and SM (stacked multicell) converters that occur without any feedback control and this process causes the clamping capacitors to reach their desired dc voltage values [7, 8]. Capacitor voltage balancing dynamics variations as a function of load current, modulation index, pulse ratio, and load frequency characteristics have been dealt with in the article [9, 10].

The open loop circuit's output is analyzed with five-level, seven-level, and nine-level configurations of FCMI and its THDs are compared with each other. A closed loop circuit is realized using the fuzzy controller using fuzzy tool box of MATLAB Simulink.

2 Flying Capacitor Multicell Inverter Analysis

2.1 *Five-Level Flying Capacitor Muticell Inverter*

Five-level FCMI is shown in Fig. 1. It has four cells with two switches in each cell which are complimentary to each other. The capacitors C_2, C_3, and C_4 are called flying capacitors and their voltages are V_{c2}, V_{c3}, and V_{c4}, respectively, these voltages are controlled to be regulated at 3 $V_{dc}/4$, $V_{dc}/2$ and $V_{dc}/4$, respectively. Table 1 lists the switch combination used to synthesize the five levels of output voltage V_{an} and the corresponding states of flying capacitors. Charging of a capacitor is indicated by a '+', discharging by a '−' while 'NC' indicates neither charging nor discharging [1]. From the Table 1, it is evident that there are multiple switch combinations for V_{an} as $+V_{dc}/4$, 0 and $-V_{dc}/4$ and these redundancies are helpful in flying capacitor voltage balancing.

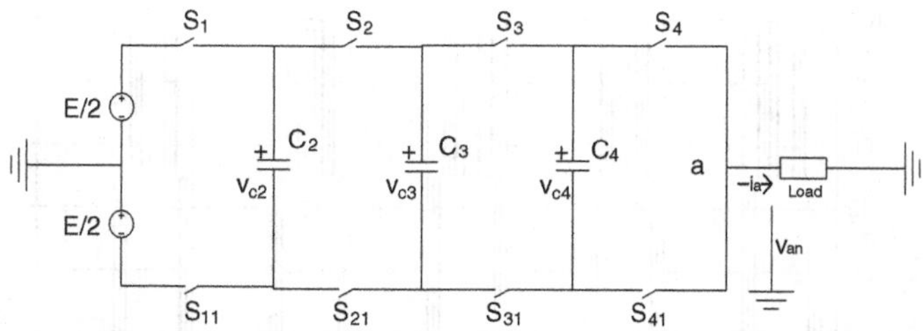

Fig. 1 Five-level FCMI

Table 1 Switching scheme for five-level FCMI

S_1	S_2	S_3	S_4	C_2	C_3	C_4	V_{an}
ON	ON	ON	ON	NC	NC	NC	$+V_{dc}/2$
ON	ON	ON	OFF	NC	NC	+	$+V_{dc}/4$
ON	ON	OFF	ON	NC	+	−	
ON	OFF	ON	ON	+	−	NC	
OFF	ON	ON	ON	−	NC	NC	
OFF	OFF	ON	ON	NC	−	NC	0
OFF	ON	OFF	ON	−	+	−	
OFF	ON	ON	OFF	−	NC	+	
ON	OFF	OFF	ON	+	NC	−	
ON	OFF	ON	OFF	+	−	+	
ON	ON	OFF	OFF	NC	+	NC	
ON	OFF	OFF	OFF	+	NC	NC	$-V_{dc}/4$
OFF	ON	OFF	OFF	−	+	NC	
OFF	OFF	ON	OFF	NC	−	+	
OFF	OFF	OFF	ON	NC	NC	−	
OFF	OFF	OFF	OFF	NC	NC	NC	$-V_{dc}/2$

The output voltage for five-level FCMI is shown in Fig. 2. The output is obtained by using phase disposition pulse-width modulation (PD-PWM) scheme. This modulation technique provides proper flying capacitor voltage balancing and gives satisfactory performance when it is used for closed-loop control. Hence, the same technique is extended for higher levels of FCMIs which are presented in the subsections that follow.

The FFT analysis window for five-level FCMI is shown in Fig. 3 and it can be inferred from the figure that the THD for five-level FCMI is 26.88 %. The THD value can be reduced by designing a proper filter circuit.

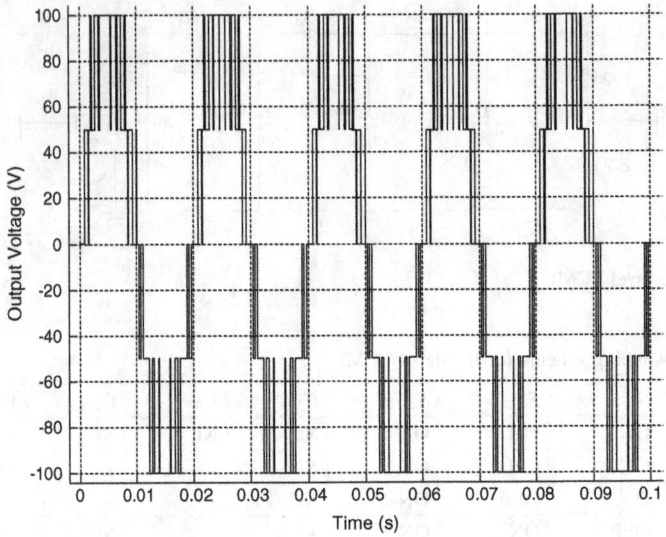

Fig. 2 Output voltage across the load for five-level FCMI

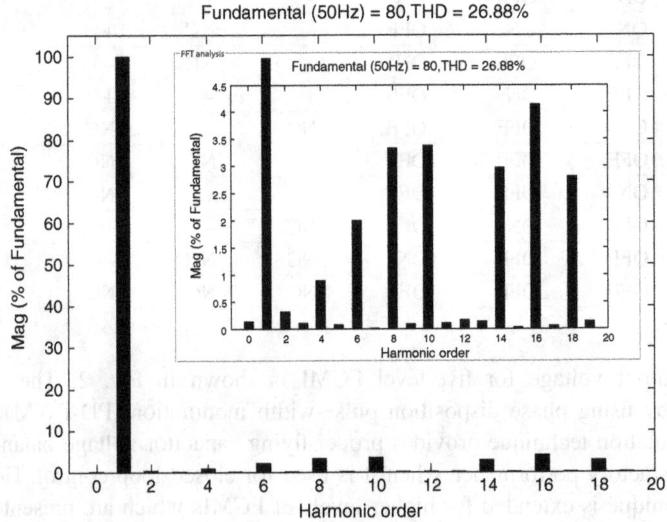

Fig. 3 FFT analysis window for five-level FCMI

2.2 *Seven-Level Flying Capacitor Muticell Inverter*

The Seven-level FCMI can be designed by adding two more cells to the existing five-level FCMI with two more flying capacitors. The simulated output for the seven-level FCMI is shown in Fig. 4, the switching sequence is created by

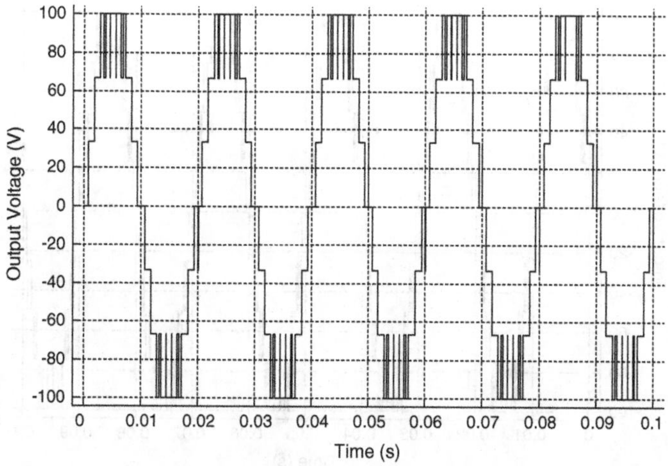

Fig. 4 Output voltage across the load for seven-level FCMI

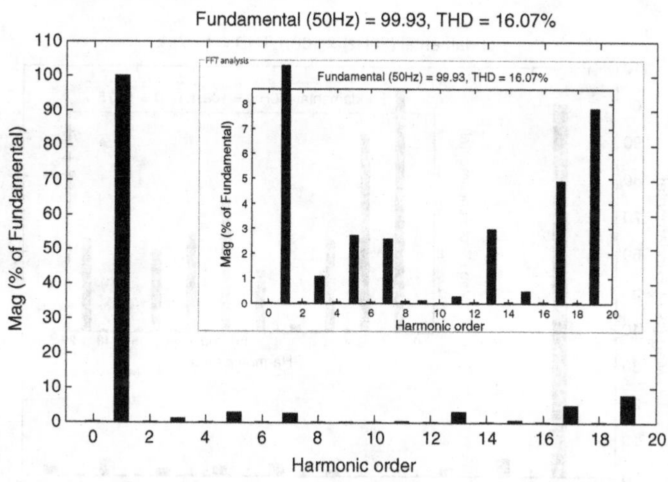

Fig. 5 FFT analysis window for seven-level FCMI

extending the switching scheme specified in Table 1. The FFT analysis window for the seven-level FCMI is shown in Fig. 5 from which it can be inferred that the THD value is 16.07 %.

2.3 Nine-Level Flying Capacitor Muticell Inverter

The nine-level FCMI can be designed by adding two more cells and two more capacitors with the seven-level FCMI. The load output voltage for this

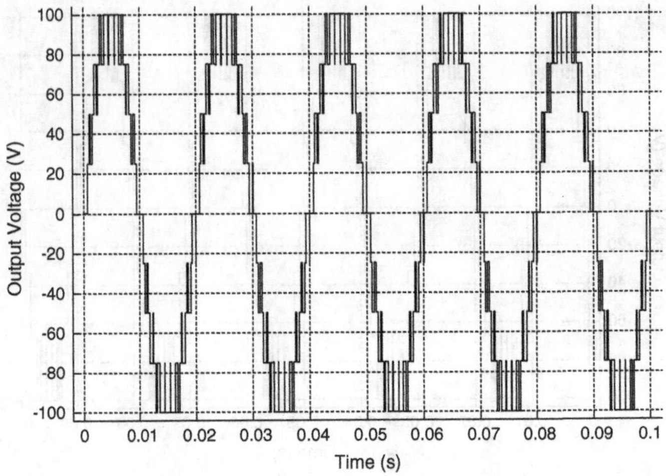

Fig. 6 Output voltage across the load for nine-level FCMI

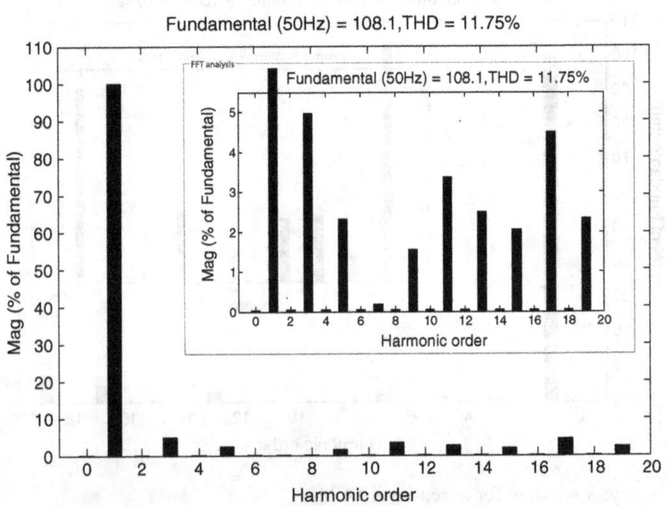

Fig. 7 FFT analysis window for nine-level FCMI

configuration of inverter is shown in Fig. 6. From the figure it can be observed that there is a nine-level created at the output, the switching sequence can be obtained by extending the switching sequence created for seven-level FCMI.

The FFT analysis for this configuration of inverter is shown in Fig. 7. From which it can be inferred that the THD for nine-level FCMI is 11.75 %. The THD for five-level configuration is 26.88 % and for seven-level it is 16.07 % the rate of reduction in THD between five-level and seven-level is 10.81 %, while the THD for

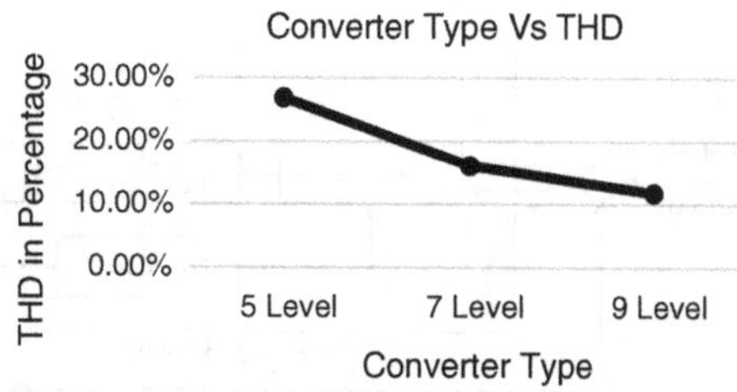

Fig. 8 Analysis of THD for different levels FCMIs

nine-level configuration is 11.75 % the rate of reduction in THD between seven-level and nine-level is 4.32 %.

Figure 8 shows the analysis of THD for various levels of FCMIs, from the figure it can be inferred that the rate of reduction in THD is steeper between five-level and seven-level as compared with the rate of reduction with seven-level and nine-level configuration.

3 Fuzzy Controller

Fuzzy logic controller is a digital control methodology that simulates human thinking by incorporating the impression inherent in all physical systems. Fuzzy logic works by turning the hard-edged world of binary control variables into "soft" grades with varying degrees of membership. Fuzzy logic is useful just because of its low specificity; i.e., for a given input, it allows a more flexible response. The fuzzy system is ideal for the control of continuous variable systems such as electric motors or positioning systems since its output is smooth and continuous. Fuzzy systems base their decisions on inputs in the form of linguistic variables. If-then rules are used to test variables which produce one or more responses depending on which rules were asserted. The response of each rule is weighed according to the confidence or degree of membership of its inputs, and the centroid of the responses is calculated to generate the appropriate output [11].

Multicell converters have large amount of switching devices in their structures. Therefore, obtaining a mathematical model for the converter is complex. Moreover, to achieve a particular state, multiple commutation states can be generated. One of the most important applications in fuzzy logic is fuzzy control. Fuzzy logic is a technique which imitates the human behavior by capturing vague information. When the control is complex and nonlinear, its mathematical model cannot be obtained and Fuzzy controls are implemented. Hence, a fuzzy control can be implemented to control a multicell converter [12].

The advantages of FLC's over the conventional controllers are: (a) it does not need an accurate mathematical model; (b) it can work with imprecise inputs; (c) it can handle nonlinearity; and (d) it is more robust than conventional nonlinear controllers [13].

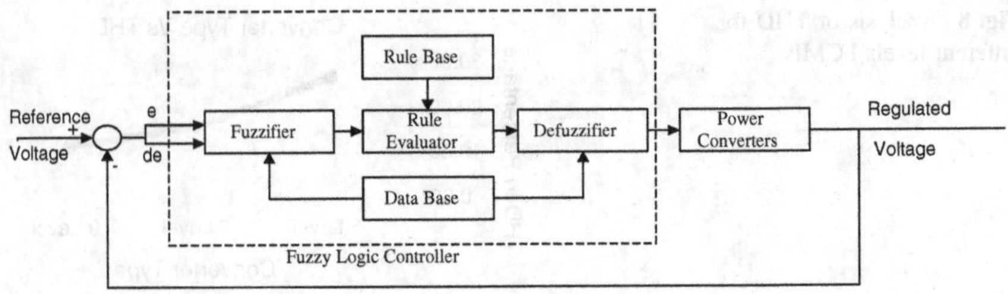

Fig. 9 Internal structure of fuzzy logic controller for power converters

The schematic diagram of a closed-loop fuzzy-controlled power converter is shown in Fig. 9. The reference voltage is compared with the output voltage to produce error signal, the error signal is fed to the fuzzy logic controller, where the calculations are done to decide the output for the fuzzy logic controller. The controller has three functional block diagrams namely, fuzzifier, rule evaluator, and defuzzifier. Fuzzification is the process of converting the numerical variables to its equivalent linguistic variables; this process is done by choosing the membership functions for the fuzzy controller. The fuzzifier finds the degree of membership functions in every linguistic variable. The rule evaluator is the decision making block in the fuzzy logic controller, it does that with the help of rule base that has been created by connecting the membership functions created using linguistic variables, and these rules are equivalent to the control gains of a conventional controller. An example for rule is as follows:

$$R_K : \text{If } E \text{ is } A_i \text{ and CE is } B_i \text{ then output is } C_i$$

where A_i, B_i, and C_i are the linguistic variables for error, change in error, and output, respectively. To evaluate this rule, fuzzy set theory is used; the basic fuzzy set operations are AND, OR, and NOT. A fuzzy set can be denoted as follows:

$$\begin{aligned} &\mu(x) \in [0, 1] \\ &A = [x, \mu(x) | x \in X] \end{aligned} \tag{1}$$

In this equation, X denotes collection of objects generically denoted by $\{x\}$, X is called the universe: if an element in the universe say x is a member of fuzzy set A then mapping is done as specified in the above equation.

Defuzzification is the reverse of fuzzification, which does the job of transforming the linguistic variables to crisp values that could be understood by the real-world. The crisp output is obtained by using the following equation:

$$\text{Output} = \sum A_i * x_i \Big/ \sum A_i \tag{2}$$

Rule base block has the linguistic control rules designed by the fuzzy logic control designer and data base block has the definition of the membership function required by the fuzzifier and defuzzifier [14].

The block diagram shown in Fig. 9 is a general block diagram for any power electronics circuit. The power converter block shown in the figure can be a single phase converter, inverter, chopper, ac voltage regulator, cycloconverter, etc, depending upon the circuit used and the output of the fuzzy logic control will be designed for that, for example if a converter is used then the fuzzy controller will control the triggering angle of the converter to control the output voltage, in case chopper is used then the duty ratio will be controlled, if an inverter is used the modulation index of the PWM signal is controlled. For cycloconverter, triggering angle along with switching sequence will be controlled; fuzzy logic controller can be designed to give two different outputs depending upon two different inputs. The inputs and the outputs are related with each other using rule base along with membership functions. Fuzzy logic controllers are capable of processing multiple inputs and sending out multiple outputs depending on the decision taken by the fuzzy controller with the membership functions and the rule base created by the fuzzy controller designer.

3.1 *Fuzzy Controller for Five-Level Flying Capacitor Multicell Inverter (FCMI)*

The fuzzy controller for five-level flying capacitor multicell converter (FCMI) is designed with the input and output membership functions Fig. 10 from which it can be observed that the range of the input membership function is different from that of the fuzzy controller designed for SCMI. The output membership function is shown in Fig. 11. The input and output membership functions are connected with the help of rule base as shown below:

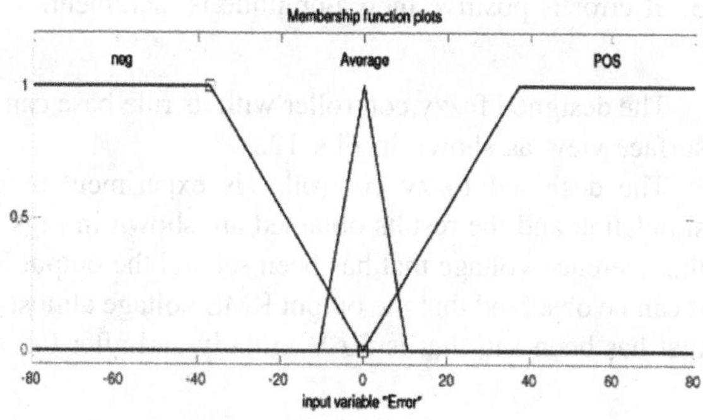

Fig. 10 Input membership functions for fuzzy controller to control five-level FCMI

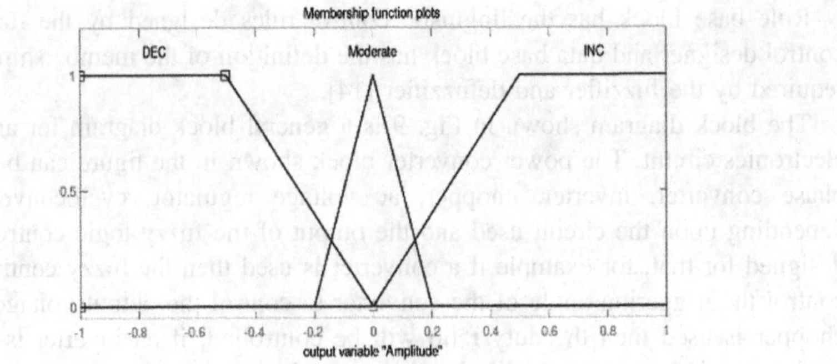

Fig. 11 Output membership functions for fuzzy controller to control five-level FCMI

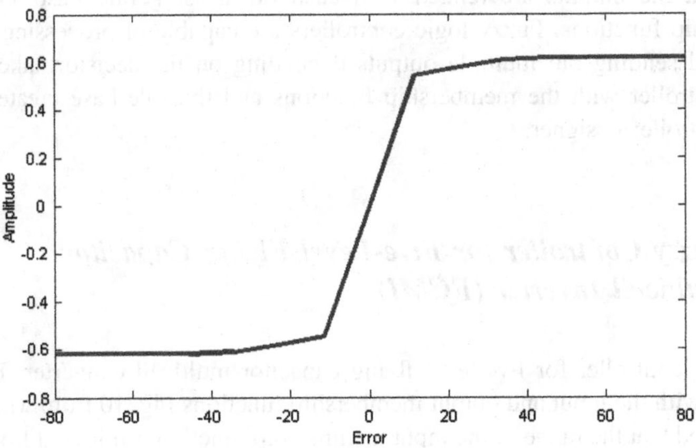

Fig. 12 Surface view of the fuzzy controller designed for five-level FCMI

1. If error is negative then amplitude is decrement
2. If error is average then amplitude is moderate
3. If error is positive then amplitude is increment.

The designed fuzzy controller with its rule base can be observed with the help of surface view as shown in Fig. 12.

The designed fuzzy controller is experimented with the five-level FCMI in simulation and the results obtained are shown in Figs. 13 and 14. Figure 13 shows the reference voltage that has been set and the output RMS voltage; from the figure it can be observed that the output RMS voltage almost follows the reference voltage that has been set, that is 50 V initially and after 0.5 s raised to 70 V. The output

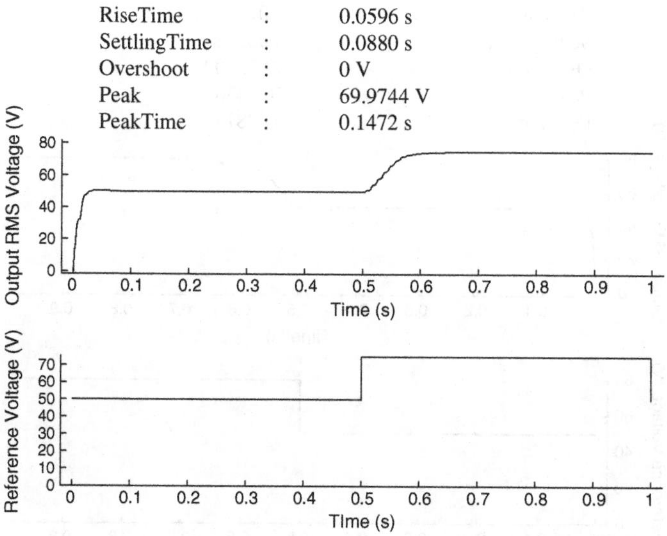

Fig. 13 Fuzzy controlled RMS voltage output for five-level FCMI

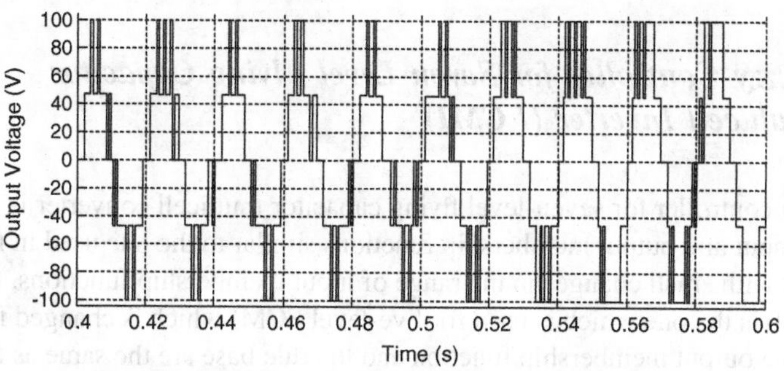

Fig. 14 Output voltage across load for fuzzy controlled five-level FCMI

RMS voltage settles after some time giving a pronounced settling time. The output load voltage for the fuzzy-controlled five-level FCMI is shown in Fig. 14. From the load voltage it can be observed that the initial part of the waveform has lesser pulse width to produce a RMS voltage of 50 V and after 0.5 s, the pulse width gradually increases to produce a RMS voltage of 70 V. The transient response for the fuzzy controlled five-level FCMI taking into consideration with reference voltage kept at 70 V is as shown below

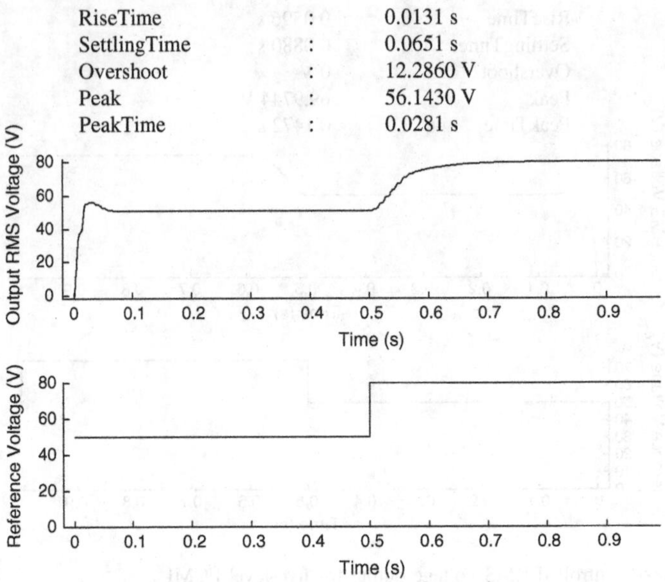

Fig. 15 Fuzzy controlled RMS voltage output for seven-level FCMI

3.2 *Fuzzy Controller for Seven-Level Flying Capacitor Multicell Inverter (FCMI)*

The fuzzy controller for seven-level flying capacitor multicell converter is designed with the input and output membership functions similar to the one used in five-level FCMI but with small changes in the range of input membership functions, the range −80 to +80 is the one which is used for five-level FCMI which is changed from −85 to +85. The output membership function and the rule base are the same as five-level FCMI. The designed fuzzy controller is experimented with simulation circuit of seven-level FCMI and the results obtained are shown in Figs. 15 and 16, respectively. Figure 15 shows the reference voltage and the corresponding RMS output voltage of the fuzzy controlled seven-level FCMI; from the output voltage it could be observed that the output RMS voltage almost follows the reference voltage but the output voltage has some settling time. Figure 16 shows the load output voltage of the fuzzy controlled seven-level FCMI; from the output waveform it could be observed that till 0.5 s pulse width the level of the voltage is low to produce the output voltage of 55 V and after 0.5 s, since the reference voltage is 80 V the output RMS voltage also settles at 80 V after some settling time. The designed fuzzy controller experimented with seven-level FCMI gives the following transient response taking into consideration the reference voltage is maintained at 50 V only.

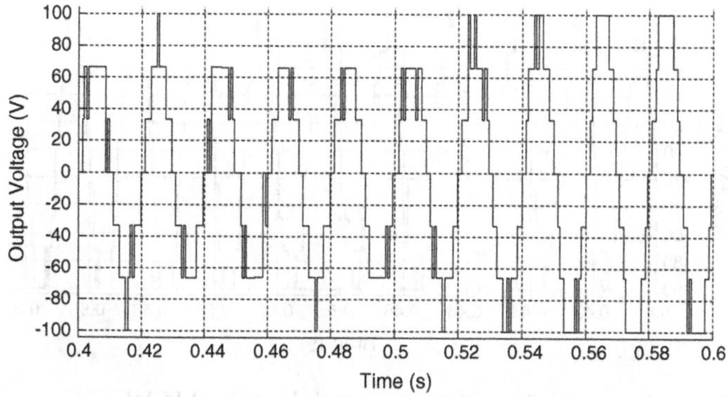

Fig. 16 Output voltage across load for fuzzy controlled seven-level FCMI

3.3 Fuzzy Controller for Nine-Level Flying Capacitor Multicell Converter (FCMI)

The fuzzy controller for nine-level FCMI is designed with the input and output membership functions similar to that of the fuzzy controller designed for seven-level FCMI. The input membership functions range is changed from −85 – +85 to −88 – +88. The designed fuzzy controller is experimented with the nine-level FCMI and the results are presented in Figs. 17 and 18. The reference voltage and the output voltage are shown in Fig. 17 from which it could be observed that the reference

RiseTime	:	0.0323s
SettlingTime	:	0.0569 s
Overshoot	:	0.1259 V
Peak	:	70.0881 V
PeakTime	:	0.0877 s

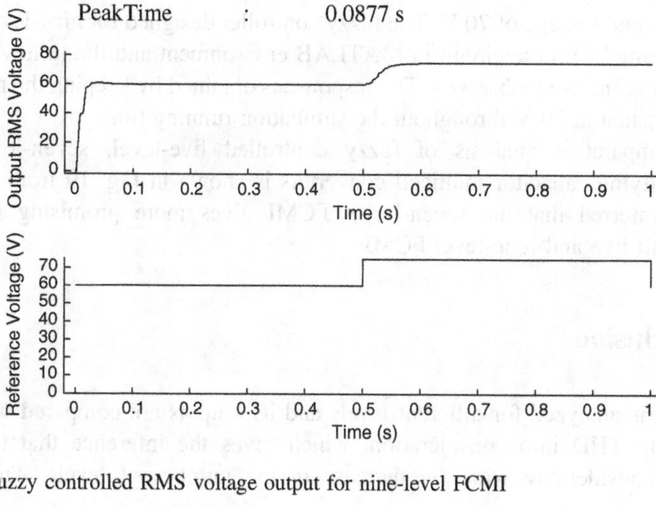

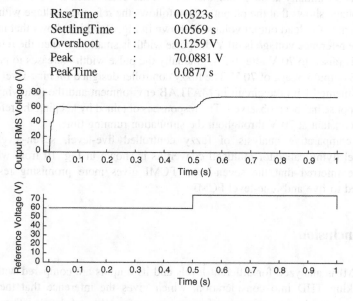

Fig. 17 Fuzzy controlled RMS voltage output for nine-level FCMI

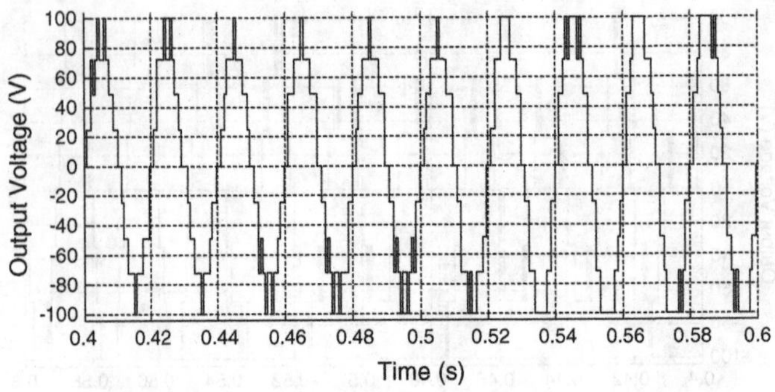

Fig. 18 Output voltage across load for fuzzy controlled nine-level FCMI

Fig. 19 Transient response comparison for fuzzy-controlled FCMIs

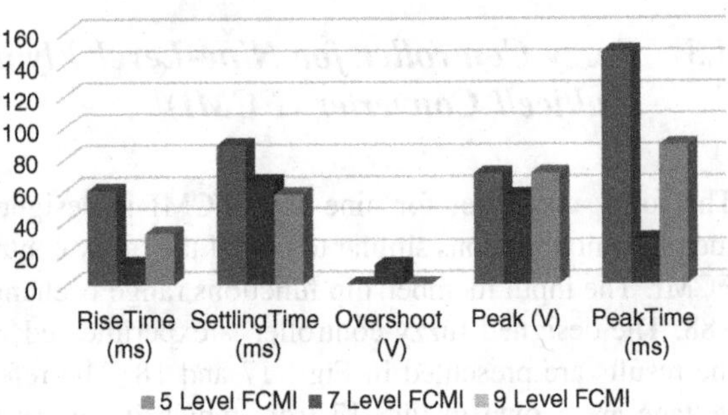

voltage is set initially at 60 V and after 0.5 s it has been raised to 70 V, the output RMS voltage shows that the output almost follows the reference voltage with some settling time. The load output voltage is shown in Fig. 18 which shows that initially when the reference voltage is 60 V the pulse width is small and when the reference voltage is raised to 70 V after 0.5 s, gradually the pulse width increases to produce the RMS output voltage of 70 V. The fuzzy controller designed for nine-level FCMI is experimented with the circuit in MATLAB environment and the following transient response has been observed. The response is obtained by keeping the reference voltage constant at 70 V throughout the simulation running time.

The comparative analysis of fuzzy controlled five-level, seven-level, and nine-level flying capacitor multicell converters is shown in Fig. 19 from which it could be inferred that the seven-Level FCMI gives more promising result as compared to five and nine-level FCMIs.

4 Conclusion

The FCMI is analyzed for different levels and its outputs are compared with each other taking THD into consideration, which gives the inference that the THD decreases considerably with the increase in the number of levels. The THDs obtained are 26.88, 16.07, and 11.75 % for five-level, seven-level and nine-level FCMI, respectively. Suitable fuzzy controllers were designed for these FCMIs and their performance has been verified. Comparison has been made between the outputs of the fuzzy-controlled FCMIs by obtaining the transient responses of the corresponding configurations. The obtained transient response is presented in the form of bar chart to make comparative analysis.

References

1. Shukla, A., Ghosh, A., Joshi, A.: Natural balancing of flying capacitor voltages in multicell inverter under PD carrier-based PWM. IEEE Trans. Power Electron. **26**(6), 1682–1693 (2011)
2. Shukla, A., Ghosh, A., Joshi, A.: Improved multilevel hysteresis current regulation and capacitor voltage balancing schemes for flying capacitor multilevel inverter. IEEE Trans. Power Electron. **23**(2), 518–528 (2008)
3. Hosseini, S.H., Khoshkbar Sadigh, A., Sharifi, A.: Estimation of flying capacitors voltages in multicell converters. In: Proceedings of 6th International ECTI Conference, vol. 1, pp. 110–113. Thailand (2009)
4. Meynard, T.A., Fadel, M., Aouda, N.: Modelling of multilevel converters. In: IEEE Trans. Ind. Electron. **44**(3), 356–364 (1997)
5. Gateau, G., Meynard, T.A., Foch, H.: Stacked Multicell Converter Properties and Design. PESC (Vancouver), vol. 3, pp. 1583–1588 (2001)
6. Song, B.M., Kim, J., Lai, J.S., Seong, K.C., Kim, H.J., Park, S.S.: A multilevel soft-switching inverter with inductor coupling. IEEE Trans. Ind. Appl. **37**(2), 628–636 (2001)
7. Sadigh, A.K., Dargahi, V., Barakati, S.M.: New Asymmetrical Cascade Multicell Converter Based on Optimized Symmetrical Modules. In: IEEE International Symposium on Industrial Electronics (ISIE), pp. 408–412 (2012). ISBN 978-1-4673-0159-6
8. Lienhardt, A.M., Gateau, G., Meynard, T.A.: Stacked multicell converter (SMC): reconstruction of flying capacitor voltages. Industrial Electronics Society, IECON. 31st Annual Conference of IEEE (2005)
9. McGrath, B.P., Holmes, D.G.: Analytical modeling of voltage balance dynamics for a flying capacitor multilevel converter. IEEE Trans. Power Electron. **23**(2), 543–550 (2008a)
10. McGrath, B.P., Holmes, D.G.: Natural current balancing of multicell current source converters. IEEE Trans. Power Electron. **23**(3), 1239–1246 (2008b)
11. Self, K.: Designing with Fuzzy Logic. IEEE Spectrum **105,** 42–44 (1990)
12. Revelo-Andrade, S., Fernández-Nava, M., Bañuelos-Sánchez, P., Uerrero-Castro, F.E.: Control algorithm of a multilevel converter based on a fuzzy inference system using MATLAB. 17th International Conference on Electronics Communications and Computers (CONIELECOMP'07) (2007)
13. Raviraj, V.S.C., Sen, P.C.: Comparative study of proportional—integral, sliding mode and fuzzy logic controllers for power converters. IEEE Trans. Ind. Appl. vol. **33**(2), (1997)
14. Ross, T.J.: Fuzzy Logic with Engineering Applications. McGraw Hill, Inc. (1997)

Influence of Double Dispersion on Non-Darcy Free Convective Magnetohydrodynamic Flow of Casson Fluid

A. Jasmine Benazir and R. Sivaraj

Abstract A numerical study on unsteady, MHD, chemically reacting, free convective, and non-Darcy flow of Casson fluid over a cone placed vertically is presented. The flow regime is influenced by double dispersion effect. The Crank–Nicolson technique is employed to solve the coupled nonlinear partial differential equations. Graphical results are obtained for various controlling parameters present in the governing equations of the problem. The graphical results are very useful to analyze the influence of various controlling parameters on Casson fluid flow. The average skin friction, rate of heat transfer coefficient, and rate of mass transfer coefficient for sundry parameters have been presented in tables below. Results indicate that enhancing the Casson fluid parameter tends to decelerate fluid flow by increasing the plastic dynamic viscosity, whereas it enhances the shear stress in the flow regime. The double dispersion effects play a vital role on sensitive controlling of energy consumptions and species concentration in a small region near to cone and plate.

Keywords Non-Darcy flow · MHD · Casson fluid · Thermal dispersion · Solutal dispersion · Chemical reaction

1 Introduction

Nowadays, non-Newtonian fluids are utilized in numerous applications of day-to-day activities such as process involved conservation of edible products, petroleum production, and preparation of syrup drugs. The distinct rheological characteristics of Casson fluid bring out this fluid to behave like non-Newtonian

A. Jasmine Benazir · R. Sivaraj (✉)
Fluid Dynamics Division, School of Advanced Sciences, VIT University,
Vellore 632014, India
e-mail: sivaraj.kpm@gmail.com

A. Jasmine Benazir
e-mail: jasminebenazir@gmail.com

M. Pant et al. (eds.), *Proceedings of Fifth International Conference on Soft Computing for Problem Solving*, Advances in Intelligent Systems and Computing 436, DOI 10.1007/978-981-10-0448-3_44

fluids. Animasaun et al. [3] have analytically studied the natural convective flow of Casson fluid by varying thermophysical properties in association with the persuit of suction and space-dependent internal heat generation. Mukhopadhyay and Vajravelu [7] have presented numerical solutions of time-dependent Casson fluid flow over a permeable sheet. Mustafa et al. [9] established a mathematical model to investigate unsteady Casson fluid flow over vertical surface which is set in motion. This theoretical problem is solved by using homotopic approach. Exponentially, oscillating and radiating Casson fluid flow subject to suction or injection toward a stretching wall by including thermal radiation effects that are investigated by Pramanik [14]. The knowledge of MHD has developed a new approach to study the behavior of flow regimes in various geometries which have practical relevance. Chamkha [4] has obtained the analytical solution for MHD convective and absorbing fluid flow over a plate which is set in a vertical motion in the presence of suction/injection. Hayat et al. [5] have presented a mathematical model to analyze the magnetohydrodynamic radiating viscoelastic Jeffrey fluid flow subject to convective boundary conditions using homotopy analysis method. Sivaraj and RushiKumar [18] have presented a numerical study to predict the flow behavior of a Walters-B model viscoelastic fluid in two different geometries by considering the variations in electric conductivity of the fluid. Srinivas et al. [19] have applied the homotopy analysis method to study the natures of hydromagnetic flow.

Kairi and Murthy [6] and Murthy [8] have used similarity solution technique to investigate the effects of thermal dispersion in addition to solutal dispersion on a power-law fluid flow under the consideration of Darcy–Forchheimer model. Rushikumar and Sivaraj [16] have obtained numerical solutions for non-Darcy flow of viscoelastic fluid in the presence of double dispersion effects and reported that the dispersion parameters have influential effects on energy consumption and mass transfer analysis. The natural convective flow past a permeable vertical wall under non-Darcian phenomenon is examined by Afify [1]. Aly and Ahmed [2] have derived a numerical scheme to analyze the free convective flow in an anisotropic cavity. Nield and Bejan [10] have provided the comprehensive mathematical aspects in association with physical aspects of convective flows in porous conduit. Pal and Mondal [12] have examined a theoretical model for the non-Darcy flow of a heat generating or absorbing fluid by considering variations of the fluid viscosity cause to experience the variations of temperature. The knowledge of chemically reacting flows is important in many scientific and engineering applications. Narayana et al. [11] adopted regular Perturbation method to analyze the chemically reacting magnetohydrodynamics flow of micropolar fluid. Pal and Mondal [13] have numerically examined the influence of chemically reacting and thermally radiating steady flow of viscous fluid with the cross diffusion effects in which they have accounted the *n*th order reaction. Rashidi et al. [15] have carried out the numerical calculation using differential transform method to study the influence of chemical reaction on fluid flow subject to a moving permeable wall in which they also have accounted the *n*th order reaction. Sivaraj and RushiKumar [17] have constructed a mathematical model to obtain analytical solution of chemically reacting flow within a channel having irregular walls as its boundary in which first order reaction is considered.

The Casson fluid model is found to be accurately applicable to many practical situations in the wings of polymer processing industries, biomechanics, etc. Hereby, authors are motivated and interested to numerically analyze a theoretical model which aims to examine the effects of thermal dispersion and solutal dispersion on chemically reacting non-Darcian flow of Casson fluid over a cone by adopting the Crank–Nicolson scheme. The impact of controlling parameters of the problem is displayed in the form of graphs and tables as well as detailed discussions are provided to validate the physical interpretations of the obtained graphical results.

2 Formulation of the Problem

Consider an unsteady and chemically reacting flow over a vertical cone which maintains a wall temperature T_w, ambient temperature T_∞, a wall concentration, C_w, and ambient concentration C_∞. The physical model of the problem is illustrated in Fig. 1. The two-dimensional system consists of the Cartesian coordinates (x, y), where x-axis is measured parallel to the surface of cone and y-axis is taken transverse to it. The magnetic field of constant intensity B_0 is applied normal to the incompressible Casson fluid non-Darcy flow saturated with porous medium. The momentum equation is influenced by thermal buoyancy and solutal buoyancy, whereas the heat and mass diffusion equations are influenced by thermal dispersion and solutal dispersion, respectively. The conservation equations that govern the physical problem can be written as

$$\frac{\partial(r\,u(t^*,x,y))}{\partial x}+\frac{\partial(r\,v(t^*,x,y))}{\partial y}=0 \tag{1}$$

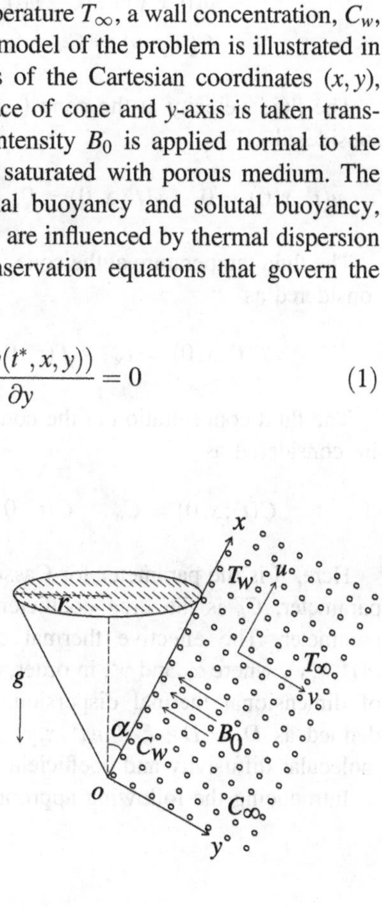

Fig. 1 Conventional form of the problem

$$\frac{\partial u(t^*,x,y)}{\partial t^*}+u(t^*,x,y)\frac{\partial u(t^*,x,y)}{\partial x}+v(t^*,x,y)\frac{\partial u(t^*,x,y)}{\partial y}=\nu\left[1+\frac{1}{\beta}\right]\frac{\partial^2 u(t^*,x,y)}{\partial y^2}$$
$$-\frac{\sigma B_0^2}{\rho}u(t^*,x,y)-\frac{C_b}{\sqrt{k_1}}u(t^*,x,y)^2-\frac{\nu}{k_1}u(t^*,x,y) \tag{2}$$
$$+g\beta_T\cos(\alpha)(T(t^*,x,y)-T_\infty)+g\beta_C\cos(\alpha)(C(t^*,x,y)-C_\infty)$$

$$\frac{\partial T(t^*,x,y)}{\partial t^*}+u(t^*,x,y)\frac{\partial T(t^*,x,y)}{\partial x}+v(t^*,x,y)\frac{\partial T(t^*,x,y)}{\partial y}=\frac{\partial}{\partial y}\left(\alpha_e\frac{\partial T(t^*,x,y)}{\partial y}\right) \tag{3}$$

$$\frac{\partial C(t^*,x,y)}{\partial t^*}+u(t^*,x,y)\frac{\partial C(t^*,x,y)}{\partial x}+v(t^*,x,y)\frac{\partial C(t^*,x,y)}{\partial y}=\frac{\partial}{\partial y}\left(D_e\frac{\partial C(t^*,x,y)}{\partial y}\right) \tag{4}$$
$$-K_R(C(t^*,x,y)-C_\infty)$$

The following initial conditions are assumed for fluid flow:

$$u(0,x,y)=0,\quad v(0,x,y)=0,\quad T(0,x,y)=T_\infty,$$
$$C(0,x,y)=C_\infty\quad \text{for all}\quad x,y \tag{5}$$

The fluid velocity at the cone ($y=0$) and far-off from the cone ($y\to\infty$) are considered as

$$u(t^*,x,0)=0,\quad v(t^*,x,0)=0,\quad u(t^*,0,y)=0,\quad u(t^*,x,y)\to 0|_{y\to\infty} \tag{6}$$

The fluid temperature at the cone ($y=0$) and far-off from the cone ($y\to\infty$) are considered as

$$T(t^*,x,0)=T_w,\quad T(t^*,0,y)=T_\infty,\quad T(t^*,x,y)\to T_\infty|_{y\to\infty} \tag{7}$$

The fluid concentration at the cone ($y=0$) and far-off from the cone ($y\to\infty$) are considered as

$$C(t^*,x,0)=C_w,\quad C(t^*,0,y)=C_\infty,\quad C(t^*,x,y)\to C_\infty|_{y\to\infty} \tag{8}$$

Here, β is the parameter for Casson, k_1 is the dimensional porous permeability parameter, C_b is the drag coefficient, and K_R is dimensional chemical reaction parameter. The effective thermal diffusivity (α_e) is defined as $\alpha_e=\alpha_0+\gamma^* d\, u(t^*,x,y)$, where α_0 and γ^*, in order, are constant thermal diffusivity and coefficient of dimensional thermal dispersion. The effective molecular diffusivity (D_e) is defined as $D_e=D+\xi^* d\, u(t^*,x,y)$, where D and ξ^*, in order, are the constant molecular diffusivity and coefficient of dimensional molecular dispersion.

Introducing the following appropriate dimensional variables:

$$X=\frac{x}{L},\quad Y=\frac{y}{L}(Gr_T)^{1/4},\quad R=\frac{r}{L},\quad r=x\sin(\alpha),$$
$$U(t,X,Y)=\frac{u(t^*,x,y)L}{\nu}(Gr_T)^{-1/2},\quad V(t,X,Y)=\frac{v(t^*,x,y)L}{\nu}(Gr_T)^{-1/4},$$
$$t=\frac{\nu t^*}{L^2}(Gr_T)^{1/2},\quad \theta(t,X,Y)=\frac{T(t^*,x,y)-T_\infty}{T_w-T_\infty},\tag{9}$$
$$\phi(t,X,Y)=\frac{C(t^*,x,y)-C_\infty}{C_w-C_\infty},\quad Gr_T=\frac{g\beta_T(T_w-T_\infty)L^3}{\nu^2}$$

By means of Eq. (9), we obtain the following dimensionless form for the Eqs. (1)–(4)

$$\frac{\partial(R\,U(t,X,Y))}{\partial X}+\frac{\partial(R\,V(t,X,Y))}{\partial Y}=0\tag{10}$$

$$\frac{\partial U(t,X,Y)}{\partial t}+U(t,X,Y)\frac{\partial U(t,X,Y)}{\partial X}+V(t,X,Y)\frac{\partial U(t,X,Y)}{\partial Y}$$
$$=\left(1+\frac{1}{\beta}\right)\frac{\partial^2U(t,X,Y)}{\partial Y^2}-\left(M+\frac{1}{K}\right)U(t,X,Y)-F_IU(t,X,Y)^2\tag{11}$$
$$+\cos(\alpha)\theta(t,X,Y)+N\cos(\alpha)\phi(t,X,Y)$$

$$\frac{\partial\theta(t,X,Y)}{\partial t}+U(t,X,Y)\frac{\partial\theta(t,X,Y)}{\partial X}+\left(V(t,X,Y)-\gamma\frac{\partial U(t,X,Y)}{\partial Y}\right)\frac{\partial\theta(t,X,Y)}{\partial Y}$$
$$=\left(\frac{1}{P_r}+\gamma U(t,X,Y)\right)\frac{\partial^2\theta(t,X,Y)}{\partial Y^2}\tag{12}$$

$$\frac{\partial\phi(t,X,Y)}{\partial t}+U(t,X,Y)\frac{\partial\phi(t,X,Y)}{\partial X}+\left(V(t,X,Y)-\xi\frac{\partial U(t,X,Y)}{\partial Y}\right)\frac{\partial\phi(t,X,Y)}{\partial Y}$$
$$=\left(\frac{1}{S_c}+\xi U(t,X,Y)\right)\frac{\partial^2\phi(t,X,Y)}{\partial Y^2}-K_r\phi(t,X,Y)\tag{13}$$

The dimensionless initial conditions for the fluid flow are

$$U(0,X,Y)=0,\quad V(0,X,Y)=0,\quad \theta(0,X,Y)=0,$$
$$\phi(0,X,Y)=0\quad\text{for all } X,\ Y\tag{14}$$

The dimensionless velocity at the cone $(Y=0)$ and far-off from the cone $(Y\to\infty)$ are considered as

$$\begin{aligned} &U(t,X,0)=0, \quad V(t,X,0)=0, \quad U(t,0,Y)=0, \\ &U(t,X,Y)\to 0|_{Y\to\infty} \end{aligned} \tag{15}$$

The dimensionless temperature at the cone ($Y = 0$) and far-off from the cone ($Y \to \infty$) are considered as

$$\theta(t,X,0)=1, \quad \theta(t,0,Y)=0, \quad \theta(t,X,Y)\to 0|_{Y\to\infty} \tag{16}$$

The dimensionless concentration at the cone ($Y = 0$) and far-off from the cone ($Y \to \infty$) are considered as

$$\phi(t,X,0)=1, \quad \phi(t,0,Y)=0, \quad \phi(t,X,Y)\to 0|_{Y\to\infty} \tag{17}$$

Where, the Hartman number (M), permeability parameter for porous medium (K), local inertia coefficient (F_I), buoyancy ratio parameter (N), Prandtl number (P_r), thermal dispersion parameter (γ), Schimdt number (S_c), solutal dispersion parameter (ξ), and chemical reation parameter (K_r) are given by:

$$\begin{aligned} &M=\frac{\sigma B_0^2L^2}{\mu(Gr_T)^{1/2}}, \quad \frac{1}{K}=\frac{L^2}{k_1(Gr_T)^{1/2}}, \quad F_I=\frac{C_bL}{\sqrt{k_1}}, \quad N=\frac{\beta_T(C_w-C_\infty)}{\beta_C(T_w-T_\infty)}, \\ &P_r=\frac{\mu C_P}{k}, \quad \gamma=\frac{\gamma^*(Gr_T)^{1/2}d}{L}, \quad S_c=\frac{v}{D}, \quad \xi=\frac{\xi^*(Gr_T)^{1/2}d}{L}, \\ &K_r=\frac{K_RL^2}{v(Gr_T)^{1/2}} \end{aligned} \tag{18}$$

The local skin Friction, local rate of heat transfer, and local rate of mass at the wall is defined as follows:

$$\begin{aligned} &\tau=-\left(1+\frac{1}{\beta}\right)\left(\frac{\partial U(t,X,Y)}{\partial Y}\right)_{Y=0}, \quad Nu=-X\left(\frac{\partial\theta(t,X,Y)}{\partial Y}\right)_{Y=0}, \\ &Sh=-X\left(\frac{\partial\phi(t,X,Y)}{\partial Y}\right)_{Y=0} \end{aligned} \tag{19}$$

Average skin Friction, rate of heat and mass transfer coefficients are:

$$\begin{aligned} &\overline{\tau}=-\left(1+\frac{1}{\beta}\right)\int_0^1\left(\frac{\partial U(t,X,Y)}{\partial Y}\right)_{Y=0}\mathrm{d}X, \quad \overline{Nu}=-\int_0^1\left[\frac{\left(\frac{\partial\theta(t,X,Y)}{\partial Y}\right)_{Y=0}}{\theta(t,X,Y)_{Y=0}}\right]\mathrm{d}X, \\ &\overline{Sh}=-\int_0^1\left[\frac{\left(\frac{\partial\phi(t,X,Y)}{\partial Y}\right)_{Y=0}}{\phi(t,X,Y)_{Y=0}}\right]\mathrm{d}X \end{aligned} \tag{20}$$

3 Result and Discussion

The nondimensional governing Eqs. (10)–(13) are coupled, nonlinear, and partial differential equations. These equations are numerically solved and subjected to the initial and boundary conditions as specified in Eqs. (14)–(17) by adopting the Crank—Nicolson method. The momentous-governing parameters involved in the expression of fluid velocity are the Casson fluid parameter (β), buoyancy ratio parameter (N), Hartmann number (M), Forchheimer number (F_I), and permeability parameter (K) which are examined with different values to predict their significant role on controlling the fluid velocity. The choice for variations of these parameters is considered as follows: $0.5 \leq \beta \leq 2$, $1 \leq N \leq 10$, $0 \leq M \leq 3$, $0 \leq F_I \leq 1$, and $1 \leq K \leq 4$ and fixed with the following values:$\beta = 0.5$, $M = 1$, $K = 1$, $F_I = 0.3$, and $N = 1$ when analyzing the significance of other pertinent parameters. Figures 2, 3, 4, and 5 illustrate the nature and significance of the above-mentioned pertinent parameters on the flow

Fig. 2 Variation of Casson fluid parameter on fluid velocity

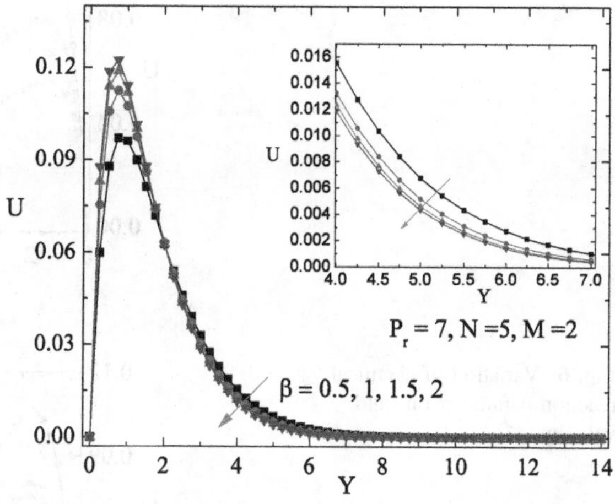

Fig. 3 Variation of magnetic field parameter on fluid velocity

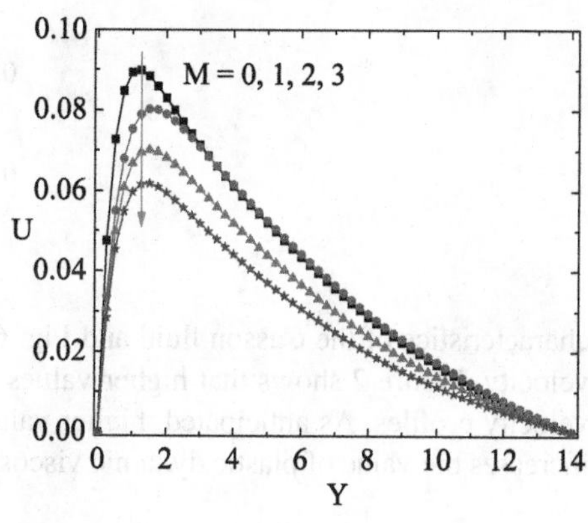

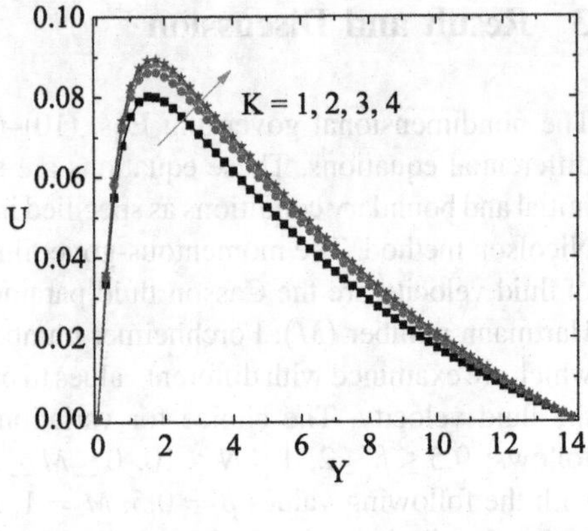

Fig. 4 Variation of permeability of porous medium on fluid velocity

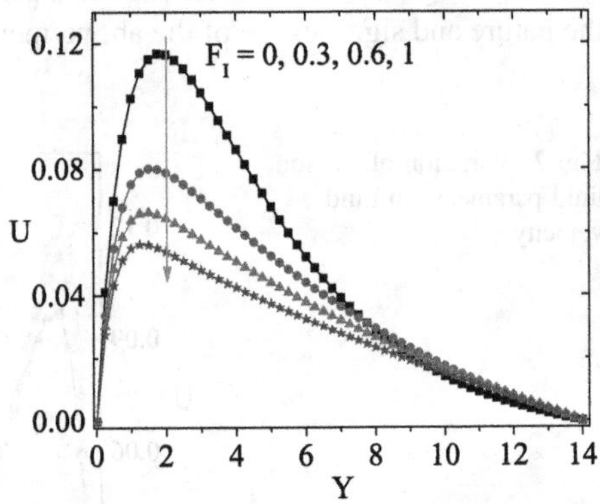

Fig. 5 Variation of local inertia coefficient on fluid velocity

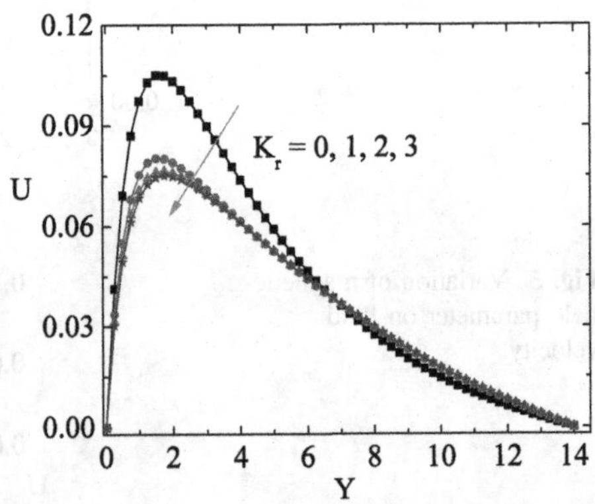

Fig. 6 Variation of chemical reaction parameter on fluid velocity

characteristics of the Casson fluid and Fig. 6 elucidates the effect of K_r on the fluid velocity. Figure 2 shows that higher values of Casson fluid parameter decrease the velocity profiles. As anticipated, higher values of β diminish the yield stress which increases the value of plastic dynamic viscosity, thereby decelerate the motion of the

Table 1 Variations of β, M, K, and F_I on $\overline{\tau}$, $\overline{Nu}$, and $\overline{Sh}$

Physical parameters	Values	$\overline{\tau}$	$\overline{Nu}$	$\overline{Sh}$
β	0.5	0.30102	0.88276	1.45056
	1	0.39706	0.96359	1.46383
	1.5	0.44772	0.99933	1.46995
M	0	0.23420	0.26062	1.44355
	1	0.15299	0.24401	1.42860
	2	0.14301	0.22971	1.42503
K	1	0.15299	0.24401	1.42860
	2	0.15869	0.25231	1.43065
	3	0.16071	0.25526	1.43138
F_I	0	0.17342	0.27434	1.43615
	0.3	0.15299	0.24401	1.42860
	1	0.13386	0.21667	1.42158

fluid. It is to be noted that when $\beta \rightarrow \infty$, the non-Newtonian behaviors disappear and the fluid purely behaves like a Newtonian fluid due to the plasticity of Casson fluid. As seen in Fig. 3, intensification in the values of Hartmann number has the penchant to slow down the velocity of Casson fluid. Reason is fortification in Hartmann number used to strengthen the Lorentz force. Figure 4 elucidates that an increase in K is used to boost the velocity of Casson fluid along the boundary layer which shows the physical situation that higher values of K used to depress the porous medium, thereby the motion of Casson fluid is accelerated. Amplifying the local inertia coefficient fortifies resistance on flow field which diminishes the velocity as shown in Fig. 5. It is clear that the absence of Forchheimer drag force in momentum boundary layer exhibits the behavior of Darcy flow. As displayed in Fig. 6, amplifying K_r results in fall off motion of Casson fluid (Tables 1 and 2).

The momentous-governing parameters involved in the expression of fluid temperature and concentration are the Prandtl number (P_r), thermal dispersion parameter (γ), Schmidt number S_c, chemical reaction parameter K_r, and solutal dispersion parameter (ξ) which are examined with different values to predict their significant role on controlling the fluid temperature and concentration. The choice for variations of these parameters is considered as follows: $0.71 \leq P_r \leq 7$, $0 \leq \gamma \leq 0.3$, $0.22 \leq S_c \leq 2$, $0 \leq K_r \leq 3$, and $0 \leq \xi \leq 0.3$ and fixed with the following values:$P_r = 0.71$, $\gamma = 0.3$, $S_c = 2$, $K_r = 1$, and $\xi = 0.3$ when analyzing the significance of other pertinent parameters. The nature and significance of the above-mentioned pertinent parameters on the characteristics of energy consumptions and mass transfer of the present model are displayed in Figs. 7, 8, 9, 10, and 11. Figure 7 illustrates that higher values of Prandtl number dilutes the fluid thermal conductivity, consequently the fluid temperature decreases for higher values of P_r. Physically, fluids with larger Prandtl numbers have higher viscosity due to less impact of the buoyancy parameter, and thus move slowly. Prandtl number variations are selected that correspond to air (P_r = 0.71), methyl chloride (P_r = 2.97),

Table 2 Variations of P_r, γ, S_c, K_r, and ξ on $\overline{\tau}$, $\overline{Nu}$, and $\overline{Sh}$

Physical parameters	Values	$\overline{\tau}$	$\overline{Nu}$	$\overline{Sh}$
P_r	0.71	0.15299	0.24401	1.42860
	4.24	0.13631	0.52409	1.41926
	7	0.13047	0.64588	1.41651
γ	0	0.67467	1.04717	1.49156
	0.15	0.67486	1.08107	1.49175
	0.3	0.67509	1.10273	1.49193
S_c	0.22	0.17818	0.27469	0.48230
	0.96	0.16230	0.25226	0.99718
	2	0.15299	0.24401	1.42860
K_r	0	0.17711	0.26538	0.45413
	1	0.15299	0.24401	1.42860
	2	0.14437	0.23912	1.96809
ξ	0	0.39790	0.33271	0.57198
	0.15	0.39813	0.33381	0.58244
	0.3	0.39836	0.33489	0.59248

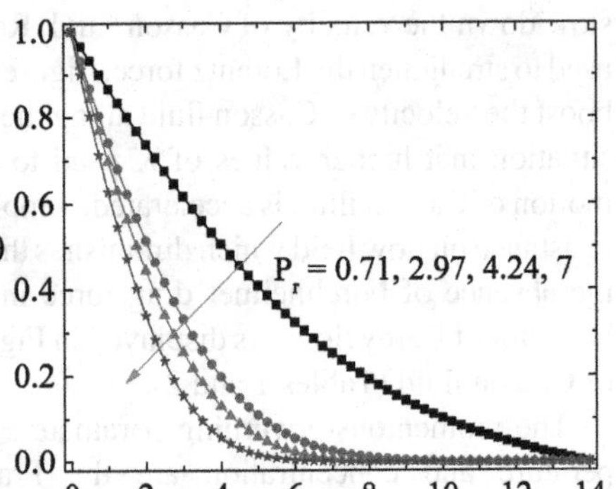

Fig. 7 Variation of Prandtl number on fluid temperature

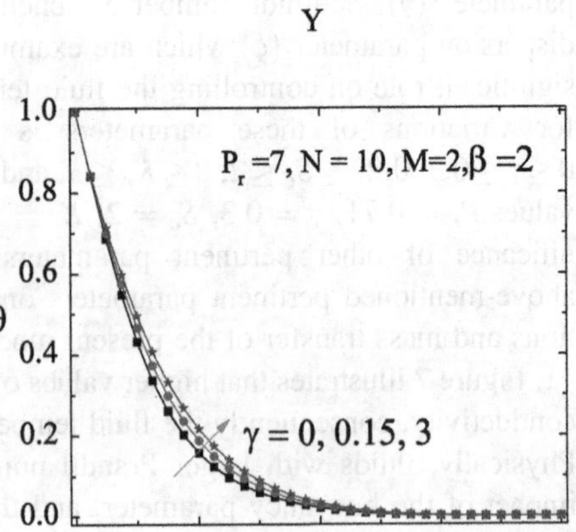

Fig. 8 Variation of thermal dispersion parameter on fluid temperature

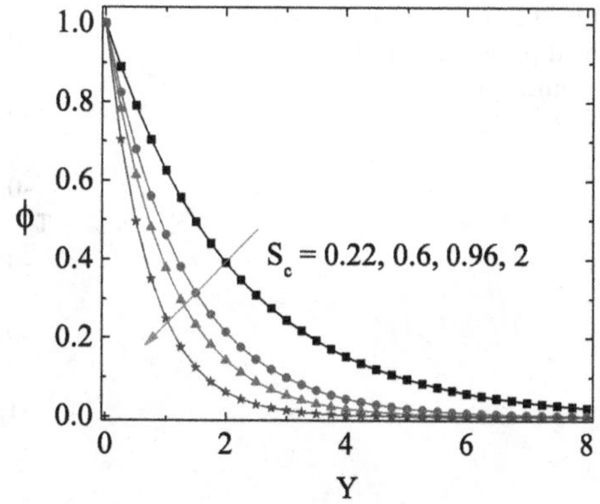

Fig. 9 Variation of Schmidt number on fluid concentration

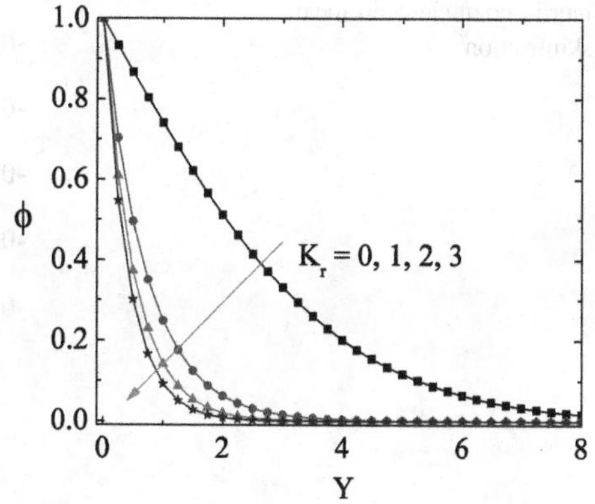

Fig. 10 Variation of chemical reaction parameter on fluid concentration

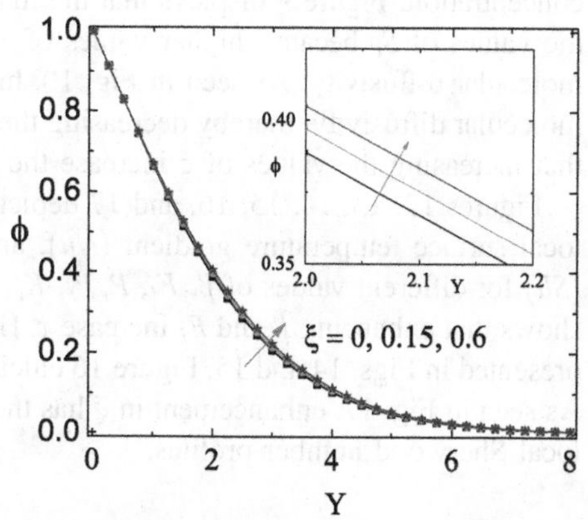

Fig. 11 Variation of solutal dispersion parameter on fluid concentration

sulfur dioxide ($P_r = 4.24$), and water ($P_r = 7$). Fluid temperature can be enhanced for escalating values of γ as shown in Fig. 8. It seems, the dispersion effects are important for sensitive control of the heat transfer in a small region near the wall.

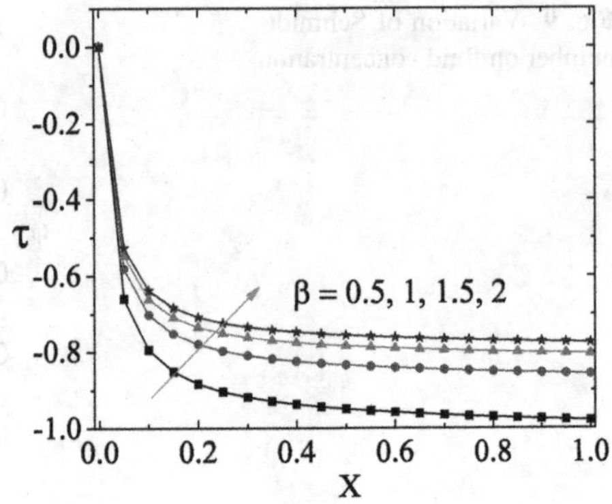

Fig. 12 Variation of Casson fluid parameter on local skinfriction

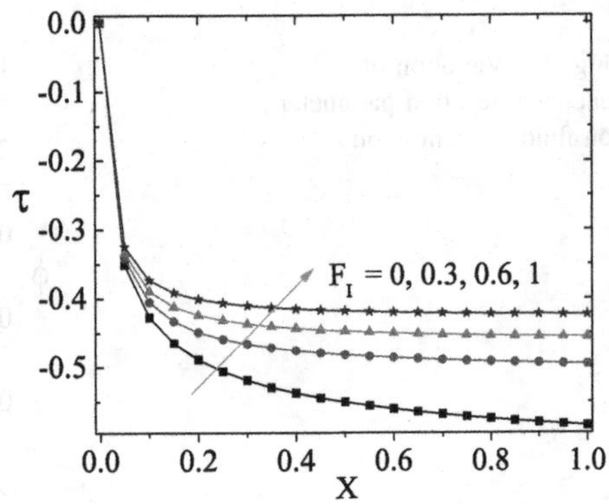

Fig. 13 Variation of local inertia coefficient on local skinfriction

Figures 9, 10, and 11 exhibit the variations of S_c, K_r, and ξ, in order, on fluid concentration. Figure 9 displays that the mass transfer is diminished for increasing the values of S_c because higher values of S_c have the tendency to dilute the fluid molecular diffusivity. As seen in Fig. 10, higher values of K_r suppresses chemical molecular diffusivity, thereby decreasing the fluid concentration. Figure 11 presents that increasing the values of ξ increase the mass transfer.

Figures 12, 13, 14, 15, 16, and 17 depict the variation of local skin friction (τ), local surface temperature gradient (Nu), and local surface concentration gradient (Sh) for different values of β, F_I, P_r, γ, K_r, and ξ, respectively. Figures 12 and 13 shows that enhancing β and F_I increase τ. Higher values of P_r and γ increase Nu as presented in Figs. 14 and 15. Figure 16 elucidates that Sh increase for increasing K_r. As seen in Fig. 17, enhancement in ξ has the tendency to increase the magnitude of local Sherwood number profiles.

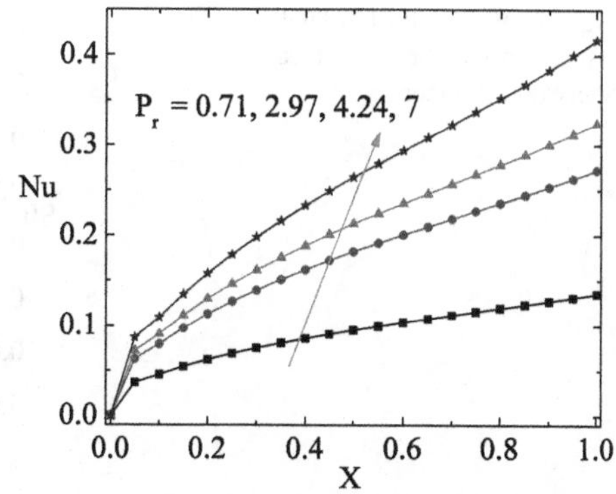

Fig. 14 Variation of Prandtl number on local Nusselt number

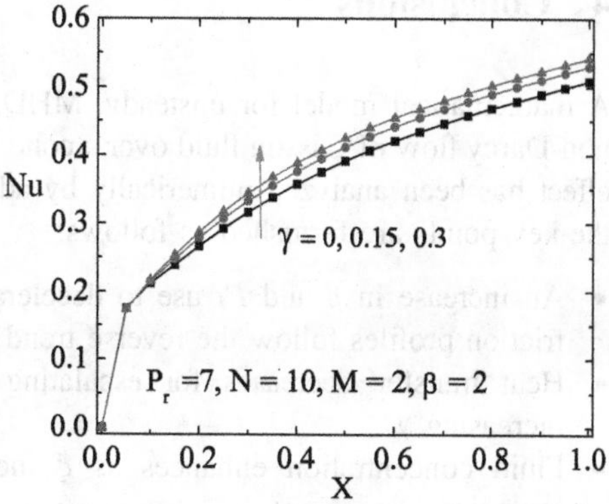

Fig. 15 Variation of thermal dispersion parameter on local Nusselt number

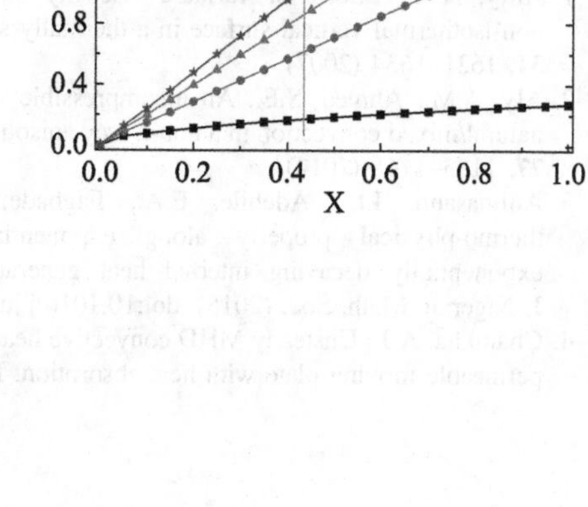

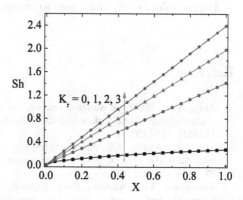

Fig. 16 Variation of chemical reaction parameter on local Sherwood number

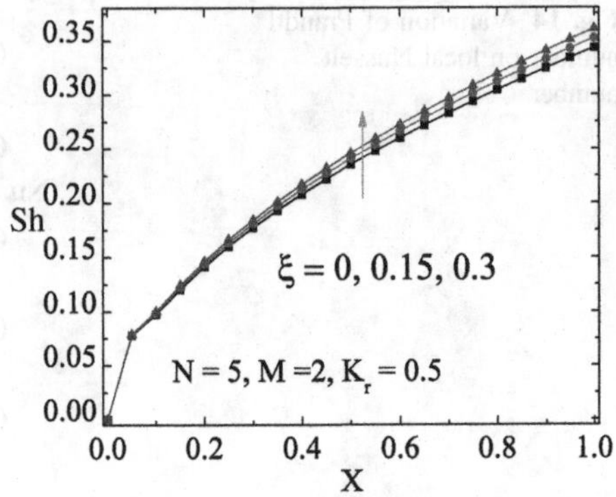

Fig. 17 Variation of solutal dispersion parameter on local Sherwood number

4 Conclusions

A mathematical model for unsteady, MHD, free convective, chemically reacting, non-Darcy flow of Casson fluid over a cone with the influence of double dispersion effect has been analyzed numerically by adopting Crank–Nicolson technique and the key points are furnished as follows:

- An increase in β and F_I use to decelerate the fluid flow conversely and skin friction profiles follow the reverse trend.
- Heat transfer decreases for escalating value of P_r while it increases for increasing γ.
- Fluid concentration enhances as ξ increases, whereas it diminishes as K_r increases.
- Higher values of P_r and γ result in accelerating the rate of heat transfer.
- Higher values of K_r and ξ lead to increase the local rate of mass transfer.

References

1. Afify, A.A.: Effects of variable viscosity on non-Darcy MHD free convection along a non-isothermal vertical surface in a thermally stratified porous medium. Appl. Math. Model. **31**, 1621–1634 (2007)
2. Aly, A.M., Ahmed, S.E.: An incompressible smoothed particle hydrodynamics method for natural/mixed convection in a non-Darcy anisotropic porous medium. Int. J. Heat Mass Transf. **77**, 1155–1168 (2014)
3. Animasaun, I.L., Adebile, E.A., Fagbade, A.I.: Casson fluid flow with variable thermo-physical property along exponentially stretching sheet with suction and exponentially decaying internal heat generation using the homotopy analysis method. J. Nigerian Math. Soc. (2015). doi:10.1016/j.jnnms.2015.02.001
4. Chamkha, A.J.: Unsteady MHD convective heat and mass transfer past a semi-infinite vertical permeable moving plate with heat absorption. Int. J. Engrg Sci. **42**, 217–30 (2004)

5. Hayat, T., Asad, S., Mustafa, M., Alsaedi, A.: MHD stagnation-point flow of Jeffrey fluid over a convectively heated stretching sheet. Comput. Fluids **108**, 179–185 (2015)
6. Kairi, R.R., Murthy, P.V.S.N.: Effect of double dispersion on mixed convection heat and mass transfer in a non-Newtonian fluid-saturated non-Darcy porous medium. J. Porous Med. **13**, 749–757 (2010)
7. Mukhopadhyay, S., Vajravelu, K.: Diffusion of chemically reactive species in Casson fluid flow over an unsteady permeable stretching surface. J. Hydrodyn. **25**, 591–598 (2013)
8. Murthy, P.V.S.N.: Effect of double dispersion on mixed convection heat and mass transfer in non-Darcy porous medium. ASME J. Heat Transfer **122**, 476–484 (2000)
9. Mustafa, M., Hayat, T., Pop, I., Aziz, A.: Unsteady boundary layer flow of a Casson fluid due to an impulsively started moving flat plate. Heat Transf. Asian Res. **40**, 563–576 (2011)
10. Nield, D.A., Bejan, A.: Convection in Porous Media. Springer-Verlag, Berlin (1992)
11. Narayana, P.V.S., Venkateswarlu, B., Venkataramana, S.: Effects of Hall current and radiation absorption on MHD micropolar fluid in a rotating system. Ain Shams Engrg. J. **4**, 843–854 (2013)
12. Pal, D., Mondal, H.: Effect of variable viscosity on MHD non-Darcy mixed convective heat transfer over a stretching sheet embedded in a porous medium with non-uniform heat source/sink. Commun. Nonlinear Sci. Numer. Simulat. **15**, 1553–1564 (2010)
13. Pal, D., Mondal, H.: Influence of chemical reaction and thermal radiation on mixed convection heat and mass transfer over a stretching sheet in Darcian porous medium with Soret and Dufour effects. Energ. Convers. Manage. **62**, 102–108 (2012)
14. Pramanik, S.: Casson fluid flow and heat transfer past an exponentially porous stretching surface in presence of thermal radiation. Ain Shams Engrg. J. **5**, 205–212 (2014)
15. Rashidi, M.M., Rahimzadeh, N., Ferdows, M., Uddin, M.J., Beg, O.A: Group theory and differential transform analysis of mixed convective heat and mass transfer from a horizontal surface with chemical reaction effects. Chem. Engrg. Commun. **199**, 1012–1043 (2012)
16. Rushikumar, B., Sivaraj, R.: MHD viscoelastic fluid non-Darcy flow over a vertical cone and a flat plate. Int. Commun. Heat Mass Transf. **40**, 1–6 (2013)
17. Sivaraj, R., RushiKumar, B.: Unsteady MHD dusty viscoelastic fluid Couette flow in an irregular channel with varying mass diffusion. Int. J. Heat Mass Transfer **55**, 3076–3089 (2012)
18. Sivaraj, R., RushiKumar, B.: Viscoelastic fluid flow over a moving vertical cone and flat plate with variable electric conductivity. Int. J. Heat Mass Transfer **61**, 119–128 (2013)
19. Srinivas, S., Reddy, A.S., Ramamohan, T.R., Shukla, A.K.: Influence of heat transfer on MHD flow in a pipe with expanding or contracting permeable wall. Ain Shams Engrg J. **5**, 817–830 (2014)

JUPred_SVM: Prediction of Phosphorylation Sites Using a Consensus of SVM Classifiers

Sagnik Banerjee, Debjyoti Ghosh, Subhadip Basu and Mita Nasipuri

Abstract One of the most important types of posttranslational modification is phosphorylation which helps in the regulation of almost all activities of the cell. Phosphorylation is the process of addition of a phosphate group to a protein after the process of translation. In this paper, we have used evolutionary information extracted from position-specific scoring matrices (PSSM) to serve as features for prediction. Support vector machine (SVM) was used the machine learning tool. The system was tested with an independent set of 141 proteins for which our system achieved the highest AUC score of 0.7327. Additionally, our system attained best results for 34 proteins in terms of AUC.

Keywords Posttranslational modification (PTM) · Support vector machine (SVM) · Position-specific scoring matrices (PSSM) · Shannon's entropy · Window consensus · Star consensus

1 Introduction

Posttranslational modification (PTM) is a biochemical modification of a protein sequence after its translation from mRNA. A protein is built using basic blocks of 20 different amino acids. The modification that takes place may be the attachment of biochemical functional groups such as acetate, phosphate, various lipids,

Sagnik Banerjee (✉)
Institute of Engineering and Management, Kolkata, India
e-mail: sagnik.banerjee@iemcal.com

Debjyoti Ghosh · Subhadip Basu · Mita Nasipuri
Jadavpur University, Kolkata, India
e-mail: debjyoti.88@gmail.com

Subhadip Basu
e-mail: subhadip@cse.jdvu.ac.in

Mita Nasipuri
e-mail: mnasipuri@cse.jdvu.ac.in

M. Pant et al. (eds.), *Proceedings of Fifth International Conference on Soft Computing for Problem Solving*, Advances in Intelligent Systems and Computing 436, DOI 10.1007/978-981-10-0448-3_45

carbohydrates, etc. Phosphorylation is the mechanism of addition of a phosphate group $\left(PO_4^{3-}\right)$ to a certain residue. Phosphorylation is essential to several biochemical processes which help in proper functioning of the cell. Annotation of residues using high-throughput experiments is extremely time-consuming. To address this issue we present a computational approach toward predicting phosphorylation sites. Studies of kinases and protein substrates are important for understanding signaling networks in cells and it helps us to develop treatments for signaling defect diseases like cancer. In eukaryotic proteins phosphorylation usually occurs on serine (S), threonine (T), tyrosine (Y) and histidine (H) residues. These phosphorylation sites have been experimentally determined by wet lab experiments like site directed mutagenesis and mass spectrometry, in either low-throughput or high-throughput manners. High-throughput methods suffer from a high false-positive and false-negative rates. Also, the cost of performing the associated experiments is quite high. Therefore, computational prediction of phosphorylation sites become increasingly important as it helps to narrow down the sites where phosphorylation may occur in a protein sequence.

There have been over more than 40 methods for prediction of phosphorylation sites described in literature since 1999. The prediction tools may be kinase-specific or nonkinase-specific. Several machine learning methods have been used for prediction of posttranslational modification sites, AMS 3.0, [3] and a 2012 update of AMS 4.0 [10]. AMS 3.0 was based only on sequence information, using artificial neural network (ANN) method. The query protein sequence was dissected into overlapping short sequence segments. Ten different physicochemical features described each amino acid.

AMS 4.0 [10] was an update to the AMS 3.0 [3]. It predicted 88 different types of single amino acid's posttranslational modifications (PTM) in protein sequences. The researchers selected the experimentally confirmed modifications from the latest UniProt and Phospho.ELM databases for training. The sequence vicinity of each modified residue was represented using amino acids' physico-chemical features encoded using high quality indices (HQI) obtaining by automatic clustering of known indices extracted from *AAINDEX* database. For this method multilayer perceptron (MLP) pattern classifiers were used. The brainstorming consensus meta-learning methodology, on the average boosted the AUC score up to 89 %, averaged over all 88 PTM types.

Apart from the AutoMotif Server (AMS) tools, there has been an array of several others. For example, neural network was used to predict glycosylation and cleavage sites. Some were also used for phosphorylation site prediction. For example, SVM was used by PhosphoSVM [6], Swaminathan's method and PredPhospho, neural networks were used by NetPhosK [8]. Despite almost a decade of research work, we are still unable to boost the precision of in silico methods which can be really useful in high-throughput context of personalized medicine. Therefore, the present research improvements concentrate on designing systems that can predict accurately with faster speed.

2 Materials and Methods

Pattern recognition is the branch of machine learning that attempts to elicit regularity in data. Classification happens to be one of the most important activities in pattern recognition. In order to classify a certain sample, we need to provide the SVM with its features. In this work we have used information extracted from position-specific scoring matrices (PSSM) as features for training and testing.

2.1 *Position-Specific Scoring Matrix*

Position-specific scoring matrix is a commonly used representation of motifs in biological sequences, to extend similarity searches, or as sequence encoding for prediction. A PSSM is constructed from a multiple sequence alignment of several sequences. The primary design objectives in developing the position-specific iterated BLAST (PSI-BLAST) program [1] were speed, ease of use, and automatic operation. In the first round PSI-BLAST will find sequence homologues, just like normal BLAST. In the second iteration of PSI-BLAST, it creates a multiple alignment of several protein sequences. From the alignment itself it figures out which residues tend to be conserved. It does this by creating a custom profile for each position in the sequence from the predesigned multiple alignments. Positions which are highly conserved are awarded high scores whereas weakly conserved positions are penalized with near zero scores. In this work, we have downloaded the set of 90 % redundancy reduced proteins from UniProt. PSSMs were generated by subjecting protein sequences up to three iterations.

2.2 *Shannon's Entropy*

Shannon's Entropy (SE) is computed from the set of 20 values that correspond to the 'weighted observed percentages rounded down' part in the PSSM. Hence, this feature, computed from the weighted observed percentages (WOP) is popularly used as a sequence conservation score to quantify how much a phosphorylation site is conserved in a protein sequence. For a given residue in a sequence, the Shannon Entropy is calculated as follows [6]:

$$\mathrm{SE} = -\sum_{i=1}^{20} p_i \log p_i \tag{1}$$

where p_i is the value in the ith column of the WOP vector.

2.3 Clustering and Classification

Clustering is a process of putting similar elements in a single group and dissimilar elements in a different group. Since the volume of negatively annotated sites is immense, we used mini batch *K*-means to cluster the data and select few elements from each cluster. Classification is the task of assigning a particular category to a certain observation. For every classification task the system needs to be trained with some data. This makes the system capable of categorizing examples. In this work we have used support vector machines, with RBF kernel, as the machine learning tool.

2.4 Star Consensus

We have used star consensus scheme to combine the results obtained for each odd sized window. In star consensus star is basically an integer. In an *n*-star consensus, a particular site is declared as positive if there are at least *n* predictors which predict that particular site as positive. In our work we have performed experiments with window sizes varying from 9 to 25. The final result was obtained by performing a star consensus of the results from the nine windows.

The entire sequence of the research work has been presented here. At first, Uniref90 was downloaded from www.uniprot.org against which the PSSMs were generated using PSI-BLAST [1]. Reviewed protein sequences which had at least one annotated site of phosphorylation were downloaded from www.uniprot.org. This resulted in 33,541 protein sequences. To remove redundancy we used CD-HIT [9] to cluster sequences at 30 % threshold. This brought down the protein count to 7242. We eliminated those sequences which had '*X*' in it. This led us to 7188 sequences. Now we segregated the test proteins. We choose 141 test proteins randomly in proportion to the number of proteins corresponding to each organism in the entire set. This left us with 7047 proteins which we used for training. PSSMs were generated for each one of 7047 proteins for three iterations against 90 % redundancy-reduced Uniprot. Data were extracted from each one of these sequences by placing a hypothetical odd sized window over a residue as depicted in Fig. 1. If that residue is phosphorylated then we treat it as positive. If the residue is one of *S*, *T*, or *Y* and no annotation is provided then we assume it to be negative. The window size was varied from 9 to 25. To reduce the number of negative data, mini batch *K*-means was applied to cluster only negative data for each window. Samples were chosen from the cluster to ensure that the number of negative and positive training

...GNK[VPPVRV**Y**GPDCVV]LM[PPEPPLsKRNPPA]LRL....

Fig. 1 Positive segment and negative segment

examples is almost equal [4]. Threefold cross-validation was done for each window and each residue. SVMs were trained with the best parameter obtained from the cross-validation experiments. Separate SVMs were designed to predict phosphorylation for each of the residues. Finally testing was done on an independent set of 141 proteins. The results obtained from the each of the 9 windows were combined using star consensus scheme.

3 Results and Discussions

In this section we present our findings when our system was tested on an independent set of 141 proteins. We trained our system with data extracted from 7047 protein sequences. For each odd sized window between 9 and 25 we trained three SVMs, one for each residue among *S*, *T*, and *Y*. Then we combined the results obtained from each window using star consensus scheme. This led us to a significant improvement in all the metrics (Fig. 2).

After obtaining the window consensus result, we compared the performance of our predictor with seven other state-of-the-art predictors. The predictors we chose for comparison were DISPHOS [12], GPS [13], Musite [7], NetPhos2.0 [5], NetPhosK1.0 [5], Phosfer [11], and PhosphoSVM [6]. For DISPHOS, NetPhos2.0, Phosfer, and PhosphoSVM we considered 0.5 as the threshold for positive data. For musite the specificity level was set to 0.95. On an average, our system performed best when compared with seven other predictors on the basis of AUC which proves that our predictor is highly robust (Fig. 3).

Below we have presented the results obtained for each residue for the metric AUC. It is clear from the graph that JUPred_SVM achieves the highest AUC for the residue *T*. Its performance in the remaining two residues is considerably good. The details about performance on other metric have been provided in the supplementary document. Musite happens to be one of those predictors for which MCC and F1 is very high but AUC is comparatively quite low which points out poor robustness of the predictor (Fig. 4).

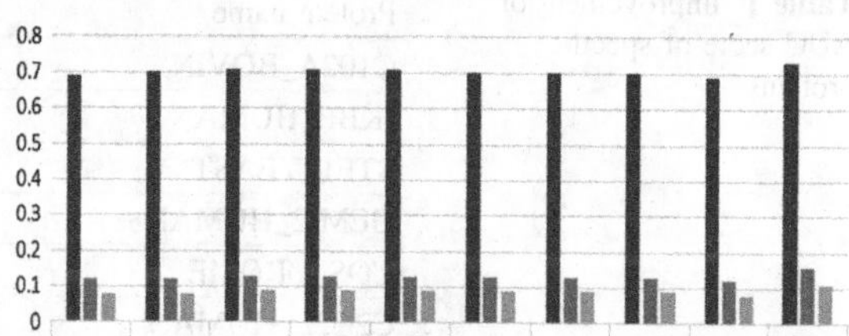

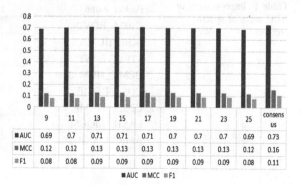

	9	11	13	15	17	19	21	23	25	consensus
AUC	0.69	0.7	0.71	0.71	0.71	0.7	0.7	0.7	0.69	0.73
MCC	0.12	0.12	0.13	0.13	0.13	0.13	0.13	0.13	0.12	0.16
F1	0.08	0.08	0.09	0.09	0.09	0.09	0.09	0.09	0.08	0.11

AUC MCC F1

Fig. 2 Distribution of results for each window and the final consensus for metrics AUC, MCC, and F1

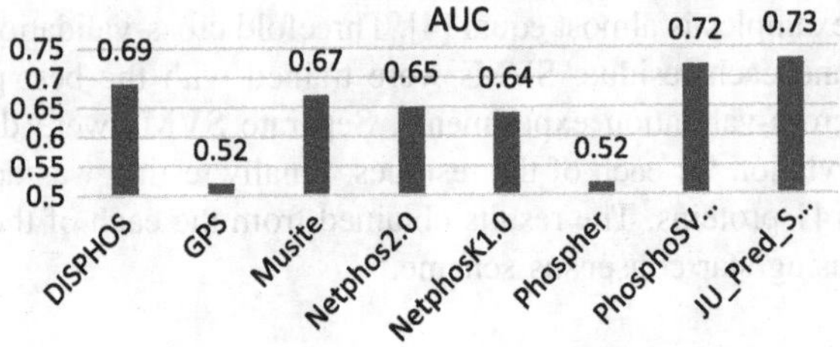

Fig. 3 Comparison of Ju_Pred_SVM with other predictors on the basis of AUC

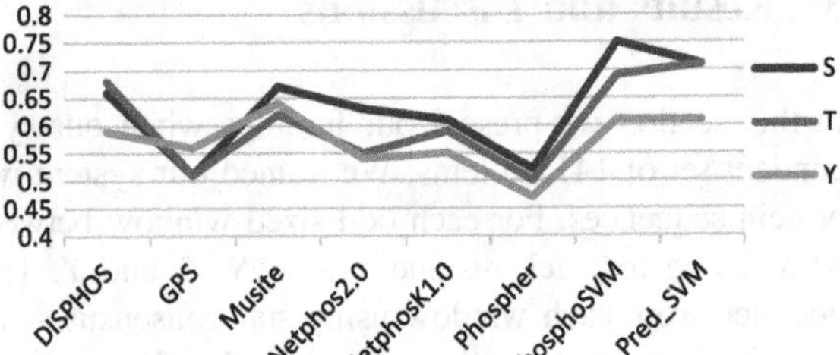

Fig. 4 Comparison of AUC for different predictors and for each residue

We further conducted an experiment on per protein basis to establish the robustness of our classifier over the others. Of 141 proteins, JUPred_SVM gives the best result for 34 protein sequences. Musite predicts 33 proteins with the highest AUC. In some proteins JUPred_SVM successfully improves the AUC count by a very large factor. The list for such proteins has been provided in Table 1. Most of the proteins in this list are of extreme importance. For instance, STF1_YEAST is essential for stabilizing the ATPase inhibitor complex, which normally dissociates when ATP is consumed externally. The GEMI2_HUMAN protein is a catalyst which assists in the assembly of small nuclear ribonucleoproteins (snRNPs), which in turn are the building blocks of the spliceosome. LAS1L_MOUSE is essential in the biogenesis of the 60S ribosomal subunit. C2C2L_HUMAN helps in positive regulation of insulin secretion involved in cellular response to glucose stimulus.

Table 1 Improvement of AUC score of specific proteins

Protein name	AUC improvement
C102A_BOVIN	0.1078
IKBE_HUMAN	0.1288
STF1_YEAST	0.2143
GEMI2_HUMAN	0.0972
COS_DROME	0.0856
C2C2L_HUMAN	0.0855
LAS1L_MOUSE	0.2313
AMRA1_MOUSE	0.0813

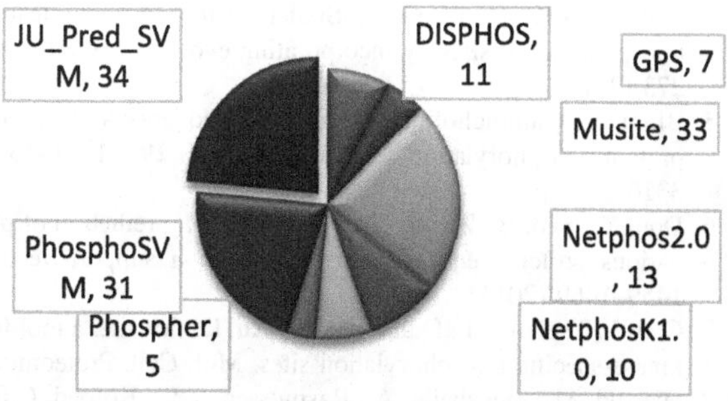

Fig. 5 Pie-chart depicting contribution of each predictor toward predicting individual protein phosphorylation ith maximum AUC score

Below we provide a pie chart which depicts the distribution of the different predictors and the number of protein sequences in which each of them could predict with the highest AUC. It is clear that the predictor JUPred_SVM stands out by successfully performing the best among all the predictors (Fig. 5).

4 Conclusion

Phosphorylation of protein regulates almost all aspects of cell life. A nonkinase-specific protein phosphorylation prediction method has been proposed in this work. This predictor attempts to predict whether a residue is phosphorylated or not from evolutionary information obtained from the PSSM. In terms of AUC our predictor performed best when tested with an independent set of 141 proteins. The AUC values were 0.711/0.705/0.614665 for S/T/Y residues. The total AUC was 0.733, better than any other predictor. Hence, big data technology [2] could be employed to harness humongous amounts of data which are often dealt in bioinformatics.

Acknowledgments Authors are indebted to CMATER, department of computer science and Engineering, Jadavpur University for providing the necessary support for carrying out this experiment.

References

1. Altschul, S.F., Madden, T.L., Schäffer, A.A., Zhang, J., Zhang, Z., Miller, W., Lipman, D.J.: Gapped BLAST and PSI-BLAST: a new generation of protein database search programs. Nucleic Acids Res. **25**(17), 3389–3402 (1997)
2. Banerjee, S., Basu, S., Nasipuri, M.: Big data analytics and its prospects in computational proteomics. In: Information Systems Design and Intelligent Applications, pp. 591–598. Springer (2015)
3. Basu, S., Plewczynski, D.: AMS 3.0: prediction of post-translational modifications. BMC Bioinformatics **11**(1), 210 (2010)

4. Biswas, A.K., Noman, N., Sikder, A.R.: Machine learning approach to predict protein phosphorylation sites by incorporating evolutionary information. BMC Bioinformatics **11**(1), 273 (2010)
5. Blom, N., Gammeltoft, S., Brunak, S.: Sequence and structure-based prediction of eukaryotic protein phosphorylation sites. J. Mol. Biol. **294**, 1351–1362 (1999). doi:10.1006/jmbi.1999.3310
6. Dou, Y., Yao, B., Zhang, C.: PhosphoSVM: prediction of phosphorylation sites by integrating various protein sequence attributes with a support vector machine. Amino Acids **46**(6), 1459–1469 (2014)
7. Gao, J., Thelen, J.J., Dunker, A.K., Xu, D.: Musite, a tool for global prediction of general and kinase-specific phosphorylation sites. Mol. Cell. Proteomics **9**(12), 2586–2600 (2010)
8. Hjerrild, M., Stensballe, A., Rasmussen, T.E., Kofoed, C.B., Blom, N., Sicheritz-Ponten, T., Larsen, M.R., Brunak, S., Jensen, O.N., Ganuneltoft, S.: (2004). Identification of phosphorylation sites in protein kinase A substrates using artificial neural networks and mass spectrometry. J. Proteome Res. **3**, 426–433. doi:10.1021/pr0341033
9. Huang, Y., Niu, B., Gao, Y., Fu, L., Li, W.: CD-HIT Suite: a web server for clustering and comparing biological sequences. Bioinformatics **26**(5), 680–682 (2010)
10. Plewczynski, D., Basu, S., Saha, I.: AMS 4.0: consensus prediction of post-translational modifications in protein sequences. Amino Acids, **43**(2), 573–582 (2012)
11. Trost, B., Kusalik, A.: Computational phosphorylation site prediction in plants using random forests and organism-specific instance weights. Bioinformatics, btt031 (2013)
12. Xue, Y., Gao, X., Cao, J., Liu, Z., Jin, C., Wen, L., Yao, X., Ren, J.: A summary of computational resources for protein phosphorylation. Curr. Protein Pept. Sci. **11**(6), 485–496 (2010)
13. Xue, Y., Liu, Z., Cao, J., Ma, Q., Gao, X., Wang, Q., Jin, C., Zhou, Y., Wen, L., Ren, J.: GPS 2.1: enhanced prediction of kinase-specific phosphorylation sites with an algorithm of motif length selection. Protein Eng. Des. Sel. **24**(3), 255–260 (2011)

Fuzzy Logic-Based Gait Phase Detection Using Passive Markers

Chandra Prakash, Kanika Gupta, Rajesh Kumar and Namita Mittal

Abstract With the advancement in technology, gait analysis plays a vital role in sports, science, rehabilitation, geriatric care, and medical diagnostics. Identification of accurate gait phase is of paramount importance. The objective of this paper is to put forward a novel approach via passive marker-based optical approach that automatically recognizes gait subphases using fuzzy logic approach from hip and knee angle parameters extracted at RAMAN lab at MNIT, Jaipur. In addition to stance phase and swing phase, the approach is capable of detecting all the subphases such as initial swing, mid swing, and terminal swing, loading response, mid stance, terminal stance and preswing. The prototype of the system provides an effective and accurate gait phase that could be used for understanding patients' gait pathology and in control strategies for active lower extremity prosthetics and orthotics. It is an automated, easy to use, and very cost-efficient yet reliable model.

Keywords Gait phase detection · Fuzzy logic · Optical based approach · Human gait analysis

1 Introduction

Over the past six decades, the quantification of gait assists health professionals to explore clinical use of gait analysis is done not only in India but worldwide. This is possible because of clinically befitting software and hardware with faster computing

Chandra Prakash (✉) · Kanika Gupta · Rajesh Kumar · Namita Mittal
Malaviya National Institute of Technology, Jaipur, India
e-mail: cse.cprakash@gmail.com

Kanika Gupta
e-mail: kanika100388@gmail.com

Rajesh Kumar
e-mail: rkumar.ee@gmail.com

Namita Mittal
e-mail: mittalnamita@gmail.com

M. Pant et al. (eds.), *Proceedings of Fifth International Conference on Soft Computing for Problem Solving*, Advances in Intelligent Systems and Computing 436, DOI 10.1007/978-981-10-0448-3_46

capability for collection and analysis of data for gait analysis. By definition, *Gait* is termed as manner of walking. Human gait is also known as bipedal and it is a method of locomotion achieved through two legs alternately to offer both support and propulsion with the condition that at least one foot should be in contact with the ground at all times [1]. While in case of running it is not necessary that at least one foot should be in contact with the ground.

The average adult takes 5000–8000 steps a day. The research associated with mechanics of human body while walking on stairs in an ascending/descending order to recognize explicit deviations in the gait pattern and determining their reasons and effects is known as gait analysis. Pathological gait identification is the most unswerving application of gait analysis [2]. It has other numerous applications, such as physical therapy, biometrics, rehabilitation, sports, science, and geriatric care [3–5]. Identification of gait phases is of vital importance in all these applications. Clinicians are able to utilize gait segmentation concept in their routine clinical practice to evaluate a patient's status, treatment, and rehabilitation for complex musculoskeletal and neurological disorders using the spatiotemporal and kinematics parameters.

Gait analysis is not a new research area. Systematic study of gait started with the description of walking principle by Leonardo da Vinci, Galileo, and Newton. In 1682 a student of Galileo, Borelli, described how balanced walking can be achieved using the concept of center of gravity of body in De Motu Animalium. Gait cycle was clearly described by Weber brothers in 1836 at Germany. Since 1960 clinical gait analysis has gained momentum. Contemporary work on clinical gait analysis has been discussed by Perry and Sutherland [2, 6].

Recognition of gait cycle phases is extensively useful to spot the time instance at which feedback should be applied for safety and effective response by the patient undergoing rehabilitation or physical therapy.

This paper implements a fuzzy-based approach for automatic detection of gait phase from hip and knee angle extracted using passive marker. This system is cheaper but efficient with respect to the other system. By comparing the normal and current gait cycle phase pattern, healthcare professionals can suggest an effective treatment. Details of the proposed system are described in the later section of this paper. The paper is planned and structured as follows: Sect. 2 covers the basics about the human gait cycle and the angle patterns followed by literature survey; Sect. 3 presents the implementation of fuzzy inference system for automatic gait phase detection; Sect. 4 discusses the results. Conclusions and future scope are discussed in Sect. 5.

2 Human Gait Cycle

Walking is a series of gait cycles. For understanding pathology, normal gait pattern is essential to detect alteration in gait. Gait is considered as stereotyped activity in both young and old healthy people.

2.1 Gait Cycle

The gait cycle is measured as the time period between two successive incidences of the recurring phenomena of walking. Gait cycle is a combined function of the lower extremity, pelvis, and spinal column.

Gait cycle begins with heel contact of either of the foot and ends with the heel contact of the same foot. Therefore, one complete gait cycle consists of two steps one of either right foot and then left, or vice versa. By convention, normal gait cycle is the time period in which heel of one foot contacts the ground when the heel contact of same foot takes place and forward propulsion of the center of gravity is involved.

A single gait cycle consists primarily of two phases: a swing phase and a stance phase [2]. In general, stance phase begins with the heel contact and ends with the toe of the same foot. The duration when the foot remains in contact with the ground is known as stance phase and accounts for approximately 60 % of the normal gait cycle. The duration when the foot is off the ground is known as swing phase and accounts for 40 % of the gait cycle. Swing phase begins with the toe off of the delete foot and ends with the heel contact of that same foot.

Stance phase and swing phase could be further segmented into eight segments and are referred as critical incidents that enable the examiners to further specify the abnormal aspects of gait. Classical gait model by Perry divides gait cycle into eight subphases (5 stance and 3 swing) [2].

Further, stance phase is divided among five subphases: initial contact, loading response, mid stance, terminal stance, and preswing whereas swing phase has three subphases: initial swing, mid swing, and terminal swing. Figure 1 shows the fundamentals of gait phases and expected interval of phases and subphases in total gait cycle.

For this study, initial contact and loading response are considered as same phase as former is an instance of loading response only. It is seen that different pathologies affect different segments of either swing or stance phase. Any abnormality suggests that there is pathology which should be identified by the examiner.

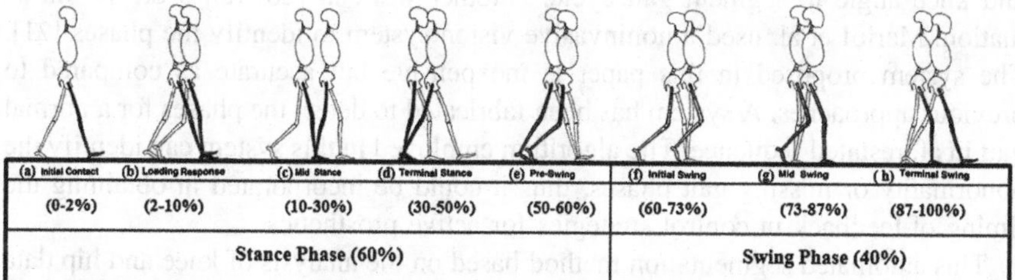

Fig. 1 Fundamental gait phases and expected interval of gait cycle [2]

2.2 Literature Survey

For quantitative gait analysis researchers have used numerous wearable sensors like gyroscopes, accelerometers, EMG sensors, force sensitive resistors (FSRs), inertial sensors, force contact sensors, foot switches, load cells, etc. [7–11]. Alternative approaches such as force plate and vision-based methods can be used to compute quantitative parameters of interest [12, 13]. All these methods have been used in gait segmentation.

Wang et al. were able to identify only initial contact, stance phase, and swing phase by using 3-axis accelerometer fixed on ankle [14]. Pappas et al. used a gyroscope attached to rear end of the shoe along with force sensitive resistors to detect heel strike (Initial Contact), stance phase, heel-off and swing phase. Only four phases were identified by this approach [7].

Computational-based techniques have also been proposed for real and precise gait phase recognition. Researchers have explored fuzzy inference system (FIS) to segment gait phase. Liu et al. used gyroscope and accelerometers to detect four gait phases using fuzzy logic due to its robustness to noise [15]. Kyoungchul et al. implemented fuzzy-based approach to detect gait phases from foot pressure patterns [16]. DeRossi, et al. have used hidden markov model for identification of six gait phases [17].

But in all these techniques, one or more sensors need to be attached to one or both legs which is not an appropriate approach as the presence of sensors, cables, or other components hinder the subject's natural motion. In contrary, vision-based analysis systems can be used to obtain gait kinematics smoothly and continuously without affecting the natural motion. The technology associated with this measurement approach has continued to change over the past decade.

A new approach using red color reflective marker is used to obtain the gait cycle and spatiotemporal and kinematic parameters [18]. This paper is the extension to this work and uses fuzzy logic for gait phase identification technique using two joint angles obtained from 2D optical system proposed at RAMAN lab, MNIT Jaipur. Other than these, time and stage variables are also considered.

Similar kind of work is reported in [19], in which author deployed hip, ankle, and knee angle to segment gait cycle. Another work in [20] relies on 3D information. Mariol et al. used a noninvasive vision system to identify the phases [21]. The system proposed in this paper is inexpensive but accurate as compared to previous approaches. A system has been fabricated to detect the phases for a normal gait in aforestated sequence. The algorithm employed in this system can identify the abnormality or missing gait phases, thus it could be incorporated in obtaining the timing of feedback in control strategies for active prosthetics.

This automated segmentation method based on the analysis of knee and hip data obtained from 2D optical system has been put to test on six healthy subjects.

3 System Architecture and Joint Angles

The proposed system consists of a digital video camera for recording and a computer for data acquisition and processing. Figure 2a illustrates the marker position. Figure 2b shows an optical motion capture system developed to detect and track the markers fastened to the subject's body at anatomical points of concern [18].

An algorithm has been developed to process video frames and knee and hip angles are obtained as shown in Fig. 2c. The fuzzy inference system (FIS) shown in Fig. 2d maps inputs (hip and knee angles) to outputs (gait phases) using a predefined set of fuzzy rules. These rules will be touched upon in the later section. Finally, the output of this system is segmented as gait cycle. MATLAB has been used to develop FIS.

In [13, 18] Prakash et al. discussed a method of obtaining joint kinematics and Spatio-temporal parameters using passive markers. An optical motion capture system is developed to detect and track the subject's joint movements in 2D space.

The data obtained from the experiment is processed using a simulation framework, which assists in kinematic and dynamic analyses of human gait. The joint coordinates obtained by tracking the five reflective markers are then used for 2D gait analysis. The data are then filtered using average filtering techniques to remove irregularity. Figure 3 shows the average knee and hip joint angle patterns for normal walking.

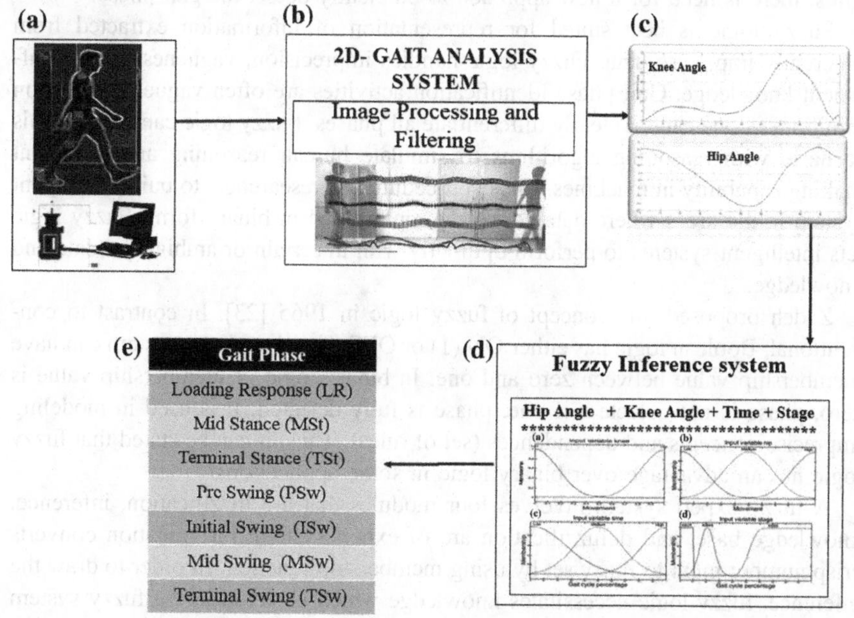

Fig. 2 Gait phase detection system methodology; **a** Marker setup and data acquisition; **b** Gait analysis system; **c** Gait kinematics extraction; **d** Fuzzy inference system for gait phase identification; **e** Phase detection

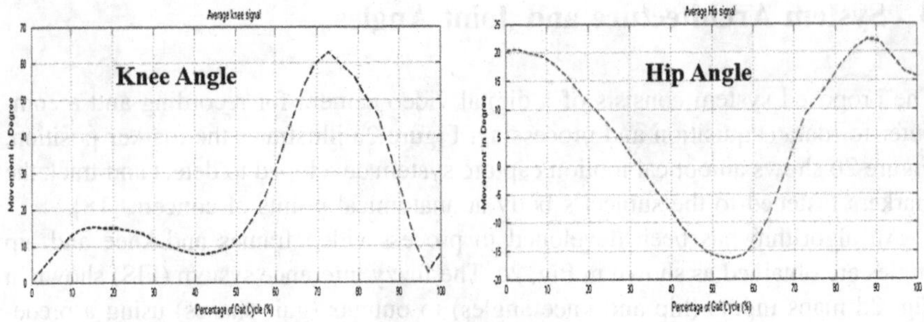

Fig. 3 Average knee and hip joint angle for one gait cycle

4 Fuzzy Inference System

At a given stance of time, kinematic parameters such as hip, ankle, and knee angle can be put to use to detect the gait phase [22]. One possible approach for gait segmentation is by setting threshold for discrete event analysis. But it has a limitation in implementation, for this change the signal should be clearly visible. Change in phases is not visible in knee and hip angle parameter as gait is not a set of isolated actions, although phases keep changing continuously and smoothly. Thus, there is need for a new approach to efficiently detect the gait phase.

Fuzzy logic is best suited for representation of information extracted from inherently imprecise data. Fuzzy logic handles imprecision, vagueness, and insufficient knowledge. Gait phase identification activities are often vague or based on intuition, as one cannot clearly differentiate all phases. Fuzzy logic can work in this scenario with reasoning algorithms to simulate human reasoning and judgment making capability in machines. These procedures let researchers to build intelligent system in the areas where data cannot be represented in binary form. Fuzzy logic lets intelligent systems to perform optimally with uncertain or ambiguous data and knowledge.

Zadeh proposed this concept of fuzzy logic in 1965 [23]. In contrast to conventional, Boolean logic has either ON (1) or OFF (0) value, fuzzy system can have membership value between zero and one. In binary logic, if membership value is zero, phase is not detected; if one, phase is fully detected. It is used in modeling imprecise concepts and dependences (set of rules). Thus, it can be stated that fuzzy logic has an advantage over binary logic in some applications.

A fuzzy expert system involves four modules that are fuzzification, inference, knowledge base, and defuzzification are of expert system. Fuzzification converts crisp number input to fuzzy set by using membership function. In order to draw the inference, fuzzy logic necessitates knowledge which is stored in the fuzzy system and provided by an expert who have experience or who knows the process of that specific domain. If-then rules suggested by expert are stored in knowledge base. Using these rules, inference engine simulates reasoning process similar to humans but output is in fuzzy form. So, there is need of conversion of fuzzy set to crisp value, i.e., defuzzification.

4.1 Input Parameter of the Fuzzy Membership Function

In this research, only two joint angles; knee and hip along with time and stage (function of time) variables have been considered. As discussed earlier hip and knee angle used has been taken from our previous work [18]. Time and stage variables have been considered along with hip and knee angles as input [21]. The membership function (MF) values for hip, knee, time, and stage input are defined based on the normative data presented in [2, 21] and is shown in Table 1.

Hip angle can be divided into three intervals (Extension, Low Flexion, and high Flexion). Knee angle movement is also divided into three intervals (Low, Medium, and High). Similarly, input variable, time is divided into low, medium, and high intervals. Stage variable is similar to time variable but not determined function of interval. Instead it has loading response (SLR), Mid stance (SMS) and swing stance (SSS) as its membership functions. In this stage input variable is employed to differentiate between the phases, mid stance and terminal swing and between loading response and initial swing. Gaussian function is used for hip, knee, and time. For stage variable, SLR uses polynomial-based Z-function, Gaussian function is used for SMS membership and S-function is used for SSS. The membership function for phase is used as triangular as shown in Fig. 4.

4.2 Fuzzy Rule Classification

Gait phase detection's If-Then rules recommended by an expert are stored in knowledge base. For defining fuzzy rules, books on gait analysis and biomechanics have been referred [1, 2, 24] but all variables are not considered in this research and shown in Table 2.

Table 1 Fuzzy membership function for inputs parameters

Input	Quantity
Hip	Extension (0 to −30), low flexion (0–15), high flexion (15–30)
Knee	Low (0–20), medium (20–40), high (40–70)
Time interval (% of gait)	Low (0–25), medium (25–75), high (75–100)
Stage	Loading response (SLR) (0–10 %), mid stance (SMS) (10–40 %), and swing stance (SSS) (40–100 %)

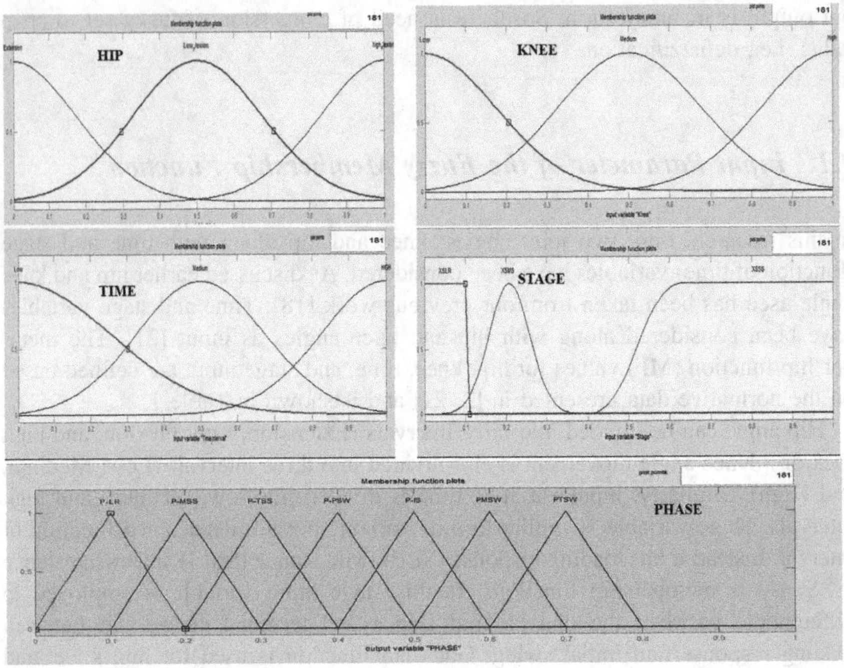

Fig. 4 Membership function for input variable: hip, knee, time and stage, and output variable phase

Table 2 Set of fuzzy rules [1, 2, 24]

	Hip	Knee	Time	Stage	Gait phase
1	High flexion	Low	Low	SLR	Loading response (LR)
2	Not high flexion	Low	Low	SMS	Mid stance (MSt)
3	Low flexion	Low	Low	SMS	Mid stance (MSt)
4	Extension	Medium	Medium	–	Terminal stance (TSt)
5	Extension	Not low	Medium	SSS	Preswing (PSw)
6	Low flexion	Not low	Medium	SSS	Initial swing (ISw)
7	High flexion	High	High	SSS	Mid swing (MSw)
8	High flexion	Low	High	SSS	Terminal swing (TSw)

The fuzzy system designed for this work is a Mamdani system. The defuzzification scheme used is centroid-based. That is, the output is just the consequence of a specific condition of the lower body section.

For example, a condition for the loading response phase in Table 2 suggests that as "If hip angle is of 20° while Knee is 18 and Time period are 18 % and Stage is also 18 % of total time, then the human motion is in the LR phase".

Fuzzy logic expresses the statement as "Hip angle is of High Flexion nature AND Knee angle has Low MF AND Time interval is also Low AND Stage is SLR then PHASE is Loading Response". Value of Membership function of the Loading Response phase is approximately equal to one. Based on these rules, gait phase can be determined at any given instance of time.

5 Experimental Result

5.1 Data Set

Five healthy young subjects were chosen for conducting the experiment. Each subject walked on a starting line six times. Before the actual trial, subject was asked to walk for three trials so that they feel comfortable with the setup. This is very important for normal data collection and subjects confirmed this in feedback. Gait data of five subjects comprising of both male and female subjects with the age group between 18 and 30 years was recorded. Markers were fastened to the clothes of the targeted subject at anatomical points of concern, i.e., shoulder, hip, knee, ankle, and toe. The Kinematics parameters of the subjects were visualized using MATLAB.

5.2 Gait Phase Identification

The result of one of the healthy subjects (age 20, weight 54) corresponding to the knee and hip is shown in Table 3 and Fig. 5. The first phase, loading response is found in the interval 0–10 %, which indicates a normal behavior occurrence of this phase.

Mid stance is detected in 10–32 %. The terminal stance phase is detected in the percentage 33–50 % and preswing is found in the range of 50–64 %. The initial swing is detected in the percentage of 64–75, mid swing is located in 75–87 %. The terminal swing phase is correctly detected in the percentage 87–100 %.

Cross validation of these results with Fig. 1 as shown in Table 3 indicates that this system reports very similar outcome with few differences but these differences are not significant. Detection of gait phase is in natural order, i.e., LR, MSt, TSt, PSw, ISw, MSw, and TSw. This shows that that subject gait pattern is free from gait

Table 3 Extracted gait phases using proposed system for a subject

Gait phase	Our result (%)	Actual values (%) [2]
Loading response (LR)	0–10	0–10
Mid stance (MSt)	10–32	10–30
Terminal stance (TSt)	33–50	30–50
Pre-Swing (PSw)	50–64	50–60
Initial swing (ISw)	64–75	60–73
Mid swing (MSw)	75–87	73–87
Terminal swing (TSw)	87–100	87–100

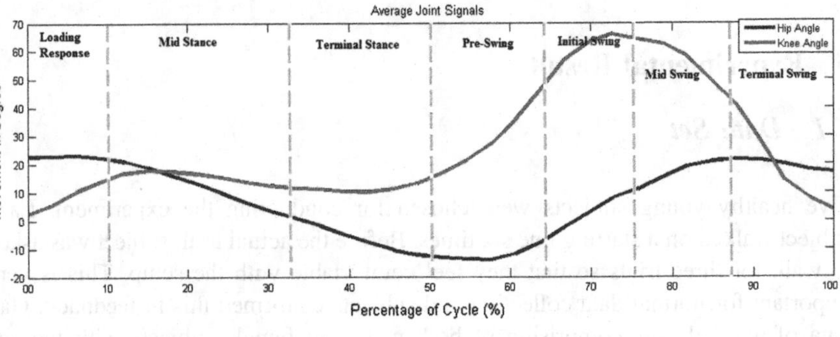

Fig. 5 Kinematics signal and phase detection

abnormalities in the subject gait pattern. The result of the fuzzy system is illustrated in Fig. 6. Based on fuzzy rules, the result follows the natural order of phase sequence and at a time only one gait phase has been detected.

Normal gait phase plots for two strides are clearly shown in Fig. 6. It can be clearly seen that the sequence of gait phases obtained is in natural order. Detected

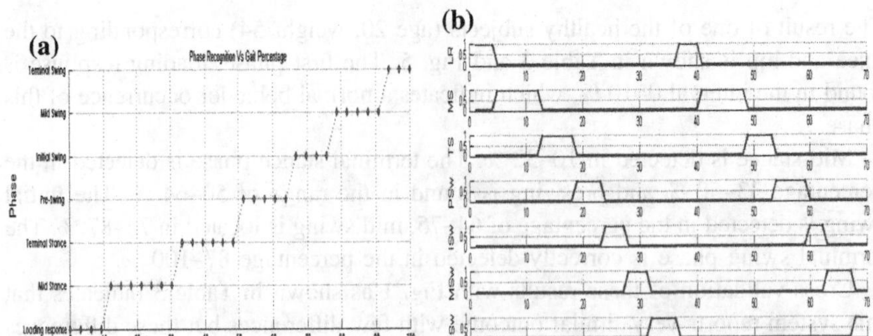

Fig. 6 **a** Gait phase and cycle percentage. **b** Normal gait phase plots for two strides

gait phases in each gait cycle had a maximum membership value of 1, stipulating that within a gait cycle, all phases were fully identified. It also indicates that a maximum mean deviation of 1.04 % with reference to normative data is spent in transition from one phase to another phase.

6 Conclusions and Future Scope

The work discussed in this paper is initial study of implementing fuzzy based techniques for gait segmentation for 2D optical-based system using passive marker. The main benefit is that these markers do not need high cost and excessive time in placing. The effect of passive makers on normal gait pattern is also very less as compared to other makers. The subjects' opinion on these marker enactments further simplified that the setup did not affect the gait performance. Findings of the experiments indicate that this technique was able to correctly segment the gait phases at very low cost using fuzzy based approach. For all the healthy subjects considered in this research identification rate in natural sequence is 100 %. This approach of gait phase detection has the potential to use in rehabilitation activities and gait analysis.

Though the method presented in the paper is cost-effective, reliability can be improved using sensors such as accelerometer, goniometer, and EMG are used. With the use of hybrid computational techniques, such as neuro-fuzzy approach, could further improve accuracy and robustness of system for the gait phase detection algorithm.

Acknowledgments The author gratefully acknowledges the support of Department of Science and Technology, India for funding this project under grant SR/S3/MERC/0101/2012.

References

1. Whittle, M.W.: Gait Analysis: An Introduction, 4th edn. Elsevier (2007)
2. Perry, A.J.: Gait Analysis: Normal and Pathological Function. Slack, NJ (1992)
3. Senanayake, B.C.M., Senanayake, S.M.: Computational intelligent gait phase detection system to identify pathological gait. IEEE Trans. Inf. Technol. Biomed. **14**(5), 1173–1179, (2010)
4. Wang, H., Wu, J., Wang, Y., Ren, L., Zhang, D., Lu, H.: Research on the lower limb gait rehabilitation. In: 2014 IEEE International Conference on Mechatronics and Automation (ICMA) pp. 1243–1247, 3–6 Aug 2014
5. Courtney, J., Paor, A.M.: A monocular marker-free gait measurement system. IEEE Trans. Neural Syst. Rehabil. Eng. **18**(4), 453–460 (2010)
6. Sutherland, D.H.: The evolution of clinical gait analysis part III—kinetics and energy assessment, Gait Posture **21**(4), 447–46 (2005)
7. Pappas, I.P., Popovic, M.R., Keller, T., Dietz, V., Morari, M.: A reliable gait phase detection system. IEEE Trans. Rehabil. Eng. **9**, 113–125 (2001)

8. Knight, J.F., Bristow, H.W., Anastopoulou, S., Baber, C., Schwirtz, A., Arvanitis, T.N.: Uses of accelerometer data collected from a wearable system. Pers. Ubiquit. Comput. **11**, 117–132 (2007)
9. Stefano, A., Burrudge, J., Yule, V., Allen, R.: Effect of gait cycle selection on EMG analysis during walking in adults and children with gait pathology. Gait Posture **20**, 92–101 (2004)
10. Bamberg, A., Benbasat, A.Y., Scarborough, D.M., Krebs, D.E., Paradiso, J.A.: Gait analysis using a shoe integrated wireless sensor system. IEEE Trans. Inf. Technol. Biomed. **12**, 413–423 (2008)
11. Bamberg, S.J.M., Benbasat, A.Y., Scarborough, D.M., Krebs, D.E., Paradiso, J.A.: Gait analysis using a shoe-integrated wireless sensor system. IEEE Trans. Inf. Technol. Biomed. **12** (4), 413–23 (2008)
12. Gouwanda, D., Senanayake, S.M.N.A.: Emerging Trends of Body—Mounted Sensors in Sports and Human Gait Analysis. In: IFMBE Proceedings, 4th Kuala Lumpur International Conference on Biomedical Engineering (2008).
13. Prakash, C., Gupta, K., Mittal, A., Kumar, R., Laxmi, V.: Passive marker based optical system for gait kinematics for lower extremity. Procedia Comput. Sci. **45**, 176–185 (2015)
14. Wang, J.-S., Lin, C.-W., Yang, Y.-T.C., Ho, Y.-J.: Walking pattern classification and walking distance estimation algorithms using gait phase information. IEEE Trans. Biomed. Eng. **59** (10), 2884–2892 (2012)
15. Liu, T., Inoue, R., Shibuta, K., Morioka, H.: Development of wearable sensor combinations for human lower extremity motion analysis, In: Proceedings of the 2006 IEEE International Conference on Robotics and Automation, Orlando, Florida, pp. 1655–1660, May 2006
16. Kong, K., Bae, J., Tomizuka, M.: Detection of abnormalities in a human gait using smart shoes, SPIE Smart Structures/NDE, Health Monitoring (2008)
17. DeRossi, S.M.M., et al.: Gait segmentation using bipedal foot pressure patterns. In: Proceedings of IV IEEE RAS/EMBS International Conference in Biomechatronics and Biomedical Robotics pp. 361–366 (2012)
18. Prakash, C., Mittal, A., Kumar, R., Mittal, N.: Identification of spatio-temporal and kinematics parameters for 2-D optical gait analysis system using passive markers. In: International Conference on Advances in Computer Engineering and Application, pp. 143–149 (2015)
19. Senanayake, C.M., Senanayake, S.M.N.A.: Evaluation of gait parameters for gait phase detection during walking. In: 2010 IEEE Conference on Multisensor Fusion and Integration for Intelligent Systems (MFI), pp. 127–132, 5–7 Sept 2010
20. MacDonald, C., Smith, D., Brower, R., Ceberio, M., Sarkodie-Gyan, T.: Determination of human gait phase using fuzzy inference. In: Proceedings of IEEE Rehabilitation Robotics, pp. 661–665 (2007)
21. Chacon-murguia, M.I., Arias-enriquez, O., Sandoval-rodriguez, R.: A fuzzy scheme for gait cycle phase detection oriented to medical diagnosis. Pattern Recogn. Lect. Notes Comput. Sci. **7914**, 20–29 (2013)
22. Senanayake, C.M., Senanayake, S.M.N.A.: Evaluation of gait parameters for gait phase detection during walking. 2010 IEEE Conference on Multisensor Fusion and Integration for Intelligent Systems (MFI), pp. 127–132, 5–7 Sept (2010)
23. Zadeh, L.A.: Fuzzy sets. Inf. Control **8**(3), 338–353 (1965)
24. Nordin, M., Frankel, V.H.: Basic Biomechanics of the Musculoskeletal System, 3rd edn., Chaps. 7–9, Lippincott Williams and Wilkins (2001)

Extraction of Retinal Blood Vessels and Optic Disk for Eye Disease Classification

V.K. Jestin and Rahul R. Nair

Abstract The retina is the important and only part of the human body from which the blood vessel information can be clearly obtained. The information about blood vessels in the retina plays an important role in the finding and efficient treatment of diseases such as glaucoma, macular degeneration, degenerative myopia, diabetic retinopathy, etc. The structure of the retinal vessels is a significant way to predict the presence of eye diseases such as hypertension, diabetic retinopathy, glaucoma, hemorrhages, retinal vein occlusion, and neovascularization. Ophthalmologists find it difficult when the diameter and turns for the retinal blood vessel or shape of the optic disk structures are complicated or a huge number of eye images are acquired to be marked by hand, all of which eventually leads to error. Therefore, an automated method for retinal blood vessel extraction and optic disk segmentation, which preserves various vessel and optic disk characteristics, is presented in this work and is attractive in computer-based diagnosis. Here, we implement a new competent method for detection of diseases using the retinal fundus image. In this anticipated work the first step is the extraction of retinal vessels by graph cut technique. The retinal vessel information is then used to calculate approximately the position of the optic disk. These results are given to an ANN classifier for the detection and classification of diseases. By robotically identifying the disease from normal images, the workload and its costs will be reduced.

Keywords Diabetic retinopathy · Hypertension · Glaucoma · Fundus images · Optic disk · ANN

V.K. Jestin (✉)
Department of ECE, Hindusthan College of Engineering and Technology, Coimbatore, Tamil Nadu, India
e-mail: jestinvk@gmail.com

R.R. Nair
Department of ECE, Believers Church Caarmel Engineering College, Ranni, Kerala, India
e-mail: rahulkottattu@gmail.com

M. Pant et al. (eds.), *Proceedings of Fifth International Conference on Soft Computing for Problem Solving*, Advances in Intelligent Systems and Computing 436, DOI 10.1007/978-981-10-0448-3_47

1 Introduction

Eye disease classification and retinal extraction is an emerging and important field in image processing. Researchers shown keen interest in extracting various features like retinal vessels, fovea, optic disk, and macula automatically from eye images [1]. The method described in this work is aimed to assist doctors for early diagnosis of various eye diseases. People use various color images to study eye diseases such as vein occlusion, DR, NPDR, etc. Fovea is a significant characteristic for detection of eye disease and has used mathematical morphological analysis on the structure of retinal vessels and the information on optic disk [2].

For diagnosis of complete diseases, assessment of retinal blood vessel is significant. It offers a lot of information conversely for easy recognition of exudates or micro aneurysms [3]. Diabetic retinopathy is of two forms, diabetes mellitus and diabetes insipidus. It is a dangerous eye disease which should be treated in the earlier stages, otherwise it could lead to lasting vision loss. The severity of the problem lies when the patients might not know about it until it reaches later stages and leads to vision loss, which becomes unavoidable [4].

In this work, graph cut technique for blood vessel segmentation is used and has implemented a preprocessing method, which consists of an efficient adaptive histogram equalization and robust distance transform. This operation improves the robustness and accuracy of the graph cut algorithm [5]. The optic disk segmentation initiates by calculating the location of the optic disk. This method uses the features of vessels to the optic disk in order to find its location. The disk area is then extracted using two methods such as MRF image reconstruction and compensation factor methods. Both methods use the feature of retinal vessels to recognize the location of the optic disk [6].

The aim of image classification method is that it allocates each input to one of the diseases pattern classes. In this work, an artificial neural network (ANN) such as backpropagation network (BPNs) is used to classify the input images into one of the output and layered feedforward is implemented such that the neurons are arranged in layers and drive signal forward and the errors are sent in backward direction. The network gets input by neurons in the input layer, and the output of the system is given by the neurons on an output layer [7]. Glaucoma is an eye disorder that affects the optic nerve, often associated with increased fluid pressure in the eye (intraocular pressure) (IOP). This disorder can be divided into two main types, "open-angle" and "closed-angle glaucoma [8].

The human eye is nearly in the shape of a sphere. Its average diameter is approximately 20 mm or more. The eye is made up of three coats, enclosing three apparent structures. The outermost layer is composed of the cornea and sclera with iris, cornea and retina is kept inside these coats.

Retinal vessels can be considered as thin elongated structures in the retina, with variation in width and length. Adaptive histogram equalization and distance transform method is used as a preprocessing method for segmentation of the blood vessel from the fundus retinal images [6].

2 Methodology

The novel work consists of four important blocks such as RGB image collection, green channel extraction and preprocessing, segmentation of blood vessels and optic disk, and diseases classification. The basic system level block diagram is shown below: RGB images are taken from the database such as DRIVE and STARE. The green channel is extracted from the RGB as it contains more information compared to other channels. Then the green channel is converted to gray image for easy processing and it undergoes preprocessing stage such as histogram equalization method for pixel equalization. Then the blood vessels are extracted using graph cut method and optic disk is also separated and eventually the diseases are classified (Fig. 1).

2.1 Vessel Segmentation by Graph Construction Method

Graph construction method is an energy-based technique for object segmentation and is mainly used for minimizing the energy obtained from the input data and also defines the relation between neighbor pixels in the given image. A set of pixels and a set of undirected edges that connect these neighboring nodes form a graph. It includes two nodes; a foreground terminal (source *S*) and a background terminal (sink *T*). Neighborhood links (*n*-links) and terminal links (*t*-links) are the two types of undirected edges. A pair of neighboring pixels $\{p, q\}$ which is available in the set of *N* (number of pixel neighbors) is connected by an *n*-link and each pixel *p* is available in the set of *P* (a set of pixels) presents two *t*-links $\{p, S\}$ and $\{p, T\}$ connecting it to each terminal [9]. Thus,

$$\varepsilon = \mathrm{NU}_{p\varepsilon P}\{\{p,S\},\{p,T\}, v = \mathrm{PU}\{S,T\}\} \tag{1}$$

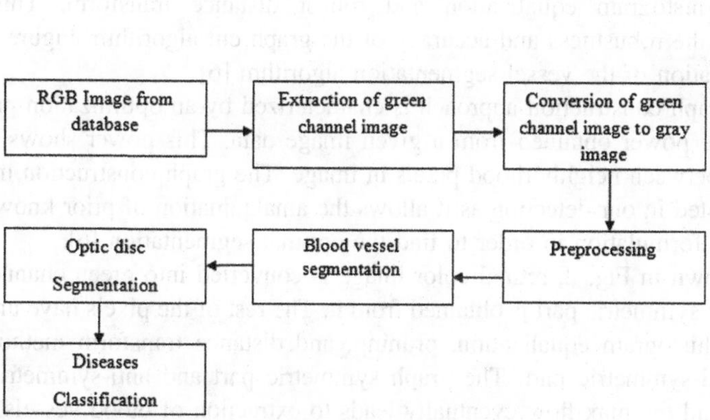

Fig. 1 Basic blocks of the system

In this segmentation method, graph construction technique is used as it incorporates the prior knowledge into the graph formation to get optimal segmentation. Assume $A = (A1, A2, Ap,. .., AP)$ which is a set of binary vector of labels assigned to each pixel p in the image P, where Ap indicate assignments to pixels p in image P and each assignment Ap is either in foreground terminal or background terminal. Thus the binary vector A can be used to obtain the segmentation of pixels and by the energy formulation of the graph the regional and boundary proprieties of vector A are derived as follows:

$$E(A) = \lambda \cdot R(A) + B(A) \tag{2}$$

where $R(A)$ against the boundary term (relationship between neighborhood pixels) $B(A)$, λ is the positive coefficient which indicates the relative importance of the regional term (likelihoods of foreground terminal or background terminal). The regional or the likelihood of the foreground terminal or background terminal is given as

$$R(A) = \sum_{p\varepsilon P} R_p(A_p) \tag{3}$$

While in the process of formulation of graph energy and its minimization in Eq. (2), the boundary term in (2) follows short edges known as the shrinking bias [10].

2.2 Extraction of Retinal Blood Vessels

Retinal vessels can be seen as a lengthy thin structure in the retina which varies in width and length. In order to fragment the blood vessel from the fundus retinal image, we have implemented a preprocessing method, which consists of an efficient adaptive histogram equalization and robust distance transform. This action improves the robustness and accuracy of the graph cut algorithm. Figure 2 shows the illustration of the vessel segmentation algorithm [6].

The graph construction approach is characterized by an optimization process to reduce the power obtained from a given image data. This power shows the relationship between neighborhood pixels in image. The graph construction method is implemented in our detection as it allows the amalgamation of prior knowledge to the graph formulation in order to find the optimal segmentation [6].

As shown in Fig. 2, retinal color image is converted into green channel image and graph symmetric part is obtained from it. The rest of the pixels have undergone adaptive histogram equalization, pruning, and distance transform method to get graph anti-symmetric part. The graph symmetric part and anti-symmetric part is used to find the max flow eventually leads to extraction of blood vessels.

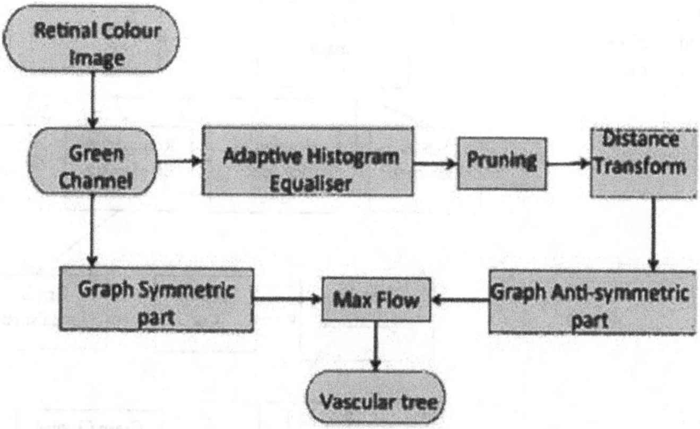

Fig. 2 Retinal vessel extraction method

2.3 Segmentation of Optic Disk

This process can be started by knowing the location of the optic disk and it uses the features of vessels to view its position.

Figure 3a, b shows the overview of image reconstruction and the compensation factor methods. The location of the optic disk is extracted by two different methods such as image reconstruction and the compensation factor methods. These methods use a common feature to find the position of the disk. The image reconstruction method can be used to eradicate the retinal vessels from the optic disk region and it is applied only on the vessel pixels to avoid changes in other structures of the image. This method is called image reconstruction. The output of the reconstruction process undergoes a segmentation method for optic disk by graph construction technique. The compensation factor process segments the optic disk using prior local intensity knowledge of the vessels.

2.4 Location Detection of Optic Disk

The location of the optic disk can be detected by segmenting the vessels from the image by iteratively moving toward the center of the disk. The vessel image is pruned via a morphological open method to eliminate thin vessels and keep the main arcade. The centroid of the arcade is calculated using the following formulation:

$$C_x = \sum_{i=1}^{K} \frac{x_i}{K} \quad C_y = \sum_{i=1}^{K} \frac{y_i}{K} \tag{4}$$

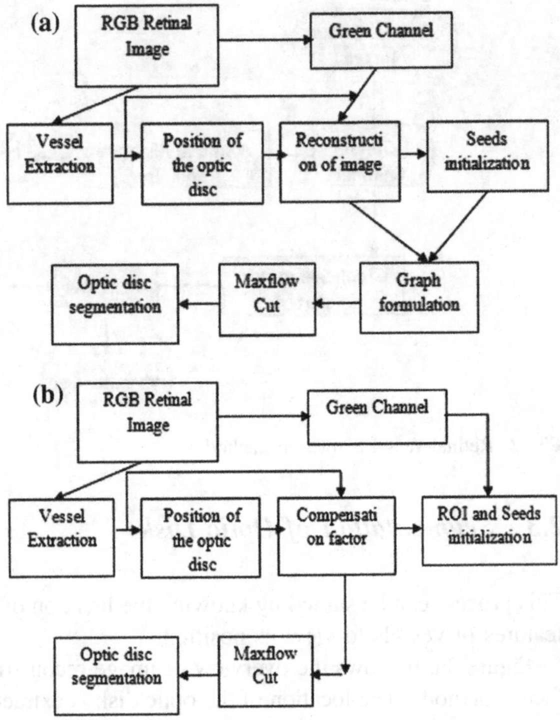

Fig. 3 **a** Image reconstruction method. **b** Compensation factor method

One percent of the brightest region is selected from the grayscale intensity of retinal image. This method finds the location of the optic disk by finding the brightest position through the most number of pixels with respect to center point. It adjusts the center point iteratively until it reaches the convergence point by reducing the distance from one center point to next one in the direction of brightest region [11].

The blood vessels are segmented using a method called as graph by introducing a compensation factor V_{ad}. The function of the graph cut algorithm usually comprises boundary and regional terms. The boundary term is used to assign weights on the edges also known as *n*-links to measure the similarity between neighboring pixels with respect to the pixel proprieties (intensity, texture, and color) and pixels with similar intensities, texture, and color have a strong link. The regional term is derived to define the likelihood of the pixel by assigning weights on the foreground or background also known as *t*-links between the image pixels and the two terminals Bg and Fg seeds. The *t*-link is defined in order to extract the blood vessels into the graph as follows:

$$S_{\text{link}} = \begin{matrix} \{-\ln P_r\left(I_p \backslash \text{Fg}_{\text{seeds}}\right) & \text{if } p \neq \text{vessel} \\ \{-\ln P_r\left(I_p \backslash \text{Fg}_{\text{seeds}}\right) + V_{\text{ad}} & \text{if } p = \text{vessel} \end{matrix} \quad (5)$$

$$T_{\text{link}} = \begin{cases} -\ln P_r\left(I_p \backslash \text{Bg}_{\text{seeds}}\right) & \text{if } p \neq \text{vessel} \\ -\ln P_r\left(I_p \backslash \text{Bg}_{\text{seeds}}\right) & \text{if } p = \text{vessel} \end{cases} \tag{6}$$

where p is the pixel in the image, Fg_{seeds} is the intensity distribution of the foreground pixels, Bg_{seeds} represents the intensity distribution of background pixels, and V_{ad} is the compensation factor given as

$$V_{\text{ad}} = \max_{p\,\varepsilon\,\text{vessel}} \left\{ -\ln P_r\left(I_p \backslash \text{Bg}_{\text{seeds}}\right) \right\} \tag{7}$$

The intensity distribution of the blood vessel pixels in the region around the optic disk makes them more likely to belong to Bg pixels than the Fg (or the optic disk pixels). Therefore, the vessels inside the disk have weak connections with neighboring pixels making them likely to be segmented by the graph cut as Bg [12].

2.5 Classification of Retinal Diseases

Disease classification is one the most important steps in this work. Segmented blood vessels and optic disk information's are given to ANN classifier for the classification of retinal diseases. The classified diseases are diabetic retinopathy, glaucoma, vein occlusion, and hypertension [13]. The classification process includes allocating each input pixels to any of the output class. Backpropagation network (BPNs) has been used in this work in order to classify the different input images. The backpropagation algorithm is a layered feedforward neural network which can send signals in forward direction and the errors are propagated in the backward direction. The network consists of 3 layers such as input layer which has 50 nodes, a hidden layer which is made up of 1–10 nodes, and an output layer having 1 node. It receives the input through its input layer process it in the hidden layer and sends the output through its output layer.

3 Results and Discussion

The method implemented in this work is tested on two public datasets, namely STARE and DRIVE, with a total of 100 images for the vessel segmentation method. The optic disk segmentation algorithm was experimented on DIARETDB1 and DIARETDB2, consisting of 129 images in total. The performances of both methods are tested against a number of alternative methods. The entire work is done with the help of MATLAB. Figures 4 and 5 show images of blood vessels segmented and the images that are manually labeled for the DRIVE and the STARE datasets, respectively.

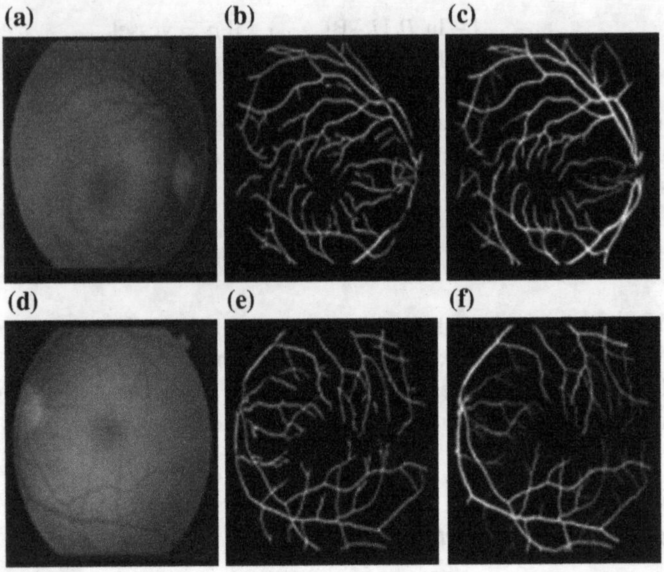

Fig. 4 DRIVE database

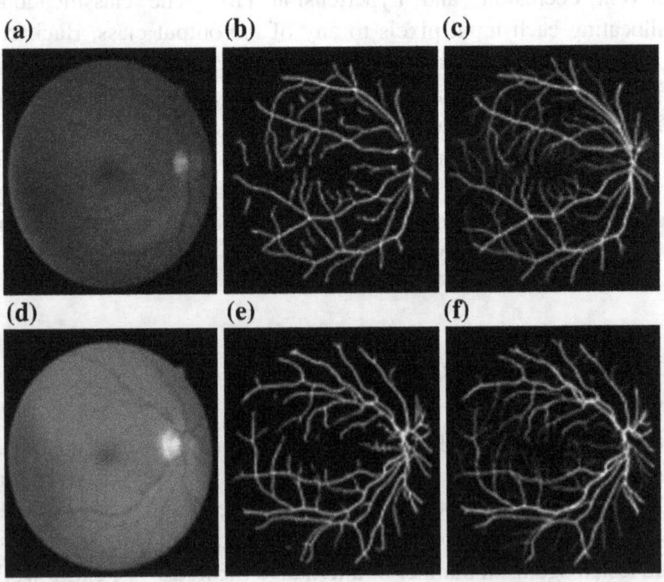

Fig. 5 STARE database

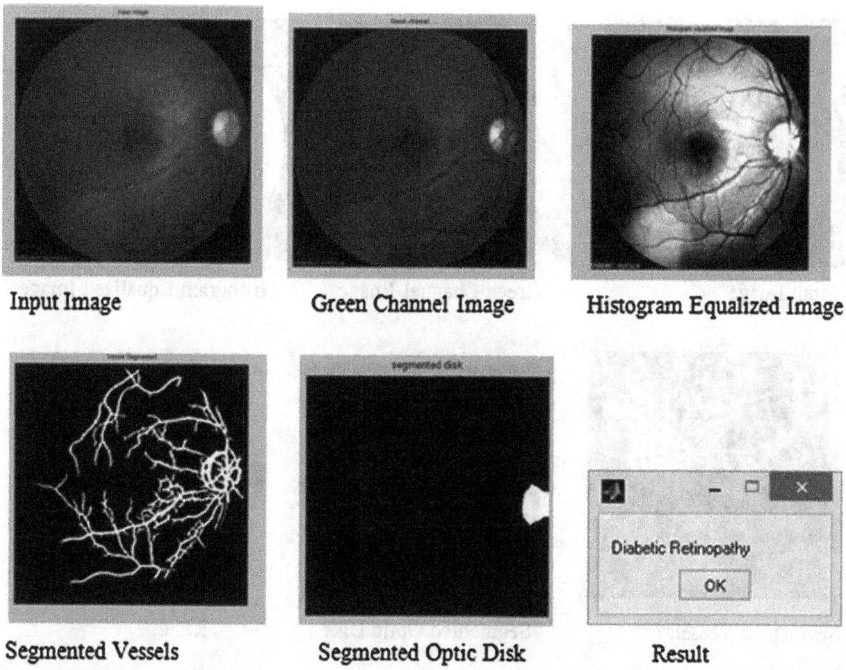

Fig. 6 Diabetic retinopathy

Figure 6 shows the processing steps for the detection of diabetic retinopathy. Input image is converted into green channel image for noise reduction. Blood vessels are segmented by the graph cut technique. MRF and compensation factor method give the optic disk. Further processing via ANN gives the effected disease. Figure 7 shows the processing steps for the detection of glaucoma.

Figure 8 shows the processing steps for hypertension detection. Figure 9 shows the processing steps for vein occlusion detection. For diabetic retinopathy (DR), the total number of correctly classified images are 17 out of 20 giving a classification accuracy of 85 %. For glaucoma, the total number of correctly classified images is 18 out of 20 giving a classification accuracy of 90 %. For hypertension, the total number of correctly classified images is 16 out of 20 giving a classification accuracy of 80 %. For vein occlusion, the total number of correctly classified images is 19 out of 20 giving a classification accuracy of 95 % and eventually leads to an overall accuracy of 87.5 % (Table 1).

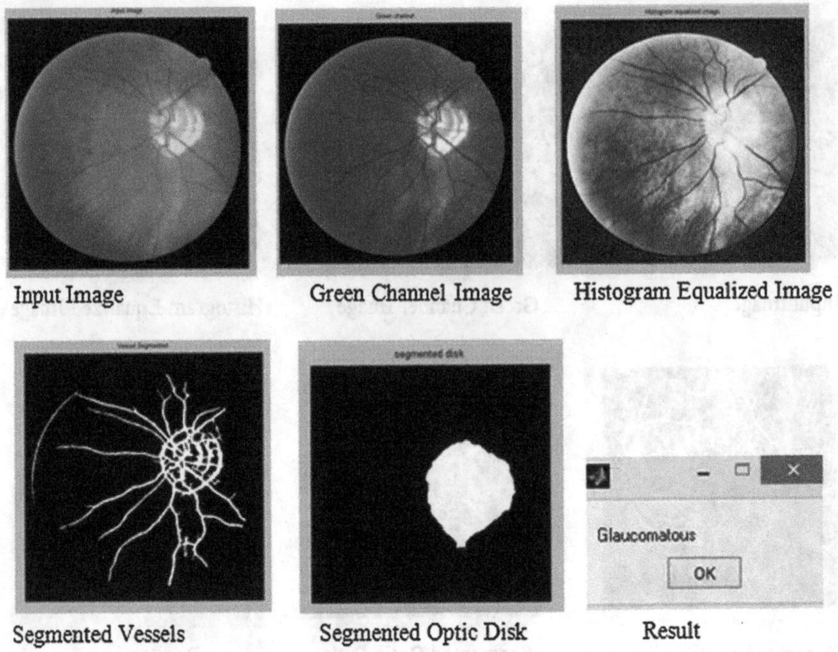

Fig. 7 Results of glaucoma

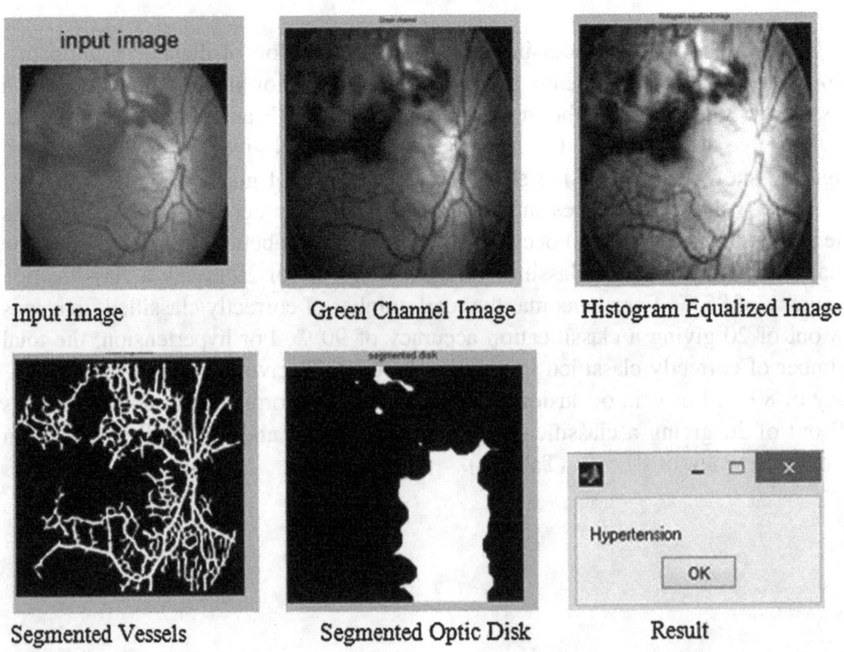

Fig. 8 Results of hypertension

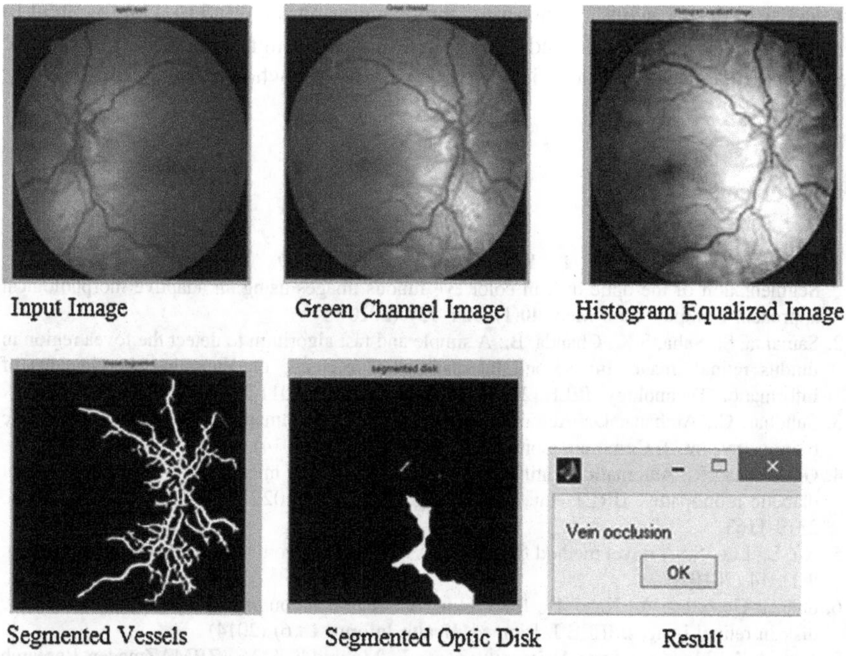

Fig. 9 Results of vein occlusion

Table 1 Performance analysis for different eye diseases

Diseases	No. of images taken	No. of images correctly classified	Classification accuracy (%)
Diabetic retinopathy	20	17	85
Glaucoma	20	18	90
Hypertension	20	16	80
Vein occlusion	20	19	95
Average result	87.5 %		

4 Conclusion

In this research work, retinal blood vessel extraction and segmentation of optic disk in retinal images have been done by combining the mechanism of image reconstruction, and compensation factor into the graph construction method.

In the second stage information extracted from retinal vessels and optic disk are given to an ANN classifier to find whether the image is infected or normal, finally it classifies the diseases. This proposed methodology can be utilized in hospitals to detect diseases occurring on the eyes by doctors easily. The future scope of this project is to detect many eye diseases, thus benefitting mankind to a large extent to be free from eye diseases leading to blindness with higher efficiency. From the results and its slotted outputs, it is clearly identify the whole concept of the work.

References

1. Welfer, D., Scharcanski, J., Kitamura, C., Pizzol, M.D., Ludwig, L., Marinho, D.: Segmentation of the optic disk in color eye fundus images using an adaptive morphological approach. Comput. Biol. Med. **40**(1), 124–137 (2010)
2. Samanta, S., Saha, S.K., Chanda, B.: A simple and fast algorithm to detect the fovea region in fundus retinal image. In: Second International Conference on Emerging Applications of Information Technology, IEEE (2011). doi:10.1109/EAIT.2011.22206
3. Sundhar, C., Archana, D.: Automatic screening of fundus images for detection of diabetic retinopathy. Int. J. Commun. Comput. Technol. **02**(1, 3) (2014)
4. Gowthaman R.: Automatic identification and classification of microaneurysms for detection of diabetic retinopathy. IJRET: Int. J. Res. Eng. Technol. **03**(02), 2321–7308 (2014). eISSN: 2319-1163
5. Xu, L., Luo, S.: A novel method for blood vessel detection from retinal images. Biomed. Eng. **9**(1), 14 (2010)
6. Salazar-Gonzalez, A., Kaba, D., Li, Y., Liu, X.: Segmentation of the blood vessels and optic disk in retinal images. IEEE J. Biomed. Health Inform. **18**(6) (2014)
7. Jain, A.K., Michigan State University, Mao, J., Mohiuddi, K.M., ZBMAZmaden Research Center.: Artificial neural networks: a tutorial
8. Kolmogorov, V., Boykov, Y.: What metrics can be approximated bygeo-cuts, or global optimization of length are and flux. In: Proceedings of 10th IEEE International Conference on Computer Vision, vol. 1, pp. 564–571 (2005)
9. Boykov, Y.Y., Jolly, M.-P.: Interactive graph cuts for optimal boundary & region segmentation of objects in N-D images. In: Proceedings of IEEE 8th International Conference on Computer Vision, vol. 1, pp. 105–112 (2001)
10. Vicente, S., Kolmogorov, V., Rother, C.: Graph cut based image segmentation with connectivity priors. In: Proceedings of IEEE Conference on Computer Vision and Pattern Recognition, vol. 1, pp. 1–8 (2008)
11. Vicente, S., Kolmogorov, V., Rother, C.: Graph cut based image segmentation with connectivity priors. In: Proceedings of IEEE Conference on Computer Vision and Pattern Recognition, vol. 1, pp. 1–8 (2008)
12. Salazar-Gonzalez, G., Li, Y., Liu, X.: Retinal blood vessel segmentation via graph cut. In: Proceedings of IEEE 11th International Conference on Control, Automation, Robotics and Vision, pp. 225–230 (2010)
13. Jestin, V.K., Anitha, J., Jude Hemanth, D.: Genetic algorithm for retinal image analysis. IJCA Special Issue on Novel Aspects of Digital Imaging Applications (DIA) (1), 48–52 (2011)

Improved Local Search in Shuffled Frog Leaping Algorithm

Tarun Kumar Sharma and Millie Pant

Abstract Shuffled frog-leaping algorithm (SFLA) is comparatively a recent addition to the family of nontraditional population-based search methods that mimics the social and natural behavior of species (*frogs*). SFLA merges the advantages of particle swarm optimization (PSO) and genetic algorithm (GA). Though SFLA has been successfully applied to solve many benchmark and real-time problems it limits the convergence speed. In order to improve its performance, the frog with the best position in each memeplexes is allowed to slightly modify its position using random walk. This process improves the local search around the best position. The proposal is named improved local search in SFLA (ILS-SFLA). For validation, three engineering optimization problems are consulted from the literature. The simulated results defend the efficacy of the proposal when compared to state-of-the-art algorithms.

Keywords Shuffled frog-leaping algorithm · SFLA · Optimization · Engineering design optimization · Metaheuristics

1 Introduction

Search methods that mimic the social and natural behavior of species have been widely accepted and used to solve nontraditional complex optimization problems. Shuffled frog leaping algorithm (SFLA) is a recent addition to the family of nature inspired algorithms, inspired by natural memetics [1]. SFLA was introduced by Eusuff and Lansey in 2003 and initially applied to optimize the pipe size and expansion of network in water distribution network design [2]. Since then SFLA

T.K. Sharma (✉)
Amity University Rajasthan, Jaipur, India
e-mail: taruniitr1@gmail.com

M. Pant
Department of Applied Science & Engineering, IIT Roorkee, Roorkee, India
e-mail: millidma@gmail.com

M. Pant et al. (eds.), *Proceedings of Fifth International Conference on Soft Computing for Problem Solving*, Advances in Intelligent Systems and Computing 436, DOI 10.1007/978-981-10-0448-3_48

and its variants have been successfully applied to solve a number of combinatorial optimization problems [3–8].

SFLA merges the advantages of particle swarm optimization (PSO) to perform local search and shuffled complex evolution algorithm [9] for global search. The brief detail of SFLA is given in Sect. 2.

Like other nature inspired metaheuristics, SFLA is limited in its convergence speed. In the present study attempt is made to accelerate basic SFLA by embedding the concept of random walk while performing local search. The proposal is tested over three engineering optimization problems, namely piston lever design; tubular column design, and three-bar truss design.

The rest of the paper is structured as follows: The modification in basic SFLA is discussed in Sect. 3 followed by a brief overview of engineering design problems in Sect. 4. Parameter settings and results are given in Sect. 5. Finally conclusions are drawn and future work are presented in Sect. 6.

2 Shuffled Frog Leaping Algorithm

Proposed by Eusuff and Lansey, SFLA proved its success when compared with GA, ant colony optimization (ACO), and the PSO [5]. SFLA is formulated on the concept of evolution of memeplexes in frogs. In essence, SLFA contains the element of both the local search method of PSO and the concept of mixing information of the shuffled complex evolution.

In SFLA, a set of frogs represents the population of possible solutions, which is partitioned into subsets called memeplexes. Different subsets have frogs from different cultures and each frog carries out a local search and the position of the worst frog is modified or updated so that the frogs can move toward optimization. When each subset evolves through a fixed number of generations or memetic evolution steps, the ideas held by the frogs within the subset are passed among subsets through shuffling process. This process of local search and shuffling of information continues until the termination criterion is satisfied.

There are four steps in SFLA:

a. Initialization Process

The initialization of a set of frogs (solutions) is similar to initialization process of other stochastic techniques, i.e., using Eq. (1). The population of frogs (P) are represented by $X_i = (x_{i1}, x_{i2, \ldots,} x_{iS})$ and then the position of each frog is generated as

$$x_{ij} = \mathrm{lb}_j + \mathrm{rand}(0, 1) \times (\mathrm{ub}_j - \mathrm{lb}_j) \tag{1}$$

for $i = 1, 2, \ldots, P$ (set of frogs); $j = 1, 2, \ldots, S$ (S-dimensional vector) and, lb_j and ub_j are the lower and upper bounds, respectively, for the dimension j.

b. Sorting and Division Process
The frogs, based on their fitness evaluations are sorted in descending order. Then the sorted population of P frogs is distributed into m subsets (memeplexes), each subset holds n frogs such that $P = m \times n$. The distribution is done such that the frog with the maximum fitness value will go into subset first, accordingly the next frog into second subset, and so on. Then X_b (best) and X_w (worst) individuals in each subset are determined.
The pictorial representation of frogs in SFLA is shown in Fig. 1.

c. Local Search Process
Worst individual position is improved using Eqs. (2) and (3):

$$D_i = \text{rand}(0, 1) \times (X_b - X_w) \quad (2)$$

$$X_w = X_w + D_i; -D_{max} \leq D_i \leq D_{max}. \quad (3)$$

where $i = 1, 2, \ldots, N_{gen}$; D is the movement of a frog, whereas D_{max} represents the maximum permissible movement of a frog in feasible domain; N_{gen} is maximum generation of evolution in each subset. The old frog is replaced if the evolution produces the better solution, else X_b is replaced by X_g (optimal solution). If no improvement is observed then a random frog is generated and replaces the old frog. This process of evolution continues till the termination criterion is met.

d. Shuffling Process
The frogs are again shuffled and sorted to complete the round of evolution. Again, follow the same four steps until the termination condition is met.

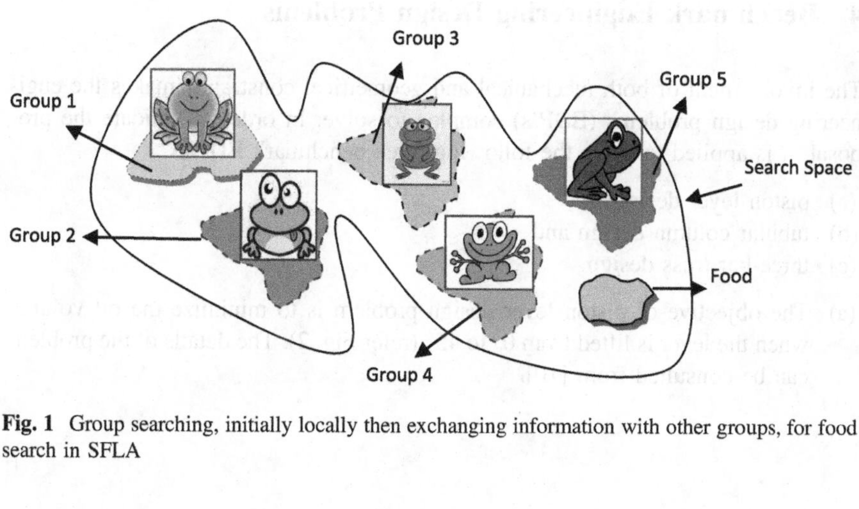

Fig. 1 Group searching, initially locally then exchanging information with other groups, for food search in SFLA

3 ILS–SFLA: Proposal

The frogs in their respective memeplexes explore the search/solution space locally. Then all the memeplexes are shuffled to find the optimal solution. If the optimal solution is not achieved then the frogs are again divided into new subsets (memplexes). This information exchange results in optimal solution. Further, it is analyzed from the local search process of basic SFLA that if difference between the position of the best frog (X_b) and the worst frog (X_w) decreases (Eqs. 2 and 3), the perturbation decreases on the position of the worst frog. This may lead to premature convergence or sometimes to stagnation while solving complex problems. In order to come through from this situation the local searching process of basic SFLA has been modified by embedding the concept of random walk. The local search is modified as follows:

$$X_b = X_b + r_n \times \lambda \tag{4}$$

where r_n is a normally distributed random number and λ is a scale factor and is given as

$$\lambda = 0.01 \times (\mathrm{ub} - \mathrm{lb}) \tag{5}$$

Further, to accelerate the searching process and widen the global search in order to balance two antagonists, exploration and exploitation the frogs individual position is modified using Eq. 6:

$$D_i = \lambda \times (X_b - X_w) \tag{6}$$

4 Benchmark Engineering Design Problems

The involvement of both mechanical and geometrical constraints makes the engineering design problems (EDP's) complex to solve. In order to validate the proposal, it is applied to solve the following three benchmark EDPs:

(a) piston lever design;
(b) tubular column design and
(c) three-bar truss design.

(a) The objective of piston lever design problem is to minimize the oil volume when the lever is lifted from 0° to 45° (refer Fig. 2). The details of the problem can be consulted from [10].

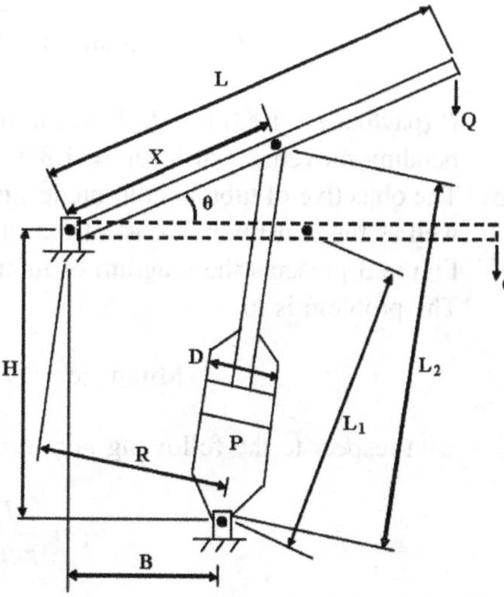

Fig. 2 An illustration of piston design

The objective function of the problem is given as

$$\text{Minimize} : f(H,B,D,X) = \frac{1}{4}\pi D^2(L_2 - L_1) \tag{7}$$

with respect to the following constraints:

$$g_1 = QL\cos\theta - RF \leq 0 \quad \text{at } \theta = 45^\circ \tag{8}$$

$$g_2 = Q(L = X) - M_{\max} \leq 0 \tag{9}$$

$$g_3 = 1.2(L_2 - L_1) - L_1 \leq 0 \tag{10}$$

$$g_4 = \frac{D}{2} - B \leq 0 \tag{11}$$

where

$$R = \frac{|-X(X\sin\theta + H) + H(B - X\cos\theta)|}{\sqrt{(X-B)^2 + H^2}}$$

$$F = \pi P D^2/4$$

$$L_1 = \sqrt{(X-B)^2 + H^2}$$

$$L_2 = \sqrt{(X\sin 45 + H)^2 + (B - X\cos 45)^2}$$

P (payload) = 1000 lbs; L (Lever arm) = 240 in.; M_{max} (maximum allowable bending movement of lever) = 1.8 × 10; and oil pressure = 1500 Ψ.

(b) The objective of tubular column design problem is to minimize the material as well as the construction cost of the structure. The problem is detailed in [10]. Figure 3 presents the diagram of tubular design.
The problem is to

$$\text{Minimize} : f(d,t) = 9.8\,\mathrm{d}t + 2d \tag{12}$$

with respect to the following nonlinear constraints:

$$g_1 = \frac{P}{\pi d t \sigma_y} - 1 \le 0 \tag{13}$$

$$g_2 = \frac{8PL^2}{\pi^3 E d t (d^2 + t^2)} - 1 \le 0 \tag{14}$$

In this study P is considered as 2500 kgf; yield stress of the material is $\sigma_y = 500\ \mathrm{kgf/cm^2}$; modulus elasticity E is given as 0.85×10^6 kgf/cm^2; and a density ρ is 0.0025 kgf/cm^3. The length of the column l is 250 cm.

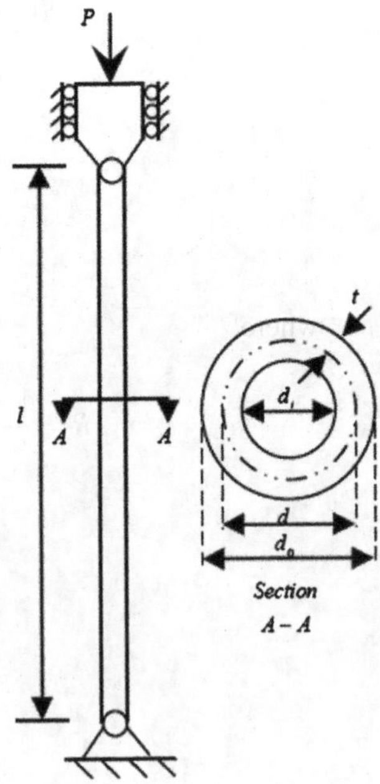

Fig. 3 An illustration of tubular column design

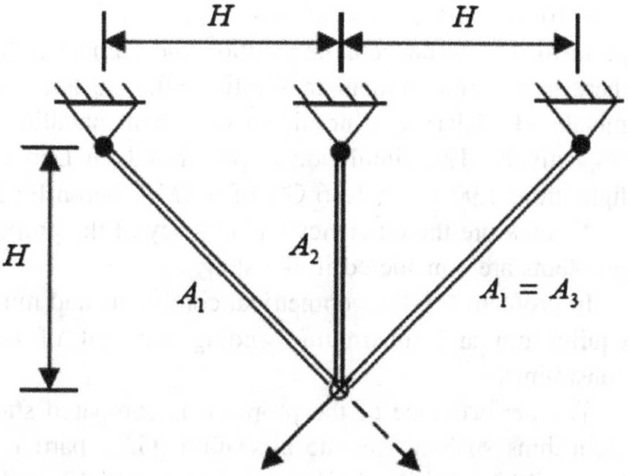

Fig. 4 An illustration of three-bar truss design

(c) The objective of three-bar truss design in Fig. 4 is to optimize the volume when the bar is subjected to a static load and stress (σ) constraints on each of the truss members.

$$\text{Minimize}: f(A_1, A_2) = (2\sqrt{2}A_1 + A_2) \times l \tag{15}$$

with respect to constraints:

$$g_1 = \frac{\sqrt{2}A_1 + A_2}{\sqrt{2}A_1^2 + 2A_1A_2}P - \sigma \le 0 \tag{16}$$

$$g_2 = \frac{A_2}{\sqrt{2}A_1^2 + 2A_1A_2}P - \sigma \le 0 \tag{17}$$

$$g_3 = \frac{1}{A_1 + \sqrt{2}A_2}P - \sigma \le 0 \tag{18}$$

where A_1, A_2 are cross-sectional areas and bounded as $0 \le A_1, A_2 \le 1$; $l = 100$ cm; $P = 2$ kN; and $\sigma = 2\,\frac{\text{kN}}{\text{cm}^2}$.

5 Optimization Parameter Settings and Results Discussion

Parameter selection is a critical issue and of much concern as it may affect the performance of heuristics. There are five parameters in SFLA: number of memeplexes (m), number of frogs (n) in each memeplexes, number of evolution or infection steps (N_{gen}) in memeplexes between two shufflings and D_{max} maximum step size allowed during evolutionary step. While testing the performance of the proposal the parameters are tuned as follows: Frog population (P) is fixed to 100;

$m = 10$; $n = 10$; $N_{gen} = 10$ and D_{max} is 100 % of variable range. 25 runs were performed to obtain reliable solution and number of iterations is fixed to 10,000. If there is no improvement in solution after 30 iterations then the process will terminate. Boundary and nonlinear constraint handling is performed using [11, 12] respectively. The simulation is performed on Deb C++ with the following configuration: 1.90 GHz, 1.86 GB of RAM Celeron(R) Dual Core.

To measure the efficiency and efficacy of the proposal, three engineering design problems are considered in the study.

In problem (a), the geometrical conditions and minimum piston stoke, the force equilibrium and maximum bending moment of the lever taken as inequality constraints.

The performance of the proposal is compared statistically with state-of-the-art algorithms such as genetic algorithm (GA), particle swarm optimization (PSO), hybrid PSO (HPSO), differential evolution (DE), and hybrid PSO with Q learning (HPSO-Q) [13]. The best values obtained for the piston parameters are as $H = 0.051$ in., $B = 2.050$ in., $D = 121.007$ in., $X = 4.109$.

The simulated results are summarized in Table 1.

The results achieved by the proposed algorithm are comparatively remarkable. Algorithm took only 13,760 NFEs, whereas for others it is not available in the literature.

The problem of tubular column has two nonlinear constraints, called buckling and yield stress and two boundary conditions d (mean diameter of the column) and t (thickness). d is within the range of 2–14 cm, whereas t lies between 0.2 and 0.8 cm. The simulated results gained by the proposal and others that are available in the literature [14, 15] are illustrated in Table 2.

From Table 2, it can be analyzed that the other two studies taken for comparison failed in achieving the feasible result because of second constraint (g_2) violation. The best, worst, mean, and standard deviation (Std. Dev.) results obtained using the proposal is 26.531, 26.579, 26.5311, and 0.0019 respectively. The best part of the proposal is that it has not violated any constraints and is able to achieve the feasible solution.

In the case of (c), for the third problem the simulated results are demonstrated in Table 3. The results of the proposal are compared with the results given in the literature for PSO, NFP, SCA, and BBA [16–19].

Table 1 Statistical simulated results of piston lever design problem

	GA	PSO	HPSO	HPSO-Q	DE	Present study
Min	161	122	162	129	159	145
Max	216	294	197	168	199	219
Mean	185	166	187	151	187	167
Std. dev.	18.2	51.7	13.4	13.4	14.2	32.8
NFE	N/A[a]	N/A	N/A	N/A	N/A	13,760

[a]Not available

Table 2 Statistical simulated results of tubular problem

	Rao [14]	Hsu and Liu [15]	Present study
d	5.44	5.4507	5.450737
t	0.293	0.292	0.292018
g_1	−0.8579	-7.8×10^{-5}	0.0000
g_2	0.0026	0.1317	0.0000
$F_{\min}$	26.5323	25.5316	26.531102

Table 3 Statistical simulated results of three-bar truss design

	PSO	NFP	SCA	BBA	Present study
A_1	0.795	0.788	0.788621	0.788786	0.788621
A_2	0.395	0.408	0.408401	0.407941	0.408273
g_1	−0.00169	*0.00082*	0	−2E−06	0
g_2	−0.26124	−0.2674	−0.26804	−0.26778	−0.26799
g_3	−0.74045	−0.73178	−0.73196	−0.73223	−0.73105
$F_{\min}$	264.3	263.68	263.8958	263.8965	263.89462

From Table 3, it can be seen that the simulated results obtained by the proposal are comparatively better and close to the best. The best result is reported in [17]. Further it took only 14,765 NFE to achieve the result in comparison to SCA [17], which took 17,600 searches to achieve the best result; at the same time SCA violated g_1 constraint.

It can be concluded that the proposal performed well to achieve comparatively good results to find feasible solutions.

6 Conclusions

In the present study, a variant of SFLA called improved local search in SFLA (ILS-SFLA) has been proposed. The proposals embed the concept of random walk. The position of the best frog (best fitness value) is updated using random walk, as this mechanism improves the local search around the best position. Later this proposal is validated on the set of three engineering design problems. The results are compared with state-of-the-art algorithms available in the literature. The simulated comparative results prove the effectiveness of modification in the basic SFLA. In the near future we will try to modify the proposal further to implement it on multiobjective optimization problems.

References

1. Eusuff, M.M., Lansey, K.E.: Optimization of water distribution network design using the shuffled frog leaping algorithm. J. Water Resour. Plann. Manag. **129**(3), 210–225 (2003)
2. Eusuff, M.M., Lansey, K.E.: Water distribution network design using the shuffled frog leaping algorithm. In: Proceedings of the 2nd World Water Congress of the International Water Association, Berlin, Germany (2001)
3. Eusuff, M.M., Lansey, K.E., Pasha, F.: Shuffled frog-leaping algorithm: a memetic meta-heuristic for discrete optimization. Eng. Optim. **38**(2), 129–154 (2006)
4. Elbeltagi, E., Hegazy, T., Grierson, D.: A modified shuffled frog-leaping optimization algorithm: applications to project management. Struct. Infrastruct. Eng.: Maint. Manag. Life-Cycl. **3**(1), 53–60 (2007) (68)
5. Elbeltagi, E., Hegazy, T., Grierson, D.: Comparison among five evolutionary-based optimization algorithms. Adv. Eng. Inform. **19**(1), 43–53 (2005)
6. Sharma, S., Sharma, T.K., Pant, M., Rajpurohit, J., Naruka, B.: Accelerated Shuffled frog-leaping algorithm. In: Proceedings of Fourth International Conference on Soft Computing for Problem Solving (SocProS-2014) Springer Berlin Heidelberg Advances in Intelligent Systems and Computing, vol. 336, pp. 181–189 (2014)
7. Naruka, B., Sharma, T.K., Pant, M., Sharma, S., Rajpurohit, J.: Differential shuffled frog leaping algorithm. In: Proceedings of Fourth International Conference on Soft Computing for Problem Solving (SocProS-2014) at NIT Silchar, Assam, Springer Berlin Heidelberg Advances in Intelligent Systems and Computing, vol. 336, pp 245–253 (2015)
8. Sharma, S., Sharma, T.K., Pant, M., Rajpurohit, J., Naruka, B.: Centroid mutation embedded shuffled frog-leaping algorithm. Procedia Comput. Sci. **46**, 127–134 (2015)
9. Duan, Q., Gupta, V.K., Sorooshian, S.: A shuffled complex evolution approach for effective and efficient global minimization. Optim. Theory Appl. **76**(3), 501–521 (1993)
10. Vanderplaats, G.N.: DOT (Design Optimization Tools) Users Manual, Version 4.20, VR&D (1995)
11. Gandomi, A.H., Yang, X.S.: Evolutionary Boundary Constraint Handling Scheme. Neural Comput. Appl. **21**(6), 1449–1462 (2012)
12. Becerra, R.L., Coello Coello, C.A.: Cultured differential evolution for constrained optimization. Comput. Methods Appl. Mech. Eng. **195**, 4303–4322 (2006)
13. Kim, P., Lee, J.: An integrated method of particle swarm optimization and differential evolution. J. Mech. Sci. Technol. **23**, 426–434 (2009)
14. Rao, S.S.: Engineering Optimization: Theory and Practice, 3rd edn. Wiley, Chichester (1996)
15. Hsu, Y.L., Liu, T.C.: Developing a fuzzy proportional-derivative controller optimization engine for engineering design optimization problems. Eng. Optim. **39**(6), 679–700 (2007)
16. Ray, T., Saini, P.: Engineering design optimization using a swarm with an intelligent information sharing among individuals. Eng. Optim. **33**(6), 735–748 (2001)
17. Tsai, J.: Global optimization of nonlinear fractional programming problems in engineering design. Eng. Optim. **37**(4), 399–409 (2005)
18. Ray, T., Liew, K.: Society and civilization: an optimization algorithm based on the simulation of social behavior. IEEE Trans. Evol. Comput. **7**(4), 386–396 (2003)
19. Park, Y.C., Chang, M.H., Lee, T.-Y.: A new deterministic global optimization method for general twice differentiable constrained nonlinear programming problems. Eng. Optim. **39**(4), 397–411 (2007)

Shuffled Frog Leaping Algorithm with Adaptive Exploration

Jitendra Rajpurohit, Tarun Kumar Sharma and Atulya K. Nagar

Abstract Shuffled frog leaping algorithm is a nature inspired memetic stochastic search method which is gaining the focus of researchers since it was introduced. SFLA has the limitation that its convergence speed decreases towards the later stage of execution and it also tends to stuck into local extremes. To overcome such limitations, this paper first proposes a variant in which a few new random frogs are generated and the worst performing frogs population are replaced by them. Experimental results show that a high number of replaced frogs does not always provide better results. As the execution progresses the optimized number of replaced frogs decreases. Based on the experimental observations, the paper then proposes another variant in which the number of replaced frogs adapts to the stage of the execution and hence provides the best results regardless of the stage of execution. Experiments are carried out on five benchmark test functions.

Keywords Shuffled frog leaping algorithm · Nature inspired computing · Stochastic search

1 Introduction

Nature inspired computing algorithms (NICA) have been gaining popularity for a few decades now. More and more problems are now being solved using algorithms that are in some way inspired by some natural phenomenon. Most NICAs are stochastic search methods. A number of NICAs have been introduced. Genetic

J. Rajpurohit (✉) · T.K. Sharma
Amity University Rajasthan, Jaipur, India
e-mail: jiten_rajpurohit@yahoo.com

T.K. Sharma
e-mail: taruniitr1@gmail.com

A.K. Nagar
Liverpool Hope University, Liverpool, UK
e-mail: nagara@hope.ac.uk

M. Pant et al. (eds.), *Proceedings of Fifth International Conference on Soft Computing for Problem Solving*, Advances in Intelligent Systems and Computing 436, DOI 10.1007/978-981-10-0448-3_49

algorithms [1] are based on natural reproduction system. Differential evolution [2] uses difference vector(s) of the participant solutions to increase diversity of the population. Particle swarm optimization [3] utilizes the way a flock of birds uses to maintain velocity and distance of each of its members. Similarly, artificial bee colony [4] and ant colony optimization [5] exploit the optimization methods used by honey bees and ants, respectively, in search for food. One of the latest such algorithms is shuffled frog leaping algorithm (SFLA) [6] which mimics the behavior of a group of frogs in a pond searching for the place with maximum food.

The remainder of this paper is structured as follows: Sect. 2 explains the working process of the basic SFLA and a brief survey of modifications found in the literature. Section 3 explains the proposed variants of the algorithm which includes adaptive exploration. Experimental setup and results are discussed in Sect. 4. Finally Sect. 5 concludes the paper.

2 Working of SFLA

SFLA combines the benefits of PSO and shuffled complex evolution (SCE). Its local search is inspired by PSO while global search process is inspired by SCE. Members of population are denoted as frogs. The basic SFLA executes according to the following steps:

Step 1: Initialization
NP (population size) frogs are initialized randomly in the feasible space.

Step 2: Sorting and creation of memeplexes
frogs are arranged in descending order of their performance based on their objective function values. Then these frogs are distributed in m memeplexes such that each memplex has equal number (n) of frogs and $m * n = \text{NP}$. This distribution is done in such a manner that frog 1 goes to memeplex 1, frog 2 goes to memeplex 2, frog m goes to memeplex m, and frog $m + 1$ goes to memeplex 1. At the end of this distribution process the best (Xb) and worst (Xw) frogs of each memeplex are identified.

Step 3: Local Search Process
Now within each memplex, Xw is moved to a new position Xw(new) calculated by Eqs. 1 and 2.

$$D_i = (Xb - Xw) * \text{rand}(0, 1) \tag{1}$$

$$Xw(\text{new}) = Xw + D_i \tag{2}$$

where i denotes a particular dimension.
If the performance of the worst frog does not improve then it is moved toward the global best frog (Xg) by replacing Xb by Xg in Eq. 1. Even if it does not improve a new random frog is generated and it replaces Xw.

The above movement of worst frog is repeated for a fixed number of times (LI) for each memeplex.

Step 4: Shuffling (Global Search) Process
This step enables the global exchange of information between frogs from all the memeplexes. Frogs from all the memeplexes are merged and sorted in descending order of their performance.
Steps 2–4 are now repeated until the termination condition is met. One execution up to Step 4 is denoted as an iteration in this paper.

Since its inception SFLA has been a primary topic for researchers in the related field. A number of improvements have been proposed and are found in the literature. Clonal selection-based SFLA [7] uses a modified clonal selection algorithm for better performing frogs and SFLA for underperforming frogs to move them toward the global best frog. Composite SFLA [8] embeds SFLA with artificial fish swarm algorithm to overcome the limitations of SFLA like slow convergence speed toward the later stage and tendency to trap into local extremes. Dichotomous search in SFLA [9] searches a better solution in forward as well as backward direction of the *Xb*. Centroid mutated SFLA [10] uses the centroid of best position of a frog and two randomly selected other frogs to find a better position for a frog. Accelerated SFLA [11] embeds a scaling factor while computing the new position to accelerate convergence and this scaling factor is computed by golden section search. Differential SFLA [12] embeds the search processes of differential evolution to maintain population diversity. Two-phase SFLA [13] improves the SFLA by embedding opposition-based learning and a scaling factor.

3 SFLA with Increased Exploration (SFLA-IE)

It can be easily observed in any run of the basic SFLA that in early stages of execution (lower values of iteration count) the global best solution improves quickly, but in the later stages its improvement either halts or becomes slow. This is because the algorithm either enters a local extreme or the difference between the member frogs becomes very small, which results in lower exploration of the search space. Successive iterations without any improvement in the best function value can be observed. This proposal aims to identify the stage when the execution has stuck into a stage of no improvement and then increasing the diversity of the population by replacing a fixed number of worst frogs by an equal number of randomly generated new frogs. For this purpose two factors are defined here:

Non-improvement factor (*I*): The number of previous successive iterations for which, if the best function value has not improved, exploration will be increased by replacing a few worst frogs by newly generated frogs.
Replacement factor (*R*): This is the number of worst frogs in the population that are being replaced by randomly generated new frogs.

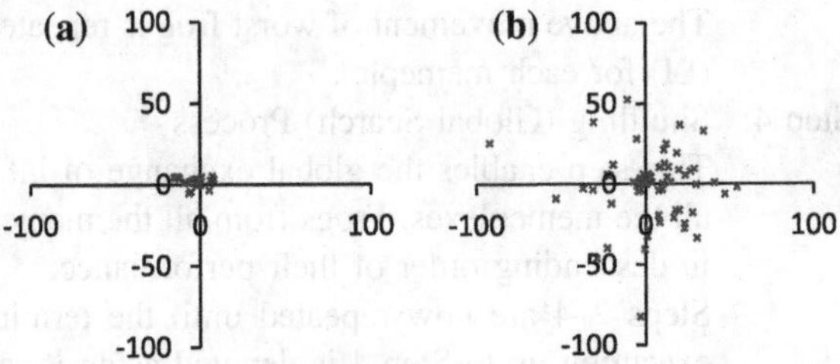

Fig. 1 Comparison of diversities of **a** SFLA and **b** SFLA-IE for $R = 20$

After each iteration, the algorithm checks the number of previous successive iterations for which the function value has not improved. If the number is equal to I, then the worst R frogs of the population are replaced by randomly generated new R frogs. This inclusion increases the diversity of the population resulting in increased convergence speed. Figure 1 shows the diversity of SFLA-IE at the first instance when new random frogs were included in the population and of SFLA after the same number of iterations.

Results of the experiments carried out show that this increased exploration at various stages of the execution increases the convergence speed and may avoid the algorithm from being stuck into a local extreme.

4 Experimental Results and Discussions

The experiments are carried out for the following 5 benchmark test functions with 10 dimensions.

1. Sphere

$$F(x) = \sum_{i=1}^{D} x_i^2$$

where $x_i \in [-100, 100]$, Minimum value $f(x) = 0$ at $x_i = (0, 0, 0\ldots)$.

2. Hyperellipsoid

$$F(x) = \sum_{i=1}^{D} x_i^2 i^2$$

where $x_i \in [-100, 100]$, minimum value $f(x) = 0$ at $x_i = (0, 0, 0\ldots)$.

3. Rastrigin

$$F(x) = \sum_{i=1}^{D} \left[x_i^2 - 10\cos(2\pi x_i) + 10\right]$$

where $x_i \in [-5.12, 5.12]$, minimum value $f(x) = 0$ at $x_i = (0, 0, 0\ldots)$.

4. Ackley

$$F(x) = -a\exp\left(-b\sqrt{\frac{1}{D}\sum_{i=1}^{D}x_i^2}\right) - \exp\left(\frac{1}{D}\sum_{i=1}^{D}\cos(cxi)\right) + a + \exp(1)$$

where $x_i \in [-32.0, 32.0]$, $a = 20$, $b = 0.2$, $c = 2\pi$, minimum value $f(x) = 0$ at $x_i = (0, 0, 0\ldots)$

5. Rosenbrock

$$\sum_{i=1}^{D-1}\left[100\left(x_{i+1} - x_i^2\right)^2 + (x_i - 1)^2\right]$$

where $x_i \in [-2.14, 2.14]$ minimum value $f(x) = 0$ at $x_i = (1, 1, 1\ldots)$.

Other settings were: Populations size (NP) = 100; number of memeplexes (m) = 10; local iterations in each memeplex (LI) = 10. All the experiments were repeated 20 times and average values of objective functions after each iteration are plotted in all the graphs. Termination condition was set to 3000 iterations. The non-improvement factor (I) was fixed to 5 for all the experiments. According to the values of replacement factor (R), the following two sets of experiments were carried out.

4.1 Constant Values of R (SFLA-IE)

For these experiments, value of R was fixed for the entire execution. Experiments were repeated for R = 10, 20, 30 and 50. Convergence graphs for sphere and hyperellipsoid functions are provided in Figs. 2 and 3 respectively.

These figures show some interesting trends. The highest value of R (50) produces the best convergence speed in the initial phases but as the execution

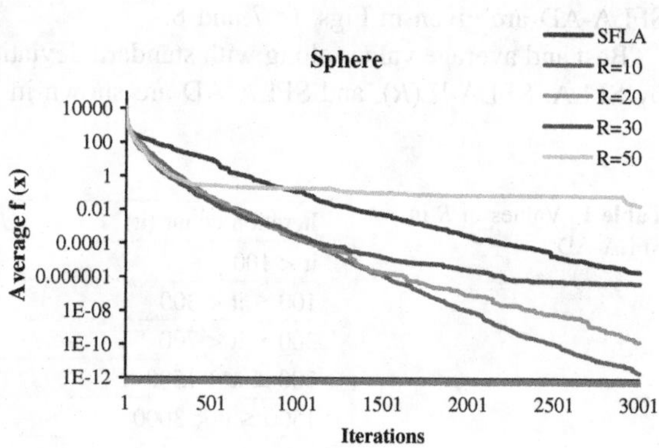

Fig. 2 Convergence graph of sphere function for *SFLA* and SFLA-IE for R = 10, 20, 30, and 50

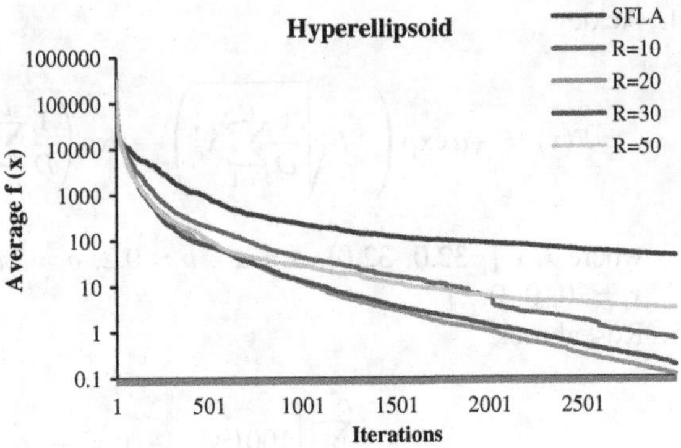

Fig. 3 Convergence graph of hyperellipsoid function for *SFLA* and SFLA-IE for $R = 10, 20, 30$, and 50

progresses, it slows down most rapidly and at one stage it even crosses the convergence speed of the basic SFLA. The lowest value taken for *R* (10), though initially has the worst convergence speed (but better than SFLA) but it maintains its convergence speed for longest of all the values of *R* and tends to cross the basic SFLA convergence curve at the end. So, it can be observed that for lower values of iteration count, higher values of *R* produce more convergence speed. But as the iteration count increases the convergence speed gets slower and this slow down is proportional to the value *R*. This observation can be used to design a variant in which the value of replacement factor (*R*) adapts to the stage of the execution. This variant is denoted as SFLA with adaptive exploration (SFLA-AD).

4.2 Adaptive Values of *R* (SFLA-AD)

By extensive experimentation, it was found that the values of *R* given in Table 1 provide almost the best convergence speed, regardless of the stage of execution the algorithm is in. Comparison of SFLA, SFLA-IE with various values of *R* and SFLA-AD is given in Figs. 4 and 5 for sphere and hyperellipsoid respectively.

Convergence graphs for the other benchmarks problems for SFLA and SFLA-AD are given in Figs. 6, 7 and 8.

Best and average values along with standard deviation for test functions obtained by SFLA, SFLA-IE(*R*), and SFLA-AD are shown in Table 2.

Table 1 Values of *R* in SFLA-AD

Iteration count (it)	Value of *R*
it < 100	50
100 ≤ it < 300	40
300 ≤ it < 700	30
700 ≤ it < 1500	20
1500 ≤ it < 3000	10

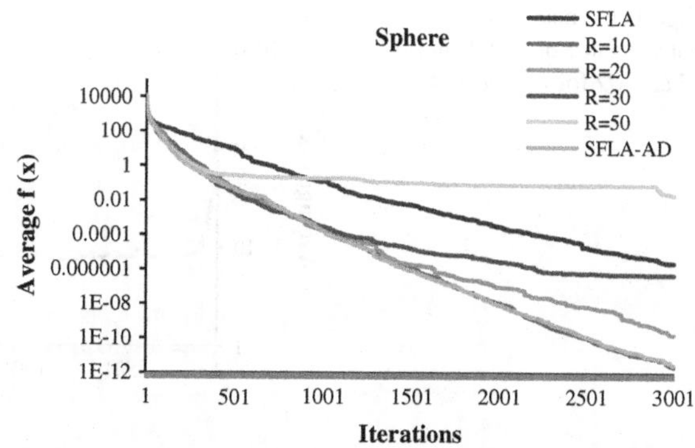

Fig. 4 Showing that *SFLA-AD* provides almost the best convergence speed at all the stages of execution (*sphere*)

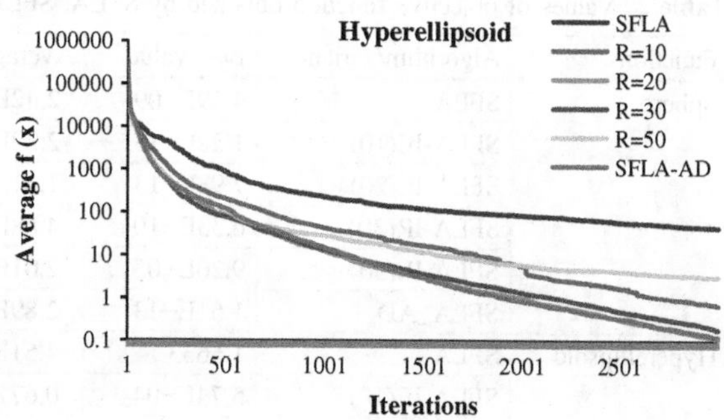

Fig. 5 Showing that *SFLA-AD* provides almost the best convergence speed at all the stages of execution (*hyperellipsoid*)

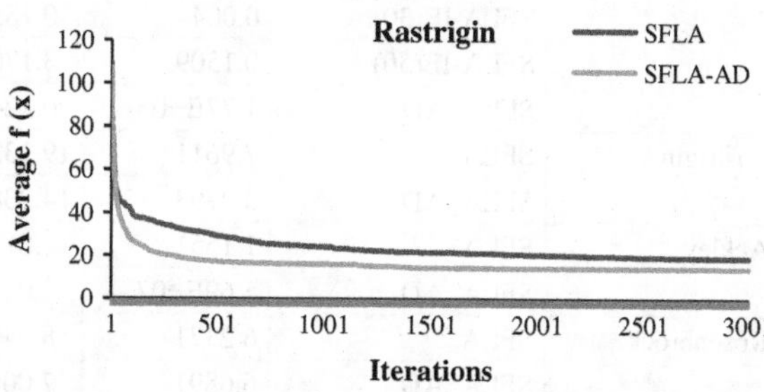

Fig. 6 Comparison of convergence of *SFLA* and *SFLA-AD* for *rastrigin*

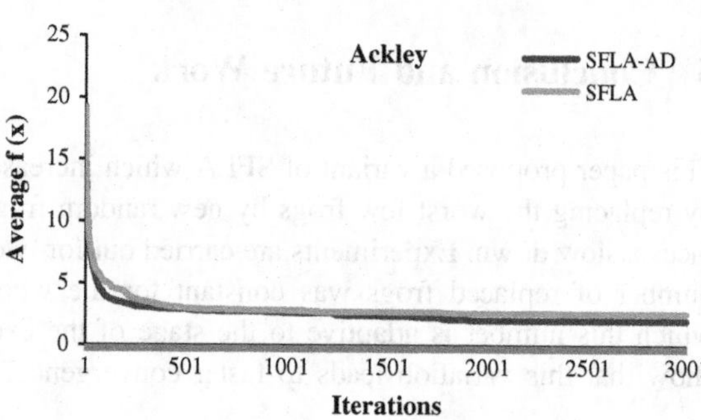

Fig. 7 Comparison of convergence of *SFLA* and *SFLA-AD* for a*ckley*

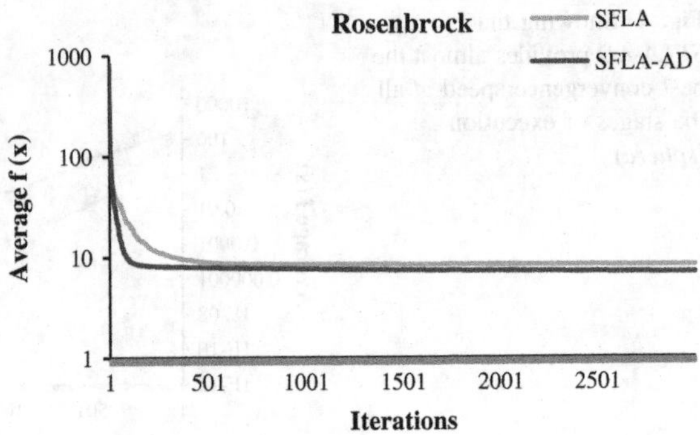

Fig. 8 Comparison of convergence of *SFLA* and *SFLA-AD* for *rosenbrock*

Table 2 Values of objective function obtained by SFLA, SFLA-IE(*R*), and SFLA-AD

Function	Algorithm/variant	Best value	Average value	Standard deviation
Sphere	SFLA	4.69E−09	2.42E−06	7.12E−06
	SFLA-IE(10)	1.22E−13	2.29E−12	1.75E−11
	SFLA-IE(20)	7.98E−13	1.61E−10	4.26E−10
	SFLA-IR(30)	6.33E−10	4.88E−07	1.36E−06
	SFLA-IE(50)	9.20E−05	2.01E−02	3.52E−02
	SFLA_AD	1.51E−13	2.89E−12	3.04E−12
Hyperellipsoid	SFLA	1.4633	4.51E+01	5.76E+01
	SFLA-IE(10)	5.74E−04	0.6775	2.2592
	SFLA-IE(20)	0.0029	0.1132	0.1157
	SFLA-IE(30)	0.0044	0.185	0.1314
	SFLA-IE(50)	0.1509	3.1703	2.7277
	SFLA_AD	1.77E−05	0.1243	0.2125
Rastrigin	SFLA	7.9611	19.1322	6.7637
	SFLA_AD	3.9798	14.2389	5.5608
Ackley	SFLA	1.1551	2.6126	0.8553
	SFLA_AD	5.69E−07	2.0214	0.8085
Rosenbrock	SFLA	6.2371	8.2401	0.7247
	SFLA_AD	6.6891	7.0067	0.1689

5 Conclusion and Future Work

This paper proposed a variant of SFLA which increases diversity of the population by replacing the worst few frogs by new random frogs whenever the convergence faces a slow down. Experiments are carried out for two schemes. First, in which the number of replaced frogs was constant for the whole execution and second, in which this number is adaptive to the stage of the execution. Experimental results show that this variation leads to faster convergence. Observations also show that this improvement in convergence speed is not uniform for all the test problems. This leads to scope of future work where the relationship of number of frogs replaced and function properties can be studied.

References

1. Goldberg, D.: Genetic Algorithms in Search, Optimization, and Machine Learning. Addison Wesley, Reading (1989)
2. Price, K., Storn, R.: Differential evolution—a simple and efficient adaptive scheme for global optimization over continuous spaces. Technical Report, International Computer Science Institute, Berkley (1995)
3. Kennedy, J., Eberhart, R.C.: Particle swarm optimization. In: Proceeding of IEEE International Conference on Neural Networks, pp. 1942–1948, Perth, Australia. IEEE Service Center, Piscataway, NJ (1995)
4. Karaboga, D., Basturk, B.: A powerful and efficient algorithm for numerical function optimization: artificial bee colony (ABC) algorithm. J. Glob. Optim. **39**, 459–471 (2007)
5. Dorigo, M., Maniezzo, V., Colorni, A.: Ant system: optimization by a colony of cooperating agents. IEEE Trans. Syst. Man Cybern. B **26**(1), 29–41 (1996)
6. Eusuff, M., Lansey, K.E.: Optimization of water distribution network design using the shuffled frog leaping algorithm. Water Resour. Plan. Manage. **129**(3), 210–225 (2003)
7. Bhaduri, A.: A clonal selection based shuffled frog leaping algorithm. In: Proceedings of IEEE International Advance Computing Conference, pp. 125–130, Patiala, India (2009)
8. Zhang, X., Hu, F., Tang J., Zou, C., Zhao, L.: A kind of composite shuffled frog leaping algorithm. In: Proceedings of Sixth International Conference on Natural Computation, pp. 2232–2235 (2010)
9. Rajpurohit, J., Sharma, T.K., Pant, M., Naruka, B., Sharma, S.: Dichotomous search in shuffled frog-leaping algorithm. In proceedings of 1st International Science and Technology Congress, pp. 131–135, Kolkata, India (2014)
10. Sharma, S., Sharma, T.K., Pant, M., Rajpurohit, J., Naruka, B.: Centroid mutation embedded shuffled frog-leaping algorithm. In: Proceedings of International Conference on Information and Communication Technologies, pp. 127–134, Kochchi, India (2014)
11. Sharma, S., Sharma, T.K., Pant, M., Rajpurohit, J., Naruka, B.: Accelerated shuffled frog-leaping algorithm. In: Proceeding of Fourth International Conference on Soft Computing for Problem Solving, pp. 181–189, Silchar, India (2014)
12. Naruka, B., Sharma, T.K., Pant, M., Sharma, S., Rajpurohit, J.: Differential shuffled frog-leaping algorithm. In: Proceeding of Fourth International Conference on Soft Computing for Problem Solving, pp. 245–253, Silchar, India (2014)
13. Naruka, B., Sharma, T.K., Pant, M., Rajpurohit, J., Sharma, S.: Two-phase shuffled frog-leaping algorithm. In: Proceeding of 3rd International Conference on Reliability, Infocom Technologies and Optimization, pp. 1–5, Noida, India (2014)

Intuitionistic Trapezoidal Fuzzy Prioritized Weighted Geometric Operator: An Algorithm for the Selection of Suitable Treatment for Lung Cancer

Kumar Vijay, Arora Hari and Pal Kiran

Abstract Lung cancer is considered as the second most common cancer and is the major cause of cancer deaths over the globe. Due to advancement in the field of medical science, different types of treatments or therapies are made available for the treatment of the disease. Multiple attribute group decision making (MAGDM) with the help of intuitionistic trapezoidal fuzzy (ITrF) information has wide applications in decision-making processes especially in the field of medical science. In this paper, we use the concept of MAGDM from a geometric point of view for selection of the most appropriate treatment from the available set of treatments for lung cancer as per the attributes. Once the disease has been diagnosed, with the help of the algorithm of intuitionistic trapezoidal fuzzy prioritized weighted geometric (ITFPWG) operators, we can select the most suitable treatment for lung cancer. Finally, we demonstrate the method by taking a hypothetical case study.

Keywords Multiple attribute group decision making (MAGDM) · Intuitionistic triangular fuzzy numbers (ITFN) · Intuitionistic trapezoidal fuzzy number (ITrFN) · Fuzzy prioritized operators · Intuitionistic trapezoidal fuzzy prioritized weighted geometric (ITFPWG) · Lung cancer

Kumar Vijay (✉)
Manav Rachna International University, Faridabad, India
e-mail: drvijaykumarsudan@gmail.com

Arora Hari · Pal Kiran
Amity University, Noida, Uttar Pradesh, India
e-mail: hdarora@amity.edu

Pal Kiran
e-mail: kiranpaldite@gmail.com

Pal Kiran
Delhi Institute of Tool Engineering, New Delhi, India

M. Pant et al. (eds.), *Proceedings of Fifth International Conference on Soft Computing for Problem Solving*, Advances in Intelligent Systems and Computing 436, DOI 10.1007/978-981-10-0448-3_50

1 Introduction

Population is growing at an exponential pace. Due to the lack of sufficient primary facilities, many lethal diseases are spreading over the globe. Patients are increasing day-by-day and the number of doctors are not sufficient to treat them. Simultaneously, the evolution of computers and many other innovative techniques tackle this problem and thus enhance the efficiency of doctors. Medical diagnosis with the help of computational techniques is an efficient tool for doctors for better decision making in the process of execution of proper treatment to patients. There are many techniques for this purpose, among these generalized fuzzy set theory is considered as an efficient tool for the prescription of suitable treatment for the disease. The concept of MAGDM has been extensively used in real-life situations. When this concept is flavored with generalized fuzzy sets, it gives more strength to the concept of decision making. Atanassov [1] proposed the concept of an intuitionistic fuzzy set for dealing with imprecise and imperfect information. The domains of both intuitionistic fuzzy sets and fuzzy sets are discrete in nature. Shu et al. [2] introduced intuitionistic triangular fuzzy numbers (ITFNs) and their operations. Wang and Zhang [3] explained that ITrFNs are the extension of ITFNs. Wang and Zhang [4] provides the better outfit of intuitionistic triangular fuzzy information for multi-criteria decision making problems. Wang and Zhang [5] developed the intuitionistic trapezoidal fuzzy weighted arithmetic averaging (ITFWAA) operator, including intuitionistic trapezoidal fuzzy weighted arithmetic averaging operator and weighted geometric averaging operator for the propose of better decision making. Wang [6] explained the expected values of ITrFN and proposed the programming method of multi-criteria decision making under ITrFN with incomplete and imprecise information. Wu and Cao [7] explains intuitionistic trapezoidal fuzzy weighted geometric (ITFWG) operator and its hybrid version. Wan [8] discussed the expected score of ITrFNs from the geometric point of view and proposed intuitionistic trapezoidal fuzzy hybrid aggregation operators. Yager [9–11] modeled the concept of prioritized operators to streamline decision making. Wei [12] elaborated the generalized concept of prioritized aggregation operators as explained by Yager [9–11] and developed some hesitant fuzzy prioritized aggregation operators. Yu [13] investigated the fact of prioritization relationship of attributes in MAGDM and developed some prioritized intuitionistic fuzzy aggregation operators. But these operators lacks over certain parameters: they cannot be used when the input is in the form of ITrFNs, having difficulty in dealing with the problems of MAGDM, where the attributes and decision makers are at different priority levels. To overrule this drawback, Zhang [14] proposed some intuitionistic trapezoidal fuzzy prioritized operators. These operators are not only compatible with the situations in which the input arguments are ITrFNs but also consider prioritization among the input arguments. In this paper, we discuss an algorithmic approach of intuitionistic trapezoidal fuzzy prioritized weighted geometric (ITFPWG) operators proposed by Zhang [14] and validate the algorithm for the selection of suitable treatment for lung cancer over certain attributes. These attributes are given in the form of ITrFNs. Once the disease is diagnosed, it is difficult for a doctor to choose the appropriate treatment from the available set of treatments. The present work helps the doctors to select the appropriate treatment for the said disease and rank them as per the attribute. For this purpose, we develop a hypothetical case study to explain the algorithm.

Some basic definitions like intuitionistic fuzzy set (IFS) given by Atanassov [1], intuitionistic trapezoidal fuzzy numbers (ITrFNs) given by Wang [3], and basic arithmetic operations of Intuitionistic trapezoidal fuzzy number (ITrFNs) given by Wang [4] have not been explained in the present text. Only mandatory notations have been considered in the preparation of this paper.

2 Preliminaries

2.1 Score Function of ITrFN

Wang [5] let $\tilde{A} = ([a_1, b_1, c_1, d_1]; \mu_{\tilde{A}}, \nu_{\tilde{A}})$ be an ITrFN, then $S(\tilde{A}) = I(\tilde{A}) \times (\mu_{\tilde{A}} - \nu_{\tilde{A}})$ is called the score function of $\tilde{A}$.
where $I(\tilde{A}) = \frac{1}{8} \times [(a_1 + b_1 + c_1 + d_1) \times (1 + \mu_{\tilde{A}} - \nu_{\tilde{A}})]$ is the expected value of ITrFN $\tilde{A}$.

2.2 Accuracy Function of ITrFN

Wang [5] let $\tilde{A} = ([a_1, b_1, c_1, d_1]; \mu_{\tilde{A}}, \nu_{\tilde{A}})$ be an ITrFN, then $H(\tilde{A}) = I(\tilde{A}) \times (\mu_{\tilde{A}} + \nu_{\tilde{A}})$ is called the accuracy function of $\tilde{A}$.

2.3 Ranking of ITrFN

Wang [5], stated that if $\tilde{A}_1$ and $\tilde{A}_2$ are two random ITrFNs, then

2.3.1 If $S(\tilde{A}_1) > S(\tilde{A}_2)$, then $\tilde{A}_1 > \tilde{A}_2$;
2.3.2 If $S(\tilde{A}_1) = S(\tilde{A}_2)$, and

If $H(\tilde{A}_1) = H(\tilde{A}_2)$, and $\tilde{A}_1 = \tilde{A}_2$;
If $H(\tilde{A}_1) > H(\tilde{A}_2)$, and $\tilde{A}_1 > \tilde{A}_2$.

2.4 *Intuitionistic Trapezoidal Fuzzy Prioritized Weighted Geometric (ITFPWG) Operator*

Yager [9] introduced the prioritized average operators (PA) under ITrF environment.

Let $\tilde{A}_i = ([a_i, b_i, c_i, d_i]; \mu_{\tilde{A}}, v_{\tilde{A}})$ $i = 1, 2, \ldots, n$ be a collection of ITrFNs, if

$$\mathrm{ITFPWG}(\tilde{A}_1, \tilde{A}_2, \ldots, \tilde{A}_n) = \tilde{A}_1^{\frac{T_1}{\sum_{j=1}^{n} T_j}} \otimes \tilde{A}_2^{\frac{T_2}{\sum_{j=1}^{n} T_j}} \otimes \cdots \otimes \tilde{A}_n^{\frac{T_n}{\sum_{j=1}^{n} T_j}}$$

where $T_1 = 1$ and $T_j = \prod_{k=1}^{j-1} S(\tilde{A}_k), (j = 2, 3, \ldots, n)$ and $S(\tilde{A})$ is the score function of $\tilde{A}$.

3 Multiple Attribute Group Decision Making (MAGDM) with Intuitionistic Trapezoidal Fuzzy Prioritized Weighted Geometric (ITFPWG) Operators

MAGDM with the help of ITrF information has been used by many researchers for the purpose of medical diagnosis. Durai [15] describe the technique with the help of algorithm and a rule base to determine diseases. Bhatla [16] used the approach for the diagnosis of heart diseases. Maryam [17] applied the same approach for the diagnosis of diabetics and asthma. In this section, we implement MAGDM under ITrF Information environment for selection of the most appropriate treatment for lung cancer.

Let $X = \{x_1, x_2, \ldots, x_m\}$ be the set of alternatives. Let $C = \{c_1, c_2, \ldots, c_n\}$ be a collection of prioritized linearly ordering attributes. The attribute c_j has a higher priority than c_k if $j<k$. Let $D = \{d_1, d_2, \ldots, d_l\}$ be the set of prioritized linearly ordering decision makers. The decision maker d_p has a higher priority than d_q, if $p<q$. Consider that each decision maker provides its own ITrF decision matrix $\tilde{\mathrm{A}}^{(k)} = \left(\tilde{A}_{ij}^{(k)}\right)_{m\times n}$ $(k = 1, 2, \ldots, l)$, where $\tilde{A}_{ij}^{(k)} = \left(\left[a_{ij}^{(k)}, b_{ij}^{(k)}, c_{ij}^{(k)}, d_{ij}^{(k)}\right]; \mu_{ij}^{(k)}, v_{ij}^{(k)}\right)$ is an ITrFN given by the decision maker $d_k \in D$, $\mu_{ij}^{(k)}$ denotes the membership number of trapezoidal fuzzy numbers $\left[a_{ij}^{(k)}, b_{ij}^{(k)}, c_{ij}^{(k)}, d_{ij}^{(k)}\right]$ on the criteria c_j,$v_{ij}^{(k)}$ denotes the non-membership number of trapezoidal fuzzy number $\left[a_{ij}^{(k)}, b_{ij}^{(k)}, c_{ij}^{(k)}, d_{ij}^{(k)}\right]$ on the criteria c_j. $\mu_{ij}^{(k)} \in [0, 1]$, $v_{ij}^{(k)} \in [0, 1]$ and $\mu_{ij}^{(k)} + v_{ij}^{(k)} \leq 1$, $i = 1, 2, 3, \ldots, m, j = 1, 2, \ldots, n$

In order to standardize decision making matrix, we may transform the ITrF decision matrix $\tilde{\mathrm{A}}^{(k)} = \left(\tilde{A}_{ij}^{(k)}\right)_{m\times n} = \left(\left(\left[a_{ij}^{(k)}, b_{ij}^{(k)}, c_{ij}^{(k)}, d_{ij}^{(k)}\right]; \mu_{ij}^{(k)}, v_{ij}^{(k)}\right)\right)_{m\times n}$ into the standardized ITrF decision matrix $\tilde{R}^{(k)} = \left(\tilde{r}_{ij}^{(k)}\right)_{m\times n} = \left(\left(\left[e_{ij}^{(k)}, f_{ij}^{(k)}, g_{ij}^{(k)}, h_{ij}^{(k)}\right];\right.\right.$ $\mu_{ij}^{(k)}, v_{ij}^{(k)}))_{m\times n}$ by the following method given by Wang [4, 5] and Zhang [18].

For cost type of criteria:

$$e_{ij}^{(k)} = \frac{\max\limits_{1\le j\le n}\left\{d_{ij}^{(k)}\right\} - a_{ij}^{(k)}}{\max\limits_{1\le j\le n}\left\{d_{ij}^{(k)}\right\} - \min\limits_{1\le j\le n}\left\{a_{ij}^{(k)}\right\}}, f_{ij}^{(k)} = \frac{\max\limits_{1\le j\le n}\left\{d_{ij}^{(k)}\right\} - b_{ij}^{(k)}}{\max\limits_{1\le j\le n}\left\{d_{ij}^{(k)}\right\} - \min\limits_{1\le j\le n}\left\{a_{ij}^{(k)}\right\}},$$

$$g_{ij}^{(k)} = \frac{\max\limits_{1\le j\le n}\left\{d_{ij}^{(k)}\right\} - c_{ij}^{(k)}}{\max\limits_{1\le j\le n}\left\{d_{ij}^{(k)}\right\} - \min\limits_{1\le j\le n}\left\{a_{ij}^{(k)}\right\}}, h_{ij}^{(k)} = \frac{\max\limits_{1\le j\le n}\left\{d_{ij}^{(k)}\right\} - d_{ij}^{(k)}}{\max\limits_{1\le j\le n}\left\{d_{ij}^{(k)}\right\} - \min\limits_{1\le j\le n}\left\{a_{ij}^{(k)}\right\}}$$

For benefit type of criteria:

$$e_{ij}^{(k)} = \frac{a_{ij}^{(k)} - \min\limits_{1\le j\le n}\left\{a_{ij}^{(k)}\right\}}{\max\limits_{1\le j\le n}\left\{d_{ij}^{(k)}\right\} - \min\limits_{1\le j\le n}\left\{a_{ij}^{(k)}\right\}}, f_{ij}^{(k)} = \frac{b_{ij}^{(k)} - \min\limits_{1\le j\le n}\left\{a_{ij}^{(k)}\right\}}{\max\limits_{1\le j\le n}\left\{d_{ij}^{(k)}\right\} - \min\limits_{1\le j\le n}\left\{a_{ij}^{(k)}\right\}},$$

$$g_{ij}^{(k)} = \frac{c_{ij}^{(k)} - \min\limits_{1\le j\le n}\left\{a_{ij}^{(k)}\right\}}{\max\limits_{1\le j\le n}\left\{d_{ij}^{(k)}\right\} - \min\limits_{1\le j\le n}\left\{a_{ij}^{(k)}\right\}}, h_{ij}^{(k)} = \frac{d_{ij}^{(k)} - \min\limits_{1\le j\le n}\left\{a_{ij}^{(k)}\right\}}{\max\limits_{1\le j\le n}\left\{d_{ij}^{(k)}\right\} - \min\limits_{1\le j\le n}\left\{a_{ij}^{(k)}\right\}}$$

In the following, we use the ITFPWG operators to develop an approach to MAGDM under ITrF environment. The algorithm follows as

Step 3.1: Transform the ITrF decision matrices $\tilde{A}^{(k)}$ into the standardized ITrF decision matrices $\tilde{R}^{(k)}$ as defined in Sect. 3.

Step 3.2: Calculate

$T^{(p)} = \left(T_{ij}^{(p)}\right)_{m\times n}, p = 2, 3, \ldots, l, i = 1, 2, 3, \ldots, m, j = 1, 2, \ldots, n,$

where $T_{ij}^{(p)} = \prod\limits_{k=1}^{p-1} S\left(\tilde{r}_{ij}^{(k)}\right), p = 2, 3, \ldots, l, i = 1, 2, 3\ldots, m, j = 1, 2\ldots, n,$

and $T_{ij}^{(1)} = 1\, i = 1, 2, 3, \ldots, m, j = 1, 2, \ldots, n$

Step 3.3: Using the ITFPWG operator Sect. 2.4:

To aggregate all the individual ITrF decision matrices $\tilde{R}^{(k)} = \left(r_{ij}^{(k)}\right)_{m\times n}$ $(k = 1, 2, \ldots, l)$ into the collective ITrF decision matrix $\tilde{R} = \left(\tilde{r}_{ij}\right)_{m\times n} =$ $\left(\left(\left[e_{ij}, f_{ij}, g_{ij}, h_{ij}\right]; \mu_{ij}, v_{ij}\right)\right)_{m\times n}$

ITFPWG operator:

$$\tilde{r}_{ij} = \text{ITFPWG}\left(\tilde{r}_{ij}^{(1)}, \tilde{r}_{ij}^{(2)}, \ldots, \tilde{r}_{ij}^{(l)}\right) = \left[\begin{array}{l} \left(\prod_{k=1}^{l}\left(e_{ij}^{(k)}\right)^{\left(\frac{T_{ij}^{(k)}}{\sum_{p=1}^{l} T_{ij}^{(p)}}\right)}, \prod_{k=1}^{l}\left(f_{ij}^{(k)}\right)^{\left(\frac{T_{ij}^{(k)}}{\sum_{p=1}^{l} T_{ij}^{(p)}}\right)}, \prod_{k=1}^{l}\left(g_{ij}^{(k)}\right)^{\left(\frac{T_{ij}^{(k)}}{\sum_{p=1}^{l} T_{ij}^{(p)}}\right)}, \prod_{k=1}^{l}\left(h_{ij}^{(k)}\right)^{\left(\frac{T_{ij}^{(k)}}{\sum_{p=1}^{l} T_{ij}^{(p)}}\right)}\right); \\ \prod_{k=1}^{l}\left(\mu_{ij}^{(k)}\right)^{\left(\frac{T_{ij}^{(k)}}{\sum_{p=1}^{l} T_{ij}^{(p)}}\right)}, 1 - \prod_{k=1}^{l}\left(1 - v_{ij}^{(k)}\right)^{\left(\frac{T_{ij}^{(k)}}{\sum_{p=1}^{l} T_{ij}^{(p)}}\right)} \end{array}\right]$$

Step 3.4: Calculate the matrix $T = \left(T_{ij}\right)_{m \times n}$ based on the following equation:

$$T_{ij} = \prod_{k=1}^{p-1} S(\tilde{r}_{ij}), T_{i1} = 1 \quad \forall i = 1, 2, 3, \ldots, m, \quad \forall j = 1, 2, \ldots, n$$

Step 3.5: Utilizing the ITFPWG operator Sect. 2.4:
To derive the collective overall preference ITrF values $\tilde{r}_1 = ([e_i, f_i, g_i, h_i]; \mu_i, v_i)$ $(i = 1, 2, \ldots, m)$ of the alternatives $x_i (i = 1, 2, \ldots, m)$

$$\tilde{r}_1 = \text{ITFPWG}(\tilde{r}_{i1}, \tilde{r}_{i2}, \ldots, \tilde{r}_{in} = (\tilde{r}_{i1})^{\left(T_{i1}/\sum_{j=1}^{n} T_{ij}\right)} \otimes (\tilde{r}_{i2})^{\left(T_{i2}/\sum_{j=1}^{n} T_{ij}\right)} \otimes \cdots (\tilde{r}_{in})^{\left(T_{in}/\sum_{j=1}^{n} T_{ij}\right)}$$

$$= \left(\begin{array}{l} \left[\prod_{j=1}^{n}(e_{ij})^{\left(T_{ij}/\sum_{j=1}^{n} T_{ij}\right)}, \prod_{j=1}^{n}(f_{ij})^{\left(T_{ij}/\sum_{j=1}^{n} T_{ij}\right)}, \prod_{j=1}^{n}(g_{ij})^{\left(T_{ij}/\sum_{j=1}^{n} T_{ij}\right)}, \prod_{j=1}^{n}(h_{ij})^{\left(T_{ij}/\sum_{j=1}^{n} T_{ij}\right)},\right]; \\ \prod_{j=1}^{n}(\mu_{ij})^{\left(T_{ij}/\sum_{j=1}^{n} T_{ij}\right)}, 1 - \prod_{j=1}^{n}(1 - v_{ij})^{\left(T_{ij}/\sum_{j=1}^{n} T_{ij}\right)}, \end{array}\right)$$

Step 3.6: Calculate the score functions Sect. 2.1 and the accuracy functions Sect. 2.2 of $\tilde{r}_i$, $\forall i = 1, 2, 3, \ldots, m$ as follows:

$$S(\tilde{r}) = \frac{1}{8} \times (e_i + f_i + g_i + h_i) \times (1 + \mu_i - \upsilon_i) \times (\mu_i - \upsilon_i)$$

$$H(\tilde{r}) = \frac{1}{8} \times (e_i + f_i + g_i + h_i) \times (1 + \mu_i - \upsilon_i) \times (\mu_i + \upsilon_i)$$

Step 3.7: Using Sect. 2.3 select the best alternative while doing ranking of all the alternatives x_i, $\forall i = 1, 2, 3, \ldots, m$

Step 3.8: End.

4 Case Study

Jemal et al. [19] in a survey in the US revealed that in the year 2014, an estimated 221,200 adults (115,610 men and 105,590 women) were diagnosed with lung cancer, which is the second most common cancer for the major cause of cancer death. With the help of experts from the same domain, we formed a knowledge base, as per the symptoms and the available set of treatments. To select the most appropriate treatment for lung cancer, the algorithm (3) follows as:

Let $x_i (i = 1, 2, \ldots, 5)$ be the available set of treatments for lung cancer. These treatments can be radiation therapy, biological therapies, personalized and targeted therapies, metastatic surgery and chemotherapy. The selection for treatment can be evaluated on the basis of the prescribed symptoms $S = (s_1, s_2, s_3, s_4)$. The common symptoms for lung cancer are coughing up blood, difficulty in breathing, bone pain, swellings on the body, etc. Also, the decision for the final treatment can be made from the panel of decision makers as $D = (d_1, d_2, d_3)$. The three ITrF decision matrix $\tilde{A}^{(k)} = \left(\tilde{A}_{ij}^{(k)}\right)_{m \times n} (k = 1, 2, 3)$ are listed in Tables 1, 2, 3, 4, 5, 6, and 7 as.

Table 1 The ITrF decision matrix $\tilde{A}^{(1)}$ provided by d_1

x_i	s_1	s_2	s_3	s_4
x_1	([2, 3, 4, 6]; 0.6, 0.3)	([4, 6, 8, 9]; 0.6,0.3)	([2, 3, 4, 6]; 0.6, 0.4)	([3, 5, 6, 9]; 0.7, 0.1)
x_2	([1, 2, 3, 6]; 0.6, 0.4)	([6, 7, 8, 9]; 0.8,0.2)	([3, 5, 6, 8]; 0.8, 0.1)	([4, 6, 8, 9]; 0.7, 0.3)
x_3	([2, 3, 4, 5]; 0.5, 0.3)	([5, 6, 7, 9]; 0.7,0.3)	([2, 3, 5, 6]; 0.6, 0.4)	([3, 5, 6, 8]; 0.7, 0.1)
x_4	([1, 2, 3, 5]; 0.6, 0.4)	([6, 7, 8, 9]; 0.8,0.1)	([3, 4, 6, 7]; 0.8, 0.1)	([4, 6, 7, 9]; 0.6, 0.3)
x_5	([2, 3, 4, 5]; 0.6, 0.3)	([5, 6, 7, 8]; 0.6,0.3)	([2, 3, 5, 9]; 0.6, 0.3)	([3, 4, 6, 8]; 0.7, 0.2)

Table 2 The ITrF decision matrix $\tilde{A}^{(2)}$ provided by d_2

x_i	s_1	s_2	s_3	s_4
x_1	([5, 6, 7, 9]; 0.8, 0.1)	([2, 3, 4, 7]; 0.7, 0.3)	([3, 4, 6, 8]; 0.5, 0.4)	([5, 6, 7, 9]; 0.7, 0.3)
x_2	([1, 4, 5, 7]; 0.5, 0.4)	([2, 4, 5, 6]; 0.5, 0.4)	([2, 5, 4, 6]; 0.7, 0.3)	([2, 6, 7, 8]; 0.8, 0.1)
x_3	([5, 6, 8, 9]; 0.8, 0.1)	([2, 3, 4, 5]; 0.6, 0.3)	([3, 4, 5, 8]; 0.5, 0.4)	([5, 6, 7, 8]; 0.7, 0.2)
x_4	([1, 4, 5, 8]; 0.5, 0.4)	([2, 4, 5, 7]; 0.5, 0.4)	([2, 3, 4, 5]; 0.6, 0.3)	([2, 5, 7, 8]; 0.8, 0.1)
x_5	([5, 6, 7, 8]; 0.8, 0.2)	([2, 3, 4, 6]; 0.6, 0.3)	([3, 4, 6, 9]; 0.5, 0.3)	([5, 6, 7, 9]; 0.7, 0.1)

Table 3 The ITrF decision matrix $\tilde{A}^{(3)}$ provided by d_3

x_i	s_1	s_2	s_3	s_4
x_1	([2, 3, 5, 7]; 0.5, 0.4)	([4, 5, 7, 8]; 0.5, 0.4)	([2, 3, 6, 8]; 0.6, 0.3)	([4, 5, 6, 8]; 0.8, 0.1)
x_2	([4, 6, 7, 9]; 0.6, 0.3)	([2, 3, 5, 6]; 0.7, 0.2)	([2, 3, 5, 6]; 0.7, 0.3)	([6, 7, 8, 9]; 0.4, 0.6)
x_3	([2, 3, 5, 6]; 0.5, 0.4)	([4, 6, 7, 8]; 0.5, 0.4)	([2, 3, 6, 8]; 0.6, 0.4)	([4, 5, 6, 7]; 0.8, 0.1)
x_4	([4, 6, 7, 8]; 0.6, 0.4)	([2, 3, 4, 5]; 0.8, 0.2)	([2, 3, 4, 6]; 0.6, 0.3)	([6, 7, 8, 9]; 0.3, 0.5)
x_5	([2, 4, 5, 6]; 0.5, 0.3)	([4, 6, 7, 8]; 0.6, 0.4)	([2, 5, 6, 8]; 0.6, 0.4)	([4–6, 8]; 0.8, 0.2)

Table 4 The ITrF decision matrix $\tilde{R}^{(1)}$ provided by d_1

x_i	s_1	s_2	s_3	s_4
x_1	([0, 0.1429, 0.2857, 0.5714]; 0.6, 0.3)	([0.2857, 0.5714, 0.8571, 1.00]; 0.6, 0.3)	([0, 0.1429, 0.2857, 0.5714]; 0.6, 0.4)	([0.1429, 0.4286, 0.5714, 1.0000]; 0.7, 0.1)
x_2	([0, 0.1250, 0.250, 0.6250]; 0.6, 0.4)	([0.6250, 0.7500, 0.8750, 1.00]; 0.8, 0.2)	([0.2500, 0.5000, 0.6250, 0.8750]; 0.8,0.1)	([0.3750, 0.6250, 0.8750, 1.00]; 0.7, 0.3)
x_3	([0, 0.1429, 0.2857, 0.428]; 0.5, 0.3)	([0.4286, 0.5714, 0.7143, 1.00]; 0.7, 0.3)	([0, 0.1429, 0.4286, 0.5714]; 0.6, 0.4)	([0.1429, 0.4286, 0.5714, 0.8571]; 0.7, 0.1)
x_4	([0, 0.1250, 0.2500, 0.50]; 0.6, 0.4)	([0.6250, 0.7500, 0.8750, 1.00]; 0.8, 0.1)	([0.2500, 0.3750, 0.6250, 0.7500]; 0.8, 0.1)	([0.3750, 0.6250, 0.7500, 1.00]; 0.6, 0.3)
x_5	([0, 0.1429, 0.2857,0.4286]; 0.6, 0.3)	([0.4286, 0.5714, 0.7143, 0.857]; 0.6, 0.3)	([0, 0.1429, 0.4286, 1.00]; 0.6, 0.3)	([0.1429, 0.2857, 0.5714, 0.8571]; 0.7, 0.2)

Table 5 The ITrF decision matrix $\tilde{R}^{(2)}$ provided by d_2

x_i	s_1	s_2	s_3	s_4
x_1	([0.4286, 0.5714, 0.7143, 1.00]; 0.8, 0.1)	([0, 0.1429, 0.2857, 0.714]; 0.7, 0.3)	([0.1429, 0.2857, 0.5714, 0.8571]; 0.5, 0.4)	([0.4286, 0.5714,0.7143, 1.0000]; 0.7, 0.3)
x_2	([0, 0.4286, 0.5714, 0.8571]; 0.5, 0.4)	([0.1429, 0.4286, 0.5714, 0.714]; 0.5, 0.4)	([0.1429, 0.2857, 0.4286, 0.7143]; 0.7, 0.3)	([0.1429, 0.7143, 0.8571, 1.0000]; 0.8, 0.1)
x_3	([0.4286, 0.5714, 0.8571, 1.00]; 0.8, 0.1)	([0, 0.1429, 0.2857, 0.428]; 0.6, 0.3)	([0.1429, 0.2857, 0.4286, 0.8571]; 0.5, 0.4)	([0.4286, 0.5714, 0.7143, 0.8571]; 0.7,0.2)
x_4	([0, 0.4286, 0.5714, 1.00]; 0.5, 0.4)	([0.1429, 0.4286, 0.5714, 0.857]; 0.5, 0.4)	([0.1429, 0.2857, 0.4286, 0.5714]; 0.6, 0.3)	([0.1429, 0.5714, 0.8571, 1.0000]; 0.8, 0.1)
x_5	([0.4286, 0.5714, 0.7143, 0.857]; 0.8, 0.2)	([0, 0.1429, 0.2857, 0.571]; 0.6, 0.3)	([0.1429, 0.2857, 0.5714, 1.0000]; 0.5, 0.3)	([0.4286, 0.5714, 0.7143, 1.0000]; 0.7, 0.1)

Using (3.1) and (3.2), we get the matrices $T^{(1)}$, $T^{(2)}$ and $T^{(3)}$ as follows:

$$T^{(1)} = \begin{bmatrix} 1 & 1 & 1 & 1 \\ 1 & 1 & 1 & 1 \\ 1 & 1 & 1 & 1 \\ 1 & 1 & 1 & 1 \\ 1 & 1 & 1 & 1 \end{bmatrix}$$

Table 6 The ITrF decision matrix $\tilde{R}^{(3)}$ provided by d_3

x_i	s_1	s_2	s_3	s_4
x_1	([0, 0.1667, 0.500, 0.8333]; 0.5, 0.4)	([0.3333, 0.5000, 0.8333, 1.000]; 0.5, 0.4)	([0, 0.1667, 0.6667, 1.000]; 0.6, 0.3)	([0.3333, 0.500, 0.6667, 1.0000]; 0.8, 0.1)
x_2	([0.285, 0.571, 0.7143, 1.00]; 0.6, 0.3)	([0, 0.1429, 0.4286, 0.5714]; 0.7, 0.2)	([0, 0.1429, 0.4286, 0.5714]; 0.7, 0.3)	([0.5714, 0.7143, 0.8571, 1.0000]; 0.4, 0.6)
x_3	([0, 0.1667, 0.50, 0.66]; 0.5, 0.4)	([0.3333, 0.6667, 0.833, 1.000]; 0.5, 0.4)	([0, 0.1667, 0.6667, 1.0000]; 0.6, 0.4)	([0.3333, 0.5000, 0.6667, 0.8333]; 0.8, 0.1)
x_4	([0.2857, 0.5714, 0.7143, 0.85]; 0.6, 0.4)	([0, 0.1429, 0.2857,0.4286]; 0.8, 0.2)	([0, 0.1429, 0.2857, 0.5714]; 0.6, 0.3)	([0.5714, 0.7143, 0.8571, 1.0000]; 0.3, 0.5)
x_5	([0, 0.33, 0.50, 0.66]; 0.5, 0.3)	([0.33, 0.6667, 0.8333, 1.00]; 0.6, 0.4)	([0, 0.5000, 0.6667, 1.00]; 0.6, 0.4)	([0.3333, 0.500, 0.6667, 1.0000]; 0.8, 0.2)

Table 7 The collective ITrF decision matrix R

x_i	s_1	s_2	s_3	s_4
x_1	([0,0.1526, 0.3010, 0.5903]; 0.6059, 0.2939)	([0, 0.4961, 0.7546, 0.9618]; 0.6098, 0.3010)	([0, 0.1458, 0.2917, 0.5785]; 0.5968, 0.3999)	([0.1831, 0.4562, 0.6005, 1.0000]; 0.7035, 0.1435)
x_2	([0, 0.1297, 0.256, 0.6310]; 0.5968, 0.399)	([0, 0.634, 0.7731, 0.906]; 0.7012, 0.2616)	([0, 0.4218, 0.5643, 0.8233]; 0.7716, 0.1592)	([0.3310, 0.6437, 0.8710, 1.0000]; 0.6899, 0.2972)
x_3	([0, 0.1481, 0.2953, 0.439]; 0.505, 0.2968)	([0, 0.4591, 0.618, 0.8743]; 0.6816, 0.3007)	([0, 0.1462, 0.4287, 0.5794]; 0.5964, 0.4000)	([0.1818, 0.4551, 0.5996, 0.8561]; 0.7042, 0.1194)
x_4	([0,0.1291, 0.255, 0.5091]; 0.597, 0.400)	([0, 0.6168, 0.755, 0.9443]; 0.6873, 0.2113)	([0, 0.3474, 0.5670, 0.7023]; 0.7463, 0.1530)	([0.3424, 0.6223, 0.7658, 1.0000]; 0.6015, 0.2902)
x_5	([0, 0.1525, 0.298, 0.4429]; 0.605, 0.2963)	([0, 0.4905, 0.645, 0.8202]; 0.600, 0.3006)	([0, 0.1508, 0.4382, 1.00]; 0.5923, 0.3005)	([0.1735, 0.3234, 0.5940, 0.8823]; 0.7043, 0.1866)

$$T^{(2)} = \begin{bmatrix} 0.0488 & 0.1323 & 0.0300 & 0.2571 \\ 0.0300 & 0.3900 & 0.3347 & 0.2012 \\ 0.0257 & 0.1900 & 0.0343 & 0.2400 \\ 0.0263 & 0.4834 & 0.2975 & 0.1341 \\ 0.0418 & 0.1254 & 0.0766 & 0.1741 \end{bmatrix}$$

$$T^{(3)} = \begin{bmatrix} 0.0197 & 0.0106 & 0.0008 & 0.0489 \\ 0.0008 & 0.0100 & 0.0368 & 0.0813 \\ 0.0109 & 0.0079 & 0.0008 & 0.0579 \\ 0.0007 & 0.0133 & 0.0207 & 0.0513 \\ 0.0129 & 0.0061 & 0.0046 & 0.0567 \end{bmatrix}$$

Using (3.3) and (3.4), we get the following matrix as

$$T = \begin{bmatrix} 1 & 0.0534 & 0.0059 & 0.0002 \\ 1 & 0.0300 & 0.0055 & 0.0012 \\ 1 & 0.0279 & 0.0036 & 0.0001 \\ 1 & 0.0264 & 0.0054 & 0.0010 \\ 1 & 0.0452 & 0.0043 & 0.0003 \end{bmatrix}$$

Using (3.5), we get the overall preference value $\tilde{r}_i = (i = 1, 2, \ldots, m)$ against the alternatives $x_i (i = 1, 2, \ldots, 5)$ which follows as

$$\tilde{r}_1 = ([0, 0.1618, 0.3153, 0.6050]; 0.6060, 0.2949)$$
$$\tilde{r}_2 = ([0, 0.1369, 0.2661, 0.6389]; 0.6005, 0.3951)$$
$$\tilde{r}_3 = ([0, 0.1527, 0.3017, 0.4484]; 0.5103, 0.2972)$$
$$\tilde{r}_4 = ([0, 0.1353, 0.2641, 0.5184]; 0.6001, 0.3946)$$
$$\tilde{r}_5 = ([0, 0.1604, 0.3090, 0.4564]; 0.6053, 0.2965)$$

Using (3.6), we calculate score functions of $\tilde{r}_i (i = 1, 2, 3, 4, 5)$ as follows: $S(\tilde{r}_1) = 0.0552$; $S(\tilde{r}_2) = 0.0322$; $S(\tilde{r}_3) = 0.0292$; $S(\tilde{r}_4) = 0.0284$; $S(\tilde{r}_5) = 0.0468$.

Finally, using (3.7), we can rank all the available alternatives x_i $(i = 1, 2, \ldots, 5)$ and then select the best alternative as $S(\tilde{r}_1) > S(\tilde{r}_5) > S(\tilde{r}_2) > S(\tilde{r}_3) > S(\tilde{r}_4)$, we have $x_1 \succ x_5 \succ x_2 \succ x_3 \succ x_4$.

Therefore, the best treatment for lung cancer is x_1.

5 Conclusion

Using ITFPWG operators, we can easily take the decision of which treatment is the most suitable from the available set of treatments for lung cancer. Also, we can rank the treatments as per their attributes. This algorithm is the most efficient tool for decision makers in handling real-life situations.

References

1. Atanassov, K.: Intuitionistic fuzzy sets. Fuzzy Sets Syst. **20**, 87–96 (1986)
2. Shu, M.H., Cheng, C.H., Chang J.R.: Using intuitionistic fuzzy sets for fault-tree analysis on printed circuit board assembly. Microelectron. Reliab. **46**, 2139–2148 (2006)
3. Wang, J.Q.: Overview on fuzzy multi-criteria decision-making approach. Control Decis. **23**, 601–606
4. Wang J.Q., Zhang Z.H.: Multi-criteria decision-making method with incomplete certain information based on intuitionistic fuzzy number. Control Decis. **24**, 226–230 (2009)
5. Wang, J.Q., Zhang, Z.: Aggregation operators on intuitionistic trapezoidal fuzzy number and its application to multi-criteria decision making problems. J. Syst. Eng. Electron. **20**, 321–326 (2009)
6. Wang, J.Q., Zhang, Z.H.: Programming method of multi-criteria decision-making based on intuitionistic fuzzy number with incomplete certain information. Control Decis. **23**, 1145–1148 (2008)
7. Wu, J., Cao, Q.W.: Same families of geometric aggregation operators with intuitionistic trapezoidal fuzzy numbers. Appl. Math. Model. **37**(1–2), 318–327 (2013)
8. Wan, S.P., Dong J.Y.: Method of trapezoidal intuitionistic fuzzy number for multi-attribute group decision. Control Decis. **25**(5), 773–776 (2010)
9. Yager, R.R.: Modeling prioritized multi-criteria decision making. IEEE Trans. Syst. Man Cybern. B. Cybern. **34**, 2396–2404 (2004)
10. Yager, R.R.: Prioritized aggregation operators. Int. J. Approximate Reasoning **48**, 63–274 (2008)
11. Yager, R.R.: Prioritized OWA aggregation. Fuzzy Optim. Decis. Making **8**, 245–262 (2009)
12. Wei, G.W.: Hesitant fuzzy prioritized operators and their application to multiple attribute decision making. Knowl.-Based Syst. **31**, 176–182 (2012)
13. Yu, X., Xu, Z.: Prioritized intuitionistic fuzzy aggregation operators. Inf. Fusion **14**, 108–116 (2013)
14. Zhang, Z.: Intuitionistic trapezoidal fuzzy prioritized operators and their application to multiple attribute group decision making. Brit. J. Math. Comput. Sci. **4**(14), 1951–1998 (2014)
15. Durai, M.S., Iyengar, N.: Effective analysis and diagnosis of lung cancer using fuzzy rules. Int. J. Eng. Sci. Technol. **2**(6), 2102–2108 (2010)
16. Bhatla, N., Jyoti, K.: A novel approach for heart disease diagnosis using data mining and fuzzy logic. In: International Journal of Computer Applications, pp. 16–21 (2012)
17. Maryam, Fazel, M.H., Mostafa.: Fuzzy rule-base expert system for evaluation possibility of fatal asthma. In: Journal of Health Informatics in Developing Countries, pp. 171–184 (2010)
18. Zhang, X., Jin, F., Liu, P.D.: A grey relational projection method for multi-attribute decision making based on intuitionistic trapezoidal fuzzy number. Appl. Math. Model. **37**(5), 3467–3477 (2013)
19. Jemal, A., Siegel, R., Ward, E., Hao, Y., Xu, J., Murray, T., Thun, M.J.: Cancer statistics. CA. **58**(2), 71–96 (2008)

Fuzzy Controller for Reversing Voltage Topology MLI

P. Ponnambalam, B. Shyam Sekhar, M. Praveenkumar, V. Surendar and P. Ravi Teja

Abstract Multilevel inverter of reversing voltage topology has emerged recently as a very important technology in the area of medium-voltage high power energy control, due to lower EMI, requirement of less number of semiconductor power devices with less blocking voltage, lower THD percentage in output voltage, and less stress on insulation. This topology overcomes the disadvantages that a normal multilevel inverter has, like increased number of components, complex power bus structure in some topologies, and voltage balancing problem at neutral point. In this paper, the multilevel inverter with reversing voltage is implemented (which was previously proposed). This topology of inverter is first simulated using MATLAB simulation in open loop, and then PWM technique is introduced to have a control over the output RMS voltage; for these topologies the THD is analyzed. Then closed-loop control is implemented using fuzzy logic. The open-loop configuration of the circuit is realized in hardware and the results are analyzed.

Keywords Multilevel inverter · Reversing voltage topology · THD · Pulse-width modulation · Fuzzy inference system · Fuzzy logic

P. Ponnambalam (✉) · B. Shyam Sekhar · M. Praveenkumar · P. Ravi Teja
SELECT, VIT University, Vellore, Tamil Nadu, India
e-mail: p.ponnambalam@gmail.com

B. Shyam Sekhar
e-mail: b.shyamsekhar@gmail.com

M. Praveenkumar
e-mail: praveen.m@vit.ac.in

P. Ravi Teja
e-mail: ravitejareddy226@gmail.com

V. Surendar
EEE Department of Kongu Engineering College, Perundurai, Erode, Tamil Nadu, India
e-mail: surendar136@gmail.com

M. Pant et al. (eds.), *Proceedings of Fifth International Conference on Soft Computing for Problem Solving*, Advances in Intelligent Systems and Computing 436, DOI 10.1007/978-981-10-0448-3_51

1 Introduction

Multilevel inverter of reversing voltage topology has been mostly used for medium or high power system applications. The general concept of multilevel inverter is that it utilizes higher number of semiconductor switches for the purpose of power conversion in small voltage steps; the multilevel inverter of reversing voltage topology has some advantages compared to other conventional power conversion topologies [1]. The smaller voltage steps lead to the production of high power quality waveforms and reduced d*v*/d*t* stress on the load. The EMI compatibility concerns are well satisfied in this topology. Another important feature of MLI is that the semiconductors are connected in series. It operates at higher voltages. The series connection was made up with the clamping diodes and it eliminates the over voltage concerns. In case the switches are not connected in series the switching can be staggered and it will reduce the switching frequency losses.

The conventional MLI power conversion circuit has the disadvantage of requiring higher number of semiconductor switches in addition to the disadvantage that it requires isolated voltage sources or series capacitors bank that could produce the smaller voltage steps. Series capacitors require voltage balancing [2]. Lower voltage rated switches are used in MLI conversion for this purpose, so the cost will not increase when compared with two-level cases. Each semiconductor switch requires gate drive circuit which will be a disadvantage. Due to the higher number of semiconductor devices the voltage balance can be addressed using redundant switching states; for a complete solution for this voltage balancing problem, we may require to create another multilevel converter [3].

In recent years, the interest has been increased on multilevel power conversion. The recent research has been involved on the unique modulation strategies and novel converter topologies. There has been so many multilevel inverter topologies emerged, but predominantly the diode clamped, capacitor clamped, and cascade and neutral-point-clamped inverters are used in many applications [4]. There are some combination of topologies, namely cascaded 3/2 multilevel inverter, which is the combination of two-level converter with three-level NPC converter which is connected in series and another series combination is three-level cascaded with five-level NPC converter namely cascaded 5/3 multilevel inverter [5]. These types of converters have their own advantages and applications.

These converters can be used in some applications like industrial drives, FACTS, and vehicle propulsion. In electrical network, STATCOM is one of the effective devices for compensating the voltage sag [6–8], and these multilevel inverters are part of STATCOM. Multilevel converters are also suitable to renewable photovoltaic energy. The power quality and efficiency concern the researchers for many applications [9]. Power quality is more important in distribution system [10], and hence most of the inverters designs would be in such a way that there is good trade-off between power quality and efficiency. Since we are concerned about power quality more number of devices or components would be used, which eventually increases the losses in the components which in turn reduces the efficiency of the converter.

Multilevel converters for low power systems have been competing with the high-frequency pulse-width modulated converters [11]. The reliability has been increased due to the improvements in the semiconductor devices and cooling systems, and mitigation of the harmonic distortion. So advanced semiconductor devices give a better performance [12]. For approaching the novel four-level inverter topology, the selection is based on a set of target, which is using minimum number of switches or minimum number of dc voltage sources. The values can be defined according to the target selection [13]. In some cases the voltage sources are not used efficiently for generating output voltage levels. For example, [14] some topologies can generate only five levels but the conventional MLI can generate nine levels with the same number of voltage sources.

In the proposed topology, symmetrical configuration of sources is used, i.e., the values in all voltage sources are same or equal. Asymmetrical topology [15] requires different voltage sources; this topology has a major drawback that there is requirement of different ratings of switches. This drawback also rises in relative or similar topologies [16–18]. Some high-frequency switches can withstand up to the maximum over all voltage which makes its application limited for the high-voltage products. A new approach [19] has been proposed named as cascaded multilevel inverter employing three-phase transformers and single dc input, which requires less number of dc supplies. So this is the disadvantage of this new approach because this approach adds many number of transformer windings which can add up to the overall volume of the circuit and the cost of the circuit also increases. There is one other topology in [20] named active neutral-point-clamped multilevel converter. For the same number of levels it requires more number of switches than the proposed topology. Some of another topologies suffer with the capacitor balancing [21–23]. This paper presents an overview of new multilevel inverter topology with fuzzy logic controller. This topology namely reversing voltage topology separates the level generation and polarity generation. For triggering the inverter circuit, the phase disposition sinusoidal pulse-width modulation is used and it can be extended to any number of levels.

2 Reversing Voltage Topology Multilevel Inverter

2.1 General Description

In conventional multilevel inverters, the power semiconductor switches are arranged in such a way for producing the high-frequency waveform in both positive and negative polarities. But for reversing voltage topology multilevel inverter to generate bipolar levels, there is no need to use all the switches. This is a hybrid multilevel topology, wherein the output voltage is divided into two parts, one is level generation part. It is the responsibility of the level generation part to generate the level in positive polarity. This part requires high-frequency switches for generating the required levels and hence these switches should have high-frequency rating. Another part is the polarity generation part; it is the responsibility of this part to generate the polarity (positive, negative) of the output voltage. This is low-frequency part, and it operates at the line frequency. This topology combines the level generation part and polarity generation part for generating the multilevel voltage output. The positive levels are generated by level generation or high-frequency switching part and another part is fed to a full-bridge inverter (polarity generation). It will generate the required polarity for the output. The block diagram of reverse voltage topology is shown in Fig. 1. The reversing voltage topology for a seven-level schematic diagram is shown in Fig. 2. It requires 10 switches and 3 isolated dc sources. The circuit shown in Fig. 2 produces required levels by the left part of the circuit, and the right part of the circuit decides the polarity of the output voltage.

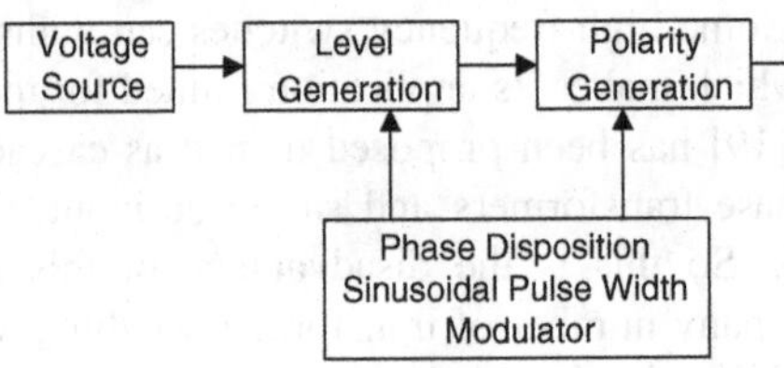

Fig. 1 Block diagram of reversing voltage topology multilevel inverter

Fig. 2 Reversing voltage topology seven-level inverter

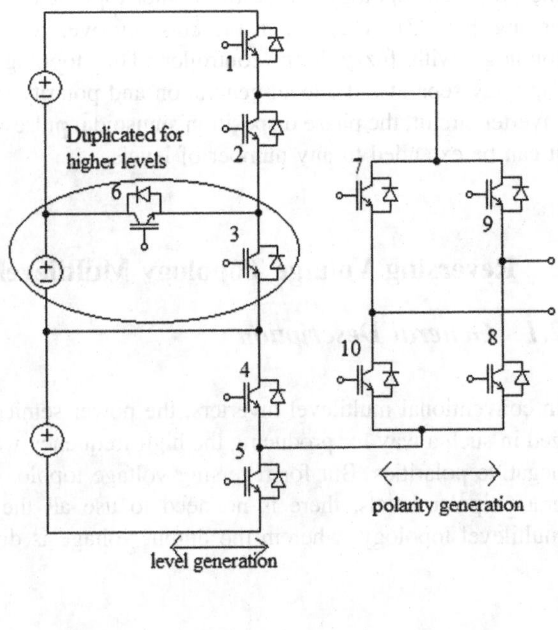

This topology easily extends to the higher voltage levels using the duplicating part specified in the circuit, which is in the middle of the circuit, so this can be easily extended to higher levels. Compared to the conventional circuits, it requires less number of components and the advantage of this topology is that it requires just three positive carrier signals. It does not require negative carrier signals, while the conventional topologies require both positive and negative carrier signals. Figure 2 shows the level generation part which generates positive levels only; it does not generate negative levels. So this topology reduces the number of carrier signals and in turn the cost of the circuit.

2.2 Switching Sequence

The switching modes of this topology are shown in Table 1. According to the table there are six possible ways to control the circuit. The power supplies are externally adjustable sources. So there is no need for voltage balancing for this work. The sequences of switches 2-3-4, 2-3-5, 2-6-5, and 1-5 are chosen for level 0–3. These sequences are shown in Fig. 3. The required levels are produced by these appropriate switching sequences. For producing the seven levels using the PD SPWM, three different carrier triangular waveform sinusoidal reference signals are used. Figure 4 shows the modulator and three carriers for SPWM.

From Fig. 3 it could be observed that there are three states that are considered for generating triggering pulses for the circuit. The first state is the 0-1-0 (Volts) and it is the lowest carrier. The second state is the 1-2-1 (Volts) and it is the middle carrier triangular wave. Final state is 3-2-3 (Volts) and it is the highest carrier wave. For each state some switching patterns are adopted to cover the voltage requirements. The pulse width can be varied by changing the sinusoidal waveform's amplitude, and this particular methodology will be used to implement the fuzzy controller.

The relation between right comparator output according to the required states and current state for switching to meet the voltage requirements happens. The right comparator here refers to the comparator output of the current state. Current also plays an important role in the efficiency of overall converter for the number of switches in the path of conducting. For example, a seven-level cascaded topology [24] has twelve switches, half of the switches conducting the inverter current in each instance.

The polarity generation part is an H-bridge inverter and works in two modes. One is forward mode and another one is reverse mode. In forward mode switches 8

Table 1 Switching sequence

Mode	Level			
	0	1	2	3
1	2, 3, 4	2, 3, 5	1, 4	1, 5
2		2, 4, 6	2, 6, 5	

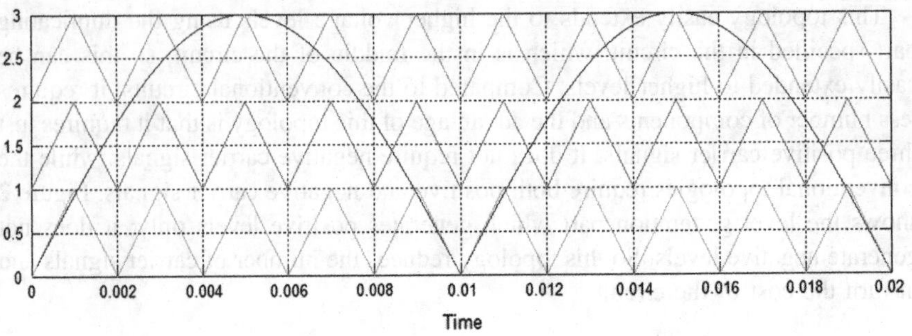

Fig. 3 Methodology for creating SPWM pulses

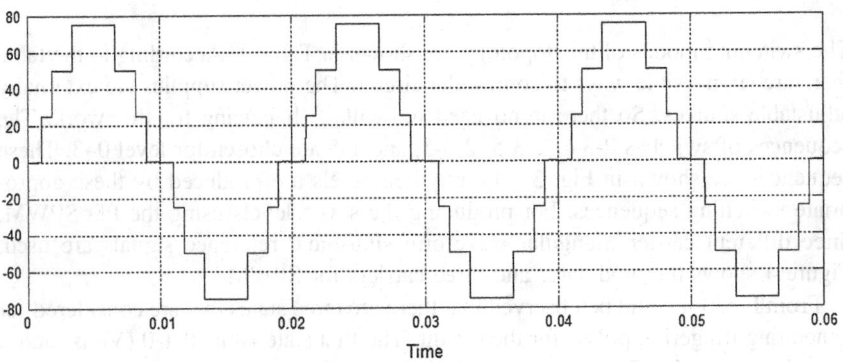

Fig. 4 Load output voltage for seven-level configuration without PWM

and 9 will conduct and this will give only the positive polarity. In reverse mode switches 7 and 10 will conduct and this will give the negative polarity. Low-frequency generation part operates at the line frequency, which changes only at the zero voltage crossings. Figure 4 shows the load output voltage of seven-level reversing voltage topology multilevel inverter without using PWM technique. But to implement fuzzy controller it is very much necessary to control the circuit with PWM Technique.

2.3 Number of Components Used

As compared to the other topologies RV topology requires less number of components and also fewer switches. So this type of topology may be used in applications where there is requirement for high-voltage power devices like in FACTS and HVDC systems (Fig. 5).

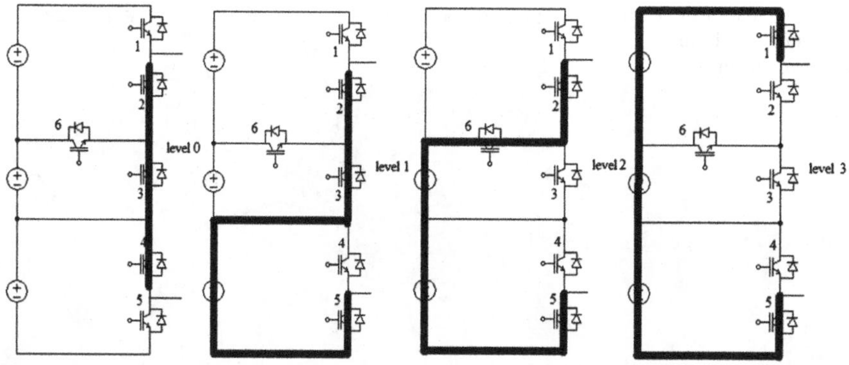

Fig. 5 Levels

2.4 Hardware Implementation

Figure 6 shows the snapshot of seven-level inverter. The inverter is controlled using dSPACE (the control signals are given through dSPACE). The control signals from the dSPACE are given to the individual switches through optocoupler TLP 250. The optocoupler acts as an isolator and it is being individually biased by separate supplies. Figure 7 shows the positive level generated from the hardware configuration which can be seen to have seven levels in the output. This level generated is given to the polarity generation circuit which just acts like a full-bridge inverter changing the polarity of the generated levels and thus giving us the result of seven-level output both in positive and negative. Figure 8 shows the total output of seven-level configuration, and this configuration can be extended to have PWM

Fig. 6 Hardware photo snap of open-loop reversing voltage topology seven-level multilevel inverter

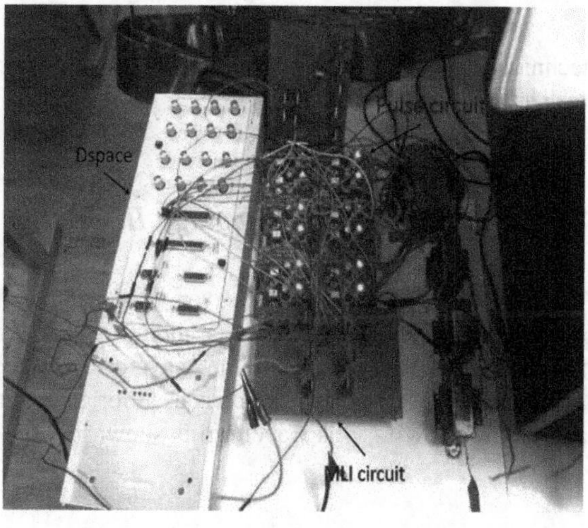

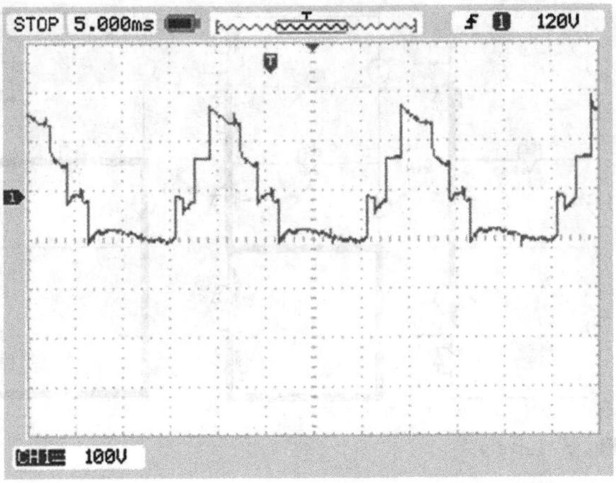

Fig. 7 Output of the level generation part in the hardware circuit

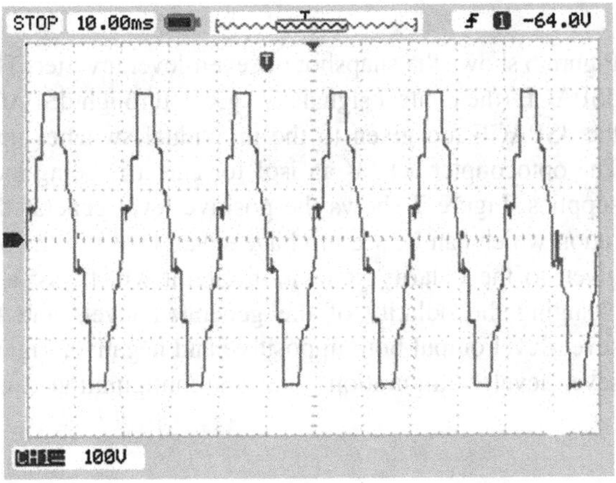

Fig. 8 Load output voltage of seven-level RV topology MLI

technique and the output voltage can be measured and fed back to dSPACE to realize the fuzzy controller.

2.5 PWM-Based RV Topology Inverters

The circuit shown in Fig. 2 is seven-level reversing voltage topology multilevel inverter for which PWM signal is applied and the voltage obtained is shown in Fig. 9. The RMS voltage of the waveform can be controlled by varying the amplitude of the sine wave, which is used for generating the PWM pulses. The FFT analysis window for this waveform is shown in Fig. 10.

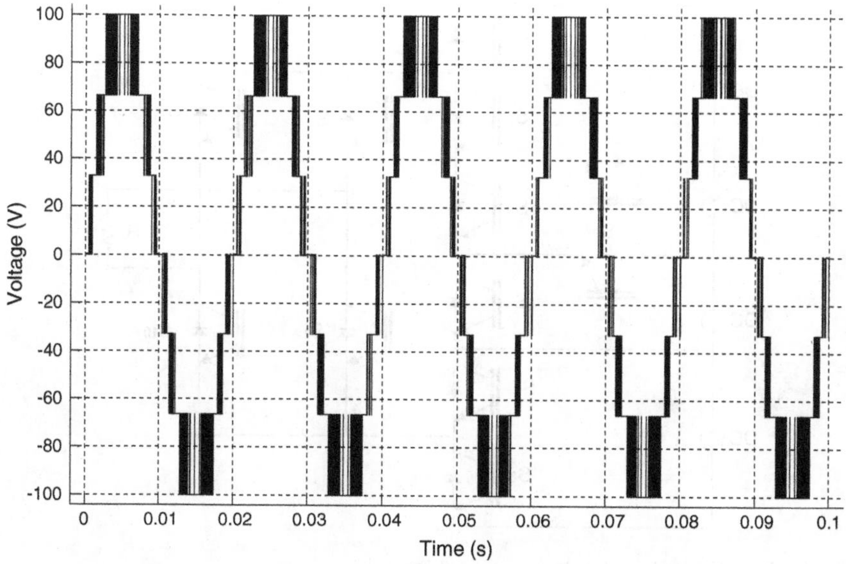

Fig. 9 RV topology inverter for seven-level configuration

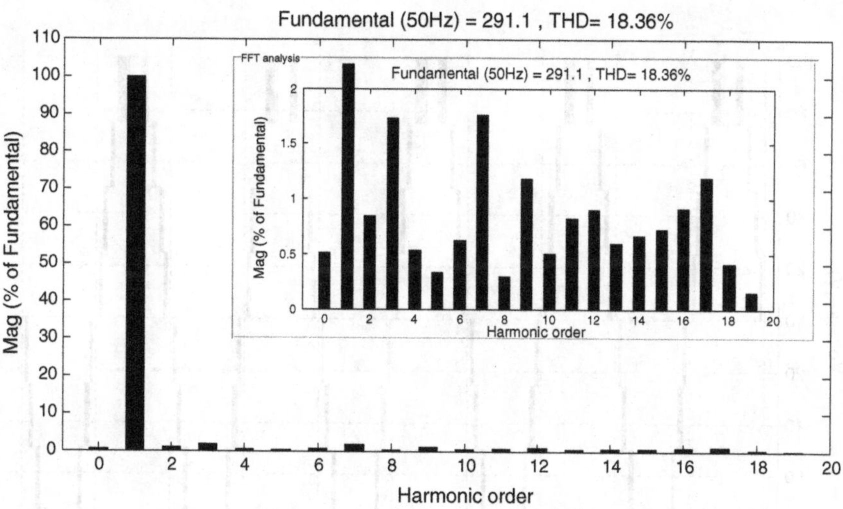

Fig. 10 FFT analysis for PWM controlled seven-level RV topology inverter

Figure 11 shows the circuit diagram for reverse voltage topology nine-level inverter; from the figure we could infer that the level is increased by 2 with increase in the number of switches by 2. The circuit working principle is similar to that of the seven-level inverter. The output load voltage of the nine-level inverter is shown in Fig. 12, and the corresponding FFT analysis window is shown in Fig. 13.

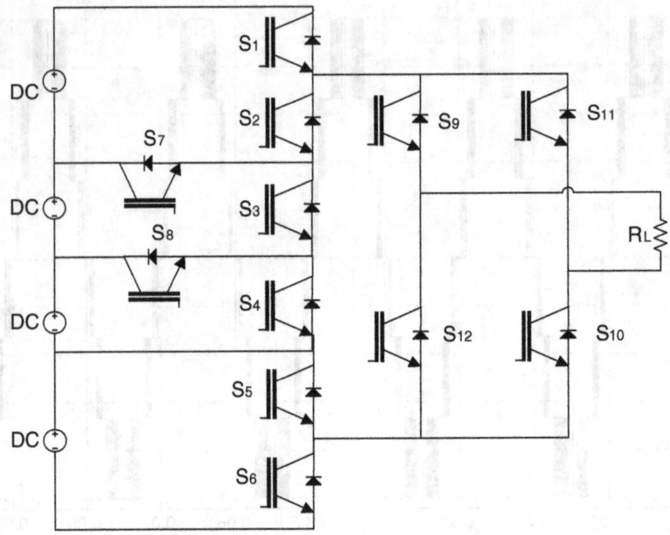

Fig. 11 Reversing voltage topology nine-level inverter

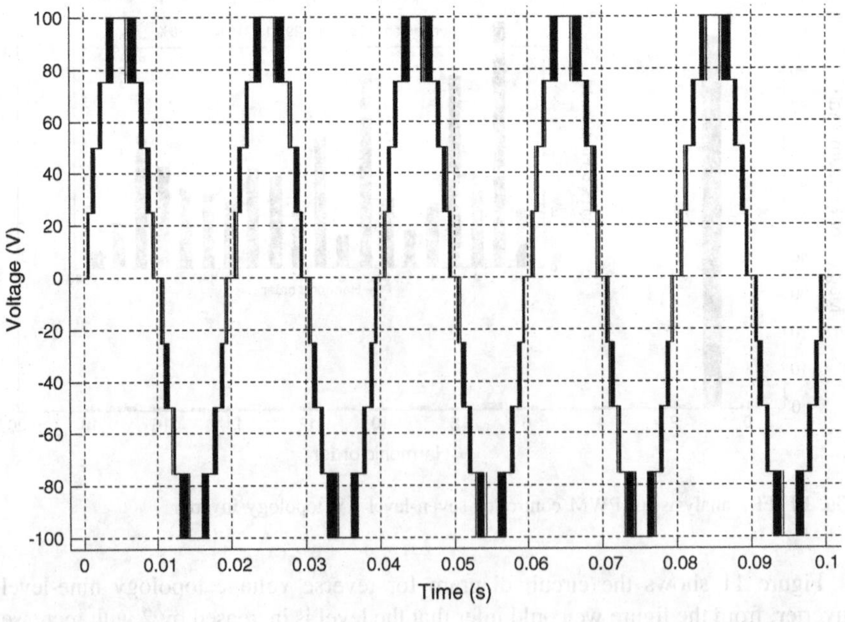

Fig. 12 Output load voltage for nine-level inverter

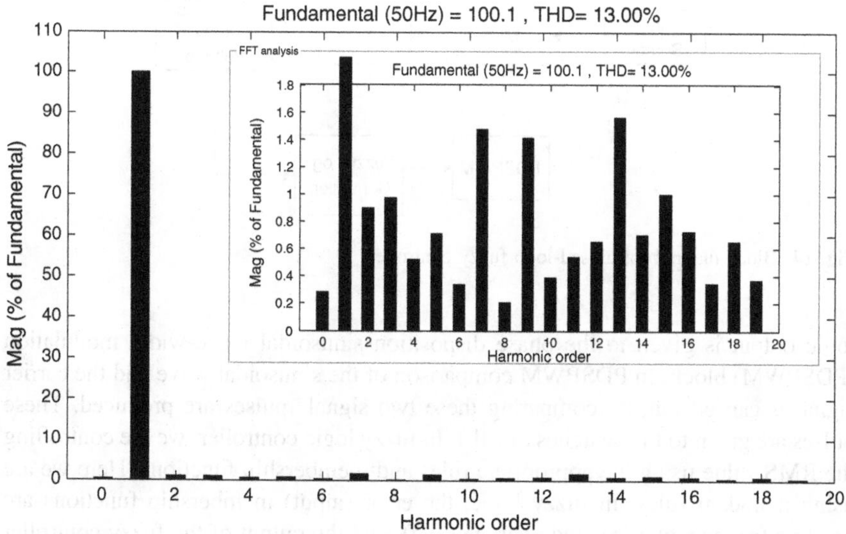

Fig. 13 FFT analysis window for nine-level RV topology inverter

The THD of nine-level RV topology inverter is found to be 13.00 %. The THD of nine-level RV topology inverter is found to be 5.36 % lesser than that of seven-level configuration. If further reduction in THD is required, then suitable filter may be designed for the circuit.

From Fig. 10 we could observe that the THD for this inverter is 18.36 %. This THD could be brought down by suitable filter design.

3 Fuzzy Controllers for RV Topology Inverters

The term "fuzzy logic" was introduced in the year of 1965 with the proposal of the fuzzy set theory. Fuzzy logic is a form of many-valued logic. It deals with reasoning that is approximate rather than fixed and exact, compared to traditional binary sets (variables may take on true or false values). Fuzzy logic variables may have a truth value that ranges between 0 and 1. Fuzzy logic has been extended to handle the concept of partial truth, where the truth value may range between completely true and completely false [25–27]. When the linguistic variables are used these degrees may be managed by specific functions.

The block diagram of closed-loop MLI is shown in Fig. 14. This block consists of voltage source, MLI, fuzzy controller, PDSPWM, and load. The voltage is supplied to the MLI, and the MLI is connected to the load. The MLI produces the required level depending on the number of switches' topology. The fuzzy logic controller takes the input from the measured output voltage of the load. The fuzzy

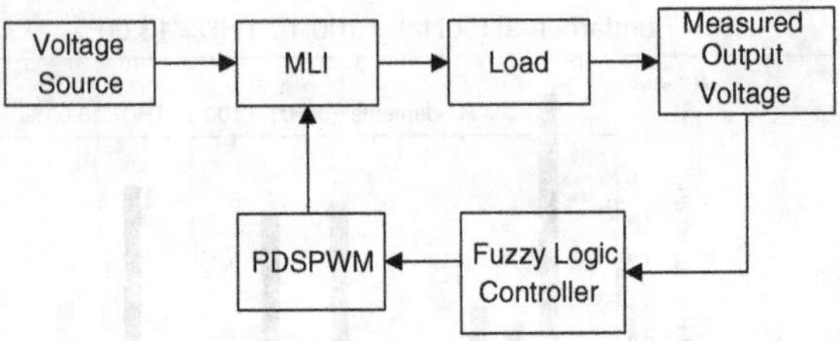

Fig. 14 Block diagram of closed-loop fuzzy controller

logic output is given to the phase disposition sinusoidal pulse-width modulation (PDSPWM) block. In PDSPWM comparison of the sinusoidal wave and the carrier signal is carried out, by comparing these two signals pulses are produced. These pulses are given to the switches of MLI. In fuzzy logic controller, we are controlling the RMS value using the appropriate rules and membership functions. Here we are using mamdani rules. In fuzzy logic, the error (input) membership functions are positive (p), negative (n), and average (avg), and the output of the fuzzy controller that is amplitude is decided by the membership function increment (inc), moderate, and decrement (dec).

Figure 15 shows the load output voltage for seven-level reversing voltage topology multilevel inverter; from the figure we can infer that the change in voltage occurs at 0.5 s, and the reference voltage and the output RMS voltage that have been used in the circuit are shown in Fig. 16. The reference voltage is kept at 45 V till 0.5 s and then it has been increased to 60 V. The output RMS voltage can be observed from Fig. 16, which gradually increases and settles at the voltage closer to the reference voltage. The load output voltage in Fig. 15 shows a sharp increase in the width of the voltage which makes the increase in voltage to reach the RMS reference voltage specified in the circuit.

Figure 17 shows the load output voltage for reversing voltage topology nine-level inverter; in this case again the reference voltage is changed from 55 to 70 V. The output RMS voltage is shown in the Fig. 18 which is found to be following the reference voltage, and this shows that the fuzzy controller design has good controllability for different reference values. When there is a change in the reference voltage, there is a change in the load voltage also which can be inferred in Fig. 17, where we can see that up to 0.5 s the output waveform is in such a way that it gives less voltage after 0.5 s and the waveform width increases to increase the output RMS voltage.

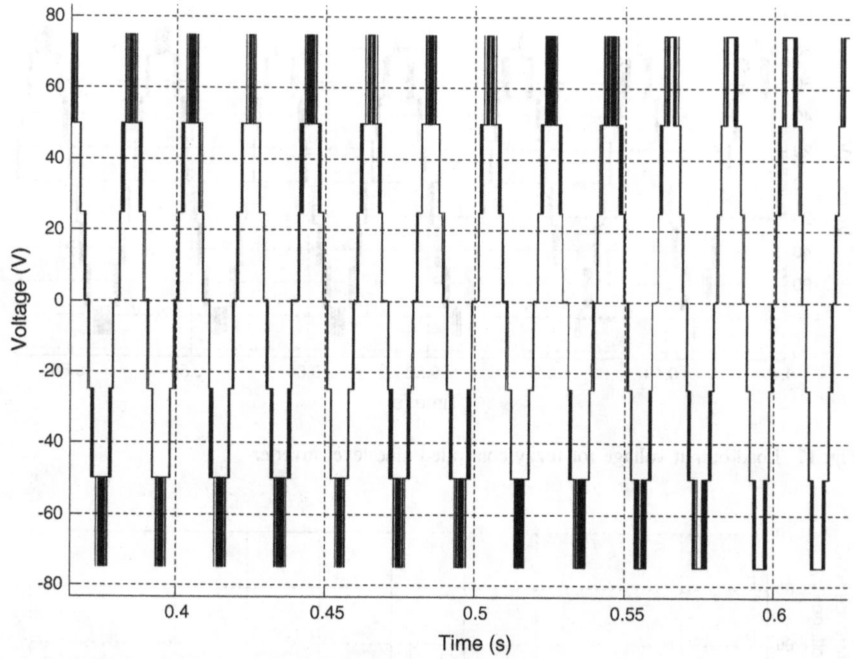

Fig. 15 Fuzzy controlled seven-level inverter load voltage

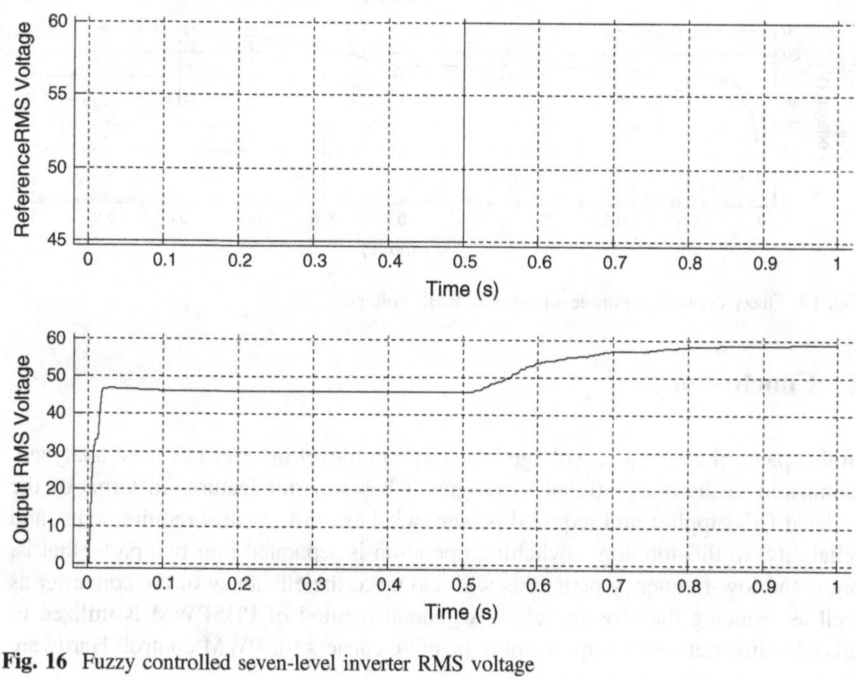

Fig. 16 Fuzzy controlled seven-level inverter RMS voltage

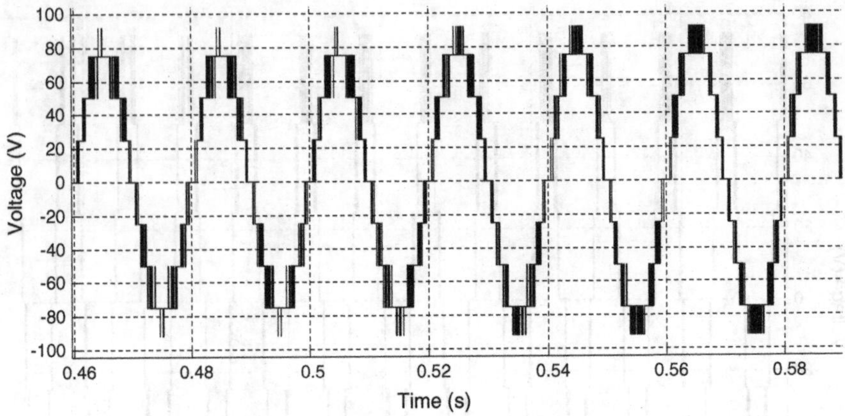

Fig. 17 Load output voltage for fuzzy controlled nine-level inverter

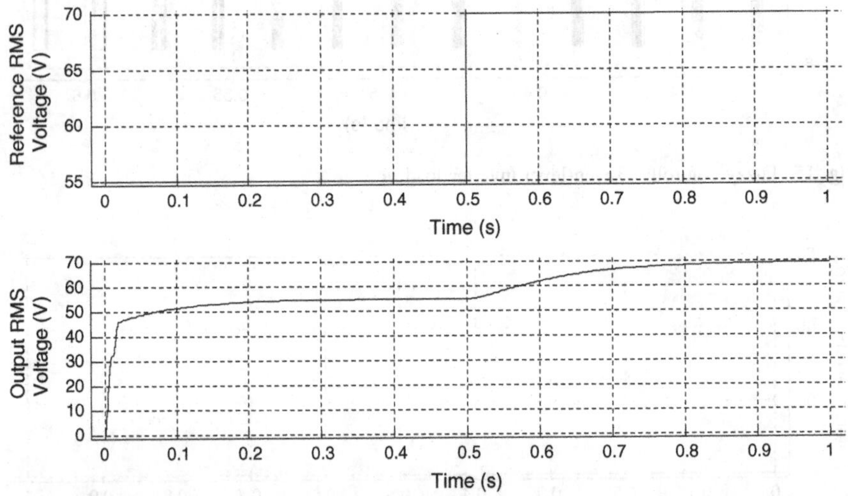

Fig. 18 Fuzzy controlled nine-level inverter RMS voltage

4 Conclusion

In this paper the reversing voltage topology multilevel inverter has been analyzed. Compared to the conventional topologies it has superior features in terms of the isolated DC supplies and required power switches, cost, control requirements, and reliability. In this topology switching operation is separated into two parts, that is, high- and low-frequency parts; this will add up to the efficiency of the converter as well as reducing the size and cost. A general method of PDSPWM is utilized to drive the inverter and it requires only positive carriers for PWM control. Hardware circuit for seven-level configuration is realized using dSPACE and the result has been presented. A closed-loop control for seven-level and nine-level inverters has been proposed and the same is realized using fuzzy controller. The result of the controlled output has been presented from which it could be found that the designed controller gives a precise output for different reference voltages.

References

1. Jang-Hwan, K., Sul, S.-K., Enjeti, P.N.: A carrier-based PWM method with optimal switching sequence for a multilevel four-leg voltage source inverter. IEEE Trans. Ind. Appl., **44**(4), 1239–1248 (2008)
2. Srikanthan, S., Mishra, M.K.: DC capacitor voltage equalization in neutral clamped inverters for DSTATCOM application. IEEE Trans. Ind. Electron. **57**(8), 2768–2775 (2010)
3. Tolbert, L.M., Peng, F.Z., Habetler, T.G.: Multilevel converters for large electric drives. IEEE Trans. Ind. Appl. **35**(1), 36–44 (1999)
4. Skvarenina, T.L.: The Power Electronics Handbook. CRC Press, Boca Raton (2002)
5. Yun, X., Zou, Y., Liu, X., He, Y.: A novel composite cascade multilevel converter. In: Proceedings 33rd IEEE IECON, pp. 1799–1804 (2007)
6. Najafi, E., Yatim, A.H.M.: A novel current mode controller for a static compensator utilizing Goertzel algorithm to mitigate voltage sags. Energy Convers. Manage. **52**(4), 1999–2008 (2011)
7. Seki, N., Uchino, H.: Converter configurations and switching frequency for a GTO reactive power compensator. IEEE Trans. Ind. Appl. **33**(4), 1011–1018 (1997)
8. Shahgholiyan, G., Haghjou, E., Abazari, S.: Improving the mitigation of voltage flicker by usage of fuzzy control in a distribution static synchronous compensator (DSTATCOM). Majlesi J. Elect. Eng. 3(2), 25–35 (2009)
9. Nakata, K., Nakamura, K., Ito, S., Jinbo, K.: A three-level traction inverter with IGBTs for EMU. In: Conference Record IEEE IAS Annual Meeting, vol. 1, pp. 667–672 (1994)
10. Jidin, A., Idris, N.R.N., Yatim, A.H.M., Sutikno, T., Elbuluk, M.E.: An optimized switching strategy for quick dynamic torque control in DTC-hysteresis-based induction machines. IEEE Trans. Ind. Electron. **58**(8), 3391–3400 (2011)
11. Daher, S., Schmid, J., Antunes, F.L.M.: Multilevel inverter topologies for stand-alone PV systems. IEEE Trans. Ind. Electron. **55**(7), 2703–2712 (2008)
12. Zambra, D.A.B., Rech, C., Pinheiro, J.R.: A comparative analysis between the symmetric and the hybrid asymmetric nine-level series connected H-bridge cells inverter. In: Proceedings Europeans Conference Power Electronics Application, pp. 1–10 (2007)
13. Babaei, E.: Optimal topologies for cascaded sub-multilevel converters. J. Power Electron. **10**(3), 251–261 (2010)
14. Mondal, G., Gopakumar, K., Tekwani, P.N., Levi, E.: A reduced switch-count five-level inverter with common-mode voltage elimination for an open-end winding induction motor drive. IEEE Trans. Ind. Electron. **54**(4), 2344–2351 (2007)
15. Beser, E., Arifoglu, B., Camur, S., Beser, E.K.: Design and application of a single phase multilevel inverter suitable for using as a voltage harmonic source. J. Power Electron. **10**(2), 138–145 (2010)
16. Ceglia, G., Guzman, V., Sanchez, C., Ibanez, F., Walter, J., Gimenez, M.I.: A new simplified multilevel inverter topology for dc-ac conversion. IEEE Trans. Power Electron. **21**(5), 1311–1319 (2006)
17. Rahim, N.A., Chaniago, K., Selvaraj, J.: Single-phase seven-level grid-connected inverter for photovoltaic system. IEEE Trans. Ind. Electron. **58**(6), 2435–2443 (2011)

18. Selvaraj, J., Rahim, N.A.: Multilevel inverter for grid-connected PV system employing digital PI controller. IEEE Trans. Ind. Electron. **56**(1), 149–158 (2009)
19. Song, S.G., Kang, F.S., Park, S.-J.: Cascaded multilevel inverter employing three-phase transformers and single dc input. IEEE Trans. Ind. Electron. **56**(6), 2005–2014 (2009)
20. Barbosa, P., Steimer, P., Steinke, J., Meysenc, L., Winkelnkemper, M., Celanovic, N.: Active neutral-point-clamped multilevel converters. In: Proceedings IEEE 36th Power Electronics Specialist Conference, pp. 2296–2301 (2005)
21. Gonzalez, S.A., Valla, M.I., Christiansen, C.F.: Analysis of a cascade asymmetric topology for multilevel converters. In: Proceedings IEEE ISIE, pp. 1027–1032 (2007)
22. Lezana, P., Rodriguez, J.: Mixed multicell cascaded multilevel inverter. In: Proceedings IEEE ISIE, pp. 509–514 (2007)
23. Stala, R.: Application of balancing circuit for dc-link voltages balance in a single-phase diode-clamped inverter with two three-level legs. IEEE Trans. Ind. Electron. **58**(9), 4185–4195 (2011)
24. Park, Y.-M., Ryu, H.-S., Lee, H.-W., Jung, M.-G., Lee, S.-H.: Design of a cascaded H-bridge multilevel inverter based on power electronics building blocks and control for high performance. J. Power Electron. **10**(3), 262–269 (2010)
25. Novak, V., Perfilieva, I., Mockor, J.: Mathematical principles of fuzzy logic Dodrecht: Kluwer Academic. ISBN 0-7923-8595-0 (1999)
26. Fuzzy logic: Stanford encyclopedia of philosophy. Stanford University. 2006-07-23. Retrieved 2008-09-30
27. Zadeh, L.A.: Fuzzy sets. inf. Control **8**(3), 338–353 (1965)

Image Quality Assessment-Based Approach to Estimate the Age of Pencil Sketch

Steven Lawrence Fernandes and G. Josemin Bala

Abstract In recent years, the increasing interest in the evaluation of biometric systems security has led to the creation of numerous and very diverse initiatives focused on this major field of research. After the occurrence of crime a skilled pencil sketch artist draws the sketches based on the description of the eyewitness. The accuracy of the skill depends on the description given by the eyewitness and the skill of the artist. After the sketch is drawn finding its age is a challenging task. In this paper we apply image quality assessment (IQA) to find the age of a pencil sketch drawn by a skilled pencil sketch artist. The database considered FGNET pencil sketch database that consists of 34 pencil sketches varying from 6 to 61 years. The IQA parameters considered are peak signal-to-noise ratio (PSNR), signal-to-noise ratio (SNR), maximum difference (MD), average departure (AD), normalized absolute error (NAE), total edge difference (TED), structural similarity index (SSI), and mean square error (MSE). The significance of this analysis is given a pencil sketch that we can quickly and effectively calculate its age and hence help the law enforcement agency to apprehend the criminals in a very short time interval. Demo version of the code along with input pencil sketches and output obtained can be downloaded from https://goo.gl/zYq3cI.

Keywords Image quality assessment · Peak signal-to-noise ratio · Signal-to-noise ratio · Maximum difference

S.L. Fernandes (✉) · G. Josemin Bala
Department of Electronics & Communication Engineering,
Karunya University, Coimbatore, India
e-mail: steva_fernandes@yahoo.com

G. Josemin Bala
e-mail: josemin@karunya.edu

M. Pant et al. (eds.), *Proceedings of Fifth International Conference on Soft Computing for Problem Solving*, Advances in Intelligent Systems and Computing 436, DOI 10.1007/978-981-10-0448-3_52

1 Introduction

Finding the age from a pencil skilled drawn from a skilled pencil sketch artist is a very challenging task [1–9]. Due to aging process, faces undergo continuous variations, and hence continuously updating the large databases will be a difficult task [10–14]. Hence, we try to find the age of the pencil sketch by applying image quality assessment (IQA) parameters. IQA attempts to assess the errors in an input pencil sketch by subjecting it to eight IQA parameters. The eight IQA parameters considered are peak signal-to-noise ratio (PSNR), signal-to-noise ratio (SNR), maximum difference (MD), average departure (AD), normalized absolute error (NAE), total edge difference (TED), structural similarity index (SSI), and mean square error (MSE) [15–24]. Each of these eight IQA parameters is defined in Table 1.

Section 2 describes the proposed system, and Sect. 3 presents the results and discussions. Section 4 draws the conclusion.

Table 1 Description of eight IQA parameters considered in the proposed system

Acronym	Name	References	Description
PSNR	Peak signal-to-noise ratio	[30]	$\mathrm{PSNR}\,(I,\hat{I}) = 10\log\left(\frac{\max(I^2)}{\mathrm{MSE(I,\hat{I})}}\right)$
SNR	Signal-to-noise ratio	[31]	$\mathrm{SNR}\,(I,\hat{I}) = 10\log\left(\frac{\sum_{i=1}^{N}\sum_{j=1}^{M}(I_{i,j})^2}{N.M.MSE(I,\hat{I})}\right)$
MD	Maximum difference	[32]	$\mathrm{MD}\,(I,\hat{I}) = \mathrm{Max}\left\lvert I_{i,j} - \widehat{I_{i,j}}\right\rvert$
AD	Average difference	[32]	$\mathrm{AD}\,(I,\hat{I}) = \frac{1}{NM}\sum_{i=1}^{N}\sum_{j=1}^{M}\left(I_{i,j} - \widehat{I_{i,j}}\right)$
NAE	Normalized absolute error	[32]	$\mathrm{NAE}\,(I,\hat{I})\frac{\sum_{i=1}^{N}\sum_{j=1}^{M}\left\lvert I_{i,j}-\widehat{I_{i,j}}\right\rvert}{\sum_{i=1}^{N}\sum_{j=1}^{M}\left\lvert I_{i,j}\right\rvert}$
TED	Total edge difference	[32]	$\mathrm{TED}\,(I,\hat{I}) = \frac{1}{NM}\sum_{i=1}^{N}\sum_{j=1}^{M}\left\lvert I_{E_{i,j}} - \widehat{I_{\mathrm{E_{i,j}}}}\right\rvert$
SSI	Structural similarity index	[29]	$\mathrm{SSI}\,(x,y) = \frac{(2\mu_x\mu_y + C_1)(2\sigma_{xy} + C_2)}{(\mu_x^2 + \mu_y^2 + C_1)(\sigma_y^2 + \sigma_y^2 + C_2)}$
MSE	Mean squared error	[29]	$\mathrm{MSE}\,(I,\hat{I}) = \frac{1}{NM}\sum_{i=1}^{N}\sum_{j=1}^{M}\left(I_{i,j} - \widehat{I_{i,j}}\right)^2$

2 Proposed System

The activity diagram and class diagram are shown in Figs. 1 and 2. The proposed system consists of the following modules:

- Query image;
- Preprocess;
- Feature extraction;
- Classification.

Query image Query image is an input pencil sketch image of which is given to the process.

Preprocess Here, we clear the noise and resize pencil sketch image. Gaussian filter is used to eliminate the noise in the pencil sketch image [25].

Feature Extraction In this process, we use image quality assessment. The input pencil sketch of measurement image I (of size $N \times M$) is refined with a low-pass

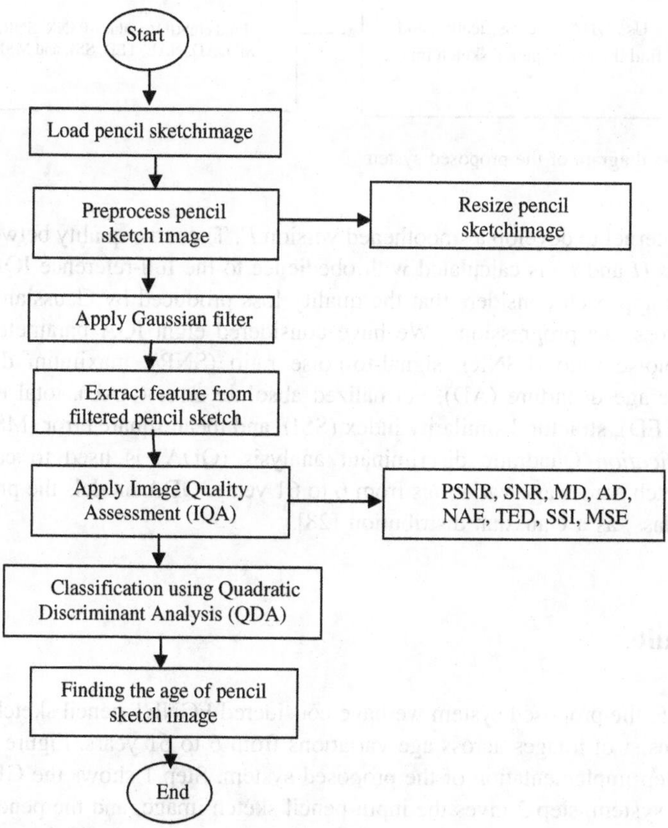

Fig. 1 Activity diagram of the proposed system

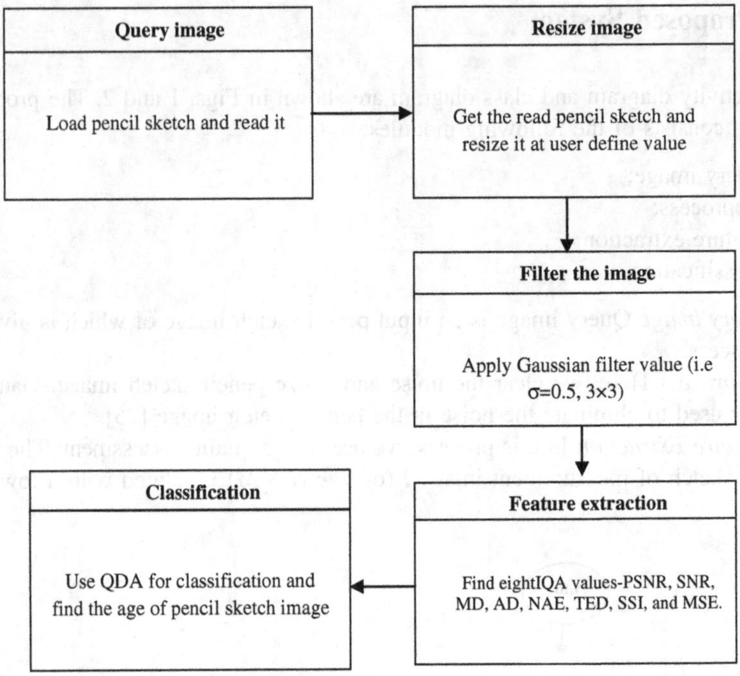

Fig. 2 Class diagram of the proposed system

Gaussian kernel to develop a smoothened version *I*ˆ. Then, the quality between both the images (*I* and *I*ˆ) is calculated with obedience to the full-reference IQA metric [26]. This approach considers that the quality loss produced by Gaussian filtering varies across age progressions. We have considered eight IQA parameters: peak signal-to-noise ratio (PSNR), signal-to-noise ratio (SNR), maximum difference (MD), average departure (AD), normalized absolute error (NAE), total edge difference (TED), structural similarity index (SSI), and mean square error (MSE) [27].

Classification Quadratic discriminant analysis (QDA) is used to categorize pencil sketch across age variations from 6 to 61 years. QDA models the propensity of each class, as a Gaussian distribution [28].

3 Results

To validate the proposed system we have considered FGNET pencil sketch images which consist of images across age variations from 6 to 61 years. Figure 3 shows step-by-step implementation of the proposed system. Step 1 shows the GUI of the proposed system, step 2 gives the input pencil sketch image, and the pencil sketch image is resized in step 3 to 256 × 256. The input pencil sketch image is then

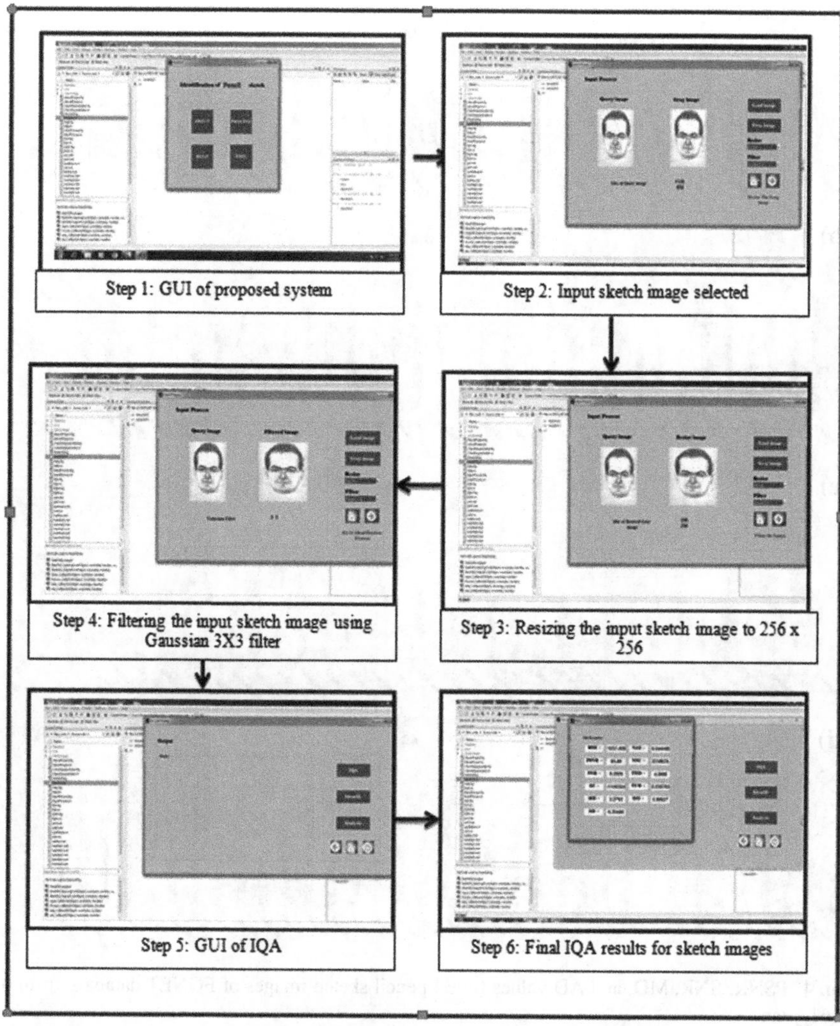

Fig. 3 Step-by-step implementation of the proposed system

filtered using 3 × 3 Gaussian filter in step 4. Step 5 shows the GUI of IQA model and the final IQA values are given in step 6 as shown in Fig. 3. Demo version of the code along with input pencil sketches and output obtained can be downloaded from https://goo.gl/zYq3cI.

The graphical representations of the various IQA parameters for 34 pencil sketch images of FGNET database from 6 to 61 years are shown in Figs. 4 and 5.

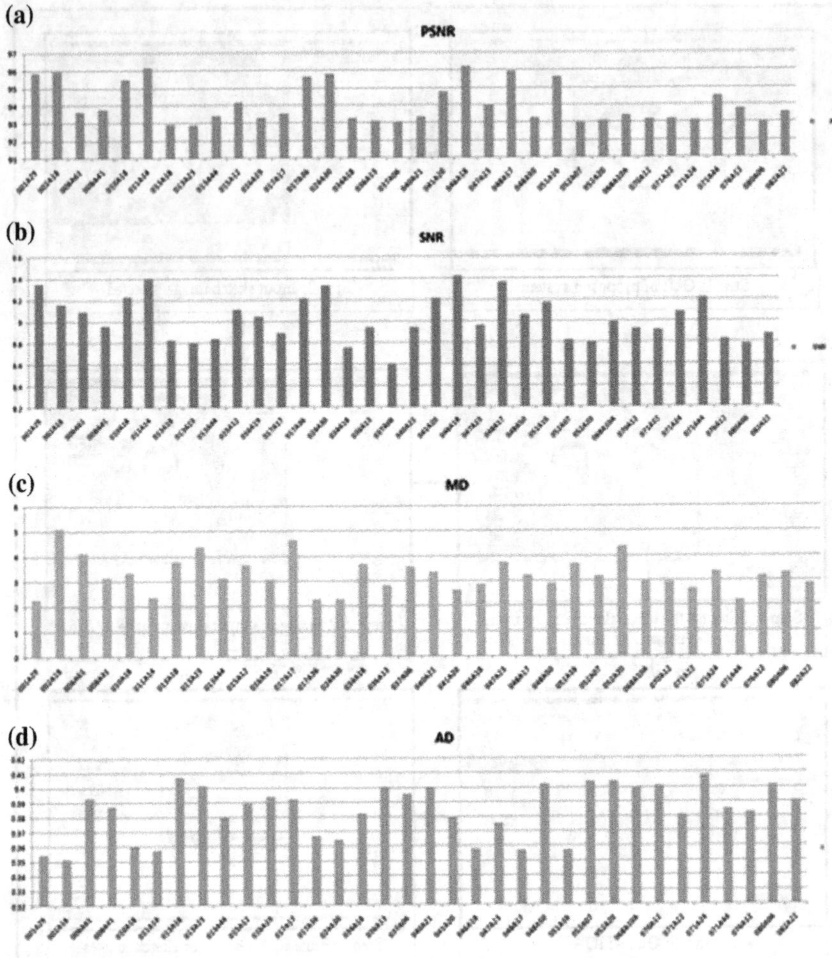

Fig. 4 PSNR, SNR, MD, and AD values for 34 pencil sketch images of FGNET database from 6 to 61 years

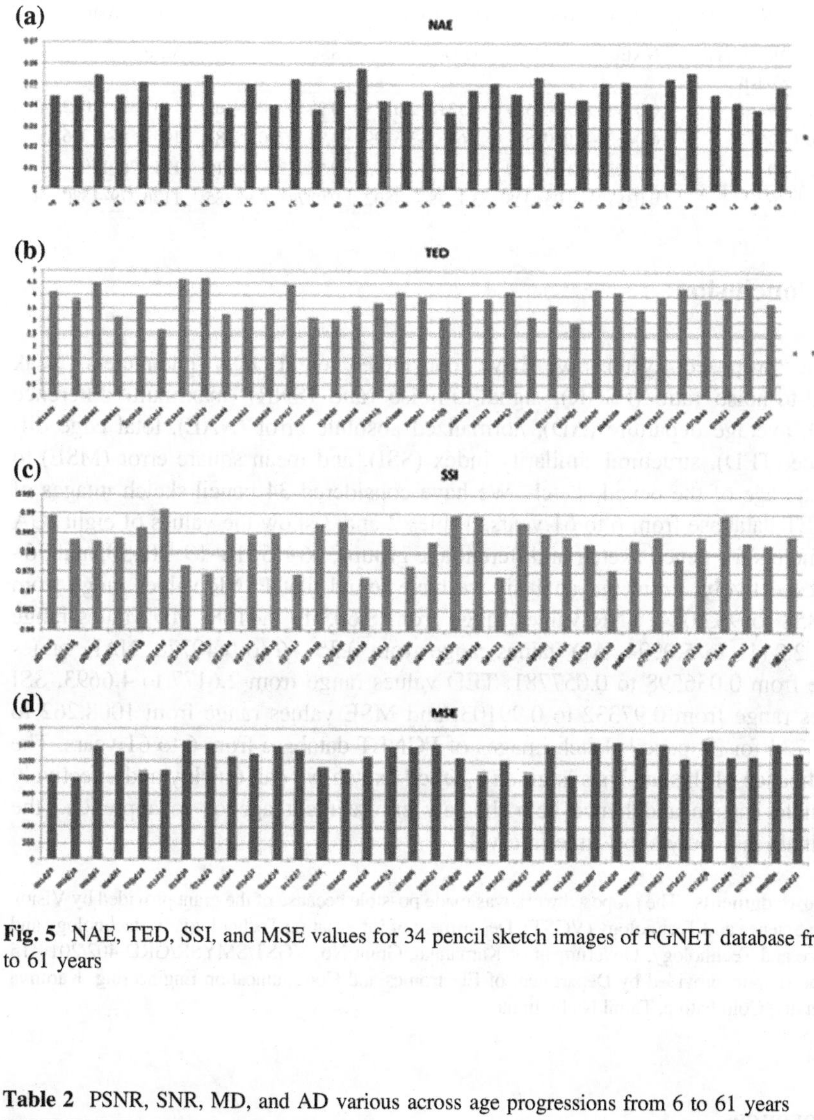

Fig. 5 NAE, TED, SSI, and MSE values for 34 pencil sketch images of FGNET database from 6 to 61 years

Table 2 PSNR, SNR, MD, and AD various across age progressions from 6 to 61 years

Years	No. of pencil sketch	PSNR	SNR	MD	AD
6–15	9	93.0415–96.1514	8.7809–9.4013	3.3124–2.3613	0.40114–0.35793
16–19	9	95.6111–93.3184	9.1697–9.0465	3.6768–3.0769	0.35713–0.39377
20–30	10	94.7923–95.7932	9.2119–9.3288	2.6594–2.2791	0.37996–0.36452
36–61	6	95.6272–93.6494	9.2118–9.091	2.2963–4.1284	0.36732–0.39311

Table 3. NAE, TED, SSI, and MSE various across age progressions from 6 to 61 years

Years	No. of pencil sketch	NAE	TED	SSI	MSE
6–15	9	0.038339–0.040549	4.5318–2.6177	0.98197–0.99103	1439.621–1082.546
16–19	9	0.04687–0.040455	3.6462–3.5309	0.98402–0.98481	1059.586–1396.451
20–30	10	0.036598–0.048399	3.1374–3.0603	0.98952–0.98819	1249.76–1106.411
36–61	6	0.038123–0.054028	3.1238–4.4802	0.98748–0.98285	1126.279–1380.956

4 Conclusion

In the proposed system we have considered eight IQA parameters: peak signal-to-noise ratio (PSNR), signal-to-noise ratio (SNR), maximum difference (MD), average departure (AD), normalized absolute error (NAE), total edge difference (TED), structural similarity index (SSI), and mean square error (MSE) to find the age of the pencil sketch. We have considered 34 pencil sketch images of FGNET database from 6 to 61 years. Tables 2 and 3 show the values of eight IQA parameters for pencil sketch in different age groups, 06–15, 16–19, 20–30, and 36–61, respectively. From our analysis we have found that PSNR values range from 92.8858 to 96.2024, SNR values range from 8.605 to 9.4168, MD values range from 2.2441 to 5.0994, AD values range from 0.35176 to 0.40778, NAE values range from 0.036598 to 0.057781, TED values range from 2.6177 to 4.6693, SSI values range from 0.97332 to 0.99103, and MSE values range from 1008.262 to 1487.734 for 34 pencil sketch images of FGNET database from 6 to 61 years. The significance of this analysis is given a pencil sketch we can quickly and effectively calculate its age and hence help the law enforcement agency to apprehend the criminals in a very short time interval.

Acknowledgments The proposed work was made possible because of the grant provided by Vision Group Science and Technology (VGST), Department of Information Technology, Biotechnology and Science and Technology, Government of Karnataka, Grant No. VGST/SMYSR/GRD-402/2014-15 and the support provided by Department of Electronics and Communication Engineering, Karunya University, Coimbatore, Tamil Nadu, India.

References

1. Jashi, P., Prakash, S.: A quality aware technique for biometric recognition. In: IEEE International Conference on Signal Processing and Integrated Networks, pp. 795–800 (2015)
2. Raghavendra, R., Raja, K.B., et al.: Automatic face quality assessment from video using gray level co-occurrence matrix: an empirical study on automatic border control system. IEEE Int. Conf. Pattern Recogn. 438–443 (2014)
3. Galbally, J., Marcel, S.: Face anti-spoofing based on general image quality assessment. IEEE Int. Conf. Pattern Recogn. 1173–1178 (2014)

4. Haque, M.A., Nasrollahi, K., et al.: Quality-aware estimation of facial landmark in video sequences. IEEE Int. Conf. Appl. Comput. Vision 678–685 (2015)
5. Chen, X., Jin, X., et al.: Learning templates for artistic portrait lightening analysis. IEEE Trans. Image Process. **24**, 608–618 (2014)
6. Lin, K., et al.: Heterogeneous feature fusion-based optimal face image acquisition in visual sensor network. IEEE Int. Conf. Instrumen. Measure. Technol. 1078–1083 (2015)
7. Jiansheng, C., Yu, D., et al.: Face image quality assessment based on learning to rank. IEEE Trans. Signal Process. Lett. **22**, 90–94 (2014)
8. Johnson, P.A., Hua, F., et al.: Comparison of quality-based fusion of face and iris biometrics. IEEE Int. Conf. Biometr. 1–5 (2011)
9. Maatta, J., Hadid, A., et al.: Face spoofing detection from single images using micro-texture analysis. IEEE Int. Conf. Biometr. 1–7 (2011)
10. Wei-Yang, L., Ming-Yang, C.: Automatic quality assessment and preprocessing for three-dimensional face recognition. IEEE Int. Conf. Inform. Security Intell. Control 266–269 (2012)
11. Abaza, A., Harrison, M.A., et al.: Quality metrics for practical face recognition. IEEE Int. Conf. Pattern Recogn. 1051–4651 (2012)
12. Lihuo, H., Dacheng, T., et al.: Sparse representation for blind image quality assessment. IEEE Int. Conf. Comput. Vision Pattern Recogn. 1146–1153 (2012)
13. Maatta, J., Hadid, A., et al.: Face spoofing detection from single images using texture and local shape analysis. IEEE Trans. Biometr. **1**, 3–1 (2012)
14. Pin, L., Haixiang, L., et al.: Facial image quality assessment based on support vector machines. IEEE Int. Conf. Biomed. Eng. Biotechnol. 810–813 (2012)
15. Alonso-Fernandez, F., Fierrez, J., et al.: Quality measures in biometric systems. IEEE Trans. Secur. Privacy **10**, 52–62 (2011)
16. Erdogmus, N., Dugelacy, J.L.: Regional confidence score assessment for 3D face. IEEE Int. Conf. Acoustics, Speech Signal Process. 1521–1524 (2012)
17. Despiegel, V., Gentric, S.: Common mistakes on face recognition based on video. IEEE Int. Conf. Biometr. Special interest Group 1–4 (2012)
18. Jianzhou, Y., Lin, S., et al.: Learning the change for automatic image cropping. IEEE Int. Conf. Comput. Vision Pattern Recogn. 971–978 (2013)
19. Bharadwaj, S., Vatsa, M., et al.: Can holistic representations be used for face biometrics quality assessment. IEEE Int. Conf. Image Process. 2792–2796 (2013)
20. Bing, X., Xinbo, G.: Visual quality assessment of the synthesized sketch. IEEE Int. Conf. Natural Comput. 317–321 (2013)
21. Haque, M.A., Nasrollahi, K. et al.: Real-time acquisition of high quality face sequences from an active pan-tilt-zoom camera. IEEE Int. Conf. Adv. Video Signal Based Surv. 443–448 (2013)
22. Xueliam, D., Qiuqi, R.: Acclimatization calculation based on image quality assessment. IEEE Int. Conf. Wireless, Mobile Multimedia Networks 284–288, (2013)
23. Lin, K., Wang, X.: Face acquiring optimization based on video sensor network. IEEE Int. Conf. Control 445–450 (2014)
24. Galbally, J., Marcel, S., et al.: Image quality assessment for fake biometric detection: application to Iris, fingerprint and face recognition. IEEE Trans. Image Process. **23**, 710–724 (2013)
25. Patil, M., Ruikar, S.D.: Super-resolution of face image extraction from a video sequence. IEEE Int. Conf. Commun. Signal Process. 1620–1624 (2014)
26. Kim, H., Lee, S.H., et al.: Investigating cascaded face quality assessment for practical face recognition system. IEEE Int. Conf. Multimedia 399–400 (2014)
27. Avcibas, I., Sankur, B., Sayood, K.: Statistical evaluation of image quality measures. J. Electron. Imag. **11**(2), 206–223 (2002)
28. Huynh-Thu, Q., Ghanbari, M.: Scope of validity of PSNR in image/video quality assessment. Electron. Lett. **44**(13), 800–801 (2008)

29. Yao, S., Lin, W., Ong, E., Lu, Z.: Contrast signal-to-noise ratio for image quality assessment. In: Proc. IEEE ICIP, pp. 397–400 (2005)
30. Eskicioglu, A.M., Fisher, P.S.: Image quality measures and their performance. IEEE Trans. Commun. **43**(12), 2959–2965 (1995)
31. Martini, M.G., Hewage, C.T., Villarini, B.: Image quality assessment based on edge preservation. Signal Process. Image Commun. **27**(8), 875–882 (2012)
32. Wang, Z., Bovik, A.C., Sheikh, H.R., Simoncelli, E.P.: Image quality assessment: from error visibility to structural similarity. IEEE Trans. Image Process. **13**(4), 600–612 (2004)

Self-Similarity Descriptor and Local Descriptor-Based Composite Sketch Matching

Steven Lawrence Fernandes and G. Josemin Bala

Abstract Composite sketching belongs to the forensic science where the sketches are drawn using freely available composite sketch generator tools. Compared to pencil sketches, composite sketches are more effective because it consumes less time. It can be easily adopted by people across different regions; moreover, it does not require any skilled artist for drawing the suspects faces. Software tool used to generate the faces provides more features which can be used by the eyewitness to provide better description, which increases the clarity of the sketches. Even the minute details of the eyewitness description can be captured with great accuracy, which is mostly impossible in pencil sketches. Now that a composite sketch is provided, it has to be identified effectively. In this paper we have analyzed two state-of-the-art techniques for composite sketch image recognition: Self-similarity descriptor (SSD)-based composite sketch recognition and local descriptors (LD)-based composite sketch recognition. SSD is mainly used for developing a SSD dictionary-based feature extraction and Gentle Boost KO classifier-based composite sketch to digital face image matching algorithm. LD is mainly used for multiscale patch-based feature extraction and boosting approach for matching composites with digital images. These two techniques are validated on FACES and IdentiKit databases. From our analysis we have found that SSD descriptor works better than LD. Using SSD method we obtained the results for FACES (ca) as 51.9 which is greater when compared to LD which gives a result of 45.8. Similarly, using SSD, values of 42.6 and 45.3 for FACES (As) and IdentiKit (As), respectively, are obtained which are much better than the values of 20.2 and 33.7 for FACES (As) and IdentiKit (As), respectively, using LD method.

Keywords Composite sketch · Pencil sketch · Self-similarity descriptor · Local descriptor

S.L. Fernandes (✉) · G. Josemin Bala
Department of Electronics & Communication Engineering,
Karunya University, Coimbatore, India
e-mail: steva_fernandes@yahoo.com

G. Josemin Bala
e-mail: josemin@karunya.edu

M. Pant et al. (eds.), *Proceedings of Fifth International Conference on Soft Computing for Problem Solving*, Advances in Intelligent Systems and Computing 436, DOI 10.1007/978-981-10-0448-3_53

1 Introduction

Law enforcement agencies prefer composite sketches which are generated by the softwares to manually produced pencil sketches. Composite sketches are easy to develop when compared to hand-drawn sketches [1]. The software-generated sketches are quite independent of artist and will be reliable across different agencies and regions [2]. The software tools used for generating composite sketches contain many predefined templates having different facial components which can be used by the eyewitness for describing the suspects. Based on narration of eye witness, the artist will develop face sketches using software. Using the obtained sketch, the existing records will be checked for a match [3]. Composite sketches in the past meant hand-drawn sketches. Artists, who use traditional pencil and paper for sketch creation, must possess a lot of talent and training for transforming the description provided by witness into a sketch. The hand-drawn sketch will not be matching the criminals exactly, since the drawn sketch might suffer from some degree of limitations such as the expertise of drawing artist, the remembering and recollecting ability of the witness, texture loss, etc. [4]. Hence, bringing out the information from the witness is a difficult task. While drawing the sketch by the artist based on the description of witness chances are there that the image may go the way that is not according to exactly what witness has seen [5]. In that case artist may have to change the sketch several times using eraser till satisfactory results are obtained to some extent. By the advancement in software technology, artists started creating composite sketches with the help of digital pencil and computer operators which leads to person's identity [6]. Some agencies where they do not have any trained forensic artist will use the computer-based systems which do not need any artistic skills. These kinds of computer-based systems generate the sketch automatically which can be handled by any ordinary person who does not possess artistic skills [7].

Composite art technique is broadly classified into two approaches:

- Self-similarity descriptor-based composite sketches recognition;
- Local descriptor-based composite sketches recognition.

It is found that dataset is created by Caucasian and Asian user, using two software tools—FACES and IdentiKit. A set of composite sketches is generated by both users using FACES toolkit [8]. In addition, the Asian user has also created another dataset using Identikit. It is found that SSD method outperforms LD method for matching composite and digital face images in both the tools FACES and IdentiKit.

2 Self-similarity Descriptor-Based Composite Sketches Recognition

Here SSD-based dictionary consisting of 50,000 images is generated using CMU multi-PIE database. Gentle Boost KO classifier is used for matching composite sketches with digital images.

Step 01: Patches of size $p \times p$ are extracted for every key point. Patches having less than $\frac{p}{2}$ pixels are discarded.

Step 02: This involves three stages with first Gaussian smoothing level being the original image followed by second level with variance = 15 and third level with variance = 21.

Step 03: From the patches corresponding to these Gaussian smoothing levels features are extracted. For each patch P_{ij}, self-similarity descriptor (SSD) S_{ij} is calculated.

Step 04: For each dictionary face image SSD features are extracted. SSD extracts these features from patches encoding the internal geometric layouts, patterns, colors, and edges. A bag-of-word is formed by the features of size 80 × 40. Each word in a bag is a histogram of size 40.

Step 05: While creating the dictionary, sketches and digital images are preprocessed and then the features are extracted from the image.

Step 06: SSD features of an image S_{ij} are matched with S^n_{ij}, according to the following equation: $\overrightarrow{d}_{in} = \coprod_{j=1}^{k} X^2(S_{ij}, S^n_{ij})$, where $\overrightarrow{d}_{in}$ represents match score histogram for nth dictionary image and the ith Gaussian smoothing level. X^2 represents the X^2 distance metric. X^2 distance metric serves the purpose of matching two histograms. $\coprod$ represents histogram operator. Histogram of $K \times 80$ scores is obtained. These scores are obtained using X^2 distance metric and S^n_{ij} is the nth dictionary image.

Step 07: 100 distinct levels are created by quantizing the above scores and the 100 levels are mapped to a real value exclusively. Now the individual score histogram obtained by repeating the process for all Gaussian smoothing level is combined together.

Step 08: The histograms obtained are joined over the entire dictionary. After computing the match score histograms for both composite and digital images, difference histogram is calculated.

Step 09: Now for the matching purpose initially SSD features from the probe sketch are obtained and matched with dictionary.

Step 10: The difference histogram obtained between probe and gallery images is passed on to Gentle Boost KO classifier, to obtain match score between probe sketch and each subject in the gallery.

Step 11: The genuine sketches are arranged in descending order of their match score and ranked list of a match for probe sketch is obtained.

3 Local Descriptor-Based Composite Sketches Recognition

In this system, Daisy descriptors are used to extract local information from the patches generated by patch-based face recognition algorithm around fiducial features. Gentle Boost KO algorithm is used to match the information obtained from the patches.

Step 01: Initially using patch-based face recognition algorithm four points one each at eyes say P_1 and P_2, one at mouth say P_3, and one at the point where centroid is formed on the face say P_4 are obtained. Centroid is formed at the tip of the nose.

Step 02: Distance between the two eye points is measured which is the Euclidian distance say D.

Step 03: At the point $P_1 = (x_1, y_1)$ a circular patch of radius $\frac{D}{4}$ is obtained.

Step 04: Now by changing the points $p_1 = \left(x_1 + \frac{D}{4}, y_1\right), p_1 = \left(x_1 + \frac{D}{2}, y_1\right), p_1 = \left(x_1 + 3\frac{D}{4}, y_1\right)\ldots$ circular patch of radius $\frac{D}{4}$ is obtained to proceed in the right until image boundary is obtained.

Step 05: Similarly, circular patch of radius $\frac{D}{4}$ is obtained by changing the points at $p_1 = \left(x_1 - \frac{D}{4}, y_1\right)), p_1 = \left(x_1 - \frac{D}{2}, y_1\right), p_1 = \left(x_1 - 3\frac{D}{4}, y_1\right), \ldots$ to proceed in the left, $p_1 = \left(x_1, y_1 + \frac{D}{4}\right), \ldots$ to proceed in the up, and $p_1 = \left(x_1, y_1 - \frac{D}{4}\right), \ldots$ to proceed in the down.

Step 06: The same process is repeated by taking the point P_2, then by taking the point P_3, and at last by taking the point P_4.

Step 07: By changing the radius to D_2 the entire algorithm is repeated and at last the radius is changed to $3D_4$ and the algorithm is again repeated.

Step 08: Three descriptors Pyramid histogram of oriented gradients (PHOG), Daisy, and extended uniform circular binary patterns (EUCLBP) are used in extracting the features for the circular facial patches.

Step 09: For efficiently matching hand-drawn sketch to digital image EUCLBP is used.

Step 10: We know that composite sketch is synthetic in nature and differs slightly from digital images. To overcome this differential factor Daisy descriptor is used.

Step 11: Pyramid Histogram of Oriented Gradients is used to extract local features for classification purpose.

Step 12: Match score S_i is obtained by comparing patch of digital image corresponding to patch of composite sketch using X^2 distance metric.

Step 13: At last verification experiments are run several times to test boosting classifier in identification mode.

4 Result and Discussions

4.1 Use of SSD Dictionary in Comparing Composite Sketches with Digital Face Images

This method performs better than existing algorithms at rank 10. Combination of three bags Bag 1, Bag 2, and Bag 3 provides better results than any other combination. The redundant and outlier features are efficiently handled by Gentle Boost KO (Table 1).

4.2 Local Descriptor-Based Composite Sketches Recognition

Here two softwares FACES and Identikit are used. Using the FACES software one Asian and one Caucasian composite sketch is prepared. Using the IdentiKit software one Asian composite sketch is prepared. FACES software yields higher accuracy in creating composite sketch than IdentiKit. Rank-10 identification accuracy of the composite sketch created by Caucasian user using FACES tool is higher when compared to Asian user using IdentiKit tool. Among the three descriptors, Daisy descriptor performs better than Enhanced Uniform Circular Local Binary Pattern (EUCLBP) and Pyramid Histogram of Oriented Gradients (PHOG). When Daisy descriptor and commercial off-the-shelf (COTS) are compared Daisy descriptor stands best. The performance is improved since the patches are extracted and then boosted (Table 2).

Table 1 Rank-10 identification accuracy (%) extracted on the extended-pattern recognition and image processing (e-PRIP) composite sketch database

Algorithm	Faces (ca)	Faces (As)	Identikit (As)
Commercial off-the-shelf	11.3 ± 2.1	7.2 ± 2.2	8.1 ± 2.1
Bag 1	45.4 ± 2.6	34.6 ± 3.1	37.4 ± 3.5
Bag 2	29.3 ± 2.9	28.0 ± 3.7	21.3 ± 4.1
Bag 3	32.0 ± 3.2	24.0 ± 3.4	26.6 ± 3.6
Equal weighted sum	33.3 ± 4.3	31.3 ± 2.8	33.6 ± 4.3
Weighted sum	48.0 ± 1.2	37.3 ± 0.8	40.0 ± 1.2
AdaBoost	46.6 ± 1.6	35.9 ± 0.7	38.6 ± 1.1
Multiscale circular weber local descriptor	23.2 ± 3.2	15.7 ± 3.0	15.4 ± 3.1
Local descriptor	32.4 ± 2.4	21.3 ± 2.1	27.6 ± 1.8
Self-similarity descriptor	51.9 ± 1.2	42.6 ± 1.2	45.3 ± 1.5

Table 2 Rank-10 identification accuracy (%) obtained on the pattern recognition and image processing composite sketch dataset

Algorithm	Faces (ca)	Faces (As)	IdentiKit (As)
Enhanced uniform circular local binary pattern	10.6	13.0	14.6
Patch—pyramid histogram of oriented gradients	29.3	4.1	21.9
Patch—Daisy	30.9	13.8	25.2
Commercial off-the-shelf	10.6	6.5	8.1
Composite and digital images matched using local descriptors	45.8	20.2	33.7

5 Conclusion

Matching composite sketches with digital photos is an interesting law enforcement problem and very limited research has been undertaken. In the first approach, self-similarity descriptor-based composite sketches recognition serves two purposes—(1) developing a SSD dictionary-based feature extraction and Gentle Boost KO classifier-based composite sketch to digital face image matching algorithm and (2) extending the PRIP sketch database by including a set of composite sketches prepared by an artist of Indian ethnicity. In the second approach, Local descriptor-based composite sketches recognition helps in multiscale patch-based feature extraction and boosting approach for matching composites with digital images. It is found that for recognizing composite sketches with digital face images, SSD method produced a much better result with the values of 51.9, 42.6, and 45.3 for FACES (Ca), FACES (As), and IdentiKit (As) tools compared to the values of 45.8, 20.2, and 33.7 for the same FACES (Ca), FACES (As), and IdentiKit (As) tools, respectively, produced by LD method. The values above clearly show that recognizing composite sketches with digital face images via SSD dictionary performs better than LD method.

Acknowledgments The proposed work was made possible because of the grant provided by Vision Group Science and Technology (VGST), Department of Information Technology, Biotechnology and Science and Technology, Government of Karnataka, Grant No. VGST/SMYSR/GRD-402/2014-15 and the support provided by Department of Electronics and Communication Engineering, Karunya University, Coimbatore, Tamil Nadu, India.

References

1. Mittal, P., Jain, A., Singh, R., Vatsa, M.: Boosting local descriptors for matching composite and digital face images. In: 20th International Conference on Image Processing, pp. 2797–2801. Melbourne, VIC (2013)
2. Mittal, P., Jain, A., Goswami, G., Singh, R., Vatsa, M.: Recognizing composite sketches with digital face images via SSD dictionary. In: IEEE International Joint Conference on Biometrics, pp. 1–6, Clearwater, FL (2014)

3. Rahim, N.N., Malek, N.A.A., Zeki, A.M., Abubakar, A.: Automatic face reconstruction system. In: 6th International Conference on Computer Science and Information Technology, pp. 208–212. Amman (2014)
4. Frowd, C.D., Park, J., McIntyre, A., Bruce, V., Pitchford, M., Fields, S., Kenirons, M., Hancock, P.J.B.: Effecting an improvement to the fitness function. How to evolve a more identifiable face. In: ECSIS Symposium on Bio-inspired Learning and Intelligent Systems for Security, pp. 3–10. Edinburgh (2008)
5. Yung-Hui, L., Savvides, M., Bhagavatula, V.: Illumination tolerant face recognition using a novel face from sketch synthesis approach and advanced correlation filters. In: IEEE International Conference on Acoustics, Speech and Signal Processing, vol. 2, p. 11. Toulouse (2006)
6. Chug, T., Bhatt, H.S., Singh, R., Vatsa, M.: Matching age separated composite sketches and digital face images. In: IEEE Sixth International Conference on Biometrics: Theory, Applications and Systems, pp. 1–6, Arlington, VA (2013)
7. Hu, H., Klare, B.F., Bonnen, K., Jain, A.K.: Matching composite sketches to face photos: a component-based approach. In: IEEE Transaction on Information Forensics and Security, vol. 8, pp. 191–204 (2013)
8. Klum, S., Hu, H., Jain, A.K., Klare, B.: Sketch based face recognition: forensic vs. composite sketches. In: IEEE International Conference on Biometrics, pp. 1–8. Madrid (2013)

Multi-objective Colliding Bodies Optimization

Arnapurna Panda and Sabyasachi Pani

Abstract Kaveh and Mahdavi proposed a new metaheuristic method in 2014 known as colliding bodies optimization (CBO). The algorithm is based on the principle of collision between bodies (each has a specific mass and velocity). The collision makes the bodies move toward the optimum position in the search space. This paper deals with the multi-objective formulation of CBO termed as MOCBO. Simulation studies on benchmark functions Schaffer N1, Schaffer N2, and Kursawe have demonstrated the superior performance of the MOCBO over multi-objective particle swarm optimization (MOPSO) and non-dominated sorting genetic algorithm II (NSGA-II). The performance analysis are carried out for the proposed and benchmark algorithms in identical platforms using response matching between obtained and true Pareto front; the convergence matric, diversity matric, and computational efficiency achieved over fifty independent runs.

Keywords Colliding bodies optimization · Multi-objective optimization · MOPSO · NSGA-II

1 Introduction

In last two decades, the multi-objective algorithms have gained popularity due to their applicability in all diversified areas of science and engineering. The survey papers [1–3] highlight the benchmark problems, research methodologies, performance indicators, and potential applications of the multi-objective algorithms. The multi-objective formulation of a problem provides a set of optimal solutions (termed as Pareto optimal), by suitably trading off between different conflicting

Arnapurna Panda (✉) · Sabyasachi Pani
School of Basic Sciences, Indian Institute of Technology Bhubaneswar,
Bhubaneswar 751007, Orissa, India
e-mail: arnapurna.math@gmail.com

Sabyasachi Pani
e-mail: spani@iitbbs.ac.in

M. Pant et al. (eds.), *Proceedings of Fifth International Conference on Soft Computing for Problem Solving*, Advances in Intelligent Systems and Computing 436, DOI 10.1007/978-981-10-0448-3_54

objectives. Each and every solution in the Pareto optimal is significant for the user (as it provides flexibility to the user to select a proper solution depending on the requirement).

The nature-inspired metaheuristics have been extensively applied to solve multi-objective optimization problems. Some popular ones which have achieved huge citations include non-dominated sorting genetic algorithm (NSGA-II) [4], multi-objective particle swarm optimization [5], multi-objective differential evolution [6], and multi-objective artificial immune system [7]. The recently developed ones are multi-objective bacterial foraging optimization [8], multi-objective artificial bee colony [9], multi-objective cat swarm optimization [10], multi-objective cuckoo search [11], multi-objective gravitational search [12], and multi-objective immunized PSO [13]. Researchers of computing and mathematics communities are developing new multi-objective nature-inspired algorithms to effectively solve the real-life engineering problems.

Observing the natural phenomenon of collision between bodies Kaveh and Mahdavi proposed a new metaheuristic method in 2014, named it as colliding bodies optimization (CBO) [14, 15]. In CBO the collision between bodies makes them move toward the optimum position in the search space. The principle of collision is governed by the coefficient of restitution, which is responsible for the elastic and inelastic collision. In order to make the algorithm popular they discussed the computer codes of CBO in [16]. They applied the CBO for solving problems of civil engineering (optimum design of truss structures [17, 18] and optimal design water distribution systems [19]). Recently, Panda and Pani [20] have applied the CBO to identify the parameters of complex Hammerstein plant.

The beauty of CBO is it deals with only one parameter, i.e., coefficient of restitution. Therefore, parameter setting is not a problem. In [20] it has been reported that the algorithm has lower computational complexity than bacterial foraging optimization for identification of complex plants. Being motivated by the advantages of CBO in this manuscript we have developed the multi-objective formulation of CBO and termed it as 'MOCBO.' On benchmark problems many multi-objective algorithms lag in convergence to true Pareto front and maintaining proper diversity among the solutions [21]. The proposed MOCBO provides accurate convergence as well as diversity compared to MOPSO and NSGA-II.

The details about the proposed algorithm are described in Sect. 2. The performance of the MOCBO is evaluated on three benchmark multi-objective functions Schaffer N1, Schaffer N2, and Kursawe. The details about these simulation studies are highlighted in Sect. 3. The comparative analyses of the results obtained by MOCBO and benchmark algorithms NSGA-II and MOPSO are reported in Sect. 4. The concluding remarks of the investigation are outlined in Sect. 5.

2 Multi-objective Colliding Bodies Optimization

2.1 Basic Definitions

Definition 1 *Multi-objective optimization*: A multi-objective problem [21, 22] can be mathematically represented as

Optimize the vector function

$$\vec{f}(\vec{x}) = [f_1(\vec{x}), f_2(\vec{x}), \ldots f_m(\vec{x})] \tag{1}$$

where $(\vec{x}) = [x_1, x_2, \ldots x_n]^T$ is a vector of decision variables $\forall\, \vec{x} \in \Omega$.

The Ω is decision space.

The $\vec{f}(\vec{x})$ consists of m objective functions such that

$$f_i : \Omega \rightarrow R, i = 1, 2, \ldots m \tag{2}$$

The R^m is the objective space.

It is important that the objectives in (1) are conflicting with each other. The improvement in one objective may lead to deterioration of another. Therefore, a single solution cannot optimize all objectives simultaneously.

Definition 2 *Pareto dominance*: Let two vectors $\vec{u} = (u_1, u_2, \ldots u_m)^T$, $\vec{v} = (v_1, v_2, \ldots v_m)^T$ and $\vec{u}, \vec{v} \in \Omega$.

The vector $\vec{u}$ is said to dominate another vector $\vec{v}$, noted by $\vec{u} \prec \vec{v}$ if:

$$\forall i \in (1, 2, \ldots m), u_i \leq v_i \text{ and } \vec{u} \neq \vec{v} \tag{3}$$

Definition 3 *Pareto Optimality*: A feasible solution $\vec{x}^* = [x_1^*, x_2^*, \ldots x_n^*]$ with $\vec{x}^* \in \Omega$ is said to be Pareto optimal solution if $\nexists\, \vec{y} \in \Omega$ such that $f(\vec{y}) \prec f(\vec{x}^*)$. The set of Pareto optimal solutions is called the **Pareto set** (PS) given by

$$PS = [\vec{x} \in \Omega | \nexists\, \vec{y} \in \Omega, f(\vec{y}) \prec f(\vec{x}^*)] \tag{4}$$

The image of PS in the objective space is called Pareto front (PF).

$$\text{PF} = [f(\vec{x}) | \vec{x} \in \text{PS}] \tag{5}$$

2.2 Concept of Colliding Bodies Optimization

The CBO is inspired by the natural phenomenon of collision between two bodies. The collision between two bodies in an isolated environment is shown in Fig. 1. During collision the total momentum of the bodies is conserved (i.e., the net

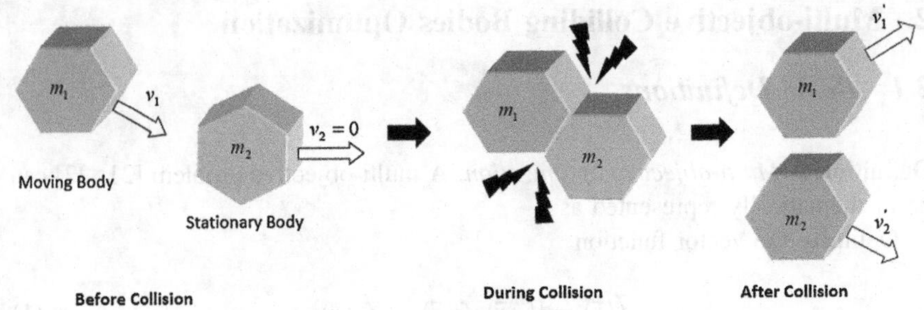

Fig. 1 Principle of collision between two bodies in an isolated environment

momentum of bodies before the collision and the momentum of all bodies after the collision becomes equal).

Consider two bodies of mass m_1 and m_2 are moving with velocities v_1 and v_2. During collision the conversion of momentum is given by

$$m_1 v_1 + m_2 v_2 = m_1 v_1' + m_2 v_2' \tag{6}$$

where v_1' and v_2' are the velocities achieved by the bodies after collision, which are mathematically expressed by

$$v_1' = \frac{(m_1 - \epsilon m_2)v_1 + (m_2 + \epsilon m_2)v_2}{m_1 + m_2} \tag{7}$$

$$v_2' = \frac{(m_2 - \epsilon m_1)v_2 + (m_1 + \epsilon m_1)v_1}{m_1 + m_2} \tag{8}$$

The ϵ is known as coefficient of restitution (CR). The CR is the ratio of relative velocity of separation to the relative velocity of approach, given by

$$\epsilon = \frac{|v_2' - v_1'|}{|v_2 - v_1|} \tag{9}$$

According to the coefficient of restitution values, there are two cases of any collision:

- Perfectly Elastic Collision ($\epsilon = 1$): In this collision there is no loss of energy. After collision, the velocity of separation for both bodies is high.
- Inelastic Collision ($\epsilon \leq 1$): In this case there is loss of energy (some of the energies get converted into other form). After collision, the velocities of separation for both bodies are low.

2.3 Details of the Proposed Algorithm

The steps of proposed multi-objective colliding bodies optimization (MOCBO) algorithm are outlined in sequel:

1. **Initialization of Positions of Bodies**

Let Z be the number of colliding bodies (CB) in the search space each with dimension D. The D represents the number of variables in the function. The CB position becomes a $Z \times D$ matrix which is initialized with random numbers in the range $[x_{\min}, x_{\max}]$, given by

$$x_i^0 = x_{\min} + \text{rand} \times (x_{\max} - x_{\min})\, i = 1, 2, \ldots, Z \tag{10}$$

where $x_{\min}$ and $x_{\max}$ are the minimum and maximum ranges of variables in the search space; rand is a random number in the range [0, 1].

2. **Evaluate the Fitness of CBs**

Substitute the position of bodies in the given multi-objective functions and evaluate the fitness.

3. **Perform Non-dominated Sorting**

The non-dominated sorting is applied to all the bodies in the search space following the definition and steps of implementation described in Sect. IV(B) of [5]. All the non-dominated solutions ($I \leq Z$) are then determined. These solutions are kept in an archive.

4. **Calculate the Mass of CBs**

The body mass of each CB is defined as

$$m_j = \frac{\frac{1}{\text{fit}(j)}}{\sum_{i=1}^{N} \frac{1}{\text{fit}(i)}}\; j = 1, 2, \ldots, Z \tag{11}$$

where

$$\text{fit}(j) = \sum_{k=1}^{D} \left|\text{fit}_j^D\right| \forall j \in [1, Z] \tag{12}$$

5. **Initial Velocities of CBs**

The masses of the CBs are arranged in ascending order. The sorted CBs are equally divided into two groups:

- Stationary CBs: The upper half of CBs represents stationary bodies and their velocities before collision is zero:

$$v_i = 0;\ \forall i = 1, 2, \ldots, \frac{Z}{2} \tag{13}$$

- Moving CBs: The lower half of CBs is movable in nature. Their velocities before collision are given by

$$v_i = x_i - x_{i-\frac{Z}{2}};\ \forall i = \frac{Z}{2} + 1, \ldots, Z \tag{14}$$

6. **Velocities After Collision**

The velocities of the stationary bodies after the collision are represented by

$$v_i' = \frac{\left(m_{i+\frac{Z}{2}} + \epsilon m_{i+\frac{Z}{2}}\right) v_{i+\frac{Z}{2}}}{m_i + m_{i+\frac{Z}{2}}};\ \forall i = 1, \ldots, \frac{Z}{2} \tag{15}$$

The velocity of movable bodies after the collision is given by

$$v_i' = \frac{\left(m_i - \epsilon m_{i-\frac{Z}{2}}\right) v_i}{m_i + m_{i-\frac{Z}{2}}};\ \forall i = \frac{Z}{2} + 1, \ldots, Z \tag{16}$$

7. **Position Update**

The positions of the stationary bodies are updated by

$$x_i^{new} = x_i + \text{rand}\ o\ v_i'\ \forall i = 1, \ldots, \frac{Z}{2} \tag{17}$$

The positions of the moving bodies are updated by

$$x_i^{new} = x_{i-\frac{Z}{2}} + \text{rand}\ o\ v_i' \forall i = \frac{Z}{2} + 1, \ldots, Z \tag{18}$$

where x_i^{new}, x_i, and v_i' are the updated position, previous position, and the updated velocity after the collision of the *i*th colliding body. The rand represents random numbers in [−1, 1] and '*o*' represents element-by-element multiplication.

8. **Computation of Coefficient of Restitution (CR)** The CR value is calculated by

$$\epsilon = 1 - \frac{\text{itr}}{\text{itr}_{\text{max}}} \tag{19}$$

where itr is the present iteration and itr_{max} is the maximum number of iteration.

9. **Termination Criteria**

The process is repeated from step 2 to 8 until a termination criterion, such as maximum iteration number, or filled archive content is satisfied.

3 Experiments on Benchmark Multi-objective Functions

3.1 Benchmark Functions

In order to evaluate the performance of the proposed MOCBO simulation studies are carried out on three benchmark unconstrained multi-objective functions. The details about these functions are outlined here:

Schaffer Function N1: This benchmark function comprise two objectives with one variable given by

$$\text{Minimize} = \begin{cases} f_1(x) = x^2 \\ f_2(x) = (x-2)^2 \end{cases} \tag{20}$$

where $x \in [-1000, 1000]$. The plot of f_1, f_2 for $x \in [-3, 3]$ is shown in Fig. 2a.

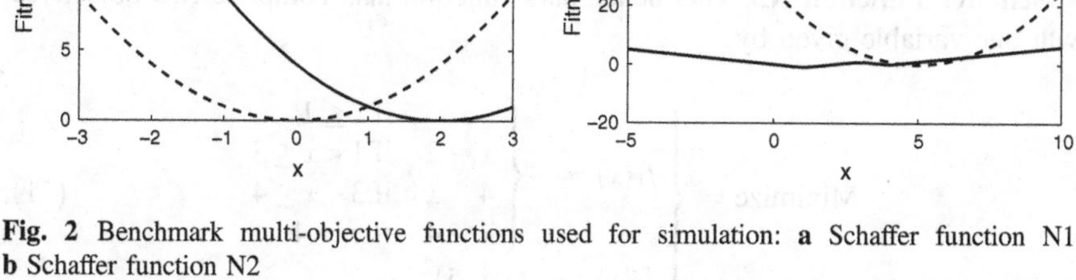

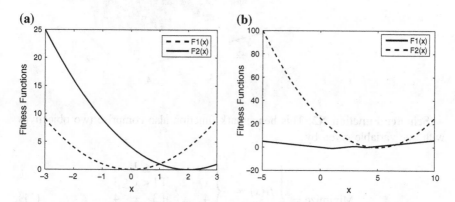

Fig. 2 Benchmark multi-objective functions used for simulation: **a** Schaffer function N1. **b** Schaffer function N2

Fig. 3 Benchmark Kursawe function used for multi-objective simulation: **a** Kursawe function F1. **b** Kursawe function F2

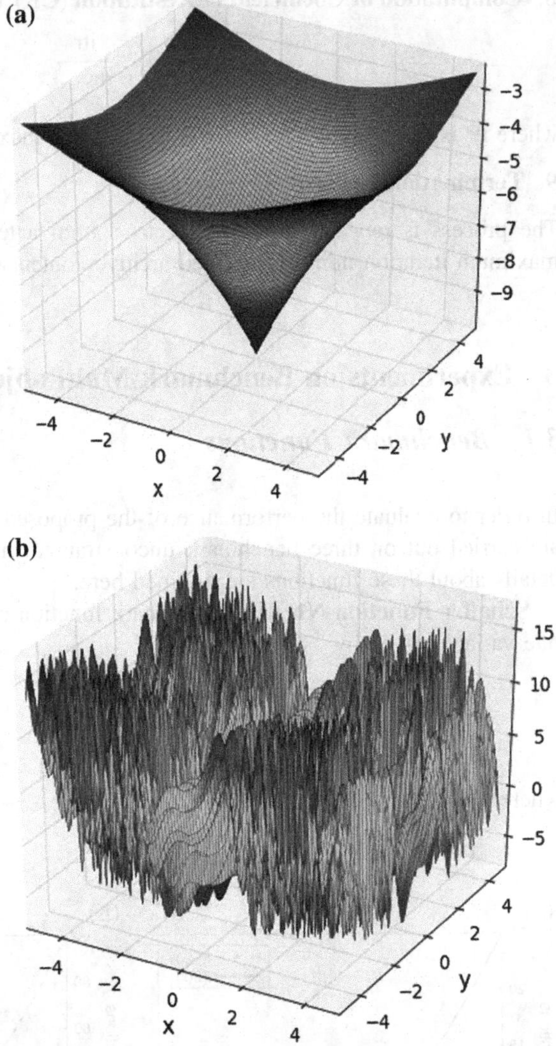

Schaffer Function N2: This benchmark function also comprise two objectives with one variable given by

$$\text{Minimize} = \begin{cases} f_1(x) = & \begin{cases} -x & \text{if } x \leq 1 \\ x-2 & \text{if } 1 < x \leq 3 \\ 4-x & \text{if } 3 < x \leq 4 \\ x-4 & \text{if } x > 4 \end{cases} \\ f_2(x) = & (x-5)^2 \end{cases} \tag{21}$$

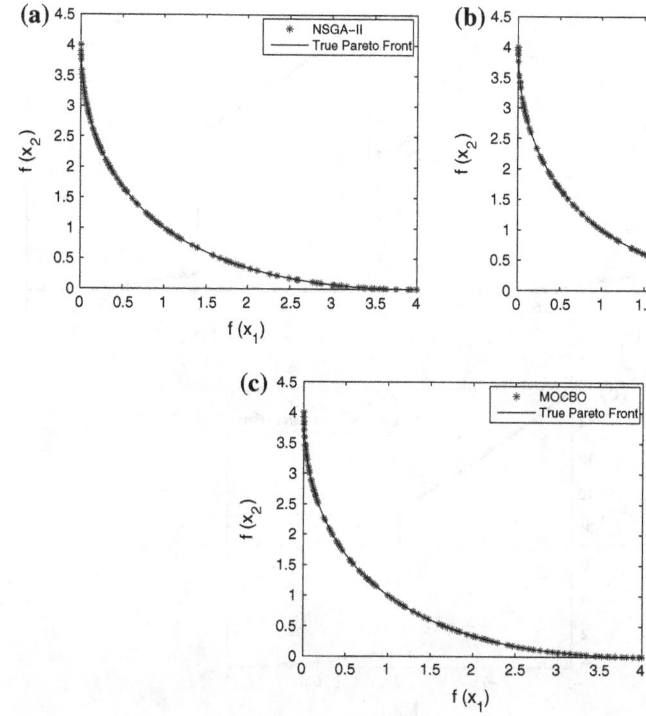

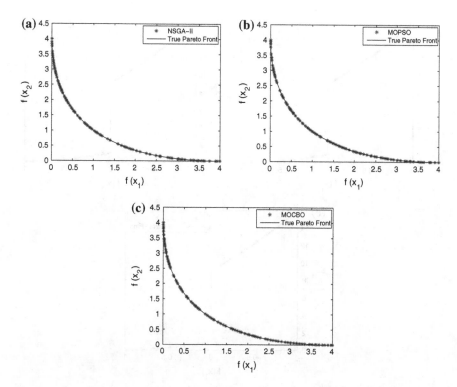

Fig. 4 Comparative results of true and obtained Pareto optimal front for Schaffer function N1 using: **a** NSGA-II. **b** Multi-objective PSO. **c** Multi-objective CBO

where $x \in [-5, 10]$. The plot of f_1, f_2 for $x \in [-5, 10]$ is shown in Fig. 2b.

Kursawe Function: This benchmark function consists of two objectives with two variables. It is unconstrained, non-convex, and disconnected in nature.

$$\text{Minimize} = \begin{cases} f_1(x) = \sum_{i=1}^{n-1} \left(-10\exp\left(-0.2\sqrt{x_i^2 + x_{i+1}^2}\right)\right) \\ f_2(x) = \sum_{i=1}^{n} \left(|x_i|^{0.8} + 5\sin(x_i^3)\right) \end{cases} \tag{22}$$

where $x \in [-5, 5]$ and $1 \leq i \leq 3$. The surface plots of $f_1(x)$ and $f_2(x)$ are shown in Fig. 3a, b, respectively.

3.2 Simulation Strategy

The simulation studies of proposed MOCBO and benchmark MOPSO and NSGA-II are carried out in a MATLAB version R2011 platform on an Intel

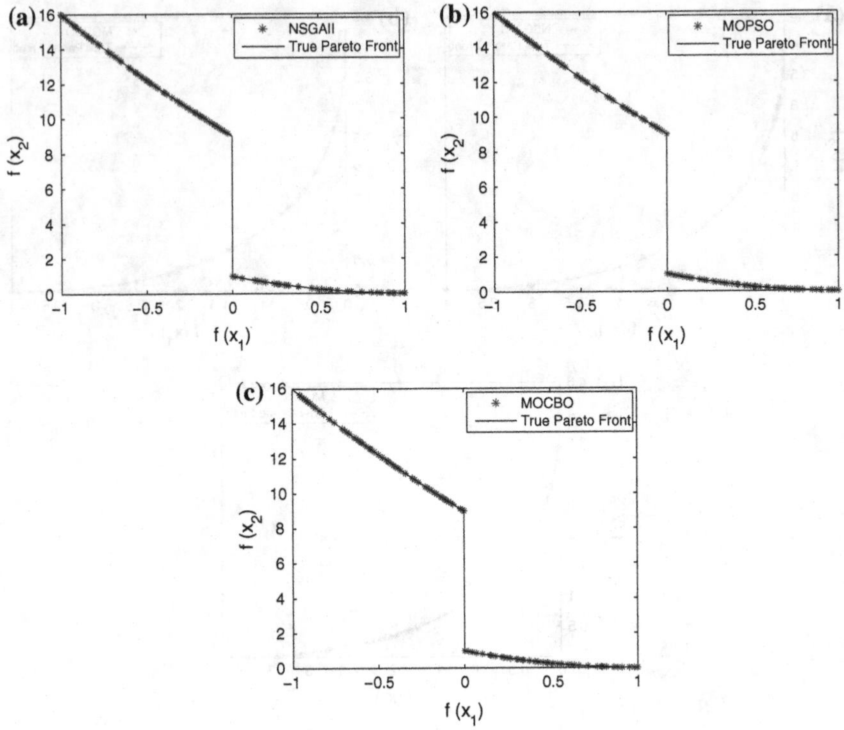

Fig. 5 Comparative results of true and obtained Pareto optimal front for Schaffer function N2 using: **a** NSGA-II. **b** Multi-objective PSO. **c** Multi-objective CBO

i7-3540M 3 GHz CPU, with an 8 GB RAM in Windows 8.1 (64-bit) environment. For all the three algorithms, the population size is taken as 100 and number of iteration is taken as 250. The rest parameters of NSGA-II and MOPSO are taken as discussed in [4, 5], respectively.

3.3 *Performance Evaluation*

Due the heuristic nature of evolutionary algorithms the results of consecutive runs usually do not match. In order to achieve proper performance evaluation, fifty independent runs of each algorithm are carried out. The results are illustrated by the best, worst, and mean performance obtained over fifty independent runs. The following performance measures are used:

1. **Plot of Pareto Fronts**: The plot of true and obtained Pareto front by the multi-objective algorithms is a visual measure about the characteristic of the Pareto front [21].

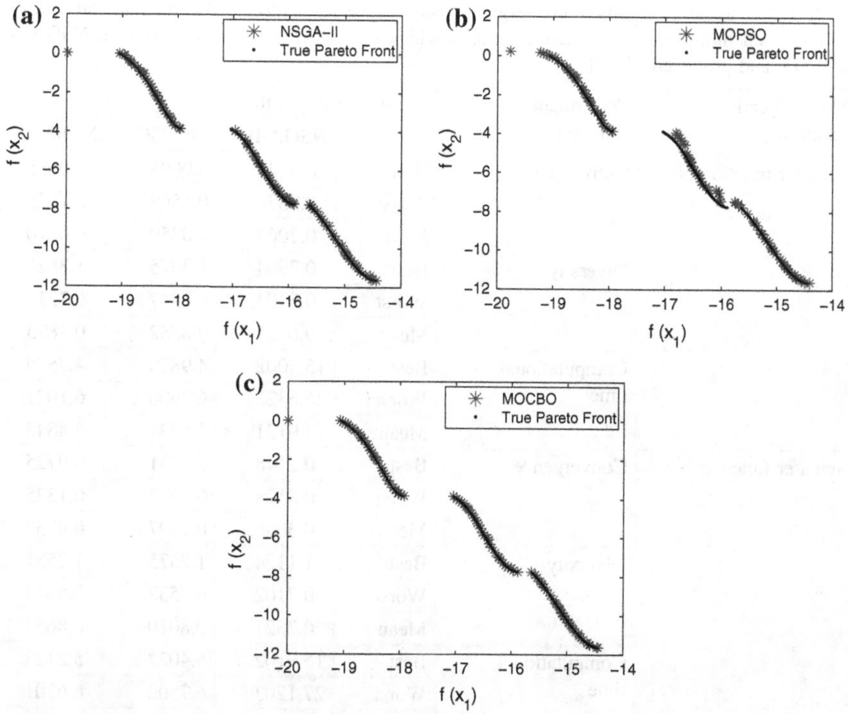

Fig. 6 Comparative results of true and obtained Pareto optimal front for Kursawe function using: **a** NSGA-II. **b** Multi-objective PSO. **c** Multi-objective CBO

2. **Convergence Metric**: It represents the convergence of the obtained Pareto point to the true Pareto front. The detailed mathematical expressions are outlined in [4].
3. **Diversity Metric**: It describes the maintenance of diversity in solutions of the Pareto optimal set. The details are given in [4].
4. **Computational Efficiency**: The best, worst, and average run time of each algorithm is considered as the measure of computational efficiency [5].

4 Result and Discussions

The true and obtained Pareto fronts (PF) obtained by the proposed MOCBO and benchmark NSGA-II and MOPSO for Schaffer N1, Schaffer N2, and Kursawe functions are shown in Figs. 4, 5, and 6, respectively. In all the three figures, it is observed that the obtained PFs lie nearer to the true fronts. The comparative results of convergence metric, diversity metric, and computational time of the algorithms

Table 1 Comparative results of convergence metric, diversity metric, and computational time on the three benchmark multi-objective functions achieved over fifty independent runs of NSGA-II, MOPSO, and proposed MOCBO

Multi-objective function	Performance function	Results	Algorithms		
			NSGA-II	MOPSO	MOCBO
Schaffer function N1	Convergence	Best	0.0148	0.0093	0.0085
		Worst	0.9578	0.1569	0.1325
		Mean	0.2069	0.0259	0.0210
	Diversity	Best	0.7904	1.3575	1.3621
		Worst	0.5104	0.6947	0.7214
		Mean	0.6425	0.8582	0.8875
	Computational time	Best	15.4002	4.9820	4.7529
		Worst	26.6423	6.2600	6.1021
		Mean	17.9121	5.5721	5.4845
Schaffer function N2	Convergence	Best	0.2148	0.1231	0.0725
		Worst	0.8578	0.1697	0.1325
		Mean	0.3096	0.1497	0.0932
	Diversity	Best	1.1124	1.2575	1.2531
		Worst	0.7102	0.7532	0.8043
		Mean	0.7821	0.8010	0.8652
	Computational time	Best	15.9102	5.4022	5.2123
		Worst	27.1203	6.7502	6.6201
		Mean	18.4201	6.0272	18.4201
Kursawe function	Convergence	Best	0.0449	0.0515	0.0349
		Worst	0.0559	0.0725	0.0559
		Mean	0.0516	0.0608	0.0512
	Diversity	Best	0.7939	0.7582	0.8648
		Worst	0.7248	0.6800	0.7203
		Mean	0.7597	0.7193	0.7709
	Computational time	Best	28.4620	2.3200	2.1024
		Worst	40.1502	3.2124	3.1520
		Mean	29.6321	2.4321	2.3727

on the three benchmark functions are reported in Table 1. From Table 1, it is observed that in all the three functions the front of MOCBO is closer to the true front (lower convergence) and maintain proper diversity among the solutions (higher diversity). It is also noted that the run time is also reported low compared to the MOPSO and NSGA-II. The run time is low because of the simplified update equations and velocity component of 50 % bodies becomes zero in each iteration (stationary bodies which reduce the calculations).

5 Conclusion

In this paper, a recently developed metaheuristic algorithm colliding body optimization is formulated to solve multi-objective problems. The multi-objective version is termed as MOCBO. Simulation studies on Schaffer N1, Schaffer N2, and Kursawe functions demonstrated the superior performance of the MOCBO over MOPSO and NSGA-II using four benchmark performance measures over fifty independent runs. Thus the proposed MOCBO is a potential candidate and in future can effectively be used to solve real-life multi-objective problems.

References

1. Zhou, A., Qu, B.Y., Li, H., Zhao, S.Z., Suganthan, P.N., Zhang, Q.: Multiobjective evolutionary algorithms: a survey of the state of the art. Swarm Evol. Comput. **1**(1), 32–49 (2011)
2. Coello Coello, C.A.: Evolutionary multi-objective optimization. A historical view of the field. IEEE Comput. Intell. Mag. **1**(1), 28–36 (2006)
3. Chiandussi, G., Codegone, M., Ferrero, S., Varesio, F.E.: Comparison of multi-objective optimization methodologies for engineering applications. Comput. Math. Appl. **63**(5), 912–942 (2012)
4. Deb, K., Pratap, A., Agarwal, S., Meyarivan, T.A.M.T.: A fast and elitist multiobjective genetic algorithm. NSGA-II: IEEE Trans. Evol. Comput. **6**(2), 182–197 (2002)
5. Coello Coello, C.A., Pulido, G.T., Lechuga, M.S.: Handling multiple objectives with particle swarm optimization. IEEE Trans. Evol. Comput. **8**(3), 256–279 (2004)
6. Xue, F., Sanderson, A.C., Graves, R.J.: Pareto-based multi-objective differential evolution. In: Proceedings of IEEE Congress on Evolutionary Computation (CEC-03), Australia, pp. 862–869 (2003)
7. Coello Coello, C.A., Corts, N.C.: Solving multiobjective optimization problems using an artificial immune system. Genet. Programm. Evolvable Mach. **6**(2), 163–190 (2005)
8. Niu, B., Wang, H., Wang, J., Tan, L.: Multi-objective bacterial foraging optimization. Neurocomputing **116**, 336–345 (2013)
9. Akbari, R., Hedayatzadeh, R., Ziarati, K., Hassanizadeh, B.: A multi-objective artificial bee colony algorithm. Swarm Evol. Comput. **2**, 39–52 (2012)
10. Pradhan, P.M., Panda, G.: Solving multiobjective problems using cat swarm optimization. Expert Syst. Appl. **39**(3), 2956–2964 (2012)
11. Yang, X.S., Deb, S.: Multiobjective cuckoo search for design optimization. Comput. Oper. Res. **40**(6), 1616–1624 (2013)
12. Hassanzadeh, H.R., Rouhani, M.: A multi-objective gravitational search algorithm. In: Proceedings IEEE Second International Conference on Computational Intelligence, Communication Systems and Networks (CICSyN), pp. 7–12 (2010)
13. Nanda, S.J., Panda, G.: Automatic clustering algorithm based on multi-objective Immunized PSO to classify actions of 3D human models. Eng. Appl. Artif. Intell. **26**(5), 1429–1441 (2013)
14. Kaveh, A., Mahdavi, V.R.: Colliding bodies optimization: a novel meta-heuristic method. Comput. Struct. **139**, 18–27 (2014)
15. Kaveh, A.: Colliding bodies optimization. In: Advances in Metaheuristic Algorithms for Optimal Design of Structures, pp. 195–232. Springer (2014)

16. Kaveh, A., Ilchi Ghazaan, M.: Computer codes for colliding bodies optimization and its enhanced version. Int. J. Optim. Civil Eng. **4**(3), 321–339 (2014)
17. Kaveh, A., Mahdavi, V.R.: Colliding bodies optimization method for optimum discrete design of truss structures. Comput. Struct. **139**, 43–53 (2014)
18. Kaveh, A., Mahdavi, V.R.: Colliding bodies optimization method for optimum design of truss structures with continuous variables. Adv. Eng. Softw. **70**, 1–12 (2014)
19. Kaveh, A., Shokohi, F., Ahmadi, B.: Analysis and design of water distribution systems via colliding bodies optimization. Int. J. Optim. Civil Eng. **4**(2), 165–185 (2014)
20. Panda, A., Pani, S.: A new model based on colliding bodies optimization for identification of Hammerstein plant. In: Proceedings of IEEE Annual India Conference (INDICON-2014), pp. 1–5 (2014)
21. Deb, K.: Multi-objective optimization using evolutionary algorithms. Wiley, New York, USA (2001)
22. Nanda, S.J., Panda, G.: A survey on nature inspired metaheuristic algorithms for partitional clustering. Swarm Evol. Comput. **16**, 1–18 (2014)

Printed Hindi Characters Recognition Using Neural Network

Vaibhav Gupta and Sunita

Abstract To recognize the Hindi characters using perceptron learning rule an algorithm is modeled and simulated in this paper. This model maps a matrix of pixels into characters on scanned images. In this paper perceptron learning rule is modeled based on mapping of input and output matrix of pixels. Perceptron learning rule uses an iterative weight adjustment that is more powerful than other learning rules. The perceptron uses threshold output function and the McCulloch–Pitts model of a neuron. Their iterative learning converges to correct weight vector, i.e., the weight vector that produces the exact output value for the training input pattern. For modeling and simulation, those Hindi characters are used which are similar to some of numeric numbers. To model and simulate the algorithm, Hindi characters are taken in form of the 5 × 3 matrix of pixels.

Keywords Perceptron learning rule · Pattern recognition · Hindi character · Character recognition · Artificial neural network · McCulloch–Pitts model · Output unit · Activation function · Test pattern

1 Introduction

Nowadays, character recognition has gained tremendous popularity in the field of pattern recognition and artificial intelligence. There is a vast growth in the technology; character recognition has significantly increased the systems' supply to meet industrial and commercial needs through the development process.

Character recognition can be classified as the recognition of optically processed characters based on quantitative or structural description of an object. This process

Vaibhav Gupta (✉) · Sunita
Instruments Research & Development Establishment,
Defence Research & Development Organization, Dehradun, India
e-mail: Vaibhav.drdo@gmail.com

Sunita
e-mail: sunnidma@yahoo.com

M. Pant et al. (eds.), *Proceedings of Fifth International Conference on Soft Computing for Problem Solving*, Advances in Intelligent Systems and Computing 436, DOI 10.1007/978-981-10-0448-3_55

involves extracting meaningful data attributes by separation of irrelevant details. The pattern recognition solutions involve many stages such as doing the measurements, preprocessing, and segmentation. Perceptron learning rule plays an important role in pattern recognition applications. One of the simplest is a single-layer network in which weights and biases could be trained to produce a correct target vector when presented with the corresponding input vector. The training technique used is called the perceptron learning rule. The perceptron gained great interest due to its ability to generalize from its training vectors and learn from initially randomly distributed connections. Perceptron can provide best solutions to overcome the problems in pattern classification. It is highly efficient, fast, and reliable networks for the problems it can solve. In addition, perceptron is very effective to understand complex networks. In this paper, modeling and simulation to recognize the Hindi printed characters are carried out using perceptron learning rule.

2 Perceptron Learning Rule Algorithm

The perceptron learning rule uses an iterative weight adjustment that is more powerful than the Hebb rule. The perceptrons use threshold output function and the McCulloch–Pitts model of a neuron. Their iterative learning converges to correct weight. The original perceptron is found to have three layers, input, weight, and output units, as shown in Fig. 1.

The input and output units have binary activation of +1 or −1 is used for the response unit. All the units have their corresponding weighted interconnections. Training in perceptron will continue until no error occurs. This net solves the problem and is also used to learn the recognition. To start the training process, initially the weights w_i and the bias b_j are set to zero. It is also essential to set the learning rate parameter α, which range between 0 and 1. Then the input is presented, and the net input is calculated by multiplying the weights with the inputs x_i and adding the result with the bias entity. Once the net input is calculated, by applying the activation function, the output of the network y_j is also obtained. This output is compared with the target t_j, where if any difference occurs, we go in for weight updation based on perceptron learning rule, else the network training is stopped.

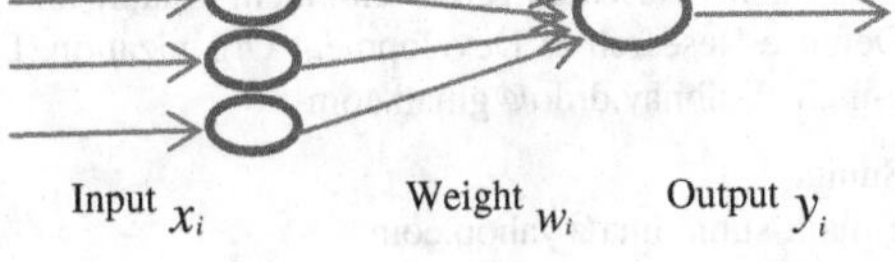

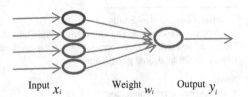

Fig. 1 McCulloch–Pitts model of a neuron

The training algorithm is as follows:

Step 1: Initialize the weights (w_i) and biases (b_i). Set the learning rate.
Step 2: When stopping condition is false, perform Steps 3–7.
Step 3: For each input training pair, do steps 4–6.
Step 4: Set identity function as an activation function for the input units x_i

$$x_i = s_i \quad \text{for } i = 1 \text{ to } n$$

Step 5: Compute the activation output of each output unit using binary step function

$$y_{-inj} = b_j + \sum_i x_i w_i \quad \begin{cases} \text{for } i = 1 \text{ to } n \\ \text{for } j = 1 \text{ to } m \end{cases}$$

$$y_j = f(y_{-inj}) = \begin{cases} 1, & \text{if} \quad y_{-inj} > \theta \\ 0, & \text{if} \quad -\theta \le y_{inj} \le \theta \\ 1, & \text{if} \quad y_{inj} < -\theta \end{cases}$$

Step 6: The weights and bias are to be updated for $j = 1$ to m and $i = 1$ to n.

$$\text{If } y_j \neq t_j \quad \text{and} \quad x_i \neq 0, \text{then}$$
$$w_{ij(\text{new})} = w_{ij(\text{old})} + \alpha t_j x_i$$
$$b_{j(\text{new})} = b_{j(\text{old})} + \alpha t_j$$
$$\text{elseif } y_j = t_j$$
$$w_{ij(\text{new})} = w_{ij(\text{old})}$$
$$b_{j(\text{new})} = b_{j(\text{old})}$$

That is, the biases and weights remain unchanged.
Step 7: Test for stopping condition.

3 Methodology to Recognize the Hindi Characters

To recognize Hindi characters उ, प, र, and त, a 5 × 3 matrix is formed to represent these characters. For any valid point it is taken as 1 and invalid point it is taken as −1. The net has to be trained to recognize all the numbers and when the test data is given, the network has to recognize the particular numbers using perception algorithm for several output classes. Input vector x_i (for i = 0–4) has 15 numbers of neurons to formulate the numbers उ, प, र, and त. Representations of all these numbers are shown in Table 1.

The weight matrices are initially set to zero. After being trained, the final weight that produces the correct output is obtained. The net has been trained to recognize

Table 1 Representation of input data x_i

उ	[1 1 0 0 0 1 1 1 0 0 0 1 1 1 0]
प	[1 1 1 1 0 1 1 1 1 0 0 1 0 0 1]
र	[1 1 1 0 0 1 1 1 0 0 1 0 0 0 1]
म	[1 1 1 0 1 1 1 1 1 0 1 1 0 0 1]
त	[1 1 1 0 0 1 1 1 1 1 0 1 1 0 1]

these numbers by giving some test data y_i using perceptron learning rule as described in Sect. 2. The number of epochs required for the training is very less.

4 Simulation Result

To simulate the algorithm, maximum numbers of epochs required to train the net are only 50. Network is trained to recognize Hindi characters उ, प, र, म, and त. Identity function and binary step function are used as an activation function. The algorithm is simulated in MATLAB.

The results are as follows:

Test pattern of character उ is recognized as उ;
Test pattern of character प is recognized प;
Test pattern of character र is recognized as र;
Test pattern of character त is recognized as त.

See Graphs 1, 2, 3, 4 and 5.

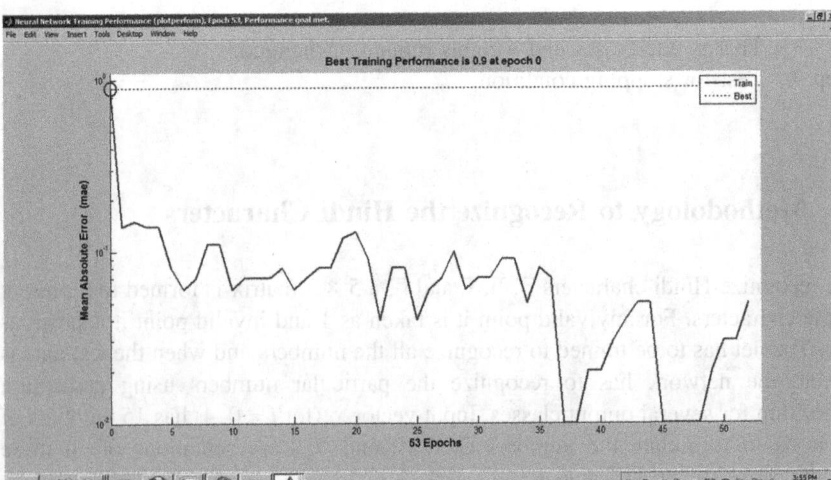

Graph 1 Training curve (epoch = 53)

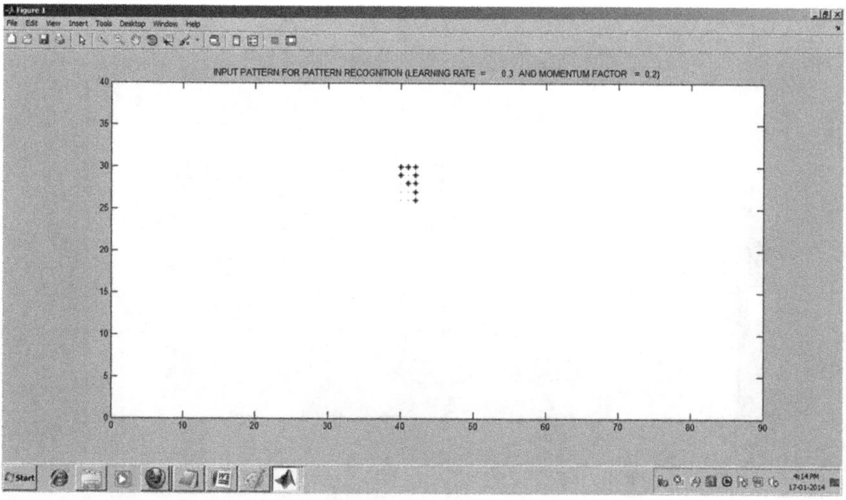

Graph 2 Input pattern for pattern recognition of प

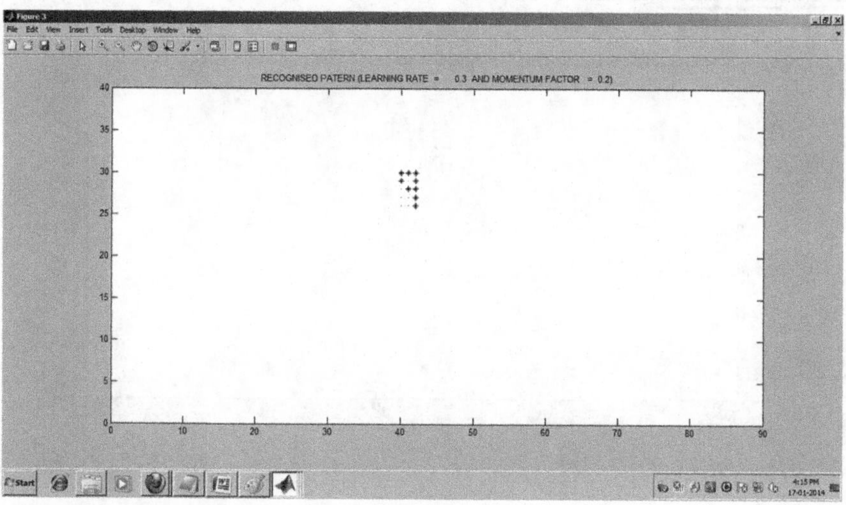

Graph 3 Recognized pattern of प

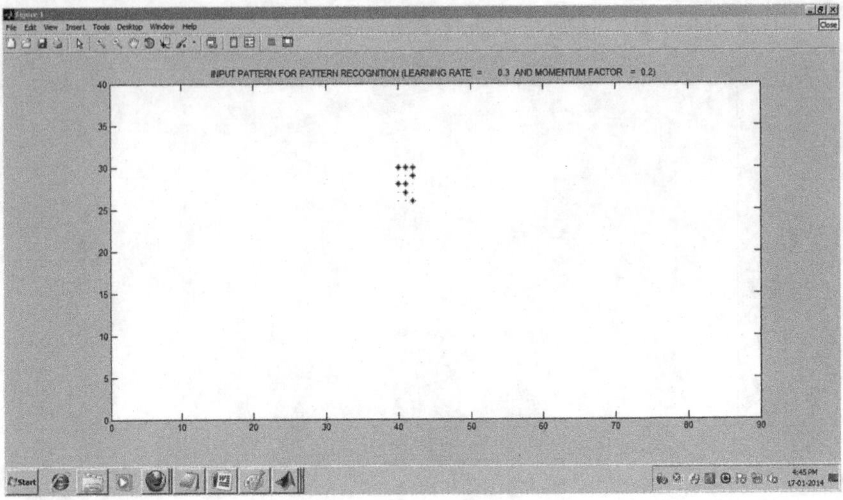

Graph 4 Input pattern for pattern recognition of र

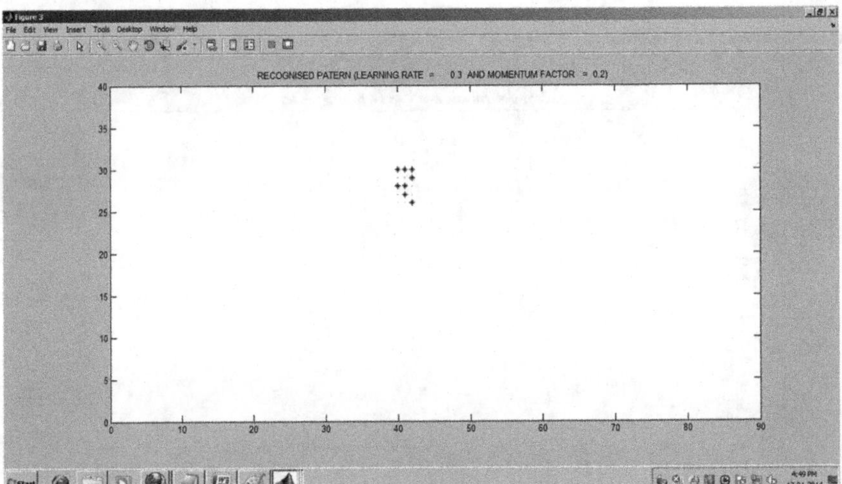

Graph 5 Recognized pattern of र

5 Conclusion

The algorithm is able to recognize all the Hindi characters उ, प, र, and त using different test data only in epoch 50. Identity function is used as an activation function for the input units and binary step function is used as an activation function for the output units.

References

1. Hirwani, A., et al.: Character recognition using multilayer perceptron. (IJCSIT) Int. J. Comput. Sci. Inf. Technol. **5**(1), 558–661 (2014)
2. Shih, F.T.: Image processing and pattern recognition fundamentals and techniques. Wiley Publications (2010)
3. El Bahi, H., Mahani, Z., Zatni, A., Saoud, S.: A robust system for printed and handwritten character recognition of images obtained by camera phone. Wseas Trans. signal Process. **11** (2015)
4. Hussain, J., Lalthlamuana: Unicode mizo character recognition system using multilayer neural network model. Int. J. Soft Comput. Eng. (IJSCE) **4**(2) (2014). ISSN: 2231-2307
5. Jayanta Kumar, B., Debnath, B., Tai-hoon, K.: Use of artificial neural network in pattern recognition. Int. J. Softw. Eng. Appl. **4**(2), 23–34 (2014)
6. Kanak, U., et al.: Handwritten character recognition system with Devanagari script. Int. J. Innov. Eng. Sci. Manage. **2**(9) (2014)
7. Vasudeva, N., Parashar, H.J., Vijendra, S.: Offline character recognition system using artificial neural network. Int. J. Mach. Learn. Comput. **2**(4) (2012)
8. Salouan, R., Safi, S., Bouikhalene, B.: Printed Arabic noisy characters recognition using the multi-layer perceptron. Int. J. Innov. Sci. Res. **9**(1), 61–69 (2014)
9. Mehta, R., Kaur, R.: Neural network classifier for isolated character recognition. Int. J. Appl. Innov. Eng. Manage. (IJAIEM) **2**(1) (2013)
10. Barve, S., Borawan, K.: Optical character recognition using artificial neural network. Int. J. Adv. Technol. Eng. Res. (IJATER) **2** (2), 139–142 (2012)
11. Kosbatwar, S.P., Pathan, S.K.: Pattern association for character recognition by back-propagation algorithm using neural network approach. Int. J. Comput. Sci. Eng. Survey (IJCSES) **3**(1) (2012)
12. Pati, S., Bhagat, H.: Character recognition system using back prorogation network. Int. J. Adv. Res. Comput. Sci. Softw. Eng. **3**(8) (2013)

Rendering Rigid Shadows Using Hybrid Algorithm

Nitin Kumar, Sugandha Agarwal and Rashmi Dubey

Abstract We present a precise and efficient hybrid algorithm for rendering rigid shadows. Our algorithm performs the combination of two major shadow rendering algorithms: shadow map and shadow volume. In our approach the first step is performed by the application of renowned shadow map algorithm which generates the shadow with aliased edges. The result generated is then passed to identify the shadow pixels. Then shadow volume algorithm is applied to generate a crisp-edged shadow of object. Shadow volume is performed only at shadow pixels to minimize the time consumption in rendering shadows. The identification of shadow pixels depends upon the hardware functionality for which a graphics processor is required. Algorithm for implementing the hybrid approach is mentioned with results in the paper.

Keywords Hybrid · Buffer · Mapping · Projection · Rendering

1 Introduction

Shadow maps and shadow volumes are two renowned means of rendering real-time shadows. Shadow maps are proficient and adaptable but they are prone to aliasing effect, whereas shadow volumes are accurate, but they posses higher fill-rate requirements and so does not scale appropriate to complex scenes.

Shadows are widely used in presenting the objects reality. The usefulness of lance Williams original shadow algorithm [1] is self-evident as Pixar's RenderMan utilizes this technique. The primary idea is apparently straightforward, yet its biggest profit

Nitin Kumar (✉) · Sugandha Agarwal · Rashmi Dubey
Computer Science Department, Amity University, Noida, Uttar Pradesh, India
e-mail: nitinaparicit@gmail.com

Sugandha Agarwal
e-mail: aga.sugandha@gmail.com

Rashmi Dubey
e-mail: rash.monu@gmail.com

M. Pant et al. (eds.), *Proceedings of Fifth International Conference on Soft Computing for Problem Solving*, Advances in Intelligent Systems and Computing 436, DOI 10.1007/978-981-10-0448-3_56

is its greatest drawback which is the discrete nature of image space computation and hence the presence of aliasing artefacts.

Rather than shadow map, which works in image space, Crow's shadow volume algorithm [2] works in object space by creating polygons to present the boundary between illuminated and shadowed regions. The shadow volume is the region covered by these polygons. A shadow query is used to verify whether a point resides within the volume.

Sen et al. [3] noticed that shadow map aliasing is just recognizable at the discontinuities between shadowed and lit regions, i.e. at the shadow shapes; on other side, shadow volumes calculate shadows accurately at each pixel but this certainty is required only at silhouettes. This analysis takes us to another technique referred to as hybrid algorithm which makes use of a slower but definite algorithm near shadow discontinuities and uses less accurate but faster algorithm at other places. In this paper we mention the hybrid algorithm for shadow rendering. The algorithm initially applies the shadow map algorithm and later applies shadow volume algorithm to generate crisp-edged rigid shadows of object.

2 Theory

A differentiation can be created between Image and Object space algorithms.

2.1 *Image Space*

The presentation of graphics in form of raster or rectangular pixels has become very popular. Raster scan displays are very flexible as they refresh the screen on the basis of the values stored in the frame buffer. Image space algorithms are very easy and accurate as they have the similar data structure as of the frame buffer. The most widely used image space method is z-buffer algorithm which is used in graphics for storing the values of z-coordinate of the displayed object.

Image space algorithms are less proper but they also give lesser constraints on scene presentation. In process, regardless of the fact that scene is presented with polygons, they can be simpler to utilize and more flexible. Image space is implemented in screen coordinate system. It determines the pixel which is visible of the object (Fig. 1).

2.2 *Object Space*

Object space algorithms provide the benefit of storing the important and related data and due to this ability the communication of the algorithm to the object is easy. The computation performed for the colour is carried only once. Object space algorithms

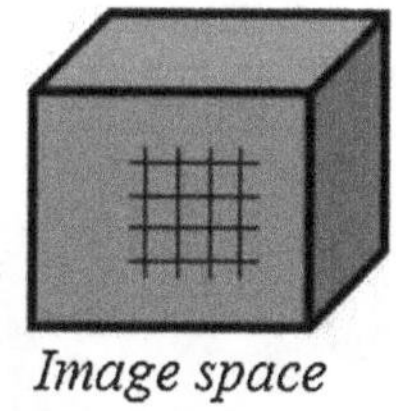

Fig. 1 Image space

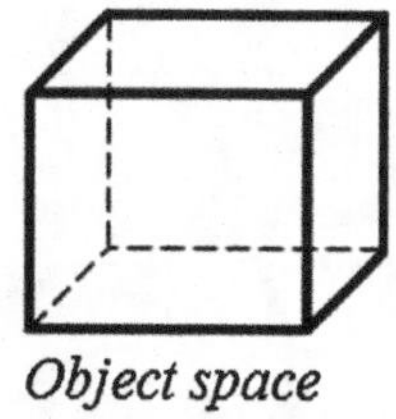

Fig. 2 Object space

also help shadow creation to increment the z-values of the 3D objects on screen. This algorithm is used in software methods as its implementation in hardware is difficult.

Object space provides more accurate shadows but in needs the accessibility to the polygonal presentation of the scene. Polygonal presentations are not present in the systems which make use of another modelling and rendering methods for example selective raycasting of objects and distance volume primitives [4]. It is presented in physical coordinate system. It determines the object part which is visible (Fig. 2).

2.3 *Soft Shadows*

Mainly, soft shadows are very less definite and blur at the edges (Fig. 3).

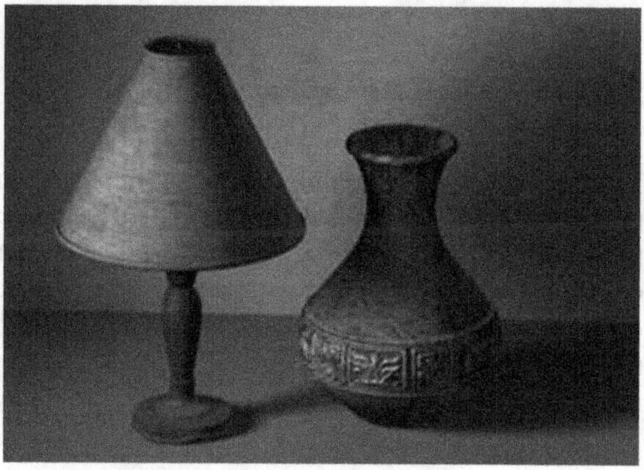

Fig. 3 Representing soft shadows of the object

Fig. 4 Representing hard shadows of object

2.4 Hard Shadows

Mainly, shadows are hard which are defined crisply and have sharp edges (Fig. 4).

3 Shadow Algorithms

Analysts have created multiple shadow algorithms over the years, few of them depend upon the classic shadow map and volume methods.

Present hard shadow techniques can be categorized into four categories:

1. Ray casting;
2. Projection;
3. Shadow volumes;
4. Shadow maps.

From these techniques, last three are used for real-time rendering.

3.1 Projection Algorithm

This method projects primitives far from the light source on the surface of other primitives. Blinn's fake shadow algorithm [5] suppresses all polygons of the object producing a shadow on the plane of other polygon. Here they can be shown in black or blended on the previously generated pixels to approximate the shadow of object on the plane. This method is basic and suits well for hardware acceleration but are not good to scale or generalize.

Normally, object space projection algorithms trim polygons into shadowed and unshadowed parts [6] in a prepass. As they work on polygon clipping, this method needs access to polygonal representation of scene.

3.2 Shadow Maps

Shadow maps were presented by Williams in 1978 [1]. This algorithm performs its operation in image space. Initially, it renders a depth map of scene from the direction of light, and depth map is then used to analyse in which samples of final image are obvious to light. Shadow maps are powerful and are backed by the graphics hardware, but they contain aliasing effect. Analysts have created many methods to address shadow map aliasing. Approaches which are based upon filtering and stochastic sampling [7] generated good antialiased shadows. As a matter of fact, this filtering requires a huge number of samples per pixel which are not only expensive as well as very crucial for real-time applications. Self-shadowing artefacts that are obtained from large filter width are totally scene dependent and are also hard to avoid. On the other hand there exist several methods to avoid aliasing by incrementing the shadow map resolution. Adaptive shadow maps [8] identify and regenerate undersampled areas at higher resolution of the shadow map. Another technique used is perspective shadow map [9] which is much easier than others. The reason behind it is that it uses an additional perspective transformation which gives the higher resolution in shadow map which resides near to the viewer. The above method is easier and works at higher speed, but on the other hand it does not eliminate the aliasing in every case. For example, Fig. 5 represents the shadows generated using the shadow map algorithm. The image presents the aliased edges of the object which is cube in the example.

Shadow map algorithm utilizes two steps to generate shadow scene. The first step is used to generate the shadow map and the other one just applies it to scene.

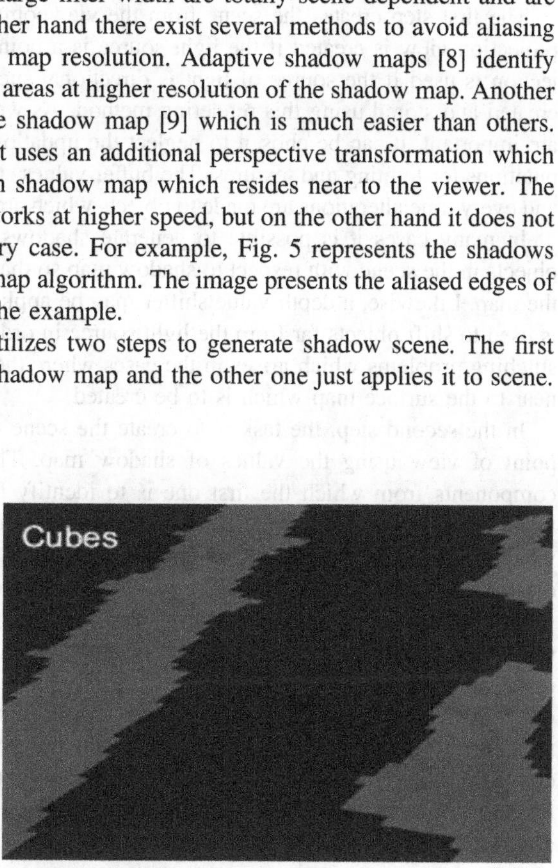

Fig. 5 Representing shadow using shadow map function

Fig. 6 Shadow map technique

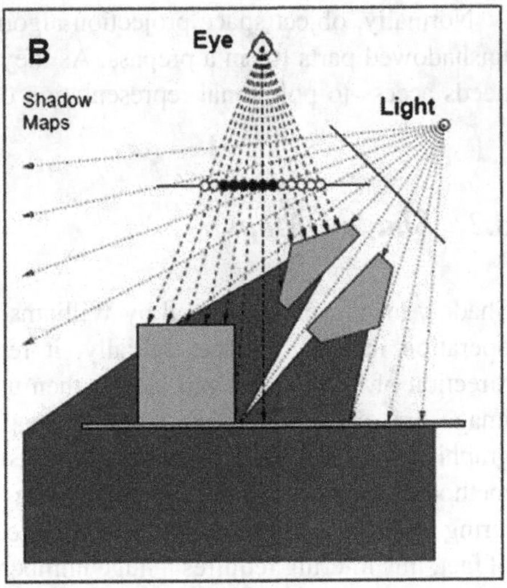

The first step creates the scene from the viewpoint of light source. Perspective projection view is created if the light source is a point, whereas orthographic projection is used if the source of light is directional such as of Sun. Depth buffer is created and stored using this rendering method. As we know that the buffer values are important, it can be chosen to neglect the updation of colour buffers and computations for lighting and textures. The buffer values stored are to be refreshed each and every time alterations are made to objects which are present in the scene (Fig. 6).

In many cases it is possible to generate shadows for only a particular set of objects in the scene with respect to shadow map so that the time taken to regenerate the map. Likewise, a depth value shifter may be applied to shadow creation which is used to shift objects far from the light source in order to provide the solution for stitching problems which arises in the cases where the value of the depth map lies near to the surface map which is to be created.

In the second step, the task is to create the scene on the basis of usual camera point of view using the values of shadow map. This task makes use of three components from which the first one is to identify the coordinates of the object which is observed from the light source. Another component is the comparison which is performed upon the coordinates in contrast to the depth map. Final component is the construction or representation of the object in shadow or light.

Testing a point of object with respect to depth map requires that its conversion must be done first from its relative position as in scene coordinates into the respective position as determined by the light. This conversion process is carried out with the help of matrix multiplication. Coordinate transformation is mainly used to identify the object position on the screen but another coordinate set is also created in order to keep the location of objects in light space. The matrix we create here to compute the world coordinates provides a value set of homogeneous coordinates which require conversion to the normalized device coordinates. It has the components (x, y, z) lying in between -1 and 1. An additional scale and bias matrix computation is being performed in implementations to convert the values from -1 to 1 into the required 0–1 values which are more regular coordinate values for the depth map.

3.3 Shadow Volumes

Virtual world of object representation is divided into two areas by the shadow volume as first, the area which lie in shadow and another which is not. This method of shadow rendering has achieved the popularity in real-time shadow generation as compared to the popular shadow mapping. Shadow volumes are used mainly because of their advantage of defining shadows accurate to the pixel. However, CPU-intensive shadow geometry creation process is required by the shadow volume. Shadow volume computation utilizes large fill time and polygons are also huge in respect of utilizing screen space as compared to shadow maps which does not carry limitations like this [10].

Shadow volume algorithm proposed by Crow [2] performs its operation in object space where enlightened and shadowed regions are represented by making polygons. These polygons capture a certain area of object which is referred as shadow volume and the evaluation of point for its presence inside the shadow volume is carried out by executing a shadow query.

Another method to accelerate the computation of volume acquired by the shadow is proposed by Heidmann [11] who uses a hardware-based stencil buffer to perform the operation of calculation. Unsuccessfully, the number of robustness issues is present in the shadow volume algorithm musing stencil buffer. Also, the algorithm proposed is unable to process the scenes which have high shadow complexity. Creating an extra geometry is involved in the process but the requirement of the large fill rate by the algorithm is the main problem. This failure is caused mainly due to two reasons. First reason is that heavy shadowed scenes which are mapped using shadow volume polygons are taking up the higher amount of screen surface, whereas the second reason is that the updations in stencil buffer are made by every shadow polygon for every rasterized pixel as the polygons in the screen space overlap.

All the techniques defined above are very useful for determining the shadows of object but they become less effective as soon as we go for the cases having heavy shadow regions. The reason for this reduced effectiveness is that the computations of pixels which exist near to the shadow are specifically those which fall in depth ranges; so in his case the pixels obtain no benefit for the optimization techniques used. Figure 7 represents the shadow of the object created using the shadow volume technique.

Lloyd et al. [12] provide another approach to compute the shadow volume which reduces the rasterization costs. The process taken by them involves the execution of query to use image space occlusions which are used for identification of blockers

Fig. 7 Shadow of the object created using the shadow volume technique

which are spread in the complete shadow, and the polygons created using shadow volume are unnecessary and can be culled. Lloyd et al. described a technique for shadow volume polygon in which screen space is made limited. This technique is performed in two ways. First, the value computations of shadow receivers are held by the depth differences obtained from the light source. Second, occlusion queries are applied to the disconnected slices obtained by the division of observer's view portion to resolve the selection of slice containing receivers. Lloyd provided a method which decreases the shadow volume polygons' number and size which eventually reduces the fill rate.

For the generation of shadow volume, a ray is fired from the light source passing from every vertex to some point from the object which is shadow casted. Volume is generated by all these fired ray projections. The determination of shadows is then performed by identifying the objects lying inside and outside of the volume; inside points are referred as shadow, whereas all the points outside the volume are light glow. In case of the modelling by polygon method, volume of the object is mainly generated by differentiating between the faces by determining whether the face is in front of the light source or lies on the back of light source. Combination of the edges which combine the front face to the back face creates the silhouette in accordance to light source. Edges which collectively generate a silhouette are expelled away from light to generate faces of shadow volume. The volume so generated must cover up the complete visible scene. In order to create a closed volume, front side and the back of this removal must be secured. On the basis of the method used to generate shadow volume, the faces of the object in the set are dependent.

Problem also exists with shadows where silhouette edges become shallow along faces. For this situation shadow created by an object over itself will be sharp, showcasing its polygonal features, while the conventional model for lighting will have a progressive change in lighting. Due to this rough shadow artefacts which lie around the silhouette edge are generated which are hard to be corrected. This problem can be minimized by incrementing the density of polygon but it will not permanently distinguish this problem. Figure 8 represents how shadow volumes are generated by the objects.

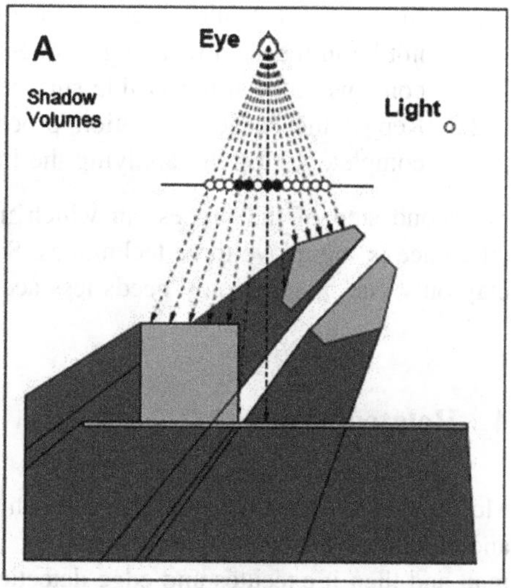

Fig. 8 Shadow volume technique

Steps involved in generating shadow volumes are as follows:

1. Search all the silhouette edge.
2. Expand all the found silhouette edge far from light source direction.
3. Generate closed volume by adding the front or back cap.

3.4 Stencil Buffer

Another buffer which is found in current graphics system along with colour and depth buffer is stencil buffer. It stores the integer value for every pixel utilizing one byte for every pixel. This buffer along with the depth buffer shares same RAM area. Usually, the stencil buffer is used for limiting the area acquired by rendering. Rendering pipeline makes use of the link between the stencil buffer and the z-buffer. Automatic increment or decrement for the stencil values assigned to every pixel can be done on the basis of the test for the pixel whether it passes or fails.

Heidmann [11] showed the concept of using the stencil buffer to generate shadows using the shadow volume algorithm for use in real-time applications fast enough to be considered. Three different techniques are defined using this concept which are depth pass, depth fail, and exclusive-or.

The process used by all these techniques is same as follows:

1. First generate the scene as it lies completely in the shadow.
2. For every light source present in the scene:

2.1 Make use of depth buffer to select the surfaces where visible surface does not lie in the shadow and generates a mask which is in stencil buffer which contains holes at the visible surface areas.
2.2 Repeat the scene generation process again considering the scene being completely glow by applying the information in stencil buffer.

Second step of the process in which mask generation is performed creates a difference in the above three techniques. Some takes only two iterations and some may only one or some only needs less accuracy in stencil buffer.

4 Related Work

McCool [13] first defines the algorithm which uses the combination of shadow map and volumes referred as hybrid algorithm. The algorithm in first generates the depth map and then it executes and edge detection technique to search for the silhouette edges. Then the algorithm regenerates the shadow volumes using the edges determined and then makes use of them to calculate the shadows in the image obtained. The benefit of McCool algorithm is that it only generates the polygons using shadow volume technique for silhouette edges that can be seen from the light source. But it has the problem of requirement to read the depth buffer again which is expensive and the polygons generated are completely rasterized. Also, the aliasing effect can generate artefacts in shadow volume generation.

Another hybrid approach is proposed by the Govindaraju et al. [14]. Their approach uses the technique of bounding the number of computation required in order to generate the precise shadow silhouette. Their approach minimizes the object processing performed by clipping algorithm. Their clipping algorithm uses the software method for performing the clipping of polygons.

5 Hybrid Algorithm

In generating our algorithm, we assume that the objects which block the light from light source to fall on the surface are polygonal, closed, and manifold. These properties help in ensuring that the implementation of the shadow volume algorithm is powerful [14].

In our algorithm, initially we apply shadow map algorithm which generates shadow of the object with aliased edges and consumes less time as compared to other shadow rendering algorithms. The result image obtained from this step is processed to identify boundary or edge shadow pixels which are aliased in order to apply the shadow volume algorithm which generates shadow with crisp edges. The assumption taken here is that the shadow boundary pixels represent a smaller part of the aggregate number of shadow polygon pixels.

Algorithm steps in detail are as follows:

1. Generating shadow map
   In generating shadow map, the light source is kept opposite to the viewing position. The shadow map algorithm is then applied which stores the depth values in buffer. The shadow map generated contains the shadow of the object with aliased edges. The shadow map is generated with small resolution as the output of this step is used to identify the shadow boundary. There lies a problem of more aliasing as the resolution is kept less but can overcome as the overall percentage is less.
2. Detecting boundary pixels of shadow
   Viewing position of the scene is kept in accordance to the viewer's direction. The technique we are using here for detecting the boundary pixels of shadow is defined by Sen [3]. All the fragments are mapped into the light space and are compared for their depth across their adjacent pixels. If the result of the comparison is found to be false, the pixel is taken as a boundary pixel. Else the pixel is not a boundary pixel.
   Detection and reduction of the boundary pixels is performed because it restricts the value of stencil fill-rate absorbed while applying shadow volume. For example, pixels lie on the back of the object with regard to light source reside in shadow and hence they are not considered as the boundary pixels. To retain the values of adjacent depth pixels in the depth buffer, z-buffering is applied. The depth buffer is retained to perform the shadow volume in next step.
3. Applying shadow volume
   Shadow volume technique here is based on stencil. This algorithm works on the basis of increasing or decreasing the values in stencil buffer depending upon the shadow volume polygon pixel depth test of either they pass or fail. In our approach we are using the method given by Everitt [15] which is depth-fail method. The main difference possessed by our approach is that we convert shadow polygon pixels and boundary pixel addresses in the frame buffer are referenced to update the stencil buffer.
   After the completion of algorithm the stencil buffer holds the nonzero values for the pixels which reside in shadow. Zero value is held for the pixels which are either not boundary or not shadowed.
4. Calculating final shadow
   The final shadow of the scene is created by drawing the scene for the updated values in stencil buffer to cast crisp shadow at the edges.

5.1 Results

Our algorithm is executed on the system with configuration of 2.3 GHz Core i5 with ATI Radeon HD 6490 M with 4 GB RAM. The resolution of the system is 1024 × 768 (Fig. 9).

5.2 Performance

Bar graph in Fig. 10 represents the comparison between the shadow map, shadow volume, and hybrid algorithm. The comparison is done on the time elapsed to generate one frame using the above algorithms. Time is considered in milliseconds.

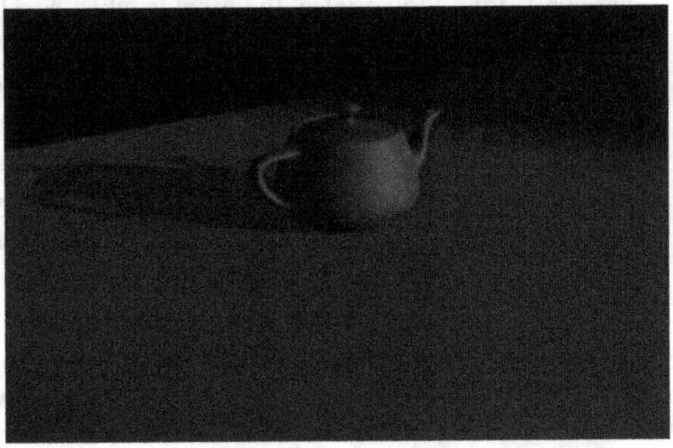

Fig. 9 Hybrid algorithm result

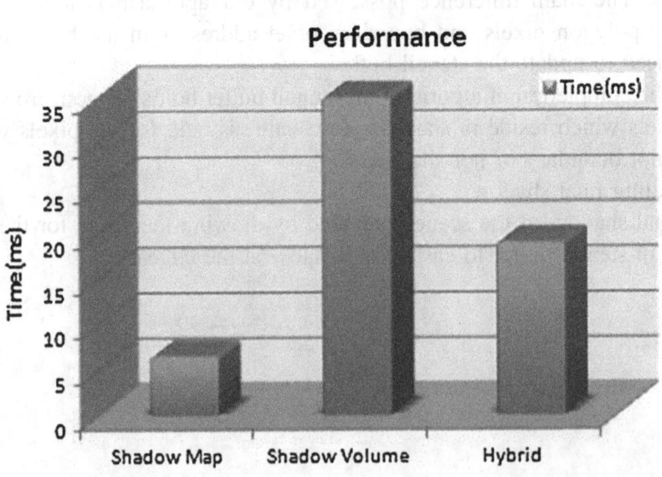

Fig. 10 Performance comparison

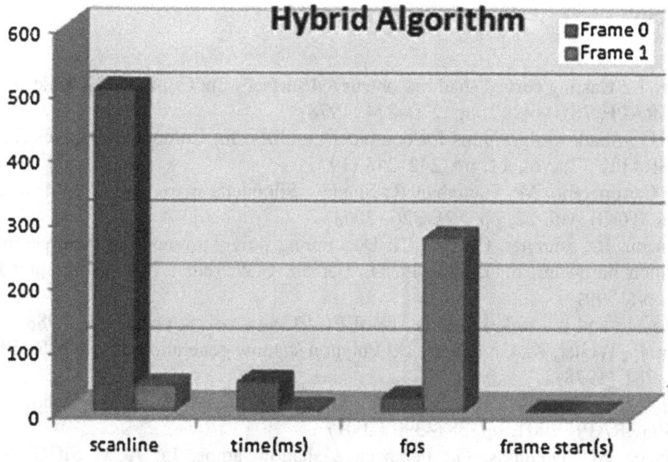

Fig. 11 Bar graph comparing the properties of two frames

The evaluation is performed on considering the rasterization and stencil updates used in the shadow volume algorithm. Both the algorithms, shadow volume and our approach hybrid, depend upon it.

Above bar graph represents the time taken by the different algorithms to render shadow of the object in the scene. It can be seen easily that the shadow map consumes least time to render and shadow volume the most. But our approach consumes the average time and generates crisp-edged shadows. Figure 11 represents the time taken by the different objects in rendering the scene using our algorithm.

6 Conclusion

Our hybrid algorithms combine two shadow rendering algorithms to create an efficient one. The main concept we have used in our algorithm is to generate the scene initially with the shadow map algorithm, and then determine the boundary pixels and at the end apply the shadow volume algorithm to generate the crisp-edged shadow.

The approach we have used decomposed the time-taking process into two parts. The first part deals with generating the shadow using the faster method but it holds the problem of aliasing in the result. The second part deals with using the slower shadow rendering method by minimizing the area to be computed and hence the overall time to compute a scene decreases.

References

1. Williams, L.: Casting curved shadows on curved surfaces. In: Computer Graphic (Proceedings of SIGGRAPH 78), vol. 12, pp. 270–274 (1978)
2. Crow, F.C.: Shadow algorithms for computer graphics. In: Computer Graphics (Proceedings of SIGGRAPH 77), vol. 11, pp. 242–248 (1977)
3. Sen, P., Cammarano, M., Hanrahan P.: Shadow Silhouette maps. In: ACM Transactions on Graphics (TOG), vol. 22, pp. 521–526 (2003)
4. Westermann, R., Sommer, O., Ertl, T.: Decoupling polygon rendering from geometry using rasterization hardware. In: Lischinski, D., Larson, G.W. (eds.) 10th Eurography Rendering Workshop (1999)
5. Blinn, J.: Me and my (fake) shadow. In: IEEE CG&A, vol. 8, pp. 82–86 (1988)
6. Atherton, P., Weiler, K., Greenberg, D.: Polygon shadow generation. In: SIGGRAPH, vol. 78, pp. 275–281 (1978)
7. Fernando, R., Fernandez, S., Bala, K., Greenberg, D.P.: Adaptive shadow maps. In: ACM SIGGRAPH 2001, pp. 387–390 (2001)
8. Stamminger, M., Drettakis, G.: Perspective shadow amps. In: ACM SIGGRAPH 2002, pp. 557–562 (2002)
9. Heidmann, T.: Real shadows, real time. In: Iris Universe, vol. 18, pp. 23–31, Nov 1991
10. Lokovic, T., Veach, E.: Deep shadow maps. In: ACM SIGGRAPH 2000, pp. 385–392 (2000)
11. Lloyd, B., Wendt, J., Govindaraju, N., Manocha, D.: CC shadow volumes. In: Proceedings of the Eurographics Symposium on rendering (2004)
12. McCool, M.D.: Shadow volume reconstruction from depth maps. In: ACM Transaction on Graphics (TOG), vol. 19, pp. 1–26 (2000)
13. Govindaraju, N.K., Lloyd, B., Yoon, S., Sud, A., Manocha, D.: Interactive shadow generation in complex environments. In: ACG Transactions on Graphics (TOG), vol. 22, pp. 501–510 (2003)
14. Everitt, C., Kilgard, M.J.: Practical and robust stencilled shadow volumes for hardware-accelerated rendering (2003)

Analysis of Role-Based Access Control in Software-Defined Networking

Priyanka Kamboj and Gaurav Raj

Abstract Lack of interoperability of traditional networking architecture reduces the network's speed, reliability, and security. Software-defined networking has decoupled control plane and data plane in order to configure the existing architecture. The OpenFlow protocol used for communication in software-defined network has been discussed in this paper [1, 2]. The main aim of the research paper is to implement role-based control model in the software-defined networking environment in order to provide more security [3, 4]. The paper also outlines about the reduction in packet loss and latency using it in software-defined network environment [5, 6].

Keywords Software-defined networking (SDN) · Open flow · Controller · Flow table · Security · Role-based access control (RBAC) · Open networking foundation (ONF) · Round-trip time (RTT)

1 Introduction

SDN is now the "next new thing" which is gaining a lot of attention and popularity for the researchers. SDN was the first commercialized work done at Berkeley and Stanford University [1, 7]. SDN has separated the control and data plane which were tightly coupled in traditional method of networking. The SDN decouples logic of traffic routing, the logic layer is replaced by a virtualized controller [8, 9]. It offers greater flexibility and efficiency and it allows to scale with the demand [10–12].

The SDN is a policy-based network approach which centrally enforces changes and a new policy is added to all the devices on the network while it also provides a tracking and auditing mechanism [13, 2, 6]. The policy-based network reduces overall

Priyanka Kamboj (✉) · Gaurav Raj
Amity School of Engineering & Technology, Amity University, Noida, India
e-mail: prinkskamboj12@gmail.com

M. Pant et al. (eds.), *Proceedings of Fifth International Conference on Soft Computing for Problem Solving*, Advances in Intelligent Systems and Computing 436, DOI 10.1007/978-981-10-0448-3_57

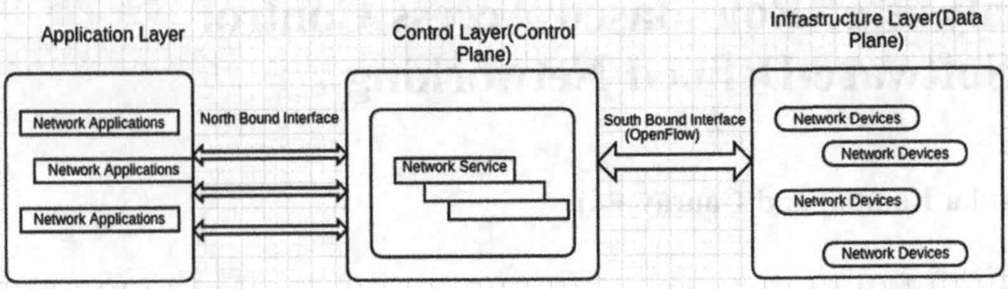

Fig. 1 Software defined networking architecture

operational expenditure of change and reduces the compliance costs of business [1, 7, 14]. The SDN is helping in providing more bandwidth, processing power, reliability, scalability, and easy programmability of the network [5]. So it is gaining a lot of attention in adopting SDN in a real-world scenario. The SDN definition was given by Open Networking Foundation (ONF) [15–18]. SDN is a three-layer architecture. The various components of the SDN have been described in the Fig. 1 [1, 7, 19, 20].

2 Role-Based Access Control

2.1 Introduction

It is a method which regulates access to resources in the network on the basis of roles of the individual users in an enterprise [3]. Roles are defined according to the responsibility and authority of the job within the enterprise. It allows users to carry out various authorized tasks by changing actions with respect to functions, relationships. In conventional access control, user access is granted or revoked on object-by-object basis [21]. In RBAC, as the needs of the enterprise emerge the new roles can be created and existing roles can be changed and discarded without the need of revising the privileges of individual user [3, 22].

It is a system which controls the user's access to resources based on the roles assigned to them. Each role is grouped with certain access rights and the user who has been entitled by the associated role can only access the resource [3, 22].

2.2 Role-Based Access Control Model

The roles are associated with the concept of user groups in access control. A role places the set of users on one side and the set of permissions are brought together on the other side. The sets of users are known as user groups. The basic concept of RBAC was introduced with the advent of multiuser and multi-application systems in 1970s [3, 21, 22].

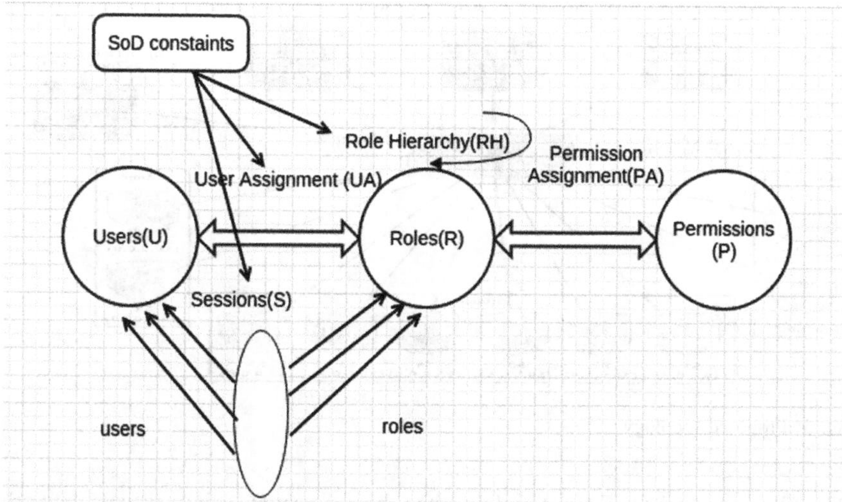

Fig. 2 Role-based access control model

RBAC model comprises of four things: users, roles, permissions, and sessions. In this model, the users (U) are employees in the organization. A role (R) is an activity or the duty which is to be performed by the users. Permissions (P) define the acceptance in order to approach a model consisting of different objects. In the session (S) the user is approaching the system with the triggered roles which are given to them [23, 21] (Fig. 2).

3 Experimental Setup

3.1 Setup of the Experiment

The already inbuilt virtual machine image is available for the Mininet software version 2.2.0.1. This image was downloaded and made to run on a Ubuntu 14.04 LTS version with VMware Workstation (version 11.0). The virtual machine was allotted 2 gigabytes of RAM and two Intel Core 5 processors. The network design was created keeping in mind similarly like an organization network [24, 25]. In order to visualize the topology created a graphical user interface, i.e., visual network description or VND is used [26, 27].

3.2 Proposed Scenario

The topology which is used for the experiment is named as "Single." In which there is a single OpenFlow switch connected to the 10 hosts and 3 resources and one

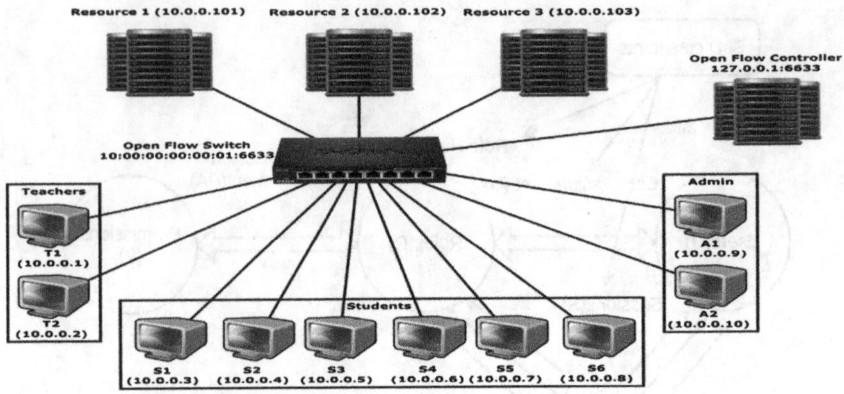

Fig. 3 Proposed topology

OpenFlow controller. This topology has been created using visual network description or VND and the custom script is written using the command line in the Mininet. Now, out of the 10 hosts, first 2 hosts are assumed to be teachers, next 6 hosts to be students, and last 2 to be admin. There are 3 resources, i.e., resource 1, resource 2, and resource 3 (Fig. 3).

A role-based access control named as RBAC module has been created which is being installed in the remote controller named as "POX." This RBAC module consists of three main files, Roles.XML, Resources.XML, and Access.XML (Fig. 4).

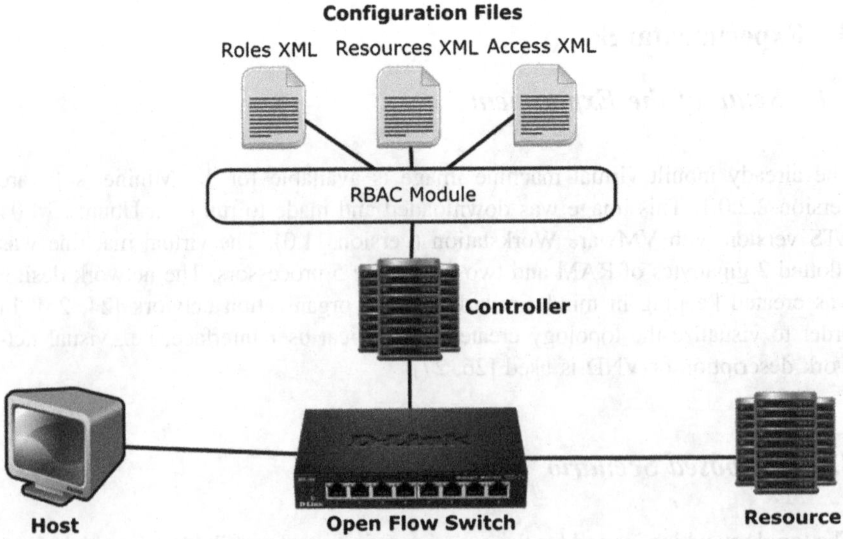

Fig. 4 Packet forwarding mechanism using RBAC

But after applying RBAC module in the OpenFlow controller whenever a host sends a request packet for a resource to the switch,if there is flow entry for the request packet in the flow table, then switch performs the action mentioned in flow table. But if there is no entry in the flow table, the switch informs the controller about the request packet. Now the RBAC module is implemented on the OpenFlow controller, so it will traverse through all the three XML files, resources.XML, roles. XML, and access.XML.

4 Experimental Results and Discussion

This section discusses about the implementation of the role-based access control in SDN in the proposed topology which is already discussed in Sect. 3.2. The RBAC module containing the three XML files is stored in the cache of the OpenFlow controller. Compared to the traditional network it is applied on each individual resource. This section will illustrate about permission Access list, running the topology on the remote POX controller.

4.1 Permission Access List

In order to apply RBAC in OpenFlow controller, the roles such as admin, i.e., A1 and A2, teacher, i.e., T1, T2, and student, i.e., S1, S2, S3, S4, S5, S6 have the permission to access which resource out of the three resources resource 1, resource 2, and resource 3 is illustrated in the table. Here, tick sign "0" depicts that the role have the permission "allow" to access the resource and the entries which do not have any value have the permission "deny" to access the resource (Table 1).

Table 1 Permission access list

	Resource 1	Resource 2	Resource 3
Teacher 1 (T1)	✓	✓	
Teacher 2 (T2)	✓	✓	
Student (S1)		✓	✓
Student (S2)		✓	✓
Student (S3)		✓	✓
Student (S4)		✓	✓
Student (S5)		✓	✓
Student (S6)		✓	✓
Admin (A1)	✓	✓	✓
Admin (A2)	✓	✓	✓

4.2 Algorithm of the Proposed Role-Based Access Control Model

Algorithm 1: Load Access file

Step 1: Parse XML file containing Access rules.
Step 2: Set resources := get elements by tag name "resource"
Step 3: For each resource in resources:
Step 4: Set name := get name of the resource.
Step 5: Set acl[name] := empty dictionary
Step 6: Set roles := from resource get elements by tag name "role"
Step 7: For each role in roles:
Step 8: Set role_name := From role get name of the role
Step 9: Set acl[name][role_name] := Access level of role on the resource.

Algorithm 2: Load Roles file

Step 1: Parse XML file containing roles of hosts.
Step 2: Set hosts := get elements by tag name "host"
Step 3: For each host in hosts:
Step 4: Set mac := get mac of the host.
Step 5: Set hostname := From host get name
Step 6: Set role := from host get the role of the host
Step 7: Set roles[mac]['hostname'] := hostname
Step 8: Set roles[mac]['role'] := role

Algorithm 3: Load Resources file

Step 1: Parse XML file containing resource present in the network.
Step 2: Set resourcexml := get elements by tag name "resource"
Step 3: For each resource in resourcexml:
Step 4: Set hostname := get name of the resource.
Step 5: Set mac := From resource get the mac address
Step 6: Set resources[mac] := hostname

Algorithm 4: _handle_PacketIn(self, event)

Step 1: Set packet := Parsed packet associated with the PacketIn event
Step 2: If checkPacketForRbac(packet) returns true, Then:
Step 3: If perms is set to 'deny', Then:
Step 4: Drop the packet.
Step 5: Return the control flow
Step 6: Set packetIn := OpenFlow packet associated with the event
Step 7: Set dpid := Dpid of the switch from which the event was fired
Step 8: Set match := Matching packets using ofp_match.from_packet
Step 9: Set match := Matching packets using ofp_match.from_packet

Step 10: Add mapping for MAC address using addMacToPort method
Step 11: If match has network protocol set as ARP request, Then:
Step 12: Process ARP request using process_arp_request method
Step 13: Else if match has network protocol set as ARP reply, Then:
Step 14: Process ARP reply using process_arp_response method.

Algorithm 5: checkPacketForRbac(self, packet)

Step 1: Set src_mac := MAC address of source host present in the packet
Step 2: Set dst_mac := MAC address of destination host present in the packet.
Step 3: If isResource(dst_mac) returns true, Then:
Step 4: Set host := Get host name using src_mac
Step 5: Set rol := Get role using src_mac
Step 6: If host is False or role is False, Then:
Step 7: Return false
Step 8: Set resource := Get resource name from dst_mac
Step 9: Set perms := Get permission using resource and role
Step 10: Return true
Step 11: Else:
Step 12: *Return false*

4.3 Analysis of the Results

The simulation of the experiment was performed for 100 instances by randomly picking the requestor and the resource. In each use case the no. of packets transmitted, no. of packets received, packet loss (in %), round-trip time (in ms) such as maximum, minimum, average, and mean deviation and the permission of either allow or deny access to the requestor of the resource were recorded and stored in the form of a table.

Figures 5 and 6 illustrate the round-trip time (in ms), i.e., minimum, average, and mean deviation taken by the role "admin" to access the resources, Resource 1, Resource 2, and Resource 3, respectively, as admin is having the permission of "allow" to access all the resources.

Figures 7 and 8 illustrate the round-trip time (in ms), i.e., minimum, average, and mean deviation taken by the role "teacher" to access the resources, Resource 1, Resource 2, respectively, as teacher is having the access permission of "deny" for the Resource 3.

Figures 9 and 10 illustrate the round-trip time (in ms), i.e., minimum, average, and mean deviation taken by the role "student" to access the resources, Resource 2, Resource 3, respectively, as student is having the access permission of "deny" for the Resource 1.

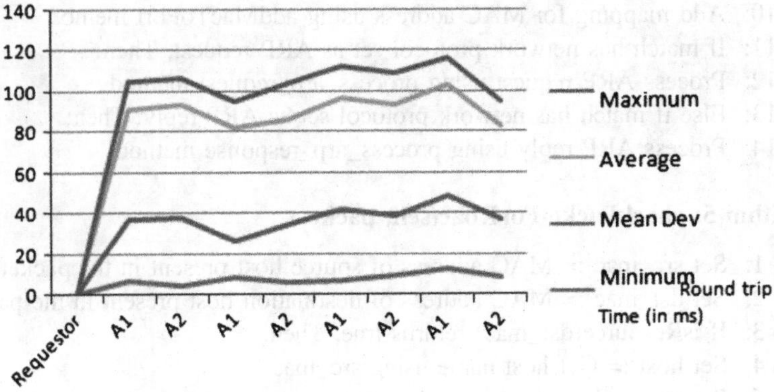

Fig. 5 Rtt for requestor admin to access Resource1

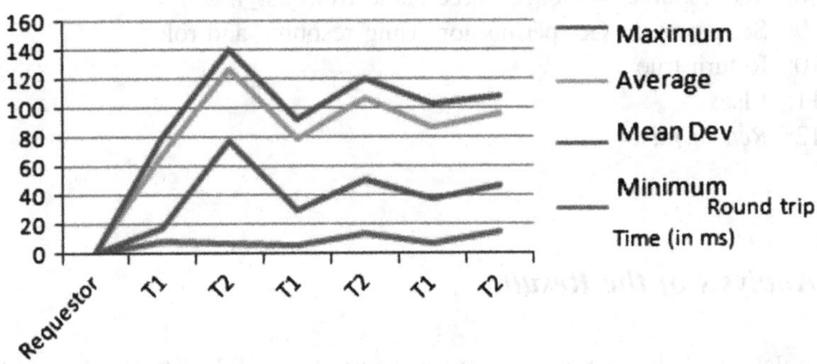

Fig. 6 Rtt for requestor admin to access Resource2

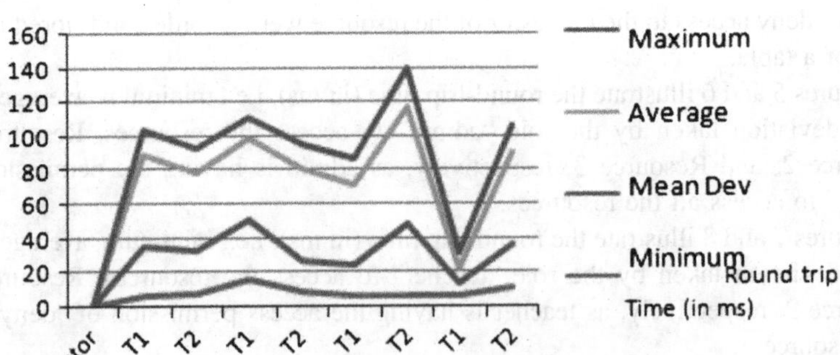

Fig. 7 Rtt for requestor teacher to access Resource1

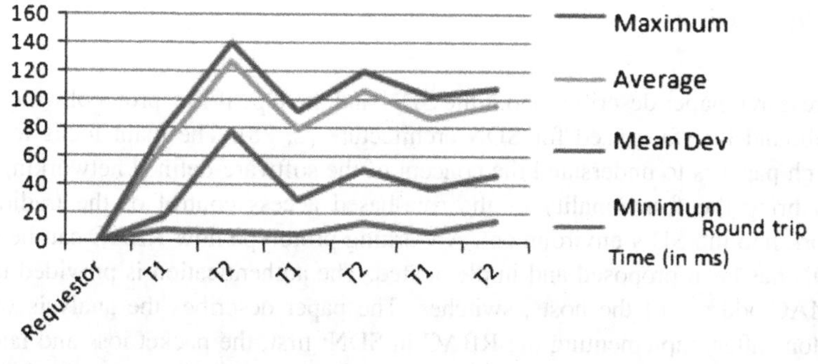

Fig. 8 Rtt for requestor teacher to access Resource2

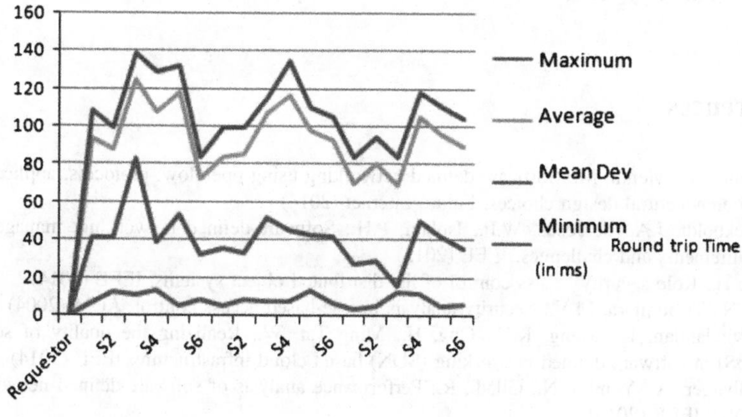

Fig. 9 Rtt for requestor student to access Resource2

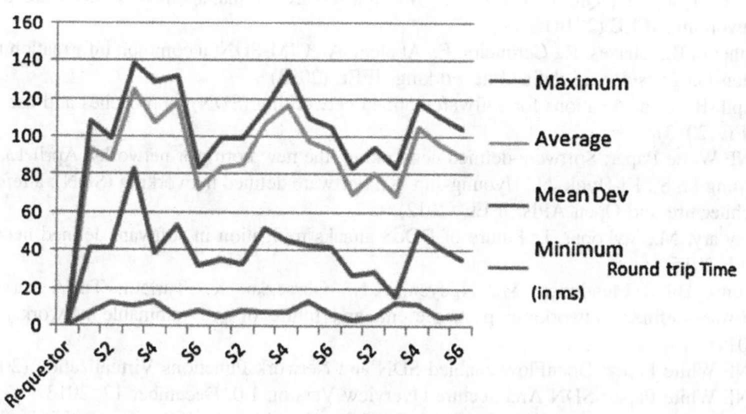

Fig. 10 Rtt for requestor student to access Resource3

5 Conclusion

This research paper describes about the SDN and the OpenFlow protocol which is a Southbound Interface used for SDN architecture [2, 28]. The main focus for this research paper is to understand the concept of the software-defined networking and try to bring the functionality of the role-based access control of the traditional network into the SDN environment. A working prototype how RBAC can be used in SDN has been proposed and implemented. The authentication is provided using the MAC address of the hosts, switches. The paper describes the analysis which was done after implementing the RBAC in SDN; first, the packet loss and latency have been reduced in SDN. Second, the controller in the SDN is centralized due to which the admin has not to monitor individual switches, resources, or the hosts in the network, they are now centrally controlled by a single entity, i.e., controller.

References

1. Braun, W., Menth, M.: Software-defined networking using openflow: protocols, applications and architectural design choices. Future Internet (2014)
2. Wickboldt, J.A., de Jesus, W.P., Isolani, P.H.: Software-defined networking: management requirements and challenges. IEEE (2015)
3. Xu, H.: Role security access control of the distributed object systems. IEEE (2014)
4. Li, N., Tripunitara, M.V.: Security analysis in role-based access control. ACM (2004)
5. Govindarajan, K., Meng, K.C., Ong, H., Ming Tat, W.: Realizing the quality of service (QoS) in software-defined networking (SDN) based cloud infrastructure. IEEE (2014)
6. Gelberger, A., Yemini, N., Giladi, R.: Performance analysis of software-defined networking (SDN). IEEE (2013)
7. Kreutz, D., Member, Ramos, F.M.V., Verissimo, P.: Software-defined networking: a comprehensive survey (2014)
8. Li, Y., Dong, L., Qu, J., Zhang, H.: Multiple controller management in software defined networking. IEEE (2014)
9. Pinheiro, B., Chaves, R., Cerqueira, E., Abelem, A.: CIM-SDN: a common information model extension for software-defined networking. IEEE (2013)
10. Kapil, B.: Considerations for software defined networking (SDN): approaches and use cases. IEEE (2013)
11. ONF White Paper: Software-defined networking: the new norm for networks, April 13, 2012
12. Myung-ki, S., Ki-Hyuk, N., Hyoung-Jun, K.: Software defined networking (SDN): a reference architecture and Open APIs. IEEE (2012)
13. Vizv´ary, M., Vykopal, J.: Future of DDoS attacks mitigation in software defined networks. IEEE 2015
14. Astuto, B.N., Mendonça, M., Nguyen, X.N., Obraczka, K., Turletti, T.: A survey of software-defined networking: past, present, and future of programmable networks. IEEE (2014)
15. ONF White Paper: OpenFlow-enabled SDN and Network Functions Virtualization (2014)
16. ONF White Paper: SDN Architecture Overview Version 1.0, December 12, 2013
17. McKeown, N., Anderson Hari Balakrishnan, T.Y.: OpenFlow: enabling innovation in campus networks, March 14 2008
18. Open Networking Foundation: https://www.opennetworking.org/

19. Nakayama, H., Mori, T., Ueno, S.: An implementation model and solutions for stepwise introduction of SDN. IEICE—Asia-Pacific Network Operation and Management Symposium (APNOMS) (2014)
20. Kevin Benton, L., Camp, J., Small, C.: OpenFlow vulnerability assessment
21. Sandhu, R.S., Coyne, E.J., Feinstein, H.L., Youman, C.E.: Role-based access control models. IEEE Comput. **29**(2), 38–47 (1996)
22. Ferraiolo, D.F., Richard Kuhn, D.: Role-based access controls. In: 15th National Computer Security Conference, pp. 554–563. Baltimore MD (1992)
23. Sasaki, T., Hatano, Y., Sonoda, K.: Load distribution of an openflow controller for role-based network access control. IEICE (2013)
24. Mininet Walkthrough: http://mininet.org/walkthrough/
25. Mininet Walkthrough: https://github.com/mininet/mininet/wiki/Documentatin
26. POX controller APIS: https://openflow.stanford.edu/display/ONL/POX+Wiki#POXWiki-POXAPIs
27. Visual Network Description: http://www.ramonfontes.com/vnd/#app=ac1d&c52eselectedIndex=0&630e-selectedIndex=1
28. Zerrik, S., Bakhouya, M., Gaber, J.: Towards a decentralized and adaptive software-defined networking architecture. IEEE (2014)

Analysis and Comparison of Regularization Techniques for Image Deblurring

Deepa Saini, Manoj Purohit, Manvendra Singh, Sudhir Khare and Brajesh Kumar Kaushik

Abstract Image deblurring or deconvolution problems are referred as inverse problems which are usually ill-posed and are quite difficult to solve. These problems can be optimized by the use of some advanced statistical methods, i.e., regularizers. There is, however, a lack of comparisons between the advanced techniques developed so far in order to optimize the results. This paper focuses on the comparison of two algorithms, i.e., augmented Lagrangian method for total variation regularization (ALTV) and primal-dual projected gradient (PDPG) algorithm for Beltrami regularization. It is shown that primal-dual projected gradient Beltrami regularization technique is better in terms of superior image quality generation while taking relatively higher execution time.

Keywords Augmented lagrangian · Beltrami regularization · Projected gradient restoration · Total variation

1 Introduction

Currently, image processing is a widespread area of research having applications in transmission and encoding, medical, remote sensing, satellite imagery and defense. Although the present image capturing systems are capable of providing high quality and magnification, substantial problems still appear in the captured images. These images may contain some form of blurring and degradation caused by the nature of optical device, camera or object motion, and atmospheric factors. In this context, image restoration techniques attempt to recover a high-quality image from the

Deepa Saini · B.K. Kaushik
Electronics and Communication Engineering Department,
Indian Institute of Technology Roorkee, Roorkee, Uttarakhand, India
e-mail: deepa.saini90@gmail.com

Manoj Purohit (✉) · Manvendra Singh · Sudhir Khare
Instruments Research and Development Establishment, Dehradun 248008, India
e-mail: manoj_irde@yahoo.co.in

M. Pant et al. (eds.), *Proceedings of Fifth International Conference on Soft Computing for Problem Solving*, Advances in Intelligent Systems and Computing 436, DOI 10.1007/978-981-10-0448-3_58

low-quality image employing one or more algorithms. Traditionally, significant efforts have been attempted for the restoration of blurred images. Some previous approaches to image restoration include, least squares filtering [1], Wiener filtering [2], neural network approach [3], Lucy Richardson deconvolution [4], and adaptive sparse domain selection [5]. Most of the methods, however, restore the image when a prior information about the point spread function (PSF) is known. In contrary, if the PSF is unknown, it is estimated by utilizing an iterative approach either in spatial or in frequency domain [6].

Generally, the image restoration problems are inverse problems which correspond to the deconvolution of the degraded image and the PSF to achieve a good-quality image. In this scenario, the challenging problem is that the original image and PSF both are unknown and need to be determined. This problem, hence, admits many or possibly an infinite number of solutions and identified as an ill-posed problem. Mathematically, an ill-posed problem is one that has the following properties: (i) The solution may not exist (ii) The solution is not unique (one-to-many) (iii) The solution's behavior does not change continuously with the initial conditions (unstable). Accordingly, some advanced statistical techniques have been introduced in image restoration. In order to obtain the solution for such problems, the statistical methods provide a principled means to estimate the uncertain or missing information. In this approach, regularizers minimize an appropriately chosen objective function, i.e., measure of the restoration error. Consequently, a constrained ill-posed problem renders an unconstrained optimization problem. Previously, some standard regularization techniques included truncated singular value decomposition (TSVD) [7], Tikhonov regularization [8], and total variation (TV) regularization [9–11]. Among these techniques, TV regularization is the most efficient way for image restoration. Unlike other techniques that tend to smoothen the edges without retaining important information, the TV regularization preserves the edge information of the restored image. According to the TV minimization technique, a problem having infinite number of solutions has high total variation. In this context, the reduction of total variation yields a close match to the solution by preserving important details and removing unwanted solutions. Thus, the regularization is obtained by minimizing the quadratic energy. However, significant problems arise in the TV regularization approach as it is restricted by non-differentiability of the penalty term and the staircasing effect.

Alternatively, there is a different line of research that started with Beltrami regularization, which is considered to be midway solution between staircasing and edge preserving. In general, this approach is based on a geometric methodology of image representation using gradient descent equations, proposed by Sochen et al. [12]. The elucidation for this method is twofold: first, the establishment of images as a Riemannian manifold and second, the metric tensor to measure image regularity. Subsequently, the metric tensor controls and stabilizes the nonlinear diffusion process by alternating the flow of forward and backward diffusions [13]. In this approach, the regularization is performed by minimizing the surface area of the image. However, the primary difficulty in this method is the optimization of the Beltrami regularizers. These problems at hand lead us to the improved algorithms for both regularizers. Taking into account the above problems of both regularization schemes, this paper emphasizes on the comparison of the recent advancement of two innovative algorithms, i.e., augmented Lagrangian method for total variation regularization (ALTV) and primal-dual projected gradient (PDPG) algorithm for Beltrami regularization.

This paper is organized in six sections, including the current introductory Sect. 1. In Sect. 2, problem of image degradation is formulated. Further, Sect. 3 evaluates the augmented Lagrangian method for total variation regularization whereas Sect. 4 concisely describes the primal-dual projected gradient algorithm for Beltrami regularization. In Sect. 5, both the algorithms are implemented and compared in regard to their performance. Finally, the work is concluded in Sect. 6.

2 Problem Formulation

In the spatial domain, the standard formulation of an imaging system can be expressed as [14]

$$I_d(x,y) = h * I(x,y) + n(x,y) \tag{1}$$

where $I(x,y) \in \mathbb{R}^{mn\times 1}$ ensures a vector representing the ideal image of size $m \times n$, $I_d(x,y) \in \mathbb{R}^{mn\times 1}$ corresponds to the observed degraded and blurred image, $n(x,y) \in \mathbb{R}^{mn\times 1}$ represents a vector denoting the noise, h is PSF or blur kernel and $*$ denotes the convolution. To deal with the spectral domain, (1) is intuitively defined as

$$I_d(u,v) = H(u,v)I(u,v) + n(u,v) \tag{2}$$

where $H \in \mathbb{R}^{mn\times mn}$ is a linear transformation matrix representing the convolution operation popularly known as the optical transfer function (OTF). The objective of image restoration is to recover the original image (I) from the degraded image (I_d) (Fig. 1).

Deconvolution refers to the case where the blur to be removed is linear and shift invariant. In particular, the deconvolution of the image I corresponds to the Fourier division of the degraded image to the PSF. In this context, if there is little or no prior information of PSF, there is no guarantee that it will result in a stable and unique solution. This results in an ill-posed problem and regularizers can solve this problem by replicating some a priori properties.

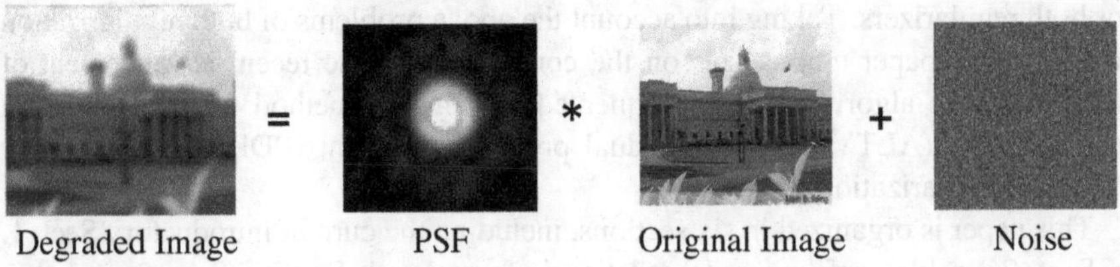

Fig. 1 Data model for the imaging system

3 Augmented Lagrangian Method for Total Variation (ALTV)

The ALTV algorithm is an advanced technique of total variation regularization introduced by Chan et al. [14]. The modification in this algorithm tends to update the half quadratic penalty parameter ρ_r to achieve faster convergence. The simplified block diagram of this approach is shown in Fig. 2. In this technique, the optimization is done by transforming the original unconstrained minimization problem to the constrained minimization. Accordingly, the augmented Lagrangian method is followed to find the solutions and the algorithm is formulated in terms of operator splitting iterations. The key idea of operator splitting method is to establish an auxiliary variable to iteratively revise the original function so that it can be minimized using simple steps per iteration.

Moreover, the alternating direction method of multipliers (ADMM) is used to iteratively solve the convex optimization subproblems. Typically, the original image I is estimated by minimizing the cost function of total variation minimization problem consisting of restoration error measure and a regularization term, noting that

$$\min_{I} \frac{\mu}{2}\|\mathrm{HI} - I_0\|^2 + \|I\|_{\mathrm{TV}} \tag{3}$$

Here, first term corresponds to the data fidelity or fit to data term estimated using the least square fit. It particularly measures the closeness of restored image to the original image I for a given PSF. μ is the regularization parameter which depends on the blurring level and governs the applied regularization. $\| \cdot \|$ and $\| \cdot \|^2$ are conventional vector1-norm and vector1-norm square, respectively. The normalized total variation of solution I is represented as $\| \mathrm{I} \|_{\mathrm{TV}}$ and it can be simplified as

$$\| I \|_{\mathrm{TV}} = \| DI \| = \sum_i \sqrt{[D_x I]_i^2 + [D_y I]_i^2} \tag{4}$$

Fig. 2 Functional block diagram of ALTV

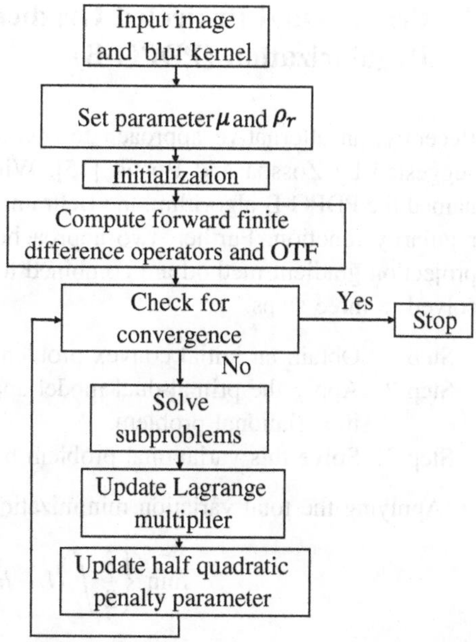

where operators D_x, D_y correspond to the forward finite-difference operators along the horizontal and vertical directions, respectively. By introducing intermediate variables u in (3), the problem involves

$$\min_{I,u} \frac{\mu}{2} \| \mathrm{HI} - I_0 \|^2 + \|u\| \quad \text{subject to}\, u = DI \tag{5}$$

where $u = [u_x^T \; u_y^T] I$ and u subproblems can be expressed as follows [15]:

$$I = f^{-1}\left[\frac{f\;[\mu H^T I_0 + \rho_r D^T u - D^T y]}{\mu |f\;[H]|^2 + \rho_r(|f\;[D_x]|^2 + |f\;[D_y]|^2)}\right] \tag{6}$$

where f denotes the Fourier transform operatorand

$$u = \max\left\{v_x - \frac{1}{\rho_r},\; 0\right\}.\; \frac{v_x}{v} \tag{7}$$

where $v = \max\left\{\sqrt{|v_x|^2 + |v_y|^2},\; \in\right\}$ and $\in = 10^{-6}$.

4 Primal-Dual Projected Gradient Beltrami Regularization (PDPG-B)

Recently, an alternative approach to optimize the Beltrami regularizer had been suggested by Zossoa and Bustin [15]. With this procedure, which is henceforth named the PDPG-B algorithm, the Beltrami energy is minimized considering as the regularity function. Further, two approaches, i.e., the primal-dual model and the projection gradient method are combined to optimize the solution. The problem is solved in three steps.

Step 1: Obtain an initial convex problem.
Step 2: Apply the primal-dual model approach to transform the initial problem in variational problem.
Step 3: Solve this variational problem by gradient projection-type method.

Applying the total variation minimization problem to the Beltrami energy

$$\min \left\{ \frac{\lambda}{2} \int (I - I_0)^2 + E_{\text{Bel}} \right\} \tag{8}$$

Here, λ is balancing parameter between fidelity and regularity. The Beltrami energy functions as

$$E_{\text{Bel}} = \int_{\Omega} \sqrt{1 + \beta^2 |\nabla I|^2 \mathrm{d}x} \tag{9}$$

Here, β is the aspect ratio of the Beltrami embedding. More specifically, the discrete equivalent of (8) is defined as

$$\min_{I} \left\{ \frac{\mu}{2} (I - I_0)^2 + \sum_{\Omega} \sqrt{1 + \beta^2 |\nabla I|^2} \right\} \tag{10}$$

After applying primal-dual problem, the Beltrami-regularized deconvolution problem can be formulated as

$$\min_{I \in \mathbb{R}^{m \times n}} \max_{\varphi \in X} \left\{ \frac{\mu}{2} (\mathrm{HI} - I_0)^2 + \sum_{\Omega} -I \operatorname{div}\varphi + \frac{\sqrt{\beta^2 - |\varphi|^2}}{\beta} \right\} \tag{11}$$

where φ is dual variable.

In particular, the implicit gradient descent is defined as

$$I^{k+1} = I^k + r_2 \left(\frac{\operatorname{div}\varphi^{k+1}}{\lambda} - h^s * (\mathrm{HI}^{k+1} - I_0) \right) \tag{12}$$

where r_2 is gradient descent step size.

After updating the primal variable through gradient descent and solving (12), the update of I is as follows:

$$I^{k+1} = f^{-1}\left\{\frac{f\{I^k + \frac{r_2}{\lambda}\operatorname{div}\varphi^{k+1}\} + r_2 Hf\{I_0\}}{1 + r_2|H|^2}\right\} \tag{13}$$

5 Results and Comparison

For the evaluation of both algorithms, seven standard test images (taken from image processing library of MATLAB by MathWorks Inc.) in the intensity range of [0, 256] are blurred by a shift-invariant Gaussian blur kernel of size 9 × 9 with different standard deviation. Both the algorithms are implemented in Matlab on a

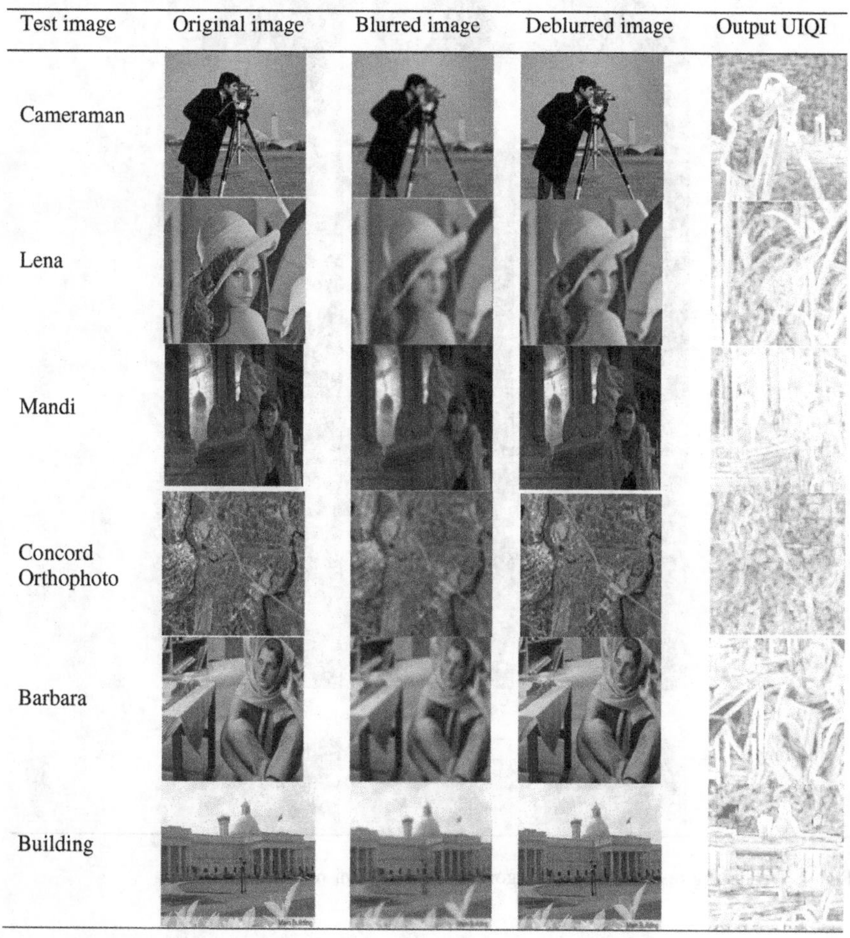

Test image
Original image
Blurred image
Deblurred image
Output UIQI
Cameraman
Lena
Mandi
Concord
Orthophoto
Barbara
Building

Fig. 3 Deblurring results of ALTV algorithm

system with 8 GB RAM and 64 bit processor running 64 bit Windows 8. The implementation of ALTV algorithm involves the regularization parameter $\mu = 10000$ while the initial value for the half quadratic penalty parameter $\rho_r = 2$ [14]. This algorithm converges in 20 iterations. The results of ALTV algorithm are shown in Fig. 3. The universal image quality index (UIQI) is also computed for output deblurred images. A similar series of test images have been applied to the PDPG-B algorithm with the value of balancing parameter between fidelity and regularity $\lambda = 0.075$, the aspect ratio of Beltrami embedding $\beta = 1$ and gradient descent parameter $r = 0.2$ [15]. The results of PDPG-B algorithm are shown in Fig. 4. This algorithm converges in 2000 iterations.

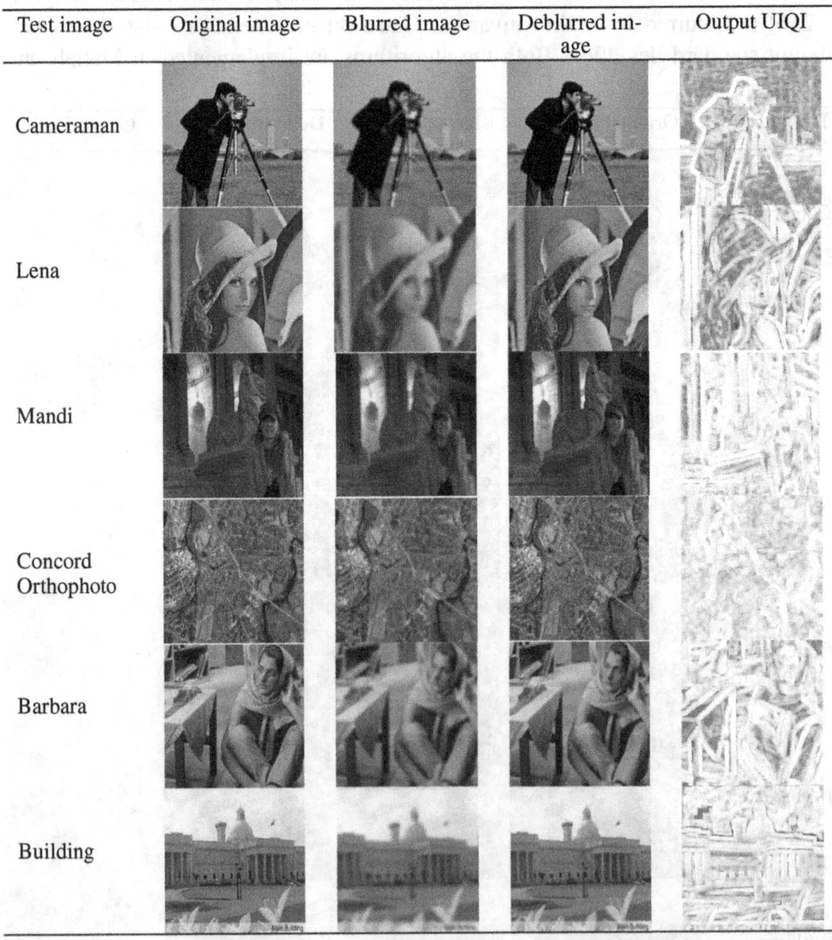

Fig. 4 Deblurring results of PDPG algorithm for Beltrami regularization

Especially, both algorithms are compared mathematically in terms of mean square error (MSE) and peak signal-to-noise ratio (PSNR) which are based on the pixel intensity difference between original and output images. Structural similarity index metric (SSIM) and UIQI are the human visual system (HVS) approach of measuring image quality. The comparison of PSNR and MSE is shown in Table 1, while the similarity between the results of both algorithms with the original image is encapsulated in Table 2. Finally, the speeds of both the algorithms are compared in Fig. 5.

The results demonstrate a slight improvement in PSNR, SSIM, and UIQI for the PDPG-B algorithm than the ALTV algorithm. The Beltrami model, however, requires more number of iterations for the convergence due to the fact that the explicit finite-difference schemes require small time steps and hence, the numbers of iterations are large. In this context, the execution timing performance may improve while updating the gradient descent parameter. However, another bottleneck of Beltrami regularization is that it is employed only on gray-scale images. The Lagrangian method can also work for the spatially variant blur; however, this work is restricted to the spatially invariant blur kernel.

Table 1 PSNR and MSE comparison of the blurred and restored images

PRM	Method	Cameraman	Lena	Mandi	Concord Orthophoto	Barbara	Building
MSE	Blurred	30.45	29.59	19.56	68.23	44.51	26.47
	ALTV	19.48	15.35	8.26	55.60	26.45	16.50
	PDPG-B	19.12	15.07	7.93	55.40	26.13	16.32
PSNR	Blurred	33.32	33.45	35.25	29.82	31.67	33.93
	ALTV	35.26	36.30	38.99	30.71	33.94	35.58
	PDPG-B	35.01	36.65	38.99	30.73	34.02	35.99

Table 2 SSIM and UIQI comparison of the blurred and restored images

PRM	Method	Cameraman	Lena	Mandi	Concord Orthophoto	Barbara	Building
SSIM	Blurred	0.778	0.742	0.865	0.7431	0.728	0.999
	ALTV	0.815	0.801	0.899	0.731	0.800	0.690
	PDPG-B	0.826	0.814	0.903	0.795	0.801	0.715
UIQI	Blurred	0.508	0.455	0.765	0.757	0.657	0.604
	ALTV	0.555	0.567	0.878	0.804	0.752	0.486
	PDPG-B	0.559	0.572	0.881	0.813	0.753	0.498

Fig. 5 Comparison of the execution time of both algorithms

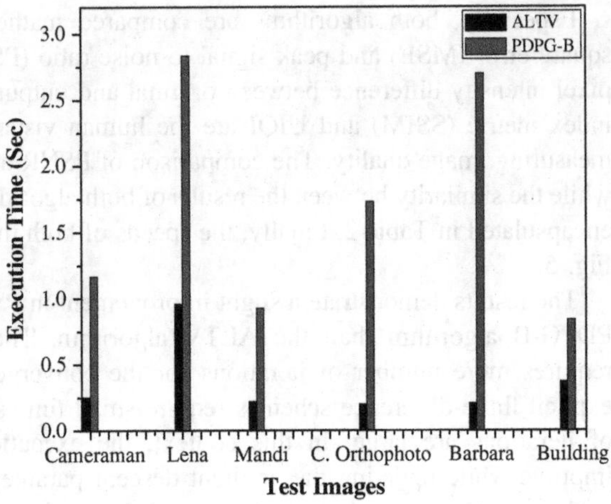

6 Conclusion

Usually, the deblurring problems are ill-posed problems and therefore, it is quite difficult to find the solution for such problems. Therefore, some advanced statistical method can be used to optimize the solutions. Henceforth, two regularization techniques, i.e., the augmented Lagrangian method for total variation regularization and the primal-dual projected gradient Beltrami algorithm for image deblurring are evaluated, implemented, and compared in this paper. The performance comparison of both algorithms reveals that the former one is faster while the performance is slightly better for the latter one. Future work in this field may focus on the combination/alteration of both algorithms to get better performance and high speed for the processing of natural images.

References

1. Helstrom, C.W.: Image restoration by the method of least squares. J. Opt. Soc. Am. A **57**, 297–303 (1967)
2. Hillery, A.D., Chin, R.T.: Iterative wiener filters for image restoration. IEEE Trans. Signal Process. **39**, 1892–1899, (1991)
3. Petersen, M.E., de Ridder, D., Handels, H.: Image processing with neural networks-A review. Pattern Recogn. Soc. **35**, 2279–2301 (2002)
4. Fish, D.A., Brinicombe, A.M., Pike, E.R.: Blind Deconvolution by means of the richardson-lucy algorithm. J. Opt. Soc. Am. A **12**, 58–65 (1995)
5. Dong, W., Zhang L., Shi, G., Wu, X.: Image deblurring and super-resolution by adaptive sparse domain selection and adaptive regularization. IEEE Trans. Image Process. **20** (2011)
6. Kundur, D., Hatzinakos, D.: Blind image deconvolution. IEEE Signal Process. Mag. 1053–588 (1996)
7. Ruhe. A. (ed.).: BIT numerical mathematics In: Hansen, P.C. (ed.) The Truncated SVD As A Method For Regularization, vol. 27, pp. 534–553. Springer (1987)

8. Ying, L., Xu, D., Liang, Z.-P.: On tikhonov regularization for image reconstruction in parallel MRI. In: Annual International Conference of the IEEE EMBS. San Francisco, CA, USA (2004)
9. Li, Y., Santosa, F.: A computational algorithm for minimizing total variation in image restoration. IEEE Trans. Image Process. **5**, 987–995 (1996)
10. Getreuer, P.: Total Variation deconvolution using split bregman. Image Process. Line **2**, 158–174 (2012)
11. Rudin, L.I., Osher, S., Fatemi E.: Nonlinear total variation based noise removal algorithms. Physica D. **60**, 259–268 (1992)
12. Sochen, N., Deriche, R., Lopez-Perez, L.: Variational beltrami flows over manifolds. Int. Conf. Image Process. Proc. IEEE, **1**(I), 861–864 (2003)
13. Sommer, G., Zeevi, Y.Y. (ed.).: Algebraic frames for the perception-action cycle. In: Sochen, N.A., Gilboa, G., Zeevil Y.Y.: Color Image Enhancement by a Forward-and-Backward Adaptive Beltrami Flow. Lecture Notes in Computer Science, vol. 1888, pp. 319–328. Springer, Berlin, Heidelberg, New York (2000)
14. Chan, S.H., Khoshabeh, R., Gibson, K.B., Gill P.E., Nguyen T.Q.: An Augmented lagrangian method for total variation video restoration. IEEE Trans. Image Process. **20**, 3097–3111 (2011)
15. Zossoa, D., Bustin, A.: A primal-dual projected gradient algorithm for efficient beltrami regularization. Comput. Vis. Image Underst. (2014)

An Approach to Solve Multi-objective Linear Fractional Programming Problem

Suvasis Nayak and A.K. Ojha

Abstract In this paper, an approach of hybrid technique is presented to derive Pareto optimal solutions of a multi-objective linear fractional programming problem (MOLFPP). Taylor series approximation along with the use of a hybrid technique comprising both weighting and ϵ-constraint method is applied to solve the MOLFPP. It maintains both priority and achievement of possible aspired values of the objectives by the decision maker (DM) while producing Pareto optimal solutions. An illustrative numerical example is discussed to demonstrate the proposed method and to justify the effectiveness, the results so obtained are compared with existing fuzzy max–min operator method.

Keywords Multi-objective linear fractional programming · Taylor series approximation · Hybrid method · Fuzzy programming · Pareto optimal solution

1 Introduction

Fractional programming belongs to the class of mathematical programming and deals with the optimization of a function existing in the form of ratio of two linear or nonlinear functions. Some common and practical instances of objectives belonging to the category of fractional programming are profit/cost, cost/time, output/employee, inventory/sale, risk assets/capital, debt/equity, and so forth. A linear fractional programming problem (LFPP) developed by Martos [1] optimizes

Suvasis Nayak (✉) · A.K. Ojha
School of Basic Sciences, Indian Institute of Technology Bhubaneswar, Bhubaneswar, India
e-mail: sn14@iitbbs.ac.in

A.K. Ojha
e-mail: akojha@iitbbs.ac.in

M. Pant et al. (eds.), *Proceedings of Fifth International Conference on Soft Computing for Problem Solving*, Advances in Intelligent Systems and Computing 436, DOI 10.1007/978-981-10-0448-3_59

an objective function (existing in form of ratio of two affine functions, i.e., linear plus constant) subject to a set of linear constraints as

$$\max \frac{\sum_{j=1}^{n} c_j x_j + \alpha}{\sum_{j=1}^{n} d_j x_j + \beta}$$

subject to

$x \in \Omega$, i.e., a set of linear constraints.

In numerous real-world decision-making situations, several fractional objectives are encountered which need to be optimized simultaneously restricted within a common set of constraints. Such mathematically modeled optimization problems are well known as multi-objective fractional programming or ratio optimization problem. Research activities in this field has been considerably accelerated since a few years because of its wide range of application in numerous important fields like engineering, economics, information theory, finance, management science, marine transportation, water resources, corporate planning, and so forth.

Charnes and Cooper [2] proposed a variable transformation technique to solve an LFPP with an additional constraint and a variable. Stancu-Minasian [3] discussed various methods and theoretical concepts on fractional programming. Costa [4] proposed an algorithm to solve MOLFPP which goes on dividing the no-dominated region to find the maximum value of the weighted sum of the objectives. Mishra [5] used weighting method to solve a bi-level LFPP. Toksar [6] proposed an approach to solve a fuzzy MOLFPP where membership functions are linearized using Taylor series expansion. Ojha and Ota [7] used a hybrid method to solve a multi-objective geometric programming problem. Valipour et al. [8] proposed parametric approach with weighting sum method to solve a MOLFPP. Ojha and Biswal [9] used ϵ-Constraint method to produce a set of Pareto optimal solutions. [10–12] discussed basic concepts and many methods of solution to a multi-objective optimization problem.

This paper is organized as follows: Sect. 2 interprets basics of multi-objective optimization and some techniques for its solution. The concept of fuzzy programming with details of fuzzy max–min operator method is discussed in Sect 3. Section 4 describes the proposed method to solve a MOLFPP. A numerical example with its solution and some remarks are incorporated in Sects. 5 and 6 contains the concluding part.

2 Multi-objective Optimization Problem

A multi-objective optimization problem (MOOP) can be mathematically stated as follows:

$$\max f(x) = (f_1(x), f_2(x), \ldots, f_k(x)) \quad \text{subject to} \quad x \in \Omega \tag{1}$$

where Ω is the nonempty compact feasible region. Usually, a single optimal solution does not exist to satisfy all the objectives simultaneously with their best individual optimality level which generates the concept of Pareto optimal solutions.

Definition *(See [11])* $x^* \in \Omega$ is a Pareto optimal solution of MOOP (1) if there does not exist another feasible solution $\bar{x} \in \Omega$ such that $f_i(\bar{x}) \leq f_i(x^*) \forall i$ and $f_j(\bar{x}) < f_j(x^*)$ for at least one j.

A set of Pareto optimal solutions comprising the most preferred optimal (best compromise) solution that satisfies all the objective functions with best possibility, can be generated using an appropriate method.

2.1 Weighting Sum Method

In this method [10, 11] nonnegative and normalized weights are assigned to all the objectives with smaller or bigger values regarding their importance at the DM as bigger and smaller weights represent greater and lesser importance of the objective, respectively. Finally, their weighted sum is optimized on the constraint feasible region to generate a set of Pareto optimal solutions by varying the weights. In other words, it scalarizes a vector optimization that converts a multi-objective to single-objective optimization problem. Mathematically, it can be stated for the MOOP (1) as

$$\max \sum_{i=1}^{k} w_i f_i(x) \quad \text{subject to} \quad x \in \Omega, \quad w_i \geq 0, \quad \sum_{i=1}^{k} w_i = 1 \tag{2}$$

Convexity of the feasible region guarantees Pareto optimality of the solution generated due to this method whereas concavity does not.

2.2 ϵ-Constraint Method

This method [10, 11, 13] is based on preference which considers an objective as the best prioritized one and converts the rest objectives with their goals as constraints, i.e., it maximizes an objective and simultaneously maintains minimum acceptability level for other objective functions. Mathematically, it can be stated as

$$\begin{gathered} \max f_s(x), \quad s \in \{1,2,\ldots,k\} \\ \text{subject to} \\ f_i(x) \geq \varepsilon_i, \quad i = 1,2,\ldots,s-1, \quad s+1,\ldots,k \\ x \in \Omega, \quad \varepsilon_i^L \leq \varepsilon_i \leq \varepsilon_i^U \end{gathered} \tag{3}$$

where $f_s(x)$ is the best prioritized function and ϵ_i^L, ϵ_i^U are, respectively, the relative minimum, maximum values of the objective $f_i(x)$ with respect to other objectives. Substituting different values of $\epsilon_i \in \left[\epsilon_i^L, \epsilon_i^U\right]$, a set of Pareto optimal solutions of (1) can be generated by solving (3).

2.3 *Hybrid Method*

This method [10] considers both priority of objective functions by assigning weights 'w_i' and achievement of minimum aspired objective values by the DM simultaneously. Mathematically, it can be stated as

$$\begin{gathered} \max \sum_{i=1}^{k} w_i f_i(x) \\ \text{subject to} \\ f_i(x) \geq \bar{f}_i, \quad x \in \Omega \, i = 1,2,\ldots,k \end{gathered} \tag{4}$$

where $\bar{f}_i$ is the aspired value for the objective $f_i(x)$.

3 Fuzzy Programming

Zadeh developed fuzzy set theory in 1965 that transforms imprecise information into precise mathematical form. Zimmermann [14] proposed fuzzy max–min operator method to solve various multi-objective optimization problems which is based on the concept of Bellman and Zadeh [15]. Each fuzzy set is associated with a membership function whose domain and range are the set of decision variables and [0, 1], respectively. An appropriate membership function is selected to determine the optimal solution. According to Zimmermanns fuzzy technique, best preferred optimal solution of a MOOP can be obtained using the following steps:

Step-1: Determine the best and worst values, i.e., the aspired $\left(f_i^{\max}\right)$ and acceptable $\left(f_i^{\min}\right)$ values, respectively, for each objective function $f_i(x), i = 1, \ldots, k$ satisfying $f_i^{\min} \leq f_i(x) \leq f_i^{\max}$, for $x \in \Omega$.

Step-2: Define following linear fuzzy membership function $\mu_i(x)$ for each objective function $f_i(x)$ to derive the best preferred optimal solution of the MOOP as

$$\mu_i(x) = \begin{cases} 0, & f_i(x) \leq f_i^{\min} \\ \frac{f_i(x) - f_i^{\min}}{f_i^{\max} - f_i^{\min}}, & f_i^{\min} \leq f_i(x) \leq f_i^{\max} \\ 1, & f_i(x) \geq f_i^{\max} \end{cases} \tag{5}$$

Step-3: Construct the following crisp model to obtain the optimal solution as follows:

$$\begin{gathered} \text{Max}\left\{ \min_{1 \leq i \leq k} \mu_i(x) \right\} \\ \text{subject to} \\ x \in \Omega \end{gathered} \tag{6}$$

Step-4: The above crisp model can be transformed into an equivalent mathematical programming problem as follows:

$$\begin{aligned} &\text{Max } \beta \\ &\text{subjectto} \\ \mu_i(x) = &\frac{f_i(x) - f_i^{\min}}{f_i^{\max} - f_i^{\min}} \geq \beta, \quad i = 1, 2, \ldots, k \\ &x \in \Omega \end{aligned} \tag{7}$$

where 'β' is an auxiliary variable assumed as the value of $\min_{1 \leq i \leq k} \mu_i(x)$. The constraints $\mu_i(x) \geq \beta$ can be replaced by $f_i(x) - \beta\left(f_i^{\max} - f_i^{\min}\right) \geq f_i^{\min}, i = 1, 2, \ldots, k$ for simplicity.

Step-5: Solve the above maximization problem to obtain the best preferred optimal solution of the MOOP and evaluate the optimal objective values at this solution.

4 Proposed Method to Solve MOLFPP

A MOLFPP can be mathematically formulated as

$$\max f_i(x) = \frac{\sum_{j=1}^{n} N_{ij}x_j + \alpha_i}{\sum_{j=1}^{n} D_{ij}x_j + \beta_i}$$

$$\text{subject to}$$

$$x \in \Omega = \{x \in R^n | Ax(\leq, =, \geq)b, x \geq 0\}$$

$$\text{where} \quad x = (x_j) \in R^n, N_{ij}, D_{ij}, \alpha_i, \beta_i \in R, b = (b_t) \in R^m \tag{8}$$

$$A = (a_{tj}) \in R^{m \times n}, \quad t = 1, 2, \ldots, m$$

$$\text{Assume that,} \sum_{j=1}^{n} D_{ij}x_j + \beta_i > 0 \quad \text{for each} \quad x \in \Omega \quad \text{and} \quad i = 1, 2, \ldots, k$$

Solution approach Maximize the fractional objective function $f_i(x)$ for each 'i' separately on the feasible region of constraints Ω to obtain their individual optimal solutions.

$$\max \frac{\sum_{j=1}^{n} N_{ij}x_j + \alpha_i}{\sum_{j=1}^{n} D_{ij}x_j + \beta_i}$$

$$\text{subject to} \sum_{j=1}^{n} a_t j x_j(\leq, =, \geq)b_t, \quad t = 1, 2, \ldots, m \text{ and } x_j \geq 0$$

Variable transformation technique [2] can be applied to obtain individual optimal solutions for each fractional objectives. For the kth objective function $f_k(x)$, this technique is mathematically interpreted as

$$\max \sum_{j=1}^{n} N_{kj}y_j + \alpha_k z$$

$$\text{subject to}$$

$$\sum_{j=1}^{n} D_{kj}y_j + \beta_k z = 1 \tag{9}$$

$$\sum_{j=1}^{n} a_{tj}y_j(\leq, =, \geq)b_t$$

$$y_j \geq 0, z > 0$$

If (y_j, z) is the optimal solution of (5) then $(x_j) = \left(\frac{y_j}{z}\right)$ is considered as the individual optimal solution of the kth objective function $f_k(x)$.

Let $X_1^*, X_2^*, \ldots, X_k^*$ be the individual optimal solutions of the objective functions $f_1(x), f_2(x), \ldots, f_k(x)$, respectively, where $X_i^* = \left(x_i^{(j)*}, j = 1, 2, \ldots, n\right), i = 1, 2, \ldots, k.$

Evaluate relative minimum and maximum objective values, i.e., ϵ_i^L and ϵ_i^U, respectively, for the ith objective function as

$$\varepsilon_i^L = \min_{1 \le j \le k} f_i\left(X_j^*\right)$$

$$\varepsilon_i^U = \max_{1 \le j \le k} f_i\left(X_j^*\right) = f_i(X_i^*)$$

Approximate each fractional objective function $f_i(x)$ by its first-order Taylor's series expansion evaluated at its individual optimal solution X_i^* as

$$f_i(x) \approx f_i(X_i^*) + \sum_{j=1}^{n} \left(x_j - x_i^{(j)*}\right) \frac{\partial f_i(X_i^*)}{\partial x_j}, \quad \text{for} \quad i = 1, 2, \ldots, k$$

As each fractional objective functions get transformed into nonfractional linear functions, the MOLFPP is converted into a single-objective linear programming problem (SOLPP) using the Hybrid method, i.e., the combination of both weighting sum and ϵ-constraint methods can be mathematically formulated as

$$\max \sum_{i=1}^{k} w_i \left[f_i \pi(X_i^*) + \sum_{j=1}^{n} \left(x_j - x_i^{(j)*}\right) \frac{\partial f_i(X_i^*)}{\partial x_j} \right]$$

$$\text{subject to}$$

$$f_i(X_i^*) + \sum_{j=1}^{n} \left(x_j - x_i^{(j)*}\right) \frac{\partial f_i(X_i^*)}{\partial x_j} \ge \varepsilon_i, \quad i = 1, 2, \ldots, k \tag{10}$$

$$\sum_{j=1}^{n} a_t j x_j (\le, =, \ge) b_t, \quad t = 1, 2, \ldots, m$$

$$x_j \ge 0, \quad w_i > =0, \quad \sum_{i=1}^{k} w_i = 1, \quad \varepsilon_i \in \left[\varepsilon_i^L, \varepsilon_i^U\right]$$

Substituting different values of ϵ_i, i.e., minimum aspired objective values to be achieved by the objectives $f_i(x)$ for different weights w_i, a set of Pareto optimal solutions can be generated by solving the above problem (7) from which the DM can choose the most preferred optimal solution by comparing the objective values.

5 Numerical Example

Consider the following multi-objective linear fractional programming problem:

$$\max f_1(x) = \frac{2x_1 + 3x_2 - 4x_3 + 3}{x_1 + x_2 + 2x_3 + 1}$$
$$f_2(x) = \frac{5x_1 - 2x_2 + x_3 + 1}{2x_1 + 3x_2 + x_3 + 2}$$

subject to

$$\Omega = \begin{cases} x_1 + x_2 + x_3 \geq 1 \\ 3x_1 - x_2 - 2x_3 \leq 2 \\ 2x_1 + x_2 + 3x_3 \leq 5 \\ x_1, x_2, x_3 \geq 0 \end{cases}$$

Individual optimal solutions of the objectives are obtained as

$$\max_{x \in \Omega} f_1(x) = 3 \quad \text{at} \quad X_1^* = (0, 1, 0)$$
$$\max_{x \in \Omega} f_2(x) = 1.5072 \quad \text{at} \quad X_2^* = (1.2308, 0, 0.8462)$$

Using the criteria of the proposed method, relative minimum and maximum values of the objectives are obtained as

$$\epsilon_1 \in [\epsilon_1^L, \epsilon_1^U] = [-1.8464, 3] \quad \text{and} \quad \epsilon_2 \in [\epsilon_2^L, \epsilon_2^U] = [-0.1010, 1.5072]$$

Fractional objectives are approximated by linear functions as

$$f_1(x) \approx -0.5x_1 - 5x_3 + 3$$
$$f_2(x) \approx 0.3741x_1 - 1.2287x_2 - 0.0956x_3 + 1.1277$$

The multi-objective LFPP is transformed into the following single-objective LPP using the proposed method as

$$\max(-0.5w_1 + 0.3741w_2)x_1 \\ - 1.2287w_2x_2 + (-5w_1 - 0.0956w_2)x_3 + (3w_1 + 1.1277w_2)$$

subject to

$$-0.5x_1 - 5x_3 + 3 \geq \varepsilon_1$$
$$0.3741x_1 - 1.2287x_2 - 0.0956x_3 + 1.1277 \geq \varepsilon_2$$
$$x_1 + x_2 + x_3 \geq 1, \quad 3x_1 - x_2 - 2x_3 \leq 2, \quad 2x_1 + x_2 + 3x_3 \leq 5$$
$$x_1, x_2, x_3 \geq 0$$
$$w_1 + w_2 = 1,\ w_1, w_2 \geq 0,\ \epsilon_1^L \leq \epsilon_1 \leq \epsilon_1^U,\ \epsilon_2^L \leq \epsilon_2 \leq \epsilon_2^U$$

Table 1 Pareto optimal solutions of the MOLFPP

w_1	w_2	ϵ_1	ϵ_2	x_1^*	x_2^*	x_3^*	f_1^*	f_2^*
1	0	>2.2	>1.2	No	–	Feasible	–	Solution
		2.2	1.2	0.7661	0.1694	0.0645	2.3165	1.1099
		2.1094	1.1746	0.7620	0.1901	0.0479	2.3940	1.0810
		2.0188	1.1491	0.7578	0.2109	0.0313	2.4728	1.0524
		1.9282	1.1237	0.7537	0.2316	0.0147	2.5529	1.0244
		1.6563	1.0474	0.7165	0.2835	0	2.6418	0.9374
0.8	0.2		Same	As	Above	With	Weights	(1,0)
0.2	0.8	>2.2	>1.2	No	–	Feasible	–	Solution
		2.2	1.2	0.7707	0.1463	0.0829	**2.2319**	**1.1429**
		2.1094	1.1746	0.7752	0.1242	0.1006	2.1521	1.1751
		2.0188	1.1491	0.7796	0.1021	0.1183	**2.0735**	**1.2079**
		1.9282	1.1237	0.7840	0.0800	0.1360	1.9963	1.2414
		1.6563	1.0474	0.7973	0.0137	0.1890	1.7724	1.3460
0	1		Same	As	Above	With	Weights	(0.2, 0.8)

Substituting different weights 'w_i' on priority basis of objectives and aspired objective values ϵ_i, a set of Pareto optimal solutions (x_1^*, x_2^*, x_3^*) are generated in the following Table 1 by solving the above problem.

5.1 Result Due to Fuzzy Method

The relative minimum and maximum values of the objectives are the worst (acceptable) and best (aspired) values, respectively, i.e.,

$$-1.8464 = \epsilon_1^L = f_1^{\min} \le f_1(x) \le f_1^{\max} = \epsilon_1^U = 3$$
$$-0.1010 = \epsilon_2^L = f_2^{\min} \le f_2(x) \le f_2^{\max} = \epsilon_2^U = 1.5072$$

For approximated linear objectives Since fractional objectives are approximated by linear objectives, constructing the linear membership functions as defined in Sect. 3 (step-2) and solving the corresponding optimization problem due to fuzzy max–min operator method as defined in Sect. 3 (step-4), the best preferred optimal solution is obtained as $x^* = (x_1^*, x_2^*, x_3^*) = (0.7718, 0.1411, 0.0871)$. The values of the given fractional objective functions at this solution are $f_1(x^*) = 2.2130$ and $f_2(x^*) = 1.1504$.

For given fractional objectives Constructing membership functions with the given fractional objectives and solving the corresponding crisp model as defined in Sect. 3, the best preferred optimal solution is obtained as $x^* = (x_1^*, x_2^*, x_3^*) = (0.7790, 0.1052, 0.1159)$. The values of the given fractional objective functions at this solution are $f_1(x^*) = 2.0842$ and $f_2(x^*) = 1.2033$.

Remark 1 Ascertaining the weights 'w_i' and changing the aspired objective values 'ϵ_i' in the range $[\epsilon_i^L, \epsilon_i^U]$, a set of Pareto optimal solutions (x_1^*, x_2^*, x_3^*) are generated which are same for the weights $(1, 0), (0.8, 0.2)$ and $(0.2, 0.8), (0, 1)$. f_1^* and f_2^* are the values of the fractional objectives evaluated at (x_1^*, x_2^*, x_3^*). Since, fractional objectives are approximated by linear functions, $f_i^* \succcurlyeq \epsilon_i (i = 1, 2)$ are satisfied where '$\succcurlyeq$' represents "greater than or approximately equal to." As $f_i^* \in [\epsilon_i^L, \epsilon_i^U]$ for each (x_1^*, x_2^*, x_3^*), it shows the correctness of the proposed method. DM can choose one Pareto optimal solution from Table 1 as the most preferred optimal solution on priority basis as required in the practical decision-making situation. If DM is still unsatisfied, more Pareto optimal solutions (x_1^*, x_2^*, x_3^*) can be generated by substituting more aspired objective values 'ϵ_i' within the specified range $[\epsilon_i^L, \epsilon_i^U]$.

Remark 2 The highlighted objective values in Table 1 obtained due to the proposed method are considerably closer to the objective values obtained due to fuzzy max–min operator method using both cases of fractional and approximated linear objective functions. Besides this, using the proposed method DM obtains a set of solutions to choose the best preferred optimal solution as per the requirement of the system but in case of fuzzy, there is no choice to choose. It justifies the use of the proposed method.

6 Conclusions

This paper comprises conversion of fractional objectives into linear functions using Taylor's series approximation and a hybrid method that combines the ideas "priority of objectives" and "achievement of possible aspired objective values" of both weighting sum and ϵ-constraint methods, is implemented to generate a set of Pareto optimal solutions of a MOLFPP. As the proposed method offers numerous options to DM, we can select one solution as the most preferred optimal solution. "LINGO" and "MATLAB" softwares are used for computational works in the numerical example. Comparison of the results so obtained with existing fuzzy max–min operator method ensures the effectiveness of the proposed method.

Acknowledgments Authors are grateful to the Editor and anonymous referees for their valuable comments and suggestions to improve the quality of presentation of the paper.

References

1. Martos, B.: Hyperbolic programming. Publ. Res. Inst. Math. Sci. **5**, 386–407 (1960)
2. Charnes, A., Cooper, W.W.: Programming with linear fractional functionals. Naval Res. logist. Q. 9, 181–186 (1962)
3. Stancu-Minasian, I.M.: Fractional programming: Theory, Methods and Applications. Kluwer Academic Publishers (1997)
4. Costa, J.P.: Computing non-dominated solutions in MOLFP. Eur. J. Oper. Res. **181**, 1464–1475 (2007)
5. Mishra, S.: Weighting method for bi-level linear fractional programming problems. Eur. J. Oper. Res. **183**, 296–302 (2007)
6. Toksar, M.D.: Taylor series approach to fuzzy multiobjective linear fractional programming. Inform. Sci. **178**, 1189–1204 (2008)
7. Ojha, A.K., Ota, R.R.: A hybrid method for solving multi-objective geometric programming problem. Int. J. Math. Oper. Res. **7**, 119–137 (2015)
8. Valipour, E., Yaghoobi, M.A., Mashinchi, M.: An iterative approach to solve multiobjective linear fractional programming problems. Appl. Math. Model. **38**, 38–49 (2014)
9. Ojha, A.K., Biswal, K.K.: Multi-objective geometric programming problem with ϵ-constraint method. Appl. Math. Model. **38**, 747–758 (2014)
10. Collette, Y., Siarry, P.: Multiobjective optimization: principles and case studies. Springer (2003)
11. Miettinen, K.M.: Nonlinear multiobjective optimization. Kluwer Academic Publisher (2004)
12. Ehrgott, M.: Multicriteria optimization. Springer (2005)
13. Haimes, Y.Y., Ladson, L.S., Wismer, D.A.: On a Bicriterion formulation of problems of integrated system identification and system optimization. IEEE Trans. Syst., Man, Cybern., Syst. **1**, 296–297 (1971)
14. Zimmermann, H.-J.: Fuzzy programming and linear programming with several objective functions. Fuzzy Sets Syst. **1**, 45–55, (1978)
15. Bellman, R.E., Zadeh, L.A.: Decision-making in a fuzzy environment. Mang. Sci. **17**, B-141 (1970)

A Fuzzy AHP Approach for Calculating the Weights of Disassembly Line Balancing Criteria

Shwetank Avikal, Sanjay Sharma, J.S. Kalra, Deepak Varma and Rohit Pandey

Abstract Disassembly of outdated and previously consumed product takes place in field of remanufacturing, recycling, reusing and disposal. The disassembly lines have become the first choice for disassembly of the product that has been consumed previously. Disassembly line should be designed and balanced properly so that it can work as efficiently as possible. There are many different criteria in the disassembly lines for selecting the parts those are to be removed. The problem of disassembly line balancing is based on these different criteria. In this paper, the weights of these criteria have been evaluated. A fuzzy analytical hierarchy process (fuzzy AHP)-based approach has been applied to calculate the weight of each criterion. With the help of the weight of these criteria, the tasks can be assigned to workstations with different precedence constraint and cycle time limit.

Keywords Product disassembly · Line balancing · Fuzzy theory · AHP

1 Introduction

The demands of new product are ever-increasing that are having more values than the existing similar types of products. This in turn has shortened the life cycle of products and these discarded products are continuously filling the available landfill space. These discarded products may be hazardous or valuable materials. It has been felt by the producers as well as by consumers to evolve a viable solution to utilize these products. So it is need of the time to disassemble the existing products and disassembly line has emerged as one of the solutions to this problem.

The main objective of product recovery is known as minimizing the amount of industrial and domestic waste going for landfills. The parts and material can be

Shwetank Avikal (✉) · Sanjay Sharma · J.S. Kalra · Rohit Pandey
Mechanical Engineering Department, Graphic Era Hill University, Dehradun, India
e-mail: shwetank.avikal@gmail.com

Deepak Varma
Centre of Transportation Engineering, IIT Roorkee, Rooekee, India

M. Pant et al. (eds.), *Proceedings of Fifth International Conference on Soft Computing for Problem Solving*, Advances in Intelligent Systems and Computing 436, DOI 10.1007/978-981-10-0448-3_60

extracted from the post-consumed product with the help of reuse, recycling and remanufacturing and the goal can be achieved [1]. Disassembly has proved its role in recovery of important material and parts by allowing their selective separation [2].

In this paper, fuzzy AHP-based technique has been used to calculate the weights of each criterion of disassembly line balancing problems. In Sect. 2, the relative literature has been reviewed. In Sect. 3 the fuzzy AHP technique has been described. The problem has been formulated in fourth section. In Sect. 5, the proposed technique has been described. Section 6 represents the calculation of weight, while the conclusion and discussion in Sect. 7.

2 Literature Review

The problem of assignment of tasks to disassembly workstations is known as disassembly line balancing problem (DLBP). In DLBP, precedence relations between the tasks must be considered and some measure of effectiveness such as minimization of cycle time, number of workstations, etc., must be optimized.

Gungor and Gupta [3–5] have described the introduction of DLBP and they proposed an algorithm for solving the problem of disassembly line balancing. Tiwari et al. [6] have suggested a Petri net-based problem-solving approach to find the disassembly strategy of a product. McGovern and Gupta [1] have presented a solution approach that was based on greedy model and 2-opt heuristic for solving the problem. McGovern and Gupta [7] have also proposed and compared many combinatorial optimization techniques for disassembly problems. McGovern and Gupta [8] have stated that DLBP is an NP hard problem and they also proposed a genetic algorithm-based approach for solving DLBP. Altekin et al. [9] have developed an algorithm that was based on a mixed integer programming algorithm for solving profit-oriented DLBP. Koc et al. [10] have suggested two exact solution techniques for disassembly line balancing problems with AND/OR precedence relation between tasks. Ding et al. [11] have proposed a multi-objective DLBP and solved this by an ant colony algorithm. Yeh [12] has proposed a modified simplified swarm optimization (SSO) method to solve disassembly sequencing problems. Karadag and Turkbey [13] have proposed a genetic algorithm (GA)-based approach for solving multi-objective optimization of a stochastic DLBP. Kalayci and Gupta [14] proposed an artificial bee colony (ABC)-based methodology for solving sequence-dependent disassembly line balancing problem.

In the present case, it seems difficult to select a task from a number of tasks for assignment to the disassembly workstation. Therefore, one can find a number of approaches for solving task assignment problem with the help of different types of problem-solving techniques. However, these techniques do not vender due consideration to view of labour, supervisor, customers, environment activities, researchers and public reviewer. Because of these issues, this problem has been taken as multi-criteria decision-making problem and F-AHP technique has been applied for finding the weight of each disassembly criterion.

3 Fuzzy AHP Approach

Saaty [15] has defined AHP as a hierarchical decomposing decision technique for solving the complex multi-criteria decision-making (MCDM) problem. In traditional AHP, the pairwise comparisons for each level with respect to each criterion are conducted by a nine-point scale. This traditional method is not capable to deal uncertainties of the decision-making problem and the actual problem carries a lot of uncertainties. To deal these problems and overcome the effect of uncertainties, fuzzy set theory has been used. In the present problem, fuzzy set theory described by Singh et al. [16] has been selected for the calculations.

4 Problem Formulation

The algorithm has been proposed for calculating the weights of each criterion of the disassembly task. With the help of these weights the tasks can be assigned to the disassembly workstation to minimize cycle time, number of workstations and maximize line efficiency. In the present problem, disassembly line balancing problem has been taken as a multi-criteria decision-making problem. Three well-known criteria have been selected for solving this complex DLBP, i.e. part demand, disassembly time and part hazardous.

5 Proposed Technique

The proposed technique has been developed for calculating the weights of different disassembly criterion. The proposed technique incorporates the views of different stackholders related to disassembly environment for solving the task assignment problems. The steps of proposed technique are following:

The proposed fuzzy AHP-based approach consists of three steps (S): (1) first one is data collection and (2) second one is the application of fuzzy AHP. In order to express a preference for the different criteria, weights (W) are allocated in Step 3.

6 Calculation of Weights of Criteria

The weights of each criterion can be calculated by fuzzy AHP after forming the decision tree (Fig. 1). For calculating the same, the people related to the disassembly environment have been asked to make pairwise comparison between each

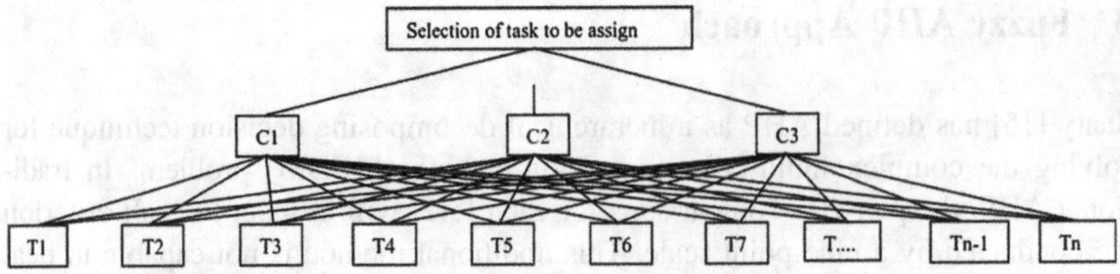

Fig. 1 The decision hierarchy for selection of tasks for assignment

Table 1 The pairwise comparison matrix for criteria

Criteria	Part removal	Part demand	Part
C1	1	0.375	0.1458
C2	3	1	0.75
C3	7	3	1

Table 2 Results obtained by AHP

Criteria	Weights (w)	λ_{max}, CI, RI	CR
C1	0.0908	$\lambda_{max} = 3.093$	0.0805
C2	0.2493		
C3	0.6599	CI = 0.0467	

criteria using the scale proposed by Satti [15]. The pairwise comparison matrix has been shown in Table 1. The weights of each criteria calculated by fuzzy AHP are presented in Table 2.

The applied Fuzzy AHP has been calculated that criterion C3 has the maximum weight as compared to other criterion. Consistency ratio of the pairwise comparison matrix is calculated as 0.0805 < 0.1. So the weights are proved to be consistent and they are ready to be used in disassembly process.

7 Conclusion and Future Work

However, the use of disassembly lines, the need of generating to perfectly balance the disassembly line has been increased. In the present problem, the weights of all criteria have been calculated by fuzzy AHP method and using these weights tasks can be assigned to disassembly workstations. In the process assignment of tasks to disassembly workstations, different types of ranking method can be used to find out the final ranking of the tasks on the bases of their criteria and their calculated weights. With the help of final ranking, the tasks can be assigned to disassembly workstation on the bases of their precedence constraints and cycle time limit.

References

1. McGovern, S.M., Gupta, S.M.: 2-Opt heuristic for the disassembly line balancing problem. In: Proceedings of the SPIE International Conference on Environmentally Conscious Manufacturing III, Providence, Rhode Island, pp. 71–84, 1994 (2003)
2. Gupta, S.M., Taleb, K.N.: Scheduling disassembly. Int. J. Prod. Res. **32**(8), 1857–1866 (1994)
3. Gungor, Gupta, S.M.: Disassembly line in product recovery. Int. J. Prod. Res. **40**(11), 2569–2589 (2002)
4. Gungor, Gupta, S.M.: Disassembly line balancing. In: Proceedings of the 1999 Annual Meeting of the Northeast Decision Sciences Institute, Newport, Rhode Island, March 24–26, pp. 193–195 (1999b)
5. Gungor, Gupta, S.M.: Issues in environmentally conscious manufacturing and product recovery: a survey. Comput. Ind. Eng. **36**(4), 811–853 (1999a)
6. Tiwari, M.K., Sinha, N., Kumar, S., Rai, R., Mukhopadhyay, S.K.: A petri net based approach to determine strategy of a product. Int. J. Prod. Res. **40**(5), 1113–1129 (2001)
7. McGovern, S.M., Gupta S.M.: Combinatorial optimization methods for disassembly line balancing. In: Proceedings of the 2004 SPIE International Conference on Environmentally Conscious Manufacturing, Philadelphia, Pennsylvania, pp. 53–66 (2004)
8. McGovern, S.M., Gupta, S.M.: Combinatorial optimization analysis of the unary NP-complete disassembly line balancing problem. Int. J. Prod. Res. **45**(18–19), 4485–4511 (2007)
9. Altekin F.T., Kandiller, L., Ozdemirel, N.E., Profit-oriented disassembly-line balancing. Int. J. Prod. Res. **46**, 2675–2693 (2008)
10. Koc, A., Sabuncuoglu, I., Erel, E.: Two exact formulations for disassembly line balancing problems with task precedence diagram construction using an AND/OR graph. IIE Trans. Oper. Eng. **41**, 866–881 (2009)
11. Ding, L.P., Feng, Y.X., Tan, J.R., Gao, Y.C.: A new multi-objective ant colony algorithm for solving the disassembly line balancing problems. Int. J. Adv. Manuf. Technol. **48**, 761–771 (2010)
12. Yeh, W.-C.: Simplified swarm optimization in disassembly sequencing problems with learning effects. Comput. Oper. Res. **39**(9), 2168–2177 (2012)
13. Karadag, A.A., Turkbey, O.: Multi-objective optimization of stochastic disassembly line balancing with station paralleling. Comput. Ind. Eng. **65**(3), 413–425, (2013)
14. Kalayci, C.B., Gupta, S.M.: Artificial bee colony algorithm for solving sequence-dependent disassembly line balancing problem. Expert Syst. Appl. **40**(18), 7231–7241, (2013)
15. Saaty, T.L., The Analytic Hierarchy Process. McGraw-Hill, New York, (1980)
16. Singh, A.R., Mishra, P.K., Jain, R., Khurana, M.K.: Robust strategies for mitigating operational and disruption risks: a fuzzy AHP approach. Int. J. Multicriteria Decis. Mak. **2**(1), 1–28 (2012)

An Efficient Compression of Encrypted Images Using WDR Coding

Manoj Kumar and Ankita Vaish

Abstract This paper presents a novel scheme for the compression of encrypted images through which we can efficiently compress the encrypted images without compromising either the compression efficiency or the security of the encrypted images. In the encryption phase, content owner encrypts the original image using pseudorandom numbers which are derived from a secret key. Then, the channel provider without the knowledge of secret key can compress the encrypted image. For compression, encrypted image is decomposed into subimages and each of these subimages is compressed independently using quantization and wavelet difference reduction coding technique. Then the compressed data obtained from all the subimages is regarded as the compressed bit stream. At receiver side, a reliable decompression and decryption technique is used to reconstruct the image from compressed bit stream. To evaluate the performance, the proposed technique has passed through a number of test cases such as compression ratio (CR) and peak signal-to-noise ratio (PSNR). All the analysis and experimental results clearly show that the proposed encryption-then-compression technique is reckon secure and shows good compression performance. To show the efficiency of proposed work it is compared with a well-known scheme on compression of encrypted images and experimental results show better compression performance with improved image quality.

Keywords Compression ratio · Image compression · Image encryption · Peak signal-to-noise ratio · Wavelet difference reduction

---

Manoj Kumar · Ankita Vaish (✉)
Babasaheb Bhimrao Ambedkar University, Lucknow, India
e-mail: av21lko@gmail.com

Manoj Kumar
e-mail: mkjnuiitr@gmail.com

M. Pant et al. (eds.), *Proceedings of Fifth International Conference on Soft Computing for Problem Solving*, Advances in Intelligent Systems and Computing 436, DOI 10.1007/978-981-10-0448-3_61

1 Introduction

In today's world, the growing technology has brought tremendous changes in our lives in a number of ways. As most of the web information is now preferred to represent as multimedia data such as audio, video, and image rather than text therefore, it is necessary to have reliable methods for efficient storage and fast transmission of these multimedia data. At the same time, the security of transmitting information has arisen the need of a joint work on encryption and compression. Compression of encrypted images has brought a lot of concern in the field of present work. The conventional way for the transmission of redundant data is to first compress the data and then encrypt the compressed data for security aspect. For better illustration, let us take a practical scenario in which a content owner say Alice wants to efficiently transmit image I to recipient say Bob over an insecure bandwidth-constrained communication channel provider say Charlie. Traditionally, it could be performed as follows: Original image I is compressed into I_c, and then I_c is encrypted into I_e using secret key. Encrypted image I_e is then forwarded to Charlie, who simply passes it to Bob. As Bob is an authorized user therefore, sequential decryption and decompression of encrypted image to be performed by Bob to get the reconstructed image I' is shown in Fig. 1a.

The compression-then-encryption (CTE) paradigm explained above accomplishes the need of many secure transmission scenarios. However, in some cases, it is necessary to protect the images prior to compression. Therefore, for such cases CTE paradigm needs to be revised as shown in Fig. 1b. Alice is concerned to protect the privacy of her data from everyone through encryption. On the other hand, to maximize network utilization, Charlie has a paramount interest in compressing all network traffic. As Alice does not believe on Charlie therefore, the secret key that was used to encrypt the original image data is kept secret from Charlie. Further, the task of compression is carried out by Charlie on encrypted data, without the knowledge of secret key. Now, to get the reconstructed image from the encrypted compressed data, joint decompression and decryption is performed by Bob. Figure 1b shows the encryption-then-compression (ETC) method in which Alice first encrypted the image then the encrypted data is compressed without compromising either the compression efficiency or security of images. At the receiver side, joint decompression and decryption is performed to get the original image content. The problem of compressing encrypted images has attracted a lot of interest in the current research work [1–6]. At first, compression of encrypted data seems to be impractical because there is a lack of redundant information in encrypted data therefore, the compression of encrypted data became a challenging task. However, lossless compression of encrypted image is possible through the use of Slepian–Wolf coding [1], when some side information is available at decoder side but unavailable at encoder side. Johnson et al. [6] have shown that the compression of encrypted data is possible through the use of coding

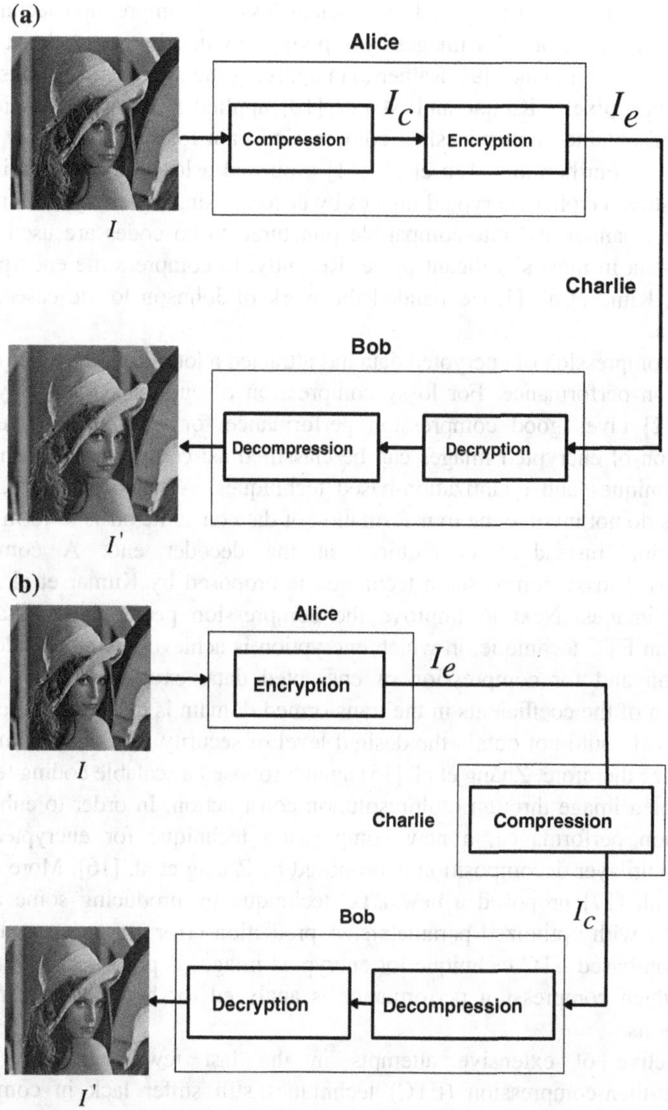

Fig. 1 **a** Traditional compression-then-encryption technique. **b** Encryption-then-compression technique

with side information principles theoretically. In [6] Johnson et al. have also shown practical schemes that can be used to losslessly compress the encrypted binary images. Further, Schonberg et al. [7, 8] probe the problem of compressing encrypted data when source statistics is not known and the sources have memory.

Lazzeretti and Barni [9] proposed an efficient lossless compression technique for encrypted gray scale or color images by applying low-density parity-check (LDPC) codes in various bit planes [9]. Rather than applying the approach of Johnson et al. [6] on image pixels, Kumar and Makur [10] applied it to the prediction error domain and obtained better lossless compression gain for encrypted gray scale or color images. Furthermore, Liu et al. [11] proposed a lossless compression technique for stream cipher encrypted images by decomposing the encrypted image in a progressive manner and rate-compatible punctured turbo codes are used to compress the data in most significant plane. Recently, to compress the encrypted data losslessly, Klinc et al. [12] expanded the work of Johnson to the case of block cipher data.

Lossy compression of encrypted data has attracted a lot of interest to improve the compression performance. For lossy compression of encrypted data, Wyner–Ziv theorem [2] gives good compression performance for Gaussian source. Lossy compression of encrypted images can be classified as: compressive sensing (CS)-based techniques and quantization-based techniques. At encoder end, CS-based techniques do not involve the transformation of data on some basis to form a sparse representation instead it is required at the decoder end. A compressive sensing-based lossy compression technique is proposed by Kumar et al. [13] for encrypted images. Next to improve the compression performance Zhang [14] proposed an ETC technique, in which encryption is achieved using pseudorandom permutation and for compression of encrypted data, excessive rough and fine information of the coefficients in the transformed domain is discarded. The scheme given in [14] could not obtain the desired level of security and better compression performance therefore, Zhang et al. [15] again proposed a scalable coding technique for encrypted image through multiresolution construction. In order to enhance the compression performance, a new compression technique for encrypted image through multilayer decomposition is proposed by Zhang et al. [16]. More recently, Zhang et al. [17] proposed a new ETC technique by producing some auxiliary information with optimized parameters. A prediction error clustering and random permutation-based ETC technique for encrypted images is proposed by Zhou et al. [18] in which compression performance is analyzed for both lossless and lossy compressions.

Irrespective of extensive attempts in the last few years, the existing encryption-then-compression (ETC) techniques still suffer lack in compression performance when compared to the state-of-the-art work on compression of unencrypted images. In this paper, a wavelet difference reduction (WDR) coding-based ETC technique is proposed through which good compression performance can be achieved without compromising the security of images. In the proposed work, at first, an original image is encrypted using pseudorandom number which is generated using a seed termed as key (k). In the compression phase, the encrypted image is decomposed into subimages and each of these subimages is compressed independently using quantization and WDR coding.

The rest of the paper is organized as follows: Sect. 2 gives a brief introduction to the WDR coding. The proposed encryption-then-compression (ETC) technique is discussed in Sect. 3. Experimental results and analysis are reported in Sect. 4. Conclusions that are drawn from the proposed work are given in Sect. 5.

2 Overview of Wavelet Difference Reduction (WDR)

Wavelet difference reduction (WDR) algorithm is given by Tian and Wells [19–21]. This algorithm follows few simple steps. First of all, discrete wavelet transform (DWT) is applied to an image. Now for wavelet coefficients $w(i)$, a bit plane encoding-based WDR algorithm is carried out. Basically, it consists of four parts as shown in Fig. 2. The first one is initialization part: in this phase, an initial threshold T_0 is selected in such a way that the value of all the transformed coefficients is less than the value of T_0 and at least one transform value has magnitude of $\frac{T_0}{2}$. In the next phase, threshold T_0 is changed to $T = T_k$, where $T_k = \frac{T_{k-1}}{2}$. Furthermore, the significance part results in new significant values $(w(i))$ which satisfy the condition $T \leq |w(i)| \leq 2T$. After that WDR method is used to encode the index of significant transform values, for each index values a binary expansion relative to threshold T_0 is calculated. The loop shown in Fig. 2 indicated the way to encode the significant transform values by bit plane encoding [21].

For an example, when threshold is $T_1 = 32$ and the significant values of wavelet coefficients are $w(1) = +48$, $w(2) = -38$, $w(7) = +42$, $w(14) = +44$, $w(36) = -57$. Here, 1, 2, 7, 14, 36 represent the indices of the significant values. Instead of working on these indices WDR works on their successive differences: 1, 1, 5, 7, 22 so-called wavelet difference reduction. The latter list of indices is generated by keeping the first number as the starting index and each successive number indicates the number of steps required to reach the next index. Further, the binary expansion of these successive differences is calculated: $(1)_2$, $(1)_2$, $(101)_2$, $(111)_2$, and $(10110)_2$. As the most significant bit (MSB) for each of these binary expansions is always 1 therefore, we can drop this bit and to generate symbol stream, signs of the significant transform values can be used instead of separators. When this MSB is dropped, we obtain the binary expansion that remains as the reduced binary expansion. For the taken example, the output of the WDR significance pass will be string of symbols + – +01 + 11–0110.

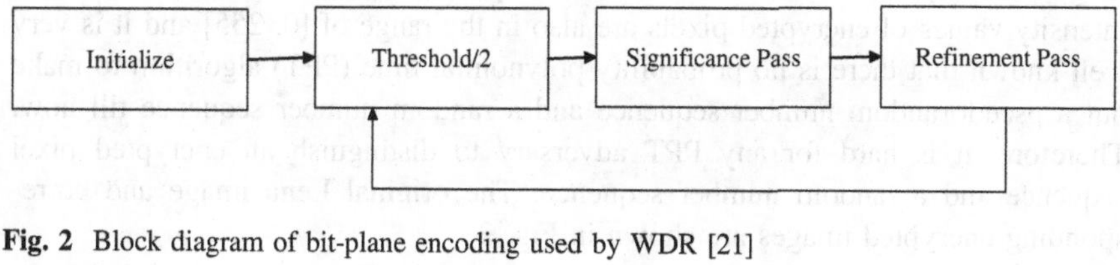

Fig. 2 Block diagram of bit-plane encoding used by WDR [21]

After the significance pass, refinement pass reduces the error by refining the already quantized values (w_q), i.e., the refinement pass refines the precision of old quantized transformed values w_q, which satisfy $|w_q| \geq 2T$. Each refined value is better approximated to original transform value.

To reconstruct an image above steps are reversed and an approximated wavelet coefficients are obtained, an inverse wavelet transform is then applied to obtain the reconstructed image.

3 Proposed Encryption-then-Compression Scheme

This section simply focuses on some motivating factors regard to the proposed image encryption-then-compression technique. At the sender end, an input image I of size $P \times Q$ is used as an input image, content owner encrypts the original image using pseudorandom number. The encrypted image is passed to channel provider who simply decomposes the encrypted image into subimages and then compression is performed on each of these subimages independently. The detailed steps are as follows:

3.1 Encryption of Image

In this phase, original image (I) is encrypted using pseudorandom numbers, that are generated by pseudorandom number generator (PRNG) sequences [17], initial seed used to generate the PRNG sequences is termed as secret key (k). For an image (I) of size $P \times Q$ where, P and Q represent the number of rows and columns in the image, the total number of pixels is $N(N = P \times Q)$. Therefore, for encryption content owner generates N pseudorandom numbers within [0, 255] and performs one-by-one addition modulo 256 to create an encrypted image.

$$E(i,j) = \mathrm{mod}[I(i,j) + S(i,j), 256], \quad 1 \leq i \leq P, \; 1 \leq j \leq Q \tag{1}$$

Here, $I(i,j)$ represents the original gray values of pixels at position (i,j) and $S(i,j)$ represents the pseudorandom numbers generated by a PRNG within [0, 255] at (i,j). The encrypted data obtained by performing addition modulo 256 to original image is represented by $E(i,j)$. As the intensity values of I and S are within the range [0, 255] and addition modulo 256 is performed over them therefore, the intensity values of encrypted pixels are also in the range of [0, 255] and it is very well known that there is no probability polynomial time (PPT) algorithm to make out a pseudorandom number sequence and a random number sequence till now. Therefore, it is hard for any PPT adversary to distinguish an encrypted pixel sequence and a random number sequence. The original Lena image and corresponding encrypted images are shown in Fig. 3.

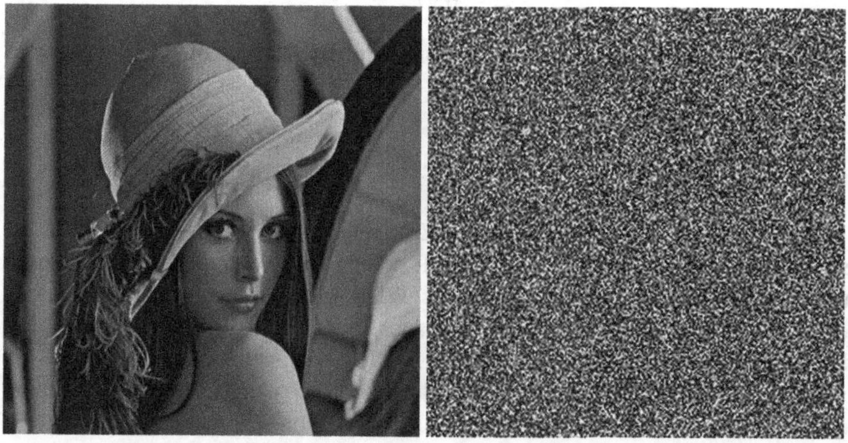

Fig. 3 **a** Original image, **b** encrypted image

3.2 Compression of Encrypted Image

In the compression phase, encrypted data is compressed by channel provider, who simply accomplished this by decomposing the encrypted data into subimages as

$$D_1(i,j) = E(2i, 2j), \quad 1 \leq i \leq P/2,\ 1 \leq j \leq Q/2 \tag{2}$$

$$D_2(i,j) = E(2i, 2j-1), \quad 1 \leq i \leq P/2,\ 1 \leq j \leq Q/2 \tag{3}$$

$$D_3(i,j) = E(2i-1, 2j), \quad 1 \leq i \leq P/2,\ 1 \leq j \leq Q/2 \tag{4}$$

$$D_4(i,j) = E(2i-1, 2j-1), \quad 1 \leq i \leq P/2,\ 1 \leq j \leq Q/2 \tag{5}$$

a step size η is used to quantize the subimage D_1 as given below:

$$C_1(i,j) = \text{floor}\left(\frac{D_1(i,j)}{\eta}\right), \quad 1 \leq i \leq P/2,\ 1 \leq j \leq Q/2 \tag{6}$$

where

$$\eta = \frac{256}{\alpha}. \tag{7}$$

α is a quantizing parameter and it is shared by the encoder and decoder. Due to this quantization, the values of C_1 lie in the range $[0, (\alpha - 1)]$. Clearly,

$$0 \leq C_1(i,j) \leq \alpha - 1 \tag{8}$$

Hence, the binary sequence used to express the data amount of $C_1(i,j)$ is denoted by B_1 as

$$B_1 = \frac{N}{4}\log_2(\alpha). \tag{9}$$

For the other subimages, (D_2, D_3, D_4) encoder compresses each of these encrypted subimages independently. WDR coding-based compression technique is applied on each of these subimages. As WDR is a wavelet-based compression technique therefore, it can compress images more efficiently without too much degrading its visual quality. As a result of WDR coding compressed bit streams are obtained. Further, collect the compressed bit streams obtained from other subimages and the value of α. The number of bits used to represent the other subimages is represented by B_2, B_3, B_4, respectively. Therefore, the total amount of compressed data is

$$B = B_1 + \frac{N}{4}(B_2 + B_3 + B_4) \tag{10}$$

Hence, the compression ratio, i.e., the ratio between the compressed data and the original image data is expressed as

$$CR = \frac{B}{8N} \tag{11}$$

3.3 *Image Reconstruction*

At receiver side, the original image content can be recovered with the help of compressed bit streams, secret key (k), and parameter α used by the encoder. With the help of compressed bit stream B_1 and parameter α, the values of $C_1(i,j)$ can be retrieved. Further $C_1(i,j)$ is decrypted to form a subimage (R_1) as

$$R_1(i,j) = \mathrm{mod}[C_1(i,j) \cdot \eta - S(2i,2j), 256] + \frac{\eta}{2}, \quad 1 \leq i \leq \frac{P}{2}, \; 1 \leq j \leq \frac{Q}{2} \tag{12}$$

where $\eta = \frac{256}{\alpha}$ and the value of $S(2i,2j)$ is derived from the secret key (k). With the help of compressed bit streams B_2, B_3, B_4, the other subimages can be reconstructed by applying the inverse of wavelet difference reduction coding followed by decryption. Suppose the subimages obtained from the compressed bit streams B_2, B_3, B_4 are R_2, R_3, R_4, respectively. Therefore, with the help of decrypted subimage R_1 and all the three subimages R_2, R_3, R_4 obtained from inverse WDR followed by decryption process, reconstructed image (R) is obtained.

$$R(2i, 2j) = R_1(i,j), \quad 1 \le i \le P/2, \ 1 \le j \le Q/2 \tag{13}$$

$$R(2i, 2j-1) = R_2(i,j), \quad 1 \le i \le P/2, \ 1 \le j \le Q/2 \tag{14}$$

$$R(2i-1, 2j) = R_3(i,j), \quad 1 \le i \le P/2, \ 1 \le j \le Q/2 \tag{15}$$

$$R(2i-1, 2j-1) = R_4(i,j), \quad 1 \le i \le P/2, \ 1 \le j \le Q/2 \tag{16}$$

Clearly, R represents the reconstructed image of size $P \times Q$ obtained from all the subimages.

4 Experimental Results and Discussion

The proposed ETC technique is implemented using Matlab 7.12.0. To evaluate the performance of proposed work, 8-bit grayscale images like Lena, Baboon, Peppers, Boat, Barbara, Cameraman, Fruits, etc., each of size 512 × 512 are used. The test images which have been used are the standard images and are most frequently used in many applications of image processing. After encrypting the original image, the encrypted data is compressed by setting maximum number of loops at encoder side with different values of α. For an instance, when Lena image is compressed using parameter $\alpha = 32$ and at WDR part of proposed scheme maximum number of loops have been set to 12. In this case, we have achieved compression ratio (CR) 0.42, with the help of the compressed data, secret key (k), and parameter α, we can retrieve the content of reconstructed image with 40.30 dB PSNR value. When we decreased the maximum number of loops at WDR on same value of α we have achieved 36.95 dB PSNR value with 0.23 CR. The proposed scheme is implemented on various test images and the CR and PSNR values in dB are calculated for each of the test images as shown in Table 1 for which we have taken $\alpha = 32$ and

Table 1 CR and PSNR values in dB for various test images

Images	α	Maximum number of loops	CR	PSNR
Lena	32	12	0.42	40.30
Boat	32	12	0.33	39.42
Baboon	32	12	0.54	37.07
Peppers	32	12	0.40	40.78
Fruit	32	12	0.50	43.97
Phantom	32	12	0.40	41.97
Cameraman	32	12	0.10	33.97
Barbara	32	12	0.30	38.97

Table 2 CR (bpp) and PSNR (dB) values of proposed technique on different parameter values

Image	α	Maximum number of loops	CR	PSNR
Lena	32	12	0.42	40.30
	32	10	0.23	36.95
	32	8	0.15	35.11
Lena	16	12	0.39	34.65
	16	10	0.19	32.65
	16	8	0.13	30.73
Lena	8	12	0.36	27.67
	8	10	0.16	27.11
	8	8	0.09	26.79
Lena	4	12	0.32	25.77
	4	10	0.13	25.09
	4	8	0.07	24.99

maximum number of loops are set to 12. Table 2 shows the CR and PSNR values in dB for the test image Lena, the value of compressed data and quality of reconstructed image are based on the parameter α and maximum number of loops set at WDR part of proposed work. For a given value of α, the decrease in the maximum number of loops at WDR part of proposed scheme has significantly improved the compression performance with slightly degrading the quality of reconstructed image. The compressor always tries to reduce the size of transmitting information with an acceptable level of distortion in the quality of reconstructed image.

The original test images and their corresponding reconstructed images are shown in Figs. 4 and 5, respectively. It is evident from Figs. 4 and 5 that the human visual

(a) (b) (c) (d) (e) (f) (g) (h)

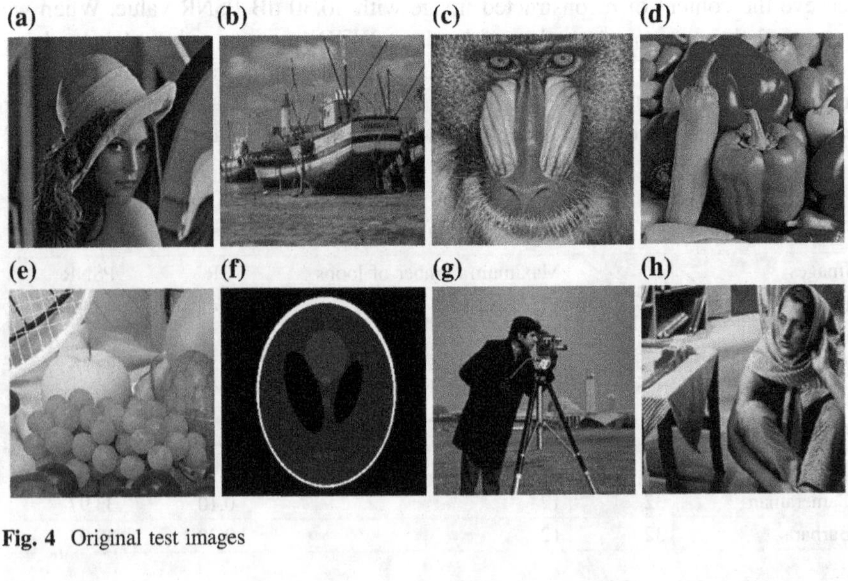

Fig. 4 Original test images

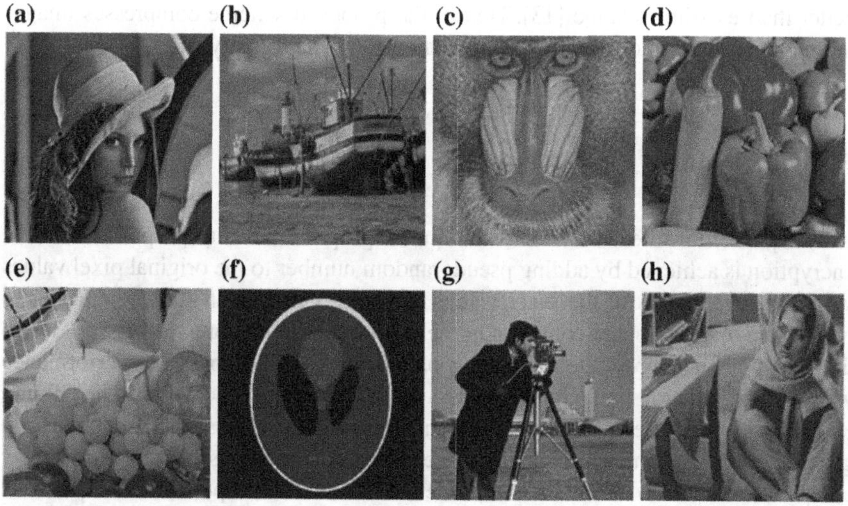

Fig. 5 Reconstructed images: **a** PSNR = 40.30, **b** PSNR = 39.42, **c** PSNR = 37.07, **d** PSNR = 40.78, **e** PSNR = 43.97, **f** PSNR = 41.97, **g** PSNR = 33.79, **h** PSNR = 38.89

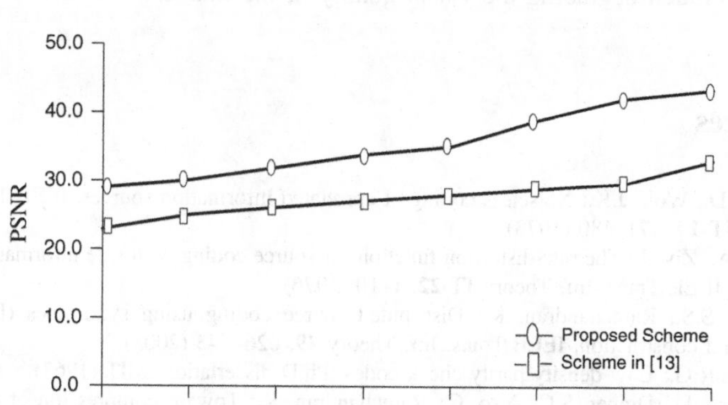

Fig. 6 Graph between CR and PSNR values in dB

system cannot make out any difference between the original and reconstructed images therefore, our scheme is reckon good without too much degrading the visual quality of reconstructed images. For a given CR, variation in the values of PSNR for Lena image is shown in Fig. 6. At the same time, in Fig. 6 the proposed scheme is also compared with scheme proposed by Kumar et al. [13] in which compressive sensing is introduced to perform lossy compression of encrypted image. It is evident from Fig. 6 that the proposed scheme outperforms the scheme proposed in [13] both in terms of CR and PSNR. The PSNR values in the proposed scheme are much better than existing scheme [13]. Hence, the proposed scheme compresses images more efficiently without too much degrading its visual quality.

5 Conclusions

In this paper, an effective ETC technique is proposed. In the proposed scheme, encryption is achieved by adding pseudorandom number to the original pixel values which simply masks all the pixel values and makes it difficult for any attacker to predict the original image content. On the other hand, in compression phase, due to the selection of appropriate quantizing parameter good compression performance with less degradation in the quality of reconstructed image is achieved. The CR and PSNR values are dependent on different values of compression parameters. In general, we can say that the quality of reconstructed image is better even for low compression rate. To demonstrate the effectiveness of proposed technique it is compared with an existing scheme and it shows better compression performance with improved image quality. Proposed scheme has an advantage of using WDR as a coding technique due to which we are obtaining high compression performance without too much degrading the visual quality of the image.

References

1. Slepian, D., Wolf, J.K.: Noiseless coding of correlated information sources. IEEE Trans. Inf. Theory **IT-19**, 471–480 (1973)
2. Wyner, A., Ziv, J.: The rate-distortion function for source coding with side information at the decoder. IEEE Trans. Inf. Theory **IT-22**, 1–10 (1976)
3. Pradhan, S.S., Ramchandran, K.: Distributed source coding using syndromes (DISCUS): design and construction. IEEE Trans. Inf. Theory **49**, 626–643 (2003)
4. Gallager, R.G.: Low density parity check codes. Ph.D dissertation, MIT (1963)
5. Schonberg, D., Draper, S.C., Yeo, C., Ramchandran, K.: Toward compression of encrypted images and video sequences. IEEE Trans. Inf. Forensics Secur. **3**(4), 749–762 (2008)
6. Johnson, M., Ishwar, P., Prabhakaran, V. M., Schonberg, D., Ramchandran, K.: On compressing encrypted data. IEEE Trans. Signal Process. **52**(10), 2992–3006 (2004)
7. Schonberg, D., Draper, S.C., Ramchandran, K.: On compression of encrypted images. Proc. IEEE Int. Conf. Image Process. **128**(3), 112–124 (1987)
8. Schonberg, D., Draper, S.C., Ramchandran, K.: On blind compression of encrypted correlated data approaching the source entropy rate. In: Proceedings of 43rd Annual Allerton Conference, Allerton, IL, USA (2005)
9. Lazzeretti, R., Barni, M.: Lossless compression of encrypted gray-level and color images. In: Proceedings of 16th European Signal Processing Conference, pp. 1–5 (2008)
10. Kumar, A., Makur, A.: Distributed source coding based encryption and lossless compression of gray scale and color images. In: Proceedings of MMSP, pp. 760–764 (2008)
11. Liu, W., Zeng, W., Dong, L., Yao, Q.: Efficient compression of encrypted gray scale images. IEEE Trans. Signal Process. **19**(4), 1097–1102 (2010)
12. Klinc, D., Hazay, C., Jagmohan, A., Krawczyk, H., Rabin, T.: On compression of data encrypted with block ciphers. IEEE Trans. Inf. Theory **58**(11) 6989–7001 (2012)

13. Kumar, A., Makur, A.: Lossy compression of encrypted image by compressing sensing technique. In: Proceedings of TENCON2009 IEEE Region 10 Conference, vol. C-23, no. 1, pp. 1–6 (2009)
14. Zhang, X.: Lossy compression and iterative reconstruction for encrypted image. IEEE Trans. Inf. Forensics Secur. **6**(1), 53–58 (2011)
15. Zhang, X., Feng, G., Ren, Y., Qian, Z.: Scalable coding of encrypted images. IEEE Trans. Image Process. **21**(6), 3108–3114 (2012)
16. Zhang, X., Sun, G., Shen, L., Qin, C.: Compression of encrypted images with multilayer decomposition. IEEE Trans. Signal Process. **78**(3), 1–13 (2013)
17. Zhang, X., Ren, Y., Shen, L., Qian, Z., Feng, G.: Compressing encrypted images with auxiliary information. IEEE Trans. Multimedia **16**(5), 1327–1336 (2014)
18. Zhou, J., Liu, X., Rao, K.R.: Designing an efficient image encryption-then-compression system via prediction error clustering and random permutation. IEEE Trans. Inf. Forensics Secur. **9**(1), 39–50 (2014)
19. Tian, J., Wells, R.O., Jr.: Image data processing in the compressed wavelet domain. In: Yuan, B. Tang, X. (eds.) Proceedings of 3rd International Conference on Signal Processing, vol. 147, no. 2, pp. 978–981 (1996)
20. Vaish, A., Kumar, M.: WDR coding based image compression technique using PCA. In: IEEE International Conference of Signal Processing and Communication, pp. 356–361 (2015)
21. Tian, J., Wells, R.O., Jr.: Embedded image coding using wavelet difference reduction. In: Topiwala, P. (ed.) Wavelet Image and Video Compression. Kluwer Academic Publication, pp. 289–301 (1998)

SVD-Based Fragile Reversible Data Hiding Using DWT

Manoj Kumar, Smita Agrawal and Triloki Pant

Abstract In today's growing world of digital technology, access to the multimedia content is very easy and for some sensitive applications such as medical imaging, military system, legal problems, it is very essential to not only reinstate the original media without any loss of information but also to increase content's security. Reversible data hiding is an approach to extract the information embedded covertly as well as the host image. In this paper, we have proposed a novel hybrid reversible watermarking scheme based on DWT and SVD. In this scheme, we have provided double layer of security by utilizing the multiresolution property of wavelet and strong features of SVD. In the proposed scheme, watermark is embedded into the singular values of all high-frequency subbands obtained by wavelet decomposition of the original image and at the time of extraction, watermark bits are used along with singular vectors to obtain the original image. Our scheme provides high security even after the extraction of watermark, without knowing the extraction algorithm, original image cannot be recovered in its entirety. The proposed scheme is tested on various test images and the obtained results after applying different performance metrics such as PSNR and UIQI show the effectiveness of the proposed scheme.

Keywords Reversible watermarking · Discrete wavelet transform · Singular value decomposition · PSNR · UIQI

Manoj Kumar · Smita Agrawal (✉)
Babasaheb Bhimrao Ambedkar University, Lucknow, UP, India
e-mail: smita.bbau@gmail.com

Manoj Kumar
e-mail: mkjnuiitr@gmail.com

Triloki Pant
Indian Institute of Information Technology, Allahabad, Allahabad, UP, India
e-mail: trylukky@gmail.com

M. Pant et al. (eds.), *Proceedings of Fifth International Conference on Soft Computing for Problem Solving*, Advances in Intelligent Systems and Computing 436, DOI 10.1007/978-981-10-0448-3_62

1 Introduction

There is always a story behind every picture, but how would anyone know whether the picture is tampered to change the story being told? Here comes the role of watermarking. The science of covertly hiding and conveying information is gaining a lot of importance and significance due to the very fast growing Internet and network technology. Digital watermarking is a process of secretly hiding information in multimedia content to deal with tampering, copying, piracy, and other such type of issues. But in this process, the pixel values are changed in such a way that they cannot be reversed to get back the original information which is not tolerable in some very sensitive imaging application areas, such as military field, medical imaging, legal matters, etc. This not only requires content authentication or protection but also the original information for their specific tasks. Reversible watermarking, a special type of digital watermarking, has come into existence for such type of highly sensitive applications where even a small loss of information is not acceptable. For example, a radiologist, who is doing diagnosis on the basis of the electronically received record of the patient which includes a CT image, may wrongly diagnose the presence or absence of a particular disease, even if there is a minor change in information. So, in such scenarios, where every bit of information is important, reversible watermarking is useful which not only authenticates the content but also recovers original data bit by bit. Due to this, reversible watermarking is also called "lossless" or "invertible" data hiding. Reversible watermarking is a type of fragile watermarking in which watermark is altered or destroyed if any modification is made in watermarked content. Therefore, being a subset of fragile watermarking, reversible watermarking is not robust against any signal processing attacks such as cropping, compression, etc.

Many reversible watermarking techniques have been proposed since Barton [1] established the concept of reversible watermarking in 1997. The work done in the reversible watermarking can be divided mainly into spatial, transform, and compressed domains. Honsinger et al. [2] proposed a reversible watermarking scheme in which they claimed to extract not only the embedded information but also to recover the exact original image. Their scheme was based on modulo 256 addition and suffered from salt-and-pepper noise. Since then many techniques based on different concepts such as difference expansion [3–8], histogram modification [9–12], prediction error-based embedding [13, 14], compression-based embedding [15–17], embedding in transform domain [18, 19] or combination of these [20], have been proposed.

Although, in literature a lot of work have been done in the field of irreversible data hiding using SVD [21–23], but very less work is reported in the field of lossless data hiding [24]. There are various properties of SVD, such as multiresolution, orthogonal subspaces, singular value distribution, etc., that have been exploited in many simple digital watermarking schemes but rarely in reversible watermarking. In simple digital watermarking, SVD plays an eminent role due to its various properties. The SVD contains the maximum energy in few coefficients and has the capability to adjust the variations in an image [25]. Most of the developed SVD-based watermarking techniques are based on the strength of singular values that specify the luminance of an image. Other than this, intrinsic algebraic image properties are represented by singular values. Therefore, small variation in singular values does not have remarkable influence on the watermarked image quality [25]. So, the SVD-based digital watermarking techniques have either used the largest singular values or the lowest singular values to embed the watermark either using quantization or using addition. In the field of reversible watermarking, these properties are still unexploited. Yoo et al. [24] proposed a reversible watermarking algorithm in which data for authentication was produced from XORing of the watermark image and singular values of one block image.

SVD provides some interesting algebraic and structural properties of an image. Use of SVD along with DWT can play a prominent role in reversible watermarking techniques, as the combination of salient features of these two provides more security in invertible data hiding. In our proposed work, we utilize the unexploited concept of SVD along with DWT to propose a novel reversible watermarking technique. Here, we are trying to use the properties of SVD in reversible data hiding.

2 Image Decompositions

2.1 Discrete Wavelet Transform (DWT)

2-D discrete wavelet transform is a widely used transform in image processing. DWT is based on the concept of wavelets. It is localized both in frequency and time domains. This reveals spatial and frequency aspects simultaneously. It is used for analyzing an image at different resolutions in different frequency components. Due to multiresolution property, features that may go unnoticed at one resolution, may be easily detected at another. Multiresolution analysis comprises image pyramid and subband coding theory.

For obtaining 2-D wavelet decomposition, 1-D DWT can be applied on image first in horizontal and then in vertical direction using different filters. 2-D DWT decomposes the image into two parts: approximation and detailed parts. Approximation part contains one low-frequency subband, LL and detailed part contains three high-frequency subbands, LH, HL, and HH. Approximation part can be further decomposed into four subbands. Decomposed subbands can be used to reconstruct the original image using inverse DWT.

High-frequency subbands of wavelet decomposition have Laplace distribution and this property can be utilized for the data embedding [26].

DWT of image $f(x,y)$ of size $M \times N$ is defined as [27]

$$W_{\varphi}(j_0,m,n) = \frac{1}{\sqrt{MN}} \sum_{x=0}^{M-1} \sum_{y=0}^{N-1} f(x,y)\phi_{j_0,m,n}(x,y) \tag{1}$$

$$W_{\psi}^{i}(j,m,n) = \frac{1}{\sqrt{MN}} \sum_{x=0}^{M-1} \sum_{y=0}^{N-1} f(x,y)\psi_{j,m,n}^{i}(x,y) \tag{2}$$

where $W_{\varphi}(j_0,m,n)$ defines approximation part of image $f(x,y)$ and $W_{\psi}^{i}(j,m,n)$ defines horizontal, vertical, and diagonal parts.

For the given Eqs. (1) and (2), inverse DWT is defined as

$$\begin{aligned} f(x,y) &= \frac{1}{\sqrt{MN}} \sum_{m} \sum_{n} W_{\varphi}(j_0,m,n)\phi_{j_0,m,n}(x,y) \\ &+ \frac{1}{\sqrt{MN}} \sum_{i=H,V,D} \sum_{j=j_0}^{\infty} \sum_{m} \sum_{n} W_{\psi}^{i}(j,m,n)\psi_{j,m,n}^{i}(x,y) \end{aligned} \tag{3}$$

2.2 *Singular Value Decomposition (SVD)*

SVD is a well-known technique of linear algebra to decompose a matrix into a set of linearly independent components such that each of them has its own energy contribution. It is a numerical method used to diagonalize the matrices in numerical analysis [28, 29], however, it can be seen from some other points of views also. On one hand, it can be seen as a method for changing correlated variables into a set of uncorrelated ones that better uncover the various relationships among the original data elements. On the other hand, SVD is a method for finding and ordering the dimensions along which data points show the most variation [30].

SVD decomposes a matrix A into matrices U, S, and V such that $A = USV^T$ (T represents transpose), where U and V contain singular vectors and S is a diagonal matrix containing singular values. The SVD of a digital image I of size $m \times n$ with $m \geq n$ can be defined as

$$I = USV^T \tag{4}$$

where U is a matrix of size $m \times m$ of left singular vectors, S is diagonal matrix of size $m \times n$ containing singular values in decreasing order, and V is matrix of size $n \times n$ containing right singular vectors.

Every singular value deals with the luminance of an image while the singular vectors specify the geometrical structure of the image [31]. Therefore, any change in U and V can change the geometrical structure of image but any small change in singular values would not affect the image in terms of its perceptibility.

3 Proposed Scheme

SVD is a very useful transformation that has been widely used in image processing and still has very much potential to be used in various areas. In the field of irreversible watermarking, various techniques have been proposed using SVD. In [21], authors proposed a digital watermarking technique based on SVD that exploits both singular values and singular vectors for embedding the watermark. In 2013, Benhocine et al. [22] gave technique for embedding the watermark based on SVD. In [23], authors have proposed optimal watermarking technique using particle swarm optimization and SVD. In this scheme, singular values of chosen detail subband are modified by multiple scaling factors and singular values of watermark are embedded. Therefore, it is evident that SVD has been used widely for developing irreversible watermarking techniques. It opens a spectrum of applications of SVD in reversible data hiding. So, we have proposed a reversible watermarking technique using unexploited concept of SVD that we have discussed in this section.

Our proposed reversible watermarking scheme utilizes the properties of not only DWT but also of SVD. We have utilized singular values of an image to hide the data in wavelet coefficient such that at the time of extraction, along with watermark, exact original image can also be recovered. Our scheme provides double layer of security: First at the SVD level and again at the level of embedding of watermark in image. The values of watermark are added to the singular values of wavelet coefficients and again, these watermark bits are embedded in the image. Thus, for extraction, one has to extract the watermark bits and using these bits along with singular vectors, original image can be recovered. If anyone finds out the watermark bits from watermarked image, even then original image cannot be recovered without knowing the exact extracting algorithm.

The mathematical theory behind our proposed scheme is as follows:

Let A be a matrix of size $m \times n$, then SVD of matrix A is

$$A = USV^T$$

Now, $US_1V^T = A_1$ (say), where $S_1 = S_{ii} + W_i$; $W \in \{0, 1\}$

Again,

$$V' = V_{ij}^T * W_i$$

(scalar multiplication of W_i with the ith row of V^T)

$$E = UV'$$

where E is the matrix of extra terms which came due to the addition of string W to singular values S_{ii}

Now, $A_1 - E = A$ (original matrix)

Following is the explanation of above theory for a 2×2 matrix A. Let us consider the singular value decomposition of a matrix A be $A = USV^T$ where

$$U = \begin{pmatrix} a & b \\ c & d \end{pmatrix}, S = \begin{pmatrix} e & 0 \\ 0 & f \end{pmatrix}, V = \begin{pmatrix} g & h \\ i & j \end{pmatrix} \tag{5}$$

Therefore,

$$A = USV^T = \begin{pmatrix} aeg + bfh & aei + bfj \\ ceg + dfh & cei + dfj \end{pmatrix} \tag{6}$$

Now, let us consider a watermark $W = (w_1, w_2)$ to be added to the diagonal elements of the diagonal matrix S. After the addition of watermark in singular values, the matrix S becomes

$$S_1 = \begin{pmatrix} e + w_1 & 0 \\ 0 & f + w_2 \end{pmatrix} \tag{7}$$

Now, let

$$A_1 = US_1V^T$$

$$= \begin{pmatrix} a & b \\ c & d \end{pmatrix} \begin{pmatrix} e + w_1 & 0 \\ 0 & f + w_2 \end{pmatrix} \begin{pmatrix} g & i \\ h & j \end{pmatrix}$$

$$A_1 = \begin{pmatrix} aeg + \underline{aw_1g} + bfh + \underline{bw_2h} & aei + \underline{aw_1i} + bfj + \underline{bw_2j} \\ ceg + \underline{cw_1g} + dfh + \underline{dw_2h} & cei + \underline{cw_1i} + dfj + \underline{dw_2j} \end{pmatrix} \tag{8}$$

It can be analyzed from above that there are extra terms (underlined) in every element of A_1 due to the watermark bits added to the singular values, which includes the values from U and V also. So, we have adjusted it in such a way that we get original value of A.

Now, using watermark W, matrix A_2 is computed such that

$$A_2 = V_{ij}^T * W_i \tag{9}$$

(scalar multiplication of W_i with the ith row of V^T)
Now,

$$A_3 = UA_2 = \begin{pmatrix} aw_1g + bw_2h & aw_1i + bw_2j \\ cw_1g + dw_2h & cw_1i + dw_2j \end{pmatrix}$$

$$\begin{aligned} B &= A_1 - A_3 \\ &= \begin{pmatrix} aeg + aw_1g + bfh + bw_2h & aei + aw_1i + bfj + bw_2j \\ ceg + cw_1g + dfh + dw_2h & cei + cw_1i + dfj + dw_2j \end{pmatrix} \\ &\quad - \begin{pmatrix} aw_1g + bw_2h & aw_1i + bw_2j \\ cw_1g + dw_2h & cw_1i + dw_2j \end{pmatrix} \\ &= \begin{pmatrix} aeg + bfh & aei + bfj \\ ceg + dfh & cei + dfj \end{pmatrix} \\ &= A \end{aligned}$$

It is evident from above example that the matrix A (original image) can be recovered from A_1 (watermarked image). It shows that by an adjustment of singular vectors and watermark, the original image can be recovered even after addition of watermark in its singular values.

Now, we explain embedding and extracting algorithm used in the proposed scheme.

3.1 Embedding Algorithm

1. Apply the discrete wavelet transform on the original image I and decompose I into four subbands, LL, LH, HL, and HH.
2. Apply SVD on each subband except LL.
3. Embed watermark bits in singular values of each subband such that

$$S_1(i,j) = \begin{cases} S(i,j) + W(i); & \text{if } i = j \\ S(i,j); & \text{otherwise} \end{cases}$$

4. Reconstruct each high-frequency subband as US_1V^T.
5. Apply inverse DWT to obtain the intermediate watermarked image.
6. Now embed watermark bits in image as follows:
   a. Choose any column(s) or row(s) of the image, notably last column.
   b. Here values are in double data type; extract integer part, left shift it. Make least significant bit (LSB) as 1 if bit to be embedded is 1 otherwise keep LSB as it is. Combine integer part with fraction part and obtain final watermarked image.

3.2 *Extracting Algorithm*

3.2.1 Extracting Watermark

1. Consider the watermarked image.
2. Consider the last column(s) or row(s) as used in embedding procedure and extract the integer part from it.
3. Number will be in $2p$ or $2p + 1$ form where p is the pixel value. Convert it into binary.
4. If LSB is 1, extract watermark security key bit as 1, subtract the LSB from the number, divide the resultant by 2, and construct values by putting together obtained quotient and fraction part otherwise set watermark bit as 0 and to obtain the values, divide the number by 2 and put together with its fractional part.

   a. For example, if the value from last column is obtained as 23.9504 then binary representation of integer part 23 would be 00010111. Here LSB is 1, so 1 is stored as watermark bit, and 1 is subtracted from the number and resultant is 22. After dividing 22 by 2 and putting the quotient with the fractional part of the number, 11.9504 is obtained which is the pixel value after extracting the watermark bit.

5. Repeat Step 2 by the number of times of the length of the watermark to extract the watermark.

Following are the steps to recover original image using extracted watermark:

3.2.2 Recovering Original Image

1. Apply DWT on the image left after the extraction of watermark.
2. Apply SVD on each subband except LL to obtain the values U', S', and V'.
3. Use extracted watermark for recovering original image. Make a matrix B_1 by scalar multiplication of W_i with the ith rows of V'^T, such that

$$B_1 = V'^T_{ij} W_i \tag{10}$$

Now,

$$B_2 = U' B_1 \tag{11}$$

Apply this procedure on each high-frequency subband.
Subtract B_2 from original values of that subband, i.e.,

$$B = (U'S'V'^T) - B_2; \tag{12}$$

4. Apply Inverse DWT on subbands obtained from Step 3.
5. The image obtained from the Step 4 is the original image.

4 Results and Discussion

The proposed algorithm has been applied on various test images using MATLAB. The images used for testing the proposed scheme are shown in Fig. 1. The proposed scheme can be applied on many levels of decompositions of DWT and therefore security will be increased but it may increase the computational complexity. At first level of wavelet decomposition of image of size $N \times N$, four subbands of size $N/2 \times N/2$ would be obtained. The maximum embedding that can be done in one subband using our proposed scheme is the total number of singular values of that subband, i.e., $N/2$. Thus, total embedding capacity at first level of decomposition would be $3N/2$ when embedding is done in detailed part, i.e., in three high-frequency subbands. If the image is of size 512×512, for one-level decomposition, the no. of embedded bits $= (3 \times 512)/2 = 768$. The embedding capacity at the k-level decomposition $= 3N \sum 1/2^k \rightarrow 3N$ as $k \rightarrow \infty$. Therefore, the maximum embedding capacity would converge toward $3N$ for any level of decomposition using DWT.

Being the subset of fragile watermarking technique, proposed reversible data hiding scheme cannot be tested against various signal processing attacks such as compression, cropping, etc.

In this scheme, we are dealing with double data type as after decomposition wavelet subbands are in double data type and again SVD is applied on the same data.

Two layers of embedding provide higher security such that after extracting watermark at first level, if extraction algorithm is not known, the original image cannot be recovered in its entirety.

4.1 Performance Metrics

4.1.1 Peak Signal-to-Noise Ratio (PSNR)

For measuring the distortion and visual quality, peak signal-to-noise ratio (PSNR) is used. It is used to measure distortion between original and watermarked images. Higher value of PSNR indicates better visual quality of the watermarked image.

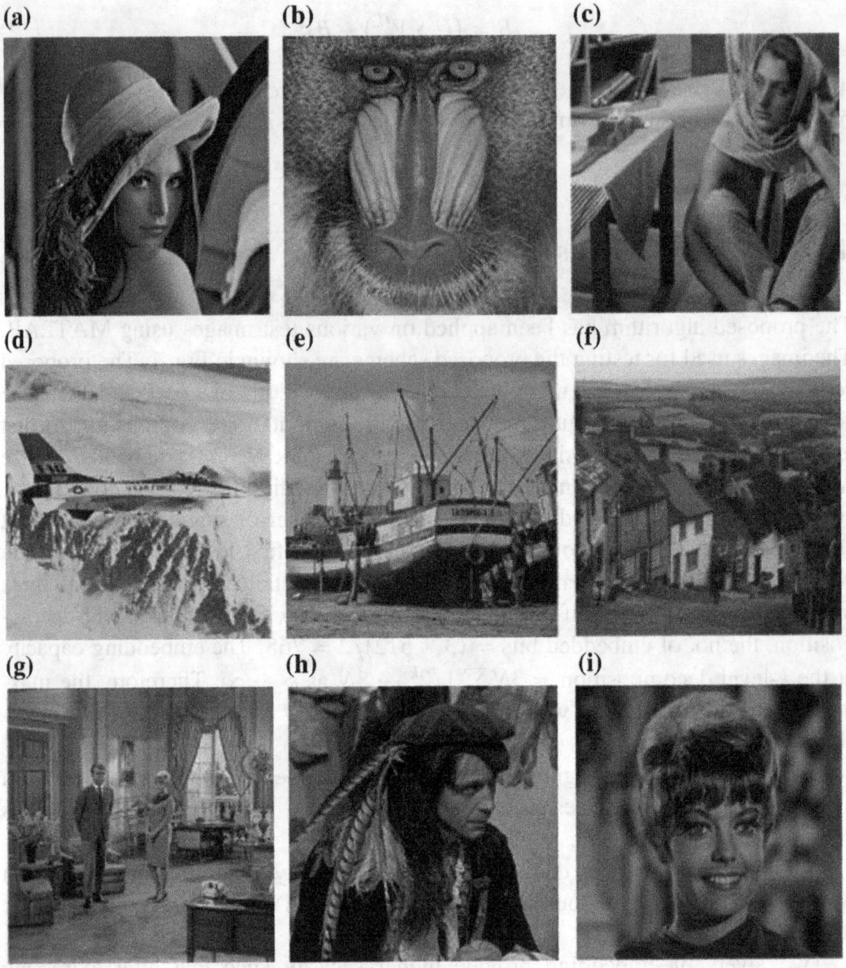

Fig. 1 Various standard test images of size 512 × 512. **a** Lena. **b** Mandrill. **c** Barbara. **d** Airplane. **e** Boat. **f** Goldhill. **g** House. **h** Man. **i** Zelda

$$\mathrm{PSNR} = 10 \times \log_{10}\left(\frac{255 \times 255}{\mathrm{MSE}}\right) \tag{13}$$

where mean squared error (MSE) is defined as

$$\mathrm{MSE} = \frac{1}{\mathrm{mn}}\sum_{i=0}^{m-1}\sum_{j=0}^{n-1}[I(i,j) - W(i,j)]^2$$

where I is the original image and W is the watermarked image.

Table 1 PSNR and UIQI values for various test images at different capacity using proposed scheme

Images	Capacity (no. of bits)	PSNR (db)	UIQI
Lena	256	43.1746	0.9992
	512	39.2699	0.9984
	768	38.6632	0.9974
Mandrill	256	42.9822	0.9996
	512	40.0038	0.9987
	768	38.5014	0.9981
Barbara	256	42.0653	0.9992
	512	39.2615	0.9985
	768	38.0083	0.9976
Airplane	256	38.5510	0.9990
	512	35.0109	0.9980
	768	34.9578	0.9975
Boat	256	41.5999	0.9993
	512	39.8263	0.9982
	768	38.0002	0.9978
Goldhill	256	42.7617	0.9993
	512	42.3330	0.9988
	768	41.0620	0.9982
House	256	40.2097	0.9992
	512	38.5040	0.9985
	768	36.0883	0.9977
Man	256	42.5304	0.9991
	512	39.0324	0.9982
	768	35.4231	0.9972
Zelda	256	43.7914	0.9991
	512	41.1350	0.9983
	768	40.4958	0.9974

Computed PSNR values for different test images at different capacities have been shown in Table 1. These PSNR values are obtained when the maximum embedding capacity is utilized (all watermark bits as 1). It is evident from the PSNR values at different capacity that our scheme provides good visual quality of watermarked image.

4.1.2 Universal Image Quality Index (UIQI)

Universal image quality index (UIQI) is an index designed for measuring image distortion as a combination of three factors—loss of correlation, contrast distortion, and luminance distortion [32]. The UIQI between original image I and watermarked image W is defined as

$$Q = \frac{4\sigma_{IW}\mu_I\mu_W}{(\sigma_I^2+\sigma_W^2)(\mu_I^2+\mu_W^2)} \tag{14}$$

where μ_I, μ_W, σ_I^2, σ_W^2 are the averages and variances of *I* and *W*, respectively, and σ_{IW} is the covariance between *I* and *W*.

The dynamic range of *Q* is [−1, 1]. The values of UIQI for proposed scheme are shown in Table 1. For proposed scheme, the values of UIQI between original images and watermarked images are almost near to 1 which shows the effectiveness of the proposed scheme.

5 Conclusions

A new scheme on reversible watermarking is presented in this paper, to deal with the extraction of watermark and also the recovery and security of the original image. The scheme proposed in this paper is based on the concept of discrete wavelet transform and singular values of high-frequency subbands. Proposed scheme is tested on various images and the experimental results are obtained which are visually good and PSNR values are also high. The high PSNR values at maximum embedding capacity indicate that our proposed scheme is good enough to produce high-quality watermarked images. UIQI values for proposed scheme also show that the degradation is almost negligible in watermarked images.

Proposed scheme provides a strong reversible feature of singular values that is used in proposed extraction algorithm to get back the original singular values for recovering the original image exactly. Two layers of embedding and salient algebraic property of the singular values provide very strong aspect of security. Properties of SVD are still not exploited to its fullest in the area of lossless data hiding and more work can be done on this aspect in future. Location map can be used to eliminate the restriction of using specific column(s) or row(s) at the time of watermark embedding in image.

References

1. Barton, J.M.: Method and apparatus for embedding authentication information within digital data. U.S. Patent No. 5646997 (1997)
2. Honsinger, C.W., Jones, P.W., Rabbani, M., Stoffel, J.C.: Lossless recovery of an original image containing embedded data. U.S. Patent No. 6278791. (2001)
3. Alattar, A.M.: Reversible watermark using difference expansion of triplets. In: Proceedings of International Conference on Image Processing (ICIP 2003), pp. 501–504, Sept 2003
4. Alattar, A.M.: Reversible watermark using difference expansion of quads. In: Proceedings of IEEE International Conference on Acoustics, Speech, and Signal Processing (ICASSP '04), pp. 377–380, May 2004

5. Lee, C.C., Wu, H.C., Tsai, C.S., Chu, Y.P.: Adaptive lossless steganographic scheme with centralized difference expansion. Pattern Recogn. **41**, 2097–2106 (2008)
6. Lin, C.C., Yang, S.P., Hsueh, N.L.: Lossless data hiding based on difference expansion without a location map. In: Proceedings of Congress on Image and Signal Processing (CISP'08), pp. 8–12 (2008)
7. Thodi, D.M., Rodríguez, J.J.: Expansion embedding techniques for reversible watermarking. IEEE Trans. Image Process. **16**, 721–730 (2007)
8. Tian, J.: Reversible data embedding using a difference expansion. IEEE Trans. Circuits Syst. Video Technol. **13**, 890–896 (2003)
9. Ni, Z., Shi, Y., Ansari, N., Su, W.: Reversible data hididng. IEEE Trans. Circuits Syst. Video Technol. **16**(3), 354–362 (2006)
10. Tai, W., Yeh, C., Chang, C.: Reversible data hiding based on histogram modification of pixel differences. IEEE Trans. Circuits Syst. Video Technol. **19**(6), 906–910 (2009)
11. Vleeschouwer, C.D., Delaigle, J.F., Macq, B.: Circular interpretation of histogram for reversible watermarking. In: IEEE 4th Workshop on Multimedia Signal Processing, pp. 345–350 (2001)
12. Vleeschouwer, C.D., Delaigle, J.F., Macq, B.: Circular interpretation of bijective transformations in lossless watermarking for media asset management. IEEE Trans. Multimedia **5**, 97–105 (2003)
13. Thodi, D.M., Rodríguez, J.J.: Prediction-error based reversible watermarking: In: Proceedings of International Conference on Image Processing, pp. 1549–1552 (2004)
14. Leung, H.Y., Cheng, L.M., Liu, F., Fu, Q.K.: Adaptive reversible data hiding based on block median preservation and modification of prediction errors. J. Syst. Softw. **86**, 2204–2219 (2013)
15. Fridrich, J., Goljan, M., Du, R.: Invertible authentication. In: Proceedings of SPIE Security and Watermarking of Multimedia Content, San Jose, U.S.A, pp. 197–208 (2001)
16. Fridrich, J., Goljan,M., Du, R.: Lossless data embedding—new paradigm in digital watermarking. EURASIP J. Appl. Signal Process. 185–196, (2002)
17. Celik, M.U., Sharma, G., Tekalp, A.M., Saber, E.: Lossless generalized-LSB data embedding. IEEE Trans. Image Process. **14**(2), 253–256 (2005)
18. Lee, S., Yoo, C.D., Kalker, T.: Reversible Image Watermarking Based on Integer-to-Integer Wavelet Transform. IEEE Trans. Inf. Forensics Secur. **2**(3), 321–330 (2007)
19. Xuan, G., Yang, C., Zheng, Y., Shi, Y.Q., Ni, Z.: Reversible data hiding based on wavelet spread spectrum. In: Proceedings of IEEE International Workshop on Multimedia Signal Processing (MMSP2004), Siena, Italy, pp. 211–214, Sept 2004
20. Alattar, A.M.: Reversible watermark Using the Difference Expansion of a Generalized Integer Transform. IEEE Trans. Image Process. **13**, 1147–1156 (2004)
21. Chang, C., Tai, P., Lin, C.: SVD-based digital watermarking scheme. Pattern Recogn. Lett. **26**, 1577–1586 (2005)
22. Benhocine, A., Laouamer, L., Nana, L., Pasc, A.C.: New images watermarking scheme based on singular value decomposition. J. Inf. Hiding Multimedia Signal Process. **4**(1), 9–18 (2013)
23. Loukhaoukha, K., Nabti, M., Zebbiche, K.: A robust SVD-based image watermarking using a multi-objective particle swarm optimization. OPTO-Electron. Rev. **22**(1), 45–54 (2014)
24. Yoo, K.S., Cho, O.H., Lee, W.H.: Reversible watermarking for authentication using singular value decomposition. In: Proceedings of International Conference on Electronics, Information and Communications, pp. 875–878 (2008)
25. Sadek, R.A.: SVD based image processing applications: state of the art, contributions and research challenges. Int. J. Adv. Comput. Sci. Appl. **3**(7), 26–34 (2012)
26. Li, C.T.: Emerging digital forensics applications for crime detection, prevention and security. IGI-Global, NY (2013)
27. Gonzalez, R.C., Woods, R.E.: Digital Image Processing, 3rd edn. Pearson Publication, New Delhi, India (2013)
28. Andrews, H.C., Patterson, C.L.: Singular value decompositions and digital image processing. IEEE Trans. Acoust. Speech Signal Process. **24**, 26–53 (1976)

29. Kamm, J.L.: SVD-based methods for signal and image restoration. Ph. D. Thesis (1998)
30. Baker, K.: Singular value decomposition tutorial, pp. 1577–1586 (2013)
31. Ganic, E., Zubair, N., Eskicioglu, A.M.: An optimal watermarking scheme based on singular value decomposition. In: Proceedings of IASTED International Conference on Communication, Network, and Information Security (CNIS 2003), pp. 85–90. Uniondale, NY, Dec 2003
32. Wang, Z., Bovik, A.C.: A universal image quality index. IEEE Signal Process. Lett. **9**(3), 81–84 (2002)

Static Economic Dispatch Incorporating UPFC Using Artificial Bee Colony Algorithm

S. Sreejith, Velamuri Suresh and P. Ponnambalam

Abstract Static economic dispatch is a real-time problem in power system network. Here, the real power output of each generating unit is calculated with respect to forecasted load demand over a time horizon while satisfying the system constraints. This paper explains the impact of unified power flow controller (UPFC) in static economic dispatch (SED) using artificial bee colony (ABC) algorithm. UPFC is a converter (shunt and series)-based FACTS device, which can control all the parameters in a transmission line individually or simultaneously. ABC algorithm that imitates the foraging behavior of honey bees is used as an optimization tool. The impact of UPFC in reducing the generation cost, loss, and improving voltage profile, power flow are demonstrated. The studies are carried out in an IEEE 118 bus test system and a practical South Indian 86 bus utility.

Keywords Economic dispatch · Artificial bee colony · Unified power flow controller · Voltage source converter

1 Introduction

Economic dispatch of generating units is one of the significant functions of contemporary energy management system. The static economic dispatch problem (SED) can be formulated as a constrained optimization problem which reduces the

S. Sreejith (✉) · Velamuri Suresh · P. Ponnambalam
School of Electrical Engineering, Vellore Institute of Technology,
Vellore, Tamil Nadu, India
e-mail: sreejith.s@vit.ac.in

Velamuri Suresh
e-mail: velamuri.suresh@vit.ac.in

P. Ponnambalam
e-mail: ponnambalam.p@vit.ac.in

M. Pant et al. (eds.), *Proceedings of Fifth International Conference on Soft Computing for Problem Solving*, Advances in Intelligent Systems and Computing 436, DOI 10.1007/978-981-10-0448-3_63

total generation cost within committed units satisfying system equality and inequality constraints. In conventional methods the cost curves of power generators are usually assumed to be quadratic and monotonically increasing functions. A variety of nonlinearities are present in the cost curves of modern generating units due to valve point loading. This results in inaccurate assumptions and the results in approximate solutions. On other hand, the evolutionary methods such as differential evolution (DE), particle swarm optimization (PSO), genetic algorithms (GA), and bacterial foraging (BF) are free from these convexity assumptions and perform better since they have excellent parallel search capability. Therefore, the above methods are particularly popular for solving nonlinear and nonconvex optimization problems. In dynamic programming [1] the shape of cost curves is unrestricted, but it consumes more time and suffers from dimensional issues. Optimization methods like dynamic programming [1], genetic algorithm [2–4], evolutionary programming [3], and particle swarm optimization [5–7] solve nonconvex optimization problems in a faster rate and efficient manner. ABC algorithm is used efficiently [8] for solving constrained optimization problems [9], so ABC algorithm is used as an optimization tool in the proposed methodology. It is a well-known fact that the consumption of power is increasing day by day. Simultaneously, there is a slow growth in the generation sector also. The growth in the power generation should be met with the necessary infrastructures added to the transmission system. The transmission system should be capable of handling the maximum power flow through the transmission lines without exceeding the permissible MVA limits. In order to use the existing transmission corridor efficiently, by transmitting maximum power through the transmission lines, FACTS (flexible AC transmission systems) devices are suitably incorporated into the power system network. Installation of FACTS devices in the network will improve the real power flow by providing reactive power support to the lines. In 1988, Hingorani [10] initiated the concept of FACTS devices with their applications. The UPFC is one of the multipurpose devices of the FACTS family which is used for improving power system stability and damping power system oscillations [11–13]. A current injected UPFC model of UPFC for improving power system stability is discussed in [14]. In [15] PSO algorithm is applied to locate the optimal position of UPFC considering the cost of installation and system loadability. The optimal location is identified by calculating the maximum system loadability (MSL) index.

2 Economic Dispatch Problem

The economic load dispatch is used to allocate the active power demand among the committed generating units. This is carried out by satisfying the system and unit constraints. In reality, the load demand changes with respect to time. Thereby, the entire dispatch period is divided into number of subintervals and static economic dispatch problem (EDP) is employed for each interval. The objective of EDP is to reduce the fuel cost simultaneously satisfying the constraints. The objective function is given as follows:

$$F(P_i) = \sum_{i=1}^{N} a_i + b_i \cdot P_i + c_i \cdot P_i^2 \tag{1}$$

subjected to the following constraints:

2.1 System Constraint

2.1.1 Real Power Constraint

To satisfy the load demand, the sum of the system load and the transmission line losses must be equal to the total generation. The power balance constraint is

$$\sum_{i=1}^{N} P_i - P_D - P_L = 0 \tag{2}$$

Kron's loss formula is used to find the approximate value of the losses, which is given by

$$P_L = \sum_{i=1}^{N} \sum_{i'=1}^{N} P_i B_{ii} P_i + \sum_{i=1}^{N} B_{io} P_i + B_{oo} \tag{3}$$

2.2 Unit Constraint

2.2.1 Generation Capacity Limit

The active power generated is limited by minimum and maximum power limits.

$$P_{i,\min} \le P_i \le P_{i,\max} \tag{4}$$

where F—Fuel cost function, a_i, b_i, c_i—Generator cost coefficients of ith unit, N—Number of generating units, P_i—Generation power output of unit i, $P_{i,\min}$ and $P_{i,\max}$—Minimum and maximum real power outputs of unit i, P_D—Total load demand, P_L—Total losses, and B_{00}, B_{i0}, B_{ii}—B loss coefficients.

2.3 UPFC Constraints

$$V_{vR\min} \leq V_{vR} \leq V_{vR\max} \quad (5)$$

$$V_{cR\min} \leq V_{cR} \leq V_{cR\max} \quad (6)$$

$$\delta_{vR\min} \leq \delta_{vR} \leq \delta_{vR\max} \quad (7)$$

$$\delta_{cR\min} \leq \delta_{cR} \leq \delta_{cR\max} \quad (8)$$

where

$V_{vR\min}, V_{vR\max}$ Minimum and maximum amplitudes of shunt voltage source
$V_{cR\min}, V_{cR\max}$ Minimum and maximum amplitudes of series voltage source
$\delta_{vR\min}, \delta_{vR\max}$ Minimum and maximum phase angles for shunt voltage sources
$\delta_{cR\min}, \delta_{cR\max}$ Minimum and maximum phase angles for series voltage sources.

3 Implementation of ABC Algorithm for Static Economic Dispatch

ABC algorithm is a metaheuristic algorithm for optimizing numerical problems [8, 9]. The performance of ABC algorithm is significant in solving various researches and engineering problems. The main components of ABC model are employed bees, food sources, onlooker or unemployed bees, and the dancing area. The employed bees fly round the search space and find the food sources using their experience. After completing their search process, the information on food source is shared to the unemployed bees which are in the hive. The information is shared by waggle dancing performed in the dancing area. By this dance, the information such as distance, direction, and the quality of the food sources are shared. The distance of the food source depends on the duration of the dance. The dancing time period will give the distance from the current position. In this work, the ABC algorithm is applied to determine the output power of all generation units for a specified load demand at a particular time horizon (T) to reduce the total generation cost. Here, solution of the problem is given by the position of the food source. The nectar amount will determine the quality of the solution. The steps for the proposed work are as follows:

Step 1 Specify the system and network parameters (generation power limits, ramp limits cost coefficients). Initialize the ABC algorithm parameters and the termination criteria.

Step 2 Initialize the population with m solutions in the solution space (food source positions), where m is the size of population. The solutions are represented by a

D-dimensional vector individually, where D is the total number of parameters which is to be optimized. In this work D is the number of generators. The elements of each solution denoted as x_{ij} is the real power output of generators which are distributed uniformly between their limits. This is given in (9).

$$P_{ij} = P_{j\min} + \text{rand}(0,1) * (P_{j\max} - P_{j\min}) \tag{9}$$

For each interval within the scheduling time, an initial population can be generated as in (10).

$$M = \begin{bmatrix} P_{11} & P_{12} \ldots & P_{1N} \\ P_{21} & P_{22} \ldots & P_{2N} \\ \cdot & \cdot & \cdot \\ \cdot & \cdot & \cdot \\ P_{m1} & P_{m2} \ldots & P_{mN} \end{bmatrix} \tag{10}$$

where P_{ij} is the real power output of the jth generator for the ith individual, N is the generating units. The size of employed bees will be half of the colony size.

Step 3 Fitness function evaluation

The fitness value evaluated for each food source in the colony is carried out using (11).

$$\text{Fitness} = \text{A} * [1 - \%\text{Cost}] + \text{B} * [1 - \%\text{Error}] \tag{11}$$

$$\text{Error} = \left| \sum_{i=1}^{N} P_i - P_L - P_D \right| \tag{12}$$

$$\%\text{Error} = \frac{\text{String error} - \text{Min error}}{\text{Max error} - \text{Min error}} \tag{13}$$

where

Stringcost	Generation cost of individual string,
Mincost	Minimum value of objective function,
Maxcost	Maximum value of objective function,
Stringerror	Error of individual string in meeting the power balance,
Minerror	Minimum error of constraints, and
Maxerror	Maximum error of constraints.

Minerror, Maxerror, and Stringerror are determined using (12). In (12) the individual with the highest fitness value will have the lowest cost. The best fitness value within the individuals and the cost corresponding to that are determined. The parameters accountable for the minimum cost are memorized. The cycle count is set and the following steps are carried out till it reaches the termination criteria MCN

Step 4 Position modification and site selection

The position of employed bees is modified for searching a new food source. The new food source is calculated by changing the value of any one of the *D* parameters (position of old food source) selected while all other parameters are kept unchanged. For the proposed problem this can be represented as in (14).

$$P_{ij} = P_{ij} + \phi_{ij} * (P_{ij} - P_{kj}) \tag{14}$$

where *j* is the index of the randomly generated selected parameter. If the n solution resulting from the modified position value violates the constraints, they are set to the maximum limits. Using (11) the fitness value of the new source is calculated. Here greedy selection method is employed as the selection mechanism. If the fitness of the new position is less, then the old population is retained.

Step 5 Recruiting onlooker bees for selected sites

After the completion of the search process by the employed bees they will share the information regarding the food sources to the onlooker bees. The onlooker bee evaluates the nectar information and finds out a food source depending on the fitness value probability. Roulette wheel selection technique is employed to place the onlookers onto the food source sites.

Step 6 Position modification by onlookers

The onlookers position is modified and the nectar amount of the candidate source is checked. Greedy selection mechanism is applied to select between the old and the new positions here. The nector information is evaluated by the onlooker bee and selects a food source with a probability. Roulette wheel selection technique is employed to place the onlookers onto the food source sites.

Step 7 Abandon sources

If there is no improvement in a solution representing a food source for a specified number of trials, then that food source will be abandoned. The scout bee discovers a new food source (*Xi*).

Step 8 The best solution achieved so far is memorized and the cycle count is incremented.

Step 9 Termination of the process

If the cycle count reaches the maximum number of cycles (MCN) the process is terminated. Otherwise, go to Step 4. The best fitness value and its corresponding food source are retained once the termination criteria are reached. This is the optimum output power of generating units for that time horizon.

Step 10 The count is incremented and the Steps from 2 to 9 are repeated for the time intervals. The total generation cost is computed for all the subintervals of the total time period *T*.

4 Mathematical Modeling of UPFC

Unified power flow controller comprises of two voltage source converters (VSC), one of them in series with the transmission line and the another in parallel with the line. Figure 1 shows the equivalent circuit of UPFC. Using UPFC, the active, reactive power flow and the magnitude of voltage in the UPFC terminals can be controlled individually or simultaneously. The magnitude of the inverter output V_{CR} decides the voltage regulation. The phase angle of the output voltage δ_{cR} decides the mode of power flow control in UPFC. The shunt converter may generate or absorb reactive power to provide independent control of voltage magnitude in its terminal in addition to its supportive role for active power exchange between series converter and the line. Figure 1 consists of two voltage sources (shunt and series) and a real power constraint equation that links the two VSC. The power flow equations for a three-phase UPFC is given as

$$E_{vR} = V_{vR}(\cos\delta_{vR} + j\sin\delta_{vR}) \tag{15}$$

$$E_{cR} = V_{cR}(\cos\delta_{cR} + j\sin\delta_{cR}) \tag{16}$$

The active power and reactive power equations at bus i are

$$\begin{aligned} P_i = {} & V_i^2 G_{ii} + V_i V_j\big(G_{ij}\cos(\theta_i - \delta_j) + B_{ij}\sin(\theta_i - \delta_j)\big) \\ & + V_i V_{cR}\big(G_{ij}\cos(\theta_i - \delta_{cR}) + B_{ij}\sin(\theta_i - \delta_{cR})\big) \\ & + V_i V_{cR}(G_{vR}\cos(\theta_i - \delta_{cR}) + B_{vR}\sin(\theta_i - \delta_{cR})) \end{aligned} \tag{17}$$

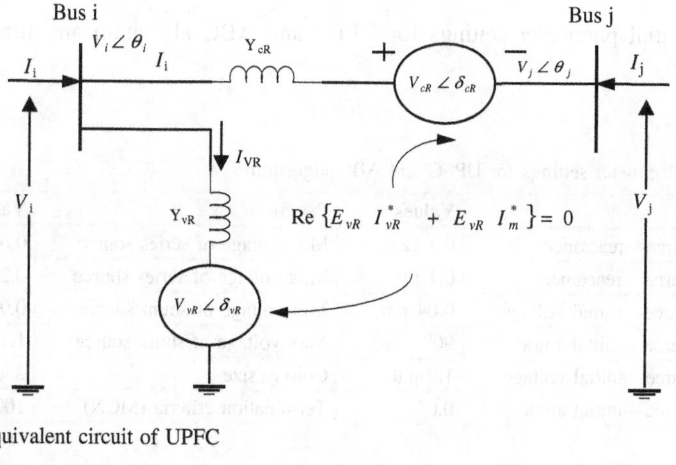

Fig. 1 Equivalent circuit of UPFC

$$Q_i = -V_i^2 B_{ii} + V_i V_j \big(G_{ij} \sin(\theta_i - \delta_j) + B_{ij} \cos(\theta_i - \delta_j)\big) + V_i V_{cR} \big(G_{ij} \sin(\theta_i - \delta_{cR}) + B_{ij} \cos(\theta_i - \delta_{cR})\big) + V_i V_{cR} (G_{vR} \sin(\theta_i - \delta_{cR}) + B_{vR} \cos(\theta_i - \delta_{cR})) \quad (18)$$

$$P_j = V_j^2 G_{jj} + V_j V_i \big(G_{ji} \cos(\theta_j - \delta_i) + B_{ji} \sin(\theta_j - \delta_i)\big) + V_j V_{cR} \big(G_{jj} \cos(\theta_j - \delta_{cR}) + B_{jj} \sin(\theta_j - \delta_{cR})\big) \quad (19)$$

The active power and reactive power equations at bus j are

$$Q_j = -V_j^2 B_{jj} + V_j V_i \big(G_{ji} \sin(\theta_j - \delta_i) + B_{ji} \cos(\theta_j - \delta_i)\big) + V_j V_{cR} \big(G_{jj} \sin(\theta_j - \delta_{cR}) + B_{jj} \cos(\theta_j - \delta_{cR})\big) \quad (20)$$

In a series converter,

$$P_{cR} = V_{cR}^2 G_{jj} + V_{cR} V_i \big(G_{ij} \cos(\delta_{cR} - \delta_i) + B_{ij} \sin(\delta_{cR} - \delta_i)\big) + V_{cR} V_j \big(G_{jj} \cos(\delta_{cR} - \delta_j) + B_{jj} \sin(\delta_{cR} - \delta_j)\big) \quad (21)$$

$$Q_{cR} = -V_{cR}^2 B_{jj} + V_{cR} V_i \big(G_{ij} \sin(\delta_{cR} - \delta_i) - B_{ij} \cos(\delta_{cR} - \delta_i)\big) + V_{cR} V_j \big(G_{jj} \sin(\delta_{cR} - \delta_j) - B_{jj} \cos(\delta_{cR} - \delta_j)\big) \quad (22)$$

Also in a shunt converter,

$$P_{vR} = -V_{vR}^2 G_{vR} + V_{cR} V_i (G_{vR} \cos(\delta_{vR} - \delta_i) + B_{vR} \sin(\delta_{vR} - \delta_i)) \quad (23)$$

$$Q_{vR} = V_{vR}^2 B_{vR} + V_{cR} V_j \big(G_{vR} \sin(\delta_{vR} - \delta_j) - B_{vR} \cos(\delta_{vR} - \delta_j)\big) \quad (24)$$

The initial parameter settings for UPFC and ABC algorithm are furnished in Table 1.

Table 1 Parameter settings for UPFC and ABC algorithm

Parameter	Values	Parameter	Values
Shunt source—reactance	0.1 Ω	Min voltage of series source	0.001 p.u.
Series source—reactance	0.1 Ω	Max voltage of series source	0.2 p.u.
Series source—initial voltage	0.04 p.u.	Min voltage of shunt source	0.9 p.u
Series source—initial angle	90°	Max voltage of shunt source	1.1 p.u.
Shunt source—initial voltage	1.0 p.u.	Colony size	250
Shunt source—initial angle	0.0°	Termination criteria (MCN)	100

5 Results and Discussion

5.1 SED Without UPFC in IEEE 118 Bus System

In this case, static economic dispatch is done without incorporating UPFC in IEEE 118 bus system. The data for the test system are obtained from https://www.ee.washington.edu/research/pstca/pf118/pg_tca118bus.htm. The dispatch is carried out for one hour and the real power dispatch is given in Table 2. The generation cost without FACTS devices is $99699.05. The voltage profile is shown in Fig. 2. It is evident that the voltages in certain buses are below 1.0 p.u.

See Table 2.

Table 2 SED without UPFC in IEEE 118 bus system

Generator	Dispatch	Generator	Dispatch	Generator	Dispatch	Generator	Dispatch
G4 (MW)	14.90	G34(MW)	30.00	G69(MW)	176.74	G92(MW)	167.33
G6(MW)	30.00	G36(MW)	47.61	G71(MW)	40.14	G99(MW)	300.00
G8(MW)	30.00	G40(MW)	17.07	G72(MW)	30.00	G10(MW)	218.71
G10(MW)	300.00	G42(MW)	30.00	G73(MW)	8.43	G103(MW)	20.00
G12(MW)	300.00	G46(MW)	100.00	G74(MW)	13.40	G104(MW)	68.70
G15(MW)	20.26	G49(MW)	210.67	G76(MW)	59.34	G105(MW)	100.00
G18(MW)	100.00	G54(MW)	175.34	G77(MW)	47.78	G107(MW)	20.00
G19(MW)	30.00	G55(MW)	100.00	G80(MW)	253.19	G110(MW)	50.00
G24(MW)	30.00	G56(MW)	100.00	G82(MW)	100.00	G111(MW)	100.00
G25(MW)	300.00	G59(MW)	97.22	G85(MW)	30.00	G112(MW)	58.52
G26(MW)	350.00	G61(MW)	200.00	G87(MW)	223.80	G113(MW)	100.00
G27(MW)	30.00	G62(MW)	33.15	G89(MW)	200.00	G116(MW)	50.00
G31(MW)	30.00	G65(MW)	420.00	G90(MW)	17.50	Generation cost ($)	99699.05

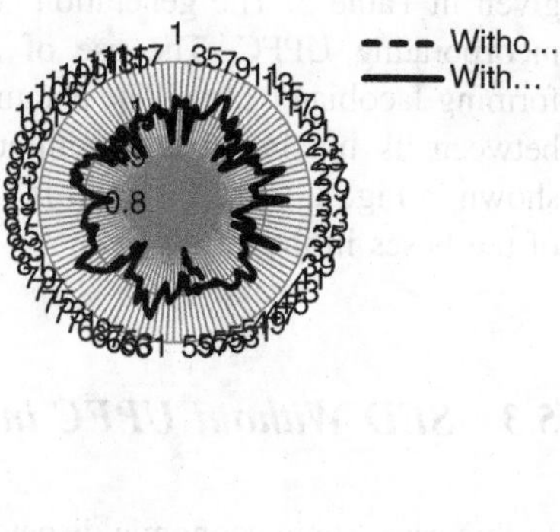

Fig. 2 Voltage profile without and with UPFC

Table 3 SED incorporating UPFC devices in 118 bus system

Generator	Dispatch	Generator	Dispatch	Generator	Dispatch	Generator	Dispatch
G4(MW)	22.62	G34(MW)	30.00	G69(MW)	300.00	G92(MW)	300.00
G6(MW)	5.74	G36(MW)	100.00	G71(MW)	80.00	G99(MW)	240.23
G8(MW)	30.00	G40(MW)	30.00	G72(MW)	24.88	G10(MW)	300.00
G10(MW)	196.04	G42(MW)	30.00	G73(MW)	30.00	G103(MW)	20.00
G12(MW)	300.00	G46(MW)	38.90	G74(MW)	20.00	G104(MW)	100.00
G15(MW)	30.00	G49(MW)	250.00	G76(MW)	100.00	G105(MW)	100.00
G18(MW)	71.25	G54(MW)	228.68	G77(MW)	53.64	G107(MW)	17.21
G19(MW)	30.00	G55(MW)	100.00	G80(MW)	169.98	G110(MW)	50.00
G24(MW)	28.15	G56(MW)	100.00	G82(MW)	41.54	G111(MW)	100.00
G25(MW)	179.74	G59(MW)	87.63	G85(MW)	30.00	G112(MW)	89.63
G26(MW)	350.00	G61(MW)	200.00	G87(MW)	156.45	G113(MW)	37.59
G27(MW)	30.00	G62(MW)	30.22	G89(MW)	171.35	G116(MW)	50.00
G31(MW)	30.00	G65(MW)	190.71	G90(MW)	20.00	Generation cost ($)	95924.00
G32(MW)	100.00	G66(MW)	420.00	G91(MW)	36.15		

Table 4 UPFC parameters in IEEE 118 bus system

Parameters	Values
V_{vR} (p.u.)	1.03
V_{cR} (p.u.)	0.58
$\delta_{vR}(°)$	−6.33
$\delta_{cR}(°)$	−94.12

5.2 SED with UPFC in IEEE 118 Bus Test System

Here, economic dispatch is carried out by incorporating UPFC in the test system. The initial parameter settings for UPFC are given in Table 1. The real power dispatch of generators incorporating FACTS devices is given in Table 3.

UPFC will inject power as required and maintains the power flow in the system, thus reducing the losses and burden of generators. The dispatch is very economical compared to the results without UPFC. The generation cost obtained with UPFC is given in Table 2. The generation cost is reduced compared to the results without incorporating UPFC. The use of ABC algorithm eliminates the complexity of forming Jacobian matrix. The parameters of UPFC are given in Table 4, which is in between its maximum and minimum limits. The voltage profile with UPFC is shown in Fig. 2. Form Fig. 2, it is evident that UPFC improves the voltage profiles of the buses in the test system.

5.3 SED Without UPFC in South Indian 86 Bus System

In this case, static economic dispatch is carried out in South Indian 86 bus system without incorporating UOFC. The line data, bus data, and generation coefficients

Table 5 Dispatch and cost with UPFC in SED (South Indian 86 bus)

Devices/generator	Without FACTS devices	With UPFC	Devices/generator	Without FACTS devices	With UPFC
G1	122.10	117.34	G10	106.21	88.641
G2	125.83	127.59	G11	75.779	64.401
G3	107.97	107.84	G12	89.164	74.766
G4	104.68	99.420	G13	109.85	100.15
G5	75.600	71.562	G14	135.30	94.036
G6	101.67	95.266	G15	127.98	101.64
G7	121.89	91.518	G16	132.35	110.53
G8	96.992	111.76	G17	95.103	93.466
G9	112.90	112.12	Generation cost without UPFC (₹)		359826.0
			Generation cost with UPFC(₹)		359294.3

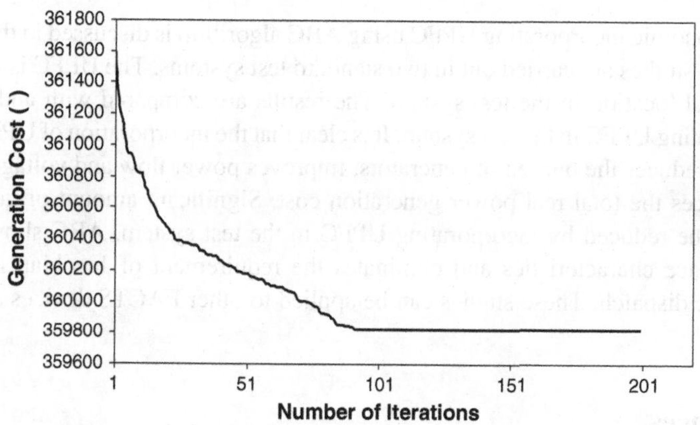

Fig. 3 Cost convergence for South Indian 86 bus system without UPFC

for the test system are obtained from [16]. The dispatch is carried out for 1 h. The real power dispatch of generating units is given in Table 5. The generation cost without FACTS devices is ₹36188.00. The graph of generation cost without FACTS device is shown in Fig. 3, which shows the better convergence property of ABC algorithm.

5.4 *SED with UPFC in South Indian 86 Bus System*

Here, economic dispatch is done by incorporating all FACTS individually in the test system. UPFC is incorporated and SED is done for a time horizon of one hour and the generator dispatch with FACTS devices are given in Table 5. The FACTS devices will inject power as required and maintain the power flow in the system, thus reducing the losses and burden of generators. Because of this, the dispatch is very economical compared to the case without UPFC. The generation cost obtained with UPFC is given in Table 5. The generation cost is reduced in this case compared to the results without incorporating FACTS devices. UPFC is the multi-converter device which has the combined shunt and series compensation has the generation cost of ₹359294.3, which is less compared to the case without UPFC. With the incorporation of UPFC, the power flow can be controlled without violating the economic optimal dispatch such that thermal limits of the transmission lines are not exceeded, losses and generation costs are minimized. The installation cost of UPFC is not considered in this paper.

6 Conclusion

Static economic incorporating UPFC using ABC algorithm is discussed in this paper. Two case studies are carried out in two standard test systems. The UPFC is placed in its optimal location in the test system. The results are compared with and without incorporating UPFC in the test system. It is clear that the incorporation of UPFC in the network reduces the burden of generators, improves power flow and voltage profile, and reduces the total real power generation cost. Significant amount of generation cost can be reduced by incorporating UPFC in the test system. ABC shows better convergence characteristics and eliminates the requirement of Jacobian matrix in economic dispatch. These studies can be applied to other FACTS devices also.

References

1. Shoults, R.R.: A dynamic programming based method for developing dispatch curves when incremental heat rate curves are non-monotonically increasing. IEEE Trans. Power Syst. **1**, 10–16 (1986)
2. Walter, D.C., Sheble, G.B.: Genetic algorithm solution of economic load dispatch with valve point loading. IEEE Trans. Power Syst. **8**, 1325–1332 (1993)
3. Sinha, N., Chakrabarti, R., Chattopadhyay, P.K.: Evolutionary programming techniques for economic load dispatch. IEEE Trans. Evol. Comput. **7**(1), 83–94 (2003)
4. Damousis, I. G., Bakirtzis, A.G., Dokopolous, P.S.: Network constrained economic dispatch using realcoded genetic algorithms. IEEE Trans. Power Syst. **18**(1), 198–205 (2003)
5. Selvakumar, A.I., Thanushkodi, K.: A new particle swarm optimization solution to nonconvex economic dispatch problems. IEEE Trans. Power Syst. **22**(1), 42–51 (2007)
6. Chaturvedi, K.T., Pandit, M., Srivastava, L.: Self-organizing hierarchical particle swarm optimization for nonconvex economic dispatch. IEEE Trans. Power Syst. **23**(3), 1079–1087 (2008)
7. Park, J.B., Jeong, Y.W., Shin, J.R.: An improved particle swarm optimization for nonconvex economic dispatch problems. IEEE Trans. Power Syst. **25**(1), 156–166 (2010)
8. Karaboga, D., Basturk, B.: Artificial Bee Colony (ABC) Optimization Algorithm for Solving Constrained Optimization Problems, vol. 4529, pp. 789–798. Springer (2007)

9. Karaboga, D., Basturk, B.: On the performance of artificial bee colony (ABC) algorithm. Appl. Soft Comput. **8**, 687–697 (2007)
10. Hingorani, N.G., Gyugyi, L.: Understanding FACTS: Concepts and Technology of Flexible AC Transmission Systems. IEEE press (2000)
11. Makombe, T., Jenkins, N.: Investigation of a unified power flow controller. In: Generation Transmission and Distribution IEE Proceedings, vol. 146, no. 4, pp. 400–408 (1999)
12. Xie, H., Xu, Z., Lu, Q., Song, Y.H., Yokoyama, A., Goto, M.: Integrated linear and nonlinear control of unified power flow controllers for enhancing power system stability. Electr. Power Compon. Syst. **31**, 335–347 (2003)
13. Ilango, G.S., Nagamani, C., Sai, A.V.S.S.R., Aravindan, D.: Control algorithms for control of real and reactive power flows and power oscillation damping using UPFC. Electr. Power Syst. Res. **79**, 595–605 (2009)
14. Meng, Z.J., So, P.L.: A current injection UPFC model for enhancing power system dynamic performance. In: Power Engineering Society Winter Meeting, vol. 2, pp. 1544–1549, 2000. IEEE (2000)
15. Saravanan, M., Mary Raja Slochanal, S., Venkatesh, P., Prince Stephen Abraham, J.: Application of particle swarm optimization technique for optimal location of FACTS devices considering cost of installation and system loadability. Electr. Power Syst. Res. **77**, 276–283 (2007)
16. Transmission element in southern region at a glance 2008, PowerGrid Corporation of India Ltd, Bangalore-560009

Edge Preservation Based CT Image Denoising Using Wavelet and Curvelet Transforms

Manoj Kumar and Manoj Diwakar

Abstract Computed tomography (CT) is a well-known medical radiological tool to diagnose the human body. Radiation dose is one of the major factors, which affects the quality of CT images. High radiation dose may improve the quality of image in terms of reducing noise, but it may be harmful for the patients. Due to low radiation dose, reconstructed CT images are noisy. To improve quality of noisy CT image, a postprocessing method is proposed. The goal of proposed scheme is to reduce the noise as much as possible by preserving the edges. The scheme is divided into two phases. In first phase, wavelet transform based denoisng is performed using bilateral filtering and thresholding. In second phase, a method noise thresholding based on curvelet transform is performed using the outcome of first phase. The proposed scheme is compared with existing methods. From experimental evaluation, it is observed that the performance of proposed scheme is superior to existing methods in terms of visual quality, PSNR and image quality index (IQI).

Keywords Image denoising · Bilateral method · Thresholding · Wavelet transform · Curvelet transform

1 Introduction

CT examination is widely used in medical science for detection of diseases such as lung cancer. High Radiation dose from clinical CT scanning is an increasing risk of cancer in the patients [1, 2]. Low-dose CT imaging may produce noisy images and degrade the diagnostic performance [3, 4]. Thus, there is a need to develop the techniques which can control the noise in low-dose CT scan images. It is observed

Manoj Kumar (✉) · Manoj Diwakar
Department of Computer Science, Babasaheb Bhimrao Ambedkar University, Lucknow, India
e-mail: mkjnuiitr@gmail.com

Manoj Diwakar
e-mail: manoj.diwakar@gmail.com

M. Pant et al. (eds.), *Proceedings of Fifth International Conference on Soft Computing for Problem Solving*, Advances in Intelligent Systems and Computing 436, DOI 10.1007/978-981-10-0448-3_64

that the CT images are generally degraded due to additive Gaussian noise. Denoising of an image is a popular method in image processing with a challenge to recover visually accepted image and preserve the details in terms of edges and texture. However, it is almost impossible to recover the visually accepted image without losing any details because of hardware/software faults, low radiation, finite computation, low data absorption and data loss during acquisition [5, 6]. The presence of noise may degrade the important features and reduce the clarity of edges in low contrast images [7].

Various postprocessing techniques have been proposed to denoise the images. In spatial domain, statistical linear filtering methods are popular to denoise the images but these blur the edges which may not be tolerable for medical images [8]. Some nonlinear filters in spatial domain are very popular for noise reduction and edge preservation such as bilateral filter [9, 10] and non-local means (NLM) filter [11]. Bilateral filter is a non-iterative, local filtering method which provides edge-preserved smoothing data [12]. NLM filter provides denoised images in terms of sharp edges based on self-similarity approach [13]. Edge-preserved denoising filters such as NLM and bilateral filters in spatial domain are providing good results but as the noise increases, very small edges in medical images may lose at the time of denoising. One other way for edge preservation and noise reduction is wavelet-based denoising specially for medical images. Several techniques have been produced to suppress the noise for preserving the edges in medical images for example multiscale wiener filtering method for low-dose CT images [14], image denoising algorithm via best wavelet packet base using Wiener cost function [15] and many other functions. Thresholding is one popular technique to reduce the noise in wavelet domain. Thresholding has been performed in various strategies where coefficients are thresholded based on coefficient dependency or inter-scale dependency such as ideal spatial adaptation via wavelet shrinkage [16]. Thresholding is one of the important tools for denoising. VISUShrink [17, 18] is a non-adaptive universal threshold, which depends only on number of samples and known to find smoothed images because its threshold choice can be large due to its dependence on the number of pixels in the images. SUREShrink [19, 20] uses a hybrid of the universal and the SURE [Steins Unbiased Risk Estimator] thresholds, and performs better than VISUShrink. BayesShrink [21, 22] minimizes the Bayes risk estimator function assuming generalized Gaussian approximation and thus finds adaptive threshold value. Recently, a new adaptive wavelet packet thresholding function [23] is used to denoised images using optimal linear thresholding function where noise is reduced and edges are preserved. Medical images can be compromised with small noise but not with loosing of medical relevant details.

In order to resolve the problem, we propose a scheme to reduce the noise and provide maximum edge information from noisy CT images using the combination of wavelet and curvelet transform. The paper is organized as follows. The related theories of the proposed scheme are introduced in Sect. 2. In Sect. 3, the proposed method is presented in details. Experimental results, including discussions and comparison with other denoising methods, are given in Sect. 4. Finally, conclusions are summarized in Sect. 5.

2 Preliminaries

This section covers two related concepts, including curvelet transform and bilateral filter which are in Sects. 2.1 and 2.2 respectively.

2.1 Curvelet Transform

The curvelet transform is based on multiscale ridgelets transform which provides several features in compare to wavelets and steerable pyramids such as more multidirectional, more multiresolution and co-localized the shape of frames in both time and frequency domain [24]. In curvelet transform, the signals are decomposed into several subbands where different orientations and positions are obtained at different scales in both frequency and time domain. It also obeys the scaling rule for the purpose fine scalability. The curvelet transform is a combination of wavelet and ridgelet transform. An image is isolated using 2D wavelet transform where spatial partitioning is used to divide each scale into blocks. Large and small scale wavelet components are partitioned into large and small block size respectively. To obtain curvelet coefficients, ridgelet transform is applied over the each block [25]. Image edges obtained through ridgelet are efficiently good and look like straight line over the certain scales. Curvelet transform is a localized ridgelet transform and extract more sharp edges such as curves.

The four major stages of discrete curvelet transform can be summarized over the function $f(x_1, x_2)$ as below:

(a) Subband decomposition: An image f is filtered into subbands by *à* trous algorithm [25] as below,

$$f \mapsto (P_0 f, \Delta_1 f, \Delta_2 f, \ldots) \tag{1}$$

where $P_0 f$ is a bank of low-pass filter and $\Delta_1, \Delta_2 \ldots$ are the bandpass (high-pass) filters.

(b) Smooth Partitioning: it is defined that each subband is smoothly windowed into a set of dyadic squares.

$$h_Q = (w_Q \cdot \Delta_S f) \tag{2}$$

where, w_Q is collection of smooth windowing function localized near dyadic squares Q_S, $Q \in Q_S$ and $S \geq 0$.

(c) Renormalization: Each dyadic square is further renormalized to the unit scale of [0, 1] × [0, 1] using the following function

$$g_Q = (T_Q^{-1} \cdot h_Q) \tag{3}$$

where, $(T_Q f)(x_1, x_2) = 2^S f(2^S x_1 - k_1, 2^S x_2 - k_2)$ is the renormalizing operator.

(d) Ridgelet analysis: For each $a > 0$, $b \in \Re$ and $\theta \in [0, 2\pi])$ in a given function $f(x_1, x_2)$, discrete ridgelet coefficients can be defined as:

$$\Re_f(a, b, \theta) = \int f(x_1, x_2)\Psi_{a,b,\theta}(x_1, x_2)\mathrm{d}x_1\mathrm{d}x_2 \tag{4}$$

where, a, b, and θ are the scale, location, and the orientation parameters, respectively, and Ψ is the wavelet function as, $\Psi_{a,b,\theta}(x) = a^{1/2}\Psi(\frac{x_1 \cos\theta + x_2 \sin\theta - b}{a})$. Discrete ridgelet transform is obtained through radon transform, can be analyzed as

$$\Re_f(a, b, \theta) = \int \wp_f(t, \theta)a^{1/2}\Psi\left(\frac{t - b}{a}\right)\mathrm{d}t \tag{5}$$

where, t is variable and $\wp_f$ is the Radon transform, which can be defined as below,

$$\wp_f(t, \theta) = \int f(x_1, x_2)\delta(x_1 \cos\theta + x_2 \sin\theta - t)\mathrm{d}x_1\mathrm{d}x_2 \tag{6}$$

where, δ is Dirac distribution.
2D fast discrete curvelet transform (2D FDCT) can be implemented via USFFT or wrapping methods. Both are used to translate the curvelets at each scale and angle but the way of spatial grid is differ. 2D FDCT via wrapping is implemented using parabolic scaling, anisotropic law, tight framing, and wrapping. The implementation is simple, fast, and less redundant.

2.2 *Bilateral Filtering*

Bilateral filter is a local, nonlinear and non-iterative edge preserving filter, introduced by Tomasi and Manduchi [9]. It has two filter kernels, first is a spatial filter kernel which behaves like a classical low-pass filter and other is an edge stoping function which attenuates the filter kernel when the intensity difference between the pixels is large. In bilateral method, spatial filter kernel is used to get the geometric closeness between two pixels. Similarly, edge stoping function is used for gray-level similarity between two pixels. Both filter kernels are based on Gaussian distribution, where the weights of filter depend on not only Euclidean distance but also on the distance in gray space. The advantage of bilateral filter is that it provides smooth and edge-preserved images using neighboring pixels. The Bilateral filter output $B_f(p)$ at a pixel location p is calculated as follows:

$$B_f(p) = \frac{1}{D} \sum_{q \in S} E_{\sigma_S}(\|p - q\|) \cdot E_{\sigma_r}(|B(p) - B(q)|)B(q) \tag{7}$$

where, $E_{\sigma_S}(\|p-q\|) = e^{-\frac{\|p-q\|^2}{2\sigma_S^2}}$ is a geometric closeness function,

$E_{\sigma_r}(|B(p) - B(q)|) = e^{-\frac{|B(p)-B(q)|^2}{2\sigma_r^2}}$ is a gray-level similarity/edge-stopping function,

$D = \sum_{q \in S} E_{\sigma_S}(\|p - q\|) \cdot E_{\sigma_r}(|B(p) - B(q)|)$ is a normalization constant, $\|p - q\|$ is the Euclidean distance between p and q. S is a spatial neighborhood of p. σ_S and σ_r control the behavior of bilateral filtering. The value of σ_S is chosen based on desired amount of low-pass filtering. Similarly, the value of σ_r is chosen based on desired amount of an edge to be preserved.

3 Proposed Methodology

With this assumption that the CT images are corrupted by Gaussian noise with zero mean and different variances, a scheme is proposed which combines the advantages of wavelet and curvelet transforms.

Let the noisy image $Y(i,j)$ be expressed as

$$Y(i,j) = W(i,j) + \eta(i,j) \tag{8}$$

where, $W(i,j)$ is a noiseless image and $\eta(i,j)$ is an additive Gaussian noise.

The block diagram of the proposed scheme is shown in Fig. 1. In our proposed scheme, wavelet transform is used to decompose the input image into low and high-frequency subband. Low-frequency subband is filtered using bilateral filtering. To denoise high-frequency subbands, a threshold value is estimated and threholding is performed. The threshold value can be estimated as,

$$\lambda = \frac{\sigma_\eta^2}{\sigma_W} \tag{9}$$

The variance σ_W^2 of noiseless image can be extracted as,

$$\sigma_W^2 = \max(\sigma_Y^2 - \sigma_\eta^2, 0) \tag{10}$$

where $\sigma_Y^2 = \frac{1}{b}\sum_{i=1}^{b} Y_i^2$ and b represents the pixel number in a selected block.

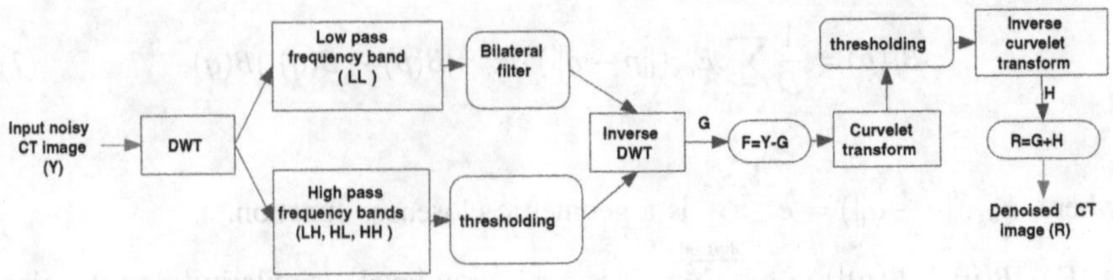

Fig. 1 Proposed scheme

The noise variance (σ_η^2) using robust median estimation method [26] can be estimated as,

$$\sigma_\eta^2 = \left[\frac{\text{median}(|Y(i \cdot j)|)}{0.6745}\right]^2 \tag{11}$$

where, $Y(i,j) \in HH$ subband.

To apply thresholding, soft thresholding method [27] is used which is defined as,

$$G := \begin{cases} \text{sign}(Y)(|Y| - \lambda), & |Y| > \lambda \\ 0, & \text{Otherwise} \end{cases} \tag{12}$$

The outcome of inverse wavelet reconstruction comes as denoised image (G). This denoised image is further processed to extract the missing edge information of the input noisy image. The original noisy CT image (Y) is subtracted from denoised image (G). To recover the maximum details such as edges, curvelet transform based thresholding is performed over subtracted image (F). The noise variance (σ_γ^2) of curvelet coefficient can be estimated as

$$\sigma_\gamma^2 = \sqrt{\frac{\sum_{i=1}^{N}\sum_{j=1}^{N} F_\gamma^{\text{Curvelet}} \cdot F_\gamma^{\text{Curvelet}^*}}{N^2}} \tag{13}$$

After estimating noise variance, curvelet coefficients are thresholded using following denoised function:

$$R_\gamma^{\text{Curvelet}} := \begin{cases} F_\gamma^{\text{Curvelet}}, & |F_\gamma^{\text{Curvelet}}| \geq K\sigma\sigma_\gamma \\ 0, & \text{Otherwise} \end{cases} \tag{14}$$

where, $F_\gamma^{\text{Curvelet}^*}$ is the complex conjugate of $F_\gamma^{\text{Curvelet}}$, σ is the noise variance of input subtracted noisy image (F) and K is the noise control scale dependent value.

Inverse curvelet transform gives an outcome of curvelet denoised image (H) which is added with denoised image (G) to get maximum extracted edges and denoised image (R).

The proposed scheme can be summarized with the following steps:

Step 1: Perform discrete wavelet transform (DWT) on input noisy CT image (Y) to obtain approximation and detail parts.
Step 2: For each decomposition level, apply bilateral filtering using Eq. (7) over the low-frequency subband.
Step 3: Denoise the image using following steps:

- (i) Estimate noise variance using Eq. (11)
- (ii) Estimate threshold value using Eq. (9)
- (iii) Apply thresholding on detail parts using Eq. (12)

Step 4: Apply inverse discrete wavelet transform (Inverse DWT) to obtain denoised image (G).
Step 5: Subtract original noisy CT image (Y) to denoised image (G) as, $F = Y - G$
Step 6: Apply curvelet transform on F and perform following steps:

- (i) Calculate noise variance (σ_γ^2) using Eq. (13)
- (ii) Perform thresholding using Eq. (14)

Step 7: Apply Inverse curvelet transform to achieve the output image (H).
Step 8: Perform $R = G + H$, to achieve the final denoised image with improved edges.

4 Results and Discussion

The experimental evaluation is performed on low-quality CT images with size 512 × 512. The CT scanned test images shown in Fig. 2a–c is obtained from public access database (https://eddie.via.cornell.edu/cgibin/datac/logon.cgi), and CT scanned test image shown in Fig. 2d is obtained from a diagnosis Center. The proposed image denoising method is applied to all test images corrupted by additive Gaussian white noise at four different noise levels (σ): 10, 20, 30, and 40.

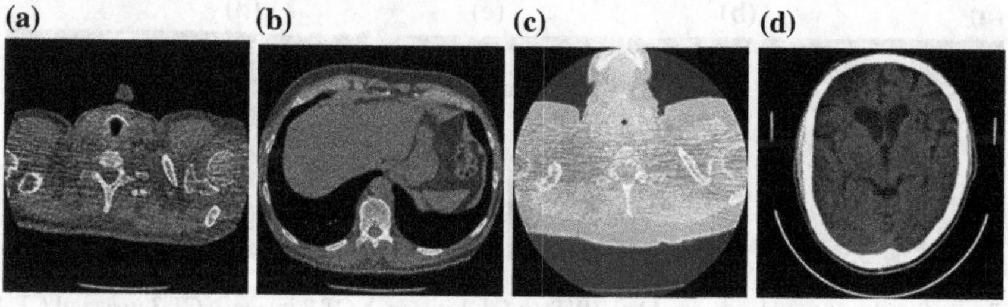

Fig. 2 Original CT image data set. **a** CT 1 image, **b** CT 2 image, **c** CT 3 image, **d** CT 4 image

Figure 2a–d are considered as CT images 1, 2, 3, and 4, respectively. Figure 3a–d are showing noisy test image data set with (σ) = 20. Over the noisy input images, 2D DWT is performed to get low and high-frequency subbands. Low-frequency subband is filtered using bilateral method. In our results, several parameters for bilateral filtering are used such as patch size is 10 × 10, σ_S is 1.2 and σ_r is 0.13. In high-frequency subbands, threshold value is estimated and thresholded for their respective subband as discussed in proposed methodology. Inverse DWT, gives an reconstructed denoised image. But from experimental evaluations (where we subtract the original noisy CT image to denoised image), it is observed that the edges are missing as the level of noise increases. To recover that, curvelet transform via wrapping is used to extract more edge information. Over the subtracted image (F), curvelet-based denoising as discussed in proposed methodology have been performed. To obtain final denoised image (R), subtracted denoised image (H) and denoised image (G) are added. To validate the superiority of proposed scheme, the performance is compared with existing methods. The existing methods for comparison are wavelet thresholding (WT) [21], wavelet thresholding and bilateral filtering (WTBF) [10] and curvelet-based thresholding (CBT) [25]. Figures 4a–d, 5a–d, 6a–d, and 7a–d are showing the results of wavelet thresholding (WT), wavelet thresholding and bilateral filtering (WTBF), curvelet-based thresholding (CBT), and proposed scheme, respectively. For CT images (1–4), PSNR and IQI are measured for proposed scheme as well as existing methods with noise level $\sigma = [10, 20, 30, 40]$.

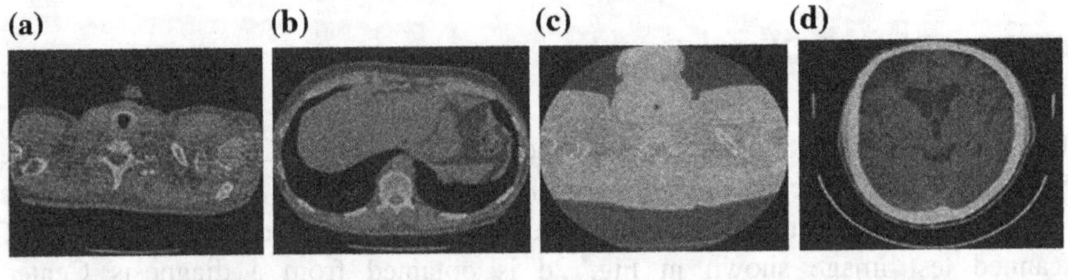

Fig. 3 Noisy CT image data set (σ = 20). **a** CT 1 image, **b** CT 2 image, **c** CT 3 image, **d** CT 4 image

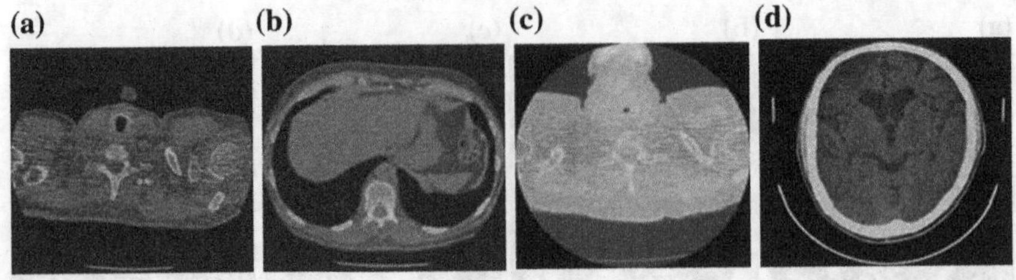

Fig. 4 Results of wavelet thresholding (WT). **a** CT 1 image, **b** CT 2 image, **c** CT 3 image, **d** CT 4 image

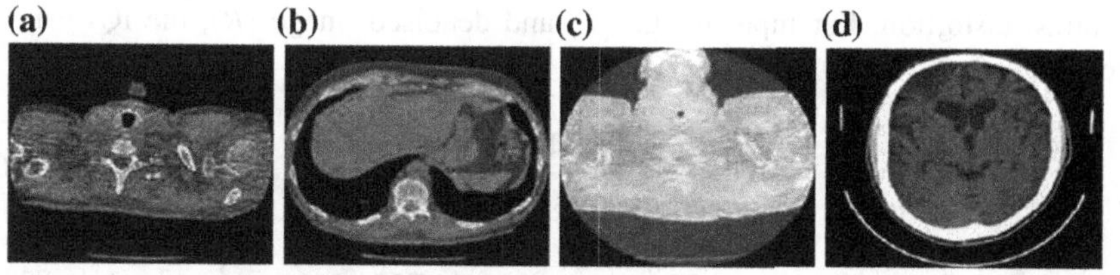

Fig. 5 Results of wavelet thresholding and bilateral filtering (WTBF). **a** CT 1 image, **b** CT 2 image, **c** CT 3 image, **d** CT 4 image

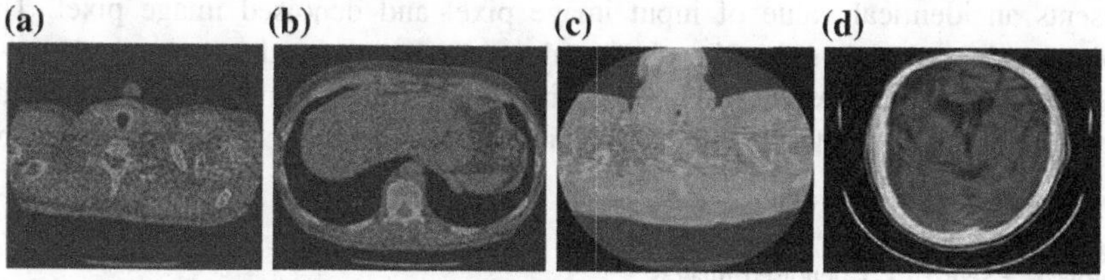

Fig. 6 Results of curvelet-based denoising (CBT). **a** CT 1 image, **b** CT 2 image, **c** CT 3 image, **d** CT 4 image

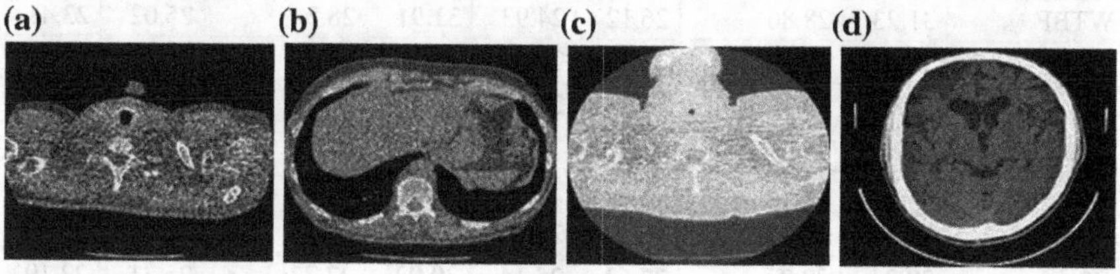

Fig. 7 Results of proposed scheme. **a** CT 1 image, **b** CT 2 image, **c** CT 3 image, **d** CT 4 image

Peak signal-to-noise ratio is an important factor to evaluate denoising performance. The high PSNR value represents more similarity between the denoising and original image than lower PSNR value. For input image (Y) and denoised image (R), the PSNR can be expressed as:

$$\mathrm{PSNR} = 10 \times \log_{10}\left(\frac{255 \times 255}{\mathrm{MSE}}\right) \tag{15}$$

where, Mean Square Error (MSE) is defined as-

$$\mathrm{MSE} = \frac{1}{mn}\sum_{i=0}^{m-1}\sum_{j=0}^{n-1}[Y(i,j) - R(i,j)]^2$$

Image quality index (IQI) is one of the important factors to analyze the performance of image denoising in terms of correlation, luminance distortion and contrast distortion. For input image (Y) and denoised image (R), the IQI can be defined as

$$\mathrm{IQI} = \frac{4\sigma_{YR}\bar{Y}\bar{R}}{(\sigma_Y^2 + \sigma_R^2)[(\bar{Y})^2 + (\bar{R})^2)]} \tag{16}$$

where, $\bar{Y} = \frac{1}{N}\sum_{i=1}^{N} Y_i$, $\bar{R} = \frac{1}{N}\sum_{i=1}^{N} R_i$, $\sigma_Y^2 = \frac{1}{N-1}\sum_{i=1}^{N}(Y_i - \bar{Y})^2$, $\sigma_R^2 = \frac{1}{N-1}\sum_{i=1}^{N}$ $(R_i - \bar{R})^2$ and $\sigma_{YR} = \frac{1}{N-1}\sum_{i=1}^{N}(Y_i - \bar{Y})(R_i - \bar{R})$.

The quality of image index range lies between 1 to −1. The best value 1 represents an identical value of input image pixel and denoised image pixel. The lowest value −1 shows that the pixels values are uncorrelated.

Tables 1 and 2, respectively, show the PSNR (in dB) and IQI values of the denoised images relative to their original images using proposed and existing

Table 1 PSNR of CT denoised images

σ	10	20	30	40	10	20	30	40
Input image	CT1	512 × 512			CT2	512 × 512		
WT	32.41	29.03	26.72	23.12	30.34	28.62	25.41	22.69
WTBF	31.23	28.86	26.12	24.93	**31.91**	28.34	25.02	23.41
CBT	32.11	29.06	27.35	25.01	31.73	28.08	25.94	22.51
Proposed	**32.92**	**29.16**	**27.73**	**25.33**	31.83	**28.76**	**26.03**	**23.83**
Input image	CT3	512 × 512			CT4	512 × 512		
WT	31.62	**29.75**	27.02	25.13	31.52	28.35	**26.65**	22.42
WTBF	30.57	28.12	26.93	24.36	30.14	28.75	26.35	23.10
CBT	30.34	29.23	**27.63**	25.14	30.03	27.72	25.31	22.19
Proposed	**31.97**	29.41	27.45	**25.19**	**31.73**	**28.91**	26.23	**23.31**

Table 2 IQI of CT denoised images

σ	10	20	30	40	10	20	30	40
Input image	CT1	512 × 512			CT2	512 × 512		
WT	0.9946	0.9826	0.9156	0.8943	0.9952	0.9812	0.9187	0.8692
WTBF	0.9958	0.9863	0.9212	0.8623	**0.9996**	0.9823	0.9223	0.8732
CBT	0.9994	0.9882	0.9287	0.8713	0.9993	0.9891	0.9272	0.8791
Proposed	**0.9996**	**0.9894**	**0.9313**	**0.9041**	0.9994	**0.9895**	**0.9283**	**0.8871**
Input image	CT3	512 × 512			CT4	512 × 512		
WT	0.9982	**0.9812**	0.9234	0.8712	0.9983	0.9893	**0.9423**	0.8699
WTBF	0.9985	0.9702	0.9282	0.8713	0.9989	0.9785	0.9267	0.8755
CBT	0.9991	0.9731	**0.9402**	0.8793	0.9994	0.9892	0.9142	0.8795
Proposed	**0.9992**	0.9390	0.9683	**0.8981**	**0.9998**	**0.9899**	0.9409	**0.8898**

methods. The best values amongst all the methods are represented in bold. The results shown in tables demonstrate that in most of the cases, the proposed method is superior to all other methods. From experimental results, it is observed that the visual quality of proposed scheme gives better denoising near the edges while other methods give noisy results on the edges.

5 Conclusions

A postprocessing approach is investigated to reduce noise in CT images and extracted more edge information by taking the advantage of wavelet and curvelet transforms. To provide smooth and edge-preserved image, adaptive bilateral based wavelet thresholding is used. Using method noise concept, the edges are more extracted using curvelet-based thresholding. Final outcomes of proposed scheme are excellent in terms of noise reduction and edge preservation. The PSNR and IQI values of proposed scheme are better in comparison of existing methods. Apart from PSNR and IQI, the visual quality of proposed scheme is also better in terms of clinically relevant details. Experimental results demonstrate that our proposed method: (i) effectively eliminate the noise in CT images, (ii) preserve the edge and structural information, and (iii) retain clinically relevant details. The proposed scheme can be more enhanced in future by optimizing the variation of similar patches using soft computing methods such as genetic algorithms and neural networks.

References

1. Huda, W., Scalzetti, E.M., Levin, G.: Technique factors and image quality as functions of patient weight at abdominal CT. J. Radiol. **217**(2), 430–435 (2000)
2. Boone, J.M., Geraghty, E.M., Seibert, J.A., Wootton-Gorges, S.L.: Dose reduction in pediatric CT: a rational approach. J. Radiol. **228**(2), 352–360 (2003)
3. Siegel, M.J., Schmidt, B., Bradley, D., Suess, C., Hildebolt, C.: Radiation dose and image quality in pediatric CT: effect of technical factors and phantom size and shape. J. Radiol. **233** (2), 515–522 (2004)
4. Li, T., Li, X., Wang, J.: Nonlinear sinogram smoothing for low-dose x-ray CT. IEEE Trans. Nucl. Sci. **51**(5), 2505–2513 (2004)
5. Candes, E.J., Romberg, J., Tao, T.: Robust uncertainty principles: exact signal reconstruction from highly incomplete frequency information. IEEE Trans. Inf. Theory **52**(2), 489–509 (2006)
6. Samuel, J.L., Emil, Y.S., Xiaochuan, P.: Accurate image reconstruction from few-view and limited-angle data in diffraction tomography. J. Opt. Soc. Am. A **25**(7), 1772–1782 (2008)
7. Kim, D., Ramani, S., Fessler, J.A.: Accelerating X-ray CT ordered subsets image reconstruction with Nesterov first-order methods. In: Proceedings of Fully Three-Dimensional Image Reconstruction in Radiology and Nuclear Medicine, pp. 22–25 (2013)

8. Kachelriess, M., Knaup, M., Bockenbach, O.: Hyperfast parallel-beam and cone-beam backprojection using the cell general purpose hardware. J. Med. Phys. **34**(4), 1474–1486 (2007)
9. Tomasi, C., Manduchi, R.: Bilateral filtering for gray and color images. In: Proceedings on Computer Vision, Bombay, pp. 839–846 (1998)
10. Wenxuan, S., Jie, L., Minyuan, W.: An image denoising method based on multiscale wavelet thresholding and bilateral filtering. Wuhan Univ. J. Nat. Sci. **15**(2), 148–152 (2010)
11. Buades, A., Coll, B., Morel Song, J.M.: A review of image denoising algorithms, with a new one. SIAM J. Multiscale Model. Simul. **4**(2), 490–530 (2005)
12. Shreyamsha Kumar, B.K.: Image denoising based on non-local means filter and its method noise thresholding. SIViP **7**(6), 1211–1227 (2013)
13. Mustafa, Z.A., Kadah, Y.M.: Multi resolution bilateral filter for MR image denoising. In: Proceedings of 1st Middle East Conference on Biomedical Engineering (MECBME), Sharjah, pp. 180–184 (2011)
14. Li, K., Zhang, R.: Multiscale wiener filtering method for low-dose CT images. In: Proceedings of IEEE Biomedical Engineering and Informatics, pp. 428–431 (2010)
15. Li, Y., Yi, X., Xu, J., Li, Y.: Wavelet packet denoising algorithm based on correctional wiener filtering. J. Inf. Comput. Sci. **10**(9), 2711–2718 (2013)
16. Chang, S.G., Yu, B., Vetterli, M.: Adaptive wavelet thresholding for image denoising and compression. IEEE Trans. Image Process. **9**(9), 1532–1546 (2000)
17. Marpe, D., Cycon, H.L., Zander, G., Barthel, K.U.: Context-based denoising of images using iterative wavelet thresholding. In: Proceedings of SPIE on Visual Communications and Image Processing, pp. 907–914 (2002)
18. Cristobal, G., Cuesta, J., Cohen, L.: Image Filtering and denoising through the scale transform. In Proceedings of the International Symposium on Time-Frequency and Time-Scale Analysis, Pittsburgh, pp. 617–620 (1988)
19. Donoho, D.L., Johnstone, I.M.: Ideal spatial adaptation via wavelet shrinkage. Biometrika **81**(3), 425–455 (1994)
20. Donoho, D.L.: Denoising by soft thresholding. IEEE Trans. Inf. Theory **41**(3), 613–627 (1995)
21. Chang, S.G., Yu, B., Vetterli, M.: Spatially adaptive thresholding with context modeling for image denoising. IEEE Trans. Image Process. **9**(9), 1522–1531 (2000)
22. Donoho, D.L., Johstone, I.M.: Adapting to unknown smoothness via wavelet shrinkage. J. Am. Stat. Assoc. **90**(432), 1200–1224 (1995)
23. Fathi, A., Naghsh-Nilchi, A.R.: Efficient image denoising method based on a new adaptive wavelet packet thresholding function. IEEE Trans. Image Process. **21**(9) 3981–3990 (2012)
24. Donoho, D.L., Duncan, M.R.: Digital Curvelet Transform: Strategy, Implementation and Experiments. Standford University, Standford (1999)
25. Bhadauria, H.S., Dewal, M.L.: Performance evaluation of curvelet and wavelet based denoising methods on brain computed tomography images. IEEE International Conference on Emerging Trends in Electrical and Computer Technology (ICETECT), pp. 666–670 (2011)
26. Liu, X., Tanaka, M., Okutomi, M.: Single image noise level estimation for blind denoising. IEEE Trans. Image Process. **22**(12), 5226–5237 (2013)
27. Borsdorf, A., Raupach, R., Flohr, T., Hornegger, J.: Wavelet based noise reduction in CT-images using correlation analysis. IEEE Trans. Med. Imaging **27**(12), 1685–1703 (2008)

Critical Analysis of Clustering Algorithms for Wireless Sensor Networks

Santar Pal Singh, Kartik Bhanot and Sugam Sharma

Abstract The scientific and industrial community increased their attention on wireless sensor networks (WSNs) during the past few years. WSNs are used in various critical applications like disaster relief management, combat field reconnaissance, border protection, and security observation. In such applications a huge number of sensors are remotely deployed and have cooperatively worked in unaccompanied environments. The disjoint groups are formed from these sensor nodes and such nonoverlapping groups are known as clusters. Clustering schemes have proven to be effective to support scalability. In this paper, authors have reported a detailed analysis on clustering algorithms and have outlined the clustering schemes in WSNs. We also make a comparative analysis of clustering algorithms on the basis of different parameters like cluster stability, cluster overlapping, convergence time, failure recovery, and support for node mobility. Moreover, we highlight the various issues in clustering of WSNs.

Keywords Wireless sensor network · Clustering algorithms · Convergence · Scalability

1 Introduction

Latest developments in wireless communication and low-cost sensor technology have enabled the emergence and evolution of sensor networks as new paradigm of computer networking [1]. The WSNs were initially motivated by military appli-

S.P. Singh (✉)
Electronics and Computer Discipline, DPT, IIT Roorkee, Rookee 247667, India
e-mail: spsingh78@gmail.com

Kartik Bhanot
ICT Department, MIT, Manipal 576104, Karnatka, India
e-mail: ks7795170077@gmail.com

Sugam Sharma
Centre for Survey Statistics and Methodology, Iowa State University, Ames, IA 50010, USA
e-mail: sugam.k.sharma@gmail.com

M. Pant et al. (eds.), *Proceedings of Fifth International Conference on Soft Computing for Problem Solving*, Advances in Intelligent Systems and Computing 436, DOI 10.1007/978-981-10-0448-3_65

cations but nowadays, sensor networks are used in various civilian application areas like: monitoring, tracking, control, automation, and healthcare applications. A WSN is a collection of huge number of low-cost, tiny wireless nodes enabled with sensing, processing, and transmitting capabilities [2, 3]. The WSNs communication architecture is shown in Fig. 1. The key components of a node are: a microsensor to detect the desired event, a microprocessor to process the information, memory to store the information, a battery, and a transceiver to communicate with rest of the network [3]. Sensor nodes communicate together by many wireless strategies and these strategies are administered by routing protocols.

Based on network structure, WSN routing schemes can be divided into flat structure routing, hierarchy-based routing, and location-based routing. Flat structure routing algorithms are relatively effective in small scale networks. However, they are unattractive in large scale networks because resources constraints, and more data processing power and bandwidth are required. In hierarchical routing, nodes execute different tasks and are typically grouped into clusters on the basis of specific requirements. Routing is a very critical technology in WSNs. Clustering routing protocols are very effective in context with network topology, data aggregation, and energy optimization.

In this paper, authors opt to categorize clustering algorithms for wireless sensor networks. In Sect. 2, clustering phenomenon has been discussed. Section 3, focuses on the classification of clustering algorithms for WSNs. In Sect. 4, authors compare various popular clustering algorithms on the basis of certain attributes. In Sect. 5, some challenges in clustering have been discussed. Finally, Sect. 6 concludes the paper.

2 Clustering

Earlier peoples have relied on wired sensors, for tasks such as temperature monitoring and monitoring life signs in hospitals. WSNs provide unforeseen applications in new area of design [1, 5, 6]. Since the primary advantage of WSNs is the ability

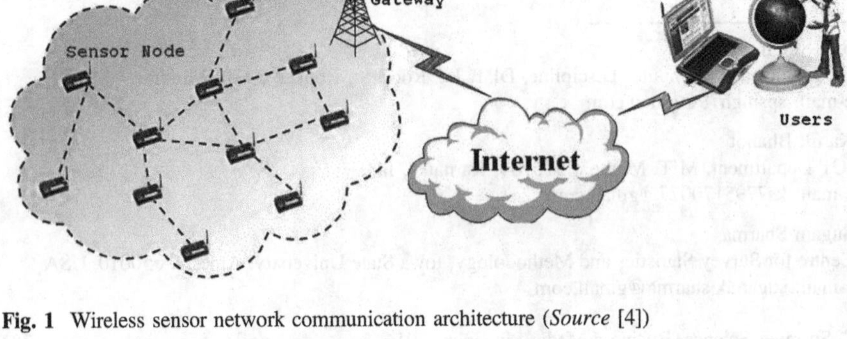

Fig. 1 Wireless sensor network communication architecture (*Source* [4])

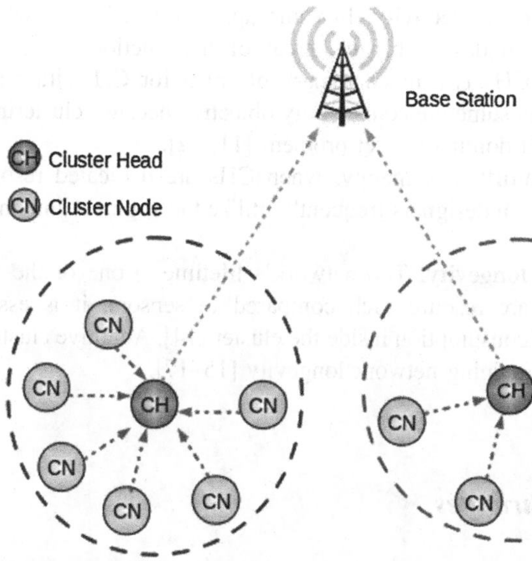

Fig. 2 Clustered wireless sensor network (*Source* [7])

to deploy them in an ad hoc fashion, as it is not feasible to arrange these nodes into groups' pre-deployment. Due to this reason, there has been much research in ways of creating these organizational structures. The architecture of clustered wireless sensor networks has been shown in Fig. 2.

Clustering is the vital branch of the organizational structure. Sensor node is the multifunctional and core component of a WSN. Sensor node can take on several roles in network like sensing, storage, processing, and routing. Clusters are the WSN's organizational units. Clusters can simplify the communication tasks. Cluster heads (CHs) are the organizational head of a cluster. Base station (BS) is at the upper layer of the hierarchical WSN. The clustering phenomenon plays an essential role in the network, so we have to discuss the objective and attributes of the clustering phenomenon.

2.1 Clustering Objectives

Clustering objectives vary according to applications. The following focuses on popular objectives of clustering:

- **Load balancing**: Allocation of nodes between the clusters is the basic objective wherever CHs carry out information processing or cluster management duties [8]. Load balancing is urgent issue in WSNs in which CH is selected from the existing nodes [9, 10].

- **Fault tolerance and connectivity**: In some applications, WSNs set in harsh environments cause nodes to bigger threat of malfunction and/or damage. Conveying backup CHs is a much remarkable plan for CH failure recovery. When few sensors presume the connectivity objective because clustering as one of various connected dominating set problem [11, 12].
- **Minimal cluster count**: It is mainly, when CHs are dedicated resource-rich nodes [13]. The system designers frequently utilize the slightest amount of such nodes.
- **Maximal network longevity**: The network's lifetime is one of the key concerns. When CHs are resource-rich compared to sensors, it is essential to optimize the energy consumption inside the cluster [14]. Adaptive clustering is a feasible option for attaining network longevity [15–17].

2.2 Clustering Attributes

This section identifies and focuses on the following attributes: cluster count, cluster topology, intercluster connectivity, and scalability that are used to classify and discriminate the clustering algorithms in WSNs.

- **Cluster properties**: Often clustering techniques try hard to attain some characteristics of the formed clusters. Taxonomy of cluster properties is shown in Fig. 3.
- **Cluster head capabilities**: Network model affects the clustering method; especially the capabilities of nodes. Certain attributes [18] of CH node are the distinguishing factors in clustering techniques which are: node type, mobility, and role. Taxonomy of CH capabilities are shown in Fig. 4.
- **Clustering process**: The organization of the complete clustering method and individuality of an algorithm differs considerably among clustering approaches. The attributes are found relevant: methodology, CH selection, node grouping objective, algorithmic complexity. Taxonomy of clustering process is shown in Fig. 5.

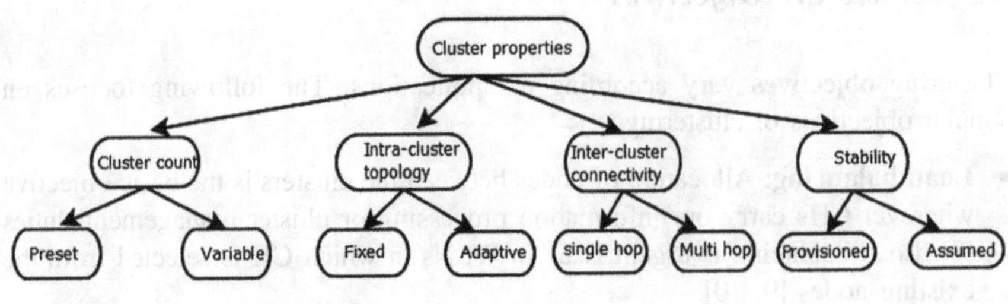

Fig. 3 Taxonomy of cluster properties

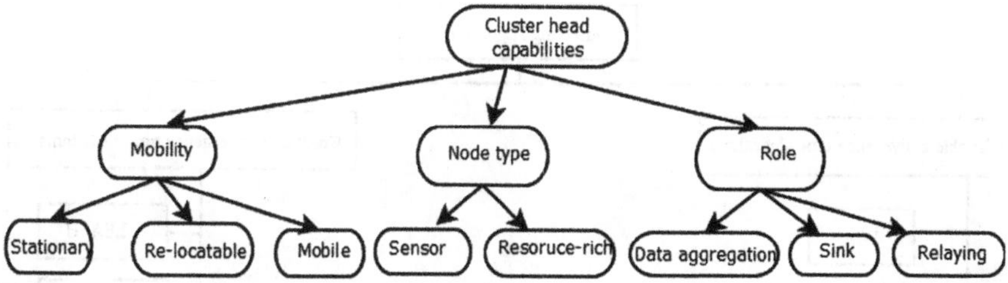

Fig. 4 Taxonomy of CH capabilities

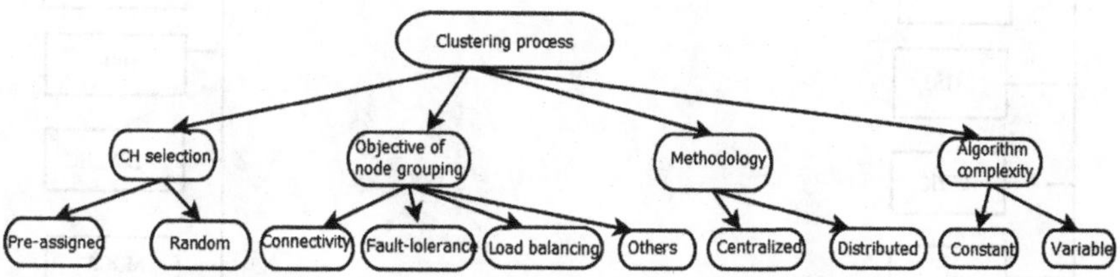

Fig. 5 Taxonomy of clustering process

3 Classification of Clustering Algorithms

Clustering algorithms play a pivotal role in attaining application specific goals. On the basis of convergence time of the algorithms, cluster routing algorithms are classified into two broad categories as shown in Fig. 6. The protocols reported under various schemes are briefly discussed in this section.

3.1 *Variable Convergence Time Algorithms*

The convergence of clustering algorithms depends on time. Some of the algorithms have convergence time of the order of *n*, i.e., O(n) where *n* represents nodes in the network. Hence, it is convenient to employ such algorithms to small scale networks. The popular variable convergence time algorithms are: LCA [19], RCC [20], CLUBS [21], GS^3 [22], EEHC [6], AC [23], and HCC [24] as shown in Fig. 6.

3.2 *Constant Convergence Time Algorithms*

The clustering algorithms that converge totally in a predetermined number of iterations are called constant convergence time clustering algorithms. These

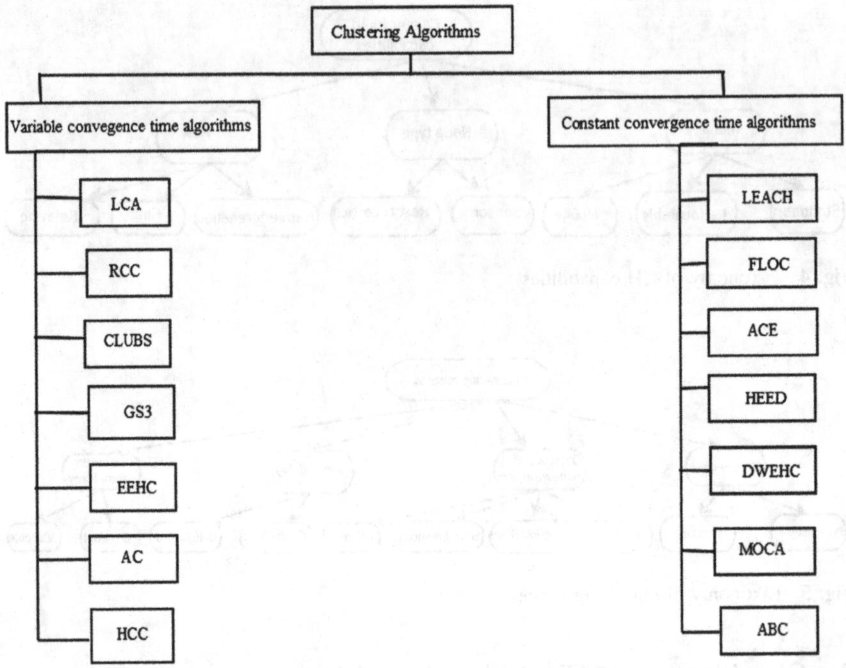

Fig. 6 Classification of clustering algorithms in WSNs

algorithms generally pursue a localized approach in which nodes execute the algorithms autonomously and their membership decision depends on state of their neighbors and their own states [25]. The popular constant convergence time clustering algorithms are: LEACH [26], FLOC [27], ACE [28], HEED [29], DWEHC [30], MOCA [31], and ABC [32] as shown in Fig. 6.

4 Comparison

Here in this section, we have performed the comparison of the clustering algorithms based on certain parameters. Table 1 summarizes the comparison between clustering algorithms on the basis of cluster properties. Table 2 summarizes the comparison between clustering algorithms on the basis of CH capabilities. Table 3 summarizes the comparison between clustering algorithms on the basis of clustering process. Table 4 summarize the comparison between some popular clustering algorithms in WSNs.

Table 1 Summary of comparison between clustering algorithms based on clustering properties

Clustering approaches	Cluster count	Intracluster topology	Intercluster connectivity	Stability
LCA	Variable	Fixed	Single-hop	Provisioned
RCC	Variable	Adaptive	Single-hop	Provisioned
CLUBS	Variable	Fixed	Multi-hop	Assumed
GS^3	Preset	Adaptive	Multi-hop	Provisioned
EEHC	Variable	Adaptive	Single-hop	Assumed
LEACH	Variable	Fixed	Single-hop	Provisioned
FLOC	Variable	Fixed	Single-hop	Provisioned
ACE	Variable	Adaptive	Single-hop	Provisioned
HEED	Variable	Fixed	Multi-hop	Assumed
DWEHC	Variable	Adaptive	Single-hop	Provisioned
MOCA	Variable	Fixed	Multi-hop	Assumed

Table 2 Summary of comparison between clustering algorithms based on CH capabilities

Clustering approaches	Mobility	Node type	Role
LCA	Mobile	Sensor	Data aggregation
RCC	Mobile	Sensor	Relaying
CLUBS	Relocatable	Sensor	Relaying and data aggregation
GS^3	Relocatable	Resource-rich	Relaying
EEHC	Stationary	Sensor	Relaying and data aggregation
LEACH	Stationary	Sensor	Relaying
FLOC	Relocatable	Sensor	Relaying and data aggregation
ACE	Relocatable	Sensor	Relaying and data aggregation
HEED	Stationary	Sensor	Relaying and data aggregation
DWEHC	Stationary	Sensor	Relaying and data aggregation
MOCA	Stationary	Sensor	Relaying and data aggregation

5 Challenges for Clustering in WSNs

The much research work has been done to query the challenges of clustering schemes in WSNs and to improve the characteristics of clustering algorithms, but there are still several issues needed to be addressed for the efficient use of clustering algorithms for WSNs, which are as follows:

- ***Role of cluster heads***: The normal sensor nodes (SNs) in cluster only pass on and relay their data to the relevant CHs, while CHs receive data from all the cluster members, perform processing function on data, and transmit the data to the base station (BS). Therefore, energy at CHs might be exhausted rapidly than SNs. Once the CH runs out of energy, it is no longer active, and all cluster nodes

Table 3 Summary of comparison of clustering algorithms based on clustering process

Clustering algorithms	Methodology	Node grouping objective	CH selection	Algorithm complexity
LCA	Distributed	Connectivity	Random	Variable
RCC	Hybrid	Stability and simplicity	Random	Variable
CLUBS	Distributed	Management and scalability	Random	Variable
GS^3	Distributed	Scalability and fault tolerance	Pre-assigned	Variable
EEHC	Distributed	Energy saving	Random	Variable
LEACH	Distributed	Energy saving	Random	Constant
FLOC	Distributed	Fault tolerance and scalability	Random	Constant
ACE	Distributed	Load balancing and scalability	Random	Constant
HEED	Distributed	Energy saving	Random	Constant
DWEHC	Distributed	Energy saving	Random	Constant
MOCA	Distributed	Connectivity and overlapping	Random	Constant

Table 4 Summary of comparison between popular clustering algorithms

Clustering approaches	Node mobility	Cluster overlapping	Location awareness	Failure recovery	Cluster stability	Convergence time
LCA	Possible	No	Mandatory	Yes	Medium	Variable
RCC	Yes	No	Mandatory	Yes	Medium	Variable
CLUBS	Possible	High	Not mandatory	Yes	Medium	Variable
GS^3	Possible	Low	Mandatory	Yes	Medium	Variable
EEHC	No	No	Mandatory	N/A	Low	Variable
AC	Yes	No	Mandatory	Yes	Low	Variable
HCC	Possible	Low	Not mandatory	Yes	Medium	Variable
LEACH	Fixed BS	No	Not mandatory	Yes	Medium	Constant
FLOC	Possible	No	Not mandatory	Yes	High	Constant
ACE	Possible	Low	Not mandatory	Yes	High	Constant
HEED	Stationary	No	Not mandatory	N/A	High	Constant
DWEHC	Stationary	No	Mandatory	N/A	High	Constant
ABC	No	No	Mandatory	Yes	High	Constant
MOCA	Stationary	Yes	Not mandatory	N/A	High	Constant

lose communication ability. Hence, it is necessary to rotate the role of CHs among nodes. So this issue is crucial for optimal network lifetime.

- ***Optimal cluster size***: The size of cluster is a significant parameter. When the cluster size is reduced, each cluster consumes less power. However, the number of CH will then be improved, so the resulting network formed by these CHs will be more complex. The less number of CHs will form a simpler backbone network. Yet that would require larger cluster size, so the power consumption in each cluster becomes higher. There is a tradeoff between number of CHs and cluster size. The majority of the clustering algorithms presume the network is structured into equal sized clusters but then there are unequal loads on CHs. So the open issues are how to choose optimum cluster size without node location awareness and how to utilize knowledge of the sinks location for competent cluster formation.
- ***Optimum mode of communication between SNs and CHs***: In clustered sensor networks, nodes can use either single-hop or multi-hop communication to transmit their data to their relevant CH. Single-hop communication between SNs and CHs might not be the optimum choice when propagation loss index for intracluster communication is large. So the open issue is, how to achieve optimum communication between SNs and CHs.

6 Conclusion

The WSNs have fascinated huge interest from research and industrial community over past several years. The rising list of both civil and military applications can utilize WSNs for improved efficiency; especially in unfriendly and inaccessible terrain. Examples include disaster relief, border protection, battle field, and security surveillance. In these applications a large number of sensors are expected, requiring careful architecture and management of network. To support scalability, grouping of nodes to form clusters is a trendy move toward WSNs. In this work, we surveyed the status of research and addressed some challenges in the clustering schemes. This paper classifies the taxonomy of relevant attributes of clustering techniques. In this work, we compare the various clustering algorithms on the basis of some parameters such as cluster properties, CH capabilities, and clustering process and represent them in tabular form. On the basis of comparison between different schemes, it is clear that clustering algorithms are useful in performance improvement of WSNs. This paper will be very useful for the research groups who are interested in the development, modification, or optimization of routing algorithms for WSNs.

Acknowledgments The authors would like to acknowledge ministry of human resource and development (MHRD) for providing financial support for this work under research scholar's grant.

References

1. Akyildiz, I.F., Su, W., Sankarasubramaniam, Y., Cyirci, E.: Wireless sensor networks: a survey. Comput. Netw. **38**(4), 393–422 (2002)
2. Romer, K., Mattern, F.: The design space of wireless sensor networks. IEEE Wirel. Commun. **11**(6), 54–61 (2004)
3. Hu, F., Siddiqui, W., Cao, X.: SPECTRA: secure power-efficient clustered—topology routing algorithm in large-scale wireless micro-sensor networks. Int. J. Inf. Technol. **11**(2), 95–118 (2005)
4. http://www.cs.sjtu.edu.cn/~yzhu/nrl/images/sensor_net.jpg
5. Meyer, S., Rakotonirainy, A.: A survey on research on context-aware homes. Workshop on wearable, invisible, context-aware, ambient, pervasive and ubiquitous computing, Adelaide, Australia (2003)
6. Bandyopadhyay, S., Coyle, E.J.: An energy efficient hierarchical clustering algorithm for wireless sensor networks. IEEE INFOCOM (2003)
7. http://vlssit.iitkgp.ernet.in
8. Gupta, G., Younis, M.: Load-balanced clustering in wireless sensor networks. In: Proceedings of the International Conference on Communication (ICC 2003), Anchorage, Alaska (2003)
9. Younis, O., Fahmy, S.: HEED: A hybrid, energy-efficient, distributed clustering approach for Ad Hoc sensor networks. IEEE Trans. Mob. Comput. **3**(4), 366–379 (2004)
10. Younis, M., Akkaya, K., Kunjithapatham, A.: Optimization of task allocation in a cluster-based sensor network. In: Proceedings of the 8th IEEE Symposium on Computers and Communications (ISCC'2003), Antalya, Turkey (2003)
11. Garcia, F., Solano, J., Stojmenovic, I.: Connectivity based k-hop clustering in wireless networks. Telecommun. Syst. **22**(1), 205–220 (2003)
12. Fernandess, Y., Malkhi, D.: K-clustering in wireless ad hoc networks. In: Proceedings of the 2nd ACM International Workshop on Principles of Mobile Computing (POMC'02), Toulouse, France, (2002)
13. Oyman, I.E., Ersoy, C.: Multiple sink network design problem in large scale wireless sensor networks. In: Proceedings of the IEEE International Conference on Communications (ICC 2004), Paris (2004)
14. Younis, M., Youssef, M., Arisha, K.: Energy-aware management in cluster-based sensor networks. Comput. Netw. **43**(5), 649–668 (2003)
15. Dasgupta, K., Kukreja, M., Kalpakis, K.: Topology-aware placement and role assignment for energy-efficient information gathering in sensor networks. In: Proceedings of 8th IEEE Symposium on Computers and Communication (ISCC'03), Kemer-Antalya, Turkey (2003)
16. Moscibroda, T., Wattenhofer, R.: Maximizing the lifetime of dominating sets. In: Proceedings of the 19th IEEE International Parallel and Distributed Processing Symposium (IPDPS'05), Denver, Colorado (2005)
17. Khanna, R., Liu, H., Chen, H.: Self-organization of sensor networks using genetic algorithms. In: Proceedings of the 32nd IEEE International Conference on Communications (ICC'06), Istanbul, Turkey (2006).
18. Abbasi, A.A., Younis, M.: A survey on clustering algorithms for wireless sensor networks. Comput. Commun. **30**, 2826–2841(2007)
19. Baker, D.J., Ephremides, A.: The architectural organization of a mobile radio network via a distributed algorithm. IEEE Trans. Commun. **29**(11), 1694–1701(1981)
20. Xu, K., Gerla, M.: A heterogeneous routing protocol based on a new stable clustering scheme. In: Proceedings of IEEE Military Communication Conference (MILCOM), CA, USA (2002)
21. Nagpal, R., Coore, D.: An algorithm for group formation in an amorphous computer. In: Proceedings of 10th International Conference on Parallel and Distributed Systems (PDCS 98), Las Vegas, USA (1998)

22. Zhang, H., Arora, A.: GS^3: scalable self-configuration and self-healing in wireless networks. In Proceedings of 21st ACM Symposium on Principle of Distributed Computing (PODC), Montreal, Canada (2002)
23. Lin, C.R., Gerla, M.: Adaptive clustering for mobile wireless networks. IEEE J. Sel. Areas Commun. **15**(7), 1265–1275 (1997)
24. Banerjee, S., Khuller, S.: A clustering scheme for hierarchical control in multi-hop wireless networks. In: Proceedings of 20th Joint Conference of the IEEE Computer and Communication Societies (INFOCOM), Anchorage, AK (2001)
25. Lui, Xu-Xun.: A survey on clustering routing protocols in wireless sensor networks. Sens. 2012(12), 11113–11153 (2012)
26. Heinzelman, W.B., Chandrakasan, A.P., Balakrishnan, H.: Application specific protocol architecture for wireless micro-sensor networks. IEEE Trans. Wireless Netw. **1**(4), 660–670 (2002)
27. Demirbas, M., Arora, A., Mittal, V.: FLOC: a fast local clustering service for wireless sensor networks. In: Proceedings of Workshop on Dependability Issues in Wireless Ad Hoc Networks and Sensor Networks, Florence Italy (2004)
28. Chan, H., Perrig, A.: ACE: an emergent algorithm for highly uniform cluster formation. In: Proceedings of the 1st European Workshop on Sensor Networks (EWSN), Berlin, Germany (2004)
29. Younis, O., Fahmy, S.: HEED: A Hybrid, Energy-Efficient, Distributed clustering approach for Ad Hoc sensor networks. IEEE Trans. Mob. Comput. **3**(4), 366–379 (2004)
30. Ding, P., Holliday, J., Celik, A.: Distributed energy efficient hierarchical clustering for wireless sensor networks. In: Proceedings of IEEE International Conference on Distributed Computing in Sensor Systems (DCOSS), Marina Del Ray, CA (2005)
31. Youssef, A., Younis, M., Youssef, M., Agarwala, A.: Distributed formation of overlapping multi-hop clusters in wireless sensor networks. In: Proceedings of 49th Annual IEEE Global Communication Conference (Globecom), San Francisco, CA (2006)
32. Wang, K., Ayyash, S.A., Little, T.D.C., Basu, P.: Attribute based clustering for information dissemination in wireless sensor networks. In: Proceedings of 2nd Annual IEEE Communication Society Conference on Sensor and Ad Hoc Communication and Networks (SECON), Santa Clara, CA (2005)

Nelder-Mead and Non-uniform Based Self-organizing Migrating Algorithm

Dipti Singh and Seema Agrawal

Abstract Self-organizing migrating algorithm (SOMA) is a novel approach capable to solve almost all type of functions. SOMA is highly effective evolutionary optimization technique and has proved its efficiency in solving many real-life applications. This paper presents a new optimization technique M-NM-SOMA to solve global optimization problems. In the proposed algorithm, SOMA is hybridized with Nelder-Mead method as crossover operator and non-uniform mutation operator in order to avoid premature convergence and keep the diversity of the population. The main feature of this algorithm is that it works for very low population size. To authenticate the efficiency of the proposed algorithm, it is tested on 17 benchmark test problems taken from the literature and the obtained results are compared with the results of other existing algorithms. Numerical and graphical results show that M-NM-SOMA has better global search ability and is very efficient, reliable, and accurate in comparison with other algorithms.

Keywords Self-organizing migrating algorithm · Nelder-Mead crossover operator · Non-uniform mutation · Particle swarm optimization · Global optimization

1 Introduction

A broad class of population-based algorithms for solving global optimization problems has been developed till date. Some of them are genetic algorithms (GA) [1], differential evolution (DE) [2], particle swarm optimization (PSO) [3], ant colony optimization (ACO) [4], and self-organizing migrating algorithm (SOMA)

Dipti Singh
Department of Applied Sciences, Gautam Buddha University, Greater Noida, India
e-mail: diptipma@rediffmail.com

Seema Agrawal (✉)
Department of Mathematics, S.S.V. College, Hapur, India
e-mail: seemagrwl7@gmail.com

M. Pant et al. (eds.), *Proceedings of Fifth International Conference on Soft Computing for Problem Solving*, Advances in Intelligent Systems and Computing 436, DOI 10.1007/978-981-10-0448-3_66

[5], etc. Among the above mentioned algorithms, SOMA is comparatively a new comer to the class of population-based stochastic search technique capable of handling all type of functions. SOMA can be classified as an evolutionary algorithm, regardless of the fact that no evolution takes place, i.e., no new generations of individuals are created during the search; only the positions of the individuals in the search space are changed during a generation called 'migration loop'. The main features of this algorithm are:

(i) It works efficiently for very low population size.
(ii) It quickly converges to global optimal solutions.

Despite the fact of several attractive features, sometimes SOMA may converge prematurely and the solution may trap to local optima and this situation arises with the increase of dimensionality. As a result, there is diversity loss in the population. To maintain the diversity mechanism, SOMA can be hybridized with local search techniques or other population-based techniques. Hybridization is a grouping of two or more algorithms, in which one seeks a promising region within the large solution space expected to contain global minima, and the other makes use of the search domain to find the best solution rapidly and more precisely. Several attempts have been made earlier to hybridize population-based techniques with other existing approaches [6–12]. First variant of SOMA was developed by Deep and Dipti which is the hybridization of GA and SOMA [13].

Recently, Dipti and Seema developed a number of variants of SOMA, named SOMAQI, SOMA-M, and M-SOMAQI [14–16]. In this paper, a novel variant of SOMA (M-NM-SOMA) based on Nelder-Mead crossover operator and non-uniform mutation operator is proposed. The performance of M-NM-SOMA has been evaluated on the set of 17 benchmark problems and the comparison of it is made with standard PSO and SOMA.

The paper is structured in the following manner: In Sect. 2, preliminaries are presented. M-NM-SOMA is presented in Sect. 3. In Sect. 4, the experimental results are shown. Finally, the paper concludes with Sect. 5 depicting the outcome of the current study.

2 Preliminaries

2.1 Self-Organizing Migrating Algorithm

Zelinka and Lampinen [17] first introduced SOMA, which is inspired by the collective behavior of intelligent creatures. This algorithm travels in migration loops and in each migration loop, active individual (individual having worst fitness value) travels a finite distance toward leader (individual having best fitness value) in N (path length/step size) moves of defined length (move size). This path is perturbed randomly by perturbation parameter (PRT) which is defined in the range

[0, 1]. Perturbation vector (PRT) controls the perturbation and is created before an individual proceeds toward leader in the following manner:

$$\begin{matrix} \text{PRTVector}_j = 1 \text{ if } \text{rnd}_j < \text{PRT}, \\ \text{PRTVector}_j = 0, \text{otherwise} \end{matrix}, \quad j = 1, 2, 3 \ldots n$$

where rnd_j is uniformly distributed random number in (0, 1) and n is the number of decision variables.

More information regarding SOMA can be obtained from [18].

2.2 Nelder-Mead Crossover Operator

The Nelder-Mead simplex search method is a direct search method, originally proposed by Spendley et al. and later modified by Nelder and Mead [19]. First of all, a population is initialized and a simplex is created using (n + 1) points (n: the number of variables of a function) chosen arbitrarily from the population. In each migration, the worst point in the simplex is selected first. Then, a new simplex is formed from the old simplex through a sequence of elementary geometric transformations (reflection, contraction, expansion). After each transformation, the current worst point is replaced by a better one. In the proposed algorithm, Nelder-Mead simplex search method is used as a linear Nelder-Mead crossover operator which creates a new point using two out of three randomly chosen points from population.

The computational steps of NM crossover operator method are as follows:

Step1: choose parameters $\gamma > 1, \beta > 0$;

Step2: create an initial simplex with randomly chosen three vertices; find x_h (the worst point), x_l (the best point), x_g (next to the worst point); calculate their function values f_r, f_l, f_g; the worst point x_h is reflected with respect to the centroid (x_c) of other two points;

$$\begin{aligned} & x_r = 2x_c - x_h. \text{ (reflection)} \\ & \text{if } f_r < f_l \\ & x_{new} = (1+\gamma)x_c - \gamma x_h. \text{(expansion)}. \\ & \text{else if } f_r >= f_h \\ & x_{new} = (1-\beta)x_c + \beta x_h. \text{ (contraction)}. \\ & \text{else if } f_g < f_r < f_h \\ & x_{new} = (1+\beta)x_c - \beta x_h. \text{ (contraction)}. \end{aligned} \tag{1}$$

calculate f_{new} and replace x_h by x_{new}.

Step3: this process continues until termination criterion is satisfied.

2.3 Non-uniform Mutation Operator

Non-uniform mutation operator was proposed by Michalewicz [20] to decrease the weakness of random mutation in the real-coded GA. Non-uniform mutation randomly selects one solution x_k from the population and its value is created according to the following rule:

$$\begin{aligned} x'_k &= x_k + (ub_k - x_k) \cdot T \text{ if } \gamma \prec 0.5 \\ \text{Or} \quad x'_k &= x_k + (x_k + lb_k) \cdot T \text{ if } \gamma \geq 0.5 \end{aligned} \quad (2)$$

where $T = \left(\mu\left(1 - \frac{t}{t_{\max}}\right)\right)^b$ with γ and μ two uniformly distributed random numbers in the interval [0, 1], lb_k and ub_k are the lower and upper bound of x_k, $b > 0$ is a parameter determining the degree of uniformity, t is a migration number, and t_{max} the maximum number of migrations allowed to run. Non-uniform mutation has fine-tuning capabilities to achieve high precision.

3 Proposed Hybrid M-NM-SOMA Algorithm

In this section, a variant of SOMA, M-NM-SOMA has been proposed which is the hybridization of SOMA with Nelder-Mead crossover operator and non-uniform mutation operator. The convergence of standard SOMA is so fast that all other individuals move closer to the best individual very quickly. This causes the population diversity decrease and leads to the premature convergence. To overcome the above problems, SOMA is hybridized with Nelder-Mead crossover operator and non-uniform mutation operator to maintain the diversity among the solutions in the search space.

3.1 Methodology of Hybridization

First, the population is initialized randomly spread over the search domain. At each migration the individuals having highest fitness value as leader and having least fitness value as active are selected. Now the active individual travels a finite distance towards leader in N moves of defined length. Among the positions created, the best position is selected and replaces the active individual if it is better than active individual. Now, leader and active individuals are selected again from the population and a new point is created using Nelder-Mead crossover operator using Eq. (1). This new point is accepted only if it is better than active individual and is replaced with active individual. Then leader and active individuals are selected again from the population and a new point is created using non-uniform mutation using Eq. (2). This new point is accepted only if it is better than active individual and is replaced with active individual. The process is continued until some termination criterion is satisfied.

4 Experimental Results

The presented algorithm M-NM-SOMA is programmed using C++ and is executed on a Pentium III PC. M-NM-SOMA is used to obtain the results of 17 benchmark problems taken from the literature. All the problems are of minimization with minimum value 0. The seventeen problems with initialization range are given in Table 1. M-NM-SOMA is probabilistic technique and relies a lot on the generation of random numbers; therefore 30 trials of each problem are carried out. A run is measured to be a success if the solution obtained is within 1 % of the preferred precision. The termination criterion of the proposed algorithm in either a run is a success or a preset number of migrations (10,000) are performed.

In order to make a comparative analysis of M-NM-SOMA with SOMA and standard PSO, various performance measures are considered like mean objective function value to check the efficiency and reliability, average number of function evaluations to check the convergence speed, and one more measure success rate is also considered.

The main parameters of M-NM-SOMA are population size, PRT, move size, and path length. The population size is taken as ten for all the problems. PRT parameter varies from 0.1 to 0.9 depending on the problem. The other parameters, move size and path length are taken as 0.31 and 3. Trials for the 17 problems are performed for dimensions (dim) n = 30, 50 and 100.

Table 2 shows successful runs of a total of 30 runs, corresponding to M-NM-SOMA, PSO, and SOMA. Results show that M-NM-SOMA is best in all 17 problems for dim 30 and 100 and it is best in 16 problems for dim 50.

Table 3 shows the average number of function evaluations corresponding to M-NMSOMA, PSO, and SOMA. Results show that M-NM-SOMA is best in 16 problems for all the three dim 30, 50, and 100. Hence on the basis of results, we can say that M-NM-SOMA shows better convergence accuracy.

Table 4 shows the mean objective function value corresponding to M-NM-SOMA, PSO, and SOMA. Results show that M-NM-SOMA is best in 17 problems for dim 30 and is best in 16 problems for dim 50 and 100. Hence, M-NM-SOMA is *most reliable and efficient.* The problems which could not be solved by the particular algorithm is given the symbol (*) at the corresponding entries. The best results are highlighted in bold characters.

Figures 1, 2 and 3 show the mean best objective function value curves for selected benchmark problems and from the figures it is very clear that M-NM-SOMA converges very fast. Hence the presented algorithm M-NM-SOMA shows its superiority over other algorithms PSO and SOMA.

Table 1 Benchmark functions

S. no.	Name	Function	Range
1	Ackley	$20\exp\left(-0.02\sqrt{\frac{1}{n}\sum_{i=1}^{n}x_i^2}\right)-\exp\left(\frac{1}{n}\sum_{i=1}^{n}\cos(2\pi x_i)\right)+20+e$	[−30,30]
2	Cosine mixture	$0.1n+\sum_{i=1}^{n}x_i^2-0.1\sum_{i=1}^{n}\cos(5\pi x_i)$	[−1, 1]
3	Exponential	$1-\left(\exp\left(-0.5\sum_{i=1}^{n}x_i^2\right)\right)$	[−1, 1]
4	Griewank	$1+\frac{1}{4000}\sum_{i=1}^{n}x_i^2-\prod_{i=1}^{n}\cos\left(\frac{x_i}{\sqrt{i}}\right)$	[−600, 600]
5	Levy and Montalvo-1	$\frac{\pi}{n}\left(10\sin^2(\pi y_1)+\sum_{i=1}^{n}(y_i-1)^2[1+10\sin^2(\pi y_{i+1})]+(y_n-1)^2\right), y_i=1+\frac{1}{4}(x_i+1)$	[−10, 10]
6	Levy and Montalvo-2	$0.1\left[\left(\sin^2(3\pi x_1)+\sum_{i=1}^{n-1}(x_i-1)^2[1+\sin^2(3\pi x_{i+1})]+(x_n-1)^2[1+\sin^2(2\pi x_n)]\right)\right]$	[−5, 5]
7	Rastrigin	$10n+\sum_{i=1}^{n}[x_i^2-10\cos(2\pi x_i)]$	[−5.12, 5.12]
8	Rosenbrock	$\sum_{i=1}^{n-1}\left[100(x_{i+1}-x_i^2)^2+(x_i-1)^2\right]$	[−30, 30]
9	Schewefel-3	$\sum_{i=1}^{n}\lvert x_i\rvert+\prod_{i=1}^{n}\lvert x_i\rvert$	[−10, 10]
10	Step function	$\sum_{i=1}^{n}\left(x_i+{}^1/_2\right)^2$	[−100, 100]
11	Dejong's function with noise	$\sum_{i=1}^{n-1}(i+1)x_i^4+\mathrm{rand}(0,1)$	[−1.28, 1.28]
12	Sphere	$\sum_{i=1}^{n}x_i^2$	[−5.12, 5.12]
13	Axis parallel hyper ellipsoid	$\sum_{i=1}^{n}ix_i^2$	[−5.12, 5.12]
14	Ellipsoidal	$\sum_{i=1}^{n}(x_i-i)^2$	[−n, n]
15	Brown3	$\sum_{i=1}^{n-1}\left((x_i^2)^{(x_{i+1}^2+1)}+(x_{i+1}^2)^{(x_i^2+1)}\right)$	[−1, 4]
16	New function	$\sum_{i=1}^{n}(0.2x_i^2+0.1x_i^2\sin(2x_i))$	[−10, 10]
17	Cigar	$x_1^2+100.00\sum_{i=2}^{n}x_i^2$	[−10, 10]

Table 2 Successful runs of M-NM-SOMA, SOMA and PSO for Dim. 30, 50 and 100

P. no.	Dimension	No. of successful runs out of 30		
		PSO	SOMA	M-NM-SOMA
1	30	21	03	**30**
	50	06	0	**30**
	100	0	0	**30**
2	30	25	18	**30**
	50	05	0	**30**
	100	0	0	**30**
3	30	30	30	**30**
	50	29	30	**30**
	100	0	30	**30**
4	30	14	03	**29**
	50	09	07	**30**
	100	0	5	**30**
5	30	09	30	**30**
	50	02	28	**30**
	100	0	28	**30**
6	30	27	23	**30**
	50	22	19	**30**
	100	0	05	**22**
7	30	0	0	**10**
	50	0	0	**13**
	100	0	0	**13**
8	30	0	02	**25**
	50	0	0	**28**
	100	0	0	**20**
9	30	11	30	**30**
	50	01	30	**30**
	100	0	0	**30**
10	30	27	30	**30**
	50	17	30	**30**
	100	0	8	**30**
11	30	0	29	**30**
	50	08	30	**30**
	100	0	12	**30**
12	30	30	30	**30**
	50	30	30	**30**
	100	01	27	**30**
13	30	22	30	**30**
	50	13	30	**30**
	100	0	30	**30**

(continued)

Table 2 (continued)

P. no.	Dimension	No. of successful runs out of 30		
		PSO	SOMA	M-NM-SOMA
14	30	10	29	**30**
	50	06	**29**	28
	100	03	0	**08**
15	30	28	30	**30**
	50	18	30	**30**
	100	0	01	**30**
16	30	15	27	**30**
	50	02	26	**30**
	100	0	0	**30**
17	30	09	30	**30**
	50	07	30	**30**
	100	0	22	**30**

Table 3 Average number of function evaluations of M-NM-SOMA, SOMA, and PSO for Dim. 30, 50, and 100

P. no.	Dimension	Average no. of function evaluations of successful runs		
		PSO	SOMA	M-NM-SOMA
1	30	142,777	47,818	**5861**
	50	180,730	90,010	**5824**
	100	200,020	180,010	**5490**
2	30	141,465	20,023	**2897**
	50	189,732	90,010	**2992**
	100	200,020	180,010	**3391**
3	30	126,190	13,702	**1963**
	50	176,762	27,044	**2068**
	100	200,020	84,144	**2383**
4	30	143,084	43,534	**4168**
	50	172,335	72,521	**4195**
	100	200,020	153,290	**4198**
5	30	106,784	15,471	**14,163**
	50	143,400	27,558	**26,755**
	100	200,020	80,829	**79,537**
6	30	131,592	22,045	**16,636**
	50	165,766	46,238	**23,359**
	100	200,020	150,842	**65,635**

(continued)

Table 3 (continued)

P. no.	Dimension	Average no. of function evaluations of successful runs		
		PSO	SOMA	M-NM-SOMA
7	30	200,020	36,010	**22,091**
	50	200,020	90,010	**10,555**
	100	200,020	180,010	**6730**
8	30	200,020	180,010	**53,014**
	50	200,020	180,010	**69,596**
	100	200,020	180,010	**97,221**
9	30	129,625	35,712	**1574**
	50	157,040	78,148	**1942**
	100	200,020	180,010	**2110**
10	30	136,301	36,372	**4021**
	50	168,018	70,822	**4252**
	100	200,020	180,010	**4568**
11	30	200,020	25,285	**6105**
	50	191,830	64,604	**9691**
	100	200,020	172,555	**11,851**
12	30	125,305	22,924	**2278**
	50	15,537	44,712	**2467**
	100	199,600	129,308	**2931**
13	30	129,417	26,786	**3886**
	50	163,240	57,088	**3391**
	100	200,020	141,010	**3622**
14	30	94,340	**31,748**	32,540
	50	125,224	**70,444**	79,807
	100	173,520	**180,010**	200,010
15	30	129,973	40,526	**3055**
	50	168,554	75,885	**3055**
	100	200,020	178,372	**3323**
16	30	114,530	22,478	**3580**
	50	149,360	42,675	**3350**
	100	200,020	180,010	**3875**
17	30	137,506	25,036	**4945**
	50	168,194	50,138	**5115**
	100	200,020	145,313	**5428**

Table 4 Mean objective function value of M-NM-SOMA, SOMA, and PSO for Dim. 30, 50, and 100

P. no.	Dimensions	Mean of objective function value of successful runs		
		PSO	SOMA	M-NM-SOMA
1	30	0.00975	0.00905	**0.000836**
	50	0.00995	3.413	**0.000819**
	100	20.016	9.06	**0.000940**
2	30	0.00937	0.00817	**0.000678**
	50	0.00965	0.5119	**0.000722**
	100	36.364	2.336	**0.000881**
3	30	0.00940	0.00810	**0.000556**
	50	0.00970	0.00940	**0.000741**
	100	0.999	0.00810	**0.000891**
4	30	0.00960	0.00758	**0.000803**
	50	0.00967	0.00895	**0.000855**
	100	0.0552	0.00940	**0.000852**
5	30	0.00892	0.00708	**0.000918**
	50	0.00951	0.00785	**0.000944**
	100	5.023	0.00758	**0.00658**
6	30	0.00886	0.00884	**0.000932**
	50	0.00912	0.00901	**0.00828**
	100	4.327	0.00926	**0.00917**
7	30	22.866	17.32	**0.00851**
	50	76.179	35.06	**0.00765**
	100	345.606	118.156	**0.00881**
8	30	5.38	0.0330	**0.00952**
	50	35.175	276.092	**0.00976**
	100	195.04	1284.4	**0.00939**
9	30	0.00967	0.00832	**0.00709**
	50	0.00943	0.00995	**0.00774**
	100	23.244	1.078	**0.00803**
10	30	0.00954	**0.00811**	0.00859
	50	0.00983	**0.00882**	0.00894
	100	*	0.0488	**0.00860**
11	30	0.0385	0.00703	**0.00651**
	50	0.00981	0.00868	**0.00710**
	100	*	0.00969	**0.00780**
12	30	0.00992	0.00815	**0.00805**
	50	0.00960	0.00884	**0.00881**
	100	0.00999	0.00771	**0.00735**
13	30	0.00950	0.00807	**0.000666**
	50	0.00980	0.00820	**0.00789**
	100	*	0.00921	**0.00884**

(continued)

Table 4 (continued)

P. no.	Dimensions	Mean of objective function value of successful runs		
		PSO	SOMA	M-NM-SOMA
14	30	0.00956	0.00867	**0.00797**
	50	0.00976	0.00859	**0.00815**
	100	0.00945	6.862	**0.00565**
15	30	0.00938	0.00814	**0.000830**
	50	0.00978	0.00845	**0.000781**
	100	*	0.00984	**0.000956**
16	30	0.00964	0.00851	**0.000876**
	50	0.00994	0.00821	**0.000907**
	100	116.265	2.466	**0.000953**
17	30	0.00965	0.00834	**0.000766**
	50	0.00970	0.00885	**0.000721**
	100	*	0.00744	**0.000889**

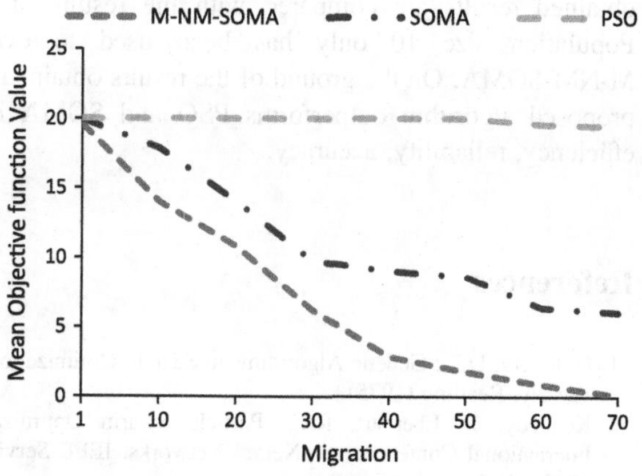

Fig. 1 Convergence graph of Ackley function for dim. 30

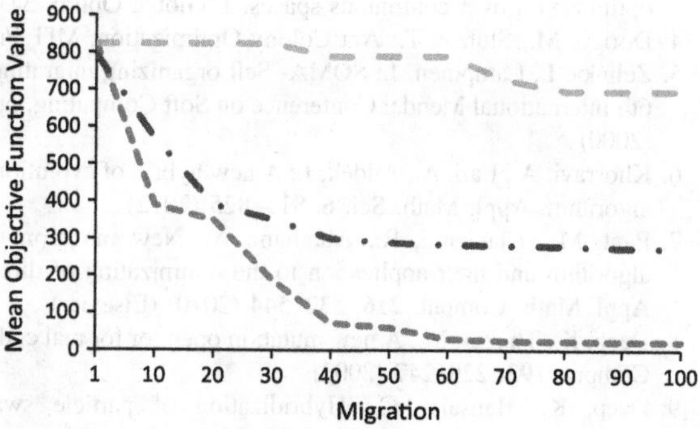

Fig. 2 Convergence graph of Rastrigin function for dim. 50

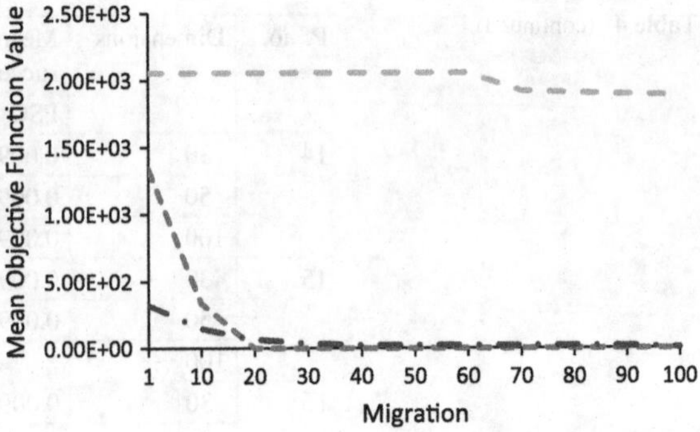

Fig. 3 Convergence graph of Dejong function with noise for dim. 100

5 Conclusions

In this paper, a new variant of SOMA, M-NM-SOMA has been proposed. The proposed algorithm is evaluated on 17 unconstrained benchmark problems and obtained results are compared with the results of standard PSO and SOMA. Population size 10 only has been used to evaluate the performance of M-NM-SOMA. On the ground of the results obtained, it can be concluded that the proposed algorithm outperforms PSO and SOMA in terms of population size, efficiency, reliability, accuracy.

References

1. Goldberg, D.E.: Genetic Algorithms in Search, Optimization and Machine Learning. Addison Wesely, Reading (1975)
2. Kennedy, J., Eberhart, R.C.: Particle Swarm Optimization. In: Proceedings of IEEE International Conference on Neural Networks, IEEE Service Center, Piscataway, pp. 1942–1948 (1995)
3. Storn, R., Price, K.: Differential evolution—a simple and efficient heuristic for global optimization over continuous spaces. J. Global Optim. 341–359 (1997)
4. Dorigo, M., Stützle, T.: Ant Colony Optimization, MIT Press, 2004
5. Zelinka, I., Lampinen, J.: SOMA- Self organizing migrating algorithm. In: proceedings of the 6th International Mendel Conference on Soft Computing, pp. 177–187, Brno, Czech, Republic (2000)
6. Khosravi, A., Lari, A., Addeh, J.: A new hybrid of evolutionary and conventional optimization algorithm. Appl. Math. Sci. **6**, 815–825 (2012)
7. Pant, M., Thangaraj, R., Abraham, A.: New mutation schemes for differential evolution algorithm and their application to the optimization of directional over-current relay settings. Appl. Math. Comput. **216**, 532–544 (2010) (Elsevier)
8. Deep, K., Thakur, M.: A new mutation operator for real coded genetic algorithms. Appl. Math. Comput. **193**, 229–247 (2007)
9. Deep, K., Bansal, J.C.: Hybridization of particle swarm optimization with quadratic approximation. Opsearch **46**, 3–24 (2009) (Springer)

10. Xing, L.N., Chen, Y.W., Yang, K.W.: A novel mutation operator base on immunity operation. European Journal of Operational Research, vol. 197, pp. 830–833 (2009)
11. Deep, K., Das, K.N.: Performance improvement of real coded genetic algorithm with Quadratic approximation based hybridization. Int. J. Intell. Defence Support Syst. **2**, 319–334, (2010) (Inderscience Publisherss)
12. Esmin, A.A.A., Matwin, S.: A hybrid particle swarm optimization algorithm with genetic mutation. Int. J. Innovative Comput. Inf. Control **9**, 1919–1934 (2013)
13. Deep, K., Dipti: A new hybrid self organizing migrating genetic algorithm for function optimization. In: IEEE Congress on Evolutionary Computation, pp. 2796–2803 (2007)
14. Singh, D., Agrawal, S., Singh, N.: A novel variant of self organizing migrating algorithm for function optimization. In: Proceedings of the 3rd international conference on soft computing for problem solving. Advances in intelligent and Soft Computing, vol. 258, pp. 225–234 (2013)
15. Singh, D., Agrawal, S.: Hybridization of self organizing migrating algorithm with mutation for global optimization. In: Proceedings of the international conference on mathematical sciences (ICMS), Elsevier, pp. 605–609 (2014)
16. Singh, D., Agrawal, S.: Hybridization of self organizing migrating algorithm with quadratic approximation and non uniform mutation for function optimization. In: Proceedings of 4th International Conference on Soft Computing for Problem solving, Advances in Intelligent Systems and Computing, Springer, vol. 335, pp. 373–387 (2014)
17. Zelinka, I., Lampinen, J.: SOMA—Self organizing migrating algorithm. In: Proceedings of the 6th International Mendel Conference on Soft Computing, pp. 177–187, Brno, Czech, Republic (2000)
18. Zelinka, I.: SOMA—Self Organizing Migrating Algorithm. In: Onwubolu, G.C., Babu, B.V. (eds.) New optimization techniques in engineering, Springer, Berlin (2004)
19. Nelder, J.A., Mead, R.A.: A simplex method for function minimization. Comput. J. **7**, 308–313 (1965)
20. Michalewicz, Z.: Genetic algorithms + Data structures = Evolution programs, 3rd edn. Springer (1996)

Surrogate-Assisted Differential Evolution with an Adaptive Evolution Control Based on Feasibility to Solve Constrained Optimization Problems

Mariana-Edith Miranda-Varela and Efrén Mezura-Montes

Abstract This paper presents an adaptive evolution control based on the feasibility of solutions, which is used with the nearest-neighbor regression surrogate model, to approximate the objective function value and the sum of constraint violation when solving constrained numerical optimization problems. The search algorithm used is the "differential evolution with combined variants" (DECV) and the constraint-handling technique adopted is the set of feasibility rules. The approach is compared against one state-of-the-art algorithm that employs the same surrogate model with an adaptive evolution control, as well. Twenty-four well-known test problems are solved in the experiments. From the obtained results, it is found that the evolution control based on the feasibility of solutions reduces the number of evaluations in the expensive model, particularly in problems with inequality constraints.

Keywords Evolutionary algorithms · Constrained numerical optimization problems · Surrogate models

1 Introduction

Real-world optimization problems can be expressed through expensive and complex models. Nowadays, a good alternative to solve them is the use of meta-heuristics, such as evolutionary algorithms (EAs). The EAs work with a set of solutions, named population, which is evaluated at each generation. This evaluation is a disadvantage when the model of the problem is complex and expensive.

The aforementioned drawback has been tackled with surrogate models (also called *approximate models* or *metamodels* [6, 19]), which can approximate the original

M.-E. Miranda-Varela (✉) · E. Mezura-Montes
Artificial Intelligence Research Center, University of Veracruz, Xalapa, Veracruz, Mexico
e-mail: memiranda.v@gmail.com

E. Mezura-Montes
e-mail: emezura@uv.mx

M. Pant et al. (eds.), *Proceedings of Fifth International Conference on Soft Computing for Problem Solving*, Advances in Intelligent Systems and Computing 436, DOI 10.1007/978-981-10-0448-3_67

expensive model with a lower computational cost [6]. Moreover, this approximation is less complex than the original model. Different approaches have been coupled with EAs. Such combination is known as a surrogate-assisted evolutionary algorithm [8]. A benefit of this approximation is to save evaluations in the expensive model and, subsequently, the computational cost of the optimization process is reduced.

Several models have been employed to approximate the objective function. Examples of surrogate models are as follows: nearest-neighbor (NN) regression model, radial basis function (RBF), neural networks (NNs), polynomial models (PM), also called response surface methodology (RSM), kriging, also called Gaussian process, and support vector machines (SVM). An overview of the aforementioned models can be found in [4, 7].

Figure 1 shows the required elements (stars) to assemble a surrogate model within an EA. Surrogate models are linked to a set of solutions, named *training set*. These solutions are evaluated in the original model and their information provides knowledge about the search space. Furthermore, they are used either to build (*offline training* stage) or to update (*online training* stage) the training set. After each online training, the quality of the approximation increases [8].

In order to update the training set, it is important to choose the most appropriate solution, because a surrogate model may introduce a false optimum. Therefore, both models, the approximate and the original one, must be used together in the search process. This process is called *evolution control*, also known as *model management*.

Two categories of evolution control are considered [21]: fixed evolution control and adaptive evolution control. In the first one, the surrogate model is used by predefined individuals (individual-based) or at preestablished generations (generation-based). On the other hand, the adaptive evolution control adjusts the use of a surrogate model according to its accuracy.

In the individual-based control, three are the most common strategies to select the solutions to be evaluated in the original model [8, 17]: the first one consists of choosing the best solution according to the surrogate model. This strategy is the most popular and has been applied by Elsayed et al. [2], Regis [13–15] and Shi [20].

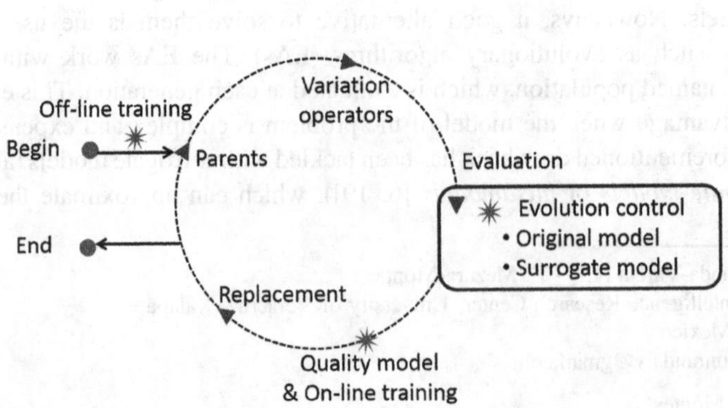

Fig. 1 Triangles represent EA elements and stars indicate the elements of a surrogate model in an EA

The second one is based on selecting a random solution. Fonseca [3] has used this criterion. Finally, the third strategy refers to pick the worst solution obtained by the surrogate model. The idea of this procedure allows to explore farther areas of the training set [8]. Other strategies are based on clusters [5].

Runarsson in [17] proposed an adaptive evolution control, which uses a sequential technique in order to update the training set. This task is performed at each generation, where at least one expensive model evaluation is carried out. But, in the worst case all the populations are evaluated in the expensive model. In [18], the number of evaluations in the worst case was reduced to a quarter of the population. From the above literature review it is noticed that none of the previous evolution controls use the feasibility of the population as a condition to begin approximating the solutions through a surrogate model.

This work precisely proposes an evolution control that employs the feasibility of solutions as a criterion to determine which solutions will be evaluated either in the original or in the surrogate model, when solving constrained numerical optimization problems (CNOP). The NN regression model is one of the most simple surrogate models [17] and it is adopted in this work. The differential evolution with combined variants (DECV) [11], an algorithm recently proposed that showed very competitive results for solving CNOPs, is used as the search algorithm, while a set of well-known test functions is solved in the experiments. The final results obtained and a number of evaluations in the original model made by the proposed approach are compared against the approach in [18].

The paper is organized as follows: the problem of interest is stated in Sect. 2. The proposed surrogate-assisted DECV is explained in Sect. 3. The experiments and results are shown in Sect. 4. Finally, the conclusions and future work are found in Sect. 5.

2 Problem Statement

In this work, the tackled problem is the general nonlinear programming problem, also known as the constrained numerical optimization problem (CNOP), which is defined, without loss of generality, as follows:

Minimize

$$f(\mathbf{x}), \mathbf{x} = (x_1, x_2, \ldots, x_n) \in \mathbb{R}^n. \tag{1}$$

subject to

$$g_i(\mathbf{x}) \leq 0, i = 1, \ldots, m \tag{2}$$

$$h_j(\mathbf{x}) = 0, j = 1, \ldots, p. \tag{3}$$

where $f(\mathbf{x})$ is the objective function, m is the number of inequality constraints, p is the number of equality constraints, and $\mathbf{x} \in \mathcal{S} \cap \mathcal{F}$, $\mathcal{S} \subseteq \mathbb{R}^n$, where $\mathcal{S}$ is the search space which is bounded by lower and upper limits for each design variable, $L_i \leq x_i \leq U_i$ and the feasible region is denoted by $\mathcal{F}$.

3 Surrogate-Assisted DECV

In this work, a unique surrogate model is proposed to approximate both the objective function value and the sum of constraint violation by means of the NN regression model, which is coupled with the DECV algorithm. In the following sections, the NN regression model, the required elements of a surrogate model, and the description of DECV are detailed. In Sect. 3.3, the evolution control proposed in this work is explained.

3.1 NN Regression Model

NN regression model is one of the most simple surrogate models [17]. It is based on the k-nearest neighbors method (k-NN) and belongs to the memory-based learners [3]. The prediction of new values (Eq. 4) is given by the average of the k-closest stored solutions.

$$\hat{f}(\mathbf{x}_i) = \frac{\sum_{j=1}^{k} f(\mathbf{y}_j)}{k}, \mathbf{y}_j \in D \tag{4}$$

where $\mathbf{x}_i$ is the point that will be approximated; D is the training set, its elements are evaluated in the original function (f); $\mathbf{y}_j$ represents the nearest points to $\mathbf{x}_i$, according to Euclidean distance and k is the number of nearest neighbors.

3.2 Training Set

The information stored in the training set (D) to approximate the objective function value and the sum of constraint violation is defined by $D = \{<x_{i,1}, \ldots, x_{i,j}, \ldots, x_{i,n}>, f(\mathbf{x}_i), \phi(\mathbf{x}_i)\}$, where $x_{i,j}$ represents the jth design variable of solution $\mathbf{x}_i$, $f(\mathbf{x}_i)$ is its corresponding objective function value and $\phi(\mathbf{x}_i)$ is its corresponding sum of constraint violation, which is calculated as $\phi(\mathbf{x}_i) = \sum_{j=1}^{m} \max(0, g_j(\mathbf{x}_i)) + \sum_{j=1}^{p} \max(0, abs(h(\mathbf{x}_j)) - \delta)$, where m is the number of inequality constraints, p is the number of equality constraints, and δ is a small tolerance to convert an equality constraint into an inequality constraint.

The strategies used in offline training and online training stages are:

- *Offline training*: The training set is started with the initial population, generated at random with uniform distribution, and their opposite points [12], which are calculated as $\breve{\mathbf{x}}_i = L + U - \mathbf{x}_i$, where $\breve{\mathbf{x}}_i$ is the opposite point of $\mathbf{x}_i$, and L and U are the lower and upper limits, respectively.
- *Online training*: The information of the training set must be updated with new solutions evaluated in the original model. These solutions are selected by the evolution control (see Sect. 3.3).

3.3 Evolution Control

The evolution control proposed in this work is shown in Algorithm 3.3, which evaluates the generation (line 2) or only some solutions (line 4) in the original model. The former gives the opportunity for exploring the search space at the beginning of the process. As a result, the quality of solutions in the training set will improve, while the accuracy of the model increases. When a feasible solution is found in the population, *Choose_vector* is called to select those solutions that will be evaluated in the original model, and the remaining will be approximated by NN regression model.

Algorithm 1 Evolution_control

Require: Population P_{G-1}, offspring u, training set $train_{set}$, generations percent (per_{gen})
Ensure: Solutions that will be evaluated in the original model ind_{orig}
1. **if** there are feasible solutions in P_{g-1} **then**
2. $\quad ind_{orig}$ = *Choose_vectors*
3. **else**
4. $\quad ind_{orig} = u$

The details of the function *Choose_vector* are in Algorithm 3.3, where the solutions to be evaluated in the original model (ind_{orig}) are selected. There are two cases: (1) no solutions are chosen (i.e., all solutions will be evaluated in the surrogate model) because the mean of the standard deviation of each design variable (SD) of the current population with respect to their centroid is below a small tolerance, 1E-06 in this case (i.e. the algorithm is close to converge), or (2) a set of solutions are selected according to their proximity to the best solution of the training set (b_{ts}). For such task, the centroids (mean of a set) of the training set ($train_{set}$) and the offspring set (u) are calculated (m_{ts} and m_u, respectively). After that, the distances between b_{ts} and each centroid (m_{ts} and m_u) are determined (d_{ts} and d_u, respectively). Based on the generation percentage (*per_cent*), the distance which determines the area to choose individuals from the offspring set to be evaluated in the original model is computed as follows: when *per_cent* is greater than 50 %, the minimal distance (d) between d_{ts} and d_u is considered. Otherwise, the maximum distance (d) between d_{ts} and d_u is adopted. After that, the solutions whose distance is less than d are chosen to be evaluated in the original model. Finally, the function *adjust* is called to assure that the number of solutions to be calculated in the original model is according to the current generation.

Algorithm 2 Choose_vectors

Require: Offspring u, population P_{G-1}, training set $train_{set}$, generations percent (per_gen)
Ensure: Solutions that will be evaluated in the original model ind_{orig}

if Standard deviation (*SD*) of P_{G-1} is less than 1E-06 **then**
  All solutions are evaluated in the surrogate model ($ind_{orig} = \emptyset$)
else
  Select the best solution of the training set b_{ts}
  m_{ts} = *centroid* (*train_set*) {The *centroid* calculates the mean of a population with respect their distance}
  m_u = *centroid* (u)
  Calculate the distances (d_{ts} and d_u) between b_{ts} and the centroids (m_{ts} and m_u)
  if per_gen > 50 **then**
    $d = \min(d_{ts}, d_u)$ {*min* returns the minimum value}
  else
    $d = \max(d_{ts}, d_u)$ {*max* returns the maximum value}
  Choose those solutions in u with a less distance than d
  ind_{orig} = *adjust* (ind_{orig}, *per_gen*)

The *adjust* function controls the number of solutions to be evaluated in the original model ($n_{ind_{orig}}$), based on the percentage of generations already computed and $n_{ind_{orig}}$. Regarding the percentage of generations, four intervals are considered: [0, 25], (25, 50], (50, 75], and (75, 100]. In the first interval, the lower number allowed is 10 % of the population. For the second interval the 25 % of the population is considered. In the third interval, the upper value permitted is 25 % of the population. Finally, the last interval considers 10 % of the population. When $n_{ind_{orig}}$ is out of the limit, the distance (d) is resized according to the product of *SD* and the upper limit of each interval in opposite order, i.e., 1, 0.75, 0.5, and 0.25, respectively.

3.4 Quality Model Measurement

The most common quality model measure is the mean square error (MSE), where $\hat{y}$ is the approximated value, y is the value of the original function, and n_{err} is the number of elements used to calculate the error. MSE is simple and can be used in a generation-based model [8]. In this work, such measure is applied as indicated in Eq. 5.

$$MSE = \frac{\sum_{i=0}^{n_{err}} (\hat{y}_i - y_i)^2}{n_{err}}. \tag{5}$$

3.5 DECV

Differential evolution (DE) is a simple, but highly competitive EA proposed by Storn & Price [22], to solve numerical optimization problems. DE has a mechanism to generate multiple search directions (Eq. 6), based on the distribution of solutions in the current population [11]. DE works as follows: each individual (vector) i (called target vector at the reproduction step) of the population at generation G ($\mathbf{x}_{i,G}$), generates one offspring ($\mathbf{u}_{i,G}$), called *trial vector*, which is computed by using a mutation and a crossover operators. The mutation operator requires three randomly chosen vectors ($r_0 \neq r_1 \neq r_2 \neq i$), which are substituted in Eq. 6, where F > 0. As a result, a mutant vector ($\mathbf{v}_i$) is obtained.

$$v_{i,j,G} = x_{r_0,j,G} + F(x_{r_1,j,G} - x_{r_2,j,G}). \tag{6}$$

After the mutation, the trial vector is generated by the combination of the target and the mutant vectors. The crossover is detailed in Eq. 7.

$$u_{i,j,G+1} = \begin{cases} v_{i,j,G} & \text{if } (rand_j \leq Cr) \text{ or } (j = jrand) \\ x_{i,j,G} & \text{otherwise} \end{cases}. \tag{7}$$

where *CR* is a parameter defined by the user (crossover parameter) which controls the number of design variables to be copied from the mutant vector into the trial vector. *rand* is a function that returns a real number between 0 and 1. *jrand* $\in [1, n]$ ensures that at least one element of the mutant vector is copied to the trial vector, then the trial vector is not an exact copy of the target vector.

Finally, a trial vector replaces the target vector for the next generation if it is better, according to the fitness value. This process is given in Eq. 8.

$$\mathbf{x}_{i,G+1} = \begin{cases} \mathbf{u}_{i,G} & \text{if } (f(\mathbf{u}_{i,G}) \leq f(\mathbf{x}_{i,G})) \\ \mathbf{x}_{i,G} & \text{otherwise} \end{cases}. \tag{8}$$

The above described DE variant is called DE/rand/1/bin. However, there are other variants [10], where one of them is DE/best/1/bin, which uses the best vector of the population to generate the mutant vector, i.e., it replaces the r_0 vector in Eq. 6.

A previously published empirical study of DE to solve CNOPs [11] found that DE/rand/1/bin generates more diverse search directions. Therefore, it promotes the exploration of the search space, while DE/best/1/bin is able to promote the exploitation of promising regions of the search space. Based on those findings, in [11] a new variant called DECV was proposed, which takes advantage of the exploration of the DE/rand/1/bin variant and the exploitation of the DE/best/1/bin variant. The search process begins with the first variant (DE/rand/1/bin), but when there is a 10 % of feasible solutions in the population, it is switched to the second variant (DE/best/1/bin) (see [11] for more details).

Based on the fact that CNOPs are tackled in this work, a constraint-handling technique is required to bias the solutions to the feasible region of the search space. In this case, the feasibility rules proposed by Deb [1] were adopted.

3.6 SA-DECV

Algorithm 3.6 details the proposal of a surrogate-assisted evolutionary algorithm by using DECV (SA-DECV) to tackle CNOPs. Text in *italics* indicates the necessary elements to assemble the NN regression model, which were explained in previous sections.

Algorithm 3 The proposal uses the following elements: Opposition-Based solutions to improve the off-line training, the variation operators of DECV in order to generate new individuals, and the evolution control based on feasibility, that selects the vectors to be evaluated in the original model. Text in *italics* represents the components required to assemble a surrogate model in DECV. The constraint-handling technique is utilized in Selection function (step 12).

---

1. *P* (Initial population)
2. Evaluate population
3. *P* = Opposition_Based_Optimization
4. $train_{set}$ = *Off_line_training*
5. g = 1
6. **while** g < MAX_GEN **do**
7. u = Operators_DECV
8. ind_{orig} = *evolution_control*
9. Evaluate the current population (Surrogate and/or original model)
10. MSE_T = *Measure_model_quality*
11. $train_{set}$ = *On_line_training*
12. P = Selection
13. g = g + 1
14. Return best individual

---

4 Experiments and Results

One objective of using a surrogate model is the reduction of evaluations in the original model. For that reason, in this work the convergence and the computational cost are analyzed with the following performance measures:

- The feasibility probability, *FP*, is the number of feasible trials (*f*) divided by the total number of tests or independent runs performed (*t*), as $FP = f/t$.
  A feasible trial is an independent run where, at least, one feasible solution (which satisfied all constraints of the problem) was generated.
- The probability of convergence, *P*, is calculated by the ratio of the number of successful trials (*s*) to the total number of tests or independent runs performed (*t*), as $P = s/t$.
  A successful trial is an independent run where the best feasible solution found $f(\mathbf{x})$ is close to the best known feasible value or feasible optimum solution $f(\mathbf{x}^*)$. The closeness is given by $f(\mathbf{x}^*) - f(\mathbf{x}) \leq \varepsilon$, where ε is a small tolerance, the value used for ε is 1E-04.
- The average number of function evaluations AFES is calculated by averaging the number of evaluations required on each successful trial to reach the vicinity of the best known feasible value or feasible optimum solution, as $AFES = 1/s \sum_{i=1}^{s} EVAL_i$, where $EVAL_i$ is the number of evaluations required to reach the vicinity of the best known feasible value or optimum solution in the successful trial *i*.
- The successful performance *SP* is defined by *P* and *AFES*. It measures the speed and reliability of an algorithm, as $SP = AFES/P$, where a lower value means a better performance.

The algorithm was tested in a well-known set of test functions, which are described in [9]. Thirty independent runs were performed for each test problem. The SA-DECV's performance was compared against the proposal called ASRES [18].

The parameters for the SA-DECV were those suggested in [11]: population size $NP = 90$, $CR = 1$, $F = 0.9$, tolerance for the equality constraints $\delta = 1E - 04$. Regarding the surrogate model, the size of the training set was 2 * *NP*, the number of elements used to calculate the error was $n_{err} = 5\,\%\,NP$ and the number of nearest neighbors was $k = n$ (dimensionality of the problem). The maximum number of evaluations in the original model was set to *Max_FES* = 240,000, which is used to fix the number of generations ($MAX\_GEN = Max\_FES/NP$).

Table 1 Statistical results (*B* Best, *Md* Median and *SD* Standard Deviation) obtained by the surrogate-assisted DECV (SA-DECV) and the approximate SRES (ASRES) in the benchmark problems

	Prob.	SA-DECV	ASRES	Prob.	SA-DECV	ASRES
B	g01	−15	−15	g12	−1	−1
Md	**−15**	**−15**	**−15**	**−1**	**−1**	**−1**
SD		8.04E−01	0.00E+00		0.00E+00	0.00E+00
B	g02	−0.713	−0.739	**g13**	0.453	0.054
Md	**−0.804**	−0.555	**−0.7**	**0.054**	0.965	**0.054**
SD		8.63E−02	2.10E−02		1.60E−01	1.40E−01
B	g03	−0.333	−0.998	g14	−47.24	−47.765
Md	**−1.001**	−0.023	**−0.995**	**−47.765**	−45.42	**−47.763**
SD		6.89E−02	1.60E−03		1.29E+00	3.30E−03
B	g04	−30665.539	−30665.539	g15	961.715	961.715
Md	**−30665.539**	**−30665.539**	**−30665.539**	**961.715**	**961.715**	**961.715**
SD		2.22E−11	0.00E+00		3.89E−01	4.60E−13
B	g05	5126.497	5126.497	g16	−1.905	−1.905
Md	**5126.497**	5152.172	**5126.497**	**−1.905**	**−1.905**	**−1.905**
SD		8.63E+01	2.30E−12		4.51E−16	1.10E−15
B	g06	−6961.814	−6961.814	g17	8872.62	8853.54
Md	**−6961.814**	**−6961.814**	**−6961.814**	**8853.54**	8945.78	**8853.54**
SD		0.00E+00	1.90E−12		7.75E+01	8.50E−01
B	g07	24.409	24.306	g18	−0.866	−0.866
Md	**24.306**	25.593	**24.307**	**−0.866**	−0.864	**−0.866**
SD		2.14E+01	1.60E−03		9.73E−02	5.30E−02
B	g08	−0.096	−0.096	g19	35.404	32.665
Md	**−0.096**	**−0.096**	**−0.096**	**32.656**	55.808	**32.79**
SD		4.23E−17	5.10E−17		1.71E+01	1.60E−01
B	g09	680.630	680.630	g21	193.732	–
Md	**680.630**	680.633	**680.630**	**193.725**	**208.444**	–
SD		1.17E+00	2.20E−06		9.75E+01	–
B	g10	7260.041	7049.408	g23	−308.803	−348.816
Md	**7049.248**	8138.902	**7053.856**	**−400.055**	−103.823	**−106.585**
SD		9.05E+02	1.90E+01		1.68E+02	1.40E+02
B	g11	0.75	0.75	g24	−5.508	−5.508
Md	**0.75**	**0.75**	**0.75**	**−5.508**	**−5.508**	**−5.508**
SD		1.14E−16	1.10E−16		2.71E−15	0.00E+00

Values in **boldface** indicate the global optimum (below problem name) and the best median value

ASRES uses a (μ, λ) evolution strategy as search algorithm. Furthermore, it employs the stochastic ranking [16] as constraint-handling technique, which is also utilized in its adaptive evolution control.

The statistical values of the final results obtained by SA-DECV and ASRES on the set of benchmark problems are summarized in Table 1. Problems g20 and g22 were excluded because no feasible solutions were found by any variant. In the case of test problem g21, only SA-DECV reported feasible solutions. The aforementioned problems were not included in the statistical validation, which was made with the 95 %-confidence Wilcoxon test using the medians of each problem. This validation showed significant differences between SA-DECV and ASRES.

Based on the best solutions, SA-DECV and ASRES reached the global optimum in twelve problems (g01, g04, g05, g06, g08, g09, g11, g12, g15, g16, g18, and g24). SA-DECV found the best known solution in nine test problem with only inequality constraints (g01, g04, g06, g08, g09, g12, g16, g18 and g24), and in three test problems with equality constraints (g05, g11 and g15). In other test problems with equality constraints, SA-DECV reported feasible solutions but they were far of the feasible global optimum.

The performance measure values (FP, P and SP) and the statistical values of the number of evaluations in the original model (function evaluations—FEs) obtained by SA-DECV and ASRES are summarized in Table 2. Test problems g02, g03, g10, g19, g21, and g23 were excluded because no successful trials were found by comparing both the algorithms.

The overall results for the FP measure show that SA-DECV and ASRES were able to reach the feasible region in 16 test problems g01, g04, g05, g06, g07, g08, g09, g11, g12, g13, g14, g15, g16, g17, g18, and g24. Based on the results of the P measure, SA-DECV and ASRES were capable to find a feasible solution in the vicinity of the feasible global optimum in seven test problems g04, g06, g08, g11, g12, g16 and g24. ASRES was able to reach the global optimum in test problems g07, g13, g14 and g17, while SA-DECV found feasible solutions but they remained away from the best known feasible value. Despite the fact that, based on the P value, ASRES was capable to reach the best feasible solution in test problems g01, g04, g06, g08, g09, g12, g16, g18, and g24, SA-DECV was able to match such performance but with a lower number of evaluations in the expensive model. These problems only have inequality constraints. Finally, derived on the results of the SP measure, SA-DECV reported the lowest SP values in seven test problems: g01, g04, g06, g08, g12, g16, and g24. All previous problems only have inequality constraints. In summary, SA-DECV was able to provide good results, with a lower number of evaluations in the original expensive model for problems with only inequality constraints.

Table 2 Statistical results (*B* Best, *Md* Median and *SD* Standard Deviation) for the FEs to obtain a successful trial and the performance measures (FP, S and SP) required, using the surrogate-assisted DECV (SA-DECV) and the approximate SRES (ASRES)

Alg.	Prob.	FEs			FP	P	SP	Prob.	FEs			FP	P	SP
		B	Md	SD					B	Md	SD			
SA-DECV	g01	**6683**	**7972**	805	**1**	0.83	**9745**	g12	**363**	**1139**	283	**1**	**1**	**1062**
ASRES		31,698	34,911	2581	**1**	**1**	35,406		1476	3101	584	**1**	**1**	2996
SA-DECV	g04	**3801**	**5113**	1013	**1**	**1**	**5356**	g13	–	–	–	**1**	0	–
ASRES		13,616	15,041	706	**1**	**1**	15,104		**9936**	**11,459**	624	**1**	**0.84**	**13,422**
SA-DECV	g05	58,688	69,418	10,396	0.96	0.1	69,1943	g14	–	–	–	**1**	0	–
ASRES		**17,526**	**19,188**	885	**1**	**1**	**19,281**		**84,631**	**92,820**	11,581	**1**	**0.08**	**1,160,256**
SA-DECV	g06	**6094**	**7335**	527	**1**	**1**	**7333**	g15	21,915	46,105	14,091	0.96	0.63	75,231
ASRES		7875	9542	918	**1**	**1**	9603		**7015**	**8473**	620	**1**	**1**	**8519**
SA-DECV	g07	–	–	–	**1**	0	–	g16	**7728**	**10,149**	1353	**1**	**1**	**10,304**
ASRES		**75,890**	**76,782**	918	**1**	**0.08**	**959,775**		13,638	16,341	1470	**1**	**1**	16,179
SA-DECV	g08	**396**	**676**	139	**1**	**1**	**699**	g17	–	–	–	**1**	0	–
ASRES		603	1007	229	**1**	**1**	1027		**17,817**	**19,296**	9743	**1**	**0.76**	**28,277**
SA-DECV	g09	**8377**	**9831**	1347	**1**	0.2	50,803	g18	**32,751**	**32,751**	–	**1**	0.03	982,530
ASRES		26,972	30,241	2535	**1**	**1**	**30,618**		36,502	41,015	3214	**1**	**0.92**	**44,391**
SA-DECV	g11	2653	10,478	4907	**1**	**1**	10,613	g24	**2491**	**3444**	468	**1**	**1**	**3430**
ASRES		**1701**	**2800**	550	**1**	**1**	**2792**		2909	3648	391	**1**	**1**	3638

Values in **boldface** indicate the best results

5 Conclusions and Future Work

This paper proposed an adaptive evolution control based on feasibility, added to a DE variant called SA-DECV, to solve constrained numerical optimization problems. The results and number of evaluations were compared against those obtained by one state-of-the-art method called ASRES. Such results suggested that SA-DECV reached the global optimum with a lower number of evaluations in the original model, in presence of only inequality constraints. For problems with equality constraints, SA-DECV reported feasible solutions but not necessarily close to the global optimum. Considering the fact that the best performance of SA-DECV was achieved in problems with only inequality constraints, other constraint-handling techniques will be considered. Moreover, other surrogate models (RBF, SVM) will be analyzed.

Acknowledgments The first author acknowledges support from the Mexican National Council for Science and Technology (CONACyT) through a scholarship to pursue graduate studies at University of Veracruz. The second author acknowledges support from CONACyT through project No. 220522.

References

1. Deb, K.: An efficient constraint handling method for genetic algorithms. In: Computer Methods in Applied Mechanics and Engineering, pp. 311–338 (1998)
2. Elsayed, S.M., Ray, T., Sarker, R.A.: A surrogate-assisted differential evolution algorithm with dynamic parameters selection for solving expensive optimization problems. In: IEEE Congress on Evolutionary Computation (CEC), pp. 1062–1068 (2014)
3. Fonseca, L.G., Barbosa, H.J.C., Lemonge, A.C.C.: A similarity-based surrogate model for enhanced performance in genetic algorithms. OPSEARCH **46**(1):89–107. Springer (2009)
4. Forrester, A.I.J., Keane, A.J.: Recent advances in surrogate-based optimization. Progress in Aerospace Sciences, **45**(1–3):50–79 (2009)
5. Gräning, L., Jin, Y., Sendhoff, B.: Efficient evolutionary optimization using individual-based evolution control and neural networks: A comparative study. In: Proceeding of European Symposium on Artificial Neural Networks (ESANN), pp. 27–29 (2005)
6. Hüsken, M., Jin, Y., Sendhoff, B.: Structure optimization of neural networks for evolutionary design optimization. Soft Comput. J. **9**(1), 21–28 (2005)
7. Jin, Y.: A comprehensive survey of fitness approximation in evolutionary computation. Soft Comput. **9**(1):3–12 (2005)
8. Jin, Y.: Surrogate-assisted evolutionary computation: Recent advances and future challenges. Swarm Evol. Comput. **1**(1):61–70 (2011)
9. Liang, J.J., Runarsson, T., Mezura-Montes, E., Clerc, M., Suganthan, P., Coello-Coello, C.A., Deb, K.: Problem definitions and evaluation criteria for the CEC 2006 special session on constrained real-parameter optimization. Technical report, Nanyang Technological University, Singapure (2005)
10. Mezura-Montes, E., Velázquez-Reyes, J., Coello-Coello, C.A.: A comparative study of differential evolution variants for global optimization. In: Cattolico, M. (ed.) GECCO, pp. 485–492 (2006)
11. Mezura-Montes, E., Miranda-Varela, M.E., Gómez-Ramón, R.C.: Differential evolution in constrained numerical optimization: an empirical study. Inf. Sci. **180**(22):4223–4262 (2010)

12. Rahnamayan, S., Tizhoosh, H.R., Salama M.M.A.: Opposition-based differential evolution. IEEE Trans. Evol. Comput. **12**(1):64–79 (2008)
13. Regis, R.G., Shoemaker, C.A.: A stochastic radial basis function method for the global optimization of expensive functions. INFORMS J. Comput. **19**(4):497–509 (2007)
14. Regis, R.G.: Stochastic radial basis function algorithms for large-scale optimization involving expensive black-box objective and constraint functions. Comput. Oper. Res. **38**(5):837–853 (2011)
15. Regis, R.G.: Evolutionary programming for high-dimensional constrained expensive black-box optimization using radial basis functions. IEEE Trans. Evol. Comput. **18**(3):326–347 (2014)
16. Runarsson, T.P., Yao, X.: Stochastic ranking for constrained evolutionary optimization. IEEE Trans. Evol. Comput. **4**:284–294 (2000)
17. Runarsson, T.P.: Constrained evolutionary optimization by approximate ranking and surrogate models. In: Parallel Problem Solving from Nature—PPSN VIII, Lecture Notes in Computer Science, Springer, pp. 401–410 (2004)
18. Runarsson, T.P.: Approximate evolution strategy using stochastic ranking. In: IEEE Congress on Evolutionary Computation, CEC 2006, pp. 745–752 (2006)
19. Sasena, M.J., Papalambros, P., Goovaerts, P.: Exploration of metamodeling sampling criteria for constrained global optimization. Eng. Optim. **34**(3):263–278 (2002)
20. Shi, L., Rasheed, K.: Asaga: An adaptive surrogate-assisted genetic algorithm. In: Ryan, C., Keijzer, M. (eds.) GECCO, pp. 1049–1056 (2008)
21. Shi, L., Rasheed, K.: A survey of fitness approximation methods applied in evolutionary algorithms. In: Hiot, L.M., Ong, Y.S., Tenne, Y., Goh, C.K. (eds.) Computational Intelligence in Expensive Optimization Problems, Adaptation Learning and Optimization, chapter 1, vol. 2, pp. 3–28. Springer, Berlin, Heidelberg (2010)
22. Storn, R., Price, K.: Differential evolution—A simple and efficient adaptive scheme for global optimization over continuous spaces (1995)

PSO-TVAC-Based Economic Load Dispatch with Valve-Point Loading

Parmvir Singh Bhullar and Jaspreet Kaur Dhami

Abstract In this paper, an effective and reliable variant of particle swarm optimization with time-varying acceleration coefficients (PSO-TVAC) technique is proposed for the economic load dispatch problem considering valve-point loading effect. Main objective of the economic load dispatch (ELD) problem is to minimize the fuel cost by allocating the generation of the committed units subjected to the equality and inequality constraints. Equation of the economic dispatch objective function gets modified with the addition of a new parameter which represents the effect of valve-point loading. While exploring the use of PSO and its other variants to the economic dispatch problem, a number of research works have not considered the transmission losses properly. This paper represents the usefulness of the proposed PSO-TVAC algorithm to significantly reduce the fuel cost while taking into account the effect of transmission losses along with the non-convex characteristic due to valve loading. The results have been demonstrated for three-generator and ten-generator test systems.

Keywords Economic load dispatch · Valve-point loading · Particle swarm optimization · Time-varying acceleration coefficients · Transmission losses

1 Retroscope

ED problem is a very well known and heavily researched problem in the field of power system. Operating cost of the thermal power plant is reduced by proper allocation of the amount of power to the committed units that subject to the equality and inequality constraints [1].

P.S. Bhullar (✉) · J.K. Dhami
B.B.S.B.E. College, Fatehgarh Sahib, India
e-mail: parmvirsinghbhullar@yahoo.com

J.K. Dhami
e-mail: jaspreet.dhami@bbsbec.ac.in

M. Pant et al. (eds.), *Proceedings of Fifth International Conference on Soft Computing for Problem Solving*, Advances in Intelligent Systems and Computing 436, DOI 10.1007/978-981-10-0448-3_68

Traditionally, Lagrangian multipliers method is used to solve the ED problem with the units of monotonically increasing piecewise linear cost functions. Practically, input–output characteristics of generating units are highly nonlinear due to other constraints like valve-point loading and multiple fuel effects, etc. Such problems can be solved using either deterministic or stochastic algorithms. The deterministic methods such as lambda iteration, best first, hill climbing, gradient methods, and others depend on the simplicity-based concept that assumes the cost curves of ED problem's generator which are convex. However, due to the addition of the practical system constraints, such as the valve-point effects, the generator cost curves are no longer convex hence the ED problem becomes a non-convex constrained minimization problem. Thus, there is a large emphasis on adopting stochastic algorithms to solve the economic dispatch problem effectively. Benevolent results have been reported during the past few years and several methods like genetic algorithm (GA) [2], evolutionary programming (EP) [3], tabu search (TS) [4], simulated annealing (SA) [5], differential evolution (DE) [6], etc., have been successfully implemented for practical ED problems thus significantly improving the existing results of the problem in terms of decreased fuel cost.

In this paper, the bioinspired algorithm PSO-TVAC is proposed in solving the non-convex economic dispatch problem. Moreover, the proper tuning of this method is also affecting the results. So all the points have been considered up to the best level into this paper. The simulation results performed on standard practical test systems of three and ten-generator systems are presented and discussed with transmission losses into consideration and with proper tuning. In next section, formulation of economic load dispatch problem is discussed and in third section PSO-TVAC algorithm and its implementation is discussed. In ensuing section, test problems as well as the experimental results are reported.

2 ED Problem Formulations

The aim of the economic dispatch problem is to minimize the fuel cost of generation [7]. The dispatch problem can be stated mathematically as follows:

$$\text{Minimize} \quad F(P_{gi}) = \sum_{i=1}^{NG} F(P_{gi}) \tag{1}$$

$$F(P_{gi}) = a_i P_{gi}^2 + b_i P_{gi} + C_i \tag{2}$$

subject to the equality real power balance constraints is

$$\sum_{i=1}^{NG} P_{gi} = P_D + P_L \tag{3}$$

$$P_L = \sum_{i=1}^{n}\sum_{j=1}^{n} P_i B_{ij} P_j + \sum_{i=1}^{n} B_{oi} P_i + B_{oo}^2 \quad (4)$$

The inequality constraint of limits on the generator outputs is

$$P_{gi}^{\min} \le P_{gi} \le P_{gi}^{\max} \quad (5)$$

where

P_{gi} is the decision variable, i.e., real power generation
P_d is the real power demand
NG is the number of generation plants
$P_{gi}^{\min}$ is the lower permissible limit of real power generation
$P_{gi}^{\min}$ is the upper permissible limit of real power generation
$F(P_{gi})$ is the operating fuel cost of the ith plant of real power generation
a_i, b_i, and c_i are the cost coefficients of the *i*th generator
P_L represents the transmission losses
B_{ij} is the *ij*th element of loss coefficient symmetric matrix B
B_{io} is the *i*th element of the loss coefficient vector
B_{oo} is loss coefficient constant.

However, the cost function of a generator is not always differentiable due to valve-point effects and/or change of fuels as the valve-point effects introduce ripples in the heat-rate curve [8]. The fuel cost function with valve-point loadings of the generators is as shown in Fig. 1:

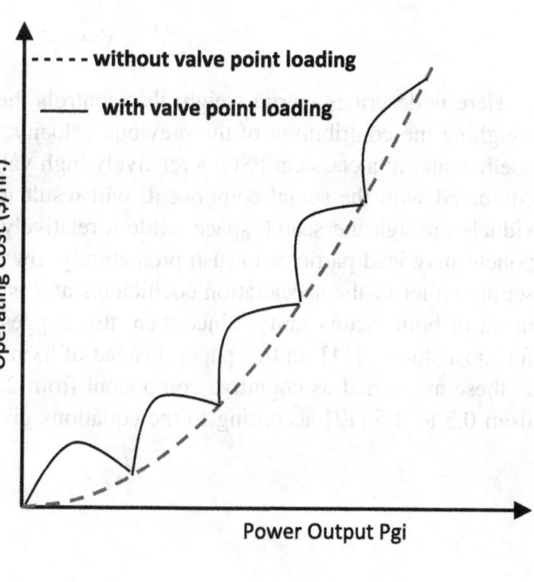

Fig. 1 Operating cost characteristics with valve-point loading

Mathematically, economic dispatch problem considering valve-point loading is defined as minimize operating cost

$$F = \sum_{i=1}^{NG} \left(c_i + b_i P_{gi} + a_i P_{gi}^2\right) + \left|e_i \sin\left(f_i\left(P_{gi}^{\min} - P_{gi}\right)\right)\right| \quad (6)$$

where a_i, b_i, c_i, e_i, and f_i are the fuel cost coefficients of generator i with valve-point loading effects.

3 Proposed Algorithm

3.1 Introduction

This section portrays the proposed algorithm which is one of the variants of PSO. PSO is a well researched artificial intelligence technique for optimization and search based on the principle of social behavior of animals [9]. PSO is very good at finding good enough solutions for a large domain of problems, such as constrained optimization problems, multiobjective optimization problems, etc. The PSO technique involves the initialization of scattered random, i.e., initial solutions in the problem space. Each particle updates its position using the velocity of particle. Both position and velocity are updated in such a manner based upon their own experience and the experience of its neighbors [10]. The most important factor in PSO algorithm is the setting or adjusting the value of parameters given in Eq. (7):

$$v_i^{(t+1)} = wv_i^t + c_1 \text{rand}\left(p_i^{lb} - x_i^t\right) + c_2 \text{rand}\left(p_i^{gb} - x_i^t\right) \quad (7)$$

$$x_i^{t+1} = x_i^t + v_i^{t+1} \quad (8)$$

Here w describes inertia weight that controls the momentum of the particle by weighing the contribution of the previous velocity. C_1 and C_2 are the acceleration coefficients. In a classical PSO, a relatively high value of the cognitive component, compared with the social component, will result in excessive wandering of individuals through the search space while a relatively high value of the social component may lead particles to rush prematurely toward a local optimum. Moreover, setting either of the acceleration coefficients at 2 is suggested in order to make the mean of both factors unity. Since then, this suggestion has been extensively used for most studies [11]. In this paper, instead of fixing the acceleration coefficients at 2, these are varied as cognitive component from 2.5 to 0.5 and social component from 0.5 to 2.5 [12] according to the equations given below:

$$C_1 = (C_{1f} - C_{1i})\frac{\text{iter}}{\text{MAX.ITR}} + C_{1i} \tag{9}$$

$$C_2 = (C_{2f} - C_{2i})\frac{\text{iter}}{\text{MAX.ITR}} + C_{2i} \tag{10}$$

C_{1f}, C_{2f}, C_{1i} and C_{2f} are final and initial values of acceleration coefficients, respectively.

p_i^{lb} and p_i^{gb} are the particles' personal and global best values, respectively. A flowchart showing general algorithm of PSO is drawn in Fig. 2.

3.2 *Implementation of PSO-TVAC to ELD Problem*

In this paper, an algorithm is developed to solve a constrained and unconstrained ED problem using PSO-TVAC to obtain solution of high quality. The algorithm is utilized mainly to determine the optimal allocation of power among the committed units and thus minimizing the total generation cost.

For the implementation of PSO-TVAC algorithm to solve the ED problem, mentioned steps should be followed:

Step 1: initialization of particles by randomly initializing generation of each generator according to the limit of each unit. These initial individuals must be feasible candidate solutions that satisfy the practical operation constraints. 100 numbers of particles are assumed in the proposed algorithm.

Step 2: Each set of initial solution in the space should satisfy the Eq. (3). If any combination does not satisfy the constraints then they are set according to the power balance equation:

$$P_d = P_D + P_L - \sum_{\substack{i=1 \\ i \neq d}}^{NG} P_{gi} \tag{11}$$

Step 3: The objective is to be evaluated for each individual, i.e., fitness of each individual is to be calculated.

Step 4: Compare fitness of each individual with its pbest. The best fitness value among the pbests is denoted as gbest.

Step 5: Modify the member velocity 'v' of each individual P_g, according to Eq. (7) where C_1 and C_2 are to be adjusted according to the Eqs. (9) and (10) and 'w' is set at 0.7.

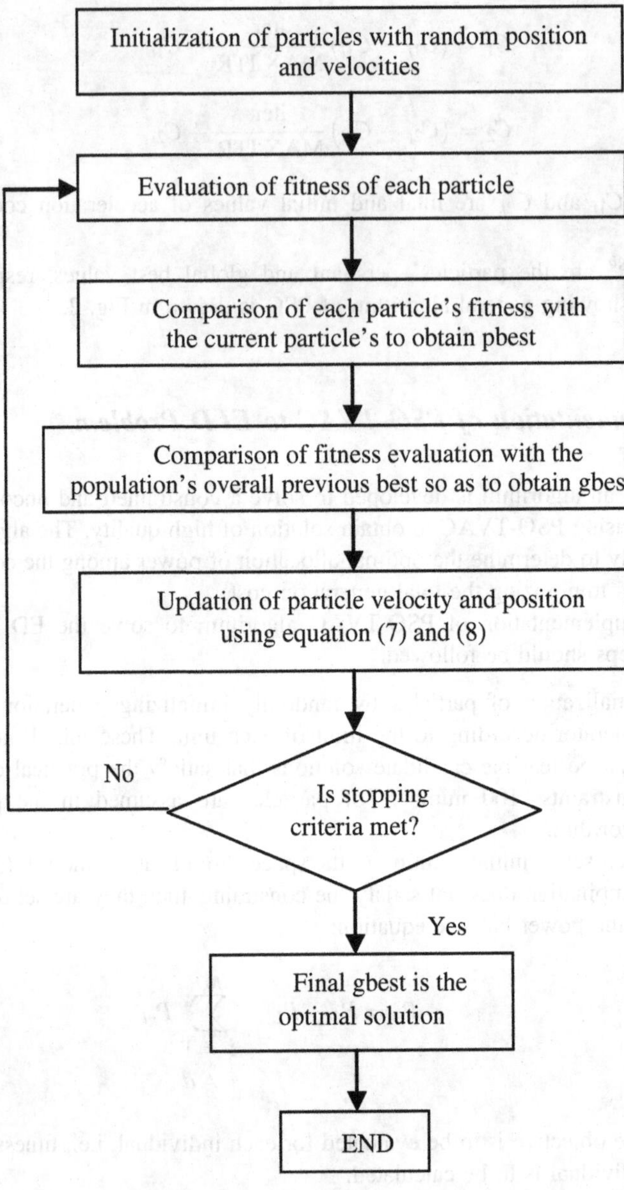

Fig. 2 Flowchart showing general PSO algorithm

Step 6: The velocity components constraint occurring in the limits from the following conditions are checked.

$$V_{\min} = -0.5 * P^{\min}$$
$$V_{\max} = +0.5 * P^{\max}$$

Step 7: Modify the member position 'x' of each individual P_g according to Eq. (8).

Step 8: If the fitness value of each individual is better than the previous pbest value, the current value to be set is pbest. If the best pbest is better than gbest, the value to be set is gbest.

Step 9: If the number of iterations reaches the maximum, then go to step 10. Otherwise, go to step 3. Numbers of iterations are assumed to be 80 in proposed algorithm.

Step 10: The individual that generates the final gbest is the optimal generation power of each unit with the minimum total generation cost.

4 Simulation Results

In this section, the results of ELD after the implementation of proposed algorithm based on PSO-TVAC are discussed. The program is implemented in MATLAB environment 7.5.0. The main objective is to minimize the cost of generation of plants considering valve-point loading effects. The performance is evaluated with and without considering transmission losses using two standard generator test systems, i.e., Three-generator test system and ten-generator test system. A reasonable B-loss coefficient matrix of the power system network has been employed to calculate the transmission loss. Results are compared and presented with different tables for valve and without valve-loading effects. For the same data, minimized cost using conventional PSO algorithm [13] is also compared with the proposed PSO-TVAC method.

4.1 Case I—Three-Generator Test System

Electric power system of three generators has been studied. In this case, the load demand, P_D is taken as 850 MW. Data are taken from standard three-generator test system [14].

Results are compared in Table 1 without valve-point loading using PSO-TVAC algorithm. It is clear from the results that operating cost is increased when transmission losses are included.

Table 1 Results for three-generator test system using PSO-TVAC (without valve loading effect)

No.	P_g (min.) MW	P_g (max.) (MW)	PSO-TVAC	
			Without losses (MW) run 1	With losses (MW) run 2
1	100	600	406	350
2	100	400	327	350
3	50	200	116	154.4
$\sum_{i=1}^{3} P_{gi}$	850 MW (load demand)		850	854.48
Total generation cost ($/h)			8194.9	8243.7

Table 2 Results for three-generator test system using PSO-TVAC (with valve loading effect)

No.	P_g (min.) (MW)	P_g (max.) (MW)	PSO-TVAC	
			Without losses (MW) run 3	With losses (MW) run 4
1	100	600	325	301.2
2	100	400	375	365.2
3	50	200	150	188
$\sum_{i=1}^{3} P_{gi}$	850 MW (load demand)		850	854.48
Total generation cost ($/h)			8208.7	8271.5

Table 3 Comparison of operating costs without losses

Algorithm used	Without valve loading	With valve loading
PSO	8196.9	8217.3
PSO-TVAC	8194.9	8208.7

In Table 2 results are discussed for three-generator test system using PSO-TVAC with valve loading effects.

In Table 3, using conventional PSO and PSO-TVAC, a comparison is made between the optimized costs, without considering transmission losses, including valve-point loading.

4.2 Case II—Ten-Generator Test System

In this case, the load demand, P_D is taken as 2000 MW. Data is taken from standard ten-generator test system [15].

Results are compared in Table 4 without valve-point loading using PSO-TVAC algorithm. It is clear from the results that cost has been increased when transmission losses are considered.

In Table 5 optimized costs with valve-point loading are compared for transmission losses and without them.

Table 4 Results for ten-generator test system using PSO-TVAC (without valve loading effect)

No.	P_g (min.) (MW)	P_g (max.) (MW)	PSO-TVAC	
			Without losses (MW) run 1	With losses (MW) run 2
1	10	55	51.5	53.2
2	20	80	76.8	76.2
3	47	120	117	120
4	20	130	103	112
5	50	160	100	111
6	70	240	95.8	125
7	60	300	298	222
8	70	340	337	304
9	135	470	469	468
10	150	470	359	469
$\sum_{i=1}^{10} P_{gi}$	2000 MW (load demand)		2006.6	2059.9
Total generation cost ($/h)			107,480	110,670

Table 5 Results for tem-generator test system using PSO-TVAC (with valve loading effect)

No.	P_g (min.) (MW)	P_g (max.) (MW)	PSO-TVAC	
			Without losses (MW) run 3	With losses (MW) run 4
1	10	55	53.8	52.4
2	20	80	78.9	77
3	47	120	109	117
4	20	130	125	128
5	50	160	98.8	122
6	70	240	90.4	122
7	60	300	298	297
8	70	340	330	265
9	135	470	468	455
10	150	470	351	467
$\sum_{i=1}^{10} P_{gi}$	2000 MW (load demand)		2002.9	2102
Total generation cost ($/h)			107,380	113,400

Table 6 Comparison of operating costs without losses

Algorithm used	Without valve loading	With valve loading
PSO	107,390	107,620
PSO-TVAC	107,480	107,380

In Table 6 comparison of conventional PSO and PSO-TVAC is shown for optimized costs, without considering transmission losses and including valve-point loading.

5 Conclusion

In this paper, a new variant of classic PSO technique is successfully implemented and some important conclusions can be easily made by observing the numerous comparisons made in the result tables between the new and old version of the optimization algorithm. As the major concern is the minimization of the operating cost of the thermal power plant, taking into account both the transmission losses and valve-point loading effect while fulfilling other system constraints, the objective is achieved up to a satisfactory level by the use of proposed PSO-TVAC technique.

References

1. Wood, A.J., Wollenberg, B.F.: Power Generation, Operation and Control, 2nd edn. Wiley, New York (1996)
2. Walters, D.C., Sheble, G.B.: Genetic algorithm solution of economic dispatch with valve point loading. IEEE Trans. Power Syst. **8**(3), 1325–1332 (1993)
3. Sinha, N., Chakrabarti, R., Chattopadhyay, P.K.: Evolutionary programming techniques for economic load dispatch. IEEE Trans. Evol. Comput. **7**(1), 83–94 (2003)
4. Min, L.W., Sheng, C.F., Tong, T.M.: An improved tabu search for economic dispatch with multiple minima. IEEE Trans. Power Syst. **17**, 108–112 (2002)
5. Wong, K.P., Fung, C.C.: Simulated annealing based economic dispatch algorithm. In: IEEE Proceedings C: Generation, Transmission and Distribution, vol. 140, pp. 509–515 (1993)
6. Khamsawang, S., Jiriwibhakorn, S.: Solving the economic dispatch problem by using differential evolution. Int. Sci. Index **3**(4), 550–554 (2009)
7. Dhillon, J.S., Kothari, D.P.: Power System Optimization, 2nd edn. Prentice Hall of India (2006)
8. Chen, C.H., Yeh, S.N.: Particle swarm optimization for economic power dispatch with valve-point effects. In: Transmission & Distribution Conference and Exposition, Caracas, pp. 1–5 (2006)
9. Kennedy, J., Eberhart, R.: Particle swarm optimization. In: Proceedings of IEEE International Conference on Neural Networks, Perth, Australia, vol. 4, pp. 1942–1948 (1995)
10. Eberhart, R., Kennedy, J.: A new optimizer using particle swarm theory. In: Proceedings of 6th Symposium on MicroMachine and Human Science, Nagoya, Japan, pp. 39–43 (1995)
11. Suganthan, P.N.: Particle swarm optimizer with neighbour-hood operator. In: Proceedings of the IEEE International Congress on Evolutionary Computation, vol.3, pp. 1958–1962 (1999)
12. Ratnaweera, A., Halgamuge, S.K.: Self-Organizing hierarchical particle swarm optimizer with time-varying acceleration coefficients. IEEE Trans. Evol. Comput. **8**(3), 240–255 (2004)
13. Bhullar, P.S., Dhami, J.K.: Particle swarm optimization based economic load dispatch with valve point loading. Int. J. Eng. Res. Technol. **4**(5), 1064–1070 (2015)
14. Singh, N., Kumar, Y.: Economic load dispatch with valve point loading effect and generator ramp rate limits constraint using MRPSO. Int. J. Adv. Res. Comput. Eng. Technol. (IJARCET). **2**(4), 1472–1477 (2013)
15. Manteaw, E.D., Odero, N.A.: Combined economic and emission dispatch solution using ABC_PSO hybrid algorithm with valve point loading effect. Int. J. Sci. Res. Publ. **2**(12), 1–9 (2012)

A Heuristic Based on AHP and TOPSIS for Disassembly Line Balancing

Shwetank Avikal

Abstract Disassembly lines have become one of the most suitable ways for the disassembly of large products or small products in large quantities for efficient working of disassembly line; its design and balancing is prudent. In disassembly lines, task assignment in appropriate schedule is necessary for designing and balancing the line. In this paper, a heuristic based on multi criteria decision-making (MCDM) technique has been proposed for assignment of tasks to the disassembly workstations. In the proposed heuristic, Analytical Hierarchy Process (AHP) and the Technique for Order Preference by Similarity to Ideal Solution (TOPSIS) has been used for the prioritizing of task for the assignment to workstations. The proposed heuristic has been compared to other heuristic and it has been found that it performs well and gives sufficiently better results.

Keywords Analytical hierarchy process (AHP) · Heuristic · TOPSIS · Line balancing · MCDM · Product disassembly

1 Introduction

Ecological reflections and strengthening legislation are introducing greater emphasis on the recovery and recycling of all the materials of manufactured products. Amplified public consciousness in environmental-related issues has led to a rising concern with the environmental implications of product design, its manufacturing process, and reutilization of used and outdated produced. The discussed issue has also led a number of countries to execute a stricter legislation, such as product "take back" laws those have been implemented in European countries and the "recyclability" laws in Japan, and to place responsibility of safely disposal of products at the end of their life cycles on the people related to their primary developer and user.

Shwetank Avikal (✉)
Mechanical Engineering Department, Graphic Era Hill University, Dehradun, India
e-mail: shwetank.avikal@gmail.com

M. Pant et al. (eds.), *Proceedings of Fifth International Conference on Soft Computing for Problem Solving*, Advances in Intelligent Systems and Computing 436, DOI 10.1007/978-981-10-0448-3_69

The paper is structured as follows. Section 2 deals with a brief literature review. Sections 3 and 4 describes the computational process of AHP and TOPSIS one by one, and Sect. 5 defines formulation deal problem. Section 6 shows the proposed heuristic and a numerical example selected from the literature have been solved in Sect. 7. In the end, Sect. 8 deals with the conclusion of present research work.

2 Literature Review

The basic disassembly line balancing problem (DLBP) may be described as the task should be assigned to the assignment in such a way that the precedence relations between the tasks are satisfied and some other measure of effectiveness is optimized.

Recently, [1] have presented an approach based on Ant Colony Optimization (ACO) for disassembly sequencing for multiple objectives. McGovern and Gupta [2] have explained that the problem of disassembly line balancing as a NP hard problem and they also suggested a genetic algorithm-based solution approach for obtaining optimal solutions for DLBP. Agrawal and Tiwari [3] have suggested a collaborative ACO algorithm for solving stochastic mixed-model, U-shaped disassembly line balancing problem and sequencing problem. They tested the proposed approach on the problems developed by Design of Experiment techniques. They also performed Analysis of Variance to find out the impact of various factors on the objective. Altekin et al. [4] have proposed a mixed integer programming algorithm for the solving of DLBP. Koc et al. [5] have suggested two exact formulations for disassembly line balancing problems with task precedence diagram construction using AND/OR precedence graph. Ding et al. [6] have proposed a multi-objective disassembly line balancing problem and applied an ant colony algorithm for its solution. Karadag and Turkbey [7] have proposed a genetic algorithm (GA)-based approach for solving multi-objective optimization of a stochastic DLBP. Kalayci and Gupta [8] have proposed an artificial bee colony (ABC)-based methodology for solving sequence-dependent disassembly line balancing problem.

A large number of research methodology have been reported in the literature for balancing the disassembly line perfectly such as: heuristics approaches, mathematical model, meta-heuristics, and other solving techniques. A number of factors on which disassembly line depends such as part demand, part removal time, part hazard, etc., have been considered for the developing of these solution techniques. There is a strong need to incorporate the experiences and views of labors, supervisors, customers, environment activists, researchers, and public reviewers. In the present work, the disassembly line balancing problem has been considered as a multi-objective decision-making (MCDM) problem. An AHP and TOPSIS method-based heuristic has been developed and attempt has been made to balance the disassembly line.

3 AHP Technique

AHP is a research method to support rational decision-making on several qualitative factors [9]. It is an outstanding tool for solving complex multi criteria decision-making (MCDM) problems. This technique is also known as a method of eigenvector. This technique compares all alternatives to other alternatives alternative and develops a pairwise comparison matrix. AHP technique suggested by Lee et al. [10] has been taken for the calculation.

4 TOPSIS Method

Hwang and Yoon proposed TOPSIS method of ranking the alternatives in 1981. It ranks different alternatives on the basis of their closeness to a zenith point and a nadir point those are also known as hypothetical positive ideal alternative and hypothetical negative ideal alternative. Both alternatives have been shown in the TOPSIS as a point in this space [11–13]. According to [13] the procedure of the standard TOPSIS method involves mainly six steps. The method proposed by [13] has been considered for the calculation.

5 Problem Formulation

The algorithm has been proposed for calculating the weights of each criterion of the disassembly task. With the help of these weights, the tasks can be assigned to the disassembly workstation to minimize cycle time, number of workstations, and maximize line efficiency. In the present problem, disassembly line balancing problem has been taken as a multi criteria decision making problem. Three well known criteria have been selected for solving this complex DLBP, i.e., Part demand, Disassembly time, and Part hazardous.

In the case of DLBP, selection of tasks for the assignment on workstation is a big task. Therefore, a lot of assignment techniques have been proposed for selecting tasks for assignment to the workstation with the use of different priority rules and other techniques. However, these techniques do not vender due consideration to view of labor, supervisor, customers, environment activities, researchers, and public reviewer. Because of these issues, the disassembly line balancing problem has been taken as MCDM problem. This MCDM problem have been solved by AHP and TOPSIS methods, as to prioritize the tasks for the assignment to the disassembly workstation.

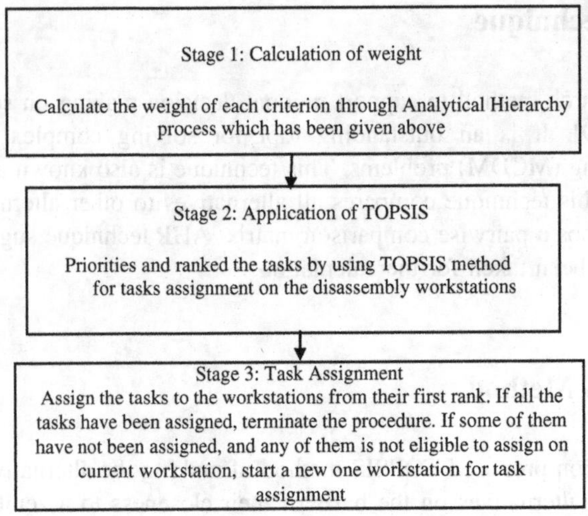

Fig. 1 The AHP- and TOPSIS-based heuristic approach

6 Proposed Heuristic

The proposed heuristic has been based on multi criteria decision-making techniques. This heuristic has been divided into three stages such as: calculation of weights of all criterions, implementation of TOPSIS method for calculating the rank of the task, and the last one is assignment of the tasks to the disassembly workstations. The stages of proposed heuristics have shown in Fig. 1.

7 Computational Example

The algorithm proposed in the article has been tested on different disassembly line balancing problems to calculate its performance. Here, the proposed heuristic has been applied to find a solution to the disassembly line balancing problem presented by McGovern and Gupta [14]. The objective of this problem is to complete disassembly. The precedence diagram of the problem is given in Fig. 2 and the numerical data have been given in Table 1.

7.1 Calculation of Weights of Criteria

With the help of decision hierarchy tree (Fig. 3), the weight of each criterion has been calculated by AHP. The pairwise comparison matrix of AHP has been given in Table 2 and the weights calculated by AHP have been given in Table 3.

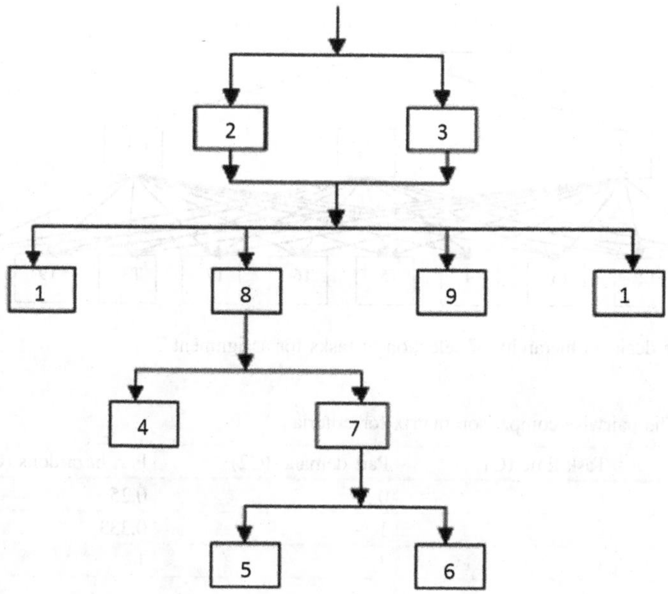

Fig. 2 Precedence diagram of proposed example

Table 1 Date of knowledge-based example from literature

Task no.	Processing time	Hazardous (Y/N)	Part demand
01	14	0	00
02	10	0	500
03	12	0	00
04	18	0	00
05	23	0	00
06	16	0	485
07	20	1	295
08	36	0	00
09	14	0	360
10	10	0	00

In the disassembly line balancing, it has been found by AHP that criteria C3 is most important among others. Consistency ratio of the pairwise comparison matrix is calculated as $0.015558 < 0.1$. So, the weights are shown to be consistent and they are used in the selection process.

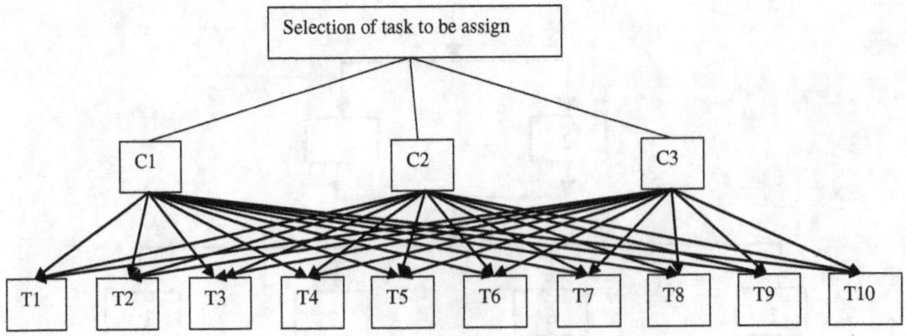

Fig. 3 The decision hierarchy of selection of tasks for assignment

Table 2 The pairwise comparison matrix for criteria

Criteria	Task time (C1)	Part demand (C2)	Part hazardous (C3)
C1	1	0.5	0.25
C2	2	1	0.333
C3	4	3	1

Table 3 Results (criterion weight) obtained by AHP

Criteria	Weights (w)	λmax, CI, RI	CR
C1	0.137299	λmax = 3.018047	0.015558
C2	0.239432	CI = 0.009024	
C3	0.623269		

7.2 Determination of Final Rank

TOPSIS has been used to calculate the final ranks of parts those are to be removed with the use of criterion weight calculated by AHP. The part's data (Table 3) available in the example is a crisp data and this data has been normalized in Table 4 for calculation and assignment of tasks.

The weighted decision matrix has been formed by using Step. No two, and has been shown in Table 5. The positive ideal solution (PI) and the negative ideal solution (NI) for the discussed problem have been calculated by Step no. three and given in Tables 6 and 7. Euclidean distance method (step 4) has been used to calculate the separation distance of each tasks from PI and the NI and given in Table 9.

Step 5 has been used for calculating the relative closeness or similarity degree of each task from the ideal solution, and step 6 has been used for calculating the final ranks of tasks on the basis of relative closeness (Table 8).

Table 4 Normalized data of selected example [14]

Tasks	C1	C2	C3
1	0.080924855	0	0
2	0.057803468	0.262467	0
3	0.069364162	0	0
4	0.104046243	0	0
5	0.132947977	0	0
6	0.092485549	0.393701	0
7	0.115606936	0.154856	1
8	0.208092486	0	0
9	0.080924855	0.188976	0
10	0.057803468	0	0

Table 5 Weighted decision data matrix

Tasks	C1	C2	C3
1	0.011110881	0	0
2	0.007936344	0.062843	0
3	0.009523612	0	0
4	0.014285419	0	0
5	0.018253591	0	0
6	0.01269815	0.094265	0
7	0.015872687	0.037077	0.623269
8	0.028570837	0	0
9	0.011110881	0.045247	0
10	0.007936344	0	0
Smax	0.028570837	0.094265	0.623269
Smin	0.007936344	0	0

Table 6 Positive ideal solution for all tasks

Tasks	C1	C2	C3
1	0.00030485	0.008886	0.388464
2	0.000425782	0.000987	0.388464
3	0.000362797	0.008886	0.388464
4	0.000204073	0.008886	0.388464
5	0.000106446	0.008886	0.388464
6	0.000251942	0	0.388464
7	0.000161243	0.00327	0
8	0	0.008886	0.388464
9	0.00030485	0.002403	0.388464
10	0.000425782	0.008886	0.388464

Table 7 Negative ideal solution for all tasks

Tasks	C1	C2	C3
1	1.00777E-05	0	0
2	6.29856E-05	0.003949	0
3	9.06992E-05	0	0
4	0.000195456	7.9E-05	0.150905
5	0.000317831	9.75E-07	0.150905
6	0.000152161	0.00729	0.150905
7	0.000245505	0.000795	0.055133
8	0.000810222	7.9E-05	0.150905
9	0.000117917	0.002047	0.150905
10	6.04522E-05	1.07E-05	0

Table 8 Separation distance, their relative closeness to PI and NI, and final rank

Tasks	S+	S−	RI	Rank
1	0.663886	0.001842	0.002767	10
2	0.656603	0.06964	0.09589	7
3	0.6639	0.005527	0.008256	9
4	0.66386	0.429993	0.393099	5
5	0.663835	0.429906	0.39306	6
6	0.655639	0.439862	0.401517	2
7	0.063661	0.227982	0.781715	1
8	0.663808	0.430233	0.393251	4
9	0.657889	0.432727	0.396773	3
10	0.663916	0.006058	0.009043	8

7.3 Task Assignment to Workstations

The final ranks of the parts have been calculated by AHP and TOPSIS method. With the use of this rank, the tasks can be assigned to workstations. The cycle time for solving the problem has been taken as 40 s and in the solution, only five disassembly workstations are required for the assignment of task to workstations. The task assignment has been shown in Table 9. Here, we can see for the maximum utilization of cycle time, the proposed approach is capable to reduce the cycle time up to 37 s.

Table 9 Task assignment to workstations

Workstation	Task number	Task time	Idle time
01	02	10	30
	03	12	18
	09	14	04
02	08	36	04
03	07	20	20
	06	16	04
04	04	18	22
	10	10	12
05	05	23	17
	01	14	03

8 Conclusions and Discussion

It has been seen that the proposed approach can improve the utilization of disassembly line because of the use of multi criteria decision-making techniques that have been discussed in current work. The proposed heuristic has been tested in terms of station time (CT) and it is found that it can reduce station time in a valuable quantity for the full disassembly of selected problem example. It includes several significant disassembly criteria and different associated factors into task assignment practice and also provides better outcomes as equated to the approach suggested by McGovern and Gupta [14]. The heuristic suggested in current work has been demonstrated with suitable example selected from literature and compared result proves that significant enhancement in the performance has been reported as matched with other heuristic.

References

1. McGovern, S.M., Gupta, S.M.: Ant colony optimization for disassembly sequencing with multiple objectives. Int. J. Adv. Manuf. Technol. **30**, 481–496 (2006)
2. McGovern, S.M., Gupta, S.M.: Combinatorial optimization analysis of the unary NP-complete disassembly line balancing problem. Int. J. Prod. Res. **45**(18–19), 4485–4511 (2007)
3. Agrawal, S., Tiwari, M.K.: A collaborative ant colony algorithm to stochastic mixed-model U-shaped disassembly line balancing and sequencing problem. Int. J. Prod. Res. **46**(6), 1405–1429 (2008)
4. Altekin, F.T., Kandiller, L., Ozdemirel, N.E.: Profit-oriented disassembly-line balancing. Int. J. Prod. Res. **46**(10), 2675–2693
5. Koc, A., Sabuncuoglu, I., Erel, E.: Two exact formulations for disassembly line balancing problems with task precedence diagram construction using an AND/OR graph. IIE Trans. Oper. Eng. **41**, 866–881 (2009)
6. Ding, L.P., Feng, Y.X., Tan, J.R., Gao, Y.C.: A new multi-objective ant colony algorithm for solving the disassembly line balancing problems. Int. J. Adv. Manuf. Technol. **48**, 761–771

7. Karadag, A.A., Turkbey, O.: Multi-objective optimization of stochastic disassembly line balancing with station paralleling. Comput. Indus. Eng. **65**(3), 413–425 (2013)
8. Kalayci, C.B., Gupta, S.M.: Artificial bee colony algorithm for solving sequence-dependent disassembly line balancing problem. Expert Syst. Appl. **40**(18), 7231–7241 (2013)
9. Saaty, T.L.: On polynomials and crossing numbers of complete graphs. J. Combin. Theor. **10**, 183–184 (1971)
10. Lee, S., Kim, W., Kim, Y.M., Oh, K.J.: Using AHP to determine intangible priority factors for technology transfer adoption. Expert Syst. Appl. **39**, 6388–6395 (2012)
11. Chamodrakas, I., Leftheriotis, I., Martakos, D.: In-depth analysis and simulation study of an innovative fuzzy approach for ranking alternatives in multiple attribute decision making problems based on TOPSIS. Appl. Soft Comput. **11**, 900–907 (2011)
12. Hwang, C.L., Yoon, K.: Multiple attribute decision making: methods and applications. Springer, Heidelberg (1981)
13. Lin, M.C., Wang, C.C., Chen, M.S., Chang, C.A.: Using AHP and TOPSIS approaches in customer-driven product design process. Comput. Indus. **59**, 17–31 (2008)
14. McGovern, S.M., Gupta, S.M.: 2-Opt heuristic for the disassembly line balancing problem. In: Proceedings of the SPIE International Conference on Environmentally Conscious Manufacturing III, Providence, RI, pp. 71–84, 2003.

Use of Intuitionistic Fuzzy Time Series in Forecasting Enrollments to an Academic Institution

Bhagawati Prasad Joshi, Mukesh Pandey and Sanjay Kumar

Abstract Fuzzy time series (FTS) forecasting models are widely applicable when the information is imprecise and vague. The concept of fuzzy set (FS) is generalized to intuitionistic fuzzy set (IFS) and proved that it is more suitable and powerful tool to deal with real life problems under uncertainty as compared to FSs theory. In this study, first we extended the definitions of FTS to the IFSs and proposed the notion of intuitionistic FTS. Further, the presented concept of intuitionistic FTS is applied to develop a forecasting model under uncertainty. Then, it is applied to the benchmark problem of the historical enrollments data of University of Alabama and the obtained results are compared with the results obtained by existing methods to show its effectiveness as compared to FTS.

Keywords Fuzzy time series · Intuitionistic fuzzy sets · Intuitionistic fuzzy time series

1 Introduction

The theory of FSs [1] is successfully implemented by Song and Chissom and developed the FTS models [2, 3], and applied it to the student enrollments of university of Alabama [3, 4]. A simplified method for time series forecasting by using the arithmetic operations is presented by Chen [5] and found the results of

B.P. Joshi (✉)
Department of Applied Sciences, Seemant Institute of Technology, Pithoragarh, India
e-mail: bpjoshi.13march@gmail.com

Mukesh Pandey
Department of Computer Science and Engineering, Seemant Institute of Technology, Pithoragarh, India
e-mail: mukesh3iov@gmail.com

Sanjay Kumar
Department of Mathematics, Statistics & Computer Science,
G.B. Pant University of Agriculture & Technology, Pantnagar, India
e-mail: skruhela@hotmail.com

M. Pant et al. (eds.), *Proceedings of Fifth International Conference on Soft Computing for Problem Solving*, Advances in Intelligent Systems and Computing 436, DOI 10.1007/978-981-10-0448-3_70

higher accuracy. In [6], Huarng introduced a heuristic model for FTS forecasting with the help of increasing and decreasing relations to enhance the forecasting results. A forecasting method based on high-order FTS was proposed in [7]. Lee and Chou [8] modified the Chen's method [5] and improved the outputs. After that, many researchers [9–12] etc. did significant contribution in the theory of FTS.

In [13], Atanassov extended the concept of FS, and defined the concept of IFS by incorporating the degree of uncertainty to FSs. The theory of IFS is more suitable and powerful tool to deal with real life problems under uncertainty and vagueness as compared to FSs theory. Due to the handling property of IFS with uncertainty Joshi and Kumar [14–16] proposed FTS forecasting model based on IFS. The forecasted values obtained under IFS [15] are close to the actual ones as compared to other forecasting methods, which show the effectiveness of incorporating IFSs to the FTS models. Thus, in this paper, the concept of intuitionistic FTS is proposed by studying the definitions of FTS presented by Song and Chissom. Based on the presented concept an intuitionistic FTS forecasting model is proposed and implemented to the benchmark problem of the Alabama's University enrollments data. The forecasted values are compared with other models in terms of mean square error, average forecasting error and the close trend with actual ones to show the superiority and effectiveness of the proposed method over existing methods. So, the concept of intuitionistic FTS can be used as a powerful tool to forecast in real life situation.

2 Preliminaries

Here, some basic definitions of FTS are reviewed and presented that can be used to introduce the concept of intuitionistic FTS model.

Definition 1 A FS is a class of elements having a continuous grade of membership. Let $U = \{u_1, u_2, u_3,\ldots, u_n,\}$ is the Universe of discourse in which u_i are the possible linguistic values of U, then a FS A_i in U is defined by

$$A_i = \mu_{A_i}(u_1)/u_1 + \mu_{A_i}(u_2)/u_2 + \mu_{A_i}(u_3)/u_3 + \cdots + \mu_{A_i}(u_n)/u_n \tag{1}$$

where μ_{A_i} is the membership function of the FS A_i, such that $\mu_{A_i} : U \to [0, 1]$

Definition 2 If $Y(t)$ ($t = \ldots,0, 1, 2, 3,\ldots$), is a subset of U on which FSs $f_i(t)$, ($i = 1, 2, 3,\ldots$) are defined and $F(t)$ is the grouping of f_i, then $F(t)$ is defined as FTS on $Y(t)$.

Definition 3 If $F(t)$ is happening only by $F(t - 1)$ and is represented by $F(t-1) \to F(t)$; then there exists a fuzzy relationship between $F(t)$ and $F(t - 1)$ and it can be expressed as the fuzzy logical relational equation

$$F(t) = F(t-1) o R(t, t-1), \tag{2}$$

Here, "*o*" is max–min composition operator. Then, this relation R is said to be 1st-order model of $F(t)$. Further if fuzzy relation $R(t, t-1)$ of $F(t)$ does not depend upon time t, that is to say for different times t_1 and t_2, $R(t_1, t_1 - 1) = R(t_2, t_2 - 1)$, then $F(t)$ is called a time invariant FTS.

Definition 4 If $F(t)$ is happening by more FSs, $F(t-n)$, $F(t-n+1)$,…, $F(t-1)$, the fuzzy relationship is denoted by $A_{i_1}, A_{i_2}, A_{i_3}\ldots, A_{i_n} \rightarrow A_j$ and is called nth-order FTS model, here,

$$F(t-n) = A_{i_1}, F(t-n+1) = A_{i_2}, \ldots, F(t-1) = A_{i_n}. \quad (3)$$

3 Intuitionistic FTS

In [15], Joshi and Kumar applied the concept of IFS theory to FTS and introduced a forecasting model which is more powerful as compared to existing models. Motivated by this and the FTS proposed by Song and Chissome, here we presented the concept of an intuitionistic FTS.

Definition 5 An IFS is a class of objects with a continuous grade of membership and non-membership. If U is the Universe of discourse with $U = \{u_1, u_2, u_3, \ldots, u_n,\}$, where u_i are possible linguistic values of U, then an IFS I_i of U is defined by

$$I_i = \langle \mu_{I_i}(u_1), \nu_{I_i}(u_1) \rangle / u_1 + \langle \mu_{I_i}(u_2), \nu_{I_i}(u_2) \rangle / u_2 + \cdots + \langle \mu_{I_i}(u_n), \nu_{I_i}(u_n) \rangle / u_n \quad (4)$$

where μ_{I_i} and ν_{I_i} be the membership function and non-membership function of the IFS I_i, such that $\mu_{I_i} : U \rightarrow [0, 1]$ and $\nu_{I_i} : U \rightarrow [0, 1]$, respectively, and $0 \leq \mu_{I_i} + \nu_{I_i} \leq 1$. The term $\pi_{I_i} = 1 - \mu_{I_i} - \nu_{I_i}$ is called the degree of non-determinacy and remains indeterministic due to the hesitation of the decision-maker.

Definition 6 Let $Y(t)$ ($t = \ldots, 0, 1, 2, 3, \ldots$), be a subset of U on which the IFSs $f_i(t)$, ($i = 1, 2, 3, \ldots$) are defined. Each $f_i(t)$ is comprising of two parts one is membership function and other is non-membership function. If $F(t)$ denotes the collection of all f_i, then $F(t)$ is called an intuitionistic FTS on $Y(t)$.

Definition 7 If $F(t)$ is happening only by $F(t-1)$ and is denoted by $F(t-1) \rightarrow F(t)$; then there is an intuitionistic fuzzy logical (IFL) relationship between $F(t)$ and $F(t-1)$ and it can be expressed as the IFL relational equation: $F(t) = F(t-1)\ o\ R(t, t-1)$, here, "*o*" is Max–Min composition operator for membership function and Min–Max composition operator for non-membership function. The relation R is called 1st-order model of $F(t)$. Further if the IFL relation $R(t, t-1)$ of $F(t)$ does not depend upon time t, that is to say for different times t_1 and t_2, $R(t_1, t_1 - 1) = R(t_2, t_2 - 1)$, then $F(t)$ is called a time invariant intuitionistic FTS. But if $R(t, t - 1)$ is time dependent, i.e., for different times t_1 and t_2, if $R(t_1, t_1 - 1)$ may be different from $R(t_2, t_2 - 1)$ for any time t then $F(t)$ is called time-variant intuitionistic FTS.

Definition 8 The IFL relationships, which have the same left hand sides, can be grouped together to form IFL relationship groups. For example, for the identical left hand side A_i such grouping can be depicted as follows:

$$\left.\begin{array}{ccc} A_i & \rightarrow & A_{j_1} \\ A_i & \rightarrow & A_{j_2} \\ \ldots & \ldots & \ldots \end{array}\right\} \Rightarrow A_i \rightarrow A_{j_1}, A_{j_2}, \ldots \tag{5}$$

Definition 9 If $F(t)$ is a time invariant intuitionistic FTS and caused by only $F(t - 1)$, then the IFL relationship $F(t-1) \rightarrow F(t)$; is called a first order IFL relationship. If $F(t)$ is caused by more IFSs, $F(t - n)$, $F(t - n + 1)$,…, $F(t - 1)$, the IFL relationship is represented by following expressions:

$$f_{i_1}, f_{i_2}, f_{i_3}, \ldots, f_{i_n} \rightarrow f_j,$$

here, $F(t-n) = f_{i_1}, F(t-n+1) = f_{i_2}, \ldots, F(t-1) = f_{i_n}$. This relationship is called nth-order intuitionistic FTS model.

Definition 10 If $F(t)$ is happening by more IFSs, $F(t - n)$, $F(t - n + 1)$,…, $F(t - 1)$, then the intuitionistic fuzzy relationship is represented by $F(t-n), F(t-n+1), \ldots, F(t-1) \rightarrow F(t)$ and is called the one-variable nth-order intuitionistic FTS forecasting model. If $F(t)$ is happening by $(F_1(t - 1), F_2(t - 1))$, $(F_1(t - 2)\ F_2(t - 2))$,…, $(F_1(t - n)\ F_2(t - n))$, then this IFL relationship is represented by

$$(F_1(t-n), F_2(t-n)), (F_1(t-n+1), F_2(t-n+1)), \ldots,$$
$$(F_1(t-1), F_2(t-1)) \rightarrow F(t)$$

and is called the two-variables nth-order intuitionistic FTS forecasting model, where $F_1(t)$ and $F_2(t)$ are called the main factor and the secondary factor intuitionistic FTS, respectively.

In the similar way, m-variables nth-order IFL relationship is defined as

$$(F_1(t-n), F_2(t-n), \ldots, F_m(t-n)), (F_1(t-n+1), F_2(t-n+1), \ldots, F_m(t-n+1)), \ldots,$$
$$(F_1(t-1), F_2(t-1), \ldots, F_m(t-1)), \rightarrow F(t)$$

Here, $F_1(t)$ is called the main factor and $F_2(t)$, $F_3(t)$,…, $F_m(t)$ are called secondary factor intuitionistic FTS.

4 Proposed Method for Intuitionistic FTS Forecasting

First, we define the Universe set U based on the available range of time series data with the rule $U = [D_{\min} - D_1, D_{\max} + D_2]$ where D_1 and D_2 are two proper positive numbers and $D_{\min}$, $D_{\max}$ be minimum and maximum of the available time series data, respectively, and partition set U into equal length of intervals: $u_1, u_2, \ldots, u_m$. Different methods of partitioning the Universe of discourse are available in literatures; here, we are not considering this issue. Define IFSs $A_1, A_2, \ldots, A_m$ by overlapping these partition intervals such that $U = \bigcup_{i=1}^{m} A_i$ and obtain the intuitionistic fuzzification of the student enrollments. Then, establish the IFL relationships by the following rule: "If A_i, A_j be the intuitionistic fuzzy production of month n and A_k, A_l be the intuitionistic fuzzified production of month $n + 1$, then the IFL relation is denoted as $A_i, A_j \rightarrow A_k, A_l$. Here A_i, A_j is called the current state and A_k, A_l is the next state."

Then, we grouped these IFL relationships and formed IFL relationship groups. Here, we are proposing a simple forecasting model by using intuitionistic FTS that is why we are not defining the membership and non-membership degrees for each IFS. Some notations are presented below to define rules for forecasting

$[*A_j]$ is corresponding interval u_j
$U[*A_j]$ is the upper bound of interval u_j

For an IFL group relation $A_i, A_j \rightarrow A_k, A_l$:

A_i, A_j be the intuitionistic fuzzified enrollments of year n
A_k, A_l be the intuitionistic fuzzified enrollments of year $n + 1$
$F_{k,l}$ is the forecasted enrollments of the year $n + 1$

Then,

$$F_{k,l} = \frac{1}{2}\left(U[*A_k] + U[*A_l]\right).$$

For an IFL groups relation $A_i, A_j \rightarrow A_{k_1}, A_{l_1}$, $A_i, A_j \rightarrow A_{k_2}, A_{l_2}, \ldots$, $A_i, A_j \rightarrow A_{k_p}, A_{l_q}$:

Then

$$F_{k,l} = \frac{1}{(p+q)}\left(\sum_{i=1}^{p} U[*A_{k_i}] + \sum_{j=1}^{q} U[*A_{l_j}]\right). \tag{6}$$

As we know that the forecasting accuracy of a method is generally calculated in terms of mean square error (MSE) and in terms of average forecasting error (AFE). Lower the MSE or AFE, better the forecasting method. The MSE and AFE are defined by following expressions:

$$\text{Mean Square Error} = \frac{\sum_{i=1}^{n} (\text{actual}_i - \text{forecasted}_i)^2}{n} \tag{7}$$

$$\text{Forecasting Error} = \frac{|\text{forecasted} - \text{actual}|}{\text{actual}} \times 100 \tag{8}$$

$$\text{Average forecasting Error (in \%)} = \frac{\text{sum of forecasting error}}{\text{number of errors}} \tag{9}$$

5 Forecasting of Student Enrollments

The proposed method is demonstrated by considering the benchmark problem of the historical enrollments data of University of Alabama and the step wise results are placed below

Step 1: Universe of discourse $U = [13000, 20000]$ is defined for enrollments at University of Alabama by taking $D_1 = 55$ and $D_2 = 663$.

Step 2: Partition universe of discourse into the following 14 intervals.

$u_1 = [13000, 13500]$,	$u_2 = [13500, 14000]$,	$u_3 = [14000, 14500]$,
$u_4 = [14500, 15000]$,	$u_5 = [15000, 15500]$,	$u_6 = [15500, 16000]$,
$u_7 = [16000, 16500]$,	$u_8 = [16500, 17000]$,	$u_9 = [17000, 17500]$,
$u_{10} = [17500, 18000]$,	$u_{11} = [18000, 18500]$,	$u_{12} = [18500, 19000]$,
$u_{13} = [19000, 19500]$,	$u_{14} = [19500, 20000]$,	

Step 3: Partition universe of discourse into the following 13 IFSs.

$A_1 = [13000, 14000]$,	$A_2 = [13500, 14500]$,	$A_3 = [14000, 15000]$,
$A_4 = [14500, 15500]$,	$A_5 = [15000, 16000]$,	$A_6 = [15500, 16500]$,
$A_7 = [16000, 17000]$,	$A_8 = [16500, 17500]$,	$A_9 = [17000, 18000]$,
$A_{10} = [17500, 18500]$,	$A_{11} = [18000, 19000]$,	$A_{12} = [18500, 19500]$,
$A_{13} = [19000, 20000]$,		

Step 5: Using the proposed algorithm for intuitionistic fuzzification in Sect. 4, time series data of enrollments are intuitionistic fuzzified and presented in Table 1.
IFL relationships and IFL relationship groups of enrollments are placed in Tables 2 and 3.

Step 6: On the basis of IFL relationship groups of enrollments the forecasted output is computed using Eq. (6) placed in Table 4 along with results obtained by other methods.

Table 1 Actual and fuzzified enrollments of University of Alabama

Year	Actual enrollments	Intuitionistic fuzzified enrollments	Year	Actual enrollments	Intuitionistic fuzzified enrolments
1971	13,055	A_1	1982	15,433	A_4, A_5
1972	13,563	A_1, A_2	1983	15,497	A_4, A_5
1973	13,867	A_1, A_2	1984	15,145	A_4, A_5
1974	14,696	A_3, A_4	1985	15,163	A_4, A_5
1975	15,460	A_4, A_5	1986	15,984	A_5, A_6
1976	15,311	A_4, A_5	1987	16,859	A_7, A_8
1977	15,603	A_5, A_6	1988	18,150	A_{10}, A_{11}
1978	15,861	A_5, A_6	1989	18,970	A_{11}, A_{12}
1979	16,807	A_7, A_8	1990	19,328	A_{12}, A_{13}
1980	16,919	A_7, A_8	1991	19,337	A_{12}, A_{13}
1981	16,388	A_6, A_7	1992	18,876	A_{11}, A_{12}

Table 2 IFL relationships of student enrollments

$A_1 \to A_1, A_2$	$A_1, A_2 \to A_1, A_2$	$A_1, A_2 \to A_3, A_4$
$A_3, A_4 \to A_4, A_5$	$A_4, A_5 \to A_4, A_5$	$A_4, A_5 \to A_5, A_6$
$A_5, A_6 \to A_5, A_6$	$A_5, A_6 \to A_7, A_8$	$A_7, A_8 \to A_7, A_8$
$A_7, A_8 \to A_6, A_7$	$A_6, A_7 \to A_4, A_5$	$A_4, A_5 \to A_4, A_5$
$A_4, A_5 \to A_4, A_5$	$A_4, A_5 \to A_4, A_5$	$A_4, A_5 \to A_5, A_6$
$A_5, A_6 \to A_7, A_8$	$A_7, A_8 \to A_{10}, A_{11}$	$A_{10}, A_{11} \to A_{11}, A_{12}$
$A_{11}, A_{12} \to A_{12}, A_{13}$	$A_{12}, A_{13} \to A_{12}, A_{13}$	$A_{12}, A_{13} \to A_{11}, A_{12}$

Table 3 IFL relationship groups of student enrollments

$A_1 \to A_1, A_2$		
$A_1, A_2 \to A_1, A_2$	$A_1, A_2 \to A_3, A_4$	
$A_3, A_4 \to A_4, A_5$		
$A_4, A_5 \to A_4, A_5$	$A_4, A_5 \to A_5, A_6$	
$A_5, A_6 \to A_5, A_6$	$A_5, A_6 \to A_7, A_8$	
$A_6, A_7 \to A_4, A_5$		
$A_7, A_8 \to A_7, A_8$	$A_7, A_8 \to A_6, A_7$	$A_7, A_8 \to A_{10}, A_{11}$
$A_{10}, A_{11} \to A_{11}, A_{12}$		
$A_{11}, A_{12} \to A_{12}, A_{13}$		
$A_{12}, A_{13} \to A_{12}, A_{13}$		$A_{12}, A_{13} \to A_{11}, A_{12}$

Step 7: The superiority of the developed model is checked with existing methods. In order to do so, MSE and AFE have been calculated and are placed in Table 5. Further, the forecasted values obtained by the developed model are closer to the actual ones as compared to other methods and depicted in Fig. 1.

Table 4 Forecasted enrollments by proposed and existing methods

Year	Actual	Proposed model	Joshi and Kumar [15]	Lee and Chou [8]	Huarng Heuristic [6]	Chen model [5]	Song and Chissom [3]
1972	13,563	13,750	14,250	–	–	–	–
1973	13,867	14,250	14,246	–	–	–	–
1974	14,696	14,250	14,246	14,568	14,000	14,000	14,000
1975	15,460	15,500	15,491	15,654	15,500	15,500	15,500
1976	15,311	15,500	15,491	15,654	15,500	16,000	16,000
1977	15,603	16,250	15,491	15,654	16,000	16,000	16,000
1978	15,861	16,250	16,345	15,654	16,000	16,000	16,000
1979	16,807	17083.3	16,345	16,197	16,000	16,000	16,000
1980	16,919	17083.3	15,850	17,283	17,500	16,833	16,813
1981	16,388	15,250	15,850	17,283	16,000	16,833	16,813
1982	15,433	15,500	15,850	16,197	16,000	16,833	16,789
1983	15,497	15,500	15,450	15,654	16,000	16,000	16,000
1984	15,145	15,500	15,450	15,654	15,500	16,000	16,000
1985	15,163	15,500	15,491	15,654	16,000	16,000	16,000
1986	15,984	16,250	15,491	15,654	16,000	16,000	16,000
1987	16,859	17083.3	16,345	16,197	16,000	16,000	16,000
1988	18,150	18,750	17,950	17,283	17,500	16,833	16,813
1989	18,970	19,250	18,961	18,369	19,000	19,000	19,000
1990	19,328	19,000	18,961	19,454	19,000	19,000	19,000
1991	19,337	19,000	18,961	19,454	19,500	19,000	19,000
1992	18,876	19,250	18,961	–	19,000	19,000	–

Table 5 Comparison of MSE and AFE of proposed method with existing methods

Model	Proposed	[15]	[8]	[6]	[5]	[3]
MSE	169085.4	175559.6	240,047	239,483	439,421	458,438
AFE	2.049743	2.069264	2.49977	3.21735	2.48254	3.34967

6 Conclusion and Future Remarks

In this paper, we extended the definitions of FTS to the IFSs and proposed the notion of intuitionistic FTS. Based on the proposed IFTS, a forecasting model is presented under uncertainty. Then, the developed model has been implemented to the benchmark problem of the historical enrollments data of University of Alabama and the results are compared with the results of existing methods to show its superiority and effectiveness as compared to FTS. Further, the intuitionistic FTS forecasting model gives more accurate values in terms of close trend to the actual ones (see Fig. 1), MSE and AFE (see Table 5) in compare to other existing methods.

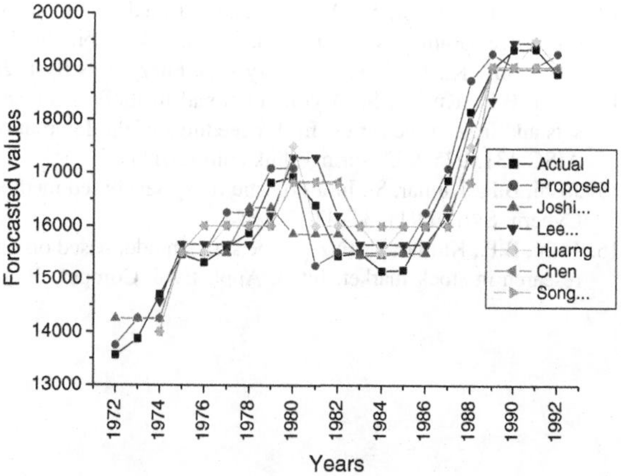

Fig. 1 Comparison of proposed method with other methods

IFSs theory is more powerful tool to deal with real life problems under uncertainty and vagueness. Earlier [14–16] observed that time series forecasting models under IFSs are superior and more efficient as compared to forecasting models under FSs, so we are focusing on this issue and it is proved here. In future, we will try to improve the presented results and implement it to forecast other instruments. Further, we will work on develop some new forecasting models using the presented notion of IFTS.

References

1. Zadeh, L.A.: Fuzzy set. Inf. Control **8**, 338–353 (1965)
2. Song, Q., Chissom, B.: Fuzzy time series and its models. Fuzzy Sets Syst. **54**, 269–277 (1993)
3. Song, Q., Chissom, B.: Forecasting enrollments with fuzzy time series—Part I. Fuzzy Sets Syst. **54**, 1–9 (1993)
4. Song, Q., Chissom, B.: Forecasting enrollments with fuzzy time series—Part II. Fuzzy Sets Syst. **64**, 1–8 (1994)
5. Chen, S.M.: Forecasting enrollments based on fuzzy time series. Fuzzy Sets Syst. **81**, 311–319 (1996)
6. Huarng, K.: Heuristic models of fuzzy time series for forecasting. Fuzzy Sets Syst. **123**, 369–386 (2001)
7. Chen, S.M.: Forecasting enrollments based on high-order fuzzy time series. Cybern. Syst. **33**, 1–16 (2002)
8. Lee, H.S., Chou., M.T.: Fuzzy forecasting based on fuzzy time series. Int. J. Comput. Math. **81** (7), 781–789 (2004)
9. Singh, S.R.: A simple method of forecasting based on fuzzy time series. Appl. Math. Comput. **186**, 330–339 (2007)
10. Liu, H.T.: An improved fuzzy time series forecasting method using trapezoidal fuzzy numbers. Fuzzy Optim. Decision Making **6**, 63–80 (2007)
11. Liu, H.T., Wei, N.C., Yang, C.G.: Improved time-variant fuzzy time series forecast. Fuzzy Optim. Decision Making **8**, 45–65 (2009)

12. Joshi, B.P., Kumar, S.: A computational method for fuzzy time series forecasting based on difference parameters. Int. J. Model. Simul. Sci. Comput. **4**(1), 1250023-1-1250023-12 (2013)
13. Atanassov, K.: Intuitionistic fuzzy sets. Fuzzy Sets Syst. **20**, 87–96 (1986)
14. Joshi, B.P., Kumar, S.: A computational method of forecasting based on intuitionistic fuzzy sets and fuzzy time series. In: Proceedings of the International Conference on SocProS 2011, AISC 131, 925–932. springerlink.com (2011)
15. Joshi, B.P., Kumar, S.: Intuitionistic fuzzy sets based method for fuzzy time series forecasting. Cybern. Syst. **43**(1), 34–47 (2012)
16. Joshi, B.P., Kumar, S.: Fuzzy time series model based on intuitionistic fuzzy sets for empirical research in stock market. Int. J. Appl. Evol. Comput. **3**(4), 71–84 (2012)

An Improved Privacy-Preserving Public Auditing for Secure Cloud Storage

Mukund N. Kulkarni, Bharat A. Tidke and Rajeev Arya

Abstract Distributed computing or Cloud Computing could be a net fundamentally based rising and rapidly developing model. Inside which client will store their insight remotely and revel in the on-interest top quality applications and administrations from a mutual pool of configurable processing assets, while not the weight of local source and upkeep. Hence, accuracy and security of information could be a prime concern. Physical ownership of the outsourced information is limited to clients. Making certain trustworthiness could be a troublesome undertaking, especially for clients with confined figuring assets. Additionally, client should be prepared to utilize the distributed storage on the grounds that it is local stockpiling, without apprehension concerning the wish to review its honesty. In this way, allowing public auditability for conveyed stockpiling is of basic hugeness; so, cloud tenants or cloud users will use a third party auditor (TPA) to envision the uprightness of saved data and be clear. To solidly present a decent outsider evaluator, the inspecting strategy should usher in no new vulnerabilities toward client learning protection, and present no further online weight to client.

Keywords Cloud computing · Cloud storage · Privacy · Privacy-preserving data storage · Privacy-preserving · Public auditability · Cloud tenant · Third party auditor (TPA)

M.N. Kulkarni · B.A. Tidke
Flora Institute of Technology, Khopi, Pune, India
e-mail: donkulkarni1@gmail.com

B.A. Tidke
e-mail: batidke@gmail.com

Rajeev Arya (✉)
Electronics & Computer Discipline, IITR SRE Campus, IIT Roorkee, Roorkee, India
e-mail: rajeev.arya.ism@gmail.com

M. Pant et al. (eds.), *Proceedings of Fifth International Conference on Soft Computing for Problem Solving*, Advances in Intelligent Systems and Computing 436, DOI 10.1007/978-981-10-0448-3_71

1 Introduction

Cloud has been notional in light of the way that the front line data advancement arrangement for endeavors. Distributed computing is immerging creating innovation used in business, IT commercial ventures which offer administrations like system access, assets, base, stage, and quick asset snap according to client request [1]. The client will be ready to get to the administrations at whatever time, wherever on-interest through net. In cloud computing, the information of client is unified to the distributed storage. Cloud storage could be an encapsulation of arranged online stockpiling amid which the data is keep at virtualized pools of capacity. Power meaning of cloud computing as: "Cloud computing could be a model for authorizing advantageous, on-interest system access to a mutual pool of configurable processing assets (e.g., systems, servers, stockpiling, applications, and administrations) that may be apace provisioned and free with ostensible administration exertion or administration supplier association" [2]. A few clients from remote area use benefits perpetually accordingly there could emerge a few issues like learning security, information uprightness, element upgrades. Every time it is unattainable for client to imagine the data is being steady that is keep at distributed storage. In this way client persistently needs that cloud server ought to should keep up information uprightness and protection. Cloud administration suppliers square measure the different substances that store information and supply administrations to the client. The wellbeing and information respectability issues emerge attributable to taking after reasons:

(a) The classes of intruders like inner and outside and their capacity of hostile the cloud.
(b) The protection risks associated with the cloud, and where important issues with assaults and countermeasures.
(c) Rising cloud security dangers.

Some different issues like absence of guiding and experience, unapproved auxiliary use, multifaceted nature of regulative consistence, absence of client administration, tending to transborder data stream limitations, lawful vulnerability, constrained uncovering to the data availability, area of data, exchange and maintenance, data security, and uncovering of breaches [3, 4]. The cloud server stores awesome measure of data of learning of data that does not supply or ensure on information trustworthiness and consistency. This drawback is tended to and settled by giving open reviewing for secure cloud.

To guarantee the data security and trustworthiness and to scale back online weight it is of significance to adjust open reviewing administration for distributed storage, so client may depend on outsider reviewer (TPA) to review the data. TPA will be the inspecting strategy in the interest of the client. The TPA organization has abilities and experience that may sporadically check the honesty of the data keep in cloud. The client doesn't have the abilities that the TPA has. The TPA check the rightness of information kept in cloud in the interest of client and keep up the uprightness of learning. Facultative open inspecting administration can assume a critical part for protection learning security and amp; minimizing the data hazard from programmers.

2 Literature Review

The general public auditability, i.e., "provable data possession or ownership" (PDP) may be a model for ensuring ownership of learning documents on untrusted stockpiles. The subject uses the RSA-based for the most part homomorphic non-direct authenticators for evaluating outsourced information and proposes indiscriminately testing various pieces of the document. Be that as it may, the overall public auditability in their subject requests the straight blend of examined squares presented to outside examiner. Once utilized straightforwardly, the convention isn't undeniably security defensive, thus may spill client learning information to the evaluator. Juels et al. [5] illustrate a "proof of retrievability" (PoR) model, wherever spot-checking and mistake remedying codes zone unit wont to ensure every ownership and retrievability of learning documents on remote file administration frameworks. Be that as it may, the measure of review difficulties a client will perform may be a secured priori, and open auditability is not bolstered in their principle subject. In spite of the fact that they depict a basic Merkle Hash tree development as open PoRs, this methodology singularly works with scrambled data. Dodis et al. [6] gives a study on totally diverse variations of PoR with individual auditability. Shacham et al. [7] style partner degree enhanced PoR subject built with full evidences of security inside of the security model delineated in [8]. Sort of alike the improvement in [9], they utilize publically unquestionable homomorphic non-straight authenticators that territory unit designed from indisputably secure BLS marks. Bolstered the rich BLS development, a minimal and open certain subject is acquired. Once more, their methodology doesn't bolster security defensive examining for a comparative reason as [9]. The propose allowing a TPA to remain focused capacity genuine by starting encoding (the learning the information the information) then causation assortment of preprocessed symmetric-keyed hashes over the scrambled information to the reviewer. The reviewer checks each the respectability of the information record and in this way the server's ownership of a previously dedicated coding key. This topic singularly works for scrambled records and it experiences the evaluator statefullness and delimited utilization, which can without a doubt envoy online weight to clients once the keyed hashes zone unit drained. The dynamic rendition of the past PDP subject, abuse singularly normal key cryptography, however, with a delimited scope of reviews. Think about an undifferentiated from backing for halfway element information stockpiling in an exceedingly circulated circumstance with additional element of learning blunder confinement. In an exceedingly resulting task, Wang et al. [10] propose to blend BLS-based Homomorphic Linear Authenticator with Merkle Hash Tree to bolster every open verification and full information elements. For all intents and purposes in the

Table 1 Examination of remote respectability varification plans

Scheme	Computation		Communication	Privacy	Dynamic	Probability of detection
	Server	Verifier				
PDP [12]	$O(t)$	$O(t)$	$O(1)$	Yes	No	$1-(1-\rho)^t$
CPDP [7]	$O(t+s)$	$O(t+s)$	$O(t+s)$	No	No	$1-(1-\rho)^{ts}$
DPDP [13]	$O(t \log n)$	$O(t \log n)$	$O(t \log n)$	No	No	$1-(1-\rho)^t$
Audit [1, 14]	$O(t \log n)$	$O(t \log n)$	$O(t \log n)$	Yes	Yes	$1-(1-\rho)^t$
IPDP [15, 16]	$O(ts)$	$O(t+s)$	$O(t+s)$	Yes	Yes	$1-(1-\rho)^{ts}$

meantime built up a skip records based for the most part subject to adjust evident information ownership with full elements support. In any case, the check in these two conventions needs the direct blend of examined pieces even as [9, 11], thus does not bolster security defensive inspecting. While all the on top of plans offer routes for practical examining and evident certification on the accuracy of remotely keep information, none of them meet every one of the necessities for security defensive open reviewing in distributed computing. A ton of essentially, none of those plans think about bunch examining, which might enormously scale back the calculation cost on the TPA once reiteration with a larger than average scope of review assignment (Table 1).

3 Problem Statement

3.1 System Model

Consider a cloud data stockpiling administration including three totally diverse elements, as delineated in Fig. 1 the cloud client (CU), which have extraordinary

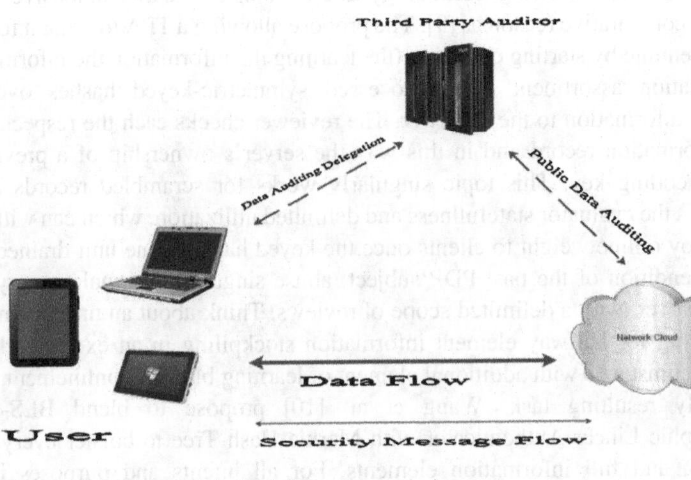

Fig. 1 A typical architecture of cloud data storage service

arrangement of data records to be keep inside of the cloud. The cloud service provider (CSP) and cloud server (CS), that is inspect by cloud service provider to deliver saved data control and has crucial to organizer space and calculation assets. The external auditor (TPA), who has expertise and talents that cloud tenants do not have and is trusty to survey the cloud network storage administration honest for the client upon challenge. Cloud tenants believe the cloud server for a large saved data and preservation. They will further strongly move with the cloud server to get to and overhaul their keep data for differed application capacities. To abstain from squandering the calculation asset also in light of the fact that the online weight, cloud clients may place to third gathering inspector for making certain the saved large data integrity of their outsourced data, while want to stay their data individual from outsider evaluator. Consider the presence of a semi-credible cloud server as will. In particular, in a large segment of your time it carries on accordingly and doesn't get sidetracked from the recommended convention execution. In any case, for his or her own edges the cloud server may disregard to stay or intentionally erase from time to time got to data records that fit in with standard cloud clients. Also, the cloud server may plan to disguise the data corruption created by server hacks or Byzantine failure to keep up name. Expect the outsider reviewer, who is within the matter of inspection, solid and independent, thus has no momentum to cooperate with one of two the cloud server or the clients all through the auditing system. In any case, it put away the client if the outsider evaluator may take in the outsourced data when the review. To admire the cloud server to answer to the task assigned to TPA's, the cloud tenant will sign a declaration allowing audit rights to the TPA's open key, and each one reviews from the TPA square measure recorded against such a testimonial.

3.2 *Existing System*

Most of the TPA systems are based on the two basic schemes, first is MAC-based solution and second is HLA-based solution.

3.2.1 MAC-Based Solution

The data blocks are uploaded with their MACs to the cloud servers and also the matching secret key are sent to the third party auditor by the user. To see the correctness of knowledge, the blocks are haphazardly retrieved with the MACs and third party auditor verifies the correctness by victimization the key. Some draw-backs during this approach areas third party auditor retrieve knowledge blocks it violates the condition of privacy conserving. This approach has high computation and communication complexities. To avoid giving access of knowledge blocks to third party auditor the verification may involve of equality checking. Message authentication code keys are chosen haphazardly by the cloud user, MACs for the complete record are precomputed, and also the verification data that's the keys and also the MACs are shared with the third party auditor. The keys shared with the cloud server. Third party auditor can evoke a recent keyed mac for comparison when, during this approach privacy-preserving is achieved. However, it is sure shortcomings

(a) The amount of times file will be audited is restricted as a result of the amount of secret keys should be fastened a priori. Once all potential secret keys are used user has got to retrieve full knowledge to cipher and send new MACs to TPA.
(b) Third party auditor has got to uphold and inform states between audits, as there are sizable amount of audit requests from multiple users. Third party auditor conjointly has got to keep track of the MAC keys that are discovered to metallic element.
(c) The dynamic knowledge is not distributed expeditiously.

3.2.2 HLA-Based Solution

Like MACs HLA are confirmation data. Not like MACs HLAs is combined from a linear combination of the information blocks. HLAs is accustomed verify the integrity of knowledge blocks while not downloading the information blocks. HLA permits economical information auditing and consumes solely constant information measure. But HLA technique could reveal user information info to third party auditor as he will merely solve a system of linear equations employed in the HLA technique. Therefore, it violates the privacy conserving guarantee.

3.3 Design Goal

To enable privacy-preserving public auditing for cloud knowledge storage below the higher than mentioned model, the projected protocol style ought to attain the subsequent security and performance guarantee.

i. *Public auditability*: to allow third party auditor to check and certify the flawlessness of the cloud data on interest while not saving a duplicate of the entire data or introducing extra online burden to the cloud tenants or subscribers.
ii. *Storage correctness*: to assure that there's no cheating by cloud server that may pass the TPA's audit while not actually storing users' data perfect.
iii. *Privacy-preserving*: to confirm that the third party reviewer or auditor cannot extract users' data content from the information collected throughout the auditing procedure.

iv. *Lightweight*: to allow third party auditor to complete reviewing or auditing with least communication and computation overhead.

4 Proposed System

In the anticipated framework, a third party auditor or reviewer who can perform reviewing task to ensure data respectability of outsourced data on clients' solicitation. Client can produce information and can exchange the data documents with the information on cloud server. TPA can perform the reviewing on clients ask for and can get the information from the client. The topic can contain four algorithms.

a. *KeyGen*: It will be started by the client for creating keys.
b. *SigGen*: Is applied by the cloud tenant to create metadata.
c. *GenProof*: Is race to produce the evidence of information uprightness by the cloud server.
d. *VerifyProof*: Is rush to review the confirmation by the TPA.

General society examining framework can be composed in two stages: Setup and Review.

Setup: The user prepares the public and secret parameters of the framework by executing KeyGen, and preprocesses the information record Fl by applying SigGen to produce the reviewing metadata. The client stores the information record Fl at the cloud server, erases its nearby duplicate and send the verification metadata to TPA for review. As a major aspect of prepreparing, the client may modify the data file Fl by expanding it or including extra metadata to be put away at server.

Review: The TPA issues a challenge message to the cloud server to verify that the cloud server has held the information record or file Fl legitimately at the season of the review. The cloud server will get a reaction message from an element of the put away information record or file Fl by executing GenProof. Utilizing the confirmation metadata, the TPA checks the reaction through VerifyProof.

4.1 Definition and Framework

MD5: Is a hash calculation used to decide the trustworthiness of a record by giving a 128 bits advanced mark. MD5 computerized mark is similar to a unique finger impression for a record; changing only one single byte in a document will bring about an alternate MD5 hash esteem. MD5 hashes can be utilized to list documents on a record framework and after that decide at a later date that the documents have not been adjusted at all, for instance on the off chance that somebody broke into a framework and changed framework records.

SHA: Stands for secure Hash Algorithm. SHA-256 and SHA-512 are novel hash capabilities computed with 32-bit and 64-bit phrases. They use specific shift quantities and constants, but their constructions are or else virtually identical, differing most effective within the quantity of rounds in hash algorithm.

To attain cozy segregation for privacy-preserving in cloud computing, this paper suggests to integrate the homomorphic linear authenticator with random covering technique. In the protocol, the linear combo of sampled blocks in the server's response is included with randomness produced the server. With random protecting, the TPA now not has all of the vital expertise to build up a proper crew of linear equations and thus cannot derive the person's information, no matter how many linear combinations of the same set of file blocks can be gathered. On the other hand, the correctness proof of the block authenticator pairs can still be applied in a brand new manner. The Proposed design makes use of a public key based HLA, to furnish the auditing protocol with public auditability.

4.2 Scheme Details

The proposed scheme is as follows:

Setup Phase: The cloud tenant or user or subscriber executes KeyGen to create the general public and secret parameters. Exceptionally, the client picks an arbitrary marking key pair (spk, ssk), an irregular $x \leftarrow Z_p$, an irregular point of interest $u \leftarrow G_1$, and figures $v \leftarrow g^x$. The key parameter is sk = (*x*, ssk) and the public parameters are pk = (spk, *v*, *g*, *u*, *e*(*u*, *v*)).

Given an information file Fl = {*mi*}, the cloud tenant or cloud user executes SigGen to figure authenticator $\sigma_i \leftarrow (H(W_i) \cdot u_{mi})^x \in G_1$ for every last *i*. Right here W_i = name ǁ *i* and name is picked by using the cloud user uniformly at random from Z_p because the identifier of file Fl. The set of authenticators by using $\phi = \{\sigma_i\}$ $1 \leq i \leq n$. The last a part of SigGen is for making certain the uprightness of the particular file identifier title. One basic way to deal with do this is to compute *t* = name ǁ SSigssk(name) because the file tag for Fl, where SSigssk(name) is the signature on title below the private key ssk. For straightforwardness, we expect the TPA knows the number of blocks *n*. The individual then sends Fl along with the verification metadata (ϕ, *t*) to the server and deletes them from regional storage.

Review Phase: The TPA first fetches the file tag *t*. With respect to the system described in the Setup phase, the TPA checks the signature SSigssk(name) via spk, and quits by aborting FALSE if the checks failed. Otherwise, the third party auditor recovers name.

Now, it comes to the "critical" portion of the reviewing procedure. To create the challenge message for the review "chal", the third outsider examiner or reviewer picks a arbitrary c-component subset $I = \{s_1 \ldots s_c\}$ of set [1, *n*]. For every component $i \in I$, the TPA also likewise picks an arbitrary value v_i. Make a sliding window of little duration '*t*'. There is a thread T_i for every time window *t*. This window will be moved over the enquiry range. The message "chal" determines the positions of the blocks required to be examined. The third party reviewer sends chal = $\{(I, v_i)\}_{i \in I}$ to the server. After receiving challenge chal = $\{(I, v_i)\}_{i \in I}$, the server executes *GenProof* to create a reaction evidence of information stockpiling rightness. In particular, the server picks an irregular component $r \leftarrow Z_p$ and calculates $R = e(u, v)^r \in G_T$. The operation will be connected over the whole dataset all the while utilizing multithreading. Every thread will contains fixed number blocks to accomplish a deliberate stratified testing of file blocks. Let μ' signify the straight blend of examined pieces in every thread T_i specified in chal: $\mu' = \sum_{i \in I} v_i m_i$. To dazzle μ' with r, the server computes: $\mu = r + \gamma\mu' \bmod p$, where $\gamma = h(R) \in Z_p$. In the interim, the server likewise ascertains a totaled authenticator $\sigma = \Pi_{i \in I} \sigma_i^{vi} i \in G_1$. It then sends $\{\mu, \sigma, R\}$ as the reaction evidence of capacity rightness to the TPA. With the reaction, the TPA runs *VerifyProof* to approve.

5 Performance Evaluation

We now check the execution of the proposed security keeping open examining plans to demonstrate that they are absolutely lightweight. We will point of convergence on the expense of the effectivity of the privateness-safeguarding convention. The test is done making utilization of JDK and J2EE servlets on a Windows 7 operating system with an Intel Core i5 2410M processor, 4 GB of memory and a 5400 rpm Western Digital 500 GB Serial ATA HDD. The code makes use of aws sdk for java version 1.8. J2EE is utilized to create servlets for communication between cloud server and cloud user or client and TPA which is likewise accessible as a web transporter for cloud user. DES encryption is decided for encryption of information for capacity.

For top assurance of inspection, the extra cost for privacy preservation maintaining assurance on the server aspect would be negligible against the total server computation for response new release. On the inspector part in the wake of accepting the check demand from CU (cloud user) the pieces of the document are partitioned into the gatherings of 4 (four). The number of threads can be raised dynamically as the number blocks are elevated. The time required for each block creation, encryption of each and every block and hash tag creation may range and it's is determined by the info which is contained within the file to be saved at cloud house.

Table 2 shows time required for blocks creation, encryption and hash tag creation for different file sizes of .pdf extension having text, images, tables, and variety of data.

As this is a systematic stratified sampling utilized on the other hand of the consecutive examining the proficiency will enhance sensibly. The size of 4 blocks are selected for every thread as to take capabilities of most common desktop computer processor and hardware which comprises two cores and two or four threads. This multithreading will contract the examining and check time for the

Table 2 Time required for blocks creation, encryption, and hash tag creation for different files

File size in kb	No. of blocks	Blocks creation in ms	Encryption in ms	Hash tag creation in ms	File size in kb	No. of blocks	Blocks creation in ms	Encryption in ms	Hash tag creation in ms
51	10	604	102	12	154	10	2014	107	10
	20	978	122	10		20	1372	145	8
	30	930	242	14		30	2363	460	9
	40	1185	146	22		40	1401	167	13
	50	1591	218	15		50	2038	244	21
	60	1184	340	36		60	2044	348	24
	70	1788	400	13		70	2146	398	14
100	10	1523	59	10	200	10	2261	133	6
	20	1178	275	29		20	1684	120	12
	30	3085	1037	14		30	1981	127	7
	40	1368	319	13		40	1830	674	14
	50	1686	255	25		50	2578	348	15
	60	1793	336	18		60	2848	299	27
	70	1763	1371	25		70	4325	430	24

given task. The test is finished on the pdf files which include not handiest textual content, tables and portraits. At that point this can likewise be accomplished on every single sort of document to make the pieces and get well it by method for the procedure of retrievability. The execution is done with the help of a cellular 2G/3G broadband knowledge connection. With the cell broadband the time required is additionally different at whatever point, the reason is the supply of data transfer capacity. The execution time for every operation could also be reduced if we use a committed leased excessive pace internet connection. Tables 3, 4 and 5 thinks about the execution time for a record of the proposed framework and the framework proposed in [1].

The accompanying figure demonstrates the time correlation of the proposed framework and the framework with consecutive testing system proposed in [1] (Fig. 2).

Table 3 Comparison of auditing time between proposed scheme and scheme in [1] for 51 kb file

No. of blocks	Time for auditing for proposed scheme in ms	Time for auditing for scheme [1] in ms
10	25,239	51,525
20	12,020	56,865
30	26,433	76,713
40	34,244	159,991
50	73,782	179,361
60	30,507	1,504,912
70	45,969	283,440
80	32,797	183,038
90	66,631	376,727
100	211,091	404,282

Table 4 Comparison of auditing time between proposed scheme and scheme in [1] for 100 kb file

No. of blocks	Time for auditing for proposed scheme in ms	Time for auditing for scheme [1] in ms
10	30,614	86,822
20	16,622	52,461
30	37,864	166,697
40	41,053	114,261
50	32,714	101,740
60	21,577	165,519
70	42,720	155,704
80	28,302	161,476
90	40,385	181,270
100	68,432	199,857

Table 5 Comparison of auditing time between proposed scheme and scheme in [1] for 154 kb file

No. of blocks	Time for auditing for proposed scheme in ms	Time for auditing for scheme [1] in ms
10	14,338	35,081
20	15,057	48,069
30	28,788	80,114
40	18,580	151,069
50	56,096	182,791
60	24,556	246,533
70	48,536	245,221
80	85,762	374,652
90	51,103	185,348
100	64,035	404,913

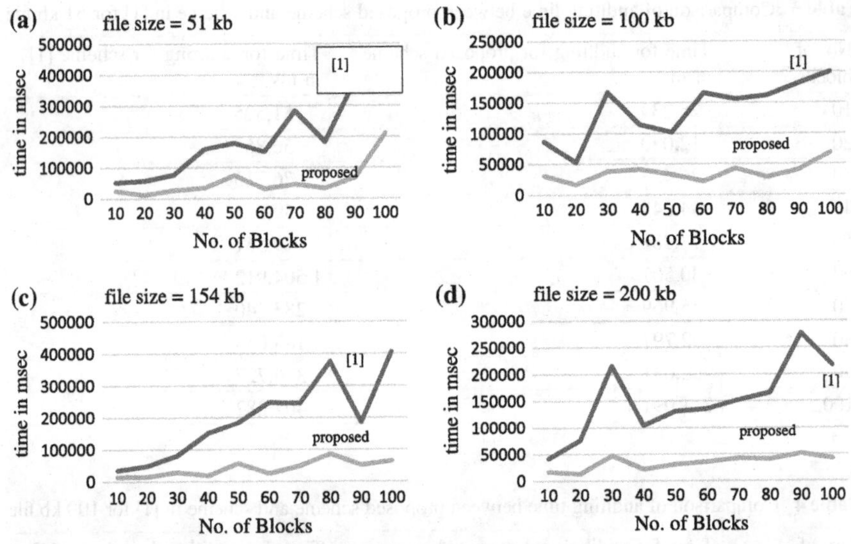

Fig. 2 Charts for comparison between proposed protocol and [1]

6 Conclusion

In this paper, we advocate an expanded methodology for protection safeguarding in distributed computing strategy for information stockpiling security. It use the homomorphic straight authenticator and irregular securing to ensure that the TPA would not be taught any abilities with respect to the learning content material put away on the cloud server for the time of the effective reviewing method, which no more just disposes of the weight of cloud individual from the repetitive and potentially costly inspecting endeavor, additionally helps the clients' stress of their outsourced information spillage. Given that TPA might just simultaneously control several review sessions from exceptional clients for their outsourced information records, we have the capacity to additional drag out our privateness-keeping open inspecting convention directly into a multiuser domain, the spot the TPA can take an interest in more than one examining obligations in a bunch technique for better effectivity. It requires less time for pursuit operation this suggests the methodology is quick fast. The early reaction is conceivable as a predetermined home window recognizes the vindictive undertaking; an alarm is produced and in like manner need not to stay up for the entire information look. The division of the inquiry control gives a more noteworthy oversees over the danger discovery and counter-active action. Our preparatory examination will presumably be performed on Amazon S3 simple stockpiling extra to explore the quick proficiency of the configuration on both the cloud and the examiner side. The information may be saved money on Amazon S3 Storage administration.

References

1. Wang, C.: Privacy-preserving public auditing for storage security in cloud computing. In: Proceedings IEEE INFOCOM 10, March 2010
2. Mell, P., Grance, T.: Draft NIST Working Definition of Cloud Computing (2009)
3. Pearson, S.: Privacy, security and trust in cloud computing. In: Privacy and Security for Cloud Computing (2012)
4. Cloud Security Alliance: Top Threats to Cloud Computing
5. Juels, A., Burton, J., Kaliski, S.: PORs: proofs of retrievability for large files. In: Proceedings ACM Conference Computer and Communication Security (CCS '07), pp. 584–597, Oct. 2007
6. Dodis, Y., Vadhan, S.P., Wichs, D.: Proofs of retrievability via hardness amplification. Proc. Theory Cryptograph. Conf. Theor. Cryptograph. (TCC), pp. 109–127 (2009)
7. Shacham, H., Waters, B.: Compact proofs of retrievability. In: Proceedings International Conferrence Theory and Application of Cryptography and Information Security: Advances in Cryptography (Asiacrypt), vol. 5350, pp. 90–107, Dec 2008
8. Marium, S., Nazir, Q., Ahmed, A., Ahthasham, S., Mirza, A.M.: Implementation of EAP with RSA for enhancing the security of cloud computing
9. Kiran Kumar, K., Padmaja, K., Radha Krishna, P.: Automatic protocol blocker for privacy-preserving public auditing in cloud computing. Int. J. Comput. Sci. Technol. **3**, pp. 936–940. ISSN. 0976-8491 (Online). ISSN: 2229-4333 (Print), March 2012
10. Cong, W., Sherman S.-M., Chow, Q.W., Kui, R., Wenjing, L.: Privacy-preserving public auditing for secure cloud storage. IEEE Trans. Comput **62**(2), 2013
11. Shrinivas, D.: Privacy-preserving public auditing in cloud storage security. Int. J. Comput. Sci. Inf. Technol. **2**(6), 2691–2693. ISSN: 0975-9646 (2011)
12. Ateniese, G., Burns, R.C., Curtmola, R., Herring, J., Kissner, L., Peterson, Z.N.J., Song, D.X.: Provable data possession at untrusted stores. In: Proceedings of the 14th ACM conference on computer and communications security (CCS'07), pp. 598–609. ACM (2007)
13. Erway, C.C., Küpçü, A., Papamanthou, C., Tamassia, R.: Dynamic provable data possession. In: Proceedings of the 16th ACM Conference on Computer and Communications Security (CCS'09), pp. 213–222. ACM (2009)

14. Wang, Q., Wang, C., Ren, K., Lou, W., Li, J.: Enabling public auditability and data dynamics for storage security in cloud computing. IEEE Trans. Parallel Distrib. Syst. **22**(5), 847–859 (2011)
15. Zhu, Y., Hu, H., Ahn, G., Yu, M.: Cooperative provable data possession for integrity verification in multi-cloud storage. IEEE Trans. Parallel Distrib. Syst. **23**(12) 2231–2244 (2012)
16. Zhu, Y., Wang, H., Hu, Z., Ahn, G.J., Hu, H., Yau, S.S.: Dynamic audit services for integrity verification of outsourced storages in clouds. In: Proceedings of the 2011 ACM Symposium on Applied Computing (SAC'11), pp. 1550–1557. ACM (2011)

A Simulation Study with Mobility Models Based on Routing Protocol

Arvind Kumar Shukla, C.K. Jha and Rajeev Arya

Abstract Mobility is an inherent character of wireless Ad Hoc Networks. These networks are characterized by node mobility and requires infrastructure. In the earlier decade, an important amount of study was dedicated to develop mobility models appropriate for estimating the show of Wireless Ad Hoc Network. Simulation is an important mechanism for the indication of fresh concepts in Wireless Ad Hoc networking. The origin of Ad Hoc networks depends on the kind of protocol for estimation, which is previously to be accomplished in a real-world situation. In pre estimation of protocol to support Ad Hoc network, the protocol must be according to common environment. The protocol is uncontrolled within broadcast area, where sub buffer rooms are specified for storage space of communication, information transfer model, and realistic movement of node. Accessing of network is basic purpose to operate complex networking system. The existing mobility model is distinguished from synthetic in term of their experimental and statistical characteristics. The main objective of this paper is to define different mobility models in context to devise more suitable choice for performance evaluation using routing protocols. In this paper, a relative analysis of mobility models existing, are discussed on a variety of simulation setting parameter like packet delivery rate (PDR), Throughput, Average End to End Delay.

Keywords ns2.35 · Routing protocol · Ad Hoc network · Mobility models · BONNMOTION 2.0

A.K. Shukla (✉) · C.K. Jha
Department of AIM & ACT, Banasthali Vidyapith, Vidyapith, Rajasthan, India
e-mail: toarvindshukla@gmail.com

C.K. Jha
e-mail: ckjha1@gmail.com

Rajeev Arya
Electronics and Communication Discipline, DPT, IIT Roorkee, Roorkee, India
e-mail: rajeev.arya.iit@gmail.com

M. Pant et al. (eds.), *Proceedings of Fifth International Conference on Soft Computing for Problem Solving*, Advances in Intelligent Systems and Computing 436, DOI 10.1007/978-981-10-0448-3_72

1 Introduction

Wireless and self configurable behavior of Ad Hoc networks makes them compatible for many scenarios such as mobile battlefield, campus, disaster relief, sensing and monitoring, etc. The mobility of nodes is the key aspect of mobile Ad Hoc networks, and the performance of Ad Hoc networks needs to be study in existence of mobility [1–4]. "The real-life mobility models can be very difficult depending on the procedure objectives of mobile nodes that are a part of the independent scheme. The more difficult the mobility model is, the more difficult it is to model for additional information requires to be integrated. The simulations are applied to study the impact of the mobility in Wireless Ad Hoc network [2, 5]. The node mobility of Wireless Ad Hoc networks sources the network topology to modify with the moment, and Wireless Ad Hoc network performances should be animatedly readjusted to such modifies. So, the networking and protocol presentation of Wireless Ad Hoc networks have very much influenced by the incidence of topology adjusts. The performance of networks can vary significantly with unusual mobility models [3, 6]. Additionally, by unstable dissimilar parameters of a known mobility model, Wireless Ad Hoc network performances are consequence by a vast area. The choice of a mobility model may need an infrastructure, traffic model over the wireless Ad Hoc network which significant and purpose performances. Network simulators emerged as the most general method of evaluating the performance of the large complex networking system. However, for system concerning mobile nodes, the association of the nodes has a significant control on the simulation results. These are collected from location of mobile nodes known to communicate with one another over a general wireless channel [1, 3, 5]. A mobility model effort to mimic the progress of genuine mobile nodes to vary the velocity and the route through moment. Mobility model that exactly symbolizes the individuality of the sensor nodes in Ad Hoc networks has the key to study whether an agreed protocol is in use in the exacting type of mobile area [1–3]. The feasible moves toward for modeling of the mobility model are of two categories: syntactic and traces. The traces offer those mobility models that are realistic in real-life schemes. In trace-based model, the whole thing is deterministic. Though, mobile Ad Hoc networks have so far to be arranging extensively to recognize the traces concerning a large number of members and an acceptably long inspection time. In deficiency of traces, the syntactic mobility model that has been projected to position for the schedule of mobile nodes practically in Ad Hoc networks. The syntactic mobility models can to be classified support on the explanation of the mobility model in Ad Hoc networks [1, 7]: entity mobile schedule and collection mobile actions. In the previous case, the mobility models effort to expect mobile's cross model interchange place at a known point of moment under different network state. In this article, we cover measured the effect of mobility model for the presentation of routing protocols DSR, (Reactive Protocol). For test purposes, we have chosen two mobility scenarios: Simple Human Mobility Model and Manhattan models. These two Mobility Models are chosen randomly and signify the chance of its practical purpose in the future. Performance assessment has also been manner across unstable node density and the number of hops. In the previous test performed by some researcher's exposed exponential range of performance of the mobility models using routing protocols, node density, and span of information booths. In the second section, we will study about the routing protocols for Wireless Ad Hoc wireless networks can be right for use in our study. In the third section, we will study mobility models used for simulation. In the Fourth Section, we will run numerous simulation scenarios for mobility models in the network. We will present the outcome and calculate the performance of each tested mobility models using protocols, based on the simulation results in the fifth section the parameters are used for performance analysis result for simulations and to conclude, in a sixth division we will finish the paper.

2 Protocol Used for Simulative Study

2.1 Dynamic Source Routing Protocol (DSR)

It is an on-demand routing intended to confine the bandwidth devoted through packets within Ad Hoc networks by reducing the cyclic renew communication essential in table driven approach. As its name tells, this protocol use source routing algorithm to determine routes the main difference between this and further on-demand protocol is that it has inspired less and therefore do not require cyclic hello packet broadcast, which are applied by a node to notify its neighbors of its occurrence. The primary approach of protocol throughout the route formation phase is to set up a route by flood route demand packets (RREQ) in the wireless network. The target node, on receiving a route demand packet, reacts by transmitting a route, respond packet (RREP) back to source, which carry the route navigate by the route demand packet obtain [8, 9]. This is called route discovery and is one of its two major phases along with route maintenance. For the Route invention and maintenance, source node that does not leada path to the target. When it has data packets to be sent to that target, it floods a RREQ throughout the wireless network. Every node, upon getting a RREQ, rebroadcasts the packet to its neighbors if it have not onward it previously, provided that the node is not the target node and that the packet's moment to live (TTL) counter has not been surpassed. On arrival the path reply, the target node should have a path to the source node. If the path is in the target node's path cache, the path would be applied. Or else, the node will reverse the path based on the route record in the RREP message header. Nodes can also learn about the neighboring path traverse by data packets if operate in the promiscuous. This path supply is also applied during the route manufacture phase in the mode that if an intermediary node that receive a RREQ has a path to the target node in its own path cache, and after that its reply to the source node by transmitting the RREP with the complete path information from the source node to the target node. As every node can odd the additional with a RREQ, it seems that loops might be produced as well as multiple communications of the similar RREQ, for example by a focus node that receives it through many paths. To avoid this, each RREQ carries a cycle number created by the source node and the route it has traveled. A node, upon getting a RREQ, checks the cycle number of the packet previous to on warding it; therefore it is forwarded only if it is not a replacement RREQ. In the episode of serious communication, the route preservation Phase is starting where by the path error packets (RERR) are created on a node. That node sends to the others so they will eliminate the route that applies that hop, so every route, including it is reduced at that point [3, 6, 10]. Over again, the Route detection Phase is initiated to decide the mainly viable path.

3 Mobility Models Used in Simulative Study

These segments present the mobility models applied in the simulation study are obtainable, evaluated and explained Simple human Mobility Model and Manhattan Mobility Model. Mobility models can be differentiated according to their compute of how two nodes have needed into their movement. If two sensor nodes are moving in the same way, then they have a maximum spatial dependence and a compute of how magnitude, current velocity and direction are connected to preceding velocity. Sensor nodes having similar velocity have a high sequential need. Descriptions of mobility models are given below.

3.1 Simple Human Mobility Model

Till now The easier mobility model was introduced by Greede et al. [11], called SHMM. It defines a realistic indoor scenario as any organization or campus the permanent nodes carry split nodes to access particular location. Then nodes are grouped that demonstrate familiar welfare after which additional nodes are considered as guest nodes. According to specified grid individual node are considered featuring a mobility model based on trace [10.20.21.22]. It is similar to city section or Manhattan model in case of easiness. The earlier models are more sufficient to other model that is the main divergence.

3.2 Manhattan Mobility Model

This model was given by Bai et al. [12] and Bai and Helmy [13]. In this model, nodes movement route is redefined. The arguments *–u* and *–v* set the block among the route. It is similar to city section mobility model. It may not be define a target point to be reached. A novel direction is chosen from the existing ones, each time it reaches a novel crossing, to any distributive node. Speed may be changed as a divided schematic method, or according to scenario constraints [14–16]. It classifies the related model of nodes on the avenues and it may valuate in model association in an urban area. The collection of horizontal and Vertical Avenue is the area [4, 8] the relative model of mobile nodes in account of defined by maps followed by Manhattan mobility. It follows a grid road topology. This model was proposed for urban scenario where streets are pre-behavioral. The node moves in horizontal and vertical way on urban map, in this mobility model. The Manhattan Mobility Model utilizes an anticipated approach in the estimation of node movements, because, at every intersection in route to keep moving in the similar way. Though this model affords flexibility for the nodes to modify the way, it imposes geographic limitations on node mobility.

4 Simulations

The wireless networks analyzed have been carried out applying Network Simulator ns2.35 and its related tools for simulation and study of analysis. We select a Linux platform, i.e., UBUNTU 12.04 LTS, as Linux recommend a number of programming improvement tools that can be applied through the simulation procedure. We have produced mobility scenarios of Mobility Model is applying BONNMOTION2.0; *they can be included into TCL scripts. Random traffic links of*

Table 1 Performance parameters

Parameter	Value
Channel type	Wireless channel
Simulator	NS 2 (Version 2.35)
Protocols	DSR
Simulation duration	500 s
Number of nodes	25, 50, 75
Transmission range	250 m
Movement model	Simple human, mahattam
MAC layer protocol	802.11
Pause time (s)	15 ± 4 s
Maximum speed	25
Minimum speed	0.5
Packet rate	4 packet/s
Traffic type	CBR (constant bit rate)
Data payload	512 bytes/packet
Max of CBR connections	10, 20, 40
Environment Size	700 m * 700 m
Channel type	Wireless channel

CBR can be set up among mobile nodes applying a traffic-scenario creator script [3, 6]. BONNMOTION is java supported tool for creating mobility scenario for several mobility models, developed by University of Bonn, Germany (Table 1).

5 Performance Parameters and Results Analysis

5.1 Performance Parameters

The organization of routing protocols is through the following important quality of services (QoS) metrics for usual procedures:

5.1.1 Packet Delivery Ratio (PDR)

It has classified in [17, 18] as the fraction among the amount of packets created with the application layer. It has the fraction of data packets send to the target to those created from the starting point. It is estimated by separating the amount of packets obtained by target throughout the packet initiated from the source.

$$\text{PDF} = (\text{Pr}/\text{Ps}) * 100,$$
$$\text{Pr} = \text{total Packet obtain}$$
$$\text{Ps} = \text{the total Packet transmit.}$$

5.1.2 Throughput

It has the standard amount of messages effectively send per unit time number of bits delivered per second [18].

Throughput = (Total received packets/total simulation time) kbits/s
N = number of data sources.

5.1.3 Average End to End Delay

It has described as the time in use for a packet to be broadcast across an Ad Hoc from basis to target.

$$\text{D} = (\text{Tr} - \text{Ts}),$$
$$\text{Tr} = \text{receive Time}$$
$$\text{Ts} = \text{sent Time}$$

5.2 Result Analysis

In the case of presentation investigation we have measured presentation parameters. In Figs. 1, 2, and 3 the analysis is focusing on evaluating the show on routing overhead, and packet delivery ratio throughput. The results also evaluated with two mobility models that we had selected. The result will explain the show of mobility models among respect to protocols that have been chosen. Under dissimilar mobility model.

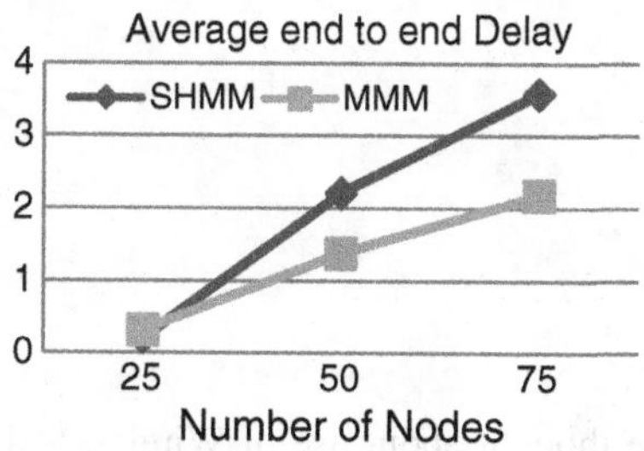

Fig. 1 Average end to end delay versus number nodes for mobility models

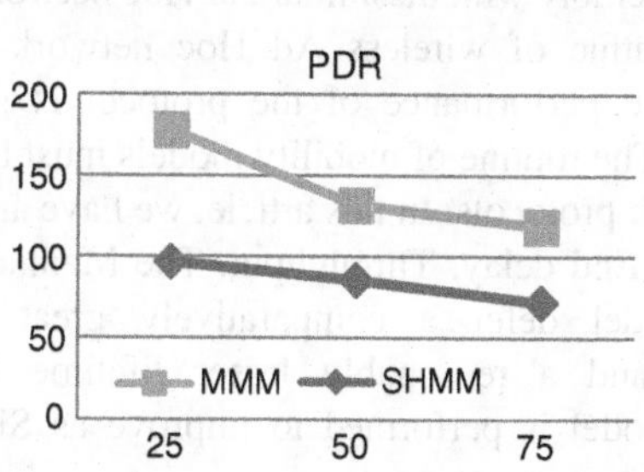

Fig. 2 Packet delivery ratio versus number nodes for mobility models

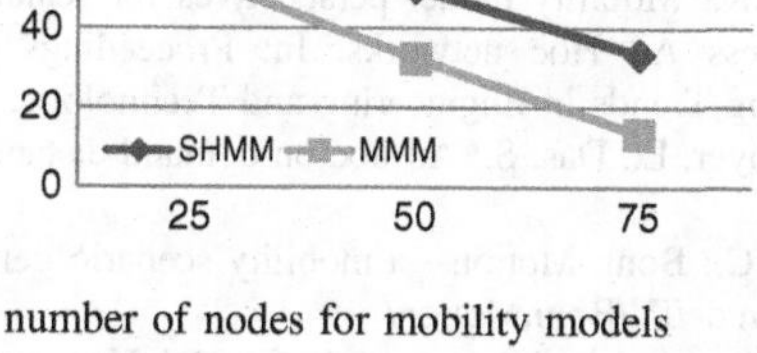

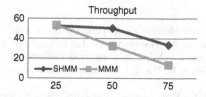

Fig. 3 Throughput versus number of nodes for mobility models

5.2.1 Average End to End Delays

See Fig. 1.

5.2.2 Packet Delivery Ratio (PDR)

See Fig. 2.

5.2.3 Throughput

See Fig. 3.

6 Conclusion

In this section, we have thought about use of Wireless Ad Hoc protocols such as DSR for wireless Ad Hoc networking simulation. We have considered Manhattan and Simple Human Mobility Model as mobility models. The practice of a mobility model may vary considerably with dissimilar Ad Hoc network protocols. The above Figures shows the routine of wireless Ad Hoc network mobility models with routing. As shown, the performance of the protocol is greatly affected by the mobility of the nodes. The routine of mobility models must be calculated among the help of Ad Hoc network protocols. In this article, we have analysis three parameters PDR, Average End to End delay, Throughput. The Manhattan model and Simple Human Mobility Model defer a comparatively great amount of hops for minimum-hop paths and a reasonably lesser lifetime for stable paths. The Manhattan mobility model is performed to improve as Simple Human Mobility Model.

References

1. Kulkarni, S.A. Rao, G.R.: Mobility model perspectives for scalability and routing protocol performances in wireless Ad Hoc networks. In: Proceedings of the First International Conference on Emerging Trends in Engineering and Technology, pp. 176–181, July 2008
2. Perkins, C., Belding-Royer, E., Das, S.: Ad-hoc on-demand distance vector (AODV) routing. In: IETF (2003)
3. Gerharz, M., de Waal, C.: Bonn Motion—a mobility scenario generation tool. University of Bonn. www.cs.uni-bonn.de/IV/BonnMotion/
4. Camp T., et al.: A survey of mobility models for Ad Hoc network research. In: Wirel. Commun. Mob. Comput. Spec. issue Mob. Ad Hoc Netw. Res. Trends Appl. **2**, 483–502 (2002)

5. Johnson, D., Hu, Y., Maltz, D.: The dynamic source routing protocol (DSR) for mobile ad-hoc networks for IPv4. In: IETF (2007)
6. Sargolzaey, H., Moghanjoughi, A.A., Khatun, S.: A review and comparison of reliable unicast routing protocols for mobile Ad-Hoc networks. Int. J. Comput. Sci. Netw Secur. **9**(1), 186–196 (2009)
7. Camp, T., et al.: A survey of mobility models for Ad Hoc network research. In: WCMC (2002)
8. Bai, F., Helmy, A.: A survey of mobility modeling and analysis in wireless Adhoc networks. In: Wireless Ad Hoc and Sensor Networks, Kluwer Academic Publishers (2004)
9. Perkins, C.E., Bhagwat, P.: Highly dynamic destination sequenced distance vector routing (DSDV) for mobile computers. Comput. Commun. Rev. 234–244, (1994)
10. Kumar, S., Sharma, S.C., Suman, B.: Mobility metrics based classification and analysis of mobility model for tactical network. Int. J. Next-Gen. Netw. **2**(3), 39–51 (2010)
11. Greede, A., Allen, S.M., Whitaker, R.M.: A simple human mobility model for opportunistic networks. In: 9th Annual Postgraduate Symposium on the Convergence of Telecommunications, Networking and Broadcasting, Liverpool, UK (2008)
12. Bai, Narayanan, S., Helmy, A.: Important: A framework to systematically analyze the impact of mobility on performance of routing protocols for adhoc networks. In: INFOCOM 2003. Twenty-Second Annual Joint Conference of the IEEE Computer and Communications Societies. IEEE, vol. 2, pp. 825–835 (2003)
13. Bai, F., Helmy, A.: A survey of mobility models. In: Wireless Adhoc Networks. CiteSeerX—Scientific Literature Digital Library and Search Engine (2008)
14. Iyer, A., Rosenberg, C., Karnik, A.: What is the right model for wireless channel interference. IEEE Trans. Wirel. Commun. **8**(5), 2662–2671 (2009)
15. Liu, J., Jiang, X., Nishiyama, H. Kato, N.: On the delivery probability of two-hop relay MANETs with erasure coding. IEEE Trans. Commun. (2013)
16. Andrews, J.G., Haenggi, M., Jindal, N.: A primer on spatial modeling and analysis in wireless networks. IEEE Commun. Mag., **48**(11), 156–163, (2010)
17. Singh M, Singh D.: Impact and performance of mobility models in wireless Adhoc networks. In: Fourth International. Conference on Computer Sciences and Co vergence Information Technology (2009) 978-0-7695-3896
18. Broch. J. et al.: A performance comparison of ad-hoc multi-hop wireless net works routing protocols. In: Proceedings IEEE/ACM MOBICOM (1998)

Genetic-Based Weighted Aggregation Model for Optimization of Student's Performance in Higher Education

Preeti Gupta, Deepti Mehrotra and Tarun Kumar Sharma

Abstract Most of the real-life problems are optimization problems, where the aim is to develop a model that optimizes certain output criteria. Education domain though a nonprofit sector intends to optimize its functioning by adopting procedures that tend toward knowledge building. Increasing the student's performance has always been an area of interest among these education organizations. The paper exemplifies the usage of Binary encoded genetic algorithm to model student's performance in a course pertaining to higher education. It gives significance to the variables identified responsible, for affecting the performance of the students in the course under study. Adopting such knowledge-based activities may help the organizations to eventually establish themselves as a Knowledge Centric Higher Education Organization.

Keywords Genetic algorithm · Higher education · Population · Knowledge · Selection · Mutation

1 Introduction

Optimization is a procedure which is performed iteratively by comparing various solutions till an optimal or a favorable solution is derived. Numerous problems in real world aim to acquire optimization by maximizing or minimizing certain output criteria [1].

Preeti Gupta (✉)
Amity School of Engineering & Technology, Amity University Rajasthan, Jaipur, India
e-mail: preeti_i@rediffmail.com

Deepti Mehrotra
Amity School of Engineering & Technology, Amity University Uttar Pradesh, Noida, India
e-mail: mehdeepti@gmail.com

T.K. Sharma
Amity Institute of Information Technology, Amity University Rajasthan, Jaipur, India
e-mail: taruniitr1@gmail.com

M. Pant et al. (eds.), *Proceedings of Fifth International Conference on Soft Computing for Problem Solving*, Advances in Intelligent Systems and Computing 436, DOI 10.1007/978-981-10-0448-3_73

Genetic Algorithm (GA) is a stochastic search process which is based on the procedure of natural selection and biological evolution. The technique draws its inspiration from the Darwin's theory of survival of the fittest [2]. Under specific selection criteria GA works with a population composed of numerous individuals to acquire a status where "fitness" is maximized. Apart from extensive work carried out by researchers like John Holland, David Goldberg, the work of De Jong was much appreciated as it exhibited the usefulness of the GA for function optimization.

As GA deals with number of dimensions (continuous or discrete) with extreme cost surface, it optimizes while searching for optimum solutions. Moreover it does not require derivative information about the problem. So genetic algorithms produce quality results in situations where traditional optimization techniques fail.

There are two encoding representations in GA, binary and continuous. Though both schemes follow the same procedure to model genetic recombination, one uses encoded binary string for representing variables and works with the binary strings to maximize the fitness, while the other works with the continuous variables to maximize the fitness. For the optimization task, the presented work uses binary encoding method since GA originated with a binary string variable representation. The work models the student performance problem in higher education domain using binary GA.

The paper is organized as follows. Section 2 establishes the higher education scenario. Section 3 elaborates the elements of GA in brief and applies GA procedure to the problem under consideration. Section 4 summarizes the importance of the technique for developing knowledge centric higher education organizations. Finally, the conclusions are drawn and presented in Sect. 5 along with future directions.

2 Higher Education Scenario

With the dawn of privatization in the field of education, number of private universities and colleges are on its growth. The management of these institutions is in constant search to excel in the existing processes. Over the years, analyzing the performance of students in the end term examination for a subject has attracted various researchers to carry out work in this direction. The performance of the students depends on number of factors. Curriculum designing which has been an important in-house academic process in an educational organization is boosted by identifying effective factors in deciding the performance of the students. In the work [3] the authors have considered number of independent attributes affecting the result of students in the end term examination for Analog Electronics and are described in Table 1. Further the attributes identified by experiential learning were evaluated and graded on the basis of higher information gain as information gain increases with the average purity of the subsets that a variable produces. Strategy adopted was to choose attribute with greatest information gain.

Table 1 Information gain of independent attributes

Attributes	Description	InfoGain
ContinuousEvaluationMarks	Continuous evaluation of the student in terms of class test, quiz, assignment (Max. marks 30)	0.496774
BaseSubMarks	Marks obtained by the students in the prerequisite subject in the previous semester	0.74
Prac_orient_for_the_sub	Practical orientation toward that subject	0.41
Attendance	Attendance in the course	0.69
SGPA_prev	SGPA in previous semester	0.665656

It was witnessed that the variable *BaseSubMarks* has the highest information gain and hence has the maximum effect on the performance of the student in the subject of Analog Electronics.

The presented work starts with an initial population of random members. Characteristics of the students associated with scoring high marks in the end term examination of the subject under scrutiny are represented with the binary sequence. If an attempt is made to obtain the students with the highest marks, then only a few of the merit holder students are to be considered. Traits of high scorer students must be determined. From this population of high scoring students, random selection of two individuals is to be made to create students having high scoring qualities. The new generated students have a significant likelihood of being high scorers as their parents have traits that make them high scoring. These new generated students replace two students that did not score high enough. To maintain the diversity of the population enough students' offsprings are created to bring the population back to its initial size. Repeating on this process leads to a student with high scoring intellect [4].

3 Elements of Binary Genetic Algorithm

The Genetic Algorithm very similar to any other optimization technique works with variables, the cost function, and the cost. It culminates work by testing for convergence. Primarily two types of representations are there in GA [5]. (1) Binary Coded (2) Real Coded.

For evaluating the fitness value Binary-Coded (genotype) GAs must decode a chromosome into a real value (phenotype) whereas real-coded GAs operates on the actual real value phenotype. To complete one generation of a GA, crossover and mutation are applied to the whole population of solutions after reproduction. Different criteria are used to terminate the program, such as maximum number of generations, desired accuracy, etc. (Fig. 1).

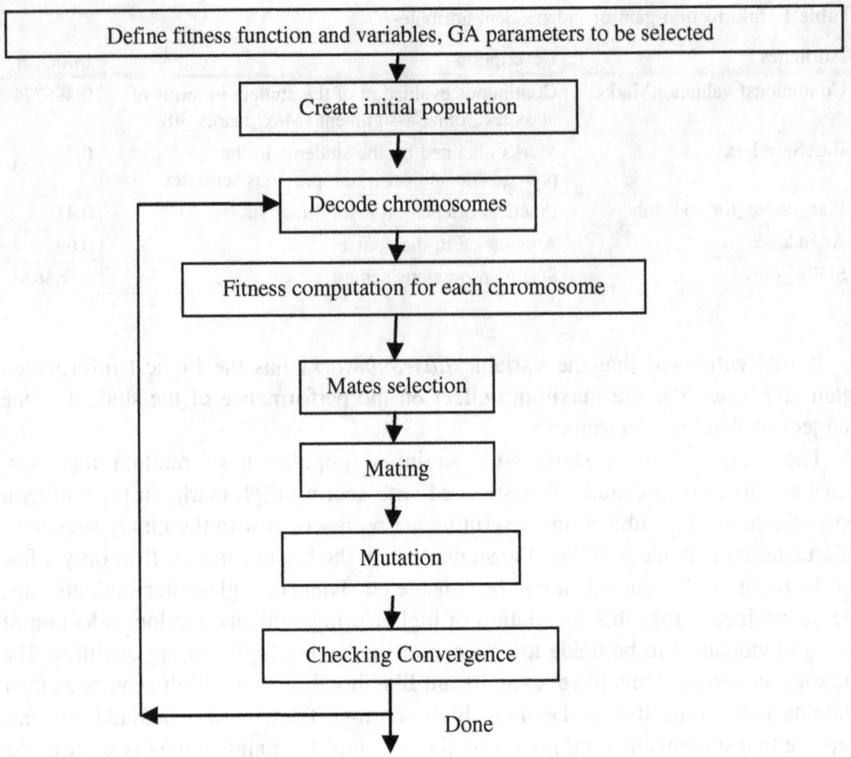

Fig. 1 Flowchart depicting binary GA

3.1 Optimization Parameter Settings

The following optimization parameters are taken into consideration and are shown in Table 2.

Table 2 Optimization parameter setting for the work

Parameters	Details	Parameters	Details	Parameters	Details
GA type	Simple	Selection	Roulette wheel	Number of generations	10
Chromosome encoding	Binary	Cross-over operator	Binary	Fitness function	Weighted aggregation
Size of population	10	Mutation type/rate	Bit flip/0.06	Termination criteria	After 100 iterations, on observing result is not improving

3.2 Selection of Variables

A fitness function yields output from a set of input attribute depicted by a chromosome. The term fitness is extensively used to depict the output of the objective function in the GA. Fitness implies a maximization problem. The objective is to adjust the output in some desirable way finding the suitable values for the input variables. GA begins by defining an array of variable values to be optimized in the form of chromosome. If N variables are used to define a chromosome (an N-dimensional optimization problem) given by p_1, p_2,…. then the chromosome is represented as an N element row vector. Chromosome = [p_1, p_2, p_3,… p_N]. Each chromosome has a fitness found by evaluating the fitness function F, at p_1, p_2, ... The defined problem is a 5 variable problem as described in Table 1.

3.3 Encoding and Decoding of Variables

Since binary digits are used to represent variables, continuous values are to be converted into binary, and vise versa. The binary encoded GA works with binary digits. The variable x is depicted by a sequence of bits that is N_{gene} long. If $N_{\text{gene}} = 3$ with limits defined by $0 \le x \le 7$, then 8 possible values can be represented using 3 bits. [2]. The encoding of the variables is in Table 3.

The study defines a chromosome that has five variables, each encoded with $N_{\text{gene}} = 3$ bits. The scheme under consideration is working with binary encodings, but the fitness function often requires continuous variables. Hence before evaluating fitness function, chromosome must be decoded. There is a need to map genotype representation (binary string) to phenotype representation (candidate solution) through a mapping function, when ever binary encoded GA is used. This can be done by using the mapping function given through Eq. (1).

$$d(ub, lb, l, c) = [(ub - lb)decode(c)]/2^l + lb \tag{1}$$

where,

lb lower bound
ub upper bound
l chromosome length (in bits)
c chromosome.

3.4 The Population and Fitness Function Computation

Group of chromosomes is known as population. There is a need to randomly select initial population and apply the fitness function on the genotype.

Table 3 Encoding of variables

Attribute–ContinuousEvaluationMarks (maximum marks 30)			Attribute–SGPA_Prev (maximum SGPA 10)		
Levels	Marks between	Encoding	Levels	SGPA between	Encoding
A	0–3	000	A	0–2	000
B	4–7	001	B	2.1–4	001
C	8–11	010	C	4.1–5	010
D	12–15	011	D	5.1–6	011
E	16–19	100	E	6.1–7	100
F	20–23	101	F	7.1–8	101
G	24–27	110	G	8.1–9	110
H	28–30	111	H	9.1–10	111
Attribute–BaseSubMarks (maximum marks 100)			Attribute–attendance (maximum attendance 100 %)		
Levels	Marks between	Encoding	Levels	Attendance between (%)	Encoding
A	0–20	000	A	Below 65	000
B	21–40	001	B	65.1–70	001
C	41–50	010	C	70.1–75	010
D	51–60	011	D	75–80	011
E	61–70	100	E	80.1–85	100
F	71–80	101	F	85.1–90	101
G	81–90	110	G	90.1–95	110
H	91–100	111	H	95.1–100	111
Attribute–Prac_orient_for_the_sub (Maximum Marks 100)					
Levels	Marks between	Encoding			
A	0–20	000			
B	21–40	001			
C	41–50	010			
D	51–60	011			
E	61–70	100			
F	71–80	101			
G	81–90	110			
H	91–100	111			

A fitness function quantifies the optimality of a solution (chromosome) so that that particular solution may be ranked against all the other solutions. It depicts the closeness of a given 'solution' to the desired result. Most functions are stochastic and designed so that a small proportion of less fit solutions are selected. This helps keep the diversity of the population large, preventing premature convergence on poor solutions [6].

The Fitness function for optimizing end term performance in the subject of Analog Electronics is a five variable function and is given by Eq. (2).

Table 4 Fitness evaluation

Student	Genotype	Fitness	Student	Genotype	Fitness
Student 1	100101100100101	12.67	Student 6	110101101010011	12.01
Student 2	011100011011011	9.36	Student 7	110110110001011	12.72
Student 3	000010100011100	7.99	Student 8	101101100011111	13.2
Student 4	011011101100101	11.69	Student 9	010011011010100	8.02
Student 5	111110110111110	18.49	Student 10	001010010000100	4.84

$$F(v_1, v_2, v_3, v_4, v_5) = 0.4 * v_1 + 0.66 * v_2 + 0.74 * v_3 + 0.69 * v_4 + 0.41 * v_5 \quad (2)$$

where,
v_1 ContinuousEvaluationMarks
v_2 SGPA_Prev
v_3 BaseSubMarks
v_4 Attendance
v_5 Prac_orient_for_the_sub

The coefficients of the variables are the corresponding information gain. The fitness values are compiled in Table 4.

3.5 Selection and Reproduction

For reproduction chromosomes are selected from the population of parents for performing cross-over and to produce off springs. According to Darwin's principle of survival of the fittest, the best one should survive and create new and healthy off springs. All the reproduction operators in GA rely on the idea that above average strings are picked from the current population and multiple copies are inserted in the mating pool in a probabilistic manner.

For selection the study uses Roulette–wheel selection technique [7]. Strings selected from the mating pool have probability proportional to the fitness.

The probability of an individual selection is given by the ratio of its fitness to the sum of fitness of all individuals of the current population. The relation is given in Eq. (3).

$$P(h_i) = F(h_i) / \sum F(h_j) \quad (3)$$

where h_i is the ith individual of current population and size of population is N. Spin wheel (N) times to select (N) individuals, j value goes from 1 to N. Ten parents are selected using the Roulette Wheel selection procedure and is shown in Table 5.

Table 5 On applying Roulette wheel selection

Student	Genotype	Fitness	Student	Genotype	Fitness
Student 1	100101100100101	12.67	Student 4	011011101100101	11.69
Student 1	100101100100101	12.67	Student 5	111110110111110	18.49
Student 1	100101100100101	12.67	Student 5	111110110111110	18.49
Student 3	000010100011100	7.99	Student 5	111110110111110	18.49
Student 4	011011101100101	11.69	Student 9	010011011010100	8.02

3.6 *Cross-Over*

After the selection phase the population is enhanced with healthy individuals. Selection is followed by cross-over. The idea behind crossover is that the new chromosome may be better than both of the parents, if it takes the best characteristics from each of the parents. The aim of cross-over is to search the parameter space for maximal information preservation as parent strings are sequences of good strings selected during reproduction.

The study implements single point crossover [8]. It includes selecting the cross-over site at random along the length of the mated string and exchanging the bits next to the cross site Table 6 depicts the same.

3.7 *Mutation*

After crossing over the strings are subjected to mutation. In the study, mutation of a bit involves flipping it, i.e., transforming 0–1 and 1–0 with a small mutation

Table 6 Creating offspring using single-point crossover

Student	Genotype	Student	Genotype
Student 1 (Parent)	100|101100100101	Student 1 (Parent)	100|101100100101
Student 3 (Parent)	000|010100011100	Student 4 (Parent)	011|011101100101
Off spring 1	100010100011100	Off spring 3	100011101100101
Off spring 2	000101100100101	Off spring 4	011101100100101
Student	Genotype	Student	Genotype
Student 1(Parent)	100|101100100101	Student 4 (Parent)	011|011101100101
Student 5 (Parent)	111|110110111110	Student 5 (Parent)	111|110110111110
Off spring 5	100110110111110	Off spring 7	011110110111110
Off spring 6	111101100100101	Off spring 8	111011101100101
Student	Genotype		
Student 5(Parent)	111|110110111110		
Student 9 (Parent)	010|011011010100		
Off spring 9	111011011010100		
Off spring 10	010110110111110		

Table 7 Fitness evaluation after applying mutation on off springs

Student	Genotype	Fitness	Student	Genotype	Fitness
Off spring 1	100010100011100	9.59	Off spring 6	111101100101101	14.56
Off spring 2	000101100100101	11.07	Off spring 7	111110110111110	18.49
Off spring 3	101011101100101	12.49	Off spring 8	111011101000101	10.53
Off spring 4	011111100100101	13.59	Off spring 9	111011001010100	8.54
Off spring 5	100110110111010	15.65	Off spring 10	010110010111110	13.53

Table 8 Population at $t = 0$

Student	Genotype	Fitness	Student	Genotype	Fitness
Student 1	100101100100101	12.67	Student 6	110101101010011	12.01
Student 2	011100011011011	9.36	Student 7	110110110001011	12.72
Student 3	000010100011100	7.99	Student 8	101101100011111	13.2
Student 4	011011101100101	11.69	Student 9	010011011010100	8.02
Student 5	111110110111110	18.49	Student 10	001010010000100	4.84

probability. It is important in the search that diversity be preserved. The mutation operator preserves the diversity of population. Mutation probabilities are smaller in the population hence can conclude that mutation is considered as a secondary mechanism of GA. In the study, the probability is 0.06. The off springs after mutation is shown in Table 7 with muted bit in bold.

3.8 The Next Generation

After all the steps are over the population at $t = 0$, which is depicted in Table 8 is replaced by population represented in Table 9.

The population at $t = 1$ is thus

Table 9 Population at $t = 1$

Student	Genotype	Fitness	Student	Genotype	Fitness
Off spring 1	100010100011100	9.59	Off spring 6	111101100101101	14.56
Off spring 2	000101100100101	11.07	Off spring 7	111110110111110	18.49
Off spring 3	101011101100101	12.49	Off spring 8	111011101000101	10.53
Off spring 4	011111100100101	13.59	Off spring 9	111011001010100	8.54
Off spring 5	100110110111010	15.65	Off spring 10	010110010111110	13.53

Table 10 Population at $t = 9$

Student	Genotype	Fitness	Student	Genotype	Fitness
Off spring 1	100110110111010	15.65	Off spring 6	100011111010101	12.19
Off spring 2	100010100011100	9.59	Off spring 7	101100110111110	16.37
Off spring 3	001110110111010	14.45	Off spring 8	111011001010101	8.95
Off spring 4	100111110111010	16.31	Off spring 9	100110010110110	13.64
Off spring 5	101110110101010	14.67	Off spring 10	010110111111010	15.59

Similarly working through few more generations, at $t = 9$ the population is depicted through Table 10.

3.9 Convergence

Whether an acceptable solution is reached or a set number of iterations are exceeded drive the progress of generations. In absence of mutation after a while all the chromosomes and their fitness value would become the same. At this time the algorithmic procedure should be stopped [9]. Population mean and maximum fitness are important population statistics that most GAs keep track of. For the study under consideration, after ten generations the average global maximum is found to be 17.8.

4 Importance of the Study

As optimizing the end term performance in the course was the ultimate aim. Iterating through the generations and after the stopping criteria is met the characteristics required to achieve maximum performance in the course can be identified by decoding the bit sequences. Moreover, the following knowledge (effect of one variable on the other) can be put to best use for proper curriculum planning, designing adequate lesson plans and evaluation criteria and adoption of suitable pedagogical techniques for improving the overall performance of the course [10].

5 Conclusion and Future Scope

Usage of GA, as an optimization tool will help to dig out patterns and establish relationships between variables that are not visible openly. This can later be put to best of advantage for developing a knowledge-based education system. The study implements simple GA with a single objective, i.e., identifying characteristics of students related to maximization of end term result in a higher education course.

The study can be extended to implement multi objective optimization in many real life scenarios. Many real life data mining problems such as classification problems having multiple conflicting objectives that are to be optimized simultaneously to obtain optimum solutions. For solving multi objective optimization problems, there is a need of working with multi objective evolutionary algorithms (MOEAs). One such variant of MOEA is Multi Objective Genetic algorithm (MOGA) that can be used to model real life scenarios that are usually multi objective.

References

1. Simon, D.: Evolutionary Optimization Algorithm Biologically-Inspired and Population-Based Approaches to Computer Intelligence, Wiley Pub. (2013)
2. Haupt, R.L., Haupt, S.E.: Practical Genetic Algorithms. Wiley (2004)
3. Gupta, P., Mehrotra, D., Sharma, T.K.: Identifying knowledge indicators in Higher Education Organization. Proc. Comput. Sci. **46**, 449–456 (2015)
4. YilDiz, O., Bal, A., Gulsecen, S., Kentli, F.D.: A genetic –fuzzy based mathematical model to evaluate distance education students' academic performance. Proc. Social Behav. Sci. **55**, 409–418 (2012)
5. McCall, J.: Genetic algorithms for modeling and optimization. J. Comput. Appl. Math. **184**, 205–222 (2005)
6. Xilin, Z., Shiming, L.: Optimization model of higher education resources allocation based on genetic algorithm. Manage. Sci. Eng. **7**(3),76–80 (2013)
7. Kumar, J.: Blending Roulette wheel selection & rank selection in genetic algorithms. Int. J. Mach. Learn. Comput. **2**(4), 365–369 (2012)
8. Mendes, J.M.: A comparative study of crossover operators for genetic algorithms to solve the job shop scheduling problem. WSEAS Trans. Comput. 4 **12**, 164–173 (2013)
9. Greenhalgh, D., Marshall, S.: Convergence criteria for genetic algorithms. SIAM J. Comput. **30**(1), 269–282 (2000)
10. Gupta, P., Mehrotra, D.: Effective curriculum development through rule induction in knowledge centric higher education organization. IET digital Library, 9.09 (2013)

Improved Convergence Behavior by Using Best Solutions to Enhance Harmony Search Algorithm

Ali Maroosi and Ravie Chandren Muniyandi

Abstract Harmony search is an emerging meta-heuristic optimization algorithm inspired from music improvisation processes, and able to solve different optimization problems. In the previous studies harmony search is improved by information of the best solution. This increases speed of coverage to the solution but chance of immature coverage to the local optimum increases by this way. Thus, this study uses information from the *p* of the best solutions to accelerate coverage to optimal solution while avoiding immature coverage. Simulation results show the proposed approach applied for different numerical optimization problems has better performance than previous approaches.

Keywords Harmony search · Optimization algorithm · Best solutions

1 Introduction

The harmony search (HS), inspired by the improvisation processes for music, was introduced by Geem et al. [1]. In HS, combination of existing solutions is selected iteratively to enhance the solution quality. The HS has several advantages. It uses information from all solutions in the harmony memory to generate a new solution, whereas the traditional optimization techniques, such as the genetic algorithms, only consider two parent solutions. Moreover, it has few mathematical requirements and derivative information is not needed, because it uses the stochastic random

A. Maroosi · R.C. Muniyandi (✉)
Center for Software Technology and Management, Faculty of Information Science and Technology, National University of Malaysia, Bangi 43600, Selangor, Malaysia
e-mail: ravie@ukm.edu.my

A. Maroosi
e-mail: ali.maroosi@gmail.com

A. Maroosi
Department of Computer Engineering and IT, University of Torbat-e-Heydarieh, Torbat-e-Heydarieh, Khorasan, Iran

M. Pant et al. (eds.), *Proceedings of Fifth International Conference on Soft Computing for Problem Solving*, Advances in Intelligent Systems and Computing 436, DOI 10.1007/978-981-10-0448-3_74

search [2–4]. The HS has been successfully applied to various areas, including the binary coded optimization problems [5], the reaction kinetic parameter estimation [6], the power economic load dispatch [7], the cost minimization [8], the damage detection [9], the feature selection [10], the machine learning [11], and the classification [12].

In comparison with other meta-heuristic algorithms, the HS algorithm requires fewer mathematical requirements and may be easily implemented for solving various types of optimization problems in engineering. With this feature, derivative information is not needed, since it uses stochastic random search [13]. The benefits of HS and its implementations in various areas have attracted more interest in recent times. However, the standard HS has constant parameters that affect accuracy and speed of convergence to achieve the optimal solution. Different variants of the HS have been devised to improve the optimization, including an ensemble of parameter sets, which can self-adaptively select the best control parameters during the evolution process. Mahdavi et al. [13] and Pan et al. [14] introduced a self-adaptive global-best HS algorithm and demonstrated that it was more effective in finding better solutions than several other HS algorithms. Omran and Mahdavi [15] established a new global-best HS, enthused by the notion of particle swarm optimization algorithms, and evaluated its efficiency in numerical and integer programming problems.

Zhao et al. [16] used the HS algorithm for the dynamic multi-swarm particle swarm optimizer (DMS-PSO) that improved the performance compared to the standard DMS-PSO and HS. A differential approach was proposed by Zhao et al. [16], inspired by the differential evolution (DE) algorithm to improve the HS. Arul et al. [17] and Coelho et al. [17] proposed a chaotic differential harmony search algorithm, where the pitch adjustment operator in the HS algorithm was modified as that of the DE operator. The chaotic sequences using the logistic map were used to generate the values of the DE parameter. Similarly, many other studies have implemented the adaptive selection schemes for choosing and tuning the HS parameters by incorporating the fuzzy c-means algorithm, clonal selection algorithm, and artificial immune system among others.

Most of the studies tried to improve harmony search by adaptive tuning of its parameters and used information of the best solution to improve HS. In this paper, a new algorithm is introduced which information of the *p* best solutions to improve HS. Since the introduced algorithm works with the *p* best solutions, it is called an improved best harmony search (IBHS). This research is organized as follows. Section 2 describes the Harmony search algorithm. The proposed harmony search algorithm is described in detail in Sect. 3. Simulation of proposed harmony search algorithm is reported in Sects. 4 and 5 summarizes the contributions of this research.

2 Harmony Search

In the HS, each solution is called the harmony and is signified via a n-dimensional vector. At the beginning of the algorithm, a population of harmony vectors are randomly created in the search space and accumulated in the harmony memory (HM) [1]. A new harmony is then created based on all the existing harmonies in the HM by applying a harmony memory consideration rate (HMCR) and a pitch adjustment rule. The newly generated harmony is compared to the existing harmonies and replaces the worst one if it has a better state [1]. The algorithm repeats until it meets a given termination criterion. The three basic steps of the algorithm are described below in details [1, 18].

Step 1. Initializing the problem and algorithm parameters: the optimization problem is defined as: Minimize $f(x)(x = (x_1, x_2, \ldots, x_n)$ subject to $x_L \leq x_i \leq x_U$, $(i = 1, 2, \ldots, n)$, where $f(x)$ is the desired objective function, x is the set of each decision variable, n is the number of decision variables, and x_i is the ith decision variables, where x_L and x_U are the lower and upper limits.

In the HS, each individual, x is considered as one harmony H, and the decision variable x_i, considered as a pitch harmony H_i. The lower and upper limits of the variables (x_L and x_U, respectively) define the feasible range of the pitch, H_L and H_U, respectively. The HS parameters, harmony memory size (HMS), harmony memory considering rate (HMCR), pitch adjusting rate (PAR), and the termination criterion are set in this step. All the solution vectors are stored in a memory location, called the harmony memory (HM).

Step 2. Initializing the harmony memory: The HMS randomly generated solution vectors are stored in the HM.

$$\mathrm{HM} = \begin{bmatrix} H_1^1 & H_2^1 & \cdots & H_n^1 \\ H_1^2 & H_2^2 & \cdots & H_n^2 \\ \vdots & \vdots & \vdots & \vdots \\ H_1^{\mathrm{HMS}-1} & H_2^{\mathrm{HMS}-1} & \cdots & H_n^{\mathrm{HMS}-1} \\ H_1^{\mathrm{HMS}} & H_2^{\mathrm{HMS}} & \cdots & H_n^{\mathrm{HMS}} \end{bmatrix}$$

Step 3. Improvising a new harmony: a new harmony vector, $H^{\mathrm{new}} = (H_1^{\mathrm{new}}, H_2^{\mathrm{new}}, \ldots, H_n^{\mathrm{new}})$ is generated. The value of each decision variable is chosen randomly from all the existing corresponding values in the HM. The HMCR, which can vary between 0 and 1, is the probability of choosing one value from the existing values stored in the HM, and $(1 - \mathrm{HMCR})$ is the probability of random generation of one value from the possible range of values. Each new decision variable has a pitch adjustment process with the PAR probability. The algorithm is as follows:

1. for ($i = 1$ to n) do
2. if ($r_1 < HMCR$) then
3. $H_i^{new} = H_i^j; j \in (1,2,...,HMS)$
4. if ($r_2 < PAR$) then
5. $H_i^{new} = H_i^{new} \pm r_3 \times bw; j \in (1,2,...,HMS), r_1, r_2, r_3 \in (0,1)$
6. endif
7. else
8. $H_i^{new} = H_L + r_4 \times (H_U - H_L), r_4 \in (0,1)$
9. endif
10. endfor

where bw is an arbitrary band width.

Step 4. Updating the harmony memory: the new harmony is compared to the existing harmonies in the HM and replaces the position of the worst one if it has a better fit value.

Step 5. Termination criterion: the algorithm continues until it meets the termination criterion, which could be simply the number of improvisations (NI).

3 Improved Best Harmony Search

In this study, information of the p best harmonies in HM according to the fitness of individuals is used to improve new harmony. In the first iterations, those p harmonies are not similar and a new harmony uses the best harmonies that belong to different spaces of solutions (exploration). In the last iterations the p harmonies have near values to the optimal harmony. Therefore, in the last iterations the new solution is improved by information from solutions near to optimal solution (exploitation). The p best harmonies are called pBest and its consideration rate, pBestR, increases during the iterations:

$$p\text{BestR} = c_1 \times (\text{Iter}/\text{Iter}_{\max})$$

where c_1 is a positive constant less than one, the Iter is the current iteration and $\text{Iter}_{\max}$ is the maximum iteration.

Information of the new harmony is improved by information of the best solution in previous approaches or by pBest solutions in this study. In the case that this new harmony has better fitness, it replaces the worst harmony in the HM. Previous algorithms need more exploitation during the last iterations of algorithms. When a new harmony has better fitness than the worst harmony in HM, it replaces it. This causes solutions to move toward the best solution, rather than improve the previous best solution. Whereas, in the proposed approach the best solution is improved by a rate called BestSolR when the iteration is larger than some α percent of maximum iteration (Iter > $\alpha \times \text{Iter}_{\max}$). The new improved best solution ($H^{\text{new(Best)}}$) replaces the best solution (H^{Best}) if it has better fitness than the previous best solution, in contrast to previous approaches which consider only the worst solution. This prevents other solutions moving toward the best solution and diversity remains high during the computation. The BestSolR increases with increasing number of iterations

$$\text{BestSolR} = c_2 \times (\text{Iter}/\text{Iter}_{\max})$$

where c_2 is a positive constant less than one. The algorithm of the IBHS is as follows:

if ($r_0 < BestSolR$ and $Iter > \alpha \times Iter_{\max}$)then
    for (i = 1 to n) do
        $H_i^{new(Best)} = H_i^{Best} \pm r_0' \times bw;\ r_0, r_0' \in (0,1)$
    endfor
else
    for (i = 1 to n) do
        if ($r_1 < HMCR$) then
            $H_i^{new} = H_i^j;\ j \in (1,2,\ldots,HMS)$
            if ($r_2 < BestR$) then
                $H_i^{new} = H_i^{new} + (H_i^{pBest} - H_i^{new}) \times bw; r_1, r_2 \in (0,1)$
            elseif ($r_3 < PAR$) then
                $H_i^{new} = H_i^{new} \pm r_3 \times bw; r_1, r_3 \in (0,1)$
            end if
        else
            $H_i^{new} = H_L + r_4 \times (H_U - H_L), r_4 \in (0,1)$
        endif
    endfor
endif

If $H^{\text{new(Best)}}$ has a better fitness value than the current best harmony, H^{Best}, it replaces it. If not, H^{new} is compared with the existing harmonies in the HM and replaces the worst one if it has a better fitness value. Other steps of IBHS are the same as conventional HS.

4 Simulation Results and Discussion

Ten (F1 to F10) numerical optimization problems from the CEC05 benchmark functions [19] has been chosen to compare the performance of the proposed approach with the selected previous approaches as follows:

- original harmony search (HS) [3] (Lee and Geem 2005),
- improved harmony search (IHS) [4] (Mahdavi et al. 2007),
- differential harmony search (DHS) [20] (Chakraborty et al. 2009),
- population-based harmony search (HSpop) [21] (Mukhopadhyay et al. 2008),
- improved global-best harmony search (IGHS) [22] (El-Abd 2013).

The bias of the fitness for the CEC05 benchmark functions are subtracted from the final fitness results. The simulations were conducted on a computer with Intel core i7, 3.6 GHz processor, four cores, with 8 GB of RAM, running MATLAB.

The parameters for each algorithm in these simulations are presented in Table 1. In all experiments, the termination criterion is 30,000× dimension. The initial HM is generated randomly based on the variable ranges of each test function. Simulation results were averaged over 30 runs in 10, 30, and 50 dimensions.

The convergence behavior of the proposed algorithm (IBHS) and other algorithms for F2 and F4 are shown in Figs. 1 and 2, respectively. IBHS has better convergence to the solution for F2, where all other algorithms converged too quickly, which implies IBHS also produces a better solution, which clearly improves if allowed further iterations. For both F2 and F4, IBHS has similar or inferior performance than other algorithms for low iterations, but shows better performance as the number of iterations increases. Unlike the other approaches, PAR is high for IBHS in the first iterations and decreases linearly, bw is also high in the first iterations therefore the diversity of the algorithm will be high. In the last iterations the best solution improved by rate c_2 to exploit the optimum solution.

Table 1 Parameters setting for the proposed and previous algorithms

Algorithms	HMS	HMCR	PAR	bw	Others
HS	40	0.9	0.3	0.01	–
HS_{pop}	40	0.99	0.5	Standard deviation of population	–
HIS	40	0.95	$PAR_{min} = 0.01$ $PAR_{max} = 0.99$	$bw_{min} = 0.0001$ $bw_{max} = (H_U - H_L)/20$	–
DHS	40	0.9	–	–	$F \in \text{rand}(1, 0)$
IGHS	40	0.99	$PAR_{min} = 0.01$ $PAR_{max} = 0.99$	$bw_{min} = 0.0001$ $bw_{max} = (H_U - H_L)/20$	–
Proposed IBHS	40	0.99	$PAR_{min} = 0.01$ $PAR_{max} = 0.99$	Standard deviation of population	$c_1 = 0.1$, $c_2 = 0.5$ $\alpha = 0.7$

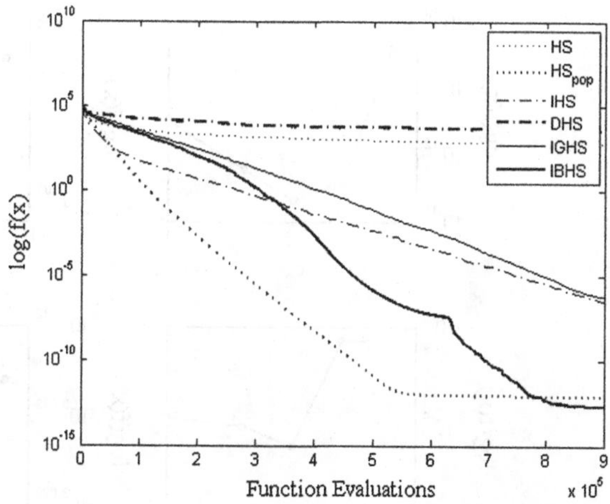

Fig. 1 Convergence behavior of proposed algorithm (IBHS) and other algorithms for F2: shifted schwefel 1.2

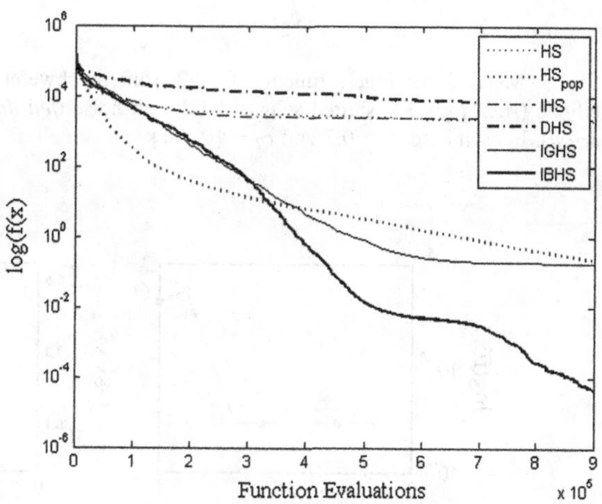

Fig. 2 Convergence behavior of proposed algorithm (IBHS) and other algorithms for F4: shifted schwefel 1.2 with noise

Figures 3 and 4 show the effects of the parameters c_2 and α for the proposed IBHS algorithm for functions F2: Shifted Schwefel 1.2, F3: Shifted Rotated Elliptic, F4: Shifted Schwefel 1.2 with Noise, and F5: Schwefel 2.6. When c_2 is small, the algorithm does not have the opportunity to exploit the solution. Alternately, when c_2 is large, the algorithm converges to a premature solution. The optimum value for c_2 is approximately 0.5.

Similar behavior is evident for α. When it is small, the algorithm tries to improve information of the best solution in the first iterations, which causes premature convergence for functions that need more diversity and exploration times such as F3: Shifted Rotated Elliptic (right top), and F5: Schwefel 2.6 (right down) in Fig. 4. When α is large, the algorithm does not have sufficient time to exploit the solution performs poorly. The optimum value of α is approximately 0.7.

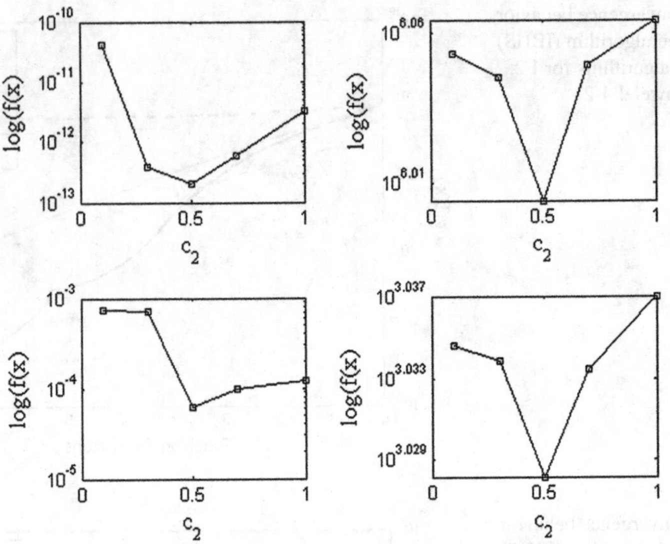

Fig. 3 Mean of the fitness function for F2: shifted schwefel 1.2 (*left top*), F3: shifted rotated elliptic (*right top*), F4: shifted schwefel 1.2 with noise (*left down*), and F5: Schwefel 2.6 (*right down*). In each case, $\alpha = 0.7$ and c_2 = {0.1–1}

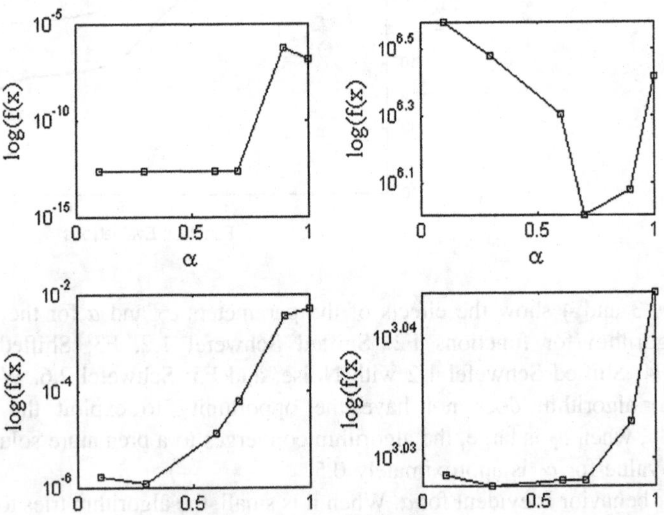

Fig. 4 Mean of the fitness function for F2: shifted schwefel 1.2 (*left top*), F3: shifted rotated elliptic (*right top*), F4: schwefel 1.2 with noise (*left down*), and F5: schwefel 2.6 (*right down*). In each case $c_2 = 0.5$ and α = {0.1–1}

5 Conclusion

The results showed that the proposed IBHS algorithm with adaptive parameter has better performance compared to previous approaches. With this approach, solutions for optimization problems by using HS could be further enhanced.

In future work, we would incorporate IBHS with a parallel framework to improve the performance of the algorithms and to investigate the use Graphics Processing Units as a parallel tool for exploiting the parallelism of the framework.

Acknowledgments This work has been supported by the Science Fund of the MOSTI–Ministry of Science, Technology and Innovation (Malaysia; Grant code: 01-01-02-SF1104).

References

1. Geem, Z.W., Kim, J.H., Loganathan, G.: A new Heuristic optimization algorithm: harmony search. Simulation **76**(2), 60–68 (2001)
2. Coelho, L.D.S., Mariani, V.C.: An improved harmony search algorithm for power economic load dispatch. Energy Conv. Manag. **50**(10), 2522–2526 (2009)
3. Lee, K.S., Geem, Z.W.: A new meta-heuristic algorithm for continuous engineering optimization: harmony search theory and practice. Comput. Meth. Appl. Mech. Eng. **194** (36), 3902–3933 (2005)
4. Mahdavi, M., Fesanghary, M., Damangir, E.: An improved harmony search algorithm for solving optimization problems. Appl. Math. Comput. **188**(2), 1567–1579 (2007)
5. Wang, L., Yang, R., Xu, Y., Niu, Q., Pardalos, P.M., Fei, M.: An improved adaptive binary harmony search algorithm. Inf. Sci. **232**, 58–87 (2013)
6. Ma, S., Dong, Y., Sang, Z., Li, S.: An improved aea algorithm with harmony search (Hsaea) and its application in reaction kinetic parameter estimation. Appl. Soft Comput. **13**(8), 3505–3514 (2013)
7. Coelho, L.D.S., Bernert, D.L.D.A., Mariani, V.C.: Chaotic differential harmony search algorithm applied to power economic dispatch of generators with multiple fuel options. In: 2010 IEEE Congress on Evolutionary Computation (CEC), pp. 1–5 (2010)
8. Kaveh, A., Ahangaran, M.: Discrete cost optimization of composite floor system using social harmony search model. Appl. Soft Comput. **12**(1), 372–381 (2012)
9. Miguel, L.F.F., Miguel, L.F.F., Kaminski, Jr, J., Riera, J.D.: Damage detection under ambient vibration by harmony search algorithm. Expert Syst. Appl. **39**(10), 9704–9714 (2012)
10. Diao, R., Shen, Q.: Feature selection with harmony search. IEEE Trans. Syst. Man Cybern. **42** (6), 1509–1523 (2012)
11. Landa-Torres, I., Ortiz-Garcia, E.G., Salcedo-Sanz, S., Segovia-Vargas, M.J., Gil-Lopez, S., Miranda, M., Leiva-Murillo, J.M., Del Ser, J.: Evaluating the internationalization success of companies through a hybrid grouping harmony search—extreme learning machine approach. IEEE J. Selected Topics Signal Process. **6**(4), 388–398 (2012)
12. Kulluk, S., Ozbakir, L., Baykasoglu, A.: Training neural networks with harmony search algorithms for classification problems. Eng. Appl. Artif. Intell. **25**(1), 11–19 (2012)
13. Mahdavi, M., Fesanghary, M., Damangir, E.: An Improved Harmony Search Algorithm for Solving Optimization Problems. Applied Mathematics and Computation. 188(2), 1567–1579 (2007).

14. Pan, Q.-K., Suganthan, P.N., Tasgetiren, M.F., Liang, J.J.: A self-adaptive global best harmony search algorithm for continuous optimization problems. Appl. Math. Comput. **216** (3), 830–848 (2010)
15. Omran, M.G., Mahdavi, M.: Global-best harmony search. Appl. Math. Comput. **198**(2), 643–656 (2008)
16. Zhao, S.-Z., Suganthan, P.N., Pan, Q.-K., Fatih Tasgetiren, M.: Dynamic multi-swarm particle swarm optimizer with harmony search. Expert Syst. Appl. **38**(4), 3735–3742 (2011)
17. Arul, R., Ravi, G., Velusami, S.: Chaotic self-adaptive differential harmony search algorithm based dynamic economic dispatch. Int. J. Electr. Power Energy Syst. **50**, 85–96 (2013)
18. Enayatifar, R., Yousefi, M., Abdullah, A.H., Darus, A.N.: Lahs: a novel harmony search algorithm based on learning automata. Commun. Nonlinear Sci. Numer. Simul. **18**(12), 3481–3497 (2013)
19. Suganthan, P.N., Hansen, N., Liang, J.J., Deb, K., Chen, Y.-P., Auger, A., Tiwari, S.: Problem definitions and evaluation criteria for the Cec 2005 special session on real-parameter optimization. KanGAL Report (2005)
20. Chakraborty, P., Roy, G.G., Das, S., Jain, D., Abraham, A.: An improved harmony search algorithm with differential mutation operator. Fund. Inform. **95**(4), 401–426 (2009)
21. Mukhopadhyay, A., Roy, A., Das, S., Abraham, A.: Population-variance and explorative power of harmony search: an analysis. In: Third International Conference on Digital Information Management ICDIM 2008, pp. 775–781 (2008)
22. El-Abd, M.: An improved global-best harmony search algorithm. Appl. Math. Comput. **222**, 94–106 (2013)

MVM: MySQL Versus MongoDB

Purva Grover and Rahul Johari

Abstract The Literature survey exhibits lack of quality research work in field of database(s) when it comes to task of performing comparison between real-world database entities. In this research paper, we have compared and contrast between the two open-source RDBMS (Relational database management system): MySQL and MongoDB. Comparison between two databases was done on basis of database operations, such as insertion, deletion, selection, projection et al. It is true that selection of the database in application depends entirely on database operations and we observed that in some database operations and applications MySQL performed better than MongoDB whereas in some applications MongoDB resulted in better performance. For the evaluation and analysis we obtained the real-time traces of diabetic dataset comprising of 100,000 records with 51 columns and put it to test for efficiency and performance to both the RDBMS and in the end the database operation execution time was recorded and analyzed.

Keywords MySQL · MongoDB · Queries · Performance of mySQL and mongoDB · Syntax of queries

Purva Grover (✉) · Rahul Johari
University School of Information and Communication Technology,
Guru Gobind Singh Indraprastha University, Drawka Sector 14, Delhi, India
e-mail: purva.usict.00416@ipu.ac.in

Rahul Johari
e-mail: rahul@ipu.ac.in
URL: http://ipu.ac.in/

M. Pant et al. (eds.), *Proceedings of Fifth International Conference on Soft Computing for Problem Solving*, Advances in Intelligent Systems and Computing 436, DOI 10.1007/978-981-10-0448-3_75

1 Introduction

1.1 Computer Applications

In most of the computer application, we require the database to archive the data for the future use and analysis. For these applications the success of the application are highly depended on the following activities: (1) Insertion of a record; (2) Selection of a particular record; (3) Querying database (4) Creation of Summary tables (5) Deletion of a particular record et al. The users of the applications are mostly concerned with these activities. In this research paper, we had taken two open-source database under GNU license viz. MySQL and MongoDB.

1.2 MySQL Databases

The word MySQL [1] can be broken down into two words My and SQL. The SQL stands for structured query language. MySQL is most widely used relational database management system. MySQL is available in two editions: (1) Open-source MySQL Community Server (2) The Proprietary Enterprise Server In this research work, we are using the Open-source MySQL Community Server for over study.

1.3 MongoDB

MongoDB [2] is the most popular NoSQL [3] database system. MongoDB comes under the document-oriented database type of NoSQL. MongoDB supports dynamic schema which makes the integration of data easier for some of the applications.

1.4 Comparisons Between MySQL and MongoDB

For making the comparisons between MySQL [1] and MongoDB [2] we are using the dataset Diabetes 130-US hospitals for years 1999–2008 Data Set—Beata Strack, Jonathan P. DeShazo, Chris Gennings, Juan L. Olmo, Sebastian Ventura, Krzysztof J. Cios, and John N. Clore, Impact of HbA1c Measurement on Hospital Readmission Rates: Analysis of 70,000 Clinical Database Patient Records, BioMed Research International, vol. 2014, Article ID 781670, 11 pages, 2014. The dataset contains the following attributes: Frequency, admission source id, number inpatient, acarbose, glimepiride pioglitazone, encounter id, time in hospital, diag 1, miglitol, metformin rosiglitazone, patient nbr, payer code, diag 2, troglitazone, metformin pioglitazone, race, medical specialty, diag 3, tolazamide, change, gender, num lab procedures, number diagnoses, examide, diabetesMed, age, num procedures, max glu serum, citoglipton, readmitted, weight, num medications, A1Cresult, insulin, admission type id, number outpatient, metformin, glyburide metformin, discharge disposition id, number emergency, repaglinide, glipizide metformin.

2 Problem Statement

Whether we choose relational database MySQL for our application or go for the NoSQL database MongoDB? We need our application to perform exceptionally well in the selection and aggregation operation. The performance of insertion and deletion should be taking medium time.

3 Related Work

In [4] author(s) had undertaken the study and analysis of the real-time disaster dataset belonging to states of India from 2010 to 2013. The data was imported from excel file into the MongoDB (NoSQL Document Database). Data was extracted from MongoDB by writing the NoSQL queries than the graphs were plotted using the extracted data. In [5] author(s) discuss about the history and the emergence of big data, how traditional DBMS could not compete with large data set and what are the issues and challenges of big data and the tools currently being used to implement and analyze the big data. In [6] author(s) gave the overview of Big data and its related components like cloud computing, distributed computing, data mining, etc. The paper introduced 3 V limitation of relational databases system which is at the roots of origin of Big Data. The paper discussed Big Data application(s), Big Data datasets and Big Data Tools. In [7] author(s) shown the analysis of large wireless network dataset which possessed real-time movement traces of the nodes. The author(s) designed relational algebra and SQL queries which were ran in the hive editor of hadoop software after importing it into hadoop software file browser against the user inputs from a web page. In [8] author(s) had beautifully contrasted between the MongoDB, NoSQL open-source database and Oracle, relational commercial database on the basis of syntaxes and query execution in milliseconds. In [9] author(s) had proposed CK-Means algorithm to analysis data stored in distributed clusters and had presented new strategy to parallelize K-Means++. They had made comparison between K-Means (existing clustering analysis algorithm), SKMeans (existing clustering analysis algorithm) and their new CK-Means algorithm. In [10] author(s) had introduced Top 10 algorithms in data mining. These top 10 algorithms covers the topics: association analysis, classification, clustering, link mining, and statistical learning of data mining. This paper describes these 10 algorithms, their impact and further research areas in algorithms.

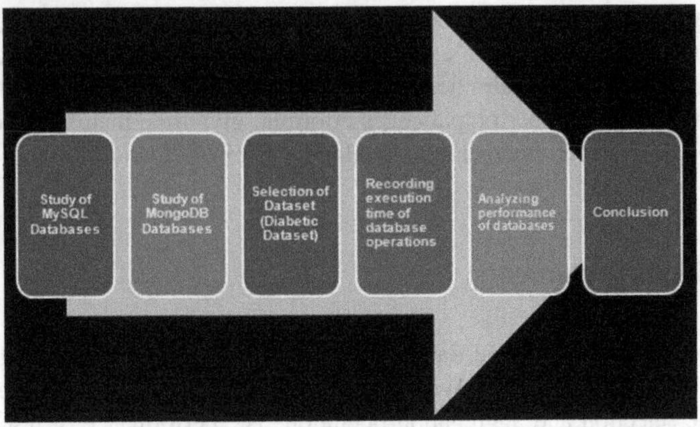

Fig. 1 Block diagram indicating the steps followed

4 Motivation and Methodology Adopted

Selecting the suitable database for the software is a big task therefore, in this research paper, we are contrasting between MySQL and MongoDB on various database operations. The flowchart in Fig. 1 gives the overview of the methodology adopted for this research paper.

5 Simulation Performed

In this research paper as discussed we had chosen two databases MySQL and MongoDB. For these two databases we had recorded the query execution time in seconds for various databases operation. Table 1 gives the execution time for creation of schema and Insertion on table. Table 2 gives the execution time for deletion in table and dropping of schema. Tables 3, 4 and 5 contains the syntax of queries of MongoDB and MySQL. Tables 6, 7 and 8 gives the execution time for 10,000, 50,000 and 100,000 records, respectively. Table 9 gives the execution time for 100,000, 50,000, 100,000 records for various number of condition. Using the values in Table 1 we had plotted the line graph in Fig. 2. Using the values in Table 3 we had plotted the line graph in Fig. 2. Using the values in Tables 6, 7 and 8 we had plotted the line and bar graphs in Figs. 3, 4, 5, 6, 7 and 8 to analysis the performance of MySQL and MongoDB.

Table 1 Creation of schema and insertion in table

Number of records	MySQL (s)		MongoDB (s)
	Creation of table	Importing data	Importing data
1	0.30	0.11	0.015
10	0.23	0.21	0.047
100	0.32	0.20	0.110
1000	0.29	0.63	0.238
5000	0.34	1.86	0.780
10,000	0.31	1.96	1.607
20,000	0.23	2.46	3.510
30,000	0.28	3.21	5.290
40,000	0.27	3.59	6.800
50,000	0.30	5.29	8.900
60,000	0.25	4.99	20.000
70,000	0.22	5.87	13.630
80,000	0.36	6.54	12.886
90,000	0.28	7.46	53.870
100,000	0.28	6.86	23.618

Table 2 Deletion in table and dropping of schema

Number of records	MySQL (s)		MongoDB (s)	
	Dropping of table	Removing of data	Dropping of collection	Removing of records
1	0.16	0.1	0.031	0.032
10	0.16	0.13	0.032	0.032
100	0.12	0.16	0.033	0.047
1000	0.12	0.16	0.032	0.203
5000	0.22	0.13	0.094	0.359
10,000	0.14	0.46	0.047	0.593
20,000	0.34	0.55	0.047	0.78
30,000	0.17	0.67	0.031	0.842
40,000	0.14	1.09	0.078	0.78
50,000	0.17	1.44	0.063	1.045
60,000	0.16	2.14	0.062	1.232
70,000	0.23	2.12	0.063	1.732
80,000	0.11	2.09	0.046	2.652
90,000	0.11	2.43	0.062	3.604
100,000	0.14	1.89	0.032	5.039

Table 3 Syntax for distinct values query

Missing values %	Column names	MySQL	MongoDB
No missing values	Gender	SELECT distinct('gender') FROM 'total10000';	db.Total10000. distinct ("gender")
1 % missing value	Diagnosis 3	SELECT distinct('diag 3') FROM 'total10000';	db.Total10000. distinct ("diag 3")
2 % missing value	Race	SELECT distinct('race') FROM 'total10000';	db.Total10000. distinct ("race")
52 % missing value	Payer code	SELECT distinct('payer code') FROM 'total10000';	db.Total10000. distinct ("payer code")
53 % missing value	Medical specialty	SELECT distinct('medicalspecialty') FROM 'total10000';	db.Total10000. distinct ("medicalspecialty")
97 % missing value	Weight	SELECT distinct('weight') FROM 'total10000';	db.Total10000. distinct ("weight")

Table 4 Syntax for summary tables query

Missing values %	Column names	MySQL	MongoDB
No missing Values	Gender	SELECT gender,Count(*) FROM 'total10000' group by gender;	db.Total10000.group({key: {gender:1},cond: {},reduce:function (curr, result) {result.total += curr. Frequency;},initial: {total: 0}})
1 % missing Value	Diagnosis 3	SELECT diag3,Count(*) FROM 'total10000' group by diag 3;	db.Total10000. group({key: {diag3: 1}, cond: {}, reduce: function(curr, result) {result.total += curr.Frequency;}, initial: {total: 0}})
2 % missing Value	Race	SELECT race,Count(*) FROM 'total10000' group by race;	db.Total10000. group({key: {race: 1}, cond: {}, reduce: function(curr, result) {result.total += curr.Frequency;}, initial: {total: 0}})

6 Result Obtained

Below are the results obtained from the above simulations:

1. Figure 2 shows that MySQL is far better than MongoDB for creation and insertion operation. For deletion operation, Till 70,000 records the performance of MongoDB is better than MySQL, but after 70,000 records the performance of MySQL is better than MongoDB.

Table 5 Syntax for selection condition query

Number of conditions	Column names	MySQL	MongoDB
1 condition	encounterid	SELECT * FROM 'total10000' WHERE 'encounterid' = "248916";	db.Total10000. find(encounterid: 248916)
2 condition	encounterid, patientnbr	SELECT * FROM 'total10000' WHERE 'encounterid' = "236316" and 'patientnbr' = "40523301";	db.Total10000. find ({encounterid: 236316,patientnbr: 40523301})
3 condition	encounterid, patientnbr, gender	SELECT * FROM 'total10000' WHERE 'encounterid' = "253380" and 'patientnbr' = "56480238" and 'gender' = "Female";	db.Total10000. find ({encounterid: 253380,patientnbr: 56480238,gender:"Female"})

Table 6 Query execution time for 10,000 records

Query executed for	MySQL (s)		MongoDB (s)	
	Distinct values	Summary tables	Distinct values	Summary tables
No missing values	0.04	0.39	0.078	0.390
1 % missing values	0.05	0.05	0.109	0.639
2 % missing values	0.05	0.03	0.078	0.406
52 % missing values	0.03	0.05	0.093	0.406
53 % missing values	0.05	0.05	0.125	0.421
97 % missing values	0.05	0.05	0.078	0.437

Table 7 Query execution time for 50,000 records

Query executed for	MySQL (s)		MongoDB (s)	
	Distinct values	Summary tables	Distinct values	Summary tables
No missing values	0.19	0.19	0.203	1.545
1 % missing values	0.09	0.14	0.359	1.716
2 % missing values	0.14	0.16	0.156	1.311
52 % missing values	0.11	0.08	0.156	1.295
53 % missing values	0.12	0.11	0.405	1.42
97 % missing values	0.14	0.16	0.375	1.28

Table 8 Query execution time for 100,000 records

Query executed for	MySQL (s)		MongoDB (s)	
	Distinct values	Summary tables	Distinct values	Summary tables
No missing values	0.23	0.25	0.25	2.434
1 % missing values	0.19	0.23	0.608	3.12
2 % missing values	0.2	0.25	0.234	2.605
52 % missing values	0.14	0.17	0.312	2.636
53 % missing values	0.22	0.23	0.811	2.855
97 % missing values	0.16	0.2	0.764	2.465

Table 9 Deletion in table and dropping of Schema

Number of records	MySQL (s)			MongoDB (s)		
	Number of conditions			Number of conditions		
	1	2	3	1	2	3
100,000	0.09	0.03	0.13	0.069	0.078	0.078
500,000	0.20	0.14	0.19	0.110	0.078	0.078
1,000,000	0.30	0.31	0.33	0.184	0.125	0.125

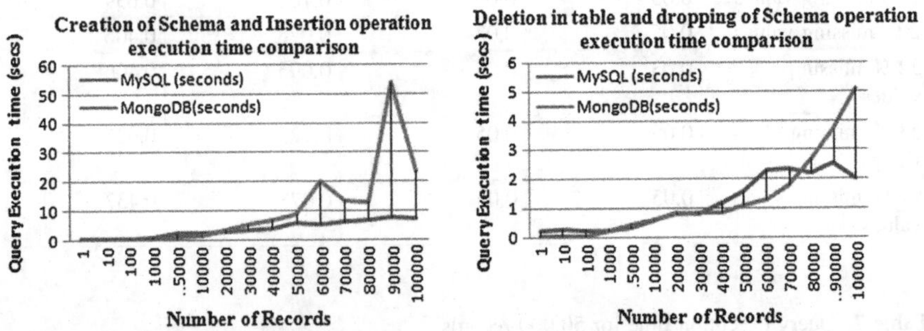

Fig. 2 MySQL and MongoDB comparisons on creation, insertion, and deletion operation

2. Figures 3 and 4 shows that the query execution time increases as the number of records increases.
3. Figure 5 shows that MySQL is better than MongoDB for distinct values query for various number of records.
4. Figure 6 shows that MySQL is better than MongoDB for summary tables queries for various number of records.
5. Figure 7 shows for MySQL selection query execution time as the number of records increases the selection query takes more time to select the records. To select the record on basis of three conditions takes more time than to select the

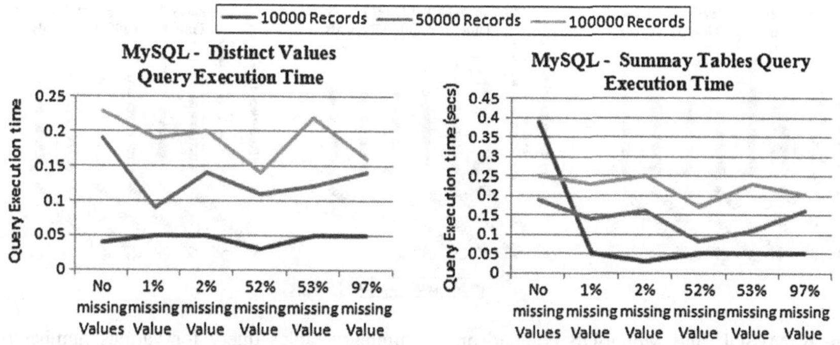

Fig. 3 MySQL database performance for various number of records

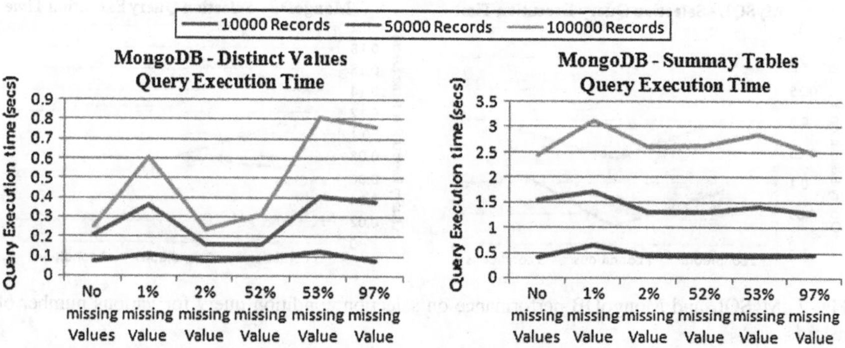

Fig. 4 MongoDB database performance for various number of records

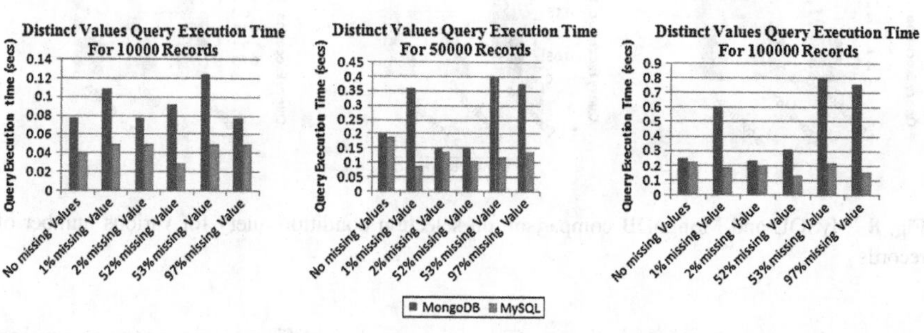

Fig. 5 MySQL and MongoDB comparison on distinct values query for various number of records

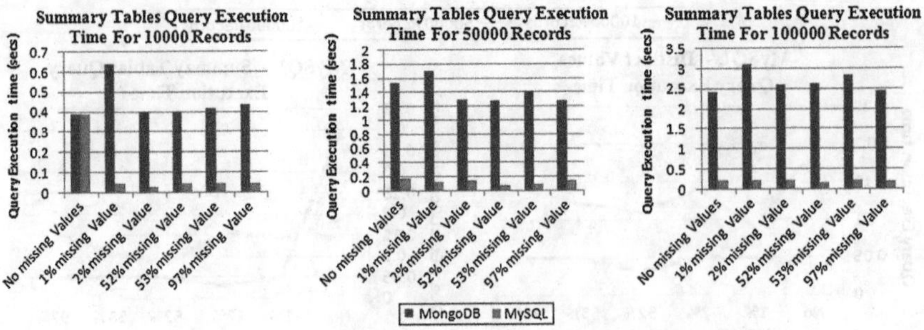

Fig. 6 MySQL and MongoDB comparison on summary tables query for various number of records

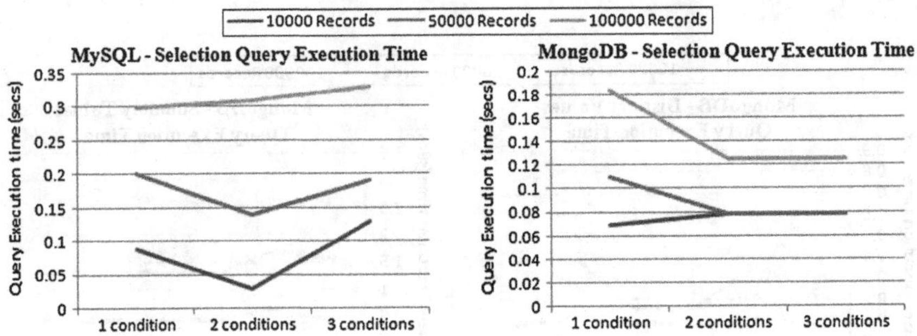

Fig. 7 MySQL and MongoDB performance on selection condition query for various number of records

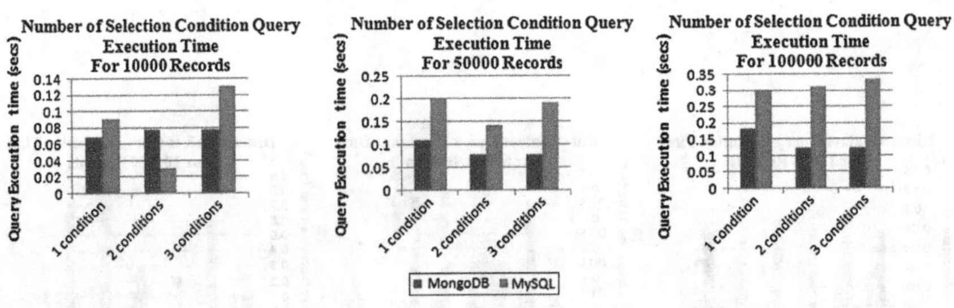

Fig. 8 MySQL and MongoDB comparison on selection condition query for various number of records

record on one or two conditions. There is very less difference in query execution time in MongoDB selection query execution time when selection is based on two and three conditions.

6. Figure 8 shows that MongoDB is better than MySQL for selection queries for various number of records.

7 Conclusion

The simulation results shows that "selection operation" is better performed in MongoDB for infinite numbers of records where as for "summary table operation" in MySQL exhibits better performance than in MongoDB. For Insertion of records there is no need for building schema in MongoDB whereas for MySQL it is necessary to built schema before inserting the records. The "Syntax for queries" is simpler and easier in MySQL as compared to MongoDB.

8 Future Work

In Future, same simulation of queries can be performed on more than 100,000 records. The simulation can also take into account more database operations. The contrast between the two open-source database technologies can be done by building the web application.

References

1. http://dev.mysql.com/downloads/mysql
2. https://www.mongodb.org/
3. https://en.wikipedia.org/wiki/NoSQL
4. Grover, P., Johari, R.: Disaster Big Data AnalysiS WIPS Symposium. In 3rd International Conference on Big Data Analytics (BDA), Delhi University, Springer, Dec 2014
5. Bhardwaj, V., Johari, R.: Big Data Analysis: Issues and Challenges. In IEEE International Conference on Electrical, Electronics, Signals, Communication and Optimization (EESCO), VIIT, Visakhapatnam, Andhra Pradesh, Jan 2015
6. Grover, P., Johari, R.: BCD: Big Data, Cloud Computing and Distributed Computing. In IEEE Global Conference on Communication Technologies (GCCT-2015) Kanyakumari, TamilNadu, April 2015
7. Bhardwaj, V., Johari, R.; Priti bhardwaj query execution evaluation in wireless network using MyHadoop. In 4th IEEE International Conference on Reliability, Infocom Technologies and Optimization (ICRITO 2015), AMITY University, Sept 2015
8. Boicea, A., Radulescu, F.: Laura Ioana Agapin: MongoDB vs Oracle—database comparison: In: 2012 Third International Conference on Emerging Intelligent Data and Web Technologies (2012)
9. Esteves, R.M., Hacker, T., Rong, C.: A new approach for accurate distributed cluster analysis for Big Data: competitive K-Means. Int. J. Big Data Intell. **1**(1/2), 50–64 (2014)
10. Wu, C., Kumar, V., Quinlan, J.R., Ghosh, J., Yang, Q., Motoda, H., McLachlan, G.J., Ng, A., Liu, B., Yu, P.S., Zhou, Z., Steinbach, M., Hand, D.J., Steinberg, D.: In: 1 Top 10 Algorithms in Data Mining. Springer, London (2007)

A Novel Single Band Microstrip Antenna with Hexagonal Fractal for Surveillance Radar Application

Shailendra Kumar Dhakad, Neeraj Kumar, Ashwani Kr. Yadav, Shashank Verma, Karthik Ramakrishnan and Jyotbir Singh

Abstract This paper aims at developing a microstrip patch antenna with hexagonal fractal pattern for a ground-based surveillance radar. The antenna performs in the *X*-band in between 8 and 9.5 GHz, applicable for short-range search. The ground plane has been varied in the design to observe its effect in the gain and VSRW parameters of the antenna. The final antenna design works at 9.2 GHz (*X*-band range) with a return loss of −28.63 dB after various stages of slotting in the ground plane depicting the effect of modification of ground plane parameters in the design.

Keywords Hexagonal · Fractal · Surveillance radar · Return loss · Voltage standing wave ratio (VSWR) · *X*-band

S.K. Dhakad (✉) · Shashank Verma · Karthik Ramakrishnan · Jyotbir Singh
Department of Electronics Electrical and Electronics, BITS Pilani KK Birla Goa Campus, Sancoale, India
e-mail: skdhakad@goa.bits-pilani.ac.in

Shashank Verma
e-mail: shashank3959@gmail.com

Karthik Ramakrishnan
e-mail: karthik.r2203@gmail.com

Jyotbir Singh
e-mail: f2012381@goa.bits-pilani.ac.in

Neeraj Kumar · A.Kr. Yadav
ASET, Amity University Rajasthan, Jaipur, India
e-mail: nkumar@jpr.amity.edu

A.Kr. Yadav
e-mail: ashwaniy2@gmail.com

M. Pant et al. (eds.), *Proceedings of Fifth International Conference on Soft Computing for Problem Solving*, Advances in Intelligent Systems and Computing 436, DOI 10.1007/978-981-10-0448-3_76

1 Introduction

Microstrip antennas are becoming increasingly popular for the ease in their fabrication and varied applications. Their application [1, 2] is particularly useful in mechanically rugged and light weight subsystems especially for radars demanding a low profile. Aircraft, missiles, ships, tanks may make use of these antennas for diverse mobile applications. *X*-band is well suited for short-range search [3, 4]. Good target resolution can be achieved when frequency band allocations are wide by use of narrow pulses with wide emission bandwidth.

Symmetric fractal patterns [5–8] increase the surface current path length. This is in contrast to the conventional square patch antenna which has a higher resonant frequency. This modification [9, 10] has been added to the design to achieve the desired characteristic.

X-band frequency band has frequency range of 8.0–12.0 GHz, being used in weather monitoring, defence communication, air traffic control, space communication, maritime tracking, and high resolution imaging.

The effect of addition of slots [11, 12] in the ground plane, on bandwidth parameters and on gain, is analyzed in this paper. Optimization of the design for implementing the 9.2 GHz *X*-band antenna design for surveillance radar systems and examining the advantages of adding these slots has been further discussed in this paper toward the end.

2 Antenna Configuration and Design

The proposed antenna is shown in Fig. 1. It is etched on an economical FR4-epoxy substrate which has a thickness = 1.6 mm, permittivity = 4.4, and loss tangent = 0.002. In order to increase the bandwidth, fractals are made on the hexagonal patch design. An edge-based rectangular feed line of 50 Ω, with dimensions 22 mm × 1.66 mm, is used to feed the system.

Table 1 depicts the dimensions of the antenna proposed. When the length of edge of the patch is changed, it results in shifting of the operating frequency, with increasing the length of the patch corresponding to a decrease in frequency, and vice versa.

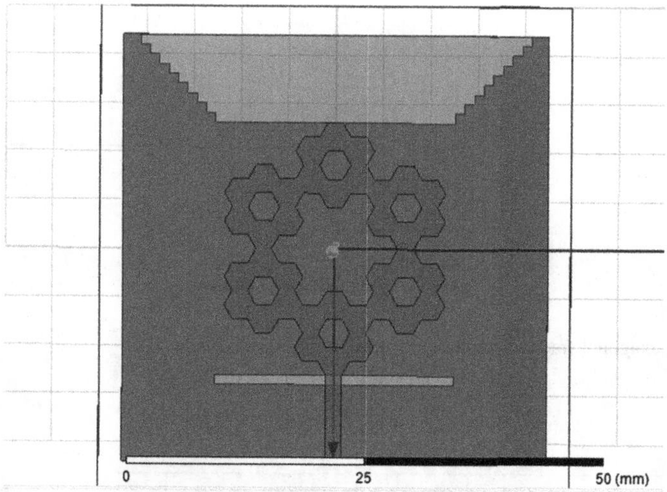

Fig. 1 Top view of Microstrip antenna

Table 1 Antenna parameters

Parameters	Value (mm)
Substrate length	44
Substrate width	44
Hor. Slot length	25.33
Hor. Slot width	1
Step of ladder	1
Hexagon length	1.666

3 Results and Discussion

The proposed fractal slotted hexagonal patch antenna has been simulated in Ansoft high frequency structure simulator (HFSS v13.0). The impedance bandwidth, VSWR, gain, and radiation patterns of the antenna are measured. To investigate the effects of different parameters of the proposed antenna, a parametric study has been done. Five prototypes of the proposed antenna, with different slot shapes in its ground plane are characterized as follows (Figs. 2, 3, 4, 5 and 6):

a. S_{11} *Parameter Rectangular Plots*

Design 1 (Fig. 7).
Design 2 (Fig. 8).
Design 3 (Figs. 9, 10 and 11).

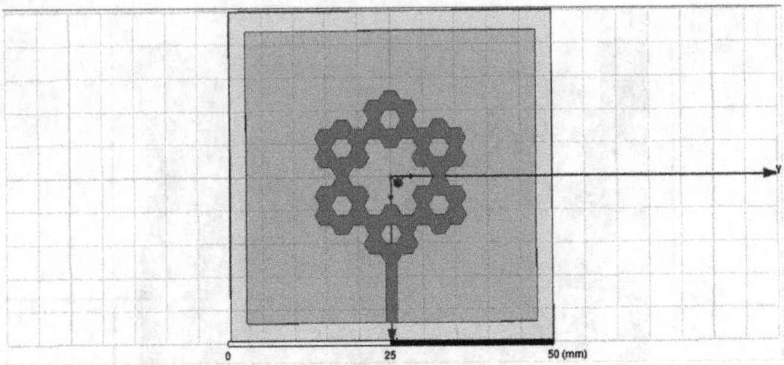

Fig. 2 A hexagonal fractal antenna with no slots in ground plane

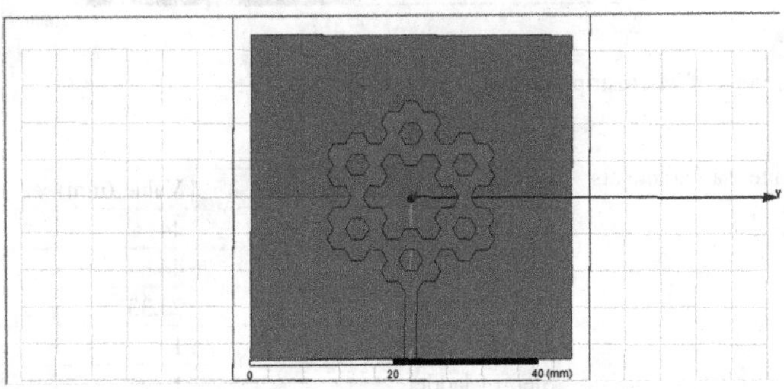

Fig. 3 A hexagonal fractal antenna with only half ground plane

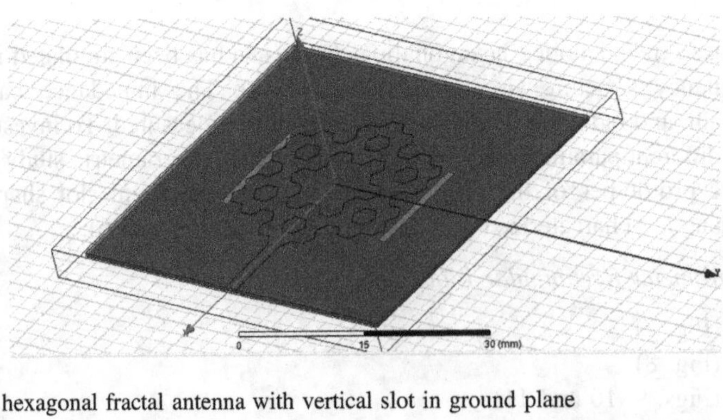

Fig. 4 A hexagonal fractal antenna with vertical slot in ground plane

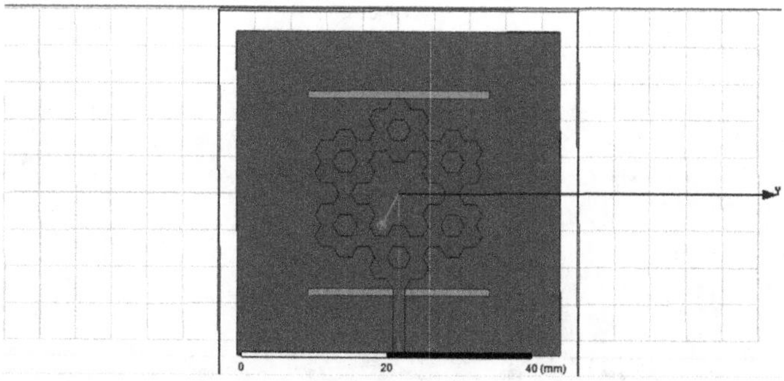

Fig. 5 A hexagonal fractal antenna with horizontal slots in ground plane

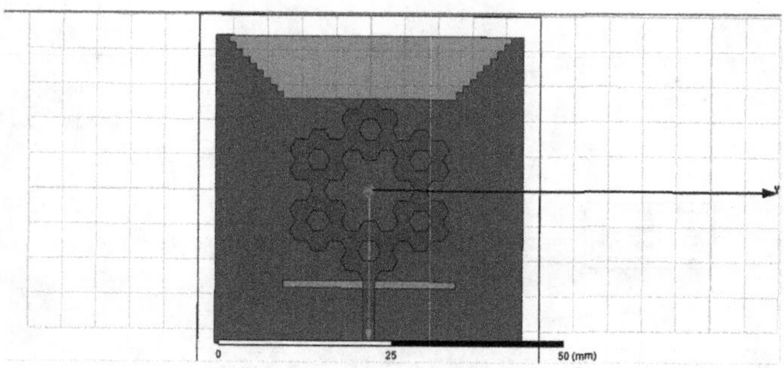

Fig. 6 A hexagonal fractal antenna with a horizontal slot and a step pattern slot in the ground plane

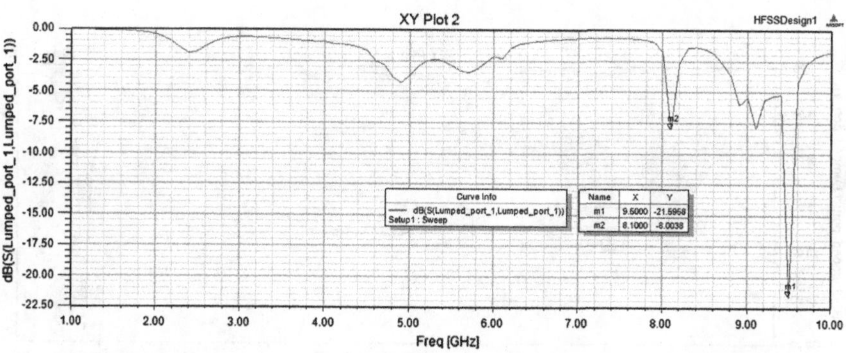

Fig. 7 Return loss result for designed hexagonal fractal antenna with no slots in ground plane

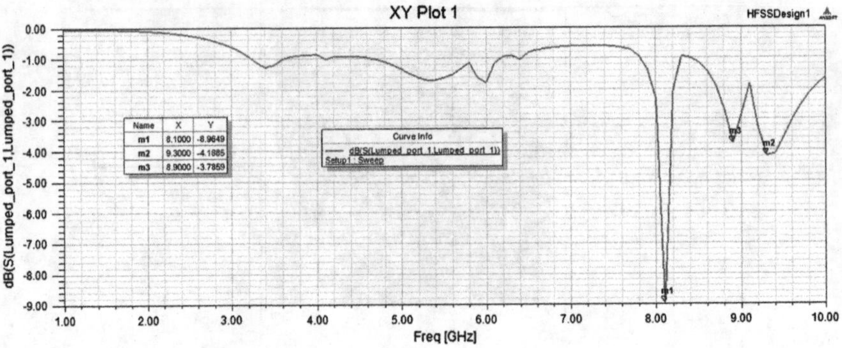

Fig. 8 Output return loss for hexagonal fractal antenna with only half ground plane

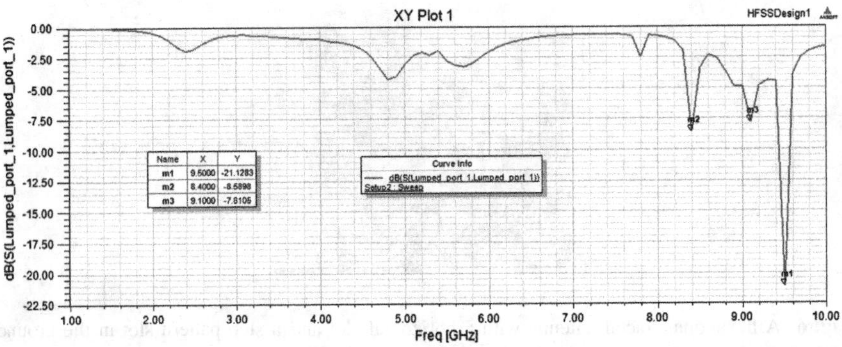

Fig. 9 Return loss for hexagonal fractal antenna with vertical slot in ground plane

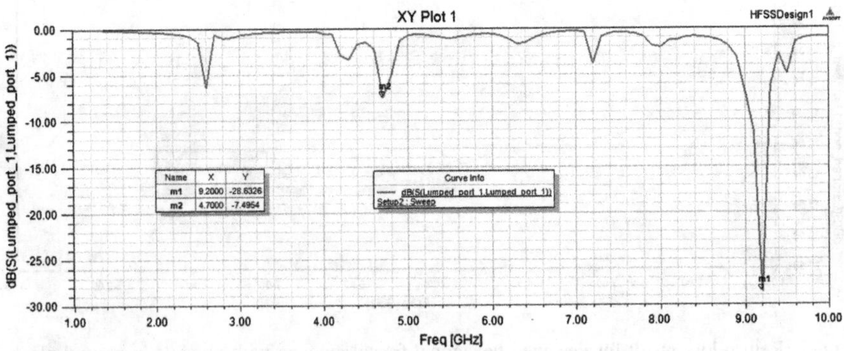

Fig. 10 Return loss for fractal antenna with horizontal slots in ground plane

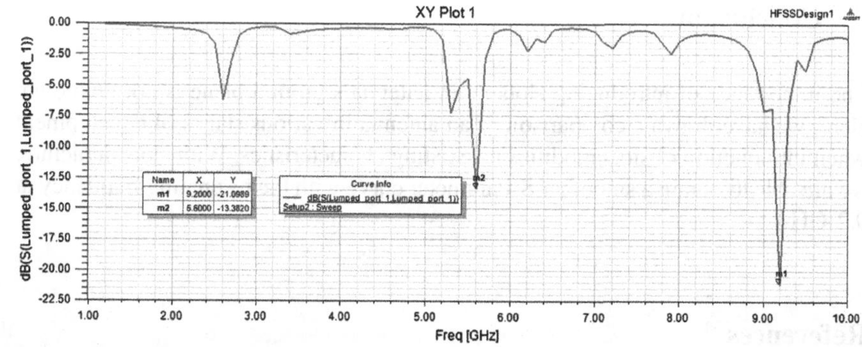

Fig. 11 Return loss for fractal antenna with a horizontal slot and a step pattern slot in the ground plane

(a) *With the insertion of slots it is observed that the fundamental resonant frequency is lowered.* It is observed that Design 1, which has no slots in the ground plane has a frequency of 9.5 GHz. With the insertion of slots in Designs 2, 3, 4, and 5, the resonant frequency decreases to a lower value. Hence, we can get lower ranges of frequency with the addition of slots in the ground plane.

(b) *With insertion of slots, gain of the antenna improves.* The first design which has no slots in its ground plane has a gain of −21.95 dB and as the slots are inserted in a specific manner the gain improves in some cases and reduces for the others. A total of 4 more design have been made where the gain increases for Design 4 and 5 and reduces for the others.

(c) *The range of the radar improves with insertion of slots in the ground plane.* As the ground plane slots are varied, the range of the surveillance varies and we are able to detect a much diverse range hence improving our application in radar technology.

(d) The VSWR parameter is a measure that numerically describes how well the antenna is impedance matched to the surveillance radar system it is connected to. *In the proposed antenna, The VSWR value reduces and approaches 1 as the slots are introduced.* Where the impedance of the radio and transmission line is well matched to the antenna's impedance (Table 2).

Table 2 Performance comparison of the proposed antenna with different slots length in ground plane

	Frequency (GHz)	Gain (dB)	VSWR
Design 1	9.5	−16.95	1.18
Design 2	8.1	−8.96	2.10
Design 3	9.43	−20.65	1.16
Design 4	9.2	−22.63	1.07
Design 5	9.2	−25.89	1.02

4 Conclusion

The importance of introducing slots in an antenna's ground plane is not unknown. Here, we have drawn a comparison of an antenna in various stages of development when introduction of slots modifies its existing characteristics. With slots in picture, we are able to obtain a smoother S-Parameter curve with the resonating frequency at 9.2 GHz.

References

1. Maryam, M., Ghobadi, C., Nourinia, J., Poorahmadazar, J.: Small monopole antenna with modified slot ground plane for UWB applications. 20th Iranian Conference on Electrical Engineering (ICEE2012) (2012)
2. Chuang, H.-R.: A 3-8-GHz broadband planar triangular sleeve monopole antenna for UWB communication. In: 2007 IEEE Antennas and Propagation International Symposium, May 2007
3. Liang, J., Chiau, C.C., Chen, X., Parini C.G.: Study of a printed circular disc monopole antenna for UWB systems. IEEE Trans. Antennas Propag. **53**(11), 3500–3504 (2005)
4. Immoreev, I., Fedotov, D.V.: Ultra wideband radar systems: advantages and disadvantage. In: IEEE Conference of Ultra Wideband Systems and Technologies (UWBWST) Digestive Technology Papers, pp. 201–205, May 2002
5. FCC, Washington, DC, USA, First report and order on ultra-wideband technology (2002)
6. Jensen, M.A., Wallace, J.W.: A review of antennas and propagation for MIMO wireless communication. IEEE Trans. Antennas Propag. **52**(11), 2810–2824 (2004)
7. Ramavath Ashok, K., Choukiker, Y.K., Behera, S.K.: Design of hybrid fractal antenna for UWB application. 2012 International Conference on Computing Electronics and Electrical Technologies (ICCEET), (2012)
8. Lin, C.-C., Kan, Y.-C., Kuo, L.-C., Chuang, H.-R.: A planar triangular monopole antenna for UWB communication. IEEE Microwave Wirel. Compon. Lett. **15**(10), 624–626 (2005)
9. Chung, K., Hong, S., Choi, J.: Ultrawide-band printed monopole antenna with band-notch filters. Microw. Antennas Propag. **1**(2), 518–522 (2007)
10. Amiri, Sh., Ojaroudi, N., Geran, F., Ojaroudi, M.: A novel and compact monopole antenna with band-stop performance for UWB applications. In: 2012 20th Telecommunications Forum (TELFOR) (2012)
11. Ojaroudi, M., Yazdanifard, S., Ojaroudi, N., Naser-Moghaddasi, M.: Small square monopole antenna with enhanced bandwidth by using inverted T-Shaped slot and conductor-backed plane. IEEE Trans. Antennas Propag. **59**(2), 670–674 (2011)
12. Li, P., Liang, J., Chen, X.: Study of printed elliptical/circular slot antennas for ultra wideband applications. IEEE Trans. Antennas Propag **54** (6), 1670–1675 (2006)

Optimization of Hyperspectral Images and Performance Evaluation Using Effective Loss Algorithm

Srinivas Vadali, G.V.S.R. Deekshitulu and J.V.R. Murthy

Abstract An effective lossy algorithm for compressing hyperspectral images using singular value decomposition (SVD) and discrete cosine transform (DCT) has been proposed. A hyperspectral image consists of a number of bands where each band contains some specific information. This paper suggests compression algorithms that compress the hyperspectral images by considering image data, band by band and compress each band employing SVD and DCT. The compression performance of the resultant images is evaluated using various objective image quality metrics.

Keywords Hyperspectral images · Effective loss algorithm · Singular value decomposition · DCT

1 Introduction

Hyperspectral images are obtained from airborne to space-borne sensors, thereby; transmit to base station for processing, while onboard systems exhibit limited storage and power. Compression algorithms with high performance and low

Srinivas Vadali (✉) · J.V.R. Murthy
Department of Computer Science and Engineering, University College of Engineering, Jawaharlal Nehru Technological University Kakinada, Kakinada, Andhra Pradesh, India
e-mail: vadalisrinivas16@gmail.com
URL: http://www.springer.com/aisc

J.V.R. Murthy
e-mail: mjonnalagedda@gmail.com

G.V.S.R.Deekshitulu
Department of Mathematics, University College of Engineering, Jawaharlal Nehru Technological University Kakinada, Kakinada, Andhra Pradesh, India
e-mail: dixitgvsr@hotmail.com

M. Pant et al. (eds.), *Proceedings of Fifth International Conference on Soft Computing for Problem Solving*, Advances in Intelligent Systems and Computing 436, DOI 10.1007/978-981-10-0448-3_77

complexity are needed to compress image data [1]. The compression techniques are classified as lossless and lossy. The lossy compression achieves higher compression ratios when compared to lossless compression algorithms [2]. Various image compression algorithms are found in the literature for both lossy and lossless compression techniques. Cheng[] proposed an improved version of EZW (Embedded Image Coding using Zero trees of Wavelet Co-efficient) algorithm for compressing the AVIRIS images (Airborne Visible/Infrared Imaging Spectrometer) to perform compression ratio for both lossless and lossy compression [1]. A Joint KLT (Karhunen–Loeve Transform) Lasso algorithm for compressing hyperspectral images was proposed by Simplice et al. [2]. Cheng introduced a compression method that uses a hybrid transformation which includes integer Karhunrn–Loeve transformation (KLT) and integer discrete wavelet transformation (DWT) [3]. A low-complexity compression algorithm for hyperspectral images based on distributed source coding multi-linear was suggested by Nian and Wan [4]. A new lossy image compression technique which uses singular value decomposition (SVD) and wavelet difference reduction was discussed in [5]. SVD is numerical technique which is used to diagonalize matrices [6]. Medical images were compressed by discrete cosine transform (DCT) spectral similarity strategy in [7]. A hybrid approach for image compression using neural networks, vector quantization and DCT is suggested in [8] to achieve good energy compaction. In [9], DCT is employed to compress the colour images, while the fractal image compression is employed to evade the repetitive compressions of analogous blocks. A modified coding framework based on H.264/SVC is employed in [10]. The same can be applied in the compression of hyperspectral images. Various other methods also have been suggested in the literature, but they are found to be complex and generally not recommended for hyperspectral image compression. In this paper, it is proposed two lossy image compression algorithms, namely SVD and DCT to compress the hyperspectral images by extracting band by band. SVD is used in image compression to achieve high quality with less computational complexity [11]. DCT is another popular compression technique, to exhibit high correlated signals with respect to performance of optimum transform KLT [7]. The remainder of the paper is organized as follows: While Sect. 2 discusses how SVD works followed by DCT in Sect. 3, Section IV presents experimental results followed by conclusion in Section V.

1.1 Hyperspectral Image Compression Using SVD

Singular value decomposition (SVD) is a technique implemented on various image processing applications [12]. SVD can be executed on any arbitrary, square, rectangular, reversible or non-reversible matrix of $M \times N$ size. SVD is a numerical technique that performs diagonalization on matrices. The pixel values are stored in a matrix to perform operation. However, SVD compression offers good image quality, though compression ratio is quite low. The transformed image is not compressed, but the data is created by product expansion using singular values. Initially, singular value maintains maximum information and gradually reduces with advancement of subsequent singular values. As inferred above, initial singular values are adopted to portray the image, than by snubbing subsequent singular values. Upon comparing reconstructed image with original image, there is no adequate information loss [6]. Finer the initial singular values, the better the quality and worse the compression, vice versa. The purpose of (SVD) is factor matrix A into

$$A = \mathrm{USA}^T \tag{1}$$

where U and V are orthonormal matrices and S is a diagonal matrix which contains singular values. The singular values are appearing in descending order along the main diagonal of S.

The block diagram in Fig. 1 shows how SVD works, on each band of the Hyperspectral images.

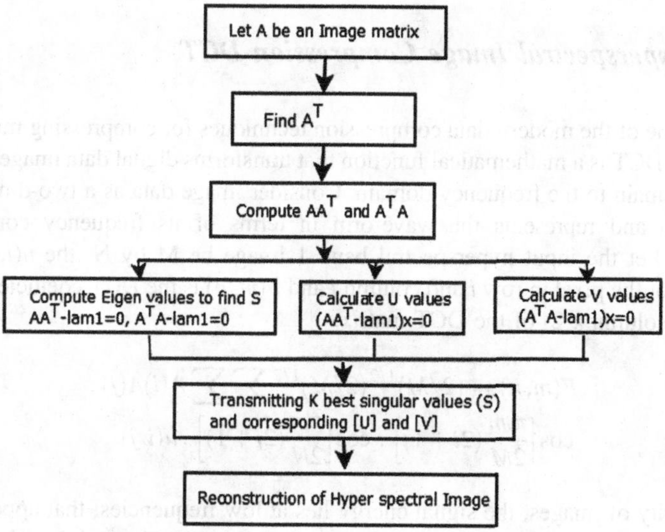

Fig. 1 Block diagram for SVD image compression

```
CIR= multibandread('paris.lan',[512,512,7], 'unit=>unit8',128, 'bil',
'ieee-le', {'Band','Direct',[1]});
imwrite(CIR,'D:/band1.jpg');          //Band 1 image

Then the SVD is computed as follows
i= imread('D:/band1.jpg');  //Read Band 1 image
[u d v] = svd(i);           //Compute SVD

The first K larger values are included in the resultant compressed image as follows
k=input('enter k');                  //include first larger K values
for j=1:k
b=b+d(j,j)*u(i,j));
end
imshow(b);                       //display the compressed image
```

Fig. 2 Block diagram for SVD image compression

The SVD algorithm has been implemented in MATLAB by appending *K* larger values (and dropping the smaller values), to yield final compressed values/image. The following function is used in MATLAB to read the hypercube band by band (Fig. 2).

1.2 Hyperspectral Image Compression DCT

DCT is one of the modern data compression techniques for compressing multimedia data. The DCT is a mathematical function that transforms digital data image from the spatial domain to the frequency domain. Consider image data as a two-dimensional waveform and represents the waveform in terms of its frequency components [13, 14]. Let the input hyperspectral band 1 image be M by N, the $u(i, j)$ is the intensity of the pixel in row i and column j and $F(m, n)$ is the DCT coefficient in row $k(1)$ and column $k(2)$ of the DCT matrix.

$$F(m,n) = (2/M)^{1/2}(2/M)^{1/2}\sum\sum A(i)A(j)\,.\\ \cos\left[\frac{\pi.m}{2M}(2i+1)\right].\cos\left[\frac{\pi.n}{2N}(2j+1)\right].u(i,j) \tag{2}$$

for majority of images, the signal energy lies at low frequencies; that appear in the upper left corner of the DCT. Compression is achieved by lower right values which constitute to higher frequencies, with slight visible distortion. The DCT input is an

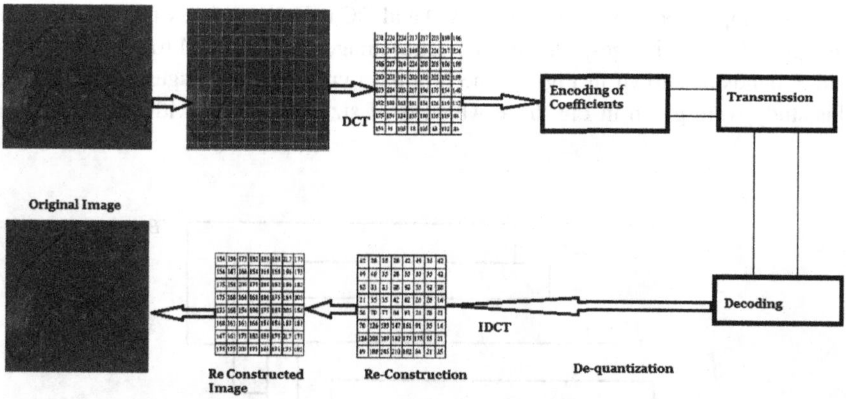

Fig. 3 Block diagram for DCT image compression

Fig. 4 Block diagram for DCT image compression

```
//Read the Hyper-spectral image to be compressed
        T = dctmtx(8);
        B = blkproc(I,[8 8],'P1*x*P2',T,T');
            //Create a 8*8 matrix for mask
        B2 = blkproc(B,[8 8],'P1.*x',mask);
        I2 = blkproc(B2,[8 8],'P1*x*P2',T',T);
        imshow(I2);
```

8 by 8 array of integers. Array contains pixels where each pixel grey scale varies from 0 to 255, i.e., 8 bits [15].

The block diagram for DCT based hyperspectral image compression is given Fig. 3 with details of typical pixel values. The image is divided into 8 × 8 blocks resulting in 64 pixels.

The following MATLAB code segment is used to compress each band of the hyperspectral image using DCT (Fig. 4).

2 Experimental Results

The hyperspectral image is compressed using SVD and DCT, quality of image is evaluated by different image quality metrics. Landsat images of.lan format are used for the analysis of hyperspectral images. The.lan file extension is associated with the ERDAS image geospatial data authoring software for Microsoft Windows Operating Systems and it stores it as a raster image. The Landsat images considered in this study were Limbe.lan, Port Moresby.lan, Tokyo.lan, Addis Ababa.lan and Madang.lan. The hypercube of each of images is analysed band by band, and each band is compressed separately using SVD and DCT. Both are lossy methods, which are applied on 2D images. The hyperspectral image consists of M rows, N columns and T bands as shown in Fig. 5. The original hyperspectral images considered in this study were given in Fig. 6a–e which is of size 512 × 512 with seven bands.

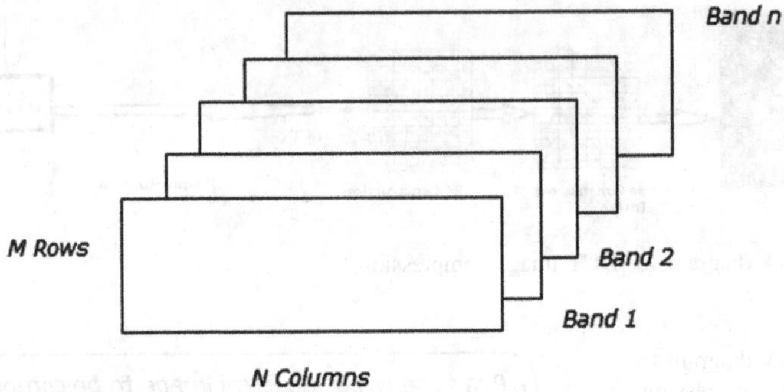

Fig. 5 Block diagram for DCT image compression

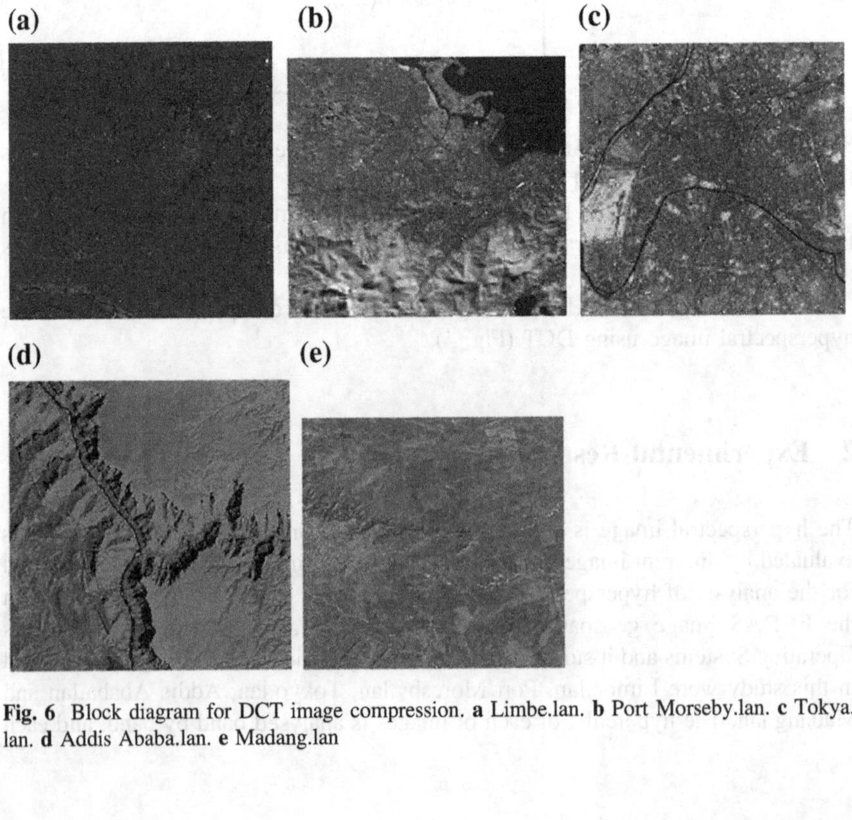

Fig. 6 Block diagram for DCT image compression. **a** Limbe.lan. **b** Port Morseby.lan. **c** Tokya.lan. **d** Addis Ababa.lan. **e** Madang.lan

2.1 *Image Quality Measures Used in This Study*

The compressed images are evaluated using performance metrics, such as mean square error (MSE), peak signal-to-noise ratio (PSNR), normalized cross correlation (NCC), structural content (SC), maximum difference (MD) and normalized absolute error (NAE). Table 1 compares size of images before and after the SVD and DCT compression. Table 2 lists values of various quality measures that are estimated after the SVD and DCT compression.

Mean Square Error (MSE) Mean square error measures the error with respect to the centre image values, i.e. the mean of the pixel values of the image, and by averaging the sum of squares of the error between the two images.

$$\mathrm{MSE} = \frac{1}{\mathrm{MN}} \sum \sum (u(i,j) - v(i,j))^2 \quad (3)$$

where $u(i, j)$ and $v(i, j)$ represent two images of size $M \times N$, u is the original image and v is the reconstructed image. A lower value of MSE signifies lesser error in the reconstructed image [16, 17].

Peak Signal-to-Noise Ratio (PSNR) A PSNR of zero can be obtained if the image is completely white and other is completely black (or vice versa). Higher values imply better image [16, 17]. It is measured in decibels (db).

$$\mathrm{PSNR} = 20 \log 10 \left(\frac{L^2}{\frac{1}{\mathrm{MN}} \sum \sum (u(i,j) - v(i,j))^2} \right) \quad (4)$$

where L is the number of grey levels in the image.

Table 1 Comparision of image size

Name of the hyperspectral image Size-1793		Original image size	
Limbe.lan(Band-1)	46 KB	41 KB	24 KB
Port Morseby.lan(Band-1)	26 KB	25 KB	17 KB
Tokyo.lan(Band-1)	32 KB	29 KB	19 KB
Addis Ababa.lan(Band-1)	28 KB	26 KB	19 KB
Manding.lan(Band-1)	37 KB	36 KB	23 KB

Table 2 Image quality measures

Limbe.lan		PortMoresby.lan		Tokyo.lan	
SVD	DCT	SVD	DCT	SVD	DCT
35.81	54.48	6.51	10.3	9.6	15
32.59	30.76	40	37.96	38.0	36.0
0.99	0.99	0.99	0.99	0.99	0.99
1.002	1.004	1.001	1.002	1.02	1.034
51.0	95.0	29.0	61.0	29.0	97.0
0.047	0.047	0.026	0.028	0.031	0.037

Normalized Cross Correlation (NCC) NCC gives the structural content of the image [17, 18].

$$\mathrm{NCC} = \sum\sum u(i,j)v(i,j) / \sum\sum u(i,j)^2 \tag{5}$$

Structural Content (SC) Measures the image similarity of an $M \times N$ matrix image, the formula for finding structural content variation factor is:

$$\mathrm{SC} = \sum\sum (u(i,j) / \sum\sum v(i,j)) \tag{6}$$

where SC is the structural content factor of two images, $u(i, j)$ pixel values of original image before compression and $v(i, j)$ pixel values of degraded image after compression [18].

Maximum Difference (MD) Measures to find the difference between the original and compressed image [18]. The higher the difference, the lesser will be the image quality.

$$\mathrm{MD} = (|u(i,j) - v(i,j)|) \tag{7}$$

Normalized Absolute Error (NAE) To find the difference between the original and reconstructed image [18]. Lesser the error, higher will be the quality.

$$\mathrm{NAE} = \sum\sum |u(i,j) - v(i,j)| / \sum\sum |v(i,j)| \tag{8}$$

The original size of Hyperspectral image is 1793 KB. When decomposed band by band, there is a decrease in file size of resultant image. Refering Table 2, the MSE and MD values vary significantly, but the values of PSNR, NCC, SC and NAE are not affected.

3 Conclusion

This paper introduced two methods for band-by-band compression of hyperspectral images using SVD and DCT techniques. Although these techniques achieve compression for images in each band with substantial reduction in size, the study found that not all the suggested objective quality metrics perform well in terms of estimating the quality of the compressed hyperspectral images. Moreover, the quality of image significantly depends on the type of the application in which the image is used. Further methods could be explored to combine the band wise compression of hyperspectral images and to achieve better quality along with the desired compression.

References

1. Cheng, K., Dill, J.: Hyper spectral images lossless compression using the 3D binary EZW algorithm. In: Proceedings of the SPIE 8655, Image Processing: Algorithms and Systems XI, 865515, Feb 19, 2013. doi:10.1117/12.2002820
2. Alissou, S.A., Zhang, Y.: Hyper spectral data compression using lasso algorithm for spectral decorrelation. In: Proceedings of the SPIE 9124, Satellite Data Compression, Communications, and Processing X, 91240A, May 22, 2014. doi:10.1117/12.2053265
3. Cheng, K., Dill, J.: An improved EZW Hyper spectral Image compression. J. Comput. Commun. **2**, 31–36. doi:10.4236/jcc.2014.22006
4. Nian, Y., He, M., Wan, J.: Low-Complexity compression algorithm for hyper spectral images based on distributed source coding. Math. Prob. Eng. **2013**, Article ID 825673, 7 pp. (2013)
5. Anbarjafari, G. et al.: Lossy image compression using singular value decomposition and wavelet difference reduction. Digital Signal Processing (Impact Factor: 1.92). Sep 2013. doi:10.1016/j.dsp.2013.09.008
6. Jayaraman, S., Sakirajan, S.E., Veera Kumar, T.: Digital image processing. Tata McGraw-Hill Education Private Ltd (2009)
7. Wu, Y-G., Tai, S-C.: Medical image compression by discrete cosine transform spectral similarity strategy. IEEE Trans. Inf. Technol. Biomed. **5**(3), 236, 243 (2001)
8. Mohamed Zorkany, E.l.: A hybrid image compression technique using neural network and vector quantization with DCT. Adv. Intell. Syst. Comput. **233**, 233–24 (2014)
9. Rawat, C.S., Meher, S.: A hybrid image compression scheme using DCT and fractal image compression. Int. Arab J. Inf. Technol. **10**(6), 553–555 (2013)
10. Balaji, L., Thyagharajan, K.: H.264/SVC Mode decision based on mode correlation and desired mode list. Int. J. Autom. Comput. **11**(5), 510–516 (2008). ISSN:1476–8186
11. Kahu, S., Rahate, R.: Image compression using singular value decomposition. Int. J. Advancements Res. Technol. **2**(8), (2013)
12. Sadek, R.A.: SVD based image processing applications: State of the Art, contributions and research challenges. Int. J. Adv. Comput. Sci. Appl. **3**(7), (2012)
13. Watson, A.B.: Image compression using the discrete cosine transform. Math. J. **4**(1), 81–88 (1994)
14. Zhou, X.H.: Research on DCT-based image compression quality. Cross Strait Quad-Regional Radio Sci. Wireless Technol. Conf. (CSQRWC) **2**, 1490–1494 (2011)
15. Cabeen, K., Gent, P.: Image compression and the discrete cosine trans form, Math 45, College of the Redwoods
16. Maruthi, R., Sankarasubramanian, K.: Assessing the blurred image quality using some uni-variate and bi-variate measures, IJCECA-SERC-DST `ISSN 0974-4983, Spring Edition 2010, pp. 32–38, vol. 02, Issue 03, Scientific Engineering Research Corporation
17. Naidu, V.P.S., Raol, J.R.: Pixel level image fusion using wavelets and PCA. Defence Sci. J. **58**(3), 338–352 (2008)
18. Desai, D., Kulkarni, L.: A quantitative comparative study of analytical and iterative reconstruction technique. Int. J. Image Process. (IJIP) **4**(4), (2010)

Comparative Study of Bakhshālī Square Root Method with Newton's Iterative Method

Nidhi Handa, Tarun Kumar Gupta and S.L. Singh

Abstract The study is aimed at comparing the convergence of Newton's iterative method and Bakhshālī square root (BSR) method. It is shown that BSR procedure naturally leads to a superfast computation of the square root problems under study. It was then concluded that of the two methods considered, BSR method is the most effective. It has been shown by comparing Newton's iterative method with BSR method with a suitable example.

Keywords Bakhshālī manuscript (BM) · Newton's formula · Āryabhaṭa's digital square root method

1 Introduction

Bakhshālī manuscript (BM) is the name given to the mathematical work written on birch bark and was found in 1881 A.D. in the village of Bakhshālī, approximately 80 km northeast of Peshawar (then in India, now in Pakistan). The manuscript was edited with an elaborate introduction by a British scholar G.R. Kaye and published in 1927. Another edition has been brought out more recently by a Japanese scholar Takao Hayashi as his doctoral study (under David Pingree, USA) was granted the

Nidhi Handa (✉)
Department of Mathematics, Gurukula Kangri Vishwavidyalaya,
Haridwar 249404, Uttrakhand, India
e-mail: nidhi_6744@yahoo.com

T.K. Gupta
Department of Mathematics and Statistics, Gurukula Kangri Vishwavidyalaya,
Haridwar 249404, Uttrakhand, India
e-mail: tarunkumar.iitr2008@gmail.com

S.L. Singh
Emeritus Fellow, UGC (Ex.), Department of Mathematics and Statistics,
Gurukula Kangri Vishwavidyalaya, Haridwar, India

M. Pant et al. (eds.), *Proceedings of Fifth International Conference on Soft Computing for Problem Solving*, Advances in Intelligent Systems and Computing 436, DOI 10.1007/978-981-10-0448-3_78

Ph.D. degree in USA and the book was published in 1995 in Groningen (Europe). R.C. Gupta says that it is Indian mathematics, however, India is not involved [2, 6].

According to the Bombay Govt. Gazette of 13th August 1881 [5]. Subsequently, the manuscript was sent to Dr. Rudolf Hoernle the head of the Calcutta schools for examination and publication. Dr. Hoernle published some parts of it in the Indian Antiquary pages 33–48 and 275–279 Bombay 1888. Later, the whole manuscript, as is the fate of almost every old Indian manuscript, was taken to England and is now the property of the Bodlian library Oxford. G.R. Kaye has edited this Bakhshālī manuscript in three parts, the first two parts are published in 1927 and the third part in 1933 on behalf of the Archaeological survey of India, New Imperial series Vol. XLIII [10].

The manuscript is written in Sāradā characters on leaves of birch bark (birch tree is known as the bhūrja tree) and consists of some 70 leaves (folios). According to Datta, the work is not a treatise on mathematics in the true sense but a running commentary on some earlier original text. Thus, we must distinguish between [3, 4].

- The date of the original treatise consisting of the Sūtras and examples only;
- The date of the commentary which presents BM work and which consists of rules, examples, solution of the examples, verification, etc.;
- The date of the present copy of the manuscript which may be quite late and involves many scribes.

According to Plofker, in mathematical notation the BM is unique and extremely precious as a source of direct information about how medieval Sanskrit mathematics was actually written. Unsurprisingly, numerals are invariably in decimal place value form with zero represented by a round dot [8].

1.1 Location of Bakhshālī [11]

See Fig. 1.

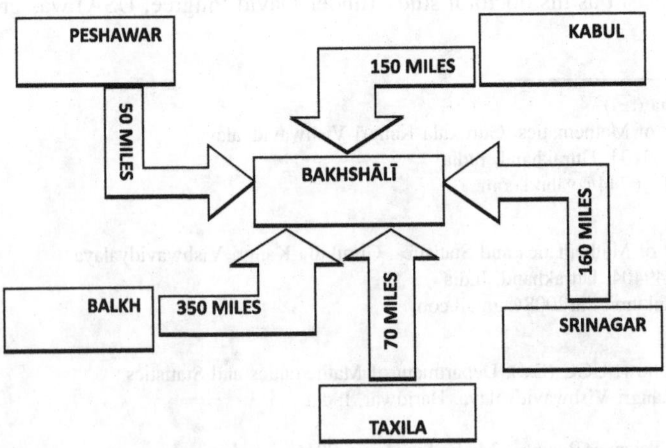

Fig. 1 Mapping of Village Bakhshālī

2 Folio 56

Original Sanskrit verses on Bakhshālī's iterative process [11].

अकृते श्लिष्टकृत्यूनात् शेषच्छेदो द्विसंगुणः।
तद्वर्गदलसंश्लिष्टहृतिः शुद्धिकृतिः क्षयः।

akṛte ślişṭa kṛtyūnā śeṣacchedodvisaṃguṇaḥ |
tadvargadala saṃśliṣṭahṛti śuddhikrti kṣayaḥ |

The above sūtra has been translated by G.R. Kaye as-

The mixed surd is lessened by the square portion and the difference divided by twice that. The difference is divided by the quantity and half that squared is the loss.

Indeed, English translation of this rule given by Kaye is discarded as being wrong and meaningless.

Now, Prof K. Ramasubramanian (IIT-Mumbai) presents the meanings of few other words appearing in the above verse [9].

2.1 *Glossary of Technical Terms [9]*

Sanskrit text and English translation with mathematical notation are as follow:

śliṣṭakṛti (Approximate Square root 'A'), *tadvargadala* (Half of the Square of that) $\frac{1}{2}\left(\frac{b}{2A}\right)^2$, *saṃśliṣṭahṛti* (Division by the composite that is $\div\left(A+\frac{b}{2A}\right)$, *kṣayaḥ* (Subtraction of that), *śuddhikrti* (the refined square root), *kṛtyūnā* (dividing by approx. square root), krti (square), *śeṣacchedodvisaṃguṇaḥ* (Divisor of $\frac{b}{A}$ multiplied by 2).

After one has deciphered this recipe, it turns to be equivalent to the following formula:

$$\sqrt{N}=\sqrt{A^2+b}\approx A+\frac{b}{2A}-\frac{\left(\frac{b}{2A}\right)^2}{2\left(A+\frac{b}{2A}\right)}$$

Where: ($A^2 < N$ and as close as possible to N)

3 Difference between Newton–Raphson Formula (Newton's Iterative Formula) and Bakhshālī's Iterative Formula

1. Newton's method has a quadratic convergence whereas Bakhshālī's iterative formula has a quartically convergence (for proofs [1, 7]).
2. Newton's formula of convergence provided the initial approximation x_0 which is chosen sufficiently close to the root [7].
   The accuracy depends upon how close we choose the initial value A to be, or how small b is. However, even if b is large, successive iterates would lead to the exact value.
3. The Bakhshālī square root (BSR) algorithm often skips some of the steps encountered in the Newton's iterative formula.
4. The solution is reached after many steps using Newton's iterative formula, which is in contrast to a single step in the case of the BSR formula.
5. BSR method ingeniously avoids large numbers in the calculations than N–R method.

3.1 *Comparison between* N–R *Iterative Method and Bakhshālī Square Root Method Using an Illustrative Example*

An example is given to facilitate the comparison of the *N–R* method with the BSR method.

3.1.1 Newton's Iterative Method

Example Find an iterative formula to find $\sqrt{N}$ where N is a positive number and hence find $\sqrt{83}$ correct to nine decimal places.

Solution: We can derive the following result from the Newton's iterative formula. Let us consider $x^2 - N = 0$. Taking $f(x) = x^2 - N$ we have $f'(x) = 2x$. Then, Newton's formula gives $x_{n+1} = x_n - \frac{f(x_n)}{f'(x_n)}$ we get $x_{n+1} = \frac{1}{2}\left[x_n + \frac{N}{x_n}\right]$ which is required Newton's iterative formula. Now, for $N = 83$, the root lies in the interval (9, 9.5). Take the initial approximation as $x_0 = 9$. Using Newton's iterative formula, we get $x_{n+1} = \frac{x_n^2 + 83}{2x_n}$, $n = 0, 1, 2\ldots$ starting with $x_0 = 9$, we obtain $x_1 = 9.111111111$, $x_2 = 9.110433604$, $x_3 = 9.110433579$, $x_4 = 9.110433579$. Hence, the correct root of 83 is 9.110433579.

3.1.2 Bakhshālī's Iterative Method

Let N = 83. The integer which is closet to N is 9. Hence, we choose A = 9 then taking b = 2. Now, using Bakhshālī formula $\sqrt{N} = \sqrt{A^2+b} \approx A + \frac{b}{2A} - \frac{\left(\frac{b}{2A}\right)^2}{2\left(A+\frac{b}{2A}\right)}$ we get $\sqrt{83} = 9.110433604$. This is correct up to seven decimal places.

It seems that in place of scientific calculator for computation of higher digits we can take help of Āryabhaṭa's digital square root method. It can be said that there are so many square root methods in pre-modern Indian mathematics so that we choose any of them [1, 8].

4 The Bakhshālī Algorithm to Compute the Square Root of a Non Square Number

There are six steps of the Bakhshālī algorithm given below.

Suppose, we wish to find the solution of $f(x) = x - \sqrt{N} = 0$ (where: $N > 0$) then the steps are as follows:

Step I: Make an initial guess 'A' at the square root by picking any number less than N.

Step II: Now, we express N as $N = A^2 + b$ or $b = N - A^2$.

Step III: Compute the ratio 'p', of b and twice the initial guess 'A' that is $p = \frac{b}{2A}$ (i.e. "the divisor of remainder is multiplied by two").

Step IV: To calculate the sum of initial guess and the ratio p that is $q = A + p$.

Step V: Compute your next guess 'A' as $A = q - \frac{p^2}{2A}$.

Step VI: Go back to step II using the guess just computed. Steps II through VI are repeated several times until the implementor sees fit.

The following computer program based on the Bakhshālī's iterative method can be used for finding square root of N (where N is a non square number).

4.1 *Computer Program Based on Bakhshālī's Iterative Method*

```
/*Write a program for BAKHSHALI SQUARE ROOT FORMULA*/
#include <iostream>
using namespace std;
int main()
{
float x,p,A,N,b = 1.0;
```

```
cout << "enter the value of A=";
cin>>A;
N = (A*A)+b;
cout << "Bakhshali Square Root Algorithm" << endl<<endl;
while (b<11)
{
p = b/(2*A);
x = float(A + p − ((p*p)/(2*(A + p))));
cout.precision(16);
cout<<"f(x) = x-sqrt(N) = 0: square root of ("<< N <<") is" << x << endl<<endl;
N++;
b++;
}
system("pause");
return 0;
}
```

OUTPUT—*for Original output see Appendix*

5 Generalization of Bakhshālī's Iterative Formula

$$\text{Let } f(x) = x^2 - N = 0 = \begin{cases} A + \dfrac{b_i}{2A} - \dfrac{\left(\frac{b_i}{2A}\right)^2}{2\left(A + \frac{b_i}{2A}\right)} \\ b_i = i(i \text{ vary from 1 to 10}) \\ \text{Where: } A^2 < N \text{ and as close as possible to } N \end{cases}$$

For each value of *N*, ten successive values are being generated. Further, the ten integers have been considered less than or equal to for *N*.

Comparing the both columns (in Table 1), we find that Bakhshālī's approximation is correct up to five or six places of decimals; and hence the results are near enough. Further, we are given some iteration graphs of BSR and *N–R* method.

5.1 Iteration Graphs of BSR and N–R Method

From Figs. 2, 3, 4 and 5, it is clear that BSR method contains single iteration whereas *N–R* method contains four or five iterations. So that BSR method rapidly converges than *N–R* method. It is also seen that BSR method nearly coincides to *N–R* method.

Table 1 Convergence of BSR and *N–R* methods in the calculation of roots of $f(x) = x^2 - N = 0$

$82 \leq N \leq 91$	Bakhshālī square root method	No. of iteration in BSR method	Newton's iterative method	No. of iteration in *N–R* method
82	9.05538514	1	9.055385138	4
83	9.110433604	1	9.110433579	4
84	9.165151515	1	9.16515139	4
85	9.219832736	1	9.219544457	4
86	9.273619428	1	9.273618495	5
87	9.327380952	1	9.327379053	4
88	9.380834977	1	9.380831519	4
89	9.433986928	1	9.433981132	4
90	9.486842105	1	9.48683298	4
91	9.539405685	1	9.539392014	4

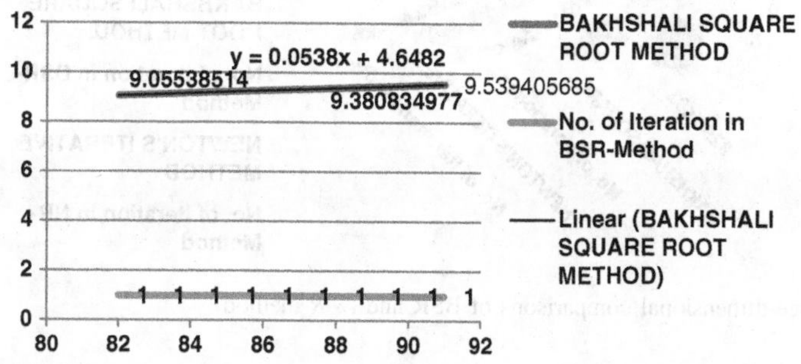

Fig. 2 Two-dimensional iteration graph of BSR-method

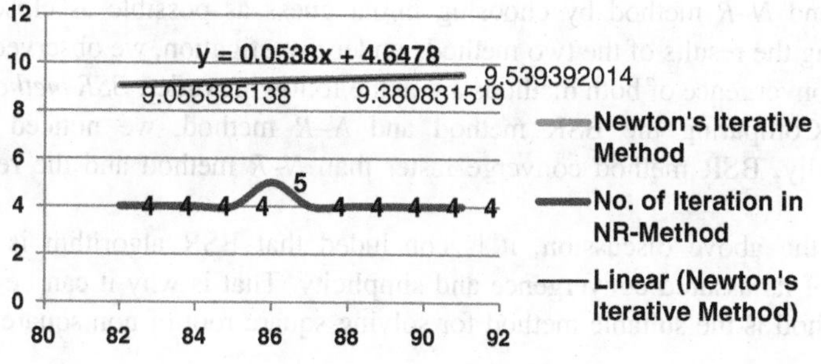

Fig. 3 Two-dimensional iteration graph of *N–R* method

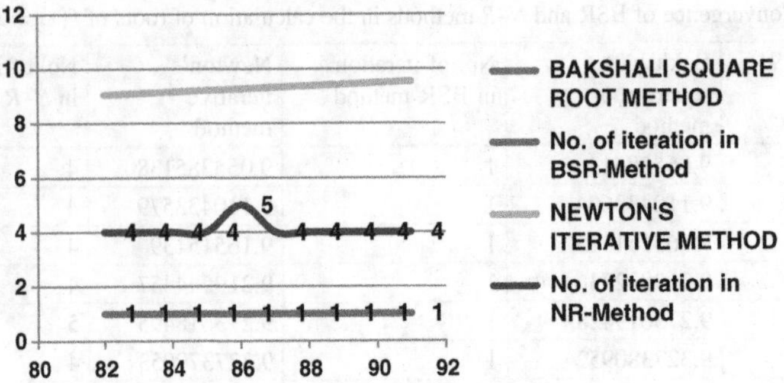

Fig. 4 Two-dimensional comparisons of BSR and *N–R* method

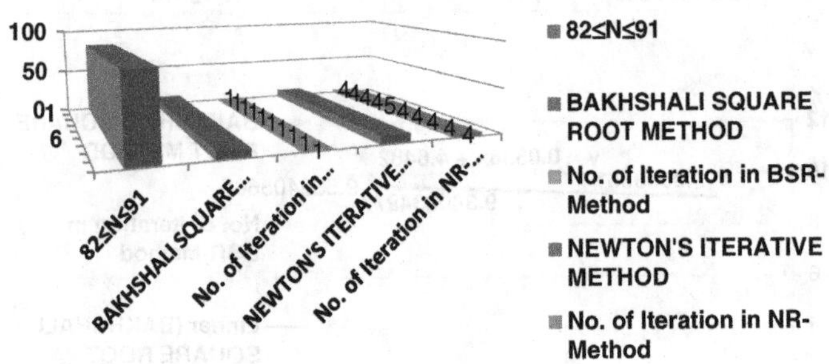

Fig. 5 Three-dimensional comparisons of BSR and *N–R* method

6 Results and Discussion

We have calculated the square root of nonsquare number from 82 to 91 using BSR method and *N–R* method by choosing initial guess as possible as closest to *N*. Comparing the results of the two methods under investigation, we observed that the rates of convergence of both methods are in the following order: *BSR method* > *N–R method*. Comparing the BSR method and *N–R* method, we noticed that the undoubtedly, BSR method converge faster than *N–R* method and the results are accurate.

From the above discussion, it is concluded that BSR algorithm is the best because of its assured convergence and simplicity. That is why it can be said that BSR method is the suitable method for solving square root of non square number.

7 Concluding Remarks

The concept of BSR algorithm can be used as an important tool worldwide in numerical analysis if this concept is introduced.

It appears that in future with the help of computer application of Bakhshālī's iterative method, a new theory can be developed in numerical analysis as a strong tool.

BSR method is computationally simpler than other square rooting method. One advantage of BSR method over *N*–*R* method is that it converges fast. The best thing for BSR method is that it consumes less time and takes single iteration to find the root than the *N*–*R* method. Newton's iterative method has the drawback of more complicated calculations.

Acknowledgments The author thanks Prof. S.L. Singh and Prof. K. Ramasubramanian for carefully going through the drafts and for stimulating discussions.

Appendix

Original output of executed program based on BSR method is given below.

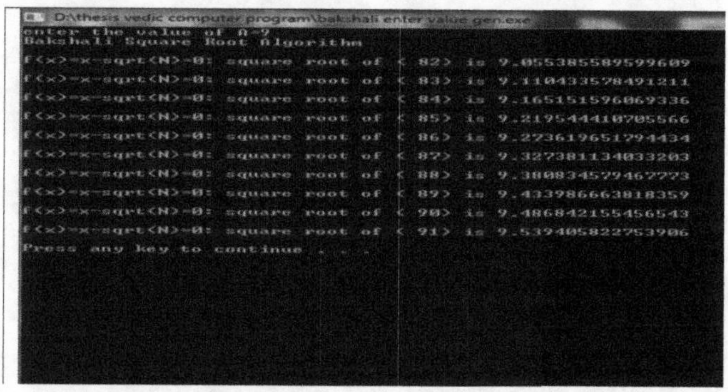

References

1. Bailey, D.H., Borwein, J.M.: Ancient Indian square roots: an exercise in forensic paleo-mathematics. Manuscript, pp. 3–5 (2011)
2. Channabasappa, M.N.: On the square root formula in the Bakhshālī Manuscript. IJHS **11**(2), 112–124 (1975)
3. Datta, B., Singh, A.N.: History of hindu mathematics, Part II. Asia Publishing House (1962)
4. Datta, B.: The Bakhshālī Mathematics. Bull. Calcutta Math. Soc. **21**(1–60), 4–6 (1929)

5. Gurjar, L.V.: Ancient Indian Mathematics and Vedha. Ideal Book Service, pp. 49–75. Poona, India (1947)
6. Gupta, R.C.: Some equalization problems from the Bakhshālī Manuscript. IJHS **21**(1), 51–61 (1986)
7. Jain, M.K., Iyengar, S.R.K., Jain, R.K.: Numerical methods for scientific and engineering computation, 1st edn. New Age Pub., Delhi (1984)
8. Plofker, K.: Mathematics in India. Princeton University Press, Princeton, NJ (2009). The Genre of Medieval Mathematics, Mathematics in India, CHOM 7, reprinted by HBA, pp. 157–162 (2012)
9. Ramasubramanian, K.: Lecture based on "The Bakhshālī Manuscript" (from NPTEL Course on Mathematics in India)
10. Shirali, S.A.: The Bakhshālī square root formula. Resonance **17**(09), 884–894 (2012)
11. Sarasvati, S.S.P., Dr Jyotishmati, U.: The Bakhshālī Manuscript, Dr. Ratna Kumari Svadhyaya Sansthan, Allahabad. http://www.scribd.com/doc/75830711/TheBakhshālī-Manuscript

Dynamic Resource Allocation for Multi-tier Applications in Cloud

Raghavendra Achar, P. Santhi Thilagam, Meghana, B. Niha Fathima Haris, Harshita Bhat and K. Ekta

Abstract Increasing demand for computing resources and widespread adaption of service-oriented architecture has made cloud as a new IT delivery mechanism. Number of cloud providers offer computing resources in the form of virtual machines to the cloud customers based on business requirements. Load experienced by the present business applications hosted in cloud are dynamic in nature. This creates a need for a mechanism which allocates resources dynamically to the applications in order to minimize performance degradations. This paper presents a mechanism which dynamically allocates the resources based on load of the application using vertical and horizontal scaling. Cloud environment is set up using Xen cloud platform and multi-tier web application is deployed on virtual machines. Experimental study conducted for various loads show that proposed mechanism ensures the response time is within the acceptable range.

Keywords Cloud computing · Virtualization · Multi-tier application · Resource allocation

---

Raghavendra Achar (✉) · P. Santhi Thilagam
National Institute of Technology Karnataka, Surathkal, India
e-mail: raghunitk@gmail.com

P. Santhi Thilagam
e-mail: santhisocrates@gmail.com

Meghana · B. Niha Fathima Haris · Harshita Bhat · K. Ekta
St. Joseph Engineering College, Mangaluru, India
e-mail: meghana.kadri@gmail.com

B. Niha Fathima Haris
e-mail: nihaharis92@gmail.com

Harshita Bhat
e-mail: harshyc.c@gmail.com

K. Ekta
e-mail: ekta2010.s@gmail.com

M. Pant et al. (eds.), *Proceedings of Fifth International Conference on Soft Computing for Problem Solving*, Advances in Intelligent Systems and Computing 436, DOI 10.1007/978-981-10-0448-3_79

1 Introduction

Cloud computing [1] is becoming one of the most prominent computing paradigm where virtualized IT resources are delivered to the cloud customers using pay-as-you-go basis. Cloud resources are offered using three delivery models namely software as a service (SaaS), platform as a service (PaaS) and infrastructure as a service (IaaS). In SaaS, a software application is made available to the requester. Salesforce is an example for SaaS. Application development platform is made available by the PaaS provider. Microsoft Azure is one of the example for PaaS. In IaaS, an infrastructure is provided as a service to the requester. Some of the providers who offers IaaS are Amazon and Rackspace. Virtualization [2, 3] is the technology which realize the vision of utility computing. Some of the hypervisors used in virtualization are Xen, KVM, VMware, etc. Cloud provides computing resources in the form of virtual machine (VM), which is a abstract machine runs on physical machine. The requester using the virtual machine will have the feel of working on physical machine. VM live migration is one of the widely used technique for dynamic resource allocation in a cloud environment. Live VM migration [4–6] refers to the process of moving a running virtual machine between different physical machines with minimum down time. Due to pay-as-you-go nature, many business applications are moved to the cloud. However, static allocation of resources to the applications results in overutilization and underutilization of resources. Overloaded servers results in performance degradation of applications and under loaded servers results in inefficient resource utilization. In this paper, we present a mechanism which dynamically allocates a resource based on the load using horizontal and vertical scaling.

The rest of this paper is organized as follows. Section 2 reviews the related work on resource allocation. Section 3 defines the problem. Section 4 presents methodology. Section 5 describes experiments and evaluation, and Sect. 6 presents the conclusion.

2 Related Work

There have been some work toward resource allocation in cloud. Paper [7] presents dynamic resource scaling architecture in cloud environment. The architecture consists of front-end load balancer to route the user requests to the web applications which are deployed on virtual machines. The proposed work focuses on increasing the resource utilization with minimum number of virtual machines. Dynamic resource scaling algorithm is proposed which distribute the load among web applications based on number of active user sessions. Paper [8] proposes a dynamic provisioning mechanism for multi-tier web applications. Here, queuing model is used to determine the amount of resources required for each tier in the multi-tier web application. In this, combination of both reactive and predictive techniques are used, which determine when additional resources are provisioned in order to meet the response time requirements. Paper [9] proposes a resource allocation mechanism which focuses on minimizing the cost of infrastructure and to run the applications with minimum SLA violation. The experiment is conducted using eucalyptus private cloud environment. Authors in [10] proposes a provisioning mechanism using genetic algorithm. Proposed system consist of a component called resource manager. This component is responsible for dynamically allocating resources by considering the time, costs and availability of physical resources. Experiment is conducted using open nebula cloud environment. Paper [11] presents a resource allocation policy to minimize the number of rejection of resource requests. The proposed policy is implemented in haizea resource lease manager. The experiment conducted shows the acceptance of leases and efficient utilization of resources as compared to the existing resource allocation policies in haizea. Authors in [12] proposed a novel dynamic resource provisioning architecture based on distributed decisions. The proposed approach is illustrated by considering cloud scenario where each physical server independently makes decision based on workload characteristics, current state of the system, and provisioning optimizer in order to provision the resource based on requirements. Authors in [13] proposed architecture for provisioning resources to multi-tier applications with minimum overhead. Hybrid queuing model is used to determine the amount of resources to be allocated to each tier of multi-tier web applications. Based on this, novel provisioning mechanism is proposed. Author also proposed meta-heuristic solutions based on users various performance requirements of the application. In this work, we have considered both horizontal and vertical scaling for efficient resource allocation.

3 Problem Statement

In the world of cloud computing, dynamic resource allocation is an interesting issue which is open for research. The success of this rising model is dependent on the effectiveness of techniques used to allocate the resources in a most optimal way. Thus to achieve this, the following scenario has been taken up as the problem statement.

Let $C = \{S_1, S_2, \ldots, S_n\}$, where C is a cloud and $S_1, S_2, \ldots, S_n$ are the servers. Let $S_j = \{v_{j1}, v_{j2}, \ldots, v_{jl}\}$, where $v_{j1}, v_{j2}, \ldots, v_{jl}$ are the virtual machines in the server S_j. Let $v_{ji} = \{v\_id_{\mathrm{ji}}, v\_cpu_{\mathrm{ji}}, v\_ram_{\mathrm{ji}}\}$, where v_id_{ji} is the VM Id, v_cpu_{ji} is the speed of the processor and v_ram_{ji} is the RAM size of the VM. Let $\mathrm{CPU}_{\mathrm{max}_j}$ be the changeable maximum CPU that can be allocated to the virtual machine v_{ji}. Let W be the multi-tier application running in different virtual machines and R be the response time of the application. The problem is to allocate the right amount of resources to the application with the help of horizontal and vertical scaling to ensure that the response time R is within the acceptable range.

4 Methodology

This section describes the methodology used in order to achieve the specific objectives. Presented work is based on XCP (Xen Cloud Platform) and Credit Scheduler. We brief about Xen, Credit Scheduler followed by system architecture, interactions, and algorithms.

4.1 Xen and Credit Scheduler

Xen [2] is a hypervisor which allows the execution of multiple virtual machines on a single physical machine. It is responsible for CPU scheduling of the all virtual machines running on the hardware. Xen not only abstracts the hardware for the virtual machines but also controls the execution of virtual machines. Domain U (DOM_U) are the virtual machines which has no direct access to physical hardware on the machine. Domain O(DOM_O) is a virtual machine running on the Xen hypervisor which has special rights to access physical I/O resources as well as interact with other virtual machines (Domain U). All Xen virtualization environments require Domain O to be running before any other virtual machines can be started. The credit scheduler [3] is a proportional fair share CPU scheduler, which is the default scheduler for Xen hypervisor. In credit scheduler each domain is assigned weight and cap. The values for weight can be varied from 1 to 65535 and the default value is 256. A domain with weight 512 will get double cpu usage as with weight of 256.

4.2 System Architecture

Figure 1 shows the system architecture. The architecture consists of n number of servers and cloud controller. Each server is having a hypervisor like Xen to run multiple virtual machines. Requester requests for computing resources through cloud controller. The required computing resources are provided in the form of virtual machine.

Fig. 1 System architecture

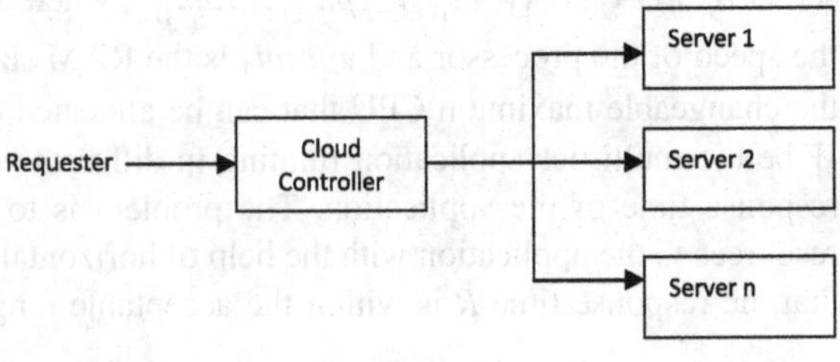

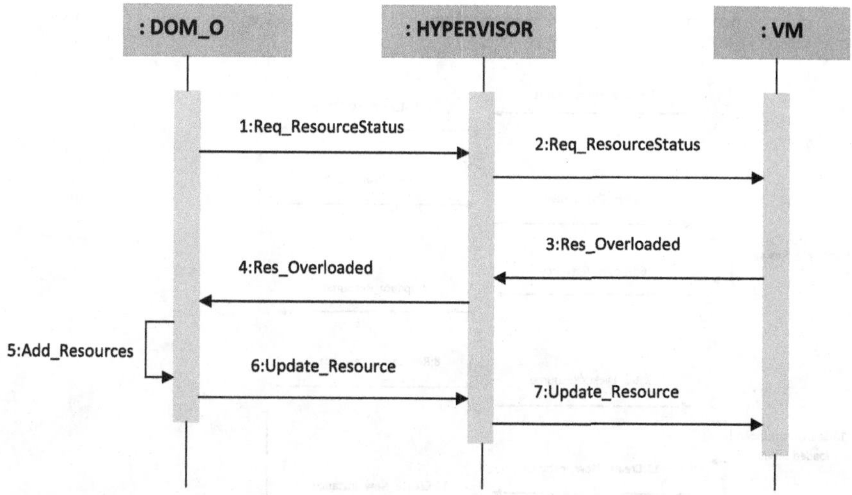

Fig. 2 Sequence of activities to scale up the resources

Following scenario in Fig. 2 depicts the interaction among various components when applications are running in a virtual machine. During the interaction, the DOM_O sends a Req_ResourceStatus message to the virtual machine, requesting the details of resource usage. If the resources granted to the virtual machine is overutilized (over loaded), the virtual machine sends a Res_Overloaded message back to the DOM_O via the hypervisor. The DOM_O makes use of the existing scaling methods to indicate the level to which the resources must be scaled up. The information provided by the DOM_O is then sent to the hypervisor which performs the job of scaling up of the resources, the result of which is reflected in the operation of the virtual machine.

Below in Fig. 3 depicts the scenario where the virtual machine is in need of resources. The DOM_O sends a Req_ResourceStatus message to the virtual machine through the hypervisor requesting the details of resource utilization. The virtual machine sends an Res_Overloaded message back to the DOM_O stating that resources are overloaded. The DOM_O searches for resources that are not used or wasted by other virtual machine. When resources are found, the resources are allocated to the virtual machine as per requirements (scaling up). In case, if there is no enough resources that can be provided by the current server, new instance of virtual machine is created in the suitable target server. The DOM_O identify the suitable target server and create a new VM instance in the target server. The server to which the virtual machine is instantiated is decided based on the selection criteria.

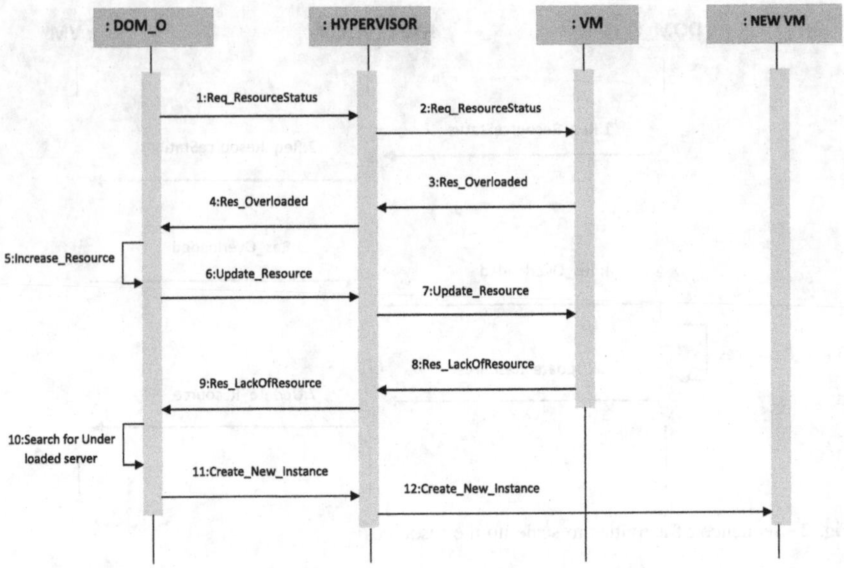

Fig. 3 Sequence of activities to instantiate new VM in the target server

4.3 Predicting the Resource Requirements

In order to predict the additional resource required by the application, we use exponential smoothing model. The exponential smoothing model takes previous values and predicts the new value. It accomplishes this by computing the weighted average of the two values. The formula is

$$F_{t+1} = \alpha D_t + (1 - \alpha) F_t$$

Here D_t is the observed value and F_t is the predicted value at time t. We use 0.5 as a α value for the exponential smoothing model.

4.4 Dynamic Resource Allocation Algorithm

Dynamic resource allocation algorithm verifies the utilization of CPU are in the defined limits. Based on utilization, algorithm initially scales the resources in an upward or downward manner based on resource availability. If the resources required for scaling are not available in the current server, the algorithm creates a new instance of VM on the suitable server. Proposed algorithm dynamically allocates the resource using vertical and horizontal scaling. Xen credit scheduler is used to scale the resource vertically. Xen API is used to get and control the information about virtual machines. The vertical scaling is done by changing the cap value of credit scheduler. When multiple user requests arrive, the algorithm compares the CPU utilization of the virtual machine with various threshold values. Based on this, algorithm proceeds to increase the cap value until the maximum limit has reached. Subsequently, a new virtual machine is instantiated.

Algorithm 1 Dynamic Resource Allocation

```
while true do
  for each server S in the pool do
    for each vm instance i running in S do
      cpu_pred ← next cpu usage using prediction algorithm
      cap ← the cap value of instance i
      if cap == 0 then
        cap ← 20
      end if
      if cap == 20 then
        if cpu_pred > 70 then
          cap ← 50
        else if cpu_pred < 30 then
          remove instance i
          remove instance i from load balancer
        end if
      else if cap == 50 then
        if cpu_pred > 70 then
          cap ← 100
        else if cpu_pred < 30 then
          cap ← 20
        end if
      else if cap == 100 then
        if cpu_pred > 70 then
          Instantiate a new vm
          Add this vm instance to the load balancer
        else if cpu_pred < 30 then
          cap ← 50
        end if
      end if
      Wait for duration T
    end for
  end for
end while
```

Algorithm initially set the cap value to 20 and predicts the next CPU utilization using prediction algorithm (exponential smoothing). When the cap value is 20 and predicted value is more than 70 %, the cap value is changed to 50 (i.e., vertical scaling). If predicted value is less than 30 % newly created web server instance will be removed. If cap value is 50 and predicted value is more than 70 %, the cap value is changed to 100 (i.e., vertical scaling). If predicted value is less than 30 % the cap value will be reduced to 20. Similarly if cap value is 100 and predicted value is more than 70 %, new VM instance is created (i.e., horizontal scaling). If predicted value is less than 30 % the cap value will be reduced to 50. After duration *T* the process will be repeated.

5 Experiments and Evaluation

Inorder to verify the proposed algorithm, we conducted a experiment using three servers having configuration Intel i3 2.93 GHz processor and 2 GB RAM as shown in Fig. 4. All three servers installed with Xen Cloud Platform (XCP 1.6.1) [14] which contains a hypervisor to run multiple virtual machines. NFS is configured on ubuntu 13.04 machine having configuration Intel Core 2 Duo CPU T6500 @ 2.10 GHz processor, 4 GB RAM, and 500 GB storage. The set up also consists of XenCenter enabled system to monitor the virtual machines. We have implemented algorithms in shell script using Xen API. Virtual Machines are created on each server with suitable operating system.

To evaluate the proposed algorithm, Multi-tier web application is created with four virtual machines as shown in Fig. 5. Two virtual machines installed with windows XP is configured as a webservers using apache tomcat server. One virtual machine installed with MySQL acts as a back-end database server. Virtual machine installed with ubuntu is configured as a load balancer using apache http server and mod jk connector. Load balancer is used to route the request to appropriate web server. The web server query a back-end database to retrieve the information stored in database. Virtual machine installed with Windows Server 2008 is configured as a

Fig. 4 Experimental set up of cloud architecture

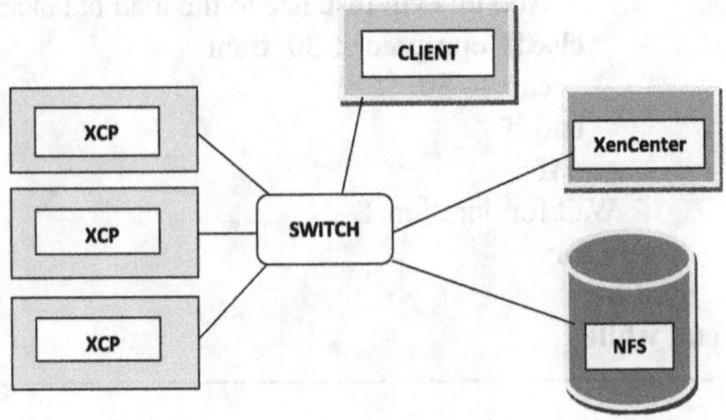

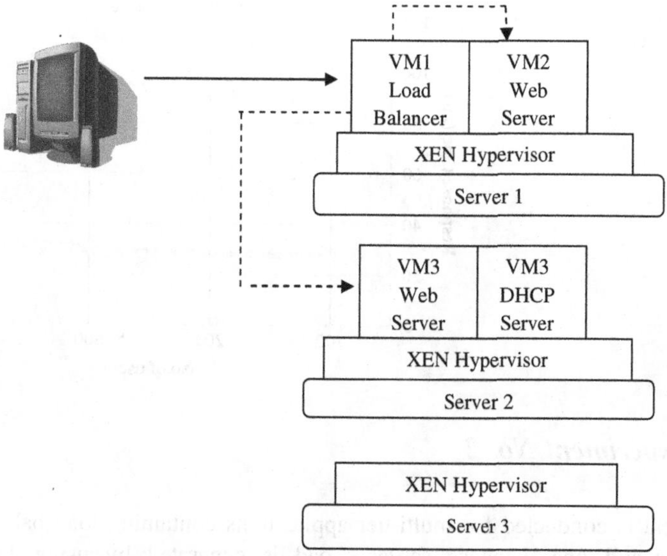

Fig. 5 Multi-tier applications in virtual machines

DHCP (Dynamic Host Configuration Protocol) server which assigns IP addresses to the newly instantiated virtual machines.

As shown in Fig. 5 request to the web application was made via load balancer which route the request to the two back-end servers in a round-robin fashion. The aim here is to balance the load among the virtual machines that host a particular web application. This is done via the load balancer. In order to generate load, we use of Apache JMeter [15], which sends multiple HTTP requests to the load balancer. The load balancer then routes these requests to the available virtual machines. In order to verify the performance of a applications number of experiments conducted by increasing the resources horizontally and vertically. We have used the response time of the virtual machines as metric to measure the performance of applications running in cloud.

5.1 *Experiment No. 1*

Experiment conducted with one webserver running a simple web application. Load is generated on the web applications by increasing the number of requests in the range 100, 200, 500, and 1000 respectively using JMeter and response time is measured. Same experiment is repeated by scaling the resources vertically by increasing the cap value. Figure 6 shows the response time versus number of users. It is observed that scaling resource vertically results in decreasing response time.

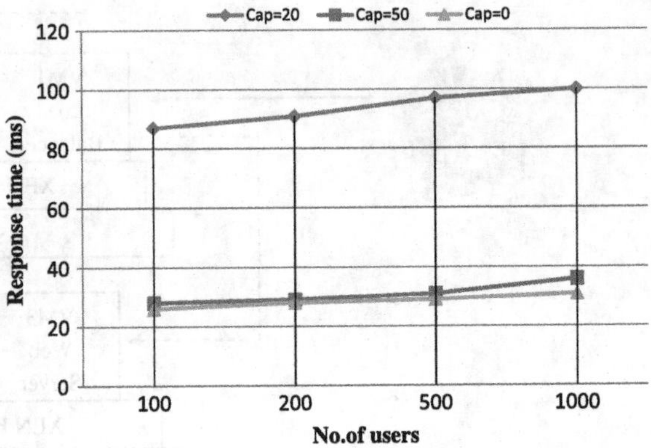

Fig. 6 Response time versus number of users

5.2 Experiment No. 2

Experiment is conducted for multi-tier applications containing load balancer, one webserver and one database server. Load is generated by using JMeter by increasing the number of request in the range 100, 200, 500, and 1000 respectively. Response time is measured. The experiment is repeated by increasing the web servers and configuring that with load balancer. The load generated in this case is sent to the load balancer which in turn handles these requests in a round-robin fashion. Figure 7 shows the response time versus number of users. The performance of applications running in virtual machines improved in terms of response time by increasing the resources horizontally.

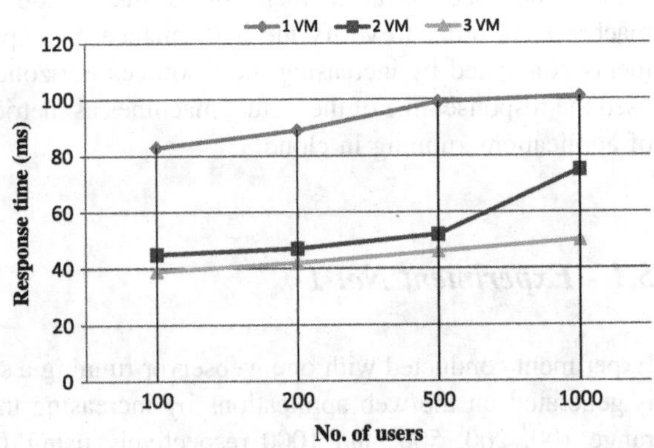

Fig. 7 Response time versus number of users

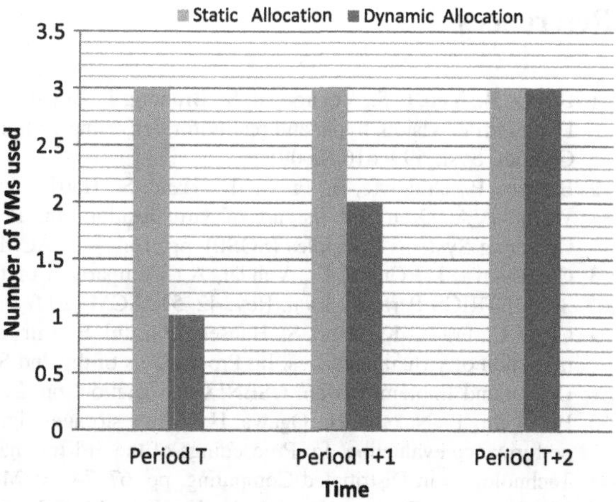

Fig. 8 Resource used by static allocation versus dynamic allocation

5.3 *Experiment No. 3*

The previous experiment conducted (Experiment No. 2) shows that increasing the number of VM instance results in better response time. Applications running in the cloud are dynamic in nature. Static resource allocation results in under utilization of resources whenever the application is lightly loaded, over utilization of resources whenever there is a heavy load. There is a need to allocate the resources on the fly. Algorithm 1 presented above allocates the resources based on the needs. To evaluate the proposed algorithm, we created one web server connected to the load balancer. We created Xen server pool and executed our algorithm in a master server. Graph below shows the resource utilization by the application (i.e., no. of virtual machines) versus time. It is observed that proposed algorithm allocated resources only when it is needed (Fig. 8).

6 Conclusion

In this paper, we have presented dynamic resource allocation algorithm to improve the performance of multi-tier applications in terms of response time. Proposed algorithm use both horizontal and vertical scaling of resources to improve the performance. Experiment is conducted on Xen cloud platform. A multi-tier web application is developed and deployed on virtual machines with a front-end load balancer. Apache JMeter is used to generate the load. Experiments are repeated for various loads. It is observed that dynamic resource allocation ensures the response time is always within the acceptable range. The proposed algorithm maximizes resource utilization in each virtual machine and minimizes the total number of deployed instances.

References

1. Buyya, R., Yeo, C.S., Venugopal, S., Broberg, J., Brandic, I.: Cloud computing and emerging IT platforms: vision, hype, and reality for delivering computing as the 5th utility. Future Gen. Comput. Syst. 599–616 (2009)
2. Barham, P., Dragovic, B., Fraser, K., Hand, S., Harris, T., Ho, A., Neugebauer, R., Pratt, I., Warfield, A.: Xen and the art of virtualization. In: Proceedings of the Symposium on Operating Systems Principles (SOSP), pp. 164–177. ACM (2003)
3. Cherkasova, L., Gupta, D., Vahdat, A.: Comparison of the three CPU schedulers in Xen. SIGMETRICS Perform. Eval. Rev. 42–51. ACM (2007)
4. Clark, C., Fraser, K., Hand, S., Hansen, J.G., Jul, E., Limpach, C., Pratt, I., Warfield, A.: Live migration of virtual machines. In: Proceedings of the 2nd Symposium on Networked Systems Design and Implementation, USENIX Association, pp. 273–286. ACM (2005)
5. Hirofuchi, T., Nakada, H., Ogawa, H.: A live storage migration mechanism over wan and its performance evaluation. In: Proceedings of the 3rd International Workshop on Virtualization Technologies in Distributed Computing, pp. 67–74. ACM (2009)
6. Hirofuchi, T., Ogawa, H., Nakada, H., Itoh, S., Sekiguchi, S.: A live storage migration mechanism over WAN for relocatable virtual machine services on clouds. In: Proceedings of the 9th International Symposium on Cluster Computing and the Grid, pp. 460–465. IEEE (2009)
7. Chieu, T.C., Mohindra, A., Karve, A.A.: Scalability and performance of web applications in a compute cloud. In: Proceedings of Eighth International Conference on e-Business Engineering, pp 317–323. IEEE (2011)
8. Urgaonkar, B., Shenoy, P., Chandra, A., Goyal, P.: Dynamic provisioning of multi-tier internet applications. In: Proceedings of Second International Conference on Autonomic Computing, pp. 217–228. IEEE (2005)
9. Haitao, Y., Jing, B., Bo, H.L., Xudong, C., Ming, T.: SLA-based virtualized resource allocation for multi-tier web application in cloud simulation environment. In: Proceedings of International Conference on Industrial Engineering and Engineering Management, pp. 1681–1685. IEEE (2012)
10. Apostol, E., Baluta, I., Gorgoi, A., Cristea, V.: Efficient manager for virtualized resource provisioning in cloud systems. In: Proceeding of International Conference on Intelligent Computer Communication and Processing, pp. 511–517. IEEE (2011)
11. Amit, N, Sanjay, C., Gaurav, S.: Policy based resource allocation in IaaS cloud. In: Future Generation Computer System, vol. 28, Issue 1, pp. 94–103 (2012)
12. Chieu, T.C., Hoi, C.: Dynamic resource allocation via distributed decisions in cloud environment. In: Proceedings of Eighth International Conference on e-Business Engineering, pp. 125–130. IEEE (2011)
13. Zhiliang, Z., Jing, B., Haitao, Y., Ying, C.: SLA based dynamic virtualized resources provisioning for shared cloud data centers. In: Proceedings of 4th International Conference on Cloud Computing, pp. 630–637. IEEE (2011)
14. XCP Download: http://www.xen.org/download/xcp/index.html. February 2013
15. JMeter Download: http://jmeter.apache.org/index.html. February 2015

Comparison of Image Restoration and Segmentation of the Image Using Neural Network

B. Sadhana, Ramesh Sunder Nayak and B. Shilpa

Abstract In present day almost all of the image restoration method suffer from weak convergence properties. Also for Point Spread Function (PSF), some methods make restrictive assumptions. Some original images restrict algorithms portability to many applications. Current situation is using deburring filters, images are restored without the information of blur and its value. In this paper, method of artificial intelligence is implemented for restoration problem in which images are degraded by a blur function and corrupted by random noise. This methodology uses back propagation network with gradient decent rule which consists of three layers and uses highly nonlinear back propagation neuron for image restoration to get a high quality of restored image and attains fast neural computation, less complexity due to the less number of neurons used and quick convergence without lengthy training algorithm. The basic performance of the neural network based restoration along with segmentation of the image is carried out.

Keywords Image restoration · Deblurring · BPN · Blur parameter · Point spread functions · Noise to signal ratio

1 Introduction

Restoring image means either to limit the degradations made during capture or completely remove it. Almost all of the naturally captured images may have undergone some or other sort of degradations. The assumption made here is that,

B. Sadhana (✉) · R.S. Nayak · B. Shilpa
IS&E, Canara Engineering College, Benjanapadavu, Mangaluru
574219, Karnataka, India
e-mail: sadhana26rai@gmail.com

R.S. Nayak
e-mail: ramesh.nayak.spi@gmail.com

B. Shilpa
e-mail: shilpanarsha@rediffmail.com

M. Pant et al. (eds.), *Proceedings of Fifth International Conference on Soft Computing for Problem Solving*, Advances in Intelligent Systems and Computing 436, DOI 10.1007/978-981-10-0448-3_80

using mathematical models, the degradation process is described. In this paper, a neural network approach is introduced to implement image restoration and segmentation used in image processing techniques. A neural network based on back propagation neurons is used for the same blur parameter identification. Different types of blurs or noises are considered Gaussian, Motion and Disk [1]. Using back propagation neural network, the parameters of concerned operator are identified. Once the type of noise or blur is identified, using restoration or deblurring method, the image can be restored. Here the preliminary work considered is conservative image restoration. Between deconvolution and BPN, comparison is made. The work has been extended along with the segmentation of the image using neural networks. By employing neural networks, it can suppress image blurring by learning a sufficient number of data sets to indicate a relationship between image blurring and the corresponding magnetic nanoparticles MNP location which exist at the boundary of the field free point, are also detected. In order to overcome these problems, it is possible to use a new reconstruction method using neural networks [2]. Some techniques have image restoration using interpolation, in painting and denoising the noisy image bi-iterative method. The main task of image restoration is to capture a noisy image and estimating the original image [3]. For applications which restore standard images (frame based), Hessian matrix is found quite efficient. In frame based image restoration with balanced regularization, our numerical simulations illustrate efficiency [4].

2 Framework for Image Deblurring Using Back Propagation Neural Network

Firstly, the image can be selected from multi source to initiate the processing. After image is been selected, pre-processing step is been done and image is tested for noises and blur that are predominant and uses filters which is suited for removing the noise and blur to enhance the image for the best output for next process. Using BPN, training is done for the parameters which are extracted from noise type. The network is simulated to restore the image.

3 Denoising

3.1 Filtering Process

Denoising is the process of removing the unwanted noise from the image with the help of filters. Different types of filters [5] are applied for extraction of the image. For modifying else enriching, the image filtering techniques are used [6]. With

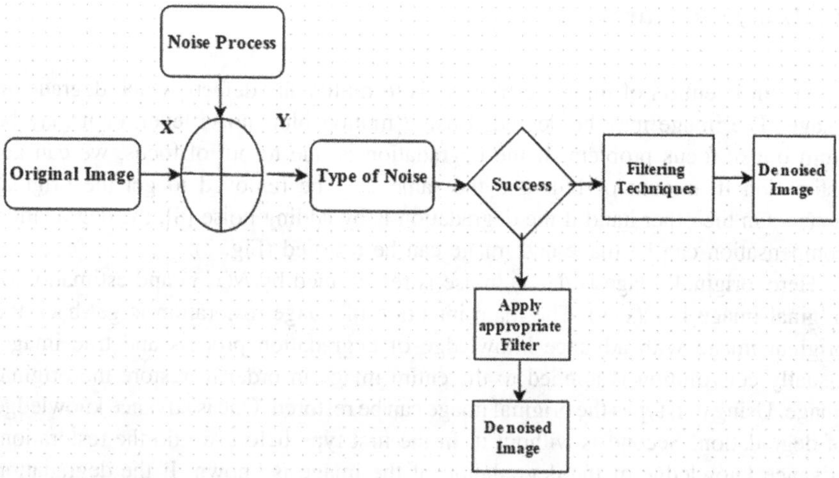

Fig. 1 Denoising model

filters, we can remove certain features or emphasize some features. It's a neighbourhood operation. Algorithms are applied to determine values of the output pixels from the values of the neighbourhood pixels. One-pixel neighbor may be other pixel. Both pixels location is relative to each other.

Figure 1 show the denoising model which deals with noise removal from the input image. Here the model takes *X* as original image and noise gets added during the processing of the image. *Y* is the input noisy image which is taken as the input image for the noise estimation model. Once the type of the noise is found then the appropriate filter for is used to remove the noise from the image. If the image contains combination of noises then filtering techniques such as average filter, Gaussian filter, median filter, laplacian filter, high pass filter, low pass filter, bilateral filters are used.

3.2 Performance Analysis

The performance analysis is done for all the filters and it is tabulated. Performance can be analyzed using parameters like MSE and PSNR. One more method is to get the noisy image [7] by subtracting the original image and the filtered image. The subtracted output image will give the amount of noise present after filtering. PSNR and MSE value for all the filters with respect to noisy image and filtered image is calculated in order to find the efficiency of filtering technique.

4 Image Restoration

The main intention of image restoration is to nullify the defects which degrade an image. The image may be degraded using motion blur, noise level, or it may be from out of focus problem. If the degradation is due to out of focus, we can get estimation if amount of blur and the same may be removed to get the original image. On the other hand if the degradation is by adding noise [8], the degradation compensation can be made and image can be restored (Fig. 2).

Here, original image is *f*(*x*, *y*), noise is represented by *N*(*x*, *y*) and estimation of original image is *f*^(*x*, *y*). The main intention of image restoration is get back the original image with advance knowledge of degradation process and true image. Usually convolution is applied to the entire image in order to restore the original image. Using two types the original image can be restored. One is advance knowledge of degradation. Second is without it. In the first type before we do the restoration, advance knowledge of the degradation of the image is known. If the degradation knowledge is known previously as due to relative motion only the degradation happened, then the modeling determines motion and speed of the motion [9].

4.1 Blind Deconvolution

When there is no clue about the blur and noise, blind deconvolution algorithm is used. This algorithm almost brings back the image and also the point spread function (PSF). User may use their function to pass the PSF constraints. To get the degradation, the input image is made blur with the help of convolution using low pass LSI filter (h). For the resulting image Gaussian Noise in included. For Power Spectrum Equalization (PSE), h is made to have zero phases. Using zero phase filter, it is common to model usual degradations (camera misfocus, atmospheric turbulence, etc.) 'h' value may not available often, in this case different methods can be tried to restore the image. But success rate is little here. The process is Blind Deconvolution, since without knowing the value of blur function, the prediction of

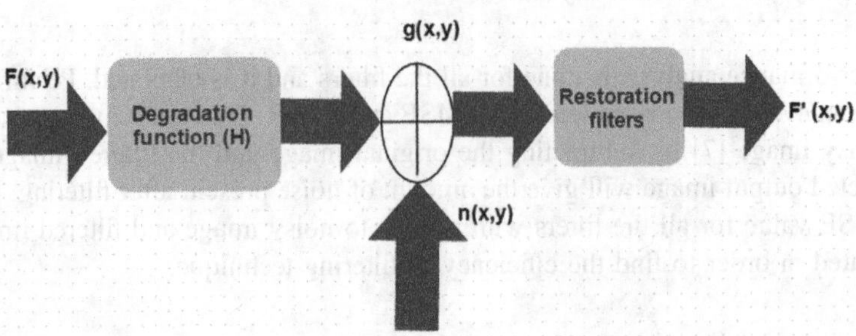

Fig. 2 Degradation/Restoration model

h value is tried in deconvolution. Homomorphism ideas used here. Here, convolution with noise formed as degradation. When considering frequency domain, multiplication is treated as convolution. Logarithm of multiplication is added when additive noise is ignored.

4.2 *Restoration Using Neural Networks*

The main problem of current restoration procedure is about poor convergence properties. The algorithms arrive at local minima. Also these is more computational work. Hence they are considered less in image related applications. Even some methods have limited assumption in point spread function. Also some algorithms are just not suitable for different applications. Without having the knowledge of blur and its properties sometimes deblurring filters are used [8]. The original machine intelligent solution of the blur and blur parameters identification problem is presented in this paper which is handled by BPN. "Figure 3" shows the neural network with three layers which are input layer, hidden layer and output layer.

Back propagation neurons may be helpful for detecting blur and its parameters [10]. Using single layered neural network, the type of distorting operator can be identified. The values of the parameters can be obtained using likewise neural networks. After the type of blur and its parameter is been identified from the image, it is restored back by same neural network.

4.2.1 Parameter Estimation

Blur parameters are fed as a training input to the adopted BPN network [11]. It takes the blur parameters from the blur patterns of the selected image. Point spread

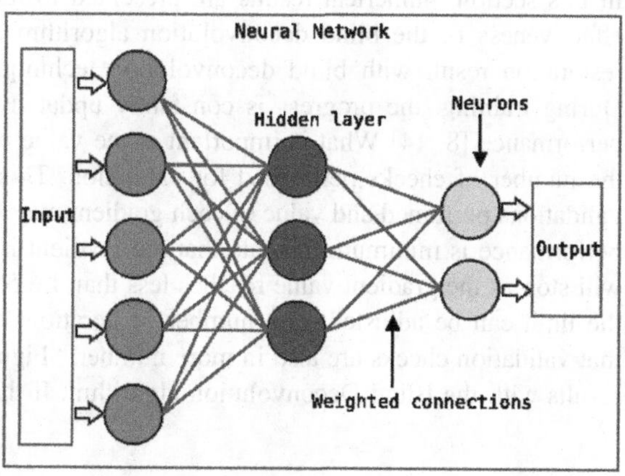

Fig. 3 Neural network trains the neurons

function is the main reason for the blur's PSF and it is a degree to which an optical will blur the point of light. The blurred spot of the single point is called the point spread function. The noise level will also be removed by the appropriate filter according to the blur identified.

- Gaussian {0.0113, 0.0838, 0.0113, 0.0838, 0.6193, 0.0838, 0.0113, 0.0838, 0.0113}
- Disk {0, 0, 0, 0.0012, 0.0050, 0.0063, 0.0050, 0.0012}
- Motion {0.1111, 0.1111, 0.1111, 0.1111, 0.1111, 0.1111, 0.1111, 0.1111, 0.1111}.

4.2.2 Back Propagation Neural Network

Widrow-Hoff learning rule is termed most for back propagation. This is actually useful in multiple neural networks and also for nonlinear differentiable transfer functions [12]. To train a network, input vectors as well as target vectors are used. This will continue till a corresponding function is found out. Here specific input vectors are linked with corresponding output vectors. Networks that are fluctuating values, sigmoid layer, and also which has linear output layer can approximate any function, provided the discontinuity is finite. There are three layers in Back Propagation Neural Networks [13]. One among them is input, other is hidden and the third one is output layer. The training data is given as input for first. That is input layer. The performance of the methods can be computed with respect to PSNR and MSE.

5 Experimental Results

In this section, numerical results are presented to illustrate the efficiency and the effectiveness of the blind deconvolution algorithm and the neural network. The restoration result with blind deconvolution technique is shown in the "Fig. 4". During training, the progress is constantly updated in the training window. The performance [8, 14] What is important is the value of gradient performance. Also the number of checks performed for validation. This value, that is the number of validations perfumed and value of each gradient may limit the training. The training performance is minimum indicate that the gradient is very minimum. The training will stop as the gradient value reaches less than Le-5. With the help of parameters, the limit can be adjusted. The number of iterations which are successful indicate that validation checks are also in more number. "Figure 5" shows the experimental results with the Blind Deconvolution algorithm. If this number reaches the default

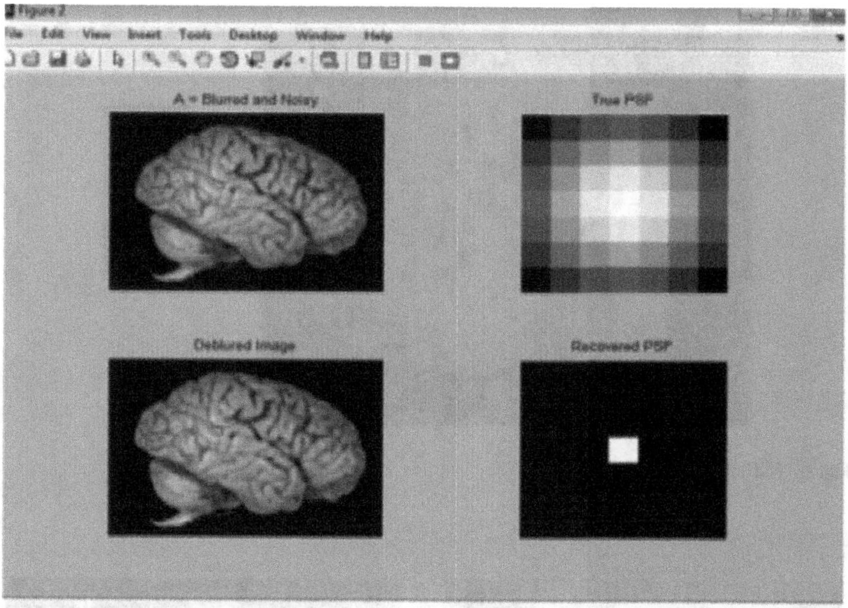

Fig. 4 Results with blind deconvolution

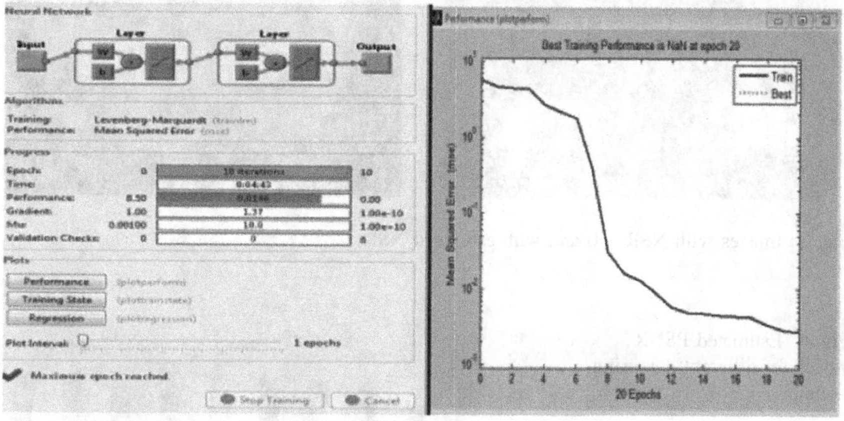

Fig. 5 Results with neural network

value, the training will be stopped. “Figures 6” and “7” shows the PSF of the image along with NSR = O, estimated NSR.

From the experimental results it’s cleared that the restoration or denoising with the neural network is efficient technique with infinity PSNR and zero MSE and it can be shown graphically with the experimental results in “Figs. 8” and “9”. For the

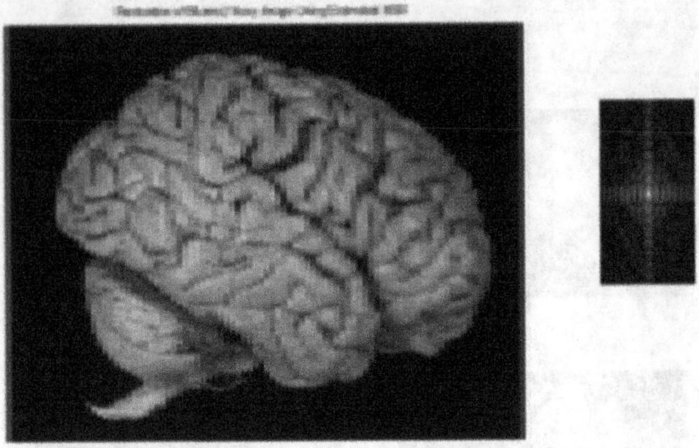

Fig. 6 PSF of the image

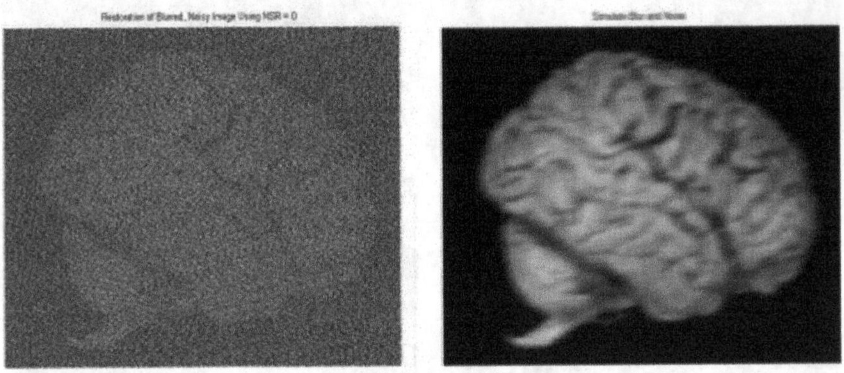

Fig. 7 Images with NSR = 0 and with estimated NSR

Fig. 8 Estimated PSNR values for different denoising techniques

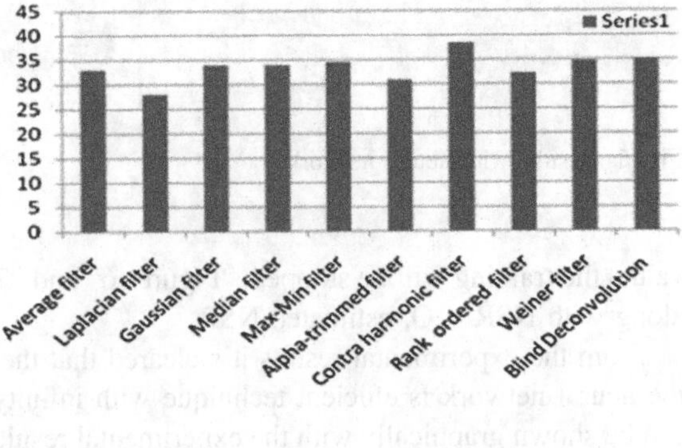

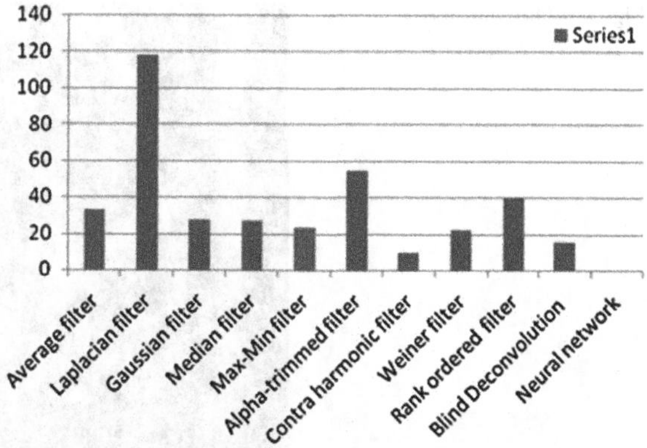

Fig. 9 Estimated MSE values for different denoising techniques

good and efficient restoration technique or algorithm the PSNR [6] should be very large value which is usually calculated in terms of db. Similarly MSE value and should be approximately equals to zero.

6 Skin Color Segmentation Process Using Neural Network

In image processing, the segmentation is considered as important. Almost all methods of segmentation are parametric. The parameters considered plays important role in segmentation [15]. To separate optic disc region and cup from interior surface picture of an eye need to be more accurate and efficient. In case of supervised segmentation, a priori knowledge is needed for successful segmentation. So, nonparametric and unsupervised segmentation method is used when a priori information is not available [12]. From the segmented output which is shown in Fig. 10 it can be noticed that all skin regions are found brighter than the non-skin region. Skin may not be there in all the considered regions. This will indicate that the region is having the skin color. But analogues to this, the process also show up area that which is not of skin color. Such area need to be neglected in face detection process. Compared to other parts, the brightness of the skin is more, hence they can be segmented from rest. It will be difficult to have a single threshold number for different person's skin to process. Figure 11 shows gradient magnitude of the image.

Adaptive thresholding is conducted after learning as "decreasing the threshold value will eventually increase the segmented area. Figures 12 and 13 shows segmented and adaptive thresholding of the image. As the skin region recognized reach to 100 %, the segmented region will keeps decreasing. But eventually the segmented region will increase when the threshold value is little and along with the skin area, non-skin area are also included.

Fig. 10 Original image

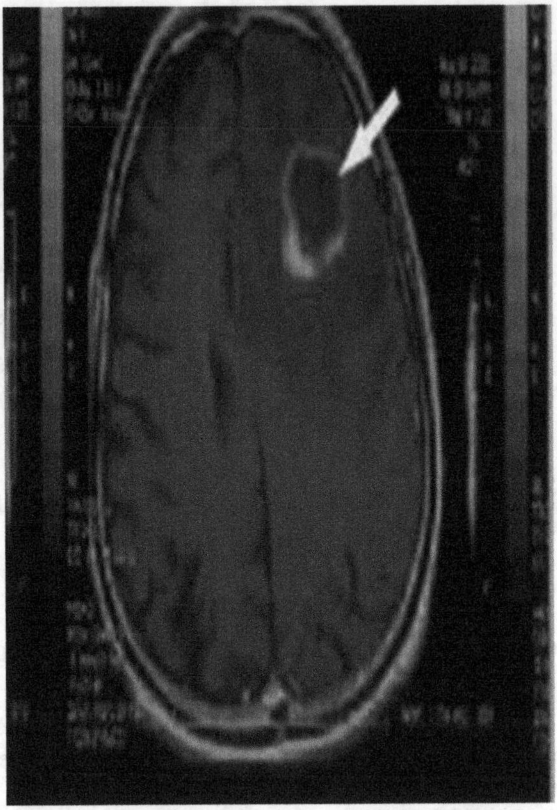

Fig. 11 Gradient magnitude of the image

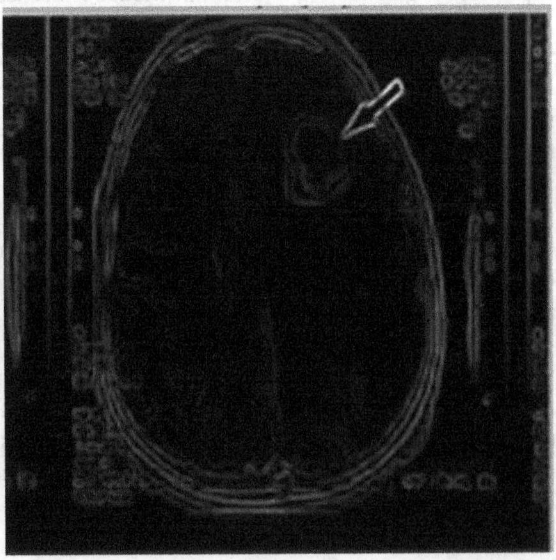

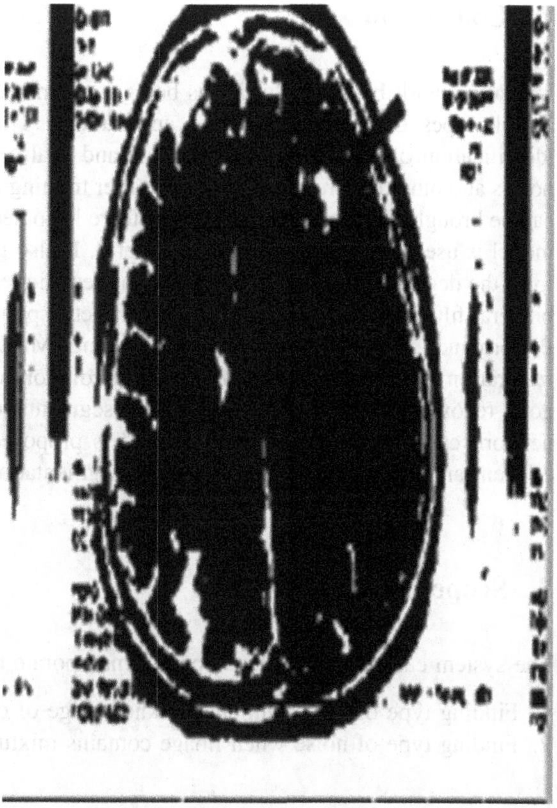

Fig. 12 Segmented image

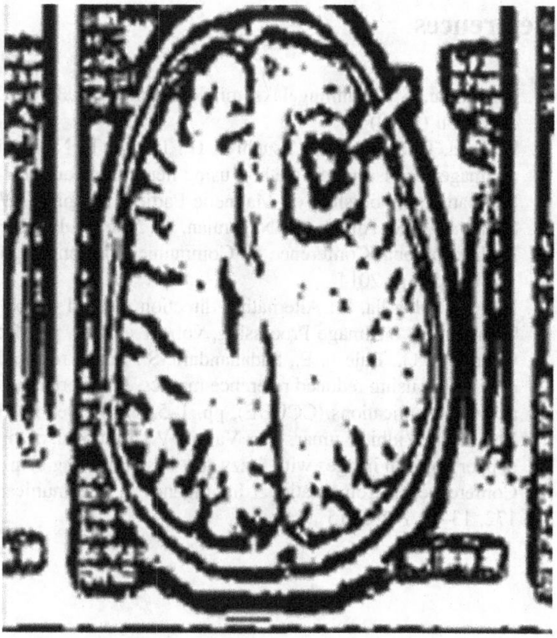

Fig. 13 Adaptive thresholding

7 Conclusion

Neural network based deblurring is being implemented and it will be more useful for all types of image processing applications. This attempt is very useful in identification of what type of a blur it is and it also helps working with different noises and other restoration techniques. After training of neural network, the images can be brought to original form. For this there is no need of knowledge about which model is used to blur the image to corrupt it. It also gives the original information from the degraded image. Different filtering techniques are used and concluded that bilateral filtering and hybrid filter works in better passion for these type of images. Performance is calculated using PSNR and MSE parameters. Finally image restoration is carried out using blind deconvolution and PSF values are varied for good recovery of the original image. The segmentation of the image using neural network can be carried out. The method is proposed with the aim of providing efficient and effective restoration and the segmentation.

8 Scope for Future Work

The system can be further developed to incorporate the following:

1. Finding type of noise without the knowledge of original noiseless image.
2. Finding type of noise when image contains mixture of noise.

References

1. Annadurai, S., Shanmugalakshmi, R.: Fundamentals of Digital Image Processing. Pearson education (2009)
2. Hatsuda, T., Shimizu, S., Tsuchiya, H., Takagi, T., Noguchi, T., Ishihara, Y.: A basic study of an image reconstruction method using neural networks for magnetic particle imaging. In: 5th International Workshop on Magnetic Particle Imaging (IWMPI), pp. 1–1, 26–28 Mar 2015
3. Sankaran, K.S., Ammu, G., Nagarajan, V.: Non local image restoration using iterative method. In: International Conference on Communications and Signal Processing (ICCSP), pp. 1740–1744, 3–5 Apr 2014
4. Xie, S., Rahardja, S.: Alternating direction method for balanced image restoration. In: IEEE Transactions on Image Processing, vol. 21, no. 11, pp. 4557–4567, Nov 2012
5. Coumar, S.O., Rajesh, P., Sadanandam, S.: Image restoration using filters and image quality assessment using reduced reference metrics. In: International Conference on Circuits, Controls and Communications (CCUBE), pp. 1–5, 27–28 Dec 2013
6. Rai, C.S.; Vidhi, Kumar, M.; Varun, V.: Removal of high density Gaussian and salt and pepper noise in images with fuzzy rule based filtering using MATLAB. In: IEEE International Conference on Computational Intelligence & Communication Technology (CICT), pp. 166–172, 13–14 Feb 2015

7. Aizenberg, I.N., Butakoff, C., Karnaukhov; Nikolay, V.N., Merzlyakov, S., Milukova, O.: Blurred image restoration using the type of blur and blur parameter identification on the neural network. In: Proceedings of SPIE 4667, Image Processing: Algorithms and Systems, 460, May 23 2002
8. Chandler, D., Hemami, S.: VSNR: A wavelet-based visual signal to noise ratio for natural images. In: IEEE Transactions on Image Processing, vol. 16, no. 9, pp. 2284–2298, Sept 2007.
9. Aizenberg, I., Bregin, T., Butakoff, C., Karnaukhov, V., Merzlyakov, N., Milukova, O.: Type of blur and blur parameters identification using neural network and its application to image restoration. In: Dorronsoro, J.R. (eds.) Lecture Notes in Computer Science, vol. 2415. Springer-Verlag, Berlin, Heidelberg, New York (2002)
10. Khare, C., Nagwanshi, K.K.: Implementation and analysis of image restoration techniques. Int. J. Comput. Trends Technol. 2011. ISSN: 2231-2803
11. Wang, Y., Zhang, W., Fu, W.: Back Propogation(BP)-neural network for tropical cyclone track forecast. In: 19th International Conference on Geoinformatics, pp. 1–4, 24–26 June 2011
12. Issac, A., Parthasarthi, M., Dutta, M.K.: An adaptive threshold based algorithm for optic disc and cup segmentation in fundus images. In: 2nd International Conference on Signal Processing and Integrated Networks (SPIN), 2015, pp. 143–147, 19–20 Feb 2015
13. Fernandez-Redondo, M., Hernandez-Espinosa, C.: A comparison among weight initialization methods for multilayer feedforward networks. In: Proceedings of the IEEE-INNS-ENNS International Joint Conference on Neural Networks, 2000, IJCNN 2000, vol. 4, pp. 543–548 (2000)
14. Yadav, P.: Color image noise removal by modified adaptive threshold median filter for RVIN. In: International Conference on Electronic Design, Computer Networks & Automated Verification (EDCAV), pp. 175–180, 29–30 Jan 2015
15. Sammouda, R.S., Xinning, Wang, Basilion, J.P.: Hopfield Neural Network for the segmentation of Near Infrared Fluorescent images for diagnosing prostate cancer. In: 6th International Conference on Information and Communication Systems (ICICS), pp. 111–118, 7–9 Apr 2015

Performance Evaluation of PCA and ICA Algorithm for Facial Expression Recognition Application

Manasi N. Patil, Brijesh Iyer and Rajeev Arya

Abstract In everyday interaction, our face is the basic and primary focus of attention. Out of many human psycho-signatures, the face provides a unique identification of a person by the virtue of its size, shape, and different expressions such as happy, sad, disgust, surprise, fear, anger, neutral, etc. In a human computer interaction, facial expression recognition is an interesting and one of the most challenging research areas. In the proposed work, principle component analysis (PCA) and independent component analysis (ICA) are used for the facial expressions recognition. Euclidean distance classifier and cosine similarity measure are used as the cost function for testing and verification of the images. Japanese Female Facial Expression (JAFFE) database and our own customized database are used for the analysis. The experimental result shows that ICA provides improved facial expression recognition in comparison with PCA. The PCA and ICA provides detection accuracy of 81.42 and 94.28 %, respectively.

Keywords Cosine similarity measure · Eigen faces · Euclidean distance classifier · Feature extraction · Human computer interface (HCI) · ICA · PCA

M.N. Patil (✉) · Brijesh Iyer
Department of Electronics and Telecommunication Engineering,
Dr. Babasaheb Ambedkar Technological University,
Lonere 402103, Raigad, Maharashtra, India
e-mail: mansipatil261@gmail.com

Brijesh Iyer
e-mail: brijeshiyer@dbatu.ac.in

Rajeev Arya
IIT, Roorkee 247667, India
e-mail: rajeevarya.ism@gmail.com

M. Pant et al. (eds.), *Proceedings of Fifth International Conference on Soft Computing for Problem Solving*, Advances in Intelligent Systems and Computing 436, DOI 10.1007/978-981-10-0448-3_81

1 Introduction

Human face is the basic trait which differentiates one person to another by means of its shape, texture, and facial expressions (FE). Of all these traits, facial expression plays a vital role in human identification as it is very unique and related with the psychological condition of the human being. The automatic detection of human FE may be very useful for law enforcement agencies, surveillance, and security applications. Humans may easily recognize facial expressions and emotion of other person. However, it is difficult for a machine to automatically recognize the expression and mood of the person. Owing to this fact, this area is always very challenging and attracted the interest of researchers and academicians to find out viable solutions. For effective HCI, the systems must be intelligent to detect and analyze the input query so that it can automatically recognize the FE such as happy, sad, disgust, fear, neutral, surprise, anger, etc. From early 1970s, the research in the detection and analysis of human FE was initiated. Ekman and Friesen had firstly reported the research in the area of human FE detection and analysis [1, 2]. Ever since then, the detection and analysis of human FE acquired a hot spot in many day-to-day applications such as biometric identity authentication, ATM access security, crime detection, computer security, emotion recognition, etc.

Techniques such as hidden markov model (HMM), artificial intelligence, neural network-based approach, and PCA had been used for the human FE analysis. All these methods suffer from high false detection, huge computational requirement, large training time or a combination of them. PCA is widely used in face and facial expressions recognition application for dimensionality reduction and feature extraction. It reduces large dimensionality data space into smaller dimensionality feature space [3–5]. Based on the concept of eigen faces, PCA represents each face image as a linear combination of eigen faces. Human FE may be recognized by comparing the characteristics feature vectors of the face to those of known face images in the training set. ICA is mostly used for facial expression recognition due to its ability to extract local features. ICA is a generalization of PCA and works on the principle of blind source separation (BSS) technique. ICA extracts the individual signal from the mixture of different signals. It assumes that the input image is a linear mixture of train images and it then reduces the statistical dependencies of data to produce statistically independent basis images and coefficients [6–10].

In the present work, PCA is used for the reduction of dimensions and ICA is applied on reduced PCA subspace to find statistically independent basis images for the corresponding facial expression image feature extraction. The performance evaluation is carried out using Euclidean distance classifier and cosine similarity measure for matching and testing of the images. ICA results are compared with the results obtained using PCA on the same datasets. The rest of the paper is organized as follows: Section 2 describes the PCA methodology whereas ICA approach is detailed in Sect. 3. The results are discussed in Sect. 4 of the paper. Conclusions and future scope is given in Sect. 5.

2 PCA Methodology

2.1 Overview

PCA finds a linear projection of high-dimensional data into a lower dimensional subspace. It is used to find the eigenvectors to represent the face image within the entire image space. These eigenvectors define the subspace of face images, which is called as *Face Space*. Further, human face may be represented as linear combination of weighted eigenvectors called eigenfaces. The database of facial expression images is divided into training set and test set. Each image of size $N \times N$ in training set may be represented as column vector of size N^2. Let there are M images in training dataset and they may be represented as $X_1, X_2, X_3, \ldots X_M$.

1. Turn each image in training set into column vector and form a column matrix L. Each column of L represents a training image.
2. The average face image (mean image) may be calculated as

$$m = \frac{1}{M}\sum_{i=1}^{M} X_i \tag{1}$$

3. Subtract mean image vector m from each column of L (i.e., each train image) to form mean-centered matrix A.

$$\Phi_i = X_i - m \tag{2}$$

4. The matrix A is set of all the mean-centered images and given by

$$A = [\Phi_1 \Phi_2 \ldots \Phi_M] \tag{3}$$

5. The covariance matrix of mean-centered image is given by, $C = A \cdot A^T$ with a dimension of $N^2 \times N^2$. To determine the N eigen values and eigen vectors requires a huge amount of calculations which may be minimized by dimensionality reduction approach. Hence, covariance matrix $C = A \cdot A^T$ is estimated which is much smaller matrix of dimension $M \times M$.
6. Do singular value decomposition (SVD) on C and Calculate eigen values λ_i and eigen vectors V_i of covariance matrix $C = A^T \cdot A$ such that

$$A^T \cdot AX_i = \lambda_i \cdot X_i \tag{4}$$

$$AA^T \cdot AX_i = \lambda_i (A \cdot X_i) \tag{5}$$

Here eigen vectors corresponding to $A \cdot A^T$ may be calculated with reduced dimensionality where λ_i is eigen value and $A \cdot X_i$ eigen vector denoted by U_i. It is seen that U_i resembles facial images which looks ghostly and these images are called as Eigen faces.

7. Select Eigen vectors with large eigenvalues and obtain corresponding eigenfaces. Energy mainly locates in subspace constituted by first few eigenvectors with large eigenvalues. Hence, eigenvectors with less eigenvalues may be neglected, i.e., select first k eigenvectors such that $k < M$ and calculate corresponding k eigenfaces. Therefore, great compression can be achieved with improved computational efficiency.
8. Project each image in the training set onto this face space of eigenfaces and calculate feature vectors also called as weight vectors as-

$$W_k = U^T \cdot (X_k - m) \quad \text{where} \tag{6}$$

where $k = 1, 2, 3, \ldots, M$ and $(X_k - m)$ represents mean-centered image. Hence by using Eq. (6), projection of each train image may be obtained as W_1 for 1st image, W_2 for 2nd image, W_3 for 3rd image, and so on.

2.2 *Facial Expression Recognition Process*

- Read the test image of person of same size $N \times N$ and convert test image to column vector T.
- Subtract mean image of training set from test image and project this mean-subtracted test image on face space (eigen faces) to obtain a weight vector w given as-

$$W_k = U^T \cdot (T - m) \tag{7}$$

- Calculate Euclidean distance between projected test image and projected train images.

$$\epsilon_k^2 = \|W - W_k\|^2 \text{ where, } k = 1, 2, 3, \ldots, M \tag{8}$$

- The train image with minimum value of Euclidean distance. If the test image closely represents the face image, it is assumed to fall in the same training set as that of the face image and the expression of test image is the expression corresponding to that training image.

3 ICA Methodology

3.1 *Overview*

In a task such as facial expression recognition, most of the facial expression information is contained in the high-order relationships among the image pixels. PCA decorrelates second-order statistics, whereas ICA minimizes both second-order as well as high-order dependencies in the input. ICA is generalization of PCA which separates the high-order moments of the input in addition to the second-order moments. ICA algorithm maximizes the mutual information between the input and the output. In the present work, PCA is used for preprocessing and then ICA algorithm is applied on the reduced PCA subspace to find statistically independent basis images.

3.2 Preprocessing with PCA

1. Centering: Subtract mean image vector m from each training image to form mean-centered matrix A.

$$X_c = X_i - m \tag{9}$$

   The matrix A is set of all the mean-centered images.

$$A = [X_{c1}X_{c2}\ldots XC_{CM}] \tag{10}$$

2. Whitening: Whitening linearly transforms the observation vector such that its components are uncorrelated and having unit variance. Let X_w is whitened vector then it satisfies the following Eq. (11)

$$E\{X_w \cdot X_w^T\} = I \tag{11}$$

   Calculate covariance matrix $C = A^T \cdot A$ of reduced dimensions and using SVD calculate Eigenvalue matrix D and corresponding orthonormal Eigenvector matrix V. The observation vector is whitened and calculate true Eigen vectors of original covariance matrix of higher dimensions $C = A \cdot A^T$

$$\tilde{X} = D^{-0.5} V^T A^T \tag{12}$$

3.3 ICA Representations for Facial Expression Analysis

In the proposed work, the role of ICA is to extract features from the whitening preprocess. Our aim is to find out a set of statistically independent basis images. From linear input mixture model, $X = AS$; where A is an unknown full rank $M \times M$ mixing matrix (Fig. 1).

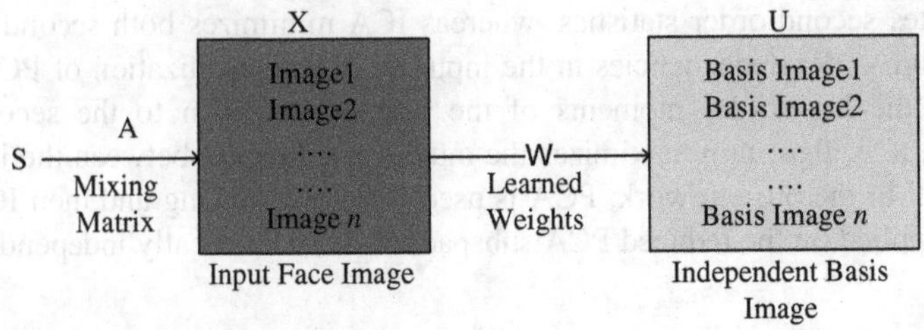

Fig. 1 Calculation of statistically independent basis images

The face images in the row vector of X are linear mixture of an unknown set of statistically independent source image S. Each row of X is a face image vector, which are concatenated to a $1 \times M^2$ dimensional row vector, x_i. The algorithm needs to find both A and S using X. After calculating matrix A, we can calculate its inverse, $W = A^{-1}$ to obtain U mixing matrix as $U = WX$. Let P_m denote the matrix containing the first m principle component axes in its columns. ICA is performed on P_m^T to produce a matrix of m independent source images in the rows of U. The coefficients b for the linear combination of basis images in U that comprised the face images in X was determined as follows.

The PC representation of the set of zero-mean images in X based on P_m is defined as $R_m = X \cdot P_m$. Further, the coefficients of ICA for training images can be obtained as $B = R_m \cdot W^{-1}$. Hence, the rows of $B = R_m \cdot W^{-1}$ contained the coefficients for the linear combination of statistically independent sources U.

The PCA coefficients for input test image is obtained by $R_{\text{test}} = X_{\text{test}} W^{-1}$and the coefficients of ICA for test image can be obtained as $B_{\text{test}} = R_{\text{test}} W^{-1}$.

Each face image is comprised as a linear combination of coefficients b and independent basis images u. The Corresponding face image is given as $b_1 * u_1 + b_2 * u_2 + \cdots + b_n * u_n$.

3.4 Facial Expression Recognition Process

Facial expression recognition performance is evaluated using Euclidean distance classifier and cosine similarity measure. Coefficient vectors in each test set were assigned the class label of the coefficient vector in the training set that was most similar as evaluated by the cosine of the angle between them.

$$C = \frac{B_{\text{test}} \cdot B_{\text{train}}}{\|B_{\text{test}}\| \cdot \|B_{\text{train}}\|} \tag{13}$$

4 Experimental Results

The performance of the PCA and ICA for facial expression recognition is verified using JAFFE database [11]. It consist of number of individuals = 10; total number of images in database = 213 with image resolution = 256 × 256 pixels (gray scale images); No. of facial expressions = 7 facial expressions (angry, disgust, fear, happy, neutral, sad, and surprise) posed by 10 Japanese female models.

Figure 2 shows sample images from JAFFE. Each person has three to four images of each expression. 143 images are used for training and 70 images are used for testing. There are 10 test images of each expression. The PCA and ICA algorithms are tested on various facial expressions

4.1 Preprocessing Steps

We crop the image and take face portion and resize the image to size 227 × 200. The advantage of these preprocessing steps is reduced computation time, increased speed of recognition and less memory for storage. For JAFFE database the original size of image in database is 256 × 256; cropped image size = 130 × 160, and resized it to 60 × 75. Figure 3a shows original image of size 256 × 256. Figure 3b shows cropped image of size 130 × 160 and Fig. 3c shows resized image of size 60 × 75. Figure 4 shows the mean face image of all training images of JAFFE database.

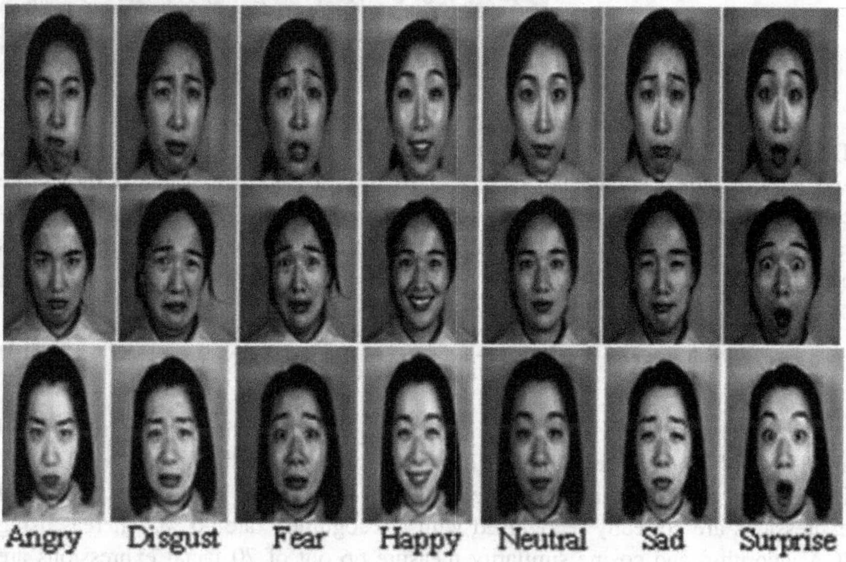

Fig. 2 Sample images from JAFFE database

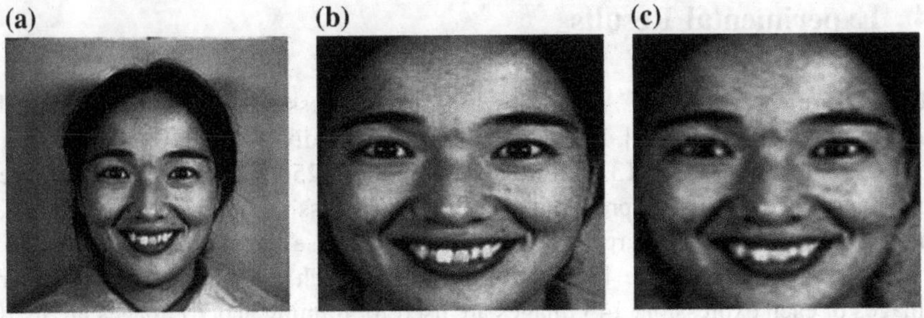

Fig. 3 **a** Original image; **b** Cropped image and **c** Resized image

Fig. 4 Mean face image

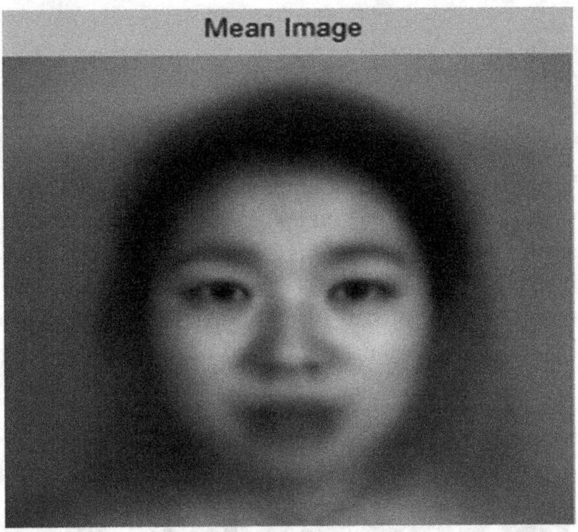

Tables 1 and 2 shows the confusion matrix for 7 basic facial expressions using PCA and ICA in which euclidean distance is used as a classifier. Table 3 shows the confusion matrix for ICA using cosine similarity measure. The recognition accuracy (RA) of PCA and ICA for facial expression recognition on JAFFE database is estimated as:

$$\text{RA} = \frac{\text{No. of test image expressions correctly recognized}}{\text{Total no. of test images}}. \quad (14)$$

Using PCA, for JAFFE test dataset, the image expressions correctly recognized are 57 out of 70. The facial expression recognition rate is observed to be 81.42 %. Using ICA algorithm and Euclidean distance classifier, 63 out of 70 facial expressions are correctly recognized with a recognition rate 90 %, whereas using ICA algorithm and cosine similarity measure 66 out of 70 facial expressions are correctly recognized and facial expression recognition rate observed is 94.28 %.

Table 1 Confusion matrix for FER using PCA (Euclidean distance)

Expression	Anger	Disgust	Fear	Happy	Sad	Surprise	Neutral	Accuracy rate (%)
Anger	7	2	0	0	1	0	0	70
Disgust	1	9	0	0	0	0	0	90
Fear	0	0	8	0	1	0	1	80
Happy	0	0	0	10	0	0	0	100
Sad	0	0	2	1	6	0	1	60
Surprise	0	0	0	0	0	8	2	80
Neutral	0	0	0	1	0	0	9	90
Average accuracy rate (%)								81.42

Table 2 Confusion matrix for FER using ICA (Euclidean distance)

Expression	Anger	Disgust	Fear	Happy	Sad	Surprise	Neutral	Accuracy rate (%)
Anger	10	0	0	0	0	0	0	100
Disgust	1	9	0	0	0	0	0	90
Fear	0	0	10	0	0	0		100
Happy	0	0	0	10	0	0	0	100
Sad	0	0	0	2	6	1	1	60
Surprise	0	1	0	1	0	8	0	80
Neutral	0	0	0		0	0	10	100
Average accuracy rate (%)								90

Table 3 Confusion matrix for FER using ICA (cosine similarity measure)

Expression	Anger	Disgust	Fear	Happy	Sad	Surprise	Neutral	Accuracy rate (%)
Anger	10	0	0	0	0	0	0	100
Disgust	0	9	0	1	0	0	0	90
Fear	0	0	10	0	0	0	0	100
Happy	0	0	0	9	0	0	1	90
Sad	0	0	0	1	9	0	0	90
Surprise	0	0	0	1	0	9	0	90
Neutral	0	0	0	0	0	0	10	100
Average accuracy rate (%)								94.28

It is observed that the RA is sensitive to the selection of Eigen values and corresponding Eigen vectors since the facial expression recognition rate is directly proportional to the number of feature vectors. Using PCA and ICA, we get 100 % of recognition rate for few expressions for 70 test images. However, as number of images increases the recognition rate decreases. ICA provides better results using

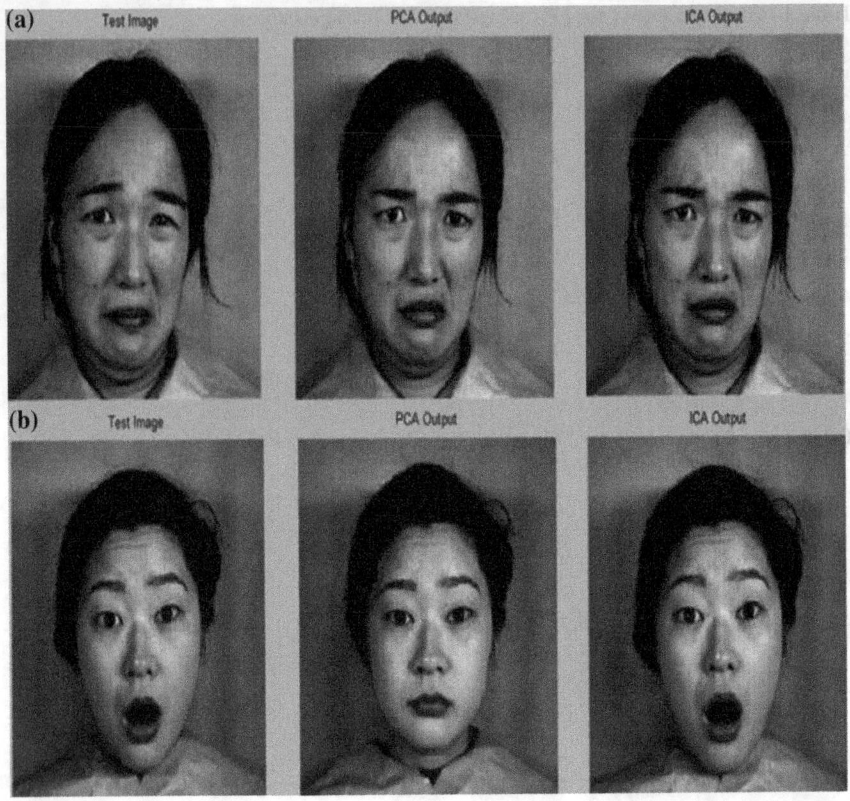

Fig. 5 **a** Recognition success output expression match of PCA and ICA. **b** Recognition failure output expression PCA mismatch and ICA match

cosine similarity measure rather than Euclidean distance classifier. PCA is faster than ICA to recognize different expressions. However, ICA is superior in terms of detection accuracy.

Figure 5a, b shows the output of PCA and ICA expression match for success as well as failure case, respectively. From Fig. 5a, it may be seen that both PCA and ICA correctly recognized disgust expression from JAFFE database. Figure 5b depicts that PCA fails and ICA successfully recognizes the surprise expression from JAFFE database. Figure 6 shows facial expression recognition rate comparison between PCA and ICA using Euclidean distance as a cost function and ICA using Cosine similarity as a cost function for different expressions.

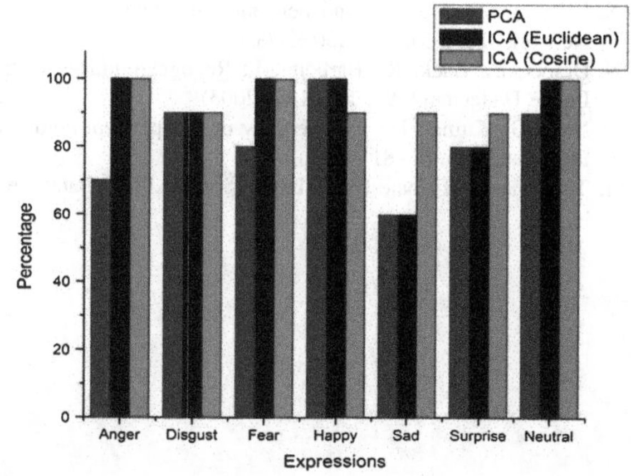

Fig. 6 Comparision of PCA and ICA results for test images

5 Conclusions

In this paper, performance evaluation of PCA and ICA methodologies for facial expression recognition has been carried out. The novelty of the present work is use of PCA for preprocessing (i.e., dimensionality reduction) of database images and ICA for feature extraction along with Euclidean distance and cosine similarity measure cost functions. Experimental results show that ICA gives better accuracy and lower error rates for facial expression recognition than PCA algorithm. ICA representations are designed to maximize information transmission in the presence of noise and thus they may be more robust to variations such as lighting conditions, changes in hair, make-up, etc. In future, an adaptive methodology may be developed depending on the specific facial expression of the human being.

References

1. Ekman, P., Friesen, W.V.: Constant across cultures in the face and emotion. Jr. Pers. Soc. Psychol. **17**(2), 124–129 (1971)
2. Ekman, P., Priesen, W.: Facial Action Coding System: A Technique for the Measurement of Facial Movements. Consulting Phychologists Press, Palo Alto, CA (1978)
3. Turk, A., Alex, P.: Face recognition using eigenfaces. In: IEEE Computer Society Conference on Computer Vision and Pattern Recognition (1991)
4. Garg, A., Choudhary, V.: Facial expression recognition using principal component analysis. Int. J. Sci. Eng. Res. Technol. (2012)
5. Meher, S., Maben, P.: Face recognition and facial expression identification using PCA. In: IEEE International Advanced Computing Conference, pp. 1093–1098 (2014)
6. Zia Uddin, Md., Lee, J., Kim, T.: An enhanced independent component-based human facial expression recognition from video. IEEE Trans. Consumer Electron. **55**(4), 2216–2224 (2009)
7. Stewart, M., Javier, B., Movellan, R., Sejonowski, T.: Face recognition by independent component analysis. IEEE Trans. Neural Networks **13**(6), 1450–1464 (2002)

8. Hyvarinen, A., Oja, E.: Independent component analysis: algorithm and applications. Neural Networks **13**(4–5), 411–430 (2000)
9. Draper, B., Baek, K., Bartlett, M.: Recognizingfaces with PCA and ICA. Comput. Vision Image Understand. **91**, 115–137 (2003)
10. Naik, G., Kumar, D.: An overview of independent component analysis and its applications. Informatica **35**, 63–81 (2011)
11. The Japanese Female Facial Expression (JAFFE) Database: http://www.kasrl.org/jaffe.html

Modified PSO Algorithm for Controller Optimization of a PFC Ćuk Converter-Fed PMBLDCM Drive

Rinku K. Chandolia and Sanjeev Singh

Abstract This paper presents a modified particle swarm optimization (PSO) algorithm for the selection of controller parameter for a Ćuk converter-fed PMBLDCM drive. The main objective of the proposed algorithm for controller optimization is to achieve power factor correction (PFC) at AC mains of the PMBLDCM drive. The PSO is modified to achieve the above objectives for a PMBLDC motor rated as 1.01 KW, 3000 rpm, 310 V, and 3.2 Nm. The complete drive is designed, modeled, and its performance is simulated in MATLAB-Simulink. The simulated results of the drive are presented to demonstrate the desired power quality at AC mains along with the desired speed and torque for the searched values of the controller parameters.

Keywords Power Factor Correction (PFC) · PMBLDCM · Power Quality (PQ) · Ćuk converter · Controller optimization · PSO

1 Introduction

Permanent magnet brushless DC motors (PMBLDCM) have high efficiency, high torque, low volume, high reliability, and low maintenance. Due to these features, it is used in the field of speed control applications such as military and aerospace appliances, as position control in machine tools, robotics, and high-precision servos. Due to the absence of brushes, PMBLDC motors are safer to work in the environment where there exists danger of explosion. Advancement in power electronics led its applications in the field of electric vehicles (EVs) and hybrid electric vehicles (HEVs), treadmills, robotics, hard disk of computer drives, air conditioner compressors, fans, freezers, and plenty of other applications [1].

R.K. Chandolia (✉) · S. Singh
EIE Department, SLIET Longowal, Sangrur, Punjab, India
e-mail: rkcrinku@gmail.com

S. Singh
e-mail: sschauhan@sliet.ac.in

M. Pant et al. (eds.), *Proceedings of Fifth International Conference on Soft Computing for Problem Solving*, Advances in Intelligent Systems and Computing 436, DOI 10.1007/978-981-10-0448-3_82

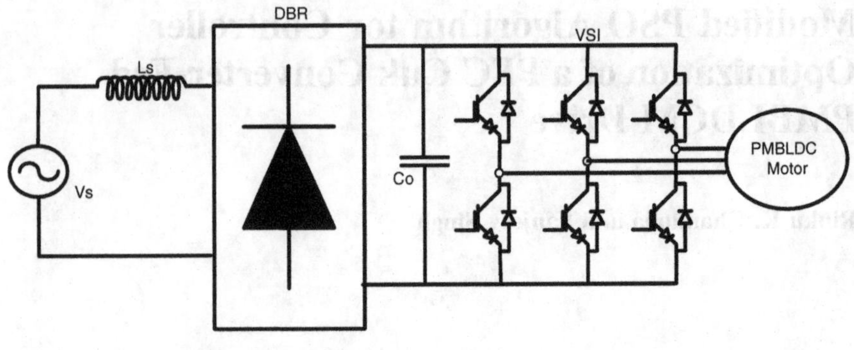

Fig. 1 Single phase DBR-VSI-fed PMBLDC motor

The PMBLDCMs are controlled using voltage source inverter (VSI) fed from AC mains through a diode bridge rectifier (DBR) and a capacitive filter at DC link as shown in Fig. 1. Due to uncontrolled charging and discharging of DC link capacitor, DBR draws a peaky current from the AC mains as shown in Fig. 2. This pulse-shaped current results in poor power quality at AC mains having total harmonic distortion (THDi) 86.46 %, crest factor (CF) 2.3347, power factor (PF) 0.7209, and displacement power factor (DPF) 0.9530. The situation becomes more severe when many such drives are employed simultaneously. These drives are operated from utility supply; therefore, they should conform to international PQ standards. There are various international standards for electrical power quality out of which IEC-61000-3-2 presents the limits of current harmonics for drives [2].

Power factor correction (PFC) and power quality (PQ) improvement are synonymous terms that represent reduced total harmonic distortion (THD) in current as well as voltage at AC mains, power factor (PF) near unity value, and crest factor (CF) as per pure sine wave [3].

To get PQ improvement or PFC as per international standards, a DC–DC converter is used in between DBR and VSI with PFC control scheme to obtain reduction of THD, improvement in PF and CF, etc. The Ćuk converter is proposed

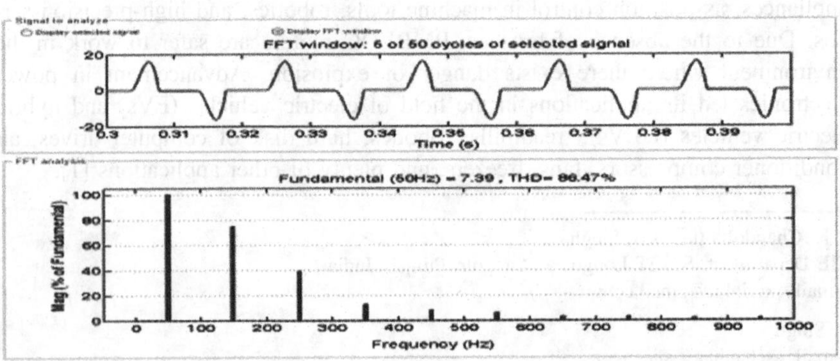

Fig. 2 Current waveform and its harmonic spectra at AC mains for PMBLDCM drive

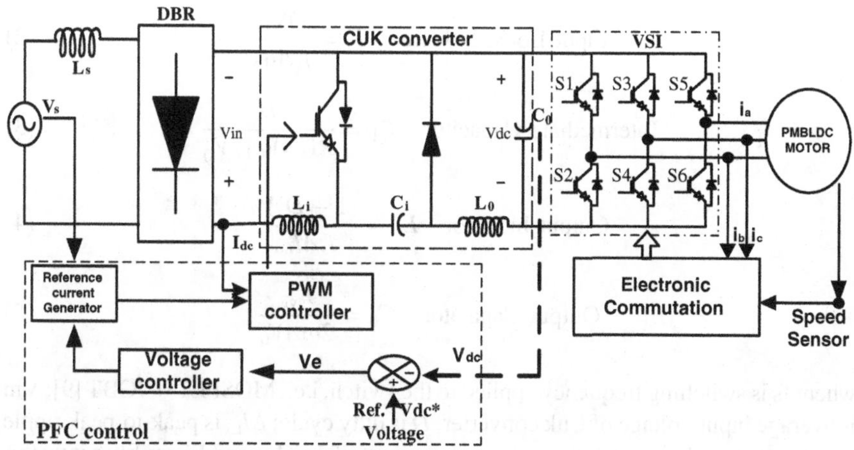

Fig. 3 Block diagram of PFC-controlled Ćuk converter-fed PMBLDC motor drive

in this paper as it has advantages like continuous input and output currents, small output filter, and better efficiency over buck–boost and SEPIC converters [4, 5].

Figure 3 shows the PFC control scheme for PMBLDCM drive which uses current multiplier control and operates in continuous conduction mode of the converter. The voltage controller used is a simple proportional–integral (PI) controller and requires optimum selection of the controller parameters Kp and Ki.

Selection of PI controller parameters is not simple because Kp and Ki are conflicting in nature. If Kp is higher than Ki, it results in high overshoot and oscillation during transient conditions; whereas, if Kp is lower than Ki, it results in sluggish response of the drive. Thus, optimization techniques based on random search methods like PSO, genetic algorithms, and ANN, etc., are used by various authors for various applications [6–8]. This paper proposes a modified PSO algorithm to search the optimum gains for power quality controller of a Ćuk converter-fed PMBLDCM drive.

2 Design of Ćuk Converter for PFC of PMBLDCM Drive

Figure 3 shows the block diagram of Ćuk converter-fed PMBLDCM drive. The Ćuk converter is one of the buck–boost converters with capacitive energy transfer, therefore, reduced EMI as compared to others. The design parameters of Ćuk converter are calculated using following equations [4]:

$$\text{Output Voltage} \quad V_{dc} = \frac{DV_{in}}{1 - D} \tag{1}$$

$$\text{Input Boost Inductor} \quad L_i = \frac{\mathrm{DV_{in}}}{f_s \Delta I_{\mathrm{Li}}} \tag{2}$$

$$\text{Intermediate Capacitor} \quad C_1 = \frac{D}{\mathrm{R}f_s \Delta V_{C1}/V_0} \tag{3}$$

$$\text{Output Inductor} \quad L_0 = \frac{(1-D)V_0}{f_s \Delta I_{L0}} \tag{4}$$

$$\text{Output Capacitor} \quad C_0 = \frac{I_{\mathrm{avg}}}{2\omega \Delta V_0} \tag{5}$$

where fs is switching frequency applied to the switch, i.e., MOSFET or IGBT [9]; Vin is average input voltage of Ćuk converter; D is duty cycle; ΔI_{Li} is peak-to-peak ripple current of I_{Li}; ΔI_{Lo} is peak-to-peak ripple current of I_{Lo}; Lo and Li are filter inductor and input boost inductor, respectively; C_0 and C_1 are filter capacitor and intermediate energy transfer capacitor, respectively; I_{av} is Average DC output current.

3 PFC Control Scheme

PFC control scheme in current multiplier approach mainly consists of three parts, i.e., voltage controller, reference current generator, and PWM controller. The modeling equations [4] for these components are summarized below:

3.1 Voltage Controller

$$\text{Voltage Error} \quad V_e(k) = V^*_{\mathrm{dc}}(k) - V_{\mathrm{dc}}(k) \tag{6}$$

PI controller output at kth instant

$$I_c(k) = I_c(k-1) + K_p\{V_e(k) - V_e(k-1)\} + K_i V_e(k) \tag{7}$$

3.2 Reference Current Generator with PWM Controller

$$i^*_d = I_c(k) u_{\mathrm{vs}} \tag{8}$$

$$\text{Unit template} \quad u_{\mathrm{Vs}} = \frac{v_d}{u_m}; \quad v_d = |v_s|; \quad v_s = V_{\mathrm{sm}} \sin \omega t \tag{9}$$

$$\text{PWM Controller} \quad \Delta i_d = i_d^* - i_d \tag{10}$$

$$\text{If} \quad k_d \Delta i_d > m_d(t) \quad \text{then} \quad S = 1 \tag{11}$$

$$k_d \Delta i_d \le m_d(t) \quad \text{else} \quad S = 0 \tag{12}$$

The voltage controller consists of a proportional-integral (PI) controller which is used to control the DC link voltage. Its output is multiplied to get the current multiplier control for power quality improvement at AC mains of the PMBLDCM drives. Therefore, the controller parameters are very important for effective control system.

4 PSO Algorithm

Conventional PSO technique is a search-based technique in which populations of particles are randomly generated to suit the objective function. In this paper, it is ensured that the particles are generated in all the search space with modification in the algorithm and applied to the multiobjective problem. These randomly generated particles move in search space using velocity parameter. New particles are generated using previous iteration and velocity [7, 8, 10]. The velocity is also varied as the search progresses to facilitate exploration and exploitation. The step-by-step method of the modified PSO is explained below:

Step 1: Generate population of particles in search space to cover entire area using the following equation:

$$\begin{aligned} X(j+N*(i-1),1) &= X\text{min} + (j-1)*p + \text{rand}(1); \\ X(j+N*(i-1),2) &= X\text{min} + (j-1)*p + \text{rand}(1); \\ p &= (X_{\max} - X_{\min})/N \end{aligned} \tag{13}$$

where i and j varies from 1 to N, which is any integer number.

Step 2: Initialize pbest and gbest with their position. Set initial velocity of agent (particles).

Step 3: Calculate the personal best of particle for particular iteration.
If fitness(p) is better than fitness(pbest), then pbest = p; otherwise, retain previous pbest.

Step 4: Calculate gbest that is suggested by all the particles for particular iteration.

$$\text{gbest} = \min(\text{pbest}); \tag{14}$$

Step 5: Update the velocity of particles using the following equation:

$$V(i^{\text{th}}, j^{\text{th}}) = \text{WV}(i^{\text{th}-1}, j^{\text{th}-1}) + c1 * r1 * (\text{pbestp}(i^{\text{th}}, j^{\text{th}}) - X(i^{\text{th}}, j^{\text{th}})) + c2 * r2 * (\text{gbestp}(i, j) - X(i^{\text{th}}, j^{\text{th}})); \tag{15}$$

In this equation, w is inertia of the previous velocity, $c1$ and $c2$ are acceleration constants, and $r1$ and $r2$ are randomly generated numbers.

Step 6: Update position of particles using the following equation:

$$X(i^{\text{th}+1}, j^{\text{th}+1}) = X(i^{\text{th}}, j^{\text{th}}) + V(i^{\text{th}}, j^{\text{th}}); \tag{16}$$

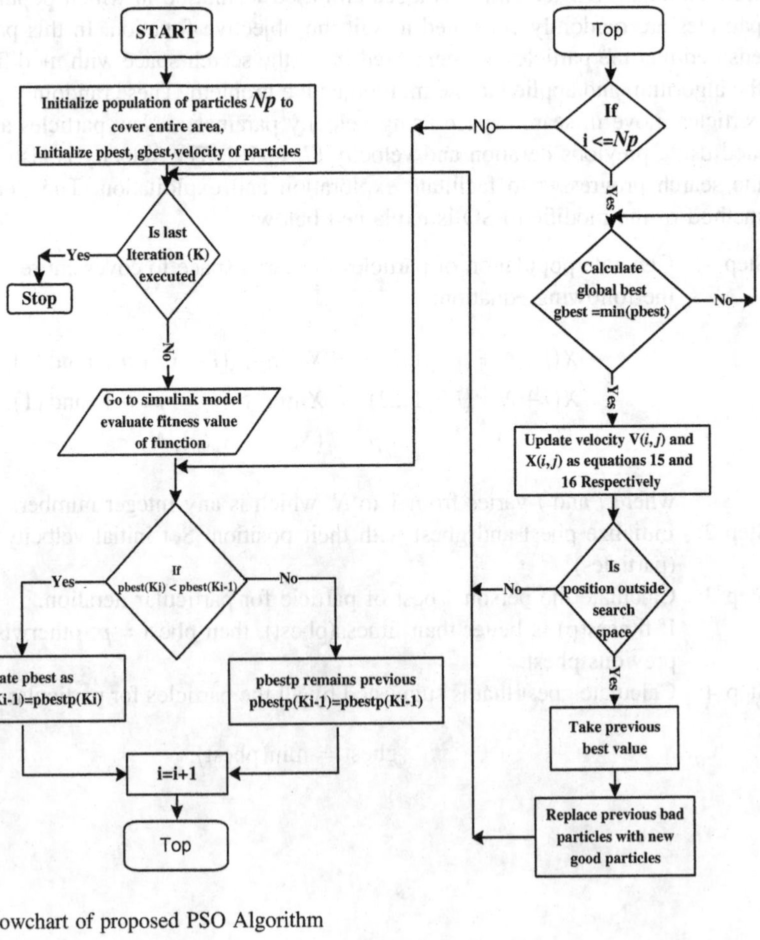

Fig. 4 Flowchart of proposed PSO Algorithm

If the position of particles lies outside the search space region, then take the previous best value as per fitness value of function.

Step 7: Go to Step 3 and repeat until all iterations are not executed. Stop and give gbest as optimal solution for define function (Fig. 4).

Using the above steps, the optimum values of Kp and Ki are searched so that the objective function is minimized.

$$\text{Minimize} \quad \left\{0.15\left(\frac{1}{\text{PF}}\right) + 0.50 \times \text{THD} + \frac{\text{del}V}{320} \times 100 \times 0.35\right\}$$
$$\text{Subject to} \quad \left.\begin{matrix} \text{PF} \geq 0.999 \\ \text{THD} \leq 5\,\% \\ \text{del}V_0 \leq 3\,\% \end{matrix}\right\} \text{in Steady State...} \tag{17}$$

5 Performance Evaluation of PMBLDCM Drive

The complete PMBLDCM drive as per the rated data shown in Table 1 is modeled in MATLAB-Simulink environment and called in the PSO program for optimization of the controller parameters for desired results. The minimization of objective function defined in Eq. (17) considers near unity PF, reduction of THD, and ΔV. The obtained performance of the drive is presented in the form of crest actor (CF), displacement power factor (DPF), total harmonic distortion (THDi) at AC mains at the rated torque, and different running conditions.

5.1 Controller Parameter Optimization

To cover complete search space, 25 particles are generated and applied to the PMBLDCM drive. These particles carry their best fitness and go toward global best fitness in search space in successive iteration. Velocity of particles is changed sequentially after 10 iterations as 1, 0.85, 0.7, and 0.55. Initially, generated positions of particles are shown in Fig. 5a. The behavior of particles in successive iteration is shown in Fig. 5b–d, respectively.

Table 1 Rated values of PMBLDC Motor

Motor parameters	Value	Unit
Rated torque, speed, power	3.2, 3000, 1.01	Nm, rpm, kW
Inertia, Resistance, Inductance	2.2, 3.58, 9.13	kg, cm^2, mH, Ω
Back emf constant K_b, poles	0.418, 4	Vs/rad,

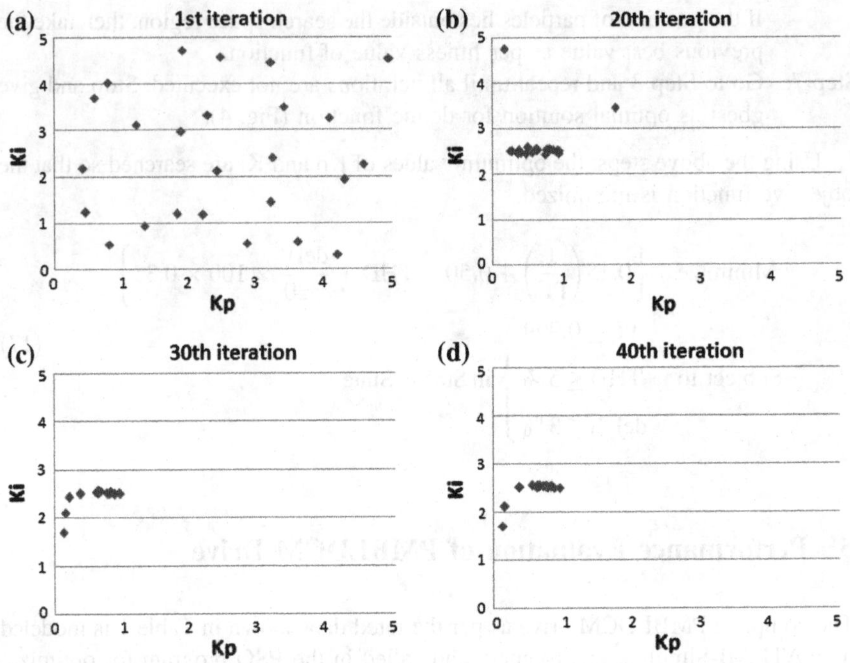

Fig. 5 Behavior of swarm particles in search space under successive iteration

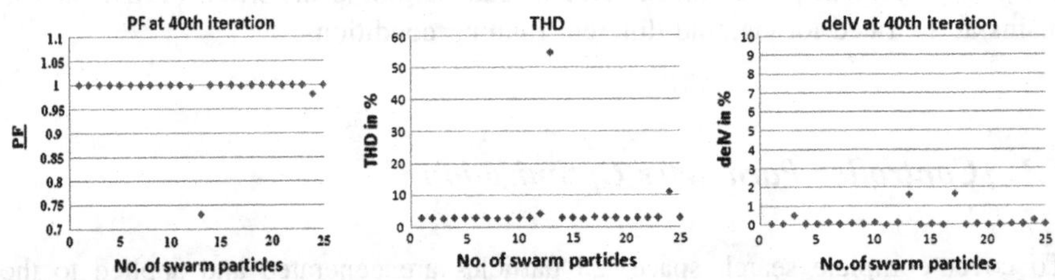

Fig. 6 Swarm particles evaluated value of PF, THD, and ΔV at 40th iteration

After 40th iteration, more than 20 swarm particles come close to gbest point and give global best position as gbestp (global best position). The gbestp is nothing but PI controller (Kp, Ki) value which is (0.3796, 2.5080) as shown in Fig. 5d.

At 40th iteration, all 25 particles have individual best position, the PF, THD, and delV values for these best positions are shown in Fig. 6, which show that nearly all particles have more than 0.999 PF, less than 5 % of THD, and less than 3 % of variation in output voltage (delV). This concluded that almost all particles are close to gbestp, as discussed in earlier section.

5.2 Performance of the PMBLDCM Drive with Optimum Controller

The results obtained for the performance of PMBLDCM drive with optimized PFC controller are shown in Figs. 7, 8, and 9, respectively. Using optimized controller, the input current becomes sinusoidal and THD is reduced to 2.13 % from 86.47 % as shown in Fig. 7.

Figure 8 shows the zero-crossing of input current (I_s) with supply voltage (V_s) and PF at AC mains. The zero-crossing comes at the same time since the power factor (PF) increases to 0.9995 from 0.7209.

Figure 9 shows the transient and steady state response of the drive for input current (I_s), supply voltage (V_s), DC link voltage (V_{dc}), phase A current (I_a), torque (T), and speed (N) at running speed of 2000 rpm. The reference DC link voltage (V_{dc}) is achieved within 0.07 s and the reference speed is achieved within 0.1 s. The torque ripple of the drive remains within limits.

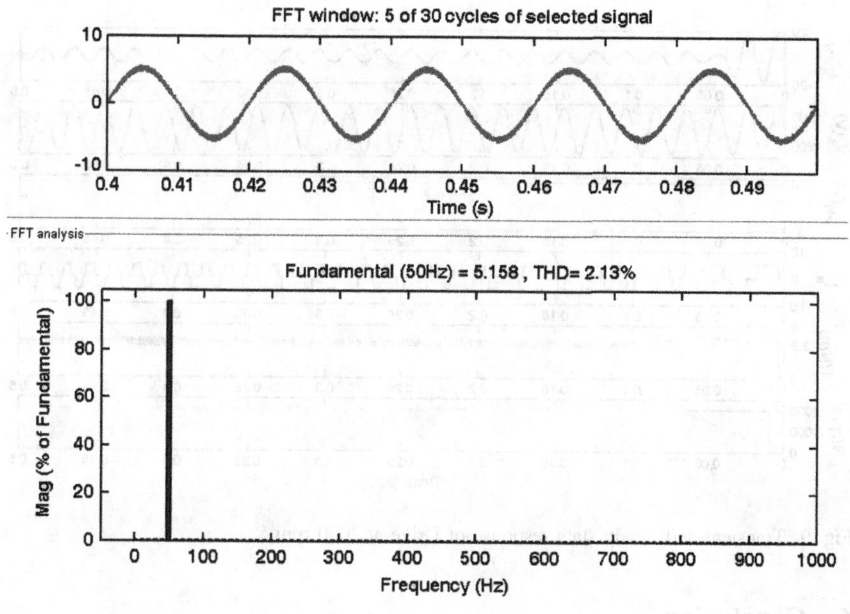

Fig. 7 Current (I_s) waveform at AC mains and its THD at 2000 rpm

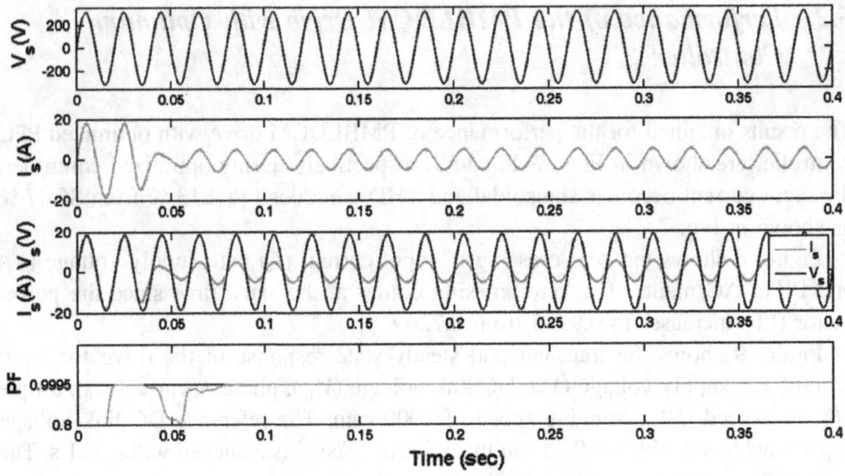

Fig. 8 Zero crossing of I_s, V_s, and PF response at AC mains

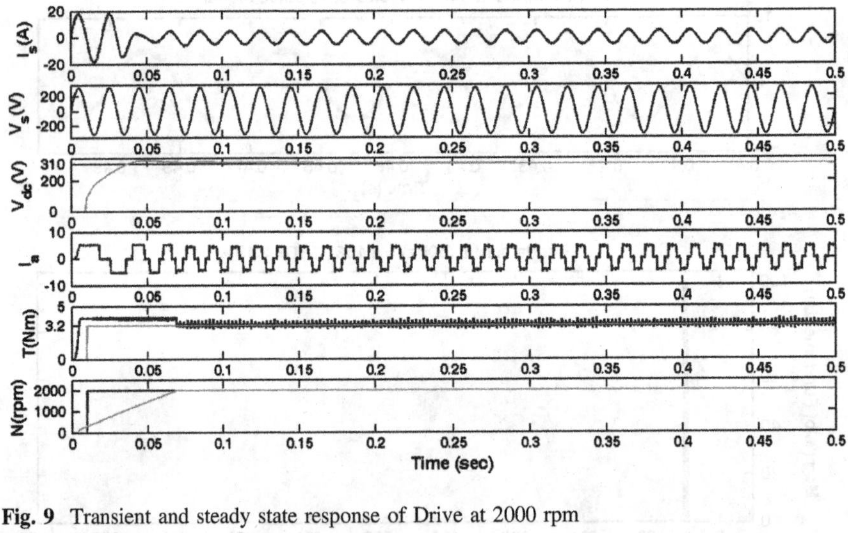

Fig. 9 Transient and steady state response of Drive at 2000 rpm

6 Conclusion

The PSO algorithm is used after modification for optimization of controller parameters for a Ćuk converter to result power factor correction at AC mains of a PMBLDCM drive. The complete drive is modeled and simulated in MATLAB-Simulink and the obtained results have been presented for power quality improvement at AC mains in terms of PF, THDi, and ΔV as 0.9995, 2.13 %, and

0.015 V, respectively, which is well within IEC standards. With the optimized controller, the PMBLDCM drive has shown excellent PQ improvement with desired speed control at rated torque. On the basis of this investigation, it is concluded that this controller optimization algorithm shall be useful for the PMBLDCM drive in varied range of loading and operating conditions.

References

1. Krishnan, R.: Permanent Magnet Synchronous and Brushless DC Motor Drives, pp. 457–473. Prentice Hall (2001)
2. 'Limits for Harmonic Current Emissions (Equipment input current ≤16 A per phase).' International Standard IEC-61000-3-2 (2000)
3. Singh, B., Singh, S.: State-of-Art on permanent magnet brushless DC motor drives. J. Power Electron. **9**(1), 1–17 (2009)
4. Singh, B., Singh, S.: Single phase PFC topologies for permanent magnet brushless DC motor drives. IET Power Electron. **3**(2), 147–175 (2010)
5. Singh, S., Singh, B.: A voltage controlled PFC Ćuk converter based PMBLDCM drive for air-conditioners. IEEE Trans. Ind. Appl. **48**(2), 832–838 (2012)
6. Popadic, B., Dumnic, B., Milicevic, D., Katic, V., Corba, Z.: Tuning methods for PI controller—Comparison on a highly modular drive. IEEE (2013)
7. Singh, S., Singh, B.: Particle swarm optimization for power quality improvement of a 12-Pulse Rectifier-Chopper Fed LCI-Synchronous motor drive. Int. J. Int. Syst. Technol. Appl. (IJISTA) **11**(3/4), 267–285 (2012)
8. Singh, S., Singh, B.: Optimized passive filter design using modified particle swarm optimization algorithm for a 12-Pulse Converter Fed LCI-Synchronous motor drive. IEEE Trans. Ind. Appl. **50**(4), 2681–2689 (2014)
9. Erickson R.W.: Fundamentals of Power Electronics, Chapman and Hall, New York (1997)
10. Panda, S., Padhy, N.P.: A PSO-based SSSC controller for improvement of transient stability performance. Int. J. Intell. Syst. Technol. **2**(I), 28–35, (2007)

Speed Controller Optimization for PMSM Drive Using PSO Algorithm

Paramjeet Singh Jamwal and Sanjeev Singh

Abstract This paper presents the use of particle swarm optimization (PSO) algorithm modified for the search of optimized gain values of speed controller for a permanent magnet synchronous motor (PMSM) drive. The PSO is modified to generate particles in complete search space and have a multiobjective problem involving speed as well as torque error as independent variables of fitness function so as to minimize these errors. The proposed algorithm is modeled and simulated in MATLAB/Simulink environment. The obtained results are presented to demonstrate the effectiveness of the modified PSO algorithm for the desired speed control of the PMSM drive.

Keywords PSO · PMSM · Speed control · Controller optimization · Simulink

1 Introduction

Permanent magnet synchronous motor (PMSM) is a three-phase synchronous motor with permanent magnet rotor and sinusoidal back emf. The control of motor requires continuous rotor position using encoders or resolvers accomplished by electronic controller [1]. In the PMSM, power output is controlled by all the three phases unlike the power output in permanent magnet brushless DC (PMBLDC) motor which is controlled by two phases. PMSM are small in size, high in efficiency, and fast in response. These characteristics of PMSM increase its popularity and find application in domestic and commercial sectors especially in medical, vehicles, and position control systems, e.g., hybrid electric vehicle [2], satellite [3], spacecraft [4], roller conveyers [5], etc. The electronic control of PMSM requires a

P.S. Jamwal (✉) · S. Singh
Electrical and Instrumentation Engineering Department, Sant Longowal Institute of Engineering and Technology, Longowal 148106, Punjab, India
e-mail: paramjeet.jamwal@gmail.com

S. Singh
e-mail: sschauhan@sliet.ac.in

M. Pant et al. (eds.), *Proceedings of Fifth International Conference on Soft Computing for Problem Solving*, Advances in Intelligent Systems and Computing 436, DOI 10.1007/978-981-10-0448-3_83

Fig. 1 Schematic diagram of DBR-VSI-fed PMSM drive

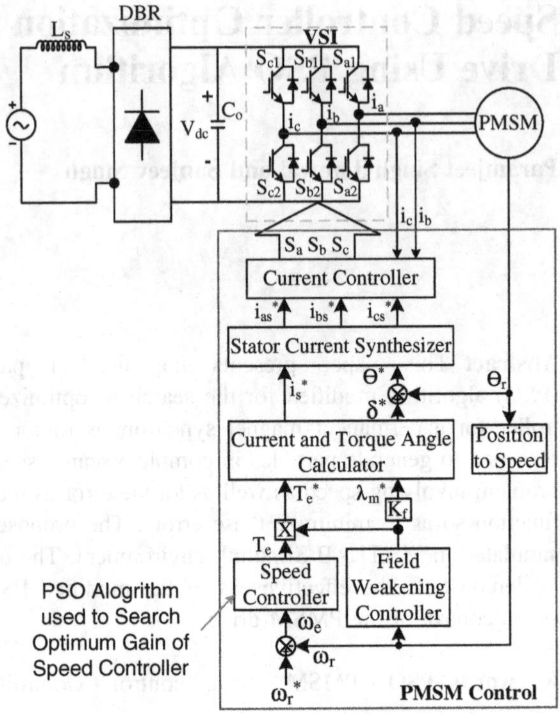

voltage source inverter (VSI) to feed the desired current as per requirement guided by the position information. Figure 1 shows the schematic diagram of the control system of a PMSM drive which depicts a diode bridge rectifier (DBR) for providing DC to VSI because usually DC supply is not available and it is obtained using DBR. The speed control is obtained using a speed controller which has a proportional-integral (PI) controller. Due to changing dynamic conditions of load, the controller parameter, i.e., proportional and integral gains K_P and K_I need to be optimized so that desired results are obtained in all operating conditions.

The particle swarm optimization (PSO) was discussed for the first time in 1995, by James Kennedy and Russell Eberhart [6]. After that, PSO has gone through several modifications to apply it in different applications; some of them are reported in the literature [7–10]. PSO stores two types of best values, personal best, which is the fitness value for each particle, and global best, which is selected out of all personal best values. PSO is applicable to nonlinear problems, in which a number of random particles are generated to initialize the population between desired ranges.

Therefore, this paper presents the swarm behaviour and application of PSO algorithm modified to obtain the desired operating conditions while finding the optimum controller parameter. This paper is presented in five main parts namely introduction, control scheme for PMSM drive, modified PSO algorithm, results for validation of proposed concepts, and conclusion.

2 Control Scheme for PMSM Drive

There are a number of control methods available for the PMSM [11] but this paper uses vector control method [12]. In this method, torque and mutual flux can be controlled separately which results in a fast and efficient control of the PMSM drive. The operation of PMSM drive, as shown in Fig. 1, depends upon the switching of VSI (S_a, S_b, and S_c) received through *PWM controller* [13], which operates on the basis of error in current, which is the difference of simulated motor current (i_{as}, i_{bs}, and i_{cs}) and reference motor current (i_{as}^*, i_{bs}^*, and i_{cs}^*). Three-phase simulated motor current is obtained through *Three-Phase Current Calculator* from two-phase current which is sensed from the stator of the motor using two sensors. Three-phase reference motor current is obtained through *Stator Current Synthesizer* from stator current (i_s^*) and stator angle (Θ_s^*). Stator angle is the sum of the rotor position (Θ_r) and torque angle (δ^*) of the motor. Stator current and torque angle are obtained through *Current and Torque Angle Calculator* from reference mutual flux linkage (λ_m^*) and reference electromagnetic torque (T_e^*). Reference mutual flux linkage is the multiplication of mutual flux linkage (λ_m) and constant of proportionality (K_f). Reference electromagnetic torque is the multiplication of mutual flux linkage and reference torque (T^*). Mutual flux linkage is obtained through *Field Weakening* controller from rotor speed (ω_r) and reference torque is obtained through PI controller from speed error (ω_e). Rotor speed is obtained through *Position to Speed* converter from sensed rotor position and speed error is the difference of reference rotor speed (ω_r^*) and rotor speed.

PI controller being the simplest controller is used as *Speed Controller* which processes the speed error and maintains the speed and torque of the drive at

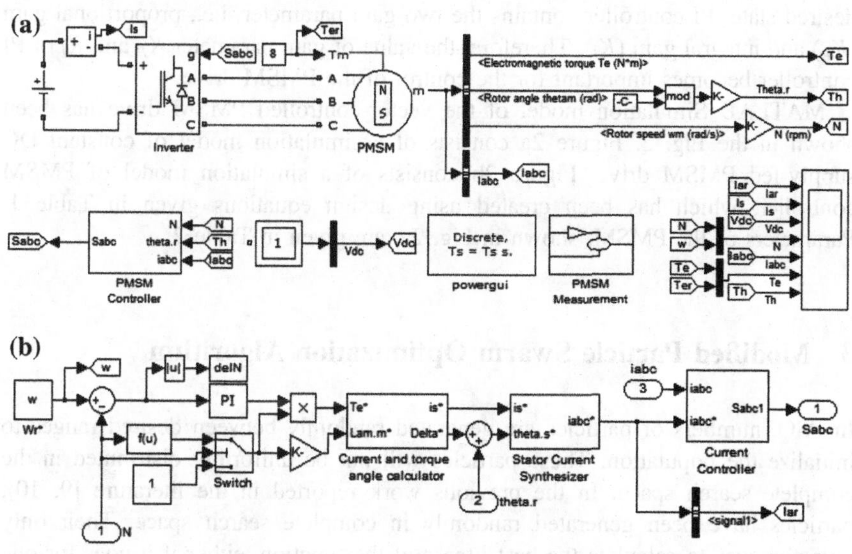

Fig. 2 Simulation model of a Constant DC supply-fed PMSM Drive and b PMSM controller

Table 1 Design equation of PMSM controller

System	Equation
Field weakening controller	$\lambda_m = \begin{pmatrix} 1 & \text{if} \quad \omega_r < \omega_r^* \\ 2 - \frac{\omega_r}{\omega_r^*} & \text{if} \quad \omega_r \geq \omega_r^* \end{pmatrix}$
PI controller	$\omega_e(k) = \omega_r(k)^* - \omega_r(k)$
	$T_e = T_k = T_{k-1} + K_P \times \{\omega_e(k) - \omega_e(k-1)\} + K_i \times \omega_e(k)$
	$T_e^* = T_e \times \lambda_m$
Current and torque angle calculator	$i_T^* = \frac{T_e^*}{\frac{3}{2}\frac{P}{2}\{\lambda_{af} + (L_d - L_q)\, i_f^*\}}$
	$i_f^* = \frac{\sqrt{\lambda_m^2 - (L_q \times i_T^*)^2} - \lambda_{af}}{L_d}$
	$i_s^* = \sqrt{i_T^{*2} + i_f^{*2}}$
	$\delta^* = \tan^{-1}\left(\frac{i_T^*}{i_f^*}\right)$
Stator current synthesizer	$\begin{bmatrix} i_a^* \\ i_b^* \\ i_c^* \end{bmatrix} = i_s^* \times \begin{bmatrix} \text{Sin } \theta_s^* \\ \text{Sin } \left(\theta_s^* - \frac{2\pi}{3}\right) \\ \text{Sin } \left(\theta_s^* + \frac{2\pi}{3}\right) \end{bmatrix}$
PWM current controller	$S_{a_1} = 1$ and $S_{a_2} = 0, \quad if\ \Delta i_a > m_s$
	$S_{a_1} = 0$ and $S_{a_2} = 1, \quad if\ \Delta i_a \leq m_s$
	$S_{b_1} = 1$ and $S_{b_2} = 0, \quad if\ \Delta i_b > m_s$
	$S_{b_1} = 0$ and $S_{b_2} = 1, \quad if\ \Delta i_b \leq m_s$
	$S_{c_1} = 1$ and $S_{c_2} = 0, \quad if\ \Delta i_c > m_s$
	$S_{c_1} = 0$ and $S_{c_2} = 1, \quad if\ \Delta i_c \leq m_s$

desired state. PI controller contains the two gain parameter, i.e., proportional gain (K_P) and integral gain (K_I). Therefore, the value of gain parameter K_P and K_I of PI controller becomes important for the control of the PMSM drive.

MATLAB Simulation model of the vector-controlled PMSM drive has been shown in the Fig. 2. Figure 2a consists of a simulation model of constant DC supply-fed PMSM drive. Figure 2b consists of a simulation model of PMSM controller, which has been created using design equations given in Table 1. Parameters of the PMSM, shown in Fig. 2a, are given in Table 2.

3 Modified Particle Swarm Optimization Algorithm

In PSO, numbers of particles are generated randomly between desired ranges to initialize the population. These particles will not be uniformly distributed in the complete search space. In the previous work reported in the literature [9, 10], particles have been generated randomly in complete search space. Their only objective was to calculate the best fitness of the function, either if it goes for one

Table 2 PMSM Parameters

Parameter	Symbol	Value
Rated torque	T	8 Nm
Maximum torque	T_m	10 Nm
Voltage	V_{dc}	300 V
Rated speed	ω_r	2000 rpm
Stator phase resistance	R_s	0.9585 Ω
d-axis inductance	L_d	0.00525 H
q-axis inductance	L_q	0.00525 H
Rotor flux linkage	λ	0.1827 V s
Inertia	J	0.0006329 kg m2
Friction factor	F	0.0003035 N m s
Pole	P	8
Constant of proportionality	K_f	0.2

iteration or goes for 100 iteration. If best value is not achieved, the function may trap into infinite loop.

$$X(j+N\times(i-1),1)=X_{\min}+\{(j-1)+\text{rand}(1)\}\times l \quad (1)$$

$$X(j+N\times(i-1),2)=X_{\min}+\{(i-1)+\text{rand}(1)\}\times l \quad (2)$$

$$l=\frac{X_{\max}-X_{\min}}{N} \quad (3)$$

where i and j varies from 1 to N, which is any integer number.

Therefore, in this paper, a modification is proposed in PSO, i.e., particles are generated in square of number (N) using Eqs. 1, 2. For example, if the user selects $N = 5$, then $N^2 = 25$ particles will be generated. Every particle is generated randomly but in a particular area of square, whose length is calculated using Eq. 3. Therefore, this method is called as *Square Particle Generation* method, which is adopted to cover the complete search space. The pictorial view of square particle generation method is shown in Fig. 3.

The selection of particles and number of iteration depends upon the range of search space and accuracy level of fitness function, respectively. A square area has

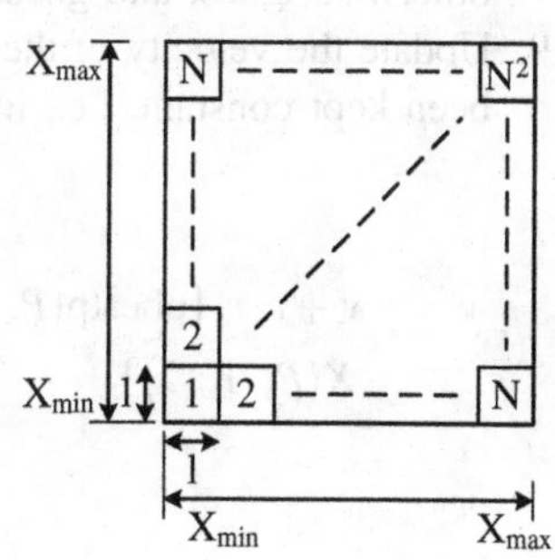

Fig. 3 Square particle generation method

been selected and particles are generated in the form of square of a number, so that it can be distributed equally in complete search space. The number of iterations are so decided that it can describe the swarm behavior along with optimum value calculation.

The steps of the modified PSO algorithm, whose flowchart is shown in Fig. 4 are as follows:

1. First of all, the user is asked to enter the number (N) of square particle (N^2), range (X_{min} and X_{max}) of the square particle, and the number of iterations (I) to be performed.
2. Then, the program generates the N^2 particles using Eqs. 1, 2, generate their velocity V randomly between 0–1, and initialize the iteration with I = 1. Initialize the particle's best value, pbest(P) and its position, pbest(P, 1:2) by putting maximum value that function cannot achieve. Similarly, initialize the particle's global best value, gbest and its position gbestp(1, 1:2) by putting maximum value that function also cannot achieve.
3. Start with the first particle by putting P = 1.
4. The position of particle P, $X(P,1) = X(1,1)$ and $X(P,2) = X(1,2)$ is assigned as the value of speed controller gain for K_P and K_I, respectively.
5. Run the simulation model of the PMSM drive, as shown in Fig. 2, for the obtained value of K_p and K_i for particle P.
6. Evaluating the fitness of the function for particle P given in Eq. 4 which is dependent of error in speed (ΔN) and error in torque (ΔT). In this function, 20 % weightage has been given to the speed and 80 % weightage to the torque and our task is to minimize it.

$$f = 0.2 \times \Delta N/2000 + 0.8 \times \Delta T/8 \tag{4}$$

7. Compare the particle's current fitness (fn) value with its personal best (pbest) value. If pbest is greater than fn, replace the fitness (pbest(P)) and position (pbestp(P, 1:2)) with fn and $X(P$, 1:2), respectively; otherwise, pbest and pbestp remains unchanged.
8. Compare the particle's current personal best (pbest) value with global best (gbest) value. If gbest is greater than pbest, replace the global gbest (gbest) and its position (gbestp(1, 1:2)) with pbest(P) and pbestp(P, 1:2), respectively; otherwise, gbest and gbestp remains unchanged.
9. Update the velocity of the particle (P) using the Eq. 5. In this algorithm, w has been kept constant, i.e., w = 1.

$$v = \text{w}v + c_1 r_1\{\text{pbestp}(P,\ 1:2) - X(P,\ 1:2)\} + c_2 r_2\{\text{gbestp}(P,\ 1:2) - X(P,\ 1:2)\} \tag{5}$$

Fig. 4 Flowchart of the modified PSO algorithm

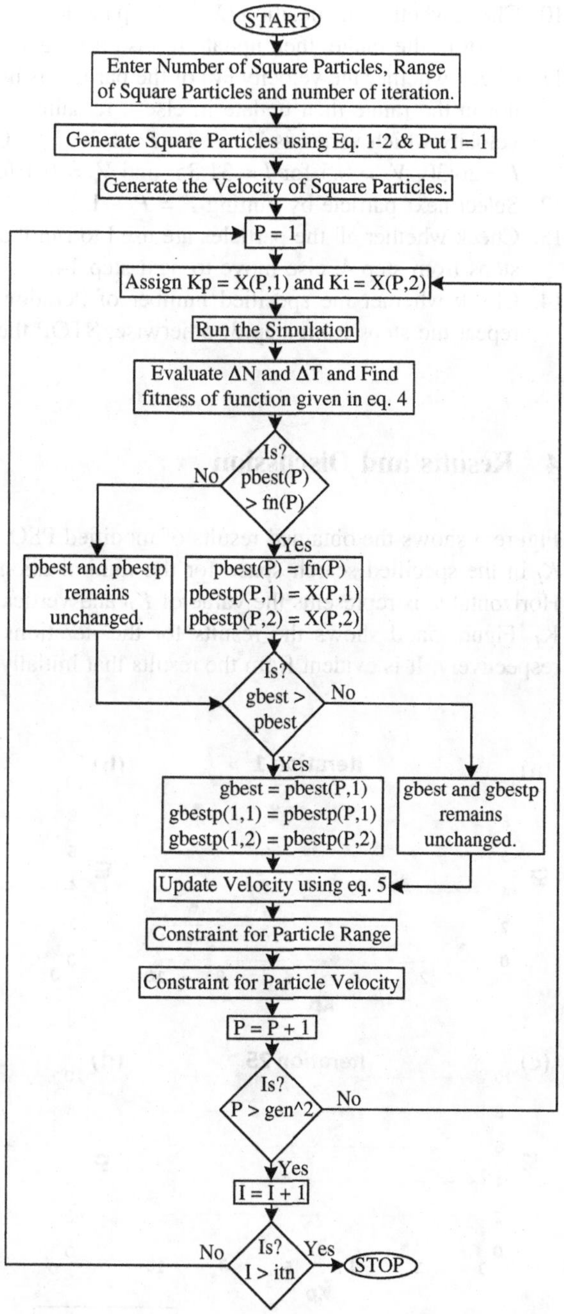

10. Check whether the position (X) of the particle is in the range ($X_{min} - X_{max}$). If it is not in the range, then update it, else it remains unchanged.
11. Check whether the velocity (V) of the particle is in the range ($-V_c$ to V_c). If it is not in the range then update it, else it remains unchanged. V_c is constraint of velocity which varies as $V_c = 1$ for $I = 1{:}20$, $V_c = 0.8$ for $I = 21{:}25$, $V_c = 0.6$ for $I = 26{:}30$, $V_c = 0.5$ for $I = 31{:}35$, and $V_c = 0.4$ for $I = 36$: more.
12. Select next particle by putting $P = P + 1$.
13. Check whether all the particles are used to run the simulation. If not repeat the steps from step 4, else move to next step 14.
14. Check whether the specified number of iteration has been performed. If not repeat the steps from step 3; otherwise, STOP the program.

4 Results and Discussion

Figure 5 shows the obtained results of modified PSO in terms of values of K_P and K_I in the specified search space for the minimization of function given in Eq. 4. Horizontal axis represents the value of K_P and vertical axis represents the value of K_I. Figure 5a–d shows the results for the iterations number 1, 10, 25, and 40, respectively. It is evident from the results that initially particles are scattered in the

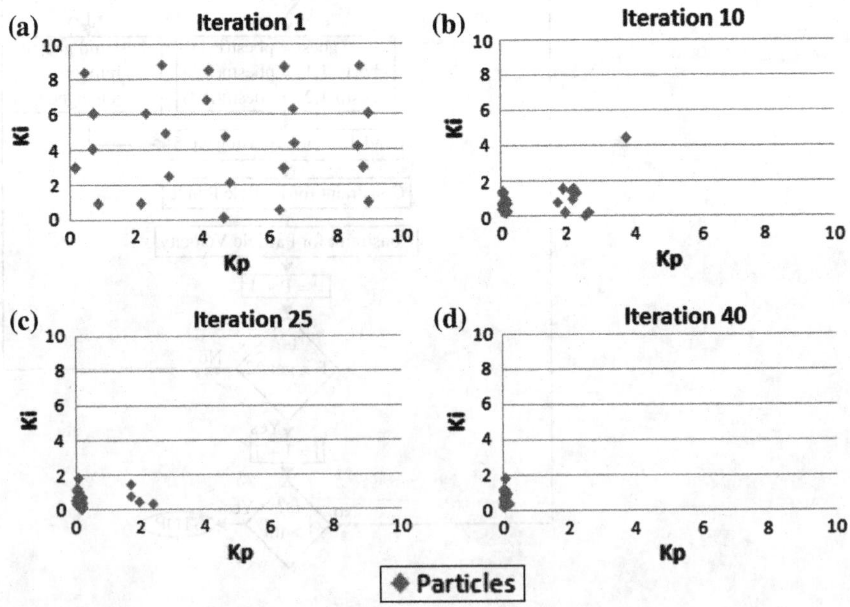

Fig. 5 Gain parameters K_P and K_I of speed controller in terms of particle position using modified PSO algorithm

complete region; thereafter, two clusters are formed between 0–2 at the end of tenth iteration. Further, the second cluster starts to merge with the first cluster by the end of the 25th iteration and it converges during the 40th iteration. The values of controller gains searched by the modified PSO algorithm are $K_P = 0.030541$ and $K_I = 1.078264$.

From top to bottom, MATLAB simulation results of the constant DC source (V_{dc}), three-phase simulated motor current (i_S), electromagnetic torque (T_e), rotor position (Θ_r), and rotor speed (N) of the PMSM drive at obtained value of K_P and K_I during torque and speed variation are shown in Fig. 6a, b, respectively.

During torque variation, initially drive has been driven at the rated torque of 8 Nm, after 0.22 s torque on the drive has been reduced to 4 Nm and after 0.43 s

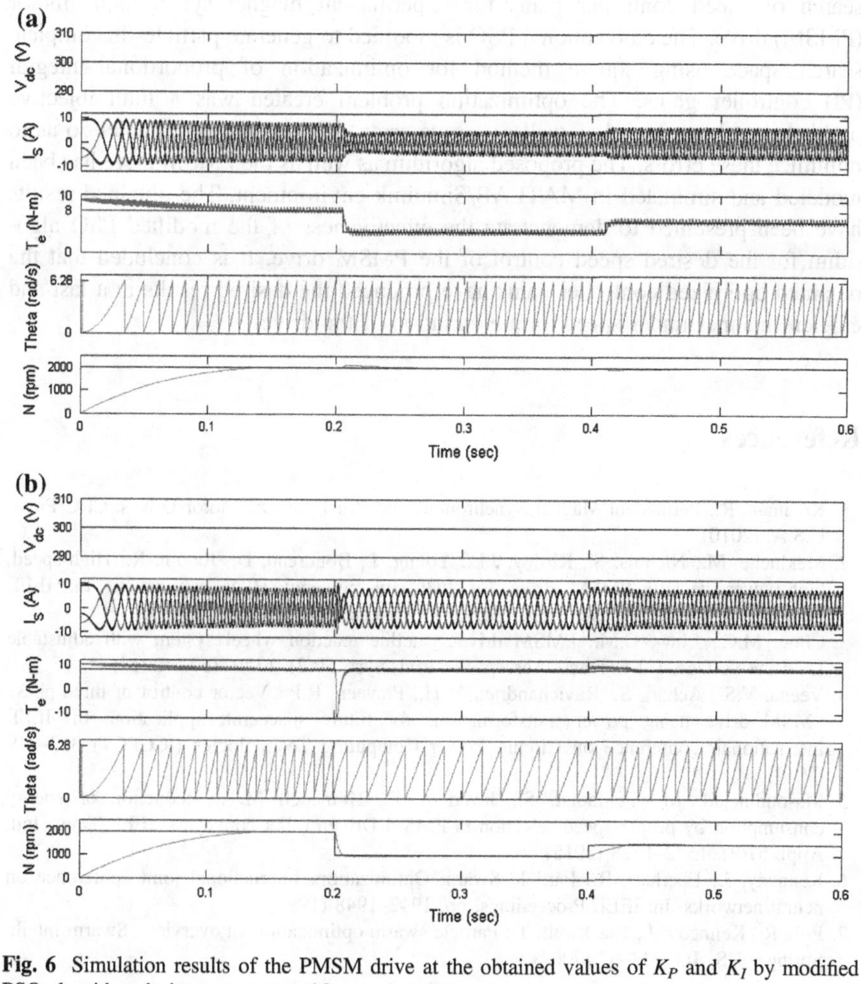

Fig. 6 Simulation results of the PMSM drive at the obtained values of K_P and K_I by modified PSO algorithm during **a** torque and **b** speed variation

torque on the drive has been increased to 6 Nm. During speed variation, initially drive has been at the rated speed of 2000 rpm, after 0.2 s speed of the drive has been reduced to 1000 rpm, and after 0.4 s speed of the drive has been increased to 1500 rpm. The controller tries to maintain the speed and torque error as minimum as possible during every variation. During different variations of torque and speed, the drive achieves its rated value in short duration which shows the effectiveness of the speed controller and its parameter.

5 Conclusion

A modified particle swarm optimization (PSO) algorithm has been presented, for search of speed controller gains for a permanent magnet synchronous motor (PMSM) drive. The conventional PSO is modified to generate particles in complete search space using square method for optimization of proportional–integral (PI) controller gains. The optimization problem created was a multiobjective problem involving speed as well as torque error as independent variables so as to minimize these errors. The proposed algorithm as well as the PMSM drive has been modeled and simulated in MATLAB/Simulink environment. The obtained results have been presented to demonstrate the effectiveness of the modified PSO algorithm for the desired speed control of the PMSM drive. It is concluded that the obtained optimized controller gains have produced the desired results in a fast and efficient manner under speed/torque variation of the drive.

References

1. Krishnan, R.: Permanent Magnet Synchronous and Brushless DC Motor Drives. CRC Press, U.S.A. (2010)
2. Mekhiche, M., Nichols, S., Kirtley, J.L., Young, J., Boudreau, D., Jodoin, R.: High-speed, high-power density PMSM drive for fuel cell powered HEV application. In: IEEE International Electric Machines and Drives Conference, pp. 658–663 (2001)
3. Chou, M.C., Liaw, C.M.: PMSM driven satellite reaction wheel system with adjustable DC-Link voltage. IEEE Trans. Aerosp. Electron. Syst. **50**(2), 1359–1373 (2014)
4. Veena, V.S., Achari, S., Ravichandran, M.H., Praveen, R.P.: Vector control of three phase PMSM drive using power transformations for future spacecraft application. In: IEEE International Conference on Circuit, Power Computing Technologies (ICCPCT) 313–319 (2014)
5. Masoudinejad, M., Feldhorst, S., Javadian, F., Homepel, M.T.: Reduction of energy consumption by proper speed selection in PMSM-Driven roller conveyers. IEEE Trans. Ind. Appl. **51**(2), 1572–1578 (2015)
6. Kennedy, J., Eberhart, R.: Particle Swarm Optimization. International joint conference on neural networks. In: IEEE Proceedings, pp. 1942–1948 (1995)
7. Poli, R., Kennedy, J., Blackwell, T.: Particle swarm optimization an overview. Swarm Intell., Springer US, **1**(1), 33–57 (2007)

8. Thangaraj, R., Pant, M., Abraham, A., Snasel, V.: Modified particle swarm optimization with time varying velocity vector. Int. J. Innovative Comput., Inf. Control (ICIC) **8**(1(A)), 201–218 (2012)
9. Singh, S., Singh, B.: Particle swarm optimization for power quality improvement of a 12-Pulse Rectifier-Chopper fed LCI-Synchronous motor drive. Int. J. Intell. Syst. Technol. Appl. (IJISTA) **11**(3/4), 267–285 (2012)
10. Singh, S., Singh, B.: Optimized passive filter design using modified particle swarm optimization algorithm for a 12-Pulse Converter fed LCI-Synchronous motor drive. IEEE Trans. Ind. Appl. **50**(4), 2681–2689 (2014)
11. Dwivedi, S.K., Laursen, M., Hansen, S.: Voltage vector based control for PMSM in industry Applications. In: IEEE International Symposium on Industrial Electronics (ISIE), pp. 3845–3850 (2010)
12. Dwivedi, S., Singh, B.: Vector control versus torque control comparative evaluation for PMSM drive. IEEE Power Electronics, Drives and Energy Systems (PEDES) & 2010 Power India, Joint International Conference, pp. 1–8 (2010)
13. Dost, P., Sourkounis, C.: On influence of various modulation schemes on a PMSM within an electric vehicle. In: IEEE Industry Applications Society Annual Meeting, pp. 1–10 (2014)

Parallelization of Simulated Annealing Algorithm for FPGA Placement and Routing

Rajesh Eswarawaka, Pavan Kumar Pagadala, B. Eswara Reddy and Tarun Rao

Abstract This paper aims to parallelize the simulated annealing algorithm used for the placement of circuit elements in the logic blocks of an FPGA. It intends to introduce the simulated annealing algorithm and the placement problem, analyzes the complexities involved, and justifies the use of simulated annealing as the algorithm for placement ahead of other algorithms. It explains the accuracy of the simulated annealing algorithm using a simple example which, also aims to explore parallelization techniques currently in use, such as parallel moves, area-based partitioning, Markov chains, and suggests possible improvements in the same using a combination of the above, using GPGPUs and investigate further the effects of move biasing. Also, the VPR (versatile placement and routing) CAD tool is introduced and key functions related to placement are explained [1]. The use of GPGPUs to achieve the required parallelism and speedup is discussed, along with the difficulties involved in implementing the same.

Rajesh Eswarawaka (✉) · P.K. Pagadala · B. Eswara Reddy · T. Rao
Dayananda Sagar College of Engineering, Bangalore, India
e-mail: rajesheminent@gmail.com
URL: http://www.springer.com/aisc

P.K. Pagadala
e-mail: pavankumarpagadala@gmail.com

B. Eswara Reddy
e-mail: eswarcsejntu@gmail.com

T. Rao
e-mail: Tarun636@gmail.com

Rajesh Eswarawaka · P.K. Pagadala · B. Eswara Reddy
Bharat Institute of Technology, Hyderabad, India

Rajesh Eswarawaka · P.K. Pagadala · B. Eswara Reddy
Jawaharlal Nehru Technological University, Hyderabad, Hyderabad, India

M. Pant et al. (eds.), *Proceedings of Fifth International Conference on Soft Computing for Problem Solving*, Advances in Intelligent Systems and Computing 436, DOI 10.1007/978-981-10-0448-3_84

Keywords Field-programmable gate array · Genetic algorithm · Genetic annealing · Parallel genetic algorithm · Simulated annealing · Stochastic tunneling

1 Introduction

Placement is the process that maps the circuit components to the logic blocks of the FPGA. The inputs required are the architecture description of the FPGA (number of CLBs, pin positions, etc.) and the net list describing the circuit. The placement process gives the exact blocks which each circuit component will occupy. Figure shows a typical FPGA whose logic blocks will be occupied by various circuit components. The challenge is to obtain an optimal placement where the placement cost function is minimized [2]. The cost function generally includes the total wirelength congestion of wire channel widths. Figure 1 highlights the difference between a random placement of blocks and the final placement after running the algorithm. It is therefore evident that the placement algorithm is critical to reduce the wirelength. Our contribution in this paper is another sequence of slicing lines which is more adequate for FPGA circuits than the traditional methods. This sequence reduces the maximum cut size and the total wirelength of the circuit. The rest of this paper is organized as follows. In the next section, we define the other algorithms used for the placement problem. In Sect. 3, we briefly describe the experimental setup. In Sect. 4, we describe the results and analyses. Future work is presented in Sect. 5. Finally, conclusions are made in Sect. 6.

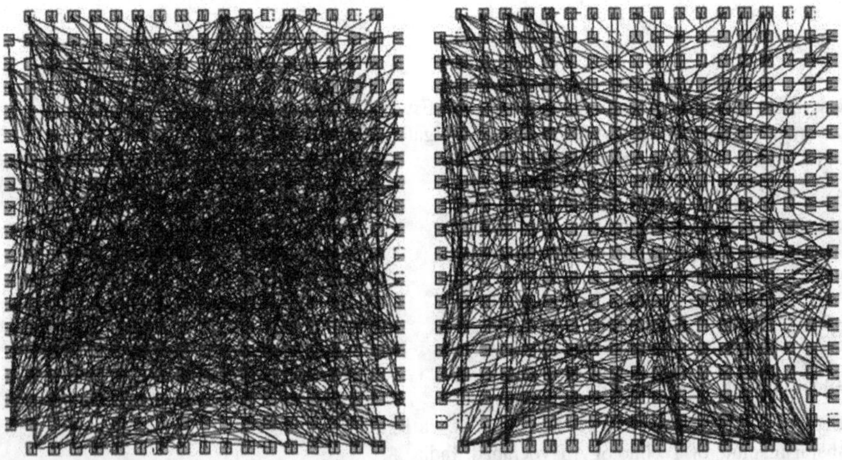

Fig. 1 Difference in congestion of wires due to application of placement algorithm

1.1 The Representation of the Placement in FPGA

Placement algorithms are NP-Complete in nature. Consider the following example circuit in the figure to illustrate the complexity. Assume that the FPGA consists of 6 × 6 = 36 configurable logic blocks (CLBs). For the circuit in consideration, eight blocks will be required to be configured. These eight blocks may be chosen in ${}^{36}C_8$ ways [3]. The placement costs may further vary depending on the way these eight blocks are configured as shown in figure. Therefore, there are ${}^{36}C_8$ placements possible, each with its own associated placement function value. In general, the number of placements possible is in the order of n^n, where n is the number of logic blocks. It would be too time consuming to check the placement costs of all the possible placements individually.

It would therefore be ideal to apply a heuristic like simulated annealing to arrive at the solution faster. Also since the placement code is sometimes bundled with the FPGA by the manufacturer, the user may not have control over it. This should not create a bottleneck in terms of overall circuit performance.

Simulated Annealing: It is a probabilistic algorithm for the global optimization problem of locating a good approximation to the global optimum of a given function in a large search space. The following are the basic elements of this algorithm [4].

1.2 Analysis

The algorithm has to be analyzed based on If $x(t)$ converges to the optimal set S^* The time taken for this to occur. The SA algorithm ensures that it is not stuck at a local minima and converges to the global minima [5, 6]. This is due to the acceptance of states with higher cost function based on the temperature (T) of the system as shown in figure. The time taken to converge at the global optima depends on the rate at which T is decreased. There is a trade-off between the quality of the solutions and the time required to compute them. A quality result generally requires a slow cooling schedule (Fig. 2).

Implementation of Simulated Annealing as the Placement Algorithm: The finite set S is the set of all placements possible. The initial placement is created by randomly placing the circuit blocks in the FPGA. The neighborhood consists of all other placements possible [7]. The cost function can be defined as either wire-length-driven, time-driven, or path-driven. After the initial placement, a certain number of moves are performed to see whether the cost is reduced or not according at a certain temperature. If the cost decreases, then the move is always accepted. If the cost increases, there is still probability for the move to be accepted which is given by $e^{-\delta(C/T)}$, where C is the change in the placement cost function the move causes and T is the temperature. The annealing schedule will determine to a large extent the quality of results obtained [8]. Even for a good annealing schedule,

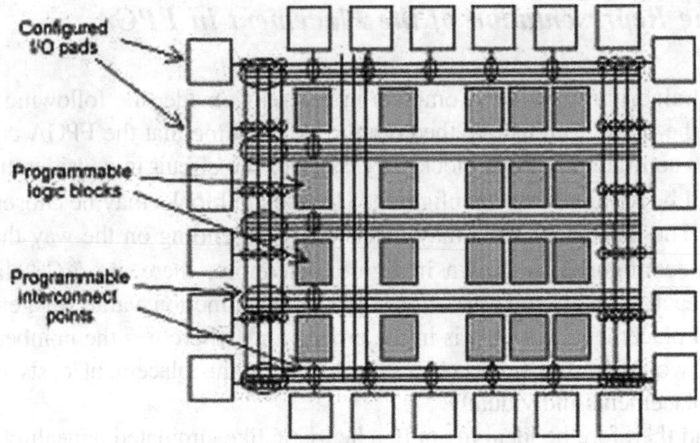

Fig. 2 Typical FPGA Island structure

millions of block swaps are evaluated at each temperature in the case of many circuits. The most time-consuming and computationally intensive part is calculating the cost caused by the swap. It is crucial to make this part as fast as possible.

Algorithm 1 Simulated Annealing Algorithm

Require: A finite set S
J is a cost function defined on S
$S^* \subset S$ is the set of global minima of J
For every $i \in S$, Set $S(i)$ represents the neighbours where $S(i) \subset (S - i)$
For every i, there is a collection of positive coefficients q_{ij}, $j \in S(i)$, such that $\sigma_{j \in S(i)}, q_{ij}$=1.
j is a member of $S(i)$ only if i is a member of $S(j)$
A non increasing function $T : N \rightarrow (0, \infty)$ (cooling schedule).
Initial state $x(0)$ is member of S
Given these elements, the algorithm as consists of a discrete-time inhomogeneous Markov Chain $x(t)$.
At any time t, if current state $x(t)$=i
j is chosen at random.
The probability that a given $j \in S(i)$ is q_{ij}.
Once j is chosen, the next state $x(t+1)$.
If $J(j)$<$J(i)$, **then** $x(t+1)$=j.
If not
then
$x(t+1)$=j with probability $exp^{[-(J(j)-J(i))/T(t)]}$
$x(t+1)$=i otherwise
end

---

2 Other Algorithms Used for Placement

2.1 *Quadratic Placement*

Quadratic placement algorithm uses the square of wirelength as the objective function. It tries to minimize the cost by solving the linear equations thus generated [9]. Although quadratic placement only considers the square of wirelength, it can efficiently finish the placement process with almost no quality lost. However, since the squared wirelength is the only factor considered in the objective function, the timing part of the placement cannot be shown in the quadratic placement (Fig. 3).

2.2 *Min-Cut Placement*

It is a partitioning-based placement method which recursively applies bipartitioning to map the net list of the circuit into the FPGA layout region. It minimizes the number of cuts of the nets and leaves the highly connected logic blocks in one partition [10]. It considers the delay of the circuit as the main parameter of the cost function. There is degradation in quality and the partitioning process is complex and requires heuristics to be performed to compute the partitions [11]. Popularly, Greedy algorithms are used in making the partitions which means there is a chance that the global minima will not be reached.

2.3 *Iterative Placement*

An iterative placement improvement algorithm takes an existing placement and tries to improve it by moving the logic cells and produces results as good as simulated annealing. Complexity of algorithm is greatest among placement algorithms [12].

Fig. 3 Hill climbing ability of simulated annealing

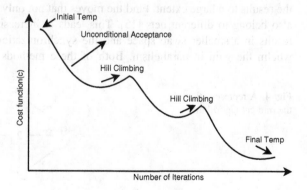

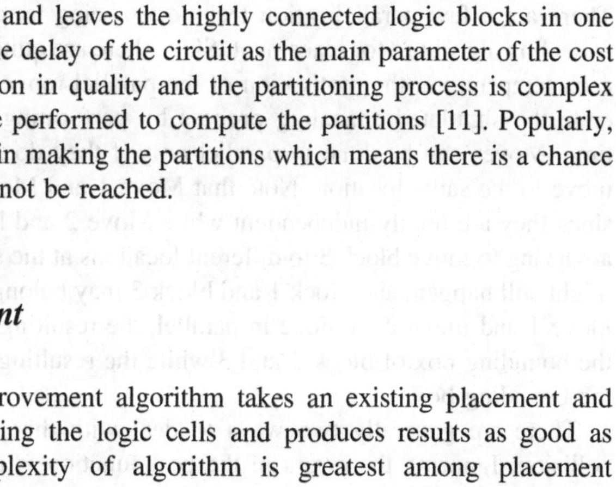

2.4 Need of Simulated Annealing

As we have seen in the previous section, there are many algorithms available to tackle the placement problem. But each of them had certain disadvantages. For simulated annealing, the advantages are: Unlike other algorithms, the placement cost can be defined as wirelength-driven, time-driven, or path-driven, or a combination of these factors. It outperforms the other placers in terms of quality of result as long as direct comparisons can be made [13]. The global minima are reached due to the hill climbing capability of the algorithm. The only drawback is the very slow nature of each move in annealing. Therefore, it is desirable that the simulated annealing algorithm be made faster using techniques like parallelization as it produces the most optimum results among available algorithms (Fig. 4).

2.5 Parallel Simulated Annealing Algorithms

There are a few parallelization techniques being used to speed up the simulated annealing algorithm for placement. Since there are quite a large number of moves at each temperature, the motivation of the parallel move approach is trying to accelerate the simulated annealing process by performing several moves at the same time. Moves can be done in parallel only if they do not move the same block or move to the same location. Note that Move 1 and Move 2 can be done in parallel since they are totally independent while Move 2 and Move 3 cannot because they are trying to move block 3 to different locations at the same time. Net cost collision might still happen, and block 1 and block 3 may belong to the same net [14]. While move 1 and move 2 are done in parallel, the resulting bounding box of move 1 is the bounding box of block 2 and 3 while the resulting bounding box of move 2 is the bounding box.

There are generally two ways to deal with the move collision and net cost collision. Ignoring the errors in the cost function is the easiest way to deal with these collisions. The problem is that it has negative effects on the accuracy of the cost function which interferes with the acceptance of moves. This adversely affects the results to a large extent. Find the moves that not only move different blocks, but also belong to different nets [15]. This leads to increasingly restricted moves and results in a smaller swap space and the synchronization overheads tend to overwhelm the gain in parallelism. Both of these methods show negative speedups.

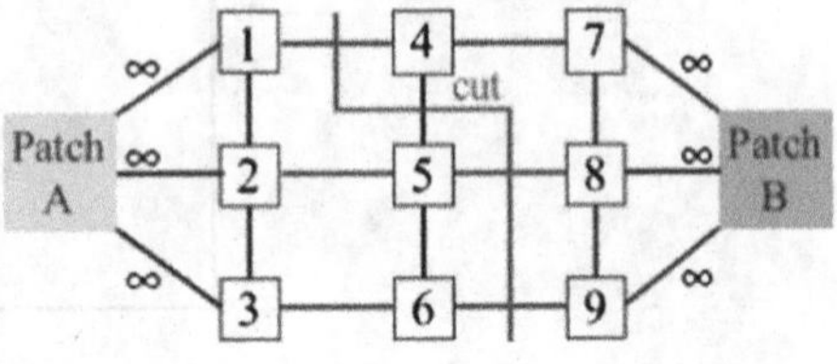

Fig. 4 A representation of the min-cut algorithm

The reason is that the overhead of synchronization outweighs the advantages of parallelization.

2.6 Area-Based Partitioning

It partitions the area of FPGA and assigns the partitioned areas to different processors. The whole circuit is partitioned into four parts, and each processor is in charge of one partition. The moves evaluated are much less restricted than the move parallel approach. However, collisions could still happen because multiple processors may move blocks belonging to the same net across the partition [16]. For example, the bounding box of block 1, 2, and 3 cannot be computed since they belong to different partitions. These errors can be tolerated because with cooling temperature, the swaps are tend to happen between nearby blocks and the chance that nets which span over two or more partitions will be chosen is minimal The experimental results show that a nonlinear speedup has been gained compared to the sequential placer and the cost does not degrade much with the increasing processors. This is due to the less synchronization requirements.

2.7 Parallel Markov Chains

The restriction on moves is removed altogether by assigning the whole FPGA to each processor. Each processor carries out simulated annealing on the whole FPGA starting with a different random state. To avoid concurrent updates to data structures, each processor does simulated annealing on a local copy of the FPGA. At periodic intervals, the results from all the processors are combined. The result of the processors can be combined. Among all the processors, take the best placement obtained by a processor as the new combined placement [17]. This approach is very efficient in terms of computation time and does well in terms of preserving the quality of solution. If we consider simulated annealing as a search path where moves are proposed and either accepted or rejected depending on a particular cost evaluation and a random seed, each search path can be viewed as a Markov Chain. This approach then essentially implements parallel Markov Chains. The speedup achieved is nearly linear and this is due to the fact that synchronization is minimum. There is only a slight degradation of results with increase in processors (Fig. 5).

2.8 Usage of GPGPUs and Methods Used

General-purpose computing on graphics processing units is the utilization of a graphics processing unit (GPU), which typically handles computation only for

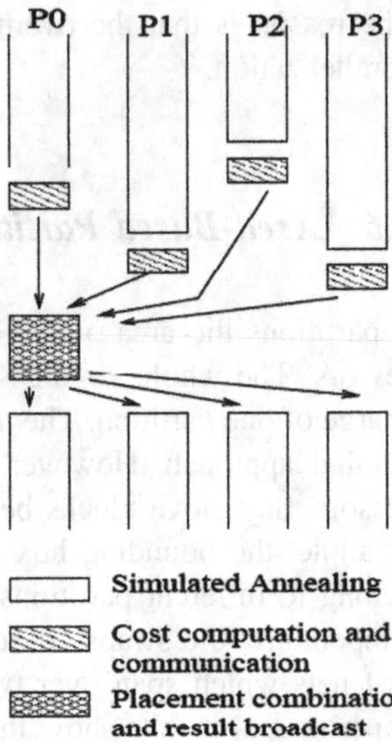

Fig. 5 A representation of the Markov Chain

computer graphics, to perform computation in applications traditionally handled by the central processing unit (CPU). Any GPU providing a functionally complete set of operations performed on arbitrary bits can compute any computable value. Additionally, the use of multiple graphic cards in one computer, or large numbers of graphics chips, further parallelizes the already parallel nature of graphics processing. GPGPUs have been used to parallelize the SA using the following techniques: Parallel Moves: Quality of results has not been very good. Area-Based Partitioning: Results have shown that parallelism has not affected the quality of results much, with a significant speedup also being achieved by Parallel Markov Chains: This has not been implemented for placement purposes [18].

2.9 *Proposed Approaches to Parallelism*

By implementing Markov Chains on GPGPUs, we will be able to increase the rate of cooling of the annealing schedule. This will provide speedup which will not involve other constraints. The number of moves performed at a given temperature can also be reduced if the number of cores is increased [18]. A combination of parallel Markov chains and area-based partitioning, where the parallel Markov chains technique is applied after partitioning the area. The speedup of the algorithm can be improved by using more cores and finding the optimum neighborhood size and cooling schedule. The cost function of placement is changed to reduce loss in quality of result due to parallelization. Move biasing can be further explored to better understand trade-offs between run time and quality of results.

3 Experimental Setup

3.1 Versatile Placement and Routing (VPR)

VPR is an FPGA CAD tool which performs placement and routing given the net list and a text file describing the FPGA architecture. VPR's output consists of the placement and routing, as well as statistics useful in assessing the utility of an FPGA architecture, such as routed wirelength, track count, and maximum net length. It is thus easy to obtain results for different benchmark circuits and compare the respective results. The placement algorithm used is Simulated Annealing. The figure describe the minimum number of tracks per channel required for a successful routing by various CAD tools on a set of 9 benchmark circuits [1]. Since VPR outperforms other existing CAD tools in terms of routing, it will be advantageous to improve the used simulated annealing placement algorithm (Fig. 6 and Table 1).

3.2 Overview of Some Important Functions of VPR

There are various functions used to implement placement in VPR. The important ones are

$try_{place()}$ This function does almost all of the placement-related work. It takes parameters like type of scheduling required, channel widths etc. User is able to choose different options like time-driven or path-driven

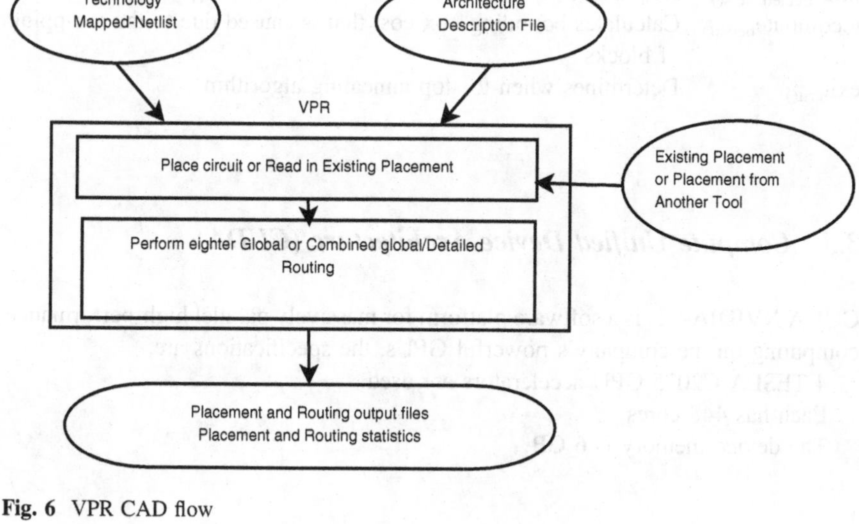

Fig. 6 VPR CAD flow

Table 1 Versatile place and route (VPR) outperforms other tools

Details	SEGA [19]	GBP [20]	OCG [21]	IKMB	SEGA	**VPR**
symml9	9	9	9	8	7	**6**
alu12	10	11	9	9	8	**8**
alu15	13	14	12	11	10	**9**
apex13	13	11	10	10	10	**8**
example18	17	13	12	11	10	**9**
k2	16	17	16	15	14	**12**
term1	9	10	9	8	8	**7**
toolarge	11	12	11	10	10	**8**
vda	14	13	11	12	12	**10**
total	112	110	99	94	89	**77**

placements. Invokes the simulated annealing algorithm to place the circuit. Cost calculations and approximations are done in this function

$\text{Try}_{\text{swap()}}$ Does the work of swapping the blocks during the placement process and picks a block and moves it to another spot. If that is occupied, it switches the same and assesses the change in cost function and decides to whether accept or reject the move

$\text{Starting}_{t()}$ Finds the starting temperature by trying one move and setting *t* accordingly high to allow all moves to be accepted

$\text{update}_{t()}$ Updates the temperature according to the selected annealing schedule; the user is allowed to specify his own schedule and we can modify to our requirements easily.

Other important Functions are

$\text{find}_{\text{affected nets()}}$ Required to check if nets affected and thus requiring updation

$\text{recompute}_{\text{bcost()}}$ Calculates bounding box cost that is caused due to the swapping of blocks

$\text{exit}_{\text{crit()}}$ Determines when to stop annealing algorithm.

3.3 Compute Unified Device Architecture (CUDA)

CUDA NVIDIA-92s is a software platform for massively parallel high-performance computing on the company's powerful GPUs, the specifications are,

4 TESLA C2075 GPU accelerators are used

Each has 448 cores

The device memory is 6 GB.

The CUDAC platform is used for programming purpose and it supports other languages, application programming interfaces, or directives-based approaches are supported, such as FORTRAN, DirectCompute, Open ACC.

3.4 Benchmark Circuits

Circuits for which the algorithms will be tested need to be chosen such that the complexity is high. This is ensured by choosing Benchmark Berkeley Logic Interchange Format (BLIF) circuits. The goal of BLIF is to describe a logic-level hierarchical circuit in textual form, thus making it useful for the purpose of placement. These BLIF circuits can be passed as input to VPR. BLIF circuits are popularly used, making analysis easy. These circuits need to be converted to .net format to be compatible with VPR and T-VPack is used for this.

4 Results and Analysis

Parallel Markov chains were implemented on the CUDA-capable machine. The number of cores used was 10. Table 2 compares serial and parallel execution times for different benchmark circuits. The parallel version of the Simulated Annealing algorithm is on an average approximately 4 times faster than the serial placement algorithm.

4.1 Comparison of Cost Functions

Table 3 compares the serial and parallel execution times for different benchmark circuits. The parallel version of the algorithm results in 5–20 % degradation of quality. This may be due to varying factors like different random seed, change in cooling schedule, number of moves per temperature etc.

Table 2 Comparative execution times

Circuit	Parallel execution time (s)	Serial execution time (s)	Speedup
clma	34.16000	143.27000	4.194
elliptic	8.38000	33.37000	3.982
ex1010	15.19000	61.6000	4.055
frisc	9.7000	38.19000	3.937
pdc	15.71000	63.94000	4.070
s38417	32.87000	135.08000	4.109
spla	13.3000	53.72000	4.031

Table 3 Comparative costs

Circuit	Parallel execution costs	Serial execution costs	Cost (%)
clma	1834.423	1587.561	15.550
elliptic	585.035	494.379	18.337
ex1010	888.893	827.748	7.387
frisc	731.946	609.494	20.091
pdc	1134.622	1051.926	7.861
s38417	1242.357	1029.044	20.730
spla	926.148	878.762	5.392

5 Future Work

The use of more cores will improve speedup. The loss in quality of result can be removed by changing the parameters like cooling schedule etc. These results have to be verified with larger circuits. Memory can be managed better to get better results. The technique can be combined with area-based partitioning to further speed up the placement process.

6 Conclusion

The various aspects of the placement algorithm were studied and the following conclusions drawn. Simulated annealing is an ideal placement algorithm in terms of quality of result obtained. The parallelization of simulated algorithm is a challenge involving identification of portions that can be parallelized, and techniques to implement them. Existing techniques can be improved using further changes in parallelization or aspects of the algorithm itself like the updating of temperature, annealing schedule etc. An ideal way to test new approaches is to use the VPR tool whose functions allow the user to modify the simulated annealing algorithm. The CUDAC platform was used to implement the parallel algorithm on the GPGPU. The parallel markov chains technique was used to speed up the algorithm. Results suggest that further improvement can be achieved.

Acknowledgments The authors thank Sri.T.V. Bala Krishna Murthy who gave the entire support to write this article in such a good manner.

References

1. Betz, V., Rose, J.: VPR: A new packing, placement and routing tool for FPGA research. In: International Workshop on Field Programmable Logic and Applications (1997)
2. Haldar, M., Nayak, A., Choudhary, A., Banerjee, P.: Parallel algorithms for FPGA placement. In: Proceedings of the Great Lakes VLSI Conference (2000)

3. Shi, X.: FPGA Placement Methodologies: A Survey. University of Alberta, Edmonton, Canada (2009)
4. Smith, M.J.S.: Application-specific integrated circuits. In: Proceedings of the VLSI Systems Series, June 1997
5. Chandy, J.A., Banerjee, P.: Parallel simulated annealing strategies for VLSI cell placement. In: 9th International Conference on VLSI Design (1996)
6. Deb, K., Agarwal, S., Pratap, A., Meyarivan, T.: A fast and elitist multi-objective genetic algorithm. IEEE Trans. Evol. Comput. **6**(2), 182–197 (2002)
7. Janaki Ram, D., Sreenivas, T.H., Ganapathy Subramaniam, K.: Parallel simulated annealing algorithms. J. Parallel Distrib. Comput. **37**, 207a¸ S212 (1996)
8. Deb, K.: Multi-objective Optimization Using Evolutionary Algorithms. Wiley, Chichester, UK (2001)
9. Choong, A., Beidas, R., Zhu, J.: Parallelizing simulated annealing-based placement using GPGPU. In: Parallelizing Simulated Annealing-Based Placement using GPGPU
10. Bertsimas, D., Tsitsiklis, J.: Simulated annealing. Stat. Sci. **8**(1), 10–15 (1993)
11. Goldberg, D.E., Deb, K.: A Comparison of selection schemes used in genetic algorithms. In Foundations of Genetic Algorithms Pages (FOGA-1) pp. 69–93. (1991)
12. Schug, A., Herges, T., Wenzel, W.: Reproducible protein folding with the stochastic tunneling method. Phys. Rev. Lett. **91**, 158102 (2003)
13. Tolley, A.J., Wyman, M.: Stochastic tunneling in DBI inflation, ArXiv e-prints, Sep 2008
14. Baumketner, A., Shimizu, H., Isobe, M., Hiwatari, Y.: Stochastic tunneling minimization by molecular dynamics: An application to heteropolymer models. Phys. A Stat. Mech. Applicat. **310**(139), 150 (2002)
15. Michalewicz: Genetic Algorithms+Data Structures Evolution Programs. Springer (1992)
16. Lin, M., Wawrzyneki, J.: Improving FPGA placement with dynamically adaptive stochastic tunneling. In: IEEE Transactions on Computer-Aided Design of Integrated Circuits and Systems, vol. 29, no. 12, Dec 2010
17. Baruch, Z., Cret, O., Pusztai, K.: Placement Algorithm for FPGA Circuits. Computer Science Department, Technical University of Cluj-Napoca, 26, BariÅ£iu St., 3400 Cluj-Napoca, Romania
18. Goldberg, D.E.: Genetic Algorithms for search, optimization, and machine learning. Addison-Wesley, Reading, MA (1989)
19. Queipo, N.V., Gil, G.F.: Multi objective optimal placement of connectively and conductively cooled electronic components on printed wiring boards. ASME Trans. J. Electron. Packag. **122**, 152–159 (2000)
20. Vose, M.D.: Simple Genetic Algorithm: Foundation and Theory. MIT Press, Aim Arbor, MI (1999)
21. Deb, K., Jain, P., Gupta, N., Maji, H.: Multi-Objective placement of electronic components using Evolutionary Algorithm. KanGAL Report No:2002006

Review of Image Acquisition and Classification Methods on Early Detection of Skin Cancer

M. Reshma and B. Priestly Shan

Abstract The word cancer is enough to send many people into a spin. However, most types of skin cancer have a very favorable prognosis. They are common and very treatable. Melanoma is the skin cancer of most concern. Minor skin cancers often appear as a spot or sore that will not heal. Melanomas may arise in a pre-existing skin mole that has become darker or changed in appearance. More often they will appear as a new mole or an unusual freckle. Nearly all skin cancers are related to excessive UV radiation. The depletion of the earth's ozone layer also appears to be increasing the risk of developing skin cancer. With melanoma, family history also seems to be a factor. Detection at the melanoma in situ stage provides the highest curable rate for melanoma. The aim of this paper is to provide the summary of all the available methods and stages of melanoma identification.

Keywords Vertical growth phase (VGP) · Epiluminescence dermatoscope · Optical coherence tomography (OCT)

1 Introduction

Melanoma is a cancer that develops in pigment cells called melanocytes. It occurs often in people with light complexion who had a high exposure to sunlight. Caught early, most melanomas can be cured with relatively minor surgery. Melanoma can be more serious than the other forms of skin cancer, because it may spread (metastasize) to other parts of the body and cause serious illness and death.

M. Reshma (✉)
Department of E&C, Sathyabama University, Chennai, India
e-mail: reshma.m03@gmail.com

M. Reshma
Department of DECS, VIAT, VTU, Muddenahalli, Bangalore, India

B. Priestly Shan
Royal College of Engineering & Technology, Akkikaru, Thrissur 680604, Kerala, India
e-mail: priestlyshan@gmail.com

M. Pant et al. (eds.), *Proceedings of Fifth International Conference on Soft Computing for Problem Solving*, Advances in Intelligent Systems and Computing 436, DOI 10.1007/978-981-10-0448-3_85

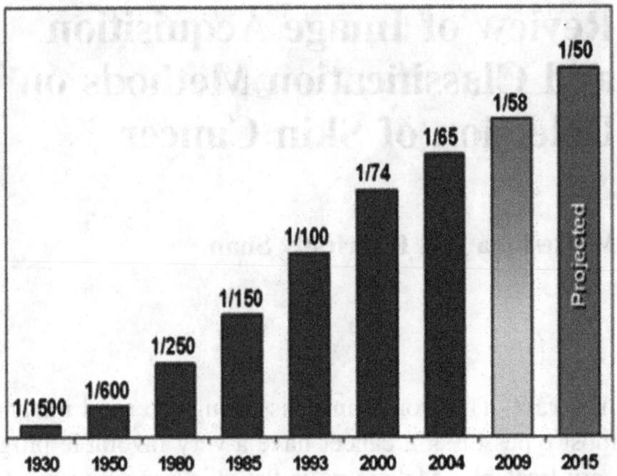

Fig. 1 Lifetime risk of an American developing invasive melanoma

The most common forms of melanoma are superficial spreading melanoma, nodular melanoma, and lentigo maligna [1]. Patients themselves are the first to detect many melanomas. Spots suspicious for melanoma show one or more of the following features (the *ABCD*s): *A*symmetry, *B*order irregularity, *C*olor changes, and a *D*iameter more than the size of a pencil eraser. Doctors diagnose melanoma by biopsy [2]. The number of malignant melanoma (MM) cases worldwide has increased faster than any other cancer in recent decades as shown in Fig. 1. Melanoma development and progression pass through several distinct stages. Primary MM may develop from precursor melanocytic nevi, which is the first step, although more than 60 % of cases are believed to arise de novo (i.e., not from a preexisting pigmented lesion) [3, 4].

Radial growth phase (RGP) of primary melanoma is the next step of MM progression. The cells in this phase are locally invasive but to the vertical growth phase (VGP) of primary lesions. In this step, melanoma cells infiltrate and invade the dermis as a large cluster of cells and exhibit metastatic potential. Metastasis to distant organs followed by overgrowth of tumor cells in affected sites is the last step of MM progression. Research in melanoma is headed in three directions: prevention, more precise diagnosis, and better treatment for advanced disease.

2 Methodology

This section describes the common steps involved in the early detection of skin cancer Fig. 2. Each stage is based on characteristics such as tumor thickness, ulceration, and the involvement of lymph nodes or organs. Once diagnosed the stages of melanoma then guide the treatment approach.

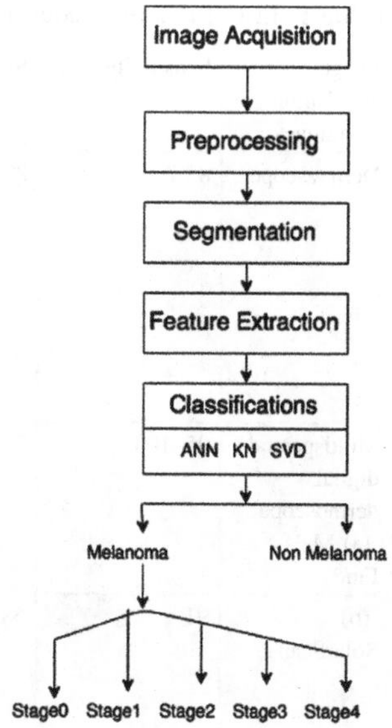

Fig. 2 Steps in melanoma detection

2.1 Image Acquisition

The first step in early inspection of skin cancer involves the acquisition of lesioned digital image. Around 1990, dermatoscopy also known as epiluminescence microscopy was the examination of skin lesions with a dermatoscope. This traditionally consists of a magnifier (typically $x10$), a non-polarized light source, a transparent plate, and a liquid medium between the instrument and the skin, which allows inspection of skin lesions unobstructed by skin surface reflections [5]. When the images or video clips are digitally captured or processed, the instrument can be referred to as a "digital epiluminescence dermatoscope." However, the effectiveness of this technique depends on experience. Digital dermatoscopy images are stored and compared to images obtained during the patient's next visit [6]. Suspicious changes in such a lesion are an indication for excision. Skin lesions, which appear unchanged over time are considered benign. Lack of training represents the most important barrier for a wider use of dermoscopy. Time consumption represented the second described disadvantage of dermoscopy. But it is an inexpensive, rapid, and safe procedure that allows a better selection of lesions requiring biopsy compared with the naked eye. The list of different acquisition techniques is given in Table 1.

In multispectral digital dermoscopy, a sequence of images are obtained using given bands of wavelengths; as the wavelength of light varies, it is able to penetrate the skin to differing depths. Multispectral imaging takes sequences of images at varying wavelengths (from 400 to 1000 nm), thus supplying information about a range of depths within the lesion in vivo. Information found at different depths are useful in differentiating between benign and malignant pigmented skin lesions.

Table 1 List of different acquisition techniques

Image acquisition techniques	Sensitivity (%)	Specificity (%)	Advantages	Disadvantages	Cost
Dermtascope	85.7	83.4	Accuracy by dermatoscopy increases melanoma pick up by 20 %. Compared with naked eye examination	Lack of training, time consumption	Inexpensive
Multispectral digital dermoscope (a) Mela Find	95–100	70–85	Multispectral sequence of images created in 3 s; Hand-held scanner	–	Expensive
(b) SolarScan	91	68	Database for comparison, session, and image-level accuracy calibration; recorded body graphic map	Requires oil immersion	Expensive
Ultrasound scanning	99	99	Information about inflammatory processes of skin in relation to nerves and vessels	Tumor thickness may be overestimated because of underlying inflammatory infiltrate; melanoma metastasis cannot be separated from that of another tumor; images can be difficult to interpret	Cost effective
Laser-based technology (a) Confocal scanning laser microscopy (CSLM)	98.2	98.9	Histopathological evaluation at bedside with similar criteria; longer wavelengths can measure up to papillary dermis; fiber-optic imaging allows for flexible hand-held devices	Poor resolution. Captures structures only to depth of 300 lm; melanomas without in situ component will likely escape detection.	Inexpensive

(continued)

Table 1 (continued)

Image acquisition techniques	Sensitivity (%)	Specificity (%)	Advantages	Disadvantages	Cost
(b) Optical l Coherence Tomography (OCT)			High-resolution cross-sectional images resembling histopathological section of skin; 4 mm scan length obtained in 4 s; higher resolution than ultrasound and greater detection depth than CSLM; Doppler and phase-resolved techniques allow visualization of vessels	Photons are scattered more than once, which can lead to image artifacts; ointment or glycerol may be needed to reduce scattering and increase detection depth; visualization of architectural changes and not single cell	Inexpensive
Electrical bioimpedance	92–100	67–80	Complete examination lasts 7 min	Electrical impedance properties of human skin vary significantly with the body location, age, gender, and season	low cost
Magnetic Resonant imaging (MRI)	–	–	Permits clear differentiation of the SC, epidermis and dermis in vivo	equipment size; acquisition time; need for specialized training; contraindicated in patients with metal implants	Expensive

This technique offers the advantage of analyzing features indiscernible to the human eye, probing up to 2 mm below the surface of the skin. Other noninvasive imaging tools are MelaFind, a multispectral digital dermoscope with a specialized imaging probe and software to assist with differentiation between early melanoma and other skin lesions. SolarScan, a robust diagnostic instrument for pigmented or partially pigmented melanocytic lesions of the skin features a 5-megapixel 3-chip digital video camera providing improved lesion monitoring. For better resolution of small lesions located near the skin surface, ultrasound scanning is a safe noninvasive method that in some settings can be used to show subtle differences between nevi and melanoma. Use of high frequency transducers lead to the tradeoff of decreased depth of penetration by the ultrasound waves, leaving the choice of the probe frequency dependent on the diameter and site of the lesion. To speedup the detection of deadly melanoma, a new laser-based technology was developed without the expense of false diagnosis and unnecessary surgery. Confocal scanning laser microscopy (CLSM) (reflectance confocal microscopy [RCM]) provides real-time in vivo visualization of subsurface skin structures at a resolution that approaches that of light microscopy. The confocal microscope uses a near-infrared laser at 830 nm operating at a power of less than 20 mW. The penetration depth of imaging is 200–500 μm, allowing visualization of the epidermis and the superficial dermis. Optical coherence tomography (OCT) is a technique that enables an examination of the skin to a depth of about 1 mm. Under OCT, melanomas demonstrate increased architectural disarray, less-defined dermoepidermal borders, and vertically oriented icicle-shaped structures not seen in nevi. Optical coherence tomography (OCT) is an emerging imaging technology based on light reflection. It provides real-time images with up to 2 mm penetration into the skin and a resolution of approximately 10 m. The combination of high-resolution and relatively high-imaging depth places OCT in the imaging-gap between ultrasound and confocal microscopy. Electrical impedance spectroscopy (EIS) is another imaging method, called electronic biopsy by SciBase AB (Stockholm, Sweden), uses a probe which painlessly penetrates the stratum corneum, measuring the overall resistance in tissues at alternating currents and frequencies. Here, the device evaluates the resistance of the cells to determine whether the cells are concerning for melanoma. Bioimpedance levels are a function of cell shape and structure, cell membranes, and the amount of water present. Based on these features, electrical impedance of cancer and benign cells are different because cancer cells typically have a different shape, size, and orientation than benign cells do. Currently, cellular magnetic resonance imaging (MRI) is a newly emerging field of imaging research that is expected to have a large impact on cancer research thought to be less useful in discriminating banal melanocytic lesions from those which deserve biopsy but may provide some future benefit in evaluating melanoma thickness, allowing better preparation for removal and treatment. MRI can provide a dynamic view of disease progression, delivers results in real time, and is free from sectioning-related artifacts. Thus, MRI may provide a more complete picture of the overall biologic process under investigation. Cellular MRI offers the opportunity for early detection of small numbers of metastatic cells and also provides the opportunity for studying micrometastatic processes in their earliest developmental stages in the target organs of interest. Cellular MRI is a young field of imaging research that combines the high resolution of micro-MRI with sensitive iron oxide-based contrast agents for cell and receptor labeling. The major disadvantages for these noninvasive examinations are the relatively high cost and patient anxiety, mainly caused by false positive results leading to biopsy of benign pigmented skin lesions. Many of these methods only propose to differentiate between malignant and benign skin lesion but do not identify early indications of lesion. Table 1 shows the advantages and disadvantages of different acquisition techniques.

2.2 Preprocessing Images

Image preprocessing can significantly increase the reliability of optical inspection. Factors responsible for degradation of image quality with different types of scanners include: *Sharpness* determines the amount of detail an image can convey. *Noise* is a random variation of image density, visible as grains in film and pixel-level variations in digital images. *Dynamic range* (or exposure range) is the range of light levels a camera can capture, usually measured in f-stops, EV (exposure value), or zones (all factors of two in exposure). *Tone reproduction* is the relationship between scene luminance and the reproduced image brightness. *Contrast*, also known as gamma, is the slope of the tone reproduction curve in a log–log space. *Color accuracy* is an important but ambiguous image quality factor. *Distortion* is an aberration that causes straight lines to curve. *Exposure accuracy* can be an issue with fully automatic cameras and with video cameras [7]. *Lens flare*, including "veiling glare" is stray light in lenses and optical systems caused by reflections between lens elements and the inside barrel of the lens. *Color moiré* is artificial color banding that can appear in images with repetitive patterns of high spatial frequencies, like fabrics or picket fences. *Artifacts* can cause significant visual artifacts, including data compression and transmission, over sharpening "halos" and loss of fine, and low-contrast detail. Preprocessing is composed of three steps [8]. It starts by using a median filter aimed at cleaning the image by eliminating certain defects, then performing a morphological closing aimed at eliminating all artifacts such as hair, and finally improving the color between the lesion and healthy skin. Figure 3 shows the sample of original and preprocessed image.

2.3 Segmentation

Image segmentation is the process of partitioning a digital image into multiple segments. Image segmentation is typically used to locate objects and boundaries (lines, curves, etc.) in images. More precisely, image segmentation is the process of assigning a label to every pixel in an image such that pixels with the same label share certain characteristics. Figure 4 shows the sample of lesion segmentation after preprocessing the image.

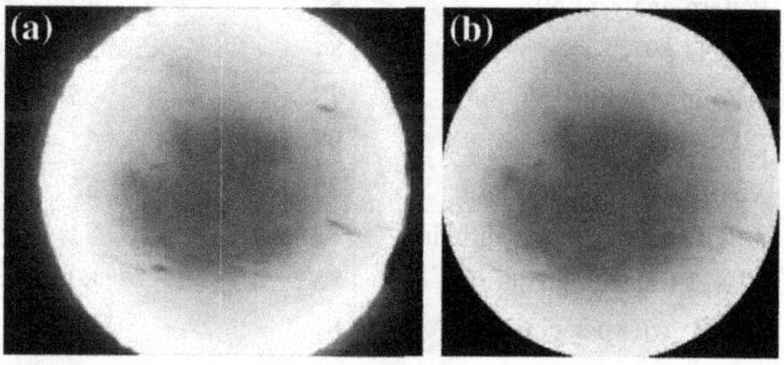

Fig. 3 Sample of **a** original and **b** preprocessed image

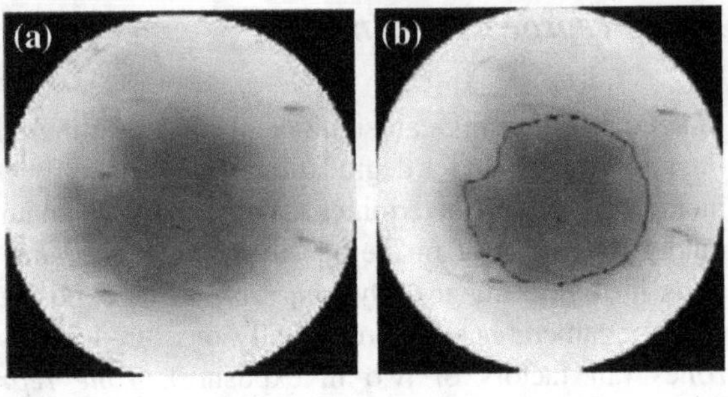

Fig. 4 Sample of **a** preprocessed image **b** segmented image

2.4 Feature Extraction

Color, texture, and edges are three basic features. Texture is a tactile or visual characteristic of a surface. Texture primitives (or texture elements, texels) are building blocks of a texture. Edges correspond to large discontinuities in the image. The texture features consist of a variation of local binary pattern (LBP) in which the strength of the LBPs is used to extract scale-adaptive patterns at each pixel, followed by the construction of a histogram. For color feature extraction, we used standard HSV histograms [9]. The extracted features are concatenated to form a feature vector for an image, followed by classification using support vector machines. Table 2 lists the different feature extraction methods employed for the early detection of skin cancer.

2.5 Classification

Wide ranges of classifiers are available and each one of them has its strengths and weaknesses. The performance of classifier depends greatly on the characteristics of

Table 2 Lists the different feature extraction methods

Feature extraction methods	Sensitivity (%)	Specificity (%)	Accuracy (%)
ABCD	3.3	97.8	–
Pattern analysis 59	85	79	71
ABCD 59	84	75	76
7-Point checklist 59	78	65	58
CASH61	98	68	–
Menzies 63	85	85	81
ABCDE descriptor validation	82	81	Results–encouraging
Wavelet Decomposition	85	80	58.44
Simple wrapper curve lets	87	79	86.57

Table 3 Comparison of different classifier methods [12]

Author	Year	Classifier	Results
Nikhil Cheerla, Debbie Frazier.	2014	Neural network classifier	
Ru-Song et al.	2012	Combined neural network classification and data training	
Binamrata Baral, Sandeep Gonnade, Toran Verma	2014	ANFIS: Neural network (a) ANFIS Classifier (b) Optimization of ANFIS Classifier [13] (c) Optimization of ANFIS Classifier	
Mahammed et al.	2014	On artificial neural network for the recognition of malignant melanoma	The classification results show an increasing true detection rate and a decreasing false positive rate.
Md. Khaled Abu et al.	2013	Backpropagation Neural network classifier	
Catarina Barata, et al.	2013	Three classifiers are considered in this work, i.e., Ada Boost, SVM, and KNN classifier.	
Rahil Garnavi, Mohammad Aldeen	2012	Support Vector Machine	

the data to be classified and there is no single classifier that works best on all given problems [10]. Various empirical tests have been performed to compare different classifier performance and to find out the relationship between characteristics of data and the classifier performance [11]. The input given to the classifier are the methods of features extraction which classifies the given data set into cancerous or noncancerous which is the last step in melanoma identification which further judges at what stage the melanoma has reached for the early detection of human survival. Table 3 shows the comparison of different classifier methods.

2.6 *Stages of Melanoma*

Unlike most other cancers, the physical thickness of the primary melanoma is directly related to its likelihood of metastases. Initially, it was observed that the extent of anatomic invasion by the primary tumor may predict the 10 year survival probability (Table 4) [14]. The Clark classification involves staging the primary lesion based on the anatomic level of invasion into the dermis or subcutaneous fat rather than based on its metric depth. According to Breslow's thickness, tumor depth is measured from the granular cell layer downward using an ocular micrometer. If the tumor is sphere shaped, the maximal thickness, as measured from

Table 4 Stages of melanoma

Stages	Anatomic location of melanoma cells	10 year survival (%)
Stage 0	Confined to epidermis	99
Stage 1	Penetrating the papillary dermis	96
Stage 2	Filling the papillary dermis	90
Stage 3	Extending into the reticular dermis	67
Stage 4	Invasion of the subcutis	26

the granular cell layer to the deepest component of the tumor, is mathematically related to the tumor volume. To date, tumor thickness remains the most powerful prognostic indicator that can be determined from evaluation of the primary melanoma itself. The anatomic stage of melanoma invasion has been reported to offer additional prognostic information in thin primary MM, as for example, tumors with less than 1 mm thickness but greater than stage 3 have a worse prognosis than lesions of the same thickness with a stage of 3 or less. Furthermore, it has been shown that a Clark's stage of III or higher is an independent predictor of positive sentinel lymph node biopsy.

3 Conclusion

From the above survey, it is evident that the methods proven to be the best for each stage depends on the more dataset used and evaluation rules applied. The ABCD criteria were intended to be a simple tool that could be implemented in daily life, a mnemonic "as easy as ABC" to alert both laypersons and healthcare professionals to the clinical features of early melanoma. However, the challenge of all of these techniques is that they still require a "good eye" to select the lesions for evaluation among the sea of lesions that are prevalent. In this paper, we reviewed various skin cancer diagnostic method among which ABCD rule based is most commonly used method and thus yields good results in the early detection of melanoma diagnosis.

References

1. Weinstock, M.: Cutaneous melanoma: Public health approach to early detection. Dermatol. Ther. **19**(1), 22–31 (2006)
2. Halpern, A., Marghoob, A., Bialoglow, T., Witmer, W., Slue, W.: Standardized positioning of patients (poses) for whole body cutaneous photography. J. Am. Acad. Dermatol. **49**(4), 593–598 (2003)
3. Massone, C., Brunasso, A.M., Campbell, T.M., Soyer, H.P.: Mobile teledermoscopy—melanoma diagnosis by one click in Seminars in cutaneous medicine and surgery, pp. 203–205 (2009)

4. Wadhawan, T., Situ, N., Lancaster, K., Yuan, X., Zouridakis, G.: SkinScan©: A portable library for melanoma detection on handheld devices in Biomedical Imaging: From Nano to Macro. In: IEEE International Symposium, pp. 133–136 (2011)
5. Boldrick, J., Layton, C., Ngyuen, J., Swtter, S.: Evaluation of digital dermoscopy in a pigmented lesion clinic: Clinician versus computer assessment of malignancy risk. J. Am. Acad. Dermatol. 56(3), 417–421 (2007)
6. Doukas, C., Stagkopoulos, P., Kiranoudis, C., Maglogiannis, I.: Automated skin lesion assessment using mobile technologies and cloud platforms. In: Engineering in Medicine and Biology Society(EMBC), 2012 Annual International Conference of the IEEE, pp. 2444–2447 (2012)
7. Grana, C. Pellacani, G., Cucchiara, R., Seidenari, S.: A new algorithm for border description of polarized light surface microscopic images of pigmented skin lesions. IEEE Trans. Med. Imaging **22**(8), 959–964 (2003)
8. Abbas, Q., Celebi, M.E., Garcia, I.F., Ahmad, W.: Melanoma recognition framework based on expert definition of abcd for dermoscopic images. Skin Res. Technol. **19**(1), e93–e102 (2013)
9. Barata, C., Ruela, M., Francisco, M., Mendonca, T., Marques, J.S.: Two systems for the detection of melanomas in dermoscopy images using texture and color features. IEEE Syst. J. **99**, 1–15 (2013)
10. Surowka, G., Grzesiak-Kopec, K.: Different learning paradigms for the classification of melanoid skin lesions using wavelets. In: Proceedings of the 29th Annual International Conference of the IEEE EMBS, pp. 3136–3139 Aug 2007
11. Marques, J.S., Barata, C., Mendonca, T.: On the role of texture and color in the classification of dermoscopy images. In: IEEE 34th EMBC, pp. 4402–4405 (2012)
12. Md. Abu Mahmoud, K., Al-Jumaily, A., Takruri, M.: Wavelet and curvelet analysis for automatic identification of melanoma based on neural network classification. Int. J. Comput. Inf. Syst. Ind. Manag. (IJCISIM). ISSN 2150–7988, **5**, 606–614 (2013)
13. Ramlakhan, K., Shang, Y.: Mobile, automated skin lesion classification system in tools with artificial intelligence (ICTAI). In: 23rd IEEE International Conference, pp. 138–141 (2011)
14. Friedman, R.J., Rigel, D.S., Kopf, A.W.: Early detection of malignant melanoma: The role of physician examination and self-examination of the skin. CA: A Cancer J. Clin. **35**(3) (May/June 1998)

Enhancement of Mobile Ad Hoc Network Security Using Improved RSA Algorithm

S.C. Dutta, Sudha Singh and D.K. Singh

Abstract The RSA algorithm is used in different communication networks in order to ensure data confidentiality. This paper proposed an improvement in RSA algorithm for the increased security enhancement in mobile ad hoc network. The proposed method is best for small messages and also applicable for much increased data security in different types of network including new generation network. For large volume of data, we have added DES with proposed RSA and getting better and secure transmission. We have used key length up to 2048 bits considering better security, computing speed, and processor condition. Key length can be increased depending upon the conditions.

Keywords Ad hoc wireless network · Mobile ad hoc network (MANET) · Public key cryptography · DES · RSA algorithm

1 Introduction

An ad hoc network is a wireless network with mobile computing devices that use wireless transmission for communication without having fixed infrastructure [1]. A mobile ad hoc network is a network that is merely comprised of mobile devices

S.C. Dutta (✉)
Department of Computer Science and Engineering,
BIT Sindri, Dhanbad 828123, India
e-mail: dutta_subhash@yahoo.com

Sudha Singh
Department of Computer Engineering, MGM College of Engineering
and Technology, Kamothe, Navi Mumbai 410209, India
e-mail: sudha_2k6@yahoo.com

D.K. Singh
Department of Electronics and Communication Engineering,
National Institute of Technology, Patna 800005, India
e-mail: dksingh_bit@yahoo.com

M. Pant et al. (eds.), *Proceedings of Fifth International Conference on Soft Computing for Problem Solving*, Advances in Intelligent Systems and Computing 436, DOI 10.1007/978-981-10-0448-3_86

without any preestablished infrastructure. These network find application in several areas like military communication, emergency situations that need quick deployment of a network, hybrid wireless network etc. Security is the major concern in rapid development of such wireless communication. To provide security, Data Encryption Standard (DES) [2], Diffie and Hellman [3], Rivest, Shamir and Adleman (RSA) [4] are the most widely used concepts.

RSA algorithm [4] was discovered by a group at MIT in 1978. Their method is

(a) Choose two prime numbers p and q (typically greater than 10^{100}).
(b) Compute $n = p \times q$ and $^{\Phi}(n) = (p - 1) \times (q - 1)$.
(c) Choose a number relatively prime to $^{\Phi}(n)$ and call it e.
(d) Find d such that $e \times d \equiv 1 \bmod {}^{\Phi}(n)$

Now divide the plain text into blocks of k (where k is the largest integer for which $2^k < n$) bits so that each plaintext P falls in the interval $0 \leq P < n$. To encrypt a message P, compute $C = P^e \bmod n$. To decrypt C, compute $P = C^d \bmod n$. The RSA algorithm robustness is ensured by the complexity of large number factorization. RSA different weaknesses of this algorithm could be observed and many attacks against it are developed successfully [5].

2 Related Work

To provide data security and confidentiality, researchers and developers are working on RSA since the last three decades in various types of networks [5–12]. Also different types of works are going on for digital signatures, RSA signatures cryptanalysis [13–18]. Developers are recommending higher key length for encryption, but time for decryption will be more in this case [19].

In our proposed RSA encryption algorithm for MANET, we specified a key length of 1024 bits and studied it for a key length of 2048 bits and higher ones.

3 Proposed Improvement of RSA Algorithm and Its Flow Model

Proposed enhanced RSA method is given below:

(a) Select two very big prime numbers p and q.
(b) Compute $n = p \times q$ and $^{\Phi}(n) = (p - 1) \times (q - 1)$.
(c) Public key component e ($e < (^{\Phi}(n)/10)$) is randomly generated and prime to $^{\Phi}(n)$. Also e should not belong to the set {3,5,7}. Conditions on e are applied to maintain the algorithm speed because we are working with keys containing large number of bits.
(d) Find d such that $e \times d \equiv 1 \bmod {}^{\Phi}(n)$

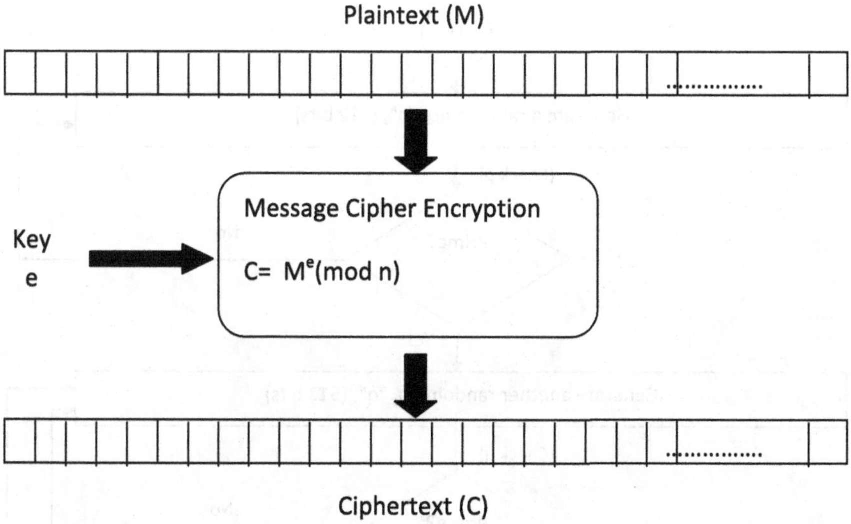

Fig. 1 RSA model

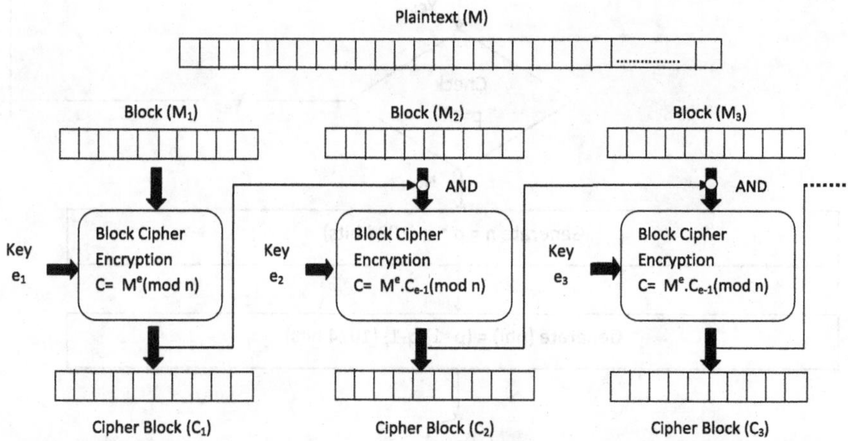

Fig. 2 Proposed RSA model

Fig. 3 Primary number generation

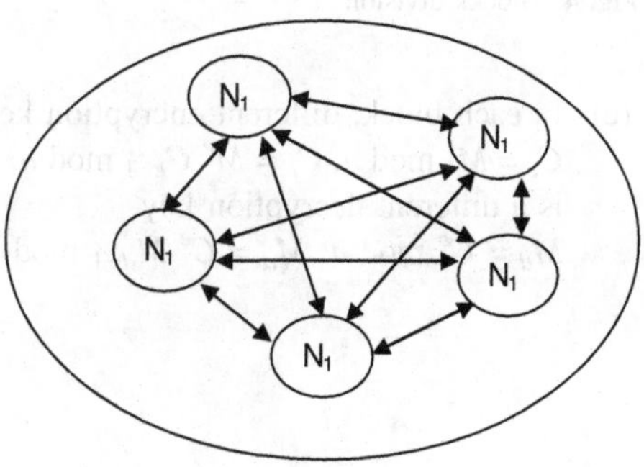

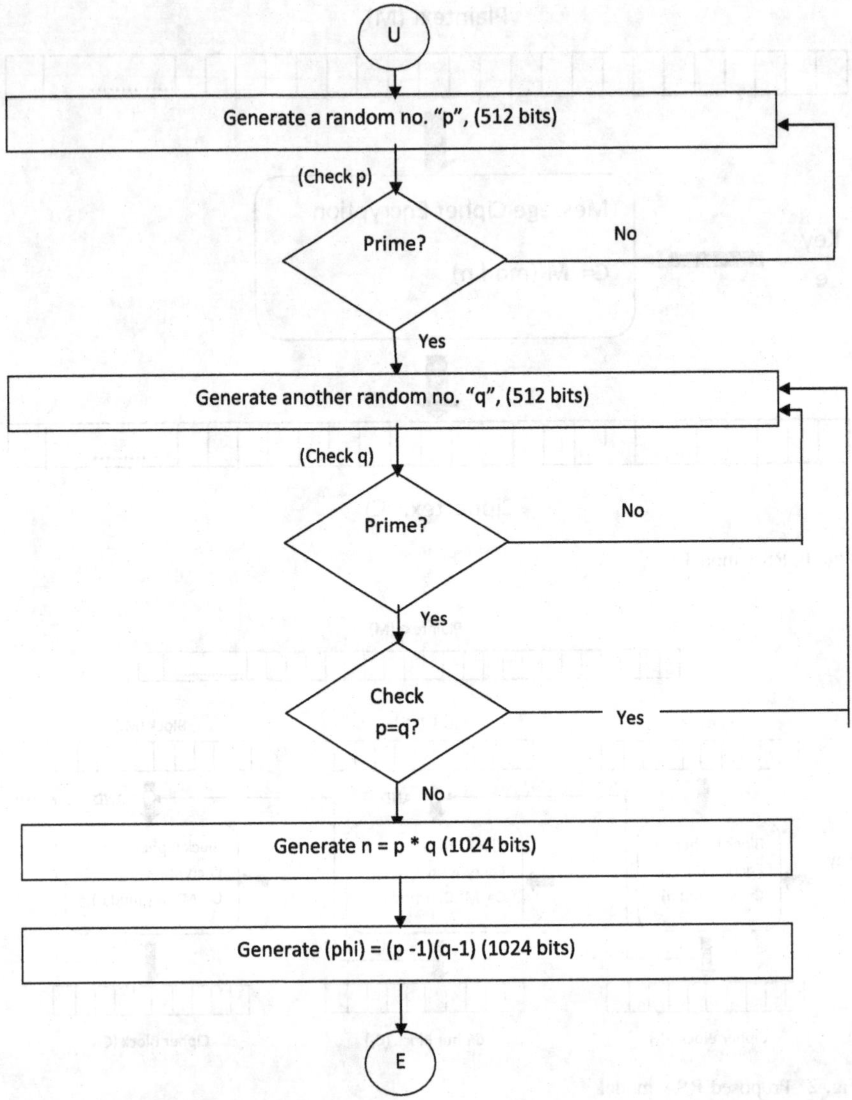

Fig. 4 Blocks division

(e) In each block, different encryption key is used.
$C_e = M^e \bmod n$; $C_e = M^e\, C_{e-1} \bmod n$, $e \geq 7$, and for each encryption key there is a different decryption key.
$M_d = C^e \bmod n$; $M_d = C^e\, M_{d-1} \bmod n$.

Example $P = 11$, $q = 17$, $n = p * q = 187$ and ${}^{\Phi}(n) = (p - 1) \times (q - 1) = 160$

Now select e such that $e < {}^{\Phi}(n)/10$, $\gcd(e, {}^{\Phi}(n)) = 1$ and e does not belongs to the set {3,5,7}.

$e_1 = 11$, $e_2 = 13, \ldots$
$d_1 = 131$, $d_2 = 37, \ldots$
$M_1 = 13$
$C_1 = M_1^{e1} \bmod n = 13^{11} \bmod 187 = 123$
$M_1 = C_1^{d1} \bmod 187 = 123^{131} \bmod 187 = 13 = M_1$.

We have to do encryption and decryption iteratively to avoid computational errors. In order to enhance the proposed RSA algorithm for security reasons, we have to fragment the message into blocks with an imposed length and apply the encryption algorithm on each block. Then proceed to generate randomly, the public key component e with conditions to maintain the algorithm speed. Algorithm should compute iteratively the encrypted message to avoid computational errors caused by large numbers; $C_e = M^e \bmod n = (M \cdot C_{e-1}) \bmod$ n, where with various conditions on e. These conditions are required to improve the computational speed and security of RSA in MANET. There should be often change in the encryption key for each input

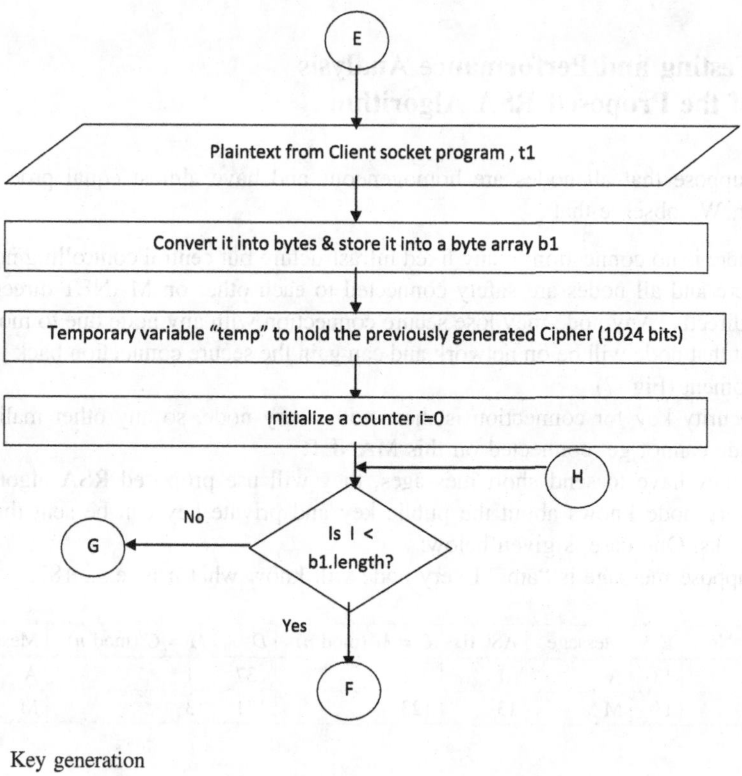

Fig. 5 Key generation

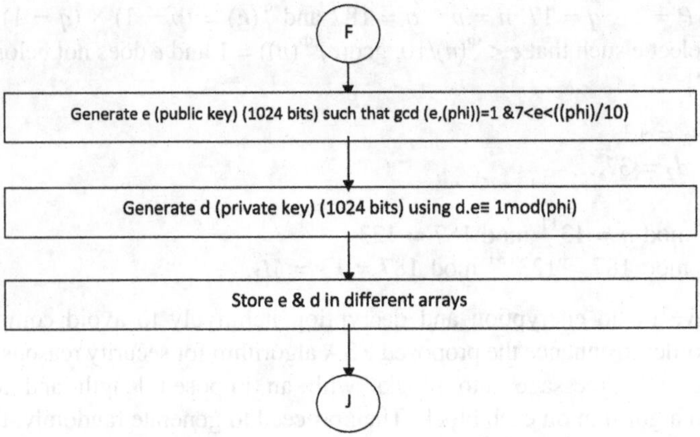

Fig. 6 Encryption

block to increase the robustness of the algorithm. Comparison between the RSA model and the proposed RSA model can be easily visualized by the Figs. 1 and 2.

Flow model of the encryption part of the proposed RSA consists of four parts: primary number generation (Fig. 3), block division (Fig. 4), key generation (Fig. 5), and encryption (Fig. 6). Flow models are given at the end of this paper.

4 Testing and Performance Analysis of the Proposed RSA Algorithm

We suppose that all nodes are homogeneous and have almost equal processing power. We observe that

1. There is no connection to any fixed infrastructure but central controlling node is there and all nodes are safely connected to each other on MANET directly or indirectly. Any node may lose secure connection with any node due to mobility but that node will be on network and can gain the secure connection back in any moment (Fig. 7).
2. Security key for connection is there with every node, so any other malicious node cannot get connected on this MANET.
3. If they have to send short messages, they will use proposed RSA algorithm. Every node knows about the public key and private key can be sent through blocks. One case is given below:
   Suppose message is "am." Every node will know what n is. e.g., 187.

Block No.	E	Message	ASCII	$C = M^e (\text{mod } n)$	D	$M = C^d (\text{mod } n)$	Message
1	13	A	1	1	37	1	A
2	11	M	13	123	131	13	M

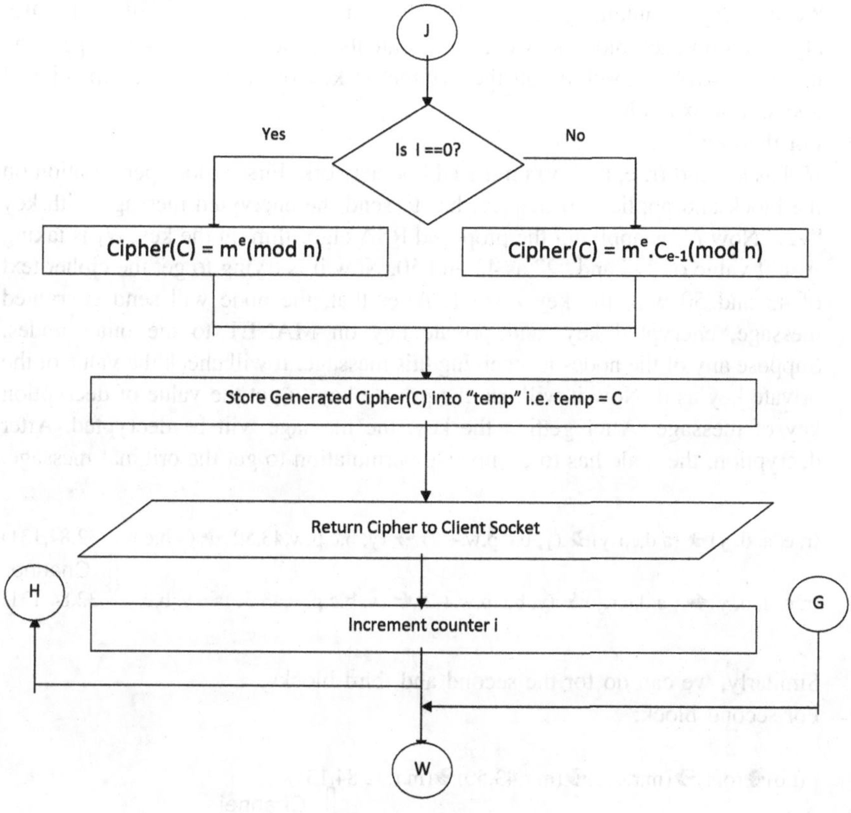

Fig. 7 An example of ad hoc wireless network

4. Due to the conditions applied on $^{\Phi}(n)$, time required for computation is as low as multiplying two array of numbers which means within a fraction of second. Encryption and decryption both are fast.
5. If we have to send a long message then also the proposed algorithm will work and it will work fast compared to all existing algorithms. But due to mobility in MANET, sometimes there will be some difficulty in locating the correct identification of block because we cannot get full features of TCP/IP in such network.
6. Solution is dividing the long messages into small blocks and using the symmetric cryptosystem on that block and for hiding symmetric key, using the proposed RSA algorithm.
7. One case is given below:
   Suppose the message is "Ready to go." Every node know what is n, e.g., 187. We are using alphabetical set for shifting and ASCII set for taking values of taking different characters. Message is divided into three blocks: B_1 is containing

"ready," B_2 is containing "to," and B_3 is containing "go." We will apply DES algorithm on each block to secure the contents of the block. Now we apply the proposed RSA algorithms on the symmetric key of each block as mentioned above. For example,
For the first block:
N_1 has to send (r, e, a, d, y) on the ad hoc network. First it does permutation on the block and applies shifting. N_1 has to send the encrypted message with key "+2." Now N_1 is applying the proposed RSA algorithm on the key. N_1 is taking ASCII value of "+" and "2" as 43 and 50. Now it is trying to get the cipher text of 43 and 50 with the key e = 11. After that, the node will send encrypted message, encrypted key, and private key on MANET to the other nodes. Suppose any of the nodes is receiving this message. It will check the value of the private key as d. Now it will use the private key to get the value of decryption key of message. After getting the key, the message will be decrypted. After decryption, the node has to do inverse permutation to get the original message.

(r, e, a, d, y)→ (a,d,e,r,y)→ (y, b,c,p,w,+,2) → (y, b,c,p,w,43,50)→ (y,b,c,p,w,32,84,131)
⇓ Channel
(r, e, a, d, y)← (a,d,e,r,y)← (y, b,c,p,w,+,2)← (y, b,c,p,w,43,50)←(y, b,c,p,w,32,84,131)

Similarly, we can do for the second and third blocks.
For second block:

(t,o)→(o, t)→(m,r,+,2)→(m,r,43,50)→(m,r,32,84,131)
⇓ Channel
(t,o)← (o, t)←(m,r,+,2)←(m,r,43,50)←(m,r,32,84,131)

For third block:

(g,o)→(e,m)→ (e,m,+,2)→(e,m,43,50)→(e,m,32,84,131)
⇓ Channel
(g,o)←(e,m)← (e,m,+,2)←(e,m,43,50)←(e,m,32,84,131)

8. We can use different RSA keys for different message blocks for better security. Since speed is good, we have used the key with 2048 bits for better security in proposed algorithm.

We successfully verified the proposed RSA using socket programming in Java. The random RSA public and private key pair with arbitrary length can be generated effectively by using the Java BigInteger. A 1024 bits RSA can be generated within 1 min on common PC platform (with processor-Intel(R) Core™2 Duo CPU T7500@ 2.20 GHz 2.20 GHz), while the encryption/decryption operation on data less than 1024 can be done within 2 s.

5 Conclusion and Future Work

This paper has proposed an improvement in RSA algorithm for the security enhancement in Mobile ad hoc network. It is slow for large volume of data but more secure in mobile ad hoc network. To increase the speed for large volume of data in such network, we have utilized the power of symmetric cryptosystem like DES with the proposed RSA algorithm. Simulation work on this work is under process for keys higher than 2048 bits.

References

1. Muthy, C.S.R., Manoj, B.S.: Adhoc Wireless Network Architecture and Protocols, 20th edn. Pearson Education Inc. (2014)
2. Stallings, W.: Cryptography and network security, 4th edn., Tenth impression. Pearson education Inc. (2011)
3. Difie, W., Hellman, M.: Multiuser cryptographic techniques. IEEE Trans. Inf. Theory (1976)
4. Rivest, R., Shamir, A., Adleman, L.: A method for obtaining digital signatures and public key cryptosystems. Commun. ACM **21**(2), 120–126 (1978)
5. Frunza, M., Scripariu, L.: Improved RSA encryption algorithm for increased security of wireless networks. In: IEEE International Symposium on Signals, Circuits and Systems, pp. 1–4 (2007)
6. Ren, W., Miao, Z.: A hybrid encryption algorithm based on DES and RSA in Bluetooth communication. In: IEEE 2nd International Conference on Modeling, Simulation and Visualization Methods, pp. 221–225 (2010)
7. Sexsena, S., Kapoor, B.: An efficient parallel algorithm for secured data communication using RSA public key cryptography method. In: IEEE International Conference on Advanced Computing, pp. 850–854 (2014)
8. Prema, G., Natarajan, S.: An enhanced security algorithm for wireless application using RSA and genetic approach. In: IEEE 4th International Conference on Computing, Communication and Networking Technologies, pp. 1–5 (2013)
9. De Costa, C., Moren, R.L., Carpinteiro, O.S., Pimenta, T.C.: A 1024 bit CMOS. In: IEEE 25th International Conference on Microelectronics, pp. 1–4 (2013)
10. Tan, X., Li, Y.: Parallel analysis of an improved RSA algorithm. In: IEEE International Conference on Computer Science and Electronics Engineering, pp. 318–320 (2012)
11. Li, Y., Liu, Q., Li, T.: Design and implementation of an improved RSA algorithm. In: IEEE Conference on E-Health Networking, Digital Ecosystems and Technologies, pp. 390–393 (2010)
12. Meshram, C.: An efficient ID based cryptographic encryption based on discrete logarithm problem and integer factorization problem. Elsevier's Inf. Process. Lett. **115**, 351–358 (2015)
13. Couvreur, C., Quisquater, J.: Fast decipherment algorithm of RSA public key cryptosystem. Electron. Lett. **18**(21), 905–907 (1982)
14. Boneh, D., Demillo, R.A., Lipton, R.J.: On the importance of checking cryptographic protocols for fault. EUROCRYPT'97, Springer-Verlag, LNCS, vol. 1233, pp. 37–51 (1997)
15. Bellcore Press Release: New threat model breaks crypto codes, Sept 1996. http://www.bellcore.com/press/ADVSRY96/facts.html
16. Cao, Y.Y., Fu, C.: An efficient implementation of RSA digital signature algorithm. In: IEEE International Conference on Intelligent Computation Technology and Automation, pp. 100–103 (2008)

17. Dhakar, R.S., Gupta, A.K., Sharma, P.: Modified RSA encryption algorithm. In: IEEE 2nd International Conference on Advanced Computing and Communication Technologies, pp. 426–429 (2012)
18. Dongjiang, L., Yandan, W., Hong, C.: The research on key generation in RSA public-key cryptosystem. In: IEEE 4th International Conference on Computational and Information Sciences, pp. 578–580 (2012)
19. Materials downloaded from www.google.com, www.wikipedia.org, http://www.wiziq.com/tutorials/java-implementation-rsa, www.jatit.org, and www.RSAlaboratories.org.

Empirical Study of Grey Wolf Optimizer

Avadh Kishor and Pramod Kumar Singh

Abstract In this paper, the authors empirically investigate performance of the grey wolf optimizer (GWO). A test suite of six non-linear benchmark functions, well studied in the swarm and the evolutionary optimization literature, is selected to highlight the findings. The test suite contains three unimodal and three multimodal functions. The experimental results demonstrate the advantages and weaknesses of the GWO. In case of unimodal problems, initially it hastens towards the optimal solution but soon slows down because of the diversity problem. A similar behaviour is seen in case of multimodal problems with a difference that because of its behaviour it easily sticks to local optima, loses its diversity and stops any further progress. The reason is that it lacks information sharing in the pack. This insight led the authors to propose a modified GWO called the modified grey wolf optimizer (MGWO). An empirical study of the proposed algorithm MGWO shows its promising performance as the obtained results are superior to the GWO for all the test functions.

Keywords Grey wolf optimizer (GWO) · Population based optimization · Crossover operator · Modified grey wolf optimizer (MGWO)

1 Introduction

Recently, the population-based optimization methods have become main-stream in solving optimization problems in the real-life applications in engineering design, information science, economics, biology, etc. [1, 2]. Some of the well-known population-based optimization methods are genetic algorithm (GA) [3], differential

Avadh Kishor (✉) · P.K. Singh
Computational Intelligence and Data Mining Research Laboratory,
ABV-Indian Institute of Information Technology and Management,
Gwalior, India
e-mail: avadhkishor133@gmail.com

P.K. Singh
e-mail: pksingh@iiitm.ac.in

M. Pant et al. (eds.), *Proceedings of Fifth International Conference on Soft Computing for Problem Solving*, Advances in Intelligent Systems and Computing 436, DOI 10.1007/978-981-10-0448-3_87

evolution (DE) [4], particle swarm optimization (PSO) [5], and artificial bee colony (ABC) algorithm [6, 7].

The grey wolf optimizer (GWO) [8] is one of the most recently proposed population-based optimization algorithms which simulates the hunting and democratic behaviour of the grey wolves in nature. It involves very few user-defined parameters and is also easy to implement. Due to its simplicity, it has attracted attention of many researchers and has been applied successfully to solve real-world problems [2, 9, 10]. Nevertheless, as suggested by the no free lunch theorem [11], no single optimization algorithm can perform equally well for all the problems. Therefore, to know how well the algorithm performs on a particular type of optimization problem or a subset of optimization problems is difficult as well as essential. Moreover, as it is very difficult to analyse this relationship theoretically owing to the probabilistic nature of these algorithms, the empirical study is a worthwhile strategy to judge the efficacy of the algorithms. In this paper, the authors too adopt the same strategy.

The authors investigate performance of the GWO on a well-known diverse set of synthetic optimization functions with twofold objectives: (a) empirical study of the GWO to ascertain its advantages and weaknesses (b) modification in the algorithm to alleviate its weaknesses with the insights gained.

The results of empirical study show that in case of unimodal problems, it initially hastens towards the optimal solution but soon slows down because of the diversity problem. A similar behaviour is seen in case of multimodal problems with a difference that because of its behaviour it easily sticks to local optima, loses its diversity and stops any further progress. The reason is that it lacks information sharing in the pack. In other words, the GWO faces the problem of premature convergence. The insights gained out of this empirical study led the authors to propose an algorithm modified grey wolf optimizer (MGWO), which improves the information sharing mechanism of the GWO and also improves its performance.

The remainder of this paper is organised as follows. In Sect. 2, a brief outline of the GWO is provided. In Sect. 3, benchmark problems, parameter setting, and empirical results and discussion for the GWO are presented. Section 4 describes the proposed algorithm. Simulation results for the proposed algorithm are presented and discussed in Sect. 5. Finally, a brief conclusion and future scope of the work is presented in Sect. 6.

2 Grey Wolf Optimizer

The GWO is a population-based meta-heuristic algorithm, which is inspired by the hunting and democratic behaviour of the grey wolves. The grey wolves generally live in packs, which comprise 5–12 members on average. There is a strict dominant hierarchy in the group that is practised by all the members. The group is led by an alpha wolf. It is followed by beta wolf, which holds the second rank (level) in the hierarchy and assists the alpha wolf in decision making. The beta wolf conveys the instructions of the alpha wolf to all the members of the pack and acknowledges the feedback to the alpha wolf. The delta wolf from the third level pack is subordinate to the alpha wolf and beta wolf; however, they dominate the omega wolves. A mathematical modelling of the GWO is described in the following subsection.

2.1 Mathematical Modelling

In this subsection, we describe the mathematical models of social hierarchy, enriching prey, search for prey, attacking prey and hunting.

Social hierarchy: The algorithm starts with a fixed number of wolves, which are assigned random positions in the search space. Alpha (α), beta (β) and delta (δ) correspond to the best, the second best, and the third best solutions respectively, and the remainder of the solutions is considered as omega (ω).

Encircling prey: Encircling behaviour of grey wolves plays a crucial role during hunting. Let $X(t)$ and $X_p(t)$ be the positions of the wolf and prey at the tth iteration (generation) in the algorithm. The mathematical model for this strategy is as follows:

$$D = |C * X_p(t) - X(t)| \tag{1}$$

$$X(t+1) = X_p(t) - A * D \tag{2}$$

where D is considered as difference vector. $C = 2r_2$ and $A = 2a * r_1 - a$. Here r_1 and r_2 are random numbers between (0,1) and $a = 2 - 2 \times t/|\text{t\_Max}|$ [8], which decreases linearly from 2 to 0; here t_Max is the maximum number of iterations in the algorithm.

Hunting: Hunting is directed by the alpha, the beta and the delta wolves. Hence, the positions of these three best solutions are saved in the pack and the omega wolves update their positions according to them. This entire hunting approach can be modelled mathematically as follows:

$$D_\alpha = |C_1 * X_\alpha - X|; D_\beta = |C_2 * X_\beta - X|; D_\delta = |C_3 * X_\delta - X| \tag{3}$$

$$X_1 = X_\alpha - A_1 * D_\alpha; X_2 = X_\beta - A_2 * D_\beta; X_3 = X_\delta - A_3 * D_\delta \tag{4}$$

$$X(t+1) = \frac{X_1 + X_2 + X_3}{3} \tag{5}$$

Attacking prey (Exploitation): In order to model this phase of the GWO, the value of "a" is gradually decreased over the course of iteration. As mentioned above, A is a random value between $-2a$ and $2a$. Therefore, when $|A| < 1$ the wolves attack the prey.

Search for prey (exploration): Grey wolves perform searching according to the positions of the alpha, beta and delta wolves. Two components of the GWO (i) A and (ii) C are responsible to carry out the search for the prey. When $|A| > 1$ in the early iterations of the algorithm, it enforces the divergence of wolves from the prey. Moreover, a random number C between (0,2) is used to balance the exploration and exploitation.

Instead of detailed description, we have outlined the major concepts and flow of the algorithm only because of space limitation. Interested readers may refer [8] for a detailed description of the algorithm. However, the pseudo-code of the GWO algorithm is described in Algorithm 1.

Algorithm 1 Pseudo code of the GWO algorithm

Initialize the position of grey wolf population (search agents) $X_i (i = 1, 2, ..n)$, a, A, *and* C
Calculate the objective function value $f(X_i)$ for each search agent and Set:
  $f(X_\alpha) \leftarrow$ the best $f(X_i)$
  $f(X_\beta) \leftarrow$ the second best $f(X_i)$
  $f(X_\delta) \leftarrow$ the third best best $f(X_i)$
for $t \leftarrow 1$ to *maximum number of iterations* **do**
  for $i \leftarrow 1$ to *number of search agents* **do**
    update X_i by (5)
  end for
  Calculate the objective function value $f(X_i)$ for each search agent and update a, A, C, X_α, X_β, *and* X_δ
end for
return X_α

3 Empirical Study of GWO

3.1 Test Functions

Six non-linear traditional benchmark functions [8] are used to evaluate performance of the GWO. Various subsets of these test functions have been utilised in number of prior studies, e.g., [12, 13]. According to their properties these functions are categorised into two groups: unimodal and multimodal. The properties and the mathematical expressions of these test functions are described below. Note that all the functions considered for this study are minimization problems and their global minimum is 0.

GROUP I: Unimodal Functions

A function containing only one optimum value as global optimum is called unimodal function and is generally used to investigate the convergence speed of the algorithm.

In this group, the first test function is the sphere function. It is the simplest problem to solve. Its initialization range is [−100,100]. Its mathematical representation is as follows.

$$f_1(x) = \sum_{i=1}^{n} x_i^2$$

where x is a real-valued vector of dimension n and x_i denotes the ith *parameter of that vector.*

The second function is the Schwefel 2.22 function. It is a very difficult problem. Due to sharply pointed corners of the contours, algorithms usually get stuck at one of these corners without any sign of further progress [14]. Its initialization range is [−10,10]. Its mathematical representation is shown below.

$$f_2(x) = \sum_{i=1}^{n} |x_i| + \prod_{i=1}^{n} |x_i|$$

The third function is a classical optimization problem called Rosenbrock function. Its global optimum is located inside a long, narrow, parabolic shaped flat valley. It is trivial to find the valley; however, finding the global optimum is very difficult. Moreover, it can also be treated scientifically as multimodal problem [13]. Its initialization range is [−30,30]. Its mathematical formulation is given below.

$$f_3 = \sum_{i=1}^{n-1} (100(x_i^2 - x_{i+1})^2 + (1 - x_i^2))$$

GROUP II: Multimodal Functions

A function having many local optimal solutions other than the global optimal solution is known as a multimodal function and is used to test the capability of the algorithms to elude from premature convergence.

In this group, the first problem is the Rastrigin function. It is a complex multimodal problem with a large number of regularly distributed local optima. When attempting to solve it, algorithms easily get trapped into a local optimum. Thus, an algorithm with better exploration ability is likely to perform better. Its initialization range is [−5.12, 5.12]. It has the following mathematical definition.

$$f_4(x) = \sum_{i=1}^{n} [x_i^2 - 10\cos(2\pi x_i) + 10]$$

The second problem is the Ackley function. It is presumably the easiest problem among all the three problems of this group and is a widely used test problem. The Ackley function contains a narrow global optimum basin and many local optima. Its initialization range is [−32, 32]. The mathematical formulation of the function is as follows.

$$f_5(x) = -20 \times e^{\left(-0.2\sqrt{\frac{1}{n}\sum_{i=1}^{n} x_i^2}\right)} - e^{\left(\frac{1}{n}\sum_{i=1}^{n}\cos(2\pi x_i)\right)} + 20 + e$$

The third problem is the Griewank Function. It has a component $\prod_{i=1}^{n}\cos(x_i/\sqrt{i})$ that provides linkages among variables, thereby making it difficult to obtain the global optimum. Moreover, Griewank function has an interesting characteristic that it is easier for higher dimensions than the lower dimensions [12]. Its initialization range is [−600, 600] and the mathematical formulation is as follows.

$$f_6(x) = \frac{1}{4000}\sum_{i=1}^{n} x_i^2 - \prod_{i=1}^{n}\cos(x_i/\sqrt{i}) + 1$$

3.2 Parameter Setting

The parameters used for experiments are as follows. For each function, three different dimension sizes 10, 20 and 30 are examined. To test whether the GWO algorithm scales well, three different population sizes (number of search agents) 30, 60 and 120 are selected for each function with dimensions 10, 20 and 30, respectively. Furthermore, the maximum number of iterations for the dimensions 10, 20 and 30 is set to 500, 1000 and 1500, respectively. A total of 30 runs is carried out for each experimental setting.

3.3 Experimental Results and Discussion

Figure 1 shows the convergence graphs for all three unimodal functions sphere, Schwefel 2.22, and Rosenbrock with three different experimental settings. Each graph demonstrates the mean objective function value profile over the 30 runs.

It is observed from the convergence graphs of Fig. 1a–f that the GWO moves quickly towards the optimum in the early iterations but as the number of iterations increases, it slows down its convergence almost to halt. This phenomenon may be due to the linearly decreasing "a" as A is also decreased with the "a" and the value of "a" decreases as the number of iterations increases. Moreover, from (2) it is obvious that when the value of A is small, the wolves move very slowly with smaller step size towards the prey. In other words, in latter stages of search procedure the GWO favours exploitation and performs search only in its vicinity by reducing its step size.

However, the performance of the Rosenbrock function as shown in Fig. 1g–i is different than the other two unimodal functions. As mentioned in Sect. 3.1 this test

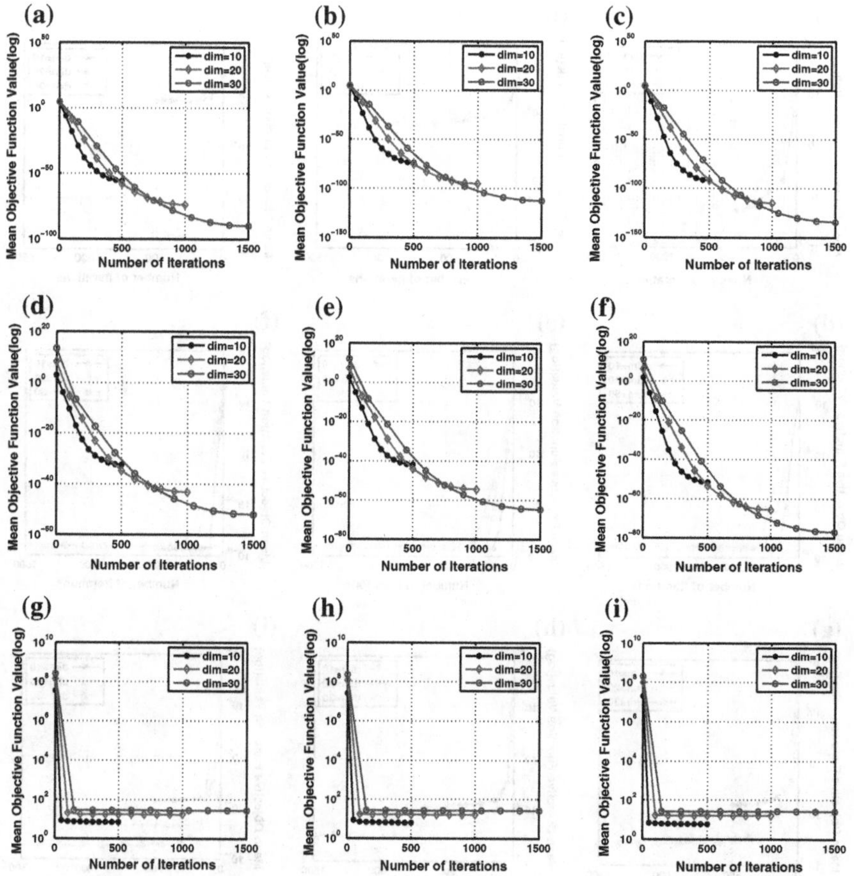

Fig. 1 Progress towards the optimum solution of GWO over three unimodal test functions. **a** Sphere with popsize = 30. **b** Sphere with popsize = 60. **c** Sphere with popsize = 120. **d** Schwefel2.22 with popsize = 30. **e** Schwefel2.22 with popsize = 60. **f** Schwefel2.22 with popsize = 120. **g** Rosenbrock with popsize = 30. **h** Rosenbrock with popsize = 60. **i** Rosenbrock with popsize = 120

function behaves like a multimodal function. Here, convergence graphs indicate that initially the GWO takes very less amount of time to locate local optimum and then stops proceeding any further. Generally, it is referred to as premature convergence. Another fascinating observation of the Rosenbrock function is that the performance of the algorithm decreases with the increasing number of function evaluations.

Figure 2 shows the convergence graphs for all the three multimodal functions Rastrigin, Griewank and Ackley with three different experimental settings. A close investigation of Fig. 2d–i reveals that the GWO finds local optima very fast and then flattens out, rapidly halting its progress. In other words, in case of multimodal

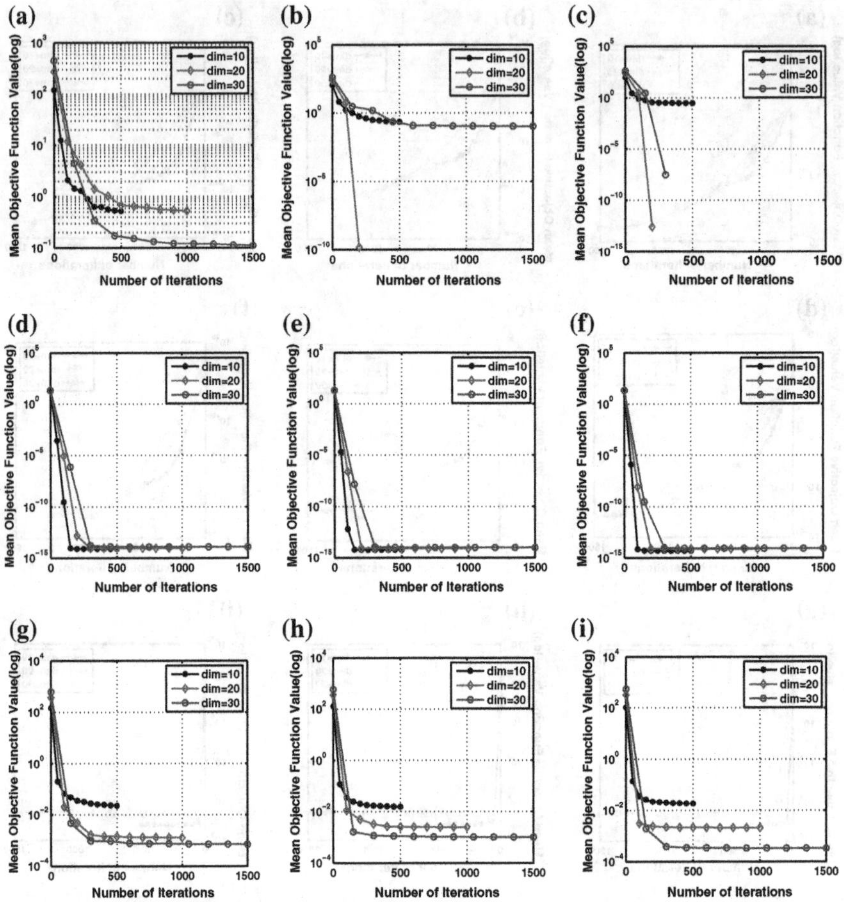

Fig. 2 Progress towards the optimum solution of GWO over three multimodal test functions. **a** Rastrigin with popsize = 30. **b** Rastrigin with popsize = 60. **c** Rastrigin with popsize = 120. **d** Ackley with popsize = 30. **e** Ackley with popsize = 60. **f** Ackley with popsize = 120. **g** Griewank with popsize = 30. **h** Griewank with popsize = 60. **i** Griewank with popsize = 120

functions, the GWO gets trapped into local optima and converges prematurely. This may be due to the following two reasons. First, by using linearly decreasing *a*, the GWO lacks exploration ability in the final stages of search (at the end of run) when it is needed to escape out local optimum. Second, the position of search agents in the GWO is updated according to the position of the alpha, beta and delta only which are located at the best, the second best and the third best positions, respectively. Thus, these best agents increase the convergence pressure and the algorithm moves quickly towards the optimum by losing its diversity that leads to premature convergence.

Table 1 Mean and the standard deviation (Std) of the best-of-run solution for 30 independent runs by GWO over all functions f1−f6

Function	Dim	Iteration	Mean best value (Standard deviation)		
			PS = 30	PS = 60	PS = 120
f1	10	500	2.47E−56(7.92E−56)	1.38E−74(3.22E−74)	1.20E−92(3.56E−92)
	20	1000	1.09E−74(4.90E−74)	8.03E−96(2.26E−95)	1.36E−115(2.76E−115)
	30	1500	8.22E−91(1.69E−90)	3.74E−113(1.50E−112)	5.62E−135(1.54E−134)
f2	10	500	4.10E−33(6.43E−33)	1.42E−42(1.89E−42)	2.96E−52(5.54E−52)
	20	1000	7.09E−44(1.20E−43)	3.83E−55(4.46E−55)	2.64E−66(2.88E−66)
	30	1500	1.38E−52(2.64E−52)	1.99E−65(3.35E−65)	8.12E−78(1.12E−77)
f3	10	500	6.78E+00(7.46E−01)	6.36E+00(6.14E−01)	5.95E+00(6.84E−01)
	20	1000	1.66E+01(4.83E−01)	1.63E+01(7.56E−01)	1.61E+01(7.00E−01)
	30	1500	2.67E+01(7.49E−01)	2.64E+01(8.62E−01)	2.60E+01(7.52E−01)
f4	10	500	5.26E−01(1.60E+00)	2.21E−01(1.21E+00)	3.09E−01(1.18E+00)
	20	1000	5.28E−01(1.63E+00)	0(0)	0(0)
	30	1500	1.11E−01(6.09E−01)	1.06E−01(5.79E−01)	0(0)
f5	10	500	7.64E−15(2.16E−15)	5.63E−15(1.70E−15)	4.91E−15(1.23E−15)
	20	1000	8.82E−15(2.02E−15)	7.64E−15(1.08E−15)	7.52E−15(1.23E−15)
	30	1500	1.17E−14(2.72E−15)	9.30E−15(2.38E−15)	8.35E−15(1.08E−15)
f6	10	500	2.24E−02(2.50E−02)	1.54E−02(1.54E−02)	1.86E−02(2.14E−02)
	20	1000	1.29E−03(3.45E−03)	2.46E−03(7.31E−03)	2.10E−03(4.94E−03)
	30	1500	7.02E−04(2.74E−03)	1.01E−03(4.04E−03)	3.35E−04(1.84E−03)

Here, Dim and PS correspond the dimension of the function and the population size, respectively

The performance of the Rastrigin function is significantly different than the other two multimodal functions used for this experiment. From Fig. 2a–c, two interesting observations of this function are (i) the average performance of the GWO on 10 dimensional problem is worse than the 20 or 30 dimensional problem, and (ii) a larger population results in slow convergence.

Table 1 demonstrates the mean and standard deviation (Std) values of the best-of-run values for 30 runs obtained by the GWO over all for the six test functions with three different experimental settings. For all the test functions, the performance of the GWO (in terms of solution quality) ameliorates as the number of potential trial solutions (population size) increases. Finally, it shows that the GWO scales well for all six test problems taken in this study.

4 The Proposed Modified GWO (MGWO)

Insights into the performance of the GWO gained from the aforementioned empirical study are as follows. Its convergence speed is reduced in the latter iterations of the runs as it favours exploration in the initial iterations and latter iterations are dedicated only to exploitation. Further, instead of sharing information among all the individuals in the pack, only three best solutions named alpha, beta and delta concentrate and share the information and pass on this information to all other individuals in the pack for further progress. In other words, the GWO lacks diversity as it concentrates only on the three best solutions. Consequently, mainly in case of multimodal problems (when many local optima are present in the search space to mislead the search), it gets stuck into a local optima, loses it diversity and faces the problem of premature convergence.

This insight led to propose a modified GWO named MGWO. For this modification, the authors introduce the crossover operator of the genetic algorithm (GA) into the original GWO. By incorporating crossover operator in the MGWO, the global search ability is improved since every member of the pack gets chance to share information with each other. It helps in maintaining necessary exploration and exploitation. Thus, it alleviates the problem of diversity and avoids premature convergence.

As depicted in Algorithm 2, all the steps in the MGWO are same as the original GWO except that a crossover operator is included as an extra operator. The crossover operator works as follows.

1. For each solution X_i in population $\mathcal{P}$, select another solution X_j randomly from $\mathcal{P}$. Note that both the solutions X_i and X_j are different.

Algorithm 2 Pseudo code of the MGWO algorithm

Initialize the position of grey wolf population (search agents) $X_i (i = 1, 2, ..n)$, a, A, *and* C
Calculate the objective function value $f(X_i)$ for each search agent and Set:
  $f(X_\alpha) \leftarrow$ the best $f(X_i)$
  $f(X_\beta) \leftarrow$ the second best $f(X_i)$
  $f(X_\delta) \leftarrow$ the third best best $f(X_i)$
for $t \leftarrow 1$ to *maximum number of iterations* **do**
  for $i \leftarrow 1$ to *number of search agents* **do**
    update X_i by (5)
  end for
  Calculate the objective function value $f(X_i)$ for each search agent and update a, A, C, X_α,X_β, *and* X_δ
  /* Crossover */
  for $i \leftarrow 1$ to *number of search agents* **do**
    Pick a solution X_j randomly from the population to perform crossover with X_i
    Generate a new solution X_i^{new} by combining X_i and X_j by (6)
    Apply greedy selection between X_i^{new} and X_i and keep fitter individual in the population
  end for
end for
return X_α

2. Produce a new solution X_i^{new} (say, offspring) by combining both the parent solutions as:

$$X_i^{\text{new}} = \text{rand}(0, 1) * X_i + \text{rand}(0, 1) * X_j \tag{6}$$

Here, rand(0,1) denotes a random number between 0 and 1.

A greedy selection is performed between the offspring X_i^{new} and the parent X_i. If the objective function value of the offspring is better, it replaces the parent. Otherwise, the parent remains unchained.

5 On the Performance of MGWO

To assess the relative performance of the MGWO, it is compared with the GWO on all the test functions listed in Sect. 3.1. The parameter setting for this comparative study is as follows. Both the dimension of the functions and population size are 30 each. The number of runs is also 30 for each function by both the algorithms. For fair comparison, the termination criteria are set to the maximum number of function evaluations (FEs). Both the algorithms terminate after 45000 FEs. It amounts to 500 iterations for the GWO and 250 iterations for the MGWO as the FEs are doubled, in comparison to GWO, in iteration in the MGWO because of the crossover operator.

Figure 3 shows convergence curves of both the algorithms. It is clearly visible that the MGWO outperforms the GWO in all the six benchmark functions. It reflects the efficacy of the incorporation of the crossover operator introduced into the original GWO.

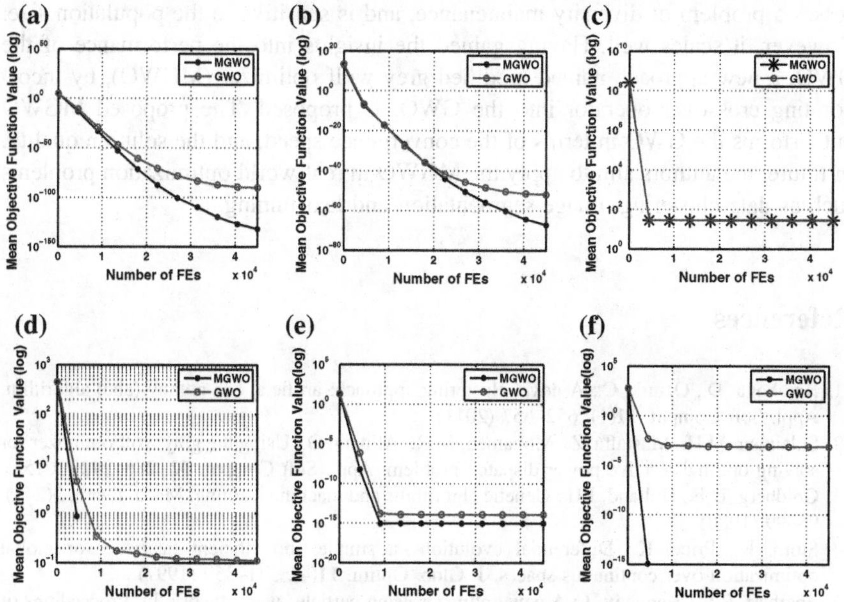

Fig. 3 Progress towards the optimum solution of MGWO over all the test functions. **a** Sphere function. **b** Schwefel2.22 function. **c** Rosenbrock Function. **d** Rastrigin function. **e** Ackley function. **f** Griewank function

Table 2 Mean and the standard deviation (Std) of the best-of-run solution for 30 independent runs by the MGWO and the GWO

Algorithm	PS	Dim	Mean best value (Standard deviation)		
			f1	f2	f3
MGWO	30	30	1.09E−132(5.32E−132)	4.55E−68(1.02E−67)	27.31407(0.937394)
GWO	30	30	8.22E−91(1.69E−90)	1.38E−52(2.64E−52)	2.67E + 01(7.49E−01)
			f4	f5	f6
MGWO	30	30	0(0)	8.88E−16(8.88E−16)	0(0)
GWO	30	30	1.11E−01(6.09E−01)	1.17E−14(2.72E−15)	7.02E−04(2.74E−03)

Perspective values for the GWO has been considered from the previous study (refer, Table 1)

Table 2 presents the mean and standard deviation values of all the six test functions achieved by the MGWO. It is interesting to see that the MGWO is far better than the GWO in terms of solution quality.

6 Conclusion

In this paper, the performance of the GWO algorithm has been extensively investigated by experimental studies of six synthetic test functions well studied in the literature. The experimental results demonstrate that it lacks exploration, processes a problem of diversity maintenance, and is sensitive to the population size. However, it scales well. Having gained the insights into the performance of the GWO, a new approach named modified grey wolf optimizer (MGWO), by incorporating crossover operator into the GWO, is proposed. The proposed MGWO outperforms the GWO in terms of the convergence speed, and the solution quality. In future, the authors aim to apply the MGWO in real-world optimization problems such as data clustering, image segmentation and text mining.

References

1. Karaboga, D., Ozturk, C.: A novel clustering approach: artificial bee colony (abc) algorithm. Appl. Soft Comput. **11**(1), 652–657 (2011)
2. Sulaiman, M.H., Mustaffa, Z., Mohamed, M.R., Aliman, O.: Using the gray wolf optimizer for solving optimal reactive power dispatch problem. Appl. Soft Comput. **32**, 286–292 (2015)
3. Goldberg, D.E., Holland, J.H.: Genetic algorithms and machine learning. Mach. Learn. **3**(2–3), 95–99 (1988)
4. Storn, R., Price, K.: Differential evolution—a simple and efficient heuristic for global optimization over continuous spaces. J. Glob. Optim. **11**(4), 341–359 (1997)
5. Eberhart, R.C., Kennedy, J.: A new optimizer using particle swarm theory. In: Proceedings of the Sixth International Symposium on Micro Machine and Human Science, vol. 1, pp. 39–43. New York, NY, 1995

6. Karaboga, D.: An idea based on honey bee swarm for numerical optimization. Technical report, Technical Report-TR06, Erciyes university, engineering faculty, computer engineering department, 2005
7. Karaboga, D., Basturk, B.: A powerful and efficient algorithm for numerical function optimization: artificial bee colony (abc) algorithm. J. Glob. Optim. **39**(3), 459–471 (2007)
8. Mirjalili, S., Mirjalili, S.M., Lewis, A.: Grey wolf optimizer. Adv. Eng. Softw. **69**, 46–61 (2014)
9. Emary, E., Zawbaa, H.M., Grosan, C., Hassenian, A.E.: Feature subset selection approach by gray-wolf optimization. In: Afro-European Conference for Industrial Advancement, pp. 1–13. Springer, 2015
10. Song, X., Tang, L., Zhao, S., Zhang, X., Li, L., Huang, J., Cai, W.: Grey wolf optimizer for parameter estimation in surface waves. Soil Dyna. Earthquake Eng. **75**, 147–157 (2015)
11. Wolpert, D.H., Macready, W.G.: No free lunch theorems for optimization. IEEE Trans. Evol. Comput. **1**(1), 67–82 (1997)
12. Whitley, D., Rana, S., Dzubera, J., Mathias, K.E.: Evaluating evolutionary algorithms. Artif. Intell. **85**(1), 245–276 (1996)
13. Huang, V.L., Suganthan, P.N., Liang, J.L.: Comprehensive learning particle swarm optimizer for solving multiobjective optimization problems. Int. J. Intell. Syst. **21**(2), 209–226 (2006)
14. Schwefel, H.P.P.: Evolution and Optimum Seeking: The Sixth Generation. John Wiley & Sons, Inc. (1993)

Evaluation of Huffman-Code and B-Code Algorithms for Image Compression Standards

Chanda, Sunita Singh and U.S. Rana

Abstract To reduce the quantity of data without excessively reducing the quality of the multimedia data is called compression. Compressed multimedia data are faster for transition and storing as compared to the original uncompressed multimedia data. For JPEG and JPEG 2000 images there are various techniques and standards for data compression. These standards consist of different functions such as color space conversion and entropy coding. Huffman codes and B-codes are normally used in the entropy coding phase.

Keywords Huffman codes · B-codes · Image compression

1 Introduction

Information, in its many forms, is a valuable commodity in today's society, and the amount of information is increasing at a phenomenal rate. As a result, the ability to store, access, and transmit information in an efficient manner has become crucial. This is particularly true in the case of digital images. A large number of bits is typically required to represent even a single digital image and with the rapid advances in sensor technology and digital electronics, this number grows larger with each new generation of products [3, 15]. Furthermore, the number of digital images created each day increases as more applications are found. In order to have efficient utilization of disk space and transmission rate, images need to be compressed.

Chanda (✉)
Uttarakhand Technical University, Dehradun, India
e-mail: chanda.nautiyal@gmail.com

Sunita Singh
IRDE (DEAL), Dehradun, India
e-mail: sunnidma@yahoo.com

U.S. Rana
DAV (PG) College, Dehradun, India
e-mail: drusrana@yahoo.co.in

M. Pant et al. (eds.), *Proceedings of Fifth International Conference on Soft Computing for Problem Solving*, Advances in Intelligent Systems and Computing 436, DOI 10.1007/978-981-10-0448-3_88

Image compression is the technique of reducing the file size of an image without compromising the image quality at an acceptable level [6, 17]. This reduction in file size saves disk/memory space and allows faster transmission of images over a medium. A wide range of common techniques to reduce the number of bits required for the representation of digital image have been developed over the years, and novel approaches continue to emerge [4, 14, 18].

In this paper we have compared two compression techniques Huffman coding and B-Coding. Entropy and average length of each code has also been calculated.

2 Huffman Code

Proposed by Dr. David A. Huffman in 1952 as "A method for the construction of minimum redundancy code," Huffman code is a technique for compressing data. Huffman's algorithm looks at the occurrence of each character and it as a binary string in an optimal way [1, 2, 12, 16]. Huffman coding is a form of statistical coding which attempts to reduce the amount of bits required to represent a string of symbols. The pixels in the image are treated as symbols. The algorithm accomplishes its goals by allowing symbols to vary in length. Shorter codes are assigned to the most frequently used symbols, and longer codes to the symbols which appear less frequently in the string (that is where the statistical part comes in). Code word lengths vary and will be shorter for the more frequently used characters.

2.1 Huffman Algorithm

- Step 1—Read the image on to the workspace of the matlab.
- Step 2—Call a function which will find the symbols (i.e., pixel value which is non-repeated).
- Step 3—Call a function which will calculate the probability of each symbol.
- Step 4—Probabilities of symbols are arranged in decreasing order.
- Step 5—The two smallest probabilities are combined by addition to form a new set of probabilities.
- Step 6—The new set of probabilities, which has one fewer probability than the original set, is again ordered according to magnitude. Equal probabilities can be ordered in any way. (e.g., in Fig. 1 0.1 obtained by combining input probabilities 0.06 and 0.04 could be placed in any three of the bottom column 2 entries.)
- Step 7—Go to step 5. Repeat this process until only two probabilities are left.

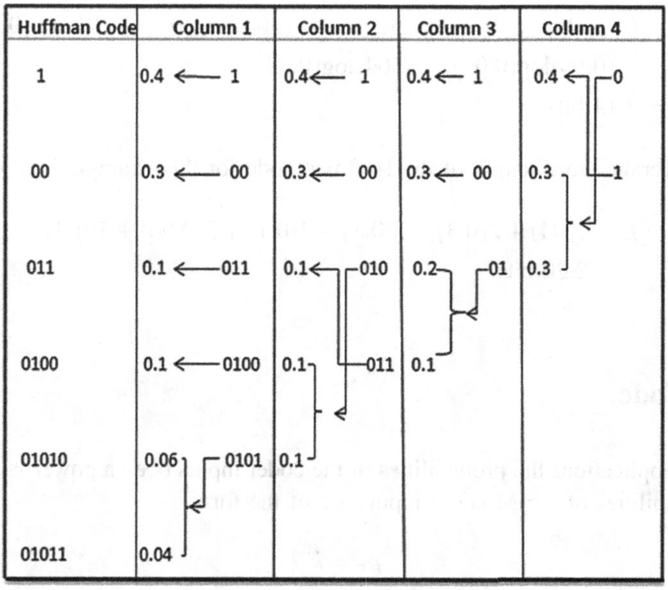

Huffman Code
Column 1
Column 2
Column 3
Column 4
1
00
011
0100
01010
01011
0.4 1
0.3 00
0.1 011
0.1 0100
0.06 0101
0.04
0.4 1
0.3 00
0.1 010
0.1 011
0.1
0.4 1
0.3 00
0.2 01
0.1
0.4 0
0.3 1
0.3

Fig. 1 Huffman coding

2.2 *How to Assign Codes?*

Code words are generated by starting at the last step and working backward [5, 7, 10]. Let us take six input probabilities as listed in column 1 of Fig. 1. We start by assigning 0 to one of the last two combined probabilities, and 1 to the other, as illustrated in Fig. 1, where we have placed a 0 to the left of the 0.6 in column 4 and a 1 to the left of the 0.4. We now proceed backward to column 3, decomposing probabilities and generating code words as we go: For example, the 0.6 in column 4 is decomposed back into the two 0.3 probabilities in column 3 and further 0.3 is decomposed into 0.2 and 0.1. The 0 associated with the 0.6 remains the first bit of each of its decomposed code words and the 1 associated with 0.4 remains the first bit of the 0.4 in column 3. A second bit, a 0 and 1, respectively, is appended to each of the code words associated with their reconstructed probabilities to obtain the code words in column 3. The same procedure is repeated to go back to column 2, and finally to the input probabilities, at which point we have a code word assigned to each input level.

It can be proved that the procedure outlined above generates a compact code. For the input probabilities listed in Fig. 1 the entropy is

$$H = (-0.4)\log(0.4) - (0.3)\log(0.3) - (0.1)\log(0.1) - (01)\log(0.1)$$
$$- (0.06)\log(0.06) - (0.04)\log(0.04)$$
$$= 2.14 \text{ bits.}$$

The average word length of the Huffman code for this example is

$$R = 1(04) + 2(0.3) + 3(0.1) + 4(0.1) + 5(0.06) + 5(0.04)$$
$$= 2.20 \text{ bits.}$$

3 B-Code

In some applications the probabilities of the coder inputs obey a power law; that is, the probabilities of the M coder inputs are of the form

$$p_k = k^{-\gamma}$$

for $k = 1, 2, \ldots, M$, and some positive constant γ [5]. The B-codes are nearly optimal for data that obey.

The B_1-code is presented in Fig. 2. Half of the bits in each code word are "continuation" bits labeled C and the other half are "information" bits. The information bits use a natural code that increases in length, as illustrated in Fig. 2. The continuation bit is, of course, either 0 or 1, but it can be determined by either of two rules: for 2-level data where each pixel is white or black, the continuation bit can be set equal to the gray level, say, $C = 0$ for black and $C = \text{I}$ for white. The other possibility is to let it alternate with each code word since its only purpose is to signify how long a code word is. For example, the sequence of code words for the sequence of inputs w_1, w_8, w_5 could be 00 10 10 11 01 00 or 10 00 00 01 11 10, where we have underlined the continuation bits. A change in the continuation bits signifies the start of a new code word. Note that the code is not instantaneous because the decoder must look ahead to the next continuation bit in order to determine whether or not the present code word has ended.

Fig. 2 B-coding

Probabilities	B_1 - Code	B_2 - Code
0.4	C0	C00
0.3	C1	C01
0.1	C0C0	C10
0.1	C0C1	C11
0.06	C1C0	C00C00
0.04	C1C1	C00C01

3.1 How to Assign Codes?

Implementation of the B_1-code is much simpler than for the Huffman code. For example, in coding run lengths the coder for the information bits is simply an up-counter that counts up by one for each new datum until the end of the run is reached. At the end of each run the counter is reset to zero and the continuation bit is flipped. Higher-order B-codes can also be constructed. A B_n-code uses n information bits for each continuation bit, as illustrated in Fig. 2 for $n = 2$.

For the set of input probabilities listed in Fig. 2, the average length of the B_1-code is

$$\begin{aligned} R &= 2(0.4) + 2(0.3) + 4(0.1) + 4(0.1) + 4(0.06) + 4(0.04) \\ &= 2.6 \end{aligned}$$

and the average length of the B_2-code is

$$\begin{aligned} R &= 3(0.4) + 3(0.3) + 3(0.1) + 3(0.1) + 6(0.06) + 6(0.04) \\ &= 3.3 \end{aligned}$$

4 Implementation of the Algorithms

The input image is a JPEG image of 46.7 kB. Read this on the workspace of Matlab. A digital image is two dimensional light intensity function $f(x, y)$ where x and y denote spatial co-ordinate and the amplitude or value of 'f' at any point (x, y) is called intensity or brightness of the image at that point. This value is taken as image symbol. Probability of each symbol has been calculated and arranged from highest to lowest probability as shown in Table 1. Entropy of the input image is calculated by the following formula

$$H = -\sum_{i=1}^{n} p_i \log p_i$$

where p = probability of symbol.

For the given input image entropy was calculated as 7.5434. Our aim is to represent the image with less number of bits. Huffman coding and B-coding algorithms are applied for the generation of codes as described in Sects. 2 and 3.

After executing the above mentioned coding algorithms, average length of each code has been calculated. Average word length using the above codes is presented in Table 2.

Table 1 Table for code length

Symbol	Prob.	Code length by Huffman code	Code length by B_1 code	Symbol	Prob.	Code length by Huffman code	Code length by B_1 code
218	0.0024	9	2	75	0.0011	10	16
220	0.0034	8	2	49	0.0015	9	16
221	0.0035	8	4	33	0.0019	9	16
223	0.0039	8	4	46	0.0016	9	16
222	0.0031	8	4	65	0.0011	10	16
224	0.005	8	4	86	0.0014	10	16
217	0.0026	9	8	101	0.0024	9	16
219	0.0028	9	8	147	0.0042	8	16
225	0.0049	8	8	190	0.001	10	16
226	0.0044	8	8	128	0.0061	7	16
227	0.005	8	8	129	0.0059	7	16
228	0.0057	7	8	98	0.0021	9	16
229	0.0053	8	8	76	0.0012	10	16
216	0.0026	9	8	50	0.0015	9	16
213	0.0026	9	8	45	0.0016	9	16
209	0.0023	9	8	64	0.0012	10	16
206	0.002	9	8	85	0.0014	10	16
203	0.002	9	8	100	0.0023	9	16
202	0.0022	9	8	99	0.0021	9	16
204	0.0018	9	8	77	0.0012	10	16
205	0.002	9	8	51	0.0014	10	16
207	0.0019	9	8	44	0.0016	9	16
214	0.0025	9	16	62	0.0012	10	16
231	0.0041	8	16	83	0.0013	10	16
233	0.006	7	16	171	0.0034	8	16
236	0.0086	7	16	32	0.002	9	16
234	0.0065	7	16	43	0.0017	9	16
232	0.0066	7	16	82	0.0013	10	16
215	0.0026	9	16	189	0.0012	10	16
212	0.0028	8	16	185	0.0014	10	16
211	0.0029	8	16	30	0.002	9	16
210	0.0025	9	16	41	0.0017	9	16
200	0.0015	9	16	61	0.0012	10	16
198	0.0013	10	16	81	0.0012	10	16
197	0.0013	10	16	136	0.0046	8	16
193	0.0009	10	16	79	0.0012	10	16
191	0.001	10	16	53	0.0013	10	16
187	0.0013	10	16	40	0.0017	9	16
183	0.0016	9	16	59	0.0012	10	16

(continued)

Table 1 (continued)

Symbol	Prob.	Code length by Huffman code	Code length by B_1 code	Symbol	Prob.	Code length by Huffman code	Code length by B_1 code
178	0.0024	9	16	97	0.002	9	16
174	0.0029	8	16	78	0.0012	10	16
170	0.0035	8	16	52	0.0013	10	16
169	0.0034	8	16	80	0.0012	10	16
162	0.0037	8	16	199	0.0014	9	16
161	0.0039	8	16	55	0.0013	10	16
157	0.0042	8	16	26	0.0023	9	16
154	0.004	8	16	36	0.0018	9	16
148	0.0043	8	16	93	0.0018	9	16
146	0.0044	8	16	92	0.0017	9	16
143	0.0044	8	16	194	0.001	10	16
142	0.0044	8	16	54	0.0013	10	16
141	0.0044	8	16	28	0.0022	9	16
140	0.0047	8	16	38	0.0018	9	16
144	0.0042	8	16	57	0.0012	10	16
145	0.0044	8	16	96	0.0019	9	16
139	0.0044	8	16	39	0.0017	9	16
137	0.0046	8	16	58	0.0012	10	16
135	0.0048	8	16	60	0.0012	10	16
138	0.0045	8	16	235	0.0064	7	16
134	0.0049	8	16	88	0.0015	9	16
133	0.0051	8	16	188	0.0011	10	16
132	0.0053	8	16	63	0.0011	10	16
125	0.0065	7	16	35	0.0019	9	16
123	0.0067	7	16	70	0.0011	10	16
121	0.0069	7	16	201	0.0017	9	16
120	0.0071	7	16	91	0.0016	9	16
119	0.0071	7	16	71	0.0011	10	16
117	0.0067	7	16	69	0.0012	10	16
118	0.007	7	16	31	0.002	9	16
116	0.0066	7	16	67	0.0011	10	16
115	0.0062	7	16	90	0.0016	9	16
114	0.0056	8	16	95	0.0019	9	16
113	0.0054	8	16	73	0.0012	10	16
122	0.0068	7	16	29	0.0021	9	16
104	0.0028	9	16	89	0.0016	9	16
94	0.002	9	16	68	0.0011	10	16
74	0.0011	10	16	237	0.0082	7	16
48	0.0015	9	16	72	0.0011	10	16

(continued)

Table 1 (continued)

Symbol	Prob.	Code length by Huffman code	Code length by B_1 code	Symbol	Prob.	Code length by Huffman code	Code length by B_1 code
34	0.0019	9	16	42	0.0017	9	16
47	0.0016	9	16	27	0.0022	9	16
66	0.0011	10	16	84	0.0013	10	16
87	0.0014	9	16	196	0.0012	10	16
102	0.0025	9	16	238	0.0082	7	16
110	0.0043	8	16	37	0.0018	9	16
111	0.0047	8	16	239	0.0096	7	16
107	0.0033	8	16	240	0.0057	7	16
105	0.0029	8	16	241	0.0127	6	16
108	0.0036	8	16	242	0.01	7	16
124	0.0066	7	16	243	0.0105	7	16
126	0.0064	7	16	56	0.0012	10	16
112	0.0051	8	16	244	0.0128	6	16
109	0.0039	8	16	245	0.0154	6	16
106	0.0032	8	16	247	0.0123	6	16
103	0.0026	9	16	248	0.0113	7	16
130	0.0057	7	16	246	0.0116	6	16
163	0.0037	8	16	250	0.0144	6	16
165	0.0036	8	16	249	0.0086	7	16
152	0.0043	8	16	251	0.0124	6	16
151	0.0041	8	16	252	0.0134	6	16
155	0.004	8	16	24	0.0024	9	16
156	0.0039	8	16	253	0.0145	6	16
159	0.0041	8	16	254	0.045	5	16
160	0.0041	8	16	255	0.0041	8	16
166	0.0035	8	16	25	0.0023	9	16
167	0.0036	8	16	23	0.0025	9	16
164	0.0036	8	16	22	0.0027	9	16
153	0.0041	8	16	21	0.0028	9	16
150	0.0042	8	16	18	0.0031	8	16
149	0.0043	8	16	14	0.0039	8	16
127	0.0064	7	16	12	0.0042	8	16
168	0.0033	8	16	19	0.0029	8	16
158	0.0041	8	16	8	0.0049	8	16
180	0.0021	9	16	7	0.0053	8	16
177	0.0025	9	16	20	0.0029	8	16
176	0.0025	9	16	17	0.0034	8	16
175	0.0028	9	16	3	0.0069	7	16
182	0.0017	9	16	6	0.0056	8	16

(continued)

Table 1 (continued)

Symbol	Prob.	Code length by Huffman code	Code length by B_1 code	Symbol	Prob.	Code length by Huffman code	Code length by B_1 code
181	0.0019	9	16	11	0.0044	8	16
186	0.0013	10	16	9	0.0046	8	16
184	0.0015	9	16	10	0.0045	8	16
179	0.0023	9	16	16	0.0036	8	16
173	0.0032	8	16	5	0.0059	7	16
172	0.0036	8	16	13	0.0042	8	16
208	0.0021	9	16	4	0.0065	7	16
230	0.0055	8	16	1	0.0078	7	16
195	0.0012	10	16	0	0.0257	5	16
192	0.0009	10	16	2	0.0074	7	16
131	0.0055	8	16	15	0.0036	8	16

Table 2 Average word length

Huffman code	B1-code
8.4346	13.0426

5 Conclusion

Compression is an important technique in the multimedia computing field. This is because we can reduce the size of data. Transmitting and storing the reduced data on the Internet and storage devices are faster and cheaper than uncompressed data. Many image and video compression standards such as JPEG, JPEG2000, MPEG-2, and MPEG-4 have been proposed and implemented [8, 9, 11, 13]. This paper focuses on implementing Huffman coding and B-coding algorithms for image compression in order to clarify their differences from different points of view such as implementation and average word length. We have explained these algorithms in detail and implemented. From implementation point of view, B-coding is easier than Huffman coding while average word length reduces using Huffman code. B-Coding needs less execution time than the Huffman coding. This means that in some applications where time is not so important we can use Huffman algorithm for better compression, while for some applications where time is important such as real-time applications, B-coding algorithm can be used.

As a future work more focus can be on reduction of average word length using the new techniques. The proposed algorithms can be experimented on different kinds of data sets like audio, video, text as till now it is restricted to images.

References

1. Anitha, S.: Lossless image compression and decompression using Huffman coding. Int. Res. J. Eng. Technol. (IRJET) **2**(1), 240–247 (2015)
2. Arti: Performance analysis of Huffman coding algorithm. Int. J. Adv. Res. Comput. Sci. Softw. Eng. **3**(5), 615–619 (2013)
3. Farid, H.: Fundamentals of Image Processing (2010)
4. Gomathi, K.V., Lotus, R.: Digital image compression techniques. IJRET: Int. J. Res. Eng. Technol. **3**(10), 285–290 (2014)
5. Gonzalez, C.R., Woods, E.R.: The encoding process. Digital Image Processing, 2nd edn., pp. 265–267. Printice Hall (2002)
6. http://www.en.wikipedia.org/wiki/Entropy_encoding
7. http://www.en.wikipedia.org/wiki/Huffman_coding
8. http://www.en.wikipedia.org/wiki/Lossless_JPEG
9. http://www.howstuffworks.com/file-compression.htm
10. http://www.huffmancoding.com, Wikipedia, the free encyclopedia
11. http://www.JPEG.org, Wikipedia, the free encyclopedia
12. Mathur, K.M., Loonker, S., Saxena, D.: Lossless Huffman coding technique for image compression and reconstruction using binary trees. Int. J. Comp. Tech. Appl. **3**(1), 76–79 (2012)
13. Nelson, M., Gailly, L.J.: The Data Compression Book, 2nd edn. (1995)
14. Padmaja, G.M., Nirupama, P.: Analysis of various image compression techniques. ARPN J. Sci. Technol. **2**(4), 371–376 (2012)
15. Rabbani, M., Jones, W.P.: Digital Image Compression Techniques (1991)
16. Shahbahrami, A., Bahrampour, R., Rostami, M., Mobarhan, M.A.: Evaluation of Huffman and arithmetic algorithms for multimedia compression standards. Int. J. Comput. Sci. Eng. Appl. (IJCSEA) **1**(4), 34–47 (2011)
17. Sharma, M.: Compression using Huffman coding. IJCSNS Int. J. Comput. Sci. Netw. Secur. **10**(5), 133–141 (2010)
18. Singh, A., Gahlawat, M.: Image compression and its various techniques. Int. J. Adv. Res. Comput. Sci. Softw. Eng. **3**(6), 650–654 (2013)

Author Index

M. Pant et al. (eds.), *Proceedings of Fifth International Conference on Soft Computing for Problem Solving*, Advances in Intelligent Systems and Computing 436, DOI 10.1007/978-981-10-0448-3

Printed in the United States
By Bookmasters